Dimitrios C. Kravvaritis, Athanasios N. Yannacopoulos
Variational Methods in Nonlinear Analysis

Also of Interest

Differential Equations
A First Course on ODE and a Brief Introduction to PDE
Antonio Ambrosetti, Shair Ahmad, 2024
ISBN 978-3-11-118524-8, e-ISBN (PDF) 978-3-11-118567-5

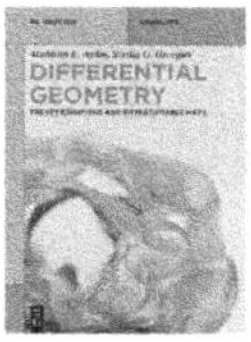

Differential Geometry
Frenet Equations and Differentiable Maps
Muhittin E. Aydin und Svetlin G. Georgiev, 2024
ISBN 978-3-11-150089-8, e-ISBN (PDF) 978-3-11-150185-7

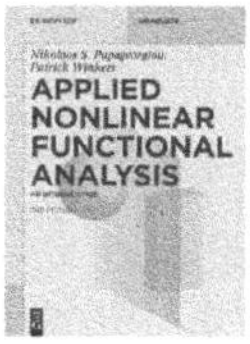

Applied Nonlinear Functional Analysis
An Introduction
Nikolaos S. Papageorgiou, Patrick Winkert, 2024
ISBN 978-3-11-128421-7, e-ISBN (PDF) 978-3-11-128695-2

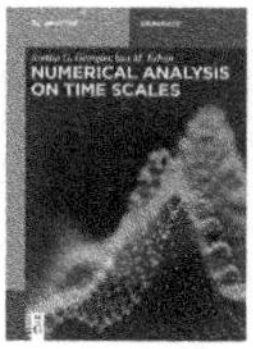

Numerical Analysis on Time Scales
Svetlin G. Georgiev, Inci M. Erhan, 2022
ISBN 978-3-11-078725-2, e-ISBN (PDF) 978-3-11-078732-0

Nonlinear Functional Analysis and Applications
Jesús Garcia-Falset, Khalid Latrach, 2023
ISBN 978-3-11-103096-8, e-ISBN (PDF) 978-3-11-103181-1
in: *De Gruyter Series in Nonlinear Analysis and Applications*
ISSN 0941-813X

Dimitrios C. Kravvaritis, Athanasios N. Yannacopoulos

Variational Methods in Nonlinear Analysis

Optimization and Partial Differential Equations Applications

2nd, revised edition

DE GRUYTER

Mathematics Subject Classification 2020
35-02, 65-02, 65C30, 65C05, 65N35, 65N75, 65N80

Authors

Prof. Dr. Dimitrios C. Kravvaritis
National Technical University of Athens
School of Applied Mathematical and
Physical Sciences
Heroon Polytechniou 9
Zografou Campus
157 80 Athens, Greece

Prof. Dr. Athanasios N. Yannacopoulos
Athens University of Economics
and Business
Statistics
Patission 76
104 34 Athens, Greece

ISBN 978-3-11-133325-0
e-ISBN (PDF) 978-3-11-133329-8
e-ISBN (EPUB) 978-3-11-133335-9

Library of Congress Control Number: 2025937603

Bibliographic information published by the Deutsche Nationalbibliothek
The Deutsche Nationalbibliothek lists this publication in the Deutsche Nationalbibliografie; detailed bibliographic data are available on the Internet at http://dnb.dnb.de.

Cover image: gt29 / iStock / Getty Images Plus
Typesetting: VTeX UAB, Lithuania

www.degruyterbrill.com
Questions about General Product Safety Regulation:
productsafety@degruyterbrill.com

DCK: To my wife Katerina and my two sons Christos and Nikos

ANY: To my fixed point Electra
and to Nikos, Jenny and Helen

Spent the morning putting in a comma and the afternoon removing it

G. Flaumbert or O. Wilde

If you can't prove your theorem, keep shifting parts of the conclusion to the assumptions, until you can

E. De Giorgi (according to Wikipedia)

Preface

This is a revised and considerably extended second edition of a book with the same title.

Nonlinear analysis is a vast field, which originated from the need of extending classical linear structures in analysis in order to treat nonlinear problems. It can be considered as general term for describing a wide spectrum of abstract mathematical techniques, ranging from geometry, topology, and analysis, which nevertheless find important applications in a variety of fields, such as mathematical modeling (e. g., in economics, engineering, biology, decision science, etc.), optimization, or in applied mathematics (e. g., integral, differential equations and partial differential equations). It is the aim of this book to select out of this wide spectrum certain aspects that the authors consider as most useful in a variety of applications, and, in particular, aspects of the theory related with variational methods, which are most closely related to applications in optimization and PDEs.

The book is intended for graduate students, lecturers, or independent researchers, who desire a concise introduction to nonlinear analysis, to learn the material independently, use it as a guideline for the organization of a course, or use it as a reference when trying to master certain techniques to use in their own research. We hope that we have achieved a good balance between theory and applications. All the theoretical results are proved in full detail, emphasizing certain delicate points. Many examples are provided, guiding the reader either towards extensions of the fundamental results or towards important applications in a variety of fields. Large sections are devoted to the applications of the theoretical results in optimization, PDEs, and variational inequalities, including hints towards a more algorithmic approach. Keeping the balance right, the volume manageable but at the same time striving to provide as wide a spectrum of techniques and subjects as possible has been a very tricky business. We hope that we have managed to present a good variety of these parts of nonlinear analysis, which find the majority of applications in the fields that are the within the core interest of this book, i. e., optimization and PDEs. Naturally, we had to draw lines in our indulgence to the beautiful subjects we decided to present (otherwise we would have a nine-volume treatise, rather than a nine-chapter book), but at the same time we had to completely omit important fields, such as degree theory, and we apologize for this, hoping to make up for this omission in the future! At any rate, we hope that our effort (which by the way was extremely enjoyable!) will make itself useful to those wishing a concise introduction to nonlinear analysis with an emphasis towards applications, and will help newcomers of any mathematical age to the field, colleagues in academia to organize their courses, and practitioners or researchers towards obtaining the methodological framework that will help them towards reaching their goal.

This book consists of ten chapters, which we think cover a wide spectrum of concepts and techniques of nonlinear analysis and their applications.

Chapter 1 collects, mostly without proofs, some preliminary material from topology, linear functional analysis, Sobolev spaces, and multivalued maps, which are considered

https://doi.org/10.1515/9783111333298-204

as necessary in order to proceed further to developing the material covered in this book. Even though proofs are not provided for this introductory material, we try to illustrate and clarify the concepts using numerous examples.

Chapter 2 develops calculus in Banach spaces and focuses on convexity as a fundamental property of subsets of Banach spaces as well as on the properties of convex functions on Banach spaces. This material is developed in detail, starting from topological properties of convex sets in Banach space, in particular their remarkable properties with respect to the weak topology, and then focuses on properties of convex functions with respect to continuity and semicontinuity or with respect to their Gâteaux or Fréchet derivatives. Extended comments on the deep connections of convexity with optimization are made. The important concepts of the lower semicontinuous envelope as well as of Γ-convergence, an important tool in the study of convergence for variational problems with multiple and diverse applications are introduced in the end of this chapter.

Chapter 3 is devoted to an important tool of nonlinear analysis, which is used throughout this book, fixed point theory. One could devote a whole book to the subject, so here we take a minimal approach, of presenting a selection of fixed point theorems, which we find indispensable. This list consists of the Banach contraction theorem, Brouwer's fixed point theorem, which, although finite dimensional in nature, forms the basis for developing a large number of useful fixed point theorems in infinite dimensional spaces, Schauder's fixed point theorem, and Browder's fixed point theorem. Special attention is paid to the construction of iterative schemes for the approximation of fixed points, and in particular the Krasnoselskii–Mann iterative scheme, which finds interesting applications in numerical analysis. We also introduce the Caristi fixed point theorem and its deep connection with a very useful tool, the Ekeland variational principle. The chapter closes with a fixed point theorem for multivalued maps.

Chapter 4 provides an introduction to nonsmooth convex functions and the concept of the subdifferential. The theory of the subdifferential as well as subdifferential calculus is developed gradually, leading to its applications to optimization. The Moreau proximity operator in Hilbert spaces is introduced, and its use in a class of popular optimization algorithms, which go under the general name of proximal algorithms is presented.

Chapter 5 is devoted to the theory of duality in convex optimization. The chapter begins with an indispensable tool, a version of the minimax theorem, which allows us to characterize saddle points for functionals of specific structure in Banach spaces. We then define and provide the properties of the Fenchel–Legendre conjugate as well as the biconjugate, and illustrate them with numerous examples. To further illustrate the properties of these important transforms, we introduce the concept of the inf-convolution, a concept of interest in its own right in optimization. Having developed the necessary machinery, we present the important contribution of the minimax approach in optimization, starting with a detailed presentation of Fenchel's duality theory, and then its generalization to a more general scheme that encompasses Lagrangian duality as well as the augmented Lagrangian method. These techniques are illustrated in terms of various examples and applications. The chapter closes by presenting an important class of numerical methods based on the duality approach and discussion of their properties.

Chapter 6 is an introduction to the calculus of variations and its applications. After a short motivation, we start with a warm up problem, and in particular the variational theory of the Laplacian. This is a nice opportunity to either introduce or remind (depending on the reader) some important properties of the Laplacian, such as the solvability and the properties (regularity etc.) of Poisson's equation and the spectral theory related to this operator. Having spent some time on this important case, we return to the general case, presenting results on the lower semicontinuity of integral functionals in Sobolev spaces, the existence of minimizers, and their connection with the Euler–Lagrange equation. We then consider the problem of possible extra regularity of the minimizer beyond that guaranteed by their membership in a Sobolev space (already established by existence). To this end, we provide an introduction to the theory of regularity of solutions variational problems, initiated by De Giorgi, and in particular present some fundamental results concerning Hölder continuity of minimizers or higher weak differentiability. We then continue with two important examples: a) the application of the calculus of variations to semilinear elliptic problems and b) quasilinear elliptic problems, in particular, problems related to the p-Laplace operator. The chapter closes with two important examples of the application of the concept of Γ-convergence to the calculus of variations: a) the homogenization of elliptic problems with fast oscillating coefficients and b) the study of the p-Laplace operator when the coefficient p is varying.

Chapter 7 considers variational inequalities, a subject of great interest in various fields, ranging from mechanics to finance or decision science. As a warm up, we revisit the problem of the Laplacian, but now related to the minimization of the Dirichlet functional on convex closed subsets of a Sobolev space, which naturally lead to a variational inequality for the Laplace operator. Then, the Lions–Stampacchia theory for existence of solutions to variational inequalities (not necessarily related to minimization problems) is developed, initially in a Hilbert space setting and with generalizations to variational inequalities of second kind. We then proceed to study the problem of approximation of variational inequalities, presenting the penalization method, as well as the internal approximation method. The chapter closes with an important class of applications, the study of variational inequalities for second-order elliptic operators. As the Lions–Stampacchia theory covers equations as a special case, we take up the opportunity to start our presentation with second-order elliptic equations, and in this way generalize certain properties that we have seen for the Laplacian in Chapter 6, to more general equations. We then move to variational inequalities and discuss various issues such as solvability, comparison principles, etc., and close the chapter with a study of variational inequalities involving the p-Laplace operator.

Chapter 8 introduces critical point theory for functionals in Banach spaces. Having understood quite well the properties of minimizers of functionals and their connection with PDEs, it is interesting to see whether extending this analysis to critical points in general may provide some useful insights. The chapter starts with the presentation of an important class of theorems related to the Ambrosseti–Rabinowitz mountain pass theorem and it generalizations (e. g., the saddle point theorem). This theory provides general criteria under which critical points of certain functionals exist. Through the Euler–Lagrange equation, such existence results may prove useful to the study of certain PDEs or variational inequalities. Our applications focus on semilinear and quasilinear PDEs, related to the p-Laplacian.

Chapter 9 introduces a very important part of nonlinear analysis, the study of monotone type operators. During the development of the material up to now, we have seen that monotonicity in some form or another has played a key role to our arguments (such as the monotonicity of the Gâteaux derivative of convex functions, or the coercivity of bilinear forms in the Lions–Stampacchia theory etc.). In this chapter, motivated by these observations, we study the monotonicity properties of operators as a primary issue and not as a derivative of other properties, e. g., convexity. As we shall see, this allows us to extend certain interesting results that we have arrived at via minimization or in general variational arguments for operators that may not necessarily enjoy such a structure. We start our study with monotone operators, then move to maximal monotone operators, and then introduce pseudomonotone operators. Our primary interest is in obtaining surjectivity results for monotone type operators, which will be subsequently used in the study of certain PDEs or variational inequalities, but along the way encounter certain interesting results, such as the Yosida approximation etc. The theory is illustrated with applications in quasilinear PDEs, differential inclusions, and variational inequalities.

The book finishes with a new addition, Chapter 10, which deals with evolution problems. This type of problems is very important both from the point of view of theoretical developments as well as from the point of view of applications. We start our cover of this topic with the theory of nonlinear semigroups, focusing on the theory of their generation by a certain type of nonlinear operators called m-dissipative operators, which are closely related to monotone operators, as well as their connection with nonlinear evolution problems. We then move to a complementary approach covering a wider range of problems, a variational approach based upon the concept of pseudomonotonicity, introduced in Chapter 9. The chapter concludes with applications of these two approaches to concrete problems of evolution equations, such as semi- and quasilinear parabolic and hyperbolic problems. Such problems arise naturally in various models from the physical sciences, image processing, and financial economics.

We wish to acknowledge the invaluable support of various people for various reasons. We must start with acknowledging the support of Dr. Apostolos Damialis, at the time Mathematics Editor at De Gruyter, whose contribution made this book a reality. Then, we are indebted to Nadja Schedensack at De Gruyter for her superb handling of all the details and her understanding.

Concerning the second edition of the book, we are indebted to Ranis Imbragimov, who kindly contacted us concerning the opportunity of a revised edition as well to Ute Skambrachts at De Gruyter, for her wonderful guidance and patience throughout the working of this revision (which took a bit longer than expected, but hopefully it is worth the waiting!). We are also grateful to the technical editorial team of De Gruyter and in particular Vilma Vaičeliūnienė and Ina Talandienė for their invaluable and constant support.

As this material has been tested on audiences at our respective institutions over the years, we wish to thank our students; they have helped us shape and organize this material through the interaction with them. We both feel compelled to mention specifically George Domazakis, whose help in proof reading and commenting some of this material as well as with technical issues related to the De Gruyter latex style file has been invaluable; we deeply thank him for that. We are also thankful to Kyriakos Georgiou for his kind offer for proof reading the manuscript and his useful comments.

There are many colleagues and friends, with whose interaction and collaboration throughout these years have shaped our interests and encouraged us. We need to mention, Professors N. Alikakos, I. Baltas, D. Drivaliaris, N. Papageorgiou, G. Papagiannis, G. Smyrlis, I. G. Stratis, A. Tsekrekos, S. Xanthopoulos, A. Xepapadeas, W. G. Weber and Anh Tuan Dang. At a personal level, ANY is indebted to Electra Petracou for making all this feasible through her constant support and devotion.

Athens, 2025

D. C. Kravvaritis, NTUA,
A. N. Yannacopoulos, AUEB

Contents

Notation

The following notation is used throughout this book:

General notation

- We will use the notation $A \subset B$ denoting both strict or nonstrict inclusion. If we want to emphasize that an inclusion is strict, we will denote it by $A \subsetneq B$. We will also use the notation $A \setminus B$ for the difference of two sets.
- $\mathbf{1}_A$ is the indicator function of a set A, defined by $\mathbf{1}_A(s) = 1$, if $s \in A$ and $\mathbf{1}_A(s) = 0$ otherwise.
- I_A is the convex indicator function of a set A, defined by $I_A(s) = 0$, is $s \in A$ and $I_A(s) = +\infty$ otherwise.
- The standard notations for the set of real numbers $\mathbb{R}$ and the set of natural numbers $\mathbb{N}$ is used. We will also use the notation $\mathbb{R}_+ = \{x \in \mathbb{R} : x \geq 0\}$.
- $\mathbb{R}^d = \{(x_1, \dots, x_d) : x_i \in \mathbb{R}, \ i = 1, \dots, d\}$.
- $x = (x_1, \dots, x_d)$ or $z = (z_1, \dots, z_d)$ is used to denote the elements of $\mathbb{R}^d$, while $|x| = (|x_1|^2 + \cdots + |x_d|^2)^{1/2}$ is used to denote the Euclidean norm for $\mathbb{R}^d$, and $x \cdot x$ the usual inner product.
- By $B(x_o, R) := \{x \in \mathbb{R}^d : |x - x_o| \leq R\}$, we denote the ball in $\mathbb{R}^d$, with center x_0 and radius R, and by $Q(x_0, R)$ we denote the cube centered at x_0 of side R.
- $\mathcal{D} \subset \mathbb{R}^d$ is a nonempty open subset of $\mathbb{R}^d$.
- A domain in $\mathbb{R}^d$ is an open and connected subset of $\mathbb{R}^d$.
- $\partial\mathcal{D}$, the boundary of $\mathcal{D}$, assumed to be sufficiently smooth, e. g., Lipschitz, unless explicitly stated otherwise. If $\partial\mathcal{D}$ is piecewise C^1 the domain $\mathcal{D}$ will be called regular.
- $\mathbb{M}_{m\times n}$ is the set of $m \times n$ matrices. If $\mathbf{A} \in \mathbb{M}_{m\times n}$, by $\mathbf{A}^T$, we denote its transpose, and in the case where $m = n$ by $\det(\mathbf{A})$, we denote its determinant.
- Throughout the most part of this book, X and Y are used for denoting generic Banach spaces. The elements of the space X are denoted by x or z (using if necessary additional demarcations, e. g., x_o, z_0, etc.), while the elements of Y are denoted by y (or using if necessary additional demarcations).
- Similarly, the duals of the above are denoted by $X^\star$ and $Y^\star$, respectively. The elements of $X^\star$ are denoted by $\mathsf{x}^\star$ or $\mathsf{z}^\star$ (using if necessary additional demarcations, e. g., $\mathsf{x}_o^\star$, $\mathsf{z}_0^\star$, etc.), while the elements of $Y^\star$ are denoted by $\mathsf{y}^\star$ (or using if necessary additional demarcations).
- The duality pairing between two spaces $X, X^\star$ is denoted by $\langle\cdot,\cdot\rangle_{X^\star,X}$, with the subscript dropped when no confusion is possible.
- Hilbert spaces are denoted by H, and the inner product by $\langle\cdot,\cdot\rangle_H$ (with the subscript dropped when no confusion may arise).
- Operators, linear or nonlinear, between two spaces X and Y are denoted by sans serif fonts, e. g., A, B, L, etc. The action of a linear operator A on an element $\mathsf{x} \in X$ is denoted by Ax, while for a nonlinear operator it is denoted by A(x).
- $f : X \to Y$ is used to denote maps in general between two topological spaces (including linear functionals), whereas F is used to denote possible nonlinear functionals $F : X \to \mathbb{R}$.
- Df is used to denote the Gâteaux or Fréchet derivative of a map $f : X \to Y$.
- If $f : \mathcal{D} \subset \mathbb{R}^d \to \mathbb{R}$, we will use the notation $\nabla f = (\frac{\partial}{\partial x_1}f, \dots, \frac{\partial}{\partial x_d}f)$ for its gradient.
- If $f = (f_1, \dots, f_d) : \mathcal{D} \subset \mathbb{R}^d \to \mathbb{R}^d$, we will use the notation $\operatorname{div} f = \frac{\partial}{\partial x_1}f_1 + \cdots \frac{\partial}{\partial x_d}f_d$ for its divergence.
- By $\frac{df}{dt}$, we denote the derivative of a map $f : [0, T] \subset \mathbb{R} \to X$ (the case $X = \mathbb{R}$ included). Please note that we denote the derivative by $\frac{df}{dt}$ and **NOT** f'! The ordinary prime is used to denote an alternative element of any set.
- $C \subset X$ is used to denote a convex subset of X.
- $\varphi : X \to \mathbb{R} \cup \{+\infty\}$ is used to denote convex functions.

https://doi.org/10.1515/9783111333298-206

- $\arg\min_{z\in C}\varphi(z) := \{x \in C : \varphi(x) = \inf_{z\in C}\varphi(x)\}$ is the set of minimizers of the function $\varphi : X \to \mathbb{R}\cup\{+\infty\}$ over $C \subset X$. By $\text{cont}(\varphi)$, we denote the set of points where φ is finite and continuous.

More specialized notation

- A linear or nonlinear operator: $\mathbf{D}(A)$ its domain, $\mathbf{Gr}(A)$ its graph, $\mathbf{N}(A)$ its kernel, $\mathbf{R}(A)$ its range, Definition 1.1.1, Section 1.1.1.
- $\mathcal{B}(X, Y)$: the space of linear bounded operators $A : X \to Y$; Definition 1.1.1, Section 1.1.1.
- $\mathcal{L}(X, Y)$: the space of linear continuous operators $A : X \to Y$; Definition 1.1.1, Section 1.1.1.
- $\mathcal{L}(X)$: the space of linear continuous operators $A : X \to X$; Definition 1.1.1, Section 1.1.1.
- $X \simeq Y$, isometric spaces; Definition 1.1.3, Section 1.1.1.
- $X^\star$: the dual space of a normed space X; Definition 1.1.12, Section 1.1.2.
- $X^{\star\star}$: the bidual space of a normed space X; Definition 1.1.20, Section 1.1.3.
- $J : X \to 2^{X^\star}$: the duality map; Definition 1.1.17, Section 1.1.2.
- $j : X \to X^{\star\star}$: canonical embedding, Theorem 1.1.23, Section 1.1.3.
- $\overline{B}_X(0,1) = \overline{B_X(0,1)} = \{x \in X : \|x\|_X \le 1\})$: the closed unit ball of X, Section 1.1.4.
- $B_X(0,1) = B_X(0,1) = \{x \in X : \|x\|_X < 1\}$: the open unit ball of X.
- $S_X(0,1) = \{x \in X : \|x\|_X = 1\}$: the unit sphere of X.
- $x_n \to x$: strong convergence in a Banach space X; Definition 1.1.29, Section 1.1.5.2.
- $\sigma(X^\star, X)$: the weak* topology on $X^\star$; Definition 1.1.33, Section 1.1.6.1.
- $X^\star_{w^\star}$ is $X^\star$ equipped with the weak* topology on $X^\star$; Definition 1.1.33, Section 1.1.6.1.
- $x_n^\star \overset{\star}{\rightharpoonup} x^\star$: weak* convergence in $X^\star$; Definition 1.1.39, Section 1.1.6.2.
- $\sigma(X, X^\star)$: the weak topology on X; Definition 1.1.43, Section 1.1.7.1.
- X_w is X equipped with the weak topology on X; Definition 1.1.43, Section 1.1.7.1.
- $x_n \rightharpoonup x$: weak convergence in X; Definition 1.1.51, Section 1.1.7.2.
- $A^\star : Y^\star \to X^\star$: the adjoint of the linear operator $A : X \to Y$; Definition 1.1.19, Section 1.1.2.
- $\text{index}(A) = \dim(\mathbf{N}(A)) - \dim(Y/\mathbf{R}(A))$: the index of a Fredholm operator; Definition 1.3.5, Section 1.3.
- $\rho(A)$: the resolvent set of a bounded linear operator A; Definition 1.3.11, Section 1.3.
- $\sigma(A)$: the spectrum of a bounded linear operator A; Definition 1.3.11, Section 1.3.
- $\mathcal{B}(\mathcal{D})$: the Borel σ-algebra on $\mathcal{D} \subset \mathbb{R}^d$, Section 1.4.
- $\mu_{\mathcal{L}}$: the Lebesgue measure on $\mathbb{R}^d$, Section 1.4. The Lebesgue measure of a set $\mathcal{D} \subset \mathbb{R}^d$ is denoted either by $\mu_{\mathcal{L}}(\mathcal{D})$ or simply by $|\mathcal{D}|$ when there is no risk of confusion.
- $L^p(\mathcal{D})$: Lebesgue space; Definition 1.4.1, Section 1.4.
- For any $p \in \mathbb{N}$, $p > 1$, we denote by p^* the conjugate $\frac{1}{p} + \frac{1}{p^*} = 1$. If $p = 1$, then $p^* = \infty$, whereas if $p = \infty$, $p^* = 1$.
- $(L^p(\mathcal{D}))^\star \simeq L^{p^*}(\mathcal{D})$.
- $\text{osc}(x_o, r) := \sup_{x\in B(x_o,r)} f(x) - \inf_{x\in B(x_o,r)} f(x)$: the oscillation of a function $f : \mathbb{R}^d \to \mathbb{R}$ on the ball $B(x_0, r)$; Definition 1.9.17, Section 1.9.4.
- $⨍_A u(x)dx := \frac{1}{|A|}\int_A u(x)dx$ for every $A \subset \mathbb{R}^d$, $u \in L^1(A)$ is the mean value of the function f on A; Definition 1.9.17, Section 1.9.4.
- $\mathcal{M}(D)$: the set of Radon measures on $\mathcal{D} \subset \mathbb{R}^d$; Definition 1.4.3, Section 1.4.
- $C_c(\mathcal{D}) := \{\phi \in C(\mathcal{D}) : \phi(x) = 0,\ \forall x \in \mathcal{D}\setminus\mathcal{D}_o,\ \mathcal{D}_o \subset \mathcal{D},\ \text{compact}\}$, Section 1.4.
- $C_c^\infty(\mathcal{D})$: the set of infinitely continuously differentiable functions on $\mathcal{D}$ with compact support, Section 1.5.
- $C^{0,\alpha}(\mathcal{D})$: the space of Hölder continuous functions $f : \mathcal{D} \subset \mathbb{R}^d \to \mathbb{R}$ with Hölder exponent $\alpha \in (0,1]$; Definition 1.9.15, Section 1.9.4.
- $W^{k,p}(\mathcal{D})$: the k-Sobolev space of order p; Definition 1.5.3, Section 1.5.
- $W^{k,p}_{\text{loc}}(\mathcal{D})$; Definition 1.5.3, Section 1.5.

- $W_0^{k,p}(\mathcal{D})$: the k-Sobolev space of order p of functions of vanishing trace on $\partial\mathcal{D}$; Definition 1.5.6, Section 1.5.
- $W^{-k,p^*}(\mathcal{D}) = (W^{k,p}(\mathcal{D}))^*$; Definition 1.5.6, Section 1.5.
- $\mathfrak{s}_p := \frac{dp}{d-p}$: the critical exponent for the Sobolev embedding $L^{\mathfrak{s}_p}(\mathcal{D}) \hookrightarrow W^{1,p}(\mathcal{D})$, $p \le d$.
- $W^{1,p,q}(I;X,Y)$: the Sobolev–Bochner spaces; Definition 1.6.9, Section 1.6.
- $W^{1,p}(I;X) := W^{1,p,p^*}(I;X,X^*)$: a special Sobolev–Bochner space; Definition 1.6.9, Section 1.6.
- $f : X \to 2^Y$: a multivalued map, Section 1.8.
- $f^u(A)$: the upper inverse of a set A for a multivalued map $f : X \to 2^Y$; Definition 1.8.1, Section 1.8.
- $f^\ell(A)$: the lower inverse of a set A for a multivalued map $f : X \to 2^Y$; Definition 1.8.1, Section 1.8.
- $DF(\mathrm{x};\mathrm{h})$: the directional derivative of $F : X \to \mathbb{R}$ at $\mathrm{x} \in X$ in the direction $\mathrm{h} \in X$; Definition 2.1.1, Section 2.1.1.
- $DF(\mathrm{x})$: the Gâteaux derivative of $F : X \to \mathbb{R}$ at $\mathrm{x} \in X$; Definition 2.1.2, Section 2.1.2. If it exists, the Fréchet derivative is denoted the same; Definition 2.1.6, Section 2.1.3.
- $C \subset X$ is in general a convex set; Definition 1.2.1, Section 1.2.1.
- $\mathrm{core}(C) = \{\mathrm{x} \in C : \forall \mathrm{z} \in S_X,\ \exists \epsilon > 0,\ \text{such that } \mathrm{x} + \delta \mathrm{z} \in C,\ \forall \delta \in [0,\epsilon]\}$ is the core or algebraic interior of a convex set; Definition 1.2.5, Section 1.2.1.
- $\mathrm{conv}\, A$ is the set of all convex combinations from $A \subset X$; Definition 1.2.16, Section 1.2.2.
- $\varphi : X \to \mathbb{R} \cup \{+\infty\}$ is a convex function; Definition 2.3.2, Section 2.3.1.
- $\mathrm{dom}\, \varphi = \{\mathrm{x} \in X : \varphi(\mathrm{x}) < +\infty\}$ is the domain of a convex function; Definition 2.3.4, Section 2.3.1.
- $\mathrm{epi}\, \varphi = \{(\mathrm{x},\lambda) \in X \times \mathbb{R} : \varphi(\mathrm{x}) \le \lambda\}$ is the epigraph of a convex function; Definition 2.3.4, Section 2.3.1.
- $I_C(\mathrm{x})$ is the indicator function for a convex set; Definition 2.3.10, Section 2.3.2.1.
- $\sigma_C : X^* \to \mathbb{R} \cup \{+\infty\}$ is the support function of a convex set $C \subset X$; Definition 2.3.13, Section 2.3.2.3.
- $\mathrm{P}_C : H \to C$ is the projection operator to the closed convex subset $C \subset H$; Definition 2.5.1, Section 2.5.
- $E^\perp$ is the orthogonal complement of closed linear subspace $E \subset H$; Definition 2.5.4, Section 2.5.
- $\partial\varphi : X \to 2^{X^*}$ is the subdifferential of a convex function; Definition 4.1.1, Section 4.1.
- $N_C(\mathrm{x})$ is the normal cone of $C \subset X$ at $\mathrm{x} \in X$; Definition 4.1.5, Section 4.1.
- $D^+\varphi(\mathrm{x},\mathrm{h})$ is the right hand side directional derivative of the function φ at $\mathrm{x} \in X$ in the direction $\mathrm{h} \in X$; Definition 4.2.6, Section 4.2.2.
- $D^-\varphi(\mathrm{x},\mathrm{h})$ is the left hand side directional derivative of the function φ at $\mathrm{x} \in X$ in the direction $\mathrm{h} \in X$; Example 4.2.9, Section 4.2.2.
- $\mathrm{prox}_\varphi : H \to H$ is the proximity operator related to the convex function $\varphi : H \to \mathbb{R} \cup \{+\infty\}$; Definition 4.4.1, Section 4.4.1.
- $\varphi_\lambda : H \to \mathbb{R}$ is the Moreau–Yosida approximation of a convex function φ; Definition 4.4.3, Section 4.4.2.
- $\varphi^* : X^* \to \mathbb{R} \cup \{+\infty\}$ is the Fenchel–Legendre conjugate of the function $\varphi : X \to \mathbb{R} \cup \{+\infty\}$; Definition 5.2.1, Section 5.2.1.
- $\varphi = \varphi_1 \Box \varphi_2$ is the inf-convolution of $\varphi_1, \varphi_2 : X \to \mathbb{R} \cup \{+\infty\}$; Definition 5.2.24, Section 5.2.4.
- $F^{**} : X \to \mathbb{R} \cup \{+\infty\}$ is the Fenchel–Legendre biconjugate of $\varphi : X \to \mathbb{R} \cup \{+\infty\}$; Definition 5.2.11, Section 5.2.2.
- $\mathbb{A}(F) := \{g : X \to \mathbb{R} : g \text{ continuous and affine, } g \le F\}$, is the set of continuous affine functions minorizing F; Definition 5.2.13, Section 5.2.2.
- F^Γ is the Γ-regularization of F; Definition 5.2.13, Section 5.2.2.
- $R(\lambda,\mathrm{A}) = \mathrm{R}_\lambda, \lambda > 0$ is the resolvent operator for the nonlinear operator A; Definition 9.4.21, Section 9.4.6.
- $\mathrm{A}_\lambda, \lambda > 0$ is the resolvent operator for the nonlinear operator A; Definition 9.4.21, Section 9.4.6.

1 Preliminaries

In this introductory chapter, we collect (mostly without proofs but with a number of illustrative examples) the fundamental notions from linear functional analysis, convexity, Lebesgue–Sobolev, Sobolev–Bochner spaces, and multivalued mappings that are essential in proceeding to the main topic of this book: nonlinear analysis and its various applications. For detailed coverage of the standard material in this chapter, we refer to [1, 4, 5, 33, 56, 142] (upon which our presentation was based) .

1.1 Fundamentals in the theory of Banach spaces

Most of this book will concern Banach spaces, i. e., complete normed spaces. In this section, we recall some of their fundamental properties.

1.1.1 Linear operators and functionals

Linear mappings between normed spaces and functionals play an important role in (linear) functional analysis.

Definition 1.1.1 (Operators). Let X, Y be two normed spaces.

(i) A map $\mathsf{A} : \mathbf{D}(\mathsf{A}) \subset X \to Y$ (not necessarily linear) is called an operator (or functional if $Y = \mathbb{R}$). We further define the sets

$$\begin{aligned} \mathbf{D}(\mathsf{A}) &:= \{\mathsf{x} \in X \,:\, \mathsf{A}(\mathsf{x}) \in Y\}, \\ \mathbf{Gr}(\mathsf{A}) &:= \{(\mathsf{x}, \mathsf{A}(\mathsf{x})) \,:\, \mathsf{x} \in \mathbf{D}(\mathsf{A})\} \subset X \times Y, \\ \mathrm{N}(\mathsf{A}) &:= \{\mathsf{x} \in X \,:\, \mathsf{A}(\mathsf{x}) = 0\}, \\ \mathbf{R}(\mathsf{A}) &:= \{\mathsf{y} \in Y \,:\, \exists\, \mathsf{x} \in X, \text{for which } \mathsf{A}(\mathsf{x}) = \mathsf{y}\} = \mathsf{A}(X), \end{aligned}$$

called the domain, graph, kernel, and range of A respectively.

(ii) If A is such that $\mathsf{A}(\lambda_1 \mathsf{x}_1 + \lambda_2 \mathsf{x}_2) = \lambda_1 \mathsf{A}(\mathsf{x}_1) + \lambda_2 \mathsf{A}(\mathsf{x}_2)$, it is called a linear operator (or linear functional, denoted by f if $Y = \mathbb{R}$).

(iii) The linear operator $\mathsf{A} : X \to Y$ is called bounded if $\sup\{\|\mathsf{A}(\mathsf{x})\|_Y \,:\, \mathsf{x} \in X,\ \|\mathsf{x}\|_X \le 1\} < \infty$. The set of bounded linear operators between X and Y is denoted by $\mathcal{B}(X, Y)$.

The set of continuous linear operators between X and Y is denoted by $\mathcal{L}(X, Y)$. In the special case where $Y = X$, we will use the simplified notation $\mathcal{L}(X)$ instead of $\mathcal{L}(X, X)$.

The following theorem collects some useful properties of linear operators and functionals over normed spaces:

Theorem 1.1.2 (Elementary properties of linear operators and functionals). *Let* $\mathsf{A} : X \to Y$ *be a linear mapping.*

https://doi.org/10.1515/9783111333298-001

(i) *The following are equivalent:* (a) A *is continuous at* 0*;* (b) A *is continuous;* (c) A *is uniformly continuous;* (d) A *is Lipschitz;* (e) A *is bounded. It therefore holds that* $\mathcal{B}(X,Y) = \mathcal{L}(X,Y)$.

(ii) *It also holds that*

$$\begin{aligned}\|A\|_{\mathcal{L}(X,Y)} &:= \sup\{\|A(x)\|_Y \,:\, x \in X,\ \|x\|_X \leq 1\}\\ &= \sup\{\|A(x)\|_Y \,:\, x \in X,\ \|x\|_X = 1\}\\ &= \sup\left\{\frac{\|A(x)\|_Y}{\|x\|_X} \,:\, x \in X,\ x \neq 0\right\}\\ &= \inf\{c > 0 \,:\, \|A(x)\|_Y \leq c\|x\|_X,\ x \in X\}.\end{aligned}$$

In fact, $\mathcal{L}(X,Y)$ *is a normed space with norm* $\|\cdot\|_{\mathcal{L}(X,Y)}$*, and if* Y *is a Banach space,* $\mathcal{L}(X,Y)$ *is a Banach space also.*

An important special class of continuous linear functionals are isomorphisms and isometries.

Definition 1.1.3 (Isomorphisms and isometries). Let X, Y be normed spaces.

(i) A linear operator $A : X \to Y$, which is 1–1 and onto (bijection), and such that A and A^{-1} are bounded is called an isomorphism. If an isomorphism A exists between the normed spaces X and Y, the spaces are called isomorphic, and this is denoted by $X \simeq Y$.

(ii) A linear operator $A : X \to Y$, which is 1–1 and onto (bijection), and such that $\|A(x)\|_Y = \|x\|_X$, for every $x \in X$, is called an isometry. If an isometry A exists between the normed spaces X and Y, the spaces are called isometric. Clearly, an isometry is an isomorphism.

An isomorphism leaves topological properties between the two spaces invariant, whereas an isometry leaves distances invariant. One easily sees that if X_1, X_2, X_3 are normed spaces and $X_1 \simeq X_2$, and $X_2 \simeq X_3$, then $X_1 \simeq X_3$, with a similar result holding for isometric spaces. This shows that the relation $X \simeq Y$ is an equivalence relation in the class of normed spaces. Furthermore, one can show that a linear 1–1 and onto (bijective) operator between two normed spaces is an isomorphism if and only if there exists $c_1, c_2 > 0$ such that $c_1\|x\|_X \leq \|A(x)\|_Y \leq c_2\|x\|_X$ for every $x \in X$, while an isometry maps the unit ball of X to the unit ball of Y, thus preserving the geometry of the spaces. As we will see quite often, when two normed spaces are isometric, we will "identify" them in terms of this isometry.

One of the fundamental results concerning linear spaces and linear functionals is the celebrated Hahn–Banach theorem. We will return to this theorem very often, engaging a number of its multiple versions, especially its versions related to separation of convex subsets of normed spaces. Here we present its analytic form and will return to its equivalent geometric forms (in terms of separation theorems) in Section 1.2, where we study convexity and its properties in detail.

In its analytic form, the Hahn–Banach theorem deals with the extension of a linear functional defined on a subspace X_0 of a (real) linear space X to the whole space.

Theorem 1.1.4 (Hahn–Banach I). *Let X_0 be a subspace of a (real) linear space X, and let $p : X \to \mathbb{R}$ be a positively homogeneous subadditive functional on X, i. e., a functional with the properties* (a) $p(\lambda x) = \lambda p(x)$ *for every* $\lambda > 0$ *and* $x \in X$, *and* (b) $p(x_1 + x_2) \le p(x_1) + p(x_2)$ *for every* $x_1, x_2 \in X$. *If* $f_0 : X_0 \to \mathbb{R}$ *is a linear functional such that* $f_0(x) \le p(x)$ *for every* $x \in X_0$, *there exists a linear functional* $f : X \to \mathbb{R}$ *such that* $f(x) = f_0(x)$ *on* X_0 *and* $f(x) \le p(x)$ *for every* $x \in X$.

As a corollary of the above, we may obtain the following version of the Hahn–Banach theorem for normed spaces (simply apply the above for the choice $p(x) = \|f_0\|_{\mathcal{L}(X_0^\star,\mathbb{R})}\|x\|_X$).

Theorem 1.1.5 (Hahn–Banach II). *If X is a (real) normed linear space, X_0 is a subspace, and $f_0 : X_0 \to \mathbb{R}$ is a continuous linear functional on X_0, then there exists a continuous linear extension $f : X \to \mathbb{R}$, of f_0 such that* $\|f_0\|_{\mathcal{L}(X_0^\star,\mathbb{R})} = \|f\|_{\mathcal{L}(X^\star,\mathbb{R})}$.

A second fundamental theorem concerning bounded linear operators in Banach spaces is the Banach–Steinhaus theorem or principle of uniform boundedness, which provides a connection between the following two different types of boundedness and convergence in the normed space $\mathcal{L}(X, Y)$.

Definition 1.1.6 (Norm and pointwise bounded families). Let X, Y be normed spaces.
(i) A family of operators $\{A_\alpha : \alpha \in \mathcal{I}\} \subset \mathcal{L}(X, Y)$ is called norm bounded if $\|A_\alpha\|_{\mathcal{L}(X,Y)} < \infty$ for every $\alpha \in \mathcal{I}$, and pointwise bounded if $\|A_\alpha(x)\|_Y < \infty$ for every $\alpha \in \mathcal{I}$ and $x \in X$.
(ii) A sequence $(A_n)_{n\in\mathbb{N}}$ is called norm convergent to $A \in \mathcal{L}(X, Y)$ if $\|A_n - A\|_{\mathcal{L}(X,Y)} \to 0$ as $n \to \infty$, and pointwise convergent to $A \in \mathcal{L}(X, Y)$ if $A_n(x) \to A(x)$ for every $x \in X$.

The Banach–Steinhaus theorem states the following:

Theorem 1.1.7 (Banach–Steinhaus). *Let X, Y be Banach spaces.*[1] *Consider a family of continuous linear mappings $\mathcal{A} := \{A_\alpha : X \to Y : \alpha \in \mathcal{I}\} \subset \mathcal{L}(X, Y)$, (not necessarilly countable). $\mathcal{A}$ is norm bounded if and only if it is pointwise bounded. The principle of course remains valid for linear functionals, upon setting $Y = \mathbb{R}$.*

In other words, if we have a family of linear operators $\{A_\alpha : X \to Y : \alpha \in \mathcal{I}\} \subset \mathcal{L}(X, Y)$ such that $\sup_{\alpha\in\mathcal{I}} \|A_\alpha x\|_Y < \infty$ for every $x \in X$, then $\sup_{\alpha\in\mathcal{I}} \|A_\alpha\|_{\mathcal{L}(X,Y)} < \infty$; i. e.,

1 Y may in fact simply be a normed space, see, e. g., the principle of uniform boundedness Theorem 14.1 [56].

there exists a constant $c > 0$ such that $\|A_\alpha x\|_Y \le c\|x\|_X$ for every $x \in X$ and $\alpha \in \mathcal{I}$, with the constant c being independent of both x and α. The Banach–Steinhaus theorem has important implications for operator theory. The following example shows us how to use it to define linear bounded operators as the pointwise limit of a sequence of bounded operators, a task often fundamental in numerical analysis and approximation theory:

Example 1.1.8 (Definition of operators as pointwise limits). Consider a sequence of linear bounded operators $(A_n)_{n\in\mathbb{N}} \subset \mathcal{L}(X, Y)$ with the following property: For every $x \in X$, there exists $y_x \in Y$ such that $A_n(x) \to y_x$ in Y. Then, these pointwise limits define an operator $A : X \to Y$ by $A(x) := y_x = \lim_n A_n(x)$ for every $x \in X$, which is linear and bounded, and

$$\|A\|_{\mathcal{L}(X,Y)} \le \liminf_{n\to\infty} \|A_n\|_{\mathcal{L}(X,Y)}. \tag{1.1}$$

By the Banach–Steinhaus theorem, there exists a $c > 0$ such that $\|A_n x\| \le c\,\|x\|$ for all $x \in X$, $n \in \mathbb{N}$. We pass to the limit as $n \to \infty$ and since $A_n(x) \to A(x)$ in Y for every $x \in X$ we have that $\|A_n(x)\|_Y \to \|A(x)\|_Y$ as $n \to \infty$, therefore, $\|A(x)\|_Y \le c\,\|x\|_X$ for every $x \in X$ from which follows that $A \in \mathcal{L}(X, Y)$. Furthermore, taking the limit inferior on the inequaltity $\|A_n(x)\|_Y \le \|A_n\|_{\mathcal{L}(X,Y)}\|x\|_X$ for every x, we have that

$$\|A(x)\|_Y = \lim_n \|A_n(x)\|_Y = \liminf_n \|A_n(x)\|_Y \le \liminf_n \|A_n\|_{\mathcal{L}(X,Y)}\,\|x\|_X, \quad \forall x \in X,$$

where upon taking the supremum over all $\|x\| \le 1$, we obtain (1.1). ◁

The other fundamental result concerning bounded linear operators between Banach spaces is the open mapping theorem, which is again a consequence of the Baire category theorem (for a proof see, e. g., [33]).

Theorem 1.1.9 (Open mapping). *Let X, Y be Banach spaces and* $A : X \to Y$ *a linear bounded operator, which is surjective (onto). Then,* A *is also open, i. e., the image of every open set in X under* A *is also open in Y.*

Example 1.1.10. Let $A : X \to Y$ be a linear bounded operator, which is onto and one to one (bijective). Then A^{-1} is a bounded operator.

By the open mapping theorem if $\mathcal{O} \subset X$ is open then $A(\mathcal{O}) \subset Y$ is open. Since $A(\mathcal{O}) = (A^{-1})^{-1}(\mathcal{O})$, it follows that A^{-1} maps open sets to open sets, hence, a) it is continuous, and being linear b) it is bounded. ◁

An important consequence of the open mapping theorem is the closed graph theorem.

Theorem 1.1.11 (Closed graph). *Let X, Y be Banach spaces and* $A : X \to Y$ *be a linear operator (not necessarily bounded).* A *is continuous (hence, bounded) if and only if* **Gr**(A) *is closed.*

Using the open mapping theorem, we can show that if **Gr**(A) is closed in $X \times Y$, the operator A is continuous, a fact which is not true for nonlinear operators.

1.1.2 The dual $X^\star$

An important concept is that of the dual space, which can be defined in terms of continuous linear mappings from X to $\mathbb{R}$ (functionals).

Definition 1.1.12 (Dual space). Let X be a normed space. The space $\mathcal{L}(X,\mathbb{R})$ of all continuous linear functionals from X to $\mathbb{R}$ is called the (topological) dual space of X, and is denoted by $X^\star$, its elements by $x^\star$ and their action on elements $x \in X$, defines a duality pairing between X and $X^\star$ by $x^\star(x) =: \langle x^\star, x\rangle_{X^\star,X}$, often denoted for simplicity as $\langle\cdot,\cdot\rangle$.

Proposition 1.1.13 (The dual as a normed space). *Let X be a normed space. The dual space $X^\star$ is a Banach space*[2] *when equipped with the norm*[3] $\|x^\star\|_{X^\star} = \sup\{|\langle x^\star, x\rangle| : \|x\|_X \le 1\}$, *where $\langle\cdot,\cdot\rangle$ is the duality pairing between the two spaces, $\langle x^\star, x\rangle = x^\star(x)$, for every $x \in X$, $x^\star \in X$.*

Often it is convenient to understand a dual space of a normed linear space in terms of its isometry to a different normed linear space. One of the most important results of this type holds in the special case where X is a Hilbert space. In this case, one has the celebrated Riesz representation theorem [33, 53].

Theorem 1.1.14 (Riesz). *Let $X = H$ be a Hilbert space with inner product $\langle\cdot,\cdot\rangle_X$. For every $x^\star \in X^*$, there exists a unique $x \in X$ (called the Riesz representative) such that $x^\star(z) = \langle x^\star, z\rangle = \langle x, z\rangle_X$ for every $z \in X$. Moreover, $\|x^\star\|_{X^\star} = \|x\|_X$.*

The mapping $i_X : X = H \to X^*$, defined for any $x \in X$ by $i_X(x) = T_x$, where $T_x \in X^*$ is the continuous linear map $z \mapsto \langle z, x\rangle_X$, a bijective isometric operator, which allows for every $x^* \in X^*$ to be written as T_x for a unique $x \in X = H$. This mapping $i_X : X \to X^*$ is the Riesz isometry, and $x = i_X^{-1}(x^\star)$ is the element in X, whose existence is guaranteed by Theorem 1.1.14 (see [53]).

Remark 1.1.15. The Riesz isometry allows us, if we wish so, to identify elements $x^\star \in X^*$ by their Riesz representatives $x = i_X^{-1}(x^\star) \in X$ (treating i_X as the identity), thus effectively identifying X with X^* (writing $X \simeq X^*$).

However, there are many important cases where we choose not to do so ([33, 53]). An interesting example is that of an evolution or Gel'fand triple. A Gel'fand triple consists of three Hilbert spaces $X_0 \hookrightarrow X \simeq X^* \hookrightarrow X_0^*$, with X_0 continuously embedded and dense in X, and normed with a different norm than the larger space X. Here we identify X with its dual (i. e., assume that the Riesz isometry $i_X : X \to X^*$ is the identity map, but we do not do the same for X_0 and X_0^*, i. e., we do not consider $i_{X_0} : X_0 \to X_0^*$ to be the identity. Such constructions play an important role in the study of PDEs and variational inequalities and will be considered in detail later on in this book. As a first example,

---

2 Even if X is not!

3 Or equivalently $\|x^\star\|_{X^\star} = \sup_{x\in X\setminus\{0\}} \frac{|\langle x^\star, x\rangle|}{\|x\|_X} = \sup_{x\in X,\ \|x\|_X=1} |\langle x^\star, x\rangle|$.

consider the sequence spaces $X_0 = \{x = (x_n)_{n\in\mathbb{N}} : \|x\|_{X_0}^2 := \sum_{n\in\mathbb{N}} w_n^2|x_n|^2 < \infty\}$ and $X_0^* = \{x^* := (x_n^*)_{n\in\mathbb{N}} : \|x^*\|_{X^*}^2 = \sum_{n\in\mathbb{N}} w_n^{-2}|x_n^*|^2 < \infty\}$ for some sequence $(w_n)_{n\in\mathbb{N}}$, $w_n > 0$ for all $n \in \mathbb{N}$, and $X = X^* = \ell^2$. In the above, $\langle x^*, x\rangle_{X^*,X} = \sum_{n\in\mathbb{N}} x_n^* x_n$.

For more general Banach spaces one may still identify the dual space with a concrete vector space.

Example 1.1.16 (The dual space of sequence spaces). Let $X = \ell^p = \{x = (x_n)_{n\in\mathbb{N}} : x_n \in \mathbb{R},\ \sum_{n\in\mathbb{N}} |x_n|^p < \infty\}$, $p \in (1,\infty)$, equipped with the norm $\|x\|_{\ell^p} = (\sum_{n\in\mathbb{N}} |x_n|^p)^{1/p}$. Its dual space is identified with the sequence space ℓ^{p^*} with $\frac{1}{p} + \frac{1}{p^*} = 1$. ◁

Similar constructions can be done for other spaces, see, e. g., Sections 1.4 and 1.5.

Since the dual space $X^\star$ is the space of continuous linear functionals on X, the Hahn–Banach theorem plays an important role in the study of the dual space. One of the many instances where this shows up is in the construction of the duality map from X to $X^\star$, the existence of which follows by an application of the Hahn–Banach theorem.

Definition 1.1.17 (The duality map). The duality map $J : X \to 2^{X^\star}$ is the set valued nonlinear map $x \mapsto J(x)$, where

$$J(x) := \{x^\star \in X^\star : \langle x^\star, x\rangle = \|x\|_X^2 = \|x^\star\|_{X^\star}^2\}.$$

As we shall see, the duality map satisfies the important property

$$\|z\| - \|x\| \geq \langle x^\star, z - x\rangle, \quad \forall\, x^\star \in J(x).$$

Example 1.1.18. For every $x \in X$ it holds that

$$\begin{aligned}\|x\|_X &= \sup\{|\langle x^\star, x\rangle| : x^\star \in X^\star,\ \|x^\star\|_{X^\star} \leq 1\}\\ &= \max\{|\langle x^\star, x\rangle| : x^\star \in X^\star,\ \|x^\star\|_{X^\star} \leq 1\}.\end{aligned}$$

The fact that the supremum is attained (for $x^\star = \frac{z^\star}{\|z^\star\|_{X^\star}}$, where $z^\star \in J(x)$) is very important, since in general the supremum in the definition of the norm $\|\cdot\|_{X^\star}$ may not be attained (unless certain conditions are satisfied for X, e. g., if X is a reflexive Banach space; see [33]). ◁

At this point we recall the notion of the adjoint operator.

Let X, Y be Banach spaces, with duals $X^\star$, $Y^\star$ respectively. We will first define the concept of the adjoint or dual operator.

Definition 1.1.19 (Adjoint of an operator). Let $A : X \to Y$ be a linear operator between two Banach spaces. The adjoint operator $A^\star : Y^\star \to X^\star$ is defined by

$$\langle A^\star(y^\star), x\rangle_{X^\star,X} = \langle y^\star, A(x)\rangle_{Y^\star,Y},$$

for every $x \in X$, $y^\star \in Y^\star$.

The adjoint or dual operator generalizes in infinite dimensional Banach spaces the concept of the transpose matrix. If X, Y are Hilbert spaces, the term adjoint operator is used, and upon the identifications $X^\star \simeq X$ and $Y^\star \simeq Y$, the definition simplifies to $\langle \mathsf{A}^\star \mathsf{y}, \mathsf{x}\rangle_X = \langle \mathsf{y}, \mathsf{A}\mathsf{x}\rangle_Y$ for every $\mathsf{x} \in X$, $\mathsf{y} \in Y$. An operator $\mathsf{A} : X \to X$, where $X = H$ is Hilbert space, such that $\mathsf{A}^\star = \mathsf{A}$ is called a self adjoint operator. For such operators, one may show that the operator norm admits the representation[4] $\|\mathsf{A}\| = \sup_{\|\mathsf{x}\|=1} |\langle \mathsf{A}\mathsf{x}, \mathsf{x}\rangle|$.

1.1.3 The bidual $X^{\star\star}$

Definition 1.1.20 (Bidual of a normed space). The dual space of $X^\star$ is called the bidual and is denoted by $X^{\star\star} := (X^\star)^\star$.

Let us reflect a little bit on the nature of elements of the bidual space $X^{\star\star}$. They are functionals from $X^\star \to \mathbb{R}$, i. e., continuous linear mappings $\mathsf{x}^{\star\star} : X^\star \to \mathbb{R}$ that take a functional $\mathsf{x}^\star : X \to \mathbb{R}$ and map it into a real number $\mathsf{x}^{\star\star}(\mathsf{x}^\star)$. This allows us to define a duality pairing between $X^\star$ and $X^{\star\star}$ by $\langle \mathsf{x}^{\star\star}, \mathsf{x}^\star\rangle_{X^{\star\star},X^\star} = \mathsf{x}^{\star\star}(\mathsf{x}^\star)$ for every $\mathsf{x}^{\star\star} \in X^{\star\star}$ and every $\mathsf{x}^\star \in X^\star$. Clearly, this is a different duality pairing than the one defined between X and $X^\star$. If we were too strict on notation, we should use $\langle\cdot,\cdot\rangle_{X^\star,X}$ for the latter and $\langle\cdot,\cdot\rangle_{X^{\star\star},X^\star}$ for the former. But doing as mentioned would make notation unnecessarily heavy; we will often just use $\langle\cdot,\cdot\rangle$ for both when there is no risk of confusion. For instance as $\mathsf{x}, \mathsf{x}^\star, \mathsf{x}^{\star\star}$ will be used for denoting elements of $X, X^\star, X^\star$ respectively, it will be clear that $\langle \mathsf{x}^\star, \mathsf{x}\rangle$ corresponds to the duality pairing between X and $X^\star$, whereas $\langle \mathsf{x}^{\star\star}, \mathsf{x}^\star\rangle$ corresponds to the duality pairing between $X^\star$ and $X^{\star\star}$

Proposition 1.1.21 (The bidual as a normed space). *Let X be a normed space. The bidual space $X^{\star\star}$ of X, is a Banach space endowed with the norm*

$$\|\mathsf{x}^{\star\star}\|_{X^{\star\star}} = \sup\{\langle \mathsf{x}^{\star\star}, \mathsf{x}^\star\rangle_{X^{\star\star},X^\star} \,:\, \|\mathsf{x}^\star\|_{X^\star} \le 1\}.$$

In the following example we construct an important linear mapping from X to $X^{\star\star}$:

Example 1.1.22. Consider any $\mathsf{x} \in X$. For any $\mathsf{x}^\star \in X^\star$ define the mapping $\mathsf{x}^\star \mapsto \langle \mathsf{x}^\star, \mathsf{x}\rangle_{X^\star,X}$, which is a mapping from $X^\star$ to $\mathbb{R}$. Call this mapping $f_\mathsf{x} : X^\star \to \mathbb{R}$. Since $|f_\mathsf{x}(\mathsf{x}^\star)| = |\langle \mathsf{x}^\star, \mathsf{x}\rangle| \le \|\mathsf{x}^\star\|_{X^\star}\|\mathsf{x}\|_X$ it is easy to see that f_x is a bounded mapping, and $\|f_\mathsf{x}\|_{\mathcal{L}(X^\star,\mathbb{R})} \le \|\mathsf{x}\|_X$. In fact, $\|f_\mathsf{x}\|_{\mathcal{L}(X^\star,\mathbb{R})} = \|\mathsf{x}\|_X$. ◁

The above example introduces the following important concept: For any $\mathsf{x} \in X$ the mapping $\mathsf{x}^\star \mapsto \langle \mathsf{x}^\star, \mathsf{x}\rangle_{X^\star,X}$ can be considered as a linear mapping $f_\mathsf{x} : X^\star \to \mathbb{R}$, which

4 If $m = \sup_{\|\mathsf{x}\|=1} |\langle \mathsf{A}\mathsf{x}, \mathsf{x}\rangle|$ it is straightforward to show that $m \le \|\mathsf{A}\|$. For the reverse inequality express $\langle \mathsf{A}\mathsf{x}, \mathsf{z}\rangle = \frac{1}{4}(\langle \mathsf{A}(\mathsf{x}+\mathsf{z}), \mathsf{x}+\mathsf{z}\rangle - \langle \mathsf{A}(\mathsf{x}-\mathsf{z}), \mathsf{x}-\mathsf{z}\rangle) \le \frac{1}{4}(|\langle \mathsf{A}(\mathsf{x}+\mathsf{z}), \mathsf{x}+\mathsf{z}\rangle| + |\langle \mathsf{A}(\mathsf{x}-\mathsf{z}), \mathsf{x}-\mathsf{z}\rangle|) \le \frac{m}{4}(\|\mathsf{x}+\mathsf{z}\|^2 + \|\mathsf{x}-\mathsf{z}\|^2) = \frac{m}{2}(\|\mathsf{x}\|^2 + \|\mathsf{z}\|^2)$ and choose $\mathsf{z} = \frac{\|\mathsf{x}\|}{\|\mathsf{A}\mathsf{x}\|}\mathsf{A}\mathsf{x}$.

by the fundamental estimate $|\langle x^\star, x\rangle_{X^\star,X}| \le \|x^\star\|_{X^\star}\|x\|_X$ is continuous. Therefore, it is a linear functional on $X^\star$ hence, an element of $(X^\star)^\star = X^{\star\star}$. Note that by definition $f_x(x^\star) = \langle x^\star, x\rangle$, and we may interpret $f_x(x^\star)$ as the duality pairing between $X^\star$ and $X^{\star\star}$ so that $f_x(x^\star) = \langle f_x, x^\star\rangle_{X^{\star\star},X^\star}$. Hence, f_x may be interpreted as this element of $X^{\star\star}$, for which it holds that

$$\langle f_x, x^\star\rangle_{X^{\star\star},X^\star} = \langle x^\star, x\rangle_{X^\star,X}, \quad \forall x^\star \in X^\star. \tag{1.2}$$

For this mapping it holds that $\|f_x\|_{X^{\star\star}} = \|x\|_X$. Now define the map $j : X \to X^{\star\star}$ by $j(x) = f_x$ for any $x \in X$. Note that by (1.2) for the map $j : X \to X^{\star\star}$ it holds that $\langle j(x), x^\star\rangle_{X^{\star\star},X^\star} = \langle x^\star, x\rangle_{X^\star,X}$ for every $x^\star \in X^\star$. Since $\|f_x\|_{X^{\star\star}} = \|j(x)\|_{X^{\star\star}} = \|x\|_X$, the mapping j is one to one and conserves the norm. If we define $Y := \mathbf{R}(j) = \{x^{\star\star} \in X^{\star\star} : \exists x \in X,\ j(x) = x^{\star\star}\}$, then obviously $j : X \to Y$ is onto, and since it is an isometry, we may identify X with $Y = \mathbf{R}(j)$. Note that in general $j : X \to X^{\star\star}$ is not onto. In terms of the mapping $j : X \to X^{\star\star}$, the vector space X can be considered as a subspace of its bidual space $X^{\star\star}$,, and for this reason the mapping is called the canonical embedding of X into $X^{\star\star}$.

The above discussion leads to the following fundamental theorem:

Theorem 1.1.23 (Canonical embedding of X into $X^{\star\star}$). *Let X be a normed space. Then there exists a linear operator $j : X \to X^{\star\star}$, with $\langle j(x), x^\star\rangle_{X^{\star\star},X^\star} = \langle x^\star, x\rangle_{X^\star,X}$ for all $x^\star \in X^\star$, such that $\|j(x)\|_{X^{\star\star}} = \|x\|_X$ for all $x \in X$. This operator is called the canonical embedding of X into $X^{\star\star}$ and is not in general surjective (onto), but allows the identification of X with a closed subspace of $X^{\star\star}$, in terms of an isomorphic isometry.*

For certain types of normed spaces, the canonical embedding may be surjective (onto).

Definition 1.1.24 (Reflexive space). If $j : X \to X^{\star\star}$ is surjective, i. e., $j(X) = X^{\star\star}$, then the space X is called reflexive.

For a reflexive space, j is an isometry between X and $X^{\star\star}$, hence X and $X^{\star\star}$ are isomorphic isometric and can be identified in terms of the equivalent relation $X \simeq X^{\star\star}$. By the Riesz isometry all Hilbert spaces are reflexive.

Example 1.1.25 (Biduals of sequence spaces). We may illustrate the above concepts using the example of sequence spaces (see Example 1.1.16). Since in the case $p \in (1,\infty)$, it holds $(\ell^p)^\star \simeq \ell^{p^\star}$ for $\frac{1}{p} + \frac{1}{p^\star} = 1$, following the same arguments as in Example 1.1.16; for $\ell^{p^\star}$, we see that $(\ell^p)^{\star\star} = (\ell^{p^\star})^\star \simeq \ell^{p^{\star\star}}$ with $\frac{1}{p^\star} + \frac{1}{p^{\star\star}} = 1$. One immediately sees that in the case where $p \in (1,\infty)$, we have that $p^{\star\star} = p$ so that $(\ell^p)^{\star\star} \simeq \ell^p$ and the spaces are reflexive. The situation is different in the cases where $p = 1$ and $p = \infty$. ◁

Reflexive Banach spaces satisfy the following fundamental properties:

Proposition 1.1.26 (Properties of reflexive spaces). *Let X be a Banach space.*
(i) *If X is reflexive, every closed subspace is reflexive.*
(ii) *X is reflexive if and only if $X^\star$ is reflexive.*

1.1.4 Different choices of topology on a Banach space: strong and weak topologies

Often the topology induced by the norm on a Banach space X, the strong topology is too strong. Let us clarify what we imply by that. Compactness, openess, closedness of sets as well as continuity and semicontinuity of mappings are topological properties, and whether they hold or not depends on the topology with which a set is endowed. As the choice of topology changes the choice of open and closed sets, it is clear that a set $A \subset X$ may be open (closed) for a particular choice of topology on X and not enjoying these properties for another choice. Similarly, whether or not a set $A \subset X$ is compact depends on whether any open covering of the set admits a finite subcover, and in turn whether a cover is open or not depends on the choice of topology. Hence, a subset $A \subset X$ may be compact for a particular choice of topology, while not being compact for some other choice. Similarly, a mapping $f : X \to Y$ is called continuous if f^{-1} maps open sets of Y to open sets of X and again whether a set is open or not depends on the choice of topology. Hence, a map may be continuous for some choice of topology but not for some other choice. Finally, a mapping $F : X \to \mathbb{R}$ is lower semicontinuous if the set $\{x \in X : F(x) \leq c\}$ is closed for any $c \in \mathbb{R}$, and clearly again this depends upon the choice of topology.

As a rule of thumb one can intuitively figure out that as we weaken a topology (i. e., if we choose a topology whose collection of open sets is a subset of the collection of open sets of the original one), then it is easier to make a given set compact. However, at the same time it makes it more difficult to make a mapping continuous. As an effect of that, if we choose a topology on X which is weaker than the strong topology, we may manage to turn certain important subsets of X, which are not compact when the strong topology is used (such as for instance the closed unit ball $\overline{B}_X(0,1) = \overline{B_X(0,1)} = \{x \in X : \|x\|_X \leq 1\}$), into compact sets. This need arises when we need to guarantee some limiting behavior of sequences in such sets, as is often the case in optimization problems.

Since one tries to juggle between compactness and continuity, one way to define a weaker topology is by choosing the coarser (weaker) topology that makes an important family of mappings continuous. If this family of mappings is a family of seminorms on X then this topology enjoys some nice properties, for example, it is a locally convex topology, which is Hausdorff (i. e., enjoys some nice separation properties). When trying this approach with X, for the choice of the family of mappings defined by the duality pairing, we see that we may obtain a weak topology, the weak topology on X, which provides important compactness properties, and furthermore complies extremely well with the property of convexity. This approach can be further applied on the Banach space $X^\star$, again for a family of mappings related to the duality pairing, and obtain a weak topology on $X^\star$, called the weak* topology, which again enjoys some very convenient properties with respect to compactness.

1.1.5 The strong topology on X

We start by recalling the strong (norm) topology on a Banach space X.

1.1.5.1 Definitions and properties

Definition 1.1.27 (Strong topology). Let X be a Banach space. The strong topology on X is the topology generated by the norm.[5] If X is endowed by the strong topology, then it satisfies the first axiom of countability, i. e., at every point the defining system of neighborhoods has a countable basis.[6]

The strong topology on X has the following important property, which will lead us to the necessity of endowing X with a different (weaker) topology:

Theorem 1.1.28. *The closed unit ball* $\overline{B}_X(0,1) := \{x \in X : \|x\|_X \le 1\}$ *is compact if and only if X is of finite dimension.*

1.1.5.2 Convergence in the strong topology

Definition 1.1.29 (Strong convergence in X). A sequence $(x_n)_{n\in\mathbb{N}} \subset X$ converges strongly to x, denoted by $x_n \to x$, if and only if $\lim_{n\to\infty} \|x_n - x\|_X = 0$.

By the reverse triangle inequality for the norm, according to which $|\,\|x_n\|_X - \|x\|_X\,| \le \|x_n - x\|_X$, we see that the norm is continuous with respect to strong convergence, in the sense that if $x_n \to x$, then $\|x_n\|_X \to \|x\|_X$, as $n \to \infty$.

The next result, connects completeness with the convergence of series in X.

Theorem 1.1.30. *Let $(X, \|\cdot\|_X)$ be a normed linear space. The space $(X, \|\cdot\|_X)$ is complete if and only if it has the following property: For every sequence $(x_n)_{n\in\mathbb{N}} \subset X$ such that $\sum_n \|x_n\|_X < \infty$, we have that $\sum_n x_n$ converges strongly.*[7]

1.1.5.3 The strong topology on $X^\star$

Since $X^\star$ can be considered as a Banach space when endowed with the norm $\|\cdot\|_{X^\star}$, we can consider $X^\star$ as a topological space endowed with the strong topology defined by the norm $\|\cdot\|_{X^\star}$. This topology enjoys the same properties as the strong topology on X,, so in complete analogy with Theorem 1.1.28, we have

---

5 i. e., the topology of the metric generated by the norm.

6 See Definition 1.9.6 in Section 1.9.1.

7 i. e., the sequence $(s_n)_{n\in\mathbb{N}} \subset X$ defined by $s_n = \sum_{i=1}^n x_i$ converges strongly to some limit $x \in X$.

Theorem 1.1.31. *The closed unit ball* $\overline{B}_{X^\star}(0,1) = \{x^\star \in X^\star : \|x^\star\|_{X^\star} \le 1\}$ *is compact if and only if* $X^\star$ *is of finite dimension.*

We may further consider the strong convergence in $X^\star$:

Definition 1.1.32 (Strong convergence in $X^\star$). A sequence $(x_n^\star)_{n\in\mathbb{N}} \subset X^\star$ converges strongly to $x^\star$, denoted by $x_n^\star \to x^\star$, if and only if $\lim_{n\to\infty} \|x_n^\star - x^\star\|_{X^\star} = 0$.

Clearly, the analogue of Theorem 1.1.30 holds for the strong topology on $X^\star$.

1.1.6 The weak* topology on $X^\star$

The strong topology on $X^\star$ is perhaps too strong as Theorem 1.1.31 shows. To regain the compactness of $\overline{B}_{X^\star}(0,1)$ in infinite dimensional spaces, we need to endow $X^\star$ with a weaker topology, called the weak* topology.

1.1.6.1 Definition and properties

Let X be a normed space, $X^\star$ its dual, and $\langle\cdot,\cdot\rangle$ the duality pairing between them.

Definition 1.1.33 (Weak* topology on $X^\star$). The topology generated on $X^\star$ by the family of mappings $\mathscr{M}^\star := \{f_x : X^\star \to \mathbb{R},\ x \in X\}$ defined for each $x \in X$ by $x^\star \mapsto f_x(x^\star) = \langle x^\star, x\rangle$ for every $x^\star \in X^\star$ is called the weak* topology on $X^\star$ and is the weaker topology on $X^\star$ for which all the mappings in the collection $\mathscr{M}^\star$ are continuous. The weak* topology will be denoted by $\sigma(X^\star, X)$, and the space $X^\star$, endowed with the weak* topology, is denoted by $X^\star_{w^\star}$.

For the construction of topologies by a set of mappings, see [33].

An alternative (and equivalent) way of defining the weak* topology on $X^\star$ is by considering the collection of seminorms on $X^\star$ defined by the collection of mappings $\{p_x : X^\star \to \mathbb{R}_+,\ x \in X\}$, where $p_x(x^\star) = |\langle x^\star, x\rangle|$, and then defining the weak topology on X as the topology generated by this family of seminorms, in the sense that it is the weakest topology making these seminorms continuous. Recall that such topologies are called locally convex topologies. The construction of a local basis for this topology is presented in the next example.

Example 1.1.34 (A local basis for the weak* topology). For any $x_0^\star \in X^\star$, fix $n \in \mathbb{N}$ finite, $\epsilon > 0$, and consider the sets of the form

$$\mathcal{O}^\star(x_0^\star; \epsilon, x_1, \dots, x_n) := \bigcap_{i=1}^{n} \{x^\star \in X : |\langle x^\star - x_0^\star, x_i\rangle| < \epsilon\}, \quad x_i \in X,$$

called $(x_1, \dots, x_n)$ semi-ball of radius ϵ centered at $x_0^\star$. The collection of sets

$$\mathfrak{B}^\star_{x_0^\star} := \{\mathcal{O}^\star(x_0^\star; \epsilon, x_1, \dots, x_n) : \forall\, \epsilon > 0,\ \forall\, \{x_1, \dots x_n\} \subset X,\ \text{finite},\ n \in \mathbb{N}\},$$

forms a basis of neighborhoods of $x_0^\star$ for the weak* topology $\sigma(X^\star, X)$, and this helps us characterize weak* open sets in $X^\star$. A subset $\mathcal{O}^\star \subset X^\star$ is weak* open if and only if for every $x_0^\star \in \mathcal{O}^\star$ there exists $\epsilon > 0$ and $x_1, \dots, x_n \in X$ such that the semi-ball $\mathcal{O}^\star(x_0^\star; \epsilon, x_1, \dots, x_n) \subseteq \mathcal{O}^\star$. ◁

The weak* topology on $X^\star$ is a weaker topology than the strong topology on $X^\star$, so the same subsets of $X^\star$ may have different topological properties depending on the chosen topology. However,

Example 1.1.35. A subset $A \subset X^\star$ is bounded if and only if it is weak* bounded.

If A is bounded (in the strong topology on $X^\star$), then for every strong neighborhood N_0, there exists $\lambda > 0$ such that $A \subset \lambda N_0$. To show that it is weak* bounded, we need to show the above for every weak* neighborhood $N_0^{w^\star}$, for a possibly different $\lambda' > 0$. Consider any weak* neighborhood of the origin $N_0^{w^\star}$, which by the fact that the weak* topology on $X^\star$ is weaker than the strong topology on $X^\star$ is also a strong neighborhood, so the result follows by the fact that A is bounded in the strong topology.

For the converse, assume that A is weak* bounded, i. e., for every weak* neighborhood $N_0^{w^\star}$ of the origin, there exists $\lambda' > 0$ such that $A \subset \lambda' N_0^{w^\star}$. Then, $\sup\{|\langle x^\star, x\rangle| : x^\star \in A\} < \infty$ for every $x \in X$, so by the Banach–Steinhaus theorem $\sup\{\|x^\star\|_{X^\star} : x^\star \in A\} < \infty$, therefore A is bounded in the strong topology. ◁

The weak* topology has the following important properties (see e. g. [33]):

Theorem 1.1.36. *The weak* topology, $\sigma(X^\star, X)$, has the following properties:*
(i) *The space $X^\star$ endowed with the weak* topology is a Hausdorff*[8] *locally convex space.*
(ii) *The weak* topology is metrizable if and only if X is finite dimensional.*
(iii) *A normed space X is separable if and only if the weak* topology restricted to the closed unit ball $\overline{B}_{X^\star}(0,1) = \{x^\star \in X^\star : \|x^\star\|_{X^\star} \le 1\}$ is metrizable.*[9]
(iv) *Every nonempty weak* open subset $A \subset X^\star$ is unbounded, in infinite dimensional spaces.*

Note that even though $X^\star$ can be turned into a normed space using $\|\cdot\|_{X^\star}$, when endowed with the weak*, $\sigma(X^\star, X)$ topology, it is no longer in general a metric space on account of Theorem 1.1.36(ii). In this sense, assertion (iii) is very important since metrizable topologies enjoy some convenient special properties, resulting from the fact that they admit a countable basis and are first countable. For example, it allows us to use weak* sequential compactness as equivalent to weak* compactness for (norm) bounded subsets of $X^\star$. However, caution is needed as the metrizability does not hold over the whole

8 i. e., a topological space such that any two distinct points have distinct neighborhoods.

9 Clearly, this result extends to any norm bounded and closed (with respect to the strong topology on $X^\star$) set of the form $A = \{x^\star \in X^\star : \|x^\star\|_{X^\star} \le c\}$ for some $c \in (0, \infty)$. Note that the weak* topology is never metrizable on the whole of $X^\star$.

of $X^\star$, thus requiring in general the use of nets to replace sequences when dealing with the weak* topology in general.

On the other hand, the weak* topology displays some remarkable compactness properties, known as the Alaoglu (or Alaoglu–Banach–Bourbaki) theorem, a fact that makes the use of this topology indispensable.

Theorem 1.1.37 (Alaoglu). *The closed unit ball in $X^\star$, $\overline{B}_{X^\star}(0,1) := \{x^\star \in X^\star \,:\, \|x^\star\|_{X^\star} \le 1\}$ is weak* compact (i. e., compact for the $\sigma(X^\star, X)$ topology).*[10] *If furthermore, X is separable, then every bounded sequence in $X^\star$ has a weak* convergent subsequence.*

Example 1.1.38 (Weak compactness of the duality map). For any $x \in X$, the duality map $J(x)$ (recall Definition 1.1.17) is a bounded and weak* compact set in $X^\star$.

This follows since for any fixed $x^\star \in J(x)$, it holds that $\|x^\star\|_{X^\star} \le \|x\|_X$ so that $J(x)$ is a bounded set. The weak* compactness then follows by Alaoglu's theorem (see Theorem 1.1.37). ◁

Even though, nets provide a complete characterization of the weak* topology, there are still important properties that may be captured using sequences.

1.1.6.2 Convergence in the weak* topology

Definition 1.1.39 (Weak* convergence). A sequence $(x_n^\star)_{n\in\mathbb{N}} \subset X^\star$ converges weak* to $x^\star$, denoted by $x_n^\star \overset{\star}{\rightharpoonup} x^\star$, if it converges with respect to the weak* topology $\sigma(X^\star, X)$, i. e., if and only if $\lim_{n\to\infty}\langle x_n^\star, x\rangle = \langle x^\star, x\rangle$ for every $x \in X$.

One may similarly define weak* convergence for a net and show that a net $\{x_\alpha^\star \,:\, \alpha \in \mathcal{I}\} \subset X^\star$ converges weak* to $x^\star$ if and only if $\langle x_\alpha^\star, x\rangle \to \langle x^\star, x\rangle$ for every $x \in X$.

The weak* convergence has the following useful properties:

Proposition 1.1.40. *Consider the sequence $(x_n^\star)_{n\in\mathbb{N}} \subset X^\star$. Then,*
(i) *If $x_n^\star \to x^\star$ then $x_n^\star \overset{\star}{\rightharpoonup} x^\star$.*
(ii) *If $x_n^\star \overset{\star}{\rightharpoonup} x^\star$, then the sequence $(\|x_n^\star\|_{X^\star})_{n\in\mathbb{N}} \subset \mathbb{R}$ is bounded, and $\|x^\star\|_{X^\star} \le \liminf_n \|x_n^\star\|_{X^\star}$, i.e, the norm $\|\cdot\|_{X^\star}$ is sequentially lower semicontinuous with respect to weak* convergence.*
(iii) *If $x_n^\star \overset{\star}{\rightharpoonup} x^\star$ and $x_n \to x$, then $\langle x_n^\star, x_n\rangle \to \langle x^\star, x\rangle$.*

Property (i) follows easily from the fact that $|\langle x_n^\star - x^\star, x\rangle| \le \|x_n^\star - x^\star\|_{X^\star}\|x\|_X$. Property (iii) follows by a rearrangement of the difference $|\langle x_n^\star, x_n\rangle - \langle x^\star, x\rangle| = |\langle x_n^\star, x_n - x\rangle + \langle x_n^\star - x^\star, x\rangle|$, and the fact that the weak* convergent sequence $(x_n^\star)_{n\in\mathbb{N}} \subset X^\star$ is norm bounded, which in turn comes as a consequence of the Banach–Steinhaus theorem (see Theorem 1.1.7).

---

10 Hence, also the set $c\overline{B}_{X^\star}(0,1) = \{x^\star \in X^\star \,:\, \|x^\star\|_{X^\star} \le c\}$. This result also implies that any bounded subset of $X^\star$ is relatively weak* compact in $X^\star$ so that bounded and weak* closed subsets of $X^\star$ are weak* compact.

Property (ii) can be seen in a number of ways, we provide one of them in the following example:

Example 1.1.41 (Weak* (sequential) lower semicontinuity of the norm $\|\cdot\|_{X^\star}$). The function $x^\star \mapsto \|x^\star\|_{X^\star}$ is weak* lower semicontinuous and weak* sequentially lower semicontinuous.

The weak* lower semicontinuity of the norm $\|\cdot\|_{X^\star}$ follows from the fact that the functions $f_x : X^\star \to \mathbb{R}$, defined by $f_x(x^\star) = \langle x^\star, x\rangle$ for every $x^\star \in X^\star$ are (by definition) continuous with respect to the weak* topology, and since $\|x^\star\|_{X^\star} = \sup\{|\langle x^\star, x\rangle| : \|x\|_X \leq 1\} = \sup\{|f_x(x^\star)| : \|x\|_X \leq 1\}$, we see that the function $\|\cdot\|_{X^\star}$ is the (pointwise) supremum of a family of continuous functions, hence, it is semicontinuous. The latter can be easily seen as follows: if $f(x) = \sup_{\alpha\in\mathcal{I}} f_\alpha(x)$, with f_α continuous (or lower semicontinuous), then $\{x \in X : f(x) \leq c\} = \bigcap_{\alpha\in\mathcal{I}}\{x \in X : f_\alpha(x) \leq c\}$, which is a closed set as the intersection of the closed sets $\{x \in X : f(x) \leq c\}$.

The weak* sequential lower semicontinuity can follow by another application of the Banach–Steinhaus Theorem 1.1.7, working as in Example 1.1.8, for the sequence of operators $(A_n)_{n\in\mathbb{N}} \subset \mathcal{L}(X,\mathbb{R})$ defined by $A_n(x) = \langle x_n^\star, x\rangle$ for any $x \in X$, which converges pointwise to the operator $A \in \mathcal{L}(X,\mathbb{R})$ defined by $A(x) = \langle x^\star, x\rangle$ for every $x \in X$. ◁

Example 1.1.42 (Does every norm bounded sequence in $X^\star$ have a weak* convergent subsequence?). Can we claim that if $(x_n^\star)_{n\in\mathbb{N}} \subset X^\star$ satisfies $\sup_{n\in\mathbb{N}}\|x_n^\star\|_{X^\star} < c$ for some $c > 0$ independent of n, there exists $x_o^\star \in X^\star$ and a subsequence $(x_{n_k}^\star)_{k\in\mathbb{N}}$ such that $x_{n_k}^\star \overset{\star}{\rightharpoonup} x_o^\star$ as $k \to \infty$? The answer is no, unless X is separable.

Note that since in general (see Theorem 1.1.36) the weak* topology is not metrizable, Alaoglu's theorem, which guarantees weak* compactness of (norm) bounded subsets of $X^\star$, does not guarantee weak* sequential compactness, i. e., that any sequence $(x_n^\star)_{n\in\mathbb{N}} \subset X^\star$, such that there exists a constant $c > 0$ (independent of n) with the property $\|x_n^\star\|_{X^\star} \leq c$, for every $n \in \mathbb{N}$, has a convergent subsequence in the sense of convergence in the weak* topology on $X^\star$. This will only be true when X is separable, in which case, the weak* topology restricted on this bounded set will be metrizable. ◁

1.1.7 The weak topology on X

The strong topology on X is perhaps too strong as Theorem 1.1.28 shows. To regain the compactness of the closed unit ball $\overline{B}_X(0,1)$ in infinite dimensional spaces, we need to endow X with a weaker topology, called the weak topology.

1.1.7.1 Definitions and properties

Let X be a normed space, $X^\star$ its dual and $\langle\cdot,\cdot\rangle$ the duality pairing among them.

Definition 1.1.43 (Weak topology on X). The topology generated on X by the family of mappings $\mathscr{M} := \{f_{x^\star} : X \to \mathbb{R},\ x^\star \in X^\star\}$ defined for each $x^\star \in X$ by $x \mapsto f_{x^\star}(x) = \langle x^\star, x\rangle$

for every $x \in X$, is called the weak topology on X and is the weaker topology on X for which all the mappings in the collection $\mathcal{M}$ are continuous. The weak topology will be denoted by $\sigma(X, X^\star)$, and the space X endowed with the weak topology will be denoted by X_w.

For the construction of topologies by a set of mappings, see [33].

An alternative (and equivalent) way to define the weak topology on X is by considering the collection of seminorms on X defined by the collection of mappings $\{p_{x^\star} : X \to \mathbb{R}_+ \,,\ x^\star \in X^\star\}$ where $p_{x^\star}(x) := |\langle x^\star, x\rangle|$, and then defining the weak topology on X as the topology generated by this family of seminorms, in the sense that it is the weakest topology making these seminorms continuous. Recall that such topologies are called locally convex topologies. The construction of a local basis for this topology is presented in the next example.

Example 1.1.44 (A local basis for the weak topology). For any $x_0 \in X$, fix n as a finite integer and $\epsilon > 0$, and consider the sets of the form

$$\mathcal{O}(x_0; \epsilon, x_1^\star, \dots, x_n^\star) := \bigcap_{i=1}^{n} \{x \in X \,:\, |\langle x_i^\star, x - x_0\rangle_{X^\star, X}| < \epsilon,\ x_i^\star \in X^\star\},$$

called $(x_1^\star, \dots, x_n^\star)$ semi-ball of radius ϵ centered at x_0. The collection of semiballs

$$\mathfrak{B}_{x_0} := \{\mathcal{O}(x_0; \epsilon, x_1^\star, \dots, x_n^\star) \,:\, \forall\, \epsilon > 0,\ \forall\, \{x_1^\star, \dots x_n^\star\} \subset X^\star,\ \text{finite},\ n \in \mathbb{N}\},$$

forms a basis of neighborhoods of x_0 for the weak topology $\sigma(X, X^\star)$, and this helps us characterize the open sets for the weak topology on X (or weakly open sets). A subset $\mathcal{O} \subset X$ is weakly open if and only if for every $x_0 \in \mathcal{O}$ there exists $\epsilon > 0$ and $x_1^\star, \dots, x_n^\star \in X^\star$ such that the semiball $\mathcal{O}(x_0; \epsilon, x_1^\star, \dots, x_n^\star) \subset \mathcal{O}$. ◁

Example 1.1.45. As a simple geometrical illustration of how the semiballs may look like, consider the case where $X = \mathbb{R}^2$, $x_0 = (0, 0)$ and $x_1^\star = (1, 1)$, $x_2^\star = (1, -1)$. Then,

$$\mathcal{O}(x_0; \epsilon, x_1^\star) = \{(x_1, x_2) \,:\, |x_1 + x_2| \le \epsilon\}, \quad \text{and}$$
$$\mathcal{O}(x_0; \epsilon, x_1^\star, x_2^\star) = \{(x_1, x_2) \,:\, |x_1 + x_2| \le \epsilon,\ |x_1 - x_2| \le \epsilon\},$$

which look like strips and intersection of strips, respectively. However, this geometrical picture, which is valid in finite dimensional spaces, may be misleading in infinite dimensions (see, e. g., Theorem 1.1.46(iv)). ◁

The following theorem (see e. g. [33]) collects some fundamental results for the weak topology:

Theorem 1.1.46. *The following are true for the weak topology $\sigma(X, X^\star)$:*

(i) *The space X endowed with the weak topology $\sigma(X, X^\star)$ is a Hausdorff locally convex space.*

(ii) *The weak topology is weaker than the norm topology, i. e., every weakly open (closed) set is strongly open (closed). The weak and the strong topology coincide if and only if X is finite dimensional.*
(iii) *The weak topology is metrizable*[11] *if and only if X is finite dimensional.*
(iv) *$X^\star$ is separable if and only if the weak topology $\sigma(X, X^\star)$ on X restricted on the closed unit ball $\overline{B}_X(0,1) = \{x \in X : \|x\|_X \leq 1\}$ is metrizable. Clearly the result extends to any closed norm bounded $A \subset X$.*[12]
(v) *If X is infinite dimensional, then for any $x_o \in X$, every weak neighborhood $N^w(x_o)$ contains an affine space passing through x_o.*

Assertion (iii) indicates that we have to be cautious with the weak topology since it will not enjoy some of the convenient special properties valid for metrizable or first countable topologies. For instance it is not necessarily true that compactness coincides with sequential compactness, and in general the weak closure of a set is expected to be a larger set than the set of limits of all weakly converging sequences in the set. In this respect, Assertion (iv) is very important since it allows us to re-instate these convenient characterizations in terms of sequences for the weak topology in special cases. For example, it allows us to use sequential compactness as equivalent to compactness for bounded subsets of X in the weak topology, or it allows the characterization of weak closures of sets in terms of limits of sequences. However, caution is needed as the metrizability does not hold over the whole of X, thus requiring in general the use of nets to replace sequences when dealing with the weak topology generally. An interesting question is what kind of Banach spaces X satisfy the condition that $X^\star$ is separable? This would always hold true for instance if X is separable and reflexive, as in such a case, $X^\star$ is separable. A further example of nonreflexive X with separable $X^\star$ is the case where $X = c_0$ and $X^\star = \ell^1$ (see, e. g., [56]).

The weak topology on X is weaker than the strong topology on X so concepts like closed and open sets, compact sets, etc., vary with respect the topology chosen on X. The following example shows that as far as the concept of boundedness is concerned the choice of topology does not make a difference:

Example 1.1.47 (Weakly bounded set vs bounded set). A subset $A \subset X$ is weakly bounded if and only if it is bounded.[13]

It is easy to see that if A is bounded, then it is also weakly bounded. Since A is bounded for every (strong) neighborhood N_0, there exists $\lambda > 0$ such that $A \subset \lambda N_0$. To show that A is weakly bounded, we must show that for every weak neighborhood

11 I. e. there exists a metric that induces that weak topology.

12 A possible choice of metric d on X can be $d(x_1, x_2) = \sum_{n=1}^{\infty} 2^{-n} |\langle x_n^\star, x_1 - x_2 \rangle|$, where $(x_n^\star)_{n \in \mathbb{N}} \subset X^\star$.

13 In fact, this result holds in a more general setting than that of Banach spaces and is known as the Mackey theorem (see, e. g., Theorem 6.24 in [5]).

N_0^w, there exists $\lambda' > 0$ such that $A \subset \lambda' N_0^w$. However, any weak neighborhood N_0^w is also a strong neighborhood, so the required property holds for $\lambda' = \lambda$.

For the converse, the argument is slightly more involved. Assume that A is weakly bounded so that for every weak neighborhood N_0^w there exists $\lambda' > 0$ such that $A \subset \lambda' N_0^w$. This implies that the family of linear maps $\mathcal{A} := \{f_{\mathrm{x}} : X^\star \to \mathbb{R} : \mathrm{x} \in A\}$, defined by $f_{\mathrm{x}}(\mathrm{x}^\star) := \langle \mathrm{x}^\star, \mathrm{x}\rangle$ for every $\mathrm{x}^\star \in X^\star$, is pointwise bounded. By the Banach–Steinhaus Theorem 1.1.7, this is also norm bounded, i. e., the elements of $\mathcal{A}$ are bounded when considered as elements of $\mathcal{L}(X^\star, \mathbb{R})$, which coincides with $X^{\star\star}$. But $\|f_{\mathrm{x}}\|_{\mathcal{L}(X^\star,\mathbb{R})} = \|\mathrm{x}\|_X$ (see Example 1.1.22), which means that A is norm bounded, hence, it is strongly bounded. ◁

However, certain topological properties are significantly different if we change topologies on X. Clearly, since the weak topology is weaker than the strong (norm) topology, any weakly open set is open, however, as the following example shows, the converse is not true. Similarly for closed sets (unless these are convex, where in this case the concept of weak and strong closedness coincide, as we will prove in Proposition 1.2.12).

Example 1.1.48. In infinite dimensions, the open unit ball $B_X(0,1) = \{\mathrm{x} \in X : \|\mathrm{x}\|_X < 1\}$ has empty weak interior, hence is not open in the weak topology. In fact (see Theorem 1.1.46(v)), every nonempty weakly open subset is unbounded in infinite dimensions. Note however, that the closed unit ball $\overline{B}_X(0,1) = \{\mathrm{x} \in X : \|\mathrm{x}\|_X \le 1\}$ is also weakly closed (as a result of convexity)!

Suppose not, and consider a point x_o in the weak interior of $B_X(0,1)$. Then, there exist $\epsilon > 0$, $n \in \mathbb{N}$, and $\mathrm{x}_1^\star, \dots, \mathrm{x}_n^\star \in X^\star$ such that $\mathcal{O}(\mathrm{x}_0; \epsilon, \mathrm{x}_1^\star, \dots, \mathrm{x}_n^\star) \subset B_X(0,1)$. We claim that we may obtain a point $\mathrm{z}_0 \in X \setminus \{0\}$, such that $\langle \mathrm{x}_i^\star, \mathrm{z}_0\rangle = 0$ for all $i = 1, \dots, n$. Indeed, if not $\bigcap_{i=1}^n \mathbf{N}(\mathrm{x}_i^\star) = \{0\}$, where by $\mathbf{N}(\mathrm{x}_i^\star)$, we denote the kernel of the functional $\mathrm{x}_i^\star$ considered as a linear mapping $\mathrm{x}_i^\star : X \to \mathbb{R}$, and since for any $\mathrm{x}^\star \in X^\star$, $\{0\} \subset \mathbf{N}(\mathrm{x}^\star)$, we conclude that $\bigcap_{i=1}^n \mathbf{N}(\mathrm{x}_i^\star) \subset \mathbf{N}(\mathrm{x}^\star)$, therefore using a well known result from linear algebra[14] $\mathrm{x}^\star \in \operatorname{span}\{\mathrm{x}_1^\star, \dots, \mathrm{x}_n^\star\}$, which contradicts the assumption that $\dim(X) = \infty$. Then, for every $\lambda \in \mathbb{R}$, we have that $\mathrm{x}_o + \lambda \mathrm{z}_o \in \mathcal{O}(\mathrm{x}_0; \epsilon, \mathrm{x}_1^\star, \dots, \mathrm{x}_n^\star) \subset B_X(0,1)$, which implies that $\|\mathrm{x}_o + \lambda \mathrm{z}_0\|_X \le 1$, which is clearly a contradiction. ◁

Example 1.1.49 (The closure of the unit sphere). Assume that $\dim(X) = \infty$, and let $S_X(0,1) = \{\mathrm{x} \in X : \|\mathrm{x}\|_X = 1\}$ and $\overline{B}_X(0,1) = \{\mathrm{x} \in X : \|\mathrm{x}\|_X \le 1\}$. Then $\overline{S_X(0,1)}^w = \overline{B}_X(0,1)$. In general, not every element of $\overline{B}_X(0,1)$ can be expressed as the limit of a weakly con-

---

14 If on a linear space for the linear maps $f, f_1, \dots, f_n : X \to \mathbb{R}$ it holds that $\bigcap_{i=1}^n \mathbf{N}(f_i) \subset \mathbf{N}(f)$, then f is a linear combination of the $f_1, \dots, f_n$. This can be proved by induction. It is easy to see that the claim is true for $n = 1$. Assuming it holds for n, if $\bigcap_{i=1}^{n+1} \mathbf{N}(f_i) \subset \mathbf{N}(f)$, then it holds that $\bigcap_{i=1}^n \mathbf{N}(\phi_i) \subset \mathbf{N}(\phi)$, where $\phi_i = f_i|_{\mathbf{N}(f_{n+1})}$, $\phi = f|_{\mathbf{N}(f_{n+1})}$, $i = 1, \dots, n$. So applying the result at level n to these functions, we have that $\phi \in \operatorname{span}\{\phi_1, \dots, \phi_n\}$, i. e., for some $\lambda_1, \dots, \lambda_n \in \mathbb{R}$, we have that $\mathbf{N}(f_{n+1}) \subset \mathbf{N}(f - \sum_{i=1}^n \lambda_i f_i)$; applying the result for $n = 1$, we conclude the existence of λ_{n+1} such that $f = \sum_{i=1}^{n+1} \lambda_i f_i$, concluding the proof.

vergent sequence in $S_X(0,1)$, except in special cases. Such is, e. g., the case where $X = \ell^1$ on account of the Schur property that states the equivalence of the weak and the strong convergence.

To check this, if we take for the time being for granted that $\overline{B}_X(0,1)$ is a weakly closed set (for a proof see Proposition 1.2.12) so that $\overline{S_X(0,1)}^w \subset \overline{B}_X(0,1)$, since $S_X(0,1) \subset \overline{B}_X(0,1)$ and $\overline{S_X(0,1)}^w$ is the smallest weakly closed set containing $S_X(0,1)$. To prove the reverse inclusion, consider any $x_o \in \overline{B}_X(0,1)$, and take a weak (open) neighbourhood $N^w(x_o)$. Then, by the construction of the local basis for the weak topology (see Example 1.1.44), there exists $x_1^\star, \dots, x_n^\star \in X^\star$ and $\epsilon > 0$ such that $\mathcal{O}(x_o; \epsilon, x_1^\star, \dots, x_n^\star) := \{x \in X : \langle x_i^\star, x - x_o \rangle, \ i = 1, \dots, n\} \subset N^w(x_o)$. With the same arguments as in Example 1.1.48, there exists nonzero $z_o \in X$ such that $\langle x_i^\star, z_o \rangle = 0$ for all $i = 1, \dots, n$ so that for any $\lambda \in \mathbb{R}$, $x_o + \lambda z_o \in \mathcal{O}(x_o; \epsilon, x_1^\star, \dots, x_n^\star) \subset N^w(x_o)$. Since $\|x_o\|_X \le 1$, there exists $\lambda_o \in \mathbb{R}$ such that $\|x_o + \lambda_o z_o\|_X = 1$, therefore $x_o + \lambda_o z_o \in \mathcal{O}(x_o; \epsilon, x_1^\star, \dots, x_n^\star) \cap S_X(0,1) \subset N^w(x_o) \cap S_X(0,1)$ so that $N^w(x_o) \cap S_X(0,1) \neq \emptyset$. Since for any $x_o \in \overline{B}_X(0,1)$ and any weak neighborhood $N^w(x_o)$ of x_o we have that $N^w(x_o) \cap S_X(0,1) \neq \emptyset$, we see that $x_o \in \overline{S_X(0,1)}^w$, so that since x_o was arbitrary, we conclude that $\overline{B}_X(0,1) \subset \overline{S_X(0,1)}^w$, and the proof is complete. ◁

On the other hand, when it comes to the question of continuity of linear operators, the choice of topology does not make too much difference [33].

Theorem 1.1.50. *A linear operator $A : X \to Y$ is continuous with respect to the weak topologies on X and Y (weakly continuous) if and only if it is continuous with respect to the strong topologies on X and Y (strong or norm continuous), the same holding of course for linear functionals ($Y = \mathbb{R}$).*

Even though nets provide a complete characterization of the weak topology, there are still important properties that may be captured using sequences.

1.1.7.2 Convergence in the weak topology

Definition 1.1.51 (Weak convergence). A sequence $(x_n)_{n \in \mathbb{N}} \subset X$ converges weakly to x, denoted by $x_n \rightharpoonup x$, if it converges with respect to the weak topology $\sigma(X, X^\star)$, i. e., if and only if

$$\lim_{n \to \infty} \langle x^\star, x_n \rangle_{X^\star, X} = \langle x^\star, x \rangle_{X^\star, X}, \quad \forall\, x^\star \in X^\star.$$

One may similarly define weak convergence for a net and show that a net $\{x_\alpha : \alpha \in D\} \subset X$ converges weakly to x if and only if $\langle x^\star, x_\alpha \rangle_{X^\star, X} \to \langle x^\star, x \rangle_{X^\star, X}$ for every $x^\star \in X^\star$.

Remark 1.1.52 (The Urysohn property). Since $X^w := (X, \sigma(X, X^\star))$ is a Hausdorff topological space, we have uniqueness of weak limits. A sequence weakly converges to a point if each subsequence contains a further subsequence, which converges to this point. The same property naturally holds for the strong topology.

Example 1.1.53. Show that the weak convergence enjoys the Urysohn property. Indeed assume by contradiction that every subsequence of $(x_n)_{n\in\mathbb{N}} \subset X$ has a further subsequence that converges weakly to x but $x_n \not\rightharpoonup x$ in X. Then, there exists $\epsilon > 0$, some $x^\star \in X^\star$ and a subsequence $(x_{n_k})_{k\in\mathbb{N}}$ with the property $|\langle x^\star, x_{n_k}\rangle - \langle x^\star, x\rangle| \geq \epsilon$ for every k. However, by assumption, this subsequence has a further subsequence $(x_{n_{k_\ell}})_{\ell\in\mathbb{N}}$ such that $x_{n_{k_\ell}} \rightharpoonup x$, which clearly is a contradiction. ◁

Proposition 1.1.54 (Properties of weak convergence). *Let $(x_n)_{n\in\mathbb{N}}$ be a sequence in X. Then,*

(i) *If $x_n \to x$, then $x_n \rightharpoonup x$.*
(ii) *If $x_n \rightharpoonup x$, then the sequence $(x_n)_{n\in\mathbb{N}}$ is bounded (i. e., $\sup_{n\in\mathbb{N}} \|x_n\|_X < \infty$), and $\|x\|_X \leq \liminf_n \|x_n\|_X$, i. e., the norm is sequentially weakly lower semicontinuous.*
(iii) *If $x_n \rightharpoonup x$ and $x_n^\star \to x^\star$ in $X^\star$, then $\langle x_n^\star, x_n\rangle \to \langle x^\star, x\rangle$.*

The first claim is immediate from the fact that the weak topology is weaker (coarser) than the strong topology. The second claim may be proved in various ways; we present one possible proof in Example 1.1.55 below. The third claim follows easily by a rearrangement of $\langle x_n^\star, x_n\rangle - \langle x^\star, x\rangle = \langle x_n^\star - x^\star, x_n\rangle + \langle x^\star, x_n - x\rangle$, using the norm boundedness of the sequence $(\|x_n\|_X)_{n\in\mathbb{N}} \subset \mathbb{R}$.

Example 1.1.55 (Weak (sequential) lower semicontinuity of the norm). The function $x \mapsto \|x\|_X$ is (sequentially) lower semicontinuous for the weak topology.

The proof of this claim (stated in Proposition 1.1.54(ii)) follows by the observation that the norm can be represented as $\|x\|_X = \sup\{|\langle x^\star, x\rangle| : x^\star \in X^\star, \|x^\star\|_{X^\star} \leq 1\}$ (see Example 1.1.18). Then using the same reasoning as in Example 1.1.41, based on the remark that in the weak topology for any $x^\star \in X^\star$ fixed, the functions $f_{x^\star} : X \to \mathbb{R}$, defined by $f_{x^\star}(x) = \langle x^\star, x\rangle$, are continuous and the supremum of a family of continuous functions is lower semicontinuous (see the argument in Example 1.1.41). Based on the foregoing, we obtain the weak lower semicontinuity of the norm. For the sequential weak lower semicontinuity, we may reason once more as in Example 1.1.41, reversing the roles of x and $x^\star$. ◁

Example 1.1.56 (If $x_n \rightharpoonup x$ and $x_n^\star \overset{\star}{\rightharpoonup} x^\star$, it may be that $\langle x_n^\star, x_n\rangle \not\to \langle x^\star, x\rangle$). Consider the space $X = \ell^p$, $p \in (0,\infty)$, with dual space $X^\star = \ell^{p^\star}$, with $\frac{1}{p} + \frac{1}{p^\star} = 1$, and bidual $X^{\star\star} = (\ell^{p^\star})^\star = \ell^p$. Let $e_n = \{\delta_{in} : i \in \mathbb{N}\}$, which may be considered either as an element of X, $X^\star$ or $X^{\star\star}$. Consider the sequences $(x_n)_{n\in\mathbb{N}} = (e_n)_{n\in\mathbb{N}} \subset X$ and $(x_n^\star)_{n\in\mathbb{N}} = (e_n)_{n\in\mathbb{N}} \subset X^\star$. One may easily confirm that $x_n \rightharpoonup 0$, and $x_n^\star \overset{\star}{\rightharpoonup} 0$. Indeed, considering any $x^\star \in X^\star$, of the form $x = \{x_i : i \in \mathbb{N}, \sum_{i\in\mathbb{N}} |x_i|^{p^\star} < \infty\}$ so that $\langle x^\star, x_n\rangle = x_n \to 0$ as $n \to \infty$, since $\sum_{i\in\mathbb{N}} |x_i|^{p^\star} < \infty$, with a similar treatment for $x_n^\star \rightharpoonup 0$. However, $\langle x_n^\star, x_n\rangle = 1$ for every $n \in \mathbb{N}$ so that $\langle x_n^\star, x_n\rangle \not\to 0$. ◁

Even though the weak topology is not metrizable in general, the celebrated Eberlein–Smulian theorem (see, e. g., [4] or [104]) shows that for the weak topology the concepts of sequential compactness and compactness are equivalent, as for metrizable topologies.

Theorem 1.1.57 (Eberlein–Smulian). *Let X be a Banach space and consider $A \subset X$. The following are equivalent:*
(i) *A is compact in the weak topology $\sigma(X, X^\star)$ (weakly compact).*
(ii) *A is sequentially compact in the weak topology $\sigma(X, X^\star)$ (weakly sequentially compact).*

1.1.7.3 The weak topology for reflexive spaces

For the particular case of reflexive Banach spaces X endowed with the weak topology, one may obtain some important information, which essentially follows by the important compactness properties of the weak* topology on $X^\star$.

Before proceeding, we present in the next example the possible connections of the weak* topology on $X^\star$ with the weak topology on X.

Example 1.1.58 (The weak topology on $X^\star$). Since $X^\star$ is a normed space, with dual $X^{\star\star}$, what would happen if we tried to repeat the construction of Definition 1.1.43 for $X^\star$? Replacing X by $X^\star$ and $X^\star$ by $X^{\star\star}$ in this construction, we create thus the topology $\sigma(X^\star, X^{\star\star})$. This will be another topology on $X^\star$, called the weak topology on $X^\star$. How does this topology compare with the weak* topology $\sigma(X^\star, X)$ on $X^\star$?

In general, we may consider that $X \subset X^{\star\star}$ (in the sense of the canonical embedding of Theorem 1.1.23), which gives rise to the weak* topology $\sigma(X^\star, X)$ being weaker than the weak topology on $X^\star$ constructed here $\sigma(X^\star, X^{\star\star})$. However, if X is reflexive, then the canonical embedding is onto and $X \simeq X^{\star\star}$ so that the topologies $\sigma(X^\star, X)$ and $\sigma(X^\star, X^{\star\star})$ coincide. This means that in reflexive spaces, we may consider the weak topology on $X^\star$ as equivalent to the weak* topology on $X^\star$, thus enjoying the important compactness properties of the weak* topology ensured by Alaoglu's theorem. ◁

After the above preparation, we may state the following result (see, e. g., [33] or [104]), which provides some important information concerning compact sets in this topology, called weakly compact sets:

Theorem 1.1.59. *Let X be a Banach space. The following statements are equivalent:*
(i) *X is reflexive.*
(ii) *The closed unit ball of X, $\overline{B}_X(0,1) = \{x \in X : \|x\|_X \le 1\}$ is weakly compact.*
(ii′) *The closed unit ball of $X^\star$, $\overline{B}_{X^\star}(0,1) = \{x^\star \in X^\star : \|x^\star\|_{X^\star} \le 1\}$ is compact when $X^\star$ is endowed with the weak topology.*
(iii) *For every bounded sequence $(x_n)_{n\in\mathbb{N}} \subset X$ there exists a subsequence $(x_{n_k})_{k\in\mathbb{N}}$, which converges for the topology $\sigma(X, X^\star)$ (a weakly converging subsequence), i. e., there exists $x_o \in X$ such that $x_{n_k} \rightharpoonup x_o$ in X.*

(iv) *For every* $x^\star \in X^\star$, *the supremum in the definition of the dual norm is attained:*

$$\|x^\star\|_{X^\star} = \max\{\langle x^\star, x\rangle \;:\; \|x\|_X \le 1\}.$$

This theorem is very useful, as it allows us to extract a weakly convergent subsequence out of every bounded sequence in a reflexive Banach space, and so it will be used very often throughout this book. It plays in reflexive Banach spaces the role that the Bolzano–Weierstrass theorem plays in $\mathbb{R}$. To see how reflexivity leads to weak compactness of the closed unit ball, we may follow the reasoning of the next example.

Example 1.1.60 (Weak compactness of the closed unit ball in reflexive spaces). For a reflexive space, we know that the canonical embedding $j : X \to X^{\star\star}$ is onto so that $X \simeq X^{\star\star}$, and we can identify $X = (X^\star)^\star$. By Proposition 1.1.26, since X is reflexive so is $X^\star$. Hence, by the reasoning of Example 1.1.58, the weak* topology on $(X^\star)^\star = X$ coincides with the weak topology on $(X^\star)^\star = X$. By Alaoglu's theorem (see Theorem 1.1.37) the (norm) closed unit ball $\overline{B}_X(0,1)$ of $(X^\star)^\star = X$ is weak* compact, but since (by reflexivity) the weak* and the weak topology on $(X^\star)^\star = X$ coincide, $\overline{B}_X(0,1)$ is also weakly compact. But then the Eberlein–Smulian theorem (see Theorem 1.1.57) guarantees weak sequential compactness, from which we can conclude that every sequence in $\overline{B}_X(0,1)$ hence, any (norm) bounded sequence, admits a weakly convergent subsequence. ◁

Example 1.1.61 (The supremum in the definition of the norm is attained for reflexive spaces). We will show that if X is reflexive, then for any $x^\star \in X^\star$ fixed, there exists $x_o \in X$ such that $\|x^\star\|_{X^\star} = \langle x^\star, x_o\rangle$ with $\|x_o\|_X \le 1$.

By the reflexivity of X, it holds that the (norm) closed unit ball $\overline{B}_X(0,1)$ is weakly compact, hence from Eberlein–Smulian theorem (see Theorem 1.1.57) weakly sequentially compact. By definition $\|x^\star\|_{X^\star} = \sup\{\langle x^\star, x\rangle \;:\; x \in X,\ \|x\|_X \le 1\}$, so we construct a sequence $(x_n)_{n\in\mathbb{N}} \subset B$, with the property[15] $\langle x^\star, x_n\rangle \to \|x^\star\|_{X^\star}$. The weak sequential compactness of $\overline{B}_X(0,1)$ implies the existence of a subsequence $(x_{n_k})_{k\in\mathbb{N}}$ of the above sequence and a $x_o \in \overline{B}_X(0,1)$ such that $x_{n_k} \rightharpoonup x_o$ as $k \to \infty$. This implies that $\langle x^\star, x_{n_k}\rangle_{X^\star,X} \to \langle x^\star, x_o\rangle$. On the other hand, since the whole sequence $(x_n)_{n\in\mathbb{N}}$ satisfies $\langle x^\star, x_n\rangle \to \|x^\star\|_{X^\star}$, the same holds for the weakly converging subsequence $(x_{n_k})_{k\in\mathbb{N}}$, and we easily conclude that $\|x^\star\|_{X^\star} = \langle x^\star, x_o\rangle$. The converse statement is known as James theorem; its proof is more subtle (see, e. g., [104]). The reasoning used here will be used repeatedly in this book for providing results concerning the maximization of functionals in reflexive Banach spaces. ◁

15 The construction of this sequence follows by elementary means from the definition of the supremum as the least upper bound; for any $n \in \mathbb{N}$, $\|x^\star\|_X - \frac{1}{n}$ is not an upper bound of the quantity $\langle x^\star, x\rangle$ for $x \in B$, which implies the existence of an $x_n \in B$ such that $\|x^\star\|_X - \frac{1}{n} < \langle x^\star, x_n\rangle \le \|x^\star\|_X$ which has the required properties.

The following result (see, e. g., [62] or [140]) will be used frequently throughout this book.

Proposition 1.1.62. *Let X be a reflexive Banach space, $A \subset X$, bounded, and $x_o \in \overline{A}^w$, where by $\overline{A}^w$ we denote the weak closure of A. Then, there exists a sequence $(x_n)_{n\in\mathbb{N}} \subset A$ which is weakly convergent to x_o in X.*

Proof. Let us fix the point $x_o \in \overline{A}^w$, and let us denote by $\overline{B}_{X^\star}(0,1) = \{x^\star \in X^\star : \|x^\star\|_{X^\star} \leq 1\}$ the closed unit ball of $X^\star$.

First we will show the existence of a countable subset $A_0 \subset A$ such that $x_o \in \overline{A_0}^w$. To this end we claim that for the fixed $x_o \in \overline{A}^w$, and for any choice of integers n, m, and every $(x_1^\star, \dots, x_m^\star) \in (\overline{B}_{X^\star}(0,1))^m = \overline{B}_{X^\star}(0,1) \times \cdots \times \overline{B}_{X^\star}(0,1)$, we can find $x \in A$ such that $|\langle x_k^\star, x - x_o\rangle| < \frac{1}{n}$ for all $k = 1, \dots, m$. Indeed, since $x_o \in \overline{A}^w$, by definition, for any weak neighborhood $N^w(x_o)$ it holds that $N^w(x_o) \cap A \neq \emptyset$, hence, recalling the construction of weak neighborhoods the claim holds. This can be rephrased as that for any $(x_1^\star, \dots, x_m^\star) \in (\overline{B}_{X^\star}(0,1))^m$ it holds that $(x_1^\star, \dots, x_m^\star) \in \bigcup_{x\in A} V_{n,m}(x)$, where

$$\begin{aligned} V_{n,m}(x) &= \left\{(x_1^\star, x_2^\star, \dots, x_m^\star) \in (\overline{B}_{X^\star}(0,1)^\star)^m \ : \ |\langle x_k^\star, x - x_o\rangle| < \frac{1}{n},\ k = 1, \dots, m\right\} \\ &\subset (\overline{B}_{X^\star}(0,1))^m, \end{aligned}$$

which are clearly open sets (for the weak topology). By this rephrasal, we conclude that $\{V_{n,m}(x) : x \in A\}$ constitutes an open cover for $(\overline{B}_{X^\star}(0,1))^\star$, which is weakly compact by Alaoglu's theorem, hence it must admit a finite subcover. Therefore, there exists a finite subset of A, which we denote by $S_{n,m} \subset A$, such that $(\overline{B}_{X^\star}(0,1)^\star)^m \subset \bigcup_{x\in S_{mn}} V_{m,n}(x)$. As this holds for every n, m we may define $A_0 = \bigcup_{n,m\in\mathbb{N}} S_{n,m}$, which is a countable subset of A, and has the sought for property that $x_o \in \overline{A_0}^w$. In order to show that $x_o \in \overline{A_0}^w$, we must show that for every weak neighborhood $N^w(x_o)$ of x_o, it holds that $N^w(x_o) \cap A_0 \neq \emptyset$. Indeed, let $N^w(x_o)$ be any weak neighborhood of x_o. By the construction of weak neighborhoods on X, there exists n, m and $(x_1^\star, \dots, x_m^\star) \in (\overline{B}_{X^\star}(0,1))^m$ such that $\bigcap_{i=1}^m \{x \in X : |\langle x_i^\star, x - x_o\rangle < \frac{1}{n}\} \subset N^w(x_o)$. Since $(x_1^\star, \dots, x_m^\star) \in (\overline{B}_{X^\star}(0,1))^m \subset \bigcup_{x\in S_{n,m}} V_{n,m}(x)$, there exists $x \in S_{n,m}$, for which $(x_1^\star, \dots, x_m^\star) \in V_{n,m}(x)$, i. e., by the definition of $V_{n,m}(x)$, there exists $x \in S_{n,m} \subset A_0 = \bigcup_{n,m} S_{n,m}$, with the property $|\langle x_i^\star, x - x_o\rangle < \frac{1}{n}$ for all $i = 1, \dots, m$, which means that $x \in \bigcap_{i=1}^m \{x \in X : |\langle x_i^\star, x - x_o\rangle| < \frac{1}{n}\} \subset N^w(x_o)$, therefore $x \in \mathcal{N}'_{x_o}$. In summary, we have shown that for every weak neighborhood $N^w(x_o)$ of x_o, there exists a $x \in A_0$ such that $x \in N^w(x_o)$ so that $N^w(x_o) \cap A_0 \neq \emptyset$. Since $N^w(x_o)$ is arbitrary, we have that for every open neighborhood $N^w(x_o)$ of x_o it holds that $N^w(x_o) \cap A_0 \neq \emptyset$, which implies that $x_o \in \overline{A_0}^w$.

Now, let X_0 be the smallest closed linear subspace of X which contains A_0 and x_o. Then, X_0 is a separable and reflexive Banach space. By the Hahn–Banach theorem, each functional $x_0^\star \in X_0^\star$ can be extended to a functional $x^\star \in X^\star$. Therefore, x_o lies in the weak closure of A_0 with respect to the weak topology on X_0. Since A_0 is bounded and X_0 is separable and reflexive, the weak X_0 topology on A_0 is metrizable. Thus, there exists a sequence $(x_n)_{n\in\mathbb{N}} \subset A_0$ such that $x_n \rightharpoonup x_o$ in X_0. Since $X^\star \subset X_0^\star$, we conclude that $x_n \rightharpoonup x$ in X. □

As is natural, certain functions that are continuous with respect to the strong topology may no longer be continuous in the weak topology, as for example the function $x \mapsto \|x\|_X$, but this important function remains sequentially lower semicontinuous with respect to the weak topology. Furthermore, the weak topology has some rather convenient properties with respect to convexity, such as for instance that convex sets, which are closed under the strong topology, are also closed under the weak topology, but those will be treated in detail in Section 1.2.2.

1.2 Convex subsets of Banach spaces and their properties

Convexity plays a fundamental role in analysis. Here we review some of the fundamental concepts that will be very frequently used in this book. For detailed accounts of convexity, we refer to e. g. [5, 20], or [28].

1.2.1 Definitions and elementary properties

We now introduce the concept of a convex subset of X.

Definition 1.2.1. Let X be a Banach space and $C \subset X$. The set C is called convex if it has the property

$$tx_1 + (1-t)x_2 \in C, \quad \forall x_1, x_2 \in C, \forall t \in [0,1].$$

Notation 1.2.2. We will also use the notations $[x_1, x_2] = \{tx_1 + (1-t)x_2 \,:\, t \in [0,1]\}$, $(x_1, x_2] = \{tx_1 + (1-t)x_2 \,:\, t \in (0,1]\}$, and $[x_1, x_2) = \{tx_1 + (1-t)x_2 \,:\, t \in [0,1)\}$ to simplify the exposition. With this notation, C is convex if for any $x_1, x_2 \in C$ it holds that $[x_1, x_2] \subset C$.

Example 1.2.3. Let $\mathcal{D} \subset \mathbb{R}^d$, and take $X = L^p(\mathcal{D})$ for $p \in [1,\infty]$. Then each element $x \in X$ is identified with a function $u : \mathcal{D} \to \mathbb{R}$ such that $\|u\|_{L^p(\mathcal{D})} < \infty$. If ψ is any element of $X = L^p(\mathcal{D})$, the set $C := \{u \in L^p(\mathcal{D}) \,:\, u(x) \geq \psi(x) \text{ a.e.}\}$ is a convex subset of X. ◁

The following topological properties of convex sets (see, e. g., [5]) are useful:

Proposition 1.2.4 (Intersections, interior and closure of convex sets). *In what follows, $C \subset X$ will be a convex set:*

(i) *Let $\{C_\alpha \,:\, \alpha \in \mathcal{I}\}$ be an arbitrary family of convex sets. Then, $\bigcap_{\alpha\in\mathcal{I}} C_\alpha$ is also convex.*
(ii) *If $x \in \operatorname{int}(C)$ and $z \in C$, then $[x, z) \subset \operatorname{int}(C)$.*

(iii) *If* $x_1 \in \operatorname{int}(C)$ *and* $x_2 \in \overline{C}$, *then* $[x_1, x_2) \in \operatorname{int}(C)$.
(iv) *The interior* $\operatorname{int}(C)$ *and closure* $\overline{C}$ *are convex sets.*
(v) *If* $\operatorname{int}(C) \neq \emptyset$, *then* $\operatorname{int}(\overline{C}) = \operatorname{int}(C)$ *and* $\overline{\operatorname{int}(C)} = \overline{C}$.

A useful concept is that of the algebraic interior.

Definition 1.2.5 (Algebraic interior). The algebraic interior or core of a set is defined as

$$\operatorname{core}(C) = \{x \in C : \forall z \in S_X, \ \exists \epsilon \text{ such that } x + \delta z \in C, \ \forall \delta \in [0, \epsilon]\}.$$

For a convex set C, the core can equivalently be defined as

$$\operatorname{core}(C) = \left\{x \in C : X = \bigcup_{\lambda > 0} \lambda(C - x)\right\}.$$

Clearly $\operatorname{int}(C) \subset \operatorname{core}(C)$. However, there are important situations in which $\operatorname{int}(C) = \operatorname{core}(C)$. As we shall see when studying convex functions, it is important to identify the interior of the domain of certain convex sets (for example, the effective domain of a convex function, which has important effects on the continuity of the function in question). On the other hand, it is much easier to check if a point is in the core of a convex set. Therefore, knowing when the two sets coincide may often be useful. However, before doing that we will need another definition.

Definition 1.2.6 (Convex series and convex series closed (CS) sets). An infinite sum of the form $\sum_{n=1}^{\infty} \lambda_n x_n$, where $\lambda_n \in [0, 1]$ with $\sum_{n=1}^{\infty} \lambda_n = 1$ is called a convex series. A set C is called convex series closed (or CS- closed) if for every convex series $\sum_{n=1}^{\infty} \lambda_n x_n = x$, with $x_n \in C$, it holds that $x \in C$.

It is straighforward for the reader to check that since X is a Banach space, any convex series of elements on a bounded subset $A \subset X$ (which is by construction Cauchy) is convergent. CS-closedness of C, guarantees that the limit is in C. Clearly, every CS-closed set is convex. Every closed convex set C is CS-closed, but the converse does not always hold true, as CS-closed sets may not be closed. Therefore, CS-closedness is often a useful generalization of closedness. The following proposition (see [28]) is very useful:

Proposition 1.2.7. *Let* $C \subset X$ *be a convex set. Then,*
(i) *If* $\operatorname{int}(C) \neq \emptyset$ *then* $\operatorname{int}(C) = \operatorname{core}(C)$.
(ii) *If* C *can be expressed as the countable union of closed sets* $C = \bigcup_n A_n$, *then* $\operatorname{int}(C) = \operatorname{core}(C)$.
(iii) *If* C *is CS-closed (see Definition* 1.2.6*), then* $\operatorname{core}(C) = \operatorname{int}(C) = \operatorname{int}(\overline{C})$.

1.2.2 Separation of convex sets and its consequences

Convex sets enjoy separation properties related to the various forms of the Hahn–Banach theorem (see, e. g., [33]). We state a form of this theorem, which will be used very often in this book.

Definition 1.2.8 (Closed hyperplanes and half spaces). Let X be a normed linear space.
(i) The set $H_{x^\star,\alpha} \subset X$ defined by $H_{x^\star,\alpha} := \{x \in X : \langle x^\star, x\rangle = \alpha\}$ for some $x^\star \in X^\star \setminus \{0\}$ and $\alpha \in \mathbb{R}$, is called a closed hyperplane.
(ii) The set $H_{x^\star \le \alpha} := \{x \in X : \langle x^\star, x\rangle \le \alpha\}$ is called a closed half-space.

Using the above notions, we may present the geometric version of the Hahn–Banach theorem, which essentially offers information concerning the containtment of convex sets in half-spaces.

Theorem 1.2.9 (Strict separation). *Suppose C_1 and C_2 are two disjoint, convex subsets of a normed space X, where C_1 is compact and C_2 is closed. Then, there exists a closed hyperplane that strictly separates C_1 and C_2, i. e., there exists a functional $x^\star \in X^\star \setminus \{0\}$ and an $\alpha \in \mathbb{R}$ such that*

$$\langle x^\star, x_1\rangle < \alpha, \quad \forall x_1 \in C_1, \quad \textit{and} \quad \langle x^\star, x_2\rangle > \alpha \quad \forall x_2 \in C_2.$$

A convenient corollary of the separation theorem arises when one of the two convex sets degenerates to a point.

Proposition 1.2.10. *Let $C \subset X$ be a nonempty closed convex subset of a normed linear space. Then, each $x_o \notin C$ can be strictly separated from C by a closed hyperplane, i. e., there exists $(x^\star, \alpha) \in (X^\star \setminus \{0\}) \times \mathbb{R}$ such that*

$$\langle x^\star, x_o\rangle > \alpha, \quad \textit{and} \quad \langle x^\star, x\rangle < \alpha, \quad \forall x \in C.$$

This proposition[16] essentially states that a nonempty closed convex set C is contained in a (suitably chosen) closed half space, whereas any point $x_o \notin C$ in its complement. In fact, this result can be rephrased to convey that a nonempty closed convex set can be expressed in terms of the intersection of all the closed half-spaces $H_{x^\star \le \alpha}$ that contain it, $C = \bigcap_{C \subset H_{x^\star \le \alpha}} H_{x^\star \le \alpha}$, an observation that is important in its own right.

A final reformulation of the separation theorem, which is often useful, is the following version:

16 Whose proof follows directly from Theorem 1.2.9 by considering the compact set $C_1 = \{x_o\}$ and the closed set $C_2 = C$.

Proposition 1.2.11. *Let C_1, C_2 be two nonempty disjoint convex susets of a normed linear space such that C_1 has an interior point. Then, C_1 and C_2 can be separated by a closed linear hyperplane, i. e., there exists $(x^\star, \alpha) \in (X^\star \setminus \{0\}) \times \mathbb{R}$ such that*

$$\langle x^\star, x_1 \rangle \leq \alpha, \quad \forall x_1 \in C_1, \quad \textit{and} \quad \langle x^\star, x_2 \rangle \geq \alpha \quad \forall x_2 \in C_2.$$

As a result of the separation theorem, convex sets have a number of remarkable and very useful properties. For example, for convex sets, the concepts of weak and strong closedness coincide, meaning that if a convex set $C \subset X$ is closed in the strong topology of X, it is also closed in the weak topology of X. Clearly, the same is true for closed linear subspaces.

Proposition 1.2.12. *A closed convex set C is also weakly closed.*

Proof. Assume that C is strongly closed. We will show that it is also weakly closed. This is equivalent to showing that its complement $C^c = X \setminus C$ is open in the weak topology. Consider any $x_o \in C^c$. Since C is closed, and $x_o \notin C$, by an application of the Hahn–Banach theorem (in the form of Proposition 1.2.10), there exists $x_o^\star \in X^\star$ and $\alpha \in \mathbb{R}$ such that $\langle x_o^\star, x_o \rangle < \alpha < \langle x_o^\star, z \rangle$ for any $z \in C$. Define the set $V = \{x \in X \,:\, \langle x_o^\star, x \rangle < \alpha\}$, which is open in the weak topology. Observe that $x_o \in V$, so $N^w(x_o) := V$ is a (weak) neighborhood of x_o. By the strict separation $V \cap C = N^w(x_o) = \emptyset$, which implies $V = N^w(x_o) \subset C^c$. So for every $x_o \in C^c$, we may find a (weak) neighborhood $N^w(x_o)$ such that $N^w(x_o) \subset C^c$, which means that C^c is weakly open.[17] □

Remark 1.2.13. An alternative way to phrase the above statement is that for a convex set C in a Banach space X, the strong and the weak closure of C, denoted by $\overline{C}$ and $\overline{C}^w$, respectively, coincide. Since the weak topology is coarser than the strong, it is clear that $\overline{C} \subseteq \overline{C}^w$. It remains to show the opposite inclusion. Since $\overline{C}$ is (by definition) strongly closed and convex, by Proposition 1.2.12 it is also weakly closed. Therefore, $\overline{C}^w \subseteq \overline{C}$ leading to the stated result.

Remark 1.2.14. A weakly closed set is also weakly sequentially closed.[18] By that observation, the result stated in Proposition 1.2.12 is very useful and will be employed quite often in the following setting: If a convex set $C \subset X$ has the property that for any sequence $(x_n)_{n\in\mathbb{N}} \subset C$ such that $x_n \to x$ in X it holds that $x \in C$, then it also has the property that for any sequence $(\bar{x}_n)_{n\in\mathbb{N}} \subset C$ such that $\bar{x}_n \rightharpoonup \bar{x}$, it holds that $\bar{x} \in C$.

An important side result of Proposition 1.2.12 is the following weak compactness result (the reader may also wish to revisit Example 1.1.48):

17 Since C^c is a weak neighborhood of each of its points, $C^c = \bigcup_{x_o \in C^c} N^w(x_o)$ so C^c is weakly open, as the arbitrary union of weakly open sets.

18 In fact a closed set is sequentially closed for any topology.

Proposition 1.2.15. *Let X be a reflexive Banach space and $C \subset X$ a convex, closed, and bounded set. Then C is weakly compact.*

Proof. By Proposition 1.2.12, since C is convex and closed, it is also weakly closed. Since C is bounded, there exists $\lambda > 0$ such that $C \subset \lambda B_X(0,1)$, where $\overline{B}_X(0,1)$ is the closed unit ball in X. This implies that $C = C \cap \lambda\overline{B}_X(0,1)$. Since X is reflexive by the Eberlein–Šmulian theorem, $B_X(0,1)$ is weakly compact, hence $\lambda\overline{B}_X(0,1)$ enjoys the same property. Therefore, since $C = C \cap \lambda\overline{B}_X(0,1)$, it is expressed as the intersection of the weakly closed set C and the weakly compact set $\lambda\overline{B}_X(0,1)$; it is weakly compact. □

An important corollary of the above proposition is Mazur's lemma. In order to prove it, we need the concept of the convex hull of a set.

Definition 1.2.16 (Convex hull). Let $A \subset X$. The convex hull of A denoted by $\operatorname{conv} A$ is the set of all convex combinations from A:

$$\operatorname{conv} A = \left\{ \mathrm{x} \,:\, \exists \mathrm{x}_i \in A,\ t_i \in [0,1],\ i = 1,\dots,n,\ \sum_{i=1}^{n} t_i = 1, \text{ and } \mathrm{x} = \sum_{i=1}^{n} t_i \mathrm{x}_i \right\}.$$

The convex hull of A, $\operatorname{conv} A$, is the smallest convex set, including A. The smallest closed convex set, including A, is called the closed convex hull of A and is denoted by $\overline{\operatorname{conv}} A$. It can be easily seen that $\overline{\operatorname{conv} A} = \overline{\operatorname{conv}} A$.

Proposition 1.2.17 (Mazur). *Let X be a Banach space, and consider a sequence $(\mathrm{x}_n)_{n\in\mathbb{N}} \subset X$ such that $\mathrm{x}_n \rightharpoonup \mathrm{x}$ in X. Then, there exists a sequence $(\mathrm{z}_n)_{n\in\mathbb{N}} \subset X$ consisting of (finite) convex combinations of terms of the original sequence such that $\mathrm{z}_n \to \mathrm{x}$ (strongly). The sequence $(\mathrm{z}_n)_{n\in\mathbb{N}}$ can be constructed so that for each $n \in \mathbb{N}$, the term z_n is either* (i) *a (finite) convex combination of the terms $\{\mathrm{x}_n, \mathrm{x}_{n+1}, \dots\}$ or* (ii) *a (finite) convex combination of the terms $\{\mathrm{x}_1, \mathrm{x}_2, \dots, \mathrm{x}_n\}$.*

Proof. The proof is based on Proposition 1.2.12. Since for every $n \in \mathbb{N}$ we have that $\mathrm{x}_n \in C := \operatorname{conv}\{\mathrm{x}_1, \mathrm{x}_2, \dots\}$, and because $\mathrm{x}_n \rightharpoonup \mathrm{x}$, we have that $\mathrm{x} \in \overline{C}^w$. But C is convex, so that by Proposition 1.2.12, it holds that $\overline{C}^w = \overline{C}$, hence $\mathrm{x} \in \overline{C}$, so that there exists a sequence $(\mathrm{z}_n)_{n\in\mathbb{N}} \subset C$, strongly converging to x. To show claim (i), fix any $n \in \mathbb{N}$. Clearly, $\mathrm{x}_k \in C_n := \operatorname{conv}\{\mathrm{x}_n, \mathrm{x}_{n+1}, \dots\}$ for any $k \geq n$. Since $\mathrm{x}_k \rightharpoonup x$ as $k \to \infty$, it holds that $\mathrm{x} \in \overline{C_n}^w$. But C_n is a convex set so that by Proposition 1.2.12, $\overline{C_n}^w = \overline{C_n}$ for every n, therefore, $\mathrm{x} \in \overline{C_n}$. The strong closure is characterized as the set of limits of strongly convergent sequences, so there exists a sequence $\{\mathrm{z}_k \,:\, k \in \mathbb{N}\} \subset C_n$ such that $\mathrm{z}_k \to \mathrm{x}$ as $k \to \infty$. Since our result holds for any n, for each k pick $n = n(k) = k$, which leads to the desired result. Claim (ii) follows similarly: □

Proposition 1.2.18 (Mazur). *If X is a Banach space, $A \subset X$ compact, then $\overline{\operatorname{conv} A}$ is compact.*

Proof. We briefly sketch the proof.[19] If $A = \{x_1, \dots, x_n\}$ is finite, then we may consider $\operatorname{conv} A$ as the image of the map $(t_1, \dots, t_n) \mapsto \sum_{i=1}^n t_i x_i$, where $(t_1, \dots, t_n)$ takes values in the compact set $\{(t_1, \dots, t_n) : t_i \in [0,1], \sum_{i=1}^n t_i = 1, i = 1, \dots, n\}$. Since this map is continuous, it follows that $\operatorname{conv} A$ is compact, therefore $\overline{\operatorname{conv} A}$ is also compact.

If A is not finite but compact, for any $\epsilon > 0$, there exists a finite set $A_\epsilon = \{x_1, \dots, x_n\} \subset A$ such that $A \subset \bigcup_{i=1}^n \overline{B}(x_i, \epsilon)$. Note that $\bigcup_{i=1}^n \overline{B}(x_i, \epsilon) \subset \operatorname{conv} A_\epsilon + \overline{B}(0, \epsilon)$, which by the above argument is closed and convex. Therefore, $A \subset \operatorname{conv} A_\epsilon + \overline{B}(0, \epsilon)$, and since $\operatorname{conv} A_\epsilon + \overline{B}(0, \epsilon)$ is a closed convex set containing A and $\overline{\operatorname{conv} A}$ is the smallest closed convex set containing A, it must hold that

$$\overline{\operatorname{conv} A} \subset \operatorname{conv} A_\epsilon + \overline{B}(0, \epsilon). \tag{1.3}$$

Since A_ϵ is a discrete set, we know from the first part of the proof that $\operatorname{conv} A_\epsilon$ is compact, so there exists a finite set $A_1 := \{z_1, \dots, z_m\} \subset \operatorname{conv} A_\epsilon$ such that $\operatorname{conv} A_\epsilon \subset \bigcup_{i=1}^m \overline{B}(z_i, \epsilon)$. Noting that $\bigcup_{i=1}^m \overline{B}(z_i, \epsilon) = A_1 + \overline{B}(0, \epsilon)$ and combining the previous inclusion with (1.3), we see that

$$\overline{\operatorname{conv} A} \subset A_1 + \overline{B}(0, 2\epsilon), \tag{1.4}$$

which together with the observation that $A_1 \subset \operatorname{conv} A_\epsilon \subset \overline{\operatorname{conv} A}$ allows us to conclude that $\overline{\operatorname{conv} A}$ is totally bounded, hence compact. □

Remark 1.2.19. A similar result, known as the Krein–Smulian weak compactness theorem, holds for weak compactness. According to that, if $A \subset X$ be a weakly compact set, then $\overline{\operatorname{conv} A}^w$ is also weakly compact. For a proof see, e. g., [104] p. 254 or [56] p. 164.

A natural question arising at this point is concerning the behavior of convex sets of $X^\star$ in the weak* topology. In general, Proposition 1.2.12 is not true, i. e., a convex $A \subset X^\star$ which is closed in the strong topology of $X^\star$ is not necessarily closed in the weak* topology of $X^\star$, unless of course X is reflexive. One may construct counterexamples for the general case, where X is a nonreflexive Banach space. A general criterion for weak* closedness of convex subsets $A \subset X^\star$, is the Krein–Smulian theorem, according to which A is weak* closed if and only if $A \cap r\overline{B}_{X^\star}(0,1)$ is weak* closed for every $r > 0$ (such sets are often called bounded weak* closed). If X is a separable Banach space, then one may show that a convex set $A \subset X^\star$ is bounded weak* closed if and only if it is weak* sequentially closed. Therefore, for separable X, a convex set $A \subset X^\star$ is weak* closed if and only if it is weak* sequentially closed. This characterization can be quite useful in a number of cases.

19 The reader can find more details in [140], or in a slightly more general framework in [5] or [96].

1.3 Compact operators and completely continuous operators

Compact operators is an important class of operators that plays a fundamental role in both linear and nonlinear analysis. We review here some of the fundamental properties of compact operators (see, e. g., [56]).

Definition 1.3.1 (Compact and completely continuous operators).
(i) Let X and Y be Banach spaces. An operator $\mathsf{A} : X \to Y$ (not necessarily linear) is called compact if it maps bounded sets of X into relatively compact sets of Y (i. e., into sets whose closure is a compact set).
(ii) Let X and Y be Banach spaces. An operator $\mathsf{A} : X \to Y$ is called completely continuous if the image of every weakly convergent sequence in X under A converges in the strong (norm) topology of Y, i. e., if $x_n \rightharpoonup x$ in X implies $\mathsf{A}(x_n) \to \mathsf{A}(x)$ in Y, (i. e., $\|\mathsf{A}(x_n) - \mathsf{A}(x)\|_Y \to 0$).

Example 1.3.2 (Finite rank operators are compact). Consider a linear operator $\mathsf{A} \in \mathcal{L}(X, Y)$ such that $\dim(\mathbf{R}(\mathsf{A})) < \infty$. Such an operator, called a finite rank operator, is compact.

This follows easily from the definition since closed and bounded sets in finite dimensional spaces are compact. ◁

In general, the classes of compact operators and completely continuous operators are not comparable. However, the following general results hold:

Theorem 1.3.3. *Let X, Y, Z be Banach spaces and $\mathsf{A} : X \to Y$, $\mathsf{B} : Y \to Z$ be linear operators.*
(i) *If A is a compact operator, then it is completely continuous.*
(ii) *If X is reflexive, and A is completely continuous, then A is compact.*
(iii) *(Schauder) A bounded operator A is compact if and only if A^* is compact.*
(iv) *The product $\mathsf{BA} : X \to Z$ is compact if one of the two operators is compact and the other is bounded.*
(v) *Consider a sequence of compact operators $(\mathsf{A}_n)_{n\in\mathbb{N}} \subset \mathcal{L}(X, Y)$ that converges in the strong operator topology to an operator $\mathsf{A} : X \to Y$ such that $\mathsf{A} \in \mathcal{L}(X, Y)$. Then A is compact.*

An important class of compact operators are integral operators.

Example 1.3.4 (Certain integral operators are compact). Consider a compact interval $[a, b] \subset \mathbb{R}$ and the Banach space

$$X = C([a, b]) = \{\phi : [a, b] \to \mathbb{R},\ \phi \text{ continuous}\},$$

endowed with the norm $\|\phi\| = \sup_{t\in[a,b]} |\phi(t)|$. An element $x \in X$ is identified with a continuous function $\phi : [a, b] \to \mathbb{R}$. We now consider the integral operator $\mathsf{A} : C([a, b]) \to C([a, b])$ defined by

$$\mathsf{A}\phi(t) = \int_a^b K(t,\tau)\phi(\tau)d\tau,$$

where $K : [a,b]\times[a,b] \to \mathbb{R}$ is continuous, and we use the simpler notation $\mathsf{A}\phi$, instead of $\mathsf{A}(\phi)$, to denote the element of $C([a,b])$, which is the image of $\phi \in C([a,b])$ under A. Then A is a linear and compact operator. The result can be extended for integral operators on $X = C(\mathcal{D})$ in the case where $\mathcal{D} \subset \mathbb{R}^d$ is a suitable domain, defined by its action on any $\phi \in C(\mathcal{D})$ by $\mathsf{A}\phi(x) = \int_{\mathcal{D}} K(x,z)\phi(z)dz$ for any $z \in \mathcal{D}$.

The linearity of the operator follows by standard properties of the integral. In order to show compactness, consider the bounded set $U \subset X$, of functions $\phi \in X$ such that $\sup_{t\in[a,b]} |\phi(t)| < c$ for some constant c. Then, the set $\mathsf{A}(U)$ is bounded, as can be seen by the estimate

$$\left|\int_a^b K(t,\tau)\phi(\tau)d\tau\right| \le (b-a) \max_{t_1,t_2\in[a,b]} |K(t_1,t_2)| \|\phi\| < c'$$

for a suitable constant c'. We will show that $\mathsf{A}(U)$ is equi-continuous, hence relatively compact in $X := C([a,b])$ by the Arzelá–Ascoli theorem (see Theorem 1.9.13). Indeed, since K is continuous on the compact set $[a,b] \times [a,b]$, it is also uniformly continuous, hence for every $\epsilon' > 0$ there exists $\delta > 0$ such that

$$|K(t_1,\tau) - K(t_2,\tau)| < \epsilon' \quad \text{for all } t_1, t_2, \tau \in [a,b] \text{ with } |t_1 - t_2| < \delta. \tag{1.5}$$

Then,

$$|\mathsf{A}\phi(t_1) - \mathsf{A}\phi(t_2)| \le \int_a^b |K(t_1,\tau) - K(t_2,\tau)||\phi(\tau)|d\tau \le (b-a)\epsilon' c, \tag{1.6}$$

for any $\phi \in U$. Consider any $\epsilon > 0$; let $\epsilon' = \frac{\epsilon}{(b-a)c}$, and choose $\delta > 0$ such that (1.5) holds. Then from (1.6), we can conclude that for all $t_1, t_2 \in [a,b]$ with $|t_1 - t_2| < \delta$ and all $\phi \in U$, it holds that $|\mathsf{A}\phi(t_1) - \mathsf{A}\phi(t_2)| < \epsilon$, hence $\mathsf{A}(U)$ is equi-continuous, and by the Arzelá–Ascoli theorem (see Theorem 1.9.13), we have the compactness of the operator A. ◁

Definition 1.3.5 (Fredholm operators). A bounded linear operator $\mathsf{A} : X \to Y$ is called a Fredholm operator if (a) $\mathbf{R}(\mathsf{A})$ is a closed subset of Y, (b) $\dim(\mathbf{N}(\mathsf{A})) < \infty$, and (c) $\dim(Y/\mathbf{R}(\mathsf{A})) < \infty$. The quotient space $Y/\mathbf{R}(\mathsf{A})$ is often called the co-kernel of the operator A, and is denoted by coker(A). The difference $\text{index}(\mathsf{A}) := \dim(\mathbf{N}(\mathsf{A})) - \dim(Y/\mathbf{R}(\mathsf{A}))$ is called the Fredholm index of A.

Example 1.3.6 (Finite dimensional operators are Fredholm). If $\dim(X) < \infty$ and $\dim(Y) < \infty$, then any $\mathsf{A} \in \mathcal{L}(X,Y)$ is Fredholm with $\text{index}(\mathsf{A}) = \dim(X) - \dim(Y)$. This is a well known result from linear algebra.

Example 1.3.7 (Fredholm operators related to integral operators). In the framework of Example 1.3.4, consider the operator $\mathsf{B} : X \to X$ defined by $\mathsf{B} = \mathrm{I} - \mathsf{A}$, which acts on any function $\phi \in C(\mathcal{D})$ by $\mathsf{B}\phi(x) = \phi(x) - \int_{\mathcal{D}} K(x,z)\phi(z)dz$. This is a Fredholm operator.

Fredholm operators have some interesting properties with respect to duality, as this is reflected in the properties of the adjoint (dual) operator, as well as some useful connections with compact operators.

Theorem 1.3.8 (Properties of Fredholm operators). *Let* $\mathsf{A} : X \to Y$ *be a bounded linear operator between two Banach spaces* X, Y. *Then,*
(i) A *is Fredholm if and only if* $\mathsf{A}^\star$ *is Fredholm.*
(ii) *If* A *is Fredholm, then* $\dim(\mathbf{N}(\mathsf{A}^\star)) = \dim(Y/\mathbf{R}(\mathsf{A}))$ *and* $\dim(X^\star/\mathbf{R}(\mathsf{A}^\star)) = \dim(\mathbf{N}(\mathsf{A}))$.
(iii) A *is a Fredholm operator if and only if there exists a bounded linear operator* $\mathsf{B} : X \to Y$ *such that the operators* $I_X - \mathsf{BA} : X \to X$ *and* $I_Y - \mathsf{AB} : Y \to Y$ *are compact.*
(iv) *If* X, Y, Z *are Banach spaces and* $\mathsf{A} : X \to Y$, $\mathsf{B} : Y \to Z$ *are Fredholm operators, then their composition* $\mathsf{BA} : X \to Z$ *is also a Fredholm operator with* $\mathrm{index}(\mathsf{BA}) = \mathrm{index}(\mathsf{A}) + \mathrm{index}(\mathsf{B})$.
(v) *If* $\mathsf{A}, \mathsf{B} : X \to Y$ *are a Fredholm and a compact operator respectively, then* $\mathsf{A} + \mathsf{B}$ *is Fredholm and* $\mathrm{index}(\mathsf{A} + \mathsf{B}) = \mathrm{index}(\mathsf{A})$.
(vi) *If* $\mathsf{A}, \mathsf{B} : X \to Y$ *are a Fredholm and a bounded operator, respectively, then there exists* $\epsilon_0 > 0$ *such that* $\mathsf{A} + \epsilon\mathsf{B}$ *is Fredholm for* $\epsilon < \epsilon_0$ *and* $\mathrm{index}(\mathsf{A} + \mathsf{B}) = \mathrm{index}(\mathsf{A})$.

Theorem 1.3.8 in the special case where $X = Y$ can lead to a very useful result, known as the Fredholm alternative, concerning the solvability of linear operator equations, of the form $(I - \mathsf{A})\mathsf{x} = \mathsf{z}$, where $\mathsf{A} : X \to X$ is a linear compact operator, and $\mathsf{z} \in X$ is given.

Theorem 1.3.9 (Fredholm alternative). *Let* $\mathsf{A} \in \mathcal{L}(X)$ *be a compact operator. Then, upon defining the annihilator of a set* $A \subset X$ *as*

$$A^\perp := \{\mathsf{x}^\star \in X^\star \,:\, \langle \mathsf{x}^\star, \mathsf{x}\rangle = 0,\ \forall\, \mathsf{x} \in A\},$$

we have that: $\dim(\mathbf{N}(I - \mathsf{A})) = \dim(\mathbf{N}(I - \mathsf{A}^\star)) < \infty$, *and* $\mathbf{R}(I - \mathsf{A})$ *is closed with* $\mathbf{R}(I - \mathsf{A}) = \mathbf{N}(I - \mathsf{A}^\star)^\perp$. *In particular,* $\mathbf{N}(I - \mathsf{A}) = \{0\}$ *if and only if* $\mathbf{R}(I - \mathsf{A}) = X$.

In particular also, the Fredholm alternative states that *either* the operator equation $(I-\mathsf{A})\mathsf{x} = \mathsf{z}$ admits a unique solution for any $\mathsf{z} \in X$, *or* the homogeneous equation $(I-\mathsf{A})\mathsf{x} = 0$ admits n linearly independent solutions, and then the inhomogeneous equation admits solutions only as long as $\mathsf{z} \in \mathbf{N}(I - \mathsf{A}^\star)^\perp$. Then, the solvability condition of the operator equation $(I-\mathsf{A})\mathsf{x} = \mathsf{z}$ is related to the existence of nontrivial solutions to problem $(I-\mathsf{A})\mathsf{x} = 0$ or the invertibility of the operator $I - \mathsf{A}$. The latter problem is called an eigenvalue problem.

Example 1.3.10 (Solvability of integral equations). In the multidimensional framework of Example 1.3.4, consider the integral equation

$$\phi(x) - \lambda \int_{\mathcal{D}} K(x,z)\phi(z)dz = \psi(x),$$

for any $x \in \mathcal{D}$, where $\psi \in C(\mathcal{D})$ is a known function, and $\lambda \in \mathbb{R}$ is a known constant. Then, on account of the results of Example 1.3.4, we may apply the Fredholm alternative and conclude that this problem admits a unique solution for any $\psi \in C(\mathcal{D})$, as long as $\lambda \in \mathbb{R}$ is such that the homogeneous problem $\phi(x) - \lambda \int_{\mathcal{D}} K(x,z)\phi(z)dz = 0$ admits only the trivial solution $\phi(x) = 0$ for all $x \in \mathcal{D}$. This tells us that the solvability of the problem occurs only for special types of ψ, if λ is such that $\phi(x) - \lambda \int_{\mathcal{D}} K(x,z)\phi(z)dz = 0$ admits nontrivial solutions, i. e., if λ coincides with eigenvalues of the integral operator. A characterization of such ψ requires the study of the dual operator. ◁

As problems of the form $(I - \mathsf{A})\mathsf{x} = 0$ play an important role in linear functional analysis, we close this section by recalling some fundamental facts concerning the spectrum of linear operators in Banach spaces. Since it is well known from linear algebra that real matrices may well have complex eigenfunctions, the same is anticipated to hold in infinite dimensions, and we therefore, in general, need to consider complex valued Banach spaces, in which the concept of the norm is generalized so that $\|\lambda \mathsf{x}\| = |\lambda| \, \|\mathsf{x}\|$ for all $\mathsf{x} \in X$ and $\lambda \in \mathbb{C}$, so as to treat this problem in its full generality (see [104]).

Definition 1.3.11 (The resolvent set and the spectrum). Let $\mathsf{A} : X \to X$ be a bounded linear operator.

(i) The resolvent set of A is the set $\rho(\mathsf{A}) := \{\lambda \in \mathbb{C} : (\mathsf{A} - \lambda I) : X \to X \text{ is bijective}\}$.
(ii) The spectrum of A is the set $\sigma(\mathsf{A}) := \mathbb{C} \setminus \rho(\mathsf{A})$.
(iii) A $\lambda \in \mathbb{C}$ is called an eigenvalue of A if $\mathbf{N}(\mathsf{A} - \lambda I) \neq \{0\}$. In general, there are elements of $\sigma(\mathsf{A})$ that are not eigenvalues.

Theorem 1.3.12 (Spectrum of compact operators). *Let X be an infinite dimensional Banach space and $\mathsf{A} \in \mathcal{L}(X)$ be a compact operator. Then $0 \in \sigma(\mathsf{A})$ and $\sigma(\mathsf{A})$ consists either* (a) *only of* 0, *or* (b) *of* 0 *and a finite set of eigenvalues, or* (c) *of* 0 *and a countable set of eigenvalues converging to* 0.

A lot more can be said when $X = H$, a separable Hilbert space, and $\mathsf{A} : H \to H$ has some rather special properties, in particular that $\mathsf{A}^* = \mathsf{A}$ (see, e. g., [33]).

Theorem 1.3.13. *Let $X = H$ be a separable infinite dimensional Hilbert space, and let $\mathsf{A} : H \to H$ be a linear self adjoint compact operator. Then, there exists a Hilbert basis $(\mathsf{x}_n)_{n\in\mathbb{N}}$, composed of eigenvectors of* A.

Example 1.3.14 (Construction of a basis using eigenvectors of an operator). Construct a basis for H consisting of the eigenvectors of the compact self adjoint operator[20] A, assuming for simplicity that A is also positive definite, i. e., $\langle \mathsf{A}\mathsf{x}, \mathsf{x}\rangle > 0$ for every $\mathsf{x} \in H \setminus \{0\}$.

20 This is essentially the proof of Theorem 1.3.13.

For a self-adjoint operator[21] $\sigma(A) \neq \{0\}$ (unless $A = 0$), so let $(\lambda_n)_{n\in\mathbb{N}}$ be the countable set of (distinct) eigenvalues of A, setting also $\lambda_0 = 0$, and define the closed linear subspaces $E_n = \mathrm{N}(\lambda_n I - A)$, $n = 0, 1, \ldots$, which can easily be seen to be orthogonal ($E_0 = \mathrm{N}(A) = \{0\}$ by the positive definite property,[22] whereas E_n, $n = 1, \ldots$ are finite dimensional). Let $E = \mathrm{span}(\bigcup_{n=1}^{\infty} E_n)$, and note that both E and $E^{\perp}$ are invariant under the action of A. Consider the operator $A_0 = A|_{E^{\perp}} : E^{\perp} \to E^{\perp}$, which is in turn a self-adjoint compact operator, for which[23] $\sigma(A_0) = \{0\}$, and this in turn shows that $A_0 = 0$, i. e., $Ax = 0$ for every $x \in E^{\perp}$, hence $E^{\perp} \subset \mathrm{N}(A) = \{0\}$, so $E^{\perp} = \{0\}$, which implies the density of[24] E in H. Since each of the E_n, $n \in \mathbb{N}$ is finite dimensional, form a basis on H by taking the union of the bases of each E_n. This basis allows us to express the action of the operator on any element $x = \sum_{n=1}^{\infty} x_n$, $x_n \in E_n$, as $Ax = \sum_{n=1}^{\infty} \lambda_n x_n$ (as the finite rank operators $A_k : H \to \bigcup_{n=1}^{k} E_n$, defined by $A_k x = \sum_{n=1}^{k} \lambda_n x_n$ converge in the operator norm to A since $\|A_k - A\| \leq \sup_{n\geq k+1} |\lambda_n| \to 0$, as $k \to \infty$). It is often convenient, by counting the eigenvalues with multiplicity, so that they are not necessarily distinct anymore, and denoting the new sequence by $(\lambda_n')_{n\in\mathbb{N}}$ to reorganize this basis as a basis $(x_n')_{n\in\mathbb{N}}$ with the property that $Ax_n' = \lambda_n' x_n'$, allowing for more straightforward eigenvector expansions. ◁

1.4 Lebesgue spaces

A fundamental example of Banach spaces of great importance in applications are Lebesgue and Sobolev spaces. As Lebesgue spaces are usually familiar to anyone that has had a first course in measure theory or functional analysis, we will here only recall some very fundamental facts concerning Lebesgue spaces (for an excellent account see [142]). We also assume familiarity of the reader with the fundamentals of Lebesgue's theory of integration. Even though, Lebesgue spaces can be defined on general measure spaces and for vector valued functions, these extensions will not be needed for most of the material covered in this book, with minor exceptions. So, in this section, we contain ourselves to an exposition of the theory for functions $u : \mathcal{D} \subset \mathbb{R}^d \to \mathbb{R}$, or when needed to functions $v : \mathcal{D} \subset \mathbb{R}^d \to \mathbb{R}^m$, where $\mathcal{D}$ is an open subset of $\mathbb{R}^d$. The more general case will be sketched where and if needed.

Let $\mathcal{D} \subset \mathbb{R}^d$ be a bounded open subset, and let $\mu_{\mathcal{L}}$ be the Lebesgue measure on $\mathcal{D}$. We will denote by $\mathcal{B}(\mathcal{D})$ the Borel σ-algebra on $\mathcal{D}$ (i. e., the smallest σ-algebra containing all open subsets of $\mathcal{D}$), and from now on, we will restrict our attention to Borel measurable

---

21 As can be seen, for instance, by the observation that for self-adjoint operators the spectral norm $\rho(A) := \sup\{|\lambda| : \lambda \in \sigma(A)\} = \|A\|$, see also [33] for an alternative proof.

22 For a general operator E_0 may be infinite dimensional.

23 Suppose not, then there exists $x \in E^{\perp} \setminus \{0\} \subset H$ and $\lambda \neq 0$ such that $A_0 x = \lambda x$ i. e. λ is one of the eigenvalues of A, say λ_n, and $x \in E^n \subset E$. Hence, $x \in E \cap E^{\perp}$, therefore $x = 0$ contradiction.

24 One could obtain the same result omitting the positive definite property, by defining $E = \bigcup_{n=0}^{\infty} E_n$ and proceeding as above to conclude that $E^{\perp} \subset \mathrm{N}(A) \subset E$, hence $E^{\perp} = \{0\}$.

functions on $\mathcal{D}$ (i. e., functions $u : \mathcal{D} \to \mathbb{R}$ with the property that for any Borel measurable set $I \subset \mathcal{B}(\mathbb{R})$, it holds that $u^{-1}(I) \in \mathcal{B}(\mathcal{D})$), hereafter simply called measurable. We will also use the notation $|A|$ for $\mu_{\mathcal{L}}(A)$ for every $A \in \mathcal{B}(\mathcal{D})$. For simplicity, we will denote the Lebesgue integral of such a function over any $A \subset \mathcal{D}$ by $\int_A u dx = \int_{\mathcal{D}} u \mathbf{1}_A \, dx$. Moreover, whenever we use the notation $u = v$ or $u \le v$ a. e., we mean that $u(x) = v(x)$ or $u(x) \le v(x)$, respectively, for x a. e. in $\mathcal{D}$. Similarly, $u_n \to u$ a. e. implies $u_n(x) \to u(x)$, for x a. e. in $\mathcal{D}$.

Definition 1.4.1 (Lebesgue spaces). For $1 \le p \le \infty$, the function spaces[25]

$$L^p(\mathcal{D}) := \Big\{ u : \mathcal{D} \to \mathbb{R} \; : \; \text{measurable with} \int_{\mathcal{D}} |u|^p dx < \infty \Big\}, \quad p \in [1, \infty),$$

$$L^\infty(\mathcal{D}) := \Big\{ u : \mathcal{D} \to \mathbb{R} \; : \; \text{measurable with} \operatorname*{ess\,sup}_{\mathcal{D}} |u| < \infty \Big\}$$

are called the Lebesgue spaces of order p.

We will also define the spaces $L^p_{\text{loc}}(\mathcal{D})$, $1 \le p \le \infty$ as the spaces of all functions $u : \mathcal{D} \to \mathbb{R}$ such that $u \mathbf{1}_K \in L^p(\mathcal{D})$ for every compact subset $K \subset \mathcal{D}$.

It is easily seen that these function spaces are vector spaces. In fact, they can be turned into Banach spaces with a suitable norm.

Theorem 1.4.2. *The spaces $L^p(\mathcal{D})$ equipped with the norms*

$$\|u\|_{L^p(\mathcal{D})} := \Big\{ \int_{\mathcal{D}} |u|^p dx \Big\}^{1/p}, \quad p \in [1, \infty),$$

$$\|u\|_{L^\infty(\mathcal{D})} := \operatorname*{ess\,sup}_{\mathcal{D}} |u|,$$

are Banach spaces, with $L^2(\mathcal{D})$ being a Hilbert space. In particular, the L^p norms for $p \in [1, \infty]$ satisfy the Hölder inequality, according to which if $u_1 \in L^p(\mathcal{D})$, $u_2 \in L^{p^*}(\mathcal{D})$, with the conjugate exponent p^* defined by $\frac{1}{p} + \frac{1}{p^*} = 1$, it holds that

$$\|u_1 u_2\|_{L^1(\mathcal{D})} \le \|u_1\|_{L^p(\mathcal{D})} \|u_2\|_{L^{p^*}(\mathcal{D})},$$

where the case $p = \infty$ is included with $p^* = 1$. As a simple consequence of this inequality, we see that if $|\mathcal{D}| < \infty$, then $L^q(\mathcal{D}) \subset L^p(\mathcal{D})$ if $p < q$ and

$$\|u\|_{L^p(\mathcal{D})} \le |\mathcal{D}|^{1/p - 1/q} \|u\|_{L^q(\mathcal{D})}.$$

25 Recall that for a measurable function u, the essential supremum is defined as $\operatorname{ess\,sup}_{\mathcal{D}} u = \inf\{c \in \mathbb{R} : \mu\{x \in \mathcal{D} : u(x) > c\} = 0\}$. In most cases, we will use the simplified notation sup, implicitly meaning the essential supremum.

Using the Hölder inequality, we can see that as long as $|\mathcal{D}| < \infty$ an $L^q(\mathcal{D})$ function can be considered as in $L^p(\mathcal{D})$ if $p < q$, in the sense that the identity map $I : L^q(\mathcal{D}) \to L^p(\mathcal{D})$ is continuous. For various reasons, it may be convenient to consider the "outmost" case of $L^1(\mathcal{D})$ as subset of a larger space, the space of Radon measures. Connecting measures with integrable functions is rather natural. One of the ways to envisage that is to recall the Radon–Nikodym theorem, according to which if a measure ν on $\mathcal{B}(\mathcal{D})$ is absolutely continuous to the Lebesgue measure on[26] $\mathcal{B}(\mathcal{D})$, then there exists a function $u \in L^1(\mathcal{D})$ such that $\nu(A) = \int_A u dx$ for every $A \in \mathcal{B}$. In this sense, one may consider some measures on $\mathcal{B}(\mathcal{D})$ (i. e., those absolutely continuous to the Lebesgue measure) as elements of $L^1(\mathcal{D})$, and therefore try to embed $L^1(\mathcal{D})$ in a larger space of signed measures.

We introduce the following definition:

Definition 1.4.3 (Radon measures on $\mathbb{R}^d$). An outer measure ν on $D \subset \mathbb{R}^d$ is called a Radon measure if it is Borel regular, i. e., if for every $A \subset D \subset \mathbb{R}^d$, there exists a Borel set B such that $\nu(A) = \nu(B)$, and it is finite on compact sets. The set of signed Radon measures on $D \subset \mathbb{R}^d$ will be denoted by $\mathcal{M}(D)$ and can be turned into a Banach space with the total variation norm defined by $\|\nu\|_{\mathcal{M}(D)} := |\nu| := \sup\{\sum_i |\nu(D_i)| : \{D_i\}$ countable partition of $D\}$.

The space of Radon measures is related via duality with the space of continuous functions, as the following theorem asserts (see, e. g., Theorems 19.54 and 19.55 of [142]):

Theorem 1.4.4. *Let $\mathcal{D} \subset \mathbb{R}^d$ be a bounded open set, and consider the Banach space*

$$C_c(\mathcal{D}) := \{\phi \in C(\mathcal{D}) : \phi(x) = 0,\ \forall x \in \mathcal{D} \setminus \mathcal{D}_0,\ \mathcal{D}_0 \subset \mathcal{D},\ \textit{compact}\},$$

the space of continuous functions on $\mathcal{D}$ with compact support, equipped with the norm $\|\phi\|_{C_c(\mathcal{D})} = \sup_{x\in\mathcal{D}} |\phi(x)|$. Then, $(C_c(\mathcal{D}))^* \simeq \mathcal{M}(\mathcal{D})$ is the following sense:

(i) *For every continuous linear mapping $f : C_c(\mathcal{D}) \to \mathbb{R}$, there exists a unique signed Radon measure $\nu \in \mathcal{M}(\mathcal{D})$ such that $f(\phi) = \int_{\mathcal{D}} \phi d\nu$ for every $\phi \in C_c(\mathcal{D})$ and $\|f\|_{(C_c(\mathcal{D}))^*} = |\nu|$, where $|\nu|$ is the total variation of the Radon measure ν and*
(ii) *If for each $\nu \in \mathcal{M}(\mathcal{D})$, we consider the bounded linear functional $j_\nu : C_c(\mathcal{D}) \to \mathbb{R}$ defined by $j_\nu(\phi) = \int_{\mathcal{D}} \phi d\nu$, then the mapping $\nu \mapsto j_\nu$ is a linear isometry from $\mathcal{M}(\mathcal{D})$ to $(C_c(\mathcal{D}))^*$.*

The same results hold upon replacing $C_c(\mathcal{D})$ by $C(\overline{\mathcal{D}})$ equipped with the usual norm $\|\phi\|_{C(\overline{\mathcal{D}})} := \sup_{x\in\overline{\mathcal{D}}} |\phi(x)|$ and $\mathcal{M}(\mathcal{D})$ by $\mathcal{M}(\overline{\mathcal{D}})$.

The above theorem provides an alternative definition for the total variation norm in terms of $\|\nu\|_{\mathcal{M}(D)} = \sup\{\int_{\mathcal{D}} \phi d\nu : \phi \in C_c(\mathcal{D}),\ \|\phi\|_{C_c(\mathcal{D})} \le 1\}$. Via the connection of the two spaces, we may also define a convenient weak* topology on $\mathcal{M}(D)$ and a useful notion

26 In the sense that for any $A \subset \mathcal{D}$ such that $\mu_{\mathcal{L}}(A) = 0$, it holds that $\nu(A) = 0$.

of weak* convergence. It also allows us to consider the following inclusions: $L^q(\mathcal{D}) \subset L^p(\mathcal{D}) \subset \mathcal{M}(\mathcal{D})$ for $1 \leq p < q \leq \infty$ (and $\mu(\mathcal{D}) < \infty$). This inclusion will be very useful when considering convergence of sequences in Lebesgue spaces.

We are now ready to present some fundamental results concerning the characterization of the dual spaces of Lebesgue spaces.

Theorem 1.4.5 (Duals of Lebesgue spaces). *Let $\mathcal{D} \subset \mathbb{R}^d$ be open and bounded.*

(i) *If $p \in (1,\infty)$, then $(L^p(\mathcal{D}))^\star \simeq L^{p^\star}(\mathcal{D})$, where $p \in (1,\infty)$ and $\frac{1}{p} + \frac{1}{p^\star} = 1$, in terms of the duality pairing $\langle u, v\rangle_{(L^p(\mathcal{D}))^\star, L^p(\mathcal{D})} = \int_{\mathcal{D}} uv dx$. In particular, the spaces $L^p(\mathcal{D})$ are reflexive and separable for $p \in (1,\infty)$.*

(ii) *$(L^1(\mathcal{D}))^\star \simeq L^\infty(\mathcal{D})$ in terms of $\langle u, v\rangle_{(L^1(\mathcal{D}))^\star, L^1(\mathcal{D})} = \int_{\mathcal{D}} uv dx$, and $L^1(\mathcal{D})$ is separable but (on account of* (iii) *below) $L^1(\mathcal{D})$ is not reflexive.*

(iii) *$(L^\infty(\mathcal{D}))^\star \subset L^1(\mathcal{D})$, with the inclusion being strict. $L^\infty(\mathcal{D})$ is neither separable nor reflexive.*

The result that $(L^\infty(\mathcal{D}))^\star \subsetneq L^1(\mathcal{D})$ (with the inclusion being strict) means that there exist continuous linear mappings $f : L^\infty(\mathcal{D}) \to \mathbb{R}$ that may not be represented as $f(u) = \int_{\mathcal{D}} vu dx$ for some $v \in L^1(\mathcal{D})$ (depending only on f) for every $u \in L^\infty(\mathcal{D})$. The full characterization of the space $(L^\infty(\mathcal{D}))^\star$ requires the space of finite additive bounded variation measures, but this will not be needed here.

Example 1.4.6 (The duality map for Lebesgue spaces). Recall the definition of the duality map (see Definition 1.1.17). Letting $X = L^p(\mathcal{D})$, $1 < p < \infty$, so that $X^\star = L^{p^\star}(\mathcal{D})$, (which is strictly convex so that, by Theorem 2.6.17, the dualiity mapping is single valued). Then it is easy to verify that for any $u \in L^p(\mathcal{D})$, we have that $(\mathrm{J}(u))(x) = \|u\|^{2-p}_{L^p(\mathcal{D})} |u(x)|^{p-2} u(x) \times \mathbf{1}_{\{u(x)\neq 0\}}$ for x a. e. in $\mathcal{D}$, where $p^\star$ is the conjugate exponent of p. The reader can verify that the definition is satisfied for this choice. In the special case, where $p = 2$, in which X is a Hilbert space, the above result simplifies to $\mathrm{J}(u) = u$.

In the case where $p = 1$, the situation differs, and for any $u \in L^1(\mathcal{D})$, we have that $\mathrm{J}(u) = \{v \in L^\infty(\mathcal{D}) : v = \|u\|_{L^1(\mathcal{D})} w_o\}$, where $w_o \in L^\infty(\mathcal{D})$ is any function of the form

$$w_o(x) = \begin{cases} 1, & \text{on } \{x \in \mathcal{D} : u(x) > 0\} \\ w, & \text{on } \{x \in \mathcal{D} : u(x) = 0\} \\ -1, & \text{on } \{x \in \mathcal{D} : u(x) < 0\}, \end{cases}$$

where w is any element of $L^\infty(\mathcal{D})$ such that $\|w\|_{L^\infty(\mathcal{D})} < 1$. This is a multi-valued (or set-valued) map. ◁

Functions in Lebesgue spaces can be approximated by sequences of continuous or even smooth functions as the following density result shows:

Theorem 1.4.7 (Density results for Lebesgue spaces). *The space*

$$C_c(\mathcal{D}) := \{\phi \in C(\mathcal{D}) \,:\, \phi(x) = 0,\ \forall x \in \mathcal{D} \setminus \mathcal{D}_o,\ \mathcal{D}_o \subset \mathcal{D},\ \textit{compact}\}$$

is dense in $L^p(\mathcal{D})$ for $p \in [1, \infty)$. The same holds for

$$C_c^\infty(\mathcal{D}) := \{\phi \in C^\infty(\mathcal{D}) \,:\, \phi(x) = 0,\ \forall x \in \mathcal{D} \setminus \mathcal{D}_o,\ \mathcal{D}_o \subset \mathcal{D},\ \textit{compact}\}.$$

The characterization of the duals of various Lebesgue spaces allows us to use the general concepts of strong, weak, and weak* convergence that have been provided in Sections 1.1.5.3, 1.1.6, 1.1.7. The definitions are straightforward adaptations of the general definitions, provided in these sections using the characterization of the dual spaces given in Theorem 1.4.5. Since we consider $L^1(\mathcal{D})$ as a subset of $\mathcal{M}(\overline{\mathcal{D}})$, we will also require a concept of convergence in this space, which will follow from the characterization of $\mathcal{M}(\overline{\mathcal{D}})$ as a dual space (see Theorem 1.4.4), which will correspond to weak* convergence. This discussion leads to the following definitions, which are collected in concrete form for the convenience of the reader:

Definition 1.4.8 (Various modes of weak and strong convergence). The following modes of weak convergence are frequently used:
(i) If $(\nu_n)_{n\in\mathbb{N}} \subset \mathcal{M}(\mathcal{D})$, then $\nu_n \overset{\star}{\rightharpoonup} \nu$ if $\int_{\mathcal{D}} \phi d\nu_n \to \int_{\mathcal{D}} \phi d\nu$ for every $\phi \in C_c(\mathcal{D})$.
(ii) If $(u_n)_{n\in\mathbb{N}} \subset L^p(\mathcal{D})$, $p \in [1, \infty)$, then $u_n \rightharpoonup u$ if $\int_{\mathcal{D}} u_n v dx \to \int_{\mathcal{D}} uv dx$ for every $v \in L^{p^\star}(\mathcal{D})$, where $p^\star$ is the conjugate exponent of p.
(iii) If $(u_n)_{n\in\mathbb{N}} \subset L^\infty(\mathcal{D})$, then $u_n \overset{\star}{\rightharpoonup} u$ if $\int_{\mathcal{D}} u_n v dx \to \int_{\mathcal{D}} uv dx$ for every $v \in L^1(\mathcal{D})$.
(iv) If $(u_n)_{n\in\mathbb{N}} \subset L^p(\mathcal{D})$, $p \in [1, \infty]$, then $u_n \to u$ in $L^p(\mathcal{D})$ if $\|u_n - u\|_{L^p(\mathcal{D})} \to 0$.
(v) For any sequence of functions $(u_n)_{n\in\mathbb{N}}$, then $u_n \to u$, a. e. (i. e., $u_n(x) \to u(x)$, x - a. e. in $\mathcal{D}$) if $\mu_{\mathcal{L}}\{x \in \mathcal{D} \,:\, u_n(x) \not\to u(x)\} = 0$.

The following results are classical (see, e. g., [142]) and are collected here for the convenience of the reader:

Theorem 1.4.9. *The following hold.*
(i) *Fatou's Lemma: If $u_n \ge 0$ and $\liminf_n u_n = u$ a. e., then $\int_{\mathcal{D}} u dx \le \liminf_n \int_{\mathcal{D}} u_n dx$.*
(ii) *Monotone convergence: If $u_n \le u_{n+1}$ a. e. for every $n \in \mathbb{N}$ and $u_n \to u$ a. e., then $\lim_n \int_{\mathcal{D}} u_n dx = \int_{\mathcal{D}} u dx$.*
(iii) *Lebesgue's dominated convergence: If $u_n \to u$ a. e., and there exists $c > 0$ such that $\|u_n\|_{L^1(\mathcal{D})} < c$ for every $n \in \mathbb{N}$, then $\lim_n \int_{\mathcal{D}} u_n dx = \int_{\mathcal{D}} u dx$.*
(iv) *If $u_n \to u$ in $L^p(\mathcal{D})$, $p \in [1, \infty)$, then there exists a subsequence $(u_{n_k})_{k\in\mathbb{N}}$ such that $|u_{n_k}| \le h$, a. e., where $h \in L^p(\mathcal{D})$, with $u_{n_k} \to u$ a. e. On the other hand, if $(u_n)_{n\in\mathbb{N}} \subset L^p(\mathcal{D})$, $u \in L^p(\mathcal{D})$, $p \in [1, \infty)$ and $u_n \to u$ a. e. and $\|u_n\|_{L^p(\mathcal{D})} \to \|u\|_{L^p(\mathcal{D})}$, then $u_n \to u$ in $L^p(\mathcal{D})$.*

The weak* convergence of measures will turn out to be a very useful tool as it satisfies some very useful lower semicontinuity and compactness properties. To anticipate its use, consider the following line of argumentation: Since $L^p(\mathcal{D})$ are reflexive for $p \in (1,\infty)$, we know from abstract arguments based on the use of Theorem 1.1.59 that bounded sequences in $L^p(\mathcal{D})$ admit a subsequence having a weak limit in the same space. Similarly, by using the general abstract arguments for the weak* topology, the same is true in $L^\infty(\mathcal{D})$ with respect to the weak* convergence. This argument however, cannot be extended to bounded sequences in $L^1(\mathcal{D})$, where, by the nonreflexivity, it may well be that bounded sequences in $L^1(\mathcal{D})$ may not admit subsequences with weak limits in the same space. To save the situation, we may always consider a bounded sequence in $L^1(\mathcal{D})$ as a bounded sequence in the larger space $\mathcal{M}(\mathcal{D})$ (or $\mathcal{M}(\overline{\mathcal{D}})$), and then use the weak* compactness of measures in order to obtain a limit, which may be a measure rather than an $L^1(\mathcal{D})$ function. In order to guarantee that the limit of a uniformly bounded sequence in $L^1(\mathcal{D})$ is indeed in $L^1(\mathcal{D})$, rather than in the larger space $\mathcal{M}(\mathcal{D})$, we need a more restrictive condition, that of uniform integrability (or equi-integrability), which requires that

$$\forall\, \epsilon > 0,\ \exists \delta > 0\ \forall\, E \subset \mathcal{D} \text{ with } |E| < \delta;\ \text{it holds that } \int_E |u_n| dx < \epsilon,\ \forall\, n \in \mathbb{N}. \tag{1.7}$$

Note that for any sequence $(u_n)_{n\in\mathbb{N}} \subset L^1(\mathcal{D})$ such that $\sup_{n\in\mathbb{N}} \|u_n\|_{L^1(\mathcal{D})} < \infty$, property (1.7) holds, but, in general, $\delta = \delta(\epsilon, n)$, whereas uniform integrability requires that the $\delta = \delta(\epsilon)$ only, i. e., the same choice of δ will work for the whole family, hence uniform integrability is a more restrictive condition. In the case of bounded $\mathcal{D} \subset \mathbb{R}^d$ considered here, we have an alternative, sometimes easier to verify, equivalent criterion for uniform integrability: the De la Vallée-Pousin criterion, according to which a sequence $(u_n)_{n\in\mathbb{N}} \subset L^1(\mathcal{D})$ is uniformly integrable if and only if there exists an increasing function $\gamma : \mathbb{R}_+ \to \mathbb{R}_+$ with $\lim_{t\to\infty} \frac{\gamma(t)}{t} = \infty$ such that $\sup_n \int_{\mathcal{D}} \gamma(|u_n|) dx < \infty$.

In the following theorem, we collect some useful weak compactness results for Lebesgue spaces and measures:

Theorem 1.4.10 (Weak* compactness for Lebesgue spaces and measures). *Let $\mathcal{D} \subset \mathbb{R}^d$ be a bounded open set.*

(i) *If $(\mu_n)_{n\in\mathbb{N}} \subset \mathcal{M}(\mathcal{D})$ is a bounded sequence (in terms of the total variation norm), then there exists a measure $\mu \in \mathcal{M}(\mathcal{D})$ and a subsequence $(\mu_{n_k})_{k\in\mathbb{N}} \subset \mathcal{M}(\mathcal{D})$ such that $\mu_{n_k} \overset{\star}{\rightharpoonup} \mu$ in $\mathcal{M}(\mathcal{D})$, as $k \to \infty$. Furthermore, for any $A \subset \mathcal{D}$ open, it holds that $\mu(A) \le \liminf_k \mu_{n_k}(A)$.*
(ii) *If $(u_n)_{n\in\mathbb{N}} \subset L^1(\mathcal{D})$ is a bounded sequence, then there exists a function $u \in L^1(\mathcal{D})$ and a subsequence $(u_{n_k})_{k\in\mathbb{N}}$ such that $u_{n_k} \rightharpoonup u$ in $L^1(\mathcal{D})$ if and only if $(u_n)_{n\in\mathbb{N}}$ is uniformly integrable or, equivalently, if and only if there exists an increasing function $\gamma : \mathbb{R}_+ \to \mathbb{R}_+$ with $\lim_{t\to\infty} \frac{\gamma(t)}{t} = \infty$ such that $\sup_n \int_{\mathcal{D}} \gamma(|u_n|) dx < \infty$.*

(iii) *If* $(u_n)_{n\in\mathbb{N}} \subset L^p(\mathcal{D})$, $p \in (1,\infty)$, *is a bounded sequence, then there exists a function* $u \in L^p(\mathcal{D})$ *and a subsequence* $(u_{n_k})_{k\in\mathbb{N}}$ *such that* $u_{n_k} \rightharpoonup u$ *in* $L^p(\mathcal{D})$.
(iv) *If* $(u_n)_{n\in\mathbb{N}} \subset L^\infty(\mathcal{D})$ *is a bounded sequence, then there exists a function* $u \in L^\infty(\mathcal{D})$ *and a subsequence* $(u_{n_k})_{k\in\mathbb{N}}$ *such that* $u_{n_k} \overset{\star}{\rightharpoonup} u$ *in* $L^\infty(\mathcal{D})$.

Assertions (i), (iii), and (iv) of the above theorem follow from the general properties of weak* convergence and weak convergence in reflexive spaces, whereas (ii) is a consequence of the Dunford–Pettis weak compactness criterion for $L^1(\mathcal{D})$ combined with the De la Vallée-Poussin criterion for uniform integrability.

The weak compactness results allow us to guarantee, under boundedness conditions, the existence of weakly converging subsequences. Clearly, weak convergence does not imply strong convergence, nor does it guarantee the existence of an a. e. converging subsequence (as does the strong convergence), unless some extra conditions are met. The following proposition (see, e. g., Corollaries 2.49 and 2.58, [80]) summarizes certain such cases:

Proposition 1.4.11. *Let* $p \in (1,\infty)$
(i) *If* $(u_n)_{n\in\mathbb{N}} \subset L^p(\mathcal{D})$ *and* $\|u_n\|_{L^p(\mathcal{D})} < c$ *for every* $n \in \mathbb{N}$ *with* $p \in (1,\infty)$, *then if* $u_n \to u$ *a. e., it holds that* $u \in L^p(\mathcal{D})$ *and* $u_n \rightharpoonup u$ *in* $L^p(\mathcal{D})$.
(ii) *If* $u_n \rightharpoonup u$ *in* $L^p(\mathcal{D})$ *and* $\|u_n\|_{L^p(\mathcal{D})} \to \|u\|_{L^p(\mathcal{D})}$, *then* $u_n \to u$ *in* $L^p(\mathcal{D})$.

A very useful result is the following:

Proposition 1.4.12 (Vitali). *Let* $p \in [1,\infty)$ *and* $(f_n)_{n\in\mathbb{N}}$ *be a sequence of measurable functions on* $\mathcal{D} \subset \mathbb{R}^d$ *(with* $\mathcal{D}$ *not necessarily bounded). Then* $f_n \to f$ *in* $L^p(\mathcal{D})$ *if and only if*
(i) $f_n \to f$ *in measure.*
(ii) *The sequence* $(|f_n|^p)_{n\in\mathbb{N}}$ *is equi-integrable.*
(iii) *For every* $\epsilon > 0$, *there exists* $A \subset \mathcal{D}$ *such that* $|A| < \infty$ *and* $\int_{\mathcal{D}\setminus A} |f_n|^p dx < \epsilon$ *for every* n.

Remark 1.4.13. Conditions (ii) and (iii) are always satisfied if there exists a function $g \in L^p(\mathcal{D})$ such that $|f_n| \le g$ a. e. for every n. Condition (iii) is always true if $|\mathcal{D}| < \infty$.

Operators between Lebesgue spaces will play an important role in this book.

Example 1.4.14 (Integral operators between Lebesgue spaces). Consider the operator $\mathsf{A} : L^2(\mathcal{D}) \to L^2(\mathcal{D})$ acting on any $\phi \in L^2(\mathcal{D})$ by

$$\mathsf{A}\phi(x) = \int_{\mathcal{D}} K(x,z)\phi(z)dz, \quad \text{a.e, } x \in \mathcal{D},$$

where $K : \mathcal{D}\times\mathcal{D} \to \mathbb{R}$ is a continuous function. Then A is a compact linear operator. The same result applies if the kernel function K is $L^2(\mathcal{D}\times\mathcal{D})$. ◁

One may easily extend the above definitions and properties to Lebesgue spaces for vector valued function $u : \mathcal{D} \subset \mathbb{R}^d \to \mathbb{R}^m$, by defining the Lebesgue spaces $L^p(\mathcal{D}; \mathbb{R}^m)$ as follows: Since any $u : \mathcal{D} \subset \mathbb{R}^d \to \mathbb{R}^m$ can be represented as $u = (u_1, \ldots, u_m)$, where $u_i : \mathcal{D} \subset \mathbb{R}^d \to \mathbb{R}$, $i = 1, \ldots, m$, we may define

$$L^p(\mathcal{D}; \mathbb{R}^m) = \{u : \mathcal{D} \to \mathbb{R}^m \ : \ u_i \in L^p(\mathcal{D}),\ i = 1, \ldots, m\}, \quad p \in [1, \infty].$$

A suitable norm for such spaces can be $\|u\|_{L^p(\mathcal{D};\mathbb{R}^m)} = \{\sum_{i=1}^m \|u_i\|^p_{L^p(\mathcal{D})}\}^{1/p}$ for $p \in [1, \infty)$ or $\|u\|_{L^\infty(\mathcal{D});\mathbb{R}^m} = \max_{i=1,\ldots,m} \|u_i\|_{L^\infty(\mathcal{D})}$, but other suitable choices of equivalent norms are possible. Unless absolutely necessary, we will use the simplified notation $L^p(\mathcal{D})$ for $L^p(\mathcal{D}; \mathbb{R}^m)$.

1.5 Sobolev spaces

Sobolev spaces are fundamental in the study of variational problems and in partial differential equations. We recall some important facts concerning Sobolev spaces (for a detailed account see, e. g., [1, 33] or [98]).

As before, let $\mathcal{D} \subseteq \mathbb{R}^d$ be a nonempty open set, and let $C_c^\infty(\mathcal{D})$ be the set of infinitely continuously differentiable functions on $\mathcal{D}$ with compact support.

Definition 1.5.1. The Sobolev spaces $W^{1,p}(\mathcal{D})$. $1 \leq p < \infty$, are defined as

$$W^{1,p}(\mathcal{D}) := \Big\{u \in L^p(\mathcal{D}) \ : \ \exists\, v_i \in L^p(\mathcal{D}), \text{ such that } \int_{\mathcal{D}} u \frac{\partial \phi}{\partial x_i} dx = -\int_{\mathcal{D}} v_i \phi dx,$$
$$\forall \phi \in C_c^\infty(\mathcal{D}),\ i = 1, \ldots, d\Big\}.$$

The functions v_i are called the weak partial derivatives of u in the direction x_i, and are denoted by $v_i = \frac{\partial u}{\partial x_i}$. We will use the notation $\nabla u = (\frac{\partial u}{\partial x_1}, \ldots, \frac{\partial u}{x_d})$ for the gradient of u, defined in the weak sense.

We will also define the spaces $W^{1,p}_{\text{loc}}(\mathcal{D})$, $1 \leq p \leq \infty$ as the spaces of all functions $u : \mathcal{D} \to \mathbb{R}$ such that $u\mathbf{1}_K \in W^{1,p}(\mathcal{D})$ for every compact subset $K \subset \mathcal{D}$.

In a similar fashion, as for the gradient, we may define weak versions of other operators that are encountered in vector calculus, for example, the divergence operator, which acts on vector fields $w = (w_1, \ldots, w_d)$ as

$$\nabla \cdot w = \operatorname{div} w = \sum_{i=1}^d \frac{\partial w}{\partial x_i} \tag{1.8}$$

with all partial derivatives interpreted in the weak sense.

The weak derivatives are best understood in the sense of distributions, using the concept of duality for the space $C_c^\infty(\mathcal{D})$, and can be shown to enjoy a number of convenient properties of the classical derivative, such as the Leibnitz rule. One may define higher weak derivatives in the same fashion, i. e., using the integration of parts formula for a function that is infinitely smooth. For example, we may define the functions v_{ij} in terms of the integration by parts formula $\int_{\mathcal{D}} v_{ij}\phi dx = \int_{\mathcal{D}} u \frac{\partial^2 \phi}{\partial x_j \partial x_i} dx$ for every $\phi \in C_c^\infty(\mathcal{D})$ and understand the function v_{ij} as the weak second partial derivative of u with respect to x_i and x_j, denoting it by $v_{ij} = \frac{\partial^2 u}{\partial x_j \partial x_i}$. Higher weak derivatives will be compactly denoted using the convenient multi-index notation $\alpha = (\alpha_1, \ldots, \alpha_d) \in \mathbb{N}^d$, and upon defining $|\alpha| = \sum_{i=1}^d \alpha_i$, we will denote $\partial_x^\alpha u := \frac{\partial^{|\alpha|}}{\partial x_1^{\alpha_1} \cdots \partial x_d^{\alpha_d}} u$. When $\alpha = (0, \ldots, 0)$, we have that $\partial_x^\alpha u = u$, and when $\alpha = e_i = (\delta_{ij} : j = 1, \ldots, d)$, we have that $\partial_x^\alpha u = \frac{\partial u}{\partial x_i}$.

Example 1.5.2. If $u \in W^{1,2}(\mathcal{D})$, then $u^- := \max(-u, 0) \in W^{1,2}(\mathcal{D})$ also, and $\nabla u^- = -\nabla u \mathbf{1}_{u<0}$. Similarly for $u^+ := \max(u, 0)$, with $\nabla u^+ = \nabla u \mathbf{1}_{u>0}$. ◁

We can then define higher-order Sobolev spaces recursively as follows:

Definition 1.5.3 (The Sobolev spaces $W^{k,p}(\mathcal{D})$). The Sobolev spaces $W^{k,p}(\mathcal{D})$, $1 < p \leq \infty$, are the spaces[27]

$$W^{k,p}(\mathcal{D}) := \left\{ u \in W^{k-1,p}(\mathcal{D}) \; : \; \frac{\partial u}{\partial x_i} \in W^{k-1,p}(\mathcal{D}),\; i = 1, \ldots, d \right\}.$$

We will also define the spaces $W^{k,p}_{\text{loc}}(\mathcal{D})$, $1 \leq p \leq \infty$ as the spaces of all functions $u : \mathcal{D} \to \mathbb{R}$ such that $u \mathbf{1}_K \in W^{k,p}(\mathcal{D})$ for every compact subset $K \subset \mathcal{D}$.

The Sobolev spaces can be turned into Banach spaces with a suitable norm.

Theorem 1.5.4 ($W^{k,p}(\mathcal{D})$ are Banach spaces). *Let $\mathcal{D} \subset \mathbb{R}^d$ be open and bounded.*

(i) *The Sobolev spaces $W^{1,p}(\mathcal{D})$, $p \in [1, \infty]$ when endowed with one of the two equivalent norms,*

$$\|u\|_{W^{1,p}(\mathcal{D})} := \left\{ \|u\|^p_{L^p(\mathcal{D})} + \sum_{i=1}^d \left\| \frac{\partial u}{\partial x_i} \right\|^p_{L^p(\mathcal{D})} \right\}^{1/p}, \quad p \in [1, \infty),$$

$$\begin{aligned} \|u\|_{W^{1,\infty}(\mathcal{D})} &:= \max\left\{ \|u\|_{L^\infty(\mathcal{D})}, \left\| \frac{\partial u}{\partial x_1} \right\|_{L^\infty(\mathcal{D})}, \ldots, \left\| \frac{\partial u}{\partial x_d} \right\|_{L^\infty(\mathcal{D})} \right\} \\ &= \max_{\alpha,\, |\alpha| \leq 1} \|\partial_x^\alpha u\|_{L^\infty(\mathcal{D})}, \quad p = \infty, \end{aligned}$$

or

$$\|u\|_{W^{1,p}(\mathcal{D})} = \|u\|_{L^p(\mathcal{D})} + \sum_{i=1}^d \left\| \frac{\partial u}{\partial x_i} \right\|_{L^p(\mathcal{D})}, \quad p \in [1, \infty]$$

27 Equivalently $u \in W^{k,p}(\mathcal{D})$ if and only if $u \in L^p(\mathcal{D})$, and for every α with $|\alpha| \leq k$, there exists $v_\alpha \in L^p(\mathcal{D})$ such that $\int_{\mathcal{D}} u \partial_x^\alpha \phi dx = (-1)^{|\alpha|} \int_{\mathcal{D}} v_\alpha \phi dx$ for every $\phi \in C_c^\infty(\mathcal{D})$.

are Banach spaces. In the special case, $p = 2$*, the space* $W^{1,2}(\mathcal{D})$ *is a Hilbert space with inner product*

$$\langle u_1, u_2\rangle_{W^{1,2}(\mathcal{D})} = \langle u_1, u_2\rangle_{L^2(\mathcal{D})} + \sum_{i=1}^{d}\left\langle \frac{\partial u_1}{\partial x_i}, \frac{\partial u_2}{\partial x_i}\right\rangle_{L^2(\mathcal{D})}.$$

$W^{1,p}(\mathcal{D})$ *are separable for* $p \in [1,\infty)$ *but not for* $p = \infty$.

(ii) *Similarly, the Sobolev spaces* $W^{k,p}(\mathcal{D})$, $p \in [1,\infty]$ *endowed with any of the equivalent norms,*

$$\|u\|_{W^{k,p}(\mathcal{D})} := \Big(\sum_{|\alpha|\leq k} \|\partial_x^\alpha u\|_{L^p(\mathcal{D})}^p\Big)^{1/p}, \quad p \in [1,\infty),$$
$$\|u\|_{W^{k,p}(\mathcal{D})} := \max_{\alpha,\, |\alpha|\leq k} \|\partial_x^\alpha u\|_{L^\infty(\mathcal{D})}, \quad p = \infty,$$

or

$$\|u\|_{W^{k,p}(\mathcal{D})} := \sum_{|\alpha|\leq k} \|\partial_x^\alpha u\|_{L^p(\mathcal{D})}, \quad p \in [1,\infty],$$

are Banach spaces. When $p = 2$*, the space* $W^{k,p}(\mathcal{D})$ *is a Hilbert space with inner product*

$$\langle u_1, u_2\rangle_{W^{k,2}(\mathcal{D})} = \sum_{|\alpha|\leq k} \langle \partial_x^\alpha u_1, \partial_x^\alpha u_2\rangle_{L^2(\mathcal{D})}.$$

$W^{k,p}(\mathcal{D})$ *are separable for* $p \in [1,\infty)$ *but not for* $p = \infty$.

The following density result, clarifies the nature of functions in Sobolev spaces:

Theorem 1.5.5 (Meyers–Serrin). *If* $\mathcal{D} \subset \mathbb{R}^d$ *is open, then* $C^\infty(\mathcal{D}) \cap W^{k,p}(\mathcal{D})$ *is dense in* $W^{k,p}(\mathcal{D})$*, so that* $W^{k,p}(\mathcal{D})$ *can be defined as the completion of* $C^\infty(\mathcal{D})$ *in the norm* $\|\cdot\|_{W^{k,p}(\mathcal{D})}$.

We will also use the following subsets of the Sobolev spaces $W^{k,p}(\mathcal{D})$:

Definition 1.5.6. The Sobolev space $W_0^{1,p}(\mathcal{D})$ is defined as the closure of the set $C_c^\infty(\mathcal{D})$ with respect to the norm $\|\cdot\|_{W^{1,p}(\mathcal{D})}$. Similarly, the Sobolev space $W_0^{k,p}(\mathcal{D})$ is defined as the closure of the set $C_c^\infty(\mathcal{D})$ with respect to the norm $\|\cdot\|_{W^{k,p}(\mathcal{D})}$. We also define the spaces $W^{-1,p^*}(\mathcal{D}) := (W^{1,p}(\mathcal{D}))^*$ and $W^{-k,p^*}(\mathcal{D}) := (W^{k,p}(\mathcal{D}))^*$, where $\frac{1}{p} + \frac{1}{p^*} = 1$. Theorem 1.5.4 holds for the Sobolev spaces $W_0^{k,p}(\mathcal{D})$.

The spaces $W_0^{k,p}(\mathcal{D})$ can be considered as consisting of these functions in $W^{k,p}(\mathcal{D})$, which "vanish" on $\partial\mathcal{D}$, where this can be made precise once the concept of trace for a Sobolev space is defined.

The duals of the Sobolev spaces $W^{k,p}(\mathcal{D})$ play an important role in many applications. The elements of these spaces are distributions. The following result sheds some light to the nature of the elements of these spaces:

Proposition 1.5.7 (Riesz representation in Sobolev spaces). *Let $p \in [1, \infty)$, and let $p^\star$ be so that $\frac{1}{p} + \frac{1}{p^\star} = 1$. Then,*

(i) *For any $u \in (W^{1,p}(\mathcal{D}))^\star$ there exist $v_0, v_1, \dots, v_d \in L^{p^\star}(\mathcal{D})$ such that*

$$\langle u, w \rangle_{(W^{1,p}(\mathcal{D}))^\star, W^{1,p}(\mathcal{D})} = \int_{\mathcal{D}} \left(v_0 w + \sum_{i=1}^{d} v_i \frac{\partial w}{\partial x_i} \right) dx, \quad \forall\, w \in W^{1,p}(\mathcal{D}). \tag{1.9}$$

The spaces $W^{1,p}(\mathcal{D})$ are reflexive for $p \in (1, \infty)$.

(ii) *For any $u \in (W^{k,p}(\mathcal{D}))^\star$, there exists a family $\{v_\alpha : |\alpha| \le k\} \subset L^{p^\star}(\mathcal{D})$ such that*

$$\langle u, w \rangle_{(W^{k,p}(\mathcal{D}))^\star, W^{k,p}(\mathcal{D})} = \sum_{0 \le |\alpha| \le k} \int_{\mathcal{D}} v_\alpha \partial_x^\alpha w dx, \quad \forall\, w \in W^{k,p}(\mathcal{D}). \tag{1.10}$$

The spaces $W^{k,p}(\mathcal{D})$ are reflexive for $p \in (1, \infty)$.

It is important to note that the above representation *does not* imply that $(W^{1,p}(\mathcal{D}))^\star \simeq L^{p^\star}(\mathcal{D}; \mathbb{R}^{d+1})$ (for a discussion see, e. g., Remark 10.42 in [98])! Although we will be more specific for $(W_0^{k,p}(\mathcal{D}))^\star$, the full characterization of $(W^{k,p}(\mathcal{D}))^\star$ in general will not concern us here (see, e. g., [1, 33] or [98]), but the result of Proposition 1.5.7 allows us to characterize sufficiently weak convergence in $W^{k,p}(\mathcal{D})$.

Example 1.5.8 (Weak convergence in $W^{k,p}(\mathcal{D})$). Consider a sequence $(u_n)_{n \in \mathbb{N}} \subset W^{k,p}(\mathcal{D})$ and some $u \in W^{k,p}(\mathcal{D})$, $p \in [1, \infty)$. Then, $u_n \rightharpoonup u$ in $W^{k,p}(\mathcal{D})$ if and only if $u_n \rightharpoonup u$ and $\partial_x^\alpha u_n \rightharpoonup \partial_x^\alpha u$, $1 \le |\alpha| \le k$, in $L^p(\mathcal{D})$. By allowing $0 \le |\alpha| \le k$, we can simply use the notation $\partial_x^\alpha u_n \rightharpoonup \partial_x^\alpha u$, $0 \le |\alpha| \le k$, in $L^p(\mathcal{D})$ to include weak convergence of both the function and its derivatives.

Assume that $u_n \rightharpoonup u$ in $W^{k,p}(\mathcal{D})$ so that $\langle w, u_n - u \rangle_{(W^{k,p}(\mathcal{D}))^\star, W^{k,p}(\mathcal{D})} \to 0$ for all $w \in (W^{k,p}(\mathcal{D}))^\star$. Then for any $v \in L^{p^\star}(\mathcal{D})$, and any α such that $0 \le |\alpha| \le k$ set $v_\alpha = v$, and consider the element $w_\alpha \in (W^{k,p}(\mathcal{D}))^\star$ generated by v_α, i. e., $w_\alpha(u) = \int_{\mathcal{D}} v_\alpha \partial_x^\alpha u dx$, for any $u \in W^{k,p}(\mathcal{D})$. Since $u_n \rightharpoonup u$ in $W^{k,p}(\mathcal{D})$, it holds that $\langle w_\alpha, u_n - u \rangle_{(W^{k,p}(\mathcal{D}))^\star, W^{k,p}(\mathcal{D})} \to 0$, which implies that $\int_{\mathcal{D}} v_\alpha \partial_x^\alpha (u_n - u) dx = \int_{\mathcal{D}} v \partial_x^\alpha (u_n - u) dx \to 0$ for any $v \in L^{p^\star}(\mathcal{D})$, hence $\partial_x^\alpha u_n \rightharpoonup \partial_x^\alpha u$, $|\alpha| \le k$, in $L^p(\mathcal{D})$.

For the converse, consider that $u_n \rightharpoonup u$ and $\partial_x^\alpha u_n \rightharpoonup \partial_x^\alpha u$, $0 \le |\alpha| \le k$, in $L^p(\mathcal{D})$, so that for every $v \in L^{p^\star}(\mathcal{D})$ and every α such that $0 \le |\alpha| \le k$, it holds that $\langle v, \partial_x^\alpha (u_n - u) \rangle_{L^{p^\star}(\mathcal{D}), L^p(\mathcal{D})} \to 0$. Consider now any $w \in (W^{k,p}(\mathcal{D}))^\star$. By Proposition 1.5.7, there exists a collection $\{v_\alpha : 0 \le |\alpha| \le k\} \subset L^{p^\star}(\mathcal{D})$ such that $\langle w, u_n - u \rangle_{(W^{k,p}(\mathcal{D}))^\star, W^{k,p}(\mathcal{D})} = \sum_{0 \le |\alpha| \le k} \langle v_\alpha, \partial_x^\alpha (u_n - u) \rangle_{L^{p^\star}(\mathcal{D}), L^p(\mathcal{D})} \to 0$. ◁

Remark 1.5.9 (What is the nature of the elements of $(W^{k,p}(\mathcal{D}))^\star$ and $(W_0^{k,p}(\mathcal{D}))^\star = W^{-k,p^\star}(\mathcal{D})$?). For the characterization of $W^{-k,p^\star}(\mathcal{D})$, we need to define the space of distributions $\mathscr{D}(\mathcal{D})^\star$ on $\mathcal{D}$, which is the set of functionals $f : C_c^\infty(\mathcal{D}) \to \mathbb{R}$, endowed with the weak* topology according to which $f_n \rightharpoonup f$ in $\mathscr{D}(\mathcal{D})^\star$ if $\langle f_n, \phi\rangle \to \langle f, \phi\rangle$ for every $\phi \in C_c^\infty(\mathcal{D})$. We should note that to define the space of distributions properly, we need to define the space of test functions $\mathscr{D}(\mathcal{D})$, which is the space $C_c^\infty(\mathcal{D})$, endowed with a proper topology τ. Distributions can be considered as generalizations of either functions or measures, in the sense that for every $\mu \in \mathcal{M}(\mathcal{D})$ and $u \in L^p(\mathcal{D})$, we may define distributions $T_\mu, T_u : \mathscr{D}(\mathcal{D}) \to \mathbb{R}$ by $\langle T_\mu, \phi\rangle = \int_\mathcal{D} \phi d\mu$ and $\langle T_u, \phi\rangle = \int_\mathcal{D} u\phi dx$ for every $\phi \in \mathscr{D}(\mathcal{D})$, respectively. We may also define derivatives of distributions, as the distributions $\partial^\alpha T_{v_\alpha}$ defined by $\partial^\alpha T_{v_\alpha}(\phi) = (-1)^{|\alpha|} \int_\mathcal{D} v_\alpha \partial_x^\alpha \phi dx$ for every $\phi \in \mathscr{D}(\mathcal{D})$.

With these definitions at hand, by restricting the action of each element $u \in (W^{k,p}(\mathcal{D}))^\star$ to $\mathscr{D}(\mathcal{D})$ and using Proposition 1.5.7, we have that $\langle T_v, \phi\rangle = \sum_{0\le|\alpha|\le k}(-1)^{|\alpha|} \times \langle \partial^\alpha T_{v_\alpha}, \phi\rangle$ for every $\phi \in \mathscr{D}(\mathcal{D})$, where $v = \{v_\alpha : 0 \le |\alpha| \le k\} \subset L^{p^\star}(\mathcal{D})$ is the family of functions referred to in Proposition 1.5.7. Conversely, for any family of functions $v = \{v_\alpha : 0 \le |\alpha| \le k\} \subset L^{p^\star}(\mathcal{D})$, the distribution $T_v : \mathscr{D}(\mathcal{D}) \to \mathbb{R}$, defined as above, can be extended (but not uniquely!) to an element $u \in (W^{k,p}(\mathcal{D}))^\star$. On the other hand, its extension to an element $u \in (W_0^{k,p}(\mathcal{D}))^\star$ is unique, therefore, allowing us to identify $W^{-k,p^\star}(\mathcal{D})$ with the subspace of distributions of the form $\sum_{0\le|\alpha|\le k}(-1)^{|\alpha|}\partial^\alpha T_{v_\alpha}$ for any $v = \{v_\alpha : 0 \le |\alpha| \le k\} \subset L^{p^\star}(\mathcal{D})$.

The fact that Sobolev functions admit integrable weak derivatives incites useful integrability and continuity properties. These properties can be presented in terms of embedding theorems, which go by the general name of Sobolev embeddings. We will say that X is continuously embedded in Y and use the notation $X \hookrightarrow Y$ when $X \subset Y$ and the identity mapping $I : X \to Y$ is continuous, whereas if $I : X \to Y$ is compact, we will call it a compact embedding and use the notation $X \overset{c}{\hookrightarrow} Y$. Clearly, if $X \hookrightarrow Y$, there exists $c > 0$ such that $\|u\|_Y \le c\|u\|_X$ for every $u \in X \subset Y$. The embedding theorems often require restrictions on $\partial\mathcal{D}$ the boundary of the domain $\mathcal{D}$.

Definition 1.5.10 (Lipschitz domain). A domain $\mathcal{D} \subset \mathbb{R}^d$ is called Lipschitz if $\partial\mathcal{D}$ is Lipschitz, i. e., it can be expressed (at least locally) in terms of Lipschitz graphs. In particular, a domain such that $\partial\mathcal{D}$ is bounded is called Lipschitz if for every $x_o \in \partial\mathcal{D}$ there exists a neighborhood $N(x_o)$ and local coordinates $z = (z', z_d) \in \mathbb{R}^{d-1} \times \mathbb{R}$ with $z = 0$ at $x = x_o$, a Lipschitz function $f : \mathbb{R}^{d-1} \to \mathbb{R}$ and $r > 0$, such that

$$\mathcal{D} \cap N(x_o) = \{(z', z_d) \in \mathcal{D} \cap N(x_o) : z' \in Q_{d-1}(0, r),\ z_d > f(z')\}.$$

The assumption of Lipschitz domains will be made throughout this book.

Theorem 1.5.11 (Sobolev embeddings). *Let $\mathcal{D} \subset \mathbb{R}^d$ be a bounded open subset with Lipschitz boundary. Then, for any $0 \le k_2 \le k_1$, we have the following continuous embeddings:*

(i) *If* $(k_1 - k_2)p_1 < d$, *then* $W^{k_1,p_1}(\mathcal{D}) \hookrightarrow W^{k_2,p_2}(\mathcal{D})$ *for* $1 \leq p_2 \leq \frac{dp_1}{d-(k_1-k_2)p_1}$, *with the embedding being compact if* $1 \leq p_2 < \frac{dp_1}{d-(k_1-k_2)p_1}$. *In particular, if* $k_1 p_1 < d$, *then* $W^{k_1,p_1}(\mathcal{D}) \hookrightarrow L^{p_2}(\mathcal{D})$ *for* $1 \leq p_2 \leq \frac{dp_1}{d-k_1p_1}$, *with the embedding being compact if* $1 \leq p_2 < \frac{dp_1}{d-k_1p_1}$.

(ii) *If* $(k_1 - k_2)p_1 = d$, *then* $W^{k_1,p_1}(\mathcal{D}) \hookrightarrow W^{k_2,p_2}(\mathcal{D})$ *for* $1 \leq p_2 < \infty$, *with the embedding being compact. In particular, if* $k_1 p_1 = d$, *then* $W^{k_1,p_1}(\mathcal{D}) \hookrightarrow L^{p_2}(\mathcal{D})$ *for* $1 \leq p_2 \leq \infty$, *with the embedding being compact.*

(iii) *If* $(k_1 - k_2)p_1 > d$ *with* $\frac{d}{p_1} \notin \mathbb{N}$ *and* $(k_1 - k_2 - 1)p_1 < d < (k_1 - k_2)p_1$, *then* $W^{k_1,p_1}(\mathcal{D}) \hookrightarrow C^{k_2,\gamma}(\overline{\mathcal{D}})$ *for* $0 < \gamma \leq (k_1 - k_2) - \frac{d}{p_1}$, *where* $C^{k_2,\gamma}$ *is the space* k_2 *times differentiable functions with all derivatives up to order* k_1 *in* $C^{0,\gamma}$, *the space of Hölder continuous functions with exponent* γ *(see Section 1.9.4). If* $\frac{d}{p_1} \notin \mathbb{N}$, *then* $W^{k_1,p_1}(\mathcal{D}) \hookrightarrow C^{k_2,\gamma}(\overline{\mathcal{D}})$ *for* $0 < \gamma < 1$, *with the embeddings being compact if* $(k_1 - k_2)p_1 > d \geq (k_1 - k_2 - 1)p_1$ *and* $0 < \gamma < (k_1 - k_2) - \frac{d}{p_1}$.

The results remain true for $W_0^{k,p}(\mathcal{D})$ *for arbitrary domains.*

Remark 1.5.12. An easy to memorize result is that (under the stated conditions on $\mathcal{D}$) if $p < d$, then $W^{1,p}(\mathcal{D}) \hookrightarrow L^{\mathfrak{s}_p}(\mathcal{D})$, with $\mathfrak{s}_p := \frac{dp}{d-p}$, the critical Sobolev exponent, which provides the largest exponent for the embedding to hold. Naturally, the embedding holds for every $L^p(\mathcal{D})$ with $p \leq \mathfrak{s}_p$, however, the embedding is compact only for $p < \mathfrak{s}_p$. Moreover, $W^{1,p}(\mathcal{D}) \hookrightarrow C(\overline{\mathcal{D}})$ for $p > d$. The compact embeddings guaranteed by the Rellich–Kondrachov theorem may provide useful information, as it allows us to ensure that a bounded sequence in a reflexive Sobolev space, admits a strongly convergent subsequence in a properly selected Lebesgue space (hence the whole sequence by Theorem 1.3.3 or by the Urysohn property) and hence a further subsequence converging a. e..

The above theorem combines a number of results obtained independently by various authors (including Sobolev himself) into an easily (?) memorable form. The starting point for such results is the so called Gagliardo–Nirenberg–Sobolev inequality, according to which

$$\|u\|_{L^{\mathfrak{s}_p}(\mathbb{R}^d)} \leq c\|\nabla u\|_{L^p(\mathbb{R}^d)}, \quad \forall\, u \in C_c^1(\mathbb{R}^d),\ \mathfrak{s}_p := \frac{dp}{d-p},\ 1 \leq p < d. \tag{1.11}$$

This important inequality allows us to obtain embedding theorems for the space of Sobolev functions defined over the whole of $\mathbb{R}^d$, denoted by $W^{k,p}(\mathbb{R}^d)$. Then, given a domain $\mathcal{D}$, using the concept of an extension operator for $\mathcal{D}$, i. e., a continuous linear operator $\mathsf{E} : W^{k,p}(\mathcal{D}) \to W^{k,p}(\mathbb{R}^d)$ such that $\mathsf{E}u(x) = u(x)$ a. e. in $\mathcal{D}$; the embedding is extended to $W^{k,p}(\mathcal{D})$. Naturally, the existence of an extension operator depends on the domain $\mathcal{D}$, with the domains admitting such an operator called extension domains. Lipschitz domains can be shown to be extension domains (see [1]). The compact embedding results are due to Rellich and Kondrachov, whereas the embedding into the space of continuous functions is due to Morrey (see, e. g., [1] for details). These embedding results are based on inequalities between the norms of the derivatives of functions, initially obtained over the whole of $\mathbb{R}^d$, and then modified by proper extension (or restriction) arguments. Since most of these inequalities can be inferred from the embeddings, we do not reproduce them explicitly here for the sake of brevity. We only provide an important inequality, the Poincaré inequality, which is very useful in applications.

Theorem 1.5.13 (Poincaré inequality).

(i) *Let $\mathcal{D} \subset \mathbb{R}^d$ be a bounded domain.*[28] *Then, there exists $\mathbf{c}_{\mathcal{P}} > 0$ depending only on p and $\mathcal{D}$ such that*

$$\mathbf{c}_{\mathcal{P}} \|u\|_{L^p(\mathcal{D})} \leq \|\nabla u\|_{L^p(\mathcal{D})}, \quad \forall\, u \in W_0^{1,p}(\mathcal{D}),\ p \in [1, \infty).$$

(ii) *Let $\mathcal{D} \subset \mathbb{R}^d$ be a domain with Lipschitz boundary, and let $A \subset \mathcal{D}$ be a Lebesgue measurable set with positive measure, with the possibility that $A = \mathcal{D}$. Then, if $u_A = \fint_A u\,dx := \frac{1}{|A|}\int_A u\,dx$ is the average of u over A, there exists a constant $\widehat{\mathbf{c}}_{\mathcal{P}} > 0$ dependent only on p, $\mathcal{D}$, A such that*

$$\widehat{\mathbf{c}}_{\mathcal{P}} \|u - u_A\|_{L^p(\mathcal{D})} \leq \|\nabla u\|_{L^p(\mathcal{D})}, \quad \forall\, u \in W^{1,p}(\mathcal{D}),\ p \in [1, \infty).$$

The constant in the Poincaré inequality is very important, as it carries geometric information on the domain $\mathcal{D}$. We will also see that it can be characterized in terms of the solution of eigenvalue problems for operators related to the Laplace operator. For the time being, we show that by using the Poincaré inequality, we may establish an equivalent norm for the Sobolev space $W_0^{1,p}(\mathcal{D})$ using only the $L^p(\mathcal{D})$ norm of the weak gradient of u.

Example 1.5.14 (An equivalent norm for $W_0^{1,p}(\mathcal{D})$). If $\mathcal{D} \subset \mathbb{R}^d$ is a bounded domain, the norms $\|u\|_{W^{1,p}(\mathcal{D})}$ and $\|\nabla u\|_{L^p(\mathcal{D})}$ are equivalent norms for $W_0^{1,p}(\mathcal{D})$, $p \in [1, \infty)$.

It suffices to prove the existence of two constants $c_1, c_2 > 0$ such that $c_1\|\nabla u\|_{L^p(\mathcal{D})} \leq \|u\|_{W_0^{1,p}(\mathcal{D})} \leq c_2\|\nabla u\|_{L^p(\mathcal{D})}$ for all $u \in W_0^{1,p}(\mathcal{D})$.

Recall that $\|u\|^p_{W^{1,p}(\mathcal{D})} = \|u\|^p_{L^p(\mathcal{D})} + \sum_{i=1}^d \|\frac{\partial u}{\partial x_i}\|^p_{L^p(\mathcal{D})}$, whereas $\|\nabla u\|_{L^p(\mathcal{D})} = (\int_{\mathcal{D}} |\nabla u(x)|^p dx)^{1/p}$, where $|\nabla u|$ denotes the Euclidean norm of ∇u, so that $\|\frac{\partial u}{\partial x_i}\|^p_{L^p(\mathcal{D})} \leq \|\nabla u\|^p_{L^p(\mathcal{D})}$ for every $i = 1, \ldots, d$, from which we see that $\|u\|^p_{W^{1,p}(\mathcal{D})} \leq \|u\|^p_{L^p(\mathcal{D})} + d\|\nabla u\|^p_{L^p(\mathcal{D})}$. By the Poincaré inequality, $\|u\|^p_{L^p(\mathcal{D})} \leq \mathbf{c}_{\mathcal{P}}^{-p}\|\nabla u\|^p_{L^p(\mathcal{D})}$, so that combining this with the above, we conclude that $\|u\|_{W_0^{1,p}(\mathcal{D})} \leq c_1 \|\nabla u\|_{L^p(\mathcal{D})}$ for $c_1 = (\mathbf{c}_{\mathcal{P}}^{-p} + d)^{1/p}$.

28 Recall that a domain is an open and connected set. Actually, it suffices that it has finite width, i. e., that it lies between two parallel hyperplanes.

On the other hand, all norms in $\mathbb{R}^d$ are equivalent, hence there is a constant $c' > 0$ such that for all $a = (a_1, \ldots, a_d) \in \mathbb{R}^d$ it holds that $\sum_{i=1}^d |a_i|^p \geq c' (\sum_{i=1}^d |a_i|^2)^{p/2}$. Applying this to the gradient ∇u, it follows that

$$\|u\|^p_{W_0^{1,p}(\mathcal{D})} \geq \sum_{i=1}^d \left\| \frac{\partial u}{\partial x_i} \right\|^p_{L^p(\mathcal{D})} = \int_{\mathcal{D}} \sum_{i=1}^d \left| \frac{\partial u}{\partial x_i} \right|^p dx \geq c' \int_{\mathcal{D}} |\nabla u|^p dx,$$

from which we obtain that $c_1 \|\nabla u\|_{L^p(\mathcal{D})} \leq \|u\|_{W_0^{1,p}(\mathcal{D})}$ holds for $c_1 = (c')^{1/p}$. ◁

Example 1.5.15 (A useful inequality [83]). Let $\mathcal{D}$ be a bounded domain with Lipschitz continuous boundary and $u \in W^{1,p}(\mathcal{D})$, $p < d$, which vanishes on a set A_o of positive measure. Then,

$$\|u\|_{L^{s_p}(\mathcal{D})} \leq 2\widehat{\mathbf{c}}_{\mathcal{P}}(\mathcal{D}) \left(\frac{|\mathcal{D}|}{|A_o|} \right)^{\frac{1}{s_p}} \|\nabla u\|_{L^p(\mathcal{D})},$$

where $\widehat{\mathbf{c}}_{\mathcal{P}}(\mathcal{D}) > 0$ is a constant independent of A_o. This inequality is useful in estimating the size of level sets of functions.

We start by combining the Poincaré inequality of Theorem 1.5.13(ii) for the choice $A = A_o$ with the Sobolev embedding. Defining the function $v = u - u_{A_o} = u$, we have by the Sobolev embedding that $\|v\|_{L^{s_p}(\mathcal{D})} \leq c(\|\nabla v\|_{L^p(\mathcal{D})} + \|v\|_{L^p(\mathcal{D})})$, and using the Poincaré inequality on the right hand side, we obtain an inequality of the form $\|v\|_{L^{s_p}(\mathcal{D})} \leq c' \|\nabla v\|_{L^p(\mathcal{D})}$ for an appropriate constant $c' > 0$ depending on $\widehat{\mathbf{c}}_{\mathcal{P}}$. This may be sufficient for many applications, but it would be better if we could get some further information on the dependence of the constant on the measure of set A_o. This can be obtained as follows: We first work as above setting $A = \mathcal{D}$ in the Poincaré inequality of Theorem 1.5.13(ii) to obtain an inequality of the form

$$\|u - u_{\mathcal{D}}\|_{L^{s_p}(\mathcal{D})} \leq \widehat{\mathbf{c}}_{\mathcal{P}}(\mathcal{D}) \|\nabla u\|_{L^p(\mathcal{D})}, \tag{1.12}$$

for an appropriate constant $\widehat{\mathbf{c}}_{\mathcal{P}}(\mathcal{D}) > 0$. Then, since $A_o \subset \mathcal{D}$ (and is of positive measure), clearly, $\|u - u_{\mathcal{D}}\|_{L^{s_p}(\mathcal{D})} \geq \|u - u_{\mathcal{D}}\|_{L^{s_p}(A_o)} = |u_{\mathcal{D}}|\, |A_o|^{\frac{1}{s_p}}$, hence $|u_{\mathcal{D}}| \leq |A_o|^{-\frac{1}{s_p}} \|u - u_{\mathcal{D}}\|_{L^{s_p}(\mathcal{D})}$. By the triangle inequality

$$\|u\|_{L^{s_p}(\mathcal{D})} \leq \|u - u_{\mathcal{D}}\|_{L^{s_p}(\mathcal{D})} + |u_{\mathcal{D}}| |\mathcal{D}|^{\frac{1}{s_p}} \leq 2 \left(\frac{|\mathcal{D}|}{|A_o|} \right)^{\frac{1}{s_p}} \|u - u_{\mathcal{D}}\|_{L^{s_p}(\mathcal{D})},$$

where we also used the obvious estimate $|A_0| \leq |\mathcal{D}|$. The result follows by applying the Poincare-type inequality (1.12) obtained above. ◁

We now consider the behavior of functions in Sobolev spaces on the boundary $\partial\mathcal{D}$ of the domain $\mathcal{D}$ and consider the problem of restricting a Sobolev function on $\mathcal{D}$. Functions that are elements of Sobolev spaces are elements of a Lebesgue space, hence it may be that there does not exist continuous representatives of them. Trace theory is concerned with assigning boundary values to such functions. In partial differential equations, there are instances where we need to know the value of a function or of its normal derivative at $\partial\mathcal{D}$. This is related to two mappings, which are called traces.

Definition 1.5.16 (Trace operators). Let $\mathcal{D} \subset \mathbb{R}^d$ be a domain, $\partial\mathcal{D}$ its boundary, and $n(x)$ the normal vector at $x \in \partial\mathcal{D}$. Consider a function $u : \overline{\mathcal{D}} \to \mathbb{R}$.

(i) The map $\gamma_0 : u \mapsto u|_{\partial\mathcal{D}}$ is called the trace map. This map "assigns" boundary values to the function u.

(ii) The map $\gamma_1 : u \mapsto \frac{\partial}{\partial n}u = \nabla u \cdot n|_{\partial\mathcal{D}}$. This map "assigns" to u its normal derivative on the boundary.

Clearly, the above is not yet a proper definition since we have not properly defined for which type of functions nor sufficient properties of $\partial\mathcal{D}$ such that these maps make sense. For example, γ_0 can be defined for functions $u \in C(\overline{\mathcal{D}})$, and γ_1 can be defined for functions in $C^1(\overline{\mathcal{D}})$, but is this true for generalized functions or elements of Sobolev spaces? Furthermore, the answer to this question depends also on the properties of the domain $\mathcal{D}$, or rather its boundary $\partial\mathcal{D}$. The proposition that follows asserts that as long as $\mathcal{D}$ has certain regularity properties and the function belongs to a certain class of Sobolev spaces, then there exists a map that can be considered as an extension of the map assigning boundary values to this function in the case where the function is continuous or continuously differentiable.

Proposition 1.5.17 (Trace spaces). *Let $\mathcal{D}$ be a bounded domain and consider $p \in [1, \infty]$.*

(i) *If $\mathcal{D}$ is Lipschitz and $sp > 1$, then there exists a unique linear and continuous trace operator $\gamma_0 : W^{s,p}(\mathcal{D}) \to L^p(\partial\mathcal{D})$ such that $\gamma_0 u = u|_{\partial\mathcal{D}}$ for any $u \in \mathscr{D}(\partial\mathcal{D})$.*

(ii) *If $\mathcal{D}$ is Lipschitz, the range of the map γ_0, which is called the trace space $H^{1/2}(\partial\mathcal{D}) := \gamma_0(W^{1,2}(\mathcal{D}))$, is a Banach space when endowed with the norm*

$$\|u\|^2_{H^{1/2}(\partial\mathcal{D})} = \int\limits_{\partial\mathcal{D}} |u(x)|^2 d\mu(x) + \int\limits_{\partial\mathcal{D}}\int\limits_{\partial\mathcal{D}} \frac{|u(x) - u(z)|^2}{|x - z|^{d+1}} d\mu(x) d\mu(z),$$

where μ is the Hausdorff measure on $\partial\mathcal{D}$.

(iii) *If $\mathcal{D}$ is Lipschitz, there exists a linear continuous map $f : H^{1/2}(\partial\mathcal{D}) \to W^{1,2}(\mathcal{D})$ such that $\gamma_0(f(u)) = u$ for any $u \in W^{1,2}(\mathcal{D})$, with the property $\|f(u)\|_{W^{1,2}(\mathcal{D})} \le c\|u\|_{H^{1/2}(\partial\mathcal{D})}$, for every $u \in H^{1/2}(\partial\mathcal{D})$, for a constant $c = c(\mathcal{D})$.*

(iv) *If $\mathcal{D}$ is Lipschitz, the injection $H^{1/2}(\partial\mathcal{D}) \overset{c}{\hookrightarrow} L^2(\partial\mathcal{D})$ is compact.*

(v) *If $\mathcal{D}$ is C^1 and $sp > 1 + p$, then there exists a unique linear and continuous trace operator $\gamma_1 : W^{s,p}(\mathcal{D}) \to L^p(\partial\mathcal{D})$ such that $\gamma_1 u = \frac{\partial}{\partial n}u$ for any $u \in \mathscr{D}(\partial\mathcal{D})$.*

The Sobolev spaces $W^{k,p}_0(\mathcal{D})$ admit alternative characterizations in terms of trace maps. For example,

Proposition 1.5.18. *Let $\mathcal{D}$ be a bounded Lipschitz domain. Then,*

$$W_0^{1,p}(\mathcal{D}) = \{u \in W^{1,p}(\mathcal{D}) \, : \, \gamma_0 u = 0, \ a.e.\}.$$

The above can be interpreted as the statement that the Sobolev space $W_0^{1,p}(\mathcal{D})$ consists of the functions in $W^{1,p}(\mathcal{D})$, whose values vanish on $\mathcal{D}$ (in the sense of traces). A similar characterization can be obtained for other spaces $W_0^{k,p}(\mathcal{D})$, where now the normal trace may be involved. For example

$$W_0^{2,p}(\mathcal{D}) = \{u \in W^{2,p}(\mathcal{D}) \, : \, \gamma_0 u = \gamma_1 u = 0, \ a.e.\},$$

where of course now the domain will have to be so that the normal trace operator can be defined, i. e., C^1. Such characterizations are very useful when Sobolev spaces are applied to the solution of boundary value problems.

We close this section by a generalization of the classical integration by parts formulae for functions in Sobolev spaces, which importantly lead to Green's formula, which is extremely useful for applications.

Proposition 1.5.19. *Let $\mathcal{D} \subset \mathbb{R}^d$ be a bounded domain with Lipschitz boundary $\partial\mathcal{D}$.*

(i) *For any $u \in W^{1,1}(\mathcal{D})$,*

$$\int_{\mathcal{D}} \frac{\partial u}{\partial x_i}(x) u(x) dx = \int_{\partial\mathcal{D}} u(s) n_i(s) ds,$$

or in terms of vector fields

$$\int_{\mathcal{D}} \nabla \cdot u(x) dx = \int_{\partial\mathcal{D}} u(s) \cdot n(s) ds.$$

(ii) *For any $u \in W^{1,p}(\mathcal{D})$ and $v \in W^{1,p^*}(\mathcal{D})$, it holds that*

$$\int_{\mathcal{D}} \frac{\partial u}{\partial x_i}(x) v(x) dx = \int_{\partial\mathcal{D}} u(x) v(s) n_i(s) ds - \int_{\mathcal{D}} u(x) \frac{\partial v}{\partial x_i}(x) dx.$$

(iii) *For any $u \in W^{2,2}(\mathcal{D})$ and $v \in W^{1,2}(\mathcal{D})$, the first Green formula holds:*

$$\int_{\mathcal{D}} \nabla u(x) \cdot \nabla v(x) dx = \int_{\partial\mathcal{D}} \frac{\partial u}{\partial n}(s) v(s) ds - \int_{\mathcal{D}} \Delta u(x) v(x) dx,$$

where $\Delta u(x) = \sum_{i=1}^d \frac{\partial^2 u}{\partial x_i^2}(x)$ is the Laplacian.

In the above, n is the outer unit normal vector field on $\partial\mathcal{D}$, and the value of u and its derivatives on $\partial\mathcal{D}$ is understood in the sense of trace.

1.6 Lebesgue–Bochner and Sobolev–Bochner spaces

In certain cases (e. g., in evolution equations), the need arises to consider functions from intervals of $\mathbb{R}$ to some Banach space X, i. e., mappings $t \mapsto z = x(t) \in X$. So as to leave notation as intuitive as possible, we will denote such functions by $f = x(\cdot) : [0,T] \subset \mathbb{R} \to X$ (keeping x in the notation as an indicator of the Banach space in which the function takes values). Such functions can be integrated over the Lebesgue measure on $\mathbb{R}$ yielding a vector valued integral called the Bochner integral. Since $f = x(\cdot)$ takes values on X, measurability issues can be challenging. The theory of the Lebesgue integral, Lebesgue spaces, and Sobolev spaces, can be generalized for functions taking values in Banach spaces. We will use the notation $I = [0,T]$ or its open version depending on context.[29]

We collect in this section some useful definitions and facts concerning Banach space valued functions (see, e. g., [69] or [128]).

Definition 1.6.1 (Strong and weak measurability).
(i) $f : I \subset \mathbb{R} \to X$ is called strongly measurable if there exists a sequence $(s_n)_{n\in\mathbb{N}}$ of simple functions (i. e., of functions $s_n : I \subset \mathbb{R} \to X$ of the form $s_n(t) = \sum_{i=1}^{M(n)} x_{i,n} \mathbf{1}_{A_{i,n}}(t)$ for every $t \in I$, where $M(n)$ is finite, $x_{i,n} \in X$ and $A_{i,n} \subset I$ measurable sets) such that $s_n(t) \to f(t)$ in X a. e.
(ii) $f : I \subset \mathbb{R} \to X$ is called weakly measurable if for any $x^\star \in X^\star$ the function $\phi : I \subset \mathbb{R} \to \mathbb{R}$, defined by $\phi(t) := \langle x^\star, f(t)\rangle$, for every $t \in I$ is measurable.

These two concepts are connected with the celebrated Pettis theorem.

Theorem 1.6.2 (Pettis). *If X is a separable Banach space, a function f is strongly measurable if and only if it is weakly measurable.*

We may now define the Bochner integral, which is a useful vector valued extension of the Lebesgue integral.

Definition 1.6.3 (Bochner integral).
(i) The Bochner integral for a simple function $s : I \subset \mathbb{R} \to X$ of the form $s(t) = \sum_{i=1}^{M} x_i \mathbf{1}_{A_i}(t)$ is defined as $\int_I s(t)dt = \sum_{i=1}^{M} x_i \mu_{\mathcal{L}}(A_i)$, where $\mu_{\mathcal{L}}(A_i)$ is the Lebesgue measure of $A_i \subset I$.
(ii) If $f : I \subset \mathbb{R} \to X$ is a strongly measurable function and $(s_n)_{n\in\mathbb{N}}$, is a sequence of simple functions such that $s_n(t) \to f(t)$ in X, a. e. in I, we define the Bochner integral $\int_I f(t)dt := \lim_{n\to\infty} \int_I s_n(t)dt$. A strongly measurable function $f : I \subset \mathbb{R} \to X$ is called Bochner integrable if there exists a sequence of simple functions $(s_n)_{n\in\mathbb{N}}$ such that $\|s_n(t) - f(t)\| \to 0$ a. e. in I and $\int_I \|s_n(t) - f(t)\|dt \to 0$, with the last integral understood as a Lebesgue integral.

29 i. e., whether we consider the function as an element of $C([0,T];X)$ or as an element of a Lebesgue space $L^p((0,T);X)$.

If a function f is Bochner integrable, the value of the integral does not depend on the choice of the approximating sequence. Furthermore, the Pettis measurability theorem may be used to extend the class of integrable functions. Clearly, the Bochner integral satisfies the properties of the Lebesgue integral, i. e., linearity, additivity, and the fact that null sets for the Lebesgue measure do not contribute to the value of the integral. The following theorem collects the most important properties of the Bochner integral:

Theorem 1.6.4 (Properties of Bochner integral).
(i) *f is Bochner integrable if and only if the real valued function $t \mapsto \|f(t)\|$ is Lebesgue integrable and $\| \int_I f(t)dt\| \leq \int_I \|f(t)\|dt$.*
(ii) *If $\mathsf{L} : X \to X$ is a bounded linear operator and $f : I \subset \mathbb{R} \to X$ is Bochner integrable, then so is $\mathsf{L} \circ f : I \subset \mathbb{R} \to X$ and $\mathsf{L}(\int_I f(t)dt) = \int_I \mathsf{L}f(t)dt$.*
(iii) *If $f : I \subset \mathbb{R} \to X$ is Bochner integrable, then $\int_I \langle x^\star, f(t)\rangle dt = \langle x^\star, \int_I f(t)dt\rangle$ for every $x^\star \in X^\star$.*
(iv) *If $\mathsf{A} : D(\mathsf{A}) \subset X \to Y$ is a closed linear operator and $f : I \subset \mathbb{R} \to X$ is a Bochner integrable function taking values a..e in $D(\mathsf{A})$ and such that $\mathsf{A} \circ f : I \subset \mathbb{R} \to Y$ is Bochner integrable, then $\int_I (\mathsf{A}f)(t)dt \in D(\mathsf{A})$ and $\mathsf{A}\int_I f(t)dt = \int_I (\mathsf{A}f)(t)dt$.*
(v) *If $f : I \subset \mathbb{R} \to X$ is (locally) Bochner integrable, then $\lim_{h\to 0} \frac{1}{h} \int_{[t,t+h]} f(\tau)d\tau = f(t)$, a. e. $t \in I$.*

The Bochner integral can be used for the definition of an important class of Banach spaces, called Lebesgue-Bochner spaces. These play an important role in the study of evolution equations.

Definition 1.6.5 (Lebesgue–Bochner spaces).
(i) For $p \in [1, \infty)$, we define the spaces

$$L^p(I;X) := \left\{f : I \subset \mathbb{R} \to X \,:\, f \text{ measurable and } \int_I \|f(t)\|_X^p dt < \infty\right\},$$

which are Banach spaces (called Lebesgue–Bochner spaces) when equipped with the norm $\|f\|_{L^p(I;X)} = \{\int_I \|f(t)\|_X^p dt\}^{1/p}$.
(ii) In the case $p = \infty$, we define

$$L^\infty(I;X) := \left\{f : I \subset \mathbb{R} \to X \,:\, f \text{ measurable and } \operatorname*{ess\,sup}_{t\in I} \|f(t)\|_X < \infty\right\},$$

which is also a Banach space when equipped with the norm $\|f\|_{L^\infty(I;X)} = \operatorname{ess\,sup}_{t\in I} \|f(t)\|_X < \infty$.

Lebesgue–Bochner spaces satisfy versions of the Hölder inequality, in particular (a) $hf \in L^1(I;X)$ if $h \in L^p(I;\mathbb{R})$ and $f \in L^{p^*}(I;X)$ or (b) $\langle f_1, f_2\rangle \in L^1(I;\mathbb{R})$ if $f_1 \in L^{p^*}(I;X^\star)$ and $f_2 \in L^p(I;X)$, with $p \in [1,\infty]$ and $\frac{1}{p} + \frac{1}{p^*} = 1$. These results follow easily from the standard Hölder inequality and Theorem 1.6.4. One may also provide embedding theorems similar to the standard versions as well as extensions.

Theorem 1.6.6 (Lebesgue–Bochner embeddings). *If $p_1, p_2 \in [1, \infty]$ with $p_1 \leq p_2$ and $X \hookrightarrow Y$ (with continuous embedding), then $L^{p_2}(I; X) \hookrightarrow L^{p_1}(I; Y)$ (with continuous embedding). One may also consider the case where $X = Y$.*

The characterization of the dual space $L^p(I; X)$ depends on the properties of the Banach space X. For example, if X is reflexive and $p \in (1, \infty)$, then $(L^p(I; X))^* \simeq L^{p^*}(I; X^*)$, where as usual $\frac{1}{p} + \frac{1}{p^*} = 1$, the duality pairing being $\langle f_1, f_2 \rangle_{(L^p(I;X))^*, L^p(I;X)} = \int_I \langle f_1(t), f_2(t) \rangle_{X^*, X} dt$.

One may further define Sobolev spaces in this setting. We start with the definition of the generalized derivative for functions $f : I \subset \mathbb{R} \to X$.

Definition 1.6.7. Let $f : I \subset \mathbb{R} \to X$ be a vector valued function. The distributional derivative $\frac{df}{dt} \in \mathcal{L}(C_c^\infty(I; \mathbb{R}); Y)$ is the bounded linear operator defined by the integration by parts formula $\int_I \frac{df}{dt}(t)\phi(t)dt = -\int_I f(t)\frac{d\phi}{dt}(t)dt$ for every $\phi \in C_c^\infty(I; \mathbb{R})$.

Note that we may need to choose Y different than X and such that $X \subset Y$, in order for $\frac{df}{dt}$ to become a bounded linear operator. The following concrete example illustrates this phenomenon:

Example 1.6.8. Consider the function $f : [0, 1] \to X := L^2((-\pi, \pi))$ defined by $f(t) := \psi(t, \cdot)$, where $\psi(t, x) = \sum_{n \in \mathbb{N}} a_n e^{-n^2 t} \sin(nx)$ for every $t \in [0, 1]$, $x \in [-\pi, \pi]$. This function is well defined for any $(a_n)_{n \in \mathbb{N}} \subset \ell^2$. Differentiate formally the Fourier series with respect to t to get $\frac{df}{dt}(t) = \frac{\partial}{\partial t}\psi(t, \cdot)$, which gives for any $(t, x) \in [0, 1] \times [-\pi, \pi]$ that $\frac{\partial}{\partial t}\psi(t, x) = -\sum_{n \in \mathbb{N}} n^2 a_n e^{-n^2 t} \sin(nx)$, which for $(a_n)_{n \in \mathbb{N}} \subset \ell^2$ is no longer a function in $X = L^2((-\pi, \pi))$ but rather an element of $Y = W^{-1,2}((-\pi, \pi)) = (W_0^{1,2}((-\pi, \pi)))^*$. ◁

We may now define a general class of Sobolev–Bochner spaces.

Definition 1.6.9 (Sobolev–Bochner spaces). For the Banach spaces $X \subset Y$, we define the Sobolev–Bochner space

$$W^{1,p,q}(I; X, Y) = \left\{ f : I \subset \mathbb{R} \to X, \text{ with } f \in L^p(I; X), \ \frac{df}{dt} \in L^q(I; Y) \right\}.$$

This is a Banach space when equipped with the norm $\|f\|_{W^{1,p,q}(I;X,Y)} = \|f\|_{L^p(I;X)} + \|\frac{df}{dt}\|_{L^q(I;Y)}$.

We will also use the notation $W^{1,p}(I; X) := W^{1,p,p^*}(I; X, X^*)$.

The Sobolev–Bochner spaces enjoy some useful embedding theorems.

Theorem 1.6.10 (Sobolev–Bochner embedding). *Let $p, q \geq 1$ and $X \hookrightarrow Y$ (with continuous embedding). Then, $W^{1,p,q}(I; X, Y) \hookrightarrow C(I; Y)$ (with continuous embedding).*

An important special case in Definition 1.6.9 is when $Y = X^\star$, the dual space of X. In this case, one needs a construction called an evolution triple (or Gel'fand triple).

Definition 1.6.11. An evolution triple (or Gel'fand triple) is a triple of spaces $X \hookrightarrow H \hookrightarrow X^\star$, where X is a separable reflexive Banach space, H is a separable Hilbert space identified with its dual (called the pivot space), and the embedding $X \hookrightarrow H$ is continuous and dense.

Example 1.6.12. An example of an evolution triple is $W_0^{k,p}(\mathcal{D}) \hookrightarrow L^2(\mathcal{D}) \hookrightarrow W^{-k,p^\star}(\mathcal{D}) = (W_0^{k,p}(\mathcal{D}))^*$ for $2 \le p < \infty$. ◁

The continuous and dense embedding $X \hookrightarrow H$ leads to the continuous and dense embedding $H \hookrightarrow X^\star$. Indeed, for every fixed $\mathrm{h} \in H$, one may construct $\mathrm{x}_h^\star \in X^\star$, defined by $\langle \mathrm{x}_h^\star, \mathrm{x}\rangle_{X^\star,X} = \langle h, \iota(\mathrm{x})\rangle_H$ for every $\mathrm{x} \in X$, where $\iota : X \to H$ is the continuous embedding operator of X into H. We then consider the mapping $\iota^\star : H \simeq H^\star \to X^\star$, defined by $\mathrm{h} \mapsto \mathrm{x}_\mathrm{h}^\star$, which is linear, continuous, and injective.[30] The embedding $H \hookrightarrow X^\star$ is understood in terms of the mapping $\iota^\star$. The density of H into $X^\star$ follows by the reflexivity of X. In such cases, we will consider the Sobolev–Bochner spaces $W^{1,p,p^\star}(I;X,X^\star)$, where as usual $p \in (1,\infty)$ and $\frac{1}{p} + \frac{1}{p^\star} = 1$.

The following embeddings hold in an evolution triple:

Theorem 1.6.13. *Let $X \hookrightarrow H \hookrightarrow X^\star$ be an evolution triple. Then*

(i) *$W^{1,p,p^\star}(I;X,X^\star) \hookrightarrow C(I;H)$ with continuous and dense embedding for $p \in (1,\infty)$. Furthermore, for any $f_1, f_2 \in W^{1,p,p^\star}(I;X,X^\star)$ and $0 \le s \le t \le T$, the following integration by parts formula holds:*

$$\begin{aligned}&\langle f_1(t), f_2(t)\rangle_H - \langle f_1(s), f_2(s)\rangle_H\\&= \int_s^t \left(\left\langle \frac{df_1}{dt}(\tau), f_2(\tau)\right\rangle_{X^\star,X} + \left\langle \frac{df_2}{dt}(\tau), f_1(\tau)\right\rangle_{X^\star,X}\right) d\tau.\end{aligned} \tag{1.13}$$

(ii) *If $X \overset{c}{\hookrightarrow} H$, then for any $p \in (1,\infty)$, it holds that $W^{1,p,p^\star}(I;X,X^\star) \overset{c}{\hookrightarrow} L^p(I;H)$ with $\overset{c}{\hookrightarrow}$ denoting compact embedding.*

Proof. We sketch the proof. We start by the integration by parts formula. For any $f_1, f_2 \in C^1(I;X)$, formula (1.13) holds by standard calculus. Then it can be extended to any $f_1, f_2 \in W^{1,p}(I;X)$ by noting that $C^1(I;X) \subset W^{1,p,q}(I;X,Y)$ densely, as long as $p, q \ge 1$ and $X \subset Y$ continuously, so that $C^1(I;X)$ is dense in $W^{1,p}(I;X)$. □

An interesting remark is the following: Absolutely continuous functions on finite dimensional spaces enjoy some sort of weak differentiability, in the sense that for ev-

30 By the density of X into H; if $\mathrm{x}_\mathrm{h}^\star = 0$, then by its definition $\langle \mathrm{h}, \iota(\mathrm{x})\rangle_H = 0$ for every $\mathrm{x} \in X$, and since $X \hookrightarrow H$ densely, we conclude that $\mathrm{h} = 0$.

ery $f : I \to \mathbb{R}^d$ absolutely continuous, there exists a function $g : I \to \mathbb{R}^d$ such that $f(t) = \int_0^t g(\tau)d\tau$, and this function g can be considered as the derivative a. e. on I of the function f. A particularly interesting example of that is the case where f is a Lipschitz continuous function (this is essentially the celebrated Rademacher theorem for the a. e. differentiability of Lipshchitz functions). An alternative way to see this is in terms of the Radon–Nikodym theorem, by associating the function f with a vector valued measure μ on $\mathbb{R}^d$ enjoying continuity properties (as a result of the absolute continuity of f) and associating the Radon–Nikodym derivative of μ with the function g.

An important question is whether the same result is true for the absolutely continuous functions $f : I \to X$, where X is a general Banach space. As before, this is related to the Radon–Nikodym property for the Bochner integral. A notable feature and difference of the Bochner integral vs the Lebesgue integral is that unlike for the Lebesgue integral (and vector valued measures on $\mathbb{R}^d$), in which the Radon–Nikodym property holds generally, for the Bochner integral, whether the Radon-Nikodym property holds or not depends on the properties of the Banach space X. Hence, analogues of the Rademacher theorem for Banach space, valued absolutely continuous functions will hold for specific cases of Banach spaces only. Importantly, reflexive spaces X enjoy this property.

The discussion concerning the Radon–Nikodym property and its connections with the properties of vector measures defined on it, as well as functions with values on this space are quite deep, and this is not the right point to present it (we refer the interested reader, e. g., to [77] Chapter 11). For our needs here, we state the following:

Proposition 1.6.14. *Let X be a reflexive Banach space. Then any absolutely continuous function $f : I \to X$ is a. e. differentiable, and its derivative $g = \frac{df}{dt}$ is a measurable function satisfying*

$$\langle x^*, f(t)\rangle = \langle x^*, f(0)\rangle + \int_0^t \left\langle x^*, \frac{df}{dt}(\tau)\right\rangle d\tau, \quad \forall x^* \in X^*.$$

The property described in Proposition 1.6.14 is equivalent to the property that vector valued measures on X enjoy, the Radon–Nikodym property (see, e. g., Theorem 11.15 in [77]). Moreover, other Banach spaces may enjoy this property, for example, spaces with separable duals.

1.7 Linear semigroups of operators and evolution problems

An important concept is that of linear semigroups of operators. These are directly related to linear evolution problems. The theory of linear semigroups and their connection with evolution problems is a rich field, which has been intensely studied in the literature. Many excellent expositions of this theory exist. Our treatment is motivated by, and based mainly upon, the classic references [33, 75, 93, 111].

In this section, we collect (without proof) some essential results on linear semigroups to pave the way for our treatment of nonlinear evolution problems in Chapter 10. We note that when we get to this point, the reader can attain the proof of the results quoted in this section as particular examples of the theory of nonlinear semigroups (although the theory of linear semigroups precedes and has been developed independently of its nonlinear counterpart).

Definition 1.7.1. Let X be a Banach space and $\{\mathsf{S}(t)\}_{t\geq 0}$ a family of bounded linear operators on X. We say that $\{\mathsf{S}(t)\}_{t\geq 0}$ is a C_0-semigroup, if the following hold:
(i) $\mathsf{S}(0) = I$, the identity operator on X.
(ii) $\mathsf{S}(t+s) = \mathsf{S}(t)\mathsf{S}(s)$ for all $t, s \geq 0$.
(iii) For every $\mathsf{x} \in X$, $\mathsf{S}(t)\mathsf{x} \to \mathsf{x}$, as $t \to 0^+$.

Property (ii) is called the semigroup property. On the other hand, property (iii) implies continuity with respect to t in the strong operator topology. If $\|\mathsf{S}\|_{\mathcal{L}(X)} \leq 1$, then $\{\mathsf{S}(t)\}_{t\geq 0}$ is called a contraction semigroup.

A linear semigroup is connected with a linear operator, called its infinitesimal generator.

Definition 1.7.2. Let $\{\mathsf{S}(t)\}_{t\geq 0}$ be a C_0-semigroup on a Banach space X. Define an operator $\mathsf{A} : \mathbf{D}(\mathsf{A}) \subset X \to X$ as

$$\mathsf{A}\mathsf{x} := \lim_{t\to 0^+} \frac{1}{t}(\mathsf{S}(t)\mathsf{x} - \mathsf{x}),$$

with domain

$$\mathbf{D}(\mathsf{A}) = \{\mathsf{x} \in X \ : \ \text{the above limit exists in } X\}.$$

The operator A is called the *infinitesimal generator* (or simply generator) of the semigroup.

Note that if A is the generator for a semigroup $\{\mathsf{S}(t)\}_{t\geq 0}$, then we always have (at least) $0 \in \mathbf{D}(\mathsf{A})$. Furthermore, it follows from the definition that generators are always linear operators on $\mathbf{D}(\mathsf{A})$.

Through the concept of the generator, linear semigroups are connected with linear evolution problems. This connection is very important in theory and applications, and is one of the main applications of the theory of operator semigroups. We now formulate this connection explicitly:

Let X be a Banach space, and consider the following initial value problem:

$$\begin{aligned} &\frac{d\mathsf{x}}{dt}(t) = \mathsf{A}\mathsf{x}(t) + f(t), \quad 0 < t < T, \\ &\mathsf{x}(0) = \mathsf{x}_0, \end{aligned} \tag{1.14}$$

where $\mathsf{A} : \mathbf{D}(\mathsf{A}) \subset X \to X$ is a linear operator (not necessarily bounded), and $f : [0, T] \to X$ is a given function. Then, we have the following:

Theorem 1.7.3. *If* $\mathsf{A} : \mathbf{D}(\mathsf{A}) \subset X \to X$ *(with* $(\mathsf{A}, \mathbf{D}(\mathsf{A}))$ *as in Definition* 1.7.2*) is the generator for a* C_0*-semigroup* $\{\mathsf{S}(t)\}_{t\geq 0}$ *on* X*, and if* $\mathsf{x}_0 \in \mathbf{D}(\mathsf{A})$*, then the function*

$$\mathsf{x}(t) = \mathsf{S}(t)\mathsf{x}_0 + \int_0^t \mathsf{S}(t-\tau)f(\tau)d\tau, \tag{1.15}$$

is a classical solution of (1.14)*, in the sense that* $\mathsf{x} \in C([0,T]; \mathbf{D}(\mathsf{A})) \cap C^1((0,T); X)$*, with* $\mathsf{x}(0) = \mathsf{x}_0$*, and* x *satisfies* (1.14) *pointwise for* $t > 0$*, if* $f \in W^{1,1}((0,T); X)$ *or* $f \in L^1((0,T); \mathbf{D}(\mathsf{A}))$.

The representation (1.15), which resembles the variation of constants formula, still makes sense if $\mathsf{x}_0 \in X$ and $f \in L^1((0,T); X)$. In this case, we call (1.15) the mild solution of (1.14). The concept of the mild solution (which is clearly weaker than that of classical solution) will play an important role in the study of nonlinear evolution problems (see Chapter 10). Moreover, the above results can be extended to the time interval $(0, \infty)$.

An important question is which type of linear operators can be generators of C_0-semigroups.

Theorem 1.7.4 (Hille–Yosida). *A linear unbounded operator* $\mathsf{A} : \mathbf{D}(\mathsf{A}) \subset X \to X$ *on a Banach space* X *is the infinitesimal generator of a contraction semigroup on* X *if and only if*
(i) A *is closed and densely defined.*
(ii) *For each* $\lambda > 0$, $R(\lambda, \mathsf{A}) := (I - \lambda\mathsf{A})^{-1} : X \to X$ *is a bounded linear operator.*
(iii) *For each* $\lambda > 0$, $\|R(\lambda, \mathsf{A})\|_{\mathcal{L}(X)} = \|(I - \lambda\mathsf{A})^{-1}\|_{\mathcal{L}(X)} \leq 1$.

Condition (ii) *can be interpreted as* $\mathbf{R}(I - \lambda\mathsf{A}) = X$ *for all* $\lambda > 0$ *or* $(0, \infty) \subset \rho(\mathsf{A})$ *(recall Definition* 1.3.11*).*

An alternative general characterization is in terms of the class of dissipative operators.

Definition 1.7.5 (Dissipative operators). Let $\mathsf{A} : \mathbf{D}(\mathsf{A}) \subset X \to X$ be a linear operator.
(i) A is called dissipative if

$$\forall\, \mathsf{x} \in \mathbf{D}(\mathsf{A})\ \exists\, \mathsf{x}^* \in \mathsf{J}(\mathsf{x}) : \langle \mathsf{x}^*, \mathsf{A}\mathsf{x}\rangle \leq 0 \iff \|(I - \lambda\mathsf{A})\mathsf{x}\| \geq \|\mathsf{x}\|, \quad \forall\, \lambda > 0.$$

(ii) A is called m-dissipative if it is dissipative, and $\mathbf{R}(I - \lambda\mathsf{A}) = X$ for some (hence for all) $\lambda > 0$.
(iii) A is called maximal dissipative if any dissipative extension of A coincides with it.
(iv) A is called accretive (resp. m-accretive / maximal accretive) if $-\mathsf{A}$ is dissipative (resp. m-dissipative/maximal dissipative).

We then have the celebrated Lumer–Phillips theorem:

Theorem 1.7.6 (Lumer–Phillips). *A closed linear operator* $\mathsf{A} : \mathbf{D}(\mathsf{A}) \subset X \to X$ *with dense domain generates a continuous semigroup of contractions if and only if* A *is dissipative and* $\mathbf{R}(I - \lambda\mathsf{A}) = X$*, for all* $\lambda > 0$*, i. e., is m-dissipative.*

The concepts of dissipativity and accretivity or monotonicity are different sides of the same coin, as (loosely speaking), A is accretive/monotone if and only if −A is dissipative. Which one we use is simply a matter of convention or convenience, as—clearly—working with one allows us to translate the results to the other by a simple change of sign (for example, the Hille–Yosida or Lumer–Philips theorems can be reformulated in terms of −A if A is accretive/monotone). Monotonicity seems quite a natural formulation in problems related or generalizing minimization of convex functions (as the gradient of such functions enjoys monotonicity properties), whereas dissipativity (on account of the theory of linear semigroups) seems to be a natural formulation in problems related to evolution equations whose more convenient statement is as $\frac{dx}{dt} = \mathsf{A}x$, rather than $\frac{dx}{dt} + \mathsf{A}x = 0$. Dissipativity and monotonicity are important concepts in nonlinear analysis as well, and we will consider them in detail here. For the above reasons, and in order to familiarize the reader with both formulations, we will use both terminologies. So, in Chapter 9, we will study monotone type operators, and in Chapter 10, where evolution problems are studied, we will use dissipative operators (keeping always in mind that these results can be stated using accretive operators upon a sign change).

1.8 Multivalued maps

Given two metric spaces X, Y, a multivalued (or set valued) map is a mapping from X to the power set of Y, i. e., a map from points in X to subsets of Y. Multivalued maps will be denoted by $f : X \to 2^Y$. This is in contrast to single valued maps $f : X \to Y$, which map points to points, and can be considered as a special case of the above, where the image of each point is a singleton. We will encounter very often multivalued maps in this book (for example, we have already introduced the duality map), therefore we need to introduce some fundamental notions related to them (for more detailed accounts see, e. g., [5] or [88]).

We introduce the following definitions:[31]

Definition 1.8.1 (Multivalued maps). Let X, Y be two metric spaces and $f : X \to 2^Y$ a multivalued map. We define the following concepts for f:

31 Note that the terminology in multivalued maps is not standard, and many authors use different terminology and notation for the same concepts. Here we mostly follow the terminology in [5], which we find as more intuitive.

(i) The domain of definition $\mathbf{D}(f) := \{x \in X : f(x) \neq \emptyset\}$, and the graph $\mathbf{Gr}(f) := \{(x, y) \in X \times Y : y \in f(x)\}$.
(ii) The image of a set $A \subset X$, denoted by $f(A) := \bigcup_{x \in A} f(x)$, and the upper and lower inverse of a set $B \subset Y$ denoted by $f^u(B) := \{x \in X : f(x) \subset B\}$ and $f^\ell(B) := \{x \in X : f(x) \cap B \neq \emptyset\}$, respectively.
(iii) A selection from f is a single valued map $f_s : X \to Y$ satisfying $f_s(x) \in f(x)$ for all $x \in X$.
(iv) f is called closed (resp. compact) valued if for every $x \in X$ the set $f(x) \subset Y$ is closed (resp. compact), and closed if $\mathbf{Gr}(f) \subset X \times Y$ is a closed set.

Definition 1.8.2 (Upper and lower semicontinuity). Let X and Y be metric spaces. A multivalued map $f : X \to 2^Y$ is called
(i) Upper semicontinuous at $x \in X$ if for any open set $\mathcal{O} \subset Y$ such that $f(x) \subset \mathcal{O}$, we can find an open neighborhood $N(x) \subset X$ such that $f(N(x)) \subset \mathcal{O}$. We say that f is upper semicontinuous if it is upper semicontinuous at every point $x \in X$.
(ii) Lower semicontinuous at $x \in X$ if for every open set $\mathcal{O}$ such that $\mathcal{O} \cap f(x) \neq \emptyset$ we can find an open neighborhood $N(x) \subset X$ such that $f(x) \cap \mathcal{O} \neq \emptyset$ for all $x \in N(x)$. If f is lower semicontinuous at every point $x \in X$ it is called lower semicontinuous.

The above definition essentially means that the values of upper (lower) semicontinuous functions must not explode (implode) locally. For single valued maps upper and lower semicontinuity, in the sense of Definition 1.8.2, coincide with continuity, so some care has to be taken not to confuse the concept of upper and lower semicontinuity for single valued maps with that for multivalued maps.

Example 1.8.3. We give some simple examples.
(a) Lower but not upper semicontinuous: $f_1 : \mathbb{R} \to 2^{\mathbb{R}}$, defined by $f_1(t) = [1, 4]$, for $t \neq 0$ and $f_1(0) = (1, 4)$ is not upper but is lower semicontinuous at $t = 0$.
(b) Upper and lower semicontinuous: $f_2 : \mathbb{R} \to 2^{\mathbb{R}}$, defined by $f_2(t) = (1, 4)$, for $t \neq 0$ and $f_2(0) = [1, 4]$ is upper and lower semicontinuous at $t = 0$.
(c) Upper but not lower semicontinuous: $f_3 : \mathbb{R} \to 2^{\mathbb{R}}$, defined by $f_3(t) = (1, 4)$, for $t \neq 0$ and $f_3(0) = (1, 5)$ is upper but not lower semicontinuous at $t = 0$. ◁

We now give some useful equivalent characterizations for upper and lower semicontinuity of multivalued maps in metric spaces.

Theorem 1.8.4. *Let X, Y be metric spaces and $f : X \to 2^Y$ be a multivalued (or set valued) map.*
(i) *The following are equivalent:* (a) *f is upper semicontinuous;* (b) *$f^u(\mathcal{O})$ is open for every open $\mathcal{O} \subset Y$;* (c) *$f^\ell(C)$ is closed for every closed $C \subset Y$;* (d) *if $x \in X$, $(x_n)_{n \in \mathbb{N}} \subset X$, $x_n \to x$ and $\mathcal{O} \subset Y$ is an open set such that $f(x) \subset \mathcal{O}$, then there exists $n_o \in \mathbb{N}$ (depending on $\mathcal{O}$) such that $f(x_n) \subset \mathcal{O}$ for every $n \geq n_o$.*

(ii) *The following are equivalent:* (a) *f is lower semicontinuous;* (b) *$f^{\ell}(\mathcal{O})$ is open for every open $\mathcal{O} \subset Y$;* (c) *$f^{u}(C)$ is closed for every closed $C \subset Y$;* (d) *if $x \in X$, $(x_n)_{n\in\mathbb{N}} \subset X$, $x_n \to x$ and $y \in f(x)$, then we can find $y_n \in f(x_n)$, $n \in \mathbb{N}$ such that $y_n \to y$ in Y.*

Proof. (i). To show that (a) is equivalent to (b), we reason as follows: If f is upper semicontinuous and $\mathcal{O} \subset Y$ is open, then $f^{u}(\mathcal{O})$ is a neighborhood of every $x \in f^{u}(\mathcal{O})$. But a set that is a neighborhood of each of its points is open.[32] To show that (b) is equivalent to (c), note that for any $B \subset Y$, we have $f^{\ell}(Y \setminus B) = X \setminus f^{u}(B)$.

(ii) We only show the equivalence between (a) and (d). Let f be lower semicontinuous at x. Furthermore, let $x_n \to x$ and $y \in f(x)$. For every $k \in \mathbb{N}$, let $B(y, \frac{1}{k})$ denote the open ball with radius $\frac{1}{k}$ and center y. Since f is lower semicontinuous at x, there exists for every k a neighborhood V_k of x such that for every $z \in V_k$, we have $f(z) \cap B(y, \frac{1}{k}) \neq \emptyset$. Let $\{n_k : k \in \mathbb{N}\}$ be the subsequence of natural numbers such that $n_k < n_{k+1}$ and $x_n \in V_k$ if $n \geq n_k$. For n with $n_k \leq n < n_{k+1}$, we choose y_n in the set $f(x_n) \cap B(y, \frac{1}{k})$. So, the sequence constructed in this fashion is $y_n \to y$.

Conversely, assume that for every sequence $(x_n)_{n\in\mathbb{N}}$ such that $x_n \to x$ and any $y \in f(x)$, there exists a sequence $y_n \in f(x_n)$, $n \in \mathbb{N}$ such that $y_n \to y$. Suppose that f is not lower semicontinuous at x. Then, there exists an open set $\mathcal{O}'$ with $\mathcal{O}' \cap f(x) \neq \emptyset$ such that every neighborhood V of x contains a point z so that $f(z) \cap \mathcal{O}' = \emptyset$. Therefore, there exists a sequence $(x_n)_{n\in\mathbb{N}}$ converging to x with $f(x_n) \cap \mathcal{O}' = \emptyset$ for $n \in \mathbb{N}$. Let $y \in \mathcal{O}' \cap f(x)$. Then there exists a sequence $(y_n)_{n\in\mathbb{N}}$ with $y_n \in f(x_n)$ converging to y. For n large enough, we have $y_n \in \mathcal{O}'$. Thus, $f(x_n) \cap \mathcal{O}' \neq \emptyset$, a contradiction. □

Remark 1.8.5. In (ii), an equivalent statement could be (d′) if $x \in X$, $(x_n)_{n\in\mathbb{N}} \subset X$, $x_n \to x$, and $\mathcal{O} \subset Y$ is an open set such that $f(x) \cap \mathcal{O} \neq \emptyset$; then there exists $n_o \in \mathbb{N}$ (depending on $\mathcal{O}$) such that $f(x_n) \cap \mathcal{O} \neq \emptyset$ for every $n \geq n_o$. Note that for both (i) and (ii), the equivalence of (a), (b), and (c) above, does not require X, Y to be metric spaces. Additionally, the equivalence of (d) will hold in a more general framework than that of metric spaces, as long as sequences are replaced by nets.

We collect some useful properties of upper semicontinuous functions.

Proposition 1.8.6. *Let X, Y be metric spaces and $f : X \to 2^{Y}$ a multivalued map.*
(i) *If f is upper semicontinuous and has nonempty closed values (i. e., $f(x) \subset Y$ is closed for every $x \in X$), then $\mathbf{Gr}(f) \subset X \times Y$ is a closed set (i. e., f is closed).*
(ii) *If f is upper semicontinuous and has compact values, then for any compact set $K \subset X$ it holds that $f(K) \subset Y$ is compact.*
(iii) *If f has nonempty compact valued and for every $((x_n, y_n))_{n\in\mathbb{N}} \subset \mathbf{Gr}(f)$ with $x_n \to x$ there exists a subsequence $(y_{n_k})_{k\in\mathbb{N}}$ with $y_{n_k} \to y \in \mathbf{Gr}(f)$, then it is upper semicontinuous. The converse is also true.*

32 If A is a set which is a neighbourhood of each of its points, then $A = \bigcup \mathcal{O}'$ where $\mathcal{O}'$ are open sets such that $\mathcal{O}' \subset A$, hence its is open.

(iv) *If f has nonempty closed values, is closed, and is locally compact, in the sense that for any* $x \in X$ *we may find a neighborhood* $N(x)$ *such that* $\overline{f(N(x))} \subset Y$ *is nonempty and compact, then f is upper semicontinuous.*

Proof. (i) Suppose not, and consider a sequence $((x_n, y_n))_{n\in\mathbb{N}} \subset \mathbf{Gr}(f)$ such that $(x_n, y_n) \to (x, y)$, with $y \neq f(x)$. Then, since metric spaces are Hausdorff, there are two open sets $\mathcal{O}_1, \mathcal{O}_2 \subset Y$ such that $y \in \mathcal{O}_1, f(x) \subset \mathcal{O}_2$, and $\mathcal{O}_1 \cap \mathcal{O}_2 = \emptyset$. Since $y_n \to y$, and f is upper semicontinuous, from Theorem 1.8.4(i) (and in particular the equivalence of (a) and (d) using the open set $\mathcal{O}_2$), there exists $n_2 \in \mathbb{N}$ such that $f(x_n) \subset \mathcal{O}_2$ for all $n \geq n_2$. On the other hand, since $y_n \to y$, for the open set $\mathcal{O}_1 \subset Y$, there exists $n_1 \in \mathbb{N}$ such that $y_n \in \mathcal{O}_1$ for $n > n_1$. Choosing $n_o = \max(n_1, n_2)$, we have for $n > n_o$ that $f(x_n) \subset \mathcal{O}_2$, while $y_n \in \mathcal{O}_1$, and while by construction (since $(x_n, y_n) \in \mathbf{Gr}(f)$), it holds that $y_n \in f(x_n)$ for all n. This is in contradiction with $\mathcal{O}_1 \cap \mathcal{O}_2 = \emptyset$.

(ii) Let $\{\mathcal{O}_i,\ i \in I\}$ be an open cover of $f(K)$. If $x \in K$, the set $f(x)$, which is compact, can be covered by a finite number of $\mathcal{O}_i$, i. e., there exists a finite set $I_x \subset I$ such that $f(x) \subset \bigcup_{i\in I_x} \mathcal{O}_i = \mathcal{O}_x$. Since $\mathcal{O}_x$ is open, by Theorem 1.8.4(ii), $f^u(\mathcal{O}_x)$ is open, and it is obvious that $x \in f^u(\mathcal{O}_x)$. Therefore, $\{f^u(\mathcal{O}_x)\}_{x\in X}$ is an open cover of K. Since by assumption K is compact, there exist $x_1, \ldots, x_m$ such that $K \subset \bigcup_{n=1}^m f^u(\mathcal{O}_{x_n})$. Clearly, $\{\mathcal{O}_i\}_{i\in I_{x_n}}$, $n = 1, 2, \ldots, m$ is a finite subcover of $\{\mathcal{O}_i\}_{i\in I}$.

(iii) Let f be upper semicontinuous on X, $x \in X$ and $(x_n)_{n\in\mathbb{N}}$ be any sequence of X such that $x_n \to x$. The set $K = \{x, x_1, x_2, \ldots\}$ is compact, and the restriction of f on K is upper semicontinuous. Thus, by (ii), the set $f(K)$ is compact, and hence the sequence $(y_n)_{n\in\mathbb{N}}$ has a convergent subsequence $(y_{n_k})_{k\in\mathbb{N}}$. Let $y_{n_k} \to y$, as $k \to \infty$. Assume that $y \notin f(x)$. Then, there exists a closed neighborhood $\bar{U}$ of $f(x)$ with $y \notin \bar{U}$. Since f is upper semicontinuous at x, for n large enough $f(x_n) \subset \bar{U}$. Thus, $y_n \in \bar{U}$ for n large enough, and hence $y \in \bar{U}$, which is a contradiction.

For the converse, assume not. Then, there exists an open neighborhood U of $f(x)$ such that every neighborhood V of x contains a point z with $f(z) \not\subset U$. It then follows that there exists a sequence $x_n \to x$ and $y_n \in f(x_n)$ with $y_n \notin U$. By assumption, there exists a converging subsequence $(y_{n_k})_{k\in\mathbb{N}}$ with $\lim_k y_{n_k} \in f(x)$. But $y_{n_k} \notin U$ for all n, therefore $\lim_k y_{n_k} \notin U$, so $\lim_k y_{n_k} \notin f(x)$, which is a contradiction.

(iv) It suffices to show that for any closed $C \subset Y$, the set $f^\ell(C) \subset X$ is closed. Consider a sequence $(x_n)_{n\in\mathbb{N}} \subset f^\ell(C)$. Since f is locally compact, we may choose an open neighborhood $N(x) \subset X$ such that $\overline{f(N(x))}$ is compact. Hence, if $(y_n)_{n\in\mathbb{N}} \subset Y$ is such that $y_n \in f(x_n)$ for every $n \in \mathbb{N}$, there exists a subsequence $(y_{n_k})_{k\in\mathbb{N}}$ with the property $y_{n_k} \to y$ and $y \in \overline{f(N(x))}$. Clearly, $(x, y) \in \mathbf{Gr}(f) \cap (X \times C)$. Hence, $x \in f^\ell(C)$, and $f^\ell(C) \subset X$ is closed. □

Michael's selection theorem

The following important theorem due to Ernest Michael [105] (see also [5, 14, 107], or [143]) guarantees the existence of continuous selections:

Theorem 1.8.7 (Michael). *Let X be a compact metric space,*[33] *Y a Banach space, and $f : X \to 2^Y$ a lower semicontinuous set-valued map with closed and convex values. Then, f has a continuous selection.*

In order to prove this theorem, we need the following lemma:

Lemma 1.8.8. *Let X be a compact metric space, Y a Banach space, and $f : X \to 2^Y$ a lower semicontinuous set-valued map with convex values. Then, for each $\epsilon > 0$, there exists a continuous function f_ϵ such that $f_\epsilon(\mathrm{x}) \in f(\mathrm{x}) + B_Y(0,\epsilon)$ for all $\mathrm{x} \in X$.*

Proof. For every $\mathrm{x} \in X$, we choose $\mathrm{y_x} \in f(\mathrm{x})$, and denote by $B := B_Y(0,1)$. Since f is lower semicontinuous, the set $f^{\ell}(\mathrm{y_x} + \epsilon B)$ is open, and so $\{f^{\ell}(\mathrm{y_x} + \epsilon B) : \mathrm{x} \in X\}$ is an open cover of X. Since X is compact, it can be covered by a finite collection of such sets $\{f^{\ell}(\mathrm{y}_{\mathrm{x}_i} + \epsilon B) : i = 1,2,\dots,n\}$. Let us consider a partition of unity $\{\psi_i : i = 1,2,\dots,n\}$ associated with this cover (see Theorem 1.9.14). We claim that our desired map is

$$f_\epsilon(\mathrm{x}) = \sum_{i=1}^{n} \psi_i(\mathrm{x})\mathrm{y}_{\mathrm{x}_i}, \quad \forall \mathrm{x} \in X.$$

Clearly, f_ϵ is continuous, since the functions ψ_i are continuous. Let

$$I(\mathrm{x}) = \{i = 1,\dots,n : \psi_i(\mathrm{x}) > 0\}.$$

It is not empty since $\sum_{i=1}^{n} \psi_i(\mathrm{x}) = 1$. When $i \in I(\mathrm{x})$, it is easy to see that

$$\mathrm{y}_{\mathrm{x}_i} \in f(\mathrm{x}) + \epsilon B.$$

Since the set $f(x) + \epsilon B$ is convex, we conclude that $f_\epsilon(\mathrm{x}) \in f(x) + \epsilon B$. □

Proof of Theorem 1.8.7. By induction, we shall construct a sequence of continuous maps $g_n : X \to Y$, $n \in \mathbb{N}$ such that

$$g_n(\mathrm{x}) \in f(\mathrm{x}) + \frac{1}{2^n}B, \quad \forall \mathrm{x} \in X,\ n \in \mathbb{N}, \tag{1.16}$$

and

$$\|g_{n+1}(\mathrm{x}) - g_n(\mathrm{x})\|_Y < \frac{1}{2^{n-1}}, \quad \forall \mathrm{x} \in X,\ n \in \mathbb{N}. \tag{1.17}$$

For $n = 1$, we apply Lemma 1.8.8 with $\epsilon = \frac{1}{2}$, and see that $g_1(\mathrm{x}) \in f(x) + \frac{1}{2}B$. For the induction step, suppose that we have $g_1, g_2, \dots, g_n$, satisfying (1.16) and (1.17). We consider the set valued map

33 The theorem still holds if X is a paracompact space, i. e., a space where every open cover has a finite open refinement (e. g., a subset of a metric space or a compact subset of a topological space), see [107].

$$\phi_n(x) = f(x) \cap \left(g_n(x) + \frac{1}{2^n} B \right), \quad \forall x \in X.$$

From (1.16), we see that ϕ_n has nonempty values, which are convex. Moreover, ϕ_n is also lower semicontinuous. Indeed, if $x_m \to x$ and if $y \in \phi_n(x)$, by the lower semicontinuity of f there exists $y_m \in f(x_m)$ converging to y. By the construction of g_n, we have[34] $\|y_m - g_n(x_m)\|_Y < 2^{-n}$, and so $y_m \in \phi_n(x_m)$. Then, from Lemma 1.8.8, we can find $g_{n+1} : X \to Y$, a continuous function, such that

$$g_{n+1}(x) \in \phi_n(x) + \frac{1}{2^{n+1}} B, \quad \forall x \in X,$$

hence

$$g_{n+1}(x) \in g_n(x) + \frac{1}{2^n} B + \frac{1}{2^{n+1}} B \subset g_n(x) + \frac{1}{2^{n-1}} B, \quad \forall x \in X.$$

This completes the induction. Since $\sum_{n=1}^{\infty} \frac{1}{2^{n-1}} < \infty$, from (1.17) it follows that $(g_n)_{n \in \mathbb{N}}$ is a Cauchy sequence in $C(X; Y)$. So, we can find $g \in C(X; Y)$ such that $g_n \to g$ uniformly in X. Because f has closed values from (1.16) we deduce that $g(x) \in f(x)$ for all $x \in X$. Hence g is a continuous selection of f. □

Example 1.8.9 (Differential inclusions). Let $f : A \subset \mathbb{R} \times \mathbb{R}^d \to 2^{\mathbb{R}^d}$ be a lower semicontinuous set-valued map into the set of nonempty, closed, and convex subsets of $\mathbb{R}^d$, and consider the initial value problem $\frac{dx}{dt}(t) \in f(t, x(t))$ with initial condition $x(t_0) = x_0$. By Michael's selection theorem (see Theorem 1.8.7), there is a continuous selection $f_s : A \to \mathbb{R}^d$. So it is sufficient to consider the classical initial value problem $\frac{dx}{dt}(t) = f_s(t, x(t))$, with the same initial condition, which, by the classical Peano theorem, has a C^1 solution $x = x(t)$ in a neighborhood of t_0 (local solution). Then, by continuation arguments, we may obtain global results. ◁

1.9 Appendix

1.9.1 Elements of topology

In this section, we collect some elementary notions from topology that will be needed in this book. For more details, see any of the excellent texts dedicated to the subject, e. g., [70, 92, 134], and [141].

34 In fact, one can show more generally that if $f : X \to 2^Y$ is lower semicontinuous and $g : X \to Y$ is continuous, that $x \mapsto f(x) \cap (g(x) + U)$, where U is an open set is lower semicontinuous, see, e. g., [5].

Definition 1.9.1 (Topological space). Let X be a set.

(i) A collection τ of subsets of X is called a topology on X if it satisfies the following properties:
   (a) $\emptyset$ and X belong in τ.
   (b) Any union of elements of τ is an element of τ.
   (c) Finite intersections of elements of τ are elements of τ.

(ii) The elements of τ are called open sets, whereas their complements are called closed sets.

(iii) The pair (X, τ) is called a topological space.

Definition 1.9.2. A neighborhood U of $x \in X$ is a subset $U \subset X$ that contains an open set V, including x, ($x \in V \subset U$). The set of neighborhoods of x is denoted by $N(x)$ (often called the neighborhood system at x).

Though U is not necessarily open, we can always restrict ourselves to the case of open U, and require neighborhoods to be open. We will then denote by $N(x)$ the set of all open neighborhoods of x in X. As the concept of open depends on the choice of topology τ, neighborhoods and their structure depend the choice of topology.

Definition 1.9.3. A topological space in which for any two distinct points $x_1, x_2 \in X$, there exist a neighborhood U_1 of x_1 and a neighborhood U_2 of x_2 such that $U_1 \cap U_2 = \emptyset$ is called Haussdorff.

In a metric space setting (X, d), where the topology τ contains the open sets defined in terms of the open balls $B(x, r) = \{z \in X \,:\, d(x, z) < r\}$, an open set U is a set such that for every $x \in U$ there exists a ball $B(x, \epsilon) \subset U$. In this setting, a set U is a neighborhood of x if there exists an open ball $B(x, \epsilon) \subset U$. This characterization can be applied, e. g., for a Banach space X endowed with the strong topology, but some care must be taken when the space X is endowed with say the weak or the weak-$*$ topology.

With the above general definition of open and closed sets, many of the familiar concepts in the context of metric spaces can be generalized in the general context of topological spaces. For example,

Definition 1.9.4. Let (X, τ) be a topological space and $A \subset X$.

(i) x is an interior point of A in X if there exists an open set U of X such that $x \in U \subset A$,

(ii) The interior of A, denoted $\text{int}(A)$, is the set of all interior points of A (equiv., the largest open set contained in A, equiv., the union of all open sets contained in A).

(iii) The closure of A, denoted $\overline{A}$, is the smallest closed set containing A (equiv., the intersection of all closed sets containing A).

Clearly, the notions of interior and closure of a set depend on the chosen topology. Other convenient characterizations are, e. g., that $x \in \overline{A}$ if and only if every open set U containing x intersects A, or in terms of cluster points (see Definition 1.9.7). If A' is the set of all cluster points of A, then $\overline{A} = A \cup A'$, and a set is closed if and only if it contains all its cluster points.

Definition 1.9.5. Let (X, τ) be a topological space.
(i) A set $A \subset X$ is called dense in X if $\overline{A} = X$.
(ii) (X, τ) is called separable if it has a countable dense subset A.

A fundamental paradigm of separable space is $\mathbb{R}$ with the standard topology, in which case $A = \mathbb{Q}$, the rationals. Other important examples are certain L^p spaces (for $p \in [1, \infty)$) or related spaces, e. g., Sobolev spaces. For example, $L^p(\mathbb{R})$ is separable for $p \in [1, \infty)$ and a countable dense set is $A = \{\sum_{i=n}^{N} c_n \mathbf{1}_{[a_n, b_n)} : N > 0, a_n, b_n, c_n \in \mathbb{Q},\ a_n < b_n\}$ [85]. In the more general case of $L^p(I)$, $I \subset \mathbb{R}$, measurable, $p \in [1, \infty)$, a countable dense set is $A_I = \{f\mathbf{1}_A : f \in A\}$, where A is a countable dense set of $L^p(\mathbb{R})$, while generalizations for $L^p(E)$, $E \subset \mathbb{R}^d$ are possible (see, e. g., [85]). Another important example for separable spaces are Banach spaces, which admit a Schauder basis, i. e., a sequence $(b_n)_{n \in \mathbb{N}} \subset X$ such that for any $\mathrm{x} \in X$ there exists a unique sequence $(x_n)_{n \in \mathbb{N}} \subset \mathbb{R}$ so that $\mathrm{x} = \sum_{n \in \mathbb{N}} x_n b_n$, where convergence is considered in the norm topology. Clearly, a Banach space X with a Schauder basis $(b_n)_{n \in \mathbb{N}} \subset X$ is separable since $A = \{\sum_{n=1}^{N} c_n b_n : N > 0, c_n \in \mathbb{Q}\}$ is a countable dense subset of X (note however that not every separable Banach space admits a Schauder basis). As a final example, consider the Hilbert space $L^2((0,1))$ which admits the Fourier basis $(b_n)_{n \in \mathbb{Z}} \subset L^2((0,1))$ with $b_n(x) = e^{i2\pi nx} = \cos(2\pi nx) + i\sin(2\pi nx)$.

An important observation, needed in several applications, is the following: If X is a separable Banach space, then we can construct a sequence $(X_n)_{n \in \mathbb{N}}$ of finite dimensional spaces X_n such that $X_n \subset X_{n+1} \subset X$ and $\bigcup_{n \in \mathbb{N}} X_n$ is dense, is X (i. e. $\overline{\bigcup_{n \in \mathbb{N}} X_n} = X$). Moreover, for any $\mathrm{x} \in X$, we can construct a sequence $(\mathrm{x}_n)_{n \in \mathbb{N}} \subset X$ such that $\mathrm{x}_n \in X_{\ell_n}$ for all $n \in \mathbb{N}$, where $(\ell_n)_{n \in \mathbb{N}} \subset \mathbb{R}$ is an increasing sequence, with $\mathrm{x}_n \to \mathrm{x}$ in X (strong convergence).

Definition 1.9.6. Let (X, τ) be a topological space.
(i) A subset $B \subset \tau$ is a basis for τ if every member of τ is the union of members of B.
(ii) A subset $B_N \subset N(\mathrm{x})$ is a basis of neighborhoods at x if for every $U \in N(\mathrm{x})$ there is $V \in B_N$ such that $V \subset U$.
(iii) A topological space (X, τ) satisfies the first axiom of countability if it has a countable local basis, i. e., at every point $\mathrm{x} \in X$ there is a countable basis $(B_n)_{n \in \mathbb{N}}$ of neighborhoods at x. While the basis $(B_n)_{n \in \mathbb{N}}$ is not necessarily monotone, it can be chosen to be decreasing by setting $U_n = \bigcap_{i=1}^{n} B_i$ so that we obtain a countable basis $(U_n)_{n \in \mathbb{N}}$ with the property $U_{n+1} \subset U_n$ such that for any open set U containing x there exists $N \in \mathbb{N}$ so that $U_N \subset U$. Such spaces are often called first countable.
(iv) A topological space (X, τ) satisfies the second axiom of countability if it has a countable basis.

Any metrizable space satisfies the first axiom of countability. For example, a Banach space equipped with the strong topology is first countable, however, when equipped with the weak topology it is not. First countable spaces enjoy some nice properties, primarily associated with the fact that in such spaces many topological properties can be characterized in terms of sequences and their limits. Second countability is a stronger property than first countability, in the sense that a second countable space is first countable. In a second countable space, any open covering contains a countable subcover (spaces with this property are often called Lindelöf spaces). A second countable space is separable (but not necessarily the other way around).

Definition 1.9.7. Let (X, τ) be a topological space.

(i) A sequence $(x_n)_{n\in\mathbb{N}} \subset X$ converges to x in the topological space (X, τ) if for every open neighborhood $U \in N(x)$ there exists an $N \in \mathbb{N}$ such that $x_n \in U$ for all $n > N$. In this case, x is called limit of $(x_n)_{n\in\mathbb{N}}$.
(ii) x is a cluster point (or accumulation point) of $(x_n)_{n\in\mathbb{N}}$ if each neighborhood of x contains at least a term of $(x_n)_{n\in\mathbb{N}}$, distinct from x.

The limit of any converging subsequence of $(x_n)_{n\in\mathbb{N}}$ is a cluster point of $(x_n)_{n\in\mathbb{N}}$, with the reverse holding in first countable spaces. In general, topological spaces, a cluster point of a sequence, is not necessarily the limit of a subsequence. In some cases, sequences are insufficient, and the more general concept of nets is required. The notion of limit and cluster point can be extended to nets $(x_\alpha)_{\alpha\in\mathcal{I}}$, i. e., mappings from a nonempty directed set $\mathcal{I}$ to X (a directed set is a set equipped with a reflexive, transitive relation $\succeq$ such that for any $\alpha_1, \alpha_2 \in \mathcal{I}$ there exists $\alpha_3 \in \mathcal{I}$ with $\alpha_3 \succeq \alpha_1$ and $\alpha_3 \succeq \alpha_2$). Then, the items in Definition 1.9.7 can be extended for nets, verbatim, replacing the $\geq$ relation with $\succeq$.

The notion of compactness in its various forms plays an important role in analysis.

Definition 1.9.8. Let (X, τ) be a topological space and $A \subset X$.

(i) A is called compact if every open cover of A has a finite subcover.
(ii) A is called sequentially compact if every sequence in A has a convergent subsequence (with limit in A).

Other notions of compactness are available, e. g., limit point compactness, in which every infinite subset of A has a cluster point etc. Various connections between these notions exist, for example, a compact set is also limit point compact. Moreover, in metrizable spaces (i. e., spaces in which the topology is generated by a metric) the notions of compactness, sequential compactness, and limit point compactness are equivalent.

Definition 1.9.9. A family $\mathcal{A}$ of sets has the finite intersection property if the intersection of the members of each finite subfamily of $\mathcal{A}$ is nonempty.

Proposition 1.9.10. *A metric space X is compact if and only if every family $\mathcal{C}$ of closed sets with the finite intersection property having a nonempty intersection.*

Finally, we recall the celebrated Baire theorem:

Theorem 1.9.11 (Baire). *Let (X, d) be a complete metric space.*

(i) *If $(C_n)_{n\in\mathbb{N}}$ is a sequence of closed sets such that $\bigcup_{n=1}^{\infty} C_n = X$, there exists $n_0 \in \mathbb{N}$ such that*[35] $\mathrm{int}(C_{n_0}) \neq \emptyset$.

(ii) *If $(\mathcal{O}_n)_{n\in\mathbb{N}}$ is a sequence of open dense subsets of X, then $\bigcap_n \mathcal{O}_n \neq \emptyset$ (countable intersections of open dense sets are nonempty).*

1.9.2 Spaces of continuous functions and the Arzelá–Ascoli theorem

Let (X, τ_X) and (Y, τ_Y) be two topological spaces. A function $f : X \to Y$ is called continuous if f^{-1} maps open (closed) sets of Y to open (closed) sets of X.

Given any two metric spaces X and Y, we may define the space of continuous functions.

$$C(X; Y) := \{f : X \to Y \ : f \text{ continuous}\}.$$

An important theorem, which is very useful in applications is the Arzelá–Ascoli theorem. We will first need the following definitions:

Definition 1.9.12 (Equicontinuity and pointwise (pre) compactness). Let X, Y be metric spaces and $A \subset C(X, Y)$.

(i) A is called equicontinuous if for every $\epsilon > 0$ there exists $\delta > 0$ such that for every $\mathrm{x}_1, \mathrm{x}_2 \in X$ with $\mathrm{d}_X(\mathrm{x}_1, \mathrm{x}_2) < \delta$ we have $\mathrm{d}_Y(f(\mathrm{x}_1), f(\mathrm{x}_2)) < \epsilon$ for all $f \in A$.

(ii) A is called pointwise compact if for every $\mathrm{x} \in X$, the set $A(\mathrm{x}) = \{f(\mathrm{x}) \ : \ f \in A\}$ is a compact subset of Y.

(iii) A is called pointwise precompact if for every $\mathrm{x} \in X$, the set $\overline{A(\mathrm{x})} = \overline{\{f(\mathrm{x}) \ : \ f \in A\}}$ is a compact subset of Y (i. e., $A(\mathrm{x})$ has compact closure).

Theorem 1.9.13 (Arzelá–Ascoli). *Let (X, d_X) be a compact metric space. A subset $A \subset C(X, Y)$ is precompact if and only if it is pointwise precompact and equicontinuous.*

If in the above A is closed, then it is compact. In the special case, where $Y = \mathbb{R}^d$, by the Heine-Borel theorem, we may exchange pointwise precompactness with boundedness of A.

1.9.3 Partitions of unity

Theorem 1.9.14 (Partition of unity). *Let (X, τ) be a topological space and $K \subset X$ a compact subset of X. Let $\{U_i \ : \ i = 1, \dots, n\}$ be a finite covering of K with open sets. Then, there exists*

35 At least one of these closed sets has nonempty interior, i. e., there exists an $\mathrm{x} \in C_{n_0}$ and an $\epsilon > 0$ such that $B(\mathrm{x}, \epsilon) \subset C_{n_0}$.

continuous functions $\psi_i : X \to [0,1]$, $i = 1, \dots, n$, *with compact support* $\operatorname{supp}(\psi_i) \subset U_i$ *and such that* $\sum_{i=1}^{n} \psi_i(x) = 1$ *for every* $x \in K$.

1.9.4 Hölder continuous functions

Definition 1.9.15 (Hölder continuity). A function $f : \mathcal{D} \subset \mathbb{R}^d \to \mathbb{R}$ is called Hölder continuous of exponent $\alpha \in (0,1]$ if there exists $c > 0$ such that $|f(x_1) - f(x_2)| \le c|x_1 - x_2|^\alpha$ for every $x_1, x_2 \in \mathcal{D} \subset \mathbb{R}^d$. The case $\alpha = 1$ corresponds to Lipschitz continuity.

Definition 1.9.16 (Hölder spaces). We define the α-Hölder norm as

$$\|u\|_{C^{0,\alpha}(\overline{\mathcal{D}})} := \|u\|_{C(\overline{\mathcal{D}})} + \sup_{x_1,x_2 \in \mathcal{D}, x_1 \neq x_2} \left(\frac{|u(x_1) - u(x_2)|}{|x_1 - x_2|^\alpha} \right),$$

and denote the corresponding space of all functions such that the above norm is finite by $C^{0,\alpha}(\mathcal{D})$, the space of Hölder continuous functions of exponent α.

The Hölder space $C^{k,\alpha}(\overline{\mathcal{D}})$ consists of all functions $u \in C^k(\overline{\mathcal{D}})$, for which the norm

$$\|u\|_{C^{k,\alpha}(\overline{\mathcal{D}})} := \sum_{|m| \le k} \|\partial^k u\|_{C(\overline{\mathcal{D}})} + \sum_{|m|=k} \sup_{x_1,x_2 \in \mathcal{D}, x_1 \neq x_2} \left(\frac{|\partial^k u(x_1) - \partial^k u(x_2)|}{|x_1 - x_2|^\alpha} \right)$$

is finite, where we use the multi-index notation.

The Hölder spaces equipped with above norms are Banach spaces.

Definition 1.9.17 (Oscillation and mean value of a function). Let $f : \mathbb{R}^d \to \mathbb{R}$ be a given function. For any $r > 0$, we define the oscillation

$$\operatorname{osc}(x_o, r) := \sup_{x \in B(x_o, r)} f(x) - \inf_{x \in B(x_o, r)} f(x),$$

and if it is locally integrable, the mean value

$$⨍_{B(x_0,r)} f(x)dx := \frac{1}{|B(x_0, r)|} \int_{B(x_0,r)} f(x)dx.$$

Theorem 1.9.18. *A function* $f : \mathbb{R}^d \to \mathbb{R}$ *is Hölder continuous with exponent* $\alpha \in (0,1]$ *if and only if for every* $x_o \in \mathbb{R}^d$, $r > 0$ *there exists a constant* $c > 0$ *such that* $⨍_{B(x_o,r)} |f(x) - f(x_o)|dx < c\,r^\alpha$. *Alternatively, if* $\operatorname{osc}(x_o, r) < c\,r^\alpha$ *for some* $\alpha \in (0,1]$, *then* f *is Hölder continuous with Hölder exponent* α.

Proof. We sketch the proof. The direct implication is immediate. For the converse implication, assume that the claim is not true. Then, for every $c > 0$ there exist $x_{o,1}, x_{o,2} \in \mathbb{R}^d$ such that $|f(x_{o,1}) - f(x_{o.2})| \ge c|x_{o,1} - x_{o,2}|^\alpha$. Since by the triangle inequality for any $x \in \mathbb{R}^d$, we have that $|f(x) - f(x_{o,1})| + |f(x) - f(x_{o,2})| \geq |f(x_{o,1}) - f(x_{o,2})|$; calculating the mean value integrals, we have upon setting $r = 2|x_{o,1} - x_{o,2}|$ that

$$\fint_{B(x_{o,1},r)} |f(x) - f(x_{o,1})| dx + \fint_{B(x_{o,2},r)} |f(x) - f(x_{o,2})| dx \geq |f(x_{o,1}) - f(x_{o,2})|$$
$$\geq c|x_{o,1} - x_{o,2}|^{\alpha} = 2^{\alpha} c r^{\alpha},$$

which implies that $\fint_{B(x_{o,i},r)} |f(x) - f(x_{o,i})| dx \geq 2^{\alpha-1} c r^{\alpha}$ for either $i = 1$ or $i = 2$, which leads to a contradiction. The oscillation criterion follows immediately by the mean value criterion. □

1.9.5 Lebesgue points

Definition 1.9.19 (Lebesgue points). A Lebesgue point of an integrable function $f : \mathcal{D} \subset \mathbb{R}^d \to \mathbb{R}$ is a point $x_o \in \mathcal{D}$ such that $\lim_{r\to 0} \fint_{B(x_o,r)} |f(x) - f(x_o)| dx = 0$.

Any point of continuity is a Lebesgue point. The following theorem is very important:

Theorem 1.9.20 (Lebesgue points for L^1_{loc} functions). *If $f \in L^1_{\mathrm{loc}}(\mathbb{R}^d)$, then a. e. $x_o \in \mathbb{R}^d$ is a Lebesgue point for f.*

2 Differentiability, convexity, and optimization in Banach spaces

In this chapter, we consider the various notions of derivatives for mappings between Banach spaces, with special emphasis on the case of mappings taking values in $\mathbb{R}$, connecting derivatives with optimization problems. After a brief account of the general theory of optimization, focusing on the concepts of coercivity and semicontinuity, we then focus on convex functions in Banach spaces, and discuss their properties with respect to continuity, lower semicontinuity, differentiability, and, most importantly, optimization. Moreover, an important operator, the projection operator on convex and closed subsets of Hilbert spaces, is introduced and studied in detail. In addition, we consider various notions of the geometry of Banach spaces related to convexity, introducing the concept of strictly, locally, and uniformly convex Banach space and study the properties of the duality map between such spaces and their duals. Finally, we discuss the concept of Γ-convergence, an important notion of variational convergence, which allows for the convergence of minimizers for sequences of variational problems.

Throughout the whole of this chapter, unless otherwise specified, X will be a Banach space.

2.1 Differentiability in Banach spaces

Let X be a Banach space and $F : X \to \mathbb{R}$ be a functional (possibly nonlinear). In this section, we are going to introduce and discuss various concepts of differentiability, which are generalizations of the usual concept of the derivative and are often used in calculus in Banach spaces (see, e. g., [47]).

2.1.1 The directional derivative

Definition 2.1.1 (Directional derivative). The directional derivative of F at $\mathrm{x} \in X$ along the direction $\mathrm{h} \in X$ is the limit (whenever it exists):

$$DF(\mathrm{x};\mathrm{h}) = \lim_{\epsilon \to 0} \frac{F(\mathrm{x} + \epsilon\,\mathrm{h}) - F(\mathrm{x})}{\epsilon}.$$

It is not necessary that the operator, defined by $\mathrm{h} \mapsto DF(\mathrm{x};\mathrm{h})$, is a linear operator. In fact, this may not hold, even in finite dimensional spaces. As an illustrative example of that, consider the functional $F : \mathbb{R}^2 \to \mathbb{R}$, defined by $F(x_1, x_2) = \frac{x_2^{2r+1}}{(x_1^2+x_2^2)^r}$ for $(x_1, x_2) \neq (0,0)$ and $F(0,0) = 0$, for which the operator defined by $\mathrm{h} \mapsto DF(0;\mathrm{h}) = \frac{h_2^{2r+1}}{(h_1^2+h_2^2)^r}$, is clearly not a linear operator.

https://doi.org/10.1515/9783111333298-002

2.1.2 The Gâteaux derivative

If the operator defined by $h \mapsto DF(x; h)$ is a linear operator, then we may introduce the concept of Gâteaux (or weak) derivative.

Definition 2.1.2 (Gâteaux derivative). A functional F is called Gâteaux (or weakly) differentiable at $x \in X$ if $DF(x; h)$ exists for any direction $h \in X$ and the operator $h \mapsto DF(x; h)$ is linear and continuous. Then, by the Riesz representation theorem (see Theorem 1.1.14), there exists some $DF(x) \in X^\star$ such that

$$\lim_{\epsilon \to 0} \frac{F(x + \epsilon h) - F(x)}{\epsilon} =: DF(x; h) = \langle DF(x), h \rangle, \quad \forall h \in X. \tag{2.1}$$

The element $DF(x)$ is called the Gâteaux derivative of F at x.

If $X = \mathbb{R}^d$, then the Gâteaux derivative coincides with the gradient ∇F. It should also be clear by the above definition that if the Gâteaux derivative exists, it is unique. It should also be evident that if F is Gâteaux differentiable at $x \in X$ with Gâteaux derivative $DF(x) \in X^\star$, then

$$\lim_{\epsilon \to 0} \frac{1}{\epsilon} \left| F(x + \epsilon h) - F(x) - \langle DF(x), \epsilon h \rangle \right| = 0, \quad \forall h \in X. \tag{2.2}$$

For this reason, $DF(x; h) = \langle DF(x), h \rangle$ is often called the Gâteaux differential.

Furthermore, it is straightforward to extend the above definition so as to define Gâteaux differentiability and the Gâteaux derivative for a functional F on an open subset $A \subset X$; we simply have to ensure that the above definition holds for every $x \in A$. If the functional F is not defined over the whole Banach space X but only on a subset $A \subset X$, not necessarily open, then we must take some care when applying the above definitions to make sure that we only consider points x in the interior of A.

Example 2.1.3 (Differentiability of norm related functionals in Hilbert space). Let $X = H$ be a Hilbert space, and consider the functionals $F_1, F_2 : H \to \mathbb{R}$, defined by $F_1(x) = \|x\| = \sqrt{\langle x, x \rangle}$ and $F_2(x) = \frac{1}{2}\|x\|^2 = \frac{1}{2}\langle x, x \rangle$, respectively, where $\langle \cdot, \cdot \rangle$ is the inner product of the Hilbert space. Then, using the properties of the inner product, we easily find that

$$F_2(x + \epsilon h) - F_2(x) = \frac{1}{2}\langle x + \epsilon h, x + \epsilon h \rangle - \frac{1}{2}\langle x, x \rangle = \epsilon \langle x, h \rangle + \frac{\epsilon^2}{2}\|h\|^2,$$

so that dividing by ϵ and passing to the limit as $\epsilon \to 0$ yields that

$$DF_2(x; h) = \langle x, h \rangle, \quad \forall h \in X,$$

so that by the Riesz representation theorem (see Theorem 1.1.14), we conclude that $DF_2(x) = x$. Note that this result holds true, even if we choose $x = 0$.

On the other hand,

$$\begin{aligned}F_1(\mathsf{x}+\epsilon \mathsf{h}) - F_1(\mathsf{x}) &= \|\mathsf{x}+\epsilon \mathsf{h}\| - \|\mathsf{x}\| \\ &= \langle \mathsf{x}+\epsilon \mathsf{h}, \mathsf{x}+\epsilon \mathsf{h}\rangle^{1/2} - \langle \mathsf{x},\mathsf{x}\rangle^{1/2} \\ &= \left(\langle \mathsf{x},\mathsf{x}\rangle + 2\epsilon\langle \mathsf{x},\mathsf{h}\rangle + \epsilon^2\|\mathsf{h}\|^2\right)^{1/2} - \langle \mathsf{x},\mathsf{x}\rangle^{1/2} \\ &= \|\mathsf{x}\|\left(1+2\epsilon\frac{\langle \mathsf{x},\mathsf{h}\rangle}{\|\mathsf{x}\|^2} + \epsilon^2\frac{\|\mathsf{h}\|^2}{\|\mathsf{x}\|^2}\right)^{1/2} - \|\mathsf{x}\|,\end{aligned}$$

so that

$$DF_1(\mathsf{x};\mathsf{h}) = \lim_{\epsilon\to 0}\frac{1}{\epsilon}\left\{\|\mathsf{x}\|\left(1+2\epsilon\frac{\langle \mathsf{x},\mathsf{h}\rangle}{\|\mathsf{x}\|^2} + \epsilon^2\frac{\|\mathsf{h}\|^2}{\|\mathsf{x}\|^2}\right)^{1/2} - \|\mathsf{x}\|\right\} = \frac{1}{\|\mathsf{x}\|}\langle \mathsf{x},\mathsf{h}\rangle,$$

which leads to the conclusion that

$$DF_1(\mathsf{x}) = \frac{1}{\|\mathsf{x}\|}\mathsf{x},$$

by using the Riesz isomorphism. If $\mathsf{x} = 0$, then $F_1(\epsilon\mathsf{h}) - F_1(0) = |\epsilon|\|\mathsf{h}\|$, and therefore the norm is not differentiable at $\mathsf{x} = 0$. ◁

These results have to be modified in the case where X is a Banach space, where in fact the Gâteaux differentiability of the norm is related to geometric properties of the Banach space, such as for strict convexity (see Section 4.2.4).

We now present an example concerning the Gâteaux differentiability of functionals related to the norm in a special, yet very important class of Banach spaces, Lebesgue spaces.

Example 2.1.4 (Differentiability of L^p norms). Let $\mathcal{D} \subset \mathbb{R}^d$ be an open set, and consider the Lebesgue spaces $X = L^p(\mathcal{D})$, $2 < p < \infty$. Any element $\mathsf{x} \in X$ is identified with a p-integrable function $u : \mathcal{D} \to \mathbb{R}$. We will also use the notation v for any direction h of this function space, where $v \in L^p(\mathcal{D})$. Then if $F : X \to \mathbb{R}$ is defined by

$$F(u) = \|u\|^p_{L^p(\mathcal{D})} = \int_{\mathcal{D}} |u(x)|^p dx,$$

we have that

$$DF(u) = p|u|^{p-2}u = p|u|^{p-1}\operatorname{sgn}(u),$$

which clearly belongs to $X^\star = L^{p^\star}(\mathcal{D})$, where $p^\star$ is such that $1/p + 1/p^\star = 1$.

To prove the above claim, we need to consider the limit

$$\lim_{\epsilon\to 0}\frac{F(u+\epsilon v) - F(u)}{\epsilon} = \lim_{\epsilon\to 0}\frac{\int_{\mathcal{D}} |u(x)+\epsilon v(x)|^p dx - \int_{\mathcal{D}} |u(x)|^p dx}{\epsilon}.$$

Recall an elementary calculus fact; if $\psi : \mathbb{R} \to \mathbb{R}$ with $\psi(s) = |a + bs|^p$, $a, b \in \mathbb{R}$, then, for $p \in (1, \infty)$,

$$\frac{d\psi}{ds}(s) = pb|a + bs|^{p-1} \operatorname{sgn}(a + bs) = pb|a + bs|^{p-2}(a + bs).$$

Using that, along with the dominated convergence theorem[1] to interchange the limit with the integral, we obtain

$$\begin{aligned}\lim_{\epsilon \to 0} \frac{F(u + \epsilon v) - F(u)}{\epsilon} &= \int_{\mathcal{D}} \lim_{\epsilon \to 0} \frac{|u(x) + \epsilon v(x)|^p - |u(x)|^p}{\epsilon} dx \\ &= \int_{\mathcal{D}} p|u(x)|^{p-2} u(x) v(x) dx.\end{aligned}$$

The right hand side is identified as $\langle p|u|^{p-2}u, v\rangle_{L^{p^*}(\mathcal{D}), L^p(\mathcal{D})}$, which completes the proof of our claim. Recalling our indentification of x with functions $u \in L^p(\mathcal{D})$, we note that we have only considered the case where $\mathrm{x} \neq 0$ (i. e., $u(x) \neq 0$ a. e. in $\mathcal{D}$), since the norm itself is not differentiable at $\mathrm{x} = 0$. ◁

The Gâteaux derivative satisfies some properties similar to the derivative in finite dimensional spaces, such as the mean value theorem.

Proposition 2.1.5 (Mean value theorem). *Let $F : X \to \mathbb{R}$ be a Gâteaux differentiable functional. Then,*

$$\big|F(\mathrm{x}_1) - F(\mathrm{x}_2)\big| \le \sup_{t \in [0,1]} \big\|DF(t\mathrm{x}_1 + (1-t)\mathrm{x}_2)\big\|_{X^*} \|\mathrm{x}_1 - \mathrm{x}_2\|. \tag{2.3}$$

Proof. Consider the real valued function $\phi : \mathbb{R} \to \mathbb{R}$, defined by $\phi(t) := F(\mathrm{x}_1 + t(\mathrm{x}_2 - \mathrm{x}_1))$, $t \in [0, 1]$. A simple calculation (using the Gâteaux differentiability of F at x_1 in the direction $\mathrm{h} = \mathrm{x}_2 - \mathrm{x}_1$) yields,

$$\frac{d\phi}{dt}(t) = \big\langle DF(\mathrm{x}_1 + t(\mathrm{x}_2 - \mathrm{x}_1)), \mathrm{x}_2 - \mathrm{x}_1 \big\rangle.$$

We now use the mean value theorem for the real valued function ϕ according to which there exists $t_o \in [0, 1]$ such that $\phi(1) - \phi(0) = \frac{d\phi}{dt}(t_o)$, which yields

$$F(\mathrm{x}_2) - F(\mathrm{x}_1) = \big\langle DF(\mathrm{x}_1 + t_o(\mathrm{x}_2 - \mathrm{x}_1)), \mathrm{x}_2 - \mathrm{x}_1 \big\rangle,$$

from which (2.3) follows. □

---

1 The details are left to the reader. However, note that this step requires the elementary convexity inequality $|a|^p - |a - b|^p \le \frac{1}{t}[|a + tb|^p - |a|^p] \le |a + b|^p - |a|^p$, for $|t| \le 1$ (see, e. g., [99]).

2.1.3 The Fréchet derivative

The Gâteaux differentiability is not the only concept of differentiability available in Banach space.

Definition 2.1.6 (Fréchet differentiability). The functional $F : X \to \mathbb{R}$ is called Fréchet (strongly) differentiable at $x \in X$ if there exists $DF(x) \in X^*$ such that

$$\lim_{\|h\| \to 0} \frac{|F(x + h) - F(x) - \langle DF(x), h\rangle|}{\|h\|} = 0, \tag{2.4}$$

and $DF(x)$ is called the Fréchet derivative of F at x.

If a functional is Fréchet differentiable, then its Fréchet derivative is unique.

Remark 2.1.7. Note the difference with the Gâteaux derivative: if $DF(x)$ is the Gâteaux derivative of F at x, then (2.1) implies (2.2), not as strong a statement as (2.4), which in fact requires that the limit exists uniformly for all $h \in X$. To be more precise, (2.2) implies that $F(x + \epsilon h) - F(x) = \epsilon\langle DF(x), h\rangle + R(\epsilon, h)$, where R is a remainder term such that $\lim_{\epsilon\to 0} \frac{R(\epsilon,h)}{\epsilon} = 0$. However, the remainder term R depends on h in general. On the other hand, (2.4) implies that $F(x + h) - F(x) = \langle DF(x), h\rangle + R'(\|h\|)$, with R' being a remainder term, such that $\lim_{\|h\|\to 0} \frac{R'(\|h\|)}{\|h\|} = 0$. A more illustrative way of denoting R and R' is by using the Landau symbols $o(\epsilon)$ and $o(\|h\|)$, respectively. In the first case, the rate of convergence depends on the choice of direction h, whereas in the second it is uniform in h as $h \to 0$. We will sometimes use the small-o notation $F(x+h)-F(x)-\langle DF(x), h\rangle = o(\|h\|)$ to denote that $DF(x)$ is the Fréchet derivative of F at x.

As a result, if a functional is Fréchet differentiable at x, it is also Gâteaux differentiable at x, but the converse does not necessarily hold, even in finite dimensional spaces, unless the Gâteaux derivative satisfies additional conditions (see Proposition 2.1.14). As an example, consider $F : \mathbb{R}^2 \to \mathbb{R}$, defined by $F(x_1, x_2) = \frac{x_2}{x_1}(x_1^2 + x_2^2)$, if $x_1 \neq 0$ and $F(x_1, x_2) = 0$ if $x_1 = 0$, which is Gâteaux differentiable at $(0, 0)$ but not Fréchet differentiable at this point, as we can see by, e. g., choosing the sequences $h_n = (\frac{1}{n}, \frac{1}{n})$ and $h_n = (\frac{1}{n^2}, \frac{1}{n})$ (see [13]).

Fortunately, for a number of interesting cases, we have Fréchet differentiability.

Example 2.1.8 (Quadratic functions). Let $X = H$ be a Hilbert space, and let $A : H \to H$ be a linear bounded operator. Consider the functional $F : H \to \mathbb{R}$ defined by $F(x) = \frac{1}{2}\langle Ax, x\rangle$, where $\langle\cdot,\cdot\rangle$ is the inner product of H. In the special case, where $A = I$, the identity operator of this functional becomes the square of the norm. We now consider the Gâteaux derivative of F. We see that

$$\begin{aligned} F(x + \epsilon h) - F(x) &= \frac{1}{2}\langle A(x + \epsilon h), x + \epsilon h\rangle - \frac{1}{2}\langle Ax, x\rangle \\ &= \frac{1}{2}\epsilon\langle Ah, x\rangle + \frac{1}{2}\epsilon\langle Ax, h\rangle + \frac{1}{2}\epsilon^2\langle Ah, Ah\rangle \end{aligned}$$

$$= \frac{1}{2}\epsilon\langle \mathrm{h}, \mathrm{A}^*\mathrm{x}\rangle + \frac{1}{2}\epsilon\langle \mathrm{Ax}, \mathrm{h}\rangle + \frac{1}{2}\epsilon^2\langle \mathrm{Ah}, \mathrm{h}\rangle$$
$$= \epsilon\left\langle \frac{1}{2}(\mathrm{A} + \mathrm{A}^*)\mathrm{x}, \mathrm{h}\right\rangle + \epsilon^2\frac{1}{2}\langle \mathrm{Ah}, \mathrm{h}\rangle,$$

where we used the concept of the adjoint operator and the symmetry of the inner product. Using the fact that A is linear and bounded, we conclude that

$$DF(\mathrm{x}; \mathrm{h}) = \left\langle \frac{1}{2}(\mathrm{A} + \mathrm{A}^*)\mathrm{x}, \mathrm{h}\right\rangle, \quad \forall\, \mathrm{h} \in X,$$

so that, using the Riesz isomorphism,

$$DF(\mathrm{x}) = \frac{1}{2}(\mathrm{A} + \mathrm{A}^*)\mathrm{x}.$$

Note that the Gâteaux derivative of F is a symmetric operator.

We will now show that F is Fréchet differentiable with $DF(\mathrm{x}) = \frac{1}{2}(\mathrm{A} + \mathrm{A}^*)\mathrm{x}$. Indeed, retracing the steps above, we see that

$$|F(\mathrm{x} + \mathrm{h}) - F(\mathrm{x}) - \langle DF(\mathrm{x}), \mathrm{h}\rangle| = \frac{1}{2}|\langle \mathrm{Ah}, \mathrm{h}\rangle| \le c\,\|\mathrm{h}\|^2,$$

where we used the fact that A is bounded, and the constant $c > 0$ is related to the operator norm of A. Dividing by $\|\mathrm{h}\|$ and passing to the limit as $\|\mathrm{h}\| \to 0$ yields the required result (2.4). Note that assuming A is bounded (hence, continuous by linearity) is crucial for the proof of Fréchet differentiability of F. Furthermore, if $\mathrm{A} = I$, this result gives the Fréchet differentiability of the square of the norm in Hilbert space. ◁

2.1.4 C^1 functionals

Definition 2.1.9. Let $A \subset X$ be an open set. If the Fréchet (or Gâteaux) derivative of a functional $F : X \to \mathbb{R}$ exists for every $\mathrm{x} \in A \subset X$, and the mapping $\mathrm{x} \mapsto DF(\mathrm{x})$ is a continuous map, then we say that the functional F is Fréchet (or Gâteaux) continuously differentiable on A. We will use the notation $C^1(A)$ for Fréchet continuously differentiable functionals and $C^1_G(A)$ for Gâteaux continuously differentiable functionals.

Example 2.1.10 (The Dirichlet integral). Let $\mathcal{D} \subset \mathbb{R}^d$ be an open set with sufficiently smooth boundary and $X = W^{1,2}_0(\mathcal{D})$, a Sobolev space (see Definition 1.5.6), whose elements are identified with functions $u : \mathcal{D} \to \mathbb{R}$. This space is a Hilbert space when equipped with the inner product $\langle u, v\rangle := \langle \nabla u, \nabla v\rangle_{L^2(\mathcal{D})}$ (see Theorem 1.5.4 and Example 1.5.14). Consider the functional $F_{\mathcal{D}} : X \to \mathbb{R}$, defined by $F_{\mathcal{D}}(u) = \frac{1}{2}\int_{\mathcal{D}} |\nabla u(x)|^2 dx$. Then, identifying any direction $\mathrm{h} \in X$ with any function $v \in W^{1,2}_0(\mathcal{D})$, we see that

$$\frac{1}{\epsilon}(F_{\mathcal{D}}(u+\epsilon v)-F_{\mathcal{D}}(u))=\int_{\mathcal{D}}\nabla u(x)\cdot\nabla v(x)dx+\frac{\epsilon}{2}\int_{\mathcal{D}}|\nabla v(x)|^2dx,$$

and taking the limit as $\epsilon\to 0$, we obtain

$$\lim_{\epsilon\to 0}\frac{1}{\epsilon}(F_{\mathcal{D}}(u+\epsilon v)-F_{\mathcal{D}}(u))=-\int_{\mathcal{D}}\Delta u(x)v(x)dx=\langle -\Delta u, v\rangle,$$

(using the standard inner product in $L^2(\mathcal{D})$), where now $DF_{\mathcal{D}}(u)=-\Delta u$ is identified as an element of $X^\star=(W_0^{1,2}(\mathcal{D}))^*$. In this manner, we may define a mapping $\mathsf{A}:X\to X^\star$ by $u\mapsto -\Delta u$, which is the variational definition of the Laplace operator. Clearly, $F_{\mathcal{D}}$ is a symmetric bilinear form.

The operator $DF_{\mathcal{D}}(u)=-\Delta u$ is in fact the Fréchet derivative of $F_{\mathcal{D}}$. This can be checked directly from Definition 2.1.6, by noting that for any $v\in X$ we have (by Green's formula, see Proposition 1.5.19) that $\langle DF_{\mathcal{D}}(u),v\rangle=\int_{\mathcal{D}}\nabla u(x)\cdot\nabla v(x)dx$ so that

$$F_{\mathcal{D}}(u+v)-F_{\mathcal{D}}(u)-\langle DF_{\mathcal{D}}(u),v\rangle=\frac{1}{2}\int_{\mathcal{D}}|\nabla v(x)|^2dx=\frac{1}{2}\|v\|^2_{W_0^{1,2}(\mathcal{D})},$$

where, dividing by $\|v\|_{W_0^{1,2}(\mathcal{D})}$ and passing to the limit as $\|v\|_{W_0^{1,2}(\mathcal{D})}\to 0$ yields the required result (2.4).

The operator $DF_{\mathcal{D}}(u)=-\Delta u$ is continuous as a mapping from $W_0^{1,2}(\mathcal{D})$ to $W^{-1,2}(\mathcal{D})=(W_0^{1,2}(\mathcal{D}))^*$, therefore $F_{\mathcal{D}}$ is C^1. Indeed, consider any $u\in W_0^{1,2}(\mathcal{D})$, and take $w\to u$ in $W_0^{1,2}(\mathcal{D})$. Then, for any $v\in W_0^{1,2}(\mathcal{D})$,

$$|\langle DF_{\mathcal{D}}(u)-DF_{\mathcal{D}}(w),v\rangle|=\left|\int_{\mathcal{D}}\nabla(u-w)\cdot\nabla v dx\right|\le\|u-w\|_{W_0^{1,2}(\mathcal{D})}\|v\|_{W_0^{1,2}(\mathcal{D})},$$

so that

$$\begin{aligned}\|DF_{\mathcal{D}}(u)-DF_{\mathcal{D}}(w)\|_{(W_0^{1,2}(\mathcal{D}))^*}&=\sup_{v\in W_0^{1,2}(\mathcal{D})}\frac{|\langle DF_{\mathcal{D}}(u)-DF_{\mathcal{D}}(w),v\rangle|}{\|v\|_{W_0^{1,2}(\mathcal{D})}}\\&\le\|u-w\|_{W_0^{1,2}(\mathcal{D})},\end{aligned}$$

hence $DF_{\mathcal{D}}(w)\to DF_{\mathcal{D}}(u)$ in $(W_0^{1,2}(\mathcal{D}))^*$ as $w\to u$ in $W_0^{1,2}(\mathcal{D})$. ◁

2.1.5 Connections between Gâteaux, Fréchet differentiability, and continuity

The above definitions show that the C^1 property of a functional at a neighborhood $N(\mathrm{x})$ of x implies the Fréchet differentiability of F at x, and this in turn implies the Gâteaux differentiability of F at x. The reverse implications are not true in general.

Fréchet differentiability resembles the standard definition of differentiability in the sense that it is connected with continuity.

Proposition 2.1.11. *Consider a subset $A \subset X$, and let $F : A \to \mathbb{R}$ be Fréchet differentiable at a point $x \in \text{int}(A)$. Then F is continuous at x.*

Proof. Since $x \in \text{int}(A)$, there exists $\epsilon_1 > 0$ such that $x + h \in A$, as long as $\|h\| \le \epsilon_1$. Since, by assumption, (2.4) holds, for every $\epsilon > 0$ there exists an $\epsilon_2 > 0$ such that

$$|F(x + h) - F(x) - \langle DF(x), h\rangle| \le \epsilon \|h\|, \quad \text{if } \|h\| \le \epsilon_2.$$

By the triangle inequality

$$\begin{aligned}|F(x + h) - F(x)| &= |F(x + h) - F(x) - \langle DF(x), h\rangle + \langle DF(x), h\rangle| \\ &\le |F(x + h) - F(x) - \langle DF(x), h\rangle| + |\langle DF(x), h\rangle| \\ &\le |F(x + h) - F(x) - \langle DF(x), h\rangle| + \|DF(x)\|_{X^\star} \|h\|,\end{aligned}$$

where $\|DF(x)\|_{X^\star}$ is the norm of $DF(x) \in X^\star$. Let $\delta = \min(\epsilon_1, \epsilon_2)$. Then for any $\epsilon > 0$, the above estimate yields the existence of $\delta > 0$ such that,

$$|F(x + h) - F(x)| \le (\epsilon + \|DF(x)\|_{X^\star}) \|h\|, \quad \text{if } \|h\| < \delta,$$

therefore leading to the conclusion that there exists a constant $c > 0$ such that

$$|F(x + h) - F(x)| \le c \|h\|, \quad \text{if } \|h\| < \delta,$$

from which continuity of F at x follows. □

In contrast to the assurance of Proposition 2.1.11, the Gâteaux differentiability of a functional at a point does not guarantee continuity at this point for the functional, but rather a weaker property called hemicontinuity.

Definition 2.1.12 (Hemicontinuity). The functional $F : X \to \mathbb{R}$ is called hemicontinuous at $x \in X$ if the real valued function $\lambda \mapsto F(x + \lambda h)$ is continuous at $\lambda = 0$ for every $h \in X$, i. e., $\lim_{\epsilon \to 0} F(x + \epsilon h) = F(x)$ for every $h \in X$.

A hemicontinuous function needs not be continuous, even in finite dimensions, as the example $F(x_1, x_2) = \frac{x_2}{x_1}(x_1^2 + x_2^2)$ for $x_1 \neq x_2$, $F(x_1, x_2) = 0$, for $x_1 = 0$ indicates (see [13]).

Proposition 2.1.13. *Consider a subset $A \subset X$, and let $F : A \to \mathbb{R}$ be Gâteaux differentiable at a point $x \in \text{int}(A)$. Then F is hemicontinuous at x.*

Proof. Take any $h \in X$, and consider the real valued function $\phi : \mathbb{R} \to \mathbb{R}$ defined by $\phi(\lambda) := F(x + \lambda h)$. The Gâteaux differentiability of F at x implies that the function ϕ is differentiable at $\lambda = 0$, therefore it is continuous at this point. This means F is hemicontinuous at x. □

The following result allows us to pass from Gâteaux to Fréchet differentiability:

Proposition 2.1.14. *Let $F : X \to \mathbb{R}$ be Gâteaux differentiable on a neighborhood $N(\mathrm{x})$ of x (open) and $\mathrm{x} \mapsto DF(\mathrm{x})$ be continuous. Then, F is also Fréchet differentiable at x. In other words, if F is C^1_G in a neighborhood of x, then it is Fréchet differentiable at x.*

Proof. Fix $\mathrm{h} \in X$, and define $g : \mathbb{R} \to \mathbb{R}$ by $g(t) := F(\mathrm{x} + t\mathrm{h}) - F(\mathrm{x}) - t\langle DF(\mathrm{x}), \mathrm{h}\rangle$. The function g is differentiable, and an application of the mean value theorem implies that $g(1) - g(0) = \frac{dg}{dt}(t_o)$ for some $t_o \in (0,1)$. Since $\frac{dg}{dt}(t_o) = \langle DF(\mathrm{x} + t_o\mathrm{h}) - DF(\mathrm{x}), \mathrm{h}\rangle$, we obtain that $F(\mathrm{x}+\mathrm{h}) - F(\mathrm{x}) - \langle DF(\mathrm{x}), \mathrm{h}\rangle = \langle DF(\mathrm{x}+t_o\mathrm{h}) - DF(\mathrm{x}), \mathrm{h}\rangle$ and combined with the continuity of DF leads to the required result. □

Note that in general, if a functional is Fréchet differentiable at $\mathrm{x} \in X$, this does not necessarily imply that F is C^1 at x. On the other hand, if F is C^1_G at x, then by the above proposition F is also Fréchet differentiable at x; the Fréchet and the Gâteaux derivative coincide hence, and F is also continuously Fréchet differentiable at x.

2.1.6 Vector valued maps and higher order derivatives

The concepts of directional, Gâteaux and Fréchet derivatives for maps $f : X \to Y$, where X, Y are two Banach spaces, can be defined with the obvious generalizations.

Definition 2.1.15. Let $f : X \to Y$ be a map between the Banach spaces X, Y.

(i) The directional derivative at point $\mathrm{x} \in X$ along the direction $\mathrm{h} \in X$ is defined (if the limit exists) as

$$Df(\mathrm{x};\mathrm{h}) := \lim_{\epsilon\to 0} \frac{f(\mathrm{x}+\epsilon\mathrm{h}) - f(\mathrm{x})}{\epsilon}.$$

(ii) If $f : A \subset X \to Y$, $Df(\mathrm{x};\mathrm{h})$ is defined for any $\mathrm{x} \in \mathrm{int}(A)$, any direction $\mathrm{h} \in X$, and the operator $\mathrm{h} \mapsto Df(\mathrm{x};\mathrm{h})$ is linear and continuous, we will denote it by $Df(\mathrm{x}) \in \mathcal{L}(X;Y)$, and we will say that f is Gâteaux differentiable at x and define

$$Df(\mathrm{x})\mathrm{h} := \lim_{\epsilon\to 0} \frac{f(\mathrm{x}+\epsilon\mathrm{h}) - f(\mathrm{x})}{\epsilon}, \quad \forall\, \mathrm{h} \in X.$$

(iii) If the stronger result

$$\lim_{\|\mathrm{h}\|_X\to 0} \frac{\|f(\mathrm{x}+\mathrm{h}) - f(\mathrm{x}) - Df(\mathrm{x})\mathrm{h}\|_Y}{\|\mathrm{h}\|_X} = 0,$$

holds for the operator $Df(\mathrm{x}) \in \mathcal{L}(X;Y)$, then we say that f is Fréchet (or strongly) differentiable at x and $Df(\mathrm{x})$ is called the Fréchet derivative at x.

All the results stated above, such as for example Proposition 2.1.11 or the mean value theorem etc., hold subject to the necessary changes for the general case. One furthermore has the following chain rule for vector valued functions:

Proposition 2.1.16. *Suppose X, Y, Z are Banach spaces and $f_2 : X \to Y$ is Gâteaux differentiable at x and $f_1 : Y \to Z$ is Fréchet differentiable at $y = f_2(x)$. Then $f = f_1 \circ f_2 : X \to Z$ is Gâteaux differentiable at x and*

$$D(f_1 \circ f_2)(x) = Df_1(f_2(x)) \circ Df_2(x).$$

If f_2 is Fréchet differentiable at x, then so is f.

Proof. Using Definition 2.1.15 it suffices to show that

$$\lim_{\epsilon\to 0} \frac{1}{\epsilon}\|f_1(f_2(x+\epsilon h)) - f_1(f_2(x)) - \epsilon Df_1(f_2(x)) \circ Df_2(x)h\|_Z = 0, \quad \forall h \in X.$$

Adding and subtracting $Df_1(f_2(x))(f_2(x+\epsilon h) - f_2(x))$ to the quantity inside the norm and using the triangle inequality,

$$\begin{aligned}
&\frac{1}{\epsilon}\|f_1(f_2(x+\epsilon h)) - f_1(f_2(x)) - \epsilon Df_1(f_2(x)) \circ Df_2(x)h\|_Z \\
&\quad \le \frac{1}{\epsilon}\left\| f_1(f_2(x+\epsilon h)) - f_1(f_2(x)) - \epsilon Df_1(f_2(x))\left(\frac{1}{\epsilon}(f_2(x+\epsilon h) - f_2(x))\right)\right\|_Z \\
&\quad\quad + \left\| Df_1(f_2(y))\left(\frac{1}{\epsilon}(f_2(x+\epsilon h) - f_2(x) - Df_2(x)h)\right)\right\|_Z.
\end{aligned}$$

The last term in the above tends to zero, as $\epsilon \to 0$ on account of Gâteaux differentiability of f_2. To handle the first term, we multiply and divide by $\|f_2(x+\epsilon h) - f_2(x)\|_Y$ and setting $y = f_2(x)$ and $\hat{h} = f_2(x+\epsilon h) - f_2(x) \in Y$; this term is rearranged as

$$\frac{\|\hat{h}\|_Y}{\epsilon}\frac{1}{\|\hat{h}\|_Y}\|f_1(y+\hat{h}) - f_1(y) - \epsilon Df_1(y)\hat{h}\|_Z.$$

Since f_2 is Gâteaux differentiable at x, we have that $\hat{h} \to 0$ as $\epsilon \to 0$ so that by the Fréchet differentiability of f_1, we have that $\frac{1}{\|\hat{h}\|_Y}\|f_1(y+\hat{h}) - f_1(y) - \epsilon Df_1(y)\hat{h}\|_Z \to 0$. On the other hand, $\frac{\|\hat{h}\|_Y}{\epsilon} \to \|Df_2(x)\,h\|_Y$, as $\epsilon \to 0$, so the result follows. □

Higher-order derivatives may be defined in a standard fashion, e. g., by considering the derivatives of derivatives. As an example, consider second derivatives.

Definition 2.1.17. Let $f : A \subset X \to Y$ be Gâteaux differentiable at every $x \in A$, and apply Definition 2.1.15 on the map $Df(x) : A \subset X \to \mathcal{L}(X;Y)$, defined by $x \mapsto Df(x)$. If this map is Gâteaux differentiable at every $x \in A$, its Gâteaux derivative is an operator from X to $\mathcal{L}(X;Y)$, denoted by $D^2 f(x) \in \mathcal{L}(X;\mathcal{L}(X;Y))$.

If we consider a map $f : X \to Y$, which is twice Gâteaux differentiable at $x \in X$, and we take any two directions $h_1, h_2 \in X$, then $D^2f(x)h_1 \in \mathcal{L}(X;Y)$, and $(D^2f(x)h_1)h_2 \in Y$. We may thus define the map $B : X \times X \to Y$ by $(h_1, h_2) \mapsto (D^2f(x)h_1)h_2$, which is clearly a bilinear form. The special case $Y = \mathbb{R}$ is often of interest. In the case where $X = \mathbb{R}^d$ and $Y = \mathbb{R}$, $D^2f(x)$ coincides with the Hessian matrix, which is an element of $\mathbb{R}^{d\times d}$.

Taylor-type formulae hold for maps that have higher-order Gâteaux derivatives. For example:

Proposition 2.1.18. *If $F : X \to \mathbb{R}$ is twice Gâteaux differentiable, then there exists an $s \in (0,1)$ such that*

$$F(x + h) = F(x) + \langle DF(x), h\rangle + \frac{1}{2}\langle D^2F(x + s\,h)h, h\rangle.$$

Proof. If F is twice Gâteaux differentiable, then the real valued function $t \mapsto \phi(t) := F(x + t\,h)$, for every $h \in X$, is twice differentiable and application of Taylor's formula, for ϕ yields the required result. □

2.2 General results on optimization problems

Lower semicontinuity

An important concept concerning optimization is semicontinuity and, in particular, lower semicontinuity, as far as minimization is concerned. Before proceeding with that, we recall the definition of the liminf for a function $F : X \to \mathbb{R} \cup \{+\infty\}$ as

$$\liminf_{z\to x} F(z) = \sup_{U\in N(x)} \inf_{z\in U} F(z). \tag{2.5}$$

This definition can be generalized, by appropriate modification, for $F : A \subset X \to \mathbb{R} \cup \{+\infty\}$, for any accumulation point x of A.

We recall the following definitions for semicontinuous functions on X [62] (clearly local versions are possible, i. e. for specific $x \in X$):

Definition 2.2.1 (Lower semicontinuity). Let (X, τ) be a topological space. A function $F : X \to \mathbb{R} \cup \{+\infty\}$ is called

(i) Sequentially lower semicontinuous if for every $x \in X$ and for every $(x_n)_{n\in\mathbb{N}} \subset X$ such that $x_n \to x$ it holds that $F(x) \le \liminf_n F(x_n)$.

(ii) Lower semicontinuous if the sets $L_\lambda F := \{x \in X : F(x) \le \lambda\}$ are closed in X for every $\lambda \in \mathbb{R}$ (or equivalently, $\{x \in X : F(x) > \lambda\}$ open).
Two equivalent definitions for lower semicontinuity are
(ii-a) For each $x \in X$, it holds that

$$F(x) \le \liminf_{z\to x} F(z) =: \sup_{U\in N(x)} \inf_{z\in U} F(z).$$

(ii-b) For every $x \in X$ and for every $t \in \mathbb{R}$ with $t < F(x)$, there exists $U \in N(x)$ such that $t < F(z)$ for every $z \in U$.

A function $F : X \to \mathbb{R} \cup \{+\infty\}$ is called (sequentially) upper semicontinuous if $-F$ is (sequentially) lower semicontinuous.

Both definitions (i) and (ii) have a topological flavor, as both the concept of convergence as well as the concepts of closed set / open neighborhood depend on the topology with which the set X is endowed, hence leading to various versions of lower semicontinuity. For instance, if X is a Banach space, we can consider the following versions (continuing the numbering of Definition 2.2.1), which will play an important role for our purposes:

(iii) Weakly sequentially lower semicontinuous F if for every $x \in X$ and for every $(x_n)_{n\in\mathbb{N}} \subset X$ such that $x_n \rightharpoonup x$ it holds that $F(x) \leq \liminf_n F(x_n)$.

(iv) Weakly lower semicontinuous F if the sets $L_\lambda F := \{x \in X : F(x) \leq \lambda\}$ are weakly closed in X for every $\lambda \in \mathbb{R}$.

How are the various notions of lower semicontinuity related? The following proposition (see, e. g., [62]) provides some interesting connections:

Proposition 2.2.2. *The following hold:*

(i) *Let X be a Banach space and $F : X \to \mathbb{R} \cup \{+\infty\}$ (not identically equal to $+\infty$). If F is weakly lower semicontinuous, then it is also lower semicontinuous.*

(ii) *In the same context as* (i), *if F is weakly sequentially lower semicontinuous, it is also sequentially lower semicontinuous.*

(iii) *If X is any topological space, a lower semicontinuous function is also sequentially lower semicontinuous.*

(iv) *If X is first countable (e. g., a Banach space equipped with the strong topology), then the concept of lower semicontinuity and sequential lower semicontinuity coincide.*

(v) *If X is any topological space, the various definitions in Definition* 2.2.1(ii) *(i. e.,* (ii), (ii-a), *and* (ii-b)*) are equivalent.*

Proof. (i) For lower semicontinuity, it suffices to recall that any weakly closed subset of X is closed (see Theorem 1.1.46(ii)). From that, this assertion holds.

(ii) For sequential lower semicontinuity, consider any sequence $(x_n)_{n\in\mathbb{N}} \subset X$ such that $x_n \to x$. Then, it also holds that $x_n \rightharpoonup x$ in X, and by the weak sequential lower semicontinuity of F, we have that $F(x) \leq \liminf_n F(x_n)$. That implies the lower sequential semicontinuity.

(iii) Let $x \in X$ be arbitrary, and consider any sequence $(x_n)_{n\in\mathbb{N}} \subset X$ such that $x_n \to x$. For any $U \in N(x)$, there exists $N \in \mathbb{N}$ such that $x_n \in U$ for all $n > N$. Since F is lower semicontinuous at x, it holds that $F(x) \leq \sup_{U\in N(x)} \inf_{z\in U} F(z)$. Since $F(x) \geq \inf_{z\in U} F(z)$ for all $U \in N(x)$, taking the sup over all such U, we obtain that $F(x) \geq \sup_{U\in N(x)} \inf_{z\in U} F(z)$.

Hence, for a lower semicontinuous function, we obtain $F(x) = \sup_{U\in N(x)} \inf_{z\in U} F(z)$. By the definition of the supremum, for any $\epsilon > 0$, there exists $U_\epsilon \in N(x)$ such that

$$F(x) - \epsilon = \sup_{U\in N(x)} \inf_{z\in U} F(z) - \epsilon < \inf_{z\in U_\epsilon} F(z). \tag{2.6}$$

Since $x_n \to x$, for the chosen $U_\epsilon \in N(x)$, there exists $N_\epsilon \in \mathbb{N}$ such that $x_n \in U_\epsilon$ for all $n \geq N_\epsilon$. Combining that with (2.6), we obtain that

$$F(x) - \epsilon < \inf_{z\in U_\epsilon} F(z) \leq F(x_n), \quad \forall\, n > N_\epsilon,$$

and taking the liminf over n, we obtain from the above that

$$F(x) - \epsilon \leq \liminf_n F(x_n),$$

from which taking the limit $\epsilon \to 0$, we conclude that F is sequentially lower semicontinuous at x. Since $x \in X$ is arbitrary, we conclude.

(iv) We now show that if X is first countable, the converse of (iii) also holds. Suppose that F is sequentially lower semicontinuous but not lower semicontinuous. Then, by Def. 2.2.1(ii-b)), for some $x \in X$, fixed, there exists $t < F(x)$ such that $\inf_{z\in U} F(z) \leq t$, for every $U \in N(x)$. Since X is first countable (see Definition 1.9.6, Section 1.9.1), there exists a countable basis $(U_n)_{n\in\mathbb{N}}$ for $N(x)$ such that $U_{n+1} \subset U_n$. By the observation above, applied for every U_n in the basis, we can select for any $n \in \mathbb{N}$, a $x_n \in U_n$, such that $F(x_n) \leq t$. By the selection of the sequence $(x_n)_{n\in\mathbb{N}}$, we see that $x_n \to x$, and it satisfies the property $F(x_n) < t$ for all $n \in \mathbb{N}$. Taking the liminf as n tends to infinity, we obtain that $\liminf_n F(x_n) \leq t < F(x)$ (with the latter coming from the choice of t), which contradicts the assumption that F is sequentially lower semicontinuous.

(v) Def. 2.2.1(ii), (ii-a), (ii-b), are hereafter shortened to (ii), (ii-a) and (ii-b). That (ii-a) is equivalent to (ii-b) follows from the definition of the supremum and the infimum. Assume that (ii-a) holds, and set $I := \sup_{U\in N(x)} \inf_{z\in U} F(z)$. By the definition of the supremum, for every $\epsilon > 0$, there exists $U_\epsilon \in N(x)$ such that

$$I - \epsilon < \inf_{z\in U_\epsilon} F(z) \leq F(z), \quad \forall\, z \in U_\epsilon. \tag{2.7}$$

Since $\epsilon > 0$ is arbitrary, so is $t = I - \epsilon \in \mathbb{R}$, and combining (2.7) with (ii-a) we obtain (ii-b).

On the other hand, if (ii-b) holds, we apply it for $t = F(x) - \epsilon$ for any $\epsilon > 0$ and choose the corresponding $U_\epsilon \in N(x)$ guaranteed by (ii-b). Then,

$$t < F(z)\ \forall\, z \in U_\epsilon \implies t \leq \inf_{z\in U_\epsilon} F(z) \leq \sup_{U\in N(x)} \inf_{z\in U} F(z),$$

from which (ii-a) follows.

To show the equivalence of (ii) with (ii-a) and (ii-b), fix an $x \in X$ and assume that (ii-b) holds. Then, this implies that $U_{\lambda,x} := \{\lambda < F(x)\}$ is open, hence $L_{\lambda,x} := \{\lambda \geq F(x)\} = U_{\lambda,x}^c$

is closed. Hence, $L_\lambda = \bigcap_{x \in X} L_{\lambda,x}$ is closed, i. e., we get (ii). On the other hand, if (ii) holds, $U_\lambda = \{\lambda < F(x)\}$ is open. Hence, for every $x \in U_\lambda$, there exists a $U \in N(x)$ such that $U \in U_\lambda$, i. e., $\lambda < F(z)$ for all $z \in U$, and (ii-b) follows from that. □

The concept of lower semicontinuity is robust with respect to the supremum operation and the addition operation.

Example 2.2.3 (Lower semicontinuous functions related to the supremum operation).

(i) If $\{F_\alpha : \alpha \in \mathcal{I}\}$ is a family of lower semicontinuous functionals, then the functional F, defined by $F(x) = \sup_{\alpha \in \mathcal{I}} F_\alpha(x)$, is also lower semicontinuous (note that $L_\lambda F = \bigcap_{\alpha \in \mathcal{I}} L_\lambda F_\alpha$, so this is a closed set as the intersection of closed sets). The same applies if F_α are continuous for every $\alpha \in \mathcal{I}$, in which case, F is also lower semicontinuous.

(ii) Another important example of lower semicontinuous functions is $F : X \to \mathbb{R} \cup \{+\infty\}$, defined by $F(x) = \sup_{U \in N(x)} \alpha(U)$, where $\alpha : \mathcal{O} \to \mathbb{R} \cup \{+\infty\}$ is an arbitrary set function on the set $\mathcal{O}$ of open sets of X. Indeed, consider an arbitary $x \in X$ and any $U \in N(x)$. Since for any open set $U \subset X$ and any $z \in U$ it holds that $U \in N(z)$, we see that $\alpha(U) \le \sup_{U \in N(z)} \alpha(U) = F(z)$. We first take the inf over all $z \in U$ to obtain $\alpha(U) \le \inf_{z \in U} F(z)$ and subsequently the sup over all $U \in N(x)$ to obtain that $F(x) = \sup_{U \in N(x)} \alpha(U) \le \sup_{U \in N(x)} \inf_{z \in U} F(z)$, from which the lower semicontinuity of F follows (see Lemma 6.9 in [62]). ◁

A related notion is upper semicontinuity. As already mentioned in Definition 2.2.1, a function $F : X \to \mathbb{R} \cup \{+\infty\}$ is called upper semicontinuous if $-F$ is lower semicontinuous. The properties of upper semicontinuous functions can be easily inferred from the corresponding properties of lower semicontinuous functions.

Minimization problems

We now turn to a discussion of optimization problems. We only consider minimization problems, as maximization problems can be easily treated by working with $-F$.

Definition 2.2.4 (Local minima and minima). Let (X, τ) be a topological space. Consider the functional $F : A \subset X \to \mathbb{R}$.

(i) The point $x_o \in A$ is called a local minimum of F is there exists a neighborhood $N(x_o)$ of x_o such that $F(x_o) \le F(x)$ for every $x \in N(x_o) \cap A$.

(ii) The point $x_o \in A$ is called a minimum[2] of F in A if $F(x_o) \le F(x)$ for every $x \in A$.

We focus in the case where X is a Banach space. The following result, which is a variant of the Weierstrass theorem, is one of the cornerstones in optimization:

2 Or sometimes global minimum.

Theorem 2.2.5 (Weierstrass). *Let X be a reflexive Banach space, $A \subset X$ be bounded and weakly sequentially closed, and $F : A \subset X \to \mathbb{R}$ be a weakly sequentially lower semicontinuous functional. Then F admits a minimum in A, i. e., there exists $x_o \in A$ such that $F(x_o) \leq F(x)$ for every $x \in A$.*

Proof. We will use the so called direct method of the *calculus of variations*. Let $(x_n)_{n\in\mathbb{N}} \subset A$ be a minimizing sequence,[3] i. e., a sequence such that $F(x_n) \to m$, where $m = \inf_{x\in A} F(x)$. Since A is bounded, this sequence is bounded. By the reflexivity of X, using Theorem 1.1.59, we conclude that there exists a subsequence $(x_{n_k})_{k\in\mathbb{N}}$ and a $x_o \in X$ such that $x_{n_k} \rightharpoonup x_o$ in X. Since A is weakly sequentially closed, it follows that $x_o \in A$. It remains to show that x_o is such that $F(x_o) = m$, i. e., x_o is the minimizer. Indeed, since F is weakly lower sequentially semicontinuous, we have that $F(x_o) \leq \liminf_k F(x_{n_k})$, and since $(x_n)_{n\in\mathbb{N}}$ is a minimizing sequence $\liminf_n F(x_n) = \lim_k F(x_{n_k}) = m$ so that $F(x_o) \leq m$. But $m = \inf_{x\in A} F(x)$, so it follows that $F(x_o) = m$. □

Coercivity

An essential key to the proof of Theorem 2.2.5 was the fact that A was bounded, which allowed us to invoke Theorem 1.1.59 to guarantee the existence of a weak limit of a subsequence of the minimizing sequence, which is then recognized as the minimizer. Such assumptions of boundedness are often too restrictive, and they are replaced by alternative conditions on the functionals themselves that guarantee the boundedness of the minimizing sequence, even if A is unbounded. Such conditions are called coercivity conditions.

Definition 2.2.6 (Coercive functional). Let (X, τ) be a topological space. The map $F : X \to \mathbb{R} \cup \{+\infty\}$ is called coercive/sequentially coercive if the closure of $L_\lambda F := \{x \in X : F(x) \leq \lambda\}$ is countably/sequentially compact in X for all $\lambda \in \mathbb{R}$ (i. e., every sequence in $\overline{L_\lambda F}$ has a cluster point/subsequence that converges in it).

Remark 2.2.7. As compactness is related to the choice of topology on X, one may consider various criteria for coercivity. For example, if X is reflexive, the compactness of $\overline{L_\lambda}$ in the weak topology is related to the boundedness of L_λ. For example, $F : X \to \mathbb{R}$ is weakly sequentially coercive if it satisfies the condition $\lim_n F(x_n) = \infty$ for every sequence $(x_n)_{n\in\mathbb{N}} \subset X$ such that $\lim_n \|x_n\| = \infty$. The definition can be modified accordingly for $A \subset X$ as well. For more information, see [62].

Assuming coercivity of F, we have the following variant of Weierstrass's theorem:

3 This is a standard step: Since $m = \inf_{x\in A} F(x)$, for any $\epsilon > 0$, there exists $x(\epsilon) \in A$ such that $m \leq F(x(\epsilon)) < m + \epsilon$. Consider $\epsilon = \frac{1}{n}$, $n \in \mathbb{N}$, and denote the corresponding $x(\epsilon)$ by x_n. The resulting sequence has the property $m \leq F(x_n) < m + \frac{1}{n}$, and is a minimizing sequence.

Theorem 2.2.8. *Let X be a reflexive Banach space, $A \subset X$ be weakly sequentially closed, and $F : A \subset X \to \mathbb{R}$ be a weakly sequentially lower semicontinuous and sequentially coercive functional. Then F admits a minimum in A, i. e., there exists $x_o \in A$ such that $F(x_o) \leq F(x)$ for every $x \in A$.*

Proof. Assume that $m = \inf_{x \in A} F(x) < \infty$. If $(x_n)_{n \in \mathbb{N}} \subset A$ is a minimizing sequence, then by coercivity, it is necessarily bounded, and then we can follow verbatim the steps of the proof of Theorem 2.2.5. □

Example 2.2.9. Let $\mathcal{D} \subset \mathbb{R}^d$ be an open bounded domain with sufficiently smooth boundary, and consider the functional $F : X = W_0^{1,2}(\mathcal{D}) \to \mathbb{R}$, defined by $F(u) = \frac{1}{2}\int_{\mathcal{D}} |\nabla u(x)|^2 dx - \int_{\mathcal{D}} f(x)u(x)dx$, for any $u \in W_0^{1,2}(\mathcal{D})$ and a given $f \in L^2(\mathcal{D})$. The Dirichlet functional $F_D : X := W_0^{1,2}(\mathcal{D}) \to \mathbb{R}$, defined by $F_D(u) = \frac{1}{2}\int_{\mathcal{D}} |\nabla u(x)|^2 dx$, is coercive on X, as it easily follows by the Poincaré inequality (see Theorem 1.5.13), which guarantees that $\|\nabla u\|_{L^2(\mathcal{D})}$ is an equivalent norm for $X := W_0^{1,2}(\mathcal{D})$. This would not be true if we had defined the Dirichlet functional in such a way that did not allow us to use the Poincaré inequality, e. g., on the whole of $W^{1,2}(\mathcal{D})$. It is easily seen that F, which is a linear perturbation of the Dirichlet functional, is also coercive as a result of the Sobolev embedding theorem. Thus, there exists a function $u_o \in W_0^{1,2}(\mathcal{D})$, for which F admits its minimum value. This follows by a straightforward application of Theorem 2.2.8. ◁

Note that the reflexivity of the Banach space X, on which the functional is defined, plays an important role in all the above proofs. However, there are many functionals whose natural domains of definition are subsets of nonreflexive Banach spaces, e. g., $L^\infty(\mathcal{D})$, $L^1(\mathcal{D})$, or even measure spaces. In these cases, we have to use the concept of weak* convergence and apply the Banach–Alaoglu Theorem 1.1.37 concerning the weak* compactness of the unit ball of the dual space. The next theorem is an example of a result in this spirit.

Theorem 2.2.10. *Let $X \simeq Y^\star$, where Y is a separable Banach space, and let $F : X \to \mathbb{R} \cup \{+\infty\}$ be a proper (i. e., not identically equal to $+\infty$) sequentially coercive and sequentially weak* lower semicontinuous functional. Then F attains its minimum on X, and the set of minimizers is sequentially weak* compact. The result is also true when F is restricted to $A \subset X \simeq Y^\star$, as long as A is sequentially weak* closed.*

Proof. In the spirit of the direct method of the calculus of variations, consider a minimizing sequence $(x_n)_{n \in \mathbb{N}} \subset X \simeq Y^\star$. By coercivity the minimizing sequence is bounded in $X \simeq Y^\star$. Therefore, by Theorem 1.1.36(iii), there exists a subsequence $(x_{n_k})_{k \in \mathbb{N}} \subset X \simeq Y^\star$ and an element $x_o \in X \simeq Y^\star$ such that $x_n \overset{\star}{\rightharpoonup} x_o$. By the sequential weak* lower semicontinuity of F, we have that $\liminf_k F(x_{n_k}) \geq F(x_o)$, and since $(x_n)_{n \in \mathbb{N}}$ is a minimizing sequence, this implies that $\inf_{x \in X} F(x) \geq F(x_o)$ so that x_o is a minimizer. □

Remark 2.2.11. The separability of Y guarantees the metrizability of the weak* topology on bounded sets of $Y^\star \simeq X$ (see Theorem 1.1.36), therefore allowing us to use sequential characterizations of the various topological properties and in particular closedness. If Y is not separable, then the result is still true, but we must ask for weak* lower semicontinuity of F (i. e., asking that the level sets of F are weak* closed sets,[4] rather than sequential weak* lower semicontinuity[5]).

There are two important cases where we need Theorem 2.2.10 in applications: the case where $X^\star = \mathcal{M}_B(\mathcal{D})$, $Y = C_c(\mathcal{D})$ used for studying optimization problems in $L^1(\mathcal{D})$ (understood as a subspace of $\mathcal{M}_B(\mathcal{D})$) and the case where $X = L^\infty(\mathcal{D})$ and $Y = L^1(\mathcal{D})$.

Example 2.2.12. Let $X = \mathcal{M}(\mathcal{D})$, the space of Radon measures endowed with the total variation norm (see Definition 1.4.3 and Theorem 1.4.4). Consider the functional $F : X = \mathcal{M}(\mathcal{D}) \to \mathbb{R}$ defined by $F(\mu) = \|\mathsf{T}(\mu) - z\|^2_{L^2(\mathcal{D})} + \alpha\|\mu\|_{\mathcal{M}(\mathcal{D})}$, where $\mathsf{T} : \mathcal{M}(\mathcal{D}) \to L^2$ is a linear operator with the property $\|\mathsf{T}(\mu)\|_{L^2(\mathcal{D})} \leq c\|\mu\|_{\mathcal{M}(\mathcal{D})}$ for some constant $c > 0$ (independent of the choice of $\mu \in \mathcal{M}(\mathcal{D})$); $\alpha > 0$ is a known parameter, and $z \in L^2(\mathcal{D})$ is a given function. This problem admits a minimum $\mu_o \in \mathcal{M}(\mathcal{D})$. An example for the map T can be the solution map of the elliptic equation $-\Delta u = \mu$ for a measure valued source term, which is known to admit (weak) solutions $u \in W^{1,p}_0(\mathcal{D})$ for all $p \in [1, \frac{d}{d-1})$. Since $W^{1,p}_0(\mathcal{D}) \overset{c}{\hookrightarrow} L^2(\mathcal{D})$, the map defined by $\mu \mapsto \mathsf{T}(\mu) = u$, where u is the solution of the above problem, has the desired properties. Problems of this type appear in a variety of applications.

Indeed, consider a minimizing sequence $(\mu_n)_{n\in\mathbb{N}} \subset \mathcal{M}(\mathcal{D})$. Since $F(\mu_n) < c$ for some suitable constant c, it holds that $\|\mu_n\|_{\mathcal{M}(\mathcal{D})} < c$, so that by Theorem 1.4.10(i), there exists a subsequence $(\mu_{n_k})_{k\in\mathbb{N}}$ and a $\mu_o \in \mathcal{M}(\mathcal{D})$, such that $\mu_{n_k} \overset{\star}{\rightharpoonup} \mu_o$ in $\mathcal{M}(\mathcal{D})$, as $k \to \infty$. By the properties of the mapping T and the weak sequential lower semicontinuity of the norms, we conclude the existence of a minimum. ◁

First-order conditions

If F is Gâteaux differentiable, local minimizers in open subsets of X can be characterized in terms of first-order conditions related to the Gâteaux derivative DF.

Theorem 2.2.13 (First-order conditions). *Let $F : A \subset X \to \mathbb{R}$ and $\mathsf{x}_o \in \operatorname{int}(A)$ be a local minimum of F. If F is Gâteaux differentiable at x_o, then the first-order condition $DF(\mathsf{x}_o) = 0$ holds.*

Proof. Let $\mathsf{h} \in X$ be an arbitrary direction. Then, there exists $\epsilon_o > 0$ such that for each $0 < \epsilon < \epsilon_o$, it holds that $F(\mathsf{x}_o) \leq F(\mathsf{x}_o + \epsilon\,\mathsf{h})$. A simple manipulation leads to

4 And not just weak* sequentially closed!

5 If a functional F is weak* lower semicontinuous, then it is also sequentially weak* lower semicontinuous, but the converse is not true.

$$\frac{F(x_o + \epsilon h) - F(x_o)}{\epsilon} \geq 0,$$

and since F is Gâteaux differentiable at x, we have that

$$\langle DF(x_o), h \rangle \geq 0, \quad \forall h \in X.$$

Since $h \in X$ is arbitrary and X is a vector space, we may repeat the above procedure for $-h \in X$ so that we finally obtain that at the local minimum x_o it holds that

$$\langle DF(x_o), h \rangle = 0, \quad \forall h \in X,$$

which is the stated first-order condition. □

Example 2.2.14. Consider the functional defined in Example 2.1.10 by $F(u) = \frac{1}{2}\int_{\mathcal{D}} |\nabla u(x)|^2 dx - \int_{\mathcal{D}} f(x)u(x)dx$ for every $u \in W_0^{1,2}(\mathcal{D})$ and for a given $f \in L^2(\mathcal{D})$. As shown in Example 2.1.10, F admits a minimum. This functional consists of the Dirichlet functional plus a linear form. By Example 2.1.10, one can see that the Gâteaux derivative of F is $DF(u) = -\Delta u - f$ so that the first-order condition for the minimum u_o reduces to the Poisson equation $-\Delta u_o - f = 0$ with homogeneous Dirichlet boundary conditions on $\mathcal{D}$. ◁

The converse statement of Theorem 2.2.13 is clearly not true, even in finite dimensional spaces (for example critical points may not be minima). We will return to this in Chapters 5 and 8.

2.3 Convex functions

We now turn our attention to the study of convex functions from a Banach space X to $\mathbb{R}$. As we shall see, convex functions enjoy important and very useful properties with respect to differentiability, continuity, semicontinuity, and, most importantly, optimization. There are many excellent texts dedicated to convex functions (see e. g., [12, 20, 28, 74, 113], upon which our approach is based).

Notation 2.3.1. The notation $\varphi : X \to \mathbb{R}$ will be used instead of $F : X \to \mathbb{R}$ for convex maps. We will also freely move between the terms: map, functional, and function. Finally, we may allow convex functions to take the value $+\infty$, denoting that by $\varphi : X \to \mathbb{R} \cup \{+\infty\}$.

2.3.1 Basic definitions, properties, and examples

Let X be a Banach space with norm $\|\cdot\|$, $X^\star$ its dual space and $\langle \cdot, \cdot \rangle$ the duality pairing between them.

Definition 2.3.2 (Proper and convex functions). A function $\varphi : X \to \mathbb{R} \cup \{+\infty\}$ is called proper if it is not identically $+\infty$, and convex if it satisfies

$$\varphi(tx_1 + (1-t)x_2) \leq t\varphi(x_1) + (1-t)\varphi(x_2), \quad \forall x_1, x_2 \in X, \text{ and } t \in [0,1].$$

If the above inequality is strict for any $t \in (0,1)$, $x_1 \neq x_2$, the function is called strictly convex.

Definition 2.3.3 (Uniformly and strongly convex functions). A function $\varphi : X \to \mathbb{R} \cup \{+\infty\}$ is called uniformly convex if there exists an increasing function $\psi : \mathbb{R}_+ \to \mathbb{R}_+$, with $\psi(0) = 0$ (and vanishing only at 0) such that

$$\varphi(tx_1 + (1-t)x_2) \leq t\varphi(x_1) + (1-t)\varphi(x_2) - t(1-t)\psi(\|x_1 - x_2\|),$$
$$\forall x_1, x_2 \in X, \ t \in (0,1),$$

with the function ψ called the modulus of uniform convexity. In the special case, where $\psi(s) = \frac{c}{2}s^2$, $c > 0$, the function φ is called strongly convex with modulus of convexity c.

We will also need to define the following sets:

Definition 2.3.4 (Effective domain and epigraph). Consider a map $\varphi : X \to \mathbb{R} \cup \{+\infty\}$ (not necessarily convex).
(i) The effective domain of φ is defined as $\operatorname{dom}\varphi = \{x \in X : \varphi(x) < +\infty\}$.
(ii) The epigraph of φ is defined as $\operatorname{epi}\varphi = \{(x,\lambda) \in X \times \mathbb{R} : \varphi(x) \leq \lambda\}$.
(iii) The strict epigraph of φ is defined as $\operatorname{epi}_s\varphi = \{(x,\lambda) \in X \times \mathbb{R} : \varphi(x) < \lambda\}$.

Example 2.3.5. The norm of a Banach space is a convex function (as follows direcly by the triangle inequality), but it is not strictly convex, as one can see choosing $x_2 = \alpha x_1$ for any $x_1 \neq 0$, $0 < \alpha < 1$. On the other hand, if $X = H$ is a Hilbert space, the map $x \mapsto \frac{1}{p}\|x\|^p$, $p > 1$ is strictly convex, as is also the case for a large family of Banach spaces (see Theorem 2.6.5 and Remark 2.6.6). Another example of convex functional on $W_0^{1,p}(\mathcal{D})$ is the generalized Dirichlet integral $u \mapsto \frac{1}{p}\int_{\mathcal{D}} |\nabla u(x)|^p dx$, $p \in (1,\infty)$.[6] ◁

Example 2.3.6 (Convexity of the maximum functional). Let K be a compact metric space, and consider the metric space $X = C(K;\mathbb{R})$ of continuous mappings $x : K \to \mathbb{R}$, endowed with the usual sup norm $\|x\|_X := \max_{s\in K} |x(s)|$ (the max taken instead of the sup on account of the compactness of K). Define the functional $\varphi : X \to \mathbb{R}$ by $\varphi(x) = \max_{s\in K} x(s)$ (this is not the norm, as we have omitted the absolute value). This is a convex and continuous functional.

6 As can be seen directly by the convexity of the real valued function $s \mapsto \frac{1}{p}|s|^p$, or the fact that by the Poincaré inequality it is a norm.

Convexity is immediate. Indeed, take any pair $x_1, x_2 \in X = C(K, \mathbb{R})$, and form the convex combination $tx_1 + (1-t)x_2$. Then for any $s \in K$,

$$(tx_1 + (1-t)x_2)(s) = tx_1(s) + (1-t)x_2(s) \le t \max_{s\in K} x_1(s) + (1-t) \max_{s\in K} x_2(s),$$

and taking the maximum over $s \in K$ in the left-hand side leads to

$$\varphi(tx_1 + (1-t)x_2) \le t\varphi(x_1) + (1-t)\varphi(x_2),$$

which is the convexity property for φ.

For proving continuity of φ, consider any pair $x_1, x_2 \in X = C(K; \mathbb{R})$. Let $s_i \in K$ be the point on which maximum[7] for x_i is attained, i. e., $x_i(s_i) = \max_{s\in K} x_i(s)$, $i = 1, 2$. By the definition of the functional φ, we have that

$$\begin{aligned}\varphi(x_1) - \varphi(x_2) &= \max_{s\in K} x_1(s) - \max_{s\in K} x_2(s) = x_1(s_1) - \max_{s\in K} x_2(s) \le x_1(s_1) - x_2(s_1)\\ &\le \max_{s\in K} |x_1(s) - x_2(s)|.\end{aligned}$$

Interchange now the role of x_1, x_2 to get

$$\begin{aligned}\varphi(x_2) - \varphi(x_1) &= \max_{s\in K} x_2(s) - \max_{s\in K} x_1(s) = x_2(s_2) - \max_{s\in K} x_1(s) \le x_2(s_2) - x_1(s_2)\\ &\le \max_{s\in K} |x_1(s) - x_2(s)|.\end{aligned}$$

These inequalities imply that

$$|\varphi(x_1) - \varphi(x_2)| \le |x_1(s_1) - x_2(s_1)| \le \max_{s\in K} |x_1(s) - x_2(s)| = \|x_1 - x_2\|_X,$$

which is the desired continuity. ◁

Convexity of functions and convexity of sets are two concepts related through the concept of the epigraph.

Proposition 2.3.7. *A function $\varphi : X \to \mathbb{R} \cup \{+\infty\}$ is convex if and only if its epigraph,* epi φ, *is a convex set.*

Proof. Let φ be convex, and let $(x_1, \lambda_1), (x_2, \lambda_2) \in \operatorname{epi} \varphi$. Then,

$$\varphi(tx_1 + (1-t)x_2) \le t\varphi(x_1) + (1-t)\varphi(x_2) \le t\lambda_1 + (1-t)\lambda_2,$$

(since both these pairs are in the epigraph), so,

$$(tx_1 + (1-t)x_2, t\lambda_1 + (1-t)\lambda_2) \in \operatorname{epi} \varphi, \quad \forall t \in [0, 1],$$

7 Recall that K is compact, and x_1 is continuous. So, by Weierstrass, the point s_i exists.

i. e.,

$$t(x_1, \lambda_1) + (1-t)(x_2, \lambda_2) \in \operatorname{epi} \varphi, \quad \forall t \in [0,1].$$

Hence, epi φ is a convex set.

Conversely, let epi φ be convex. Since $(x_1, \varphi(x_1)), (x_2, \varphi(x_2)) \in \operatorname{epi} \varphi$, it follows that

$$\big(tx_1 + (1-t)x_2, t\varphi(x_1) + (1-t)\varphi(x_2)\big) \in \operatorname{epi} \varphi, \quad \forall t \in [0,1],$$

so

$$\varphi\big(tx_1 + (1-t)x_2\big) \le t\varphi(x_1) + (1-t)\varphi(x_2), \quad \forall t \in [0,1],$$

hence, φ is convex. □

Convex functions have the following useful properties:

Proposition 2.3.8 (Properties of convex functions).
(i) *The sum of convex functions is a convex function.*
(ii) *If $\varphi : X \to \mathbb{R} \cup \{+\infty\}$ is a convex function, then $\lambda\varphi$ is convex for any $\lambda > 0$.*
(iii) *The pointwise supremum of a set of convex functions is a convex function, i. e., if $\varphi_i : X \to \mathbb{R} \cup \{+\infty\}$ are convex for all $i \in \mathcal{I}$, then $\varphi = \sup_{i\in\mathcal{I}} \varphi_i$, defined for any $x \in X$ by $\varphi(x) := \sup_{i\in\mathcal{I}} \varphi_i(x)$, is also convex.*

Proof. (i) and (ii) follow in an elementary fashion from the definition of convex functions. For (iii), note that $\operatorname{epi} \varphi = \operatorname{epi}(\sup_{i\in\mathcal{I}} \varphi_i) = \bigcap_{i\in\mathcal{I}} \operatorname{epi} \varphi_i$. Since φ_i are convex functions, the sets epi φ_i are convex (Proposition 2.3.7) for every $i \in \mathcal{I}$, and the intersection of any family of convex sets is also convex (Proposition 1.2.4(i)). Therefore, epi φ is a convex set, and so φ is a convex function. □

We close this general discussion of convex functions with some important properties with respect to composition.

Proposition 2.3.9. *Let $\varphi : X \to \mathbb{R} \cup \{+\infty\}$ be a convex function.*
(i) *If $f : \mathbb{R} \to \mathbb{R}$ is an increasing convex function, then the composition $f \circ \varphi : X \to \mathbb{R} \cup \{+\infty\}$ is a convex function.*
(ii) *If $A : Y \to X$ be a continuous linear map, where Y is a Banach space, then $\varphi \circ A : Y \to \mathbb{R} \cup \{+\infty\}$ is a convex function.*

Proof. The proof is elementary and is left to the reader. □

2.3.2 Three important examples: the indicator, Minkowski, and support functions

There are three important examples of functions related to convex sets.

2.3.2.1 The indicator function of a convex set

Definition 2.3.10 (Indicator function of a convex set). Let $C \subset X$ be a set, and define the function $\varphi : X \to \mathbb{R} \cup \{+\infty\}$ by $\varphi(x) = I_C(x)$, where I_C is the indicator function[8] of C defined as

$$I_C(x) = \begin{cases} 0, & x \in C \\ +\infty, & x \in X \setminus C. \end{cases}$$

Since epi $\varphi = C$, it is clear, by Proposition 2.3.7, that $\varphi = I_C$ is a proper convex function if and only if $C \subset X$ is a convex set.

2.3.2.2 The Minkowski functional

Definition 2.3.11 (Minkowski functional). For any set $C \subset X$, define the functional $\varphi_C : X \to \mathbb{R}$, as $\varphi_C(x) = \inf\{t > 0 : t^{-1}x \in C\}$. This functional is called the Minkowski functional or the gauge of C.

The Minkowski functional has some very interesting properties, which allow the characterization of convex sets, as shown in the following proposition [33].

Proposition 2.3.12. *Let $C \subset X$ be an open convex set containing* 0. *Then,* (i) *C is characterized as $C = \{x \in X : \varphi_C(x) < 1\}$, and* (ii) *$\varphi_C$ is sublinear.*

Proof. (i) To show that C is characterized in terms of φ_C, let us consider first that $x \in C$. Since C is open, there exists $\epsilon > 0$ such that $B(x, \epsilon) \subset C$, hence, we also have that $(1+\epsilon)x \in C$; therefore, by the definition of φ_C, we have that $\varphi_C(x) \leq \frac{1}{1+\epsilon} < 1$. For the converse, assume that x is such that $\varphi_C(x) < 1$. By the definition of φ_C, there exists a $t \in (0, 1)$ such that $t^{-1}x \in C$. We express x in terms of a convex combination, as $x = t(t^{-1}x) + (1-t)0 \in C$, by the convexity of C.

(ii) To show that it is sublinear, consider two $x_1, x_2 \in X$ and $\epsilon > 0$. Since $\varphi_C(tx) = t\,\varphi_C(x)$ for every $t > 0$ and $x \in X$ (this is immediate), applying this property for the choice (a) $x = x_1$, $t = \frac{1}{(\varphi_C(x_1)+\epsilon)}$ and (b) $x = x_2$, $t = \frac{1}{(\varphi_C(x_2)+\epsilon)}$, we see that $\varphi_C(\frac{1}{(\varphi_C(x_i)+\epsilon)}x_i) = \frac{\varphi_C(x_i}{\varphi_C(x_i+\epsilon} < 1$ for $i = 1, 2$. Hence, by (i), $\frac{1}{(\varphi_C(x_i)+\epsilon)}x_i \in C$ for $i = 1, 2$. By the convexity of C for any $t \in [0, 1]$, we have that $t\frac{1}{(\varphi_C(x_1)+\epsilon)}x_1 + (1-t)\frac{1}{(\varphi_C(x_2)+\epsilon)}x_2 \in C$. We choose $t = \frac{\varphi_C(x_1)+\epsilon}{\varphi_C(x_1)+\varphi_C(x_2)+2\epsilon}$, which gives us that $\frac{1}{(\varphi_C(x_1)+\varphi_C(x_2)+2\epsilon)}(x_1 + x_2) \in C$, so that by (i) $\varphi_C(\frac{1}{(\varphi_C(x_1)+\varphi_C(x_2)+2\epsilon)}(x_1 + x_2)) < 1$, which is equivalent to $\frac{1}{(\varphi_C(x_1)+\varphi_C(x_2)+2\epsilon))}\varphi_C(x_1 + x_2) < 1$, which leads to $\varphi_C(x_1 + x_2) < \varphi_C(x_1) + \varphi_C(x_2) + 2\epsilon$. This is true for any $\epsilon > 0$, so passing to the limit as $\epsilon \to 0$, we obtain that $\varphi_C(x_1 + x_2) \leq \varphi_C(x_1) + \varphi_C(x_2)$. □

8 Note that the definition of the indicator function employed here is the one commonly used in convex analysis, which differs from the standard definition of the characteristic function of a set (often called indicator function too).

The proof of the geometric form of the Hahn–Banach theorem can be based on the use of the Minkowski functional along with the analytic form [33].

2.3.2.3 The support function

Definition 2.3.13 (Support function). With any convex set $C \subset X$, we may associate a function $\sigma_C : X^\star \to \mathbb{R} \cup \{+\infty\}$, defined by

$$\sigma_C(x^\star) = \sup_{x \in C} \langle x^\star, x \rangle .$$

If $x_o^\star \in X^\star, x_o^\star \neq 0$, is such that $\sigma_C(x_o^\star) = \langle x_o^\star, x_o \rangle$ for some $x_o \in X$, then $x_o^\star$ is said to support C at the point $x_o \in C$ (or equivalently is called a supporting functional at x_o).

Note that we have already encountered a particular case of the support function. In the case where $C = \overline{B}_X(0,1)$, then by the definition of the norm for $X^\star$, we have that $\sigma_{\overline{B}_X(0,1)}(x^\star) = \|x^\star\|_{X^\star}$ for every $x^\star \in X^\star$. The support function plays an important role in the theory of duality.

The convexity of the support function follows from its definition and Proposition 2.3.8, and it is easy to see that $\sigma_C(tx^\star) = t\sigma_C(x^\star)$ for every $t > 0$.

The support function of a closed convex nonempty set C characterizes the set, and allows it to be expressed as the intersection of appropriately chosen halfspaces. Using the notation of Definition 1.2.8 we have [12],

$$C = \bigcap_{x^\star \in X^\star,\, x^\star \neq 0} H_{x^\star \leq \sigma_C(x^\star)} = \{x \in X \,:\, \langle x^\star, x \rangle \leq \sigma_C(x^\star),\ \forall x^\star \in X^\star\}. \tag{2.8}$$

2.3.3 Convexity and semicontinuity

The assertion of Proposition 2.2.2(i) cannot be reversed in the general case, so we may need to distinguish between lower semicontinuity and weak lower semicontinuity. However, for convex functions this is not needed, as the two notions coincide:

Proposition 2.3.14. *Let X be a Banach space and $\varphi : X \to \mathbb{R} \cup \{+\infty\}$ a proper convex function. Then, if φ is lower semicontinuous, it is also weakly lower semicontinuous.*

Proof. It follows since the epigraph of φ is convex and closed, hence weakly closed (by Proposition 1.2.12), so we have the weak lower semicontinuity. □

Example 2.3.15 (Lower semicontinuity of indicator functions). Consider the function $\varphi = I_C$, where now $C \subset X$ is a closed convex set. It is clear that under these conditions $\varphi = I_C$ is proper, convex, and lower semicontinuous. ◁

Example 2.3.16 (Lower semicontinuity of the support function). The support function σ_C of a closed convex set (see Definition 2.3.13) is lower semicontinuous. This can be seen easily, as for any $x^\star \in X^\star$, the value of the function can be expressed as the supremum of a family of continuous functions. ◁

Example 2.3.17. A convex functional, which is lower semicontinuous, is also weakly sequentially lower semicontinuous.

One of the possible ways to show this is using the following argument based on Mazur's lemma: Take any sequence $x'_n \rightharpoonup x$, and let $c = \liminf_n \varphi(x'_n)$. There exists a subsequence, which we relabel as $(x_n)_{n\in\mathbb{N}}$ such that $\lim_n \varphi(x_n) = \liminf_n \varphi(x'_n)$, where of course $x_n \rightharpoonup x$. We must show that $\varphi(x) \le c$. By Mazur's lemma 1.2.17, there exists a sequence $(z_n)_{n\in\mathbb{N}} \subset X$ such that $z_n := \sum_{k=n}^{N(n)} a(n)_k\, x_k$, with $\sum_{k=n}^{N(n)} a(n)_k = 1$, $a(n)_k \in [0,1]$, $k = n,\dots,N(n)$, $n \in \mathbb{N}$, and $z_n \to x$. By the strong semicontinuity of φ, we have that $\varphi(x) \le \liminf_n \varphi(z_n)$. Convexity of φ implies that

$$\varphi(z_n) = \varphi\left(\sum_{k=n}^{N(n)} a(n)_k\, x_k\right) \le \sum_{k=n}^{N(n)} a(n)_k \varphi(x_k). \tag{2.9}$$

By the choice of $(x_n)_{n\in\mathbb{N}}$ for any $\epsilon > 0$, there exists $m \in \mathbb{N}$ such that $\varphi(x_n) < c + \epsilon$ for $n > m$. Using (2.9) for $n > m$ to obtain $\varphi(z_n) \le c + \epsilon$, and taking limit inferior in this, we obtain that $\liminf_n \varphi(z_n) \le c + \epsilon$. Therefore, $\varphi(x) < c + \epsilon$ for every $\epsilon > 0$, hence $\varphi(x) \le c$. ◁

Proposition 2.3.14 and argument in Example 2.3.17 imply that, as far as convex functions are concerned, we do not have to worry too much distinguishing between the various forms of lower semicontinuity since for convex functions these concepts coincide.

Example 2.3.18. The Dirichlet functional $F_{\mathcal{D}} : X := W_0^{1,2}(\mathcal{D}) \to \mathbb{R}$ (see Example 2.1.10) is convex and (by Fatou's lemma) lower semicontinuous, hence (by convexity) also weakly lower semicontinuous. ◁

Convex functions, which are lower semicontinuous, can be supported from below by affine functions as is indicated by the following proposition [12].

Proposition 2.3.19. *Let $\varphi : X \to \mathbb{R} \cup \{+\infty\}$ be a proper, convex, lower semicontinuous function. Then φ is bounded from below by an affine function, i. e., there exists $x^\star \in X^\star$ and $\mu \in \mathbb{R}$ such that*

$$\varphi(x) \ge \langle x^\star, x\rangle + \mu, \quad \forall x \in X.$$

Proof. Let $x_o \in X$ and $\lambda_o \in \mathbb{R}$ such that $\varphi(x_o) > \lambda_o$. Since $(x_o, \lambda_o) \notin \operatorname{epi}\varphi$, and $\operatorname{epi}\varphi$ is closed and convex, by the separation Theorem 1.2.9, there exists[9] a $(x_o^\star, \lambda_o^\star) \in X^\star \times \mathbb{R}$, and $\alpha \in \mathbb{R}$, such that

9 Note that if X is a Banach space, so is the space $X\times\mathbb{R}$ under the norm $\|(x,\lambda)\| = \|x\|+|\lambda|$. We note also that its dual, $(X\times\mathbb{R})^*$ can be identified with $X^\star \times \mathbb{R}$ using the duality pairing, $\langle(x^\star,\lambda^\star),(x,\lambda)\rangle = \langle x^\star, x\rangle + \lambda^\star\lambda$.

$$\langle x_o^\star, x_o \rangle + \lambda_o^\star \lambda_o < \alpha, \tag{2.10}$$

and

$$\langle x_o^\star, x \rangle + \lambda_o^\star \lambda > \alpha, \quad \forall (x, \lambda) \in \operatorname{epi} \varphi. \tag{2.11}$$

Since $(x_o, \varphi(x_o)) \in \operatorname{epi} \varphi$, we have by (2.11),

$$\langle x_o^\star, x_o \rangle + \lambda_o^\star \varphi(x_o) > \alpha. \tag{2.12}$$

It then follows from (2.10) and (2.12) that $\lambda_o^\star(\varphi(x_o) - \lambda_o) > 0$, and so $\lambda_o^\star > 0$. Now applying (2.11) to $(x, \lambda) = (x, \varphi(x)) \in \operatorname{epi} \varphi$, and dividing by $\lambda_o^\star > 0$, we obtain $\varphi(x) > \frac{\alpha}{\lambda_o^\star} + \langle -\frac{x_o^\star}{\lambda_o^\star}, x \rangle$, and upon setting $x^\star = -\frac{x_o^\star}{\lambda_o^\star}$ and $\mu = \frac{\alpha}{\lambda_o^\star}$, the claim is established. □

In fact, a little more can be stated in the same spirit as Proposition 2.3.19, allowing characterization of convex and lower semicontinuous functions in terms of the supremum of a family of affine maps. This is very useful in a number of applications, e. g., in providing lower semicontinuity results for integral functionals in the calculus of variations. This is provided in the following theorem [12, 20, 80].

Theorem 2.3.20 (Approximation of convex functions). *$\varphi : X \to \mathbb{R} \cup \{+\infty\}$ is a proper convex lower semicontinuous function if and only if*

$$\varphi(x) = \sup_{g \in \mathbb{A}(\varphi)} g(x), \quad \mathbb{A}(\varphi) := \{g : X \to \mathbb{R}, \text{ affine continuous}, \ g \leq \varphi\},$$

and the supremum is taken pointwise.

Proof. We sketch the proof. Fix any $x_o \in \operatorname{dom}(\varphi)$. To show that $\varphi(x_o) = \sup\{g(x_o) : g \in \mathbb{A}, g \leq \varphi\}$, it suffices to show that for any $\epsilon > 0$, the quantity $\varphi(x_o) - \epsilon$ is not an upper bound of the set $\{g(x_o) : g \in \mathbb{A}, g \leq \varphi\}$, i. e., that there exists an affine continuous function $g \leq \varphi$ such that $g(x_o) \geq \varphi(x_o) - \epsilon$. The existence of such a function follows from the separation argument, choosing the convex closed set $\operatorname{epi} \varphi$ and the point $(x_o, \varphi(x_o) - \epsilon) \notin \operatorname{epi} \varphi$, working similarly as in Proposition 2.3.19 above. □

In the case where $X = \mathbb{R}^d$, the supremum can be taken over a countable family of affine functions (see Proposition 4.78 in [80]), i. e., there exists a family $\{(b_n, a_n) : n \in \mathbb{N}\} \subset \mathbb{R}^d \times \mathbb{R}$ such that $\varphi(x) = \sup\{b_n \cdot x + a_n : n \in \mathbb{N}\}$.

2.3.4 Convexity and continuity

Convex functions enjoy interesting continuity properties as is shown in the following proposition [28, 29].

Proposition 2.3.21. *Let $\varphi : X \to \mathbb{R} \cup \{+\infty\}$ be a proper convex function. If φ is locally bounded at $x_0 \in \text{int}(\text{dom}\,\varphi)$ then φ is locally Lipschitz at x_0. In fact, it suffices that φ is locally bounded above at x_0.*

Proof. See Section 2.9.1 in the Appendix. □

Working as in the proof of Proposition 2.3.21, one can show for some $x \in \text{dom}(\varphi)$ the equivalence of the statements (a) φ is continuous at x; (b) φ is Lipschitz on some neighborhood of x; (c) φ is bounded on some neighborhood of x, and (d) φ is bounded above on some neighborhood of x.

Proposition 2.3.21 essentially connects the property that the interior of some level set of φ, $L_\lambda \varphi := \{x \in X : \varphi(x) \leq \lambda\}$ is nonempty with the continuity of φ. On the other hand, the domain of φ can be expressed as $\text{dom}\,\varphi = \bigcup_{\lambda \in \mathbb{R}} L_\lambda \varphi = \bigcup_{k \in \mathbb{Z}} L_k \varphi$. One could be tempted to jump to the conclusion that if $x \in \text{int}(\text{dom}\,\varphi)$ then this automatically implies the continuity of φ at x, but though this holds in finite dimensional spaces, this is not necessarily true in infinite dimensions. This happens because it is not necessarily true that $\text{int}(\text{dom}\,\varphi) \neq \emptyset$ implies the existence of some $k \in \mathbb{Z}$ such that $\text{int}(L_k \varphi) \neq \emptyset$, unless certain extra assumptions are made on the sets in question. By the Baire category Theorem 1.9.11 (Section 2.9.2), we see that closedness of the level sets can lead us to the required result. But closedness of the level sets is equivalent to lower semicontinuity of the function.

The above comments lead us to an interesting link between lower semicontinuity and continuity for convex functions.

Proposition 2.3.22. *Let $\varphi : X \to \mathbb{R} \cup \{+\infty\}$ be a proper convex lower semicontinuous function. Then,* $\text{int}(\text{dom}\,\varphi) = \text{core}(\text{dom}\,\varphi)$, *and φ is locally Lipschitz continuous on* $\text{int}(\text{dom}\,\varphi) = \text{core}(\text{dom}\,\varphi)$. *If X is finite dimensional, then lower semicontinuity of φ is not required.*

Proof. See Section 2.9.2 in the Appendix at the end of the chapter. □

2.3.5 Convexity and differentiability

Convexity is related with differentiability in the sense that if a convex function is Gâteaux differentiable, then this imposes important constraints on its derivatives [20].

Theorem 2.3.23. *Let $\varphi : X \to \mathbb{R}$ be Gâteaux differentiable in a convex open subset $C \subset X$.*
(i) *φ is convex if and only if*

$$\varphi(x_2) - \varphi(x_1) \geq \langle D\varphi(x_1), x_2 - x_1 \rangle, \quad \forall x_1, x_2 \in C. \tag{2.13}$$

(ii) *φ is convex if and only if*

$$\langle D\varphi(x_2) - D\varphi(x_1), x_2 - x_1 \rangle \geq 0, \quad \forall x_1, x_2 \in C. \tag{2.14}$$

If φ is strictly convex, the above inequalities are strict.

Proof. (i) Assume that φ is convex. Consider two points $x_1, x_2 \in C$, and take the convex combination $(1-t)x_1 + tx_2 = x_1 + t(x_2 - x_1)$. Convexity of φ implies that

$$\varphi(x_1 + t(x_2 - x_1)) \leq (1-t)\varphi(x_1) + t\varphi(x_2) = \varphi(x_1) + t(\varphi(x_2) - \varphi(x_1))$$

for all $x_1, x_2 \in C$, $t \in (0,1)$, which leads upon rearrangement to

$$\frac{\varphi(x_1 + t(x_2 - x_1)) - \varphi(x_1)}{t} \leq \varphi(x_2) - \varphi(x_1).$$

Since φ is Gâteaux differentiable at $x_1 \in X$, we may pass to the limit $t \to 0$, and interpreting $D\varphi(x_1)$ as an element of $X^\star$, leads to

$$\langle D\varphi(x_1), x_2 - x_1 \rangle \leq \varphi(x_2) - \varphi(x_1).$$

To prove the converse, suppose (2.13) holds for every $x_1, x_2 \in C$. Then, for any $t \in (0,1)$, (2.13) holds for the choices $(x_1 + t(x_2 - x_1), x_1) \in C \times C$ and $(x_1 + t(x_2 - x_1), x_2) \in C \times C$. This leads to the inequalities

$$\begin{aligned} \varphi(x_1) &\geq \varphi(x_1 + t(x_2 - x_1)) - t\,\langle D\varphi(x_1 + t(x_2 - x_1)), x_2 - x_1 \rangle, \\ \varphi(x_2) &\geq \varphi(x_1 + t(x_2 - x_1)) + (1-t)\,\langle D\varphi(x_1 + t(x_2 - x_1)), x_2 - x_1 \rangle. \end{aligned}$$

We multiply the first by $(1-t)$ and the second by t and add to obtain convexity.

(ii) Assume convexity of φ. Apply (2.13) twice, interchanging x_1 and x_2, and add to obtain (2.14).

Conversely, let (2.14) hold. An application of the mean value formula (see Proposition 2.1.18) implies that for all $x_1, x_2 \in C$ there exists $t \in (0,1)$ such that

$$\varphi(x_2) - \varphi(x_1) = \langle D\varphi(x_1 + t(x_2 - x_1)) - D\varphi(x_1), x_2 - x_1 \rangle + \langle D\varphi(x_1), x_2 - x_1 \rangle.$$

We now apply (2.14) for the pair $(x_1 + t(x_2 - x_1), x_1) \in C \times C$; we obtain

$$\langle D\varphi(x_1 + t(x_2 - x_1)) - D\varphi(x_1), x_2 - x_1 \rangle \geq 0, \quad \forall t \in (0,1).$$

Combining these two inequalities, we obtain (2.13), therefore φ is convex. □

If a convex function is twice differentiable, convexity imposes important positivity properties on its second derivative.

Theorem 2.3.24. *Let $C \subset X$ be a convex and open set and $\varphi : C \subset X \to \mathbb{R}$ be twice Gâteaux differentiable in C. The function φ is convex if and only if $D^2\varphi$ is positive definite in the sense that*

$$\langle D^2\varphi(x)h, h\rangle \geq 0, \quad \forall x \in C,\ h \in X. \tag{2.15}$$

If φ is strictly convex, the above inequality is strict.

Proof. Follows immediately by the Taylor expansion (Proposition 2.1.18). □

Proposition 2.3.25. *If a functional $\varphi : X \to \mathbb{R}$ is convex and is Gâteaux differentiable at $x \in X$, then φ is weakly lower sequentially semicontinuous at x.*

Proof. Consider a sequence $(x_n)_{n\in\mathbb{N}} \subset X$ such that $x_n \rightharpoonup x$ in X. Since φ is convex and Gâteaux differentiable at x, apply (2.13) for the choice $x_1 = x$ and $x_2 = x_n$ to obtain

$$\varphi(x_n) - \varphi(x) \geq \langle D\varphi(x), x_n - x\rangle, \quad n \in \mathbb{N}. \tag{2.16}$$

Since $x_n \rightharpoonup x$ in X, for every $x^\star \in X^\star$, it holds that $\langle x^\star, x_n - x\rangle \to 0$, as $n \to \infty$. Choose $x^\star = D\varphi(x) \in X^\star$ and take the limit inferior as $n \to \infty$ in (2.16), to obtain that

$$\liminf_n \varphi(x_n) \geq \varphi(x),$$

which guarantees the weak lower semicontinuity of φ. □

Convexity and differentiability lead to interesting inequalities for functions.

Example 2.3.26 (Convex functions with Lipschitz continuous Fréchet derivative). Assume that $X = H$ is a Hilbert space, identified with its dual, and that $\varphi : H \to \mathbb{R} \cup \{+\infty\}$ is a proper convex function, which is Fréchet differentiable with Lipschitz continuous Fréchet derivative of Lipschitz constant L. Then, for any x, z in an open convex set,

$$\begin{aligned}
&\text{(i)} \quad \varphi(z) \leq \varphi(x) + \langle D\varphi(x), z - x\rangle + \frac{L}{2}\|z - x\|^2,\\
&\text{(ii)} \quad \varphi(z) - \varphi(x) \leq \langle D\varphi(z), z - x\rangle - \frac{1}{2L}\|D\varphi(z) - D\varphi(x)\|^2,\\
&\text{(iii)} \quad \langle D\varphi(z) - D\varphi(x), z - x\rangle \geq \frac{1}{L}\|D\varphi(x) - D\varphi(z)\|^2 \quad \text{(co-coercivity).}
\end{aligned} \tag{2.17}$$

The first of these properties follows by Theorem 2.3.23, and the observation that

$$\varphi(z) - \varphi(x) = \int_0^1 \langle z - x, D\varphi(x + t(z - x))\rangle dt,$$

so that

$$\varphi(z) - \varphi(x) - \langle z - x, D\varphi(x)\rangle = \int_0^1 \langle z - x, D\varphi(x + t(z - x)) - D\varphi(x)\rangle dt \le \frac{L}{2}\|z - x\|^2,$$

where we first used the Lipschitz property of $D\varphi$ and then integrated over t.

The second follows by expressing $\varphi(z) - \varphi(x) = \varphi(z) - \varphi(x') + \varphi(x') - \varphi(x)$ for any $x' \in H$, and then estimating as

$$\begin{aligned}\varphi(z) - \varphi(x) &= \varphi(z) - \varphi(x') + \varphi(x') - \varphi(x) = -(\varphi(x') - \varphi(z)) + (\varphi(x') - \varphi(x)) \\ &\le \langle D\varphi(z), z - x'\rangle + \langle D\varphi(x), x' - x\rangle + \frac{L}{2}\|x' - x\|^2,\end{aligned} \tag{2.18}$$

where we used Theorem 2.3.23(i) for the first parenthesis and (2.17)(i) for the second. We then minimize the upper bound over $x' \in H$. The minimum of the upper bound is obtained at $x' = x - \frac{1}{L}(D\varphi(x) - D\varphi(z))$, and upon substitution in (2.18), we obtain the stated result (2.17)(ii).

The co-coercivity property follows by using the second, interchanging the role of x and z and adding. The details are left as an exercise. ◁

Example 2.3.27 (The Polyak–Lojasiewicz inequality). Let $\varphi : H \to \mathbb{R} \cup \{+\infty\}$ be Fréchet differentiable and strongly convex with modulus of convexity $c > 0$. Then, if x_o is a minimizer of φ, we have the Polyak–Lojasiewicz condition:

$$\|D\varphi(x)\|^2 \ge 2c(\varphi(x) - \varphi(x_o)), \quad \forall x \in H. \tag{2.19}$$

By simply modifying the proof of Theorem 2.3.23(i), using the definition of strong convexity (see Definition 2.3.3), and appropriately passing to the limit, taking into account the extra quadratic term, we find that for any $x_1, x_2 \in C$,

$$\varphi(x_2) \ge \varphi(x_1) + \langle D\varphi(x_1), x_2 - x_1\rangle + \frac{c}{2}\|x_2 - x_1\|^2, \quad \forall x_1, x_2 \in C.$$

We rearrange this inequality to obtain (after some algebra) that

$$\begin{aligned}\varphi(x_2) &\ge \varphi(x_1) + \langle D\varphi(x_1), x_2 - x_1\rangle + \frac{1}{2}\left\|\sqrt{c}(x_2 - x_1) + \frac{1}{\sqrt{c}}D\varphi(x_1) - \frac{1}{\sqrt{c}}D\varphi(x_1)\right\|^2 \\ &= \varphi(x_1) - \frac{1}{2c}\|D\varphi(x_1)\|^2 + \frac{1}{2}\left\|\sqrt{c}(x_2 - x_1) + \frac{1}{\sqrt{c}}D\varphi(x_1)\right\|^2 \\ &\ge \varphi(x_1) - \frac{1}{2c}\|D\varphi(x_1)\|^2 \ge \varphi(x_o) - \frac{1}{2c}\|D\varphi(x_1)\|^2,\end{aligned} \tag{2.20}$$

where the last estimate follows since x_o is a minimizer of φ. Since this is true for all $x_1, x_2 \in C$, by rearranging, we obtain (2.19). This inequality finds important applications in optimization algorithms. ◁

We will see in Chapter 4 that many of the above properties generalize for nonsmooth convex functions in terms of the subdifferential.

2.4 Optimization and convexity

The assumption of convexity allows us to deduce a number of important results regarding minimization problems as is shown in the following proposition (e. g., [20, 113]).

Proposition 2.4.1. *Let X be a reflexive Banach space and $\varphi : X \to \mathbb{R} \cup \{+\infty\}$ a convex proper and lower semicontinuous function.*
(i) *Let C be a bounded, convex, and closed subset of X. Then, φ attains its minimum on C.*
(ii) *If $C \subset X$ is convex and closed, and φ is coercive on C then it attains its minimum on C. We may allow C to be a closed subspace of X, or even X itself.*

Proof. This result is immediate by Theorems 2.2.5 and 2.2.8 and the fact that a closed convex set is also weakly closed (Proposition 1.2.12) and that a convex lower semicontinuous function is also weakly semicontinuous (Proposition 2.3.14). □

Remark 2.4.2. A convex lower semicontinuous functional is not necessarily weak* lower semicontinuous, so Proposition 2.4.1 is not necessarily true if X is nonreflexive.

Apart from simplifying existence results, convexity leads to some interesting qualitative properties, as far as minimization is concerned.

Proposition 2.4.3.
(i) *A local minimum for a convex functional defined on a convex set C is a global minimum.*
(ii) *If a functional is strictly convex, then its minimum is unique.*

Proof. (i) Assume that $x \in C$ is a local minimum, i. e., $\varphi(x_o) \leq \varphi(x)$ for every $x \in V := B(x_o, \epsilon) \cap C$, where $\epsilon > 0$ is sufficiently small. For any $z \in C$, take the convex combination $(1-t)x_o + tz = x_o + t(z - x_o)$, $t \in [0,1]$. For small enough values of t, it holds that $x_o + t(z - x_o) \in V$ and since x_o is a local minimum,

$$\varphi(x_o) \leq \varphi(x_o + t(z - x_o)),$$

and by convexity of φ, it follows that

$$\varphi(x_o + t(z - x_o)) \leq (1-t)\varphi(x_o) + t\,\varphi(z) = \varphi(x_o) + t(\varphi(z) - \varphi(x_o)).$$

Combining the above, we obtain for all positive and small enough t that

$$\varphi(x_o) \leq \varphi(x_o) + t\,(\varphi(z) - \varphi(x_o)),$$

or equivalently that $\varphi(x_o) \leq \varphi(z)$ for every $z \in C \subset X$, therefore, x_o is a global minimum.

(ii) Let $x_{0,1}, x_{0,2} \in C$ be two global minima of φ such that $x_{0,1} \neq x_{0,2}$. Consider the point $x_o = \frac{1}{2}x_{0,1} + \frac{1}{2}x_{0,2} \in C$. By strict convexity of φ, it follows that $\varphi(x_o) < \varphi(x_{0,1}) = \varphi(x_{0,2})$, which leads to contradiction. □

Example 2.4.4 (Maximum principle for convex functions). Consider a convex function φ defined on a convex subset $C \subset X$. If φ attains a global maximum at an interior point of C then, φ is constant. This result is attributed to Heinz Bauer.

Assume that φ is not constant, but it attains a global maximum at $x_o \in \text{int}(C)$. Choose $x \in C$ such that $\varphi(x) < \varphi(x_o)$ and $t \in (0,1)$ such that $z = x_o + t(x_o - x) \in C$ (this is possible since x_o is an interior point of C). That implies that $x_o = \frac{1}{1+t}z + \frac{t}{1+t}x$, and by the convexity of φ and the fact that x_o is a global maximum, we conclude that

$$\varphi(x_o) = \varphi\left(\frac{1}{1+t}z + \frac{t}{1+t}x\right) \leq \frac{1}{1+t}\varphi(z) + \frac{t}{1+t}\varphi(x) < \frac{1}{1+t}\varphi(x_o) + \frac{t}{1+t}\varphi(x_o) = \varphi(x_o),$$

hence, $\varphi(x_o) < \varphi(x_o)$, which is a contradiction. ◁

Theorem 2.4.5 (First order condition). *Let $C \subset X$, convex, and $\varphi : C \subset X \to \mathbb{R}$, Gâteaux differentiable and convex. Then $x_o \in C$ is a minimum if and only if*

$$\langle D\varphi(x_o), x - x_o \rangle \geq 0, \quad \forall x \in C. \tag{2.21}$$

If $C = X$, the first-order condition reduces to $D\varphi(x_o) = 0$.

Proof. Assume $x_o \in C$ is a minimum. Then, $\varphi(x_o) \leq \varphi(z)$ for every $z \in C$. For any $x \in C$, set $z = (1-t)x_o + t\,x = x_o + t\,(x - x_o) \in C$ for $t \in (0,1)$, and apply this inequality to obtain

$$\varphi(x_o) \leq \varphi(x_o + t\,(x - x_o)), \quad \forall x \in C,\ t > 0,$$

which yields

$$\frac{\varphi(x_o + t\,(x - x_o)) - \varphi(x_o)}{t} \geq 0, \quad t > 0,$$

and going to the limit, as $t \to 0$,

$$\langle D\varphi(x_o), x - x_o \rangle \geq 0, \quad \forall x \in C.$$

For the converse, since φ is convex and Gâteaux differentiable by (2.13), we have that

$$\varphi(x) - \varphi(x_o) \geq \langle D\varphi(x_o), x - x_o \rangle, \quad \forall x_o, x \in C.$$

Since for $x_o \in C$, it holds that $\langle D\varphi(x_o), x - x_o \rangle \geq 0$, $\forall x \in C$, we find that $\varphi(x_o) \leq \varphi(x)$ for all $x \in C$ so that x_o is a minimum. □

Remark 2.4.6. Note that for general minimization problems on convex subsets of Banach spaces, the first-order condition (2.21) takes the form of an inequality rather than an equation. Inequalities of this form are called variational inequalities and will be studied in detail in Chapter 7.

Theorem 2.4.5 can lead to useful numerical schemes for the study of optimization problems.

Example 2.4.7 (The gradient descent method). Let $\varphi : H \to \mathbb{R}\cup\{+\infty\}$ be a proper convex function on a Hilbert space with Lipschitz continuous Fréchet derivative of Lipschitz constant L. Consider the iterative scheme

$$x_{n+1} = x_n - \alpha D\varphi(x_n), \quad \alpha > 0,$$

called a gradient descent scheme. If this iterative scheme has a fixed point, i. e., if for some $x_o \in \operatorname{int}(C)$ it holds that $x_o = x_o - \alpha D\varphi(x_o)$, then x_o is a minimizer for φ. Moreover, if $\alpha = \frac{1}{L}$ and φ satisfies an inequality $\|D\varphi(x)\|^2 \geq 2c(\varphi(x) - \varphi(x_o))$ of the Polyak–Lojasiewicz-type for every $x \in H$ (see Example 2.3.27, inequality (2.19)), we have that

$$\varphi(x_n) - \varphi(x_o) \leq \left(1 - \frac{c}{L}\right)^{n-1} (\varphi(x_1) - \varphi(x_o)).$$

We prove the above claims. It is easy to see that a fixed point x_o of the gradient descent scheme (if it exists) is a critical point $D\varphi(x_o) = 0$. Hence, by convexity, x_o is a minimizer. To show the convergence estimate, we apply inequality (2.17)(i) for x_{n+1}, x_n, and taking into account the iteration scheme so that $x_{n+1} - x_n = -\alpha D\varphi(x_n)$, we see that

$$\varphi(x_{n+1}) - \varphi(x_n) \leq -\left(\alpha - \frac{L\alpha^2}{2}\right)\|D\varphi(x_n)\|^2.$$

We choose α so that $(\alpha - \frac{L\alpha^2}{2}) > 0$, e. g., $\alpha = \frac{1}{L}$. For this choice, and using also the Polya–Lojasiewicz inequality (2.19), we obtain the estimate

$$\begin{aligned}
&\varphi(x_{n+1}) - \varphi(x_n) \leq -\frac{c}{L}(\varphi(x_n) - \varphi(x_o)) \\
&\implies \varphi(x_{n+1}) - \varphi(x_o) + \varphi(x_o) - \varphi(x_n) \leq -\frac{c}{L}(\varphi(x_n) - \varphi(x_o)) \\
&\implies \varphi(x_{n+1}) - \varphi(x_o) \leq \left(1 - \frac{c}{L}\right)(\varphi(x_n) - \varphi(x_o)),
\end{aligned}$$

and the result follows by iteration. Note that we did not use the strong convexity assumption anywhere, but relied only on the Polyak–Lojasiewicz inequality, which may also hold for nonstrongly convex functions. An interesting observation is that the step size is inversely proportional to the Lipschitz constant. One may try to use varying step sizes, or even choose the step size by an appropriate optimization procedure. One way to allow for unconditional convergence would be to use the implicit version of the gradient scheme, i. e., calculating $D\varphi$ in x_{n+1} rather than x_n. This simple numerical scheme will be revisited in its various variants later on in this book to cover for cases where, e. g., smoothness is absent. ◁

2.5 Projections in Hilbert spaces

In this section, $X = H$ is a Hilbert space, identified with its dual, with inner product $\langle\cdot,\cdot\rangle$ and norm $\|\cdot\|$ such that $\|x\|^2 = \langle x, x\rangle$ for every $x \in X$.

Definition 2.5.1 (Projection operator). Let $C \subset H$ be a convex and closed set and $x \in H$. The operator $P_C : H \to C$, defined by $P_C(x) := x_o = \arg\min_{z\in C} \|x - z\|$, (i. e., where x_o is the solution of the problem $\inf_{z\in C} \|x - z\|$) is called the projection operator on C.

Definition 2.5.2 (Normal cone). With $C \subset H$ as above, for any $x \in C$, the normal cone of C at x is defined as the set $N_C(x) = \{x^\star \in H \,:\, \langle x^\star, z - x\rangle \leq 0,\ \forall z \in C\}$.

The projection operator enjoys useful properties, summarized in the following theorem (see e. g., [21, 33]:

Theorem 2.5.3 (Projection theorem and properties of projection operators). *Let H be a Hilbert space and $C \subset X$ be closed and convex. Then,*

(i) *For any $x \in H$, the minimization problem $\min_{z\in C} \|x-z\|$ has a unique solution, $x_o \in C$, characterized by the solution of the inequality*

$$\langle x - x_o, z - x_o\rangle \leq 0, \quad \forall\, z \in C. \tag{2.22}$$

An equivalent characterization is

$$x - x_o \in N_C(x_o).$$

(ii) *The projection operator $P_C : H \to C$ is single valued and nonexpansive, i. e.,*

$$\|P_C(x_1) - P_C(x_2)\| \leq \|x_1 - x_2\|, \quad \forall\, x_1, x_2 \in H.$$

Proof. (i) The proof is a straightforward application of Theorem 2.4.5, upon noting that the minimization problem defining $\min_{z\in C} \|x-z\|$ is equivalent to the minimization problem $\min_{z\in C} \|x - z\|^2$ and recalling Example 2.1.3 and Definition 2.5.2.

(ii) The operator P_C is single valued as a result of the stated non expansive property. To prove the latter we work as follows: The projection of x_1 on C satisfies

$$\langle x - P_C(x_1), z - P_C(x_1)\rangle \leq 0, \quad \forall\, z \in C.$$

Choose $z = P_C(x_2) \in C$ to obtain the inequality

$$\langle x_1 - P_C(x_1), P_C(x_2) - P_C(x_1)\rangle \le 0. \tag{2.23}$$

On the other hand, the projection of x_2 on C, satisfies,

$$\langle x_2 - P_C(x_2), z - P_C(x_2)\rangle \le 0, \quad \forall z \in C.$$

Choose $z = P_C(x_1) \in C$ to obtain the inequality

$$\langle x_2 - P_C(x_2), P_C(x_1) - P_C(x_2)\rangle \le 0. \tag{2.24}$$

Adding (2.23) and (2.24) yields

$$\langle x_1 - x_2 - (P_C(x_1) - P_C(x_2)), P_C(x_1) - P_C(x_2)\rangle \ge 0,$$

which gives

$$\langle x_1 - x_2, P_C(x_1) - P_C(x_2)\rangle \ge \|P_C(x_1) - P_C(x_2)\|^2.$$

Applying the Cauchy–Schwarz inequality on the LHS of the above and upon dividing by $\|P_C(x_1) - P_C(x_2)\|$, yields the desired result. □

Note that in general the projection operator is nonlinear. An important special case is $C = E$, where E is a closed linear subspace of X.

Definition 2.5.4. Let $E \subset H$ be a closed linear subspace. By $E^\perp$, we denote the orthogonal complement of E:

$$E^\perp := \{z \in H : \langle x, z\rangle = 0, \forall z \in E\}.$$

Proposition 2.5.5 (The projection operator on a closed subspace). *Let $E \subset H$ be a linear closed subspace. The projection operator $P_E : H \to E$ has the following properties:*

(i) *It is a linear nonexpansive operator, i. e., $\|P_E x\| \le \|x\|$ for every $x \in H$, satisfying the orthogonality condition $\langle x - P_E x, z\rangle = 0$ for all $z \in E$.*
(ii) *For every $x \in H$, there exists a unique decomposition $x = x_o + z$, where $x_o = P_E x \in E$ and $z = (I - P_E)x = P_{E^\perp} x \in E^\perp$.*
(iii) *P_E is an idempotent operator, i. e., $P_E P_E = P_E$ with the properties $N(P_E) = E^\perp$ and $P_E(H) = E$.*

Proof. (i) The projection of x on E, x_o, satisfies the inequality $\langle x - x_o, z' - x_o\rangle \le 0$ for all $z' \in E$. Take any $z \in E$, and first choose $z' = x_o + z \in E$ (by the linearity of E) which yields $\langle x - x_o, z\rangle \le 0$, and then choose $z' = x_o - z \in E$ (by the linearity of E), which yields $\langle x - x_o, z\rangle \ge 0$. Therefore, $\langle x - x_o, z\rangle = 0$ for every $z \in E$.

Consider the linear combination $x = \lambda_1 x_1 + \lambda_2 x_2 \in H$ and the projection $x_o := P_E x = P_E(\lambda_1 x_1 + \lambda_2 x_2)$. Since E is a closed linear subspace, the projection $P_E x$ is the unique element of E that satisfies

$$\langle x - x_o, z\rangle = 0, \quad \forall z \in E.$$

Now, let us calculate:

$$\begin{aligned}\langle x - (\lambda_1 P_E x_1 + \lambda_2 P_E x_2), z\rangle &= \langle \lambda_1 x_1 + \lambda_2 x_2 - (\lambda_1 P_E x_1 + \lambda_2 P_E x_2), z\rangle \\ &= \lambda_1 \langle x_1 - P_E x_1, z\rangle + \lambda_2 \langle x_2 - P_E x_2, z\rangle = 0.\end{aligned}$$

By the fact that E is a closed linear subspace, $\lambda_1 P_E x_1 + \lambda_2 P_E x_2 \in E$, and since it satisfies the above equality for any $z \in E$, we have (by the uniqueness of projections) that $P_E x = \lambda_1 P_E x_1 + \lambda_2 P_E x_2$, therefore we have linearity. The nonexpansiveness follows by Proposition 2.5.3.

(ii) Let any $x \in H$. Set $x_o = P_E x \in E$. By the definition of the projection $x_o = P_E x$ satisfies $\langle x - x_o, z\rangle = 0$ for all $z \in E$, therefore $z_0 := x - x_o \in E^\perp$. Hence, $x = x_o + z_0$ for $x_o = P_E x \in E$ and $z_0 \in E^\perp$. It remains to show that $z_0 = P_{E^\perp} x$. It suffices to show that $\langle x - z_0, w\rangle = 0$ for any $w \in E^\perp$. This follows from the fact that $x - z_0 = x_o \in E$.

(iii) The statements follow by the decomposition in (ii). □

The generalization of the concept of projection in Banach spaces is more limited for a number of reasons, among which is the possible nondifferentiability of the norm (see Example 4.3.4).

2.6 Geometric properties of Banach spaces related to convexity

In this section, we consider certain important characterizations of Banach spaces related to convexity (see, e. g., [41, 52], or [96]).

2.6.1 Strictly, uniformly, and locally uniformly convex Banach spaces

2.6.1.1 Strict convexity

Definition 2.6.1. A Banach space X is called strictly convex if its unit ball is strictly convex, i. e., if for all $x_1, x_2 \in X$, $x_1 \neq x_2$, $\|x_1\| = \|x_2\| = 1$ it holds that $\|t x_1 + (1-t) x_2\| < 1$ for all $t \in (0,1)$.

One may see that yet another equivalent way of defining strict convexity is using the midpoint, i. e., stating the definition for $t = \frac{1}{2}$ only, a condition which is easier to check (see Theorem 2.6.5).

The geometric intuition behind strictly convex Banach spaces is that if X is strictly convex, then the boundary of the unit ball does not contain any line segments. This is understood in the sense that geometrically, for any $x_1, x_2 \in S_X(0,1)$, the elements of the form $t x_1 + (1-t) x_2$, $t \in (0,1)$, can be considered as line segments, and strict convexity implies that such elements are not on $S_X(0,1)$.

Example 2.6.2 (Strict convexity depends on the choice of norm). Whether a Banach space is strictly convex or not depends on the choice of norm. For instance, $X = \mathbb{R}^d$ is strictly convex when equipped with the norms $\|x\|_p = (\sum_{i=1}^d |x_i|^p)^{1/p}$ for $p \in (1,\infty)$ but not for $p = 1$ or $p = \infty$ (to convince yourself try sketching the unit ball in $\mathbb{R}^2$ under these norms). ◁

Example 2.6.3 (Strict convexity and extreme points). Another way to state strict convexity is using the concept of extreme points. For a convex set C, a point $x \in C$ is called an extreme point of C if whenever $x = tx_1 + (1-t)x_2$ for some $t \in (0,1)$, and $x_1, x_2 \in C$, then $x = x_1 = x_2$. Taking $C = \overline{B}_X(0,1) = \{x \in X : \|x\| \leq 1\}$, the closed unit ball of X, we see that if X is strictly convex, then the extreme points of $\overline{B}_X(0,1)$ are on $S_X(0,1) = \{x \in X : \|x\|_X = 1\}$, the unit sphere of X.

First of all, we need to show that $\overline{B}_X(0,1)$ does have extreme points. Note that any $x \in \overline{B}_X(0,1)$ such that $\|x\| < 1$ cannot be an extreme point of $\overline{B}_X(0,1)$. To see this, observe that since $\|x\| < 1$, there exists $\epsilon > 0$ such that $\|x\| < \frac{1}{1+\epsilon}$, therefore $x_1 := (1+\epsilon)x \in \overline{B}_X(0,1)$. Clearly, $x_2 := (1-\epsilon)x \in \overline{B}_X(0,1)$ with $x_1 \neq x_2$, and since $x = \frac{1}{2}x_1 + \frac{1}{2}x_2$, we see that such an x is not an extreme point of $\overline{B}_X(0,1)$. Next, consider $x \in S_X(0,1)$. Suppose there exist $x_1, x_2 \in \overline{B}_X(0,1)$ such that $x_1 \neq x_2$ and $x = tx_1 + (1-t)x_2$, $t \in (0,1)$. Note that it is impossible that $\|x_1\| < 1$ and $\|x_2\| < 1$, so that it must be $\|x_1\| = \|x_2\| = 1$. But then, it is impossible that $x = tx_1 + (1-t)x_2$, $t \in (0,1)$ by strict convexity. So any point $x \in S_X(0,1)$ is an extreme point of $\overline{B}_X(0,1)$. ◁

Example 2.6.4. Any Hilbert space H is strictly convex, as can be seen by the Cauchy–Schwarz inequality. The Banach spaces $X = L^p(\mathcal{D})$ for $1 < p < \infty$ are strictly convex Banach spaces. The spaces $L^1(\mathcal{D})$ and $L^\infty(\mathcal{D})$ are not strictly convex. ◁

A useful characterization of strict convexity is given in the next theorem [41].

Theorem 2.6.5. *Let X be a Banach space. The following are equivalent:*

(i) *X is strictly convex.*

(ii) $\|x_1 + x_2\| = \|x_1\| + \|x_2\|$ *for* $x_1, x_2 \neq 0$ *if and only if* $x_1 = \lambda x_2$ *for some* $\lambda > 0$.

(iii) *For* $x_1, x_2 \in X$ *such that* $x_1 \neq x_2$ *and* $\|x_1\| = \|x_2\| = 1$, *we have that* $\|\frac{x_1+x_2}{2}\| < 1$.

(iv) *For* $1 < p < \infty$, *and* $x_1, x_2 \in X$ *such that* $x_1 \neq x_2$, *we have that*

$$\left\|\frac{x_1 + x_2}{2}\right\|^p < \frac{\|x_1\|^p + \|x_2\|^p}{2}.$$

(vi) *The problem* $\sup\{\langle x^\star, x\rangle : \|x\| \leq 1\}$ *achieves a unique solution for any*[10] $x^\star \in X^\star$, $x^\star \neq 0$.

(v) *The function* $x \mapsto \varphi(x) = \|x\|^2$ *is strictly convex.*

10 This condition is often phrased as the following: Each support hyperplane for the unit ball supports it at only one point (e. g., [104]). Recall that an $x^\star \in X^\star$, $x^\star \neq 0$, is called a support functional for $A \subset X$ at point $x_o \in X$ if $\langle x^\star, x_o\rangle = \sup\{\langle x^\star, x\rangle : x \in A\}$. The point x_o is called a support point for A, and the hyperplane $H := \{x \in X : \langle x^\star, x\rangle = \langle x^\star, x_o\rangle\}$ is called the support hyperplane for A.

Proof. We start by the proof of equivalence of (i) and (ii). Then we will show the equivalence of (i) with each of the remaining claims (iii)–(vi).

1. (i) $\Longleftrightarrow$ (ii): Assume (i) holds, and consider any $x_1, x_2 \neq 0$ such that $\|x_1 + x_2\| = \|x_1\| + \|x_2\|$. Without loss of generality, assume that $\|x_2\| \geq \|x_1\|$. Then,

$$\begin{aligned}\left\|\frac{x_1}{\|x_1\|} + \frac{x_2}{\|x_2\|}\right\| &= \left\|\left(\frac{x_1}{\|x_1\|} + \frac{x_2}{\|x_1\|}\right) - \left(\frac{x_2}{\|x_1\|} - \frac{x_2}{\|x_2\|}\right)\right\| \\ &\geq \left\|\frac{x_1}{\|x_1\|} + \frac{x_2}{\|x_1\|}\right\| - \left\|\frac{x_2}{\|x_1\|} - \frac{x_2}{\|x_2\|}\right\| = \frac{1}{\|x_1\|}\|x_1 + x_2\| - \left|\frac{1}{\|x_1\|} - \frac{1}{\|x_2\|}\right|\|x_2\| \\ &= \frac{\|x_1\| + \|x_2\|}{\|x_1\|} - \frac{\|x_2\| - \|x_1\|}{\|x_1\|\,\|x_2\|}\|x_2\| = 2,\end{aligned}$$

where we used the reverse triangle inequality and the choice of x_1, x_2. Setting $z_i = \frac{x_1}{\|x_i\|}$, $\|z_i\| = 1$, $i = 1, 2$, we see from the above that $\frac{1}{2}\|z_1 + z_2\| = 1$. But (i) (for the choice $t = 1/2$) implies that this can only hold if $z_1 = z_2$. This leads to

$$\frac{x_1}{\|x_1\|} = \frac{x_2}{\|x_2\|} \implies x_1 = \frac{\|x_1\|}{\|x_2\|}x_2,$$

i. e., (ii) holds for $\lambda = \frac{\|x_1\|}{\|x_2\|} > 0$.

For the reverse, assume that (ii) holds but (i) does not. Then, there exists $x_1 \neq x_2$, $\|x_1\| = \|x_2\| = 1$ and a $t_0 \in (0, 1)$ such that $\|t_0 x_1 + (1 - t_0)x_2\| = 1$. But,

$$\|t_0 x_1\| + \|(1 - t_0)x_2\| = t_0\|x_1\| + (1 - t_0)\|x_2\| = t_0 + (1 - t_0) = 1.$$

Hence, for the chosen x_1, x_2, we have that $\|t_0 x_1 + (1-t_0)x_2\| = \|t_0 x_1\| + \|(1-t_0)x_2\|$. Therefore, upon defining $z_1 = t_0 x_1$, $z_2 = (1 - t_0)x_2$, we obtain that $\|z_1 + z_2\| = \|z_1\| + \|z_2\|$, and by (ii), there exists $\lambda > 0$ such that $z_1 = \lambda z_2$. This implies that $t_0 x_1 = \lambda(1-t_0)x_2$, and taking norms, we obtain $t_0 = \lambda(1 - t_0)$. But then,

$$z_1 = \lambda z_2 \implies t_0 x_1 = \lambda(1 - t_0)x_2 \implies x_1 = x_2,$$

which is a contradiction. Hence, (ii) implies (i).

2. (i) $\Longleftrightarrow$ (iii): That (i) implies (iii) is immediate from the definition of strict convexity. To show that (iii) implies (i), argue by contradiction. Suppose that (iii) holds but not (i), so that (using (ii)) there exist $x_1, x_2 \in X$, without loss of generality $\|x_2\| \geq \|x_1\|$, nonparallel with $\|x_1 + x_2\| = \|x_1\| + \|x_2\|$. Define $z_i = \frac{1}{\|x_i\|}x_i$, $i = 1, 2$ such that $\|z_i\| = 1$, and $z = \frac{1}{\|x_1\|}x_2$. Then, $\|z_1 + z_2\| = \|z_1 + z - z + z_2\| \geq \|z_1 + z\| - \|z_2 - z\| = 2$, since $\|x_1 + x_2\| = \|x_1\| + \|x_2\|$. But this contradicts (i).

3. (i) $\Longleftrightarrow$ (iv): That (i) implies (iv) holds if $x_1 = \lambda x_2$, $0 < \lambda \neq 1$. Else, by the monotonicity and convexity of the real function $t \mapsto \varphi(t) = t^p$, $t \geq 0$, $1 < p < \infty$, applied to $t = \frac{1}{2}(\|x_1\| + \|x_2\|)$. Indeed, since X is strictly convex, by (ii), $\|x_1 + x_2\| < \|x_1\| + \|x_2\|$ so that,

$(\frac{1}{2}(\|x_1 + x_2\|))^p < (\frac{1}{2}\|x_1\| + \frac{1}{2}\|x_2\|)^p$, and the result follows. To show that (iv) implies (i), consider x_1, x_2 such that $\|x_1\| = \|x_2\| = 1$. Then (iv) implies (iii), which in turn implies (i).

4. (i) $\Longleftrightarrow$ (v): To show that (i) implies (v), assume on the contrary that there exist $x_1 \neq x_2$, $\|x_1\| = \|x_2\| = 1$, such that $\langle x^\star, x_1\rangle = \langle x^\star, x_2\rangle = \sup\{\langle x^\star, x\rangle \,:\, \|x\| \le 1\} = \|x^\star\|_{X^\star}$. Then $\langle x^\star, x_1 + x_2\rangle = \langle x^\star, x_1\rangle + \langle x^\star, x_2\rangle = 2\|x^\star\|_{X^\star}$. On the other hand, $\langle x^\star, x_1 + x_2\rangle \le |\langle x^\star, x_1 + x_2\rangle| \le \|x^\star\|_{X^\star}\|x_1 + x_2\|$, so combining that with the previous inequality (and dividing by $\|x^\star\|_{X^\star}$) yields that $2 \le \|x_1 + x_2\|$. But by the strict convexity of X, it follows that $\|x_1 + x_2\| < \|x_1\| + \|x_2\| = 2$, which yields a contradiction. For the converse, assume that (v) holds, but there exist $x_1, x_2 \in X$ with $\|x_1\| = \|x_2\| = 1$ such that $\|x_1 + x_2\| = 2$. Then, $\|\frac{1}{2}(x_1 + x_2)\| = 1$, and by the Hahn–Banach theorem (see Example 1.1.18), there exists $x^\star \in X^\star$ such that $\langle x^\star, \frac{1}{2}(x_1 + x_2)\rangle = \|\frac{1}{2}(x_1 + x_2)\| = 1$, from which it follows that $\langle x^\star, x_1\rangle + \langle x^\star, x_2\rangle = 2$. However, since $\langle x^\star, x_i\rangle \le 1$, $i = 1, 2$, the previous equality implies that $\langle x^\star, x_1\rangle = \langle x^\star, x_2\rangle = \|x^\star\| = 1$, which contradicts (v). We have thus shown that (v) implies (iii), which is equivalent to (i).

5. (i) $\Longleftrightarrow$ (vi): To show that (i) implies (vi), suppose that φ is not strictly convex, i. e., there exists $x_1, x_2 \in X$, $x_1 \neq x_2$ such that $\|t_o x_1 + (1 - t_o)x_2\|^2 = t_o\|x_1\|^2 + (1 - t_o)\|x_2\|^2$ for some $t_o \in (0, 1)$. On the other hand, using the triangle inequality, we have that $\|t_o x_1 + (1 - t_o)x_2\| \le t_o\|x_1\| + (1 - t_o)\|x_2\|$, so squaring, we have that

$$\begin{aligned}\|t_o x_1 + (1 - t_o)x_2\|^2 &\le t_o^2\|x_1\|^2 + 2t_o(1 - t_o)\|x_1\|\,\|x_2\| + (1 - t_o)^2\|x_2\|^2\\ &\le t_o^2\|x_1\|^2 + t_o(1 - t_o)\|x_1\|^2 + t_o(1 - t_o)\|x_2\|^2 + (1 - t_o)^2\|x_2\|^2\\ &= t_o\|x_1\|^2 + (1 - t_o)\|x_2\|^2,\end{aligned}$$

using the elementary inequality $2\|x_1\|\,\|x_2\| \le \|x_1\|^2 + \|x_2\|^2$. Comparing the first and fourth terms of the above, which are equal by assumption, so are the second and third terms, so that $\|x_1\|^2 - 2\|x_1\|\,\|x_2\| + \|x_2\|^2 = 0$, hence $\|x_1\| = \|x_2\|$. Defining $z_i = \frac{1}{\|x_i\|}x_i$ such that $\|z_i\| = 1$, $i = 1, 2$, we see that $\|t_o z_1 + (1 - t_o)z_2\| = 1$, which is a contradiction with (i).

To show that (vi) implies (i), assume that φ is strictly convex, but X is not. Then there exist $x_1, x_2 \in X$, $x_1 \neq x_2$, $\|x_1\| = \|x_2\| = 1$, and $t_o \in (0, 1)$ such that $\|t_o x_1 + (1 - t_o)x_2\| = 1$, which implies that $\|t_o x_1 + (1 - t_o)x_2\|^2 = 1 = t_o\|x_1\|^2 + (1 - t_o)\|x_2\|^2$, which contradicts (v). □

Remark 2.6.6. One may in fact replace (v) by (v′): the function $x \mapsto \varphi(x) = \|x\|^p$, $p \in (1, \infty)$ is strictly convex. Indeed, assume not, so that there exists $x_1 \neq x_2$ and $t_o \in (0, 1)$ such that $\|t_o x_1 + (1 - t_o)x_2\|^p = t_o\|x_1\|^p + (1 - t_o)\|x_2\|^p$. The triangle inequality and the monotonicity and convexity of the real valued function $s \mapsto s^p$, $s > 0$, $p \in (1, \infty)$ yields

$$\|t_o x_1 + (1 - t_o)x_2\|^p \le (t_o\|x_1\| + (1 - t_o)\|x_2\|)^p \le t_o\|x_1\|^p + (1 - t_o)\|x_2\|^p,$$

and since by assumption the left-hand side equals the right-hand side of this inequality, we see that $(t_o\|x_1\| + (1 - t_o)\|x_2\|)^p = t_o\|x_1\|^p + (1 - t_o)\|x_2\|^p$ so that defining $s = \frac{\|x_1\|}{\|x_2\|}$, we have that $(t_o s + (1 - t_o))^p - t_o s^p - (1 - t_o) = 0$. By elementary calculus,[11] we see that the only solution is $s = 1$. Hence, $\|x_1\| = \|x_2\|$, and our assumption implies that $\|t_o x_1 + (1 - t_o)x_2\|^p = \|x_1\|^p = \|x_2\|^p$, and proceeding, as in the proof above defining $z_i = \frac{1}{\|x_i\|}x_i$ such that $\|z_i\| = 1$, $i = 1, 2$, we see that $\|t_o z_1 + (1 - t_o)z_2\| = 1$, which is in contradiction with (i).

2.6.1.2 Uniform convexity

Definition 2.6.7. A Banach space X is called uniformly convex if for each $\epsilon \in (0, 2]$ and for any $\|x_1\| \leq 1$, $\|x_2\| \leq 1$, $\|x_1 - x_2\| \geq \epsilon$, there exists $\delta(\epsilon) > 0$ such that $\|\frac{x_1+x_2}{2}\| \leq 1 - \delta$.

One may immediately see from the definition that if X is uniformly convex, then it is also strictly convex, however, the converse is not necessarily true.

Example 2.6.8 (Uniform convexity depends on the choice of norm). The space $X = \mathbb{R}^d$ is not uniformly convex when endowed with the $\|\cdot\|_1$ or $\|\cdot\|_\infty$ norm (see also Example 2.6.2). ◁

Example 2.6.9 (Hilbert spaces are uniformly convex). If $X = H$ is a Hilbert space, then it is uniformly convex. This follows directly by the parallelogram identity for Hilbert spaces, from which it can be shown that for given $\epsilon > 0$, $\delta(\epsilon) = 1 - \sqrt{1 - \frac{\epsilon^2}{4}}$. ◁

Example 2.6.10. The Lebesgue spaces $X = L^p(\mathcal{D})$ for $1 < p < \infty$ are uniformly convex Banach spaces (Clarkson's theorem, see, e. g., [41] for a proof). The spaces $L^1(\mathcal{D})$ and $L^\infty(\mathcal{D})$ are not uniformly convex. The same result holds for any Lebesgue space on a more general measure μ. Similarly, for the Sobolev spaces $W^{1,p}(\mathcal{D})$ and $W_0^{1,p}(\mathcal{D})$, which are uniformly convex for $1 < p < \infty$. ◁

An alternative definition of uniform convexity, which is very useful in a number of applications, is given in the following proposition [41, 52].

Proposition 2.6.11. *A Banach space is uniformly convex if and only if for any sequences* $(x_n)_{n\in\mathbb{N}} \subset X$ *and* $(z_n)_{n\in\mathbb{N}} \subset X$ *such that* $\|x_n\| \leq 1$ *and* $\|z_n\| \leq 1$, *the property* $\|\frac{x_n+z_n}{2}\| \to 1$ *implies* $\|x_n - z_n\| \to 0$.

Proof. Assume that X is uniformly convex and that $\|\frac{x_n+z_n}{2}\| \to 1$. Suppose that $\|x_n - z_n\| \not\to 0$. This means that there exists $\epsilon > 0$ and $N \in \mathbb{N}$ such that $\|x_n - z_n\| \geq \epsilon$ for $n > N$. But by the uniform convexity of X the implication is that there exists $\delta = \delta(\epsilon)$ such that $\|\frac{x_n+z_n}{2}\| < 1 - \delta$, which contradicts the assumption $\|\frac{x_n+z_n}{2}\| \to 1$.

To prove the converse, suppose that for any sequences $(x_n)_{n\in\mathbb{N}}$, $(z_n)_{n\in\mathbb{N}}$ in the unit ball of X such that $\|\frac{x_n+z_n}{2}\| \to 1$ it holds that $\|x_n - z_n\| \to 0$ and assume that X is not uniformly convex. That implies that there exist an $\epsilon > 0$ for which (by appropriate choice of δ, e. g., $\delta = 1/n$), and the corresponding points x_n, z_n in the unit ball as in the negation

11 The function $\psi(s) = (t_o s + (1 - t_o))^p - t_o s^p - (1 - t_o)$ has a single maximum at $s = 1$ and $\psi(0) = 0$.

of Def. 2.6.7, such that $\|x_n - z_n\| \geq \epsilon$ and $1-\frac{1}{n} \leq \|\frac{x_n+z_n}{2}\| \leq 1$. So, $\|\frac{x_n+z_n}{2}\| \to 1$ but $\|x_n - z_n\| \nrightarrow 0$, a contradiction. □

Example 2.6.12 (A criterion for Cauchy sequences in uniformly convex spaces). If X is uniformly convex, a sequence $(x_n)_{n\in\mathbb{N}} \subset X$ such that $\|x_n\| \leq 1$ and $\|\frac{1}{2}(x_n + x_m)\| \to 1$, as $n, m \to \infty$ is a Cauchy sequence.

Simply apply Proposition 2.6.11 for $(x_n)_{n\in\mathbb{N}}$ and $(z_n)_{n\in\mathbb{N}}$ chosen so that $z_n = x_m$. ◁

Example 2.6.13 (In uniformly convex spaces weak convergence and convergence of norms imply strong convergence). Let X be uniformly convex, and consider a sequence $(x_n)_{n\in\mathbb{N}}$ such that $x_n \rightharpoonup x$ in X and $\|x_n\| \to \|x\|$, as $n \to \infty$. Then $x_n \to x$. This property is often called the Radon–Riesz property.

We define $z_n = \frac{1}{\|x_n\|}x_n$, $z = \frac{1}{\|x\|}x$, which are on $S_X(0,1)$. Consider any $x^\star \in X^\star$ such that $\|x^\star\|_{X^\star} = 1$ and $\langle x^\star, z\rangle = 1$ (this is always possible!). Then $\langle x^\star, \frac{1}{2}(z_n+z_m)\rangle \leq \|\frac{1}{2}(z_n+z_m)\| \leq 1$, whereas by weak convergence $\langle x^\star, z_n\rangle \to \langle x^\star, z\rangle$, so that the above inequality implies that $\|\frac{1}{2}(z_n + z_m)\| \to 1$, as $n, m \to \infty$. Then, by Example 2.6.12, we have that $(z_n)_{n\in\mathbb{N}}$ is Cauchy. Hence, $z_n \to z$ (since we already have that $z_n \rightharpoonup z$). ◁

Uniform convexity is related to reflexivity, as the following theorem shows:

Theorem 2.6.14 (Milman, Pettis). *A uniformly convex Banach space X is reflexive.*

Proof. We only sketch the proof here (see, e. g., [41] or [104] for a complete proof). By Goldstine's theorem $\overline{B}_X(0,1)$ is weak* dense in $\overline{B}_{X^{\star\star}}(0,1)$, i. e., for each $x^{\star\star} \in S_{X^{\star\star}}$, there exists a net $\{x_\alpha : \alpha \in \mathcal{I}\} \subset \overline{B}_X(0,1)$ such that $jx_\alpha \overset{\star}{\rightharpoonup} x^{\star\star}$. We consider the net $\{j(\frac{1}{2}(x_\alpha+x_\beta)) : (\alpha,\beta) \in \mathcal{I}\times\mathcal{I}\}$, for which we may show that $\|\frac{1}{2}(x_\alpha + x_\beta)\| \to 1$. So, by an argument based on Example 2.6.12, appropriately modified for nets, we conclude that the net $\{x_\alpha : \alpha \in \mathcal{I}\} \subset \overline{B}_X(0,1)$ is Cauchy, hence converging to a limit x_o, for which $jx_\alpha \to jx_o$. This implies the existence of a $x_o \in X$, for which $x^{\star\star} = jx_o$, therefore X is reflexive. □

This theorem is very useful in proving the reflexivity of classic Banach spaces, such as the Lebesgue spaces L^p, $p \in (1,\infty)$ (see, e. g., [33]).

2.6.1.3 Local uniform convexity

The concept of uniformly convex Banach spaces has been weakened in [103], where the concept of local uniform convexity has been introduced.

Definition 2.6.15 (Local uniform convexity). A Banach space X is called locally uniformly convex if for any $\epsilon > 0$ and any $x \in X$ with $\|x\| = 1$, there exists $\delta(\epsilon, x) > 0$ such that for all z with $\|z\| \leq 1$ satisfying $\|z - x\| \geq \epsilon$, it holds that $\|\frac{x+z}{2}\| \leq 1 - \delta(\epsilon, x)$.

As can be seen from the definition, a uniformly convex Banach space is also locally uniformly convex. Summarizing,

$$\text{Uniform convexity} \Longrightarrow \text{Local uniform convexity} \Longrightarrow \text{Strict convexity.} \tag{2.25}$$

It can also be proved, using similar arguments, that Proposition 2.6.11 holds in the special case that $(z_n)_{n\in\mathbb{N}}$ is the constant sequence $x_n = z$, and the assertion of Example 2.6.13 hold for locally uniformly convex Banach spaces [52].

As mentioned above, properties such as strict convexity and uniform convexity of a Banach space depend on the choice of norm. An important result is that, using an appropriate renorming, one may turn a Banach space either into a strictly convex or into a locally uniformly convex space [137].

Theorem 2.6.16 (Troyanski). *Let X be a reflexive Banach space. There exists a locally uniform convex equivalent norm $\|\cdot\|_\diamond$ such that X and $X^\star$ (equipped with the corresponding dual norm) are locally uniformly convex.*

Proof. The proof is outside the scope of the present book. We refer the interested reader to [137], or Theorems 2.9 and 2.10 in [52]. □

Geometrical properties of Banach spaces related to convexity, such as strict or uniform convexity, have deep links with properties such as the differentiability of the norm (see Chapter 4), or the properties of the duality map.

2.6.2 Convexity and the duality map

Let X be a Banach space, $X^\star$ its dual, $\langle\cdot,\cdot\rangle$ the duality pairing between them, and $J : X \to 2^{X^\star}$ defined by

$$J(x) = \{x^\star \in X^\star \,:\, \langle x^\star, x\rangle = \|x\|\,\|x^\star\|_{X^\star},\ \|x\| = \|x^\star\|_{X^\star}\},$$

the duality map (see Definition 1.1.17). This map plays a basic role in the study of the geometry of Banach spaces and in nonlinear functional analysis (see e. g.,[52]).

Some important properties of J are given in the next theorem [52].

Theorem 2.6.17. *Let X be a reflexive Banach space, and let $X^\star$ be strictly convex. Then,*

(i) *J is* (a) *single valued,* (b) *demicontinuous, i. e., for every $x_n \to x$ in X, it holds that*[12] $J(x_n) \rightharpoonup J(x)$, (c) *coercive, i. e., $\frac{\|J(x)\|_{X^\star}}{\|x\|} \to \infty$ as $\|x\| \to \infty$,* (d) *monotone, i. e., $\langle J(x_1) - J(x_2), x_1 - x_2\rangle \geq 0$ for every $x_1, x_2 \in X$, and* (e) *surjective.*

(ii) *If X is strictly convex, then J is bijective and strictly monotone, i. e., monotone and*

$$\langle J(x_1) - J(x_2), x_1 - x_2\rangle = 0, \text{ implies } x_1 = x_2.$$

Moreover, if

$$\langle J(x_n) - J(x), x_n - x\rangle \to 0, \quad \text{as } n \to \infty, \tag{2.26}$$

12 If X were not reflexive, demicontinuity would require $J(x_n) \overset{\star}{\rightharpoonup} J(x)$, see Definition 9.2.3.

then $\|x_n\| \to \|x\|$ *and* $x_n \rightharpoonup x$ *in* X. *If in addition,* X *is locally uniformly convex, then* (2.26) *implies that* $x_n \to x$ *as* $n \to \infty$.

(iii) *If* $X^\star$ *is locally uniformly convex, then* J *is continuous.*

Proof. (i) We first show that J is single valued. Let $x_i^\star \in J(x)$, $i = 1, 2$. By the definition of the duality map, we have

$$\langle x_i^\star, x\rangle = \|x\|^2 = \|x_i^\star\|_{X^\star}^2, \quad i = 1, 2,$$

so that $\|x_1^\star\|_{X^\star} = \|x_2^\star\|_{X^\star}$, and without loss of generality, let $\|x_1^\star\|_{X^\star} = \|x_2^\star\|_{X^\star} = 1$. Hence,

$$2\|x_1^\star\|_{X^\star}\|x\| \le \|x_1^\star\|_{X^\star}^2 + \|x_2^\star\|_{X^\star}^2 = \langle x_1^\star + x_2^\star, x\rangle \le \|x_1^\star + x_2^\star\|_{X^\star}\|x\|, \quad \forall x \in X.$$

This implies that

$$1 = \|x_1^\star\|_{X^\star} \le \frac{1}{2}\|x_1^\star + x_2^\star\|_{X^\star},$$

which, by the strict convexity of $X^\star$, implies that $x_1^\star = x_2^\star$ (see Theorem 2.6.5(ii)) so that J is single valued.

We now prove the demicontinuity of J. To this end, consider a sequence $(x_n)_{n\in\mathbb{N}} \subset X$ such that $x_n \to x$ in X, as $n \to \infty$. We will show that $J(x_n) \rightharpoonup J(x)$ in $X^\star$. Observe that, by the definition of the duality map and the properties of the sequence $(x_n)_{n\in\mathbb{N}}$, we have that

$$\|J(x_n)\|_{X^\star} = \|x_n\| \to \|x\|, \quad \text{as } n \to \infty. \tag{2.27}$$

Since $(J(x_n))_{n\in\mathbb{N}}$ is bounded and $X^\star$ is reflexive (see Proposition 1.1.26), there exists a subsequence, denoted the same for simplicity, and an element $x^\star \in X^\star$ such that

$$J(x_n) \rightharpoonup x^\star, \quad \text{in } X^\star \text{ as } n \to \infty.$$

By (2.27), and using the weak lower semicontinuity of the norm (see Proposition 1.1.54), we obtain

$$\|x^\star\|_{X^\star} \le \liminf_n \|J(x_n)\|_{X^\star} = \lim_n \|J(x_n)\|_{X^\star} = \|x\|. \tag{2.28}$$

Since $J(x_n) \rightharpoonup x^\star$ and $x_n \to x$, we have that $\langle J(x_n), x_n\rangle \to \langle x^\star, x\rangle$, so that from

$$\|x_n\|^2 = \langle J(x_n), x_n\rangle \to \langle x^\star, x\rangle,$$

and (2.27) we have, by the uniqueness of limits, that

$$\|x\|^2 = \langle x^\star, x\rangle. \tag{2.29}$$

This implies that

$$\|x\|^2 = \langle x^\star, x\rangle = |\langle x^\star, x\rangle| \le \|x^\star\|_{X^\star}\,\|x\|,$$

from which, dividing through by $\|x\|$, follows that

$$\|x\| \le \|x^\star\|_{X^\star}. \tag{2.30}$$

Therefore, by combining (2.28) and (2.30), $\|x\| = \|x^\star\|_{X^\star}$. Since we also have (by (2.29)) that $\langle x^\star, x\rangle = \|x\|^2$, we conclude, by the definition of the duality map, that $J(x) = x^\star$. Therefore, for the chosen subsequence, we have shown that $J(x_n) \rightharpoonup J(x)$. By the Urysohn property (see Remark 1.1.52), it finally follows that the entire sequence $(J(x_n))_{n\in\mathbb{N}}$ converges weakly to $x^\star = J(x)$. Hence, the demicontinuity of J is established.

Since, by definition, $\langle J(x), x\rangle = \|x\|^2$, coercivity of J is immediate. The monotonicity of J follows by (2.32) below. We defer the proof of surjectivity to Section 9.4.4.

(ii) We now assume further that X is strictly convex, and we prove that J is strictly monotone. Let

$$\langle J(x_1) - J(x_2), x_1 - x_2\rangle = 0. \tag{2.31}$$

By the definition of the duality map

$$\begin{aligned}\langle J(x_1) - J(x_2), x_1 - x_2\rangle &= \|x_1\|^2 - \langle J(x_2), x_1\rangle - \langle J(x_1), x_2\rangle - \|x_2\|^2\\ &\ge (\|x_1\| - \|x_2\|)^2, \quad \forall\, x_1, x_2 \in X.\end{aligned} \tag{2.32}$$

So, we express (2.31) as

$$\begin{aligned}0 &= \left\langle J(x_1) - J\left(\frac{x_1+x_2}{2}\right), \frac{x_1-x_2}{2}\right\rangle + \left\langle J\left(\frac{x_1+x_2}{2}\right) - J(x_2), \frac{x_1-x_2}{2}\right\rangle\\ &= \left\langle J(x_1) - J\left(\frac{x_1+x_2}{2}\right), x_1 - \frac{x_1+x_2}{2}\right\rangle + \left\langle J\left(\frac{x_1+x_2}{2}\right) - J(x_2), \frac{x_1+x_2}{2} - x_2\right\rangle\\ &\ge \left(\|x_1\| - \left\|\frac{x_1+x_2}{2}\right\|\right)^2 + \left(\left\|\frac{x_1+x_2}{2}\right\| - \|x_2\|\right)^2,\end{aligned}$$

where in the last estimate we have used (2.32). Hence,

$$\|x_1\| = \left\|\frac{x_1+x_2}{2}\right\| = \|x_2\|.$$

Since X is strictly convex, $x_1 = x_2$. Therefore, J is strictly monotone. Since J is strictly monotone and surjective, it is also bijective.

Now, consider a sequence $(x_n)_{n\in\mathbb{N}} \subset X$ such that

$$\langle J(x_n) - J(x), x_n - x\rangle \to 0, \quad \text{as } n \to \infty.$$

We will show that $x_n \to x$ in X.

We rewrite $\langle J(x_n) - J(x), x_n - x\rangle$ as

$$\begin{aligned}\langle J(x_n) - J(x), x_n - x\rangle &= \|x_n\|^2 - \langle J(x_n), x\rangle - \langle J(x), x_n\rangle + \|x\|^2 \\ &= (\|x_n\| - \|x\|)^2 + (\|x_n\|\,\|x\| - \langle J(x_n), x\rangle) \\ &\quad + (\|x_n\|\,\|x\| - \langle J(x), x_n\rangle),\end{aligned} \tag{2.33}$$

and observe that since $|\langle J(x_n), x\rangle| \le \|J(x_n)\|\,\|x\| = \|x_n\|\,\|x\|$ and $|\langle J(x), x_n\rangle| \le \|J(x)\|\,\|x_n\| = \|x_n\|\,\|x\|$, the second and third term on the right-hand side of (2.33) are nonnegative. Therefore, the right-hand side of (2.33) as a whole is nonnegative and since, by assumption, the left-hand side tends to 0, as $n \to \infty$, we must have

$$\|x_n\| \to \|x\|, \quad \langle J(x_n), x\rangle) \to \|x\|^2, \quad \text{and} \quad \langle J(x), x_n\rangle \to \|x\|^2 \quad \text{as } n \to \infty.$$

Since X is reflexive, by the boundedness of the sequence $(x_n)_{n\in\mathbb{N}}$, we may assume the existence of a subsequence (denoted the same for simplicity) and an element $z \in X$ such that $x_n \rightharpoonup z$ in X, and therefore

$$\langle J(x), x_n\rangle \to \langle J(x), z\rangle, \quad \text{as } n \to \infty.$$

It then follows, by the uniqueness of the limit, that

$$\langle J(x), z\rangle = \|x\|^2$$

so that since $\|x\|^2 = \langle J(x), z\rangle = |\langle J(x), z\rangle| \le \|x\|\,\|z\|$, upon dividing both sides with $\|x\|$, we obtain that

$$\|x\| \le \|z\|. \tag{2.34}$$

On the other hand, by the weak lower semicontinuity of the norm (see Proposition 1.1.54), we obtain

$$\|z\| \le \liminf_n \|x_n\| = \lim_n \|x_n\| = \|x\|. \tag{2.35}$$

Combining (2.34) with (2.35), if follows that $\|x\| = \|z\|$, therefore, by the strict convexity of X, we deduce that $x = z$. Therefore, for the chosen subsequence, we have that $x_n \rightharpoonup x$. By the Urysohn property, we may deduce convergence for the whole sequence.

Finally, let (2.26) hold and assume furthermore that X is locally uniformly convex. Since (2.26) implies that $\|x_n\| \to \|x\|$ and $x_n \rightharpoonup x$ in X, it follows from the local uniform convexity of X that $x_n \to x$, as $n \to \infty$ (see Example 2.6.13).

(iii) Let $X^\star$ be locally uniformly convex. We show that J is continuous. Consider a sequence $(x_n)_{n\in\mathbb{N}} \subset X$ such that $x_n \to x$, as $n \to \infty$. Then $\|x_n\| \to \|x\|$. By the definition of the duality map $\|J(x_n)\|_{X^\star} = \|x_n\|$ for any $n \in \mathbb{N}$ and $\|J(x)\|_{X^\star} = \|x\|$, therefore $\|J(x_n)\|_{X^\star} \to \|J(x)\|_{X^\star}$. By (i), J is demicontinuous, so $J(x_n) \rightharpoonup J(x)$ in $X^\star$. It then follows by the local uniform convexity of $X^\star$ (see Example 2.6.13) that $J(x_n) \to J(x)$, as $n \to \infty$. □

2.7 Relaxation: the lower semicontinuous envelope

Lower semicontinuity plays an important role in optimization. In the case where functionals that are not lower semicontinuous are considered, one may replace them with their lower semicontinuous envelope and consider the corresponding minimization problem, which can be thought of as a relaxation of the original problem. In this section, we present some important concepts in this direction, following essentially [62]. At this point, the reader may wish to refresh the definition of lower semicontinuity (see Definition 2.2.1 and the properties of lower semicontinuous functions in Sec. 2.2).

An important concept is that of the lower semicontinuous envelope:

Definition 2.7.1 (Lower semicontinuous envelope). Let (X, τ) be a topological space and $F : X \to \mathbb{R} \cup \{+\infty\}$ be a functional. The lower semicontinuous envelope of F is defined as

$$\operatorname{lsc}(F)(x) := \sup\{G(x) \, : \, G \leq F, \ G \text{ lower semicontinuous}\}, \tag{2.36}$$

i. e., $\operatorname{lsc}(F)$ is the maximal lower semicontinuous function below F.

An equivalent representation is

$$\operatorname{lsc}(F)(x) = \liminf_{z\to x} F(z) := \sup_{U\in N(x)} \inf_{z\in U} F(z), \tag{2.37}$$

where the supremum is taken over all subsets U in a neighborhood of x. (recall also (2.5)).

Recalling the definition of the liminf (see (2.5)) and the definition of lower semicontinuity (see Definition 2.2.1), it is elementary to see that $\operatorname{lsc}(F)$ is a lower semicontinuous function, such that $\operatorname{lsc}(F) \leq F$.

As a simple example consider $F : \mathbb{R} \to \mathbb{R}$, with $F(x) = -a\mathbf{1}_{(-\infty,0)}(x) + a\mathbf{1}_{[0,\infty)}(x)$, $a > 0$, with $\operatorname{lsc}(F)(x) = -a\mathbf{1}_{(-\infty,0]}(x) + a\mathbf{1}_{(0,\infty)}(x)$.

The equivalence of these definitions is shown below (see Proposition 3.3 in [62]).

Proposition 2.7.2. *Definitions* (2.36) *and* (2.37) *are equivalent, i. e.,*

$$\sup\{G(x) \, : \, G \leq F, \ G \text{ lower semicontinuous}\} = \liminf_{z\to x} F(z) := \sup_{U\in N(x)} \inf_{z\in U} F(z). \tag{2.38}$$

Proof. To show (2.38), we start by the LHS and note that the function G_0, defined by $G_0(x) := \sup_{U\in N(x)} \inf_{z\in U} F(z)$, is a lower semicontinuous function, which clearly satisfies $G_0 \le F$.[13] Hence, $G_0(x) \le \mathrm{lsc}(F)(x)$.

For the reverse inequality, consider any lower semicontinuous function G such that $G \le F$. By definition, at any point $x \in X$, we have $G(x) \le \sup_{U\in N(x)} \inf_{z\in U} G(z)$. Since for any $\epsilon > 0$ there exists $U_\epsilon \in N(x)$ such that $\sup_{U\in N(x)} \inf_{z\in U} G(z) - \epsilon < \inf_{z\in U_\epsilon} G(z) \le G(x)$ hence $\sup_{U\in N(x)} \inf_{z\in U} G(z) \le G(x)$. Combining these two inequalities, we have that

$$G(x) = \sup_{U\in N(x)} \inf_{z\in U} G(z) \le \sup_{U\in N(x)} \inf_{z\in U} F(z) = G_0(x).$$

We have thus shown that for any lower semicontinuous function $G \le F$, it holds that $G \le G_0$. So this inequality holds for the supremum over this set, i. e., $\mathrm{lsc}(F)(x) \le G_0(x) = \sup_{U\in N(x)} \inf_{z\in U} F(z)$ for every $x \in X$. This completes the proof. □

Clearly, as can be seen by its definition, the lower semicontinuous envelope for a function depends on the topology with which X is endowed. Moreover, when changing topology, it may be that we obtain a different lower semicontinuous envelope.

2.7.1 The lower semicontinuous envelope in the strong topology

In what follows, we consider the case where X is a Banach space endowed with the strong topology. This is a first countable space (a metric space with the metric generated by the norm of X). So the following general characterizations covers this situation:

Proposition 2.7.3. *Let X satisfy the first axiom of countability (as, e. g., when X to is a Banach space equipped with the norm topology). Then, $\Phi_0 = \mathrm{lsc}(F)$ if and only if for every $x \in X$*

(i) *For every sequence $x_n \to x$ it holds that*

$$\Phi_0(x) \le \liminf_{n\to\infty} F(x_n), \quad \textit{(liminf inequality)}.$$

(ii) *There exists a sequence $\bar{x}_n \to x$, called the recovery sequence, such that*

$$\limsup_{n\to\infty} F(\bar{x}_n) \le \Phi_0(x), \quad \textit{(limsup inequality)}.$$

Clearly, the recovery sequence satisfies $F(\bar{x}_n) \to \Phi_0(x) = \mathrm{lsc}(F)(x)$.

Proof. For the proof, see Section 2.9.3. □

---

13 This follows by the definition of the supremum since for every $\epsilon > 0$ there exists a $U_\epsilon \in N(x)$ such that $G_0(x) - \epsilon < \inf_{z\in U_\epsilon} F(z) \le F(x)$.

We note that in this context we can consider

$$\mathrm{lsc}(F)(x) = \inf\Big\{\liminf_{n\to\infty} F(x_n) \,:\, x_n \to x\Big\}.$$

2.7.2 The lower semicontinuous envelope in the weak topology

We now briefly consider the issue of how the concept of the lower semicontinuous envelope changes when X is a Banach space not endowed with the strong topology, but rather with a weak topology, in which case the convenient sequential characterization of the lower semicontinuous envelope provided by Proposition 2.7.3 no longer holds. Moreover, note that as the notion of lower semicontinuous envelope is topology-dependent, the envelope may differ with the change of topology, and may even be difficult to identify its form.

In such cases, we have to resort to the general definition (Definition 2.7.1). When working in the general framework many of the interesting properties of $\mathrm{lsc}(F)$ can be obtained by the important properties

$$\begin{aligned} \{\mathrm{lsc}(F) \le \lambda\} &= \bigcap_{t>\lambda} \overline{\{F \le t\}}, \quad \forall\, \lambda \in \mathbb{R}, \\ \mathrm{epi}(\mathrm{lsc}(F)) &= \overline{\mathrm{epi}(F)}, \end{aligned} \tag{2.39}$$

where, in the second case, the closure is taken in $X \times \mathbb{R}$. The proof of these properties is provided in Section 2.9.4 in the appendix of the chapter.

The following proposition (see Proposition 3.10 in [62]) simplifies these considerations, in the special case of convex functions:

Proposition 2.7.4. *Let X be a Banach space. If $F : X \to \mathbb{R}\cup\{+\infty\}$ is convex, then the lower semicontinuous envelope* $\mathrm{lsc}(F)$ *coincides under the strong and the weak topology.*

Proof. The convexity of F implies that $\mathrm{epi}(F)$ is a convex set, hence its strong and its weak closure coincide (see Proposition 1.2.12). Therefore, (2.39) yields that the two lower semicontinuous envelopes coincide. □

2.7.3 Relaxation and minimization

We now consider the connection between the original minimization problem and its relaxed version, i. e., that of minimizing the lower semicontinuous envelope (see, e. g., Theorem 3.8 [62]).

Proposition 2.7.5. *Let (X, τ) be a topological space and $F : X \to \mathbb{R} \cup \{+\infty\}$ be a coercive functional (see Def.* 2.2.6*). Then,*

$$\inf_{x \in X} F(x) = \min_{x \in X} \operatorname{lsc}(F)(x). \tag{2.40}$$

Moreover, every cluster point of a minimizing sequence for F is a minimizer for lsc(F).

If X is first countable, then every minimizer of lsc(F) *is the limit of a minimizing sequence for F.*

Proof. To show that the problem of minimizing lsc(F) is well posed, we need to guarantee its coercivity. This follows by the coercivity of F and the definition of lsc(F), which guarantees by (2.39) that $\{\operatorname{lsc}(F) \le \lambda\} = \bigcap_{t>\lambda} \overline{\{F \le t\}}$ (see Proposition 3.5 in [62]; see also Section 2.9.4).

Since $\operatorname{lsc}(F) \le F$, clearly for any $x \in X$,

$$\min_{x \in X} \operatorname{lsc}(F(x)) \le \operatorname{lsc}(F)(x) \le F(x),$$

and then taking the infimum over all $x \in X$, we conclude that

$$\min_{x \in X} \operatorname{lsc}(F)(x) \le \inf_{x \in X} F(x).$$

To show the reverse inequality, consider the constant function $\bar{G}(x) = \inf_{x \in X} F(x)$. This is clearly a lower semicontinuous function such that $\bar{G} \le F$. Then, for every $x \in X$,

$$\operatorname{lsc}(F)(x) = \sup\{G(x) \,:\, G \le F,\ G \text{ lower semicontinuous}\} \ge \bar{G}(x) = \inf_{x \in X} F(x),$$

and upon taking the minimum over $x \in X$, the desired reverse inequality is obtained. Combining the two inequalities, we conclude (2.40).

Consider now a minimizing sequence $(x_n)_{n\in\mathbb{N}}$ for F, and let $x_o \in X$ be a cluster point of this sequence. If x_o is the limit of a convergent subsequence $(x_{n_k})_{k\in\mathbb{N}} \subset (x_n)_{n\in\mathbb{N}}$, then, since lsc($F$) is lower semicontinuous (and lower semicontinuity implies sequential lower semicontinuity, Proposition 2.2.2(iii)) and $\operatorname{lsc}(F) \le F$, we obtain that

$$\operatorname{lsc}(F)(x_o) \le \liminf_{k\to\infty} \operatorname{lsc}(F)(x_{n_k}) \le \liminf_{k\to\infty} F(x_{n_k}) = \inf_{x \in X} F(x) = \min_{x \in X} \operatorname{lsc}(F(x)),$$

where for the last two equalities we used the fact that the chosen sequence is minimizing, and then (2.40). Hence, x_o is a minimizer for lsc(F). The proof in the general case (where x_o is a cluster point but not the limit of a converging subsequence) can be found in Theorem 3.8 in [62], and is based upon the fact that for any lower semicontinuous function G, if x_o is a cluster point of a sequence $(x_n)_{n\in\mathbb{N}}$, then $G(x_o) \le \limsup_n G(x_n)$, applied to $G = \operatorname{lsc}(F)$.

We now assume that X is first countable, so that the sequential characterization of lsc(F), as in Proposition 2.7.3, holds. Then, to show that any minimizer x_o of lsc(F) is the limit of a minimizing sequence, we use the sequential characterization of lsc(F) and take a recovery sequence $\bar{x}_n \to x_o$ such that

$$\limsup_{n\to\infty} F(\bar{x}_n) \le \mathrm{lsc}(F)(x_o) = \min_{x\in X} \mathrm{lsc}(F)(x).$$

But, combining this with (2.40) yields

$$\limsup_{n\to\infty} F(\bar{x}_n) \le \inf_{x\in X} F(x),$$

which immediately leads to the result

$$\inf_{x\in X} F(x) \le \liminf_{n\to\infty} F(\bar{x}_n) \le \limsup_{n\to\infty} F(\bar{x}_n) \le \inf_{x\in X} F(x),$$

hence $F(\bar{x}_n) \to \inf_{x\in X} F(x)$, and $(\bar{x}_n)_{n\in\mathbb{N}}$ is a minimizing sequence. □

Example 2.7.6 (Candidates for minimizers of a functional F). Suppose we wish to find the minimizer of a functional F that is coercive but not lower semicontinuous. A possible way forward can be to calculate $\mathrm{lsc}(F)$ and find its minimizers. Then, for any $x_\alpha \in \arg\min_{x\in X} \mathrm{lsc}(F)(x)$, we check whether (2.40) holds, i. e., whether $\mathrm{lsc}(F)(x_\alpha) = \inf_{x\in X} F(x) = F(x_\alpha)$. ◁

2.8 Γ-convergence

Γ-convergence is a variational type of convergence, introduced in the 1970s by De Giorgi and Franzoni, which handles the important question of the limiting behavior of a sequence of minimization problems, including the delicate issue concerning the convergence of the minimizers. Γ-convergence is a very powerful tool in variational analysis that has gained immense popularity and has been used in various and diverse applications. Classical texts on the field are [62] and [30], upon which our presentation is based (see also the concise introduction in [129] and the lecture notes [60]).

2.8.1 Definitions and examples

Following [62], we provide the following general definition for Γ-convergence:

Definition 2.8.1 (Γ-limit). Let (X, τ) be a topological space. Consider a sequence of functionals $(F_n)_{n\in\mathbb{N}}$, $F_n : X \to \mathbb{R} \cup \{+\infty\}$. We define:

(i) The lower Γ-limit of $(F_n)_{n\in\mathbb{N}}$ at a point $x \in X$ as

$$F_\Gamma^-(x) := \Gamma - \liminf_{n\to\infty} F_n(x) := \sup_{U\in N(x)} \liminf_{n\to\infty} \inf_{z\in U} F_n(z),$$

where the supremum is taken over all sets U in the set $N(x)$ of neighborhoods of x.

(ii) The upper Γ-limit of $(F_n)_{n\in\mathbb{N}}$ at a point $x \in X$ as

$$F_\Gamma^+(x) := \Gamma - \limsup_{n\to\infty} F_n(x) := \sup_{U\in N(x)} \limsup_{n\to\infty} \inf_{z\in U} F_n(z).$$

(iii) We say that $(F_n)_{n\in\mathbb{N}}$, Γ-converges to F_Γ, denoted by $F_\Gamma = \Gamma - \lim_{n\to\infty} F_n$, if

$$F_\Gamma(x) := \Gamma - \lim_{n\to\infty} F_n(x) = \Gamma - \liminf_{n\to\infty} F_n(x) = \Gamma - \limsup_{n\to\infty} F_n(x), \quad \forall x \in X.$$

By Definition 2.8.1, it is evident that the Γ-limit depends on the choice of topology, τ, for X. Moreover, there are connections between the Γ-limit and the lower semicontinuous envelope.

In the case where X is first countable (for example, a metric space endowed with the metric topology, e. g., a Banach space with the strong topology), then the above definition can be expressed in a more convenient way, involving sequences.

Definition 2.8.2 (Γ-limit, X first countable). Assume that (X, τ) is first countable and consider a sequence of functionals $(F_n)_{n\in\mathbb{N}}$, $F_n : X \to \mathbb{R} \cup \{+\infty\}$. We define:

(i) The lower Γ-limit of $(F_n)_{n\in\mathbb{N}}$ at a point $x \in X$ as

$$F_\Gamma^-(x) = \Gamma - \liminf_{n\to\infty} F_n(x) := \inf\Big\{\liminf_{n\to\infty} F_n(x_n) \ : \ x_n \to x\Big\},$$

with the infimum over the set of sequences $(x_n)_{n\in\mathbb{N}} \subset X$ such that $x_n \to x$ in the chosen topology.
An alternative characterization is as $F_\Gamma^-(x) := \Gamma - \liminf_n F_n(x)$, satisfying for every $x \in X$ the following two properties:
(i.a) For any sequence $x_n \to x$, it holds that $F_\Gamma^-(x) \le \liminf_n F_n(x_n)$.
(i.b) There exists a sequence $\bar{x}_n \to x$ such that $\liminf_n F_n(\bar{x}_n) \le F_\Gamma^-(x)$.
Clearly, $F_\Gamma^-(x) = \liminf_n F_n(\bar{x}_n)$, so $(x_n)_{n\in\mathbb{N}} \subset X$ is called the recovery sequence for F_Γ^- at $x \in X$.

(ii) The upper Γ-limit of $(F_n)_{n\in\mathbb{N}}$ at a point $x \in X$, as

$$F_\Gamma^+(x) := \Gamma - \limsup_{n\to\infty} F_n(x) := \inf\Big\{\limsup_{n\to\infty} F_n(x_n) \ : \ x_n \to x\Big\},$$

with the infimum over the set of sequences $(x_n)_{n\in\mathbb{N}}$ converging to x.
An alternative characterization is as $F_\Gamma^+(x) := \Gamma - \limsup_n F_n(x)$, satisfying for every $x \in X$ the following two properties:
(ii.a) For any sequence $x_n \to x$, it holds that $F_\Gamma^+(x) \le \limsup_n F_n(x_n)$.
(ii.b) There exists a sequence $\bar{x}_n \to x$ such that $\limsup_n F_n(\bar{x}_n) \le F_\Gamma^+(x)$.
Clearly, $F_\Gamma^+(x) = \limsup_n F_n(\bar{x}_n)$, so $(\bar{x}_n)_{n\in\mathbb{N}} \subset X$ is called the recovery sequence for F_Γ^+ at $x \in X$.

(iii) We say that $(F_n)_{n\in\mathbb{N}}$, Γ-converges to F_Γ, denoted by $F_\Gamma = \Gamma - \lim_{n\to\infty} F_n$, if

$$F_\Gamma(x) := \Gamma - \lim_{n\to\infty} F_n(x) = \Gamma - \liminf_{n\to\infty} F_n(x) = \Gamma - \limsup_{n\to\infty} F_n(x), \quad \forall x \in X.$$

An alternative characterization is as $F_\Gamma(x) := \Gamma\text{–}\lim_n F_n(x)$, satisfying for every $x \in X$ the following two properties:

(iii.a) For every $x_n \to x$, we have that

$$F_\Gamma(x) \le \liminf_{n\to\infty} F_n(x_n), \quad \text{(liminf inequality)}.$$

(iii.b) There exists a sequence $\bar{x}_n \to x$ (recovery sequence) such that

$$\limsup_{n\to\infty} F_n(\bar{x}_n) \le F_\Gamma(x), \quad \text{(limsup inequality)}.$$

Clearly, $F_\Gamma(x) = \lim_{n\to\infty} F_n(\bar{x}_n)$, hence the name recovery sequence for $(\bar{x}_n)_{n\in\mathbb{N}}$.

We can deduce Definition 2.8.2 from the general Definition 2.8.1 in the specific case where X is first countable (see Proposition 8.1 in [62]).

Proposition 2.8.3. *In the case where (X, τ) is first countable (e. g., a metric space), then Definition* 2.8.1 *reduces to Definition* 2.8.2.

Proof. For the proof see Section 2.9.5 in the appendix of the chapter. □

Γ-convergence is designed so that it performs well for the solutions of variational problems, including the convergence of minimizers, and some of its properties may look strange at first sight. Importantly, it does not coincide with pointwise convergence!

Example 2.8.4. Assume that $F_n \xrightarrow{\Gamma} F_\Gamma$. Then it does not necessarily hold that $-F_n \xrightarrow{\Gamma} -F_\Gamma$!

This follows easily by the definition. Assume for simplicity that the sequential characterization holds. If in general $F_n \xrightarrow{\Gamma} F_\Gamma$ implied $-F_n \xrightarrow{\Gamma} -F_\Gamma$, then for any sequence $(x_n)_{n\in\mathbb{N}}$ such that $x_n \to x$, the liminf inequality for $(F_n)_{n\in\mathbb{N}}$,

$$F_\Gamma(x) \le \liminf_{n\to\infty} F_n(x_n),$$

upon muliplication by -1, would imply that

$$-\liminf_{n\to\infty} F_n(x_n) = \limsup_{n\to\infty}(-F_n(x_n)) \le -F_\Gamma(x),$$

i. e., $(x_n)_{n\in\mathbb{N}}$ is a recovery sequence for $(-F_n)_{n\in\mathbb{N}}$. However, this implies that any sequence $(x_n)_{n\in\mathbb{N}}$ such that $x_n \to x$ is a recovery sequence for $(-F_n)_{n\in\mathbb{N}}$; that does not hold in general.[14] ◁

14 Alternatively, consider a sequence $(x_n)_{n\in\mathbb{N}}$, with $x_n \to x$, that is not a recovery sequence for $(F_n)_{n\in\mathbb{N}}$. Then, for with the same arguments, we see that this sequence does not satisfy the liminf inequality for $(-F_n)_{n\in\mathbb{N}}$, with the Γ-limit chosen as $-F_\Gamma$; that is a contradiction.

Example 2.8.5 (Γ-limit of constant sequence). The Γ-limit of a constant sequence of functions, $(F_n)_{n\in\mathbb{N}}$ with $F_n = F$ for all $n \in \mathbb{N}$, is not equal to F. In fact,

$$\Gamma - \lim_{n\to\infty} F_n = \operatorname{lsc}(F),$$

the lower semicontinuous envelope of F (see Definition 2.7.1). This can be seen easily using the Definition of Γ-convergence and the characterization of the lower semicontinuous envelope in Proposition 2.7.3. ◁

Example 2.8.6. A standard example used to show that Γ-convergence is not necessarily related to standard notions of convergence is in the simple case where $X = \mathbb{R}$. Consider the sequence of functions $(f_n)_{n\in\mathbb{N}}$ such that $f_n(x) = \sin(nx)$, which is known to oscillate rapidly. Even though the limit $\lim_n f_n$ does not exist, the Γ-limit exists and is the constant function $f(x) = -1$. This can be verified by the definition. It is trivial to check the liminf inequality, while a suitable recovery sequence for the limsup inequality for any $x \in \mathbb{R}$ is $\bar{x}_n = -\frac{\pi}{2n} + \frac{2\pi}{n}[\frac{nx}{2\pi}]$, where by $[\cdot]$ we denote the integer part. ◁

2.8.2 Properties of Γ-convergence

In general, the Γ-limit is not easy to identify. Moreover, as Example 2.8.5 indicates, there are connections between Γ-convergence and the notion of the lower semicontinuous envelope. Some useful properties of Γ-convergence are collected in the following proposition [62]:

Proposition 2.8.7 (Properties of Γ-convergence). *Let (X,τ) be a topological space and $(F_n)_{n\in\mathbb{N}}$ be a sequence of functionals $F_n : X \to \mathbb{R}\cup\{+\infty\}$.*

The following hold:

(i) $F_\Gamma^- := \Gamma - \liminf_{n\to\infty} F_n$ *and* $F_\Gamma^+ := \Gamma - \limsup_{n\to\infty} F_n$ *are lower semicontinuous functions. Moreover,*

$$\Gamma - \liminf_{n\to\infty} F_n = \Gamma - \liminf_{n\to\infty} \operatorname{lsc}(F_n),$$
$$\Gamma - \limsup_{n\to\infty} F_n = \Gamma - \limsup_{n\to\infty} \operatorname{lsc}(F_n).$$

(ii) *The Γ– and the pointwise limit in general satisfy the following:*

$$\Gamma - \liminf_{n\to\infty} F_n \le \liminf_{n\to\infty} F_n,$$
$$\Gamma - \limsup_{n\to\infty} F_n \le \limsup_{n\to\infty} F_n.$$

(iii) *If $F_n \to F$ uniformly then,*

$$\Gamma - \lim_{n\to\infty} F_n = \operatorname{lsc}(F).$$

If the uniform limit F is lower semicontinuous, then it coincides with the Γ-limit.

(iv) *If* $(F_n)_{n\in\mathbb{N}}$ *is equi lower semicontinuous at* $x \in X$, *i. e.,*

$$\forall \epsilon > 0,\ \exists U \in N(x)\ :\ \forall z \in U,\ \forall n \in \mathbb{N},\quad F_n(z) \geq F_n(x) - \epsilon,$$

then

$$\Gamma - \liminf_n F_n(x) = \liminf_n F_n(x), \quad \Gamma - \limsup_n F_n(x) = \limsup_n F_n(x),$$

and $(F_n)_{n\in\mathbb{N}}$ *Γ-converges to F if and only if it converges pointwise to F.*

(v) *For any two sequences of functionals* $(F_n)_{n\in\mathbb{N}}$, $(\Phi_n)_{n\in\mathbb{N}}$ *it holds that*

$$\Gamma - \liminf_n (F_n + \Phi_n) \geq \Gamma - \liminf_n F_n + \Gamma - \liminf_n \Phi_n,$$
$$\Gamma - \limsup_n (F_n + \Phi_n) \geq \Gamma - \limsup_n F_n + \Gamma - \liminf_n \Phi_n.$$

If Φ_n *are everywhere finite and* $\Phi_n \to \Phi$ *uniformly for some* $\Phi : X \to \mathbb{R}$ *everywhere finite, continuous, then the inequalities in the above become equalities, i. e.,*

$$\Gamma - \liminf_n (F_n + \Phi_n) = \Gamma - \liminf_n F_n + \Phi,$$
$$\Gamma - \limsup_n (F_n + \Phi_n) = \Gamma - \limsup_n F_n + \Phi.$$

Proof. The proof is in Section 2.9.6 in the appendix of the chapter. □

Example 2.8.8. Assume that X is a normed vector space. If $(F_n)_{n\in\mathbb{N}}$ consists of convex functions and is equibounded in a neighborhood $N(x)$ of x, i. e.,

$$\exists M > 0\ :\ |F_n(z)| \leq M, \quad \forall z \in N(x),\ n \in \mathbb{N},$$

then

$$\Gamma - \liminf_{n\to\infty} F_n(x) = \liminf_{n\to\infty} F_n(x), \quad \Gamma - \limsup_{n\to\infty} F_n(x) = \limsup_{n\to\infty} F_n(x).$$

This is an example of the application of the result in Proposition 2.8.7(iv), as in this context, $(F_n)_{n\in\mathbb{N}}$ is equi-continuous (i. e., both equi-lower semicontinuous and equi-upper semicontinuous).[15] An alternative proof in the special case where X is first countable in presented in Section 2.9.7 in the appendix. ◁

15 Recall that convex functions are locally Lipschitz; this combined with the equibounded property leads to equicontinuity.

2.8.3 Γ-convergence and optimization

We first require the following important definition:

Definition 2.8.9 (Equicoercivity). Let (X, τ) be a topological space. The sequence of functionals $(F_n)_{n\in\mathbb{N}}$, $F_n : X \to \mathbb{R} \cup \{+\infty\}$ is called equicoercive if

$$\exists\, K \subset X \text{ compact } : \inf_{x\in X} F_n(x) = \inf_{x\in K} F_n(x),\ \forall\, n \in \mathbb{N}, \quad \text{(equicoercivity)}. \tag{2.41}$$

Remark 2.8.10. An alternative way to express equicoercivity (see [62]) is by demanding that for every $\lambda \in \mathbb{R}$ there exists a compact set K_λ such that $\{F_n \le \lambda\} \subset K_\lambda$ for all $n \in \mathbb{N}$.[16]

We now have all the necessary material to present the most important result concerning Γ-convergence (see, e. g., [62]).

Theorem 2.8.11 (The fundamental theorem of Γ-convergence). *Let (X, τ) be a topological space and $(F_n)_{n\in\mathbb{N}}$ be an equicoercive sequence of functionals $F_n : X \to \mathbb{R} \cup \{+\infty\}$ such that $\inf_{x\in X} F_n(x) > -\infty$ for every $n \in N$, and assume that $F_\Gamma := \Gamma - \lim_n F_n$. Then,*
(i) *it holds that*

$$\min_{x\in X} F_\Gamma(x) = \lim_{n\to\infty} \inf_{x\in X} F_n(x);$$

(ii) *if $(x_n)_{n\in\mathbb{N}}$ is a sequence of minimizers for $(F_n)_{n\in\mathbb{N}}$ (i. e., for each n, $x_n \in \arg\min_{x\in X} F_n(x)$), then any of its cluster points is a minimizer for F_Γ, and*
(iii) *if X is first countable, every minimizer of F_Γ is the limit of a sequence of minimizers for $(F_n)_{n\in\mathbb{N}}$.*

Proof. To simplify the notation in the proof, we set $F := F_\Gamma$. Moreover, we consider the special case that X is first countable (for the general case see Theorem 7.8 and Corollary 7.20 in [62]).

(i) The equicoercivity property (2.41) plays an important role in the proof. We also note that the Γ-limit F is a lower semicontinuous function. Moreover, without loss of gen-

16 To see that this condition implies (2.41), under the extra assumption that there exists $\lambda_0 \in \mathbb{R}$ such that for all $n \in \mathbb{N}$, $\{F_n \le \lambda_0\} \neq \emptyset$, we work as follows: Apply the alternative condition to this level set and find the compact set K_{λ_0} such that $\{F_n \le \lambda_0\} \subset K_{\lambda_0}$ for all $n \in \mathbb{N}$. Set $K = K_{\lambda_0}$. We claim that K satisfies (2.41). To see this, denote $L_{\lambda_0,n} := \{F_n \le \lambda_0\} \subset K$. Then, by the inclusion of the relevant sets and the definition of $L_{\lambda_0,n}$, we have, for all $n \in \mathbb{N}$, that $\inf_K F_n \le \inf_{L_{\lambda_0,n}} F_n \le \lambda_0$. On the other hand, on $L^c_{\lambda_0,n}$, we have that $F_n > \lambda_0$, and since $K^c \subset L^c_{\lambda_0,n}$, we have that $\lambda_0 \le \inf_{L^c_{\lambda_0,n}} F_n \le \inf_{K^c} F_n$. Therefore, $\inf_K F_n \le \inf_{K^c} F_n$ for all $n \in \mathbb{N}$. This implies that $\inf_K F_n \le \inf_X F_n$. Indeed for all $\epsilon > 0$, there exists $x_\epsilon \in X$ such that $F_n(x_\epsilon) < \inf_X F_n + \epsilon$. Either $x_\epsilon \in K$ or $x_\epsilon \in K^c$. In the first case, $\inf_K F_n \le F_n(x_\epsilon) < \inf_X F_n + \epsilon$, where as in the second case $\inf_K F_n \le \inf_{K^c} F_n \le F_n(x_\epsilon) < \inf_X F_n + \epsilon$, where we also used our previous finding. In any case, taking $\epsilon \to 0$, we conclude that $\inf_K F_n \le \inf_X F_n$. Combining this with the obvious inequality $\inf_X F_n \le \inf_K F_n$, we conclude that $\inf_X F_n = \inf_K F_n$ for all $n \in \mathbb{N}$.

erality, we may assume that F_n are lower semicontinuous (else we replace F_n by $\mathrm{lsc}(F_n)$ and use Proposition 2.7.5).

We now estimate

$$\begin{aligned}\inf_{x\in X} F(x) &\overset{(1)}{\le} \min_{x\in K} F(x) \overset{(2)}{\le} \liminf_{n\to\infty} \inf_{x\in K} F_n(x) \\ &\overset{(3)}{\le} \liminf_{n\to\infty} \inf_{x\in X} F_n(x) \overset{(4)}{\le} \limsup_{n\to\infty} \inf_{x\in X} F_n(x) \overset{(5)}{\le} \inf_{x\in X} F(x).\end{aligned} \tag{2.42}$$

This clearly proves (i), as long as we can justify the intermediate steps (1)–(5) in the above estimate.

Inequality (1) is obvious since $K \subset X$; K is compact; F is lower semicontinuous, inequality (3) following from equicoercivity (see (2.41)), and inequality (4) is obvious. The missing steps are inequalities (2) and (5).

We now prove inequality (2). Since F_n are lower semicontinuous and K is compact, for each n there exists a x_n^* such that $\inf_{x\in K} F_n(x) = F_n(x_n^*)$. We may further extract a subsequence $(x_{n_k}^*)_{k\in\mathbb{N}}$ such that $\liminf_n F_n(x_n^*) = \lim_k F_{n_k}(x_{n_k}^*)$. Moreover, by the compactness of K, there exists a further subsequence of $(x_{n_k}^*)_{k\in\mathbb{N}}$, denoted the same for convenience, and a $x^* \in K$, such that $\lim_k x_{n_k} = x^*$. Upon defining the sequence $(z_n)_{n\in\mathbb{N}}$ with $z_n = x_{n_k}^*$ if $n = n_k$ and x^* otherwise, and noting that $z_n \to x^*$, using the liminf inequality for Γ-convergence, we see that $F(x^*) \le \liminf_n F_n(z_n)$. Combining all the above,

$$\begin{aligned}\inf_{x\in K} F(x) &\le F(x^*) \le \liminf_{n\to\infty} F_n(z_n) \le \liminf_{k\to\infty} F_{n_k}(z_{n_k}) \\ &= \lim_{k\to\infty} F_{n_k}(x_{n_k}^*) = \liminf_{n\to\infty} \inf_{x\in K} F_n(x),\end{aligned}$$

which is the required estimate (2). Note that in the above, we also used the obvious fact that passing to a subsequence, the liminf increases.

To prove (5), use the definition of the infimum to guarantee that for any $\epsilon > 0$ there exists a $\hat{z}_\epsilon \in X$ such that $F(\hat{z}_\epsilon) < \inf_{x\in X} F(x) + \epsilon$. Since F is a Γ-limit, we use a recovery sequence $\hat{z}_{\epsilon,n} \to \hat{z}_\epsilon$, for which $\limsup_n F_n(\hat{z}_{\epsilon,n}) \le F(\hat{z}_\epsilon)$. Hence,

$$\limsup_{n\to\infty} \inf_{x\in X} F_n(x) \le \limsup_{n\to\infty} F_n(\hat{z}_{\epsilon,n}) \le F(\hat{z}_\epsilon) < \inf_{x\in X} F(x) + \epsilon,$$

from which (5) follows by taking $\epsilon \to 0$. The far left inequality is evident since $\inf_{x\in X} F_n(x) \le F_n(\hat{z}_{\epsilon,n})$ (upon taking the $\limsup_n$ on both sides).

The proof of (i) is complete.

(ii) Let $(x_n)_{n\in\mathbb{N}}$ be a sequence of minimizers for $(F_n)_{n\in\mathbb{N}}$ and x_o a cluster point such that $x_n \to x_o$. Then

$$F_n(x_n) = \inf_{x\in X} F_n(x) = \inf_{x\in K} F_n(x),$$

where we also used the equicoercivity property (2.41). By the liminf inequality of Γ-convergence,

$$\begin{aligned} F(\mathsf{x}_o) &\le \liminf_n F_n(\mathsf{x}_n) = \liminf_n \inf_{\mathsf{x}\in K} F_n(\mathsf{x}) \\ &\overset{(3)}{\le} \liminf_n \inf_{\mathsf{x}\in X} F_n(\mathsf{x}) \overset{(4)}{\le} \limsup_n \inf_{\mathsf{x}\in X} F_n(\mathsf{x}) \overset{(5)}{\le} \inf_{\mathsf{x}\in X} F(\mathsf{x}), \end{aligned} \tag{2.43}$$

where, for the estimates in the second line, we work as in (2.42) (keeping the same numbering as in (2.42) for the ease of the reader). By (2.43), we immediately see that x_o is a minimizer for F. In the more general case, where the cluster point x_o is not a limit point, we may proceed as in the proof of Proposition 2.7.5.

(iii) We argue by contradiction. Consider any minimizer x_o of F, and assume that it is not a limit point of a sequence of minimizers $(\mathsf{x}_n)_{n\in\mathbb{N}}$ for $(F_n)_{n\in\mathbb{N}}$. That means we can find a ball around x_o, which for all (but a finite number of) elements of the sequence contains no minimizers of $(F_n)_{n\in\mathbb{N}}$. In other words, we can find $\epsilon > 0$ and an $N \in \mathbb{N}$ such that for all $n > N$, it holds that $F_n(\mathsf{z}) > \inf_{\mathsf{x}\in X} F_n(\mathsf{x}) + \epsilon$, for all z sufficiently close to x_o. We now consider a recovery sequence for x_o, $\bar{\mathsf{x}}_n \to \mathsf{x}_o$. Using the above, and replacing z by $\bar{\mathsf{x}}_n$ for sufficiently large n we obtain $F_n(\bar{\mathsf{x}}_n) > \inf_{\mathsf{x}\in X} F_n(\mathsf{x}) + \epsilon$, for large n. Taking the limit in this inequality (the limit of the RHS exists by (i) while by definition $\lim_n F_n(\bar{\mathsf{x}}_n) = F(\mathsf{x}_o)$), we obtain

$$\lim_{n\to\infty} \inf_{\mathsf{x}\in X} F_n(\mathsf{x}) + \epsilon \le \lim_{n\to\infty} F_n(\bar{\mathsf{x}}_n) = F(\mathsf{x}_o) = \min_{\mathsf{x}\in X} F(\mathsf{x}) = \lim_{n\to\infty} \inf_{\mathsf{x}\in X} F_n(\mathsf{x}),$$

where for the last part we also used (i). This is a contradiction. □

Equicoercivity plays an important role in the convergence of the sequence of minimizers $(\mathsf{x}_n)_{n\in\mathbb{N}}$ of $(F_n)_{n\in\mathbb{N}}$ to the minimizer x of the Γ-limit F. Its use is so that the sequence of minimizers $(\mathsf{x}_n)_{n\in\mathbb{N}}$ does not escape to "infinity" but rather is restricted to a compact set K such that it has a convergent subsequence whose limit can be identified as the minimizer of F.

Example 2.8.12. The family of functions $(f_n)_{n\in\mathbb{N}}, f_n(x) = 1 - e^{-(x-n)^2}$ is not equicoercive in $\mathbb{R}$, and hence the sequence of minimizers (that escape to infinity) fails to converge to a minimizer of the Γ-limit, which is the constant function $f = 1$. ◁

The following proposition provides a convenient criterion for equicoercivity (in the sense of Definition 2.8.9):

Proposition 2.8.13. *If there exists a lower semicontinuous function $\Phi : X \to \mathbb{R} \cup \{+\infty\}$ such that $F_n \ge \Phi$ for all $n \in \mathbb{N}$, then the sequence $(F_n)_{n\in\mathbb{N}}$ is equicoercive.*

Proof. The proof follows from the definitions and is left to the reader. We note that the reverse is also true (see Proposition 7.7 in [62]). □

2.8.4 Γ-convergence and weak topologies

The theory of Γ-convergence finds important application in the study of integral functionals, which are frequently encountered in the calculus of variations. We will return to this concept with concrete applications in Chapter 6. For such applications, it is particularly convenient to work in the weak topology, so it is useful to consider the concept of Γ-convergence in this context. As mentioned in the beginning of the section, the concept of the Γ-limit (like the concept of the lower semicontinuous envelope) depends on the topology chosen for X, with different topologies possibly leading to different limits. Moreover, in the case where X is first countable, the Γ-limit enjoys some nice properties, as for example the convenient representation in terms of sequences (see Definition 2.8.2). It is interesting to note that in certain important cases, in the context of Banach spaces, the Γ-limit does not change when passing from the strong to the weak topology. An important lead to such considerations is the fact that, in a Banach space setting, if the dual $X^\star$ is separable, the weak topology $\sigma(X, X^\star)$ restricted on $\overline{B}_X(0,1)$ (or in fact to any bounded $A \subset X$) is metrizable (i. e., there exists a metric d on X such that the weak topology on every norm bounded subset A of X coincides with the topology induced by the metric d); see Theorem 1.1.46.

The following proposition (see Proposition 8.10 in [62]) provides such an instance of interest, which allows for a simpler characterization of the Γ-limit in Banach spaces equipped with the weak topology.

Proposition 2.8.14. *Let X be a Banach space with separable dual $X^\star$, and let $(F_n)_{n\in\mathbb{N}}$ be a sequence of functionals $F_n : X \to \mathbb{R} \cup \{+\infty\}$ such that $F_n \geq \Phi$ for all $n \in \mathbb{N}$, where Φ is coercive in the sense that $\Phi(x) \to \infty$ as $\|x\| \to \infty$. Then, Definition 2.8.2 as well as the sequential characterization of the Γ-limit holds, replacing $x_n \to x$ by $x_n \rightharpoonup x$.*

Proof. The proof is provided in Section 2.9.8 in the appendix. □

2.9 Appendix

2.9.1 Proof of Proposition 2.3.21

Suppose that $|\varphi| < c$, on $B(x_o, 2\delta) \subset \operatorname{int}(\operatorname{dom}\varphi)$. Consider $x_1, x_2 \in B(x_o, \delta)$, $x_1 \neq x_2$, and set $z = x_2 + \frac{\delta}{\|x_1 - x_2\|}(x_2 - x_1)$, where clearly $z \in B(x_o, 2\delta)$. We express x_2 in terms of x_1 and z in terms of the convex combination

$$x_2 = \frac{\|x_1 - x_2\|}{\|x_1 - x_2\| + \delta} z + \frac{\delta}{\|x_1 - x_2\| + \delta} x_1,$$

and using the convexity of φ, we have that

$$\varphi(x_2) \leq \frac{\|x_1 - x_2\|}{\|x_1 - x_2\| + \delta} \varphi(z) + \frac{\delta}{\|x_1 - x_2\| + \delta} \varphi(x_1).$$

Subtracting $\varphi(x_1)$ from both sides, we have

$$\varphi(x_2) - \varphi(x_1) \le \frac{\|x_1 - x_2\|}{\|x_1 - x_2\| + \delta}(\varphi(z) - \varphi(x_1)) \le \frac{2c\|x_1 - x_2\|}{\|x_1 - x_2\| + \delta} \le \frac{2c}{\delta}\|x_1 - x_2\|, \tag{2.44}$$

where we used the fact that $z, x_1 \in B(x_o, 2\delta)$ for that both $|\varphi(z)| < c$, $|\varphi(x_1)| < c$, and then the obvious fact that $\|x_1 - x_2\| + \delta > \delta$. Reversing the role of x_1 and x_2 and combining the result with (2.44), we deduce that $|\varphi(x_2) - \varphi(x_1)| \le \frac{2c}{\delta}\|x_1 - x_2\|$, which is the required Lipschitz continuity property.

To show that it suffices that φ is locally bounded above at x_o, note that if $\varphi < c$ on $B(x_o, \delta) \subset \operatorname{int}(\operatorname{dom}\varphi)$, then for any $z \in B(x_o, \delta)$, we may define $2x - z \in B(x_o, \delta)$, and by convexity (upon expressing $x = \frac{1}{2}z + \frac{1}{2}(2x - z)$),

$$\varphi(x) \le \frac{1}{2}\varphi(z) + \frac{1}{2}\varphi(2x - z) \le \frac{1}{2}(\varphi(z) + c),$$

which upon rearrangement yields that $\varphi(z) \ge 2\varphi(x) - c$, hence φ is bounded below at z. Since z was arbitrary, we conclude that if φ is locally bounded above, then it is also locally bounded below, hence locally bounded.

2.9.2 Proof of Proposition 2.3.22

Proof. We first show that φ is locally Lipschitz continuous at any $x \in \operatorname{core}(\operatorname{dom}\varphi)$. Suppose that $x \in \operatorname{core}(\operatorname{dom}\varphi)$. Then, setting $L_\lambda\varphi := \{x \in X : \varphi(x) \le \lambda\}$, for any $\lambda \in \mathbb{R}$,

$$X = \bigcup_{n\in\mathbb{N}} n(\operatorname{dom}\varphi - x) = \bigcup_{n\in\mathbb{N}} n \bigcup_{k\in\mathbb{Z}} (L_k\varphi - x) = \bigcup_{(n,k)} n(L_k\varphi - x). \tag{2.45}$$

Since φ is lower semicontinuous, the sets $n(L_k\varphi - x)$ are closed, by Baire's theorem (see Theorem 1.9.11(i)), there exists $(n_0, k_0) \in \mathbb{N} \times \mathbb{Z}$ such that $\operatorname{int}(n_0(L_{k_0}\varphi - x)) \ne \emptyset$. Hence, $x \in \operatorname{int}(L_{k_0}\varphi)$, therefore, φ is bounded above by k_0 on an open neighborhood of x, and by Proposition 2.3.21, we conclude that φ is locally Lipschitz continuous at x.

We now show that $\operatorname{int}(\operatorname{dom}\varphi) = \operatorname{core}(\operatorname{dom}\varphi)$. Since for any convex set $C \subset X$ it holds that $\operatorname{int}(C) \subset \operatorname{core}(C)$ (see Section 1.2.1), it is always true that $\operatorname{int}(\operatorname{dom}\varphi) \subset \operatorname{core}(\operatorname{dom}\varphi)$. Hence, it suffices to show that $\operatorname{core}(\operatorname{dom}\varphi) \subset \operatorname{int}(\operatorname{dom}\varphi)$. Consider any $x \in \operatorname{core}(\operatorname{dom}\varphi)$. Then by (2.45) and essentially by the same arguments, we can show that $x \in \operatorname{int}(\operatorname{dom}\varphi)$. Hence, $\operatorname{core}(\operatorname{dom}\varphi) = \operatorname{int}(\operatorname{dom}\varphi)$, and since we have already proved that φ is continuous in $\operatorname{core}(\operatorname{dom}\varphi)$, it is also continuous in $\operatorname{int}(\operatorname{dom}\varphi)$.

An alternative approach is to prove that φ is continuous in $\operatorname{int}(\operatorname{dom}\varphi)$, that is, to consider any point $x \in \operatorname{int}(\operatorname{dom}\varphi)$, and then for any $z \in X$ define the function $\varphi_z : \mathbb{R} \to \mathbb{R}$ by $\varphi_z(t) := \varphi(t(z-x))$. This is a convex function. Also since $x \in \operatorname{int}(\operatorname{dom}\varphi)$, it is easy to see that $0 \in \operatorname{int}(\operatorname{dom}\varphi_z)$. So, by Proposition 2.3.21, φ_z is continuous at $t = 0$. That implies, for any $\epsilon > 0$, there exists $\delta > 0$ such that $|\varphi_z(t) - \varphi_z(0)| \le \epsilon$ for $|t| < \delta$. Hence, $\varphi(t(z - x)) <$

$\varphi(0) + \epsilon$ for $|t| < \delta$. Choosing $\lambda > \varphi(0)$, we see by this argument that there exists $t > 0$ such that $t\mathrm{z} \in L_\lambda\varphi + \mathrm{x}$ or equivalently $\mathrm{z} \in \frac{1}{t}(L_\lambda\varphi + \mathrm{x})$, with t in general depending on z. Since $\mathrm{z} \in X$ is arbitrary, repeating the above procedure for every $\mathrm{z} \in X$, we conclude that $X = \bigcup_{t>0} \frac{1}{t}(L_\lambda\varphi + \mathrm{x}) = \bigcup_n n(L_\lambda\varphi + \mathrm{x})$, and using once more the closedness of $L_\lambda\varphi$ and the Baire category theorem, we conclude continuity of φ at x using Proposition 2.3.21.

If $X \subset \mathbb{R}^d$, then it is easy to see that $\mathrm{int}(\mathrm{dom}\,\varphi) = \mathrm{core}(\mathrm{dom}\,\varphi)$. Indeed, let $e = (e_i)_{i=1,\dots,d}$ be the standard basis of $\mathbb{R}^d$, and consider any $\mathrm{x} \in \mathrm{core}(\mathrm{dom}\,\varphi)$. Then, for every $i = 1,\dots,d$, there exists $\epsilon_i > 0$ such that $\mathrm{z}_i = \mathrm{x} + \delta_i e_i \in \mathrm{dom}\,\varphi$ for every $\delta_i \in [0, \epsilon_i]$, $i = 1,\dots,d$. Take $\epsilon = \min\{\epsilon_i : i = 1,\dots,d\}$. Then $\mathrm{z} = \mathrm{x} + \delta e \in \mathrm{dom}\,\varphi$ for every $\delta \in (0, \epsilon)$. Hence, $B(\mathrm{x}, \epsilon) \subset \mathrm{dom}\,\varphi$ and $\mathrm{x} \in \mathrm{int}(\mathrm{dom}\,\varphi)$. Hence, $\mathrm{core}(\mathrm{dom}\,\varphi) \subset \mathrm{int}(\mathrm{dom}\,\varphi)$, and therefore $\mathrm{core}(\mathrm{dom}\,\varphi) = \mathrm{int}(\mathrm{dom}\,\varphi)$.

To show that φ is continuous at $\mathrm{int}(\mathrm{dom}\,\varphi) = \mathrm{core}\,(\mathrm{dom}\,\varphi)$, we can either show almost verbatim, as in the infinite dimensional case, that φ is continuous in core dom φ, and then deduce the continuity at $\mathrm{int}(\mathrm{dom}\,\varphi)$ from their equality. Alternatively, take any $\mathrm{x} \in \mathrm{int}(\mathrm{dom}\,\varphi)$, pick a $\delta > 0$ so that $\mathrm{x} + \delta e_i \in \mathrm{dom}\,\varphi$ for every $i = 1,\dots,d$, and then note that the convex hull $\mathrm{conv}\{\mathrm{x} + \delta e_1, \dots, \mathrm{x} + \delta e_d\}$ is a neighborhood of x. By the convexity of φ for any $\mathrm{z} \in \mathrm{conv}\{\mathrm{x} + \delta e_1, \dots, \mathrm{x} + \delta e_d\}$, we have that

$$\varphi(\mathrm{z}) = \varphi\left(\sum_{i=1}^{d} \lambda_i(x + \delta e_i)\right) \le \sum_{i=1}^{d} \lambda_i \varphi(\mathrm{x} + \delta e_i) \le \max\{\varphi(\mathrm{x} + \delta e_i \ : \ i = 1,\dots,d\}.$$

Hence, there exists a neighborhood of x, in which φ is bounded above by $\max\{\varphi(\mathrm{x} + \delta e_i \ : \ i = 1,\dots,d\}$, so we may conclude continuity of φ at x using Proposition 2.3.21(i). □

2.9.3 Proof of Proposition 2.7.3

We first recall that in a first countable space, the notions of lower semicontinuity and sequential lower semicontinuity coincide (see Proposition 2.2.2(iv)), so we will work with the latter. Let $\Phi_0 := \mathrm{lsc}(F)$, as in (2.36) and (2.37). For simplicity, let us consider first the case that X is a Banach space endowed with the strong topology.

(a) To show (i), we use the characterizations (2.37) of $\mathrm{lsc}(F)$, which for the particular setting becomes

$$\mathrm{lsc}(F)(\mathrm{x}) = \sup_{r>0} \inf\{F(\mathrm{z}) \ : \ \|\mathrm{z} - \mathrm{x}\| < r\}. \tag{2.46}$$

Consider any x and any sequence $\mathrm{x}_n \to \mathrm{x}$. Since Φ_0 is lower semicontinuous,

$$\Phi_0(\mathrm{x}) \le \liminf_{n\to\infty} \Phi_0(\mathrm{x}_n) \le \liminf_{n\to\infty} F(\mathrm{x}_n),$$

since by definition $\Phi_0 \le F$.

To show (ii), we will use the equivalent characterization of Φ_0, as in (2.46), and the observation that

$$\sup_{r>0} \inf\{F(z) \,:\, \|z - x\| < r\} = \sup_{n\in\mathbb{N}} \inf\left\{F(z) \,:\, \|z - x\| < \frac{1}{n}\right\}, \tag{2.47}$$

which follows from the fact that the infimum of $F(z)$ on the ball $B(x, r)$ is increasing as the radius of the ball decreases and a density argument.

By (2.46), there exists a sequence $(t_n)_{n\in\mathbb{N}}$ such that $t_n \to \Phi_0(x)$ and $t_n > \Phi_0(x)$. Combining that with (2.47), we see that, for any $n \in \mathbb{N}$,

$$t_n > \Phi_0(x) = \sup_{n\in\mathbb{N}} \inf\left\{F(z) \,:\, \|z - x\| < \frac{1}{n}\right\} \geq \inf\left\{F(z) \,:\, \|z - x\| < \frac{1}{n}\right\}.$$

By the definition of the infimum, for any fixed $n \in \mathbb{N}$, and any $\epsilon > 0$, there exists a $\bar{x}_n \in B(x, \frac{1}{n})$ such that

$$\inf\left\{F(z) \,:\, \|z - x\| < \frac{1}{n}\right\} - \epsilon > F(\bar{x}_n).$$

Summarizing the above, we get for each $n \in \mathbb{N}$ the existence of a $\bar{x}_n \in B(x, \frac{1}{n})$ (hence $(\bar{x}_n)_{n\in\mathbb{N}}$ satisfies $\bar{x}_n \to x$), such that

$$t_n > F(\bar{x}_n), \quad t_n > \Phi_0(x), \quad t_n \to \Phi_0(x).$$

We take the limsup in the above to obtain that

$$\Phi_0(x) \geq \limsup_{n\to\infty} F(\bar{x}_n), \quad \bar{x}_n \to x,$$

which shows that $(\bar{x}_n)_{n\in\mathbb{N}}$ is a recovery sequence.[17]

In the case where X is a general first countable space, the proof is similar (see Proposition 3.6. in [62]), where, for example, to prove (ii), we use the monotone countable base $(U_n)_{n\in\mathbb{N}}$ for the neighborhood system at x (such that $U_{n+1} \subset U_n$) and consider a sequence $(t_n)_{n\in\mathbb{N}} \subset \mathbb{R}$ such that $t_n \to \operatorname{lsc} F(x)$, with $t_n > \operatorname{lsc}(F)(x)$ for every $n \in \mathbb{N}$. By the definition of $\operatorname{lsc}(F)(x) = \sup_{U\in N(x)} \inf_{z\in U} F(z)$, it is easy to see that for every $n \in \mathbb{N}$, we have $t_n > \operatorname{lsc}(F)(x) \geq \inf_{z\in U_n} F(z) > F(\bar{x}_n)$ for a suitable chosen $\bar{x}_n \in U_n$.[18] By the properties of the sequence $(U_n)_{n\in\mathbb{N}}$, (for any open set U containing x there exists $N \in \mathbb{N}$ such that

17 The recovery sequence is not necessarily unique.

18 For the first inequality, we simply use the fact that the sup over all $U \in N(x)$ is greater or equal to the value of the quantity achieved for $U = U_n$. For the second inequality, we simply use the definition of the inf: Setting $\epsilon = t_n - \inf_{z\in U_n} F(z) > 0$, there exists $\bar{x}_n \in U_n$ such that $F(\bar{x}_n) < \inf_{z\in U_n} F(z) + \epsilon$, and the result follows from that.

$U_N \subset U$, and then $U_n \subset U$ for all $n > N$), we see that $\bar{x}_n \to x$. Hence since $t_n > F(\bar{x}_n)$, taking the limsup in this inequality and keeping in mind the choice of the sequence $(t_n)_{n\in\mathbb{N}}$, we conclude that $\mathrm{lsc}(F)(x) \geq \limsup_n F(\bar{x}_n)$, i. e., $(\bar{x}_n)_{n\in\mathbb{N}} \subset X$ is a recovery sequence.

2.9.4 Proof of (2.39)

We only prove the first one, as the second one is similar.

Consider any $x \in \{\mathrm{lsc}(F) \leq \lambda\}$. Then,

$$\begin{aligned}
&\mathrm{lsc}(F)(x) = \sup_{U\in N(x)} \inf_{z\in U} F(z) \leq \lambda \\
&\Longrightarrow \forall U \in N(x),\ \inf_{z\in U} F(z) \leq \lambda \\
&\Longrightarrow \forall U \in N(x),\ \forall \delta > 0,\ \inf_{z\in U} F(z) < \lambda + \delta \\
&\Longrightarrow \forall U \in N(x),\ \forall \delta > 0,\ \forall \epsilon > 0,\ \exists z_0 \in U\ :\ F(z_0) < \inf_{z\in U} F(z) - \epsilon < \lambda + \delta - \epsilon.
\end{aligned}$$

Upon relabeling $t = \lambda + \delta > \lambda$, the above yields

$$\begin{aligned}
&\forall t > \lambda,\ \forall U \in N(x),\ \forall \epsilon > 0,\ \exists z_0 \in U\ :\ F(z_0) < t - \epsilon \\
&\Longrightarrow \forall t > \lambda,\ \forall U \in N(x),\ \forall \epsilon > 0,\ U \cap \{F < t - \epsilon\} \neq \emptyset \\
&\Longrightarrow \forall t > \lambda,\ \forall \epsilon > 0,\ x \in \overline{\{F < t - \epsilon\}} \\
&\Longrightarrow \forall t > \lambda,\ x \in \overline{\{F \leq t\}} \Longrightarrow x \in \bigcap_{t>\lambda} \overline{\{F \leq t\}}.
\end{aligned}$$

Hence, $\{\mathrm{lsc}(F) \leq \lambda\} \subset \bigcap_{t>\lambda} \overline{\{F \leq t\}}$.

For the reverse inclusion, consider any $x \in \bigcap_{t>\lambda} \overline{\{F \leq t\}}$. Then,

$$\begin{aligned}
&\forall t > \lambda,\ x \in \overline{\{F \leq t\}} \\
&\Longrightarrow \forall t > \lambda,\ \forall U \in N(x),\ U \cap \{F \leq t\} \neq \emptyset \\
&\Longrightarrow \forall t > \lambda,\ \forall U \in N(x),\ \exists z_0 \in U\ :\ F(z_0) \leq t \\
&\Longrightarrow \forall \delta > 0,\ \forall U \in N(x),\ \exists z_0 \in U\ :\ F(z_0) \leq t := \lambda + \delta \\
&\Longrightarrow \forall \delta > 0,\ \forall U \in N(x),\ \inf_{z\in U} F(z) \leq t := \lambda + \delta \\
&\Longrightarrow \forall \delta > 0,\ \sup_{U\in N(x)} \inf_{z\in U} F(z) = \mathrm{lsc}(F) \leq t := \lambda + \delta \Longrightarrow x \in \{\mathrm{lsc}(F) \leq \lambda\}.
\end{aligned}$$

Hence, $\bigcap_{t>\lambda} \overline{\{F \leq t\}} \subset \{\mathrm{lsc}(F) \leq \lambda\}$.

Combining the two inclusions, we obtain the stated result.

2.9.5 Proof of Proposition 2.8.3

We show that the general Definition 2.8.1 implies Definition 2.8.2, when X is first countable (see Section 1.9.1). We only consider the case for the $\Gamma - \liminf$, leaving the rest as an exercise for the reader.

Assume that Definition 2.8.1(i) holds.

We first show that Definition 2.8.2(i-a) holds. Consider any $(x_n)_{n\in\mathbb{N}} \subset X$ such that $x_n \to x$. Take an arbitrary $U \in N(x)$. By the convergence of the sequence, we have that there exists an $N \in \mathbb{N}$ such that $x_n \in U$ for all $n > N$. Clearly, $\inf_{z\in U} F_n(z) \le F_n(x_n)$ for all $n > N$, hence taking the liminf on both sides,

$$\liminf_n \inf_{z\in U} F_n(z) \le \liminf_n F_n(x_n).$$

This is true for any $U \in N(x)$, hence, also for the supremum over all U, i. e.,

$$\Gamma - \liminf_n F_n(x) = \sup_{U\in N(x)} \liminf_n \inf_{z\in U} F_n(z) \le \liminf_n F_n(x_n),$$

i. e., the condition in Definition 2.8.2(i-a). We remark that we have not used here the assumption that X is first countable.

We now show that Definition 2.8.2(i-b) holds by constructing a recovery sequence for any $x \in X$. Take any $x \in X$. Since X is first countable (see Definition 1.9.6), there exists a sequence of neighborhoods of x, $(U_k)_{k\in\mathbb{N}} \subset N(x)$, with $U_{k+1} \subset U_k$ such that for every open set U containing x, there exists $m \in \mathbb{N}$ such that $U_m \subset U$.[19] Let $I := \Gamma - \liminf_n F_n(x) < +\infty$, and take a sequence $(\lambda_m)_{m\in\mathbb{N}} \subset \mathbb{R}$ such that $\lambda_m \to I \in \mathbb{R}$ with $\lambda_m > I$ for every $m \in \mathbb{N}$. Fix an arbitrary $U \in N(x)$, and let $M \in \mathbb{N}$ be a suitable integer such that $U_M \subset U$. For any $m > M$, we have $U_m \subset U_M \subset U$. Take an arbitrary $m > M$ and the corresponding U_m and denote $I_m := \liminf_n \inf_{z\in U_m} F_n(z)$. Note that

$$\begin{aligned}\lambda_m > I := \Gamma - \liminf_n F_n(x) &= \sup_{U\in N(x)} \liminf_n \inf_{z\in U} F_n(z) \\ &\ge \liminf_n \inf_{z\in U_m} F_n(z) =: I_m, \quad \forall m > M,\end{aligned} \tag{2.48}$$

where we used Definition 2.8.1(i) for Γ–$\liminf_n F_n$ and subsequently estimated the supremum over all $U \in N(x)$ from below by using $U = U_m$. By the properties of the liminf, fixing $m \in \mathbb{N}$, for every $\epsilon_m > 0$, we have an infinite number of terms of the sequence $(\inf_{z\in U_m} F_n(z))_{n\in\mathbb{N}}$ in $(I_m - \epsilon_m, I_m + \epsilon_m)$. For each m, we choose n_m to be the first index of this sequence that satisfies this property, making sure that $n_m > n_{m-1}$.[20] In this way,

---

19 If X is a metric space, then we can take $U_k = B(x, \frac{1}{k})$, a sequence of balls centered as x of decreasing radius.

20 This is possible since there is an infinite number of terms of the sequence that satisfies this property for each m, so that we can discard all terms until we find one that fulfills this condition.

and by appropriate choice of ϵ_m (for example, $\epsilon_m = \frac{1}{2}(\lambda_m - I_m) > 0$), we obtain for each $m \in \mathbb{N}$ the existence of a $n_m \in \mathbb{N}$ such that $n_{m-1} < n_m$ and

$$I_m < \inf_{z \in U_m} F_{n_m}(z) + \epsilon_m < I_m + 2\epsilon_m = \lambda_m. \tag{2.49}$$

By the definition of the infimum, for every $m \in \mathbb{N}$ and every $\epsilon'_m > 0$, there exists $z_m \in U_m$ such that $F(z_m) < \inf_{z \in U_m} F_{n_m}(z) + \epsilon'_m$. Choose $\epsilon'_m = \epsilon_m$ and combine the above with (2.49). This yields some $z_m \in U_m$ with the property

$$F_{n_m}(z_m) < \inf_{z \in U_m} F_{n_m}(z) + \epsilon_m < \lambda_m, \quad m > M. \tag{2.50}$$

Clearly, by the properties of $(U_m)_{m \in \mathbb{N}}$, and the fact that $U \in N(x)$ is arbitrary, we see that $z_m \to x$ as $m \to \infty$. Taking the liminf on both sides of (2.50) and recalling that $\lambda_m \to I$, we obtain

$$\liminf_m F_{n_m}(z_m) \leq \lim_m \lambda_m = I := \Gamma - \liminf_n F_n(x),$$

which is the required inequality in Definition 2.8.2(i-b), with $(z_m)_{m \in \mathbb{N}}$ as recovery sequence.

2.9.6 Proof of Proposition 2.8.7

We simplify the notation by setting $F_\Gamma^- = F^-$ and $F_\Gamma^+ = F^+$. For variety and practice, we provide some proofs in the general case and some using the sequential characterization.

(i) The proof if X is a Banach space endowed with the strong topology can proceed as follows (see e. g., [129]): We wish to show that for any sequence $z_m \to z$ it holds that $F^-(z) \leq \liminf_m F^-(z_m)$. To show this, we need to recall the definition of F^- (Def. 2.8.2(i)). We first apply it to z_m, for any $m \in \mathbb{N}$,

$$F^-(z_m) = \inf \left\{ \liminf_n F_n(x_n) \ : \ x_n \to z_m \right\};$$

so that by the definition of the infimum, using the above, for every $\epsilon_m > 0$, there exists a sequence $(x_{n,m})_{n \in \mathbb{N}}$, such that

$$F^-(z_m) + \epsilon_m > \liminf_n F_n(x_{n,m}), \quad \lim_{n \to \infty} x_{n,m} = z_m.$$

Hence, by the properties of the lim inf (there exist infinite terms of $(F_n(x_{n,m}))_{n \in \mathbb{N}}$ in the interval $(\liminf_n F_n(x_{n,m}), \liminf_n F_n(x_{n,m}) + \epsilon'))$, for every $m \in \mathbb{N}$, there exists $n = n_m$ (large enough) such that

$$F_n(x_{n,m}) < F^-(z_m) + \epsilon_m, \quad \|x_{n,m} - z_m\| < \epsilon_m.$$

Note that we can choose the n_m such that $n_m < n_{m+1}$ for all m. Taking the $\liminf_m$ on both sides of the above, yields

$$\liminf_m F_n(x_{n,m}) \le \liminf_m (F^-(z_m) + \epsilon_m) = \liminf_m F^-(z_m), \tag{2.51}$$

with the last estimate coming upon choosing $\epsilon_m \to 0$. This is part of the estimate we need. To finish the estimation, we construct the sequence $(\bar{z}_n)_{n\in\mathbb{N}}$, defined by $\bar{z}_n = x_{n,m}$, if $n = n_m$ and $\bar{z}_n = z$ otherwise. Clearly $\bar{z}_n \to z$ while $(x_{n,m})_{m\in\mathbb{N}}$ can be considered as a subsequence of $(\bar{z}_n)_{n\in\mathbb{N}}$. We now use Def. 2.8.2(i) at z to get $F^-(z) \le \liminf F_n(\bar{z}_n)$ which combined with (2.51) and $\liminf_n F_n(\bar{z}_n) \le \liminf_m F_n(x_{n,m})$ (by the properties of liminf with respect to subsequences) yields the required result.

In the general case, the proof of the lower semicontinuity relies on the fact that the function $G : X \to \mathbb{R} \cup \{+\infty\}$, defined by $G(x) = \sup_{U\in N(x)} \alpha(U)$, where α is a set function defined on a family of open subsets of X, is lower semicontinuous (see Example 2.2.3(ii)). Then, we can simply apply this result to $\alpha(U) = \liminf_n \inf_{z\in U} F_n(z)$ and $\alpha(U) = \limsup_{z\in U} F_n(z)$. To show the equalities between the Γ-limits of $(F_n)_{n\in\mathbb{N}}$ and $(\text{lsc}(F_n))_{n\in\mathbb{N}}$, it suffices to observe that $\inf_{z\in U} H(z) = \inf_{z\in U}(\text{lsc}(H))(z)$ for any function H (see [62]).

(ii) In the case where X is first countable (e. g., a Banach space endowed with the strong topology), by definition,

$$\Gamma - \liminf_n F_n(x) = \inf \left\{ \liminf_n F_n(x_n) \ : \ x_n \to x \right\} \le \liminf_n F_n(x),$$

where we choose to take an upper bound of the inf by the constant sequence $\hat{x}_n = x$. Similarly for the limsup.

In the general case, note that for every $U \in N(x)$ we have $\inf_{z\in U} F_n(z) \le F_n(x)$, and the result follows by taking $\liminf_n$ (resp. $\limsup_n$) on both sides of the above and subsequently the sup over all $U \in N(x)$.

(iii) (Sketch) We first sketch the proof if X is first countable.

Since $(F_n)_{n\in\mathbb{N}}$ converges uniformly to F, for every $\epsilon > 0$, there exists $N \in \mathbb{N}$ such that for all $x \in X$, $|F_n(x) - F(x)| < \epsilon$ for all $n > N$. We will show that, in this case, $\Phi := \text{lsc}(F)$ satisfies the liminf and the limsup inequality for Γ-convergence.

Liminf inequality: Consider any sequence $x_n \to x$. For arbitrary $\epsilon > 0$, find the corresponding N such that for all $z \in X$, $|F_n(z) - F(z)| < \epsilon$ for all $n > N$ (uniform convergence) and apply it for $z = x_n$, any element of the above sequence (for $n > N$). Then $|F_n(x_n) - F(x_n)| < \epsilon$, which implies that $-\epsilon < F(x_n) - F_n(x_n) < \epsilon$ or equiv. $F(x_n) < F_n(x_n) + \epsilon$. This is true for all $n > N$. So taking the liminf on both sides, we obtain that

$$\liminf_n F(x_n) \le \liminf_n F_n(x_n) + \epsilon.$$

On the other hand, the liminf inequality for $\Phi := \text{lsc}(F)$ yields that

$$\Phi(x) \leq \liminf_n F(x_n),$$

so that combining the two and setting $\epsilon \to 0$, we obtain the required Γ-limit liminf inequality.

Limsup inequality: Let $\bar{x}_n \to x$ be a recovery sequence for $\Phi = \text{lsc}(F)$ at point $x \in X$, i. e., a sequence such that

$$\limsup_n F(\bar{x}_n) \leq \Phi(x). \tag{2.52}$$

We apply the same reasoning as above, based on the uniform convergence of $(F_n)_{n\in\mathbb{N}}$, for any $x = \bar{x}_n$ with $n > N$, which guarantees that $|F_n(\bar{x}_n) - F(\bar{x}_n)| < \epsilon$, which implies that $-\epsilon < F_n(\bar{x}_n) - F(\bar{x}_n) < \epsilon$, equiv. $F_n(\bar{x}_n) < F(\bar{x}_n) + \epsilon$. Taking the limsup on both sides of this inequality, we obtain

$$\limsup_n F_n(\bar{x}_n) \leq \limsup_n F(\bar{x}_n) + \epsilon. \tag{2.53}$$

Combining (2.52) and (2.53), we recover the Γ-limit limsup inequality.

In the general case, we first have to note that if $F_n \to F$ uniformly, then for any open $U \subset X$, we have that $\lim_n \inf_{z\in U} F_n(z) = \inf_{z\in U} F(z)$. We then take the sup over all $U \in N(x)$ and proceed by the definitions (see Prop. 5.2 [62]).

(iv) We provide the result directly in the general case and for the liminf only. Since in general $\Gamma - \liminf_n F_n(x) \leq \liminf_n F_n(x)$, it suffices to show the reverse inequality. By the equi lower semicontinuity of $(F_n)_{n\in\mathbb{N}}$, for every $\epsilon > 0$, there exists a $U \in N(x)$ such that $F_n(x) - \epsilon < F_n(z)$ for every $z \in U$ and $n \in \mathbb{N}$. Taking the infimum over all $z \in U$, we obtain that $F_n(x) - \epsilon \leq \inf_{z\in U} F_n(z)$ for all $n \in \mathbb{N}$, and, subsequently, taking first the liminf over all n on both sides, and then estimating the resulting RHS from above by the supremum over all $U \in N(x)$, we obtain that

$$\liminf_n F_n(x) - \epsilon \leq \sup_{U\in N(x)} \liminf_n \inf_{z\in U} F_n(z) = \Gamma - \liminf_n F_n(x),$$

and the result follows since $\epsilon > 0$ is arbitrary.

(v) We provide the proof here for the case where X is first countable. For the general case, the proof follows from Definition 2.8.1 (see Ch. 6 in [62]. The first set of inequalities follows directly from the definition of $\Gamma - \liminf_n$ and $\Gamma - \limsup_n$ and the relevant inequalities for the ordinary $\liminf_n$ and $\limsup_n$. For the Γ-limsup inequality, recall the standard inequality valid for any two real sequences that $\limsup_n(a_n + b_n) \geq \limsup_n a_n + \liminf_n b_n$.[21] Then apply this for the choice $a_n = F_n(x_n)$ and $b_n = \Phi(x_n)$,

21 For any $\epsilon > 0$, there exists a $k^* \geq n$ such that $\sup\{a_k : k \geq n\} - \epsilon < a_{k^*}$. Then, $\sup\{a_k + b_k : k \geq n\} \geq a_{k^*} + b_{k^*} \geq \sup\{a_k : k \geq n\} - \epsilon + \inf\{b_k : k \geq n\}$, and since all sequences are monotone and bounded, hence convergent, by taking the limit the required inequality holds, since $\epsilon > 0$ is arbitrary.

where $x_n \to x$ is an arbitrary sequence, and then take the infimum over all such sequences to conclude.

To prove the equality we note that since $\Phi_n \to \Phi$ uniformly and Φ is continuous (hence, $\pm\Phi$ is lower semicontinuous) by (iii) $\Gamma - \lim_n \Phi_n = \Phi$ and $\Gamma - \lim_n(-\Phi_n) = -\Phi$.

We set $F_n = F_n + \Phi_n - \Phi_n$ to obtain

$$\begin{aligned}\Gamma - \liminf_n F_n &= \Gamma - \liminf_n(F_n + \Phi_n - \Phi_n) \geq \Gamma - \liminf_n(F_n + \Phi_n) + \Gamma - \liminf_n(-\Phi_n)\\ &= \Gamma - \liminf_n(F_n + \Phi_n) - \Phi. \end{aligned} \tag{2.54}$$

Moreover,

$$\Gamma - \liminf_n(F_n + \Phi_n) \geq \Gamma - \liminf_n(F_n) + \Gamma - \liminf_n(\Phi_n) = \Gamma - \liminf_n(F_n) + \Phi. \tag{2.55}$$

Thus, (2.54) yields

$$\begin{aligned}&\Gamma - \liminf_n(F_n \geq \Gamma - \liminf_n(F_n + \Phi_n) - \Phi\\ &\Longrightarrow \Gamma - \liminf_n(F_n + \Phi_n) \leq \Gamma - \liminf_n F_n + \Phi.\end{aligned}$$

which , combined with (2.55), leads to the required result for $\Gamma - \liminf$. Similarly for the other case.

2.9.7 Alternative approach to Example 2.8.8

We revisit Example 2.8.8 in the special case, where X is endowed with a topology such that the sequential characterization of Γ-convergence applies.

Convex functions are locally Lipschitz in the neighborhoods of points at which they are bounded. The latter, along with the equi-bounded property, implies that for every $x \in X$, the existence of some constant c_L and an R such that

$$|F_n(x) - F_n(z)| \leq c_L \|x - z\|, \quad \forall\, z \in B(x, R),$$

with the constants independent of n.

Consider $x_n \to x$ arbitrary. For any R and large enough n, $\|x_n - x\| \leq R$, so that by the Lipschitz property,

$$|F_n(x_n) - F_n(x)| \leq c_L \|x_n - x\| \to 0.$$

Since $\lim_n F_n(x) = F(x)$ by the pointwise convergence of $(F_n)_{n\in\mathbb{N}}$, we have that

$$
\begin{aligned}
F(\mathsf{x}) &= \liminf_n F_n(\mathsf{x}) = \liminf_n((F_n(\mathsf{x}) - F_n(\mathsf{x}_n)) + F_n(\mathsf{x}_n)) \\
&\leq \liminf_n(|F_n(\mathsf{x}) - F_n(\mathsf{x}_n)| + F_n(\mathsf{x}_n)) \\
&= \liminf_n|F_n(\mathsf{x}) - F_n(\mathsf{x}_n)| + \liminf_n F_n(\mathsf{x}_n) = \liminf_n F_n(\mathsf{x}_n),
\end{aligned}
\tag{2.56}
$$

where we also used the fact that $|F_n(\mathsf{x}) - F_n(\mathsf{x}_n)| \to 0$.

Since the sequence $\mathsf{x}_n \to \mathsf{x}$ is arbitrary, (2.56) yields

$$
\begin{aligned}
F(\mathsf{x}) &\leq \inf\left\{\liminf_n F_n(\mathsf{x}_n) \,:\, \mathsf{x}_n \to \mathsf{x}\right\} = \left(\Gamma - \liminf_n F_n\right)(\mathsf{x}) \\
&\leq \left(\Gamma - \limsup_n F_n\right)(\mathsf{x}) \overset{\text{(ii)}}{\leq} \limsup_n F_n(\mathsf{x}) = F(\mathsf{x}),
\end{aligned}
$$

hence $\Gamma - \lim_n F_n = F$.

2.9.8 Proof of Proposition 2.8.14

We only sketch the proof for the $\Gamma - \liminf$ following [62]; the other cases being similar are left as exercise for the reader.

The assumptions on X imply the existence of a metric d on X that induces the weak topology on bounded subsets of X. In particular, $\mathsf{x}_n \rightharpoonup \mathsf{x}$ if and only if $(\mathsf{x}_n)_{n\in\mathbb{N}} \subset X$ is bounded and $\mathsf{d}(\mathsf{x}_n, \mathsf{x}) \to 0$ (see Theorem 1.1.46(iv)). We will show that in this case

$$
\Gamma - \liminf_{n\to\infty} F_n(\mathsf{x}) := \sup_{U\in N(\mathsf{x})} \liminf_{n\to\infty} \inf_{\mathsf{z}\in U} F_n(\mathsf{z}) = \inf\left\{\liminf_n F_n(\mathsf{x}_n) \,:\, \mathsf{x}_n \rightharpoonup \mathsf{x}\right\}, \tag{2.57}
$$

where the LHS is the $\Gamma - \liminf$, as provided by the general Definition 2.8.1, and the RHS coincides with the $\Gamma - \liminf$, as provided by Definition 2.8.2 (with $\mathsf{x}_n \to \mathsf{x}$ replaced by $\mathsf{x}_n \rightharpoonup \mathsf{x}$). Note also that since the convergence $\mathsf{x}_n \rightharpoonup \mathsf{x}$ can be expressed in terms of a metric d, it holds that

$$
\inf\left\{\liminf_{n\to\infty} F_n(\mathsf{x}_n) \,:\, \mathsf{x}_n \to \mathsf{x}\right\} = \sup_{r>0} \liminf_{n\to\infty} \inf_{\mathsf{z}\in B(\mathsf{x},r)} F_n(\mathsf{z}), \tag{2.58}
$$

where $B(\mathsf{x}, r)$ is the open ball in (X, d) (for a proof of (2.58); see Lemma 2.9.1 below). To simplify the notation, set

$$
\begin{aligned}
\Phi_1(\mathsf{x}) &:= \sup_{U\in N(\mathsf{x})} \liminf_{n\to\infty} \inf_{\mathsf{z}\in U} F_n(\mathsf{z}), \\
\Phi_2(\mathsf{x}) &:= \sup_{r>0} \liminf_{n\to\infty} \inf_{\mathsf{z}\in B(\mathsf{x},r)} F_n(\mathsf{z}).
\end{aligned}
$$

In this notation, combining (2.57) with (2.58), we must prove that $\Phi_1(\mathsf{x}) = \Phi_2(\mathsf{x})$.

(a) We first prove that $\Phi_1(x) \le \Phi_2(x)$.

By the definition of $\Phi_1(x)$ as a supremum, for every $\epsilon_1 > 0$, there exists $U \in N(x)$ such that

$$\Phi_1(x) - \epsilon_1 < \liminf_{n\to\infty} \inf_{z\in U} F_n(z). \tag{2.59}$$

Consider now any $t \in \mathbb{R}$ such that $\Phi_2(x) < t$, and let $B = \{z \in X : \Psi(z) \le t\} \cup \{x\}$. This is a bounded set (in the norm of X by the assumptions on Ψ). By the fact that the weak topology on X on bounded sets is metrizable by the metric d, there exists a ball $B(x,r)$ in (X, d) such that $B(x,r) \cap B \subset U \cap B$. Since $B(x,r) \cap B \subset U \cap B \subset U$, we have that $\inf_U F_n \le \inf_{U\cap B} F_n \le \inf_{B(x,r)\cap B} F_n$ so that taking the liminf on both sides, $\liminf_n \inf_U F_n \le \liminf_n \inf_{B(x,r)\cap B} F_n$, and using (2.59), we obtain

$$\Phi_1(x) - \epsilon_1 < \liminf_{n\to\infty} \inf_{z\in U} F_n(z) \le \liminf_{n\to\infty} \inf_{z\in B(x,r)\cap B} F_n(z) =: I. \tag{2.60}$$

We now recall the definition of the $\liminf_n$, according to which, for any $\epsilon_2 > 0$, there exists a $N \in \mathbb{N}$ such that

$$I - \epsilon_2 < \inf_{z\in B(x,r)\cap B} F_n(z), \quad n > N \implies I < \inf_{z\in B(x,r)\cap B} F_n(z) + \epsilon_2, \quad n > N. \tag{2.61}$$

Combining (2.61) with (2.60), we obtain that

$$\Phi_1(x) - \epsilon_1 - \epsilon_2 < \inf_{z\in B(x,r)\cap B} F_n(z), \quad n > N \implies \Phi_1(x) \le \inf_{z\in B(x,r)\cap B} F_n(z), \quad n > N, \tag{2.62}$$

since $\epsilon_1, \epsilon_2 > 0$ are arbitrary.

In order to relate (2.62) to $\Phi_2(x)$, we would like to obtain an estimate for $\inf_{z\in B(x,r)} F_n(z)$, which is related to $\Phi_2(x)$. To this end, note that $B(x,r) = (B(x,r) \cap B) \cup (B(x,r) \setminus B)$, and on $B(x,r) \setminus B$, we have that $F_n \ge \Psi > t$. Hence, taking the inf, we have that

$$t \le \inf_{z\in B(x,r)\setminus B} F_n(z), \quad n \in \mathbb{N}. \tag{2.63}$$

Since (2.62) and (2.63), provide estimates for $\inf_x F_n(x)$ on two complementary subsets of $B(x,r)$, combining them, we conclude that

$$\min\{\Phi_1(x), t\} \le \inf_{z\in B(x,r)} F_n(z), \quad n > N. \tag{2.64}$$

We take the $\liminf_n$ in (2.64) to obtain

$$\min\{\Phi_1(x), t\} \le \liminf_n \inf_{z\in B(x,r)} F_n(z) \le \sup_{r>0} \liminf_n \inf_{z\in B(x,r)} F_n(z) =: \Phi_2(x). \tag{2.65}$$

Recall that we have chosen t so that $\Phi_2(x) < t$, so that (2.65) implies that

$$\Phi_1(x) \le \Phi_2(x). \tag{2.66}$$

(b) It remains to show the opposite inequality $\Phi_1(x) \geq \Phi_2(x)$.

The procedure is similar, so we only sketch it. By the definition of $\Phi_2(x)$, for any $\epsilon_1 > 0$, there exists an $r > 0$ and a ball $B(x, r)$ in (X, d) such that

$$\Phi_2(x) - \epsilon_1 < \liminf_n \inf_{z \in B(x,r)} F_n(z).$$

For any bounded set B (to be specified shortly) by the equivalence of the topologies, there exists an open set $U \in N(x)$ such that $U \cap B \subset B(x, r) \cap B$. Clearly, $U \cap B \subset B(x, r) \cap B \subset B(x, r)$ implies that $\inf_{B(x,r)} F_n \leq \inf_{B(x,r) \cap B} F_n \leq \inf_{U \cap B} F_n$. Working as above and using the properties of the liminf, for any $\epsilon_2 > 0$, we guarantee the existence of an $N \in \mathbb{N}$ such that

$$\Phi_2(x) - \epsilon_1 < \inf_{z \in B(x,r)} F_n(z) + \epsilon_2 \leq \inf_{z \in U \cap B} F_n(z) + \epsilon_2, \quad n > N.$$

By the fact that $\epsilon_1, \epsilon_2 > 0$ are arbitrary, we obtain

$$\Phi_2(x) \leq \inf_{z \in U \cap B} F_n(z), \quad n > N.$$

We now choose a $t \in \mathbb{R}$ such that $\Phi_1(x) < t$ and set $B = \{\Psi \leq t\}$, as above. We then proceed by noting that since $F_n(z) > t$ in $U \setminus B$,

$$t \leq \inf_{z \in U \setminus B} F_n(z), \quad n \in \mathbb{N},$$

so that $\min\{\Phi_2(x), t\} \leq \inf_{z \in U} F_n(z)$ for $n > N$, and passing to the $\liminf_n$,

$$\min\{\Phi_2(x), t\} \leq \liminf_n \inf_{z \in U} F_n(z) \leq \sup_{U \in N(x)} \liminf_n \inf_{z \in U} F_n(z) = \Phi_1(x),$$

where the last inequality is obvious by the definition of the supremum. Recalling that we have chosen $t \in \mathbb{R}$ such that $\Phi_1(x) > t$, we see that

$$\Phi_2(x) \leq \Phi_1(x). \tag{2.67}$$

Combining (2.66) and (2.67), we conclude. □

We finish by providing a simple lemma concerning assertion (2.58).

Lemma 2.9.1. *Assume that the convergence $x_n \to x$ can be expressed in terms of a metric* d, *as, for example, in the case that (X, d) is a metric space or the setting described in Theorem* 1.1.46*(iv). Then,*

$$\inf\left\{\liminf_n F_n(x_n) \,:\, x_n \to x\right\} = \sup_{r>0} \liminf_n \inf_{z \in B(x,r)} F_n(z), \tag{2.68}$$

where $B(x, r)$ is the ball in (X, d).

Proof. Consider any $x_n \to x$, which according to the assumptions here implies that $d(x_n, x) \to 0$. Hence, for every $r > 0$, there is a ball $B(x, r)$ in (X, d) such that $x_n \in B(x, r)$ for all $n > N$ (for some $N \in \mathbb{N}$). Then, $\inf_{z \in B(x,r)} F_n(z) \le F_n(x_n)$ for all $n > N$, and taking the $\liminf_n$ in this inequality, we obtain that

$$\liminf_n \inf_{z \in B(x,r)} F_n(z) \le \liminf_n F_n(x_n).$$

This is true for any $r > 0$, so it is also true for the supremum of the LHS over all $r > 0$. Hence,

$$\sup_{r>0} \liminf_n \inf_{z \in B(x,r)} F_n(z) \le \liminf_n F_n(x_n),$$

and since the chosen sequence $(x_n)_{n \in \mathbb{N}}$ is arbitrary, this also holds for the infimum of the RHS over the set of all sequences $x_n \to x$. This leads to the inequality

$$\sup_{r>0} \liminf_n \inf_{z \in B(x,r)} F_n(z) \le \inf\Big\{\liminf_n F_n(x_n) \,:\, x_n \to x\Big\}. \tag{2.69}$$

We now establish the reverse inequality. Consider a decreasing sequence $r_n \to 0$, and consider the balls $B(x, r_n)$ in (X, d). Then, for any $n \in \mathbb{N}$ and any $\epsilon_n > 0$, by the definition of the infimum, there exists a $z_n \in B(x, r_n)$ such that $F_n(z_n) < \inf_{z \in B(x,r_n)} F_n(z) + \epsilon_n$. Taking the *liminf*$_n$ in this inequality, and choosing the arbitrary sequence $(\epsilon_n)_{n \in \mathbb{N}}$ such that $\epsilon_n \to 0$, and observing that the sequence $(\inf_{z \in B(x,r_n)} F_n(z))_{n \in \mathbb{N}}$ is increasing, yields

$$\liminf_n F_n(z_n) \le \liminf_n \inf_{z \in B(x,r_n)} F_n(z) \le \sup_{r>0} \liminf_n \inf_{z \in B(x,r)} F_n(z). \tag{2.70}$$

By construction $d(z_n, x) \to 0$ so $z_n \to x$. Hence,

$$\inf\Big\{\liminf_n F_n(x_n) \,:\, x_n \to x\Big\} \le \liminf_n F_n(z_n), \tag{2.71}$$

and combining (2.70) with (2.71), we obtain

$$\inf\Big\{\liminf_n F_n(x_n) \,:\, x_n \to x\Big\} \le \sup_{r>0} \liminf_n \inf_{z \in B(x,r)} F_n(z). \tag{2.72}$$

By (2.69) and (2.72), we obtain the required result (2.68). □

3 Fixed point theorems and their applications

A very important part of nonlinear analysis is fixed point theory. This theory provides answers to the question of whether a map from a (subset) of a Banach space to itself (or sometimes a complete metric space to itself) admits points that are left invariant under the action of the map and is an indispensable toolbox in a variety of fields, finding many important applications, ranging from their use in abstract existence results to the construction of numerical schemes. In this chapter, we present some of the most common fixed point theorems, starting from Banach's contraction mapping principle, which guarantees existence of fixed points for strict contractions, and then gradually start removing hypotheses on the map, working our way towards more general fixed point theorems. In this short sojourn in fixed point theory, we present a) the Brower fixed point theorem (a deep topological finite dimensional result, which, nevertheless, forms the basis for many infinite dimensional fixed point theorems), then b) the Schauder fixed point theorem (which only assumes continuity of the map and convexity and compactness of the set it acts on) and its important extension, c) the Leray–Schauder alternative and then move to d) Browder's fixed point theorem for nonexpansive maps and the related e) Mann-Krasnoselskii iterative scheme for the approximation of fixed points for such maps. We then provide a useful fixed point theorem for multivalued maps and close the chapter with a very general fixed point theorem due to Caristi, which is equivalent to the extremely useful Ekeland variational principle. All fixed point theorems presented are illustrated with extensive examples from differential and integral equations, PDEs, and optimization. There are various excellent textbooks and monographs dedicated solely to fixed point theorems (see, e. g., [2] or [143]).

3.1 Banach fixed point theorem

3.1.1 The Banach fixed point theorem and generalizations

Definition 3.1.1. Let (X, d) be a metric space and[1] $A \subset X$. The map $f : A \to X$ is said to be a contraction if there exists $\varrho \in (0, 1)$ such that

$$\mathrm{d}(f(x_1), f(x_2)) \leq \varrho\, \mathrm{d}(x_1, x_2), \quad \forall\, x_1, x_2 \in A.$$

Theorem 3.1.2 (Banach). *Let $A \subset X$ be a nonempty, closed subset of a complete metric space X and $f : A \to A$ a contraction. Then f has a unique fixed point x_o. Moreover if $x_1 = x$ is an arbitrary point in A and $(x_n)_{n\in\mathbb{N}}$ a sequence, defined by*

$$x_{n+1} = f(x_n), \quad n \in \mathbb{N}, \tag{3.1}$$

1 Recall our convention of choosing $\subset$ interchangeably with $\subseteq$. It is possible that $A = X$.

https://doi.org/10.1515/9783111333298-003

then $\lim_{n\to\infty} x_n = x_o$ *and*

$$d(x_n, x_o) \leq \frac{\varrho^{n-1}}{1-\varrho} d(x_2, x_1). \tag{3.2}$$

Proof. Let $x \in A$. Then, from (3.1), we have

$$d(x_n, x_{n+1}) = d(f(x_{n-1}), f(x_n)) \leq \varrho\, d(x_{n-1}, x_n),$$

which upon iteration yields

$$d(x_n, x_{n+1}) \leq \varrho^{n-1}\, d(x_2, x_1).$$

It then follows (using also the triangle inequality) that

$$\begin{aligned} d(x_n, x_{n+m}) &\leq d(x_n, x_{n+1}) + d(x_{n+1}, x_{n+2}) + \cdots + d(x_{n+m-1}, x_{n+m}) \\ &\leq (\varrho^{n-1} + \varrho^n + \cdots + \varrho^{n+m-2})\, d(x_2, x_1) \\ &= \varrho^{n-1}(1 + \varrho + \cdots + \varrho^{m-1})\, d(x_2, x_1) \\ &\leq \frac{\varrho^{n-1}}{1-\varrho}\, d(x_2, x_1). \end{aligned}$$

Hence, $(x_n)_{n\in\mathbb{N}}$ is a Cauchy sequence, and since X is complete, there exists $x_o \in X$ such that $x_n \to x_o$. Furthermore, since A is closed, $x_o \in A$. Fixing n and letting $m \to \infty$, we obtain the estimate (3.2).

We now show the fixed point property of x_o. For any $n \in \mathbb{N}$, we have

$$\begin{aligned} d(x_o, f(x_o)) &\leq d(x_o, x_{n+1}) + d(x_{n+1}, f(x_o)) = d(x_o, x_{n+1}) + d(f(x_n), f(x_o)) \\ &\leq d(x_o, x_{n+1}) + \varrho d(x_n, x_o). \end{aligned}$$

For $n \to \infty$, we obtain that $d(x_o, f(x_o)) = 0$, i. e., $f(x_o) = x_o$.

Finally, uniqueness follows from the observation that if there exist two fixed points $x_{o,1}, x_{o,2}$, then $d(x_{o,1}, x_{o,2}) = d(f(x_{o,1}), f(x_{o,2})) \leq \varrho\, d(x_{o,1}, x_{o,2})$, and the fact that $\varrho \in (0, 1)$. □

Remark 3.1.3. All assumptions of the theorem are important. As a simple example, consider $f : \mathbb{R} \to \mathbb{R}$, defined by $f(x) = \frac{\pi}{2} + x - \operatorname{Arctan} x$, for every $x \in \mathbb{R}$. Then, $f'(x) = 1 - \frac{1}{1+x^2} < 1$ for every $x \in \mathbb{R}$, so that from the mean value theorem, for any $x_1, x_2 \in \mathbb{R}$, there exists $\xi \in (x_1, x_2)$ such that $|f(x_1) - f(x_2)| = |f'(\xi)|\, |x_1 - x_2| \leq |x_1 - x_2|$. However, f has no fixed point.

The previous remark motivates the following extension of the Banach fixed point theorem, which is due to Edelstein [71]:

Theorem 3.1.4 (Edelstein). *Let (X, d) be a compact metric space, and consider $f : X \to X$ such that $\mathrm{d}(f(x_1), f(x_2)) < \mathrm{d}(x_1, x_2)$ for every $x_1, x_2 \in X$, with $x_1 \neq x_2$. Then f has a unique fixed point in X.*

Proof. The existence of a fixed point follows by the existence (by the Weierstrass maximum theorem) of a minimum for the function $x \mapsto \mathrm{d}(x, f(x))$, at some point $x_o \in X$. This is a fixed point. For if not, then $x_o \neq f(x_o)$ and $\mathrm{d}(f(x_o), f(f(x_o)) < d(x_o, f(x_o))$, which contradicts the property of x_o as the minimizer. To show uniqueness, assume the existence of two fixed points $x_{o,1}$, $x_{o,2}$, and note that $\mathrm{d}(x_{o,1}, x_{o,2}) = \mathrm{d}(f(x_{o,1}), f(x_{o,2})) < \mathrm{d}(x_{o,1}, x_{o,2})$, a contradiction. □

The Banach fixed point theorem is one of the fundamental fixed point theorems in analysis and finds applications in practically all fields of mathematics. We briefly review here some of its important applications in nonlinear analysis.

3.1.2 Solvability of differential equations

Theorem 3.1.5 (Cauchy, Lipschitz, Picard). *Let X a Banach space and $f_o : X \to X$ a Lipcshitz map, i. e., a map such that there exists $L \geq 0$, for which $\|f_o(x_1) - f_o(x_2)\| \leq L\,\|x_1 - x_2\|$ for every $x_1, x_2 \in X$. Then for each $x_0 \in X$, the problem*

$$\begin{aligned} \frac{dx}{dt}(t) &= f_o(x(t)), \quad t \in [0, \infty), \\ x(0) &= x_0, \end{aligned} \tag{3.3}$$

has a unique solution $x \in C^1([0, \infty), X)$.

Proof. The problem (3.3) is equivalent to the following: Find $x \in C([0, \infty), X)$ such that $x(t) = x_0 + \int_0^t f_o(x(s))ds$, where the integral is understood in the Bochner sense (see Section 1.6).

For $\varrho > 0$, we consider the space $E = \{x \in C([0, \infty), X) \,:\, \sup_{t \geq 0} e^{-\varrho t}\|x(t)\| < \infty\}$. It is easy to verify the following properties:

(i) E endowed with the norm $\|x\|_E = \sup_{t \geq 0} e^{-\varrho t}\|x(t)\|$ is a Banach space.

(ii) Define the mapping $f : E \to X$ by $(f(x))(t) = x_0 + \int_0^t f_o(x(s))ds$ for any $x \in E$. Then $f : E \to E$.

(iii) It holds that $\|f(x_1) - f(x_2)\|_E \leq \frac{L}{\varrho}\|x_1 - x_2\|_E$ for every $x_1, x_2 \in E$.

Therefore, for $\varrho > L$, the mapping $f : E \to E$ is a contraction, and by the Banach fixed point theorem, it has a fixed point, which is a solution of (3.3).

Concerning the uniqueness, let x_1 and x_2 be two solutions of problem (3.3). Setting $p(t) := \|x_1(t) - x_2(t)\|$, we have, using the Lipschitz property, that $p(t) \leq L \int_0^t p(s)ds$ for $t \geq 0$, and by a standard application of Gronwall's inequality (see Section 3.7.1), we have that $p(t) = 0$ for every $t \geq 0$. □

3.1.3 Nonlinear integral equations

We consider the nonlinear integral equation

$$x(t) = \lambda \int_a^b K(t, s, x(s))ds + f_o(t), \tag{3.4}$$

where the unknown is a continuous function $x : [a, b] \to \mathbb{R}$, and $f_o : [a, b] \to \mathbb{R}$ and $K : [a, b] \times [a, b] \times [-R, R] \to \mathbb{R}$ are known functions.

Using the Banach fixed point theorem, we can show that under certain conditions (3.4) admits a unique solution.

Theorem 3.1.6. *Suppose that $f_o : [a, b] \to \mathbb{R}$ is continuous and $K : [a, b] \times [a, b] \times [-R, R] \to \mathbb{R}$ is continuous, satisfying in its domain the (uniform) Lipschitz condition $|K(t, s, x_1) - K(t, s, x_2)| \le L\,|x_1 - x_2|$. Letting $M = \max_{t,s,x} |K(t, s, x)|$, suppose that*

$$\max_{t\in[a,b]} |f_o(t)| \le \frac{R}{2}, \quad \textit{and,} \quad |\lambda| < \min\left\{\frac{R}{2M(b-a)}, \frac{1}{L(b-a)}\right\}. \tag{3.5}$$

Then, the integral equation (3.4) *admits a unique continuous solution.*

Proof. Consider $X = C([a, b])$, with elements $\mathrm{x} = x(\cdot) : [a, b] \to \mathbb{R}$, continuous. The latter is a Banach space endowed with the norm $\|\mathrm{x}\| = \max_{t\in[a,b]} |x(t)|$. Let $A = \overline{B}_X(0, R)$, be the closed ball of $C([a, b])$ of radius R. We prove that the mapping $f : X \to X$, defined by

$$(f(\mathrm{x}))(t) = \lambda \int_a^b K(t, s, x(s))ds + f_o(t), \quad \forall\, t \in [a, b],$$

maps $\overline{B}_X(0, R)$ into $\overline{B}_X(0, R)$ for a choice of R as above. Indeed, let $\mathrm{x} \in \overline{B}_X(0, R)$ so that $|x(t)| \le R$. We have

$$\begin{aligned} \mathrm{d}(0, f(\mathrm{x})) &= \max_{t\in[a,b]} \left|\lambda \int_a^b K(t, s, x(s))ds + f_o(t)\right| \\ &\le \max_{t\in[a,b]} \left|\lambda \int_a^b K(t, s, x(s))ds\right| + \max_{t\in[a,b]} |f_o(t)| \\ &\le |\lambda|\,(b-a)\,M + \frac{R}{2} \le \frac{R(b-a)M}{2M(b-a)} + \frac{R}{2} = R, \end{aligned}$$

i. e., $f(\mathrm{x}) \in \overline{B}_X(0, R)$.

Now, from the Lipschitz condition on K, we have

$$\begin{aligned}
d(f(x_1), f(x_2)) &= \max_{t\in[a,b]} \left| \lambda \int_a^b [K(t,s,x_1(s)) - K(t,s,x_2(s))]ds \right| \\
&\leq |\lambda|\, L \int_a^b |x_1(s) - x_2(s)| ds \leq |\lambda|\, L\,(b-a) \max_{s\in[a,b]} |x_1(s) - x_2(s)| \\
&= \varrho\, d(x_1, x_2),
\end{aligned}$$

where, because of (3.5),

$$\varrho = |\lambda|\, L\,(b-a) < \frac{L(b-a)}{L(b-a)} = 1.$$

Therefore, from the Banach fixed point theorem, the nonlinear integral equation (3.4) has a unique solution in $\overline{B}_X(0,R)$. □

3.1.4 The inverse and the implicit function theorems

An important application of the Banach fixed point theorem is in the proof of the implicit function theorem, which is concerning the solvability of operator equations of a more complex form than $f(x) = y$, in the sense that we now consider the solvability with respect to y of functions of the form $f(x, y) = 0$, where $x \in X$ is considered as a parameter. Clearly, one could ask the relevant question of whether we could solve $f(x, y) = 0$ for x as a function of the parameter y. We will see that the answer depends on the differentiability properties of f with respect to either one of the variables. This is the content of the implicit function theorem of Hildenbrandt and Graves ([87]; see also [47] or [143]).

These questions are addressed in the following theorem [47, 143]:

Theorem 3.1.7 (Implicit function theorem). *Let X, Y, Z Banach spaces and $U \subset X \times Y$ be an open set. Let the function $f : U \subset X \times Y \to Z$ be continuous in U and Fréchet differentiable with respect to the variable y, with continuous partial Fréchet derivative $D_y f \in C(U, \mathcal{L}(Y, Z))$. Suppose that there exists $(x_o, y_0) \in U$ such that $f(x_o, y_0) = 0$, and that the partial Fréchet derivative $D_y f$ is invertible, with bounded inverse at (x_o, y_0), i. e., $(D_y f(x_o, y_0))^{-1} \in \mathcal{L}(Z, Y)$. Then the following apply:*

(i) *There exist $r_1, r_2 > 0$ and a unique continuous function $\psi : B_X(x_o, r_1) \to B_Y(y_0, r_2)$ such that $B_X(x_o, r_1) \times B_Y(y_0, r_2) \subset U$ with the properties*

$$\psi(x_o) = y_0 \quad \textit{and} \quad f(x, \psi(x)) = 0, \quad \forall\, x \in B_X(x_o, r_1).$$

(ii) *If furthermore $f \in C^1(U, Z)$, then $\psi \in C^1(B_X(x_o, r_1), Y)$, and*

$$D\psi(x) = -(D_y f(x, \psi(x)))^{-1} \circ D_x(x, \psi(x)), \quad \forall\, x \in B_X(x_o, r_1).$$

Proof. We will work in terms of local variables $(\hat{x}, \hat{y})$ again around the point (x_o, y_o) using the notation $A := D_y f(x_o, y_o)$, which by assumption is bounded and invertible. Note that upon defining the mapping

$$\hat{y} \mapsto g_{\hat{x}}(\hat{y}) := \hat{y} - A^{-1} f(\hat{x} + x_o, \hat{y} + y_o),$$

for suitable choice of $\hat{x}$ a fixed point, $\hat{y}$ of $g_{\hat{x}}$ will correspond to a solution (x, y) of $f(x, y) = 0$ near (x_o, y_o). In terms of these local variables, U corresponds to an open set U_0 of $(0, 0)$. For every $\hat{x}$, it holds that $D_{\hat{y}} g_{\hat{x}}(\hat{y}) = I - A^{-1} D_y f(\hat{x} + x_o, \hat{y} + y_o)$, which is continuous as a mapping from U_0 to $\mathcal{L}(Y, Z)$.

(i) We can see that for every $\hat{x}$, the mappings $\hat{y} \mapsto g_{\hat{x}}(\hat{y})$ are contractions. Indeed, we have that

$$\begin{aligned} g_{\hat{x}}(\hat{y}_1) - g_{\hat{x}}(\hat{y}_2) &= \int_0^1 D_{\hat{y}} g_{\hat{x}}(t\hat{y}_1 + (1-t)\hat{y}_2)(\hat{y}_1 - \hat{y}_2) dt \\ &= \int_0^1 (I - A^{-1} D_y f(\hat{x} + x_o, (t\hat{y}_1 + (1-t)\hat{y}_2 + y_o)(\hat{y}_1 - \hat{y}_2) dt. \end{aligned} \tag{3.6}$$

By the continuity of $(x, y) \mapsto D_y f(x, y)$ on U, we have the continuity of $(\hat{x}, \hat{y}) \mapsto D_y f(\hat{x} + x_o, \hat{y} + y_o)$ on U_0. Since $A^{-1} D_y(x_o, y_o) = I$, by continuity for a chosen $\varrho \in (0, 1)$, there exist $\delta_1, \delta_2 > 0$ such that

$$\sup_{t \in [0,1]} \| I - A^{-1} D_y f(\hat{x} + x_o, t\hat{y}_1 + (1-t)\hat{y}_2 + y_o) \|_{\mathcal{L}(Y)} \le \varrho,$$

as long as $(\hat{x}, \hat{y}) \in \overline{B}_X(0, \delta_1) \times \overline{B}_Y(0, \delta_2)$. We therefore obtain by (3.6) that

$$\| g_{\hat{x}}(\hat{y}_1) - g_{\hat{x}}(\hat{y}_2) \|_Y \le \varrho \| \hat{y}_1 - \hat{y}_2 \|_Y. \tag{3.7}$$

The mapping $\hat{y} \mapsto g_{\hat{x}}(\hat{y})$ is therefore a contraction. We need to find suitable values for $\hat{x}$ such that g_x maps $\overline{B}_Y(0, r_2) \subset \overline{B}_Y(0, \delta_2)$ to itself, so that we may apply the Banach contraction mapping theorem. We note that

$$\| g_{\hat{x}}(\hat{y}) \|_Y \le \| g_{\hat{x}}(0) \|_Y + \| g_{\hat{x}}(\hat{y}) - g_{\hat{x}}(0) \|_Y \le \| A^{-1} f(\hat{x} + x_o, y_o) \|_Y + \varrho \| \hat{y} \|_Y, \tag{3.8}$$

where we used the triangle inequality, the definition of g_x and (3.7). By the continuity of f, and since $f(x_o, y_o) = 0$, we may find a r_1 such that

$$\| A^{-1} f(\hat{x} + x_o, y_o) \|_Y \le (1 - \varrho) r_2, \quad \forall \hat{x} \in \overline{B}(0, r_1).$$

Hence, by (3.8), we have that $\| g_{\hat{x}}(\hat{y}) \|_Y \le r_2$, therefore $g_{\hat{x}}$ maps $\overline{B}(0, r_2)$ to itself. By the Banach contraction theorem, for every $\hat{x} \in \overline{B}_X(0, r_1)$, the mapping $g_{\hat{x}}$ admits a fixed point $\hat{y}$, hence by the discussion above for every $x \in \overline{B}_X(x_o, r_1)$, there exists a $y \in \overline{B}_Y(y_o, r_2)$ such that $f(x, y) = 0$. We define the mapping $x \mapsto y =: \psi(x)$, where y is the above fixed point, and clearly $f(x, \psi(x)) = 0$.

In order to prove the continuity of the function ψ, for any $x_i \in B_X(x_o, r_1)$, expressed as $x_i = \hat{x}_i + x_o$, by definition (using the construction of $\psi(x_i)$ as a fixed point, we have that $\psi(x_i) = y_0 + g_{\hat{x}_i}(\psi(x_i))$, $i = 1, 2$. Noting that $f(x_i, \psi(x_i)) = 0$, $i = 1, 2$, we express $\psi(x_1) - \psi(x_2) = g_{\hat{x}_1}(\psi(x_1)) - g_{\hat{x}_1}(\psi(x_2)) + g_{\hat{x}_1}(\psi(x_2)) - g_{\hat{x}_2}(\psi(x_2))$. Using this representation, along with the triangle inequality and the contraction estimate (3.7), we obtain

$$\|\psi(x_1) - \psi(x_2)\|_Y \leq \varrho\|\psi(x_1) - \psi(x_2)\|_Y + \|g_{\hat{x}_1}(\psi(x_2)) - g_{\hat{x}_2}(\psi(x_2))\|_Y,$$

which rearranged yields that

$$\|\psi(x_1) - \psi(x_2)\|_Y \leq \frac{1}{1-\varrho}\|g_{\hat{x}_1}(\psi(x_2)) - g_{\hat{x}_2}(\psi(x_2))\|_Y. \tag{3.9}$$

Since $g_{\hat{x}}$ has by construction continuous dependence on the parameter $\hat{x}$, taking $\hat{x}_1 \to \hat{x}_2$ (and consequently $x_1 \to x_2$) by (3.9), we obtain the continuity of ψ.

(ii) The differentiability result uses similar arguments as in the proof of Theorem 3.1.8, so it is only sketched. Since $f \in C^1(U, Z)$, together with the observations that $f(x + h, \psi(x + h)) = f(x + \psi(x)) = 0$ for x and h so that we are within the domain of definition of ψ, and

$$f(x + h, \psi(x + h)) - f(x, \psi(x)) - D_x f(x, \psi(x))h - D_y f(x, \psi(x))(\psi(x + h) - \psi(x)) = o(\|h\|)$$

(using the small-o notation for simplicity) first at x_o, we obtain that

$$\psi(x_o + h) - \psi(x_o) + (D_y f(x_o, \psi(x_o)))^{-1} D_x f(x, \psi(x))h = o(\|h\|),$$

from which the claim follows at x_o. We then proceed by a continuity argument. □

By iteration of the arguments above, if $f \in C^m(U, Z)$, $1 \leq m \leq \infty$ on a neighborhood of (x_o, y_0), then $\psi \in C^m(N(x_o))$ on a neighborhood of x_o (see [143]). The proof is constructive, leading to an iterative scheme for obtaining the function ψ,[2] in terms of the fixed point scheme.

A useful corollary of the implicit function theorem is the inverse function theorem [47, 143]:

Theorem 3.1.8 (Inverse function theorem). *Let X, Y be Banach spaces, and consider $f : X \to Y$. Suppose that there exist $(x_o, y_0) \in X \times Y$ and a neighborhood $N'(x_0) \subset X$ such that $f(x_o) = y_0$ and $f \in C^1(N'(x_o); Y)$, with invertible Fréchet derivative $Df(x_o) : X \to Y$ (considered as a bounded linear operator).*

2 or the inverse of a function in the case of the inverse function theorem (see Theorem 3.1.8).

Then, the equation $f(x) = y$ admits solutions for a neighborhood of x_o, i. e., there exists $N(x_o) \subset N'(x_o)$ and $N(y_0) \subset Y$ such that $f(x) = y$ has a unique solution $x \in N(x_o)$ for every $y \in N(y_0)$. Furthermore, $f^{-1} : N(y_0) \to N(x_o)$ is C^1 with

$$Df^{-1}(y) = [Df(x)]^{-1}, \quad \text{for } y = f(x),\ \forall\, x \in N(x_o).$$

Proof. Follows directly as a corollary of Theorem 3.1.7, applied to $\bar{f}(x, y) = x - f(y)$, to retrieve the inverse function theorem for the function $y \mapsto f(y)$. Then, simply relabel. □

3.1.5 Iterative schemes for the solution of operator equations

In many cases, we need to solve operator equations of the form $f(x) = 0$ for some nonlinear operator $f : X \to Y$, where X, Y are Banach spaces. Banach's contraction theorem can prove very useful not only for guaranteeing existence of a solution, but also in constructing iterative schemes for approximating this solution. The following example provides an illustration of such schemes:

Example 3.1.9 (A Newton like method for the solution of $f(x) = 0$). Let $f : \overline{B}_X(x_o, r_o) \subset X \to Y$ for some $x_o \in X$ and $r_o > 0$, continuously differentiable and such that $Df(x_o)^{-1} \in \mathcal{L}(Y, X)$, with $\|Df(x_1) - Df(x_2)\|_{\mathcal{L}(X,Y)} \le L\|x_1 - x_2\|_X$, for every $x_1, x_2 \in \overline{B}_X(x_o, r_o)$. Assuming that $f(x) = 0$ has a unique zero $\bar{x} \in \overline{B}_X(x_o, r_o)$, we may, under certain conditions, approximate it with the Newton-like iterative method $x_{n+1} = x_n - (Df(x_o))^{-1} f(x_n)$. It is useful for applications to find a set of conditions such that the above iterative method converges to $\bar{x}$.

Consider the mapping $g : \overline{B}_X(x_o, r) \to \overline{B}_X(x_o, r)$, defined by

$$g(x) := x - (Df(x_o))^{-1} f(x), \quad \forall x \in \overline{B}_X(x_o, r),$$

for a properly chosen $r > 0$ (to be specified shortly). If we show that g has a fixed point, $\bar{x} \in \overline{B}_X(x_o, r)$, then $\bar{x}$ is the zero of f we seek, as long as[3] $r \le r_o$. We will look for a set of conditions such that g can be a contraction map, which maps $\overline{B}_X(x_o, r)$ into itself.

To check the contraction property, pick any $x_1, x_2 \in \overline{B}_X(x_o, r)$, express

$$g(x_1) - g(x_2) = x_1 - x_2 - (Df(x_o))^{-1} f((x_1) - f(x_2)),$$

and using the differentiability of f, we can write

3 By assumption, the zero of f is unique in $\overline{B}_X(x_o, r_o)$, so since a fixed point in $\overline{B}_X(x_o, r)$ is a zero of f, it must be the zero we are seeking. If the zero is not unique, it converges to one of the zeros of f in the chosen domain.

$$f(x_1) - f(x_2) = \int_0^1 \frac{d}{dt} f(x_2 + t(x_1 - x_2))dt = \int_0^1 Df(x_2 + t(x_1 - x_2))(x_1 - x_2)dt,$$

with the integral defined in the Bochner sense. Rearrange $g(x_1) - g(x_2)$ as

$$\begin{aligned} g(x_1) - g(x_2) &= (Df(x_o))^{-1} Df(x_o)(x_1 - x_2) - (Df(x_o))^{-1} f((x_1) - f(x_2)) \\ &= (Df(x_o))^{-1} [Df(x_o)(x_1 - x_2) - (f(x_1) - f(x_2)] \\ &= (Df(x_o))^{-1} \left[Df(x_o)(x_1 - x_2) - \int_0^1 Df(x_2 + t(x_1 - x_2))(x_1 - x_2)dt \right] \\ &= (Df(x_o))^{-1} \int_0^1 [Df(x_o) - Df(x_2 + t(x_1 - x_2)](x_1 - x_2)dt, \end{aligned}$$

so that now by the Lipschitz continuity of Df, we may control the size of the difference $g(x_1) - g(x_2)$. Indeed, we have that

$$\begin{aligned} &\|g(x_1) - g(x_2)\|_Y \\ &\le \|(Df(x_o))^{-1}\|_{\mathcal{L}(Y,X)} \left\| \int_0^1 [Df(x_o) - Df(x_2 + t(x_1 - x_2))](x_1 - x_2)dt \right\|_Y \\ &\le \|(Df(x_o))^{-1}\|_{\mathcal{L}(Y,X)} \int_0^1 \|[Df(x_o) - Df(x_2 + t(x_1 - x_2))](x_1 - x_2)\|_Y dt \\ &\le \|(Df(x_o))^{-1}\|_{\mathcal{L}(Y,X)} \int_0^1 \|Df(x_o) - Df(x_2 + t(x_1 - x_2))\|_{\mathcal{L}(X,Y)} \|x_1 - x_2\|_X dt \\ &\le L\|(Df(x_o))^{-1}\|_{\mathcal{L}(Y,X)} \|x_1 - x_2\|_X \int_0^1 \|x_o - x_2 - t(x_1 - x_2)\|_X dt, \end{aligned}$$

where we used the properties of the Bochner integral and the Lipschitz continuity of Df. Since $x_1, x_2 \in \overline{B}_X(x_o, r)$, by convexity $x_2 + t(x_1 - x_2) \in \overline{B}_X(x_o, r)$, hence $\|x_o - x_2 - t(x_1 - x_2)\|_X \le r$, and setting $c = \|(Df(x_o))^{-1}\|_{\mathcal{L}(Y,X)}$ to simplify notation, we see that

$$\|g(x_1) - g(x_2)\|_Y \le crL\|x_1 - x_2\|_X,$$

which is a contraction, as long as $crL < 1$.

It remains to make sure that g maps $\overline{B}_X(x_o, r)$ to itself. Consider any $x \in \overline{B}_X(x_o, r)$. Then, expressing

$$\begin{aligned} g(x) - x_o &= x - x_o - (Df(x_o))^{-1} f(x) \\ &= (Df(x_o))^{-1}[Df(x_o)(x - x_o) - f(x) + f(x_o) - f(x_o)] \\ &= (Df(x_o))^{-1}[Df(x_o)(x - x_o) - (f(x) - f(x_o))] - (Df(x_o))^{-1} f(x_o) \end{aligned}$$

and observing that

$$f(x) - f(x_o) = \int_0^1 \frac{d}{dt} f(x + t(x - x_o))dt = \int_0^1 Df(x + t(x - x_o))(x - x_o)dt,$$

so that

$$Df(x_o)(x - x_o) - (f(x) - f(x_o)) = \int_0^1 [Df(x_o) - Df(x + t(x - x_o))](x - x_o)dt,$$

we managed once more to bring into the game the difference of the Fréchet derivative at two different points, which we can control by the Lipschitz property of Df. Hence, by the above considerations, and with similar arguments as the ones we used for the contraction property, we have that

$$\begin{aligned} \|g(x) - x_o\|_X \le L\|(Df(x_o))^{-1}\|_{\mathcal{L}(Y,X)} \left(\int_0^1 \|x_o - x - t(x - x_o)\|_X dt \right) \|x - x_o\|_X \\ + \|(Df(x_o))^{-1}\|_{\mathcal{L}(Y,X)} \|f(x_o)\|_Y \end{aligned}$$

from which it follows that

$$\|g(x) - x_o\|_X \le c(r')^2 L + cc_o,$$

where $c_o = \|f(x_o)\|_Y$. In order that g maps $\overline{B}_X(x_o, r)$ to itself, it must hold that $cr^2L + cc_o \le r$.

Summarizing, since $r_o, c, c_o, L > 0$ are given by the problem, we need to make sure that r_o, c, c_o are chosen, so as there exists an $r \le r_o$ such that

$$crL < 1, \quad \text{and} \quad cr^2L + c_o \le r.$$

This problem reduces to solving a set of inequalities, which is of minor importance here. To illustrate the possibility of convergence of the scheme, consider the possibility that $r = r_o$, and choose $a < 1$ so that $cr_oL = a$. Then the second inequality becomes $ar_o + cc_o \le r$, which holds, as long as $cc_o \le (1 - a)r_o$. ◁

3.2 The Brouwer fixed point theorem and its consequences

Banach's contraction theorem is a very useful tool but rather restrictive, since the condition that f is a contraction will not hold in many cases of interest. This introduces the need for fixed point theorems that do not require that f is a contraction but only that it is continuous. Clearly, this introduces the need of further restrictions on the domain of definition of the function, such as for instance compactness. One of the first results in this direction, valid for *finite dimensional* spaces, and upon which most if not all the infinite dimensional counterparts of such results are based, is the celebrated Brouwer fixed point theorem. We will introduce here the Brouwer fixed point theorem and illustrate it with various examples and extensions. Throughout this section, we will either focus on $X = \mathbb{R}^d$, whose elements will be denoted by $x = (x_1, \dots, x_d)$, endowed with the Euclidean metric $|\cdot|$, deriving from the standard inner product $\langle x, z\rangle = x \cdot z$, or more general finite dimensional spaces.

3.2.1 Some topological notions

Before introducing the Brouwer's fixed point theorem, we need to recall some definitions.

Definition 3.2.1 (Retractions and retracts). Let X be a topological space, $A \subset X$ and $\mathfrak{r} : X \to A$ a continuous map. The map $\mathfrak{r}$ is called a retraction of X onto A if $\mathfrak{r}(x) = x$ for all $x \in A$, or equivalently, if the identity map, $I = \mathfrak{r}|_A : A \to A$ admits a continuous extension on X. The set A is called a retract of X if a retraction of X onto A exists.

Example 3.2.2. Let $X = \mathbb{R}^d$, endowed with the Euclidean norm $|\cdot|$, and $A = \overline{B}(0,R)$. A retraction for A is given by

$$\mathfrak{r}(x) = \begin{cases} x, & \text{if } |x| \le R \\ \frac{R}{|x|}x, & \text{if } |x| > R, \end{cases}$$

hence A is a retract of X. ◁

A very important result is the following "no retraction" theorem for the unit ball in $\mathbb{R}^d$. For the proof, using simplicial topology, we refer to [70].

Theorem 3.2.3 (No retraction). *The boundary* $\partial B(0,1) := \{x \in \mathbb{R}^d : |x| = 1\} = S(0,1)$ *of a d-dimensional closed ball* $\overline{B}(0,1)$ *of* $\mathbb{R}^d$ *is not a retract of* $\overline{B}(0,1)$ *for any* $d \ge 1$.

Clearly, the result holds for the d-dimensional closed ball of any radius $R > 0$ and its corresponding boundary centered at any $x_o \in \mathbb{R}^d$.

3.2.2 Various forms of the Brouwer fixed point theorem

Using the "no retraction" theorem, we may obtain Brouwer's fixed point theorem (see e. g., [66]).

Theorem 3.2.4 (Brouwer fixed point theorem I). *Let $\overline{B}(0,1) \subset \mathbb{R}^d$ be the unit closed ball and $f : \overline{B}(0,1) \to \overline{B}(0,1)$ a continuous map. Then f has a fixed point.*

Proof. Suppose that f has no fixed point. Then $f(x) - x \neq 0$ for all $x \in \overline{B}(0,1)$. We will show that then, we may construct a retraction $\mathfrak{r} : \overline{B}(0,1) \to \partial\overline{B}(0,1) = S(0,1)$, which is in contradiction with Theorem 3.2.3.

Indeed, let $\mathfrak{r}(x)$ be the point where the line segment from $f(x)$ to x intersects $S(0,1)$, i. e., $\mathfrak{r}(x) = f(x) + \lambda(x)(x - f(x))$ where $\lambda(x)$ is the positive root of $|\mathfrak{r}(x)|^2 = 1$ or

$$\lambda(x)^2|x - f(x)|^2 + 2\lambda(x)\langle f(x), x - f(x) + |f(x)|^2 = 1. \tag{3.10}$$

It then follows that

$$\lambda(x) = \frac{-\langle f(x), x - f(x)\rangle + \sqrt{\langle f(x), x - f(x)\rangle^2 + (1 - |f(x)|^2)|x - f(x)|^2}}{|x - f(x)|^2}.$$

Since $x \mapsto \lambda(x)$ is continuous, the map $x \mapsto \mathfrak{r}(x)$ would be a retraction which is a contradiction. □

Corollary 3.2.5. *Let $\overline{B}(0,R) \subset \mathbb{R}^d$ be the closed ball of $\mathbb{R}^d$ of radius R. If $f : \overline{B}(0,R) \to \overline{B}(0,R)$ is continuous, then f has a fixed point.*

Proof. We consider the mapping $g : \overline{B}(0,1) \to B(0,1)$, defined by $g(x) = \frac{f(Rx)}{R}$. Clearly, g is continuous so by Brouwer's fixed point theorem; it has a fixed point x_o, which implies that Rx_o is a fixed point for f. □

Corollary 3.2.6. *Let $f : \mathbb{R}^d \to \mathbb{R}^d$ be continuous such that $\langle f(x), x\rangle \geq 0$ for all x such that $|x| = R > 0$. Then, there exists x_o with $|x_o| \leq R$ such that $f(x_o) = 0$.*

Proof. Suppose $f(x) \neq 0$ for all $x \in \overline{B}(0,R)$. Define $g : \overline{B}(0,R) \to \overline{B}(0,R)$ by $g(x) = -\frac{R}{|f(x)|}f(x)$. This map is continuous, therefore, by Corollary 3.2.5, there exists x_o such that $x_o = g(x_o)$. We have $|x_o| = |g(x_o)| = R$. It then follows that

$$0 < R^2 = |x_o|^2 = \langle x_o, x_o\rangle = \langle x_o, g(x_o)\rangle = -\frac{R}{|f(x_o)|}\langle x_o, f(x_o)\rangle \leq 0,$$

which is a contradiction. □

Remark 3.2.7. The above result is also true if $\langle f(x), x\rangle \leq 0$ for all $x \in \mathbb{R}^d$ such that $|x| = R > 0$.

The power of Brouwer's fixed point theorem lies in the fact that it generalizes for subsets of $\mathbb{R}^d$, which are far more general than the closed ball, and in particular to compact and convex sets. A particular special case of that is the simplex, which was in fact Brouwer's original version of the theorem.

Theorem 3.2.8 (Brouwer fixed point theorem II). *Let $C \subset \mathbb{R}^d$ be a compact and convex set and $f : C \to C$ a continuous map. Then f has a fixed point.*

Proof. Let $\mathrm{P}_C : \mathbb{R}^d \to C$ be the projection on C. Since C is compact, there exists $R > 0$ such that $C \subset \overline{B}(0,R)$. Define $g : \overline{B}(0,R) \to \overline{B}(0,R)$ by $g(x) = f(\mathrm{P}_C(x))$. Since P_C is continuous, g is continuous, therefore, by Corollary 3.2.5, it has a fixed point $x_o \in \overline{B}(0,R)$, i. e., $x_o = g(x_o) \in C$. But $\mathrm{P}_C(x_o) = x_o$, and so $x_o = f(x_o)$. □

Remark 3.2.9. The conclusion of Theorem 3.2.8 holds if $f : A \to A$ is continuous, where $A \subset \mathbb{R}^d$ is homeomorphic to a compact convex set C.

Brouwer's theorem has many applications that we will see shortly. As a first application, we provide a proof of Perron–Frobenius theorem using Brouwer's theorem (see, e. g., [66]).

Example 3.2.10 (Perron–Frobenius theorem as a consequence of Brouwer's theorem). Let $\mathbf{A} = (a_{ij})_{i,j=1}^d$ be a $d \times d$ matrix with $a_{ij} \geq 0$, $i,j = 1,2,\dots,d$. Then $\mathbf{A}$ has an eigenvalue $\lambda \geq 0$ with corresponding nonnegative[4] eigenvector $x_o = (x_{o,1},\dots,x_{o,d})$.

Let $C = \{x \in \mathbb{R}^d \,:\, x_i \geq 0,\ i = 1,\dots,d,\ \sum_{i=1}^d x_i = 1\}$, which is convex and compact. If $\mathbf{A}x = 0$ for some $x \in C$, then $\lambda = 0$ is an eigenvalue, and the theorem holds. If $\mathbf{A}x \neq 0$ for every $x \in C$, then we may define the mapping $f : C \to C$, by $f(x) = \frac{\mathbf{A}x}{\sum_{i=1}^d \sum_{j=1}^d a_{ij}x_j}$ (which is well defined by our assumption) and continuous. Hence, by Brouwer's Theorem 3.2.8, f admits a fixed point $x_o = (x_{o,1},\dots,x_{o,d}) \in C$. Since $f(x_o) = x_o$, this implies that $\mathbf{A}x_o = \lambda x_o$, with $\lambda = \sum_{i=1}^d \sum_{j=1}^d a_{ij}x_{o,j} > 0$. ◁

We now give a result, which is equivalent to Brouwer's fixed point theorem (see [143]).

Theorem 3.2.11 (Knaster–Kuratowski–Mazurkiewicz (KKM) lemma). *Let $\Sigma_d = \mathrm{conv}\{e_1,\dots, e_d\}$ be the $(d-1)$-dimensional simplex generated by the elements e_i, $i = 1,\dots,d$, and let $A_1, A_2, \cdot, A_d \subset \Sigma_d$, be closed sets with the property that, for each set $I \subset \{1,2,\dots,d\}$, the inclusion*[5]

$$\mathrm{conv}\{e_i,\ i \in I\} \subset \bigcup_{i\in I} A_i, \tag{3.11}$$

holds. Then, $A_1 \cap A_2 \cap \cdots \cap A_d \neq \emptyset$.

4 i. e., satisfying $x_{o,i} \geq 0$, $i = 1,\dots,d$.

5 Recall our convention of using $\subset$ and $\subseteq$ interchangeably.

Proof. For $i = 1, 2, \dots, d$, we set $\psi_i(x) = \operatorname{dist}(x, A_i)$, the distance between x and A_i, and

$$f_i(x) := \frac{x_i + \psi_i(x)}{1 + \psi_1(x) + \cdots + \psi_d(x)}, \quad x = (x_1, x_2, \dots, x_d) \in \Sigma_d. \tag{3.12}$$

Since the mapping $x \mapsto \psi_i(x) := \operatorname{dist}(x, A_i)$ is Lipschitz for any $i = 1, \dots, d$, it is obvious that $f = (f_1, \dots, f_d) : \Sigma_d \to \Sigma_d$ is continuous. By Brouwer's fixed point theorem (see also Remark 3.2.9), f has a fixed point $x_o \in \Sigma_d$. From (3.11), we have that $x_o \in A_1 \cup A_2 \cup \cdots \cup A_d$, which implies that $\psi_i(x_o) = 0$ for one $i \in \{1, 2, \dots, d\}$. Since $f_i(x_o) = x_{o,i}$, the denominator in (3.12) must be 1. It then follows that $\psi_1(x_o) = \cdots = \psi_d(x_o) = 0$, i. e., $x_o \in A_1 \cap \cdots \cap A_d$. □

3.2.3 Brouwer's theorem and surjectivity of coercive maps

We have already encountered the notion of a coercive functional in Section 2.2, and have seen its connection with optimization problems. We will recall this notion here (in the special context of maps $f : \mathbb{R}^d \to \mathbb{R}^d$) and investigate its connection to surjectivity properties. This theme will be encountered later on in this book, in the context of nonlinear operators between Banach spaces (see Chapter 9).

Definition 3.2.12 (Coercive map). A mapping $f : \mathbb{R}^d \to \mathbb{R}^d$ is called coercive if $\lim_{|x| \to \infty} \frac{\langle f(x), x \rangle}{|x|} = +\infty$.

Coercive maps enjoy important surjectivity properties.

Proposition 3.2.13. *Let $f : \mathbb{R}^d \to \mathbb{R}^d$ be a continuous and coercive mapping. Then f is surjective, i. e., the equation $f(x) = y$ has a solution for every $y \in \mathbb{R}^d$.*

Proof. For $y \in \mathbb{R}^d$, we define $g(x) = f(x) - y$. Since f is coercive, for every $M > 0$, there exists $R > 0$ such that if $|x| > R$, then $\langle f(x), x \rangle > M\,|x|$. Choosing $M = |y|$, we obtain for $|x| = R$,

$$\langle g(x), x \rangle = \langle f(x) - y, x \rangle \geq \langle f(x), x \rangle - |y|\,|x| > 0.$$

From Corollary 3.2.6, there exists x such that $g(x) = 0$, i. e., $f(x) = y$. □

3.2.4 Application of Brouwer's theorem in mathematical economics

Brouwer's theorem and the KKM lemma has found important applications in mathematical economics and in particular to the theory of general equilibrium and the theory of games (see, e. g., [65, 112]).

A dominant model in mathematical economics is that of general equilibrium, in which a number of goods or assets are available in a market, and their prices are to be determined by the forces of demand and supply. Assume a very simple model of this form, the so called Walras exchange economy, in which a number of agents (who are not big enough economic units to affect prices directly) meet in the market place with initial endowments of d goods. These agents will exchange goods at a particular price (yet to be determined), so that the demand for each good matches its total supply (as determined by the total initial endowment of the agents in these goods). There is no production in this simplified economy, simply exchange of already produced goods, and the basic idea behind this simplified model is that agents may be endowed with bundles of goods, which are not matching their needs, and they try to meet their needs by exchanging goods with others who may have different endowments and different needs. The rate of exchange in this simplified economy defines the prices of the goods. The crucial question is whether such a decentralized scheme may function and produce a price in which total demand matches the total supply for every good. Such a price will be called a equilibrium price. We will show how the existence of such an equilibrium price can be obtained with the use of the KKM lemma. This simplified model finds important applications in economics and in finance (where it is often referred to as the Arrow–Debreu model for asset pricing).

To make the above conceptual model more concrete, assume that there are d different goods in the market and J agents, each one endowed with a quantity of the d goods $w_j = (w_{j,1}, \dots, w_{j,d}) \in \mathbb{R}^d_+, j = 1, \dots, J$. Each agent's actual needs in these goods are $x_j = (x_{j,1}, \dots, x_{j,d}) \in \mathbb{R}^d_+$. In order to reach this goal, agent j must exchange part of his/her initial endowment w_j with some other agent j', whose initial endowment is not necessarily in accordance to his/her goals. It is not necessary that goods are exchanged at an one to one basis, i. e., one unit of good i is not necessarily exchanged for one unit of good j, a fact that allows us to introduce a price p_i for good i in terms of the exchange rate with a fixed good (called the numeraire). The quantity x_j is the demand of agent j for the goods, whereas the quantity w_j is the supply of agent j for the goods. The total demand vector for the goods is $D = \sum_{j=1}^J x_j$, whereas the total supply vector for the goods is $S = \sum_{j=1}^J w_j$. Since there is no production in the economy, it must be that $D \leq S$ (meaning that this inequality holds componentwise for the vectors). If this inequality is strict, then certain goods in this economy are left unwanted. Since the total demand and supply depend on the prices of the goods, the important question is whether there exists a price vector $p = (p_1, \dots, p_d) \in \mathbb{R}^d_+$ for the goods so that $D(p) = S(p)$. Then, we say that the market clears, and the corresponding price is called an equilibrium price. An important aspect of this theory is the assumption that both the demand as well as the supply function are determined in a self consistent fashion in terms of a constrained preferences and profit maximization problem, respectively. For example, the preference maximization problem (the solution of which determines the preferred individual demand of each agent) must be solved under the individual constraints that $\sum_{i=1}^d p_i x_{j,i} = \sum_{i=1}^d p_i \cdot w_{j,i}$, where w_j is the initial endowment, and $p = (p_1, \dots, p_d)$ is the price vector (common to all agents). Since the constraints are positively homogeneous, we may consider the problem as restricted to the simplex Σ_d, by rescaling all prices p_i to $p_i' = \frac{p_i}{\sum_{k=1}^d p_k}$. As the total demand is the sum of all individual demands, the above remark applies to the total demand as well. Moreover, under sufficient conditions on the preferences, the demand function can be shown to be continuous. Similarly for the supply function. In what follows, we will assume the price as restricted to the simplex Σ_d.

Before moving to the proof of existence of such an equilibrium price, we must introduce an important identity called Walras' law, upon which the existence is based. Define the function $g : \Sigma_d \subset \mathbb{R}^d_+ \to \mathbb{R}^d$ by $g(p) = D(p) - S(p)$. This function is called the excess demand function and yields the excess total demand for each good as a function of price (note that excess demand may be negative if a good is not sufficiently wanted). It can be shown that under fairly weak assumptions, this function is continuous. Walras' law states that

$$\langle g(p), p\rangle := g(p) \cdot p = 0, \quad \forall\, p \in \Sigma_d \subset \mathbb{R}^d_+. \tag{3.13}$$

This law comes from the very simple observation that, typically, agent j will choose his/her demand by maximizing some utility function, describing his/her preferences, but subject to the individual constraint $\langle p, x_j\rangle = \langle p, w_j\rangle$, which states that the agent will choose a new consumption bundle $x_j \in \mathbb{R}^d_+$ to satisfy his/her needs, which comes at a price $\langle p, x_j\rangle$ that must be affordable to the agent. The agent only has his/her initial endowment $w_j \in \mathbb{R}^d_+$ to trade, which is of market value $\langle p, w_j\rangle$. Behavioral assumptions on the agents (nonsatiation) imply that the agent will spend all his/her initial endowment in order to reach his/her consumption goal. Adding the individual constraints over all the agents, leads to (3.13).

Here we illustrate an answer to the following related question: If the excess demand function is such that $\sum_{i=1}^d p_i g_i(p) \leq 0$ for all $p = (p_1, \dots, p_d) \in \Sigma_d \subset \mathbb{R}^d_+$ (which clearly holds since $D \leq S$), does there exist a $p_o = (p_{o,1}, \dots, p_{o,d}) \in \Sigma_d \subset \mathbb{R}^d_+$ such that $g_i(p_o) \leq 0$ for all $i = 1, \dots, d$? One can easily see that by Walras' law (3.13), $\langle g(p_o), p_o\rangle = \sum_{i=1}^d g_i(p_o) p_{o,i} = 0$, and since $g_i(p_o) \leq 0$ for every $i = 1, \dots, d$, it holds that $g_i(p_o) = 0$ for every $i = 1, \dots, d$; hence, the market clears at this price.

The following theorem, because of its applicability in mathematical economics, is known as the existence theorem of general equilibrium, and is due in various versions to Gale, Nikaido, and Debreu:

Theorem 3.2.14. *Let $g : \Sigma_d \to \mathbb{R}^d$ be continuous with $\langle p, g(p)\rangle \leq 0$ for all $p \in \Sigma_d$. Then, there exists $p_o \in \Sigma_d$ with $g_i(p_o) \leq 0$ for all $i = 1, \dots, d$.*

Proof. The sets $A_i = \{p \in \Sigma_d \,:\, g_i(p) \leq 0\}$, $i = 1, \dots, d$ are closed and satisfy (3.11). Indeed, denoting by $\{e_i \,:\, i = 1, \dots d\}$ the standard basis of $\mathbb{R}^d$, if there exists $I \subset \{1, \dots, d\}$ and a $p \in \operatorname{conv}\{e_i,\ i \in I\}$ with $g_i(p) > 0$ for all $i \in I$, then, for this p,

$$\langle p, g(p)\rangle = \sum_{i \in I} p_i g_i(p) > 0,$$

which contradicts our assumption for g. Therefore, from the KKM lemma (see Theorem 3.2.11), there exists $p_o \in A_1 \cap \cdots \cap A_d$. □

Note that this theorem requires that g is defined on Σ_d, the unit simplex of $\mathbb{R}^d$, rather than $\mathbb{R}^d_+$, so it may not be applied directly to the excess demand function. However, one may perform a scaling argument, using $p' = \frac{1}{\sum_{i=1}^d p_i} p \in \Sigma_d$ noting that the demand function is homogeneous in p of degree 0 (as follows directly by its definition in terms of an optimization problem). Upon this transformation, we apply Theorem 3.2.14 and obtain the required result.

Apart from the theory of general equilibrium, Brouwer' fixed point theorem finds other fundamental applications in mathematical economics, such as for instance in the theory of games (see, e. g., [112]).

3.2.5 Failure of Brouwer's theorem in infinite dimensions

The following proposition shows that an extension of the Brower fixed point theorem in infinite dimensional spaces is not possible:

Proposition 3.2.15 (Kakutani). *Let H be an infinite dimensional separable Hilbert space. Then, there exists a continuous mapping $f : \overline{B}_H(0,1) \to \overline{B}_H(0,1)$ that has no fixed point.*

Proof. Let $(z_n)_{n\in\mathbb{N}_0}$ be an orthonormal basis in H. For any $x \in H$, admitting an expansion as $x = \sum_{i=0}^\infty a_i z_i$ define $p(x) = \sum_{i=1}^\infty a_{i-1} z_i$ and the map f by

$$f(x) = \frac{1}{2}(1 - \|x\|)z_0 + p(x).$$

This map is continuous and maps $\overline{B}_H(0,1)$ into itself. Indeed, for $\|x\| \le 1$, we have

$$\|f(x)\| \le \frac{1}{2}(1 - \|x\|)\|z_0\| + \|x\| \le 1,$$

where we used the observation that $\|x\|^2 = \sum_{i=0}^\infty |a_i|^2$ and $\|p(x)\|^2 = \sum_{i=1}^\infty |a_{i-1}|^2$.

Suppose that f has a fixed point $x_o \in \overline{B}_H(0,1)$, i. e., there exists $x_o = \sum_{i=0}^\infty a_i z_i \in \overline{B}_H(0,1)$ such that $f(x_o) = x_o$. Then, $\frac{1}{2}(1 - \|x_o\|)z_0 + p(x_o) = x_o$, i. e.,

$$x_o - p(x_o) = \frac{1}{2}(1 - \|x_o\|)z_0. \tag{3.14}$$

We have to examine three cases:

(i) $x_o = 0$. Then, from (3.14), we have $z_0 = 0$, which is a contradiction.

(ii) $\|x_o\| = 1$. Then, from (3.14), we have $x_o = p(x_o)$, i. e., $\sum_{i=0}^\infty a_i z_i = \sum_{i=1}^\infty a_{i-1} z_i$, which implies that $a_i = 0$ for all $i = 0, 1, \ldots$, therefore, $x_o = 0$, which, by (ii), is a contradiction.

(iii) $0 < \|x_o\| < 1$. Let $x_o = \sum_{i=1}^{\infty} a_i z_i$, where $\sum_{i=0}^{\infty} |a_i|^2 < 1$. Then, from (3.14), we have

$$\sum_{i=0}^{\infty} a_i z_i - \sum_{i=1}^{\infty} a_{i-1} z_i = \frac{1}{2}(1 - \|x_o\|) z_0,$$

which implies that $a_i = a_{i-1}$ for all $i = 1, 2, \dots$, therefore $\|x_o\|^2 = \sum_{i=0}^{\infty} |a_i|^2 = \infty$, which is a contradiction.

Therefore, all three possible cases result to a contradiction, and the claim of the proposition is proved. □

3.3 Schauder fixed point theorem and Leray–Schauder alternative

Even though, Brouwer's fixed point theorem may fail in infinite dimensional spaces, one can recover similar results by imposing certain constraints on the subsets of the infinite dimensional space, on which we require the existence of a fixed point. One of the first results in this direction was proved by Julius Schauder in the 1930s, and requires convexity and compactness of the subset $A \subset X$ with which we wish to work. Since, convexity is important, we will use the notation C for A to emphasize it.

3.3.1 Schauder fixed point theorem

We need the following lemma (see e. g. [2]):

Lemma 3.3.1. *Let X be a Banach space and $C \subset X$ a compact set. For every $\epsilon > 0$, there exists a finite dimensional subspace $X_\epsilon \subset X$ and a continuous map $g_\epsilon : C \to X_\epsilon$ such that*

$$\|g_\epsilon(x) - x\| < \epsilon, \quad \forall x \in C. \tag{3.15}$$

If, in addition, C is convex, then $g_\epsilon(x) \in C$ for all $x \in C$.

Proof. Let $\epsilon > 0$. Since C is compact, there exists a finite set $\{x_1, \dots, x_n\} \subset C$, $(n = n(\epsilon))$ such that $C \subset \bigcup_{i=1}^{n} B_X(x_i, \epsilon)$. Let $X_\epsilon = \operatorname{span}\{x_1, \dots, x_n\} \subset X$. Then $\dim X_\epsilon \le n < \infty$. Define for any $i = 1, \dots, n$,

$$\psi_i(x) = \begin{cases} \epsilon - \|x - x_i\|, & \text{if } x \in B_X(x_i, \epsilon) \\ 0, & \text{otherwise.} \end{cases}$$

We can define the map $g_\epsilon : C \to X_\epsilon$ by

$$g_\epsilon(x) = \frac{\psi_1(x)x_1 + \cdots + \psi_n(x)x_n}{\psi_1(x) + \cdots + \psi_n(x)}.$$

Note that, by compactness, for each $x \in C$, there exists an $i \in \{1, \dots, n\}$ with $\psi_i(x) > 0$. It is clear that g_ϵ is continuous since each ψ_i is continuous.

For $x \in C$, we have

$$\begin{aligned}\|g_\epsilon(x) - x\| &= \left\| \frac{\psi_1(x)x_1 + \cdots + \psi_n(x)x_n}{\psi_1(x) + \cdots + \psi_n(x)} - \frac{\psi_1(x) + \cdots + \psi_n(x)}{\psi_1(x) + \cdots + \psi_n(x)} x \right\| \\ &= \left\| \frac{\psi_1(x)(x - x_1) + \cdots + \psi_n(x)(x - x_n)}{\psi_1(x) + \cdots + \psi_n(x)} \right\| \\ &\leq \frac{\psi_1(x)\|x - x_1\| + \cdots + \psi_n(x)\|x - x_n\|}{\psi_1(x) + \cdots + \psi_n(x)} < \epsilon.\end{aligned}$$

This completes the proof. □

We are now ready to state and prove the Schauder fixed point theorem.

Theorem 3.3.2 (Schauder). *Let X be a Banach space, and let $C \subset X$ be a convex compact set. If $f : C \to C$ is continuous, then f has a fixed point.*

Proof. For any $\epsilon > 0$, let X_ϵ, g_ϵ be as in Lemma 3.3.1. Since C is convex, $g_\epsilon(C) \subset C$. Let $C_\epsilon = \mathrm{conv}\{x_1, \dots, x_n\}$. Then $C_\epsilon \subset C$. Now define $f_\epsilon : C_\epsilon \to C_\epsilon$ by

$$f_\epsilon(x) = g_\epsilon(f(x)).$$

Since C_ϵ is a compact convex set in the finite dimensional space X_ϵ, it follows from Brouwer's theorem (see also Remark 3.2.9) that f_ϵ has a fixed point $x_\epsilon \in C_\epsilon \subset C$. Thus,

$$x_\epsilon = f_\epsilon(x_\epsilon) = g_\epsilon(f(x_\epsilon)).$$

Since C is compact, we can suppose that x_ϵ has a convergent subsequence (denoted the same for simplicity), i. e., that there exists $x_o \in C$ such that $x_\epsilon \to x_o$, as $\epsilon \to 0$. Now, observe that

$$\|x_o - f(x_o)\| \leq \|x_o - x_\epsilon\| + \|x_\epsilon - f(x_\epsilon)\| + \|f(x_\epsilon) - f(x_o)\|.$$

The first term on the right-hand side tends to 0 as $\epsilon \to 0$. The same holds for the last term, by the continuity of f. In addition, we have

$$\|x_\epsilon - f(x_\epsilon)\| = \|g_\epsilon(f(x_\epsilon)) - f(x_\epsilon)\| < \epsilon$$

since $f(x_\epsilon) \in C$ by (3.15). Hence, $x_o - f(x_o) = 0$, i. e., x_o is a fixed point for f. □

In many applications, C is not compact. Instead, we may have f to be a compact map. The following corollary of the Schauder fixed point theorem is useful:

Corollary 3.3.3. *Let C be a closed, bounded and convex set in a Banach space X and $f : C \to C$ be continuous and compact. Then, f has a fixed point.*

Proof. Since $\overline{f(C)}$ is compact, so, by Proposition 1.2.18, so is $C_1 = \overline{\operatorname{conv} f(C)}$. It is clear that $C_1 \subset \overline{\operatorname{conv} C} = C$, which implies that f maps C_1 into itself. Hence, f has a fixed point. □

3.3.2 Application of Schauder fixed point theorem to the solvability of nonlinear integral equations

Schauder's fixed point theorem has numerous applications in nonlinear analysis. One of its most important applications is in the theory of nonlinear PDEs or variational inequalities (see Chapter 7). For the time being, to illustrate its use and potential, we prove some applications in terms of the solvability of nonlinear integral equations.

Example 3.3.4 (Solvability of a nonlinear integral Volterra equation). Consider the following integral equation:

$$x(t) = \int_0^t K(t,\tau) f_o(\tau, x(\tau))d\tau + g(t), \quad t \in [0,1]. \tag{3.16}$$

Here, $K : [0,1] \times [0,1] \to \mathbb{R}$, $f_o : [0,1] \times \mathbb{R} \to \mathbb{R}$ and $g : [0,1] \to \mathbb{R}$ are continuous functions. Suppose furthermore that

$$|f_o(t,x)| \le a(t) + b(t)|x|, \quad t \in [0,1],\ x \in \mathbb{R},$$

where $a, b : [0,1] \to \mathbb{R}_+$ are continuous functions. Then, equation (3.16) has a solution $x(\cdot)$ in $C([0,1])$.

We will work on the space $X = C([0,1])$, with elements $\mathrm{x} = x(\cdot) : [0,1] \to \mathbb{R}$, and consider two equivalent norms, which turn it into a Banach space, the usual supremum norm $\|\mathrm{x}\| := \sup_{t\in[0,1]} |x(t)|$, and the weighted norm $\|\mathrm{x}\|_\lambda := \sup_{t\in[0,1]}(e^{-\lambda t}|x(t)|)$, for a suitable $\lambda > 0$ to be specified shortly.[6] In order to specify λ, fix $M \in \mathbb{R}$ such that $M > \|g\|$, and choose $\lambda > 0$, such that

$$(M - \|g\|)\lambda > (1 - e^{-\lambda})(\|a\| + \|b\|M)\|K\|. \tag{3.17}$$

By the choice of M, it is always possible to find such λ. We now consider the set

$$C = \left\{\mathrm{x} \in C([0,1]) \,:\, \|\mathrm{x}\|_\lambda := \max_{t\in[0,1]} e^{-\lambda t}|\mathrm{x}(t)| \le M\right\}.$$

Since the norm $\|\cdot\|_\lambda$ is equivalent on $C([0,1])$ with the norm $\|\cdot\|$, the set C is convex, bounded, and closed. Let f be the nonlinear operator defined by the right-hand side of (3.16). Since $\|\mathrm{x}\|_\lambda \le \|\mathrm{x}\|$, we have the estimate

6 The suprema are actually maxima.

$$
\begin{aligned}
\|f(x)\|_\lambda &\le \|g\| + \max_{t\in[0,1]} e^{-\lambda t}\int_0^t |K(\tau,t)|\,|f_o(\tau,x(\tau))|d\tau \\
&\le \|g\| + \max_{t\in[0,1]} e^{-\lambda t}\int_0^t |K(\tau,t)|e^{\lambda\tau}e^{-\lambda\tau}(a(\tau)+b(\tau))|x(\tau)|d\tau \\
&\le \|g\| + \|a+b|x|\,\|_\lambda \max_{t\in[0,1]} e^{-\lambda t}\int_0^t |K(\tau,t)|e^{\lambda\tau}d\tau \\
&\le \|g\| + (\|a\| + \|b\|\,\|x\|_\lambda)\,\|K\| \max_{t\in[0,1]} e^{-\lambda t}\frac{e^{\lambda t}-1}{\lambda} \\
&= \|g\| + (\|a\| + \|b\|\,\|x\|_\lambda)\,\|K\|\frac{1-e^{-\lambda}}{\lambda}.
\end{aligned}
$$

By (3.17), f maps C into C. On the other hand, $f = \mathrm{A}\mathrm{N}_{f_o}$, where $\mathrm{A} : C([0,1]) \to C([0,1])$ is the linear operator, $\phi \mapsto \mathrm{A}\phi$ for every $\phi \in C([0,1])$, with $\mathrm{A}\phi(t) = \int_0^t K(t,\tau)\phi(\tau)d\tau$ for every $t \in [0,1]$, and $\mathrm{N}_{f_o} : C([0,1]) \to C([0,1])$ is the Nemytskii operator, determined by f_o, defined by $(\mathrm{N}_{f_o}u)(t) = f_o(t,u(t))$, for every $t \in [0,1]$ and every $u \in C([0,1])$. The operator $\mathrm{A} : C([0,1]) \to C([0,1])$ is compact (see, e. g., Example 1.3.4), whereas $\mathrm{N}_{f_o} : C([0,1]) \to C([0,1])$, is a continuous operator (by the conditions imposed on f_o). It then follows that $f : C \to C$ is compact. By Schauder's fixed point Theorem 3.3.2, f has a fixed point x, in C which is a solution of the nonlinear integral equation (3.16).

The special case where $K(\tau,t) = 1$ for every τ, t and $g(t) = x_0$ for all t, allows to obtain a solution to the initial value problem $\frac{dx}{dt}(t) = f_o(t,x(t))$ with initial condition $x(0) = x_0$ on the interval $[0,1]$. The reader can check that the approach using the Schauder fixed point theorem provides less strict conditions on the parameters than the approach using Banach's fixed point theorem. ◁

3.3.3 The Leray–Schauder principle

Sometimes it is easy to find a Banach space X, for which a given operator $f : X \to X$ is compact, but it is difficult to find an invariant ball for f. In this case, the so called Leray–Schauder principle or Leray–Schauder alternative, which allows us to obtain a fixed point by using a priori bounds for the solution of an operator equation will be useful.

We present below a version of the Leray–Schauder alternative following [2, 118]:

Theorem 3.3.5 (Leray–Schauder alternative). *Let X be a Banach space, $C \subset X$ be a closed convex set, $A \subset C$ an open bounded set, and consider a point $x_A \in A$. Suppose that $f : \overline{A} \to C$ is a continuous compact map. Then,*
(i) *either f has a fixed point in $\overline{A}$,*
(ii) *or, there exist $x \in \partial A$ and $\lambda \in (0,1)$ such that $x = \lambda f(x) + (1-\lambda)x_A$.*

Proof. Assume that (ii) does not hold, i. e., $x \neq \lambda f(x) + (1-\lambda)x_A$ for every $x \in \partial A, \lambda \in [0,1]$, and f does not have a fixed point in $\overline{A}$. Note that in the above we may include $\lambda = 0, 1$ in the condition. Define the set

$$B := \{x \in \overline{A} \,:\, \exists t \in [0,1],\ x = tf(x) + (1-t)x_A\},$$

which is non empty and closed (since f is continuous). By construction $B \cap \partial A = \emptyset$, so that by Urysohn's lemma, we may find a continuous function $\phi \in C(\overline{A}; [0,1])$, which is equal to 1 on B and vanishes on ∂A. Using this function, define $f_C : C \to C$ by

$$f_C(x) = \begin{cases} \phi(x)f(x) + (1-\phi(x))x_A, & x \in \overline{A}, \\ x_A, & x \in C \setminus \overline{A}, \end{cases}$$

which by construction is continuous. Furthermore, again by construction, $f_C(C) \subset \operatorname{conv}(\{x_A\} \cup f(\overline{A}))$, and since $f(\overline{A})$ is relatively compact (by the compactness of f) by Mazur's lemma (see Proposition 1.2.18), its closure is also compact; hence, f_C is compact. By construction, f_C maps C to itself, so that we may apply Schauder's fixed point Theorem 3.3.2 to show that $f_C : C \to C$ admits a fixed point $x_o \in C$. It is easy to see that in fact $x_o \in A$, since if not, $x_o \in C \setminus A$. Hence, by the fixed point property and the definition of f_C, it follows that $x_o = f_C(x_o) = x_A \in A$, which is a contradiction. Then, $x_o = f_C(x_o) = (1-\phi(x_o))x_A + \phi(x_o)f(x_o) \in B$ (recall the definition of the set B), and since by construction $\phi(x_o) = 1$ for $x_o \in B$. We thus conclude that $x_o = f(x_o)$, so that f admits a fixed point in A. □

A particular choice for A in the application of the Leray–Schauder alternative is $A = B_X(0,R)$ for some appropriate choice of the radius $R > 0$, and through use of *a priori* bounds preclude case (ii). A version of the Leray–Schauder alternative related to such a choice is Schaeffer's fixed point theorem. This theorem is important, as it allows us to obtain results on the solvability of $x = f(x)$ just from *a priori* bounds on solutions of the family of equations $x = \lambda f(x)$ for $\lambda \in [0,1]$ (but without actually having to prove solvability for the family!).

Theorem 3.3.6 (Schaeffer). *Let X be a Banach space and $f : X \to X$ be a continuous and compact map. If the set $S := \{x : \exists \lambda \in [0,1],\ x = \lambda f(x)\}$ is bounded, then a fixed point for f exists.*

Proof. Assume the existence of a $R > 0$ such that $\|x\| < R$ for any $x \in X$ that satisfies $x = \lambda f(x)$ for some $\lambda \in [0,1]$. Define the continuous mapping $\mathfrak{r} : X \to \overline{B}_X(0,R)$ by

$$\mathfrak{r}(x) = \begin{cases} x & \|x\| \leq R, \\ \frac{R}{\|x\|}x & \|x\| > R, \end{cases}$$

which is a retraction of X onto $\overline{B}_X(0,R)$, and define $g : \overline{B}_X(0,R) \to \overline{B}_X(0,R)$ by $g = \mathfrak{r} \circ f$. This map is continuous and compact,[7] hence, by Schauder's fixed point Theorem 3.3.2, it admits a fixed point $x_o \in \overline{B}_X(0,R)$. It remains to show that x_o is actually a fixed point of f. Note that it is not possible that $\|f(x_o\| > R$. If it did, then by the fact that x_o is a fixed point of g and its definition, $x_o = g(x_o) = \frac{R}{\|f(x_o)\|} f(x_o) = \lambda f(x_o)$, for $\lambda = \frac{R}{\|f(x)\|} f(x_o) \in (0,1)$. But then x_o is a solution for $x_o = \lambda f(x_o)$ for some $\lambda \in (0,1)$ such that $\|x_o\| = \|\frac{R}{\|f(x_o)\|} f(x_o)\| = R$, which contradicts our assumption that $\|x\| < R$ (with strict inequality). Therefore, $\|f(x_o\| \le R$, hence $x_o = g(x_o) = f(x_o)$, so that x_o is a fixed point for f. □

Example 3.3.7. Let $f : X \to X$ be a continuous compact operator such that $\sup_{x \in X} \|f(x)\| < +\infty$. Then this operator admits a fixed point.

We will use Schaeffer's fixed point Theorem 3.3.6. Let $\sup_{x \in X} \|f(x)\| = M < +\infty$. For any solution of $x = \lambda f(x)$ for $\lambda \in [0,1]$, it holds that $\|x\| = \lambda \|f(x)\| \le \lambda M$, and applying Schaeffer's fixed point theorem, the claim holds. ◁

3.3.4 Application of the Leray–Schauder alternative to nonlinear integral equations

In this section, we illustrate the use for the Leray–Schauder alternative by applying it to the solvability of nonlinear integral equations. For the sake of economy of space, we only present an application to the study of Hammerstein integral equations (see [118]). There are numerous applications of the Leray–Schauder alternative also to other types of integral equations, such as Volterra-type equations. For a detailed account, we refer the reader to the monograph of [118] and references therein.

Example 3.3.8 (Hammerstein equations). Let $\mathcal{D} \subset \mathbb{R}^d$ be bounded open set, and consider a continuous vector valued function $u : \mathcal{D} \to \mathbb{R}^m$ such that

$$u(x) = \int_{\mathcal{D}} k(x,z) f_o(u(z))dz, \quad x \in \overline{\mathcal{D}}, \tag{3.18}$$

where $k : \mathcal{D} \times \mathcal{D} \to \mathbb{R}^d$ is a given continuous function and $f_o : \mathbb{R}^m \to \mathbb{R}^m$ is a given function, which is continuous if restricted to small enough values of its variable. Equation (3.18) is a nonlinear Fredholm integral equation (a special case of the general class

7 To check that we need to show that it maps bounded subsets of $\overline{B}_X(0,R)$ to precompact sets of $\overline{B}_X(0,R)$ or equivalently that for a bounded sequence $(x_n)_{n\in\mathbb{N}} \subset \overline{B}_X(0,R)$, the sequence $(g(x_n))_{n\in\mathbb{N}} \subset \overline{B}_X(0,R)$ has a convergent subsequence. Then for the sequence $(x_n)_{n\in\mathbb{N}} \subset \overline{B}_X(0,R)$, one of the two holds, either (i) $(f(x_n))_{n\in\mathbb{N}}$ has a subsequence in $\overline{B}_X(0,R)$, or (ii) $(f(x_n))_{n\in\mathbb{N}}$ has a subsequence in $X \setminus \overline{B}_X(0,R)$. In the first case, working with this subsequence (not relabeled) $(g(x_n))_{n\in\mathbb{N}} = (f(x_n))_{n\in\mathbb{N}}$, which does have a convergent subsequence by the compactness of f. In the second case, however, for any $n \in \mathbb{N}$, we have that $g(x_n) = \frac{R}{\|f(x_n)\|} f(x_n)$, and moving along a subsequence $(x_{n_k})_{k\in\mathbb{N}}$ such that $\frac{1}{\|f(x_{n_k})\|} \to c$ and $f(x_{n_k}) \to z$ for suitable c and z, we conclude that $g(x_{n_k}) \to Rc\,z$, as $k \to \infty$. So in either case, a bounded sequence is mapped onto a sequence with a converging subsequence by g so that f is compact.

of Hammerstein integral equations) with $u \in C(\overline{\mathcal{D}}; \mathbb{R}^m)$, the unknown. Since u admits a representation as $u = (u_1, \dots, u_m)$, in terms of $u_i : \mathcal{D} \to \mathbb{R}$, we may consider (3.18) as a system of m integral equations.

To apply the Leray–Schauder alternative, we need first of all to ensure that the operator in question is continuous and compact. As such, arguments will be useful in a number of situations; we will provide a positive answer to this question for a wider class of nonlinear integral operators.

Let us define the more general nonlinear operator $f : C(\overline{\mathcal{D}}; \mathbb{R}^m) \to C(\overline{\mathcal{D}}; \mathbb{R}^m)$ by $(f(u))(x) = \int_{\mathcal{D}} K(x, z, u(z))dz$, for every $x \in \overline{\mathcal{D}}$, for $K : \mathcal{D} \times \mathcal{D} \times \overline{B}_{\mathbb{R}^m}(0, R) \to \mathbb{R}^d$ for a suitable $R > 0$. In the special case, where K is of separable form, $K(x, z, s) = k(x, z)f_o(s)$, with k and f_o as above, equation (3.18) reduces to the operator equation $\mathrm{x} = f(\mathrm{x})$, for $\mathrm{x} = u(\cdot) \in X = C(\overline{\mathcal{D}}; \mathbb{R}^m)$.

The operator f, in its more general form, is continuous and compact. Continuity follows easily by the following argument: Consider any $u_o \in C(\overline{\mathcal{D}}; \mathbb{R}^m)$, and let $\overline{B}(0, R)$ be the ball of $\mathbb{R}^m$, or radius R, centered at 0. Choose any $R > 0$ such that $\|u_o\| < R$, and fix an arbitrary $\epsilon > 0$. By continuity, we have that, when restricted to the compact set $\overline{\mathcal{D}} \times \overline{\mathcal{D}} \times \overline{B}(0, R)$, K is uniformly continuous. Hence, for the chosen $\epsilon > 0$, there exists $\delta > 0$ such that for every $u \in C(\overline{\mathcal{D}}; \mathbb{R}^m)$ with $\|u - u_o\| \le \delta$, it holds that as long as $|u(z)| \le R$, then $|K(x, z, u(z)) - K(x, z, u_o(z))| \le \epsilon$ for every $x, z \in \overline{\mathcal{D}}$. Then, one easily sees that $|f(u)(x) - f(u_o)(x)| \le \epsilon|\mathcal{D}|$ for every $x \in \overline{\mathcal{D}}$, (where $|\mathcal{D}|$ is the Lebesgue measure of $\mathcal{D}$), which implies that $\|f(u) - f(u_o)\| \le \epsilon\,|\mathcal{D}|$. Therefore, passing to the limit as $\epsilon \to 0$, we obtain the continuity of f at u_o. For compactness, we have to resort to the Ascoli–Arzela Theorem 1.9.13. Consider any bounded subset $A \subset C(\overline{\mathcal{D}}; \mathbb{R}^m)$; to show that $f(A)$ is relatively compact in $C(\overline{\mathcal{D}}; \mathbb{R}^m)$, it suffices to show that it is bounded and equicontinuous. Assume that for any $u \in A$, it holds that $\|u\| \le c$ for some $c > 0$. Boundedness is easy since for any u, such that $\|u\| \le c$, it holds that $\|f(u)\| \le \max_{\overline{\mathcal{D}}\times\overline{\mathcal{D}}\times\overline{B}(0,c)} |K(x, z, v)|\,|\mathcal{D}|$. To show the equicontinuity, we will need once more to invoke the uniform continuity of K on the compact set $\overline{\mathcal{D}}\times\overline{\mathcal{D}}\times\overline{B}(0, c)$ to note that, for any $\epsilon > 0$, there exists $\delta > 0$ such that $|K(x_1, z, u(z)) - K(x_2, z, u(z))| \le \epsilon$ for any $x_1, x_2, z \in \overline{\mathcal{D}}$ with $|x_1 - x_2| < \delta$ and $u \in A$ so that $|f(u)(x_1) - f(u)(x_2)| \le \epsilon\,|\mathcal{D}|$, leading to the equicontinuity of $f(A)$.

We now turn to the solvability of (3.18) using the Leray–Schauder alternative (Theorem 3.3.5). We make the following supplementary assumptions on the data of problem (3.18), (i) $|k(x, z)| < c_1$ for every $x.z \in \overline{\mathcal{D}}$, and that (ii) there exists a continuous nondecreasing function $\psi : [0, R] \to \mathbb{R}_+$ such that $|f_o(s)| \le \psi(|s|)$ for every $s \in \mathbb{R}^m$, where by $|\cdot|$ we denote the Euclidean norm in $\mathbb{R}^m$. Then, assuming that $c_1 \le \frac{R}{\psi(R)|\mathcal{D}|}$, we can conclude the existence of a solution to (3.18) with $\|u\| \le R$.

The proof of this claim follows by the Leray–Schauder principle in the following form: We restrict our attention to solutions of (3.18) in $C(\overline{\mathcal{D}}; \overline{B}_{\mathbb{R}^m}(0, R))$, with $R > 0$ chosen as above. We make the following claim: If we manage to show that for each $\lambda \in (0, 1)$ and any solution $u_\lambda \in C(\overline{\mathcal{D}}; \overline{B}(0, R))$, of

$$u_\lambda(x) = \lambda \int_{\mathcal{D}} k(x,z) f_o(u_\lambda(z))dz, \quad x \in \overline{\mathcal{D}}, \tag{3.19}$$

it holds that $\|u_\lambda\| < R$ (with strict inequality!), then a solution $u \in C(\overline{\mathcal{D}}; \overline{B}(0,R))$ of (3.18) exists. We emphasize the fact that we only require *a priori* bounds for the solutions of (3.19), and not an actual existence proof, and these *a priori* bounds should be in terms of the strict inequality $\|u_\lambda\| < R$. The proof of this claim is immediate by the Leray–Schauder alternative (Theorem 3.3.5), or the Schaeffer fixed point theorem (see Theorem 3.3.6 and its proof).

To conclude the proof, we check whether the condition needed for the above claim holds. Consider any solution (whether it exists or not!) of (3.19) in $C(\overline{\mathcal{D}}; \overline{B}(0,R))$. Then,

$$\begin{aligned}|u_\lambda(x)| &= \lambda \left| \int_{\mathcal{D}} k(x,z) f_o(u_\lambda(z))dz \right| \\ &\le \lambda c_1 \int_{\mathcal{D}} \psi(|u_\lambda(z)|)dz \le \lambda c_1 \psi(R)|\mathcal{D}| \le \lambda R < R,\end{aligned}$$

so that by taking the supremum over all $x \in \overline{\mathcal{D}}$, we obtain the required a priori bound $\|u_\lambda\| < R$. Therefore, by using the above claim, we can guarantee the existence of a solution of (3.18) in $C(\overline{\mathcal{D}}; \overline{B}(0,R))$. ◁

3.4 Fixed point theorems for nonexpansive maps

We have so far seen that we can get fixed points for maps, which are either contractive (Banach's fixed point theorem) or if not, continuous and compact[8] (Schauder's fixed point theorem). It would be interesting to further challenge these assumptions. For instance, what would happen if we relax compactness and contractivity simultaneously?

An answer to this question can be given for maps, which may not be strict contractions but on the other hand do not increase distances, the so called family of nonexpansive maps.

We start with the definition of nonexpansive maps.

Definition 3.4.1 (Nonexpansive map). Let X and Y be two Banach spaces and $A \subset X$. A mapping $f : A \to Y$ is called nonexpansive if

$$\|f(x_1) - f(x_2)\|_Y \le \|x_1 - x_2\|_X, \quad \forall\, x_1, x_2 \in A.$$

Example 3.4.2 (Projections are nonexpansive maps). If $C \subset H$ is a closed convex subset of a Hilbert space, the metric projection P_C is a nonexpansive map (see Theorem 2.5.3). ◁

8 Or equivalently continuous and defined on compact subsets of a Banach space.

Example 3.4.3 (Nonexpansive maps and derivatives of convex functions). Assume $\varphi : H \to \mathbb{R} \cup \{+\infty\}$ a convex function on a Hilbert space with Lipschitz continuous Fréchet derivative with Lipschitz constant L. Then the map $x \mapsto x - \tau D\varphi(x)$ is nonexpansive if $\tau \le \frac{2}{L}$.

Recall the properties of Lipschitz continuous Fréchet derivatives for convex maps (Example 2.3.26) and in particular the co-coercivity estimate (2.17)(iii),

$$\langle D\varphi(z) - D\varphi(x), z - x\rangle \ge \frac{1}{L}\|D\varphi(x) - D\varphi(z)\|^2.$$

We have that

$$\begin{aligned}
\|f(x_1) - f(x_2)\|^2 &= \|(x_1 - x_2) - \tau(D\varphi(x_1) - D\varphi(x_2))\|^2 \\
&= \|x_1 - x_2\|^2 + \tau^2\|D\varphi(x_1) - D\varphi(x_2)\|^2 - 2\tau\langle x_1 - x_2, D\varphi(x_1) - D\varphi(x_2)\rangle \\
&\le \|x_1 - x_2\|^2 + \tau\left(\tau - \frac{2}{L}\right)\|D\varphi(x_1) - D\varphi(x_2)\|^2 \le \|x_1 - x_2\|^2,
\end{aligned}$$

as long as $\tau \le \frac{2}{L}$. ◁

3.4.1 The Browder fixed point theorem

For nonexpansive maps acting on uniformly convex Banach spaces, Browder has proved an important fixed point theorem (see [36]). A related result was proved independently by Kirk, under the assumption that the Banach space is reflexive and satisfies certain normality conditions (see [95]). Importantly, these results hold for Hilbert spaces, which are known to be examples of uniformly convex Banach spaces (see Example 2.6.9).

Theorem 3.4.4 (Browder). *Let X be a uniformly convex Banach space, $C \subset X$ a nonempty bounded closed convex set, and $f : C \to C$ be a nonexpansive map. Then*

(i) *f has a fixed point.*

(ii) Fix(f), *the set of fixed points for f, is closed and convex.*

Proof. We provide here for pedagogical reasons a proof in the special case, where $X = H$ a Hilbert space (see also [36]).

(i) Let $X = H$ be a Hilbert space with inner product $\langle \cdot,\cdot\rangle_H =: \langle \cdot,\cdot\rangle$ and norm $\|\cdot\|$ such that $\|x\|^2 = \langle x, x\rangle$ for every $x \in X$.

Fix any $z_0 \in C$ and $\lambda > 1$. We define the mapping $f_\lambda : C \to C$ by

$$f_\lambda(x) := \frac{1}{\lambda}z_0 + \left(1 - \frac{1}{\lambda}\right)f(x).$$

Since

$$\|f_\lambda(x_1) - f_\lambda(x_2)\| = \left(1 - \frac{1}{\lambda}\right)\|f(x_1) - f(x_2)\| \le \left(1 - \frac{1}{\lambda}\right)\|x_1 - x_2\|,$$

the maps f_λ are contractions. Then, by Banach's contraction mapping theorem, for every $\lambda > 1$, the map f_λ has a fixed point in C, i. e., there exists $x_\lambda \in C$ such that

$$x_\lambda = \frac{1}{\lambda}z_0 + \left(1 - \frac{1}{\lambda}\right)f(x_\lambda),$$

which upon rearrangement yields

$$\|x_\lambda - f(x_\lambda)\| = \frac{1}{\lambda}\|z_0 - f(x_\lambda)\| \le \frac{2}{\lambda}\sup_{x \in C}\|x\| < c,$$

with the last estimate arising from the observation that C is bounded and $f(x_\lambda) \in C$. We select a sequence $(\lambda_n)_{n\in\mathbb{N}} \subset (1, \infty)$ such that $\lambda_n \to \infty$, and moving along this sequence, the above estimate leads us to

$$\lim_{n\to\infty}\|x_{\lambda_n} - f(x_{\lambda_n})\| = 0. \tag{3.20}$$

Consider now the sequence of fixed points $(x_{\lambda_n})_{n\in\mathbb{N}} \subset C$, which is bounded, and since by convexity C is relatively weakly compact, there exists a subsequence of elements $(x_{\lambda_{n_k}})_{k\in\mathbb{N}} \subset (x_{\lambda_n})_{n\in\mathbb{N}}$ that converges weakly to some $x_o \in C$ as $k \to \infty$. We set

$$x_t = (1-t)x_o + tf(x_o), \quad t \in (0,1). \tag{3.21}$$

Keeping in mind that $X = H$ is a Hilbert space, and its norm is expressed in terms of the inner product, and since f is nonexpansive, we deduce that

$$\begin{aligned}\langle x_t - f(x_t) - (x_{\lambda_{n_k}} - f(x_{\lambda_{n_k}})), x_t - x_{\lambda_{n_k}}\rangle &= \|x_t - x_{\lambda_{n_k}}\|^2 - \langle x_t - x_{\lambda_{n_k}}, f(x_t) - f(x_{\lambda_{n_k}})\rangle \\ &\ge \|x_t - x_{\lambda_{n_k}}\|^2 - \|f(x_t) - f(x_{\lambda_{n_k}})\|\,\|x_t - x_{\lambda_{n_k}}\| \ge 0.\end{aligned}$$

Letting $x_{\lambda_{n_k}} \rightharpoonup x_o$, (and recalling that $f(x_{\lambda_{n_k}}) \to x_{\lambda_{n_k}}$, strongly, on account of (3.20)) this inequality implies that

$$\langle x_t - f(x_t), x_t - x_o\rangle \ge 0. \tag{3.22}$$

Substituting (3.21) in (3.22) and dividing by $t > 0$,

$$\langle (1-t)x_o + tf(x_o) - f((1-t)x_o + tf(x_o)), f(x_o) - x_o\rangle \ge 0,$$

whereby letting $t \to 0$, we deduce that $\|f(x_o) - x_o\|^2 \le 0$, that is, $f(x_o) = x_o$. Hence, there exists at least a fixed point of f.

(ii) That the set of fixed point of f is closed follows from the fact that f is continuous (since it is nonexpansive). To show the convexity of the set of fixed points, we reason as follows: Let x_1, x_2 be two fixed points of f, and let $x_\lambda = (1-\lambda)x_1 + \lambda x_2$, $\lambda \in [0,1]$. We will show that x_λ is also a fixed point of f.

Indeed, we have that

$$\|f(x_\lambda) - x_1\| = \|f(x_\lambda) - f(x_1)\| \leq \|x_\lambda - x_1\| = \lambda \|x_1 - x_2\|, \tag{3.23}$$

and

$$\|f(x_\lambda) - x_2\| = \|f(x_\lambda) - f(x_2)\| \leq \|x_\lambda - x_2\| = (1-\lambda)\|x_1 - x_2\|. \tag{3.24}$$

It follows that

$$\begin{aligned}\|f(x_\lambda) - x_1\| + \|f(x_\lambda) - x_2\| &\leq \|x_1 - x_2\| = \|x_1 - f(x_\lambda) + f(x_\lambda) - x_2\| \\ &\leq \|f(x_\lambda) - x_1\| + \|f(x_\lambda) - x_2\|,\end{aligned}$$

so that

$$\|x_1 - f(x_\lambda)\| + \|f(x_\lambda) - x_2\| = \|x_1 - x_2\|.$$

Since H is a Hilbert space (hence, strictly convex), it follows[9] that

$$x_1 - f(x_\lambda) = k(f(x_\lambda) - x_2),$$

for some $k > 0$. Therefore, upon rearrangement,

$$f(x_\lambda) = (1-\mu)x_1 + \mu x_2, \quad \mu = \frac{k}{k+1}. \tag{3.25}$$

Combining (3.25) with (3.23), we conclude that $\mu \leq \lambda$, while combining (3.25) with (3.24), we conclude that $\mu \geq \lambda$, hence, $\mu = \lambda$. Therefore, (3.25) yields $f(x_\lambda) = (1-\lambda)x_1 + \lambda x_2$, from which it follows that $x_\lambda = f(x_\lambda)$, i. e., x_λ is also a fixed point of f. □

The following example shows that Theorem 3.4.4 fails in a general Banach space:

Example 3.4.5. Let $\overline{B}_{c_0}(0,1)$ be the closed unit ball of c_0, the space of all null sequences $x = (x_n)_{n\in\mathbb{N}}$, $\lim_n x_n = 0$, endowed with the norm $\|x\| = \sup_i |x_i|$. We define $f : \overline{B}_{c_0}(0,1) \to \overline{B}_{c_0}(0,1)$ by $f(x) = f(x_1, x_2, \ldots) = (1, x_1, x_2, \ldots)$. Then, $\|f(x) - f(z)\| = \|x - z\|$ for every $x, z \in c_0$. However, the equation $f(x) = x$ is satisfied only if $x = (1, 1, \ldots)$, which is not in c_0. ◁

9 See Theorem 2.6.5 (and in particular characterization (ii)) and apply it to $x_1 - f(x_\lambda)$ and $x_2 - f(x_\lambda)$.

3.4.2 The Krasnoselskii–Mann Algorithm

Given that a fixed point exists for an nonexpansive operator, how can we approximate it? The Krasnoselskii–Mann algorithm provides an iterative scheme that allows this in the special case, where $X = H$ is a Hilbert space (see, e. g., [21]). The crucial observation behind this scheme is that even though the iterative procedure for a nonexpansive operator f may not converge to a fixed point of this operator,[10] nevertheless the operator $f_\lambda(x) = (1-\lambda)x + \lambda f(x)$, for any $\lambda \in (0,1)$ when iterated, weakly converges to a fixed point.[11] In particular, we introduce the following definition:

Definition 3.4.6 (Averaged operators). A map $f_\lambda : H \to H$, which can be expressed as $f_\lambda = (1-\lambda)I + \lambda f$ for some $\lambda \in (0,1)$, where f is a nonexpansive operator is called a λ-averaged operator.[12]

Example 3.4.7 (Projections are averaged operators). The projection operator P_C on a closed and convex subset $C \subset H$, which is 1/2-averaged since $2P_C - I$, is nonexpansive. ◁

Example 3.4.8 (Derivatives of convex maps and averaged operators). If $\varphi : H \to \mathbb{R} \cup \{+\infty\}$ is a convex map with Lipschitz continuous Fréchet derivative $D\varphi$, with Lipschitz constant L, then the map $f : H \to H$, defined by $f(x) = x - \tau D\varphi(x)$, is $\frac{\tau L}{2}$-averaged, as long as $\frac{\tau L}{2} < 1$.

Indeed, express $I - \tau D\varphi = (1-\nu)I + \nu f$ so that $f = I - \frac{\tau}{\nu} D\varphi$, which is nonexpansive, as long as $\frac{\tau}{\nu} < \frac{2}{L}$, so $\nu > \frac{\tau L}{2}$, and since ν must be in $(0,1)$, we must choose $\frac{\tau}{2} < 1$ (see also Example 3.4.3). ◁

Averaged operators have many important properties, such as for example that finite compositions of averaged operators are also averaged (a proof of this useful result can be found in Section 3.7.2, Lemma 3.7.2). More details concerning averaged operators and their applications in the study of various numerical algorithms will be given in Chapter 4 (see, e. g., extensions of proximal optimization methods in Section 4.5.2).

At this point, we are going to state and prove an important result (see, e. g., [21]) concerning averaged operators, and in particular that iterations of averaged operators converge weakly to a fixed point.

Theorem 3.4.9 (Krasnoselskii–Mann). *Let $C \subset H$ be a closed convex subset of a Hilbert space and $f : C \to C$ a nonexpansive operator such that* $\mathrm{Fix}(f) \neq \emptyset$, *where* $\mathrm{Fix}(f)$ *is the set of fixed points of f. Consider a sequence $(\lambda_n)_{n\in\mathbb{N}} \subset [0,1]$ such that $\sum_{n\in\mathbb{N}} \lambda_n(1-\lambda_n) = +\infty$, and construct the sequence $(x_n)_{n\in\mathbb{N}}$ such that*

10 Even if such a fixed point exists.

11 Note that f_λ and f share the same fixed points!

12 Note that nonexpansive operators need not be averaged operators, as an example, consider $f = -I$.

$$x_{n+1} = x_n + \lambda_n(f(x_n) - x_n), \quad n \in \mathbb{N}, \tag{3.26}$$

with $x_1 \in C$, *arbitrary.*

Then, $f(x_n) - x_n \to 0$ *and* $x_n \rightharpoonup x_o$, *for some* $x_o \in \text{Fix}(f)$, *for any initial condition* x_1.

Proof. The proof follows in 4 steps.

1. We first show that the sequence, defined in (3.26), converges weakly up to subsequences to some $x \in C$.

We pick any (arbitrary) $x_o \in \text{Fix}(f)$ and consider the evolution of $x_n - x_o$. We easily see that

$$\begin{aligned} x_{n+1} - x_o &= (1-\lambda_n)(x_n - x_o) + \lambda_n(f(x_n) - x_o) \\ &= (1-\lambda_n)(x_n - x_o) + \lambda_n(f(x_n) - f(x_o)), \end{aligned} \tag{3.27}$$

where we used the fact that $f(x_o) = x_o$. We further observe that in a Hilbert space for any elements $z_1, z_2 \in H$, and $\lambda \in (0,1)$, it holds[13] that

$$\|(1-\lambda)z_1 + \lambda z_2\|^2 = (1-\lambda)\|z_1\|^2 + \lambda\|z_2\|^2 - \lambda(1-\lambda)\|z_1 - z_2\|^2,$$

which applied to $\lambda = \lambda_n$, $z_1 = x_n - x_o$, $z_2 = f(x_n) - f(x_o)$, and combined with (3.27) yields that

$$\begin{aligned} \|x_{n+1} - x_o\|^2 &= \|(1-\lambda_n)(x_n - x_o) + \lambda_n(f((x_n) - f(x_o))\|^2 \\ &= (1-\lambda_n)\|x_n - x_o\|^2 + \lambda_n\|f(x_n) - f(x_o)\|^2 - \lambda_n(1-\lambda_n)\|x_n - f(x_n)\|^2 \\ &\le \|x_n - x_o\|^2 - \lambda_n(1-\lambda_n)\|x_n - f(x_n)\|^2, \end{aligned} \tag{3.28}$$

where for the last estimate we used the fact that f is nonexpansive. This estimate shows that

$$\|x_{n+1} - x_o\|^2 \le \|x_n - x_o\|^2, \quad \text{for some } x_o \in \text{Fix}(f). \tag{3.29}$$

A sequence $(x_n)_{n\in\mathbb{N}} \subset C$ satisfying (3.29) is a called a Fejér sequence with respect to Fix(f). It can easily be seen that this sequence is bounded, and thus admits a subsequence $(x_{n_k})_{k\in\mathbb{N}} \subset C$ such that $x_{n_k} \rightharpoonup x$ for some $x \in H$. Since C is closed and convex, it is also weakly sequentially closed, hence, $x \in C$.

2. We will show that $x \in \text{Fix}(f)$. We claim (to be proved in step 3) that

$$f(x_n) - x_n \to 0. \tag{3.30}$$

If claim (3.30) holds then,

$$\|f(x) - x\|^2 = \|x_{n_k} - f(x)\|^2 - \|x_{n_k} - x\|^2 - 2\langle x_{n_k} - x, x - f(x)\rangle$$

13 By an elementary expansion of the inner products.

$$
\begin{aligned}
&= \|x_{n_k} - f(x_{n_k})\|^2 + \|f(x_{n_k}) - f(x)\|^2 + 2\langle x_{n_k} - f(x_{n_k}), f(x_{n_k}) - f(x)\rangle \\
&\quad - \|x_{n_k} - x\|^2 - 2\langle x_{n_k} - x, x - f(x)\rangle \\
&\leq \|x_{n_k} - f(x_{n_k})\|^2 + \|x_{n_k} - x\|^2 + 2\langle x_{n_k} - f(x_{n_k}), f(x_{n_k}) - f(x)\rangle \\
&\quad - \|x_{n_k} - x\|^2 - 2\langle x_{n_k} - x, x - f(x)\rangle \\
&= \|x_{n_k} - f(x_{n_k})\|^2 + 2\langle x_{n_k} - f(x_{n_k}), f(x_{n_k}) - f(x)\rangle - 2\langle x_{n_k} - x, x - f(x)\rangle \to 0,
\end{aligned}
$$

as $k \to \infty$, since (by claim (3.30)) $x_{n_k} - f(x_{n_k}) \to 0$, $x_{n_k} - x \rightharpoonup 0$ and

$$f(x_{n_k}) - f(x) = f(x_{n_k}) - x_{n_k} + x_{n_k} - f(x) \rightharpoonup 0 + x - f(x).$$

Hence, $f(x) = x$ and $x \in \text{Fix}(f)$.

3. We now show claim (3.30). To show that $f(x_n) - x_n \to 0$, we proceed as follows: Adding the estimates (3.28) over n, we conclude that

$$\sum_{k=1}^{n-1} \|x_{k+1} - x_o\|^2 \leq \sum_{k=1}^{n-1} \|x_k - x_o\|^2 - \sum_{k=1}^{n-1} \lambda_k(1-\lambda_k)\|f(x_k) - x_k\|^2,$$

which upon rearrangement leads to

$$\sum_{k=1}^{n-1} \lambda_k(1-\lambda_k)\|f(x_k) - x_k\|^2 \leq \|x_1 - x_o\|^2 - \|x_n - x_o\|^2 \leq \|x_1 - x_o\|^2.$$

Upon passing to the limit $n \to \infty$ yields $\sum_{k=1}^{\infty} \lambda_k(1-\lambda_k)\|f(x_k) - x_k\|^2 \leq \|x_1 - x_o\|^2$. On the other hand, it is easy to see that the sequence $(\|f(x_n) - x_n\|)_{n\in\mathbb{N}}$ is decreasing since (using (3.26))

$$
\begin{aligned}
\|f(x_{n+1}) - x_{n+1}\| &= \|f(x_{n+1}) - f(x_n) + (1-\lambda_n)(f(x_n) - x_n)\| \\
&\leq \|x_{n+1} - x_n\| + (1-\lambda_n)\|f(x_n) - x_n\| \\
&= \|\lambda_n(f(x_n) - x_n)\| + (1-\lambda_n)\|f(x_n) - x_n\| \\
&= \|f(x_n) - x_n\|.
\end{aligned}
$$

Combining this with the fact that $\sum_{n=1}^{\infty} \lambda_n(1-\lambda_n) = \infty$, we conclude that $\lim_n \|f(x_n) - x_n\| = 0$, hence, $f(x_n) - x_n \to 0$.

4. Furthermore, one can show that the whole sequence weakly converges to some $x \in C$. For that, it suffices to show that every weakly convergent subsequence has the same limit (recall the Urysohn property, Remark 1.1.52). Indeed, assume the existence of two subsequences, $(x_{n_k})_{k\in\mathbb{N}} \subset C$ and $(x_{m_k})_{k\in\mathbb{N}} \subset C$, such that $x_{n_k} \rightharpoonup z_1 \in C$, $x_{m_k} \rightharpoonup z_2 \in C$, with $z_1 \neq z_2$. We have already shown that it must hold that $z_1, z_2 \in \text{Fix}(f)$. By the Fejér property, it holds that $\|x_{n+1} - x_o\| \leq \|x_n - x_o\|$ for every $n \in \mathbb{N}$ and $x_o \in \text{Fix}(f)$, so $(\|x_n - x_o\|^2)_{n\in\mathbb{N}}$ converges. Expressing $\|x_n - x_o\|^2 = \|x_n\|^2 - 2\langle x_n, x_o\rangle + \|x_o\|^2$, we see that $(\|x_n\|^2 - 2\langle x_n, x_o\rangle)_{n\in\mathbb{N}}$ converges for every $x_o \in \text{Fix}(f)$. We apply this for $x_o = z_1$ and

$x_o = z_2$ so that the sequences $(\|x_n\|^2 - 2\langle x_n, z_i\rangle)_{n\in\mathbb{N}}$, $i = 1, 2$ converge, and, therefore, their difference $(\langle x_n, z_1 - z_2\rangle)_{n\in\mathbb{N}}$ converges also. Assume that $\langle x_n, z_1 - z_2\rangle \to L$ for some $L \in \mathbb{R}$, as $n \to \infty$. Passing to the subsequences, $(x_{n_k})_{k\in\mathbb{N}}$ and $(x_{m_k})_{k\in\mathbb{N}}$, and subtracting, we have that $\langle x_{n_k} - x_{m_k}, z_1 - z_2\rangle \to 0$ as $k \to \infty$, and noting that $x_{n_k} - x_{m_k} \rightharpoonup z_1 - z_2$, passing to the limit as $k \to \infty$, we conclude that $\langle z_1 - z_2, z_1 - z_2\rangle = 0$, so $z_1 = z_2$. □

Example 3.4.10 (Krasnoselskii–Mann scheme for averaged operators). The Krasnoselskii–Mann (KM) scheme shows that iterates of averaged operators converge to a fixed point of the operator (if any). This is particularly important by noting that the composition of (finitely many) averaged operators remains an averaged operator, of course with a different averaging constant. For example, if $f_i : H \to H$ are ν_i-averaged, $i = 1, 2$, then, $f_1 \circ f_2$ is ν-averaged for $\nu = \frac{\nu_1+\nu_2-2\nu_1\nu_2}{1-\nu_1\nu_2}$ (for a proof of this fact see Lemma 3.7.2 in the appendix of this chapter).

Since averaged operators are also nonexpansive, one may consider a version of the Krasnoselskii–Mann scheme for averaged operators, which allows for a modification of the convergence criterion. In particular, let f be a ν-averaged operator and consider the Krasnoselskii–Mann iteration $x_{n+1} = x_n + \rho_n(f(x_n) - x_n)$ for some $\rho_n \in (0, 1)$. Since by definition $f = (1 - \nu)I + \nu f_0$, where f_0 is a nonexpansive operator, the iteration scheme can be expressed as $x_{n+1} = x_n + \nu\rho_n(f_0(x_n) - x_n)$, which is the standard KM scheme for $\lambda_n = \nu\rho_n$. This scheme converges, as long as $\sum_{n=1}^{\infty} \rho_n(1 - \nu\rho_n) = \infty$, for $\rho_n \in (0, \frac{1}{\nu})$. ◁

Example 3.4.11 (Convergence of the gradient descent method). A nice application of the Krasnoselskii–Mann scheme is to the convergence of the explicit gradient method $x_{n+1} = x_n - \tau D\varphi(x_n)$ for the minimization of a Fréchet differentiable convex function with Lipschitz derivative.

We start by noting that the mapping $x \mapsto x - \tau D\varphi(x)$ is nonexpansive if $\tau \le \frac{2}{L}$ (see Example 3.4.3). In the interest of faster convergence of the proposed numerical scheme, we choose the higher value $\tau = \frac{2}{L}$. However, even then, the mapping $x \mapsto x - \frac{2}{L}D\varphi(x)$ is not a contraction. Hence, the convergence of its iterations cannot be guaranteed by Banach's contraction theorem. It is exactly to this point that the Krasnoselskii–Mann iteration scheme can be applied. We call this mapping f_0. Consider now any $\tau \in (0, \frac{L}{2})$, and the operator $f : H \to H$, defined by $f(x) = x - \tau D\varphi(x)$. Choosing $\lambda = \frac{\tau L}{2}$, we can express $f(x) = (1 - \lambda)x + \lambda f_0(x)$, for $\lambda \in (0, 1)$, hence, f is an averaged operator, and an application of the Krasnoselskii–Mann scheme provides the weak convergence of its iterates. The same argument covers the weak convergence of the explicit gradient method for variable step sizes τ_n. The implicit gradient method, $x_{n+1} = x_n - \tau D\varphi(x_{n+1})$, displays unconditional convergence properties, even in the absence of Lipschitz continuity of the gradient, and, importantly, even in the absence of smoothness (by replacing the gradient by the subdifferential). This theme will be discussed in detail in Section 4.5. ◁

The Krasnoselskii–Mann iterative algorithm can be extended in the case where X is a Banach space, which is uniformly convex and either has a Fréchet differentiable norm [122], or satisfies the so called Opial condition, according to which if $x_n \rightharpoonup x$, then

$\limsup_n \|x_n - x\| \leq \limsup_n \|x_n - z\|$ for every $z \in X$. The strong convergence requires stronger conditions on the operator f, even in Hilbert spaces.

3.5 A fixed point theorem for multivalued maps

A fundamental result concerning monotone single or multivalued operators (not necessarily maximal) is the Debrunner–Flor extension theorem (see e. g. [110]). This theorem finds important applications in the theory of monotone operators, and most importantly in the proof of surjectivity results (see Chapter 9). We present a proof of this result following [110].

Theorem 3.5.1 (Debrunner–Flor lemma). *Let X be a Banach space, $C \subset X$ a nonempty compact and convex set, and $f : C \to X^\star$ a continuous map. Let $M \subset C \times X^\star$ be a monotone subset, i. e., such that*

$$\langle x^\star - z^\star, x - z\rangle \geq 0, \quad \forall (x, x^\star), (z, z^\star) \in M. \tag{3.31}$$

Then, there exists $x_o \in C$ such that

$$\langle x^\star - f(x_o), x - x_o\rangle \geq 0 \quad \forall (x, x^\star) \in M. \tag{3.32}$$

Proof. Suppose that no such $x_o \in C$ exists, i. e., (3.32) has no solution. We define the set

$$U(x, x^\star) = \{z \in C : \langle x^\star - f(z), x - z\rangle < 0\}.$$

The mapping $z \mapsto \langle x^\star - f(z), x - z\rangle$ is continuous, so the set $U(x, x^\star)$ is relatively open in C. Since (3.32) has no solution, the collection of $U(x, x^\star)$, for all $(x, x^\star) \in M$, forms a cover of the set C. Since C is compact, there exists a finite subcover, i. e., there exists an $m \in \mathbb{N}$, and $(x_i, x_i^\star) \in M$, $i = 1, \ldots, m$ such that $C \subset \bigcup_{i=1}^m U(x_i, x_i^\star)$. Let $\psi_1, \ldots, \psi_m$ be a partition of unity, subordinate to this cover of C, that is, ψ_i is continuous on C with $\mathrm{supp}(\psi_i) \subset U(x_i, x_i^\star)$, and

$$\sum_{i=1}^m \psi_i(z) = 1, \quad 0 \leq \psi_i(z) \leq 1, \ \forall z \in C, \ i = 1, 2, \ldots, m. \tag{3.33}$$

Let $C_1 = \mathrm{conv}(\{x_1, x_2, \ldots, x_m\})$, the convex hull of $\{x_1, x_2, \ldots, x_m\}$. For $z \in C_1$, we consider the maps defined by $z \mapsto p(z) := \sum_{i=1}^m \psi_i(z) x_i$ and $z \mapsto q(z) := \sum_{i=1}^m \psi_i(z) x_i^\star$. The map p is continuous, and by (3.33), maps the set C_1 into itself. Note that C_1 is a finite dimensional compact convex set, which is homeomorphic to a finite dimensional ball. Thus, by Brouwer's fixed point theorem, p has a fixed point, i. e., there exists $z_0 \in C_1$ such that $p(z_0) = z_0$.

Letting $a_{ij} = \langle x_i^\star - f(z_0), x_j - z_0\rangle$, we easily see, using also (3.31), that

$$a_{ij} + a_{ji} = a_{ii} + a_{jj} - \langle x_i^\star - x_j^\star, x_i - x_j\rangle \le a_{ii} + a_{jj}. \tag{3.34}$$

Now, since $p(z_0) = z_0$, we have[14]

$$\begin{aligned} 0 = \langle q(z_0) - f(z_0), p(z_0) - z_0\rangle &= \left\langle \sum_{i=1}^m \psi_i(z_0)x_i^\star - f(z_0), \sum_{j=1}^m \psi_j(z_0)x_j - z_0 \right\rangle \\ &= \sum_{i,j=1}^m \psi_i(z_0)\psi_j(z_0)a_{ij} = \sum_{i,j=1}^m \psi_i(z_0)\psi_j(z_0)\frac{a_{ij} + a_{ji}}{2}. \end{aligned} \tag{3.35}$$

From (3.34), it follows that

$$0 \le \sum_{i,j=1}^m \psi_i(z_0)\psi_j(z_0)\frac{a_{ii} + a_{jj}}{2}. \tag{3.36}$$

If $\psi_i(z_0)\psi_j(z_0) \neq 0$, then $z_0 \in U(x_i, x_i^\star) \cap U(x_j, x_j^\star)$. By the construction of $U(x, x^\star)$, this means that $a_{ii} < 0$ and $a_{jj} < 0$, which contradicts (3.36).

Therefore, from (3.35), we conclude that $\psi_i(z_0) = 0$ for all $1, 2, \dots, m$. However, since $z_0 \in C_1 \subset C$, (3.33) is contradicted. This concludes the proof. □

Theorem 3.5.2 (Kakutani). *Let X be a reflexive Banach space and $C \subset X$ nonempty compact and convex. Let $f : C \to 2^C$ be an upper semicontinuous mapping (in the sense of Definition 1.8.2) such that $f(x)$ is a nonempty, closed, and convex set for all $x \in C$. Then, there exists $x_o \in C$ such that $x_o \in f(x_o)$.*

Proof. Suppose that $0 \notin x - f(x)$ for all $x \in C$. From the Hahn–Banach separation theorem, there exists $x^\star \in X^\star$ with $\|x^\star\|_{X^\star} = 1$, and $\delta > 0$ such that $\langle x^\star, z\rangle > \delta$ for all $z \in x - f(x)$. For any $x^\star \in X^\star$, we define

$$W(x^\star) = \{x \in C : \langle x^\star, z\rangle > 0, \ \forall z \in x - f(x)\}.$$

For each $x^\star \in X^\star$ such that $\|x^\star\|_{X^\star} = 1$, let

$$U(x^\star) = \{z \in X : \langle x^\star, z\rangle > 0\}.$$

So, $x \in W(x^\star)$ if and only if $x \in C$ and $x - f(x) \subset U(x^\star)$.

We rephrase the result of the Hahn–Banach separation theorem obtained above, using this notation: For any $x \in C$, there exists[15] $x^\star(x) \in X$, with $\|x^\star(x)\|_{X^\star} = 1$ and $\delta > 0$

14 Note that $\sum_{i=1}^m \psi_i(z_0)x_i^\star - f(z_0) = \sum_{i=1}^m \psi_i(x_i^\star - f(z_0))$ by the properties of ψ_i, $i = 1, \dots, m$, and similarly for the other sum.

15 Note that by $x^\star(x)$ here we denote the dependence of $x^\star$ on x and *not* the action of the functional $x^\star$ on x; $x^\star(x) \in X^\star$!

such that $x - f(x) + \delta B(0,1) \subset U(x^\star(x))$. Since, f is upper semicontinuous, so is $-f$ (recall we use the concept as in the sense of Definition 1.8.2). Therefore, for $0 < \epsilon < \frac{\delta}{2}$, we have $-f(z) \subset -f(x) + \frac{\delta}{2}B(0,1)$ for each $z \in (x + \epsilon B(0,1)) \cap C$. For each such z, we have

$$z - f(z) \subset x + \frac{\delta}{2}B(0,1) - f(x) + \frac{\delta}{2}B(0,1) \subset U(x^\star(x)),$$

that is, $(x + \epsilon B(0,1)) \cap C \subset W(x^\star(x))$. So every point $x \in C$ lies in the interior of some $W(x^\star(x))$. It then follows that the sets $\{\text{int}(W(x^\star(x))) \,:\, x \in C\}$ form an open cover of C. By the compactness of C, there exists a finite sub-cover $\{\text{int}(W(x_i^\star)) \,:\, i = 1, \dots, n\}$, and a partition of unity $\{\psi_i \,:\, i = 1, \dots, n\}$ subordinate to this cover. Define,

$$g(x) = \sum_{i=1}^{n} \psi_i(x) x_i^\star, \quad x \in C.$$

This function is continuous from C into $X^\star$ and

$$\langle g(x), x' \rangle = \sum_{i=1}^{n} \psi_i(x) \langle x_i^\star, x' \rangle > 0, \quad \forall x \in C, \text{ and } x' \in x - f(x). \tag{3.37}$$

Applying the Debrunner–Flor lemma (see Theorem 3.5.1) for the monotone set $M = C \times \{0\}$ and for the function $-g(x)$, there exists $x_o \in C$ such that

$$\langle g(x_o), x - x_o \rangle \geq 0, \quad \forall x \in C.$$

In particular, this is true for $x = x_o'$, where x_o' is any element of $f(x_o) \subset C$, so the element $z_o = x_o - x_o' \in x_o - f(x_o)$ satisfies the inequality $\langle g(x_o), z_o \rangle \leq 0$, which contradicts (3.37). This completes the proof. □

3.6 The Ekeland variational principle and Caristi's fixed point theorem

Caristi's fixed point theorem [39] is one of the most general fixed point theorems available, in the sense that it requires practically no properties on the mapping apart from a very general contraction like condition, called the Caristi condition. As such, it is very useful in a number of applications. Furthermore, it is equivalent to an extremely useful result (produced independently) called the Ekeland variational principle [72, 73], which we will need in a number of occasions. These important results have been proved in various ways; our approach is inspired by [47] and [107].

3.6.1 The Ekeland variational principle

The Ekeland variational principle is an important tool in nonlinear analysis, which provides approximations to minima in the absence of the conditions required for applying the Weierstrass theorem. While derived independently of Caristi's theorem, it is in fact equivalent to it (see, e. g., [79]). Here we decide to prove first the Ekeland variational principle (following the approach of [47]), and then use the Ekeland variational principle to show the Caristi fixed point theorem.

Theorem 3.6.1 (Ekeland variational principle). *Let X be a complete metric space and $F : X \to \mathbb{R}$ be a lower semicontinuous functional, which is bounded below. For every $\epsilon, \delta > 0$, if $x_\delta \in X$ satisfies $F(x_\delta) \le \inf_{x\in X} F(x) + \delta$, then there exists $x_{\epsilon,\delta} \in X$ such that*

$$\begin{aligned} F(x_{\epsilon,\delta}) &\le F(x_\delta), \\ d(x_\delta, x_{\epsilon,\delta}) &\le \frac{1}{\epsilon}, \\ F(x_{\epsilon,\delta}) &< F(z) + \epsilon\delta d(z, x_{\epsilon,\delta}), \quad \forall\, z \in X,\ z \neq x_{\epsilon,\delta}. \end{aligned} \tag{3.38}$$

Proof. The proof consists in constructing a sequence of approximations that converges to the required point $x_{\epsilon,\delta}$. We proceed in 2 steps.

1. Starting from $x_0 = x_\delta$, we construct the following sequence of sets $S_n \subset X$ and elements $x_{n+1} \in S_n$:

$$\begin{aligned} S_n &= \{z \in X : F(z) \le F(x_n) - \epsilon\delta d(z, x_n)\}, \\ x_{n+1} \in S_n &: F(x_{n+1}) - \inf_{S_n} F \le \epsilon_n, \quad \epsilon_n = \frac{1}{2}\Big(F(x_n) - \inf_{S_n} F\Big). \end{aligned} \tag{3.39}$$

The above sequence is well defined. Clearly $S_n \neq \emptyset$ since $x_n \in S_n$ for every $n \in \mathbb{N}$. Moreover, the existence of x_{n+1} arises from the very definition of the infimum, such a point exists for every choice of $\epsilon_n \ge 0$; the particular choice is made to guarantee the desirable property for $(x_n)_{n\in\mathbb{N}}$, that is, a Cauchy sequence. To see that the proposed $\epsilon_n \ge 0$, simply note that by the definition of S_n for every $z \in S_n$, it holds that $F(z) \le F(x_n) - \epsilon\delta d(z, x_n) \le F(z_n)$, hence $\inf_{S_n} F \le F(x_n)$. If it happens that $\inf_{S_n} F = F(x_n)$, then we simply set $x_{n+1} = x_n$, and the iteration stops.

The following observations hold for $(x_n)_{n\in\mathbb{N}}$:

(a) The sequence $(F(x_n))_{n\in\mathbb{N}}$ is decreasing. Indeed, by construction, since $x_{n+1} \in S_n$, we have by the definition of S_n that $0 \le \epsilon\delta d(x_{n+1}, x_n) \le F(x_n) - F(x_{n+1})$, hence the monotonicity claim. Since by assumption $(F(x_n))_{n\in\mathbb{N}}$, it is bounded below, it is convergent. Therefore, $F(x_n) \to \gamma$ for a suitable γ.

(b) By the definition of S_n, once more, we have that

$$\epsilon\delta d(x_{n+1}, x_n) \le F(x_n) - F(x_{n+1}), \quad \forall\, n \in \mathbb{N}.$$

Adding over all $n, n+1, \dots, m$, and using the triangle inequality, we obtain that

$$\epsilon\delta \mathsf{d}(\mathsf{x}_m, \mathsf{x}_n) \le F(\mathsf{x}_n) - F(\mathsf{x}_m), \quad \forall\, m > n, \tag{3.40}$$

and passing to the limit, as $n, m \to \infty$, since $F(\mathsf{x}_n) \to \gamma$, and $F(\mathsf{x}_m) \to \gamma$, we conclude that $\mathsf{d}(\mathsf{x}_n, \mathsf{x}_m) \to 0$. Hence, $(\mathsf{x}_n)_{n\in\mathbb{N}}$ is Cauchy. By the completeness of X, there exists $\mathsf{z}_0 \in X$ such that $\mathsf{x}_n \to \mathsf{z}_0$.

We claim that $\mathsf{z}_0 \in X$ is the required point.

2. Note that since F is lower semicontinuous, it holds that $F(\mathsf{z}_0) \le \liminf_n F(\mathsf{x}_n)$. We now take the limit inferior of the second relation in (3.39), as $n \to \infty$, and setting $\beta_n := \inf_{S_n} F$, we conclude that

$$F(\mathsf{z}_0) \le \gamma = \liminf_n F(\mathsf{x}_n) \le \liminf_n \beta_n. \tag{3.41}$$

We check, one by one, that z_0 satisfies the stated properties in (3.38).

(a) Since $(F(\mathsf{x}_n))_{n\in\mathbb{N}}$ is decreasing and $\mathsf{x}_0 = \mathsf{x}_\delta$, clearly $F(\mathsf{x}_n) \le F(\mathsf{x}_\delta)$, and taking the limit as $n \to \infty$, using the lower semicontinuity of F yields $F(\mathsf{z}_0) \le F(\mathsf{x}_\delta)$, which is (3.38)(a).

(b) By (3.40), setting $n = 0$, bringing the $-F(\mathsf{x}_m)$ term onto the left-hand side, taking the limit inferior as $m \to \infty$, using the lower semicontinuity of F and rearranging, we have that

$$\epsilon\delta \mathsf{d}(\mathsf{z}_0, \mathsf{x}_\delta) \le F(\mathsf{x}_\delta) - F(\mathsf{z}_0) \le F(\mathsf{x}_\delta) - \inf_X F \le \delta,$$

where the second inequality follows by the trivial estimate $F(\mathsf{z}_0) \ge \inf_X F$ and the third by the choice of x_δ. This proves (3.38)(b).

(c) Suppose that z_0 does not satisfy the third condition in (3.38). Then, there exists $\hat{\mathsf{z}} \in X, \hat{\mathsf{z}} \neq \mathsf{z}_0$, such that

$$F(\mathsf{z}_0) \ge F(\hat{\mathsf{z}}) + \epsilon\delta \mathsf{d}(\hat{\mathsf{z}}, \mathsf{z}_0). \tag{3.42}$$

On the other hand, by (3.40) fixing n arbitrary, rearranging and taking the limit inferior as $m \to \infty$, using also the lower semicontinuity of F, yields

$$\epsilon\delta \mathsf{d}(\mathsf{z}_0, \mathsf{x}_n) \le F(\mathsf{x}_n) - F(\mathsf{z}_0). \tag{3.43}$$

Combining (3.42) with (3.43), we obtain that

$$F(\hat{\mathsf{z}}) + \epsilon\delta \mathsf{d}(\hat{\mathsf{z}}, \mathsf{z}_0) \le F(\mathsf{z}_0) \le F(\mathsf{x}_n) - \epsilon\delta \mathsf{d}(\mathsf{z}_0, \mathsf{x}_n)$$

so that

$$F(\hat{\mathsf{z}}) \le F(\mathsf{x}_n) - \epsilon\delta(\mathsf{d}(\mathsf{z}_0, \mathsf{x}_n) + \mathsf{d}(\hat{\mathsf{z}}, \mathsf{z}_0)) \le F(\mathsf{x}_n) - \epsilon\delta \mathsf{d}(\hat{\mathsf{z}}, \mathsf{x}_n), \tag{3.44}$$

where for the last estimate we used the triangle inequality. Since $n \in \mathbb{N}$ is arbitrary (3.44) implies that $\hat{z} \in S_n$ for every $n \in \mathbb{N}$, therefore $\hat{z} \in \bigcap_{n\in\mathbb{N}} S_n$. Since for any $n \in \mathbb{N}$, $\hat{z} \in S_n$, clearly $F(\hat{z}) \geq \inf_{S_n} F = \beta_n$, and taking the limit inferior over n, we have that $F(\hat{z}) \geq \liminf_n \beta_n$, which combined with (3.41) yields $F(z_0) \leq F(\hat{z})$, which is in contradiction with (3.42) (since we have that $\hat{z} \neq z_0$). Therefore, z_0 satisfies also the third condition in (3.38).

Combining (a), (b), and (c) above, we conclude that z_0 has the required properties and the proof is complete. □

A very common choice when applying the Ekeland variational principle is $\epsilon = \frac{1}{\sqrt{\delta}}$.

3.6.2 Caristi's fixed point theorem

Now, in the spirit of [107], we use the Ekeland variational principle to prove a very general fixed point theorem, the Caristi fixed point theorem.

Theorem 3.6.2 (Caristi). *Let (X, d) be a complete metric space, $\phi : X \to \mathbb{R} \cup \{+\infty\}$ a proper lower semicontinuous function bounded from below, and $f : X \to 2^X \setminus \{\emptyset\}$ a multivalued map that satisfies the Caristi property,*[16]

$$\mathrm{d}(x, z) \leq \phi(x) - \phi(z), \quad \forall x \in X,\ \forall z \in f(x). \tag{3.45}$$

Then, f has a fixed point in X, i. e., there exists $x_o \in X$ such that $x_o \in f(x_o)$.

Proof. We will use the Ekeland variational principle on ϕ, with $\epsilon = \delta = 1$. This provides the existence of an $x_o \in X$ such that

$$\phi(x_o) < \phi(x) + \mathrm{d}(x, x_o), \quad \forall x \neq x_o. \tag{3.46}$$

This is the required fixed point of f. Indeed, suppose that $x_o \notin f(x_o)$. Consider any $z \in f(x_o)$ (which clearly satisfies $z \neq x_o$). Since f satisfies the Caristi property, applying it for $z \in f(x_o)$ and $x = x_o \in X$, we have that

$$\phi(z) \leq \phi(x_o) - \mathrm{d}(x_o, z). \tag{3.47}$$

Combining (3.46) (applied for $x = z \neq x_o$) and (3.47), we deduce that

$$\phi(z) \leq \phi(x_o) - \mathrm{d}(x_o, z) < \phi(z),$$

which is a contradiction. □

16 In the single valued case, a contraction map f, with contraction constant $\varrho < 1$ satisfies the Caristi property upon choosing $\phi(x) = \frac{1}{1-\varrho}\mathrm{d}(x, f(x))$. In this respect, the Caristi fixed point theorem may be considered as a generalization to the Banach fixed point theorem.

3.6.3 Applications: approximation of critical points

One of the important applications of the Ekeland variational principle is in obtaining approximate critical points for F (see e. g., [79, 91]). This can be effected by rearranging the third relation in (3.38), under the extra assumptions that X is a Banach space and F is differentiable. We must introduce the notion of Palais–Smale sequence.

Definition 3.6.3 (Palais–Smale sequences and Palais–Smale condition). Let $F : X \to \mathbb{R}$ be a $C^1(X;\mathbb{R})$ functional.[17]

(i) A sequence $(x_n)_{n\in\mathbb{N}}$ is a Palais–Smale sequence for the functional if $(F(x_n))_{n\in\mathbb{N}}$ is bounded and $DF(x_n) \to 0$ (strongly).
(ii) The functional F satisfies the Palais–Smale condition if any Palais–Smale sequence has a (strongly) convergent subsequence.

The next proposition (see e. g., [79, 91]) shows the construction of approximate critical points using the Ekeland variational principle.

Proposition 3.6.4. *Assume that X is a Banach space and $F \in C^1(X;\mathbb{R})$, bounded below, that satisfies the Palais–Smale condition. Then, F has a minimizer $x_o \in X$ such that $DF(x_o) = 0$.*

Proof. We apply the Ekeland variational principle (EVI) for the choice $\epsilon = n$ and $\delta = \frac{1}{n^2}$, $n \in \mathbb{N}$, and denote the corresponding sequences $x_\delta = z_n$ and $x_{\epsilon,\delta} = x_n$. The sequence $(z_n)_{n\in\mathbb{N}}$ is by choice a minimizing sequence, and since by the first two conditions of the EVI, $F(x_n) \le F(z_n)$, and $\|z_n - x_n\| \le \frac{1}{n}$, so is $(x_n)_{n\in\mathbb{N}}$ (hence, $(F(x_n))_{n\in\mathbb{N}} \subset \mathbb{R}$ is bounded), and if any of the two converge, they should have the same limit. By the third condition in (3.38), choosing $z = x_n + th$, for $t > 0$ and $h \in X$ arbitrary, since X is a Banach space and $F \in C^1(X;\mathbb{R})$, passing to the limit as $t \to 0^+$, we conclude that $\langle DF(x_n), h\rangle < \frac{1}{n}\|h\|$. As $h \in X$ is arbitrary, setting $-h \in X$ in the place of h next, we also have $-\langle DF(x_n), h\rangle < \frac{1}{n}\|h\|$, which allows us to conclude that $|\langle DF(x_n), h\rangle| < \frac{1}{n}\|h\|$ for every $h \in X$. So upon dividing by $\|h\|$ and taking the supremum over all $\|h\| \le 1$, we obtain that $\|DF(x_n)\|_{X^\star} < \frac{1}{n}$, from which we find that $DF(x_n) \to 0$ in $X^\star$. We have thus constructed using EVI a minimizing sequence $(x_n)_{n\in\mathbb{N}}$, i. e., $F(x_n) \to \gamma = \inf_{x\in X} F(x)$ with the extra property $DF(x_n) \to 0$. At this point, there is no guarantee that the minimizing sequence admits a convergent subsequence in order to proceed with the direct method. However, $(x_n)_{n\in\mathbb{N}}$ is a Palais–Smale sequence, and since by assumption it satisfies the Palais–Smale condition, we conclude that there exists a subsequence $(x_{n_k})_{k\in\mathbb{N}}$ and a $x_o \in X$ such that $x_{n_k} \to x_o$ in X. Clearly, x_o is a minimizer for F, while $DF(x_{n_k}) \to DF(x_o) = 0$, so x_o is a critical point of F. □

17 i. e., Fréchet differentiable with continuous derivative.

3.7 Appendix

3.7.1 The Gronwall inequality

The Gronwall inequality (see e. g., [128]) is a very useful tool in the study of differential equations. In a fairly general form it can be stated as follows:

Proposition 3.7.1. *Let g be a function such that*

$$g(t) \leq c + \int_0^t (a(\tau)g(\tau) + b(\tau))d\tau, \quad t \in [0,T],$$

for some positive valued integrable functions a, b. Then,

$$g(t) \leq \left(c + \int_0^t b(\tau)e^{-\int_0^\tau a(s)ds} d\tau \right) e^{\int_0^t a(\tau)d\tau}.$$

3.7.2 Composition of averaged operators

We show that the class of ν-averaged operators is stable under compositions (see [55]).

Lemma 3.7.2. *Let H be a Hilbert space. Suppose that $f_i : H \to H$ are ν_i-averaged, $i = 1, 2$. Then, $f_1 \circ f_2$ is ν-averaged for $\nu = \frac{\nu_1+\nu_2-2\nu_1\nu_2}{1-\nu_1\nu_2}$.*

Proof. An operator f is ν-averaged if and only if

$$\|f(x_1) - f(x_2)\|^2 \leq \|x_1 - x_2\|^2 - \frac{1-\nu}{\nu}\|(I-f)(x_1) - (I-f)(x_2)\|^2, \quad \forall x_1, x_2 \in H. \tag{3.48}$$

To see that, observe that f is ν-averaged if and only if $(1-\frac{1}{\nu})I + \frac{1}{\nu}f$ is nonexpansive. To obtain the direction of the equivalence we need, it suffices to show that if (3.48) holds, then $(1-\frac{1}{\nu})I + \frac{1}{\nu}f$ is nonexpansive, i. e.,

$$\left\| \left(1 - \frac{1}{\nu}\right)(x_1 - x_2) + \frac{1}{\nu}(f(x_1) - f(x_2)) \right\|^2 \leq \|x_1 - x_2\|^2,$$

holds for any $x_1, x_2 \in H$. This follows easily by straightforward expansion of the two conditions using the fact that the norm is generated by the inner product.

We now try to check that (3.48) holds for $f = f_1 \circ f_2$ and $\nu = \frac{\nu_1+\nu_2-2\nu_1\nu_2}{1-\nu_1\nu_2}$. We apply (3.48) twice: once for f_1 and once for f_2 to obtain

$$
\begin{aligned}
&\|f_1 \circ f_2(x_1) - f_1 \circ f_2(x_2)\|^2 \\
&\overset{((3.48)\text{ for } f_1)}{\leq} \|f_2(x_1) - f_2(x_2)\|^2 - \frac{1-\nu_1}{\nu_1}\|(I - f_1)(f_2(x_1)) - (I - f_1)(f_2(x_2))\|^2 \\
&\overset{((3.48)\text{ for } f_2)}{\leq} \|x_1 - x_2\|^2 - \frac{1-\nu_2}{\nu_2}\|(I - f_2)(x_1) - (I - f_2)(x_2)\|^2 \\
&\qquad - \frac{1-\nu_1}{\nu_1}\|(I - f_1)(f_2(x_1)) - (I - f_1)(f_2(x_2))\|^2.
\end{aligned} \tag{3.49}
$$

Moreover, observe that for any $\lambda \in \mathbb{R}$, $z_1, z_2 \in H$,

$$
\|\lambda z_1 + (1-\lambda) z_2\|^2 + \lambda(1-\lambda)\|z_1 - z_2\|^2 = \lambda\|z_1\|^2 + (1-\lambda)\|z_2\|^2. \tag{3.50}
$$

We would like to apply that to the right-hand side of the above inequality for $z_1 = (I - f_1)(f_2(x_1)) - (I - f_1)(f_2(x_2))$ and $z_2 = (I - f_2)(x_1) - (I - f_2)(x_2)$, but since $\frac{1-\nu_1}{\nu_1} + \frac{1-\nu_2}{\nu_2} \neq 1$, we multiply both sides by $\frac{1}{\tau}$, where $\tau = \frac{1-\nu_1}{\nu_1} + \frac{1-\nu_2}{\nu_2}$, and apply (3.50) for $\lambda = \frac{1-\nu_1}{\tau\nu_1}$. This yields,

$$
\begin{aligned}
&\frac{1}{\tau}\Big(\frac{1-\nu_1}{\nu_1}\|(I - f_1)(f_2(x_1)) - (I - f_1)(f_2(x_2))\|^2 \\
&\qquad + \frac{1-\nu_2}{\nu_2}\|(I - f_2)(x_1) - (I - f_2)(x_2)\|^2\Big) \\
&= \|\lambda z_1 + (1-\lambda) z_2\|^2 + \lambda(1-\lambda)\|z_1 - z_2\|^2 \\
&\geq \lambda(1-\lambda)\|z_1 - z_2\|^2 \\
&= \frac{1}{\tau}\frac{(1-\nu_1)(1-\nu_2)}{\tau\nu_1\nu_2}\|(I - f_1 \circ f_2)(x_1) - (I - f_1 \circ f_2)(x_2)\|^2,
\end{aligned}
$$

which, combined with (3.49), leads to

$$
\begin{aligned}
&\|f_1 \circ f_2(x_1) - f_1 \circ f_2(x_2)\|^2 \\
&\leq \|x_1 - x_2\|^2 - \frac{(1-\nu_1)(1-\nu_2)}{\tau\nu_1\nu_2}\|(I - f_1 \circ f_2)(x_1) - (I - f_1 \circ f_2)(x_2)\|^2,
\end{aligned} \tag{3.51}
$$

and noting further that $\frac{(1-\nu_1)(1-\nu_2)}{\tau\nu_1\nu_2} = \frac{1-\nu}{\nu}$ for $\nu = \frac{\nu_1+\nu_2-2\nu_1\nu_2}{1-\nu_1\nu_2}$. Based on the latter, we have the stated result. □

4 Nonsmooth analysis: the subdifferential

Many interesting convex functions are not smooth. This chapter focuses on the study of such functions, and in particular on the concept of the subdifferential, which is a multivalued mapping generalizing the derivative. We will develop some of its important properties, as well as its calculus rules, and in particular those related to nonsmooth optimization problems. We will close the chapter with the treatment of the Moreau proximity operator in Hilbert spaces and the study of a popular class of optimization algorithms based on the concept of the subdifferential and the proximity operator, called proximal methods. These important concepts have been covered by many authors (see, e. g., [12, 13, 20, 21, 28], or [113] on which our approach was based).

4.1 The subdifferential: definition and examples

The subdifferential is a generalization of the derivative for functions that are not necessarily smooth. While the subdifferential can be defined for any function (even though, it may not exist!), we only treat here the case of convex functions, for which the subdifferential enjoys a number of useful properties.

Definition 4.1.1 (Subdifferential). Let X be a Banach space. The subdifferential of a function $\varphi : X \to \mathbb{R} \cup \{+\infty\}$ at $x \in X$, denoted by $\partial\varphi(x)$, is the set of all subgradients of φ, i. e.,

$$\partial\varphi(x) = \{x^\star \in X^\star \ : \ \varphi(z) - \varphi(x) \geq \langle x^\star, z - x\rangle, \ \forall z \in X\}.$$

The subdifferential is in general a multi-valued map, $\partial\varphi : X \to 2^{X^\star}$, where $2^{X^\star}$ is the set of all possible subsets of $X^\star$. The domain of the subdifferential $\mathbf{D}(\partial\varphi) := \{x \in X \ : \ \partial\varphi(x) \neq \emptyset\}$. It can be seen from the definition that for $x \in X$ such that $\varphi(x) = +\infty$, $\partial\varphi(x) = \emptyset$. Redressed in the language of convex analysis, this means that if $x \notin \operatorname{dom}\varphi$, then $x \notin \mathbf{D}(\partial\varphi)$. So, clearly, the two sets are related. However, they are not equal, rather, as we will see in Theorem 4.2.25, the domain of the subdifferential $\mathbf{D}(\partial\varphi)$ is dense in $\operatorname{dom}\varphi$.

To motivate the definition of the subdifferential, we provide the following proposition, which shows that if φ is Gâteaux differentiable, then the subdifferential $\partial\varphi(x)$, is a singleton consisting only of $D\varphi(x)$:

Proposition 4.1.2 (The subdifferential of a differentiable function). *Let $\varphi : X \to \mathbb{R} \cup \{+\infty\}$ be a convex function, which is Gâteaux differentiable at $x \in X$. Then, $\partial\varphi(x) = \{D\varphi(x)\}$, and to simplify the notation in such cases, we will simply write $\partial\varphi(x) = D\varphi(x)$.*

Proof. If φ is differentiable in the Gâteaux sense at $x \in X$, by the convexity of φ, it holds that (see Theorem 2.3.23)

https://doi.org/10.1515/9783111333298-004

$$\varphi(x) - \varphi(z) \geq \langle D\varphi(x), x - z\rangle, \quad \forall z \in X.$$

So, clearly, $D\varphi(x) \in \partial\varphi(x)$, i. e., $\{D\varphi(x)\} \subseteq \partial\varphi(x)$.

It remains to show that $\partial\varphi(x) \subseteq \{D\varphi(x)\}$. To this end take any element $x^* \in \partial\varphi(x)$, and any element $x_o \in X$ of the form $x_o = x + tz$ for any $t > 0, z \in X$. Then, by the definition of the subdifferential, it holds that

$$\varphi(x + tz) - \varphi(x) \geq t\,\langle x^\star, z\rangle, \quad \forall x \in X,\ t > 0.$$

Divide by t, and pass to the limit, as $t \to 0$, so that by the differentiability of φ, we obtain that

$$\langle D\varphi(x) - x^\star, z\rangle \geq 0, \quad \forall z \in X.$$

Since X is a vector space, the same inequality holds for $-z \in X$ so that

$$\langle D\varphi(x) - x^\star, z\rangle = 0, \quad \forall z \in X,$$

therefore $D\varphi(x) - x^\star = 0$ in $X^\star$, and $\partial\varphi(x) \subseteq \{D\varphi(x)\}$. The proof is complete. □

A partial converse of Proposition 4.1.2 holds, according to which if $\partial\varphi(x)$ is a singleton and φ is continuous at x, then φ is Gâteaux differentiable at $x \in X$. For the proof of the converse, we need some more properties of the subdifferential, which will be provided later on, so this is postponed until Section 4.2.2 (see Example 4.2.10).

Before proving existence of the subdifferential (Proposition 4.2.1), we consider some examples (see, e. g., [12, 13, 20, 21, 28]).

Example 4.1.3 (Subdifferential of $|x|$). Let $X = \mathbb{R}$ and consider the convex function $\varphi : \mathbb{R} \to \mathbb{R}$ defined by $\varphi(x) = |x|$. Then

$$\partial\varphi(x) = \begin{cases} 1 & \text{if } x > 0, \\ [-1, 1] & \text{if } x = 0, \\ -1 & \text{if } x < 0, \end{cases}$$

where, e. g., $\partial\varphi(x) = 1$ means that $\partial\varphi(x) = \{1\}$. This can follow either directly from the definition[1] or by applying Proposition 4.1.2. An alternative way to express this is by stating that for $x \neq 0$; the function φ is differentiable with $D\varphi(x) = \frac{x}{|x|}$, while for $x = 0$, it

1 We are looking for $x^* \in \mathbb{R}$ such that $|z| - |x| \geq x^*(z - x)$ for every $z \in \mathbb{R}$. If $x > 0$, this inequality becomes $|z| - x \geq x^*(z - x)$ for every $z \in \mathbb{R}$, so choosing $z = x \pm \epsilon$, for any $\epsilon > 0$, we have (respectively) that $x^* \leq 1$ and $x^* \geq 1$ so that $x^* = 1$. If $x < 0$, then this inequality becomes $|z| + x \geq x^*(z - x)$ for every $z \in \mathbb{R}$. So, upon the choices $z = 0$ and $z = 2x$, we have, respectively, that $x^* \leq -1$ and $x^* \geq -1$ so that $x^* = -1$. For $x = 0$, this inequality becomes $|z| \geq x^* z$ for every $z \in \mathbb{R}$, which holds for any $x^* \in [-1, 1]$.

is not differentiable but rather subdifferentiable with $\partial\varphi(0) = [-1, 1]$. Of course, the fact that for $x \neq 0$, φ is differentiable, can also be interpreted as that φ is subdifferentiable, but the set $\partial\varphi(x)$ for $x \neq 0$ is a singleton, $\partial\varphi(x) = \{\frac{x}{|x|}\}$. Note that as a set valued map, $x \mapsto \partial\varphi(x)$ is upper semicontinuous. ◁

The above result can be generalized for the norm of any normed space. This will be done in Section 4.2.4, where it will be shown that the norm is subdifferentiable and its subdifferential coincides with the duality map for this space. Before turning to the general case, we present another example related to the norm of a Hilbert space.

Example 4.1.4 (Subdifferential of the norm in Hilbert space). Let $X = H$ be a Hilbert space, with its dual identified with the space itself $X^\star \simeq H$, and define the functional $\varphi : X \to \mathbb{R}$ by $\varphi(x) = \|x\|$ for every $x \in X$. Then

$$\partial\varphi(x) = \begin{cases} \frac{1}{\|x\|}x & x \neq 0, \\ \overline{B}_H(0,1) & x = 0, \end{cases}$$

where as before, $\partial(x) = \frac{1}{\|x\|}x$ means that $\partial\varphi(x) = \{\frac{1}{\|x\|}x\}$. Indeed, for $x \neq 0$, φ is Gâteaux differentiable with[2] $D\varphi(x) = \frac{x}{\|x\|}$, so that, by Proposition 4.1.2, $\partial\varphi(x) = \{\frac{x}{\|x\|}\}$ for $x \neq 0$. At $x = 0$, the function φ is not differentiable, however, the subdifferential exists. On the other hand, by the definition, at $x = 0$, we have that for any $x^\star \in \partial\varphi(0)$; it must hold that $\|z\| \geq \langle x^\star, z\rangle_H$, for every $z \in H$. By the Cauchy–Schwarz inequality, this holds for any $x^\star \in H$ such that $\|x\| \leq 1$, therefore $\partial\varphi(0) = \overline{B}_H(0,1)$. ◁

The subdifferential can be given a geometrical meaning in terms of the concept of the normal cone.

Definition 4.1.5 (Normal cone). Let $C \subset X$ be a convex set, and consider a point $x \in C$. The set

$$N_C(x) := \{x^\star \in X^\star \ : \ \langle x^\star, z - x\rangle \leq 0, \ \forall z \in C\}$$

is called the normal cone of C at x. The convention $N_C(x) = \emptyset$ if $x \notin C$ is used.

Note that if $X = H$ is a Hilbert space, $x \in H$ and $x_o \in C \subset H$ with C closed and convex, then $x - x_o \in N_C(x_o)$ is equivalent to $x_o = P_C(x)$. The normal cone of a linear subspace $E \subset X$ of a Banach space is $N_E(x) = E^\perp$ if $x \in E$ and $N_E(x) = \emptyset$ and if $x \in X \setminus E$ (as one may easily see by setting $z = x \pm x_o$ in the definition for arbitrary $x_o \in E$).

---

2 Simply consider the function $\psi : \mathbb{R} \to \mathbb{R}$ defined by $\psi(t) = \|x + th\|$ for any $x, h \in H$, with $x \neq 0$. It holds that $\psi(t)^2 = \|x + th\|^2$, and the right-hand side differentiable at $t = 0$, with derivative x, so that taking into account as well the continuity of ψ, the result holds.

Example 4.1.6 (Subdifferential of the indicator function of a convex set). Consider a convex set $C \subset X$ and the indicator function of $C, I_C : X \to \mathbb{R} \cup \{+\infty\}$, defined as

$$I_C(x) = \begin{cases} 0, & x \in C \\ +\infty, & x \in X \setminus C. \end{cases}$$

It is clear that I_C is a proper convex function. If C is closed, then it is also lower semicontinuous (see Example 2.3.15). Then, as follows by Definition 4.1.1, the subdifferential of I_C at x is the normal cone of C at x, i. e., $\partial I_C(x) = N_C(x), \forall x \in C$. ◁

Example 4.1.7 (Subdifferential of indicator function (equiv. normal cone) of affine subspace). Let X, Y be two Banach spaces, and consider the bounded linear operator $L : X \to Y$. For a given $y \in \mathbf{R}(L)$ let $C = \{z \in X : Lz = y\}$. Then, for any $x \in C$, we have that $\partial I_C(x) = N_C(x) = (\mathbf{N}(L))^\perp$.

Indeed, let $x^\star \in N_C(x)$. Then, by definition $\langle x^\star, z - x\rangle \leq 0$ for every $z \in C$. Since $x \in C$, let us consider $z \in C$ of the form $z = x \pm \epsilon z_0$, where $z_0 \in \mathbf{N}(L)$ (i. e., $Lz_o = 0$) and $\epsilon > 0$ arbitrary. Using these, $z \in C$ for the variational inequality defining the normal cone, we conclude that $\langle x^\star, z_0\rangle = 0$ for every $z_0 \in \mathbf{N}(L)$ so that $x^\star \in (\mathbf{N}(L))^\perp$. ◁

Example 4.1.8 (Subdifferential of an integral functional). Let $\mathcal{D} \subset \mathbb{R}^d$ be an open bounded set and $X = L^p(\mathcal{D})$ so that each $x \in X$ is identified with a function $u : \mathcal{D} \to \mathbb{R}$. Let $\varphi_0 : \mathbb{R} \to \mathbb{R}$ be a convex subdifferentiable function, and consider the functional $\varphi : X \to \mathbb{R} \cup \{+\infty\}$ defined by

$$\varphi(u) = \begin{cases} \int_{\mathcal{D}} \varphi_0(u(x))dx, & \text{if } \varphi_0(u) \in L^1(\mathcal{D}), \\ +\infty, & \text{otherwise.} \end{cases}$$

Clearly, the choice of p depends on growth conditions on the function φ_0, but we prefer to pass upon this for the time being, and assume that p is chosen so that $\varphi(u)$ is finite for any $u \in L^p(\mathcal{D})$. The functional φ defined above is a convex functional, which is subdifferentiable and

$$\partial\varphi(u) = \{v \in L^{p^*}(\mathcal{D}) \,:\, v(x) \in \partial\varphi_0(u(x)), \text{ a. e. } x \in \mathcal{D}\}.$$

Indeed, it is straightforward to see that

$$A := \{w \in L^{p^*}(\mathcal{D}) \,:\, w(x) \in \partial\varphi_0(u(x)), \text{ a. e. } x \in \mathcal{D}\} \subset \partial\varphi(u)$$

since for any $w \in A$, $\varphi_0(v(x)) - \varphi_0(u(x)) \geq w(x)(v(x) - u(x))$ a. e. $x \in \mathcal{D}$, for any $v \in L^p(\mathcal{D})$, where upon integrating over all $x \in \mathcal{D}$, we conclude that $\varphi(v) - \varphi(u) \geq \langle w, v - u\rangle$ for any $v \in L^p(\mathcal{D})$, which implies that $w \in \partial\varphi(u)$.

For the reverse inclusion, consider any $w \in \partial\varphi(u)$, i.e, any w such that $\varphi(v) - \varphi(u) \geq \langle w, v - u\rangle$ for any $v \in L^p(\mathcal{D})$, which can be redressed as

$$\int_{\mathcal{D}} (\varphi_o(v(x)) - \varphi_o(u(x)) - w(x)(v(x) - u(x))dx \geq 0, \quad \forall\, v \in L^p(\mathcal{D}), \tag{4.1}$$

from which it follows that[3] $w \in A$. ◁

4.2 The subdifferential for convex functions

We now discuss the general question of existence of the subdifferential for convex functions.

4.2.1 Existence and fundamental properties

The existence of the subdifferential of a convex function[4] at its points of continuity follows by a separation argument. This shown in the following proposition (see, e. g., [12, 20]).

Proposition 4.2.1. *Let X be a Banach space, and let $\varphi : X \to \mathbb{R} \cup \{+\infty\}$ be a proper convex lower semicontinuous function. Then φ is subdifferentiable on the interior of the domain of φ, i. e., $\mathbf{D}(\partial\varphi) = \text{int}(\text{dom}\,\varphi)$. Moreover, $\partial\phi$ displays monotonicity properties, i. e.,*

$$\langle x_1^\star - x_2^\star, x_1 - x_2 \rangle \geq 0, \quad \forall\, x_i \in \mathbf{D}(\partial\varphi),\ x_i^\star \in \partial\varphi(x_i),\ i = 1, 2. \tag{4.2}$$

Proof. Let $x_o \in \text{int}(\text{dom}\,\varphi)$. By Proposition 2.3.22, φ is continuous at x_o, and so[5] $\text{int}(\text{epi}\,\varphi) \neq \emptyset$. Moreover, $(x_o, \varphi(x_o)) \notin \text{int}(\text{epi}\,\varphi)$. Then (see Proposition 1.2.11), the point $(x_o, \varphi(x_o))$ and the set $\text{epi}\,\varphi$ can be separated, i. e., there exist $(z^\star, \beta) \in X^\star \times \mathbb{R}$ and $\alpha \in \mathbb{R}$ such that

$$\langle z^\star, x_o \rangle + \beta\varphi(x_o) \geq \alpha \geq \langle z^\star, x \rangle + \beta\lambda, \quad \forall\, (x, \lambda) \in \text{epi}\,\varphi. \tag{4.3}$$

---

3 To see this, consider any measurable $\mathcal{D}_0 \subset \mathcal{D}$, and for any $v \in L^p(\mathcal{D})$, define $\bar{v}(x) = v(x)\mathbf{1}_{\mathcal{D}_0}(x) + u(x)\mathbf{1}_{\mathcal{D}_0^c}(x)$, and apply (4.1) for the choice $\bar{v}$. This gives

$$\int_{\mathcal{D}_0} (\varphi_o\big(v(x)\big) - \varphi_o\big(u(x)\big) - w(x)\big(v(x) - u(x)\big)dx \geq 0, \quad \forall\, v \in L^p(\mathcal{D}),\ \forall\, \mathcal{D}_0 \subset \mathcal{D},$$

which by the arbitrariness of $\mathcal{D}_0$ leads to $\varphi_o(v(x)) - \varphi_o(u(x)) - w(x)(v(x) - u(x)) \geq 0$, a. e., $x \in \mathcal{D}$, and hence $w(x) \in \partial\varphi_o(u(x))$, a. e., $x \in \mathcal{D}$.

4 In principle, one may define the subdifferential using Definition 4.1.1, even for functions which are not convex.

5 By continuity of φ at x_o for any $\epsilon > 0$, there exists $\delta > 0$ such that $-\epsilon < \varphi(x) - \varphi(x_o) < \epsilon$ for all $x \in X$ such that $\|x - x_o\| < \delta$. This means that $N(x_o) = \{x \in X : \|x - x_o\| < \delta\}$ is an open neighborhood of $x_o \in \text{dom}\,\varphi$, in which $\varphi(x) < \varphi(x_o) + \epsilon$, which implies that the open set $N(x_o) \times \{r : r > \varphi(x_o) + \epsilon\} \subset \text{epi}\,\varphi$, hence $\text{int}(\text{epi}\,\varphi) \neq \emptyset$.

Since for every $\epsilon > 0$, $(x_o, \varphi(x_o) + \epsilon) \in \operatorname{epi}\varphi$, (4.3) yields $\beta\epsilon \leq 0$, so $\beta \leq 0$. But $\beta \neq 0$, since if not, then (4.3) implies that $\langle z^\star, x - x_o\rangle \leq 0$ for every $(x, \lambda) \in \operatorname{epi}\varphi$, which is turn implies that $\langle z^\star, x - x_o\rangle = 0$ for every $x \in \operatorname{dom}\varphi$, therefore $z^\star = 0$, which is a contradiction. Therefore, $\beta < 0$.

On the other hand, for any $(x, \lambda) \in \operatorname{epi}\varphi$, if it holds that $\lambda \geq \varphi(x)$, then (4.3) yields

$$\langle z^\star, x_o\rangle + \beta\varphi(x_o) \geq \alpha \geq \langle z^\star, x\rangle + \beta\varphi(x), \quad \forall x \in \operatorname{dom}\varphi. \tag{4.4}$$

Dividing (4.4) by $-\beta > 0$ and setting $x_o^\star = -\frac{z^\star}{\beta}$ yields

$$\langle x_o^\star, x_o\rangle - \varphi(x_o) \geq \langle x_o^\star, x\rangle - \varphi(x), \quad \forall x \in \operatorname{dom}\varphi,$$

which can be rearranged as

$$\varphi(x) \geq \varphi(x_o) + \langle x_o^\star, x - x_o\rangle, \quad \forall x \in \operatorname{dom}\varphi,$$

i. e., $x_o^\star \in \partial\varphi(x_o)$. This concludes the proof of the first statement.

To show (4.2), take $x_i \in \mathbf{D}(\partial\varphi)$, and consider $x_i^\star \in \partial\varphi(x_i)$, $i = 1, 2$. Then, by the definition of the subdifferential, we have that

$$\begin{aligned} \varphi(z) - \varphi(x_1) &\geq \langle x_1^\star, z - x_1\rangle, \quad \forall z \in X, \\ \varphi(z) - \varphi(x_2) &\geq \langle x_2^\star, z - x_2\rangle, \quad \forall z \in X. \end{aligned}$$

Setting $z = x_2$ in the first inequality and $z = x_1$ in the second yields

$$\begin{aligned} \varphi(x_2) - \varphi(x_1) &\geq \langle x_1^\star, x_2 - x_1\rangle, \\ \varphi(x_1) - \varphi(x_2) &\geq \langle x_2^\star, x_1 - x_2\rangle, \end{aligned}$$

and upon addition, we conclude (4.2), which is the monotonocity property of the subdifferential. □

Remark 4.2.2. The subdifferential of a convex function is nonempty at the points where φ is finite and continuous, i. e., on $\operatorname{cont}(\varphi)$. This does not require lower semicontinuity of φ.

Remark 4.2.3. Note that in infinite dimensional spaces, the interior of the domain of φ may be empty. Furthermore, even in finite dimensional spaces the subdifferential may be empty at points on the boundary of the domain of[6] φ. This calls for approximations of the subdifferential (see Section 4.2.5). Through this approximation procedure, it can also be shown that the domain of the subdifferential $\mathbf{D}(\partial\varphi)$ is dense in $\operatorname{dom}\varphi$ (see the

---

6 For example, $\varphi(x) = -x^\gamma$ for $x \in (-1, 1)$ and $\varphi(x) = +\infty$ for $|x| \geq 1$, with $\gamma < 1$, where it is easily seen that $\partial\varphi(x) = \emptyset$ for every $|x| \geq 1$.

Brøndsted–Rockafellar theorem 4.2.25). This is important for the maximal monotonicity properties of $\partial\varphi$ (Theorem 9.4.18).

As already discussed and shown in Proposition 4.1.2, the subdifferential is single valued and coincides with the Gâteaux derivative for differentiable functions. In this sense, the inequality (4.2) is a monotonicity property, which generalizes the relevant monotonicity property of the Gâteaux derivative for convex functions (see (2.14), Theorem 2.3.23).

Finally, the subdifferential of a convex function enjoys convexity and compactness properties.

Proposition 4.2.4. *Let $\varphi : X \to \mathbb{R} \cup \{+\infty\}$ be a convex proper lower semicontinuous function. The subdifferential $\partial\varphi(x)$, at any $x \in X$, at which φ is continuous, is a convex, bounded, weak* closed (hence, weak* compact) subset of $X^\star$.*

Proof. Convexity and weak*-closedeness follow easily by definition. We prove that $\partial\varphi(x)$ is bounded. Since φ is finite and continuous at x, there exists an open neighborhood of x, $N(x) = \{x' \in X : \|x - x'\| < \delta\}$ such that $|\varphi(x) - \varphi(z)| < \epsilon$ for all $z \in N(x)$ (see Proposition 2.3.21). Let $x_o \in X$ such that $\|x_o\| = 1$, and take $x' = x + t\, x_o$ for $|t| < \delta$. Clearly, $x' \in N(x)$. For any $x^\star \in \partial\varphi(x)$, applying Definition 4.1.1 for the pair (x, x'), we have that

$$\varphi(x + tx_o) \geq \varphi(x) + \langle x^\star, t\, x_o\rangle. \tag{4.5}$$

Since $x + t\, x_o \in N(x)$, by the continuity of φ, we also have that

$$\big|\varphi(x + t\, x_o) - \varphi(x)\big| \leq \epsilon. \tag{4.6}$$

Therefore, combining (4.5) and (4.6), we deduce that $|\langle x^\star, x_o\rangle| \leq \frac{\epsilon}{|t|}$. This implies, by the definition of the dual norm $\|\cdot\|_{X^\star}$, that $\|x^\star\|_{X^\star} \leq \frac{\epsilon}{|t|}$ so that the set $\partial\varphi(x) \subset X^\star$ is bounded. The weak* compactness follows from the Banach–Alaoglou theorem (see Theorem 1.1.37). □

Example 4.2.5 (Subdifferential for strongly convex functions). For uniformly convex or strongly convex functions (see Definition 2.3.3), the monotonicity properties of the subdifferential may be further enhanced. Assume, for example, that $\varphi : X \to \mathbb{R} \cup \{+\infty\}$ is strongly convex with modulus of convexity c. Then, it holds that

$$\begin{aligned}
&\text{(i)} \quad \langle x^\star, z - x\rangle \leq \varphi(z) - \varphi(x) - \frac{c}{2}\|z - x\|^2, \quad \forall\, x^\star \in \partial\varphi(x),\ z \in X,\\
&\text{(ii)} \quad \langle x_1^\star - x_2^\star, x_1 - x_2\rangle \geq c\|x_1 - x_2\|^2, \quad \forall\, x_i \in X,\ x_i^\star \in \partial\varphi(x_i),\ i = 1, 2,\\
&\text{(iii)} \quad \|x_1^\star - x_2^\star\|_{X^\star} \geq c\|x_1 - x_2\|, \quad \forall\, x_i \in X,\ x_i^\star \in \partial\varphi(x_i),\ i = 1, 2.
\end{aligned} \tag{4.7}$$

To prove (i), consider any $x \in X$ (see Remark 4.2.3) and any $x^\star \in \partial\varphi(x)$ so that by definition $\varphi(z') - \varphi(x) \geq \langle x^\star, z' - x\rangle$ for every $z' \in X$. Consider $z' = x + t(z - x) = (1 - t)x + tz$

for arbitrary $z \in X$, $t \in (0,1)$ in the above definition, which along with the fact that φ is strongly convex yields

$$\begin{aligned} t\langle x^\star, z - x\rangle &\leq \varphi((1-t)x + tz) - \varphi(x) \\ &\leq (1-t)\varphi(x) + t\varphi(z) - \frac{c}{2}t(1-t)\|x - z\|^2 - \varphi(x) \\ &= t(\varphi(z) - \varphi(x)) - \frac{c}{2}t(1-t)\|x - z\|^2, \end{aligned}$$

where upon dividing by t and passing to the limit, as $t \to 0^+$, leads to (i). To get (ii) we apply (i) first for $x_1^\star \in \partial\varphi(x_1)$ and $z = x_2$, and then, for $x_2^\star \in \partial\varphi(x_2)$ and $z = x_1$ and add, while (iii) follows by the observation that $|\langle x_1^\star - x_2^\star, x_1 - x_2\rangle| \leq \|x_1^\star - x_2^\star\|_{X^\star}\|x_1 - x_2\|$, and (ii). ◁

4.2.2 The subdifferential and the right hand side directional derivative

We have seen in Proposition 4.1.2 that if a convex function φ is differentiable, its Gâteaux derivative coincides with the subdifferential. However, even if φ is not differentiable, by convexity, we may show that the right-hand side directional derivative exists and is related to the subdifferential.

Definition 4.2.6. Let $\varphi : X \to \mathbb{R}$. The limit

$$D^+\varphi(x, h) := \lim_{t\to 0^+} \frac{1}{t}(\varphi(x + t\,h) - \varphi(x)) \tag{4.8}$$

is called the right-hand side directional derivative of the function φ at $x \in X$ in the direction $h \in X$.

The right-hand side directional derivative $D^+\varphi(x, h)$ is very similar (and related) to the Gâteaux directional derivative $D\varphi(x, h) = \lim_{t\to 0} \frac{1}{t}(\varphi(x+t\,h)-\varphi(x))$, (see Sections 2.1.1 and 2.1.2) but with a very important difference! While if φ is Gâteaux differentiable at x the mapping $h \mapsto D\varphi(x, h)$ is linear in h for every $h \in X$, in general, the mapping $h \mapsto D^+\varphi(x, h)$ is sublinear in h for every $h \in X$. In fact, if at a point $x \in X$, the mapping $h \mapsto D^+\varphi(x, h)$ is linear in h for every $h \in X$, then φ is Gâteaux differentiable at $x \in X$. The following proposition (see e. g., [20, 114]) collects these useful connections.

Proposition 4.2.7 (The right-hand side directional derivative and the subdifferential). *If $\varphi : X \to \mathbb{R} \cup \{+\infty\}$ is convex, then $D^+\varphi(x, h)$ exists (but may not be finite), the map $h \mapsto D^+\varphi(x, h)$ is sublinear. At points where φ is continuous, for any $h \in X$ it holds that*

$$D^+\varphi(x, h) = \sup_{x^\star \in\partial\varphi(x)} \langle x^\star, h\rangle, \tag{4.9}$$

and the supremum is attained at some $x^\star \in \partial\varphi(x)$.

Proof. To show the existence of the limit defining $D^+\varphi(x,h)$, we simply have to note that convexity of φ leads to monotonicity of the mapping $t \mapsto \frac{1}{t}(\varphi(x+t\,h)-\varphi(x))$ for all $x, h \in X$. To see that, take $0 < t_1 \le t_2$, and write

$$x + t_1\,h = \left(1 - \frac{t_1}{t_2}\right)x + \frac{t_1}{t_2}(x + t_2 h),$$

which is clearly a convex combination since $\frac{t_1}{t_2} \in (0,1]$. Convexity of φ implies that

$$\varphi(x + t_1\,h) \le \left(1 - \frac{t_1}{t_2}\right)\varphi(x) + \frac{t_1}{t_2}\varphi(x + t_2 h),$$

which leads to

$$\frac{1}{t_1}(\varphi(x + t_1 h) - \varphi(x)) \le \frac{1}{t_2}(\varphi(x + t_2 h) - \varphi(x)).$$

This implies the monotonicity of the mapping $t \mapsto \frac{1}{t}(\varphi(x+t\,h)-\varphi(x))$ for every $x, h \in X$, which leads to the existence of the limit (not necessarily finite) as $t \to 0^+$.

To prove that the map $h \mapsto D^+\varphi(x,h)$ is sublinear, it is enough to note that this map is convex and positively homogeneous. Convexity follows by the observation that $x + t(\lambda h_1 + (1-\lambda)h_2) = \lambda(x + t\,h_1) + (1-\lambda)(x + t\,h_2)$, so by convexity of φ, we have that

$$\varphi(x + t(\lambda h_1 + (1-\lambda)h_2)) \le \lambda\varphi(x + t\,h_1) + (1-\lambda)\varphi(x + t\,h_2),$$

where upon subtracting $\varphi(x)$ from both sides, dividing by t, and passing to the limit as $t \to 0^+$ provides the required convexity. Positive homogeneity follows by a similar argument.

To show the connection of $D^+\varphi(x,h)$ with the subdifferential, observe first that by the definition of the subdifferential, it follows that if $x^\star \in \partial\varphi(x)$, then for all $t > 0$ and $h \in X$,

$$\varphi(x + th) - \varphi(x) \ge t\langle x^\star, h\rangle,$$

so that dividing by t and passing to the limit, as $t \to 0^+$, yields that for all h and every $x^\star \in \partial\varphi(x)$, it holds that

$$D^+\varphi(x,h) \ge \langle x^\star, h\rangle, \quad \forall\, x^\star \in \partial\varphi(x),$$

which in turn implies that

$$D^+\varphi(x,h) \ge \sup_{x^\star \in \partial\varphi(x)} \langle x^\star, h\rangle.$$

If we manage to show that the reverse inequality holds as well, then (4.9) is established. For that, it is enough to show that there exists a $x_o^\star \in \partial\varphi(x)$ such that[7] $D^+\varphi(x,h) \le \langle x_o^\star, h\rangle$. To show the existence of such an element, $x_o^\star \in \partial\varphi(x)$, we will use the Hahn–Banach separation theorem. To this end, define the sets

$$A := \{(z,r) \in X \times \mathbb{R} \,:\, \varphi(z) < r\},$$
$$B := \{(x + t\,h, \varphi(x) + tD^+\varphi(x,h)) \,:\, t \ge 0\},$$

that are convex subsets of $X \times \mathbb{R}$ such that[8] $A \cap B = \emptyset$. Furthermore, A is an open set. By the Hahn–Banach separation theorem, there exists a nonzero linear functional in $X^\star \times \mathbb{R}$ represented by the pair $(z_o^\star, \lambda_o) \in X^\star \times \mathbb{R}$ such that

$$\langle z_o^\star, z\rangle + \lambda_o r \le \langle z_o^\star, x + th\rangle + \lambda_o(\varphi(x) + t\,D^+\varphi(x,h)), \quad \forall\,(z,r) \in A,\ t \ge 0.$$

Setting $z = x$, $r > \varphi(z)$, and $t = 0$ in this inequality yields that $\lambda_o \le 0$, with the possibility $\lambda_o = 0$ being excluded, as it leads to the contradiction $z_o^\star = 0$. Since $\lambda_o < 0$, we may divide the inequality by $-\frac{1}{\lambda_o}$, and setting $x_o^\star = -\frac{1}{\lambda_o} z_o^\star$ leads to

$$\langle x_o^\star, z\rangle - r \le \langle x_o^\star, x + th\rangle - (\varphi(x) + t\,D^+\varphi(x,h)).$$

By the continuity of φ, we may replace r in the above inequality by $\varphi(z)$, thus obtaining

$$\langle x_o^\star, z\rangle - \varphi(z) \le \langle x_o^\star, x + th\rangle - (\varphi(x) + t\,D^+\varphi(x,h))$$

for all $t \ge 0$, and taking the limit as $t \to 0$, we see that $x_o^\star \in \partial\varphi(x)$. Setting further $z = x$, and $t = 1$ we see that $D^+\varphi(x,h) \le \langle x_o^\star, h\rangle$, and our claim is proved. □

Remark 4.2.8. The claim of Proposition 4.2.7 (the max formula) holds at any point where φ is finite and continuous, i. e., on $\text{cont}(\varphi)$. This does not require lower semicontinuity of φ. If φ is lower semicontinuous, it holds on $\text{int}(\text{dom}\,\varphi)$.

Example 4.2.9. If at a point $x \in X$ the map $h \mapsto D^+\varphi(x,h)$ is linear, then φ is Gâteaux differentiable at x, and $D^+\varphi(x,h) = D\varphi(x;h) = \langle D\varphi(x), h\rangle$ for every $h \in X$. The converse is also true.

To see the above, define the left-hand sided directional derivative $D^-\varphi(x,h) := \lim_{t\to 0^-} \frac{1}{t}(\varphi(x + th) - \varphi(x))$, and note that $D\varphi(x,h)$ exists at a point $x \in X$ if $D^+\varphi(x,h) = D^-\varphi(x,h)$. A simple scaling argument shows that for every $h \in X$ it holds that $D^-\varphi(x,h) = -D^+\varphi(x,-h)$. This implies that $D\varphi(x,h)$ exists if

7 This is easy to see since obviously $\langle x_o^\star, h\rangle \le \max_{x^\star \in \partial\varphi(x)} \langle x^\star, h\rangle$.

8 For if not, there would be a $t \ge 0$ such that $D^+\varphi(x,h) > \frac{1}{t}(\varphi(x + t\,h) - \varphi(x))$, which is in contradiction with the fact that the function $t \mapsto \frac{1}{t}(\varphi(x + t\,h) - \varphi(x))$ is increasing.

$$D^{-}\varphi(x,h) = D^{+}\varphi(x,-h) = D^{+}\varphi(x,h), \tag{4.10}$$

and, using the fact that the functional $h \mapsto D^{+}\varphi(x,h)$ is sublinear,[9] we conclude that the mapping $h \mapsto D^{+}\varphi(x,h)$ is linear. The converse follows from the observation that if the mapping $h \mapsto D^{+}\varphi(x,h)$ is linear, then $D^{+}\varphi(x,-h) = -D^{+}\varphi(x,h)$, which implies that $D^{-}\varphi(x,h) = D^{+}\varphi(x,h)$, and the Gâteaux differentiability of φ at x follows. ◁

Example 4.2.10. If the set $\partial\varphi(x)$ is a singleton at some point $x \in X$, at which φ is continuous, then φ is Gâteaux differentiable at x. This is the converse of Proposition 4.1.2.

Suppose that φ is continuous at x and $\partial\varphi(x) = \{z^{\star}\}$ for some $z^{\star} \in X^{\star}$. Then, by Proposition 4.2.7

$$D^{+}\varphi(x,h) = \max_{x^{\star}\in\partial\varphi(x)} \langle x^{\star}, h\rangle = \langle z^{\star}, h\rangle,$$

so $h \mapsto D^{+}\varphi(x,h)$ is linear, and, by Example 4.2.9, φ is Gâteaux differentiable at x, whereas by Proposition 4.1.2, $z^{\star} = D\varphi(x)$. ◁

4.2.3 Subdifferential calculus

The subdifferential has the following properties, which are usually referred to under the name subdifferential calculus [12, 20]. Note that we use $\subset$ invariably with $\subseteq$.

Proposition 4.2.11 (Subdifferential calculus).
(i) *Let $\varphi : X \to \mathbb{R} \cup \{+\infty\}$ be a convex function. Then, $\partial\lambda\varphi(x) = \lambda\partial\varphi(x)$ for every $\lambda > 0$.*
(ii) *Let $\varphi_1, \varphi_2 : X \to \mathbb{R} \cup \{+\infty\}$ be convex functions. Then,*

$$\partial\varphi_1(x) + \partial\varphi_2(x) \subset \partial(\varphi_1 + \varphi_2)(x), \quad \forall x \in X.$$

(iii) *Let $\varphi : Y \to \mathbb{R} \cup \{+\infty\}$ be a convex function and $L : X \to Y$ a continuous linear operator. Then,*

$$L^{*}\partial\varphi(Lx) \subset \partial(\varphi \circ L)(x), \quad \forall x \in X,$$

where $L^{*} : Y^{\star} \to X^{\star}$ is the adjoint (dual) operator of L.

Proof. (i) Suppose that $x^{\star} \in \partial\varphi(x)$. Then, $\varphi(z) - \varphi(x) \geq \langle x^{\star}, z - x\rangle$ for all $z \in X$, so multiplying by $\lambda > 0$, we see that $\lambda x^{\star} \in \partial(\lambda\varphi)(x)$. This implies that $\lambda\partial\varphi(x) \subset \partial(\lambda\varphi)(x)$. To prove the opposite inclusion, let $x^{\star} \in \partial(\lambda\varphi)(x)$. Then,

9 If for a sublinear map f it holds that $-f(-h) = f(h)$, then f is linear. Indeed expressing $f(x_1) = f(x_1 + x_2 - x_2) \leq f(x_1 + x_2) + f(-x_2) = f(x_1 + x_2) - f(x_2)$, we see that $f(x_1) + f(x_2) \leq f(x_1 + x_2)$, which combined with $f(x_1 + x_2) \leq f(x_1) + f(x_2)$ leads to $f(x_1 + x_2) = f(x_1) + f(x_2)$. That $f(\lambda x) = \lambda f(x)$ for every $\lambda \in \mathbb{R}$ follows by positive homogeneity and the property that $-f(-h) = f(h)$.

$$\lambda\varphi(z) - \lambda\varphi(x) \geq \langle x^\star, z - x\rangle, \quad \forall z \in X,$$

so dividing by $\lambda > 0$ yields $\frac{x^\star}{\lambda} \in \partial\varphi(x)$, which means that $x^\star \in \lambda\partial\varphi(x)$. Therefore, $\partial(\lambda\varphi)(x) \subset \lambda\partial\varphi(x)$, and the claim is proved.

(ii) Let $z_1^\star \in \partial\varphi_1(x)$ and $z_2^\star \in \partial\varphi_2(x)$. Then,

$$\varphi_i(z) - \varphi_i(x) \geq \langle z_i^\star, z - x\rangle, \quad \forall z \in X, \ i = 1, 2.$$

Adding these inequalities, we obtain that

$$(\varphi_1 + \varphi_2)(z) - (\varphi_1 + \varphi_2)(x) \geq \langle z_1^\star + z_2^\star, z - x\rangle, \quad \forall z \in X,$$

hence $z_1^\star + z_2^\star \in \partial(\varphi_1 + \varphi_2)(x)$, which implies the stated result.

(iii) For any Banach space Z we denote by $\langle \cdot, \cdot \rangle_{Z^\star, Z}$ the duality pairing between Z and $Z^\star$. Take any $x_o^\star \in L^*\partial\varphi(Lx) \subset X^\star$. Then, there exists $y_0^\star \in \partial\varphi(Lx) \subset Y^\star$, i. e., a $y_0^\star \in Y^\star$ with the property

$$\varphi(Lz) - \varphi(Lx) \geq \langle y_0^\star, Lz - Lx\rangle_{Y^\star, Y}, \quad \forall z \in X \tag{4.11}$$

such that $x_o^\star = L^* y_0^\star$. Using the definition of the adjoint operator, we can express (4.11) in the equivalent form

$$\varphi(Lz) - \varphi(Lx) \geq \langle L^* y_0^\star, z - x\rangle_{X^\star, X} = \langle x_o^\star, z - x\rangle_{X^\star, X}, \quad \forall z \in X. \tag{4.12}$$

By the definition of $\partial(\varphi \circ L)(x)$, it is evident that $x_o^\star \in \partial(\varphi \circ L)(x)$. Indeed,

$$\begin{aligned}\partial(\varphi \circ L)(x) &= \{y^* \in Y^* \ : \ (\varphi \circ L)(z) - (\varphi \circ L)(x) \geq \langle y^*, Lz - Lx\rangle_{Y^\star, Y}, \ \forall z \in X\} \\ &= \{y^* \in Y^* \ : \ (\varphi \circ L)(z) - (\varphi \circ L)(x) \geq \langle L^* y^*, z - x\rangle_{X^\star, X}, \ \forall z \in X\},\end{aligned}$$

which, combined with (4.12), yields that $x_o^\star \in \partial(\varphi \circ L)(x)$. Therefore, the stated inclusion holds true. □

It is interesting to find conditions under which the inclusions in the addition and the chain rule for the subdifferential (Proposition 4.2.11(ii) and (iii)) can be turned into equalities. We will see that continuity of the functions involved plays an important role in this respect, and in view of Proposition 2.3.21, where continuity at a single point for a convex function implies continuity at every point in the interior of its domain, we anticipate that continuity at a single point will suffice.

Theorem 4.2.12 (Moreau–Rockafellar). *Let $\varphi_1, \varphi_2 : X \to \mathbb{R} \cup \{+\infty\}$ be convex proper lower semicontinuous functions such that* $\operatorname{int}(\operatorname{dom} \varphi_1) \cap \operatorname{dom} \varphi_2 \neq \emptyset$. *Then,*

$$\partial(\varphi_1 + \varphi_2) = \partial\varphi_1 + \partial\varphi_2.$$

The condition $\operatorname{int}(\operatorname{dom}\varphi_1) \cap \operatorname{dom}\varphi_2 \neq \emptyset$ can also be interpreted as that there exists at least one point $x_o \in \operatorname{dom}\varphi_2$, at which φ_1 is finite and continuous.

Proof. By Proposition 4.2.11(ii), we have that

$$\partial\varphi_1(x) + \partial\varphi_2(x) \subset \partial(\varphi_1 + \varphi_2)(x), \quad \forall x \in X.$$

We will now show that under the extra condition imposed, the reverse inclusion also holds. To this end, let $x_o \in \operatorname{int}(\operatorname{dom}\varphi_1) \cap \operatorname{dom}\varphi_2$, fix any $x \in X$, consider any $x^\star \in \partial(\varphi_1 + \varphi_2)(x)$, and we shall show that $x^\star \in \partial\varphi_1(x) + \partial\varphi_2(x)$, i. e., $x^\star = x_1^\star + x_2^\star$, with $x_1^\star \in \partial\varphi_1(x)$ and $x_2^\star \in \partial\varphi_2(x)$. Recall, by the properties of proper lower semicontinuous convex functions (see Proposition 2.3.22), that since $x_o \in \operatorname{int}(\operatorname{dom}\varphi_1)$, the function φ_1 is continuous at x_o.

Since $x^\star \in \partial(\varphi_1 + \varphi_2)(x)$, we have that $\varphi_1(z) + \varphi_2(z) \geq \varphi_1(x) + \varphi_2(x) + \langle x^\star, z - x\rangle$ for every $z \in X$, i. e.,

$$\varphi_1(z) - \varphi_1(x) - \langle x^\star, z - x\rangle \geq \varphi_2(x) - \varphi_2(z), \quad \forall z \in X. \tag{4.13}$$

For fixed x, define the functions φ and ψ, by $z \mapsto \varphi(z)$ and $z \mapsto \psi(z)$, where

$$\varphi(z) = \varphi_1(z) - \varphi_1(x) - \langle x^\star, z - x\rangle, \quad \text{and,} \quad \psi(z) = \varphi_2(x) - \varphi_2(z).$$

Clearly, φ is convex and continuous at x_o, while ψ is a concave function. In terms of φ and ψ, (4.13) becomes

$$\varphi(z) \geq \psi(z), \quad \forall z \in X. \tag{4.14}$$

Consider the convex sets

$$A = \{(z,r) \in X \times \mathbb{R} : \varphi(z) \leq r\} = \{(z,r) \in X \times \mathbb{R} : \varphi_1(z) - \varphi_1(x) - \langle x^\star, z - x\rangle \leq r\},$$
$$B = \{(z,r) \in X \times \mathbb{R} : \psi(z) \geq r\} = \{(z,r) \in X \times \mathbb{R} : \varphi_2(x) - \varphi_2(z) \geq r\}.$$

Since A is the epigraph of the convex function φ, which is continuous at x_o, it follows that[10] $\operatorname{int}(A) = \operatorname{int}(\operatorname{epi}\varphi) \neq \emptyset$. Furthermore, by (4.14), it is clear that $\operatorname{int}(A) \cap B = \emptyset$. Therefore, by the separation theorem, $\operatorname{int}(A)$ and B can be separated by a hyperplane. It then follows that there exists $(z_o^\star, \alpha) \in X^\star \setminus \{0\} \times \mathbb{R}$ and $\rho \in \mathbb{R}$ such that

$$\langle z_o^\star, x_1\rangle + \alpha\lambda_1 \leq \rho \leq \langle z_o^\star, x_2\rangle + \alpha\lambda_2, \quad \forall (x_1,\lambda_1) \in \operatorname{int}(A), \ \forall (x_2,\lambda_2) \in B. \tag{4.15}$$

10 Consider any $\lambda > \varphi(x_o)$. Then, $(x_o, \lambda) \in \operatorname{int}(\operatorname{epi}\varphi)$. Indeed, by the continuity of φ at x_o, for any $\epsilon > 0$ there exists $\delta > 0$ such that $|\varphi(x) - \varphi(x_o)| < \epsilon$ for every x such that $\|x - x_o\| < \delta$, which implies that $\varphi(x) < \varphi(x_o) + \epsilon$ for all $x \in B(x_o, \delta)$. By choosing ϵ appropriately, we can make sure that $\varphi(x_o) + \epsilon$ may take any value in $(\varphi(x_o), \lambda)$. This means that for every $\mu \in (\varphi(x_o), \lambda)$, we may find a ball $B(x_o, \delta)$ such that for every $x \in B(x_o, \delta)$ it holds that $\varphi(x) < \mu$. That of course means that $(x, \mu) \in \operatorname{epi}\varphi$ for every $x \in B(x_o, \delta)$, $\mu \in (\varphi(x_o), \lambda)$, therefore $(x_o, \lambda) \in \operatorname{int}(\operatorname{epi}\varphi)$.

By the definition of the sets A and B, it can be seen that $(x, \epsilon) \in \text{int}(A)$ for every $\epsilon > 0$, while $(x, 0) \in B$. An application of (4.15) for this choice leads to the observation that $\langle z_0^\star, x\rangle \le \rho \le \langle z_0^\star, x\rangle - \alpha\epsilon$ for every $\epsilon > 0$, and passing to the limit, as $\epsilon \to 0$, we conclude that $\rho = \langle z_0^\star, x\rangle$. Furthermore, picking $(x, 1) \in \text{int}(A)$ and $(x, 0) \in B$ and applying (4.15), we obtain that $\alpha \le 0$. We may also observe that α may not[11] take the value 0, hence $\alpha < 0$.

We multiply both sides of (4.15) by $-\frac{1}{\alpha} > 0$, and upon defining $x_o^\star = -\frac{1}{\alpha} z_0^\star$, we obtain that

$$\langle x_o^\star, x_1\rangle - \lambda_1 \le \langle x_o^\star, x\rangle \le \langle x_o^\star, x_2\rangle - \lambda_2, \quad \forall (x_1, \lambda_1) \in \text{int}(A),\ \forall (x_2, \lambda_2) \in B, \tag{4.16}$$

where we also used the fact that $\rho = \langle z_0^\star, x\rangle$.

Inequality (4.16) guarantees that for any $(x_2, \lambda_2) \in B$, i. e., for any (x_2, λ_2) such that $\varphi_2(x) - \varphi_2(x_2) \ge \lambda_2$, it holds that $\langle x_o^\star, x\rangle \le \langle x_o^\star, x_2\rangle - \lambda_2$. Fix any $z \in X$, and set $\lambda_{o,2} = \varphi_2(x) - \varphi_2(z)$. Clearly, the pair $(z, \lambda_{o,2}) \in B$, so (4.16) yields that $\langle x_o^\star, x\rangle \le \langle x_o^\star, z\rangle + \varphi_2(z) - \varphi_2(x)$, which can be reformulated as

$$\langle -x_o^\star, z - x\rangle + \varphi_2(x) \le \varphi_2(z), \quad \forall z \in X,$$

hence $x_2^\star := -x_o^\star \in \partial\varphi_2(x)$.

Furthermore, inequality (4.16) guarantees that for any $(x_1, \lambda_1) \in \text{int}(A)$, i. e., for any (x_1, λ_1) such that $\varphi_1(x_1) - \varphi_1(x) - \langle x^\star, x_1 - x\rangle < \lambda_1$, it holds that $\langle x_o^\star, x_1\rangle - \lambda_1 \le \langle x_o^\star, x\rangle$. Fix any $z \in X$, and set $\lambda_{o,1} = \varphi_1(z) - \varphi_1(x) - \langle x^\star, z - x\rangle + \epsilon$. For any $\epsilon > 0$, the pair $(z, \lambda_{o,1}) \in \text{int}(A)$, so (4.16) yields that

$$\langle x^\star + x_o^\star, z - x\rangle + \varphi_1(x) - \epsilon \le \varphi_1(z), \quad \forall z \in X,\ \epsilon > 0,$$

and passing to the limit, as $\epsilon \to 0$, we conclude that $x_1^\star := x^\star + x_o^\star \in \partial\varphi_1(x)$.

Combining $x^\star + x_o^\star \in \partial\varphi_1(x)$ and $-x_o^\star \in \partial\varphi_2(x)$, we conclude that $x^\star = x^\star + x_o^\star + (-x_o^\star) \in \partial\varphi_1(x) + \partial\varphi_2(x)$. This concludes the proof. □

Remark 4.2.13. The subdifferenial sum rule holds, even if φ_1, φ_2 are not lower semicontinuous, if, e. g., $\text{dom}\,\varphi_1 \cap \text{cont}(\varphi_2) \neq \emptyset$, i. e., if there exists a point where one of the functions is finite and the other is continuous.

Proposition 4.2.14. *Let $L : X \to Y$ be a continuous linear operator and $\varphi : Y \to \mathbb{R} \cup \{+\infty\}$ be a convex function, which is continuous at a point of* $\text{dom}\,\varphi \cap \mathbf{R}(L)$. *Then,*

11 If it did, observing that since $x_o \in \text{dom}\,\varphi_2$, $x_o \in B$, and since x_o is a point of continuity of φ_1, there exists an open ball $B(x_o, \delta)$ such that $B(x_o, \delta) \in \text{int}(A)$. Then applying (4.15) for $x_o \in B$ and for any $B(x_o, \delta) \subset \text{int}\,A$, we conclude that $\langle z_0^\star, x_o - z'\rangle \le 0$ for every $z' \in B(x_o, \delta)$, which implies that $z_0^\star = 0$, which contradicts $z_o^\star \neq 0$.

$$L^\star \partial\varphi(Lx) = \partial(\varphi \circ L)(x), \quad \forall x \in X,$$

where $L^\star$ *is the adjoint operator of* L.

Proof. Consider the function $\psi : X \times Y \to \mathbb{R} \cup \{+\infty\}$ defined by $\psi(x,y) := \varphi(y) + I_{\mathbf{Gr}(L)}(x,y)$, where $\mathbf{Gr}(L)$ is the graph of the operator L. Clearly, $I_{\mathbf{Gr}(L)}(x,y) = 0$ if $(x,y) \in \mathbf{Gr}(L)$ and $+\infty$ otherwise. From the definition of the subdifferential, it follows that $z^\star \in \partial\varphi(Lx)$ if and only if $(z^\star, 0) \in \partial\psi(x, Lx)$. Let $y_0 \in \operatorname{dom}\varphi \cap \mathbf{R}(L)$ be a point of continuity of φ. Then, there exists $x_o \in X$ such that $y_0 = Lx_o$, and (x_o, Lx_o) is a point of continuity of ψ. We may then apply Theorem 4.2.12 to ψ. Since,

$$\partial\psi(x,y) = \partial\varphi(y) + \partial I_{\mathbf{Gr}(L)}(x,y),$$

it follows that any $(z^\star, 0) \in \partial\psi(x, Lx)$ can be expressed as $(z^\star, 0) = (0, y_1^\star) + (x_2^\star, y_2^\star)$, with ${x_2}^\star = z^\star$ and ${y_1}^\star + {y_2}^\star = 0$, where $y_1^\star \in \partial\varphi(Lx)$ and $(x_2^\star, y_2^\star) \in \partial I_{\mathbf{Gr}(L)}(x, Lx) = N_{\mathbf{Gr}(L)}(x, Lx)$, i. e.,

$$\langle x_2^\star, z - x\rangle_{X^\star,X} + \langle y_2^\star, Lz - Lx\rangle_{Y^\star,Y} \le 0, \quad \forall z \in X. \tag{4.17}$$

We have used the notation $\langle\cdot,\cdot\rangle_X$ and $\langle\cdot,\cdot\rangle_Y$ to clarify that the first duality pairing is the duality pairing between X and $X^\star$, whereas the second is between Y and $Y^\star$.

Using the linearity of L and the definition of the adjoint operator, (4.17) yields

$$\langle x_2^\star, z - x\rangle_{X^\star,X} + \langle L^\star y_2^\star, z - x\rangle_{X^\star,X} \le 0, \quad \forall z \in X,$$

from which choosing $z = x \pm z_0 \in X$ for arbitrary $z_0 \in X$, we conclude that $x_2^\star = -L^\star y_2^\star$. Therefore, $z^\star = L^\star y_1^\star \in L^\star\partial\varphi(Lx)$. This implies

$$L^\star \partial\varphi(Lx) \supset \partial(\varphi \circ L)(x), \quad \forall x \in X,$$

which when combined with Proposition 4.2.11(iii) yields the desired equality. □

Remark 4.2.15. Since, by Corollary 2.3.22, convex functions defined on open subsets of finite dimensional spaces are continuous, subdifferential calculus simplifies considerably for the subdifferentials of finite dimensional convex functions.

Example 4.2.16 (Subdifferential of the ℓ^1 norm in $\mathbb{R}^d$). Consider $X = \mathbb{R}^d$, endowed with the norm $\|x\|_1 = |x_1| + \cdots |x_d|$, where $x = (x_1, \dots, x_d)$ and $X^\star \simeq \mathbb{R}^d$. This is a norm that finds many applications in various fields, such as optimization with sparse constraints, machine learning, etc. The subdifferential of the function $\varphi : X = \mathbb{R}^d \to \mathbb{R}$, defined by $\varphi(x) = \|x\|_1$, is

$$\partial\varphi(x) = \{x^\star \in \mathbb{R}^d \ : \ x_i^\star = \operatorname{sgn}(x_i), \text{ if } x_i \neq 0,\ x_i^\star \in [-1,1] \text{ if } x_i = 0\},$$

or equivalently the Cartesian product $\partial\varphi(x) = A_1(x_1)\times\cdots\times A_d(x_d)$, where $A_i(x_i) = \partial\phi_i(x_i)$, and $\phi_i : \mathbb{R} \to \mathbb{R}$ is defined as $\phi_i(x_i) = |x_i|$, $i = 1, \dots, d$.

There are several ways to prove this result. We will use the following observation: $x^\star = (x_1^\star, \dots x_d^\star) \in \partial\varphi(x)$ if and only if $x_i^\star \in \partial\phi_i(x_i)$ for every $i = 1, \dots, d$, where $\phi_i : \mathbb{R} \to \mathbb{R}$ is defined by $\phi_i(x_i) = |x_i|$. Indeed, assume that $x^\star = (x_1^\star, \dots x_d^\star) \in \partial\varphi(x)$, so that

$$\|z\|_1 - \|x\|_1 \geq \langle x^\star, z - x\rangle, \quad \forall z \in \mathbb{R}^d, \tag{4.18}$$

where $\langle x, z\rangle = x \cdot z$, the usual inner product in $\mathbb{R}^d$. For any $i = 1, \dots, d$, choose a z_i', and select $z = x + (z_i' - x_i)e_i$, where e_i is the unit vector in the i-th direction. Clearly, for this choice of z, we have that $\|z\|_1 - \|x\|_1 = |z_i'| - |x_i|$ and $\langle x^\star, z - x\rangle = x_i^\star(z_i' - x_i)$, so (4.18) yields $|z_i'| - |x_i| \geq x_i^\star(z_i' - x_i)$, and since $z_i' \in \mathbb{R}$ is arbitrary, we conclude that $x_i^\star \in \partial\phi_i(x_i)$. For the converse, let $x = (x_1, \dots, x_d) \in \mathbb{R}^d$, $x^\star = (x_1^\star, \dots, x_d^\star) \in \mathbb{R}^d$, and assume that $x_i^\star \in \partial\phi(x_i)$ for any $i = 1, \dots, d$, so $|z_i| - |x_i| \geq x_i^\star(z_i - x_i)$ for every $x_i \in \mathbb{R}$. Adding over all $i = 1, \dots, d$, we recover (4.18), therefore $x^\star \in \partial\varphi(x)$. Our claim now follows easily from the above observation and Example 4.1.3. ◁

4.2.4 The subdifferential and the duality map

Throughout this section, let X be a Banach space with norm $\|\cdot\|$, $X^\star$ its dual with norm $\|\cdot\|_{X^\star}$, and $\langle\cdot,\cdot\rangle$ the duality pairing between them. We will study the connection of the subdifferential of a norm related functional with the duality map $\mathrm{J} : X \to 2^{X^\star}$ (see Definition 1.1.17) and explore its connections with the geometry of Banach spaces. Our approach follows [52].

We start by defining the set valued mapping $\mathrm{J}_1 : X \to 2^{X^\star}$ as

$$\mathrm{J}_1(\mathrm{x}) = \begin{cases} \{\mathrm{x}^\star \in X^\star \,:\, \|\mathrm{x}^\star\|_{X^\star} = 1,\ \langle \mathrm{x}^\star, \mathrm{x}\rangle = \|\mathrm{x}\|\}, & \text{if } \mathrm{x} \neq 0,\\ \overline{B}_{X^\star}(0,1), & \text{if } \mathrm{x} = 0, \end{cases} \tag{4.19}$$

and the family of set valued maps $\mathrm{J}_p : X \to 2^{X^\star}$ defined by

$$\mathrm{J}_p(\mathrm{x}) = \{\mathrm{x}^\star \in X^\star \,:\, \|\mathrm{x}^\star\|_{X^\star} = \|\mathrm{x}\|^{p-1},\ \langle \mathrm{x}^\star, \mathrm{x}\rangle = \|\mathrm{x}\|^p\}, \quad p > 1, \tag{4.20}$$

where it clearly holds that $\mathrm{J}_p(0) = \{0\}$, $p > 1$. For $p = 2$, the mapping J_2 coincides with the duality map (see Definition 1.1.17),

$$\mathrm{J}_2(\mathrm{x}) = \mathrm{J}(\mathrm{x}) = \{\mathrm{x}^\star \in X^\star \,:\, \langle \mathrm{x}^\star, \mathrm{x}\rangle = \|\mathrm{x}^\star\|_{X^\star}\|\mathrm{x}\|,\ \|\mathrm{x}\| = \|\mathrm{x}^\star\|_{X^\star}\},$$

while

$$\mathrm{J}_p(\mathrm{x}) = \|\mathrm{x}\|^{p-1}\mathrm{J}_1(\mathrm{x}), \quad \forall\, \mathrm{x} \in X,\ p > 1.$$

With the use of the maps J_p, $p \geq 1$, we can generalize Examples 4.1.3 and 4.1.4 in the general setting of Banach spaces. This is done in the following proposition [52].

Proposition 4.2.17 (The subdifferential of powers of the norm). *On the Banach space X, define the maps $\varphi_p : X \to \mathbb{R}$ by $\varphi_p(x) = \frac{1}{p}\|x\|^p$, $p \geq 1$. Then, $\partial\varphi_p(x) = J_p(x)$, $p \geq 1$.*

Proof. We break the proof into two steps. In step 1, we consider the case $p = 1$, and in step 2 the case $p > 1$.

1. We let $p = 1$, and noting that the definition of J_1 will need to consider the cases $x \neq 0$ and $x = 0$ separately.

Let $x \neq 0$. We first establish that $J_1(x) \subset \partial\varphi_1(x)$. Indeed, for any $x^\star \in J_1(x)$, $x \neq 0$, using (4.19), it holds that for any $z \in X$,

$$\begin{aligned}\|z\| - \|x\| &= \|z\| - \langle x^\star, x\rangle = \|z\| - \langle x^\star, z\rangle + \langle x^\star, z - x\rangle \\ &\geq \|z\| - |\langle x^\star, z\rangle| + \langle x^\star, z - x\rangle \geq \|z\| - \|x^\star\|_{X^\star}\|z\| + \langle x^\star, z - x\rangle \\ &= (1 - \|x^\star\|_{X^\star})\|z\| + \langle x^\star, z - x\rangle = \langle x^\star, z - x\rangle,\end{aligned}$$

where, in the last inequality, we used once more (4.19). Since $z \in X$ is arbitrary, we conclude that $\|z\| - \|x\| \geq \langle x^\star, z - x\rangle$ for every $z \in X$, hence, $x^\star \in \partial\varphi_1(x)$.

For the reverse inclusion, $\partial\varphi_1(x) \subset J_1(x)$, consider any $x^\star \in \partial\varphi_1(x)$. By the definition of the subdifferential,

$$\|z\| \geq \|x\| + \langle x^\star, z - x\rangle, \quad \forall z \in X. \tag{4.21}$$

Setting first $z = 0$ and second $z = 2x$ in (4.21), we obtain, respectively, that $\langle x^\star, x\rangle \geq \|x\|$ and $\langle x^\star, x\rangle \leq \|x\|$, from which follows that $\langle x^\star, x\rangle = \|x\|$. It remains to show that $\|x^\star\|_{X^\star} = 1$. Combining (4.21) with $\langle x^\star, x\rangle = \|x\|$, we see that $\|z\| \geq \langle x^\star, z\rangle$ for every $z \in X$. This implies that[12] $\|x^\star\|_{X^\star} = 1$, from which it follows that $x^\star \in J_1(x)$, thus proving the reverse inclusion.

Let $x = 0$. If $x^\star \in J_1(0) = \overline{B}_{X^\star}(0,1)$, then clearly $\langle x^\star, z\rangle \leq \|z\|$ for every $z \in X^\star$; hence $x^\star \in \partial\varphi_1(0)$. For the reverse inclusion, $x^\star \in \partial\varphi_1(0)$ implies that $\|z\| \geq \langle x^\star, z\rangle$ for every $z \in X$, and dividing by $\|z\|$ and setting $z' = \frac{1}{\|z\|}z \in \overline{B}_X(0,1)$, we see that $1 \geq \langle x^\star, z'\rangle$, which is true for any $x^\star \in \overline{B}_{X^\star}(0,1)$.

2. We now let $p > 1$. As in step 1, we will consider the cases $x \neq 0$ and $x = 0$ separately.

Let $x \neq 0$, and assume that $x^\star \in \partial\varphi_p(x)$. Then, by definition,

$$\frac{1}{p}\|z\|^p - \frac{1}{p}\|x\|^p \geq \langle x^\star, z - x\rangle, \quad \forall z \in X. \tag{4.22}$$

Choosing $z = (1 + \epsilon)x$, for $\epsilon > 0$, (4.22) yields

---

12 Dividing both sides of $\|z\| \geq \langle x^\star, z\rangle$ by $\|z\|$, and taking the supremum, we conclude that $\|x^\star\|_{X^\star} = \sup_{z\in\overline{B}_X(0,1)}\langle x^\star, z\rangle \leq 1$. On the other hand, $\bar{x} = \frac{x}{\|x\|} \in \overline{B}_X(0,1)$ and $\langle x^\star, \bar{x}\rangle = 1$, hence $\sup_{z\in\overline{B}_X(0,1)}\langle x^\star, z\rangle = 1$.

$$\frac{1}{p}\frac{(1+\epsilon)^p-1}{\epsilon}\|x\|^p \geq \langle x^\star, x\rangle,$$

and passing to the limit, as $\epsilon \to 0^+$, we conclude that $\|x\|^p \geq \langle x^\star, x\rangle$. Choosing next $z = (1-\epsilon)x$, $\epsilon > 0$ in (4.22) and rearranging gives

$$\frac{1}{p}\frac{(1-\epsilon)^p-1}{\epsilon}\|x\|^p \geq -\langle x^\star, x\rangle,$$

and passing to the limit, as $\epsilon \to 0^+$, we conclude that $\|x\|^p \leq \langle x^\star, x\rangle$. Combining the two above inequalities, we get that $\|x\|^p = \langle x^\star, x\rangle$. Then, clearly $\|x\|^p = \langle x^\star, x\rangle = |\langle x^\star, x\rangle| \leq \|x^\star\|_{X^\star}\|x\|$, and dividing both sides by $\|x\|$, we obtain that $\|x\|^{p-1} \leq \|x^\star\|_{X^\star}$. We claim that the reverse inequality also holds, i. e., that $\|x^\star\|_{X^\star} \leq \|x\|^{p-1}$, therefore $\|x\|^{p-1} = \|x^\star\|_{X^\star}$. To prove the claim, consider any $z \in X$ such that $\|z\| \leq \|x\|$ and apply (4.22). We see that $\langle x^\star, z-x\rangle \leq 0$, and recalling that $\langle x^\star, x\rangle = \|x\|^p$, we obtain that $\langle x^\star, z\rangle \leq \|x\|^p$, and dividing both sides with $\|x\|$, setting $z' = \frac{1}{\|x\|}z$, we conclude that $\langle x^\star, z'\rangle \leq \|x\|^{p-1}$. Since $z \in X$ is arbitrary (with $\|z\| \leq \|x\|$), z' an arbitrary element, $z' \in \overline{B}_X(0,1)$. Furthermore, repeating the same argument with $-z$ in the place of z, we conclude that $|\langle x^\star, z'\rangle| \leq \|x\|^{p-1}$ for every $z' \in \overline{B}_X(0,1)$, hence, taking the supremum over all such z' and recalling the definition of $\|x^\star\|_{X^\star}$, we conclude that $\|x^\star\|_{X^\star} \leq \|x\|^{p-1}$, which is the required claim.

Hence, for any $x^\star \in \partial\varphi_p(x)$, it holds that $\langle x^\star, x\rangle = \|x\|^p$ and $\|x^\star\|_{X^\star} = \|x\|^{p-1}$, which implies that $x^\star \in J_p(x)$, and we have that $\partial\varphi_p(x) \subset J_p(x)$.

To prove the reverse inclusion, recall the elementary inequality $\frac{p-1}{p}a^p + \frac{1}{p}b^p \geq a^{p-1}b$, which is valid[13] for any $a, b > 0$ and $p > 1$. Consider any $x^\star \in J_p(x)$, i. e., $\langle x^\star, x\rangle = \|x\|^p$ and $\|x^\star\|_{X^\star} = \|x\|^{p-1}$. Observe that $\frac{1}{p}\|z\|^p - \frac{1}{p}\|x\|^p = \frac{1}{p}\|z\|^p + \frac{p-1}{p}\|x\|^p - \|x\|^p$, and using the above inequality for $a = \|x\|^p$, $b = \|z\|^p$, we obtain

$$\begin{aligned}\frac{1}{p}\|z\|^p - \frac{1}{p}\|x\|^p &\geq \|x\|^{p-1}\|z\| - \|x\|^p \\ &= \|x^\star\|_{X^\star}\|z\| - \langle x^\star, x\rangle \geq \langle x^\star, z\rangle - \langle x^\star, x\rangle = \langle x^\star, z-x\rangle,\end{aligned}$$

where in the second line, we used the fact that $x^\star \in J_p(x)$ and $\langle x^\star, z\rangle \leq \|x^\star\|_{X^\star}\|z\|$. Hence, by the fact that z is arbitrary, we conclude that for any $x^\star \in J_p(x)$, it holds that $\frac{1}{p}\|z\|^p - \frac{1}{p}\|x\|^p \geq \langle x^\star, z-x\rangle$ for every $z \in X$, i. e., $x^\star \in \partial\varphi_p(x)$, and $J_p(x) \subset \partial\varphi_p(x)$.

Let $x = 0$. If $x^\star \in \partial\varphi_p(0)$, then $\frac{1}{p}\|z\|^p \geq \langle x^\star, z\rangle$ for every $z \in X$. Consider any $x \in X$, and choosing first $z = \epsilon x$, for any $\epsilon > 0$, we get $\frac{\epsilon^{p-1}}{p}\|x\|^p \geq \langle x^\star, x\rangle$, whereas choosing $z = -\epsilon x$ for any $\epsilon > 0$, we get $\frac{\epsilon^{p-1}}{p}\|x\|^p \geq -\langle x^\star, x\rangle$, from which we conclude that $|\langle x^\star, x\rangle| \leq \frac{\epsilon^{p-1}}{p}\|x\|^p$.

---

13 This follows by a simple convexity argument. Since $\varphi(s) = -\ln(s)$ is a convex function, we have that $-\ln(\frac{p-1}{p}a^p + \frac{1}{p}b^p) \leq -\frac{p-1}{p}\ln a^p - \frac{1}{p}\ln b^p$, which upon rearragement and exponentiation leads to the stated inequality.

Passing to the limit, as $\epsilon \to 0^+$, and since $x \in X$ is arbitrary, we conclude that $\langle x^\star, x\rangle = 0$ for every $x \in X$, therefore $x^\star = 0$. Hence, $\partial\varphi_p(0) = \{0\} = J_p(0)$. □

By the converse of Proposition 4.1.2 (see Example 4.2.10), we may remark that since $J_1(x) = \partial\|x\|$ if J_1 is a single valued mapping on $X \setminus \{0\}$, then the norm will be a Gâteaux differentiable function at any point but the origin. With a similar reasoning, if J_p, $p > 1$, is a single valued mapping on X, then the function $\varphi_p(x) = \frac{1}{p}\|x\|^p$ will be Gâteaux differentiable everywhere. On the other hand, the geometry of the Banach space X (in the sense of Section 2.6) is reflected on the properties of the duality mapping, as the following proposition indicates, therefore leading to connections between the geometry of the space and the differentiability of the norm. This is done in the following proposition [52].

Proposition 4.2.18. *Let X be a Banach space, with strictly convex dual $X^\star$. Then for any $p > 1$, the mappings J_p are single valued, while J_1 is single valued in $X \setminus \{0\}$.*

Proof. Consider the case $p > 1$ first. Since $J_p(0) = \{0\}$, we only need to focus on $x \neq 0$. Assume, on the contrary, that $x_1^\star, x_2^\star \in J_p(x)$, $x_1^\star \neq x_2^\star$. By the definition of J_p, it holds that $\|x_1^\star\|_{X^\star} = \|x_2^\star\|_{X^\star} = \|x\|^{p-1}$ and $\langle x_1^\star, x\rangle = \langle x_2^\star, x\rangle = \|x\|^p$, and combining these, we obtain that

$$\langle x_1^\star + x_2^\star, x\rangle = \langle x_1^\star, x\rangle + \langle x_2^\star, x\rangle = 2\|x\|^p = 2\|x\|^{p-1}\|x\| = 2\|x_1^\star\|_{X^\star}\|x\|.$$

Hence,

$$2\|x_1^\star\|_{X^\star}\|x\| = \langle x_1^\star + x_2^\star, x\rangle \le \|x_1^\star + x_2^\star\|_{X^\star}\|x\|,$$

and dividing both sides by $\|x\|$, we obtain that

$$2\|x_1^\star\|_{X^\star} = \|x_1^\star\|_{X^\star} + \|x_2^\star\|_{X^\star} \le \|x_1^\star + x_2^\star\|_{X^\star}.$$

Combining this with the triangle inequality, we immediately see that $\|x_1^\star + x_2^\star\|_{X^\star} = \|x_1^\star\|_{X^\star} + \|x_2^\star\|_{X^\star}$, which by the strict convexity of $X^\star$, using Theorem 2.6.5, implies that $x_1^\star = \lambda x_2^\star$ for some $\lambda > 0$, and since $\|x_1^\star\|_{X^\star} = \|x_2^\star\|_{X^\star}$, it must hold that $x_1^\star = x_2^\star$, which is a contradiction. A similar argument applies for J_1. □

We introduce the concept of a smooth Banach space.

Definition 4.2.19 (Smoothness). A normed space X is called smooth at a point $x_o \in S_X(0,1)$ if there exists a unique $x_o^\star \in S_{X^\star}(0,1)$ such that $\langle x_o^\star, x_o\rangle = 1$. It is called smooth if this property holds for every $x \in S_X(0,1)$.

The Gâteaux differentiability of the norm is equivalent to the smoothness of the space, as the next proposition [52] shows.

Proposition 4.2.20. *Let X be a Banach space. Then, the norm is Gâteaux differentiable on $X \setminus \{0\}$ if and only if X is smooth.*

Proof. Note that by the positive homogeneity of the norm, it suffices to restrict attention to $S_X(0,1)$. We start with the observation that an equivalent way to express smoothness is that $x^\star \mapsto \langle x^\star, x\rangle$ achieves its maximum value at a unique $x^\star \in S_{X^\star}(0,1)$. Assume that X is smooth. Let $\varphi(x) = \|x\|$, and considering $x \in S_X(0,1)$, we note that by Proposition 4.2.7, $\partial\varphi(x) = J_1(x) = \{x^\star \in X^\star : \|x^\star\|_{X^\star} = 1, \langle x^\star, x\rangle = 1\}$, so by the smoothness of X, $\partial\varphi(x)$ is a singleton and, by the converse of Proposition 4.1.2 (see Example 4.2.10), the norm is Gâteaux differentiable. For the converse, we proceed in the same fashion. □

Using Proposition 4.2.18, we may conclude the following proposition, which connects the smoothness of X with the geometry of $X^\star$:

Proposition 4.2.21. *Let X be a Banach space with strictly convex dual $X^\star$. Then, the norm is Gâteaux differentiable on $X \setminus \{0\}$, and X is smooth.*

Proof. Since $X^\star$ is strictly convex, by Proposition 4.2.18, J_1 is single valued on $X \setminus \{0\}$, hence, by Proposition 4.2.17, the subdifferential of the function $\varphi(x) = \|x\|$ is a singleton. Therefore, by the converse of Proposition 4.1.2 (see Example 4.2.10), the norm is Gâteaux differentiable at every $x \neq 0$, hence, by Proposition 4.2.20, X is smooth. □

Remark 4.2.22 (Fréchet differentiability of the norm). Under stronger conditions, we may prove that the norm is Fréchet differentiable. In particular, if X is a Banach space with locally uniformly convex dual $X^\star$, then X has a Fréchet differentiable norm (see, e. g., [68]).

4.2.5 Approximation of the subdifferential and density of its domain

We have seen in Proposition 4.2.1 that a proper convex lower semicontinuous function is subdifferentiable at its points of continuity. Two interesting questions are (a) what happens at the points where continuity does not hold, and (b) how is the set of points in which φ is subdifferentiable, i. e., $\mathbf{D}(\partial\varphi)$, related with the set of points in which φ is bounded, i. e., $\operatorname{dom}\varphi$ (which of course, by the general theory of convex functions, is related to the set of points, in which φ is continuous). The answer to these two important questions can be addressed via an approximation procedure for the subdifferential, which requires only lower semicontinuity.

Definition 4.2.23 (ϵ-subdifferential). Let $\varphi : X \to \mathbb{R} \cup \{+\infty\}$ be a proper convex lower semicontinuous function, and consider $x \in \operatorname{dom}\varphi$. For any $\epsilon > 0$, the ϵ-subdifferential of φ at x, denoted by $\partial_\epsilon\varphi(x)$, is defined as the set

$$\partial_\epsilon\varphi(x) := \{z^\star \in X^\star : \varphi(z) - \varphi(x) + \epsilon \geq \langle z^\star, z - x\rangle, \ \forall z \in X\}.$$

Example 4.2.24 (ϵ-subdifferential of the indicator function of a convex set). Consider a convex set $C \subset X$ and the indicator function of C, $I_C : X \to \mathbb{R} \cup \{+\infty\}$. Then,

$\partial_\epsilon I_C(x) = N_{\epsilon,C}(x) := \{x^\star \in X^\star : \langle x^\star, z - x\rangle \le \epsilon, \forall z \in C\}$ for any $x \in C$. This set is a generalization of the normal cone $N_C(x)$ (see Definition 4.1.5 and compare with Example 4.1.6). ◁

How is the ϵ-subdifferential related to the subdifferential of a function? It can easily be seen that for any $x \in X$, it holds that $\partial\varphi(x) = \bigcap_{\epsilon>0} \partial_\epsilon\varphi(x)$, a fact easily proved by the definitions. Furthermore, the nonemptyness of $\partial_\epsilon\varphi(x)$ for any $x \in \operatorname{dom}\varphi$ follows by the Hahn–Banach separation theorem,[14] and working similarly as for the subdifferential, one may prove that $\partial_\epsilon\varphi(x)$ is a weak* closed and convex set (details are left as an exercise). Finally, for any $\epsilon_1 < \epsilon_2$, it holds that $\partial_{\epsilon_1}\varphi(x) \subset \partial_{\epsilon_2}\varphi(x)$, while $t\partial_{\epsilon_1}(x) + (1-t)\partial_{\epsilon_2}(x) \subset \partial_{t\epsilon_1+(1-t)\epsilon_2}\varphi(x)$ for every $\epsilon_1, \epsilon_2 > 0$, $t \in [0,1]$.

The ϵ-subdifferential of φ provides an approximation of the subdifferential in the sense of Theorem 4.2.25 below, which also guarantees the density of the domain of the subdifferential in $\operatorname{dom}\varphi$. This result was first proved in [35]. The proof presented here is the one proposed in [114] (Theorem 3.17) and slightly different than the original proof of Brøndsted and Rockafellar (which does not use the Ekeland variational principle).

Theorem 4.2.25 (Brøndsted and Rockafellar). *Let X be a Banach space, $\varphi : X \to \mathbb{R}\cup\{+\infty\}$ be a proper convex and lower semicontinuous function and set arbitrary $x \in \operatorname{dom}\varphi$, $\epsilon > 0$, $x^\star \in \partial_\epsilon\varphi(x)$. Then, for any $\lambda > 0$, there exists $x_\lambda \in X$ and $x^\star_\lambda \in \partial\varphi(x_\lambda)$ such that*

$$\begin{aligned} \|x_\lambda - x\| &\le \frac{\epsilon}{\lambda}, \\ \|x^\star_\lambda - x^\star\|_{X^\star} &\le \lambda. \end{aligned} \tag{4.23}$$

This approximation guarantees the density of $\mathbf{D}(\partial\varphi)$ in $\operatorname{dom}\varphi$.

Proof. Since $x^\star \in \partial_\epsilon\varphi(x)$, it holds that $\varphi(z) - \varphi(x) + \epsilon \ge \langle x^\star, z - x\rangle$ for every $z \in X$, which we rearrange as $\varphi(z) - \langle x^\star, z\rangle + \epsilon \ge \varphi(x) - \langle x^\star, x\rangle$. Upon defining the (proper lower semicontinuous convex) function $\psi : X \to \mathbb{R}\cup\{+\infty\}$, by $\psi(z) := \varphi(z) - \langle x^\star, z\rangle$ for every $z \in X$, we express this inequality as $\psi(z) + \epsilon \ge \psi(x)$ for every $z \in X$; hence, this is true also for the infimum of ψ over X, so $\psi(x) \le \inf_{z\in X}\psi(z) + \epsilon$.

We now apply the Ekeland variational principle,[15] on ψ (taking into account the Banach space structure of X) to guarantee the existence of a $x_\lambda \in \operatorname{dom}\varphi$ such that

$$\begin{aligned} &\lambda\|x_\lambda - x\| < \epsilon, \\ &\psi(x_\lambda) \le \psi(z) + \lambda\|x_\lambda - z\|, \quad \forall z \in X. \end{aligned} \tag{4.24}$$

We will show that this x_λ is the one we seek by constructing an appropriate $x^\star_\lambda \in \partial\varphi(x_\lambda)$.

---

14 Apply the separation theorem to the set $\operatorname{epi}\varphi$ and the point $(x, \varphi(x) - \epsilon) \notin \operatorname{epi}\varphi$.

15 If δ', ϵ' are the parameters in Theorem 3.6.1, we use the scaling $\delta' = \epsilon$ and $\frac{1}{\epsilon'} = \frac{\epsilon}{\lambda}$, so $\epsilon'\delta' = \lambda$. As ϵ is considered given, we relabel the point $x_{\epsilon',\delta'}$, chosen by Ekeland variational principle, as x_λ.

Define the function $\phi : X \to \mathbb{R}\cup\{+\infty\}$ by $\phi(z) := \lambda\|z-x_\lambda\|$ for every $z \in X$, and observe that since $\phi(x_\lambda) = 0$, the second inequality of (4.24) implies that $0 \in \partial(\psi+\phi)(x_\lambda)$. However, since ϕ is continuous, by Theorem 4.2.12, we have that $\partial(\psi + \phi)(x_\lambda) = \partial\psi(x_\lambda) + \partial\phi(x_\lambda)$, hence

$$0 \in \partial\psi(x_\lambda) + \partial\phi(x_\lambda). \tag{4.25}$$

By the definition of ψ, it follows that $\partial\psi(x_\lambda) = \partial\varphi(x_\lambda) - x^\star$. Furthermore, by subdifferential calculus and Proposition 4.2.17, or by direct computation, we have that

$$\partial\phi(x_\lambda) = \{z^\star \in X^\star \ : \ \|z^\star\| \le \lambda\} = \overline{B}_{X^\star}(0,\lambda).$$

We can therefore express (4.25) as

$$0 \in \partial\varphi(x_\lambda) - x^\star + \overline{B}_{X^\star}(0,\lambda). \tag{4.26}$$

Hence, there exists $x_\lambda^\star \in \partial\varphi(x_\lambda)$ and $z^\star \in \overline{B}_{X^\star}(0,\lambda)$ such that $0 = x_\lambda^\star - x^\star + z^\star$, which implies that $x^\star - x_\lambda^\star = z^\star$, therefore $\|x^\star - x_\lambda^\star\|_{X^\star} \le \lambda$. This $x_\lambda^\star$ is the one we seek and completes the proof of the first claim. The second claim follows by the first one and a proper choice of a sequence of λ's. □

4.3 The subdifferential and optimization

The first-order conditions may be generalized, in terms of the subdifferential, turning it into a very useful tool for nonsmooth optimization (see e. g., [12, 20, 21, 29, 113]).

Proposition 4.3.1. *Let X be Banach space and $\varphi : X \to \mathbb{R}\cup\{+\infty\}$ a proper convex lower semicontinuous function. Then, $x_o \in X$ is a minimum of φ if and only if $0 \in \partial\varphi(x_o)$.*

Proof. If x_o is a minimum of φ, then $\varphi(x_o) \le \varphi(z)$ for every $z \in X$, and this can trivially be expressed as

$$\varphi(z) - \varphi(x_o) \ge 0 = \langle 0, z - x_o\rangle, \quad \forall z \in X,$$

which implies that $0 \in \partial\varphi(x_o)$. The converse follows directly from the definition of $\partial\varphi$. □

On the other hand, by replacing the subdifferential with the ϵ-subdifferential, we obtain an "approximate" minimization result.

Proposition 4.3.2. *Let X be Banach space and $\varphi : X \to \mathbb{R}\cup\{+\infty\}$ be a proper convex lower semicontinuous function. Then, $0 \in \partial_\epsilon\varphi(x_o)$ if and only if*

$$\inf_{x\in X}\varphi(x) \ge \varphi(x_o) - \epsilon. \tag{4.27}$$

Furthermore, if x_o *is an approximate minimum satisfying* (4.27), *then, for any* $\epsilon_1, \epsilon_2 > 0$, *such that* $\epsilon = \epsilon_1\epsilon_2$, *there exist* $x \in X$, $x^\star \in \partial\varphi(x)$ *such that* $\|x - x_o\| < \epsilon_2$ *and* $\|x^\star\|_{X^\star} \leq \epsilon_1$.

Proof. If $0 \in \partial_\epsilon\varphi(x_o)$, then $\varphi(z) - \varphi(x_o) + \epsilon \geq 0$ for every $z \in X$, so upon rearranging and taking the infimum over all $z \in X$, (4.27) is proved. For the converse, since (4.27) holds, $\varphi(z) - \varphi(x_o) + \epsilon \geq 0$ for every $z \in X$, so $0 \in \partial_\epsilon\varphi(x_o)$.

For the second claim, choose $\epsilon_o > 0$ such that $\varphi(x_o) - \inf_{x\in X}\varphi(x) \leq \epsilon_o \leq \epsilon_1\epsilon_2$, and choose λ so that $\frac{\epsilon_o}{\epsilon_2} \leq \lambda \leq \epsilon_1$. For this choice of ϵ_o, it holds that $0 \in \partial_{\epsilon_o}\varphi(x_o)$. So applying the Brøndsted–Rockafellar theorem 4.2.25, we conclude the existence of $x \in X$, $x^\star \in \partial\varphi(x)$ such that $\|x - x_o\| \leq \frac{\epsilon_o}{\lambda} \leq \epsilon_2$ and $\|x^\star\|_{X^\star} \leq \lambda \leq \epsilon_1$. □

Consider now a constrained convex minimization problem of the form $\inf_{x\in C}\varphi(x)$, where $C \subset X$ is a closed and convex subset of the Banach space X, and $\varphi : X \to \mathbb{R}$ is a convex function. Instead of the original constrained problem, we may consider the unconstrained problem $\inf_{x\in X}(\varphi(x) + I_C(x))$, where the original function φ is now perturbed by the indicator function of the convex set C. By the definition of the indicator function, if $x \in C$, then the minimizer of the original and the perturbed problem coincide, whereas if $x \notin C$, then the perturbed problem does not have a solution. Therefore, in some sense, the perturbed problem is a penalized version of the original one, with the indicator function playing the role of the penalty function.

The following proposition offers a generalization of the Karusch–Kuhn–Tucker conditions for the optimization of nonsmooth convex functions under convex constraints in Banach space:

Proposition 4.3.3. *Let* $C \subset X$ *be a closed convex set,* $\varphi : X \to \mathbb{R} \cup \{+\infty\}$ *be a lower semicontinuous convex function such that* φ *is continuous at some point of* C ($\operatorname{cont}(\varphi)\cap C \neq \emptyset$) *or* $\operatorname{dom}(\varphi) \cap \operatorname{int}(C) \neq \emptyset$, *and consider the problem*

$$\inf_{x\in C}\varphi(x). \tag{4.28}$$

Then x_o *is a solution of* (4.28) *if and only if there exist* $z_1^\star \in \partial\varphi(x_o)$ *with the property* $\langle z_1^\star, z - x_o\rangle \geq 0$, *for all* $z \in C$.

Stated differently, x_o *is a solution of* (4.28) *if and only if there exist* $z_1^\star \in \partial\varphi(x_o)$ *and* $z_2^\star \in N_C(x_o)$ *(i. e., such that* $\langle z_2^\star, z - x_o\rangle \leq 0$ *for every* $z \in C$*), with the property* $z_1^\star + z_2^\star = 0$.

Proof. Let $\psi : X \to \mathbb{R}$ be defined as $\psi(x) = \varphi(x) + I_C(x)$ for any $x \in X$. Then, by subdifferential calculus $\partial\psi(x) = \partial\varphi(x) + \partial I_C(x)$ and $\partial I_C(x) = N_C(x) = \{x^\star \in X^\star : \langle x^\star, z - x\rangle \leq 0, \ \forall z \in C\}$. By Proposition 4.3.1, x_o is a solution of the constrained problem if and only if $0 \in \partial\psi(x_o)$, therefore if $0 \in \partial\varphi(x_o) + \partial I_C(x_o)$. This implies that there exist $z_1^\star \in \partial\varphi(x_o)$ and $z_2^\star \in \partial I_C(x_o) = N_C(x_o)$ (i. e., $z_2^\star \in X^\star$ such that $\langle z_2^\star, z - x_o\rangle \leq 0$, $\forall z \in C$), such that $z_1^\star + z_2^\star = 0$. □

Example 4.3.4 (The projection operator revisited). Let $X = H$, a Hilbert space, identified with its dual, $C \subset X$ be a convex closed subset of X, and consider the following approximation problem: Given $x \in X$, find $x_o \in C$ such that $\|x - x_o\| = \inf_{z \in C} \|x - z\|$. This is the problem of best approximation of any element $x \in X$ by an element $x_o \in C$. We have already encountered this problem before and recognized its solution as the projection of x on C. We will now revisit it using Proposition 4.3.3.

The approximation problem is equivalent to $\min_{z \in C} \frac{1}{2}\|x - z\|^2$. Let us define the convex function $\varphi : X \to \mathbb{R}$ by $z \mapsto \frac{1}{2}\|x - z\|^2$. By direct calculation, $\partial\varphi(z) = z - x$ for any $z \in X$, and a straightforward application of Proposition 4.3.3 leads us to the conclusion that there must be a $z_1^\star = x_o - x$ and a $z_2^\star \in X$ such that $\langle z_2^\star, z - x_o\rangle \le 0, \forall z \in C$, and $z_1^\star + z_2^\star = 0$. This leads us to a representation of x_o as $x_o - x + z_2^\star = 0$, i. e., $x_o = x - z_2^\star$ where $z_2^\star$ is as above. By expressing $z_2^\star$ as $z_2^\star = x - x_o$, we can characterize x_o by the variational inequality $\langle x - x_o, z - x_o\rangle \le 0$ for every $z \in C$, which allows us to deduce that $x_o = P_C(x)$ (see Theorem 2.5.3). It also leads to the equivalent characterization $x - x_o \in N_C(x_o)$.

The projection operator can be generalized in a Banach space context under restrictive conditions on X (e. g., if X is reflexive and X and $X^\star$ are uniformly convex), in which case, the operator assigning to each $x \in X$ the element $P_C(x) = x_o = \arg\min_{z \in C} \|z - x\|$ is well posed and single valued, and is characterized by the variational inequality $\langle -J(x_o - x), x_o - z\rangle \ge 0$ for every $z \in C$ or in equivalent form $\langle J(x - x_o), x_o - z\rangle \ge 0$ for every $z \in C$, where $J : X \to X^\star$ is the duality map. However, this operator does not share the nice properties of the projection in Hilbert space (e. g., it may not be a linear operator even when $C = E$, a closed linear subspace). Note that this formulation is also useful in the case where X is a Hilbert space but not identified with its dual. ◁

Example 4.3.5 (Optimization problems with affine constraints). Let X, Y be Banach spaces, $L : X \to Y$ be a linear continuous operator, $\varphi : X \to \mathbb{R} \cup \{+\infty\}$ be a proper lower semicontinuous function that is continuous at some point in C, and consider convex minimization problems with affine constraints of the form

$$\min_{x \in C} \varphi(x), \quad C = \{x \in X : Lx = y\}. \tag{4.29}$$

Let $x_o \in C$ be a minimizer of (4.29). In this case, the normal cone of C is $N_C(x) = (\mathbf{N}(L))^\perp$ for every $x \in C$ (see Example 4.1.7) so that the first-order condition becomes $0 \in \partial\varphi(x_o) + N_C(x_o)$, where the additivity of the subdifferential follows by the conditions on φ and L. This condition implies that $x_o \in C$ solves (4.29) if and only if there exists $z_0^\star \in (\mathbf{N}(L))^\perp$ such that $-z_0^\star \in \partial\varphi(x_o)$. If we assume furthermore, that L has closed range, then by the general theory of linear operators, we know that $(\mathbf{N}(L))^\perp = \mathbf{R}(L^\star)$, where $L^\star : Y^\star \to X^\star$ is the adjoint operator. In this case, the solvability condition reduces to the existence of some $y_0^\star \in Y^\star$ such that $-L^\star y_0^\star \in \partial\varphi(x_o)$. ◁

4.4 Regularization: Moreau–Yosida approximation

4.4.1 The proximity operator: definition and fundamental properties

For this entire section, we let $X = H$ be a Hilbert space, identified with its dual, and $\varphi : H \to \mathbb{R} \cup \{+\infty\}$ be a proper convex lower semicontinuous function.

Fix a $x \in H$, and consider the minimization problem

$$\min_{z\in H} \varphi_x, \quad \text{where, } \varphi_x := \frac{1}{2}\|z - x\|^2 + \varphi(z),$$

which by the strict convexity of φ_x admits a unique solution x_o. In terms of this solution, we may define the single valued operator $\text{prox}_\varphi : H \to H$ defined by $\text{prox}_\varphi x := x_o$. This operator is called the proximity operator or Moreau proximity operator after the fundamental contribution of [106].

Definition 4.4.1 (Proximity operator). The proximity operator $\text{prox}_\varphi : H \to H$, corresponding to function φ, is defined by

$$\text{prox}_\varphi x := x_o = \arg\min_{z\in H}\left(\frac{1}{2}\|z - x\|^2 + \varphi(z)\right).$$

The proximity operator is a generalization of the projection operator. Indeed, in the special case where $\varphi = I_C$, the indicator function of a closed convex set C, the proximity operator prox_φ coincides with the projection operator P_C. Furthermore, the family of operators $\{J_\lambda : \lambda > 0\} := \{\text{prox}_{\lambda\varphi} : \lambda > 0\}$, often called resolvent of the subdifferential of φ, plays an important role in an approximation procedure of proper convex lower semicontinuous functions, called the Moreau–Yosida approximation.

We start by stating some fundamental properties of the operator prox_φ.

Proposition 4.4.2 (Properties of the proximity operator). *Let H be a Hilbert space with inner product $\langle\cdot,\cdot\rangle$, identified with its dual, $\varphi : H \to \mathbb{R} \cup \{+\infty\}$ a proper convex lower semicontinuous function, $\text{prox}_\varphi : H \to H$ the corresponding proximity operator. Then,*

(i) *The operator prox_φ is characterized by the variational inequality*

$$\langle x - \text{prox}_\varphi x, z - \text{prox}_\varphi x\rangle \le \varphi(z) - \varphi(\text{prox}_\varphi x), \quad \forall z \in H. \tag{4.30}$$

often stated in its equivalent form

$$\langle \text{prox}_\varphi x - x, z - \text{prox}_\varphi x\rangle + \varphi(z) - \varphi(\text{prox}_\varphi x) \ge 0, \quad \forall z \in H. \tag{4.31}$$

(ii) *The operator prox_φ is (a) single valued and enjoys the monotonicity property*

$$\langle \text{prox}_\varphi x_1 - \text{prox}_\varphi x_2, x_1 - x_2\rangle \ge 0, \quad \forall x_1, x_2 \in H, \tag{4.32}$$

(b) is nonexpansive, i. e.,

$$\|\mathrm{prox}_\varphi \mathrm{x}_1 - \mathrm{prox}_\varphi \mathrm{x}_2\| \le \|\mathrm{x}_1 - \mathrm{x}_2\|, \quad \forall\, \mathrm{x}_1, \mathrm{x}_2 \in H, \tag{4.33}$$

(c) is firmly nonexpansive,[16] *i. e.,*

$$\|\mathrm{prox}_\varphi \mathrm{x}_1 - \mathrm{prox}_\varphi \mathrm{x}_2\|^2 \le \langle \mathrm{x}_1 - \mathrm{x}_2, \mathrm{prox}_\varphi \mathrm{x}_1 - \mathrm{prox}_\varphi \mathrm{x}_2 \rangle, \quad \forall\, \mathrm{x}_1, \mathrm{x}_2 \in H. \tag{4.34}$$

(iii) *The operator* prox_φ *is characterized in terms of the subdifferential of* φ, *and in particular*

$$x_o = \mathrm{prox}_\varphi \mathrm{x} \quad \textit{if and only if} \quad \mathrm{x} - \mathrm{x}_o \in \partial\varphi(\mathrm{x}_\star).$$

This condition is often expressed as

$$\mathrm{prox}_\varphi(\mathrm{x}) = (I + \partial\varphi)^{-1}(\mathrm{x}).$$

An equivalent way of stating this is that $\mathrm{Fix}(\mathrm{prox}_\varphi) = \arg\min_{\mathrm{x}\in H} \varphi(\mathrm{x})$, *where by* $\mathrm{Fix}(\mathrm{prox}_\varphi)$; *we denote the set of fixed points of the proximity operator, and by* $\arg\min_{\mathrm{x}\in H} \varphi(\mathrm{x})$ *we denote the set of minimizers of the function* φ.

(iv) *The operator* $I - \mathrm{prox}_\varphi$ *is also firmly nonexpansive, i. e.,*

$$\begin{aligned} &\big\|(I - \mathrm{prox}_\varphi)(\mathrm{x}_1) - (I - \mathrm{prox}_\varphi)(\mathrm{x}_2)\big\|^2 \\ &\quad \le \big\langle (I - \mathrm{prox}_\varphi)(\mathrm{x}_1) - (I - \mathrm{prox}_\varphi)(\mathrm{x}_2), \mathrm{x}_1 - \mathrm{x}_2 \big\rangle, \quad \forall \mathrm{x}_1, \mathrm{x}_2 \in H, \end{aligned} \tag{4.35}$$

and furthermore,

$$\big\langle \mathrm{prox}_\varphi(\mathrm{x}_1) - \mathrm{prox}_\varphi(\mathrm{x}_2), (I - \mathrm{prox}_\varphi)(\mathrm{x}_1) - (I - \mathrm{prox}_\varphi)(\mathrm{x}_2) \big\rangle \ge 0, \quad \forall\, \mathrm{x}_1, \mathrm{x}_2 \in H. \tag{4.36}$$

Proof. (i) We will use the simplified notation $\mathrm{x}_o = \mathrm{prox}_\varphi \mathrm{x}$. Assume first that x_o satisfies the variational inequality (4.31). Then,[17] for any $\mathrm{z} \in H$,

$$\begin{aligned} \varphi_{\mathrm{x}}(\mathrm{z}) - \varphi_{\mathrm{x}}(\mathrm{x}_o) &= \frac{1}{2}\|\mathrm{z} - \mathrm{x}\|^2 - \frac{1}{2}\|\mathrm{x}_o - \mathrm{x}\|^2 + \varphi(\mathrm{z}) - \varphi(\mathrm{x}_o) \\ &= \frac{1}{2}\|\mathrm{x}_o - \mathrm{z}\|^2 + \langle \mathrm{x}_o - \mathrm{x}, \mathrm{z} - \mathrm{x}_o \rangle + \varphi(\mathrm{z}) - \varphi(\mathrm{x}_o) \\ &\ge \langle \mathrm{x}_o - \mathrm{x}, \mathrm{z} - \mathrm{x}_o \rangle + \varphi(\mathrm{z}) - \varphi(\mathrm{x}_o) \ge 0, \end{aligned}$$

by (4.31). Therefore, for any $\mathrm{z} \in H$, it holds that $\varphi_{\mathrm{x}}(\mathrm{z}) \ge \varphi_{\mathrm{x}}(\mathrm{x}_o)$. Hence, x_o is the minimizer of φ_{x}, which is unique by strict convexity.

---

16 This property reminds us of similar properties for the Fréchet derivative of convex functions, which are Lipschitz continuous; see Example 2.3.26!

17 Using the identity $\frac{1}{2}\|\mathrm{x}_1\|^2 - \frac{1}{2}\|\mathrm{x}_2\|^2 = \frac{1}{2}\|\mathrm{x}_1 - \mathrm{x}_2\|^2 + \langle \mathrm{x}_1 - \mathrm{x}_2, \mathrm{x}_2 \rangle$ for every $\mathrm{x}_1, \mathrm{x}_2 \in H$, for the choice $\mathrm{x}_1 = \mathrm{z} - \mathrm{x}$, and $\mathrm{x}_2 = \mathrm{x}_o - \mathrm{x}$.

For the converse, assume that x_o is the minimizer of φ_x on H. Consider any $z \in H$; for any $t \in (0,1)$, it holds that $\varphi_x((1-t)x_o + tz) - \varphi_x(x_o) \geq 0$. By the definition of φ_x, this implies that

$$\frac{1}{2}\|(1-t)x_o + tz - x\|^2 - \frac{1}{2}\|x_o - x\|^2 + \varphi((1-t)x_o + tz) - \varphi(x_o) \geq 0,$$

which upon rearrangement yields

$$0 \leq \frac{1}{2}t^2\,\|z - x_o\|^2 + t\,\langle x_o - x, z - x_o\rangle + \varphi((1-t)x_o + tz) - \varphi(x_o). \tag{4.37}$$

The convexity of φ implies that $\varphi((1-t)x_o + tz) \leq (1-t)\varphi(x_o) + t\varphi(z)$. Hence,

$$\varphi((1-t)x_o + tz) - \varphi(x_o) \leq -t\varphi(x_o) + t\varphi(z),$$

which when combined with (4.37) yields

$$0 \leq \frac{1}{2}t^2\,\|z - x_o\|^2 + t\,\langle x_o - x, z - x_o\rangle - t\varphi(x_o) + t\varphi(z).$$

Dividing by t and passing to the limit, as $t \to 0^+$, leads to (4.31).

(ii) Consider any $x_1, x_2 \in H$, and let $z_i = \mathrm{prox}_\varphi x_i$, $i = 1,2$. Apply the variational inequality (4.31) first for the choice $x = x_1$, $z = z_2$ and second for the choice $x = x_2$, $z = z_1$ and add. The terms involving the function φ cancel, thus leading to

$$\langle z_1 - x_1, z_2 - z_1\rangle + \langle z_2 - x_2, z_1 - z_2\rangle \geq 0,$$

which is rearranged as

$$0 \leq \|z_2 - z_1\|^2 \leq \langle z_2 - z_1, x_2 - x_1\rangle. \tag{4.38}$$

This proves (4.34) and (4.32). We further estimate the right-hand side of (4.38) using the Cauchy–Schwarz inequality, and divide both sides of the resulting inequality by $\|z_2 - z_1\|$ to obtain (4.33).

(iii) This can either be seen directly from the definition of the subdifferential and the characterization of $x_o = \mathrm{prox}_\varphi x$, in terms of the variational inequality (4.31) or using the rules of subdifferential calculus.

(iv) We use the notation in (ii), and observe that

$$\begin{aligned}
\|(x_1 - z_1) - (x_2 - z_2)\|^2 &= \langle x_1 - x_2, x_1 - x_2\rangle - 2\langle x_1 - x_2, z_1 - z_2\rangle + \|z_1 - z_2\|^2 \\
&\overset{(4.34)}{\leq} \langle x_1 - x_2, x_1 - x_2\rangle - 2\langle x_1 - x_2, z_1 - z_2\rangle + \langle x_1 - x_2, z_1 - z_2\rangle \\
&= \langle x_1 - x_2, x_1 - x_2\rangle - \langle x_1 - x_2, z_1 - z_2\rangle \\
&= \big\langle (x_1 - z_1) - (x_1 - z_2), x_1 - x_2\big\rangle.
\end{aligned}$$

This is the strict nonexpansive property for $I - \text{prox}_\varphi$, i. e., (4.35). Similarly,

$$\langle (x_1 - z_1) - (x_2 - z_2), z_1 - z_2 \rangle = \langle x_1 - x_2, z_1 - z_2\rangle - \|z_1 - z_2\|^2 \overset{(4.34)}{\geq} 0,$$

which proves (4.36). □

4.4.2 The Moreau–Yosida approximation

A closely related notion to the Moreau proximity operator corresponding to some nonsmooth convex function φ is its Moreau–Yosida approximation φ_λ, $\lambda > 0$, which is a family of smooth approximations to φ. This approximation is very useful in various applications, including of course optimization. For its extension in Banach spaces, see Section 9.4.6.

Definition 4.4.3 (Moreau–Yosida approximation). The function $\varphi_\lambda : H \to \mathbb{R}$, defined by

$$\varphi_\lambda(x) = \min_{z \in H} \varphi_{x,\lambda}(z) = \varphi_{x,\lambda}(\text{prox}_{\lambda\varphi}(x)), \quad \text{where, } \varphi_{x,\lambda}(z) := \frac{1}{2\lambda}\|x - z\|^2 + \varphi(z),$$

is called the Moreau–Yosida approximation (or envelope) of φ, for which the notation $J_\lambda = \text{prox}_{\lambda\varphi} = (I + \lambda\partial\varphi)^{-1}$ (resolvent of $\partial\varphi$), is often used.

The following simple observation is interesting:

Example 4.4.4 (Behavior of $\text{prox}_{\lambda\varphi}$ as $\lambda \to 0^+$). Operator $\text{prox}_{\lambda\varphi}$ "approaches" the identity operator as $\lambda \to 0^+$. In particular, it holds that $\|\text{prox}_{\lambda\varphi}x - x\|_H \to 0$, as $\lambda \to 0^+$.

Indeed, for any $\lambda > 0$, setting $x_\lambda = \text{prox}_{\lambda\varphi}x$, we have that $x - x_\lambda \in \lambda\partial\varphi(x_\lambda)$, which implies that $\lambda\varphi(z) - \lambda\varphi(x_\lambda) \geq \langle x - x_\lambda, z - x_\lambda\rangle_H$ for every $z \in H$. Choosing $z = x$ and letting $\lambda \to 0^+$, we obtain the required result (assuming $\varphi(x) - \varphi(x_\lambda)$ finite). ◁

Example 4.4.5. It is straightforward to check that in the case where $H = \mathbb{R}$, $\varphi(x) = |x|$,

$$\text{prox}_{\lambda\varphi}x := x_{\lambda,o} = \begin{cases} x - \lambda & \text{if } x > \lambda, \\ 0 & \text{if } -\lambda \leq x \leq \lambda, \\ x + \lambda & \text{if } x < -\lambda. \end{cases}$$

Then,

$$\varphi_\lambda(x) = \begin{cases} x - \frac{\lambda}{2} & \text{if } x > \lambda, \\ \frac{x^2}{2\lambda} & \text{if } -\lambda \leq x \leq \lambda, \\ -x - \frac{\lambda}{2} & \text{if } x < -\lambda. \end{cases}$$

This can be generalized for $H = \mathbb{R}^d$ and $\varphi : H \to \mathbb{R}$ is defined by $\varphi(x) = \|x\|_{\ell^1}$. Working componentwise, for $x = (x_1, \ldots, x_d)$, if $x_o = \text{prox}_{\varphi,\lambda}x$, then for any $i = 1, \ldots, d$,

$$(\text{prox}_{\lambda\varphi}x)_i := (x_{\lambda,o})_i = \begin{cases} x_i - \lambda & \text{if } x_i > \lambda, \\ 0 & \text{if } -\lambda \leq x_i \leq \lambda, \\ x_i + \lambda & \text{if } x_i < -\lambda. \end{cases}$$

In this particular case, the proximity operator is often called the shrinkage or soft threshold operator, S_λ defined by $S_\lambda(x) := \text{prox}_{\lambda\|\cdot\|_1}(x)$. ◁

The Moreau envelope φ_λ, $\lambda > 0$ of a proper convex lower semicontinuous function φ is a convex function that enjoys some nice differentiability properties, and because of that finds important applications in nonsmooth optimization. These are collected in the following proposition (see e. g., [21]).

Proposition 4.4.6 (Properties of the Moreau envelope). *Let $\varphi : H \to \mathbb{R}\cup\{+\infty\}$ be a proper convex lower semicontinuous function, and let φ_λ be its Moreau envelope. Then,*

(i) *The function φ_λ is convex for every $\lambda > 0$, with*

$$\inf_{z\in H} \varphi(z) \leq \varphi_\lambda(x) \leq \varphi(x), \quad \forall x \in H,\ \forall \lambda > 0, \tag{4.39}$$

(ii) *It holds that $\inf_{z\in H} \varphi(z) = \inf_{z\in H} \varphi_\lambda(z)$ for every $\lambda > 0$. In particular, $x_o \in \text{argmin}_{z\in H}\varphi(z)$ if and only if $x_o \in \text{argmin}_{z\in H}\varphi_\lambda(z)$ for every $\lambda > 0$, while $\lim_{\lambda\to 0} \varphi_\lambda = \varphi$, pointwise.*

(iii) *For each $\lambda > 0$, the function φ_λ is Fréchet differentiable, with monotone and Lipschitz continuous Fréchet derivative $D\varphi_\lambda$, at any $x \in H$, and*

$$D\varphi_\lambda(x) = \frac{1}{\lambda}(x - \text{prox}_{\lambda\varphi}(x)) = \frac{1}{\lambda}(I - \text{prox}_{\lambda\varphi})(x).$$

Proof. (i) By definition, $\varphi_\lambda(x) = \inf_{z\in H} \varphi_{x,\lambda}(z) \leq \varphi_{x,\lambda}(x) = \varphi(x)$. For any $x, z' \in H$, $\lambda > 0$, it holds that $\inf_{z\in H} \varphi(z) \leq \varphi(z') \leq \varphi_{x,\lambda}(z')$. Taking the infimum over $z' \in H$, we have $\inf_{z\in H} \varphi(z) \leq \varphi_\lambda(x)$ for every $x \in H$, $\lambda > 0$, i. e., (4.39). Convexity follows from standard arguments. on the pointwise infimum of families of convex functions.

(ii) Since $\inf_{z\in H} \varphi(z) \leq \varphi_\lambda(x)$ for every $x \in H$, $\lambda > 0$, taking the infimum over $x \in H$,

$$\inf_{z\in H} \varphi(z) \leq \inf_{x\in H} \varphi_\lambda(x) = \inf_{z\in H} \varphi_\lambda(z), \quad \forall \lambda > 0. \tag{4.40}$$

Since by (4.39), $\inf_{z\in H} \varphi_\lambda(z) \leq \varphi_\lambda(x) \leq \varphi(x)$ for every $x \in H$, $\lambda > 0$, taking the infimum over $x \in H$,

$$\inf_{z\in H} \varphi_\lambda(z) \leq \inf_{x\in H} \varphi(x) = \inf_{z\in H} \varphi(z), \quad \forall \lambda > 0. \tag{4.41}$$

Therefore, by (4.40) and (4.41), we conclude that $\inf_{z\in H} \varphi_\lambda(z) = \inf_{z\in H} \varphi(z)$ for every $\lambda > 0$.

Consider now $x_o \in \text{argmin}_{x\in H}\varphi(x)$. By (4.39), applied for $x = x_o$, we have

$$\inf_{z\in H} \varphi(z) \leq \varphi_\lambda(x_o) \leq \varphi(x_o) = \inf_{z\in H} \varphi(z),$$

so $\varphi_\lambda(x_o) = \inf_{z\in H} \varphi(z) = \inf_{z\in H} \varphi_\lambda(z)$ for every $\lambda > 0$. Hence, x_o is a minimizer of φ_λ for every $\lambda > 0$.

We first show that $(\varphi_\lambda)_{\lambda>0}$ is monotone. For any $x, z \in H$ and any $\lambda_1 \le \lambda_2$, it holds that $\varphi_{\lambda_2}(x) = \inf_{z'\in H} \varphi_{x,\lambda_2}(z') \le \varphi_{x,\lambda_2}(z) \le \varphi_{x,\lambda_1}(z)$. Hence, $\varphi_{\lambda_2}(x) \le \varphi_{x,\lambda_1}(z)$ for every $z \in H$, and taking the infimum over all $z \in H$, we conclude that $\varphi_{\lambda_2}(x) \le \varphi_{\lambda_1}(x)$ for all $x \in H$. Combining that with (4.39), we have that for every $x \in H$ it holds that $\varphi_{\lambda_2}(x) \le \varphi_{\lambda_1}(x) \le \varphi(x)$. Passing to the limit, as $\lambda \to 0$, requires careful considerations (see Theorem 9.4.29).

(iii) We will use the simplified notation $z_i = \text{prox}_{\lambda\varphi}(x_i)$, $i = 1, 2$. By the definition of φ_λ, we have that $\varphi_\lambda(x_i) = \frac{1}{2\lambda}\|x_i - z_i\|^2 + \varphi(z_i)$ for any $x_i \in H$, $i = 1, 2$, so subtracting, we obtain

$$\varphi_\lambda(x_2) - \varphi_\lambda(x_1) = \frac{1}{2\lambda}(\|x_2 - z_2\|^2 - \|x_1 - z_1\|^2) + \varphi(z_2) - \varphi(z_1). \tag{4.42}$$

We now recall that (see (4.30)), that for any $x, z \in H$,

$$\langle x - \text{prox}_{\lambda\varphi}(x), z - \text{prox}_{\lambda\varphi}(x)\rangle \le \lambda\varphi(z) - \lambda\varphi(\text{prox}_{\lambda\varphi}(x)), \tag{4.43}$$

so setting $x = x_1$ and $z = z_2$, we obtain upon rearranging

$$\lambda(\varphi(z_2) - \varphi(z_1)) \ge \langle x_1 - z_1, z_2 - z_1\rangle. \tag{4.44}$$

Combining (4.42) and (4.44), we obtain that

$$\varphi_\lambda(x_2) - \varphi_\lambda(x_1) \ge \frac{1}{2\lambda}(\|x_2 - z_2\|^2 - \|x_1 - z_1\|^2 + 2\langle x_1 - z_1, z_2 - z_1\rangle). \tag{4.45}$$

We would like to further enhance this inequality by noting the identity

$$\|h_2\|^2 - \|h_1\|^2 = \|h_2 - h_1\|^2 + 2\langle h_1, h_2 - h_1\rangle,$$

which applied for $h_i = x_i - z_i$, $i = 1, 2$, yields that

$$\begin{aligned}\|x_2 - z_2\|^2 - \|x_1 - z_1\|^2 + 2\langle x_1 - z_1, z_2 - z_1\rangle &= \big\|z_2 - x_2 - (z_1 - x_1)\big\|^2 \\ &\quad + 2\langle x_2 - x_1, x_1 - z_1\rangle.\end{aligned} \tag{4.46}$$

Combining (4.45) and (4.46), we get

$$\begin{aligned}\varphi_\lambda(x_2) - \varphi_\lambda(x_1) &\ge \frac{1}{2\lambda}\big(\big\|z_2 - x_2 - (z_1 - x_2)\big\|^2 + 2\langle x_2 - x_1, x_1 - z_1\rangle\big) \\ &\ge \frac{1}{\lambda}\langle x_2 - x_1, x_1 - z_1\rangle.\end{aligned} \tag{4.47}$$

Hence, rearranging (4.47),

$$0 \le \varphi_\lambda(x_2) - \varphi_\lambda(x_1) - \frac{1}{\lambda}\langle x_2 - x_1, x_1 - z_1\rangle. \tag{4.48}$$

We now return to (4.47) once more setting $x = x_2$ and $z = z_1$, and proceed as above to obtain (after multiplication by −1) that

$$\varphi_\lambda(x_2) - \varphi_\lambda(x_1) \le \frac{1}{\lambda}\langle x_2 - x_1, x_2 - z_2\rangle. \tag{4.49}$$

Combining (4.48) with (4.49), we obtain

$$\begin{aligned} 0 &\le \varphi_\lambda(x_2) - \varphi_\lambda(x_1) - \frac{1}{\lambda}\langle x_2 - x_1, x_1 - z_1\rangle \\ &\overset{(4.49)}{\le} \frac{1}{\lambda}\big(\langle x_2 - x_1, x_2 - z_2\rangle - \langle x_2 - x_1, x_1 - z_1\rangle\big) \\ &= \frac{1}{\lambda}\big(\|x_2 - x_1\|^2 - \langle x_2 - x_1, z_2 - z_1\rangle\big) \le \frac{1}{\lambda}\|x_2 - x_1\|^2, \end{aligned} \tag{4.50}$$

since by the monotonicity of the proximity operator $\langle x_2 - x_1, z_2 - z_1\rangle \ge 0$. We now pick arbitrary $x, h \in H$, and we set in (4.50) $x_1 = x$, $z_1 = \mathrm{prox}_{\lambda\varphi}(x)$, $x_2 = x + h$, and $z_2 = \mathrm{prox}_{\lambda\varphi}(x + h)$. We divide by $\|h\|$ and pass to the limit, as $\|h\| \to 0$ to obtain that

$$\lim_{\|h\|\to 0} \frac{1}{\|h\|}\left|\varphi_\lambda(x + h) - \varphi_\lambda(x) - \left\langle h, \frac{1}{\lambda}(x - \mathrm{prox}_{\lambda\varphi}(x))\right\rangle\right| = 0,$$

which proves that φ_λ is Fréchet differentiable with $D\varphi_\lambda(x) = \frac{1}{\lambda}(x - \mathrm{prox}_{\lambda\varphi}(x))$ so that

$$\begin{aligned} \langle D\varphi_\lambda(x_1) - D\varphi_\lambda(x_2), x_1 - x_2\rangle &= \frac{1}{\lambda}\langle (x_1 - x_2) - (\mathrm{prox}_{\lambda\varphi}x_1 - \mathrm{prox}_{\lambda\varphi}x_2), x_1 - x_2\rangle \\ &= \frac{1}{\lambda}\big(\|x_1 - x_2\|^2 - \langle \mathrm{prox}_{\lambda\varphi}x_1 - \mathrm{prox}_{\lambda\varphi}x_2, x_1 - x_2\rangle\big) \\ &\le \frac{1}{\lambda}\|x_1 - x_2\|^2, \end{aligned}$$

where we also used the monotonicity of the proximity operator. The above equality, combined with the fact that $\mathrm{prox}_{\lambda\varphi}$ is nonexpansive, provides the monotonicity of $D\varphi_\lambda$ (the third term is positive). Lipschitz continuity is direct from the definition of $D\varphi_\lambda$ (see also Proposition 9.4.30). □

4.5 The proximity operator and numerical optimization algorithms

The proximity operators $\mathrm{prox}_{\lambda\varphi}$ are widely used in a class of numerical minimization algorithms called proximal point algorithms [21].

The fundamental idea behind such methods is to replace the original nonsmooth optimization problem $\inf_{x\in X} \varphi(x)$ by a smooth approximation, in particular $\inf_{x\in X} \varphi_\lambda(x)$, where φ_λ is the Moreau–Yosida approximation of φ for some appropriate value of the regularization parameter $\lambda > 0$. As shown in Proposition 4.4.6, this approximation is

Fréchet differentiable with Lipschitz continuous Fréchet derivative, hence, a minimizer, for φ_λ satisfies the first-order equation $D\varphi_\lambda(x_o) = 0$, which again, by Proposition 4.4.6, reduces to $0 = \frac{1}{\lambda}(x_o - \text{prox}_{\lambda\varphi}(x_o))$. Since $\lambda > 0$, a minimizer of φ_λ turns out to be a fixed point of the proximal operator $x_o = \text{prox}_{\lambda\varphi}(x_o)$. Hence, by Proposition 4.4.2(iii), $0 \in \partial\varphi(x_o)$, therefore a minimizer of φ (this can also be considered as an alternative proof for Proposition 4.4.6(ii)). However, the mere fact that instead of dealing with the original nonsmooth problem we deal with a smooth approximation has huge advantages, as far as numerical treatment of the problem is concerned, since the smoothness allows for better convergence results.

Proximal methods are based upon the above observation. In this section, we present the standard proximal method and its convergence, as well as a number of variants, designed for solving minimization problems of the form $\inf_{x\in X}(\varphi_1(x) + \varphi_2(x))$, consisting of a smooth and a nonsmooth part, φ_1 and φ_2, respectively. While these methods can be extended to Banach spaces, we restrict our attention here in a Hilbert space setting for simplicity.

4.5.1 The standard proximal method

As already mentioned, the main idea behind proximal point algorithms is based on the important observation that $\text{Fix}(\text{prox}_\varphi) = \arg\min_{x\in H} \varphi(x)$ (see Proposition 4.4.2(iii)) and consists in trying to approximate a minimizer of φ (and the minimum value) by a sequence $(x_n)_{n\in\mathbb{N}}$, defined by the iterative procedure

$$x_{n+1} = \text{prox}_{\lambda\varphi} x_n, \quad n \in \mathbb{N}, \tag{4.51}$$

for some appropriate choice of the parameter λ. The sequence defined in (4.51) is called a proximal sequence, and under certain circumstances, can converge to a fixed point, which if φ has a minimum belongs to the set of minimizers of φ. This observation leads to the development of a class of numerical algorithms that are very popular in optimization, called proximal methods. One interesting question that arises is concerning the choice of the parameter $\lambda > 0$. One immediately sees that λ cannot be too small since $\text{prox}_{\lambda\varphi}(x) \to x$ as $\lambda \to 0$ (see Example 4.4.4), which means that for small λ, the proximal sequence will essentially not go anywhere. On the other hand, large λ provides higher speed, but one must recall that the first-order conditions are of a local nature and λ too large may create problems with that or with the calculation of the proximal operator. It is also conceivable that one may not require a uniform λ throughout the whole iteration procedure, and choose a varying λ, depending on the neighborhood of the domain of the function that the iteration has led us (similar to the gradient method for smooth problems), leading thus to an adjustment of the "speed" of the method as we proceed.

One may also view the proximal sequence (4.51) as a natural generalization for nonsmooth problems of the gradient method used for smooth problems. If φ was smooth, one could use the implicit gradient method $x_{n+1} = x_n - \lambda D\varphi(x_{n+1})$ for some appropriate $\lambda > 0$,[18] which may be rearranged as $0 = \frac{1}{\lambda}(x_{n+1} - x_n) + D\varphi(x_{n+1})$, or equivalently choose x_{n+1} as the minimizer of the function $x \mapsto \varphi(x) + \frac{1}{2\lambda}\|x - x_n\|^2$. In the case of nonsmooth φ, this leads directly to the proximal sequence (4.51). As an alternative motivation for the proximal sequence, consider the explicit gradient method for the minimization of the smooth problem $\inf_{x\in H} \varphi_\lambda(x)$ for some $\lambda > 0$. Choosing step size λ, this would yield the iterative scheme $x_{n+1} = x_n - \lambda D\varphi_\lambda(x_n) = \text{prox}_{\lambda\varphi}(x_n)$, which is exactly (4.51).

The fundamental version of such an algorithm is the following:

Algorithm 4.5.1 (Proximal point algorithm). To find an element $x_o \in \arg\min_{x\in H} \varphi(x)$, we make the following steps:
1. Choose a sequence $(\lambda_n)_{n\in\mathbb{N}}$, $\lambda_n > 0$, $\sum_{n=1}^{\infty} \lambda_n = \infty$ and an accuracy $\epsilon > 0$.
2. For an appropriate initial point x_1, iterate (until a convergence criterion is met):

$$x_{n+1} = \text{prox}_{\lambda_n\varphi} x_n := \arg\min_{z\in X}\left(\varphi(z) + \frac{1}{2\lambda_n}\|x - z\|^2\right).$$

In the case where φ is smooth, the proximal point algorithm reduces to the implicit gradient descent scheme with variable step size. Constrained optimization may be treated by appropriately modifying φ.

The convergence of the fixed point scheme is based on the fact that the proximity operator is a nonexpansive operator (see Proposition 4.4.2(ii)). The following theorem provides a weak convergence result for the proximal algorithm (see [21] or [113]). Even though, we could provide a proof using the fact that $\text{prox}_{\lambda\varphi}$ is firmly nonexpansive involving some variant of the Krasnoselskii–Mann iteration scheme, we prefer to provide a direct proof for the basic proximal point algorithm, which we feel is more intuitive.

Proposition 4.5.2. *Let $\varphi : H \to \mathbb{R}$ be a proper lower semicontinuous convex function, and consider the proximal point algorithm*

$$x_{n+1} = \text{prox}_{\lambda_n\varphi} x_n, \quad n \in \mathbb{N}, \tag{4.52}$$

with $\sum_{n=1}^{\infty} \lambda_n = \infty$. Then, $(x_n)_{n\in\mathbb{N}}$ is a minimizing sequence for φ, i. e., $\varphi(x_n) \to \inf_{x\in H} \varphi(x)$, which converges weakly to a point in the set of minimizers of φ.

Proof. The strategy of the proof is to show that the proximal sequence enjoys the so called Fejér property with respect to the set of minimizers of φ. Given a set $A \subset H$ and a sequence $(z_n)_{n\in\mathbb{N}}$, we say that it is Fejér with respect to A if

---

18 Called implicit since we calculate the gradient at x_{n+1} rather than x_n (explicit gradient scheme), thus requiring the solution of a nonlinear equation in order to retrieve x_{n+1} from x_n. The reason we decide to undertake this cost is because the implicit scheme enjoys better convergence properties than the explicit one.

$$\|z_{n+1} - z\| \leq \|z_n - z\|, \quad \forall n \in \mathbb{N},\ z \in A. \tag{4.53}$$

As we will see, Fejér sequences with respect to weakly closed sets A converge weakly to a point $z_o \in A$. Since the set of minimizers of φ is a closed convex set, if we manage to show that the proximal sequence is Fejér with respect to arg min φ, we are done. We therefore try to show that if $x_o \in \arg\min_{x\in H} \varphi(x)$, then the proximal sequence (4.52) satisfies

$$\|x_{n+1} - x_o\| \leq \|x_n - x_o\|, \quad \forall n \in \mathbb{N},\ x_o \in \arg\min_{x\in H} \varphi(x). \tag{4.54}$$

We proceed in 3 steps:

1. Using Proposition 4.4.2(iii), we see that $x_{n+1} = \text{prox}_{\lambda\varphi}(x_n)$ implies

$$x_n - x_{n+1} \in \lambda_n \partial\varphi(x_{n+1}), \quad n \in \mathbb{N}. \tag{4.55}$$

By the definition of the subdifferential (Definition 4.1.1) this means that

$$\lambda_n\varphi(z) - \lambda_n\varphi(x_{n+1}) \geq \langle x_n - x_{n+1}, z - x_{n+1}\rangle, \quad \forall z \in H. \tag{4.56}$$

Setting $z = x_n$ in (4.56), we have that

$$\begin{aligned}\varphi(x_n) - \varphi(x_{n+1}) &\geq \frac{1}{\lambda_n}\langle x_n - x_{n+1}, x_n - x_{n+1}\rangle \\ &= \frac{1}{\lambda_n}\|x_n - x_{n+1}\|^2 \geq 0,\end{aligned} \tag{4.57}$$

hence the sequence $(\varphi(x_{n+1}))_{n\in\mathbb{N}}$ is decreasing.

Consider, now any $x_o \in A := \arg\min_{x\in H} \varphi(x)$, and apply (4.56) setting $z = x_o$, to obtain

$$\frac{1}{\lambda_n}\langle x_n - x_{n+1}, x_o - x_{n+1}\rangle \leq \varphi(x_o) - \varphi(x_{n+1}). \tag{4.58}$$

We now rearrange $\|x_{n+1} - x_o\|^2$ as follows:

$$\begin{aligned}\|x_{n+1} - x_o\|^2 &= \|x_{n+1} - x_n + x_n - x_o\|^2 \\ &= \|x_n - x_o\|^2 + 2\langle x_{n+1} - x_n, x_n - x_o\rangle + \|x_{n+1} - x_n\|^2 \\ &= \|x_n - x_o\|^2 + 2\langle x_{n+1} - x_n, x_n - x_{n+1} + x_{n+1} - x_o\rangle + \|x_{n+1} - x_n\|^2 \\ &= \|x_n - x_o\|^2 - \|x_{n+1} - x_n\|^2 + 2\langle x_{n+1} - x_n, x_{n+1} - x_o\rangle \\ &\leq \|x_n - x_o\|^2 + 2\langle x_n - x_{n+1}, x_o - x_{n+1}\rangle \\ &\leq \|x_n - x_o\|^2 + 2\lambda_n(\varphi(x_o) - \varphi(x_{n+1})),\end{aligned} \tag{4.59}$$

where, in the last estimate, we used (4.58). Since $x_o \in \arg\min_{x\in H} \varphi(x)$, clearly $\varphi(x_o) - \varphi(x_{n+1})$ is negative, while setting $\gamma = \inf_{x\in H} \varphi(x)$ we have that $\varphi(x_o) = \gamma$. In view of this notation, the estimate (4.59) becomes

$$\|x_{n+1} - x_o\|^2 \le \|x_n - x_o\|^2 - 2\lambda_n(\varphi(x_{n+1}) - \gamma), \tag{4.60}$$

where $\varphi(x_{n+1}) - \gamma > 0$. We therefore conclude that the sequence $(x_n)_{n\in\mathbb{N}}$ satisfies

$$\|x_{n+1} - x_o\| \le \|x_n - x_o\|, \quad \forall n \in \mathbb{N}. \tag{4.61}$$

Hence, it enjoys the Fejér property (4.54) with respect to the set $A = \arg\min_{x\in H} \varphi(x)$.

2. Rearranging (4.60), we get that

$$\lambda_n(\varphi(x_{n+1}) - \gamma) \le \frac{1}{2}(\|x_n - x_o\|^2 - \|x_{n+1} - x_o\|^2), \quad \forall n \in \mathbb{N},$$

and adding over all $n = 1, \dots, m$, we get that

$$\begin{aligned} \sum_{n=1}^{m} \lambda_n(\varphi(x_{n+1}) - \gamma) &\le \frac{1}{2}(\|x_1 - x_o\|^2 - \|x_{m+1} - x_o\|^2) \\ &\le \frac{1}{2}\|x_1 - x_o\|^2, \quad \forall m \in \mathbb{N}. \end{aligned} \tag{4.62}$$

As mentioned above, see (4.57), the sequence $(\varphi(x_n)-\gamma)_{n\in\mathbb{N}}$ is positive and nonincreasing. Hence, $0 \le \varphi(x_{m+1}) - \gamma \le \varphi(x_{n+1}) - \gamma$ for every $n \le m$. Using this estimate in (4.62), we get that

$$(\varphi(x_{m+1}) - \gamma) \sum_{n=1}^{m} \lambda_n \le \frac{1}{2}\|x_1 - x_o\|^2, \quad \forall m \in \mathbb{N}, \tag{4.63}$$

and taking the limit as $m \to \infty$, combined with the assumption that $\sum_n \lambda_n = \infty$ leads to the result that $\lim_m \varphi(x_m) = \gamma$, so $(x_n)_{n\in\mathbb{N}}$ is a minimizing sequence. By a standard lower semicontinuity argument, any weak accumulation point of the proximal sequence (4.52) is in $x \in \arg\min_{x\in H} \varphi(x)$

3. It remains to show that the proximal sequence has weak accumulation points. To this end, recall the Fejér monotonicity property of the proximal sequence with respect to $A = \arg\min_{x\in H} \varphi(x)$, shown in (4.61). A Fejér sequence $(x_n)_{n\in\mathbb{N}}$ is bounded, since by the monotonicity of $(\|x_n - x_o\|)_{n\in\mathbb{N}}$, we have, setting $R = \|x_1 - x_o\|$, that $x_n \in \overline{B}_H(x_o, R)$ for every $n \in \mathbb{N}$. Hence, by a weak compactness argument, there exists a subsequence $(x_{n_k})_{k\in\mathbb{N}}$ and a $x_o' \in H$ such that $x_{n_k} \rightharpoonup x$. From step 2, we already know that $x_o' \in \arg\min_{x\in H} \varphi(x)$. We will show that the whole sequence (and not just the subsequence) weakly converges to this x.

Indeed, consider any two weakly converging subsequences $(x_{n_k})_{k\in\mathbb{N}}$ and $(x_{\ell_k})_{k\in\mathbb{N}}$ of the proximal sequence (4.52) such that $x_{n_k} \rightharpoonup x_A \in \arg\min_{x\in H} \varphi(x)$, $x_{\ell_k} \rightharpoonup x_B \in \arg\min_{x\in H} \varphi(x)$, as $k \to \infty$. Clearly, by (4.61), (where x_o was an arbitrary point of $\arg\min_{x\in X} \varphi(x)$) we have that both $(\|x_n - x_A\|)_{n\in\mathbb{N}}$ and $(\|x_n - x_B\|)_{n\in\mathbb{N}}$ are decreasing and bounded below, hence convergent. Let us assume that $\|x_n - x_A\| \to K_A$, $\|x_n - x_B\| \to K_B$. We first express $\|x_n - x_A\|^2 = \|x_n - x_B\|^2 + \|x_B - x_A\|^2 + 2\langle x_n - x_B, x_B - x_A\rangle$, choose the subsequence $n = \ell_k$, and pass to the limit to obtain $K_A = K_B + \|x_B - x_A\|^2$. We then repeat the above interchanging the role of A and B, choose the subsequence $n = n_k$, and pass to the limit to obtain $K_B = K_A + \|x_A - x_B\|^2$. Combining the above, we get that $x_A = x_B$. Hence, by the Urysohn property (see Remark 1.1.52), the whole sequence weakly converges to the same limit, which is a minimizer for φ. □

Example 4.5.3 (When would the proximal algorithm converge strongly to a minimizer?). For the strong convergence of the proximal algorithm to a minimizer, we need to add the extra condition that φ is uniformly convex with modulus of convexity ψ (see Definition 2.3.3). In this case, we have a unique minimizer x_o, and furthermore the proximal sequence $x_n \to x_o$ (strongly).

The uniqueness of the minimizer follows by strict convexity. Consider now the proximal sequence for which we already know, by Proposition 4.5.2, that $x_n \rightharpoonup x_o$. By the uniform convexity of φ,

$$\varphi(tx_1 + (1-t)x_2) \leq t\varphi(x_1) + (1-t)\varphi(x_2) - t(1-t)\psi(\|x_1 - x_2\|_X), \quad \forall\, x_1, x_2 \in X,\ t \in (0,1).$$

Setting $x_1 = x_n$, $x_2 = x_o$ and $t = \frac{1}{2}$, we obtain that

$$\varphi\Big(\frac{1}{2}x_n + \frac{1}{2}x_o\Big) + \frac{1}{4}\psi(\|x_n - x_o\|) \leq \frac{1}{2}\varphi(x_n) + \frac{1}{2}\varphi(x_o). \tag{4.64}$$

Note that $\frac{1}{2}x_n + \frac{1}{2}x_o \rightharpoonup x_o$ so that by lower semicontinuity $\varphi(x_o) \leq \liminf_n \varphi(\frac{1}{2}x_n + \frac{1}{2}x_o)$. We also have that $\varphi(x_n) \to \varphi(x_o) = \inf_{x\in H}\varphi(x)$. We rearrange (4.64) to keep the modulus on the left-hand side and take limit superior on both sides of the inequality to obtain

$$\begin{aligned}\frac{1}{4}\limsup_n \psi(\|x_n - x_o\|) &\leq \limsup_n\Big(\frac{1}{2}\varphi(x_n) + \frac{1}{2}\varphi(x_o) - \varphi\Big(\frac{1}{2}x_n + \frac{1}{2}x_o\Big)\Big)\\ &= \varphi(x_o) - \liminf_n \varphi\Big(\frac{1}{2}x_n + \frac{1}{2}x_o\Big) \leq 0,\end{aligned}$$

where, from the properties of ψ (increasing and vanishing only at 0), we conclude that $\psi(\|x_n - x_o\|) \to 0$, and hence $x_n \to x_o$ (strongly).

Example 4.5.4 (Rate of convergence for strongly convex functions). If the function φ is strongly convex with modulus of convexity c (see Definition 2.3.3), then we have strong convergence of the proximal scheme (4.52) and may also obtain the rate of convergence as, e. g., $\|x_N - x_o\| \leq \prod_{k=1}^{N-1}(1+2c\lambda_k)^{-1}\|x_1 - x_o\|$, where c is the modulus of strong convexity, and λ_k is the variable step size. It is interesting to compare with the relevant estimates for the gradient descent method.

To see this, the reader may wish to recall the properties of the subdifferential for strongly convex functions (4.7) in Example 4.2.5, and in particular that

$$\langle x_1^\star - x_2^\star, x_1 - x_2 \rangle \geq c \|x_1 - x_2\|^2, \quad \forall\, x_1^\star \in \partial\varphi(x_1),\ x_2^\star \in \partial\varphi(x_2). \tag{4.65}$$

Since x_o is a minimizer so that $0 \in \partial\varphi(x_o)$ and $x_{n+1} = \mathrm{prox}_{\lambda_n \varphi}(x_n)$, which is equivalent to $-\frac{1}{\lambda_n}(x_{n+1} - x_n) \in \partial\varphi(x_{n+1})$, applying (4.65) for the choice

$$x_1^\star = -\frac{1}{\lambda_n}(x_{n+1} - x_n), \quad x_1 = x_{n+1}, \quad \text{and} \quad x_2^\star = 0, \quad x_2 = x_o,$$

we obtain

$$\langle x_{n+1} - x_n, x_o - x_{n+1} \rangle \geq c\lambda_n \|x_o - x_{n+1}\|^2. \tag{4.66}$$

Being in Hilbert space, we may express the left-hand side in terms of the sum and difference of the squares of the norms, as

$$\begin{aligned} \|x_n - x_o\|^2 &= \|x_n - x_{n+1} + x_{n+1} - x_o\|^2 \\ &= \|x_{n+1} - x_n\|^2 + \|x_{n+1} - x_o\|^2 + 2\langle x_{n+1} - x_n, x_o - x_{n+1} \rangle, \end{aligned}$$

which upon solving for $2\langle x_{n+1} - x_n, x_o - x_{n+1} \rangle$ and combined with (4.66) gives

$$\|x_n - x_o\|^2 - \|x_{n+1} - x_n\|^2 - \|x_{n+1} - x_o\|^2 \geq 2c\lambda_n \|x_o - x_{n+1}\|^2. \tag{4.67}$$

Rearranging (4.67), we obtain the inequality

$$(1 + 2c\lambda_n)\|x_{n+1} - x_o\|^2 \leq \|x_n - x_o\|^2 - \|x_{n+1} - x_n\|^2 \leq \|x_n - x_o\|^2. \tag{4.68}$$

By iterating, we obtain $\|x_N - x_o\| \leq \prod_{k=1}^{N-1}(1 + 2c\lambda_k)^{-1/2}\|x_1 - x_o\|$, which is the case of constant stepsize $\lambda_k = \lambda$ reduces to $\|x_N - x_o\| \leq (1 + 2c\lambda)^{-(N-1)/2}\|x_1 - x_o\|$, which are the stated convergence estimates.

In fact, we may do a little better using again the properties of the subdifferential for strongly convex functions, and in particular (4.7)(iii),

$$\|x_1^\star - x_2^\star\| \geq c\|x_1 - x_2\|, \quad \forall\, x_1^\star \in \partial\varphi(x_1),\ x_2^\star \in \partial\varphi(x_2), \tag{4.69}$$

to estimate the term $\|x_{n+1} - x_n\|$ in (4.68) rather than discard it. In fact, since $0 \in \partial\varphi(x_o)$ and $-\frac{1}{\lambda_n}(x_{n+1} - x_n) \in \partial\varphi(x_{n+1})$, relation (4.69) guarantees that $\|x_{n+1} - x_o\| \leq \frac{1}{c\lambda_n}\|x_{n+1} - x_n\|$ so that $-\|x_{n+1} - x_n\|^2 \leq -c^2\lambda_n^2\|x_{n+1} - x_o\|^2$. Combining this with (4.68) leads to $(1 + c\lambda_n)^2\|x_{n+1} - x_o\|^2 \leq \|x_n - x_o\|^2$, which is a better estimate, so that $\|x_{n+1} - x_o\| \leq (1 + c\lambda_n)^{-1}\|x_n - x_o\|$, and by iteration, we obtain the improved convergence estimate $\|x_N - x_o\| \leq \prod_{k=1}^{N-1}(1 + 2c\lambda_k)^{-1}\|x_1 - x_o\|$, which if $\lambda_k = \lambda$ reduces to $\|x_N - x_o\| \leq (1 + 2c\lambda)^{-(N-1)}\|x_1 - x_o\|$. ◁

Example 4.5.5. Assume that we wish to minimize the quadratic function φ, defined by $\varphi(x) = \frac{1}{2}\langle x, Ax \rangle - \langle b, x \rangle$. Of course, this is a smooth problem, the solution of which is given by the solution of the equation $Ax = b$. Let us now calculate the proximal operator for φ. The solution to the minimization problem $\min_{z\in H} \frac{1}{2\lambda}\|x - z\|^2 + \frac{1}{2}\langle x, Ax\rangle - \langle b, x\rangle$ is expressed as

$$\mathrm{prox}_{\lambda\varphi}x = \left(A + \frac{1}{\lambda}I\right)^{-1}\left(b + \frac{1}{\lambda}x\right)$$

so that the proximal algorithm provides the iterative algorithm

$$x_{n+1} = \left(A + \frac{1}{\lambda}I\right)^{-1}\left(b + \frac{1}{\lambda}x_n\right),$$

which can be expressed in the equivalent form

$$x_{n+1} = x_n + \left(A + \frac{1}{\lambda}I\right)^{-1}(b - Ax_n).$$

This will converge to a solution of $Ax = b$. When H is finite dimensional with $\dim(H) = d$, in which case A corresponds to a matrix $\mathbf{A} \in \mathbb{R}^{d\times d}$, the above procedure provides an iterative algorithm for the solution of the system of linear equations $\mathbf{A}x = b$, in which at every step we have to invert the perturbed matrix $\mathbf{A} + \frac{1}{\lambda}I$, which is often preferable from the numerical point of view, especially if we have to deal with ill-posed matrices. Also, it may be that $\mathbf{A}$ is not invertible, a situation often arising in the treatment of ill-posed problems and their regularization. ◁

Other versions of the proximal point algorithm are possible, such as for instance the inertial version

$$x_{n+1} = (1 + \theta)\mathrm{prox}_{\lambda\varphi}(x_n) - \theta x_n, \quad \theta \in (-1, 1), \tag{4.70}$$

which by similar techniques can be shown to converge. For useful information on extensions of this algorithm as well as information on its concrete applications, see, e. g., [45].

4.5.2 Splitting methods: the forward-backward and the Douglas–Rachford scheme

The proximal point algorithm, though enjoying the extremely important and desirable property of unconditional convergence, suffers a serious drawback. Often the calculation of the proximal operator is a problem as difficult as the original problem we started with. So event hough it will work nicely for functions whose proximal operators are easily or even analytically obtained (as for instance for the ℓ_1 norm), in many practical situations it is not easily applicable. One way round such problems is to split the function we wish to minimize to a sum of more than one contributions, with the criterion that at least some of the constituents have easy to calculate proximal maps. This is the basic idea behind splitting methods, which is the subject of this section.

4.5.2.1 Forward-backward algorithms

An interesting extension is related to problems of the form $\min_{x\in H}(\varphi_1(x) + \varphi_2(x))$, in which both φ_1 and φ_2 are convex lower semicontinuous functions, but one of them, say φ_1, is Fréchet differentiable with Lipschitz continuous Fréchet derivative.

Starting from the first-order condition $0 \in D\varphi_1(x) + \partial\varphi_2(x)$ for a minimizer x, we split 0 as $0 = (x - z) + (z - x)$ for some auxiliary $z \in H$, chosen such that

$$x - z = \alpha D\varphi_1(x), \quad \text{and,} \quad z - x \in \alpha\partial\varphi_2(x),$$

where $\alpha > 0$ can be considered as a regularization parameter. Bringing $-x$ to the other side on the second relation, the above splitting becomes

$$\begin{aligned} x &= z + \alpha D\varphi_1(x), \\ x &= (I + \alpha\partial\varphi_2)^{-1}(z) = \mathrm{prox}_{\alpha\varphi_2}(z). \end{aligned}$$

This shows that the minimizer x must satisfy the above compatibility conditions, and is expressed in terms of the auxiliary variable z as $z = x - \alpha D\varphi_1(x)$ (this equation may be feasible to solve to obtain x in terms of z since $D\varphi_1$ is Lipschitz), where x must solve (by eliminating z) the operator equation

$$x = f(x) := \mathrm{prox}_{\alpha\varphi_2}(x - \alpha D\varphi_1(x)),$$

i. e., x is a fixed point of the operator $f : H \to H$, defined above. We may try to approximate the solution to the operator equation $x = f(x)$ in terms of a (relaxed) iterative scheme of the form

$$x_{n+1} = x_n + \lambda_n(f(x_n) - x_n),$$

where λ_n is a relaxation parameter ($\lambda_n \in (1,2)$ corresponding to over-relaxation and $\lambda_n \in (0,1)$ corresponding to under-relaxation). This may be also expressed as the two step procedure:

$$\begin{aligned} z_n &= x_n - \alpha D\varphi_1(x_n), \\ x_{n+1} &= x_n + \lambda_n(\mathrm{prox}_{\alpha\varphi_2}(z_n) - x_n) = (1 - \lambda_n)x_n + \lambda_n \mathrm{prox}_{\alpha\varphi_2}(z_n). \end{aligned}$$

Alternatively, the first step can be understood as an explicit gradient scheme for the minimization of the smooth part (which enjoys nice convergence properties), while the second step as the inertial version of the proximal algorithm (4.70) for the nonsmooth part.

In summary, in the proposed scheme, we essentially split the function $\varphi_1 + \varphi_2$ into its smooth and its nonsmooth part, treat them separately, and then combine them in terms of a fixed point iteration. This minor modification to the standard proximal scheme has certain advantages, one of which may be that the proximal operator for φ_2 may be easier to compute than the proximal operator for the sum $\varphi_1 + \varphi_2$. It also leads to small changes to the convergence properties of the scheme, but at the same time simplifies other aspects by turning the scheme into a semi-implicit (or forward-backward) one.[19]

The above considerations lead to the following algorithm:

Algorithm 4.5.6 (Forward-backward algorithm). Let φ_i, $i = 1,2$, be convex and lower semicontinuous with φ_1 Fréchet differentiable such that $D\varphi_1$ is Lipschitz continuous with Lipschitz constant L

To find an $x_o \in \arg\min_{x \in H}(\varphi_1(x) + \varphi_2(x))$:

1. Choose $\alpha > 0$ sufficiently small and a sequence $(\lambda_n)_{n\in\mathbb{N}} \subset [0,\beta]$ such that $\sum_{n=1}^{\infty}\lambda_n(\beta - \lambda_n) = +\infty$ for a suitably chosen $\beta > 1/2$, and an accuracy $\epsilon > 0$.
2. For an appropriate starting point $(x_1, y_1) \in H$, iterate

$$\begin{aligned} z_n &= x_n - \alpha D\varphi_1(x_n), \\ x_{n+1} &= x_n + \lambda_n\big(\mathrm{prox}_{\alpha\varphi_2}(z_n) - x_n\big) = (1-\lambda_n)x_n + \lambda_n \mathrm{prox}_{\alpha\varphi_2}(z_n), \end{aligned} \tag{4.71}$$

until a convergence criterion is met.

The following proposition provides convergence results for the forward-backward algorithm (see [21] or [54]). For details on extensions of the forward-backward algorithm as well as its practical application the reader may consult [45].

Proposition 4.5.7 (Convergence of forward-backward algorithm). *Assume that* $0 < \alpha < \frac{2}{L}$, *and* $\lambda_n \in (0,1)$, $\sum_{n=1}^{\infty}\lambda_n(1-\lambda_n) = +\infty$. *Then the forward backward algorithm* (4.71) *converges weakly to a minimizer of* $\varphi_1 + \varphi_2$.

Proof. Observe that the forward-backward scheme can be expressed in one equation as

$$x_{n+1} = (1-\lambda_n)x_n + \lambda_n f(x_n), \tag{4.72}$$

where $f = f_1 \circ f_2$, with $f_2 : x \mapsto x - \alpha D\varphi_1(x)$ and $f_1 : x \mapsto \mathrm{prox}_{\alpha\varphi_2}(x)$. f_1 is nonexpansive, and f_2 is also nonexpansive, as long as $0 < \alpha < \frac{2}{L}$ (see Example 3.4.3), so that f is nonexpansive. Then (4.72) reduces to the Krasnoselskii–Mann iteration scheme, and the convergence follows from Theorem 3.4.9. □

Remark 4.5.8. An alternative approach to convergence may be obtained by noting that f is a ν-averaged operator, as a composition of two ν-averaged operators (see Lemma 3.7.2) and working as in Example 3.4.10. Indeed, working exactly as in Example 3.4.7, we may prove that f_1 is $\frac{1}{2}$-averaged, while from Example 3.4.8 we know that f_2

19 In some cases it is simpler to invert $I + \lambda_n \partial\varphi_2$ than $I + \lambda_n D\varphi_1 + \lambda_n \partial\varphi_2$. As for the terminology, the step $z_n := x_n - \lambda_n D\varphi_1(x_n)$ is called the forward step, while the step $x_{n+1} = \mathrm{prox}_{\lambda_n \varphi} z_n$ is called the backward one.

is $\frac{aL}{2}$-averaged, as long as $\frac{aL}{2} < 1$. Using Lemma 3.7.2, we have that f is ν-averaged for $\nu = \frac{\nu_1+\nu_2-2\nu_1\nu_2}{1-\nu_1\nu_2} = \frac{2}{4-aL}$. So using the iterative scheme proposed in Example 3.4.10, we may choose the sequence $(\lambda_n)_{n\in\mathbb{N}}$ so that $\sum_{n=1}^{\infty}\lambda_n(\beta-\lambda_n) = \infty$ for $\beta = \frac{1}{\nu} = 2 - \frac{aL}{2}$.

Example 4.5.9 (The projected gradient method). Consider the constrained minimization problem $\inf_{x\in C}\varphi(x)$, where $\varphi : H \to \mathbb{R}\cup\{+\infty\}$ is a convex lower semicontinuous function, with Lipschitz continuous Fréchet derivative, and $C \subset H$ is a closed convex set. We may define $\varphi_1 = \varphi$, $\varphi_2 = I_C$ and express the constrained minimization problem as the unconstrained problem $\inf_{x\in H}(\varphi_1(x) + \varphi_2(x))$. Applying the forward-backward scheme to this problem, and recalling that for this choice $\mathrm{prox}_{\varphi_2} = P_C$, the projection operation on C, we retrieve the projected gradient scheme

$$x_{n+1} = P_C\big(x_n - aD\varphi_1(x_n)\big) \tag{4.73}$$

by setting the parameter λ_n in (4.71) to $\lambda_n = 1$. This extends the gradient method (see Example 3.4.11) to constrained optimization. An application of Proposition 4.5.7 shows the weak convergence of the scheme (4.73) for $a < \frac{2}{L}$, where L is the Lipschitz constant for $D\varphi$.

Example 4.5.10 (Lasso problems). An important use of the proximal point algorithm is in the, so called, Lasso optimization problems. These are minimization problems in the Hilbert space $H = \mathbb{R}^d$, endowed with the Euclidean $|\cdot| := \|\cdot\|_2$ norm, of the form

$$\min_{x\in H}\left(\frac{1}{2}\|\mathbf{A}x - b\|_2^2 + \lambda\|x\|_1\right),$$

where $\|\cdot\|_1$ denotes the 1 norm, $\mathbf{A} \in \mathbb{R}^{m\times d}$, $b \in \mathbb{R}^m$, and $\lambda > 0$. Problems of this type find many applications in high-dimensional statistics, machine learning or image processing, and their interpretation is as finding sparse solutions to least squares problems. The proximal gradient method can be directly applied to this problem by setting $\varphi_1 = \frac{1}{2}\|\mathbf{A}x - b\|_2^2$ and $\varphi_2 = \lambda\|x\|_1$ and noting that $D\varphi_1(x) = \mathbf{A}^T(\mathbf{A}x - b)$ and $\mathrm{prox}_{\lambda\varphi_2}(x) = S_\lambda(x)$, the soft thresholding operator we have defined in Example 4.4.5. The Lasso problem can be generalized in sequence spaces, in terms of the ℓ^2 and the ℓ^1 norm in the place of $\|\cdot\|_2$ and $\|\cdot\|_1$. ◁

4.5.2.2 Douglas–Rachford algorithm

A final modification of the proximal gradient algorithm, again in the spirit of a forward-backward scheme, is the Douglas–Rachford algorithm, which is often used to minimize sums of convex lower semicontinuous functions if neither of them is differentiable.

To motivate the algorithm, consider the Fermat rule for minimization of $\varphi = \varphi_1 + \varphi_2$, which under sufficient assumptions for the Moreau–Rockafellar Theorem 4.2.12 to hold, leads to $0 \in \partial\varphi_1(x) + \partial\varphi_2(x)$. We multiply this by $\lambda > 0$, and then split 0 in terms of an auxiliary $z \in H$, as $0 = -(z - x) + (z - x) = (2x - z - x) + (z - x)$. We now choose

$$2x - z - x \in \lambda\partial\varphi_1(x), \quad \text{and,} \quad z - x \in \lambda\partial\varphi_2(x),$$

and immediately (bringing the $-x$ term to the other side in both, starting from the second) recognize them as

$$x = (I + \lambda\varphi_2)^{-1}(z) = \text{prox}_{\lambda\varphi_2}(z),$$
$$x = (I + \lambda\varphi_1)^{-1}(2x - z) = \text{prox}_{\lambda\varphi_1}(2x - z).$$

This shows that a minimizer x must satisfy the above compatibility conditions, and is expressed in terms of the auxiliary variable z as $x = \text{prox}_{\lambda\varphi_2}(z)$, where z is must solve (by eliminating x) the operator equation

$$0 = \text{prox}_{\lambda\varphi_1}(2\,\text{prox}_{\lambda\varphi_2}(z) - z) - \text{prox}_{\lambda\varphi_2}(z).$$

Adding z on each side of the above, we express z as a fixed point of the map $f : H \to H$,

$$z = f(z) := z + \text{prox}_{\lambda\varphi_1}(2\,\text{prox}_{\lambda\varphi_2}(z) - z) - \text{prox}_{\lambda\varphi_2}(z).$$

This fixed point problem can be approximated by an iterative scheme, for instance by a fixed point iteration with relaxation of the form

$$z_{n+1} = z_n + \lambda_n(f(z_n) - z_n),$$

where λ_n is a relaxation parameter, with $\lambda_n \in (1, 2)$ corresponding to overrelaxation and $\lambda_n \in (0, 1)$ corresponding to underrelaxation. The above scheme may be expressed as the two-step iterative procedure:

$$\begin{aligned} x_n &= \text{prox}_{\lambda\varphi_2}(z_n), \\ z_{n+1} &= z_n + \lambda_n(\text{prox}_{\lambda\varphi_1}(2x_n - z_n) - x_n). \end{aligned} \tag{4.74}$$

The above discussion leads to the Douglas–Rachford minimization algorithm.

Algorithm 4.5.11 (Douglas–Rachford). To find a minimizer of $\varphi_1 + \varphi_2$, we make the following steps:
1. Choose a sequence $(\lambda_n)_{n\in\mathbb{N}}$ such that $\lambda_n \in (0, 2]$ for every $n \in \mathbb{N}$, satisfying $\sum_{n\in\mathbb{N}} \lambda_n(2 - \lambda_n) = \infty$, and an accuracy $\epsilon > 0$.
2. For an appropriate (x_1, z_1), iterate (until a convergence criterion is met),

$$\begin{aligned} x_n &= \text{prox}_{\lambda\varphi_2}(z_n), \\ z_{n+1} &= z_n + \lambda_n(\text{prox}_{\lambda\varphi_1}(2x_n - z_n) - x_n). \end{aligned} \tag{4.75}$$

The convergence properties of this algorithm can be proved using the reformulation of the above scheme in terms of z (after the elimination of x), using the nonexpansive properties of the proximity operator. This is sketched in the following propostion [21].

Proposition 4.5.12 (Convergence of the Douglas–Rachford algorithm). *Under the assumption that $\lambda_n \in (0,2]$ for every $n \in \mathbb{N}$, with $\sum_{n\in\mathbb{N}} \lambda_n(2-\lambda_n) = \infty$, the Douglas–Rachford iterative scheme converges weakly to a minimizer $x_o \in \arg\min_{x\in H}(\varphi_1(x) + \varphi_2(x))$.*

Proof. Define the mappings $f_i = 2\,\mathrm{prox}_{\lambda\varphi_i} - I$, $i = 1,2$, and note that $2x_n - z_n = f_2(z_n)$, so the second step in (4.75) becomes

$$\begin{aligned} z_{n+1} &= z_n + \lambda_n\Big(\mathrm{prox}_{\lambda\varphi_1}(f_2(z_n)) - \frac{1}{2}(f_2(z_n) + z_n)\Big) \\ &= z_n + \frac{\lambda_n}{2}(f_1(f_2(z_n)) - z_n), \end{aligned} \tag{4.76}$$

where $f_1(f_2(z_n)) = (f_1 \circ f_2)(z_n)$. Note that (4.76) reduces to the "averaged" like form

$$z_{n+1} = (1-\rho_n)z_n + \rho_n f(z_n),$$

where $\rho_n = \frac{\lambda_2}{2}$ and $f = f_1 \circ f_2$.

The operators f_i, $i = 1,2$, are nonexpansive operators, hence, so is $f_1 \circ f_2$. The weak convergence of the iterative scheme to a fixed point of the operator $f_1 \circ f_2$ follows from the general theory of the Krasnoselskii–Mann iterative scheme (see Theorem 3.4.9). In fact, choosing $(\lambda_n)_{n\in\mathbb{N}}$ such that $\sum_{n\in\mathbb{N}} \frac{\lambda_n}{2}(1 - \frac{\lambda_n}{2}) = \infty$ leads to the equivalent condition $\sum_{n\in\mathbb{N}} \lambda_n(2-\lambda_n) = \infty$. It is easily seen (retracing the steps in the beginning of this subsection in reverse order) that a fixed point z of $f_1 \circ f_2$ can be used in the construction of a minimizer for $\varphi_1 + \varphi_2$, in terms of $x = \mathrm{prox}_{\lambda\varphi_2}(z)$. □

Example 4.5.13 (Convex feasibility problems). A useful application of the Douglas–Rachford scheme is in convex feasibility problems, i. e., the problem, given two closed and convex sets $C_1, C_2 \subset H$, of finding a point $x \in C_1 \cap C_2$. Defining $\varphi_i = I_{C_i}$, $i = 1,2$, this problem reduces to the problem of $\inf_{x\in H}(\varphi_1(x) + \varphi_2(x))$, which has a solution, as long as $C_1 \cap C_2 \neq \emptyset$. One may use the Douglas–Rachford algorithm to find such a point $x \in C_1 \cap C_2$. Recalling that $\mathrm{prox}_{\lambda\varphi_i} = P_{C_i}$, $i = 1,2$, the operators $f_i = 2P_{C_i} - I$, $i = 1,2$, (called reflectors in this case) and the Douglas–Rachford scheme becomes (choosing $\lambda_n = 1$ for simplicity)

$$\begin{aligned} z_{n+1} &= \frac{1}{2}(I + f_1 \circ f_2)(z_n), \\ x_{n+1} &= P_{C_2} z_{n+1}, \end{aligned}$$

and the second step can only be performed once, when a stopping criterion for the z iteration is met. By the general theory of the Douglas–Rachford scheme, if $C_1 \cap C_2 \neq \emptyset$, the iteration for $(z_n)_{n\in\mathbb{N}}$ weakly converges to some z such that $P_{C_2}(z) \in C_1 \cap C_2$, while if $C_1 \cap C_2 = \emptyset$, then $\|z_n\| \to \infty$.

Convex feasibility problems, involving more than two sets, can be reduced to the above problem with 2 sets using the so called Pierra product space formulation consisting of the sets $C_1' = C_1 \times \cdots \times C_k$ and $C_2' = \{(x, \ldots x) \in H^k \,:\, x \in H\}$ (called the diagonal). Then $x \in C_1 \times \cdots \times C_k \subset H^k$ if and only if $x \in C_1' \cap C_2'$, which is a problem of the above form. Convex feasibility problems have multiple applications ranging from medical imaging to combinatorics. One common application is in the, so called, matrix completion problem, which consists, given a partial matrix, of finding a completion having desirable properties, e.g, positive semidefinite. In such cases, the space of $n \times m$ matrices is turned into a Hilbert space using the inner product $\langle \mathbf{A}_1, \mathbf{A}_2 \rangle = \mathrm{Tr}(\mathbf{A}_1^T \mathbf{A}_2)$, and the convex sets C_i, $i = 1, \ldots, k$ correspond to the desired constraints, e. g., $C_1 = \{x = \mathbf{A} \in \mathbb{R}^{d\times d} \,:\, z^T \mathbf{A} z \geq \epsilon |z|^2,\ \forall z \in \mathbb{R}^d\}$ for positive definite matrices, etc. For more details on the matrix completion problem, the reader may consult [11]. ◁

5 Minimax theorems and duality

In this chapter, we develop the theory of convex duality, which finds many interesting applications in the theory and practice of convex optimization. We begin our development of this theory by introducing a rather general version of the minimax theorem, which allows us to answer the question of whether a functional admits a saddle point. We then move to a detailed study of the Fenchel–Legendre conjugate and the bi-conjugate for convex functions and their properties. With these tools at hand, we proceed to the main aim of this chapter, the study of duality methods in optimization, which allow us to redress constrained minimization problems in saddle point form and connect with each such problem a related maximization problem called its dual (that in many cases is easier to treat than the original problem, called the primal). We first present the theory of Fenchel duality, and then a more general framework developed by Ekeland, Temam, and others, which allows the treatment of a wide class of problems arising in various applications, including data analysis, economics and finance, signal and image processing etc. These duality techniques find important applications in numerical analysis, and the chapter closes with a treatment of numerical methods for optimization problems based on such concepts. These important issues have been treated in, e. g., [12, 21, 28, 74], and [113] on which our approach is based.

5.1 A minimax theorem

Let $F : X \times Y \to \mathbb{R}$ be a function defined on the Cartesian product of the Banach spaces X, Y. It is elementary to note that it is always true[1] that

$$\sup_{y \in Y} \inf_{x \in X} F(x, y) \le \inf_{x \in X} \sup_{y \in Y} F(x, y). \tag{5.1}$$

The interval $(\sup_{y \in Y} \inf_{x \in X} F(x, y), \inf_{x \in X} \sup_{y \in Y} F(x, y))$ is called the duality gap for F.

An interesting question is: when does the opposite inequality hold? An affirmative answer to this question would lead to equality of the two sides, i. e.,

$$\sup_{y \in Y} \inf_{x \in X} F(x, y) = \inf_{x \in X} \sup_{y \in Y} F(x, y). \tag{5.2}$$

We then say that the function F satisfies the minimax equality, with the common value called a saddle value for F. This would essentially mean that the order by which we minimize and maximize does not alter the result, a fact that can be very useful as we

1 First fix $x \in X$ and observe that $\inf_{x \in X} F(x, y) \le F(x, y)$ for any x, y, then take the supremum over $y \in Y$ to obtain $\sup_{y \in Y} \inf_{x \in X} F(x, y) \le \sup_{y \in Y} F(x, y)$ for any x, and finally take the infimum over $x \in X$ to obtain the stated inequality.

https://doi.org/10.1515/9783111333298-005

shall see for a number of optimization problems. The positive or negative answer to this question is directly related to the notion of a saddle point.

Definition 5.1.1. The point $(x_o, y_0) \in X \times Y$ is called a saddle point of F if

$$F(x_o, y) \leq F(x_o, y_0) \leq F(x, y_0), \quad \forall x \in X,\ y \in Y.$$

Upon defining the sets

$$\mathfrak{m}(y) := \arg\min_{x \in X} F(x, y),$$
$$\mathfrak{M}(x) := \arg\max_{y \in Y} F(x, y),$$

i. e., the set of minimizers of $F(\cdot, y)$ and the set of maximizers of $F(x, \cdot)$, respectively, we may provide an alternative characterization of a saddle point as follows:

If (x_o, y_0) is a saddle point of F, then x_o minimizes $F(\cdot, y_0)$, whereas y_0 maximizes $F(x_o, \cdot)$, i. e.,

$$\begin{aligned} x_o &\in \mathfrak{m}(y_0) = \Big\{x \in X\ :\ F(x, y_0) = \inf_{x \in X} F(x, y_0)\Big\}, \\ y_0 &\in \mathfrak{M}(x_o) = \Big\{y \in Y\ :\ F(x_o, y) = \sup_{y \in Y} F(x_o, y)\Big\}. \end{aligned} \tag{5.3}$$

As such, a saddle point has a convenient interpretation in terms of the fixed point of the set valued map Φ, defined by $(x, y) \mapsto \mathfrak{m}(y) \times \mathfrak{M}(x)$. In view of (5.3), a saddle point (x_o, y_0) of F has the property that $(x_o, y_0) \in \Phi(x_o, y_0)$, therefore it is a fixed point of the map Φ. This connection of a saddle point with the fixed point of a set valued map allows us to show the existence of saddle points, through the use of fixed point theorems, such as for instance the Knaster–Kuratowski–Mazurkiewicz fixed point Theorem 3.2.11.

The following characterization of a saddle point is useful:

Proposition 5.1.2. *The function $F : X \times Y \to \mathbb{R}$ admits a saddle point (x_o, y_0) if and only if F satisfies the minimax equality* (5.2), *with the supremum in the first expression and the infimum on the second being attained at x_o and y_0, respectively.*

Proof. The proof is simple but is included in the appendix of the chapter (Section 5.5.1) for completeness. □

Remark 5.1.3. An alternative way of stating the above is saying that the function $F_2 : Y \to \mathbb{R}$, defined by $F_2(y) = \inf_{x \in X} F(x, y)$, attains its supremum over Y, and the function $F_1 : X \to \mathbb{R}$, defined by $F_1(x) = \sup_{y \in Y} F(x, y)$, attains its infimum over X. In fact, it is related to the existence of a $x_o \in X$ such that $\sup_{y \in Y} F(x_o, y) = \inf_{x \in X} \sup_{y \in Y} F(x, y) = \inf_{x \in X} F_1(x)$, and to the existence of a $y_0 \in Y$ such that $\inf_{x \in X} F(x, y_0) = \sup_{y \in Y} \inf_{x \in X} F(x, y) = \sup_{y \in Y} F_2(y)$. We may therefore, find the common value (saddle value) either by fixing $x = x_o$, and then maximizing the section

$F(x_o, \cdot) : Y \to \mathbb{R}$ over Y, or by fixing $y = y_0$, and then minimizing the section $F(\cdot, y_0) : X \to \mathbb{R}$ over X.

The problem of existence of saddle points for a given function $F : X \times Y \to \mathbb{R}$ has attracted a lot of attention since Von Neumann's pioneering work on functions defined on finite sets, and its connection with the foundations of game theory, and in particular bimatrix zero sum games. Since then, there have been various generalizations and reformulations. We present here a general version of the minimax theorem due to Sion ([131] and present its proof in the spirit of [107]. The theorem requires a condition less stringent than convexity.

Definition 5.1.4 (Quasi-convex and quasi-concave function). A function $\varphi : A \subset X \to \mathbb{R} \cup \{+\infty\}$ is called
(i) Quasi-convex, if for every $\lambda \in \mathbb{R}$ the sets $\{x \in A : \varphi(x) \leq \lambda\}$ are convex.
(ii) Quasi-concave, if for every $\lambda \in \mathbb{R}$ the sets $\{x \in A : \varphi(x) \geq \lambda\}$ are convex.

A convex function is clearly quasi-convex, but the converse is not necessarily true. Likewise with concave functions.

We now state and prove a version of the minimax theorem [107]:

Theorem 5.1.5 (Minimax). *Let $C_X \subset X$ and $C_Y \subset Y$ be nonempty, compact, and convex sets. Consider the function $F : C_X \times C_Y \to \mathbb{R}$ such that*
(i) *$F(\cdot, y) : C_X \to \mathbb{R}$ is lower semicontinuous and quasi-convex[2] for any $y \in C_Y$.*
(ii) *$F(x, \cdot) : C_Y \to \mathbb{R}$ is upper semicontinuous and quasi-concave[3] for any $x \in C_X$.*

Then, the function F has a saddle point.

Proof. The proof which makes use of the Knaster–Kuratowski–Mazurkiewicz lemma (Theorem 3.2.11) is broken into 3 steps.

1. By Proposition 5.1.2, it is sufficient to show that

$$\min_{x \in C_X} \max_{y \in C_Y} F(x, y) = \max_{y \in C_Y} \min_{x \in C_X} F(x, y). \tag{5.4}$$

Fix an $x \in C_X$. Since $F(x, \cdot) : C_Y \to \mathbb{R}$ is upper semicontinuous and C_Y is compact, by the Weierstrass theorem, the maximum exists, i. e., there exists $y_0 \in C_Y$ such that $\max_{y \in C_Y} F(x, y) = F(x, y_0)$. The function $h : C_X \to \mathbb{R}$, defined by $h(x) = \max_{y \in C_Y} F(x, y)$, is lower semicontinuous. Hence, by applying the Weierstrass theorem once more, $\min_{x \in C_X} h(x) = \min_{x \in C_X} \max_{y \in C_Y} F(x, y)$ exists. By similar arguments $\max_{y \in C_Y} \min_{x \in C_X} F(x, y)$ exists. It remains to show that these two values are equal.

2 i. e., for every $y \in C_Y$ and for every $\lambda \in \mathbb{R}$, the sets $\{x \in C_X : F(x, y) \leq \lambda\}$ are convex.

3 i. e., for every $x \in C_X$ and for every $\lambda \in \mathbb{R}$, the sets $\{y \in C_Y : F(x, y) \geq \lambda\}$ are convex.

2. To this end, by (5.1) it is true that

$$\max_{y\in C_Y}\min_{x\in C_X} F(x,y) \le \min_{x\in C_X}\max_{y\in C_Y} F(x,y).$$

We will show that the strict inequality cannot hold, hence the two values are necessarily equal.

Assume per contra that it did. Then, there exists $\lambda \in \mathbb{R}$ such that

$$\max_{y\in C_Y}\min_{x\in C_X} F(x,y) < \lambda < \min_{x\in C_X}\max_{y\in C_Y} F(x,y). \tag{5.5}$$

Consider the set valued mappings $F_1, F_2 : C_X \to 2^{C_Y}$, defined by

$$\begin{aligned} x &\mapsto F_1(x) = \{y \in C_Y \,:\, F(x,y) < \lambda\},\\ x &\mapsto F_2(x) = \{y \in C_Y \,:\, F(x,y) > \lambda\}, \end{aligned}$$

whose inverse mappings are defined by

$$\begin{aligned} y &\mapsto F_1^{-1}(y) = \{x \in C_X \,:\, F(x,y) < \lambda\},\\ y &\mapsto F_2^{-1}(y) = \{x \in C_X \,:\, F(x,y) > \lambda\}. \end{aligned}$$

By the assumptions on F, we see that for every $x \in C_X$, it holds that $F_1(x)$ is open in C_Y, and $F_2(x)$ is convex and nonempty (by (5.5)).[4] On the other hand, for every $y \in C_Y$, it holds that $F_1^{-1}(y)$ is convex, and $F_2^{-1}(y)$ is open.[5]

We claim that because of these properties,

$$\text{there exists} \quad \bar{x} \in C_X \quad \text{such that} \quad F_1(\bar{x}) \cap F_2(\bar{x}) \neq \emptyset. \tag{5.6}$$

Accepting this claim for the time being, we see that this leads us to a contradiction. Indeed, let $\bar{y} \in F_1(\bar{x}) \cap F_2(\bar{x})$. Then, since $\bar{y} \in F_1(\bar{x})$ by the definition of F_1 it holds that $F(\bar{x},\bar{y}) < \lambda$, while since $\bar{y} \in F_2(\bar{x})$ by the definition of F_2, it holds that $F(\bar{x},\bar{y}) > \lambda$, which is clearly a contradiction. Hence, (5.5) does not hold, and therefore (5.4) is true.

3. It remains to verify the crucial claim (5.6). This can be done considering the set valued map $\Phi : C_X \times C_Y \to 2^{C_X\times C_Y}$ defined by

$$\Phi(x,y) = (C_X \times C_Y) \cap \big(F_2(y)^{-1} \times F_1(x)\big)^c.$$

4 Since $\lambda < \min_{x\in C_X}\max_{y\in C_Y} F(x,y) \le \max_{y\in C_Y} F(x,y)$ for every $x \in C_X$, and the maximum of $F(x,y)$ is attained in C_Y, it follows that for every $x \in C_X$ there exists a $y_o = y_o(x)$ (a maximizer of $F(x,\cdot)$) such that $\lambda < F(x,y_o)$, i. e. $y_o \in F_2(x)$.

5 $F_1(x)$ is open since $F(x,\cdot)$ is upper semicontinuous; $F_2(x)$ is convex by the quasi-concavity of $F(x,\cdot)$; $F_1^{-1}(y)$ is convex by the quasi-convexity of $F(\cdot,y)$, and $F_2^{-1}(y)$ is open by the lower semicontinuity of $F(\cdot,y)$ (recall Definition 2.2.1).

3(a) Observe that, for the map Φ, it holds that[6]

$$\bigcap_{(x,y)\in C_X\times C_Y} \Phi(x,y)=\emptyset. \tag{5.7}$$

By (5.7), there exists a finite set $\{(x_i,y_i)\in C_X\times C_Y\ :\ i=1,\dots,n\}$ such that $\operatorname{conv}\{(x_i,y_i)\ :\ i=1,\dots,n\}\not\subset\bigcup_{i=1}^n\Phi((x_i,y_i))$. The existence of such a set arises from the following argument: if such a set did not exist, then for any finite set $\{(x_i',y_i')\in C_X\times C_Y\ :\ i=1,\dots,n\}$, we would have that

$$\operatorname{conv}\{(x_i',y_i')\ :\ i=1,\dots,n\}\subset\bigcup_{i=1}^n\Phi((x_i',y_i')).$$

Hence, by the KKM lemma (Theorem 3.2.11), for any such finite set we would have that $\bigcap_{i=1}^n\Phi(x_i',y_i')\neq\emptyset$. But that implies that the family $\{\Phi((x,y)\ :\ (x,y)\in C_X\times C_Y\}$ has the finite intersection property. However, recall (see Proposition 1.9.10) that a metric space is compact if and only if every collection of closed sets with the finite intersection property has a nonempty intersection. Since $C_X\times C_Y$ is by assumption compact, this would imply that $\bigcap_{(x,y)\in C_X\times C_Y}\Phi(x,y)\neq\emptyset$, which clearly contradicts our observation that $\bigcap_{(x,y)\in C_X\times C_Y}\Phi(x,y)=\emptyset$.

3(b) We now consider the finite set $\{(x_i,y_i)\in C_X\times C_Y\ :\ i=1,\dots,n\}$ of step (a). Since $\operatorname{conv}\{(x_i,y_i)\ :\ i=1,\dots,n\}\not\subset\bigcup_{i=1}^n\Phi(x_i,y_i)$, there exist $(\lambda_1,\dots,\lambda_n)$, $\lambda_i\in[0,1]$, $\sum_{i=1}^n\lambda_i=1$ such that

$$\bar{x}:=\sum_{i=1}^n\lambda_i x_i\notin\bigcup_{i=1}^n\big((C_X\times C_Y)\cap(F_2^{-1}(y_i))^c\big),$$
$$\bar{y}:=\sum_{i=1}^n\lambda_i y_i\notin\bigcup_{i=1}^n\big((C_X\times C_Y)\cap(F_1(x_i))^c\big),$$

which in turn (upon taking complements) implies that

$$\bar{x}\in\bigcap_{i=1}^n F_2^{-1}(y_i),\quad \bar{y}\in\bigcap_{i=1}^n F_1(x_i),$$

or equivalently $\bar{x}\in F_2^{-1}(y_i)$, $\bar{y}\in F_1(x_i)$ for every $i=1,\dots,n$, which in turn implies that

$$y_i\in F_2(\bar{x}),\quad x_i\in F_1^{-1}(\bar{y}),\quad i=1,\dots,n.$$

6 Suppose not. There exists $(x_o',y_o')\in\bigcap_{(x,y)\in C_X\times C_Y}\Phi(x,y)$, i. e., $x_o'\in(F_2^{-1}(y))^c$ for every $y\in C_Y$, and $y_o'\in F_1(x)$ for every $x\in C_X$. The first one implies that $y\notin F_2(x_o')$ for every $y\in C_Y$, which means that $F_2(x_o')=\emptyset$, which is in contradiction with (5.5).

By the convexity of the sets $F_2(\bar{x})$ and $F_1^{-1}(\bar{y})$, we conclude that $\bar{y} \in F_2(\bar{x})$ and $\bar{x} \in F_1^{-1}(\bar{y})$, so $\bar{y} \in F_1(\bar{x})$ and $\bar{y} \in F_2(\bar{x})$ as required. Hence we have shown (5.6) which by step 2 concludes the proof. □

There are numerous extensions to this fundamental minimax theorem, either relaxing some of the compactness assumptions or some of the convexity assumptions. Furthermore, there is a wide variety of applications of minimax theorems, ranging from game theory and decision making to optimization, PDEs, statistics, and machine learning.

Example 5.1.6 (Game theory). Consider the interaction of two players A and E. In this setting, let $F(x, y)$ be the payoff of A if he plays strategy $x \in C_X$ while E plays strategy $y \in C_Y$, and $-F(x, y)$ be the payoff for E for the same choice. Then, any element of $\arg\max_{x\in C_X} F(x, y)$ is a strategy for A, which is a best reply to the action y chosen by E. Hence, as rational agent, he would choose among these. Likewise, any element of $\arg\max_{y\in C_Y} F(x, y)$ is a strategy for E, being one of her a best replies to the action x chosen by A. Hence, as rational agent, she would choose among them. As a result of the interaction between A and E, the outcome of the game would be strategies (x_o, y_o) that satisfy the fixed point condition $x_o \in \arg\max_{x\in C_X} F(x, y_o)$ and $y_o \in \arg\max_{y\in C_Y} F(x_o, y)$, which correspond to saddle points of the payoff function F. It is important to note here that via the restatement of a saddle point as the fixed point of a set valued map, one may generalize the notion of saddle point for functions F in more than two variables, with important consequences for the theory of noncooperative games for more than two players (see e. g. [112]). ◁

5.2 Conjugate functions

5.2.1 The Legendre–Fenchel conjugate

Notation. We will adopt the convention of denoting by F any real valued function, and by φ any real valued function that is convex. Throughout this section X is a Banach space.

Definition 5.2.1 (Fenchel–Legendre conjugate or transform). Let X be a Banach space with dual $X^\star$, and denote by $\langle\cdot,\cdot\rangle$ the duality pairing between X and $X^\star$. For a proper function $F : X \to \mathbb{R} \cup \{+\infty\}$, its Legendre–Fenchel conjugate (or Legendre–Fenchel transform) $F^\star : X^\star \to \mathbb{R} \cup \{+\infty\}$, defined by

$$F^\star(x^\star) := \sup_{x\in X}(\langle x^\star, x\rangle - F(x)), \quad \forall x^\star \in X^\star.$$

The properness of F rules out the possibility of $F^\star$ admitting the value $-\infty$, thus guaranteeing that $F^\star$ is well defined. We will assume henceforth that F is proper, unless explicitly mentioned otherwise. Conjugate functions play a very important role in nonlinear analysis as the following examples (see e. g., [12, 20, 28, 57, 74]) indicate.

Example 5.2.2. Let $X = \mathbb{R}^d$ (considered as a Hilbert space identified with its dual), and consider the quadratic function $\varphi(x) = \frac{1}{2}\langle Ax, x\rangle + \langle b, x\rangle$, where $A \in \mathbb{R}^{d\times d}$ is a positive definite symmetric matrix considered as a mapping $A : X \to X$. Then, it is easy to verify that $\varphi^\star(x^\star) = \frac{1}{2}\langle x^\star - b, A^{-1}(x^\star - b)\rangle$. ◁

Example 5.2.3. Let $X = \ell^1$, the space of summable real sequences, and set $\varphi(x) = \|x\|_{\ell_1}$. Then,

$$\varphi^\star(x^\star) = \begin{cases} 0 & \text{if } \|x^\star\|_{\ell^\infty} \le 1 \\ +\infty & \text{otherwise,} \end{cases}$$

which is the convex indicator function of the closed unit ball of ℓ^∞ (which is the dual space of ℓ^1).

This can be proved directly, but is also a special case of a more general result. For a direct proof, note that for the real valued function $\varphi_i(x_i) = |x_i|$, we can easily verify that

$$\sup_{x_i\in\mathbb{R}}(x_i^\star x_i - |x_i|) = \begin{cases} 0 & \text{if } |x_i^\star| \le 1, \\ \infty & \text{otherwise.} \end{cases}$$

Hence, by the definition of the ℓ^1 norm as $\|x\|_{\ell^1} = \sum_{i=1}^\infty |x_i|$, in combination with the above result, we easily see that

$$\varphi^\star(x^\star) = \begin{cases} 0 & \text{if } \sup_{i\in\mathbb{N}} |x_i^\star| \le 1 \\ +\infty & \text{otherwise} \end{cases} = \begin{cases} 0 & \text{if } \|x^\star\|_{\ell^\infty} \le 1 \\ +\infty & \text{otherwise.} \end{cases}$$ ◁

Example 5.2.4. If $X = \mathbb{R}$ and $\varphi_i(x_i) = \frac{1}{p}|x_i|^p$, $p > 1$, it is a simple calculus fact to see that $\varphi_i^\star(x_i^\star) = \frac{1}{p^*}|x_i^\star|^{p^*}$, where $\frac{1}{p} + \frac{1}{p^*} = 1$. It can then be seen that if $X = \ell^p$, $p > 1$, with $\varphi(x) = \frac{1}{p}\|x\|_{\ell^p}^p$, then $\varphi^\star(x^\star) = \frac{1}{p^*}\|x^\star\|_{\ell^{p^*}}^{p^*}$. Note that ℓ^{p^*} can be identified as the dual of ℓ^p. ◁

The following proposition (see e. g., [12, 20, 28, 57, 74]) collects some useful properties of Fenchel–Legendre transforms:

Proposition 5.2.5 (Properties of Legendre–Fenchel transform). *Let $(X, \|\cdot\|)$, be a Banach space with dual $(X^\star, \|\cdot\|_{X^\star})$.*

(i) *$F^\star : X^\star \to \mathbb{R}\cup\{+\infty\}$ is convex and lower semicontinuous (even if F may not be convex), that satisfies the Young–Fenchel inequality*

$$F(x) + F^\star(x^\star) \ge \langle x^\star, x\rangle, \quad \forall (x, x^\star) \in X \times X^\star. \tag{5.8}$$

(ii) *$(\lambda F)^\star(x^\star) = \lambda F^\star(\frac{x^\star}{\lambda})$ for every $\lambda > 0$.*

(iii) *Let $z \in X$, and define the translation of F by z, $F_z : X \to \mathbb{R}\cup\{+\infty\}$ as $F_z(x) := F(x - z)$ for every $x \in X$. Then, $F_z^\star(x^\star) = F^\star(x^\star) + \langle x^\star, z\rangle$.*

(iv) *If* $F_1 \leq F_2$, *then* $F_1^\star \geq F_2^\star$.
(v) *Let* $\varphi_0 : \mathbb{R} \to \mathbb{R} \cup \{+\infty\}$ *be a function such that* $\varphi_0(0) = 0$ *and* $\varphi_0(t) \geq 0$, *and define* $F : X \to \mathbb{R}_+ \cup \{+\infty\}$ *as* $F(x) = \varphi_0(\|x\|)$. *Then,* $F^\star(x^\star) = \varphi_0^\star(\|x^\star\|_{X^\star})$.
(vi) *If* $F(x) = \|x\|$, *then* $F^\star(x^\star) = I_{\overline{B}_{X^\star}(0,1)}$, *where* $\overline{B}^\star := \overline{B}_{X^\star}(0,1) = \{x^\star \in X^\star \ : \ \|x^\star\|_{X^\star} \leq 1\}$.

Proof. (i) We define the functional $\psi_x : X^\star \to \mathbb{R}$, by $\psi_x(x^\star) := \langle x^\star, x\rangle - F(x)$, and note that by definition, $F^\star(x^\star) = \sup_{x\in X} \psi_x(x^\star)$ for any $x^\star \in X^\star$. Since $F^\star$ is the pointwise supremum over the family of affine functions $x^\star \mapsto \psi_x(x^\star)$, it is convex and lower semicontinuous (see Proposition 2.3.8(iii) and Example 2.2.3, respectively). By the definition of $F^\star$, we have that

$$F^\star(x^\star) \geq \langle x^\star, x\rangle - F(x), \quad \forall (x, x^\star) \in X \times X^\star,$$

from which the Young–Fenchel inequality follows.

(ii) To calculate $(\lambda F)^\star$, we need to solve the optimization problem

$$(\lambda F)^\star(x^\star) = \sup_{x\in X^\star}(\langle x^\star, x\rangle - (\lambda F)(x)).$$

We express the quantity to be optimized as

$$\langle x^\star, x\rangle - (\lambda F)(x) = \lambda\Big(\Big\langle \frac{x^\star}{\lambda}, x\Big\rangle - F(x)\Big),$$

so

$$\sup_{x\in X}\langle x^\star, x\rangle - (\lambda F)(x) = \lambda \sup_{x\in X}\Big(\Big\langle \frac{x^\star}{\lambda}, x\Big\rangle - F(x)\Big) = \lambda F^\star\Big(\frac{x^\star}{\lambda}\Big).$$

(iii) By definition,

$$F_z^\star(x^\star) = \sup_{x\in X}(\langle x^\star, x\rangle - F_z(x)) = \sup_{x\in X}(\langle x^\star, x\rangle - F(x - z)).$$

We express the quantity to be maximized as

$$\langle x^\star, x\rangle - F(x - z) = \langle x^\star, x - z\rangle - F(x - z) + \langle x^\star, z\rangle$$

so that

$$\begin{aligned}\sup_{x\in X}(\langle x^\star, x\rangle - F(x - z)) &= \sup_{x\in X}(\langle x^\star, x - z\rangle - F(x - z) + \langle x^\star, z\rangle) \\ &= \sup_{x\in X}(\langle x^\star, x - z\rangle - F(x - z)) + \langle x^\star, z\rangle = F^\star(x^\star) + \langle x^\star, z\rangle.\end{aligned}$$

(iv) Since $F_1 \leq F_2$, we have that

$$\langle x^\star, x\rangle - F_2(x) \leq \langle x^\star, x\rangle - F_1(x), \quad \forall (x, x^\star) \in X \times X^\star.$$

We first estimate the right-hand side by its supremum so that

$$\langle x^\star, x\rangle - F_2(x) \le \langle x^\star, x\rangle - F_1(x) \le \sup_{x\in X}(\langle x^\star, x\rangle - F_1(x)) = F_1^\star(x^\star), \quad \forall (x, x^\star) \in X \times X^\star,$$

and since for any fixed $x^\star \in X^\star$

$$\langle x^\star, x\rangle - F_2(x) \le F_1^\star(x^\star), \quad \forall x \in X,$$

the inequality holds also for the supremum of the left-hand side over X. Therefore,

$$F_2^\star(x^\star) = \sup_{x\in X}(\langle x^\star, x\rangle - F_2(x)) \le F_1^\star(x^\star).$$

Since this is true for any $x^\star \in X^\star$, it follows that $F_2^\star \le F_1^\star$.

(v) Recall the definition of the dual norm as $\|x^\star\|_{X^\star} = \sup_{x\in X,\ \|x\|=1} |\langle x^\star, x\rangle|$. Based on the definition of $\varphi_o^\star : \mathbb{R} \to \mathbb{R}$, and on account of the properties of φ_o, we see that

$$\begin{aligned}\varphi_o^\star(\|x^\star\|_{X^\star}) &= \sup_{r\in\mathbb{R}}\{\|x^\star\|_{X^\star} r - \varphi_o(r)\} = \sup_{r\in\mathbb{R}_+}\{\|x^\star\|_{X^\star} r - \varphi_o(r)\} \\ &\le \sup_{r\in\mathbb{R}_+}\ \sup_{x\in X,\ \|x\|=r}\{\langle x^\star, x\rangle - \varphi_o(r)\} \le \sup_{x\in X}\{\langle x^\star, x\rangle - \varphi_o(\|x\|)\} = F^\star(x^\star),\end{aligned} \tag{5.9}$$

where, for the first line, we used the fact that $\|x^\star\|_{X^\star} r - \varphi_o(r) \le 0 = \varphi_o(0)$ when $r \le 0$, and for the second the definition of the dual norm.

We note that we can establish the reverse inequality as well. Indeed, recalling that $\langle x^\star, x\rangle \le \|x^\star\|_{X^\star}\|x\|$, we see that

$$\begin{aligned}F^\star(x^\star) &= \sup_{x\in X}\{\langle x^\star, x\rangle - \varphi_o(\|x\|)\} \le \sup_{x\in X}\{\|x^\star\|_{X^\star}\|x\| - \varphi_o(\|x\|)\} \\ &\le \sup_{r\in\mathbb{R}}\{\|x^\star\|_{X^\star} r - \varphi_o(r)\} = \varphi_o^\star(\|x^\star\|_{X^\star}),\end{aligned} \tag{5.10}$$

where for the last equality we used the definition of $\varphi_o^\star$. Combining (5.9) and (5.10), we obtain the required result.

(vi) Follows from (v) setting $\varphi_0(t) = |t|$, and noting that

$$\varphi_o^\star(t^*) = \begin{cases} 0 & \text{if } |t^*| \le 1, \\ +\infty & \text{otherwise,}\end{cases}$$

(see Example 5.2.3). □

It is clear from the definition of the Legendre–Fenchel transform that if we consider it as an operator, it is not a linear operator, i. e., in general $(\varphi_1 + \varphi_2)^\star \ne \varphi_1^\star + \varphi_2^\star$. An interesting question that arises is whether $\varphi_1^\star + \varphi_2^\star$ can actually be expressed as the Legendre–Fenchel transform of some function. This leads us to the important notion of the inf-convolution (see Section 5.2.4).

Some more examples

We close this section with some more examples concerning the Fenchel–Legendre transform (see e. g., [12, 20, 28, 57, 74]).

Example 5.2.6. Let X be a Banach space and $X^\star$ its dual. Define $\varphi : X \to \mathbb{R}$ as $\varphi(x) := \frac{1}{p}\|x\|_X^p$ for some $p \in [1,\infty)$, and set $p^\star$ such that $\frac{1}{p} + \frac{1}{p^\star} = 1$ if $p \neq 1$ and $\overline{B}_{X^\star}(0,1) = \{x^\star \in X^\star : \|x^\star\|_{X^\star} \leq 1\}$. Then,

$$\varphi^\star(x^\star) = \begin{cases} \frac{1}{p^\star}\|x^\star\|_{X^\star}^{p^\star} & \text{if } p \in (1,\infty), \\ I_{\overline{B}_{X^\star}(0,1)}, & \text{if } p = 1. \end{cases}$$

This follows easily by Prop. 5.2.5(v) and (vi). ◁

The following examples show that indicator functions and support functions are connected through the Fenchel–Legendre conjugate:

Example 5.2.7. The Legendre–Fenchel conjugate of the indicator function of the closed unit ball of X is related to the norm of its dual space, $X^\star$.

Let $B := B_X(0,1) = \{x \in X : \|x\|_X \leq 1\}$, the closed unit ball of X, and consider its indicator function $\varphi := I_B$. Then,

$$\varphi^\star(x^\star) = \sup_{x\in X}(\langle x^\star, x\rangle - I_B(x)) = \sup_{x\in B}\langle x^\star, x\rangle = \sup_{x\in X, \|x\|_X=1}\langle x^\star, x\rangle = \|x^\star\|_{X^\star}.$$ ◁

Example 5.2.8. The Legendre–Fenchel conjugate of the indicator function I_C of a closed convex set C is its support function σ_C, i. e., $(I_C)^\star = \sigma_C$.

Let $C \subset X$ be a closed convex set, and consider its indicator function $\varphi := I_C$. Then,

$$\varphi^\star(x^\star) = \sup_{x\in X}(\langle x^\star, x\rangle - I_C(x)) = \sup_{x\in C}\langle x^\star, x\rangle =: \sigma_C(x^\star),$$

where $\sigma_C : X^\star \to \mathbb{R}$ is the support function of C (see Definition 2.3.13). ◁

Example 5.2.9 (The epigraph of the Legendre–Fenchel transform). Let F be any proper function (not necessarily convex) and $F^\star$ be its conjugate function. Then, upon defining, for every $x \in X$, the family of functions $F_x : X^\star \to \mathbb{R} \cup \{+\infty\}$, by $x^\star \mapsto F_x(x^\star) := \langle x^\star, x\rangle - F(x)$, it holds that

$$\operatorname{epi} F^\star = \{(x^\star, \lambda) \in X^\star \times \mathbb{R} : F(x) \geq \langle x^\star, x\rangle - \lambda, \ \forall x \in X\} = \bigcap_{x\in X} \operatorname{epi} F_x,$$

i. e., the points of the epigraph of $F^\star$ parametrize the set of affine functions minorizing F, or equivalently, the epigraph of $F^\star$ is the intersection over all $x \in X$ of the epigraphs of the family of functions $F_x : X^\star \to \mathbb{R}$.

This follows easily by the definition of the epigraph. Take any $(x^\star, \lambda) \in \operatorname{epi} F^\star$. Then, $F^\star(x^\star) \leq \lambda$, and by the definition of $F^\star$, we have that for every $x \in X$ it holds that $\langle x^\star, x\rangle - F(x) \leq \lambda$. Hence, $(x^\star, \lambda) \in \{(x^\star, \lambda) : F(x) \geq \langle x^\star, x\rangle - \lambda, \ \forall x \in X\}$, therefore $\operatorname{epi} F^\star \subset$

$\{(x^\star,\lambda) : F(x) \geq \langle x^\star, x\rangle - \lambda, \ \forall x \in X\}$. For the reverse inclusion, consider any $(x^\star,\lambda) \in \{(x^\star,\lambda) : F(x) \geq \langle x^\star, x\rangle - \lambda, \ \forall x \in X\}$. Then, $\langle x^\star, x\rangle - F(x) \leq \lambda$ for every $x \in X$, and taking the supremum over all $x \in X$, we conclude that $F^\star(x^\star) \leq \lambda$, hence $(x^\star,\lambda) \in \operatorname{epi} F^\star$. The second claim follows by observing that $\{(x^\star,\lambda) : F(x) \geq \langle x^\star, x\rangle - \lambda, \ \forall x \in X\} = \{(x^\star,\lambda) : F_x(x^\star) \leq \lambda\} = \bigcap_{x\in X} \operatorname{epi} F_x$. ◁

Example 5.2.10 (Fenchel–Legendre transform of the Moreau–Yosida regularization). Let $X = H$ be a Hilbert space, identified with its dual, and let $\varphi : H \to \mathbb{R} \cup \{+\infty\}$ be a proper lower semicontinuous convex function and ϕ_λ its Moreau–Yosida regularization defined by $\phi_\lambda(x) = \inf_{z\in H}(\varphi(z) + \frac{1}{2\lambda}\|z - x\|_H^2)$ (see Definition 4.4.3). Then, $\varphi_\lambda^\star(x^\star) = \varphi^\star(x^\star) + \frac{\lambda}{2}\|x^\star\|_H^2$.

Using the definitions, and the fact that we may interchange the order of suprema,

$$\begin{aligned}
\varphi_\lambda^\star(x^\star) &= \sup_{x\in H}(\langle x^\star, x\rangle_H - \varphi_\lambda(x)) = \sup_{x\in H}\Big(\langle x^\star, x\rangle_H - \inf_{z\in H}\Big(\frac{1}{2\lambda}\|z - x\|_H^2 + \varphi(z)\Big)\Big) \\
&= \sup_{x\in H}\Big(\langle x^\star, x\rangle_H + \sup_{z\in H}\Big(-\frac{1}{2\lambda}\|z - x\|_H^2 - \varphi(z)\Big)\Big) \\
&= \sup_{z\in H}\sup_{x\in H}\Big(\langle x^\star, z\rangle_H - \varphi(z) + \langle x^\star, x - z\rangle_H - \frac{1}{2\lambda}\|z - x\|_H^2\Big) \\
&= \sup_{z\in H}\Big(\langle x^\star, z\rangle_H - \varphi(z) + \sup_{x\in H}\Big(\langle x^\star, x - z\rangle_H - \frac{1}{2\lambda}\|x - z\|_H^2\Big)\Big) \\
&= \sup_{z\in H}\Big(\langle x^\star, z\rangle_H - \varphi(z) + \sup_{x'\in H}\Big(\langle x^\star, x'\rangle_H - \frac{1}{2\lambda}\|x'\|_H^2\Big)\Big) = \varphi^\star(x^\star) + \frac{\lambda}{2}\|x^\star\|_H^2,
\end{aligned}$$

and our claim is proved. ◁

5.2.2 The bi-conjugate function

We continue adopting the convention of denoting by F any real valued function, and by φ any real valued function that is convex.

Definition 5.2.11 (Bi-conjugate). Let $F : X \to \mathbb{R} \cup \{+\infty\}$ be a proper function. The bi-conjugate of F is the function $F^{\star\star} : X \to \mathbb{R} \cup \{+\infty\}$, defined by

$$F^{\star\star}(x) = \sup_{x^\star\in X^\star}(\langle x^\star, x\rangle - F^\star(x^\star)).$$

Remark 5.2.12. In general, we may consider $F^{\star\star}$ as a function from $X^{\star\star}$ to $\mathbb{R}$, which is the Fenchel conjugate of $F^\star$, i. e., as $(F^\star)^\star$. What we have defined in Definition 5.2.11 can then be understood as the restriction of $(F^\star)^\star$ on X (which can be seen as a subspace of $X^{\star\star}$ using the canonical embedding $j : X \to X^{\star\star}$, defined by $\langle j(x), x^\star\rangle_{X^{\star\star},X^\star} = \langle x^\star, x\rangle_{X^\star,X}$, see Theorem 1.1.23). Under this perspective, we could define

$$F^{\star\star}(j(x)) = \sup_{x^\star \in X^\star} \left(\langle j(x), x^\star \rangle_{X^{\star\star}, X^\star} - F^\star(x^\star) \right) = \sup_{x^\star \in X^\star} \left(\langle x^\star, x \rangle_{X^\star, X} - F^\star(x^\star) \right).$$

Clearly, if X is a reflexive space, these two concepts coincide.

The bi-conjugate of a function can be understood as a convexification of a function from below. We have already seen (Theorem 2.3.20) that a proper convex lower semicontinuous function can be expressed as the supremum over the family of all affine functions that minorize it. The obvious question is what happens if we drop the assumptions of convexity and lower semicontinuity. To this end, we must first define the concept of Γ-regularization:

Definition 5.2.13 (Γ-Regularization). Let $F : X \to \mathbb{R} \cup \{+\infty\}$ be a proper function, and set

$$\mathbb{A}(F) := \{g : X \to \mathbb{R} \,:\, g \text{ continuous and affine, } g \leq F\}.$$

The Γ-regularization of F is defined as the function $F^\Gamma : X \to \mathbb{R} \cup \{+\infty\}$ such that[7]

$$F^\Gamma(x) := \sup_{g \in \mathbb{A}(F)} g(x), \quad \forall x \in X.$$

It can be seen by a straightforward application of Proposition 2.3.8(iii) that F^Γ is a convex function, even if F is not. Furthermore, as the supremum of a family of affine functionals, F^Γ enjoys lower semicontinuity properties. In this respect, F^Γ can be considered as a convex regularization of F from below, as the next proposition [12] indicates.

Proposition 5.2.14. *Let $F : X \to \mathbb{R} \cup \{+\infty\}$ be a proper function. Then, it holds that $F^{\star\star} = F^\Gamma \leq F$.*

Proof. Given the fact that $F^\Gamma \leq F$ follows from the definition of F^Γ: Fixing any $x \in X$ then for any $g \in \mathbb{A}(F)$, it holds that $g(x) \leq F(x)$, hence, taking the supremum over all $g \in \mathbb{A}(F)$, we conclude that $F^\Gamma \leq F$.

It remains to show that $F^{\star\star} = F^\Gamma$. To this end, consider any affine function $g \in \mathbb{A}(F)$. Since g is affine, it is completely characterized by an element $x^\star \in X^\star$ and a real number $c \in \mathbb{R}$ such that $\langle x^\star, x \rangle + c \leq F(x)$ for every $x \in X$. Therefore, to characterize $F^\Gamma(x)$ as the supremum of the quantity $g(x)$ when $g \in \mathbb{A}(F)$, it is equivalent to characterize the supremum of the quantity $\langle x^\star, x \rangle + c$ over all $x^\star \in X^\star$ and over all $c \in \mathbb{R}$ such that $\langle x^\star, x \rangle + c \leq F(x)$. But this last condition implies that $\langle x^\star, x \rangle - F(x) \leq -c$ for all $x \in X$, so $F^\star(x^\star) := \sup_{x \in X}(\langle x^\star, x \rangle - F(x)) \leq -c$. This means, that given an $x^\star \in X^\star$ (which characterizes the affine function g, chosen), then the real number c cannot be any real number, but rather must satisfy the constraint $c \leq -F^\star(x^\star)$. Therefore,

$$F^\Gamma(x) = \sup_{x^\star \in X^\star,\ c \leq -F^\star(x^\star)} \langle x^\star, x \rangle + c.$$

7 We use the convention that $\sup \emptyset = -\infty$.

Let us consider the constraints a bit more carefully. If $F^\star(x^\star) > -\infty$ for every $x^\star \in X^\star$, then, given $x^\star \in X^\star$, c, can go up to the maximum value $c_{\max} = -F^\star(x^\star)$, so the maximum over c of the quantity $\langle x^\star, x\rangle + c$ for fixed $x^\star$ will be $\langle x^\star, x\rangle - F^\star(x^\star)$. We then vary $x^\star$ over $X^\star$ and take the supremum of the resulting quantity. This problem is simply the problem $\sup_{x^\star \in X^\star}(\langle x^\star, x\rangle - F^\star(x^\star))$, which is nothing else but $F^{\star\star}(x)$. Summarizing the above, if $x^\star$ is such that $F^\star(x^\star) > -\infty$, then

$$F^\Gamma(x) = \sup_{x^\star \in X^\star,\ c \le -F^\star(x^\star)} (\langle x^\star, x\rangle + c) = \sup_{x^\star \in X^\star} (\langle x^\star, x\rangle - F^\star(x^\star)) = F^{\star\star}(x).$$

If $F^\star(x^\star) = -\infty$ for some $x^\star \in X^\star$, then we can see that $F^\Gamma(x) = +\infty = F^{\star\star}(x)$ for all $x \in X$. □

The following proposition [12, 20, 28, 74] provides a fundamental connection between F and $F^{\star\star}$:

Proposition 5.2.15 (Fenchel–Moreau–Rockafellar). *Let $F : X \to \mathbb{R} \cup \{+\infty\}$ be a proper function. Then, $F^{\star\star} = F$ if and only if F is convex and lower semicontinuous.*

Proof. We only prove that if F is convex and lower semicontinuous that $F = F^{\star\star}$, as the other direction is easy.

Assume that F is convex and lower semicontinuous. We will show that $F = F^{\star\star}$. To this end, we will use the representation of $F^{\star\star}$ as $F^{\star\star} = F^\Gamma$ (see Proposition 5.2.14). By the definition of F^Γ, it is clear that $F^\Gamma \le F$. To show that they are equal, it is enough to show that there does not exists any $x \in X$ for which $F^\Gamma(x) < \gamma < F(x)$ for some $\gamma \in \mathbb{R}$. Assume, the contrary, and let $z \in X$ be such a point, i. e., $F^\Gamma(z) < \gamma < F(z)$. We will then show that there exists an affine function $g \in \mathbb{A}(F)$ satisfying the property $g(z) > \gamma$, therefore $F^\Gamma(z) > \gamma$ (by the definition of $F^\Gamma(z)$ as the supremum of $\hat{g}(z)$ over all $\hat{g} \in \mathbb{A}(F)$), which contradicts the choice of z.

To establish the contradiction, we will apply the strict separation Theorem 1.2.9 to the sets $C_1 := \operatorname{epi} F$ and $C_2 := \{(z, \gamma)\}$, both considered as subsets of $Y := X \times \mathbb{R}$, which is obviously a Banach space with dual[8] $Y^\star = X^\star \times \mathbb{R}$. Since F is convex by Proposition 2.3.7, C_1 is a convex set, and since F is lower semicontinuous, it is also closed. C_2 consists of a single point, so it is trivially convex and closed. Furthermore, $C_1 \cap C_2 = \emptyset$ by the choice of z. Therefore, by the strict separation theorem, there exists $y^\star := (x^\star, c) \in Y^\star$ and $\alpha \in \mathbb{R}$ such that $\langle x^\star, z\rangle + c\gamma > \alpha$ and $\langle x^\star, x\rangle + c\lambda < \alpha$ for every $(x, \lambda) \in C_1 := \operatorname{epi} F$, i.e, for every (x, λ) such that $F(x) \le \lambda$. Clearly, if $(x, \lambda) \in \operatorname{epi} F$, then for any $\mu > \lambda$, $(x, \mu) \in \operatorname{epi} F$. Applying the separation result for the new point (x, μ), we have that $\langle x^\star, x\rangle + c\mu < \alpha$ for any $\mu > \lambda$, and letting $\mu \to \infty$ leads to the conclusion $c \le 0$.

8 Any element of $y \in Y$ is identified at the pair $(x, c_1) \in X \times \mathbb{R}$; any element $y^\star \in Y^\star$ is identified as a pair $(x^\star, c_2) \in X^\star \times \mathbb{R}$, and the duality pairing between the two spaces is expressed as $\langle y^\star, y\rangle_{Y^\star,Y} := \langle x^\star, x\rangle_{X^\star X} + c_1 c_2$.

Consider now the value $F(z)$: Either (a) $F(z) < +\infty$ or (b) $F(z) = +\infty$. Both cases lead to a contradiction, as shown by constructing a $g \in \mathbb{A}(F)$ such that $g(z) > \gamma$.

(a) Let us consider the first case, $F(z) < +\infty$. This implies that the point $(z, F(z)) \in C_1 := \operatorname{epi} F$, therefore, by the separation result, we have that

$$\langle x^\star, z\rangle + cF(z) < \alpha < \langle x^\star, z\rangle + c\gamma, \tag{5.11}$$

and since $\gamma < F(z)$, this leads to the conclusion that $c < 0$. Consider the function $g : X \to \mathbb{R}$ defined by $g(x) = \frac{\alpha}{c} - \frac{1}{c}\langle x^\star, x\rangle$, which is an affine function. However, it also satisfies the property that $g(x) \le F(x)$ for every $x \in X$, therefore[9] $g \in \mathbb{A}(F)$. Furthermore, $g(z) = \frac{\alpha}{c} - \frac{1}{c}\langle x^\star, z\rangle > \gamma$, by (5.11), so this is a required g.

(b) In the second case, $F(z) = +\infty$, and since $c \le 0$, it may either hold that $c < 0$ or that $c = 0$. If $c < 0$, we may define $g : X \to \mathbb{R}$ as above and show that $g \in \mathbb{A}(F)$ and $g(z) > \gamma$. If $c = 0$, the construction of g has to be a little different. Since F is proper, there exists $x_o \in \operatorname{dom}(F)$, i. e., $x_o \in X$ such that $F(x_o) < \infty$. Then, observe that $(x_o, F(x_o)) \in C_1 = \operatorname{epi} F$, and apply the separation result to obtain that $\langle x^\star, x_o\rangle + cF(x_o) < \alpha < \langle x^\star, z\rangle + c\gamma$, which, since $c = 0$, simplifies to $\langle x^\star, x_o\rangle < \alpha < \langle x^\star, z\rangle$. We may then define an affine function $g_1 : X \to \mathbb{R}$ such that $g_1 \in \mathbb{A}(F)$ working as in case (a) with x_o in lieu of z. We may further define $g : X \to \mathbb{R}$ by $g(x) := g_1(x) + \left(\frac{|\gamma - g_1(z)|}{\langle x^\star, z\rangle - \alpha} + 1\right)(\langle x^\star, x\rangle - \alpha)$ and observe that g is affine, and it satisfies $g(x) \le g_1(x) \le F(x)$ for every $x \in X$ so that $g \in \mathbb{A}(F)$. Furthermore, $g(z) = g_1(z) + |\gamma - g_1(z)| + (\langle x^\star, z\rangle - \alpha) > \gamma$, so this is a required g.

Therefore, in all cases, we may construct a function $g \in \mathbb{A}(F)$ such that $g(z) > \gamma$. This establishes the contradiction and the proof is complete. □

Example 5.2.16. Let C be a closed convex set, and let $\varphi = I_C$ be the indicator function of C, as defined in Example 2.3.10. We have seen in Example 5.2.8 that $I_C^\star = \sigma_C$, the support function of C. Since C is closed and convex, all the requirements of the Fenchel–Moreau–Rockafellar theorem hold, so $I_C = I_C^{\star\star} = (I_C^\star)^\star = \sigma_C^\star$, which is of course related to the dual representation of convex sets in terms of the support function (see (2.8), Section 2.3.2). ◁

Remark 5.2.17 (Can we go beyond the bi-conjugate?). An interesting question that arises is whether we need to go beyond the bi-conjucate function. The answer is no, as if we did, then $\varphi^{\star\star\star} = \varphi^\star$, so any further Legendre–Fenchel transform beyond the second will coincide with either the conjugate or the bi-conjugate function. This is easy to see, as $\varphi^{\star\star} \le \varphi$, upon conjugation leads to $\varphi^\star \le \varphi^{\star\star\star}$. On the other hand, by definition $\varphi^{\star\star\star}(x^\star) = \sup_{x\in X}(\langle x^\star, x\rangle - \varphi^{\star\star}(x))$, and since $\varphi^\star(x^\star) \ge \langle x^\star, x\rangle - \varphi^{\star\star}(x)$ for every $x \in X^\star$, taking the supremum over all $x \in X$ leads to the reverse inequality $\varphi^{\star\star\star} \le \varphi^\star$, and the desired result follows.

---

9 Indeed, if $x \in X$ is such that $F(x) < +\infty$, then since $(x, F(x)) \in \operatorname{epi} F$, the separation result yields $\langle x^\star, x\rangle + cF(x) < \alpha$, which can be restated as $g(x) := \frac{\alpha}{c} - \frac{1}{c}\langle x^\star, x\rangle < F(x)$ (recall that $c < 0$). If on the contrary $x \in X$ is such that $F(x) = +\infty$, then $g(x) := \frac{\alpha}{c} - \frac{1}{c}\langle x^\star, x\rangle < +\infty = F(x)$.

5.2.3 The subdifferential and the Legendre–Fenchel transform

Recall the Young–Fenchel inequality (see Proposition 5.2.5) according to which $\varphi(x) + \varphi^\star(x^\star) \geq \langle x^\star, x\rangle$ for every $(x, x^\star) \in X \times X^\star$. A useful result is that the subdifferentials of φ and $\varphi^\star$ can be characterized as the pairs $(x, x^\star) \in X \times X^\star$, for which equality is attained in the Young–Fenchel inequality. This is shown in the next proposition [12, 20, 28, 74]:

Proposition 5.2.18. *Let $\varphi : X \to \mathbb{R} \cup \{+\infty\}$ be a proper and convex function. Then,*
(i) $\varphi(x) + \varphi^\star(x^\star) = \langle x^\star, x\rangle$ *if and only if* $x^\star \in \partial\varphi(x)$.
(ii) *If* $x^\star \in \partial\varphi(x)$, *then* $x \in \partial\varphi^\star(x^\star)$.[10]
(iii) *If furthermore, X is reflexive and φ is lower semicontinuous, then*

$$x^\star \in \partial\varphi(x) \quad \textit{if and only if} \quad x \in \partial\varphi^\star(x^\star).$$

Proof. (i) Assume that $\varphi(x) + \varphi^\star(x^\star) = \langle x^\star, x\rangle$. We will show that $x^\star \in \partial\varphi(x)$.

By definition $\varphi^\star(x^\star) = \sup_{x\in X}(\langle x^\star, x\rangle - \varphi(x))$ so that

$$\varphi^\star(x^\star) \geq \langle x^\star, z\rangle - \varphi(z), \quad \forall z \in X. \tag{5.12}$$

Since $\varphi^\star(x^\star) = \langle x^\star, x\rangle - \varphi(x)$, we substitute that into (5.12), and this yields

$$\langle x^\star, x\rangle - \varphi(x) \geq \langle x^\star, z\rangle - \varphi(z), \quad \forall z \in X,$$

which is expressed as

$$\varphi(z) - \varphi(x) \geq \langle x^\star, z - x\rangle, \quad \forall z \in X,$$

therefore $x^\star \in \partial\varphi(x)$.

Conversely, assume that $x^\star \in \partial\varphi(x)$. Then, by the definition of the subdifferential,

$$\varphi(z) - \varphi(x) \geq \langle x^\star, z - x\rangle, \quad \forall z \in X.$$

This is rearranged as

$$\langle x^\star, z\rangle - \varphi(z) \leq \langle x^\star, x\rangle - \varphi(x), \quad \forall z \in X.$$

Therefore, taking the supremum over $z \in X$,

$$\sup_{z\in X}(\langle x^\star, z\rangle - \varphi(z)) \leq \langle x^\star, x\rangle - \varphi(x),$$

and hence, by the definition of the Legendre–Fenchel conjugate of φ, the above becomes

10 Interpreted in the sense of Remark 5.2.20 is X is non reflexive.

$$\varphi^\star(x^\star) + \varphi(x) \le \langle x^\star, x\rangle. \tag{5.13}$$

On the other hand, the Young–Fenchel inequality yields

$$\varphi(x) + \varphi^\star(x^\star) \ge \langle x^\star, x\rangle,$$

which when combined with (5.13) gives us the equality

$$\varphi(x) + \varphi^\star(x^\star) = \langle x^\star, x\rangle.$$

(ii) Let X be reflexive (see Remark 5.2.20). Since $x^\star \in \partial\varphi(x)$ by (i) it holds that

$$\varphi(x) + \varphi^\star(x^\star) = \langle x^\star, x\rangle. \tag{5.14}$$

We now pick any $z^\star \in X^\star$ and apply the Young–Fenchel inequality to the pair $(x, z^\star) \in X \times X^\star$,

$$\varphi(x) + \varphi^\star(z^\star) \ge \langle z^\star, x\rangle, \tag{5.15}$$

and subtracting (5.14) from (5.15) yields

$$\varphi^\star(z^\star) - \varphi^\star(x^\star) \ge \langle z^\star - x^\star, x\rangle, \quad \forall z^\star \in X^\star.$$

This yields that $x \in \partial\varphi^\star(x^\star)$.

(iii) By the lower semicontinuity of φ, we have that $\varphi^{\star\star} = \varphi$ (see Proposition 5.2.15). Hence, applying (i) for $\varphi^\star$, noting that since X is reflexive $X^{\star\star} \simeq X$, while by lower semicontinuity $\varphi^{\star\star} = \varphi$, the result follows. □

Remark 5.2.19. The Young–Fenchel inequality yields that

$$\varphi(x) + \varphi^\star(x^\star) - \langle x^\star, x\rangle \ge 0,$$

while $\varphi(x) + \varphi^\star(x^\star) - \langle x^\star, x\rangle = 0$ if and only if $x^\star \in \partial\varphi(x)$. Therefore, any point $(x, x^\star) \in \mathbf{Gr}(\partial\varphi)$ is a minimizer of the function $\Phi : X \times X^\star \to \mathbb{R}$ defined by $\Phi(x, x^\star) := \varphi(x) + \varphi^\star(x^\star) - \langle x^\star, x\rangle$.

Remark 5.2.20. In the general case where X is not reflexive, $\partial\varphi^\star \subset X^{\star\star}$. We may then interpret the condition $x \in \varphi^\star(x^\star)$ in Proposition 5.2.18 in terms of the canonical embedding, $j : X \to X^{\star\star}$, as $j(x) \in \partial\varphi^\star(x^\star)$ (recall that in the reflexive case $j(X) = X^{\star\star}$). The result of Proposition 5.2.18 can then be generalized with analogous arguments.

Example 5.2.21. Let $X = H$ be a Hilbert space (identified with its dual) and $C \subset H$ be a closed convex set. Define the function $\varphi : H \to \mathbb{R} \cup \{\infty\}$ by $\varphi(x) = \frac{1}{2}\|x\|^2 + I_C(x)$. Then, $\varphi^\star(x^\star) = \langle x^\star, P_C(x^\star)\rangle - \frac{1}{2}\|P_C(x^\star)\|^2$, while[11] $\partial\varphi^\star(x^\star) = P_C(x^\star)$.

11 In the sense that it is a singleton containing only the element $P_C(x^\star)$.

Indeed, by definition, we have that

$$\varphi^\star(x^\star) = \sup_{x^\star \in H}\Big(\langle x^\star, x\rangle_H - \frac{1}{2}\|x\|^2 - I_C(x)\Big) = -\inf_{x^\star \in H}\Big(I_C(x) + \frac{1}{2}\|x\|^2 - \langle x^\star, x\rangle_H\Big),$$

and $x_o = \arg\min_{z\in C}\varphi(z)$ satisfies the first–order condition $-(x_o - x^\star) \in N_C(x_o)$, (recall that $\partial I_C(x) = N_C(x)$ for every $x \in C$), which is in turn equivalent (by the definition of the normal cone) to $\langle x^\star - x_o, z - x_o\rangle_H \le 0$ for every $z \in C$, i. e., $x_o = P_C(x^\star)$, and the result follows. Furthermore, since $x \in \partial\varphi^\star(x^\star)$ if and only if $x^\star \in \partial\varphi(x)$ and as $\partial\varphi(x) = x + N_C(x)$, we have that for $x^\star - x \in N_C(x)$, which implies that $x = P_C(x^\star)$. ◁

Example 5.2.22 (Moreau decomposition). Let $X = H$ be a Hilbert space (identified with its dual), and let $\varphi : H \to \mathbb{R}\cup\{+\infty\}$ be a proper convex lower semicontinuous function with convex conjugate $\varphi^\star$. Recall the definition of the Moreau proximity operator (see Definition 4.4.1) corresponding to the function φ,

$$\mathrm{prox}_\varphi(x) := \arg\min_{z\in H}\Big(\frac{1}{2}\|z - x\|^2 + \varphi(z)\Big), \tag{5.16}$$

and the corresponding proximity operator for $\varphi^\star$, defined by

$$\mathrm{prox}_{\varphi^\star}(x) := \arg\min_{z\in H}\Big(\frac{1}{2}\|z - x\|^2 + \varphi^\star(z)\Big). \tag{5.17}$$

These proximity operators satisfy the Moreau decomposition property

$$\mathrm{prox}_\varphi(x) + \mathrm{prox}_{\varphi^\star}(x) = x. \tag{5.18}$$

This follows by the definition of the proximity operator and Proposition 5.2.18. Indeed, if $z = \mathrm{prox}_\varphi(x)$, then by the first-order condition for problem (5.2.22) (see also Proposition 4.4.2(iii)), we have that $x - z \in \partial\varphi(z)$. This, by Proposition 5.2.18, is equivalent to $z \in \partial\varphi^\star(x - z)$. By the definition of $\mathrm{prox}_{\varphi^\star}$, we have that $z' = \mathrm{prox}_{\varphi^\star}(x')$ is equivalent to $x' - z' \in \partial\varphi^\star(z')$, and setting $x' = x$ and $z' = x - z$, leads to the equivalence $z \in \partial\varphi^\star(x - z)$. Combining the above, we have that $z = \mathrm{prox}_\varphi(x)$ is equivalent to $x - z = \mathrm{prox}_{\varphi^\star}(x)$, which leads to (5.18). In the particular case where $\varphi = I_E$, where E is a closed subspace of H, and where $\varphi^\star = I_{E^\perp}$, Moreau's decomposition (5.18) reduces to the standard decomposition for orthogonal projection on subspaces, i. e., $x = P_E x + P_{E^\perp} x$. Moreau's decomposition finds useful applications in numerical algorithms for optimization. ◁

Example 5.2.23 (Subdifferential of the support function in Hilbert space). Let $X = H$ be a Hilbert space (identified with its dual), $C \subset X$ be a closed convex set, and N_C its normal cone (see Definition 4.1.5). Consider the support function of this set $C : X^\star \to \mathbb{R}$, defined by $\sigma_C(x^\star) = \sup_{x\in C}\langle x^\star, x\rangle$, which of course now, since $X^\star \simeq X$, can be considered as a function of X. It holds that

$$x^\star \in \partial\sigma_C(x) \quad \text{if and only if} \quad x \in N_C(x^\star). \tag{5.19}$$

This follows by an application of Proposition 5.2.18(iii), by noting that $I_C^\star = \sigma_C$, that X is reflexive, and that $\partial I_C(x^\star) = N_C(x^\star)$ (see Example 4.1.6). ◁

5.2.4 The inf-convolution

An important and useful notion is the infimal convolution (or epi sum) of functions.

Definition 5.2.24 (Inf-convolution). Let $\varphi_1, \varphi_2 : X \to \mathbb{R} \cup \{+\infty\}$ be two proper lower semicontinuous convex functions. The inf-convolution of these functions,[12] denoted by $\varphi_1 \Box \varphi_2 : X \to \mathbb{R} \cup \{+\infty\}$ is the function defined by

$$(\varphi_1 \Box \varphi_2)(x) = \inf_{z \in X} [\varphi_1(z) + \varphi_2(x - z)] = \inf_{\substack{x_1 + x_2 = x \\ x_1, x_2 \in X}} [\varphi_1(x_1) + \varphi_2(x_2)].$$

The inf-convolution $\varphi_1 \Box \varphi_2$ is said to be exact at x if there exists $z_0 \in X$ such that $(\varphi_1 \Box \varphi_2)(x) = \varphi_1(z_0) + \varphi_2(x - z_0)$.

Example 5.2.25 (inf-convolution of the indicator function). It is straightforward to see that $I_{C_1} \Box I_{C_2} = I_{C_1 + C_2}$, where C_1, C_2 are arbitrary closed convex sets. ◁

Example 5.2.26. An important special case of the inf-convolution is the case where $\varphi_2(x - z) = \frac{1}{2}\|x - z\|^2$. This leads to the Moreau envelope, introduced in Section 4.4, which when $X = H$ is a Hilbert space can be considered as a smooth approximation for a nonsmooth convex function φ_1. The choice $\varphi_2(x - z) = \|x - z\|$, in a general Banach space, and regardless of the convexity of φ_1, leads to a Lipschitz regularization of φ_1 (see Section 9.4.6). ◁

The inf-convolution enjoys some very interesting properties, one of which is that it is turned into a sum under the Legendre–Fenchel transform. In this respect, it has a similar property as the standard convolution under the Fourier transform. Some useful properties of the inf-convolution are collected in the following proposition (see e. g. [12]):

Proposition 5.2.27. *The following properties hold for the inf-convolution.*
(i) $\operatorname{epi}_s(\varphi_1 \Box \varphi_2) = \operatorname{epi}_s \varphi_1 + \operatorname{epi}_s \varphi_2$, *hence the alternative term epi sum.*
(ii) $\operatorname{epi} \varphi_1 + \operatorname{epi} \varphi_2 \subset \operatorname{epi}(\varphi_1 \Box \varphi_2)$ *with equality*[13] *if and only if* $\varphi_1 \Box \varphi_2$ *is exact in* $\operatorname{dom}(\varphi_1 \Box \varphi_2)$.
(iii) $\varphi_1 \Box \varphi_2$ *is also a convex function.*
(iv) $(\varphi_1 \Box \varphi_2)^\star = \varphi_1^\star + \varphi_2^\star$.

12 The definition may still make sense, even if the functions are not convex or lower semicontinuous, but then the inf-convolution may not enjoy all the properties shown in this section.

13 Recall that $\subset$ is used invariably with $\subseteq$.

(v) *If* $\operatorname{int}(\operatorname{dom}\varphi_1) \cap \operatorname{dom}\varphi_2 \neq \emptyset$, *then* $(\varphi_1 + \varphi_2)^\star = \varphi_1^\star \Box \varphi_2^\star$, *and the inf-convolution* $\varphi_1^\star \Box \varphi_2^\star$ *is exact.*[14]

Proof. (i) Consider any pair $(x, \lambda) \in X \times \mathbb{R}$ such that $(x, \lambda) \in \operatorname{epi}_s(\varphi_1 \Box \varphi_2)$. This is equivalent to stating that

$$(\varphi_1 \Box \varphi_2)(x) = \inf_{\substack{x_1+x_2=x \\ x_1,x_2 \in X}} [\varphi_1(x_1) + \varphi_2(x_2)] < \lambda,$$

which in turn is equivalent to the existence of $x_1, x_2 \in X$ such that $x_1 + x_2 = x$ and $\varphi_1(x_1) + \varphi_2(x_2) < \lambda$, which is equivalent to the existence of $\lambda_1, \lambda_2 \in \mathbb{R}$ such that $\varphi_1(x_1) \leq \lambda_1$ and $\varphi_2(x_2) < \lambda_2$, while $x_1 + x_2 = x$ and $\lambda - 1 + \lambda_2 = \lambda$. This in turn, is equivalent to the existence of $(x_1, \lambda_1) \in \operatorname{epi}_s \varphi_1$ and $(x_2, \lambda_2) \in \operatorname{epi}_s \varphi_2$ such that $x_1 + x_2 = x$ and $\lambda - 1 + \lambda_2 = \lambda$. Therefore, the conclusion follows.

(ii) The inclusion $\operatorname{epi}\varphi_1 + \operatorname{epi}\varphi_2 \subset \operatorname{epi}(\varphi_1 \Box \varphi_2)$ follows by the same reasoning as above. For the opposite inclusion, we need the exactness of $\varphi_1 \Box \varphi_2$.[15] Consider any $(x, \lambda) \in \operatorname{epi}(\varphi_1 \Box \varphi_2)$, i. e., any (x, λ) with the property $(\varphi_1 \Box \varphi_2)(x) \leq \lambda$. Since $\varphi_1 \Box \varphi_2$ is exact, there exists $z_0 \in X$ such that $(\varphi_1 \Box \varphi_2)(x) = \varphi_1(z_0) + \varphi_2(x - z_0)$. This provides the existence of $\lambda_1, \lambda_2 \in \mathbb{R}$ such that $\lambda = \lambda_1 + \lambda_2$, and $x_1 = z_0$, $x_2 = x - z_0$ such that $(x_1, \lambda_1) \in \operatorname{epi}\varphi_1$ and $(x_2, \lambda_2) \in \operatorname{epi}\varphi_2$, therefore $\operatorname{epi}(\varphi_1 \Box \varphi_2) \subset \operatorname{epi}\varphi_1 + \operatorname{epi}\varphi_2$.

Assume now that $\operatorname{epi}\varphi_1 + \operatorname{epi}\varphi_2 = \operatorname{epi}(\varphi_1 \Box \varphi_2)$. That means we can express any $(x, \lambda) \in \operatorname{epi}(\varphi_1 \Box \varphi_2)$ in terms of the pairs $(z_0, \lambda_o) \in \operatorname{epi}\varphi_1$ and $(x - z_0, \lambda - \lambda_o) \in \operatorname{epi}\varphi_2$. Apply the above for $\lambda = (\varphi_1 \Box \varphi_2)(x)$. This yields $\varphi_1(z_0) \leq \lambda_o$ and $\varphi_2(x - z_0) \leq (\varphi_1 \Box \varphi_2)(x) - \lambda_o$, so adding the two inequalities, we obtain that $\varphi_1(z_0) + \varphi_2(x - z_0) \leq (\varphi_1 \Box \varphi_2)(x)$. But recall that by definition $(\varphi_1 \Box \varphi_2)(x) = \inf_{z \in X}(\varphi_1(z) + \varphi_2(x - z))$, so it must be that $\varphi_1(z_0) + \varphi_2(x - z_0) = (\varphi_1 \Box \varphi_2)(x)$, hence, $\varphi_1 \Box \varphi_2$ is exact.

(iii) This is immediate from (i) by noting that $\operatorname{epi}_s \varphi_1$ and $\operatorname{epi}_s \varphi_2$ are convex sets.

(iv) The proof follows by the definition. Indeed, setting $\varphi = \varphi_1 \Box \varphi_2$

$$\begin{aligned}
\varphi^\star(x^\star) &= \sup_{x \in X}(\langle x^\star, x\rangle - \varphi(x)) = \sup_{x \in X}\Big(\langle x^\star, x\rangle - \inf_{z \in X}[\varphi_1(z) + \varphi_2(x - z)]\Big) \\
&= \sup_{x \in X}\Big(\langle x^\star, x\rangle + \sup_{z \in X}[-\varphi_1(z) - \varphi_2(x - z)]\Big) = \sup_{x,z \in X}(\langle x^\star, x\rangle - \varphi_1(z) - \varphi_2(x - z)) \\
&= \sup_{x,z \in X}(\langle x^\star, x - z\rangle + \langle x^\star, z\rangle - \varphi_1(z) - \varphi_2(x - z))
\end{aligned}$$

14 i. e., as follows by a trivial restatement of Definition 5.2.24, there exist $x_1^\star, x_2^\star \in X^\star$ such that $x_1^\star + x_2^\star = x^\star$ and $(\varphi_1^\star \Box \varphi_2^\star)(x^\star) = \varphi_1^\star(x_1^\star) + \varphi_2^\star(x_2^\star)$.

15 Note the subtle difference between $\inf_{z \in X}(\varphi_1(z) + \varphi_2(x - z)) < \lambda$ and $\inf_{z \in X}(\varphi_1(z) + \varphi_2(x - z)) \leq \lambda$. The first one implies the existence of a $z_0 \in X$ such that $\varphi_1(z_0) + \varphi_2(x - z_0) < \lambda$. On the other hand, if $\inf_{z \in X}(\varphi_1(z) + \varphi_2(x - z)) = \lambda$, the existence of a point $z_0 \in X$ such that $\varphi_1(z_0) + \varphi_2(x - z_0) \leq \lambda$ is not guaranteed, unless the infimum is attained, i. e., the inf-convolution is exact.

$$= \sup_{z\in H}\Big(\langle x^\star, z\rangle - \varphi_1(z) + \sup_{x\in X}(\langle x^\star, x-z\rangle - \varphi_2(x-z))\Big)$$
$$= \sup_{z\in H}(\langle x^\star, z\rangle - \varphi_1(z) + \varphi_2^\star(x^\star)) = \varphi_1^\star(x^\star) + \varphi_2^\star(x^\star),$$

where we first performed the inner supremum operation for fixed z, which yields $\varphi_2^\star(x^\star)$, and then continued the calculation.

(v) We have seen in (iv) that for the inf-convolution of two convex functions φ_1, φ_2, it holds that $(\varphi_1 \square \varphi_2)^\star = \varphi_1^\star + \varphi_2^\star$. Assume that φ_1, φ_2 are lower semicontinuous. Consider now $\varphi_1^\star$ and $\varphi_2^\star$ in the place of φ_1 and φ_2, and applying this result once more to the inf-convolution $\varphi_1^\star \square \varphi_2^\star$, we obtain that $(\varphi_1^\star \square \varphi_2^\star)^\star = \varphi_1^{\star\star} + \varphi_2^{\star\star} = \varphi_1 + \varphi_2$, where for the last step we used the Fenchel–Moreau–Rockafellar theorem 5.2.15. We now take once more the Legendre–Fenchel transform on the last formula to obtain that $(\varphi_1^\star \square \varphi_2^\star)^{\star\star} = (\varphi_1 + \varphi_2)^\star$, thus providing a formula for the Legendre–Fenchel transform of the sum of two convex functions.

Assume now that $\text{int}(\text{dom}\,\varphi_1) \cap \text{dom}\,\varphi_2 \neq \emptyset$. Then, by the Moreau–Rockafellar theorem 4.2.12, $\partial(\varphi_1 + \varphi_2) = \partial\varphi_1 + \partial\varphi_2$. Take any $x \in X$, and consider any $x^\star \in \partial(\varphi_1 + \varphi_2)(x)$. Since $\partial(\varphi_1 + \varphi_2) = \partial\varphi_1 + \partial\varphi_2$, it follows that $x^\star \in (\partial\varphi_1(x) + \partial\varphi_2(x))$ so that there exist $x_1^\star \in \partial\varphi_1(x)$ and $x_2^\star \in \partial\varphi_2(x)$ with the property $x^\star = x_1^\star + x_2^\star$.

By Proposition 5.2.18, since $x^\star \in \partial(\varphi_1 + \varphi_2)(x)$, it holds that

$$(\varphi_1 + \varphi_2)(x) + (\varphi_1 + \varphi_2)^\star(x^\star) - \langle x^\star, x\rangle = 0. \tag{5.20}$$

Similarly, since $x_i^\star \in \partial\varphi_i(x)$, $i = 1, 2$, we have that

$$\varphi_1(x) + \varphi_1^\star(x_1^\star) - \langle x_1^\star, x\rangle = 0,$$
$$\varphi_2(x) + \varphi_2^\star(x_2^\star) - \langle x_2^\star, x\rangle = 0,$$

which upon addition and keeping in mind that $x_1^\star + x_2^\star = x^\star$ leads to

$$\varphi_1(x) + \varphi_2(x) + \varphi_1^\star(x_1^\star) + \varphi_2^\star(x_2^\star) - \langle x^\star, x\rangle = 0.$$

Rearranging the above we get,

$$\varphi_1^\star(x_1^\star) + \varphi_2^\star(x_2^\star) = \langle x^\star, x\rangle - (\varphi_1 + \varphi_2)(x). \tag{5.21}$$

and combining (5.21) with (5.20), we obtain that

$$\varphi_1^\star(x_1^\star) + \varphi_2^\star(x_2^\star) = (\varphi_1 + \varphi_2)^\star(x^\star). \tag{5.22}$$

This holds for any $x \in X$, $x^\star \in \partial(\varphi_1 + \varphi_2)(x)$, $x_1^\star \in \partial\varphi_1(x)$ and $x_2^\star \in \partial\varphi_2(x)$, with the constraint $x_1^\star + x_2^\star = x^\star$. Taking the infimum of the left-hand side of (5.22) over all $x_1^\star$, $x_2^\star$ satisfying the above properties and recalling the definition of the inf-convolution, we have that

$$(\varphi_1^\star \Box \varphi_2^\star)(x^\star) = (\varphi_1 + \varphi_2)^\star(x^\star). \tag{5.23}$$

There are two subtle points to be settled before closing the proof. The first one is that the inf-convolution is defined as the infimum over all $x_1^\star, x_2^\star \in X^\star$ such that $x_1^\star + x_2^\star = x^\star$ and not only over $x_i^\star \in \partial\varphi_i(x)$ satisfying the extra constraint. The second one is that we would like (5.23) to hold for any $x^\star \in X^\star$ and not just for any $x^\star \in \partial(\varphi_1 + \varphi_2)(x)$, where $x \in X$. These two points can be addressed recalling the density of the domain of the subdifferential of any convex function in its domain (see Theorem 4.2.25). The details are left to the reader. □

The differentiability properties of the inf-convolution are quite interesting. This is displayed in the following proposition [12].

Proposition 5.2.28. *If $\varphi_1 \Box \varphi_2$ is exact at $x \in X$, i. e., there exists $z_0 \in X$ such that $(\varphi_1 \Box \varphi_2)(x) = \varphi_1(z_0) + \varphi_2(x - z_0)$, then,*

$$\partial(\varphi_1 \Box \varphi_2)(x) = \partial\varphi_1(z_0) \cap \partial\varphi_1(x - z_0).$$

Proof. Since $\varphi_1 \Box \varphi_2$ is exact at x, it is expressed as $(\varphi_1 \Box \varphi_2)(x) = \varphi_1(z_0) + \varphi_2(x - z_0)$ for some $z_0 \in X$, while by Proposition 5.2.27(iv), for any $x^\star \in X$ it holds that $(\varphi_1 \Box \varphi_2)^\star(x^\star) = \varphi_1^\star(x^\star) + \varphi_2^\star(x^\star)$.

Consider any $x^\star \in \partial(\varphi_1 \Box \varphi_2)(x)$. By Proposition 5.2.18, this is equivalent to

$$(\varphi_1 \Box \varphi_2)(x) + (\varphi_1 \Box \varphi_2)^\star(x^\star) = \langle x^\star, x\rangle,$$

which, upon using the above observations, is equivalent to

$$\varphi_1(z_0) + \varphi_2(x - z_0) + \varphi_1^\star(x^\star) + \varphi_2^\star(x^\star) = \langle x^\star, x\rangle,$$

which can be reformulated as

$$\varphi_1(z_0) + \varphi_1^\star(x^\star) + \varphi_2(x - z_0) + \varphi_2^\star(x^\star) = \langle x^\star, z_0\rangle + \langle x^\star, x - z_0\rangle. \tag{5.24}$$

Young's inequality implies that

$$\begin{aligned} \varphi_1(z) + \varphi_1^\star(x^\star) &\geq \langle x^\star, z\rangle, \quad \forall z \in X, \\ \varphi_2(x - z) + \varphi_2^\star(x^\star) &\geq \langle x^\star, x - z\rangle, \quad \forall z \in X, \end{aligned} \tag{5.25}$$

with the equality holding if and only if $x^\star \in \partial\varphi_1(z)$, and $x^\star \in \partial\varphi_2(x - z)$, respectively. Adding (5.25), we obtain

$$\varphi_1(z) + \varphi_2(x - z) + \varphi_1^\star(x^\star) + \varphi_2^\star(x^\star) \geq \langle x^\star, z\rangle + \langle x^\star, x - z\rangle, \quad \forall z \in X. \tag{5.26}$$

Combining (5.24) with (5.26), we see that

$$\varphi_1(z_0) + \varphi_1^\star(x^\star) \geq \langle x^\star, z_0\rangle, \quad \text{and} \quad \varphi_2(x - z_0) + \varphi_2^\star(x^\star) \geq \langle x^\star, x - z_0\rangle,$$

therefore $x^\star \in \partial\varphi_1(z_0) \cap \partial\varphi_2(x - z_0)$. □

Example 5.2.29 (Moreau–Yosida approximation of the support function in Hilbert space). Let $X = H$ be a Hilbert space (identified with its dual), $C \subset X$ a closed and convex subset, and $\sigma_C : X^\star \to \mathbb{R}$ the support function of C defined by $\sigma_C(x^\star) = \sup_{x\in C}\langle x^\star, x\rangle$. Then, the inf-convolution of σ_C with the function $\frac{1}{2}\|\cdot\|^2$, (i. e., the Moreau–Yosida approximation of σ_C) satisfies

$$\left(\sigma_C \square \frac{1}{2}\|\cdot\|^2\right)(x) = \inf_{z\in X}\left(\sigma_C(z) + \frac{1}{2}\|x - z\|^2\right) = \langle x, P_C(x)\rangle - \frac{1}{2}\|P_C(x)\|^2, \tag{5.27}$$

and is attained at

$$z_0 := \operatorname{prox}_{\sigma_C}(x) = x - P_C(x),$$

where $P_C(x)$ is the orthogonal projection of x on the set C, and $\operatorname{prox}_{\sigma_C}$ is the proximal operator of the support function C.

To prove that, note that z_0, the minimizer of $z \mapsto \psi(z) := \sigma_C(z) + \frac{1}{2}\|x - z\|^2$ should satisfy the first-order condition $0 \in \partial\psi(z_0)$. This implies, by the Moreau–Rockafellar theorem 4.2.12, that

$$0 \in \partial\sigma_C(z_0) + \partial\left(\frac{1}{2}\|x - z\|^2\right)(z_0) = \partial\sigma_C(z_0) + z_0 - x,$$

or equivalently that $x - z_0 \in \partial\sigma_C(z_0)$. But by Example 5.2.23 (see (5.19)), this implies that

$$z_0 \in N_C(x - z_0), \tag{5.28}$$

where $N_C(x - z_0)$ is the normal cone of C at point $x - z_0$. Recalling the definition of the normal cone (see Definition 4.1.5), we see that (5.28) is equivalent to the variational inequality,

$$\langle z_0, z - (x - z_0)\rangle \leq 0, \quad \forall z \in C. \tag{5.29}$$

If we define the new variable $z_0' := x - z_0$, we can express (5.29) as

$$\langle x - z_0', z - z_0'\rangle \leq 0, \quad \forall z \in C. \tag{5.30}$$

Recall (see Theorem 2.5.1) that the orthogonal projection of x onto C, $P_C(x)$, solves the variational inequality

$$\langle x - P_C(x), z - P_C(x)\rangle \leq 0, \quad \forall z \in C, \tag{5.31}$$

and it is unique. Comparing (5.30) and (5.31), we see that $z'_0 = P_C(x)$, therefore, leading to $z_0 := \mathrm{prox}_{\sigma_C}(x) = x - P_C(x)$ as required.

Substituting this expression for the minimizer z_0 into ψ, we see that

$$\begin{aligned}\left(\sigma_C \Box \frac{1}{2}\|\cdot\|^2\right)(x) &= \sigma_C(z_0) + \frac{1}{2}\|x - z_0\|^2 \\ &= \sigma_C(x - P_C(x)) + \frac{1}{2}\|P_C(x)\|^2.\end{aligned} \tag{5.32}$$

It remains to calculate $\sigma_C(x - P_C(x))$. By definition,

$$\sigma_C(x - P_C(x)) = \sup_{z\in C}\langle x - P_C(x), z\rangle.$$

We rearrange (5.31) as

$$\langle x - P_C(x), z\rangle \le \langle x - P_C(x), P_C(x)\rangle, \quad \forall z \in C,$$

and note that for $z = P_C(x) \in C$ the equality is attained. This means that

$$\sigma_C(x - P_C(x)) = \sup_{z\in C}\langle x - P_C(x), z\rangle = \langle x - P_C(x), P_C(x)\rangle.$$

Substituting this result in (5.32), we obtain (5.27). ◁

Example 5.2.30 (Subdifferential of the Moreau–Yosida approximation of the support function in Hilbert space). In the context of Example 5.2.29 above, we show that

$$\partial\left(\sigma_C \Box \frac{1}{2}\|\cdot\|^2\right)(x) = \{P_C(x)\},$$

where $P_C(x)$ is the projection of x on C.

As we have seen in Example 5.2.29, this inf-convolution is exact and is attained at $z_0 = x - P_C(x)$. By Proposition 5.2.28, we have that

$$\partial\left(\sigma_C \Box \frac{1}{2}\|\cdot\|^2\right)(x) = \partial\sigma_C(z_0) \cap \partial\left(\frac{1}{2}\|\cdot\|^2\right)(x - z_0) = \partial\sigma_C(z_0) \cap \{x - z_0\}.$$

Consider any $x_o \in \partial(\sigma_C \Box \frac{1}{2}\|\cdot\|^2)(x)$. Then $x_o = x - z_0 = P_C(x)$ and $x_o \in \partial\sigma_C(z_0)$. In Example 5.2.23, we have shown that $x_o \in \partial\sigma_C(z_0)$ if and only if $z_0 \in N_C(x_o)$, which is expressed as $x - P_C(x) \in N_C(P_C(x))$. By the definition of the normal cone, this is equivalent to the variational inequality $\langle x - P_C(x), z - P_C(x)\rangle \le 0$ for every $z \in C$, which is of course true (this is just the variational inequality characterizing the orthogonal projection on C, see (2.22)). From the above discussion, we conclude that any $x_o \in \partial(\sigma_C \Box \frac{1}{2}\|\cdot\|^2)(x)$ is of the form $x_o = P_C(x)$, so our claim follows. ◁

5.3 Minimax and convex duality methods

The minimax theory (introduced in Section 5.1), as well as the theory of the Fenchel–Legendre transform (introduced in Section 5.2) are closely related to convex duality methods in optimization, such as for instance Fenchel duality or Lagrangian duality.

Convex duality methods is an important general class of optimization methods, in which the original minimization problem, called the primal problem, is redressed as a saddle point problem for an extended perturbation function involving an extra set of variables called the dual variables. Then using minimax, the original problem is associated with a corresponding maximin problem, which is called the dual problem. The solution to the dual problem is related to the solution of the primal problem, and, typically, the solution of the latter can be retrieved from the solution of the former. Surprisingly, the dual problem often has better properties than its primal counterpart. Hence, their connection via convex duality is extremely useful from the point of view of constructing the solution of the primal problem either analytically or numerically.

In this section, we consider the connection between the minimax theory and convex duality methods in optimization by presenting a general framework [12, 74], which as a special case, encompasses two important cases of duality, namely Fenchel duality and Lagrangian duality.[16]

5.3.1 A general framework

Let us consider the general optimization problem

$$P := \inf_{x \in X} F_0(x), \quad \text{(Primal)} \tag{5.33}$$

where X is a Banach space, and $F_0 : X \to \mathbb{R} \cup \{+\infty\}$ is a given function.[17]

The key idea in this framework is to embed the original problem (5.33) to a whole family of parametric problems by considering a parametric family of perturbations of the original problem $F : X \times Y \to \mathbb{R} \cup \{+\infty\}$, where Y is an appropriate Banach space, which can be thought of as a parameter space, and F is chosen such that $F(x, 0) = F_0(x)$ for every $x \in X$. In terms of the perturbation function, we may express the primal problem as

$$P := \inf_{x \in X} F_0(x) = \inf_{x \in X} F(x, 0). \quad \text{(Primal)} \tag{5.34}$$

16 Clearly, Lagrangian duality and Fenchel duality precedes the general construction presented in this section, developed by Ekeland and Temam (see, e. g., [74]), which generalizes them.

17 Upon suitable choice of F_0, (5.33) can be a constrained optimization problem, i. e. choosing $F_0 = \bar{F}_0 + I_C$, for $C \subset X$.

We may now define the perturbed optimization problem

$$\inf_{x \in X} F(x, y), \quad y \in Y, \tag{5.35}$$

and the corresponding value function $V : Y \to \mathbb{R}$ by

$$V(y) := \inf_{x \in X} F(x, y), \quad y \in Y, \tag{5.36}$$

with $V(0) =: P$ being the value of the primal problem. It turns out that the parametric family of problems has some remarkable structure. We shall see that the Fenchel conjugates $F^\star : X^\star \times Y^\star \to \mathbb{R} \cup \{+\infty\}$, $V^\star : Y^\star \to \mathbb{R} \cup \{+\infty\}$, of the perturbation function F and the value function V, can be used to define another class of optimization problems, the so called dual problems

$$D := \sup_{y^\star \in Y^\star} (-F^\star(0, y^\star)), \quad \text{(Dual)},$$

the value D of which is related to the value P of the original (primal) problem through the relations $V(0) = P$ and $V^{\star\star}(0) = D$. It is important to mention that often the dual problems are much better behaved than the original problem. Furthermore, minimax theorems can be used to connect the value functions of the original and the dual problems and, through the solution of the dual problem, one can then pass to the solution of the original problem in a very elegant manner. The theory of convex conjugates plays a very important role in this task, since a careful use of this theory allows one to show that as long as the value function is continuous at 0, $V^{\star\star}(0) = V(0)$, leading to $P = D$ (strong duality), whereas an element $y_0^\star$ of $\partial V(0)$, is a maximizer for the dual problem. In the case of strong duality, the maximizer $y_0^\star$ of the dual problem is connected and may be used to construct the minimizer x_o of the primal problem, and an interpretation of the pair $(x_o, y_0^\star)$ as the saddle point of the function $L : X \times Y^\star \to \mathbb{R} \cup \{+\infty\}$, defined as

$$L(x, y^\star) = -\sup_{y \in Y} (\langle y^\star, y \rangle_{Y^\star, Y} - F(x, y)), \tag{5.37}$$

is available. The connection with the minimax theorem becomes more apparent by noting that for L, defined as above,

$$F(x, 0) = \sup_{y^\star \in Y^\star} L(x, y^\star), \quad \text{and} \quad -F^\star(0, y^\star) = \inf_{x \in X} L(x, y^\star),$$

so strong duality can be expressed in minimax form as

$$\inf_{x \in X} \sup_{y^\star \in Y^\star} L(x, y^\star) = \sup_{y^\star \in Y^\star} \inf_{x \in X} L(x, y^\star).$$

Finally, we should note that as the choice of parametric families F, in which the original problem is embedded, is not unique; the general theory we are about to present is very flexible and may contain as special cases Fenchel duality or Lagrange duality.

In building up the theory summarized above, let us first consider some elementary properties of the parametric family of minimization problems (5.35) and the value function V [12, 74].

Lemma 5.3.1 (Properties of the value function and the Lagrangian). *Consider a function $F : X \times Y \to \mathbb{R} \cup \{+\infty\}$, such that $F(x, 0) = F_0(x)$, for every $x \in X$, the primal problem $\inf_{x \in X} F_0(x) = \inf_{x \in X} F(x, 0)$ and the value function $V : Y \to \mathbb{R} \cup \{+\infty\}$, defined as $V(y) := \inf_{x \in X} F(x, y)$, as well as its conjugates $V^\star : Y^\star \to \mathbb{R} \cup \{+\infty\}$ and $V^{\star\star} : Y \to \mathbb{R} \cup \{+\infty\}$. Define further, the Lagrangian function $L : X \times Y^\star \to \mathbb{R} \cup \{+\infty\}$ by*

$$L(x, y^\star) = -\sup_{y \in Y}(\langle y^\star, y\rangle_{Y^\star, Y} - F(x, y)).$$

Then, the following hold:

(i) *If F is a convex function (in both variables), then the value function V is convex, while (regardless of the convexity of F), the conjugate functions $F^\star$, $V^\star$, $V^{\star\star}$ are connected by*

$$\begin{aligned} V^\star(y^\star) &= F^\star(0, y^\star), \quad \forall y^\star \in Y^\star, \quad \textit{and} \\ V^{\star\star}(0) &= \sup_{y^\star \in Y^\star}(-F^\star(0, y^\star)) = D \quad \textit{(Dual)}. \end{aligned} \tag{5.38}$$

(ii) *If F is a proper convex lower semicontinuous function, defining the Lagrangian function $L : X \times Y^\star \to \mathbb{R}$ as in (5.37) it holds that*

$$F(x, 0) = \sup_{y^\star \in Y^\star} L(x, y^\star), \quad \textit{and} \quad -F^\star(0, y^\star) = \inf_{x \in X} L(x, y^\star), \tag{5.39}$$

so the primal and dual problems can be reformulated in terms of

$$\begin{aligned} \inf_{x \in X} F_0(x) &= \inf_{x \in X} \sup_{y^\star \in Y^\star} L(x, y^\star), \quad \textit{(Primal)}, \\ \sup_{y^\star \in Y^\star}(-F^\star(0, y^\star)) &= \sup_{y^\star \in Y^\star} \inf_{x \in X} L(x, y^\star), \quad \textit{(Dual)}. \end{aligned} \tag{5.40}$$

Proof. See Section 5.5.2, in the appendix of the chapter. □

At this point, we recall the general properties of bi-conjugate functions (Propositions 5.2.14 and 5.2.15), according to which for a proper function V it holds that $V^{\star\star} \leq V$, and if V is convex with equality at the points where V is lower semicontinuous. Therefore, Lemma 5.3.1 already reveals an interesting fact concerning the primal and the dual problem, which can be further strengthened if V is subdifferentiable at 0, which is stated in the following proposition [12, 74]:

Proposition 5.3.2 (Weak and strong duality). *Assume that $F : X \times Y \to \mathbb{R} \cup \{+\infty\}$ is convex and lower semicontinuous.*

(i) *It holds that the value of the dual problem is less or equal to the value of the primal problem,*

$$D = \sup_{y^\star \in Y^\star} (-F^\star(0, y^\star)) \le \inf_{x \in X} F(x, 0) = \inf_{x \in X} F_o(x) = P, \quad \textit{(weak duality)},$$

or in terms of the Lagrangian,

$$D = \sup_{y^\star \in Y^\star} \inf_{x \in X} L(x, y^\star) \le \inf_{x \in X} \sup_{y^\star \in Y^\star} L(x, y^\star) = P, \quad \textit{(weak duality)},$$

and the difference $P - D \ge 0$ is called the duality gap.

(ii) *If $\partial V(0) \ne \emptyset$ (as for instance if V continuous at $y = 0$), then any $y_0^\star \in \partial V(0)$ is a solution of the dual problem and*

$$D = \sup_{y^\star \in Y^\star} (-F^\star(0, y^\star)) = \inf_{x \in X} F(x, 0) = \inf_{x \in X} F_o(x) = P, \quad \textit{(strong duality)},$$

or in terms of the Lagrangian,

$$D = \sup_{y^\star \in Y^\star} \inf_{x \in X} L(x, y^\star) = \inf_{x \in X} \sup_{y^\star \in Y^\star} L(x, y^\star) = P, \quad \textit{(strong duality)},$$

and the duality gap $P - D = 0$. If furthermore, x_o is any solution of the primal problem, then, $(x_o, y_0^\star)$ is a saddle point of the Lagrangian and

$$F(x_o, 0) + F^\star(0, y_0^\star) = 0. \tag{5.41}$$

(iii) *The following are equivalent:* (a) *$(x_o, y_0^\star)$ is a saddle point of the Lagrangian L,* (b) *x_o is a solution of the primal problem, $y_0^\star$ is a solution of the dual problem, and there is no duality gap, and* (c) *the extremality condition* (5.41) *holds.*

(iv) *If the (generalized Slater) condition*

$$\exists\, \bar{x} \in X \quad \textit{such that} \quad y \mapsto F(\bar{x}, y) \textit{ is finite and continuous at } y = 0 \tag{5.42}$$

holds, then V is continuous at 0, hence, $\partial V(0) \ne \emptyset$.

Proof. (i) Since F is a convex function, V is convex (Lemma 5.3.1(i)), and by Proposition 5.2.14, it holds that $V^{\star\star}(0) \le V(0)$, a fact which, using Lemma 5.3.1(ii)–(iii), leads to weak duality.

(ii) Consider any $y_0^\star \in \partial V(0)$. Then, by the Fenchel–Young inequality (see Proposition 5.2.18) $0 \in \partial V^\star(y_0^\star)$, so by the first equation in (5.38), it holds that

$$F^\star(0, y_0^\star) = \inf_{y^\star \in Y^\star} F^\star(0, y^\star) = -\sup_{y^\star \in Y^\star} (-F^\star(0, y^\star)). \tag{5.43}$$

Hence, $y_0^\star$ is a solution of the dual problem. By the same argument,

$$V(0) + V^\star(y_o^\star) = \langle y_o^\star, 0\rangle_{Y^\star,Y} = 0,$$

and, by (5.38), this leads to

$$\inf_{x\in X} F(x,0) - \sup_{y^\star\in Y^\star}(-F^\star(0,y^\star)) = P - D = 0, \tag{5.44}$$

hence the strong duality result. If now the primal problem admits a solution, so that there exists x_o such that $F(x_o, 0) = \inf_{x\in X} F(x, 0) = P$, then, combining (5.44) with (5.43), we obtain (5.41). Recalling (5.40), we have that if x_o is a solution of the primal problem and $y_0^\star$ is a solution of the dual problem, then

$$P = \inf_{x\in X}\sup_{y^\star\in Y^\star} L(x,y^\star) = \sup_{y^\star\in Y^\star} L(x_o,y^\star) \geq L(x_o,y^\star), \quad \forall\, y^\star \in Y^\star, \tag{5.45}$$

and

$$D = \sup_{y^\star\in Y}\inf_{x\in X} L(x,y^\star) = \inf_{x\in X} L(x,y_0^\star) \leq L(x,y_0^\star), \quad \forall\, x \in X. \tag{5.46}$$

As shown above, if the dual problem admits a solution, then there is no duality gap, $P = D$. Hence, setting $y^\star = y_0^\star$ in (5.45) and $x = x_o$ in (5.46), we see that $D \leq L(x_o, y_0^\star) \leq P = D$, therefore $L(x_o, y_0^\star) = P = D$. Then, (5.45) and (5.46) can be re-interpreted as

$$L(x_o,y^\star) \leq P = L(x_o,y_0^\star) = D \leq L(x,y_0^\star), \quad \forall\,(x,y^\star) \in X\times Y^\star,$$

i. e., $(x_o, y_0^\star)$ is a saddle point for L. Note that if V is continuous at $y = 0$, then $\partial V(0) \neq \emptyset$.

(iii) We have already shown that (b) implies (c) in (ii).

We show next that (c) implies (a): Assume that (5.41) holds for some point $(x_o, y_0^\star)$. Since by the Fenchel–Young inequality $F(x,y) + F^\star(x^\star,y^\star) \geq \langle x^\star, x\rangle_{X^\star,X} + \langle y^\star, y\rangle_{Y^\star,Y}$, for every $(x,y) \in X \times Y$ and $(x^\star,y^\star) \in X^\star \times Y^\star$, choosing $x^\star = y = 0$, we see that $F(x,0)+F^\star(0,y^\star) \geq 0$ for every $(x,y^\star) \in X\times Y^\star$, with equality at $(x_o, y_0^\star)$, which combined with (5.39) leads to

$$L(x_o,y^\star) \leq \sup_{y^\star\in Y^\star} L(x_o,y^\star) = \inf_{x\in X} L(x,y_0^\star) \leq L(x,y_0^\star), \quad \forall\,(x,y^\star) \in X\times Y^\star.$$

Choosing $y^\star = y_0^\star$ in the above, we have $L(x_o, y_0^\star) \leq L(x, y_0^\star)$ for every $x \in X$, while choosing $x = x_o$ in the above leads to $L(x_o, y^\star) \leq L(x_o, y_0^\star)$ for every $y^\star \in Y^\star$, so combining these two, we have

$$L(x_o,y^\star) \leq L(x_o,y_0^\star) \leq L(x,y_0^\star), \quad \forall\,(x,y^\star) \in X\times Y^\star,$$

so $(x_o, y_0^\star)$ is a saddle point for L.

It remains to show that (a) implies (b): Recall Proposition 5.1.2. Since $(x_o, y_0^\star)$ is a saddle point of L, x_o minimizes the mapping $x \mapsto \sup_{y\in Y^\star} L(x, y^\star) = F(x, 0) = F_o(x)$ (where we used (5.39)), so it is a solution of the primal problem. On the other hand, $y_0^\star$ maximizes the mapping $y^\star \mapsto \inf_{x\in X} L(x, y^\star) = -F^\star(0, y^\star)$ (where we used once more (5.39)), so it is a solution of the dual problem. Again, by Proposition 5.1.2, no duality gap exists.

(iv) If the generalized Slater condition (5.42) holds, then V is continuous at $y = 0$. Hence, Proposition 5.3.2(iii) can be applied. Indeed, since F is a convex function, being continuous at $y = 0$ guarantees that $F(x_0, y)$ is bounded in a neighborhood of $y = 0$, let us say $B_Y(0, \delta)$. Hence, there exists $c \in \mathbb{R}$ such that $F(x_0, y) \le c$ for every $y \in B_Y(0, \delta)$. But then $V(y) := \inf_{x\in X} F(x, y) \le F(x_0, y) \le c$ for every $y \in B_Y(0, \delta)$. Since V is a convex function, this implies continuity of V at $y = 0$, and also the fact that $\partial V(0) \neq \emptyset$. □

Example 5.3.3 (A first-order condition for the saddle point formulation). Condition (5.41) can be interpreted as an extremality condition for the saddle point formulation of the Lagrangian. In particular, by the Fenchel–Young inequality (inequality (5.8) applied to the function F and its convex conjugate $F^\star$) for any pairs, $(x, y) \in X \times Y$ and $(x^\star, y^\star) \in X^\star \times Y^\star$, we have that $F(x, y) + F^\star(x^\star, y^\star) \ge \langle x^\star, x\rangle_{X^\star,X} + \langle y^\star, y\rangle_{Y^\star,Y}$. So choosing any pair of the form $(x, 0) \in X \times Y$, $(0, y^\star) \in X^\star \times Y^\star$, we have, by the Fenchel–Young inequality, that $F(x, 0) + F^\star(0, y^\star) \ge 0$ for every $(x, y^\star) \in X \times Y^\star$. In this sense, if a point $(x_o, y_0^\star)$ satisfies condition (5.41), then it is a minimizer for the function $(x, y^\star) \mapsto F(x, 0) + F^\star(0, y^\star)$. However, by (5.39), this function coincides with $(x, y^\star) \mapsto \Phi_1(x) - \Phi_2(y^\star)$, where $\Phi_1(x) = \sup_{y^\star\in Y^\star} L(x, y^\star)$, and $\Phi_2(y^\star) = \inf_{x\in X} L(x, y^\star)$. So a point $(x_o, y_0^\star)$ satisfying condition (5.41) minimizes Φ_1 over X and maximizes Φ_2 over $Y^\star$, hence is a saddle point for the Lagrangian. Note, furthermore, that once more by the Fenchel–Young inequality, condition (5.41) can be interpreted as $(0, y_0^\star) \in \partial F(x_o, 0)$. ◁

We will show in Sections 5.3.2 and 5.3.3 below, that two well known examples of duality, Fenchel duality and Lagrange duality fit within the general framework presented in this section.

5.3.2 Fenchel duality

Let us consider two Banach spaces X, Y, a bounded linear operator $\mathsf{L} : X \to Y$, two proper convex lower semicontinuous functions $\varphi_1 : X \to \mathbb{R} \cup \{+\infty\}$, $\varphi_2 : Y \to \mathbb{R} \cup \{+\infty\}$, and the optimization problem

$$\min_{x\in X}(\varphi_1(x) + \varphi_2(\mathsf{L}x)). \tag{5.47}$$

A large number of interesting real-world applications (see examples in this section) can be redressed in this general form. Importantly, optimization problems with linear constraints can be brought in the form above using the appropriate convex indicator function in the place of φ_2 to express the linear constraint.

Problem (5.47) fits exactly in the general framework developed in Section 5.3.1, upon setting $F_o = \varphi_1 + \varphi_2 \circ L$ and considering the family of perturbations $F : X \times Y \to \mathbb{R} \cup \{+\infty\}$, defined by $F(x, y) = \varphi_1(x) + \varphi_2(Lx - y)$.[18] This particular choice of the general duality framework reduces to the celebrated Fenchel duality framework, which is well suited for the treatment of problems of the general form (5.47). While the treatment of Fenchel duality can proceed as an application of the results of Section 5.3.1, we prefer to readdress it independently, providing the reader with an alternative perspective, closer to the original treatment of the problem, which is useful and instructive in its own right.

The fundamental observation leading to Fenchel duality is that, by Proposition 5.2.15, for a convex lower semicontinuous function φ, we have that $\varphi^{\star\star} = \varphi$. Then, using the definition of $\varphi^{\star\star}$, we have that

$$
\begin{aligned}
P &:= \inf_{x\in X}(\varphi_1(x) + \varphi_2(Lx)) = \inf_{x\in X}(\varphi_1(x) + \varphi_2^{\star\star}(Lx)) \\
&= \inf_{x\in X}(\varphi_1(x) + \sup_{y^\star\in Y^\star}(\langle y^\star, Lx\rangle_{Y^\star,Y} - \varphi_2^\star(y^\star)) \\
&= \inf_{x\in X}\sup_{y^\star\in Y^\star}(\varphi_1(x) + \langle L^\star y^\star, x\rangle_{X^\star,X} - \varphi_2^\star(y^\star)),
\end{aligned}
$$

with the last problem being in a minimax form. This version of the problem is interesting in its own right, leading to interesting numerical optimization algorithms. However, one may proceed a little further under the assumption that the order of the inf and sup in the above problem may be interchanged. Under sufficient conditions for interchanging the order of the inf and the sup, we may write

$$
\begin{aligned}
P &:= \inf_{x\in X}(\varphi_1(x) + \varphi_2(Lx)) = \inf_{x\in X}\sup_{y^\star\in Y^\star}(\varphi_1(x) + \langle L^\star y^\star, x\rangle_{X^\star,X} - \varphi_2^\star(y^\star)) \\
&= \sup_{y^\star\in Y^\star}\inf_{x\in X}(\varphi_1(x) + \langle L^\star y^\star, x\rangle_{X^\star,X} - \varphi_2^\star(y^\star)) \\
&= \sup_{y^\star\in Y^\star}\Big(-\sup_{x\in X}(\langle -L^\star y^\star, x\rangle_{X^\star,X} - \varphi_1(x)) - \varphi_2^\star(y^\star)\Big) \\
&= \sup_{y^\star\in Y^\star}(-\varphi_1^\star(-Ly^\star) - \varphi_2^\star(y^\star)) = \sup_{y^\star\in Y^\star}(-\varphi_1^\star(Ly^\star) - \varphi_2^\star(-y^\star)) =: D,
\end{aligned}
$$

where we have used the definition of $\varphi_1^\star$ and the vector space structure of $Y^\star$. If this step of interchanging inf and sup can be justified, then there may be a connection between the minimization problem $\inf_{x\in X}(\varphi_1(x) + \varphi_2(Lx))$ (called the primal problem) and the maximization problem $\sup_{y^\star\in Y^\star}(-\varphi_1^\star(Ly^\star) - \varphi_2^\star(-y^\star))$ (called the dual problem). It may happen in certain cases of interest that the dual problem is easier to handle both

18 Since Y is a vector space, we may equivalently use the perturbation $F(x, y) = \varphi_1(x) + \varphi_2(Lx + y)$.

analytically and numerically (for instance if φ_2 is strongly convex, then $\varphi_2^\star$ is C^1 with Lipschitz gradient). Thus, being able to guarantee a solution to the dual problem, which allows us to obtain a solution to the primal one is certainly an appealing prospect.

A rigorous answer to the above considerations is provided in the following proposition [28]:

Proposition 5.3.4 (Fenchel duality). *Let $\varphi_1 : X \to \mathbb{R} \cup \{+\infty\}$, $\varphi_2 : Y \to \mathbb{R} \cup \{+\infty\}$ be proper convex lower semicontinuous functions, $\mathsf{L} : X \to Y$ a continuous linear operator, and consider the optimization problems*

$$P := \inf_{x \in X}(\varphi_1(x) + \varphi_2(\mathsf{L}x)) \quad \textit{(Primal)},$$

and

$$D := \sup_{y^\star \in Y^\star}(-\varphi_1^\star(\mathsf{L}^\star y^\star) - \varphi_2^\star(-y^\star)) \quad \textit{(Dual)}.$$

Then, in general,

$$D := \sup_{y^\star \in Y^\star}(-\varphi_1^\star(\mathsf{L}^\star y^\star) - \varphi_2^\star(-y^\star)) \le P := \inf_{x \in X}(\varphi_1(x) + \varphi_2(\mathsf{L}x)), \quad \textit{(weak duality)}.$$

If moreover,

$$0 \in \operatorname{core}(\operatorname{dom}\varphi_2 - \mathsf{L}\operatorname{dom}\varphi_1), \tag{5.48}$$

then

$$P := \inf_{x \in X}(\varphi_1(x) + \varphi_2(\mathsf{L}x)) = D := \sup_{y^\star \in Y^\star}(-\varphi_1^\star(\mathsf{L}^\star y^\star) - \varphi_2^\star(-y^\star)), \quad \textit{(strong duality)}, \tag{5.49}$$

and if the supremum D is finite, it is attained so that the dual problem admits a solution $y_0^\star$.

In this case, the point $x_o \in X$ is optimal for the primal problem if and only if the first order conditions hold:

$$\begin{aligned} \mathsf{L}^\star y_0^\star &\in \partial\varphi_1(x_o), \\ -y_0^\star &\in \partial\varphi_2(\mathsf{L}x_o), \end{aligned} \tag{5.50}$$

where $y_0^\star \in Y^\star$ is optimal for the dual problem.

Proof. The proof proceeds in 4 steps.

1. Weak duality is a consequence of the Young–Fenchel inequality (5.8). According to that, we have

$$\varphi_1(z) + \varphi_1^\star(z^\star) \ge \langle z^\star, z\rangle, \quad \forall\,(z, z^\star) \in X \times X^\star,$$

$$\varphi_2(y) + \varphi_2^\star(y^\star) \ge \langle y^\star, y\rangle_{Y^\star,Y}, \quad \forall\, (y,y^\star) \in Y \times Y^\star.$$

Pick any $(x, y^\star) \in X \times Y^\star$, and apply the first inequality for $(x, L^\star y^\star)$ and the second for $(Lx, -y^\star)$. This yields

$$\begin{aligned}
&\varphi_1(x) + \varphi_1^\star(L^\star y^\star) \ge \langle L^\star y^\star, x\rangle, \quad \forall\, (x,y^\star) \in X \times Y^\star,\\
&\varphi_2(Lx) + \varphi_2^\star(-y^\star) \ge \langle -y^\star, Lx\rangle_{Y^\star,Y} = -\langle L^\star y^\star, x\rangle, \quad \forall\, (x,y^\star) \in X \times Y^\star.
\end{aligned}$$

By adding the two and rearranging terms, we obtain that

$$\varphi_1(x) + \varphi_2(Lx) \ge -\varphi_1^\star(L^\star y^\star) - \varphi_2^\star(-y^\star), \quad \forall\, (x,y^\star) \in X \times Y^\star,$$

and taking the infimum over all $x \in X$ on the left-hand side and the supremum over all $y^\star \in Y^\star$ on the right-hand side yields the required weak duality result.

2. Consider now the function $\varphi : Y \to \mathbb{R}$, defined by $\varphi(y) = \inf_{x\in X}(\varphi_1(x) + \varphi_2(Lx + y))$. Clearly, $\varphi(0) = \inf_{x\in X}(\varphi_1(x) + \varphi_2(Lx)) = P$, the value of the primal problem. This function is a convex function. We claim that if condition (5.48) holds (see also Remark 5.3.5), then,

$$\varphi \text{ is continuous at } 0. \tag{5.51}$$

Let us proceed by accepting the validity of this point (which is proved in step 4). By the continuity of φ at 0 (claim (5.51)) $\partial\varphi(0) \neq \emptyset$. Let $-y_0^\star \in \partial\varphi(0)$. By the definition of the subdifferential, we have that $\varphi(y) - \varphi(0) \ge \langle -y_0^\star, y - 0\rangle$, which rearranged yields

$$\begin{aligned}
\varphi(0) &\le \varphi(y) + \langle y_0^\star, y\rangle_{Y^\star,Y}\\
&\le \varphi_1(x) + \varphi_2(Lx + y) + \langle y_0^\star, y\rangle_{Y^\star,Y}\\
&= (\varphi_1(x) - \langle y_0^\star, Lx\rangle_{Y^\star,Y}) + (\varphi_2(Lx + y) - \langle -y_0^\star, Lx + y\rangle_{Y^\star,Y})\\
&= (\varphi_1(x) - \langle L^\star y_0^\star, x\rangle) + (\varphi_2(Lx + y) - \langle -y_0^\star, Lx + y\rangle_{Y^\star,Y}), \quad \forall\, (x,y) \in X \times Y,
\end{aligned} \tag{5.52}$$

where in the first line we used the definition of φ as an infimum to enhance the second inequality. We first rewrite (5.52) as

$$\langle -y_0^\star, Lx + y\rangle_{Y^\star,Y} - \varphi_2(Lx + y) \le (\varphi_1(x) - \langle L^\star y_0^\star, x\rangle) - \varphi(0), \quad \forall\, (x,y) \in X \times Y,$$

where, by keeping $x \in X$ fixed and taking the supremum over $y \in Y$, we obtain that

$$\varphi_2^\star(-y_0^\star) \le (\varphi_1(x) - \langle L^\star y_0^\star, x\rangle) - \varphi(0), \quad \forall\, x \in X,$$

and rearranging once more as

$$\langle L^\star y_0^\star, x\rangle - \varphi_1(x) \le -\varphi_2^\star(-y_0^\star) - \varphi(0), \quad \forall\, x \in X,$$

and taking the supremum over all $x \in X$, yields (after a final rearrangement)

$$\varphi(0) \leq -\varphi_1^\star(\mathsf{L}^\star \mathsf{y}_0^\star) - \varphi_2^\star(-\mathsf{y}_0^\star).$$

Clearly,

$$\begin{aligned} P = \varphi(0) &\leq -\varphi_1^\star(\mathsf{L}^\star \mathsf{y}_0^\star) - \varphi_2^\star(-\mathsf{y}_0^\star) \leq \sup_{\mathsf{y}^\star \in Y^\star} (-\varphi_1^\star(\mathsf{L}^\star \mathsf{y}_0^\star) - \varphi_2^\star(-\mathsf{y}_0^\star)) = D \\ &\leq \inf_{\mathsf{x}\in X}(\varphi_1(\mathsf{x}) + \varphi_2(\mathsf{L}\mathsf{x})) = \varphi(0) = P, \end{aligned} \tag{5.53}$$

where we used the weak duality result, $D \leq P$. Inequality (5.53), implies that

$$-\varphi_1^\star(\mathsf{L}^\star \mathsf{y}_0^\star) - \varphi_2^\star(-\mathsf{y}_0^\star) = \sup_{\mathsf{y}^\star \in Y^\star} (-\varphi_1^\star(\mathsf{L}^\star \mathsf{y}_0^\star) - \varphi_2^\star(-\mathsf{y}_0^\star)) = \inf_{\mathsf{x}\in X}(\varphi_1(\mathsf{x}) + \varphi_2(\mathsf{L}\mathsf{x})),$$

i. e., the supremum in the dual problem (if it is finite) is attained at $\mathsf{y}_0^\star$, and the strong duality, $P = D$, holds.

3. We now prove the validity of the optimality conditions (5.50). Recalling Proposition 5.2.18, condition (5.50) implies that

$$\begin{aligned} &\varphi_1(\mathsf{x}_o) + \varphi_1^\star(\mathsf{L}^\star \mathsf{y}_0^\star) = \langle \mathsf{L}^\star \mathsf{y}_0^\star, \mathsf{x}_o \rangle, \\ &\varphi_2(\mathsf{L}\mathsf{x}_o) + \varphi_2^\star(-\mathsf{y}_0^\star) = \langle -\mathsf{y}_0^\star, \mathsf{L}\mathsf{x}_o \rangle_{Y^\star, Y} = -\langle \mathsf{L}^\star \mathsf{y}_0^\star, \mathsf{x}_o \rangle, \end{aligned}$$

so adding and rearranging, we have that

$$\varphi_1(\mathsf{x}_o) + \varphi_2(\mathsf{L}\mathsf{x}_o) = -\varphi_1^\star(\mathsf{L}^\star \mathsf{y}_0^\star) - \varphi_2^\star(-\mathsf{y}_0^\star). \tag{5.54}$$

But since $\mathsf{y}_0^\star$ is the solution of the dual problem,

$$\begin{aligned} &-\varphi_1^\star(\mathsf{L}^\star \mathsf{y}_0^\star) - \varphi_2^\star(-\mathsf{y}_0^\star) \\ &\quad = D = \sup_{\mathsf{y}^\star \in Y^\star} (-\varphi_1^\star(\mathsf{L}^\star \mathsf{y}^\star) - \varphi_2^\star(-\mathsf{y}^\star)) = \max_{\mathsf{y}^\star \in Y^\star}(-\varphi_1^\star(\mathsf{L}^\star \mathsf{y}^\star) - \varphi_2^\star(-\mathsf{y}^\star)), \end{aligned}$$

and, by the strong duality, it is also true that

$$\begin{aligned} P &= \inf_{\mathsf{x}\in X}(\varphi_1(\mathsf{x}) + \varphi_2(\mathsf{L}\mathsf{x})) \\ &= D = \sup_{\mathsf{y}^\star \in Y^\star} (-\varphi_1^\star(\mathsf{L}^\star \mathsf{y}^\star) - \varphi_2^\star(-\mathsf{y}^\star)) = \max_{\mathsf{y}^\star \in Y^\star}(-\varphi_1^\star(\mathsf{L}^\star \mathsf{y}^\star) - \varphi_2^\star(-\mathsf{y}^\star)). \end{aligned}$$

Combining the above with (5.54), we see that

$$\varphi_1(\mathsf{x}_o) + \varphi_2(\mathsf{L}\mathsf{x}_o) = \inf_{\mathsf{x}\in X}(\varphi_1(\mathsf{x}) + \varphi_2(\mathsf{L}\mathsf{x})).$$

Hence, x_o is a solution of the primal problem.

4. It only remains to verify our claim (5.51), i. e., that under condition (5.48) (see also Remark 5.3.5), the function $\varphi : Y \to \mathbb{R}$, defined by $\varphi(\mathsf{y}) = \inf_{\mathsf{x}\in X}(\varphi_1(\mathsf{x}) + \varphi_2(\mathsf{L}\mathsf{x} + \mathsf{y}))$, is continuous at 0.

Assume that $0 \in \operatorname{core}(\operatorname{dom}\varphi_2 - \mathsf{L}\operatorname{dom}\varphi_1)$ holds. We note moreover that $\operatorname{dom}\varphi = \operatorname{dom}\varphi_2 - \mathsf{L}\operatorname{dom}\varphi_1$. Assuming without loss of generality and to ease notation that[19] $\varphi_1(0) = \varphi_2(0) = 0$, we define the set

$$C = \{\mathsf{y} \in Y : \exists \mathsf{x} \in \overline{B}_X(0,1) \text{ s. t. } \varphi_1(\mathsf{x}) + \varphi_2(\mathsf{Lx} + \mathsf{y}) \le 1\}.$$

It is easy to check this is a convex set. Our strategy is to show that $Y = \bigcup_{\lambda>0} \lambda C$ so that $0 \in \operatorname{core}(C)$, while C is CS-closed. Therefore, by Proposition 1.2.7, $\operatorname{core}(C) = \operatorname{int}(C) = \operatorname{int}(\overline{C})$. Hence, since $0 \in \operatorname{core}(C)$, it also holds that $0 \in \operatorname{int}(C)$. Since by the definition of C, this implies that φ is bounded above by 1 for all y in a neighborhood of 0, Proposition 2.3.21 provides the continuity of φ at 0.

We first show that $Y = \bigcup_{\lambda\in\mathbb{R}} \lambda C$, i. e., that C is absorbing. Since $0 \in \operatorname{core}(\operatorname{dom}\varphi_2 - \mathsf{L}\operatorname{dom}\varphi_1)$, recalling the definition of the core (see Definition 1.2.5), we have that for every $\mathsf{y} \in Y$, there exists $\delta > 0$ such that $\delta\mathsf{y} \in \operatorname{dom}\varphi_2 - \mathsf{L}\operatorname{dom}\varphi_1$, which means that we may choose an $\bar{\mathsf{x}} \in \operatorname{dom}\varphi_1$ such that $\mathsf{L}\bar{\mathsf{x}} + \delta\mathsf{y} \in \operatorname{dom}\varphi_2$, which in turn implies that the sum $\varphi_1(\bar{\mathsf{x}}) + \varphi_2(\mathsf{L}\bar{\mathsf{x}} + \delta\mathsf{y})$ is finite, i. e., there exists some $c \in \mathbb{R}$ such that

$$\varphi_1(\bar{\mathsf{x}}) + \varphi_2(\mathsf{L}\bar{\mathsf{x}} + \delta\mathsf{y}) = c. \tag{5.55}$$

For any $k > 1$, we may write $\varphi_1(\frac{1}{k}\bar{\mathsf{x}}) = \varphi_1(\frac{1}{k}\bar{\mathsf{x}} + (1-\frac{1}{k})0) \le \frac{1}{k}\varphi_1(\bar{\mathsf{x}}) + (1-\frac{1}{k})\varphi_1(0)$ (by convexity), and similarly for $\varphi_2(\frac{1}{k}(\mathsf{L}\bar{\mathsf{x}} + \delta\mathsf{y})) \le \frac{1}{k}\varphi_2(\mathsf{L}\bar{\mathsf{x}} + \delta\mathsf{y}) + (1-\frac{1}{k})\varphi_2(0)$, so that using the assumption that $\varphi_1(0) = \varphi_2(0) = 0$, and adding the above, we get that

$$\varphi_1\Big(\frac{1}{k}\bar{\mathsf{x}}\Big) + \varphi_2\Big(\frac{1}{k}(\mathsf{L}\bar{\mathsf{x}} + \delta\mathsf{y})\Big) \le \frac{1}{k}(\varphi_1(\bar{\mathsf{x}}) + \mathsf{x}_2(\mathsf{L}\bar{\mathsf{x}} + \delta\mathsf{y})) \le \frac{c}{k}.$$

If $\frac{c}{k} \le 1$ and $\frac{1}{k}\|\bar{\mathsf{x}}\|_X \le 1$, then the above relation implies that there exists some $\mathsf{x} = \frac{1}{k}\bar{\mathsf{x}} \in \overline{B}_X(0,1)$, such that $\mathsf{y}' := \frac{\delta}{k}\mathsf{y}$ satisfies $\varphi_1(\mathsf{x}) + \varphi_2(\mathsf{Lx} + \mathsf{y}') \le 1$, i. e., $\mathsf{y}' := \frac{\delta}{k}\mathsf{y} \in C$. Such a k can always be chosen as $k = \max(\|\bar{\mathsf{x}}\|_X, |c|, 1)$. We conclude that for any $\mathsf{y} \in Y$, there exists a $\frac{\delta}{k}\mathsf{y} \in C$ for the appropriate choice of $k > 0$, so $Y \subset \bigcup_{\lambda>0} \lambda C$, and since trivially $\bigcup_{\lambda>0} \lambda C \subset Y$, it follows that $Y = \bigcup_{\lambda>0} \lambda C$, i. e., C is absorbing. The absorbing property of C implies that $0 \in \operatorname{core}(C)$.

We now show that C is CS-closed. Consider any sequence $(\mathsf{y}_n)_{n\in\mathbb{N}} \in C$ such that $\sum_{n=1}^{\infty} \lambda_n \mathsf{y}_n = \mathsf{y}$, $\lambda_n \ge 0$, $\sum_{n=1}^{\infty} \lambda_n = 1$. We intend to show that $\mathsf{y} \in C$. Since $\mathsf{y}_n \in C$, for every $n \in \mathbb{N}$, there exists a $\mathsf{x}_n \in \overline{B}_X(0,1)$ such that

$$\varphi_1(\mathsf{x}_n) + \varphi_2(\mathsf{Lx}_n + \mathsf{y}_n) \le 1. \tag{5.56}$$

Rewrite the partial sum in the series $\sum_{n=1}^{\infty} \lambda_n \mathsf{y}_n = \mathsf{y}$ as $\bar{\mathsf{y}}_m = \sum_{n=1}^{m} \lambda_n \mathsf{y}_n = \Lambda_m(\sum_{n=1}^{m} \frac{\lambda_n}{\Lambda_m}\mathsf{y}_n)$, where $\Lambda_m = \sum_{n=1}^{m} \lambda_n$, and consider the sequence $(\bar{\mathsf{x}}_m)_{m\in\mathbb{N}}$ with $\bar{\mathsf{x}}_m = \sum_{n=1}^{m} \frac{\lambda_n}{\Lambda_m}\mathsf{x}_n$. This

19 Otherwise, we simply work with $\bar{\varphi}_i(\mathsf{z}) = \varphi_i(\mathsf{z}) - \varphi_i(0)$.

sequence is Cauchy and converges to some $x \in \overline{B}_X(0,1)$. Multiplying (5.56) for each $n = 1, \dots, m$, with $\frac{\lambda_n}{\Lambda_m}$ and adding over all such n, using also the convexity of φ_1 and φ_2, we have that

$$\varphi_1(\bar{x}_m) + \varphi_2(L\bar{x}_m + \bar{y}_m) \leq 1,$$

and passing to the limit as $m \to \infty$ and using the lower semicontinuity of φ_1 and φ_2, we conclude that

$$\varphi_1(x) + \varphi_2(Lx + y) \leq 1.$$

Hence, $y \in C$. The proof is complete. □

We remark that Proposition 5.3.4 could follow directly from Proposition 5.3.2.

Remark 5.3.5. One may replace condition (5.48) in Proposition 5.3.4 by the alternative condition

$$L \operatorname{dom} \varphi_1 \cap \operatorname{cont}(\varphi_2) \neq \emptyset,$$

where $\operatorname{cont}(\varphi_2)$ is the set of points of continuity of φ_2, in which case lower semicontinuity for φ_1, φ_2 is not required (see [28] or [29]). To see this, we only need to modify step 4 of the above proof by using the following reasoning: If $L \operatorname{dom} \varphi_1 \cap \operatorname{cont} \varphi_2 \neq \emptyset$, then, we may choose some $\bar{y} \in L \operatorname{dom} \varphi_1 \cap \operatorname{cont} \varphi_2$. Since $\bar{y}$ is a point of continuity for φ_2, then for every $\epsilon > 0$ there exists $\delta > 0$ such that for every $y' \in \delta\overline{B}_Y(0,1)$, $\varphi_2(\bar{y} + y') \leq \varphi_2(\bar{y}) + \epsilon$. Since $\bar{y} \in L \operatorname{dom} \varphi_1$ as well, there exists $x \in \operatorname{dom} \varphi_1$ such that $\bar{y} = Lx$. Using that in the estimate we have derived from continuity of φ_2, we conclude that for every $y' \in \delta\overline{B}_Y(0,1)$, $\varphi_2(Lx + y') \leq \varphi_2(Lx) + \epsilon$, which by the boundedness of $\varphi_1(x)$ (since $x \in \operatorname{dom} \varphi_1$) leads to the result that φ is bounded above in a neighborhood of 0; hence, by Proposition 2.3.21, continuous at 0.

An important remark here is that when strong duality holds and the dual problem has a solution $y_0^\star \in Y^\star$, we can try to construct a solution to the dual problem by trying a solution x_o, as indicated by (5.50). If an $x_o \in X$ satisfying (5.50) exists, then we know that this must be a solution to the primal problem. However, there is no guarantee in general that such a $x_o \in X$ exists. This requires extra conditions on the functions $\varphi_1^\star$ and $\varphi_2^\star$ and the mapping L, which will guarantee solvability of (5.50) (as an inclusion for the unknown x_o), and thus the existence of a $x_o \in X$, which solves the primal problem.

The Fenchel duality is an important result with many implications and applications.

Example 5.3.6 (The subdifferential sum rule as a result of Fenchel duality). An interesting application of Fenchel duality is that it may be used as an alternative proof of the subdifferential sum rule (see, e. g., [20]). Let φ_1, φ_2 be two convex and lower semicontinuous functions, and consider a point $x \in \operatorname{int}(\operatorname{dom}(\varphi_1)) \cap \operatorname{dom}(\varphi_2)$. Then $\partial(\varphi_1 + \varphi_2)(x) = \partial\varphi_1(x) + \partial\varphi_2(x)$.

Consider any $x^\star \in X^\star$. We will show the existence of a $x_o^\star \in X^\star$ such that

$$(\varphi_1 + \varphi_2)^\star(x^\star) = \varphi_1^\star(x^\star - x_o^\star) + \varphi_2^\star(x_o^\star). \tag{5.57}$$

Once the existence of such a point $x_o^\star$ is established, if $x^\star \in \partial(\varphi_1 + \varphi_2)(x)$, then, by Proposition 5.2.18, we have that

$$(\varphi_1 + \varphi_2)(x) + (\varphi_1 + \varphi_2)^\star(x^\star) = \langle x^\star, x\rangle,$$

which, using (5.57), yields

$$(\varphi_1 + \varphi_2)(x) + \varphi_1^\star(x^\star - x_o^\star) + \varphi_2^\star(x_o^\star) = \langle x^\star - x_o^\star, x\rangle + \langle x_o^\star, x\rangle.$$

Using Proposition 5.2.18 again for $\varphi_1, \varphi_1^\star$ and $\varphi_2, \varphi_2^\star$, respectively, we conclude that $x^\star - x_o^\star \in \partial\varphi_1(x)$ and $x_o^\star \in \partial\varphi_2(x)$. Hence, $x^\star \in \partial\varphi_1(x) + \partial\varphi_2(x)$. By the arbitrariness of $x^\star$, we conclude that $\partial(\varphi_1 + \varphi_2)(x) \subset \partial\varphi_1(x) + \partial\varphi_2(x)$, and since the reverse inclusion is always true, the equality holds.

The Fenchel duality can be used to show the claim (5.57): Define the functions $\widehat{\varphi}_1$, $\widehat{\varphi}_2$, by $\widehat{\varphi}_1(x) = -\langle x^\star, x\rangle + \varphi_1(x)$ and $\widehat{\varphi}_2(x) = \varphi_2(x)$. Note that by this choice $\inf_{x\in X}(\widehat{\varphi}_1(x) + \widehat{\varphi}_2(x)) = -(\varphi_1 + \varphi_2)^\star(x^\star)$. An easy computation shows that for any $z^\star \in X^\star$, we have $\widehat{\varphi}_1^\star(z^\star) = \varphi_1^\star(z^\star + x^\star)$ and $\widehat{\varphi}_2^\star(z^\star) = \varphi_2^\star(z^\star)$. Applying the Fenchel duality result for $\widehat{\varphi}_1 + \widehat{\varphi}_2$ (choosing $Y = X$ and $L = I$; note that condition (5.48) holds) we get that $-(\varphi_1 + \varphi_2)^\star(x^\star) = \inf_{x\in X}(\widehat{\varphi}_1(x) + \widehat{\varphi}_2(x)) = \sup_{z^\star\in X^\star}(-\widehat{\varphi}_1(z^\star) - \widehat{\varphi}_2(-z^\star))$, we obtain the claim (5.57) for $x_o^\star = -z_0^\star$ where z_0 solves the dual problem. ◁

Example 5.3.7. Let X be a Banach space, and consider the minimization problem

$$\inf_{x\in C} \varphi(x), \quad \text{where } C \subset X, \text{ closed and convex.} \tag{5.58}$$

Problem (5.58) can be brought into the general form of Fenchel duality in two different ways.

(a) Upon defining $\varphi_1 = \varphi$ and $\varphi_2 = I_C$, problem (5.58) can be expressed as $\inf_{x\in X}(\varphi_1(x) + \varphi_2(x))$. Using the Fenchel duality theorem for $Y = X$ and $L = I$, we see that the dual problem assumes the form $\sup_{x^\star\in X^\star}(-\varphi_1^\star(x^\star) - \sigma_C(-x^\star))$, where σ_C is the support function of the set C, defined as $\sigma_C = \sup_{x\in C}(\langle x^\star, x\rangle)$, see Example 5.2.8.

(b) Upon defining $\varphi_1 = \varphi + I_C$ and $\varphi_2 = 0$. We choose $Y = X$ and $L = I$ again. Then, the dual problem is $\sup_{x^\star\in X^\star}(-\varphi_1^\star(x^\star))$. We must now calculate the function $\varphi_1^\star$. One way forward would be to recall that under appropriate assumptions (see Proposition 5.2.27(v)), the Legendre–Fenchel transform of the sum can be expressed as the inf-convolution of the Legendre–Fenchel transforms. Under such conditions, $(\varphi + I_C)^\star = \varphi^\star \square I_C^\star = \varphi^\star \square \sigma_C$, where σ_C is the support function of C. The dual problem thus becomes $\sup_{x^\star\in X^\star}(-(\varphi^\star \square \sigma_C)(x^\star))$. There are certain cases, depending either on the choice of C or the choice of φ, where this inf-convolution can be explicitly calculated, leading to a well defined and manageable dual problem. Then under strong duality, once a solution to the dual problem has been obtained using (5.50), a solution to the primal problem can be found. Two explicit such examples will be presented in Sections 5.3.4.1 and 5.3.4.2 below. ◁

5.3.3 Lagrangian duality

Another important class of perturbations can be associated with Lagrangian duality [12], which corresponds to the problem

$$\min_{x \in C} \varphi_0(x), \quad C = \{x \in X : \varphi_i \leq 0, \; i = 1, \dots, n\},$$

where the function $\varphi_0 : X \to \mathbb{R} \cup \{+\infty\}$ is convex proper lower semicontinuous, and the functions $\varphi_i : X \to \mathbb{R}$, $i = 1, \dots, n$ are convex continuous functions, so the set C is a convex and closed subset of X.

We consider the following perturbation of the original problem: Choose as parameter space $Y = \mathbb{R}^n$, identified with its dual. For any $y = (y_1, \dots, y_n) \in Y$, define the sets $C(y_i) = \{x \in X : \varphi_i(x) + y_i \leq 0\}$ and the function $F : X \times Y \to \mathbb{R}$ by

$$F(x, y) = \varphi_0(x) + \sum_{i=1}^{n} I_{C(y_i)}(x).$$

Clearly $F(x, 0) = \varphi_0(x) + I_C(x)$.

We now calculate the Lagrangian function. We have that

$$\begin{aligned} -L(x, y^\star) &= \sup_{y \in Y} (\langle y^\star, y \rangle_{Y^\star, Y} - F(x, y)) \\ &= -\varphi_0(x) + \sup_{y \in Y} \left(\langle y^\star, y \rangle_{Y^\star, Y} - \sum_{i=1}^{n} I_{C(y_i)}(x) \right) \\ &= -\varphi_0(x) + \sum_{i=1}^{n} \sup_{y_i \in \mathbb{R}} (y_i y_i^\star - I_{C(y_i)}(x)), \end{aligned}$$

which, by a careful consideration of the supremum, leads to the equivalent form

$$L(x, y^\star) = \begin{cases} \varphi_0(x) + \sum_{i=1}^{n} y_i^\star \varphi_i(x), & y^\star = (y_1^\star, \dots, y_n^\star) \in \mathbb{R}_+^n, \\ -\infty & y^\star = (y_1^\star, \dots, y_n^\star) \notin \mathbb{R}_+^n, \end{cases}$$

with the vector $y^\star = (y_1^\star, \dots, y_n^\star) \in \mathbb{R}_+^n$ usually called the vector of Lagrange multipliers. The reader can easily verify[20] that

20 This just requires a careful rearrangement of the suprema; see Section 5.5.3 in the Appendix.

$$\sup_{y^\star \in Y^\star} L(x, y^\star) = \sup_{y^\star \in \mathbb{R}^n_+} \left(\varphi_0(x) + \sum_{i=1}^n y_i^\star \varphi_i(x) \right) = \varphi_0(x) + I_C(x), \tag{5.59}$$

so the primal problem can be expressed as $\inf_{x \in X} \sup_{y^\star \in \mathbb{R}^n_+} L(x, y^\star)$, while the dual problem is of the form

$$\sup_{y^\star \in Y^\star} \inf_{x \in X} L(x, y^\star) = \sup_{y^\star \in \mathbb{R}^n_+} \inf_{x \in X} \left(\varphi_0(x) + \sum_{i=1}^n y_i^\star \varphi_i(x) \right),$$

which is probably easier to handle as it is finite dimensional. An interesting interpretation of the dual problem is to first solve the inner minimization problem, which is a parametric (parameterized by the Lagrange multipliers) unconstrained minimization problem, over all $x \in X$, obtaining a solution $x_p(y^\star)$, and then try to solve the problem of maximizing $\varphi_0(x_p(y^\star)) + \sum_{i=1}^n y_i^\star \varphi_i(x_p(y^\star))$ over $y^\star = (y_1^\star, \dots, y_n^\star) \in \mathbb{R}^n_+$. By the form of the problem, especially the part $\sum_{i=1}^n y_i^\star \varphi_i(x_p(y^\star))$, we note that unless $y^\star$ is chosen so that $\varphi_i(x_p(y^\star)) \le 0$ for every $i = 1, \dots, n$, then this last problem is not well posed. This may be interpreted as that we first solve the parametric problem for any chosen value of the Lagrange multipliers, and then choose out of these only the solutions corresponding to the values of the Lagrange multipliers, which satisfy the constraints.

The general framework of Proposition 5.3.2 can be used, and the above comments can be made more precise upon assuming that the Slater condition (5.42) holds,[21] considering the saddle point formulation, and looking for a saddle point $(x_o, y_o^\star)$ for L. Since x_o minimizes $L(x, y_o^\star)$ and $y_o^\star$ maximizes $L(x_o, y^\star)$ (and minimizes $-L(x_o, y^\star)$), the first-order conditions become

$$0 \in \partial_x \varphi_0(x_o) + \sum_{i=1}^n y_{i,o}^\star \partial_x \varphi_i(x_o),$$

$$0 \in \partial_{y^\star} \left(-\varphi_0(x_o) - \sum_{i=1}^n y_i^\star \varphi_i(x_o) \right)(y_o^\star),$$

where we use the subscript in the subdifferential to denote the variable with respect to which we take the variation; in the first one we used subdifferential calculus (along with the Slater condition). We focus on the second condition, which we express (using the definition of the subdifferential) as

$$\sum_{i=1}^n (y_{i,o}^\star - y_i^\star)\varphi_i(x_o) \ge 0, \quad \forall y_i^\star \ge 0, \; i = 1, \dots, n.$$

Choosing each time $y_j^\star = y_{j,0}^\star$ for all $j \neq i$, we see that the above gives

21 For this particular case, this reduces to the Slater condition, that is, that there exists $\bar{x}$ such that $\varphi_o(\bar{x}) < +\infty$ and $\varphi_i(\bar{x}) < 0$, $i = 1, \dots, n$.

$$(y_{i,o}^\star - y_i^\star)\varphi_i(x_o) \geq 0, \quad \forall y_i^\star \geq 0,\ i = 1, \dots, n, \tag{5.60}$$

which is clearly impossible if $\varphi_i(x_o) > 0$ (since this would clearly not hold for any choice $y_i^\star = k y_{i,o}^\star$, $k > 1$). So, for the above to hold, it clearly must be that $\varphi_i(x_o) \leq 0$ for all $i = 1, \dots, n$. In fact, with a similar argument as above, one may show that (5.60) leads to $y_{i,o}^\star \varphi_i(x_o) = 0$ for any $i = 1, \dots, n$. If $\varphi_i(x_o) < 0$, then $y_{i,o}^\star - y_i^\star \leq 0$ for all $y_i^\star \geq 0$ so that $0 \leq y_{i,o}^\star \leq y_i^\star$ for every $y_i^\star \geq 0$, which leads to the conclusion that $y_{i,o}^\star = 0$ if $\varphi_i(x_o) < 0$. We therefore conclude that the conditions for $(x_o, y_0^\star)$ to be a saddle point for the Lagrangian would be

$$\begin{gathered} 0 \in \partial_x \varphi_0(x_o) + \sum_{i=1}^n y_{i,o}^\star \partial_x \varphi_i(x_o), \\ y_{i,o}^\star \geq 0, \quad \varphi_i(x_o) \leq 0, \quad i = 1, \dots, n, \\ y_{i,o}^\star \varphi_i(x_o) = 0, \quad i = 1, \dots, n. \end{gathered} \tag{5.61}$$

The optimality conditions (5.61) are called the Karush–Kuhn–Tucker (KKT) optimality conditions; the conditions have the following important implication: If x_μ is a solution of the unconstrained parametric minimization problem P_μ : $\inf_{x \in X}(\varphi_0(x) + \sum_{i=1}^n \mu_i \varphi_i(x))$ for some choice $\mu = (\mu_1, \dots, \mu_n) \in \mathbb{R}_+^n$ (the parameter μ is the Lagrange multiplier $y^\star$) such that $\varphi_i(x_\mu) \leq 0$ and $\mu_i \varphi_i(x_\mu) = 0$ for all $i = 1, \dots, n$, then $x_\mu = x_o$, where x_o is a solution of the original constrained (primal problem). This allows us to choose the solution of the constrained minimization problem out of these solutions of the unconstrained parametric problem, which satisfy the complementarity conditions.

5.3.4 Applications and examples

We close this section by providing a number of examples where the general framework developed in the previous section may prove itself useful.

5.3.4.1 Linear programming

Let $X = \mathbb{R}^n$, $Y = \mathbb{R}^m$ and $L : X \to Y$ a bounded linear map, which in this setting can be understood as an $m \times n$ real matrix $\mathbf{A} \in \mathbb{M}_{m \times n}$ and $L^\star = \mathbf{A}^T$, the transpose of $\mathbf{A}$. Clearly, in this finite dimensional setting, we will use the identifications[22] $X^\star \simeq X \simeq \mathbb{R}^n$ and $Y^\star \simeq Y \simeq \mathbb{R}^m$. Given two vectors $c \in \mathbb{R}^n$, $b \in \mathbb{R}^m$, the canonical form of a linear programming problem is

$$\inf_{x \in X \simeq \mathbb{R}^n} \langle c, x \rangle_{\mathbb{R}^n}, \quad \text{subject to} \quad \mathbf{A}x \geq b,\ x \geq 0, \tag{5.62}$$

22 But retain for convenience of the reader the notation x, y for their elements, instead of x, y.

where the notation $x \geq 0$ means that every component of the vector x satisfies the inequality. We will apply the Fenchel duality theory to this problem and show that the dual form of (5.62) is

$$\sup_{y \in Y \simeq \mathbb{R}^m} \langle y, b \rangle_{\mathbb{R}^m}, \quad \text{subject to} \quad \mathbf{A}^T y \leq c, \; y \geq 0. \tag{5.63}$$

To bring the primal problem to the standard form for which the Fenchel duality approach can be applied, we define the convex sets

$$C_1 = \{x \in \mathbb{R}^n \, : \, x \geq 0\}, \quad C_2(b) = \{y \in \mathbb{R}^m \, : \, y \geq b\}, \quad \text{and,} \quad C_3 = \{c\},$$

and the convex functions

$$\varphi_1 : \mathbb{R}^n \to \mathbb{R}, \quad \varphi_1(x) = \langle c, x \rangle_{\mathbb{R}^n} + I_{C_1}(x),$$
$$\varphi_2 : \mathbb{R}^m \to \mathbb{R}, \quad \varphi_2(y) = I_{C_2(b)}(y).$$

With the use of these two functions, the primal problem (5.62) can be expressed as $\inf_{x \in \mathbb{R}^n}(\varphi_1(x) + \varphi_2(\mathbf{A}y))$. According to the general theory of the Fenchel duality, the dual problem will be of the form[23] $\sup_{y \in \mathbb{R}^m}(-\varphi_1^\star(\mathbf{A}^T y) - \varphi_2^\star(-y))$.

To calculate $\varphi_1^\star$, we note that $\varphi_1 = \varphi + \psi$, where $\varphi(x) = \langle c, x \rangle$ and $\psi(x) = I_{C_1}(x)$. Under the conditions of Proposition 5.2.27 (v), $\varphi_1^\star = \varphi^\star \square \psi^\star = \varphi^\star \square \sigma_{C_1}$, where we used the result that the Legendre–Fenchel transform of the indicator function of a set is the support function (see Example 5.2.8). In order to calculate $\varphi_1^\star$, note that for any $x^\star \in \mathbb{R}^n$,

$$\varphi_1^\star(x^\star) = \sup_{x \in \mathbb{R}^n}(\langle x^\star, x \rangle_{\mathbb{R}^n} - \langle c, x \rangle_{\mathbb{R}^n}) = \sup_{x \in \mathbb{R}^n} \langle x^\star - c, x \rangle_{\mathbb{R}^n} = I_{C_3}(x^\star),$$

the indicator function of the set $C_3 = \{c\}$. Introducing that into the above, we obtain

$$\varphi_1^\star(x^\star) = (I_{C_3} \square \sigma_{C_1})(x^\star) = \inf_{z \in \mathbb{R}^n}(I_{C_3}(z) + \sigma_{C_1})(x^\star - z)) = \sigma_{C_1}(x^\star - c).$$

However, because of the special form of the set C_1, we can take our calculation a little further. By definition

$$\sigma_{C_1}(x^\star - c) = \sup_{x \in C_1} \langle x^\star - c, x \rangle_{\mathbb{R}^n} = \sup_{x \geq 0} \langle x^\star - c, x \rangle_{\mathbb{R}^n} = \begin{cases} +\infty & \text{if } x^\star - c > 0 \\ 0 & \text{if } x^\star - c \leq 0 \end{cases} = I_{C_1}(c - x^\star).$$

We therefore conclude that $\varphi_1^\star(x^\star) = I_{C_1}(c - x^\star)$.

We can similarly see that

$$\varphi_2^\star(y^\star) = \sigma_{C_2(b)}(y^\star) = \sup_{y \in C_2(b)} \langle y^\star, y \rangle_{\mathbb{R}^m} = \sup_{y \geq b} \langle y^\star, y \rangle_{\mathbb{R}^m}$$

23 Where, by the choice of the spaces involved and to simplify notation, we use y in the place of $y^\star$.

$$= \sup_{y \geq b} \langle y^\star, y - b \rangle_{\mathbb{R}^m} + \langle y^\star, b \rangle_{\mathbb{R}^m} = \begin{cases} +\infty & \text{if } y^\star > 0 \\ \langle y^\star, b \rangle_{\mathbb{R}^m} & \text{if } y^\star \leq 0, \end{cases}$$

which can be conveniently rephrased as $\varphi_2^\star(y^\star) = \langle y^\star, b \rangle_{\mathbb{R}^m} + I_{C_2(0)}(-y^\star)$.

The Fenchel dual problem is therefore $\sup_{y^\star \in \mathbb{R}^m}(-I_{C_1}(c - \mathbf{A}^T y^\star) + \langle y^\star, b \rangle_{\mathbb{R}^m} - I_{C_2(0)}(y^\star))$, which coincides with problem (5.63).

The reader is encouraged to obtain the same result using Lagrangian duality, using the Lagrangian function,

$$L(x, y_a^*, y_b^*) = \langle c, x \rangle_{\mathbb{R}^n} + \langle y_a^*, b - Ax \rangle_{\mathbb{R}^m} + \langle y_b^*, -x \rangle_{\mathbb{R}^n},$$

where $y_a^* \in \mathbb{R}_+^m$ are the Lagrange multipliers corresponding to the constraint $Ax \geq b$, and $y_b^* \in \mathbb{R}_+^n$ are the Lagrange multipliers corresponding to the constraint $x \geq 0$. Rearranging the Lagrangian as

$$L(x, y_a^*, y_b^*) = \langle c - A^T y_a^* - y_b^*, x \rangle_{\mathbb{R}^n} + \langle y_a^*, b \rangle_{\mathbb{R}^m},$$

we can easily obtain the dual function

$$\mathcal{D}(y_a^*, y_b^*) = \inf_{x \in \mathbb{R}^n} L(x, y_a^*, y_b^*) = \langle y_a^*, b \rangle_{\mathbb{R}^m} - I_{\{c - A^T y_a^* - y_b^* \geq 0\}}(y_a^*, y_b^*),$$

so the dual problem

$$\sup_{(y_a^*, y_b^*) \in \mathbb{R}_+^{m+n}} \mathcal{D}(y_a^*, y_b^*)$$

only makes sense under the constraint

$$c - A^T y_a^* - y_b^* \geq 0 \iff c \geq A^T y_a^* \geq A^T y_a^* + y_b^* \geq A^T y_a^*.$$

Hence, setting $y = y_a^*$, we see that the dual problem is equivalent to (5.63).

5.3.4.2 Minimization of convex functions with linear constraints

Another general class of problems in which the Fenchel dual problem can be explicitly expressed are problems of the form

$$\inf_{x \in C \subset X} \frac{1}{2}\|x\|^2 \quad \text{subject to} \quad \mathsf{L}x = c, \tag{5.64}$$

where $X = H$, Y are Hilbert spaces (identified with their duals, with Y possibly finite dimensional), $C \subset X$ is closed and convex, $\mathsf{L} : X \to Y$ is a bounded linear operator, and $c \in Y$ is known. This is a problem involving the minimization of a convex function subject to linear constraints studied in [27] (in fact can be interpreted as a projection problem as well).

Problems of this type arise quite often, e. g., in signal processing, where x represents typically a signal (modeled as an element of some Hilbert space X), which has to be matched to some measurements c for this signal. The operator L is the so called measurement operator, and it is often the case that the space Y is a finite dimensional space. For example, the operator L may model the situation of taking a finite number of measurements from the function x. As a concrete example, one may consider X as an L^2 space. Then, by the Riesz representation, any set of N continuous measurement operators $\{M_i : X = H \to \mathbb{R} : i = 1, \dots, N\}$ may be represented in terms of the inner product with N elements $z_i \in X = H$, as $M_i(x) = \langle z_i, x\rangle_H$ for every $i = 1, \dots, N$. Hence, given a set of measurements $M_i(x) = y_i$, $i = 1, \dots, N$ collected in the vector $c = (c_1, \dots, c_N)^T \in Y \simeq \mathbb{R}^N$, we may define the operator $L : X = H \to Y \simeq \mathbb{R}^N$ by $Lx = (M_1(x), \dots, M_N(x))^T$, with its adjoint $L^\star : \mathbb{R}^N \to H$, defined by $L^\star y = \sum_{i=1}^N y_i z_i$, for every $y = (y_1, \dots, y_N)^T \in Y \simeq \mathbb{R}^N$. Then the (inverse) problem of recovering a signal x, as close as possible, in the sense of minimal norm) to a reference signal x_{ref} (taken as $x_{\text{ref}} = 0$ without loss of generality) compatible with the given measurements $Lx = c$, as well some additional convex constraints (e. g., positivity etc), described by the set C, is exactly of the general form (5.64).

To apply Fenchel duality, we rewrite the primal problem in the form $\inf_{x\in X}(\varphi_1(x) + \varphi_2(Lx))$, where $\varphi_1(x) = \frac{1}{2}\|x\|^2 + I_C(x)$, and $\varphi_2(x) = I_{C_1}(x)$, where $C_1 = \{c\}$ (a convex set). The Fenchel dual is then $\sup_{y^\star\in Y^\star}(-\varphi_1^\star(L^\star y^\star) - \varphi_2^\star(-y^\star))$. Note that since Y is a finite dimensional space, then the dual problem is a finite dimensional optimization problem, which is typically a lot easier to handle than the primal infinite dimensional problem.

In order to calculate $\varphi_1^\star$ and its subdifferential, we proceed, as in Example 5.2.21, to find that $\varphi_1^\star(x^\star) = \langle x^\star, P_C(x^\star)\rangle - \frac{1}{2}\|P_C(x^\star)\|^2$, and $\partial\varphi_1^\star(x^\star) = P_C(x^\star)$ (i. e., it is a singleton). We can also calculate explicitly $\varphi_2^\star(y^\star) = \sup_{y\in Y}(\langle y^\star, y\rangle_Y - I_{C_1}(y)) = \langle y^\star, c\rangle_Y$ so that $\partial\varphi_2^\star(y^\star) = c$ (also a singleton). We therefore conclude that the Fenchel dual problem is

$$\begin{aligned}
&\sup_{y^\star\in Y^\star\simeq Y}(-\varphi_1^\star(L^\star y^\star) - \varphi_2^\star(-y^\star)) \\
&\quad = \sup_{y^\star\in Y^\star\simeq Y}\Big(-\langle L^\star y^\star, P_C(L^\star y^\star)\rangle_X + \frac{1}{2}\|P_C(L^\star y^\star)\|^2 + \langle y^\star, c\rangle_Y\Big).
\end{aligned}$$

The condition for strong duality to hold is $0 \in \operatorname{core}(c - L\operatorname{dom}\varphi_1)$ or equivalently $c \in \operatorname{core}(L\operatorname{dom}\varphi_1)$. Under this assumption, the first-order condition for a solution of the dual problem $y_o^\star \in Y^\star$ is $0 \in \partial(\varphi_1^\star(L^\star y_o^\star) + \varphi_2^\star(-y_o^\star))$, which, using subdifferential calculus, yields $0 \in L\partial\varphi_1(L^\star y_o^\star) - c$, and by the results stated above, we have

$$c = LP_C(L^\star y_o^\star). \tag{5.65}$$

The solution of the primal problem x_o, which is what we are really after, is connected to the solution of the dual problem through the inclusions

$$\begin{aligned}
L^\star y_o^\star &\in \partial\varphi_1(x_o), \\
-y_o^\star &\in \partial\varphi_2(Lx_o).
\end{aligned} \tag{5.66}$$

We can easily calculate $\partial\varphi_1(\mathsf{x}_o) = \mathsf{x}_o + N_C(\mathsf{x}_o)$, so the first of (5.66) gives that $\mathsf{L}^\star\mathsf{y}_0^\star - \mathsf{x}_o \in N_C(\mathsf{x})$, which is equivalent to the variational inequality $\langle \mathsf{L}^\star\mathsf{y}_0^\star - \mathsf{x}_o, \mathsf{z} - \mathsf{x}_o\rangle_X \leq 0$ for every $\mathsf{z} \in C$, from which follows that

$$\mathsf{x}_o = P_C(\mathsf{L}^\star\mathsf{y}_0^\star). \tag{5.67}$$

Hence, once we have found the solution to the dual problem $\mathsf{y}_0^\star$, the solution to the primal problem can be represented in terms of (5.67). Upon combining (5.67) and (5.65), we conclude that given a $\mathsf{y}_0^\star \in Y^\star$ satisfying $c = \mathsf{L}P_C(\mathsf{L}^\star\mathsf{y}_0^\star)$, the solution of the primal problem can be expressed as $\mathsf{x}_o = P_C(\mathsf{L}^\star\mathsf{y}_0^\star)$. One can verify in a straightforward manner that this is indeed the unique solution of the primal problem (see Theorem 1, in [117]). This parameterization of the solution, in terms of the solution $\mathsf{y}_0^\star$ of the nonlinear system of equations $c = \mathsf{L}P_C(\mathsf{L}^\star\mathsf{y}_0^\star)$ (which is in fact finite dimensional since $Y \simeq Y^\star \simeq \mathbb{R}^N$), is very useful both from the theoretical as well as from the computational point of view. This system requires information on the range of the nonlinear operator $\mathsf{A} := \mathsf{L}P_C\mathsf{L}^\star : Y^\star \to Y$, which satisfies the property $\langle \mathsf{y}_1^\star - \mathsf{y}_2^\star, \mathsf{A}(\mathsf{y}_1^\star) - \mathsf{A}(\mathsf{y}_2^\star)\rangle_Y \geq 0$ for every $\mathsf{y}_1^\star, \mathsf{y}_2^\star \in Y^\star$, i. e., is monotone. Many of the useful properties regarding the solvability of this system stem from this monotonicity property. We shall return to this concept in Chapter 9.

5.3.4.3 The Lasso model via duality

Another interesting application of duality, which is of great importance in statistical machine learning is in the theoretical and numerical study of the Lasso model. This is a linear regression model with an ℓ_1 regularization term aiming for a parsimonious representation of the available data, using as few features as possible. The idea behind the model is as follows: Consider a set of data (z_i, b_i), $i = 1, \dots, N$, where $z_i \in \mathbb{R}^m$ correspond to observations of a set of features, and $b_i \in \mathbb{R}$ are characteristics that are assumed to be associated with the features in terms of a linear model of the form $b_i \simeq \langle x, z_i\rangle_{\mathbb{R}^m}$ for every $i = 1, \dots, N$, where $x \in \mathbb{R}^m$ corresponds to the linear model variables and $\langle\cdot,\cdot\rangle_{\mathbb{R}^m}$ denotes the standard inner product in the space $\mathbb{R}^m$. A more convenient way of expressing that is by setting $b = (b_1, \dots, b_N)^T \in \mathbb{R}^{N\times 1}$ and $A = [z_1, \dots, z_N]^T \in \mathbb{R}^{N\times m}$ (the matrix whose rows consists of the observations of the features vectors z_i) and express the model in compact form as $Ax \simeq b$.[24] Clearly, this is often an ill-posed problem, and approximate solutions can be obtained in terms of related minimization problems with appropriate regularization terms. The Lasso model is an example of such a problem, which has gained a lot of popularity in machine learning. According to this model, the regression coefficients $x \in \mathbb{R}^m$ are selected as solutions of the minimization problem

$$\min_{x\in\mathbb{R}^m} \frac{1}{2}\|Ax - b\|_{2,N}^2 + \lambda\|x\|_{1,m}, \quad \lambda > 0, \tag{5.68}$$

24 Note the since we work in finite dimensions we adopt the notation x rather than x. Bias effects in the model can easily be included by augmenting the matrix A with an extra column of constants.

where $\|\cdot\|_{p,k}$ denotes the ℓ^p norm in $\mathbb{R}^k$, and $\lambda > 0$ is a regularization parameter. Problem (5.68) is a nonsmooth minimization problem (due to the regularization term). The first term in (5.68) corresponds to a term controlling fidelity of the chosen model x to the data, whereas the regularization term is responsible for selecting the most economic model in terms of the number of features used for modeling the data.

Problem (5.68) is in the form of problems that are amenable to Fenchel duality methods, and in fact also very close to the type of problems studied in Section 5.3.4.2. To keep as close as possible to the notation adopted here, set $\phi_1(x) = \lambda\|x\|_{1,m}$ and $\phi_2(z) = \psi_{2,b}(z) := \psi_2(z-b)$ for $\psi_2(z) = \frac{1}{2}\|z\|_{2,N}^2$. So setting $\mathsf{L}x = Ax$ as the linear operator $\mathsf{L}: \mathbb{R}^m \to \mathbb{R}^N$, we can bring (5.68) to the standard form of the Fenchel problem. It is easy to see that $L^* = A^T$. This dictates our choice of function spaces, which is $X = (\mathbb{R}^m, \|\cdot\|_{1,m})$, whose dual is chosen as $X^* = (\mathbb{R}^m, \|\cdot\|_{\infty,m})$ and $Y = (\mathbb{R}^N, \|\cdot\|_{2,N})$, whose dual is chosen as $Y^* = Y$.

We now calculate the relevant Fenchel conjugate functions. Since $\phi_1(x) = \lambda\psi_1(x)$, where $\psi_1(x) = \|x\|_{1,m}$, using the properties of the Fenchel transform, or by direct computation, we have that $\varphi_1^\star(x^*) = \lambda\psi_1^\star(\frac{x}{\lambda})$, and since ψ_1 is a norm its Fenchel transform corresponds to the indicator function of the closed ball in the dual space. Therefore, combining the above results, we conclude that

$$\varphi_1^\star(x^*) = \lambda I_{B_{X^*}}\left(\frac{x^*}{\lambda}\right) = \lambda I\left(\left\|\frac{x^*}{\lambda}\right\|_{\infty,m} \le 1\right) = \lambda I(\|x^*\|_{\infty,m} \le \lambda),$$

where I denotes as usual the convex indicator function.

For ϕ_2, using the results for the Fenchel transform of shifted functions, we see that

$$\varphi_2^\star(z^*) = \psi_2^\star(z^*) + \langle z^*, b\rangle_{\mathbb{R}^N} = \frac{1}{2}\|z^*\|_{2,N}^2 + \langle z^*, b\rangle_{\mathbb{R}^N},$$

where $\langle\cdot,\cdot\rangle_{\mathbb{R}^N}$ is the standard inner product in $\mathbb{R}^N$.

We can now express the dual problem as

$$\begin{aligned} D &= \sup_{y^*\in Y^*}(-\varphi_1^\star(\mathsf{L}^*y^*) - \varphi_2^\star(-y^*)) \\ &= \sup_{y^*\in\mathbb{R}^N}\left(-\lambda I(\|A^Ty^*\|_{\infty,m} \le \lambda) - \langle -y^*, b\rangle_{\mathbb{R}^N} - \frac{1}{2}\|y^*\|_{2,N}^2\right) \\ &= -\inf_{y^*\in\mathbb{R}^N}\left(\lambda I(\|A^Ty^*\|_{\infty,m} \le \lambda) + \langle -y^*, b\rangle_{\mathbb{R}^N} + \frac{1}{2}\|y^*\|_{2,N}^2\right) \\ &= -\inf_{y^*\in\mathbb{R}^N}\left(\lambda I(\|A^Ty^*\|_{\infty,m} \le \lambda) + \langle y^*, b\rangle_{\mathbb{R}^N} + \frac{1}{2}\|y^*\|_{2,N}^2\right), \end{aligned}$$

using the fact that $y^* \in \mathbb{R}^N$ (that is a vector space) and the symmetries of the functions.

Recalling the nature of the convex indicator function, we see that the dual problem reduces to

$$-D := \inf_{y^* \in \mathbb{R}^N} \left(\frac{1}{2} \|y^*\|_{2,N}^2 + \langle y^*, b \rangle_{\mathbb{R}^N} \right),$$
$$\text{subject to} \quad \|A^T y^*\|_{\infty,m} \le \lambda,$$

which is an easier problem to handle since it is a smooth problem with a simple constraint to handle. This is called a box constraint and can be handled with an iterated projection method. However, the above dual problem is informative also from the qualitative point of view, as the box constraint implies that, componentwise, the allowed y_i^* have lower and upper values dictated by the choice of the regularization parameter $\lambda > 0$, and the data (through the action of the design matrix A). These box constraints on the y_i^* are translated to the choice of x (which corresponds to the model) so that certain coefficients x_i, vanish, leading to the most economic model for the data considered. Indeed, considering the first-order condition (5.50), we see that for the optimal solution

$$-y^* \in \partial\varphi_2(\mathsf{L}x) \iff -y^* = Ax - b \iff Ax = b - y^*,$$

from which we can easily retrieve the required Lasso model from y^*, and also deduce that certain coefficients x_i in the model may vanish, on account of the constraints on y^*. We can obtain a similar result through the connection of the soft threshold operator with this model.

5.3.4.4 Error bounds using duality

Consider the minimization problem $\inf_{x \in X}(\varphi_1(x) + \varphi_2(\mathsf{L}x))$ in the standard setup used for Fenchel duality but with the extra assumptions that φ_2 is a uniformly convex function with modulus of convexity $\phi : \mathbb{R}_+ \to \mathbb{R}_+$ (see Definition 2.3.3), and $\mathsf{L} : X \to Y$ is coercive in the sense that there exists $c > 0$ such that $\|\mathsf{L}x\|_Y \ge c\|x\|_X$ for every $x \in X$, so that one may define an equivalent norm $\| \cdot \|_{\mathsf{L}}$, on X, by $\|x\|_{\mathsf{L}} = \|\mathsf{L}x\|_X$. We will show that under the above assumptions, if x_o is the solution to the above minimization problem, and $x \in X$ is any approximation of it, then we have the following error bound [124]:

$$2\phi(\|x - x_o\|_{\mathsf{L}}) \le \varphi_1(x) + \varphi_2(\mathsf{L}x) + \varphi_1^\star(\mathsf{L}^\star y^\star) + \varphi_2^\star(-y^\star), \quad \forall y^\star \in Y^\star. \tag{5.69}$$

To show estimate (5.69) we can work as follows: For any $\lambda \in (0, 1)$, consider the element $z = \lambda x_o + (1 - \lambda)x \in X$. By the uniform convexity of φ_2, and the convexity of φ_1, we have

$$\varphi_2(\mathsf{L}(\lambda x_o + (1-\lambda)x)) \le \lambda\varphi_2(\mathsf{L}x_o) + (1-\lambda)\varphi_2(\mathsf{L}x) - 2\lambda(1-\lambda)\phi(\|x - x_o\|_{\mathsf{L}}),$$
$$\varphi_1(\lambda x_o + (1-\lambda)x) \le \lambda\varphi_1(x_o) + (1-\lambda)\varphi_1(x),$$

so upon adding the two and rearranging, we have

$$2\lambda(1-\lambda)\phi(\|x - x_o\|_{\mathsf{L}})$$
$$\le \lambda(\varphi_1(x_o) + \varphi_2(\mathsf{L}x_o)) + (1-\lambda)(\varphi_1(x) + \varphi_2(\mathsf{L}x)) - (\varphi_1(z) + \varphi_2(\mathsf{L}z)),$$

and since $\varphi_1(z) + \varphi_2(Lz) \geq \varphi_1(x_o) + \varphi_2(Lx_o)$ for any choice of λ, the above inequality simplifies, upon dividing by $1-\lambda$, and passing to the limit $\lambda \to 1$, to

$$\begin{aligned} 2\phi(\|x - x_o\|_L) &\leq \varphi_1(x) + \varphi_2(Lx) - (\varphi_1(x_o) + \varphi_2(Lx_o)) \\ &= \varphi_1(x) + \varphi_2(Lx) - P. \end{aligned} \tag{5.70}$$

This is already a bound for the error of the approximation, nevertheless not a very convenient one since it depends on the exact value of $P = \inf_{x \in X}(\varphi_1(x) + \varphi_2(Lx))$, which is in principle unknown. As such, it will not serve as a good, *a priori* bound.

However, one may use weak (or strong) duality in order to bring this bound into a more convenient form. Indeed, by weak duality,

$$P \geq D = \sup_{y^\star \in Y^\star} (-\varphi_1^\star(L^\star y^\star) - \varphi_2^\star(-y^\star)) \geq -\varphi_1^\star(L^\star y^\star) - \varphi_2^\star(-y^\star)$$

for every $y^\star \in Y^\star$. Combining that with (5.70), we obtain a more convenient error bound:

$$2\phi(\|x - x_o\|_L) \leq \varphi_1(x) + \varphi_2(Lx) + \varphi_1^\star(L^\star y^\star) + \varphi_2^\star(-y^\star), \quad \forall y^\star \in Y^\star, \tag{5.71}$$

which is in fact (5.69). Upon judicious choice of candidates for $y^\star$ this estimate may provide easier to obtain error bounds. For example, upon the observation that $-\langle y^\star, Lx\rangle_{Y^\star,Y} - \langle -L^\star y^\star, x\rangle_{X^\star,X} = 0$ (by the definition of the adjoint operator), the error bound (5.71) can be expressed in the equivalent but more symmetric form

$$\begin{aligned} 2\phi(\|x - x_o\|_L) \leq\ & [\varphi_1(x) + \varphi_1^\star(L^\star y^\star) + \langle L^\star y^\star, x\rangle_{X^\star,X}] \\ & + [\varphi_2(Lx) + \varphi_2^\star(-y^\star) + \langle -y^\star, Lx\rangle_{Y^\star,Y}]. \end{aligned}$$

A stricter upper bound may be obtained by minimizing the right-hand side of the above inequality over $y^\star \in Y^\star$, which nevertheless may be a problem as demanding as the original one. However, the fact that since φ_2 is uniformly convex implies differentiability properties for $\varphi_2^\star$, which may be used for further refinements for the error estimates (see [124] for further details).

5.3.4.5 The augmented Lagrangian method

Consider the perturbation function

$$F_\gamma(x, y) = \varphi_1(x) + \varphi_2(Lx + y) + \frac{\gamma}{2}\|y\|^2, \quad \gamma > 0.$$

Clearly, this is an admissible perturbation since $F_\gamma(x, 0) = \varphi_1(x) + \varphi_2(Lx)$. The Lagrangian that corresponds to this perturbation provides an interesting method, which also leads to a class of popular numerical algorithms, for the solution of the minimization problem $\inf_{x \in X}(\varphi_1(x) + \varphi_2(Lx))$, called the augmented Lagrangian method. This method has a long history, originating to the (independent) contributions of Hestenes and Powel in the late

1960's, who used this method to treat nonlinear programming problems with equality constraints, and leading later to the work of Fortin and of Ito and Kunisch, who used it to treat problems related to the standard Fenchel problem, presented in this example (see [90] and references therein).

We will assume for simplicity that we are in a Hilbert space setting, and assume that $\varphi_1 : X \to \mathbb{R} \cup \{+\infty\}$ is a convex function with Lipschitz continuous Fréchet derivative, $\varphi_2 : Y \to \mathbb{R} \cup \{+\infty\}$ is a proper lower semicontinuous function (not necessarily smooth), and $\mathsf{L} : X \to Y$ is a bounded linear operator. We will also assume that $Y^\star \simeq Y$, while this is not necessarily so for X. With this information at hand, one may proceed to calculate the Lagrangian. We have that

$$\begin{aligned} L(x, y^\star) &= -\sup_{y\in Y}(\langle y^\star, y\rangle_Y - F_\gamma(x, y)) \\ &= \varphi_1(x) + \inf_{y\in Y}\Big(-\langle y^\star, y\rangle_Y + \varphi_2(\mathsf{L}x + y) + \frac{\gamma}{2}\|y\|_Y^2\Big) \\ &= \varphi_1(x) + \inf_{y'\in Y}\Big(\varphi_2(y') + \langle y^\star, \mathsf{L}x - y'\rangle_Y + \frac{\gamma}{2}\|\mathsf{L}x - y'\|_Y^2\Big), \end{aligned}$$

where for the last line we take the infimum over $y' = \mathsf{L}x + y$. We observe that, being in Hilbert space,

$$\frac{\gamma}{2}\|\mathsf{L}x - y'\|^2 + \langle y^\star, \mathsf{L}x - y^\star\rangle_Y = \frac{\gamma}{2}\Big\|\mathsf{L}x + \frac{1}{\gamma}y^\star - y'\Big\|_Y^2 - \frac{1}{2\gamma}\|y^\star\|_Y^2,$$

so

$$\begin{aligned} L(x, y^\star) &= \varphi_1(x) + \inf_{y'\in Y}\Big(\varphi_2(y') + \frac{\gamma}{2}\Big\|\mathsf{L}x + \frac{1}{\gamma}y^\star - y'\Big\|_Y^2\Big) - \frac{1}{2\gamma}\|y^\star\|_Y^2 \\ &= \varphi_1(x) + \varphi_{2,\gamma^{-1}}\Big(\mathsf{L}x + \frac{1}{\gamma}y^\star\Big) - \frac{1}{2\gamma}\|y^\star\|_Y^2, \end{aligned}$$

where, by $\varphi_{2,\gamma}$, we denote the Moreau–Yosida envelope of φ_2 (see Definition 4.4.3). By the properties of the Moreau–Yosida envelope (see Proposition 4.4.6), the function $\varphi_{2,\gamma}$ is Lipschitz continuously Fréchet differentiable (while φ_2 is not necessarily smooth) so that the Lagrangian is smooth in both variables. The above Lagrangian depends on the regularization parameter $\gamma > 0$, and is called the augmented Lagrangian. We will use the parameterization $\lambda = \gamma^{-1}$, and denote the augmented Lagrangian by

$$L_\lambda(x, y^\star) = \varphi_1(x) + \varphi_{2,\lambda}(\mathsf{L}x + \lambda y^\star) - \frac{\lambda}{2}\|y^\star\|_Y^2.$$

Using the properties of the Moreau–Yosida envelope (Proposition 4.4.6), it is easy to see that L_γ is Fréchet differentiable with respect to both variables, and in particular

$$
\begin{aligned}
D_x L_\lambda(x, y^\star) &= D_x \varphi_1(x) + \frac{1}{\lambda} L^\star(Lx + \lambda y^\star - \mathrm{prox}_{\lambda\varphi_2}(Lx + \lambda y^\star)),\\
D_{y^\star} L_\lambda(x, y^\star) &= (Lx + \lambda y^\star - \mathrm{prox}_{\lambda\varphi_2}(Lx + \lambda y^\star)) - \lambda y^\star\\
&= Lx - \mathrm{prox}_{\lambda\varphi_2}(Lx + \lambda y^\star),
\end{aligned}
\tag{5.72}
$$

or in more compact form

$$
\begin{aligned}
D_x L_\lambda(x, y^\star) &= D_x \varphi_1(x) + \frac{1}{\lambda} L^\star (I - \mathrm{prox}_{\lambda\varphi_2})(Lx + \lambda y^\star),\\
D_{y^\star} L_\lambda(x, y^\star) &= (I - \mathrm{prox}_{\lambda x_2})(Lx + \lambda y^\star) - \lambda y^\star.
\end{aligned}
\tag{5.73}
$$

One may show [90] the equivalence of the following three statements:

(a) $(x_o, y_0^\star)$ satisfies

$$
\begin{aligned}
D_x L_\lambda(x_o, y_0^\star) &= 0,\\
D_{y^\star} L_\lambda(x_o, y_o^\star) &= 0,
\end{aligned}
\tag{5.74}
$$

(b) $(x_o, y_0^\star)$ satisfies

$$
\begin{aligned}
D_x \varphi_1(x_o) + L^\star y_0^\star &= 0,\\
y_0^\star &\in \partial\varphi_2(Lx_o),
\end{aligned}
\tag{5.75}
$$

(c) $(x_o, y_0^\star)$ is a saddle point of the augmented Lagrangian L_λ.

The equivalence of (a) and (b) can easily be seen by recalling that $\mathrm{prox}_{\lambda\varphi_2} = (I + \lambda\partial\varphi_2)^{-1}$. In particular, if (b) holds, then the second of (5.75) yields that for every $\lambda > 0$, $Lx_o + \lambda y_0^\star \in (I + \lambda\partial\varphi_2)(Lx_o)$ or equivalently that $Lx_o = \mathrm{prox}_{\lambda\varphi_2}(Lx_o + \lambda y_0^\star)$, which (by (5.72)) is the second of (5.74). On the other hand, the first of (5.75) can be expressed as

$$
\begin{aligned}
0 &= D_x \varphi_1(x_o) + \frac{1}{\lambda} L^\star(\lambda y_0^\star + 0)\\
&= D_x \varphi_1(x_o) + \frac{1}{\lambda} L^\star(\lambda y_0^\star + Lx_o - \mathrm{prox}_{\lambda\varphi_2}(Lx_o + \lambda y_o^\star)),
\end{aligned}
$$

which (by (5.72)) is the first of (5.74). Suppose that (a) holds, i. e., for some $\lambda > 0$, $(x_o, y_0^\star)$ satisfies (5.74) with these Fréchet derivatives given by (5.72). The second of (5.74) implies that $Lx_o + \lambda y_0^\star \in (I + \lambda\partial\varphi_2)(Lx_o)$, which in turn implies that $y_0^\star \in \partial\varphi_2(Lx_o)$, the second relation in (5.75). Furthermore, since the second of (5.74) implies by (5.72) that $0 = Lx_o - \mathrm{prox}_{\lambda\varphi_2}(Lx_o + \lambda y_0^\star)$, combining the first of (5.74) with this and using the first of (5.72), we obtain the first of (5.75). Thus the equivalence of (a) and (b) is complete.

To show the equivalence of (a) and (c), it is enough to recall that $(x_o, y_0^\star)$ is a saddle point for L_λ if and only if

$$
x_o \in \arg\min_{x \in X} L_\lambda(x, y_0^\star),
$$

$$y_o^\star \in \arg\min_{y^\star \in Y^\star} L_\lambda(x_o, y^\star).$$

This is equivalent to $0 = D_x L_\lambda(x_o, y_o^\star)$ and $0 = D_{y^\star} L_\lambda(x_o, y_o^\star)$, hence, the stated equivalence.

We now show that if $(x_o, y_o^\star)$ satisfies any of the equivalent conditions (a), (b), or (c) above, then x_o is a solution of the original optimization problem $\inf_{x \in X}(\varphi_1(x) + \varphi_2(Lx))$. Consider condition (b), i. e., that x_o satisfies

$$D_x\varphi_1(x_o) + L^\star y_o^\star = 0 \quad \text{for some } y_o^\star \in \partial\varphi_2(Lx_o). \tag{5.76}$$

This can be expressed as $D_x\varphi_1(x_o) = -L^\star y_o^\star \in -L^\star \partial\varphi_2(Lx_o)$, which is in turn equivalent to $0 \in D_x\varphi_1(x_o) + L^\star\partial\varphi_2(Lx_o)$, whereby, using standard subdifferential calculus, we conclude that $0 \in \partial(\varphi_1 + \varphi_2 \circ L)(x_o)$. Hence, x_o is a minimizer of the original problem.

The augmented Lagrangian is to be compared with the standard Lagrangian we would get in the case where $\gamma = 0$ (equiv. $\lambda = \infty$). This would lead to the Lagrangian function

$$\begin{aligned}
L(x, y^\star) &= -\sup_{y \in Y}(\langle y^\star, y\rangle_Y - F_0(x, y)) \\
&= \varphi_1(x) - \sup_{y \in Y}(\langle y^\star, y\rangle_Y - \varphi_2(L + y)) \\
&= \varphi_1(x) - \sup_{y' \in Y}(\langle y^\star, y'\rangle_Y - \varphi_2(y') + \langle y^\star, Lx\rangle_Y) \\
&= \varphi_1(x) - \varphi_2^\star(y^\star) + \langle y^\star, Lx\rangle_Y.
\end{aligned}$$

We claim that any saddle point of the augmented Lagrangian is also a saddle point for the standard Lagrangian, which is no longer necessarily smooth in $y^\star$, with the converse being also true.

Indeed, if (x_o, y_o) is a saddle point for L, then

$$\begin{aligned}
&D_x L(x_o, y_o^\star) = 0, \\
&0 \in \partial_{y^\star} L(x_o, y_o^\star).
\end{aligned} \tag{5.77}$$

The first equation of (5.77) reduces to $D_x\varphi_1(x_o) + L^\star y_o^\star = 0$, which is the first equation in (5.75). The second inclusion of (5.77) reduces to $0 \in Lx_o - \partial\varphi_2^\star(y_o^\star)$, which is equivalent to $Lx_o \in \partial\varphi_2^\star(y_o^\star)$, so, by the Fenchel–Young inequality, we have that $y_o^\star \in \partial\varphi_2(Lx_o)$, which is the second equation in (5.75). Hence, any saddle point of L is a saddle point of L_λ for any $\lambda > 0$. The reverse implication can be also easily checked.

How does the above construction help us in the treatment of the original problem? The important difference between the standard Lagrangian approach and the augmented Lagrangian approach is that, even though a saddle point $(x_o, y_o^\star)$ of the augmented Lagrangian is also a saddle point of the standard Lagrangian, with x_o being a solution of the original problem, the smoothness of the augmented Lagrangian in both variables makes the numerical calculation of the saddle point a much easier and efficient task than the original problem of retrieving a saddle point for the original Lagrangian.

5.4 Primal dual algorithms

Convex duality methods often form the basis for a number of interesting numerical methods for the resolution of optimization problems. For simplicity, we assume that X, Y are Hilbert spaces, with respective inner products $\langle\cdot,\cdot\rangle_X$, $\langle\cdot,\cdot\rangle_Y$, and identified with their duals, in this section. However, for the ease of the reader, we still keep a different notation for the dual spaces and their elements.

Primal dual algorithms are based on the observation that a minimization problem $\min_{x\in X} f(x)$ can be brought into a saddle point form as

$$P = \inf_{x\in X}\sup_{y^\star\in Y^\star} L(x,y^\star) = \sup_{y^\star\in Y^\star}\inf_{x\in X} L(x,y^\star) = D,$$

so a solution x_o of the original (primal) problem can be retrieved from a saddle point $(x_o, y_0^\star)$ of the Lagrangian L, with $y_0^\star$ being a solution of the dual problem. Primal dual algorithms try to address simultaneously the primal and the dual problem by treating

$$\begin{aligned} x_o &\in \arg\min_{x\in X} L(x, y_0^\star),\\ y_0^\star &\in \arg\max_{y\in Y^\star} L(x_o, y^\star), \end{aligned}$$

usually considering the first-order conditions for the above problems:

$$\begin{aligned} 0 &\in \partial_x L(x_o, y_0^\star),\\ 0 &\in \partial_{y^\star} L(x_o, y_o^\star), \end{aligned}$$

and treating them in some sort of iterative scheme, the fixed point of which (if it exists) corresponds to the desired saddle point $(x_o, y_0^\star)$. Since in many cases of interest the problems under consideration do not involve smooth functions, some regularization procedure involving, e. g., the Moreau–Yosida envelope, and leading thus to proximal methods can be employed.

Such methods are often employed for optimization problems of the standard form

$$\min_{x\in X}(\varphi_1(x) + \varphi_2(Lx)),$$

which under convexity assumptions on φ_2, taking into account that $\varphi_2 = \varphi_2^{\star\star}$, can be expressed in the saddle point form (see Sec. 5.3.2)

$$\inf_{x\in X}\max_{y^\star\in Y^\star\simeq Y}\{\langle y^\star, Lx\rangle_Y - \varphi_2^\star(y^\star) + \varphi_1(x)\}, \tag{5.78}$$

where we assume, being in Hilbert space, that $Y^\star \simeq Y$. The first-order conditions for a saddle point $(x_o, y_o^\star)$ are

$$0 \in L^\star y_o^\star + \partial\varphi_1(x_o),$$
$$0 \in Lx_o - \partial\varphi_2^\star(y_o^\star),$$

which may be further interpreted as a fixed point scheme for an appropriately chosen operator. One way to numerically treat this problem is within the class of the Uzawa type algorithms, where a proximal descent method is used for the x variable, and a standard ascent method is used for the dual $y^\star$ variable, in terms of

$$x_{n+1} \in x_n - \lambda(L^\star \bar{y^\star}_n + \partial\varphi_1(x_{n+1})), \quad \lambda > 0,$$
$$y_{n+1}^\star \in y_n^\star + \mu(L\bar{x}_n - \partial\varphi_2^\star(y_{n+1}^\star)), \quad \mu > 0,$$

where the interaction between the primal and the dual problem is through the coupling of the first with the second inclusion through the terms $\bar{x}_n$ and $\bar{y}_n^\star$. These are in general functions of $x_n, x_{n+1}, y_n^\star, y_{n+1}^\star$, which may be chosen in a variety of ways, each leading to different versions of the algorithm, subject to the constraint that the conditions $x_{n+1} = x_n$ and $y_{n+1}^\star = y_n^\star$ imply that $\bar{x}_n = x_n$ and $\bar{y^\star}_n = y_n^\star$, respectively.[25]

Recalling the definition of the proximal operators $\text{prox}_{\lambda\varphi_1} = (I + \lambda\partial\varphi_1)^{-1}$, and $\text{prox}_{\mu\varphi_2^\star} = (I + \mu\partial\varphi_2^\star)^{-1}$, we may express the above scheme in proximal form as

$$x_{n+1} = \text{prox}_{\lambda\varphi_1}(x_n - \lambda L^\star \bar{y^\star}_n), \quad \lambda > 0,$$
$$y_{n+1}^\star = \text{prox}_{\mu\varphi_2^\star}(y_n^\star + \mu L\bar{x}_n), \quad \mu > 0.$$

In fact, one may reverse the order of the primal and the dual step and rewrite the above algorithm as

$$y_{n+1}^\star = \text{prox}_{\mu\varphi_2^\star}(y_n^\star + \mu L\bar{x}_n), \quad \mu > 0,$$
$$x_{n+1} = \text{prox}_{\lambda\varphi_1}(x_n - \lambda L^\star y_{n+1}^\star), \quad \lambda > 0,$$
$$\bar{x}_n = x_n + \theta(x_n - x_{n-1}), \quad \theta \in [0,1],$$

where we have also chosen $\bar{y^\star}_n = y_{n+1}^\star$, and $\theta \in [0,1]$ is a parameter, which is used in order to interpolate $\bar{x}_n$ between x_{n-1} and x_n. In the special case, $\theta = 0$, $\bar{x}_n = x_n$, and we recover a classic algorithm proposed by Arrow and Hurwicz in the late 1950s, as modified for the specific problem at hand. As it will turn out, other choices for θ, as for example $\theta = 1$, (considered as an overrelaxation step) has superior convergence

25 This is needed for compatibility with the condition that a saddle point $(x_o, y_o^\star)$ is a fixed point of the iteration scheme.

properties. Note also the similarity with the Douglas–Rachford scheme (in the special case $\mathsf{L} = I$, see also Remark 5.4.3).

We therefore, have the following algorithm, due to Chambolle and Pock (see [44]) for the resolution of minimization problems:

Algorithm 5.4.1 (Chambolle–Pock PDHG).
1. Choose $\lambda > 0, \mu > 0, \theta \in [0,1]$.
2. Choose initial condition $(\mathsf{x}_1, \mathsf{y}_1^\star) \in X \times Y$, and set $\bar{\mathsf{x}}_1 = \mathsf{x}_1$.
3. Iterate

$$\begin{aligned}
\mathsf{y}_{n+1}^\star &= \operatorname{prox}_{\mu\varphi_2^*}(\mathsf{y}_n^\star + \mu \mathsf{L}\bar{\mathsf{x}}_n), \\
\mathsf{x}_{n+1} &= \operatorname{prox}_{\lambda\varphi_1}(\mathsf{x}_n - \lambda \mathsf{L}^\star \mathsf{y}_{n+1}^\star), \\
\bar{\mathsf{x}}_{n+1} &= \mathsf{x}_{n+1} + \theta(\mathsf{x}_{n+1} - \mathsf{x}_n)
\end{aligned} \tag{5.79}$$

until a convergence criterion is met.

Though the formal similarity of this algorithm with the Douglas–Rachford scheme can be used to obtain convergence results for the above algorithm, we prefer to present here a more direct approach. The following proposition due to [44] provides some results concerning the convergence of this algorithm for the choice $\theta = 1$. Our approach follows closely the steps in [44], while in [46], the interested reader may find a revision and generalization of some of the arguments in [44].

Proposition 5.4.2. *Assume that $\lambda\mu\|\mathsf{L}\| < 1$. Then, the sequence $(w_n)_{n\in\mathbb{N}} = ((\mathsf{x}_n, \mathsf{y}_n^\star))_{n\in\mathbb{N}} \subset X \times Y^*$, generated by* (5.79) *for $\theta = 1$, is bounded in $X \times Y^*$, and the sequence of averages $(w_N)_{N\in\mathbb{N}} = ((\bar{\mathsf{x}}_N, \bar{\mathsf{y}}_N^\star))_{N\in\mathbb{N}}$, where $\bar{\mathsf{x}}_N = \frac{1}{N}\sum_{n=1}^N \mathsf{x}_i$ and $\bar{\mathsf{y}}^\star{}_N = \frac{1}{N}\sum_{n=1}^N \mathsf{y}_i^\star$, converges weakly to a saddle point of* (5.78).

Proof. The strategy of the proof is to show that the distance of the sequence $((\mathsf{x}_n, \mathsf{y}_n^\star))_{n\in\mathbb{N}}$ generated by (5.79) with respect to any saddle point is bounded, hence, weakly convergent to some point $\bar{\mathsf{w}}_o := (\bar{\mathsf{x}}_o, \bar{\mathsf{y}}_o^\star)$ in $X \times Y^\star$. We will then identify this weak limit as a saddle point. We will assume $X^\star \simeq X$ and $Y^\star \simeq Y$, but retain the notation $Y^\star$ and $\mathsf{y}^\star$ in compliance to our general notation concerning the dual formulation of optimization problems. The proof proceeds in 3 steps.

1. In the first step, we establish a generalized monotonicity result for $((\mathsf{x}_n, \mathsf{y}_n^\star))_{n\in\mathbb{N}}$ in the form of

$$\begin{aligned}
c\Big(\frac{1}{2\lambda}\|\mathsf{x} - \mathsf{x}_1\|_X^2 + \frac{1}{2\mu}\|\mathsf{y}^\star - \mathsf{y}_1^\star\|_Y^2\Big) \geq{}& \frac{1}{2\lambda}\|\mathsf{x} - \mathsf{x}_N\|_X^2 + \frac{1}{2\mu}\|\mathsf{y}^\star - \mathsf{y}_N^\star\|_Y^2 \\
&+ c\sum_{n=1}^{N-1}[\langle \mathsf{L}\mathsf{x}_{n+1}, \mathsf{y}^\star\rangle_Y - \varphi_2^\star(\mathsf{y}^\star) + \varphi_1(\mathsf{x}_{n+1})] \\
&- c\sum_{n=1}^{N-1}[\langle \mathsf{L}\mathsf{x}, \mathsf{y}_{n+1}^\star\rangle_Y - \varphi_2^\star(\mathsf{y}_{n+1}^\star) + \varphi_1(\mathsf{x})]
\end{aligned}$$

for every $(x, y^\star) \in X \times Y^\star$ for an appropriate constant $c > 0$.

We start with the general form of the algorithm as

$$\begin{aligned} y^\star_{n+1} &= \mathrm{prox}_{\mu\varphi_2^\star}(y^\star_n + \mu L\bar{x}), \\ x_{n+1} &= \mathrm{prox}_{\lambda\varphi_1}(x_n - \lambda L^\star \bar{y^\star}), \end{aligned}$$

for an appropriate choice $\bar{x}$ and $\bar{y^\star}$, to be specified during the course of the proof.

By the definition of the proximity operator, we have that

$$\begin{aligned} &\frac{1}{\lambda}(x_n - x_{n+1}) - L^\star \bar{y} \in \partial\varphi_1(x_{n+1}), \\ &\frac{1}{\mu}(y^\star_n - y^\star_{n+1}) + L\bar{x} \in \partial\varphi_2^\star(y^\star_{n+1}), \end{aligned}$$

so by the definition of the subdifferential, we have that

$$\begin{aligned} \varphi_1(x) &\geq \varphi_1(x_{n+1}) + \frac{1}{\lambda}\langle x_n - x_{n+1}, x - x_{n+1}\rangle_X - \big\langle L(x - x_{n+1}), \bar{y^\star}\big\rangle_Y, \quad \forall\, y^\star \in Y^\star, \\ \varphi_2^\star(y^\star) &\geq \varphi_2^\star(y^\star_{n+1}) + \frac{1}{\mu}\langle y^\star_n - y^\star_{n+1}, y^\star - y^\star_{n+1}\rangle_Y + \big\langle L\bar{x}, y^\star - y^\star_{n+1}\big\rangle_Y, \quad \forall\, x \in X, \end{aligned} \tag{5.80}$$

where we used $\langle x - x_{n+1}, L^\star \bar{y^\star}\rangle_Y = \langle L(x - x_{n+1}), \bar{y^\star}\rangle_X$.

We first note that the term $\langle x_n - x_{n+1}, x - x_{n+1}\rangle$ can be expressed as

$$\langle x_n - x_{n+1}, x - x_{n+1}\rangle_X = \frac{1}{2}\|x_n - x_{n+1}\|_X^2 - \frac{1}{2}\|x - x_n\|_X^2 + \frac{1}{2}\|x - x_{n+1}\|_X^2, \tag{5.81}$$

with a similar expression for $\langle y^\star_n - y^\star_{n+1}, y^\star - y^\star_{n+1}\rangle_Y$. This can be verified by direct calculation of the norms, recalling that we are in a Hilbert space setting. Substituting (5.81) into (5.80) and adding the two inequalities, we obtain

$$\begin{aligned} 0 \geq\ & \big[\, \langle Lx_{n+1}, y^\star\rangle_Y - \varphi_2^\star(y^\star) + \varphi_1(x_{n+1}) \,\big] - \langle Lx_{n+1}, y^\star\rangle_Y + \langle L\bar{x}, y^\star - y^\star_{n+1}\rangle_Y \\ & - \big[\, \langle Lx, y^\star_{n+1}\rangle_Y - \varphi_2^\star(y^\star_{n+1}) + \varphi_1(x) \,\big] + \langle Lx, y^\star_{n+1}\rangle_Y - \langle L(x - x_{n+1}), \bar{y^\star}\rangle_Y \\ & + \frac{1}{2\lambda}\|x_n - x_{n+1}\|_X^2 - \frac{1}{2\lambda}\|x - x_n\|_X^2 + \frac{1}{2\lambda}\|x - x_{n+1}\|_X^2 \\ & + \frac{1}{2\mu}\|y^\star_n - y^\star_{n+1}\|_Y^2 - \frac{1}{2\mu}\|y^\star - y^\star_n\|_Y^2 + \frac{1}{2\mu}\|y^\star - y^\star_{n+1}\|_Y^2, \end{aligned}$$

which is further rearranged as

$$\begin{aligned} & \frac{1}{2\lambda}\|x - x_n\|_X^2 + \frac{1}{2\mu}\|y^\star - y^\star_n\|_Y^2 \\ & \geq \big[\, \langle Lx_{n+1}, y^\star\rangle_Y - \varphi_2^\star(y^\star) + \varphi_1(x_{n+1}) \,\big] - \big[\, \langle Lx, y^\star_{n+1}\rangle_Y - \varphi_2^\star(y^\star_{n+1}) + \varphi_1(x) \,\big] \\ & \quad + \frac{1}{2\lambda}\|x - x_{n+1}\|_X^2 + \frac{1}{2\lambda}\|x_n - x_{n+1}\|_X^2 + \frac{1}{2\mu}\|y^\star - y^\star_{n+1}\|_Y^2 + \frac{1}{2\mu}\|y^\star_n - y^\star_{n+1}\|_Y^2 \\ & \quad + \langle L(x_{n+1} - \bar{x}), y^\star_{n+1} - y^\star\rangle_Y - \langle L(x_{n+1} - x), y^\star_{n+1} - \bar{y^\star}\rangle_Y. \end{aligned} \tag{5.82}$$

In (5.82), we may control the sign of the first two terms upon choosing $(x, y^\star)$ to be a saddle point of the Lagrangian formulation of the optimization problem. We would like to have a similar control over the last two terms. We note that upon choosing $\bar{x}, \bar{y^\star}$ in an appropriate manner, the last two terms of (5.82) can be simplified. Choosing them as in the PDHG algorithm, i.e, $\bar{x} = 2x_n - x_{n-1}$ and $\bar{y} = y_{n+1}$, we have for the last two terms of (5.82) that

$$\begin{aligned} I &:= \langle L(x_{n+1} - \bar{x}), y^\star_{n+1} - y^\star \rangle_Y - \langle L(x_{n+1} - x), y^\star_{n+1} - \bar{y^\star} \rangle_Y \\ &\geq \langle L(x_{n+1} - x_{n-1}), y^\star_{n+1} - y^\star \rangle_Y - \langle L(x_n - x_{n-1}), y^\star_n - y^\star \rangle_Y \\ &\quad - |\langle L(x_n - x_{n-1}), y^\star_{n+1} - y^\star \rangle_Y| \\ &\geq \langle L(x_{n+1} - x_{n-1}), y^\star_{n+1} - y^\star \rangle_Y - \langle L(x_n - x_{n-1}), y^\star_n - y^\star \rangle_Y \\ &\quad - \|L\| \, \|x_n - x_{n-1}\|_X \, \|y^\star_{n+1} - y^\star_n\|_Y \\ &\geq \langle L(x_{n+1} - x_{n-1}), y^\star_{n+1} - y^\star \rangle_Y - \langle L(x_n - x_{n-1}), y^\star_n - y^\star \rangle_Y \\ &\quad - \frac{\alpha}{2} \|L\| \, \|x_n - x_{n-1}\|^2_X - \frac{1}{2\alpha} \|L\| \, \|y^\star_{n+1} - y^\star_n\|^2_Y, \end{aligned} \tag{5.83}$$

for an appropriately chosen α (to be specified shortly). Combining (5.83) with (5.82), we see that

$$\begin{aligned} &\frac{1}{2\lambda} \|x - x_n\|^2_X + \frac{1}{2\mu} \|y^\star - y^\star_n\|^2_Y \\ &\quad \geq [\langle Lx_{n+1}, y^\star \rangle_Y - \varphi_2^\star(y^\star) + \varphi_1(x_{n+1})] - [\langle Lx, y^\star_{n+1} \rangle_Y - \varphi_2^\star(y^\star_{n+1}) + \varphi_1(x)] \\ &\qquad + \frac{1}{2\lambda} \|x - x_{n+1}\|^2_X + \frac{1}{2\lambda} \|x_n - x_{n+1}\|^2_X - \frac{\alpha\|L\|}{2} \|x_n - x_{n-1}\|^2_X \\ &\qquad + \frac{1}{2\mu} \|y^\star - y^\star_{n+1}\|^2_Y + \left(\frac{1}{2\mu} - \frac{\|L\|}{2\alpha} \right) \|y^\star_n - y^\star_{n+1}\|^2_Y \\ &\qquad + \langle L(x_{n+1} - x_n), y^\star_{n+1} - y^\star \rangle_Y - \langle L(x_n - x_{n-1}), y^\star_n - y^\star \rangle_Y. \end{aligned} \tag{5.84}$$

Our aim is to turn (5.84) to a comparison between $\|x_N - x\|^2 + \|y^\star_N - y^\star\|^2$ and $\|x_1 - x\|^2 + \|y^\star_1 - y^\star\|^2$, showing that $\|x_N - x\|^2 + \|y^\star_N - y^\star\|^2$ is decreasing if $(x, y^\star)$ is chosen appropriately (e. g., if $(x, y^\star) = (x_o, y^\star_o)$ is a saddle point), so the sequence $((x_N, y^\star_N))_{N \in \mathbb{N}}$ generated by the PDHG scheme is a Fejér sequence with respect to the set of saddle points of the problem, therefore leading to weak convergence of the scheme to a saddle point $(x_o, y^\star_o)$.

To this end, observe that we could turn certain parts of (5.84) into telescopic sums, leading thus to a number of simplifications. Adding (5.84) from $n = 1$ to $N - 1$, and setting $x_0 = x_1$ to eliminate certain terms, we see that after simplification, for any $(x, y^\star) \in X \times Y^\star$, it holds that

$$
\begin{aligned}
&\frac{1}{2\lambda}\|x - x_1\|_X^2 + \frac{1}{2\mu}\|y^\star - y_1^\star\|_Y^2 \\
&\quad \geq \frac{1}{2\lambda}\|x - x_N\|_X^2 + \frac{1}{2\mu}\|y^\star - y_N^\star\|_Y^2 \\
&\qquad + \sum_{n=1}^{N-1}\left[\, \langle Lx_{n+1}, y^\star\rangle_Y - \varphi_2^\star(y^\star) + \varphi_1(x_{n+1}) \,\right] \\
&\qquad - \sum_{n=1}^{N-1}\left[\, \langle Lx, y_{n+1}^\star\rangle_Y - \varphi_2^\star(y_{n+1}^\star) + \varphi_1(x) \,\right] \\
&\qquad + \frac{1}{2\lambda}\|x_N - x_{N-1}\|_X^2 + \langle L(x_N - x_{N-1}), y_N^\star - y^\star\rangle_Y \\
&\qquad + \left(\frac{1}{2\lambda} - \frac{\alpha\|L\|}{2}\right)\sum_{n=1}^{N-2}\|x_{n+1} - x_n\|_X^2 + \left(\frac{1}{2\mu} - \frac{\|L\|}{2\alpha}\right)\sum_{n=1}^{N-2}\|y_{n+1}^\star - y_n^\star\|_Y^2 \\
&\quad \geq \frac{1}{2\lambda}\|x - x_N\|_X^2 + \frac{1}{2\mu}\|y^\star - y_N^\star\|_Y^2 \\
&\qquad + \sum_{n=1}^{N-1}\left[\, \langle Lx_{n+1}, y^\star\rangle_Y - \varphi_2^\star(y^\star) + \varphi_1(x_{n+1}) \,\right] \\
&\qquad - \sum_{n=1}^{N-1}\left[\, \langle Lx, y_{n+1}^\star\rangle_Y - \varphi_2^\star(y_{n+1}^\star) + \varphi_1(x) \,\right] \\
&\qquad + \|x_N - x_{N-1}\|_X^2 + \frac{1}{2\lambda}\|x_N - x_{N-1}\|_X^2 + \langle L(x_N - x_{N-1}), y_N^\star - y^\star\rangle_Y,
\end{aligned}
\tag{5.85}
$$

as long as α is chosen so that $\frac{1}{2\lambda} - \frac{\alpha\|L\|}{2} > 0$ and $\frac{1}{2\mu} - \frac{\|L\|}{2\alpha} > 0$.

We are now in position to conclude our estimates by further estimating

$$
\begin{aligned}
\langle L(x_N - x_{N-1}), y_N^\star - y^\star\rangle_Y &\geq -\|L\|\,\|x_N - x_{N-1}\|_X\,\|y^\star - y_N^\star\|_Y \\
&\geq -\frac{\beta\|L\|}{2}\,\|x_N - x_{N-1}\|_X^2 - \frac{\|L\|}{2\beta}\|y^\star - y_N^\star\|_Y^2,
\end{aligned}
\tag{5.86}
$$

for a suitable choice for β. Substituting (5.86) into (5.85), we obtain

$$
\begin{aligned}
\frac{1}{2\lambda}\|x - x_1\|_X^2 + \frac{1}{2\mu}\|y^\star - y_1^\star\|_Y^2 \geq \frac{1}{2\lambda}\|x - x_N\|_X^2 &+ \left(\frac{1}{2\mu} - \frac{\|L\|}{2\beta}\right)\|y^\star - y_N^\star\|_Y^2 \\
&+ \sum_{n=1}^{N-1}\left[\, \langle Lx_{n+1}, y^\star\rangle_Y - \varphi_2^\star(y^\star) + \varphi_1(x_{n+1}) \,\right] \\
&- \sum_{n=1}^{N-1}\left[\, \langle Lx, y_{n+1}^\star\rangle_Y - \varphi_2^\star(y_{n+1}^\star) + \varphi_1(x) \,\right] \\
&+ \left(\frac{1}{2\lambda} - \frac{\beta\,\|L\|}{2}\right)\|x_N - x_{N-1}\|_X^2,
\end{aligned}
$$

which upon the choice $\beta = \frac{1}{\lambda\|L\|}$ reduces to

$$\begin{aligned}\frac{1}{2\lambda}\|x - x_1\|_X^2 + \frac{1}{2\mu}\|y^\star - y_1^\star\|_Y^2 \geq \frac{1}{2\lambda}\|x - x_N\|_X^2 + (1 - \lambda\mu\|L\|^2)\frac{1}{2\mu}\|y^\star - y_N^\star\|_Y^2 \\ + \sum_{n=1}^{N-1}[\langle Lx_{n+1}, y^\star\rangle_Y - \varphi_2^\star(y^\star) + \varphi_1(x_{n+1})] \\ - \sum_{n=1}^{N-1}[\langle Lx, y_{n+1}^\star\rangle_Y - \varphi_2^\star(y_{n+1}^\star) + \varphi_1(x)].\end{aligned} \tag{5.87}$$

Since, trivially, $\|x - x_N\|_X^2 \geq (1 - \lambda\mu\|L\|^2)\|x - x_N\|_X^2$, as long as $1 - \lambda\mu\|L\|^2 > 0$ for such a choice of parameters, (5.87) yields the estimate

$$\begin{aligned}&c\Big(\frac{1}{2\lambda}\|x - x_1\|_X^2 + \frac{1}{2\mu}\|y^\star - y_1^\star\|_Y^2\Big) \\ &\geq \frac{1}{2\lambda}\|x - x_N\|_X^2 + \frac{1}{2\mu}\|y^\star - y_N^\star\|_Y^2 \\ &+ c\sum_{n=1}^{N-1}[\langle Lx_{n+1}, y^\star\rangle_Y - \varphi_2^\star(y^\star) + \varphi_1(x_{n+1})] \\ &- c\sum_{n=1}^{N-1}[\langle Lx, y_{n+1}^\star\rangle_Y - \varphi_2^\star(y_{n+1}^\star) + \varphi_1(x)], \quad \forall (x, y^\star) \in X \times Y^\star,\end{aligned} \tag{5.88}$$

for $c = (1 - \lambda\mu\|L\|^2)^{-1}$.

2. We now use (5.88) to show that the sequence $((x_n, y_n^\star))_{n\in\mathbb{N}} \subset X \times Y^*$ is weakly convergent. This requires choosing $(x, y^\star) \in X \times Y^\star$ so as to further simplify (5.88) and obtain a boundedness result for the sequence. Choosing $(x, y^\star) = (x_o, y_o^\star)$, a saddle point for the Lagrangian, and denoting the saddle value by S_o, gives rise to

$$\begin{aligned}&[\langle Lx_{n+1}, y_o^\star\rangle_Y - \varphi_2^\star(y_o^\star) + \varphi_1(x_{n+1})] \geq S_o, \quad x_{n+1} \in X, \\ &[\langle Lx_o, y_{n+1}^\star\rangle_Y - \varphi_2^\star(y_{n+1}^\star) + \varphi_1(x_o)] \leq S_o, \quad y_{n+1}^\star \in Y^\star.\end{aligned} \tag{5.89}$$

Adding (5.89) from $n = 1, \dots, N - 1$, we have

$$\begin{aligned}&\sum_{n=1}^{N-1}[\langle Lx_{n+1}, y_o^\star\rangle_Y - \varphi_2^\star(y_o^\star) + \varphi_1(x_{n+1})] \\ &\quad - \sum_{n=1}^{N-1}[\langle Lx_o, y_{n+1}^\star\rangle_Y - \varphi_2^\star(y_{n+1}^\star) + \varphi_1(x_o)] \geq 0.\end{aligned} \tag{5.90}$$

Setting now $(x, y^\star) = (x_o, y_o^\star)$ (a saddle point) in (5.88), and combining the result with (5.90), we conclude that

$$\frac{1}{2\lambda}\|x_o - x_1\|_X^2 + \frac{1}{2\mu}\|y_o^\star - y_1^\star\|_Y^2 \geq (1 - \lambda\mu\|L\|^2)\Big(\frac{1}{2\lambda}\|x_o - x_N\|_X^2 + \frac{1}{2\mu}\|y_o^\star - y_N^\star\|_Y^2\Big).$$

This is true for any N, hence, the sequence $(w_N)_{N\in\mathbb{N}} := ((x_N, y_N^\star))_{N\in\mathbb{N}}$ is bounded in $X \times Y^\star$. Therefore, there exists $\bar{w}_o := (\bar{x}_o, \bar{y}_o^\star) \in X \times Y$ such that $w_n \rightharpoonup \bar{w}_o$ in $X \times Y^\star$. We will identify $\bar{w}_o$ as a saddle point.

3. In order to identify the limit, we will use once more (5.88) for a general choice of $(x, y^\star) \in X \times Y^\star$. Consider the sequence $((\bar{x}_N, \bar{y}_N^\star))_{N\in\mathbb{N}}$ defined by $\bar{x}_N = \frac{1}{N}\sum_{i=1}^N x_i$, $\bar{y}_N^\star = \frac{1}{N}\sum_{i=1}^N y_i^\star$. Clearly, $(\bar{x}_N, \bar{y}_N^\star) \rightharpoonup (\bar{x}_o, \bar{y}_o^\star)$ in $X\times Y^\star$. One easily sees that $\sum_{n=1}^N \langle Lx_n, y^\star\rangle_Y = N\langle L\bar{x}_N, y^\star\rangle_Y$, and $\sum_{n=1}^N \langle Lx, y_n^\star\rangle_Y = N\langle Lx, \bar{y}_N^\star\rangle_Y$, while, by convexity,

$$\sum_{n=1}^N \varphi_1(x_n) \geq N\varphi_1(\bar{x}_N), \quad \text{and} \quad \sum_{n=1}^N \varphi_2^\star(y_n) \geq N\varphi_2^\star(\bar{y}_N^\star).$$

Substituting these into (5.88), we obtain

$$\begin{aligned} &[\langle L\bar{x}_N, y^\star\rangle_Y - \varphi_2^\star(y^\star) + \varphi_1(\bar{x}_N)] - [\langle Lx, \bar{y}_N^\star\rangle_Y - \varphi_2^\star(\bar{y}_N^\star) + \varphi_1(x)] \\ &\quad \leq \frac{1}{cN}\left[\frac{1}{2\lambda}(\|x - x_1\|_X^2 - \|x - x_N\|_X^2) + \frac{1}{2\mu}(\|y^\star - y_1^\star\|_X^2 - \|y^\star - y_N^\star\|_X^2)\right], \\ &\quad \forall (x, y^\star) \in X \times Y^\star, \end{aligned} \tag{5.91}$$

and passing to the limit as $N \to \infty$ (using also the weak lower semicontinuity of the convex functions φ_1, $\varphi_2^\star$), we conclude that

$$\begin{aligned} &[\langle L\bar{x}_o, y^\star\rangle_Y - \varphi_2^\star(x) + \varphi_1(\bar{x}_o)] - [\langle Lx, \bar{y}_o^\star\rangle_Y - \varphi_2^\star(\bar{y}_o^\star) + \varphi_1(x)] \leq 0, \\ &\forall (x, y^\star) \in X \times Y^\star, \end{aligned}$$

so $(\bar{x}_o, \bar{y}_o^\star)$ is a saddle point of $\langle Lx, y^\star\rangle_Y + \varphi_1(x) - \varphi_2^\star(x)$. □

As seen in the proof of the above proposition (see estimate (5.91)), the rate of convergence is $O(1/N)$ (ergodic convergence). Convergence may be faster (e. g., $O(1/N^2)$) if either φ_1 or $\varphi_2^\star$ is uniformly convex. In such cases, acceleration techniques may be employed (see, e. g., [44]). In finite dimensional spaces, one may show (strong) convergence of the whole sequence.

The Chambolle–Pock algorithm may be further extended in various directions, i. e., in the study of minimization problems of the form

$$\min_{x\in X}(\varphi_1(x) + \varphi_2(x) + \varphi_3(Lx)),$$

whose minimax (saddle point) formulation is

$$\min_{x\in X}\max_{y^\star\in Y}(\langle Lx, y^\star\rangle_Y + \varphi_1(x) + \varphi_2(x) - \varphi_3^\star(y^\star)),$$

with φ_1 proper convex, lower semicontinuous with $D\varphi_1$ Lipschitz continuous, and with φ_2, φ_3 proper lower semicontinuous convex, such that their corresponding proximal maps are easy to compute (or even in cases where φ_3 is a convex lower semicontinuous function, defined by an inf-convolution; see Definition 5.2.24). Furthermore, one may consider generalized proximal maps, in which the quadratic regularization term $\frac{1}{\lambda}\|x - z\|^2$ is replaced by a regularization term involving the Bregman function $D_\psi(x, z) := \psi(x) - \psi(z) - \langle D\psi(z), x - z\rangle$ for an appropriate continuously differentiable function (a popular choice being the entropy function).

Remark 5.4.3 (Connections with the Douglas–Rachford scheme). As already mentioned, the algorithm (5.79) can be considered as an extension of the Douglas–Rachford scheme. Indeed, recalling the definition of the Moreau proximity operator, the above scheme can be expressed, upon defining the operators A, B acting on $z = (x, y^\star)^T$, by

$$A\begin{pmatrix} x \\ y^\star \end{pmatrix} = \begin{pmatrix} \frac{1}{\lambda}I & -L^\star \\ -L & \frac{1}{\mu}I \end{pmatrix}\begin{pmatrix} x \\ y^\star \end{pmatrix},$$
$$B\begin{pmatrix} x \\ y^\star \end{pmatrix} = \begin{pmatrix} \partial\varphi_1(x) \\ \partial\varphi_2^\star(y^\star) \end{pmatrix} + \begin{pmatrix} 0 & L^\star \\ -L & 0 \end{pmatrix}\begin{pmatrix} x \\ y^\star \end{pmatrix},$$

the Uzawa scheme can be expressed as

$$0 \in A(z_{k+1} - z_k) + Bz_{k+1}.$$

5.5 Appendix

5.5.1 Proof of Proposition 5.1.2

Assume that a saddle point (x_o, y_o) exists for F. By the definition of a saddle point

$$\sup_{y \in Y} F(x_o, y) = F(x_o, y_o), \tag{5.92}$$

and

$$\inf_{x \in X} F(x, y_o) = F(x_o, y_o). \tag{5.93}$$

Consider the function $F_1 : X \to \mathbb{R}$, defined by $F_1(x) = \sup_{y \in Y} F(x, y)$. Since $\inf_{x \in X} F_1(x) \le F_1(x)$ for every $x \in X$, choosing $x = x_o$ in this inequality yields that $\inf_{x \in X} F_1(x) \le F_1(x_o)$, so using the definition of F_1, we see that

$$\inf_{x \in X} \sup_{y \in Y} F(x, y) \le \sup_{y \in Y} F(x_o, y). \tag{5.94}$$

Similarly, considering the function $F_2 : Y \to \mathbb{R}$, defined by $F_2(y) = \inf_{x \in X} F(x, y)$, we note that, trivially, $F_2(y) \le \sup_{y \in Y} F_2(y)$ for every $y \in Y$, and setting $y = y_o$ yields that $F_2(y_o) \le \sup_{y \in Y} F_2(y)$. So using the definition of F_2, we see that

$$\inf_{x\in X} F(x, y_0) \le \sup_{y\in Y} \inf_{x\in X} F(x, y). \tag{5.95}$$

Combining (5.92)–(5.95) leads to the inequality

$$\inf_{x\in X} \sup_{y\in Y} F(x, y) \le \sup_{y\in Y} \inf_{x\in X} F(x, y),$$

which, when combined with (5.1), leads to the equality

$$\inf_{x\in X} \sup_{y\in Y} F(x, y) = \sup_{y\in Y} \inf_{x\in X} F(x, y).$$

Restating all these inequalities in a single line yields

$$\begin{aligned} F(x_o, y_0) = \inf_{x\in X} F(x, y_0) \le \sup_{y\in Y} \inf_{x\in X} F(x, y) &\overset{(5.1)}{\le} \inf_{x\in X} \sup_{y\in Y} F(x, y) \\ &\le \sup_{y\in Y} F(x_o, y) = F(x_o, y_o), \end{aligned}$$

from which it follows that all these inequalities are in fact equalities:

$$F(x_o, y_0) = \inf_{x\in X} F(x, y_0) = \sup_{y\in Y} \inf_{x\in X} F(x, y) = \inf_{x\in X} \sup_{y\in Y} F(x, y) = \sup_{y\in Y} F(x_o, y).$$

That means

$$\sup_{y\in Y} \inf_{x\in X} F(x, y) = \inf_{x\in X} F(x, y_0) = F(x_o, y_0),$$

so the supremum of the function $F_2 : Y \to \mathbb{R}$, defined by $F_2(y) = \inf_{x\in X} F(x, y)$, is attained at y_0, and

$$\inf_{x\in X} \sup_{y\in Y} F(x, y) = \sup_{y\in Y} F(x_o, y) = F(x_o, y_0),$$

so the infimum of the function $F_1 : X \to \mathbb{R}$, defined by $F_1(x) = \sup_{y\in Y} F(x, y)$, is attained at x_o.

For the converse, assume that

$$\sup_{y\in Y} \inf_{x\in X} F(x, y) = \inf_{x\in X} \sup_{y\in Y} F(x, y) = F(x_o, y_0), \tag{5.96}$$

with the supremum in the first expression and the infimum on the second being attained at y_0 at x_o, respectively. Then,

$$\inf_{x\in X} \sup_{y\in Y} F(x, y) = \sup_{y\in Y} F(x_o, y) \ge F(x_o, y_0), \tag{5.97}$$

where we used successively that $\sup_{y\in Y} F(\cdot, y)$ attains the infimum at x_o, and then the fact that the supremum is an upper limit, and

$$\sup_{y\in Y}\inf_{x\in X} F(x,y) = \inf_{x\in X} F(x,y_0) \le F(x_o,y_0), \tag{5.98}$$

where we used successively that $\inf_{x\in X} F(x,\cdot)$ attains the supremum at y_0, and then the fact that the infimum is a lower limit. Combining (5.97) and (5.98) into a single inequality,

$$\inf_{x\in X}\sup_{y\in Y} F(x,y) = \sup_{y\in Y} F(x_o,y) \ge F(x_o,y_0) \ge \inf_{x\in X} F(x,y_0) = \sup_{y\in Y}\inf_{x\in X} F(x,y),$$

and using the fact that by (5.96) the far left and the far right limit are equal, we see that these are, in fact, all equalities. Hence,

$$\sup_{y\in Y} F(x_o,y) = F(x_o,y_0) = \inf_{x\in X} F(x,y_0),$$

which implies that

$$F(x_o,y) \le \sup_{y\in Y} F(x_o,y) = F(x_o,y_0) = \inf_{x\in X} F(x,y_0) \le F(x,y_0), \quad \forall\, x\in X,\ y\in Y,$$

which means that (x_o,y_0) is a saddle point of F.

5.5.2 Proof of Lemma 5.3.1

(i) For the convexity of the value function, assume that F is convex in both variables x and y. Consider $y_i \in Y$, $i = 1,2$, and their convex combination $\lambda y_1 + (1-\lambda)y_2$, $\lambda \in [0,1]$. By the definition of $V(y_i)$ in terms of an infimum (see (5.36)), we see that for every $\epsilon > 0$ there exists a $x_i \in X$ such that

$$V(y_i) \le F(x_i,y_i) < V(y_i) + \epsilon, \quad i = 1,2. \tag{5.99}$$

Since F is convex in both variables,

$$\begin{aligned} F(\lambda x_1 + (1-\lambda)x_2, \lambda y_1 + (1-\lambda)y_2) &\le \lambda F(x_1,y_1) + (1-\lambda)F(x_2,y_2) \\ &\le \lambda V(y_1) + (1-\lambda)V(y_2) + \epsilon \end{aligned} \tag{5.100}$$

where for the last inequality, we used (5.99). Since

$$V(y) = \inf_{x\in X} F(x,y) \le F(x,y),$$

for every $(x,y) \in X \times Y$, clearly

$$V(\lambda y_1 + (1-\lambda)y_2) \le F(\lambda x_1 + (1-\lambda)x_2, \lambda y_1 + (1-\lambda)y_2).$$

Therefore, combining this with (5.100) leads to

$$V(\lambda y_1 + (1-\lambda)y_2) \le \lambda V(y_1) + (1-\lambda)V(y_2) + \epsilon$$

for every $\epsilon > 0$, and passing to the limit as $\epsilon \to 0^+$, yields the required convexity result.

To prove the first of (5.38), note that, by definition,

$$\begin{aligned} V^\star(y^\star) &= \sup_{y\in Y}(\langle y^\star, y\rangle_{Y^\star,Y} - V(y)) = \sup_{y\in Y}\Big(\langle y^\star, y\rangle_{Y^\star,Y} - \inf_{x\in X} F(x,y)\Big) \\ &= \sup_{(x,y)\in X\times Y}(\langle 0, x\rangle_{X^\star,X} + \langle y^\star, y\rangle_{Y^\star,Y} - F(x,y)) = F^\star(0, y^\star), \end{aligned}$$

where we also used the definition of

$$F^\star(x^\star, y^\star) = \sup_{(x,y)\in X\times Y}(\langle x^\star, x\rangle_{X^\star,X} + \langle y^\star, y\rangle_{Y^\star,Y} - F(x,y))$$

at $x^\star = 0$.

To prove the second of (5.38), note that by definition (see Definition 5.2.11 and Remark 5.2.12),

$$V^{\star\star}(0) = \sup_{y^\star\in Y^\star}(\langle y^\star, 0\rangle_{Y^\star,Y} - V^\star(y^\star)) = \sup_{y^\star\in Y^\star}(-V^\star(y^\star)) = \sup_{y^\star\in Y^\star}(-F^\star(0, y^\star)),$$

where for the last equality we used the first of (5.38).

(ii) We will use the notation F_x for the function $F_x : Y \to \mathbb{R}\cup\{+\infty\}$ defined by fixing the value of x, i. e., by $F_x(y) = F(x,y)$, and $F_x^\star : Y^\star \to \mathbb{R}\cup\{+\infty\}$, for its convex conjugate function. Note that by the definition of the Lagrangian function

$$F_x^\star(y^\star) = \sup_{y\in Y}(\langle y^\star, y\rangle_{Y^\star,Y} - F(x,y)) =: -L(x, y^\star),$$

while the bi-conjugate $F_x^{\star\star} : Y \to \mathbb{R}$ at $y = 0$ satisfies (using the definition of the bi-conjugate),

$$F_x^{\star\star}(0) = \sup_{y^\star\in Y^\star}(-F_x^\star(y^\star)) = \sup_{y^\star\in Y^\star}(L(x, y^\star)).$$

It is straightforward to see that the function $F_x : X \to \mathbb{R}\cup\{+\infty\}$ is proper convex and lower semicontinuous. Therefore, by Proposition 5.2.15, we have that $F_x^{\star\star}(0) = F_x(0)$, which leads to the conclusion that

$$\sup_{y^\star\in Y^\star}(L(x, y^\star)) = F(x, 0) = F(x).$$

We now recall the definition of $F^\star : X^\star \times Y^\star \to \mathbb{R} \cup \{+\infty\}$ and set $x^\star = 0$ to obtain

$$F^\star(0, y^\star) = \sup_{x\in X^\star} \sup_{y\in Y} (\langle y^\star, y\rangle_{Y^\star,Y} - F(x,y))$$
$$= \sup_{x\in X^\star} \sup_{y\in Y} (\langle y^\star, y\rangle_{Y^\star,Y} - F_x(y)) = \sup_{x\in X}(-L(x, y^\star)),$$

from which it follows that

$$-F^\star(0, y^\star) = \inf_{x\in X} L(x, y^\star).$$

5.5.3 Proof of relation (5.59)

Consider first an $x \in C$, i. e., $\varphi_i(x) \le 0$ for every $i = 1, \dots, n$. If $y_i^\star > 0$ for at least one $i = 1, \dots, n$, then the term $y_i y_i^\star$ is going to become large for $y_i > 0$, in which case, it definitely holds that $\varphi_i(x) \le 0 < y_i$, and the term $I_{C(y_i)}(x) = 0$, so $\sup_{y_i\in\mathbb{R}}(y_i y_i^\star - I_{C(y_i)}(x)) = +\infty$ in this case. If on the contrary $y_i^\star < 0$ for every $i = 1, \dots, n$, then each of the terms $y_i y_i^\star$ are going to become large for $y_i < 0$. We have that $\varphi_i(x) \le 0$, but if we take y_i too negative, then $\varphi_i(x) \le y_i$ will not hold, leading to $I_{C(y_i)}(x) = +\infty$. If on the other hand y_i is in the interval $\varphi_i(x) \le y_i \le 0$, then $I_{C(y_i)}(x) = 0$, and the major role in the maximization problem is played by $y_i^\star y_i$. We therefore conclude that $\sup_{y_i\in\mathbb{R}}(y_i y_i^\star - I_{C(y_i)}(x)) = y_i^\star \varphi_i(x)$ if $y_i^\star < 0$. Collecting all possible cases, we see that if $x \in C$, then $-L(x, y^\star) = +\infty$. Hence, $L(x, y^\star) = -\infty$ for $y^\star \in \mathbb{R}^n_+$, whereas $-L(x, y^\star) = -\varphi_0(x) + \sum_{i=1}^n y_i^\star \varphi_i(x)$ for $y^\star = (y_1^\star, \dots, y_n^\star) \in -\mathbb{R}^n_+$. By changing variables to $-y^\star \in \mathbb{R}^n_+$, we conclude that $L(x, y^\star) = \varphi_0(x) + \sum_{i=1} y_i^\star \varphi_i(x)$ for $y^\star = (y_1^\star, \dots, y_n^\star) \in \mathbb{R}^n_+$, whereas $L(x, y^\star) = -\infty$ for $y^\star \notin \mathbb{R}^n_+$.

Consider next an $x \notin C$. Then, $\varphi_i(x) > 0$ for every $i = 1, \dots, n$. If $y_i^\star > 0$ for at least one $i = 1, \dots, n$, then the corresponding term $y_i^\star y_i$ will grow for positive y_i. However, for large y_i, it is definitely true that $\varphi_i(x) < y_i$ so that $I_{C(y_i)}(x) = 0$, and the dominating term is $y_i^\star y_i$, so $\sup_{y_i\in\mathbb{R}}(y_i y_i^\star - I_{C(y_i)}(x)) = +\infty$. Suppose that $y_i^\star < 0$ for every $i = 1, \dots, n$. The term $y_i^\star y_i$ is going to become larger as y_i decreases. As $\varphi_i(x) > 0$, it is never true that $\varphi_i(x) \le y_i$ for $y_i < 0$, so the term $-I_{C(y_i)}(x) = -\infty$. So we must restrict ourselves to the smallest possible positive value allowed for y_i so that $\varphi_i(x) \le y_i$ holds. This is clearly $\varphi_i(x)$ so that $\sup_{y_i\in\mathbb{R}}(y_i y_i^\star - I_{C(y_i)}(x)) = y_i^\star \varphi_i(x)$. With arguments similar as above (i. e., changing variables to $-y^\star$), we find exactly the same result as in the case where $x \in C$.

6 The calculus of variations

An important part of nonlinear analysis is dedicated to the calculus of variations. The calculus of variations studies problems related to the minimization of integral functionals, defined in appropriate function spaces. It is a field that has been associated historically with many important applications in the physical sciences. Philosophical aspects aside, the calculus of variations is very useful, as important partial differential equations (PDEs) can be expressed as the first-order minimization conditions for suitable integral functionals. We may then infer qualitative or quantitative information of the behavior of the solutions of these equations from the theory of minimization of the corresponding functional. On the other hand, many important models, ranging from economics or decision science to engineering, signal and image processing etc., can be expressed in terms of integral functionals (see, e. g., [43] or [48]), so the study of minimization of integral functionals is a problem of interest in its own right (irrespective of its connections with PDEs). In this chapter, we present certain important aspects of the theory of the calculus of variations, starting with an investigation of the Dirichlet functional and its connection with the Poisson equation, then providing results concerning lower semicontinuity of general integral functionals, existence of minimizers, their connection with the solution of the Euler–Lagrange PDE and the further regularity of these minimizers. Having developed a general theory for integral functionals, we present some applications of the calculus of variations to the study of semilinear elliptic PDEs, involving the Laplace operator and quasilinear elliptic PDEs, involving the p-Laplace operator. Finally, we provide some important applications of the general theory of Γ-convergence, developed in Chapter 2, in the calculus of variations, and in particular in the study of homogenization problems and the behavior of the eigenvalues of the p-Laplacian as a function of the exponent p. There are many authoritative monographs dedicated to the calculus of variations and Γ-convergence (see, e. g., [61, 80, 83] and [12, 30, 62] on which our approach was based).

6.1 Motivation

One of the main motivations for the study of the calculus of variations is its connection with partial differential equations (PDEs). An important class of PDEs can be expressed as the first-order conditions for the minimization of integral functionals in Sobolev spaces, therefore, using the well established theory for minimization of functionals in Banach spaces, one may infer useful information concerning the solvability and the properties of the solutions of these PDEs. This sums up the general strategy of utilization of the methodology of the calculus of variations to the study of PDEs.

https://doi.org/10.1515/9783111333298-006

Example 6.1.1 (Laplace equation). Let $\mathcal{D} \subset \mathbb{R}^d$ be a bounded domain[1] with sufficiently smooth boundary $\partial\mathcal{D}$, let $f : \mathcal{D} \to \mathbb{R}$ be a given function, and for an unknown function $u : \mathcal{D} \to \mathbb{R}$, consider the Poisson equation

$$\begin{aligned} -\Delta u(x) &= f(x), \quad x \in \mathcal{D}, \\ u(x) &= 0, \quad x \in \partial\mathcal{D}. \end{aligned} \tag{6.1}$$

Let us also define the functional $F : X \to \mathbb{R}$, where $X = W_0^{1,2}(\mathcal{D})$, as

$$F(u) := \int_{\mathcal{D}} \left(\frac{1}{2}|\nabla u(x)|^2 - f(x)\,u(x)\right)dx, \quad \text{for every } u \in W_0^{1,2}(\mathcal{D}). \tag{6.2}$$

Using a simple extension of the arguments in Example 2.1.10, one can easily recognize (6.1) as the first-order condition for the minimization of the convex functional (6.2) in $W_0^{1,2}(\mathcal{D})$ (see also Example 2.2.14). Having at this point enough technical tools to treat the problem of minimizing F, and given the connection between solutions of the Poisson equation (6.1) and the minimizers of F, one is tempted to transfer this experience to the solvability and the properties of solutions for the PDE (6.1). Naturally, we may reverse our reasoning and, given knowledge concerning the solvability of the PDE (6.1), try to infer knowledge on the properties of minima of the functional F. ◁

The discussion in Example 6.1.1 can be extended to other equations, and importantly to nonlinear elliptic equations. The following example motivates this:

Example 6.1.2 (Nonlinear elliptic equations). Within the general framework of Example 6.1.1, we now start from the opposite end and consider a functional $F : X \to \mathbb{R}$, where X is an appropriate Sobolev space on $\mathcal{D}$, defined as

$$F(u) := \int_{\mathcal{D}} f(x, u(x), \nabla u(x))dx, \tag{6.3}$$

for every $u \in X$, where $f : \mathcal{D} \times \mathbb{R} \times \mathbb{R}^d \to \mathbb{R}$ is a function assumed to be sufficiently smooth in all its variables. We will use the notation $f(x, y, z)$ to denote the values of the function f, where $x \in \mathcal{D} \subset \mathbb{R}^d$, $y = u(x)$ is the value of the function $u : \mathcal{D} \to \mathbb{R}$ at the selected point $x \in \mathcal{D}$ and $z = \nabla u(x) \in \mathbb{R}^d$, the gradient of u at the selected point $x \in \mathcal{D}$. We will also denote by $\frac{\partial}{\partial y} f : \mathcal{D} \times \mathbb{R} \times \mathbb{R}^d \to \mathbb{R}$ and $\frac{\partial}{\partial z} f = \nabla_z f : \mathcal{D} \times \mathbb{R} \times \mathbb{R}^d \to \mathbb{R}^d$, the derivatives of the function f with respect to the variables y and z, respectively.

We now perform formal calculations in order to obtain the Gâteaux derivative of F and try to characterize a critical point of F. Choosing any function $v \in X$, standard calculations yield

1 Recall that a domain is an open connected set.

$$\frac{F(u+\epsilon v)-F(u)}{\epsilon}$$
$$=\int_{\mathcal{D}}\left(\frac{\partial}{\partial y}f(x,u(x),\nabla u(x))\,v(x)+\frac{\partial}{\partial z}f(x,u(x),\nabla u(x))\cdot\nabla v(x)\right)dx+O(\epsilon),$$

where we have used the Taylor expansion of the function f, and, by $O(\epsilon)$, we denote terms of order ϵ or higher. Integrating by parts and assuming that v vanishes on $\partial\mathcal{D}$, we obtain

$$\frac{F(u+\epsilon v)-F(u)}{\epsilon}$$
$$=\int_{\mathcal{D}}\left[\frac{\partial}{\partial y}f(x,u(x),\nabla u(x))-\nabla\cdot\left(\frac{\partial}{\partial z}f(x,u(x),\nabla u(x))\right)\right]v(x)dx+O(\epsilon).$$

Taking the limit as $\epsilon\to 0$, we formally obtain

$$DF(u;v)$$
$$:=\int_{\mathcal{D}}\left[\frac{\partial}{\partial y}f(x,u(x),\nabla u(x))-\nabla\cdot\left(\frac{\partial}{\partial z}f(x,u(x),\nabla u(x))\right)\right]v(x)dx, \tag{6.4}$$

so by the argumentation of Theorem 2.2.13, and the fact that v is arbitrary, the critical point(s) u of F satisfy the equation

$$\frac{\partial}{\partial y}f(x,u(x),\nabla u(x))-\nabla\cdot\left(\frac{\partial}{\partial z}f(x,u(x),\nabla u(x))\right)=0. \tag{6.5}$$

This is a PDE, called the Euler–Lagrange equation. Depending on the choice of the function f, we may obtain a large variety of nonlinear PDEs as the Euler–Lagrange equation. Of course, the above formal calculations should be made rigorous, especially the parts in which interchange of limits and integrals are involved, and this requires specific smoothness conditions on the function f and the proper choice for the function space X. Then, once the PDE in consideration has been recognized as the Euler–Lagrange equation for a functional F, it remains to check whether the functional in question indeed acquires a minimum u_o, which must be the solution of the relevant Euler–Lagrange equation, hence of the original PDE. ◁

This chapter addresses these questions, but in a manner which is rather more general in scope, than the original PDE motivation. In particular, it addresses the problem of well posedness of minimization problems for a wide class of integral functionals $F: X\to\mathbb{R}$ of the form

$$F(u):=\int_{\mathcal{D}}f(x,u(x),\nabla u(x))dx, \tag{6.6}$$

where $\mathcal{D} \subset \mathbb{R}^d$ is a bounded domain of sufficiently smooth boundary; $u : \mathcal{D} \to \mathbb{R}$ is a function such that the integral (6.6) is well defined; $\nabla u = (\frac{\partial u}{\partial x_1}, \dots, \frac{\partial u}{\partial x_d})$ its gradient; $f : \mathcal{D} \times \mathbb{R} \times \mathbb{R}^d \to \mathbb{R}$ is a given function, focusing on the case where $X = W^{1,p}(\mathcal{D})$ or $X = W_0^{1,p}(\mathcal{D})$. As already mentioned, functionals of the form (6.6) are connected with nonlinear PDEs, but their study is of interest in its own right, as integral functionals of this kind arise in a number of interesting applications.

6.2 Warm up: variational theory of the Laplacian

We start our treatment of the calculus of variations with the study of the Dirichlet and the Rayleigh quotient functionals, which are closely connected to the Laplacian and with linear problems, such as the Poisson equation and the Laplace eigenvalue problem. Even though, these problems are linear problems, they are not to be frowned upon in a book on nonlinear analysis, as they provide important background and an indispensable toolbox in the study of nonlinear problems.

6.2.1 The Dirichlet functional and the Poisson equation

Let $\mathcal{D} \subset \mathbb{R}^d$ be a bounded domain (i. e., an open and connected set) with smooth enough boundary.[2]

Proposition 6.2.1. *The Dirichlet functional $F_\mathcal{D} : W_0^{1,2}(\mathcal{D}) \to \mathbb{R}$, defined by $F_\mathcal{D}(u) = \frac{1}{2}\int_\mathcal{D} |\nabla u(x)|^2 dx$, is strictly convex and weakly sequentially lower semicontinuous.*

Proof. Convexity follows easily by the convexity of the function $z \mapsto |z|^2$, $z \in \mathbb{R}^d$, and strict convexity follows by the identity

$$\begin{aligned}
&\left|t\nabla u_1(x) + (1-t)\nabla u_2(x)\right|^2 + t(1-t)\left|\nabla u_1(x) - \nabla u_2(x)\right|^2 \\
&\quad = t\left|\nabla u_1(x)\right|^2 + (1-t)\left|\nabla u_2(x)\right|^2,
\end{aligned}$$

which is valid a. e. in $\mathcal{D}$ for every $t \in [0,1]$ upon integration.

Weak lower semicontinuity is extremely simple for this case (because of its special form), so one could argue directly by noting that if $u_n \rightharpoonup u$ in $W_0^{1,2}(\mathcal{D})$, then it is bounded, and since $W_0^{1,2}(\mathcal{D})$ is a Hilbert space, there exists a subsequence (denoted the same) such that $u_n \rightharpoonup u$ and $\nabla u_n \rightharpoonup \nabla u$ in $L^2(\mathcal{D})$ for some $u \in W_0^{1,2}(\mathcal{D})$. Along this sequence write (for a. e. $x \in \mathcal{D}$),

$$\left|\nabla u_n(x)\right|^2 = \left|\nabla u_n(x) - \nabla u(x) + \nabla u(x)\right|^2$$

2 Typically Lipschitz boundary is sufficient, unless more regularity is required, will be stated explicitly.

$$= |\nabla u(x)|^2 + 2(\nabla u_n(x) - \nabla u(x)) \cdot \nabla u(x) + |\nabla u_n(x) - \nabla u(x)|^2$$
$$\geq |\nabla u(x)|^2 + 2(\nabla u_n(x) - \nabla u(x)) \cdot \nabla u(x),$$

so upon integration,

$$F_{\mathcal{D}}(u_n) \geq F_{\mathcal{D}}(u) + \int_{\mathcal{D}} (\nabla u_n(x) - \nabla u(x)) \cdot \nabla u(x) dx. \tag{6.7}$$

Since $\nabla u_n \rightharpoonup \nabla u$ in $L^2(\mathcal{D})$, we have that $\lim_n \int_{\mathcal{D}} (\nabla u_n(x) - \nabla u(x)) \cdot \nabla u(x) dx \to 0$, so taking the limit inferior on both sides of (6.7), we conclude that $\liminf_n F_{\mathcal{D}}(u_n) \geq F_{\mathcal{D}}(u)$, hence, the required lower semicontinuity result. □

Remark 6.2.2. Clearly, the strategy used for the above lower semicontinuity result cannot be easily extended to more general integral functionals, however, convexity may prove helpful. In this general approach, we will employ the Fatou lemma for proving lower semicontinuity, upon first passing from weak to strong convergence using the Mazur lemma, and then extracting an appropriate subsequence, which converges a. e. We will study this approach in detail in Section 6.3.

We now consider the problem of minimization of perturbations of the Dirichlet functional over $X = W_0^{1,2}(\mathcal{D})$.

Proposition 6.2.3. *Consider any $f \in L^2(\mathcal{D})$, and define the functional $F : W_0^{1,2}(\mathcal{D}) \to \mathbb{R}$ by*

$$F(u) := F_{\mathcal{D}}(u) - \int_{\mathcal{D}} f(x) u(x) dx. \tag{6.8}$$

This functional has a unique minimizer in $W_0^{1,2}(\mathcal{D})$, which can be obtained as the solution of the Poisson equation

$$\begin{aligned} -\Delta u &= f, \quad \text{in } \mathcal{D}, \\ u &= 0, \quad \text{on } \partial\mathcal{D}, \end{aligned} \tag{6.9}$$

understood in the weak sense, i. e.,

$$\int_{\mathcal{D}} \nabla u(x) \cdot \nabla v(x) dx = \int_{\mathcal{D}} f(x) v(x) dx, \quad \forall v \in W_0^{1,2}(\mathcal{D}). \tag{6.10}$$

Proof. The existence of a minimizer follows by the arguments of the direct method of the calculus of variations (Theorem 2.2.5). Consider a minimizing sequence $(u_n)_{n \in \mathbb{N}}$ for F, which is bounded in $W_0^{1,2}(\mathcal{D})$ because of the Poincaré inequality. Indeed, along the minimizing sequence,

$$F(u_n) = \int_{\mathcal{D}} |\nabla u_n(x)|^2 dx - \int_{\mathcal{D}} f(x) u_n(x) dx \leq c \tag{6.11}$$

for some appropriate constant c, and since

$$\begin{aligned} F(u_n) &\geq \int_{\mathcal{D}} |\nabla u_n(x)|^2 dx - \left| \int_{\mathcal{D}} f(x) u_n(x) dx \right| \\ &\geq \int_{\mathcal{D}} |\nabla u_n(x)|^2 dx - \left(\int_{\mathcal{D}} |u_n(x)|^2 dx \right)^{1/2} \left(\int_{\mathcal{D}} |f(x)|^2 dx \right)^{1/2} \\ &\geq \int_{\mathcal{D}} |\nabla u_n(x)|^2 dx - \frac{\epsilon^2}{2} \int_{\mathcal{D}} |u_n(x)|^2 dx - \frac{1}{2\epsilon^2} \int_{\mathcal{D}} |f(x)|^2 dx \\ &\geq \left(\mathbf{c}_{\mathcal{P}}^2 - \frac{\epsilon^2}{2} \right) \int_{\mathcal{D}} |u_n(x)|^2 dx - \frac{1}{2\epsilon^2} \int_{\mathcal{D}} |f(x)|^2 dx, \end{aligned} \tag{6.12}$$

where $\epsilon > 0$ is arbitrary, and we used subsequently the Cauchy–Schwarz inequality and the Poincaré inequality. Combining (6.11) and (6.12), we see that

$$\left(\mathbf{c}_{\mathcal{P}}^2 - \frac{\epsilon^2}{2} \right) \int_{\mathcal{D}} |u_n(x)|^2 dx \leq c + \frac{1}{2\epsilon^2} \int_{\mathcal{D}} |f(x)|^2 dx,$$

and choosing ϵ so that $\frac{\epsilon^2}{2} < \mathbf{c}_{\mathcal{P}}^2$, we have that $(u_n)_{n\in\mathbb{N}}$ is bounded in $L^2(\mathcal{D})$, and returning to (6.11), we conclude that $(\nabla u_n)_{n\in\mathbb{N}}$ is also bounded in $L^2(\mathcal{D})$. Therefore, $(u_n)_{n\in\mathbb{N}}$ is bounded in $W_0^{1,2}(\mathcal{D})$.

Since $(u_n)_{n\in\mathbb{N}}$ is bounded in $W_0^{1,2}(\mathcal{D})$, which is a Hilbert space (hence reflexive), there exists a subsequence $(u_{n_k})_{k\in\mathbb{N}}$, and a $u \in W_0^{1,2}(\mathcal{D})$, such that $u_{n_k} \rightharpoonup u$ in $W_0^{1,2}(\mathcal{D})$. This weak limit is the required minimizer, by the weak lower sequential semicontinuity of the functional F (which is a linear perturbation of F_D, whose weak lower sequential semicontinuity has been established in Proposition 6.2.1.

The Gâteaux derivative of the functional F is $DF(u) = -\Delta u - f$ (see, e. g., Example 2.1.10). So the first-order condition becomes (6.9) or its equivalent weak form (6.10). The equivalence between the Poisson equation and its weak form arises by multiplying (6.9) with a sufficiently smooth test function, integrating by parts and recalling the density of smooth functions in the Sobolev space $W_0^{1,2}(\mathcal{D})$ (recall Definition 1.5.6). Uniqueness of the minimizer follows by strict convexity, however, this does not necessarily imply the uniqueness of solutions of the Poisson equation (as a critical point is not necessarily a minimum). □

Having obtained the results concerning the existence of a unique minimizer to the functional F defined in (6.8) and its connection with the Poisson equation (6.9), we may obtain important information on the Poisson PDE.

Proposition 6.2.4. *The Poisson equation* (6.9) *admits a unique solution* $u \in W_0^{1,2}(\mathcal{D})$ *for every* $f \in L^2(\mathcal{D})$. *Upon defining the linear continuous operator* $\mathsf{A} := -\Delta : W_0^{1,2}(\mathcal{D}) \to (W_0^{1,2}(\mathcal{D}))^\star = W^{-1,2}(\mathcal{D})$, *by*

$$\langle \mathsf{A}u, v\rangle_{L^2(\mathcal{D})} = \langle -\Delta u, v\rangle_{L^2(\mathcal{D})} := \langle \nabla u, \nabla v\rangle_{L^2(\mathcal{D})}, \quad \forall\, v \in W_0^{1,2}(\mathcal{D}),$$

it holds that the inverse operator $(-\Delta)^{-1} : L^2(\mathcal{D}) \to L^2(\mathcal{D})$ is well defined, single valued, self adjoint, positive definite, and compact.

Proof. The existence of solutions to the Poisson equation (6.9) has already been shown in Proposition 6.2.3. To show uniqueness, assume two solutions, u_1, u_2 of (6.9), and note that $w := u_1 - u_2$ solves $\Delta w = 0$ in $\mathcal{D}$ with homogeneous Dirichlet conditions. Take the weak form of (6.9) with w as a test function. An integration by parts implies that $\|\nabla w\|_{L^2(\mathcal{D})} = 0$, and using the Poincaré inequality and the argumentation of Example 1.5.14, we conclude that $w = 0$, hence, we have uniqueness.

The inverse operator $\mathsf{T} := (-\Delta)^{-1}$ is connected to the variational solution of the Poisson problem (6.9), in the sense that for any $f \in L^2(\mathcal{D})$; we define $\mathsf{T}f = u$, where u is the solution of problem (6.9) for the right-hand side f. By the above arguments, T is a single-valued linear operator from $L^2(\mathcal{D})$ to $W_0^{1,2}(\mathcal{D}) \overset{c}{\hookrightarrow} L^2(\mathcal{D})$. This operator is bounded, as one can easily infer by using the weak form (6.10) setting as test function $v = \mathsf{T}f$, and using the Cauchy–Schwarz inequality on the right-hand side and the Poincaré inequality on the left-hand side (details are left as an exercise). Using the boundedness of this operator and the compact embedding of $W_0^{1,2}(\mathcal{D}) \overset{c}{\hookrightarrow} L^2(\mathcal{D})$ (guaranteed by the Rellich–Kondrachov embedding, see Theorem 1.5.11), we conclude that $\mathsf{T} : L^2(\mathcal{D}) \to L^2(\mathcal{D})$ is a compact operator. The remaining properties follow easily from the weak form of (6.9). □

The compactness of the inverse operator $(-\Delta)^{-1} : L^2(\mathcal{D}) \to L^2(\mathcal{D})$ is of fundamental importance, as one may now resort to the well understood Fredholm theory for compact operators in Hilbert spaces and their spectral properties (see Theorem 1.3.13, in conjunction with Theorem 1.3.12) in order to fully specify the spectral properties of the Laplacian operator $-\Delta$. This follows from the simple observation that the eigenvalues of the operator $\mathsf{A} := -\Delta$ are clearly related to the eigenvalues of $\mathsf{T} := (-\Delta)^{-1}$, since any $\lambda \in \mathbb{R}$ is an eigenvalue of A if and only if λ^{-1} is an eigenvalue of T, and since T is a compact operator, there exists abundant useful information concerning its eigenvalues (see e. g., [33, 76]).

We close this section with some important properties of the solution of the Poisson equation, related to the maximum principle for the Laplace operator (see, e. g., [42, 76]).

Proposition 6.2.5 (Weak maximum principle).

(i) *Suppose $u \in W^{1,2}(\mathcal{D})$ satisfies $-\Delta u \geq 0$ in the weak sense, i. e., that*

$$\langle (-\Delta)u, v\rangle_{(W_0^{1,2}(\mathcal{D}))^*, W_0^{1,2}(\mathcal{D})} \geq 0, \quad \forall\, v \in W_0^{1,2}(\mathcal{D}),\ v \geq 0 \ a.\,e.$$

If $u^- := \max(-u, 0) \in W_0^{1,2}(\mathcal{D})$, then $u \geq 0$ a. e.

(ii) *Suppose $u \in W^{1,2}(\mathcal{D})$ satisfies $-\Delta u \leq 0$ in the weak sense, i. e., that*

$$\langle(-\Delta)u, v\rangle_{(W_0^{1,2}(\mathcal{D}))^\star, W_0^{1,2}(\mathcal{D})} \leq 0, \quad \forall\, v \in W_0^{1,2}(\mathcal{D}),\ v \geq 0 \text{ a. e.}$$

If $u^+ := \max(u, 0) \in W_0^{1,2}(\mathcal{D})$, then $u \leq 0$ a. e.

Proof. We only prove (i) as (ii) follows from (i) by changing u to $-u$.

Recall (Example 1.5.2) that if $u \in W^{1,2}(\mathcal{D})$, then $u^- \in W^{1,2}(\mathcal{D})$ also, and that $\nabla u^- = -\nabla u \mathbf{1}_{u<0} + 0\mathbf{1}_{u\geq 0}$. As $u^- \geq 0$, we have that $\int_{\mathcal{D}}(-\Delta u)(-u^-)dx \leq 0$, so by integration by parts

$$0 \geq \int_{\mathcal{D}} \nabla u \cdot \nabla(-u^-)dx = \int_{\mathcal{D}} |\nabla u^-|^2 dx.$$

Therefore, $\nabla u^- = 0$ a. e. in $\mathcal{D}$, and since $u^- \in W_0^{1,2}(\mathcal{D})$, we conclude that $u^- \equiv 0$ a.e in $\mathcal{D}$, hence, $u \geq 0$ a. e. in $\mathcal{D}$. □

A useful corollary of the weak maximum principle is the following comparison principle:

Proposition 6.2.6 (Comparison principle). *Let $f_1, f_2 \in L^2(\mathcal{D})$, and $u_i \in W_0^{1,2}(\mathcal{D})$ be the weak solution of*

$$-\Delta u_i = f_i, \quad i = 1, 2.$$

If $f_1 \geq f_2$ a. e. in $\mathcal{D}$, then $u_1 \geq u_2$ a. e. in $\mathcal{D}$.

Proof. Simply set $u = u_1 - u_2$ and $f = f_1 - f_2 \geq 0$ a. e. in $\mathcal{D}$. The claim can be generalized if $f_1, f_2 \in (W_0^{1,2}(\mathcal{D}))^\star = W^{-1,2}(\mathcal{D})$. □

More regular solutions satisfy the strong maximum principle [42].

Proposition 6.2.7 (Strong maximum principle).

(i) *Suppose $u \in W^{1,2}(\mathcal{D}) \cap C(\mathcal{D})$ satisfies $-\Delta u \geq 0$ in the weak sense, i. e., that*

$$\langle(-\Delta)u, v\rangle_{(W_0^{1,2}(\mathcal{D}))^\star, W_0^{1,2}(\mathcal{D})} \geq 0, \quad \forall\, v \in W_0^{1,2}(\mathcal{D}),\ v \geq 0.$$

If $u^- = \max(-u, 0) \in W_0^{1,2}(\mathcal{D})$ and $u \not\equiv 0$, then $u > 0$.

(ii) *Suppose $u \in W^{1,2}(\mathcal{D}) \cap C(\mathcal{D})$ satisfies $-\Delta u \leq 0$ in the weak sense, i. e., that*

$$\langle(-\Delta)u, v\rangle_{(W_0^{1,2}(\mathcal{D}))^\star, W_0^{1,2}(\mathcal{D})} \leq 0, \quad \forall\, v \in W_0^{1,2}(\mathcal{D}),\ v \geq 0.$$

If $u^+ = \max(u, 0) \in W_0^{1,2}(\mathcal{D})$ and $u \not\equiv 0$, then $u < 0$.

Proof. We only prove (i) as (ii) follows from (i) changing u to $-u$. The assumption that $\mathcal{D}$ is connected plays an important role in the proof. We recall the fundamental topological notion of connectedness: Since $\mathcal{D}$ is connected, the only non empty subset of $\mathcal{D}$ that is at the same time open and closed is $\mathcal{D}$ itself (see, e. g., [70]).

By the weak maximum principle (Proposition 6.2.5), it holds that $u \geq 0$ in $\mathcal{D}$. Define the set $A = \{x \in \mathcal{D} : u(x) > 0\}$, which since $u \in C(\mathcal{D})$ and $u \not\equiv 0$ is open and non empty. If we show that is it also closed, then $A = \mathcal{D}$, and we are done.

Consider the function $v : \mathcal{D} \to \mathbb{R}$, defined by $v(x) = |x|^{-\beta} - R^{-\beta}$. It is straightforward (but requires tedious algebra) to show that choosing the parameter β appropriately, the function v satisfies $-\Delta v \leq 0$ in the annulus $A_{0,\rho,R} := \{x \in \mathbb{R}^d : \rho < |x| < R\}$, and vanishes on its outer boundary.

To show that A is closed, consider any sequence $(x_n)_{n\in\mathbb{N}} \subset A$ such that $x_n \to x_o \in \mathcal{D}$. Choose R such that $B(x_o, R) \subset \mathcal{D}$, and choose N large enough so that $|x_N - x_o| < R$. Since $x_N \in A$, we have that $u(x_N) > 0$, which implies that we may find $\rho \in (0, R)$ and $\epsilon > 0$ such that $u(x) \geq \epsilon$ for $|x - x_N| = \rho$. Consider the annulus $A_{x_N,\rho,R} = \{x \in \mathbb{R}^d : \rho < |x - x_N| < R\}$ centered at x_N, and using the function v defined above shifted by x_N, define the function $w : A_{x_N,\rho,R} \to \mathbb{R}$ by $w(x) = u(x) - \epsilon\rho^\beta v(x - x_N)$, which satisfies $w \in W^{1,2}(A_{x_N,\rho,R}) \cap C(\overline{A_{x_N,\rho,R}})$ and $-\Delta w \geq 0$. By the properties of v, and since $u \geq 0$, we see that $w(x) \geq 0$ for $|x - x_N| = R$. Hence, $w^-(x) = 0$ on the outer boundary of $A_{x_N,\rho,R}$, while $w \geq 0$ on $A_{x_N,\rho,R}$. Similarly for the inner boundary. Applying the weak maximum principle for w in $A_{x_N,\rho,R}$, we conclude that $w(x) \geq 0$ for every $x \in A_{x_N,\rho,R}$. Therefore, $u(x_o) \geq \epsilon\rho^\beta v(x_o - x_N) > 0$ so that $x \in A$. □

Example 6.2.8. Minor modifications of the above arguments can be used to establish similar results for the equation

$$\begin{aligned} -\Delta u + \lambda u &= f, && \text{in } \mathcal{D}, \\ u &= 0, && \text{on } \partial\mathcal{D}, \end{aligned}$$

for $\lambda > 0$, or even for $\lambda > -\lambda_1$ where $\lambda_1 = \inf_{u \in W_0^{1,2}} \frac{\|\nabla u\|_{L^2(\mathcal{D})}}{\|u\|_{L^2(\mathcal{D})}}$, the first eigenvalue of $-\Delta$. This equation is related to the minimization of the integral functional

$$F(u) = \frac{1}{2}\int_{\mathcal{D}} |\nabla u(x)|^2 dx + \frac{\lambda}{2}\int_{\mathcal{D}} u(x)^2 dx - \int_{\mathcal{D}} f(x)u(x)dx,$$

on $W_0^{1,2}(\mathcal{D})$. The details are left as an exercise. ◁

The strong maximum principle provides important information on solutions of either the Poisson equation or of inequalities related to the Poisson equation but requires the additional property that $u \in C(\mathcal{D})$ (or even in some cases that $u \in C(\overline{\mathcal{D}})$), which we cannot know *a priori*, as all we have established so far concerning the solvability of the problem in Proposition 6.2.4 is that $u \in W_0^{1,2}(\mathcal{D})$. However, as we shall see in the next section, solutions of Poisson type equations enjoy important regularity properties.

6.2.2 Regularity properties for the solutions of Poisson type equations

Using the variational formulation, we have shown the existence of weak solutions for equations of the form

$$-\Delta u + \lambda u = f, \tag{6.13}$$

with Dirichlet boundary conditions for suitable values of λ (see Example 6.2.8). We now consider the question of regularity, i. e., can u get any better than $W_0^{1,2}(\mathcal{D})$? The answer is yes, as long as f enjoys sufficient regularity itself.

We start with the following important property of the Laplace operator, which essentially states that if the Laplacian of a function is k times differentiable (in the weak sense) the function itself gains two more weak derivatives, i. e., is $k+2$ times differentiable.

Proposition 6.2.9. *Suppose that $\mathcal{D} \subset \mathbb{R}^d$ is open and bounded and $u \in W_{\text{loc}}^{1,2}(\mathcal{D})$.*
(i) *If $\Delta u \in W_{\text{loc}}^{k,2}(\mathcal{D})$ for some $k > 0$, then $u \in W_{\text{loc}}^{k+2,2}(\mathcal{D})$.*
(ii) *If $\Delta u \in C^\infty(\mathcal{D})$, then $u \in C^\infty(\mathcal{D})$.*

Proof. The idea of the proof is sketched. To prove (i), we first note that this is true in the whole of $\mathbb{R}^d$, a fact that can be easily shown using the Fourier transform. We then use a localization (or rather extension) argument. To this end, consider any choice of sets $\mathcal{D}_o$ and $\mathcal{D}_o'$ such that $\mathcal{D}_o' \subset\subset \mathcal{D}_o \subset\subset \mathcal{D}$, and consider a function $\psi \in C_c^\infty(\mathbb{R}^d)$ such that $\psi \equiv 1$ on $\mathcal{D}_o'$ and $\operatorname{supp}\psi \subset \mathcal{D}_0$. Define the function $v = \psi u$, which is a function on $\mathbb{R}^d$. One may calculate $\Delta v = \Delta(\psi u) = \psi\Delta u + u\Delta\psi + 2\nabla u \cdot \nabla\psi \in W_{\text{loc}}^{k,2}(\mathcal{D})$ by the properties of Δu, u, and ψ (and an integration by parts argument). But then by the first step $v = \psi u \in W^{k+2,2}(\mathbb{R}^d)$, and that combined with the fact that $\mathcal{D}_o$ and $\mathcal{D}_o'$ are arbitrary, we get that $u \in W_{\text{loc}}^{k+2,2}(\mathcal{D})$. To prove (ii), we use an induction argument and the fact that $C^\infty(\mathcal{D}) = \bigcap_{m>0} W^{m,2}(\mathcal{D})$. □

The above result can be readily used to show for example that if $u \in W_{\text{loc}}^{k,2}(\mathcal{D})$ is a solution of (6.13) for $f \in W_{\text{loc}}^{k,2}(\mathcal{D})$, then $u \in W_{\text{loc}}^{k+2,2}(\mathcal{D})$. A similar argument as the above allows us to conclude that for any solution of (6.13), if $f \in W_{\text{loc}}^{m,p}(\mathcal{D})$ ($f \in C^\infty(\mathcal{D})$, respectively) and $u \in W_{\text{loc}}^{n,p}(\mathcal{D})$ for some $m \geq 0$, $n \in \mathbb{Z}$, $p \in (1,\infty)$, then $u \in W_{\text{loc}}^{m+2,p}(\mathcal{D})$ ($u \in C^\infty(\mathcal{D})$ respectively).

The following theorem, in the spirit of a more general result due to Stampacchia (see, e. g., [132]) provides L^p estimates for the solution of (6.13):

Theorem 6.2.10 (L^p estimates). *Assume that $d > 2$ and that $f \in L^p(\mathcal{D})$ for some $p > \frac{d}{2}$. Then, any weak solution of* (6.13)*, with $\lambda > 0$, is in $L^\infty(\mathcal{D})$ and $\|u\|_{L^\infty(\mathcal{D})} \leq c\|f\|_{L^p(\mathcal{D})}$ for some constant c (which does not depend on u and f).*

Proof. To show that $u \in L^{\infty}(\mathcal{D})$, it suffices to show that $|u| \leq c$, a. e., for some appropriate constant $c > 0$. This in turn is equivalent to showing that there exists a constant $c > 0$ such that $\mu_{\mathcal{L}_d}(\{x \in \mathcal{D} : |u(x)| > c\}) = 0$. The proof proceeds in 6 steps.

1. For any $k \in \mathbb{R}$, let us define $A_k := \{x \in \mathcal{D} : |u(x)| > k\}$, and set $\psi(k) := \mu_{\mathcal{L}_d}(A_k) =: |A_k|$. Clearly, for any $k_1 \leq k_2$, we have that $A_{k_2} \subset A_{k_1}$ so that $\psi(k_2) \leq \psi(k_1)$, i. e., ψ is nonincreasing. We will show that there exists a k^* such that $\psi(k^*) = 0$.

2. We recall an inequality due to Stampacchia according to which if $\phi : \mathbb{R}_+ \to \mathbb{R}_+$ is a nonincreasing function such that

$$\phi(k_2) \leq c\frac{\phi(k_1)^{\delta}}{(k_2 - k_1)^{\gamma}}, \quad \text{for every } 0 < k_1 < k_2, \tag{6.14}$$

where $\delta > 1$, $\gamma > 0$, then there exists $k^\star > 0$ such that $\phi(k^\star) = 0$, and in fact,

$$\phi(k^\star) = 0, \quad \text{for } k^\star = c^{\frac{1}{\gamma}}\phi(0)^{\frac{\delta-1}{\gamma}}2^{\frac{\delta}{\delta-1}}. \tag{6.15}$$

The proof of this inequality proceeds by considering the sequence $k_n = k^\star(1 - 2^{-n})$, for which one can show by induction that $\phi(k_n) \leq \psi(0)2^{-\frac{\gamma}{(\delta-1)}n}$, and then using monotonicity, we get

$$0 \leq \phi(k^\star) \leq \liminf_n \phi(k_n) \leq \lim_n \phi(0)^{\delta-1}2^{-\frac{\gamma}{(\delta-1)}n} = 0,$$

and the result follows.

3. We will try to show that the function ψ defined in step 1 satisfies a Stampacchia, like inequality of the form (6.14), so that applying the results of step 2, we obtain by (6.15) that ψ vanishes for some sufficiently large k^*.

To show that ψ satisfies an inequality of the form (6.14), we proceed as follows: We define the cutoff function $G_k(s) = (s-k)\mathbf{1}_{s\geq k} + (s+k)\mathbf{1}_{s\leq -k}$ and consider the composition $v_k = G_k \circ u$, which vanishes on A_k^c. This function is the composition of a globally Lipschitz function with a function in $W_0^{1,2}(\mathcal{D})$, hence, $v_k \in W_0^{1,2}(\mathcal{D})$ with $\nabla v_k = \mathbf{1}_{A_k}\nabla u$. We will use v_k as test function in the weak formulation of (6.13) in order to gain information on the behavior of the solution u of the set A_k.

Since $v_k \equiv 0$ on A_k^c and

$$\int_{\mathcal{D}} \nabla u(x) \cdot \nabla v_k(x)dx = \int_{A_k} \nabla u(x) \cdot \nabla u(x)dx = \int_{A_k} \nabla v_k(x) \cdot \nabla v_k(x)dx,$$

the weak form of (6.13) with the chosen test function yields

$$\int_{A_k} |\nabla v_k(x)|^2 dx + \lambda \int_{A_k} u(x)v_k(x)dx = \int_{A_k} f(x)v_k(x)dx. \tag{6.16}$$

We observe that

$$\int_{A_k} u(x)v_k(x)dx = \int_{\{u\leq -k\}} u(x)(u(x)+k)dx + \int_{\{u\geq k\}} u(x)(u(x)-k)dx \geq 0,$$

so since $\lambda > 0$, the weak form (6.16) implies

$$\int_{A_k} |\nabla v_k(x)|^2 dx \leq \int_{A_k} f(x)v_k(x)dx. \tag{6.17}$$

We will use the estimate (6.17) to obtain the required estimates for the measure of the set A_k.

4. Since we have information on the value of v_k on A_k but not on ∇v_k, we will try to substitute the term involving the integral of $|\nabla v_k|^2$ on A_k in (6.17) with a term involving the integral of a suitable power of v_k on the same set. To this end, we will use the Sobolev embedding Theorem 1.5.11 according to which $L^{\mathfrak{s}_2}(\mathcal{D}) \hookrightarrow W_0^{1,2}(\mathcal{D})$ for $\mathfrak{s}_2 = \frac{2d}{d-2}$. Hence, there exists a constant $c_1 > 0$ such that $\|v\|_{L^{\mathfrak{s}_2}(\mathcal{D})} \leq c_1\|v\|_{W_0^{1,2}(\mathcal{D})}$, which upon rearrangement yields

$$\int_{A_k} |\nabla v_k(x)|^2 dx \geq c_2\Big(\int_{A_k} |v_k(x)|^{\mathfrak{s}_2} dx\Big)^{2/\mathfrak{s}_2} \tag{6.18}$$

for a suitable constant $c_2 > 0$. Combining (6.18) with (6.17), we have that

$$\Big(\int_{A_k} |v_k(x)|^{\mathfrak{s}_2} dx\Big)^{2/\mathfrak{s}_2} \leq c_3 \int_{A_k} f(x)v_k(x)dx \tag{6.19}$$

for a suitable constant $c_3 > 0$. Estimate (6.19) involves only v_k, but it would be better suited for our purposes if it involved only the $L^{\mathfrak{s}_2}(A_k)$ norm of v_k. To manage that, we use Hölder's inequality on the right-hand side of (6.19) to retrieve the $L^{\mathfrak{s}_2}(A_k)$ norm for v_k. Using this strategy in (6.19) and denoting by $\mathfrak{s}_2^\star = \frac{2d}{d+2}$ the conjugate exponent for $\mathfrak{s}_2$, we have that

$$\Big(\int_{A_k} |v_k(x)|^{\mathfrak{s}_2} dx\Big)^{2/\mathfrak{s}_2} \leq c_3\Big(\int_{A_k} |f(x)|^{\mathfrak{s}_2^\star}\Big)^{1/\mathfrak{s}_2^\star}\Big(\int_{A_k} |v_k(x)|^{\mathfrak{s}_2} dx\Big)^{1/\mathfrak{s}_2},$$

where upon dividing both sides with the $L^{\mathfrak{s}_2}(A_k)$ norm for v_k, and raising to the power $\mathfrak{s}_2$, we obtain the estimate

$$\int_{A_k} |v_k(x)|^{\mathfrak{s}_2} dx \leq c_4\Big(\int_{A_k} |f(x)|^{\mathfrak{s}_2^\star} dx\Big)^{\frac{\mathfrak{s}_2}{\mathfrak{s}_2^\star}} \tag{6.20}$$

for an appropriate constant $c_4 > 0$, where $\frac{\mathfrak{s}_2}{\mathfrak{s}_2^\star} = \frac{d+2}{d-2}$.

Note that the right-hand side of estimate (6.20) allows us to involve $|A_k|$, the measure of the set A_k into the game. Expressing $|f|^{\mathfrak{s}_2} = 1\,|f|^{\mathfrak{s}_2}$, and using the Hölder inequality once more with exponents $p_1 = \frac{p}{p-\mathfrak{s}_2^\star}$ and $p_1^\star = \frac{p}{\mathfrak{s}_2^\star}$, we have

$$\begin{aligned}\int_{A_k} |f(x)|^{\mathfrak{s}_2^\star}\,dx &\le |A_k|^{\frac{p-\mathfrak{s}_2^\star}{p}} \left(\int_{A_k} |f(x)|^p\,dx\right)^{\frac{\mathfrak{s}_2^\star}{p}} \\ &\le |A_k|^{\frac{p-\mathfrak{s}_2^\star}{p}} \left(\int_{\mathcal{D}} |f(x)|^p\,dx\right)^{\frac{\mathfrak{s}_2^\star}{p}} = |A_k|^{\frac{p-\mathfrak{s}_2^\star}{p}}\,\|f\|_{L^p(\mathcal{D})}^{\mathfrak{s}_2^\star}.\end{aligned} \tag{6.21}$$

Combining (6.20) and (6.21), we obtain the estimate

$$\int_{A_k} |v_k(x)|^{\mathfrak{s}_2}\,dx \le c_5 |A_k|^{\frac{\mathfrak{s}_2(p-\mathfrak{s}_2^\star)}{\mathfrak{s}_2^\star p}}, \tag{6.22}$$

where $c_5 > 0$ is a suitable constant depending on $\|f\|_{L^p(\mathcal{D})}$ (which by assumption is finite).

5. We now show that we may use the estimate (6.22) to obtain a Stampacchia-like inequality for the function ψ, defined by $\psi(k) = |A_k|$ for every $k \in \mathbb{R}$.

Consider any $k_1 < k_2$. Clearly $A_{k_2} \subset A_{k_1}$ and $v_{k_1} := G_{k_1} \circ u \ge k_2 - k_1$ on A_{k_2}. This allows for the estimate

$$\int_{A_{k_1}} |v_{k_1}|^{\mathfrak{s}_2}\,dx \ge \int_{A_{k_2}} |v_{k_1}|^{\mathfrak{s}_2}\,dx \ge (k_2 - k_1)^{\mathfrak{s}_2} |A_{k_2}|. \tag{6.23}$$

Writing (6.22) for the level $k = k_1$ and combining the result with (6.23), we obtain the following comparison between the measures of the superlevel sets at two different values of k, as

$$(k_2 - k_1)^{\mathfrak{s}_2} |A_{k_2}| \le c_5 |A_k|^{\frac{\mathfrak{s}_2(p-\mathfrak{s}_2^\star)}{\mathfrak{s}_2^\star p}}. \tag{6.24}$$

We therefore conclude that the measure of the superlevel sets of the solution of (6.13) at two different levels $k_1 < k_2$ satisfies the inequality (6.24).

6. Upon defining $\delta := \frac{\mathfrak{s}_2(p-\mathfrak{s}_2^\star)}{\mathfrak{s}_2^\star p} = \frac{p(d+2)-2d}{(d-2)p}$, $\gamma = \mathfrak{s}_2 = \frac{2d}{d-2}$ and $\psi(k) := |A_k|$, we express (6.24) as

$$\psi(k_2) \le c_5 \frac{\psi(k_1)^\delta}{(k_2 - k - 1)^\gamma}, \quad \forall\, k_1 < k_2,$$

which is a Stampacchia-like inequality. Since by the choice of $p > \frac{d}{2}$ and $d > 2$, we have $\delta > 1$ and $\gamma > 0$, by step 1, we have the existence of a $k^\star$ such that $\psi(k) = |A_k| = 0$ for every $k > k^\star$, which is the required result that allows us to obtain the L^∞ bounds. Note the L^∞ bound depends on the constant c_5, which in turn depends on $\|f\|_{L^p(\mathcal{D})}$. □

Note that since we are in a bounded domain $\mathcal{D}$, the $L^\infty(\mathcal{D})$ bound implies $L^r(\mathcal{D})$ bounds for every $r > 0$. This remark leads to the following proposition:

Proposition 6.2.11. *Let $d > 2$, and assume that $f \in L^p(\mathcal{D})$ for some $p > \frac{d}{2}$. Then, any weak solution u of* (6.13), *with $\lambda > 0$, satisfies $u \in L^\infty(\mathcal{D}) \cap C(\mathcal{D})$.*

Proof. We approximate any $f \in L^p(\mathcal{D})$ with a sequence $(f_n)_{n\in\mathbb{N}} \subset C_c^\infty(\mathcal{D})$ such that $f_n \to f$ in $L^p(\mathcal{D})$ (see Theorem 1.4.7), and we define u_n to be the solution of $-\Delta u_n + \lambda u_n = f_n$. By Proposition 6.2.9, we have that $u_n \in C_c^\infty(\mathcal{D})$, and by linearity setting $v_n = u - u_n$ and $\bar{f}_n = f - f_n \in L^p(\mathcal{D})$, we see that v_n satisfies $-\Delta v_n + \lambda v_n = \bar{f}_n$. Applying Theorem 6.2.10 to the last equation, we see that $\|v_n\|_{L^\infty(\mathcal{D})} \le c\|f - f_n\|_{L^p(\mathcal{D})}$ so that passing to the limit as $n \to \infty$, we have that $\|u - u_n\|_{L^\infty(\mathcal{D})} \to 0$, and the result follows. □

Similar estimates can also be obtained for unbounded domains $\mathcal{D}$ at the expense of more technical arguments (see, e. g., [42] and references therein).

6.2.3 Laplacian eigenvalue problems

Let $\mathcal{D}$ be a bounded domain, and consider two related minimization problems:

$$\min_{u\in W_0^{1,2}(\mathcal{D})} \int_{\mathcal{D}} |\nabla u(x)|^2 dx, \quad \text{subject to} \quad \int_{\mathcal{D}} |u(x)|^2 dx = 1, \tag{6.25}$$

and

$$\min_{u\in W_0^{1,2}(\mathcal{D})\setminus\{0\}} F_R(u) := \min_{u\in W_0^{1,2}(\mathcal{D})\setminus\{0\}} \frac{\int_{\mathcal{D}} |\nabla u(x)|^2 dx}{\int_{\mathcal{D}} |u(x)|^2 dx}. \tag{6.26}$$

The first problem is a constrained optimization problem, which is related to the minimization of the Dirichlet functional on the surface of the unit ball in $L^2(\mathcal{D})$, denoted by $S_{L^2(\mathcal{D})}$. The second problem is an unconstrained optimization problem over the whole of $W_0^{1,2}(\mathcal{D}) \setminus \{0\}$, of the functional $F_R : W_0^{1,2}(\mathcal{D}) \setminus \{0\} \to \mathbb{R}$, defined by $F_R(u) = \frac{\int_{\mathcal{D}} |\nabla u(x)|^2 dx}{\int_{\mathcal{D}} |u(x)|^2 dx}$, called the Rayleigh–Courant–Fisher functional.

To see the connection between the two problems, note that for any $u \in W_0^{1,2}(\mathcal{D}) \setminus \{0\}$, it holds that $w := \frac{u}{\|u\|_{L^2(\mathcal{D})}} \in W_0^{1,2}(\mathcal{D}) \cap S_{L^2(\mathcal{D})}$ and $F_R(u) = F_R(w) = \int_{\mathcal{D}} |\nabla w(x)|^2 dx$. This means that (6.25) and (6.26) share the same solutions. Furthermore, it is interesting to note that the first-order condition for local minima of any of these problems reduces to an eigenvalue problem for the Laplacian,

$$\begin{aligned} -\Delta u &= \lambda u, \quad \text{in } \mathcal{D}, \\ u &= 0, \quad \text{on } \partial\mathcal{D}, \end{aligned} \tag{6.27}$$

or its equivalent weak form

$$\int_{\mathcal{D}} \nabla u(x) \cdot \nabla v(x) dx = \lambda \int_{\mathcal{D}} u(x)v(x)dx, \quad \forall\, v \in W_0^{1,2}(\mathcal{D}), \tag{6.28}$$

where both the value of $\lambda \in \mathbb{R}$ (eigenvalue) and the function $u \in W_0^{1,2}(\mathcal{D}) \setminus \{0\}$ (eigenfunction) are to be determined. Based on our observation in the previous section that $(-\Delta)^{-1} : L^2(\mathcal{D}) \to L^2(\mathcal{D})$ is a compact operator (Proposition 6.2.4) and that if μ_n is an eigenvalue of $(-\Delta)^{-1}$, then $\lambda_n = \mu_n^{-1}$ is an eigenvalue of $-\Delta$, one can, resorting to the spectral theory of compact operators on Hilbert spaces (see Theorems 1.3.12 and 1.3.13), show that the spectrum is countable so that there exists a sequence of eigenvalues $\lambda_n \to \infty$. Importantly, recalling formulation (6.26), we see that the first eigenvalue $\lambda_1 > 0$ has a variational representation as

$$\lambda_1 = \min_{u \in W_0^{1,2}(\mathcal{D}) \setminus \{0\}} \frac{\int_{\mathcal{D}} |\nabla u(x)|^2 dx}{\int_{\mathcal{D}} |u(x)|^2 dx} = \min_{u \in W_0^{1,2}(\mathcal{D}) \setminus \{0\}} F_R(u), \tag{6.29}$$

i. e., λ_1 corresponds to the minimum of the Rayleigh quotient. Another interesting observation is that this result is connected with the Poincaré inequality, in the sense that the first eigenvalue of the Laplacian provides the best estimate for the Poincaré constant.

We collect these observations, along with some other fundamental properties of the eigenvalues and eigenfunctions of the Laplacian, in the following proposition (see, e. g., [12] or [76]):

Proposition 6.2.12. *Assume that $\mathcal{D} \subset \mathbb{R}^d$ is a bounded, simply connected, and regular*[3] *open set. Problems* (6.25) *and* (6.26) *are equivalent and admit (nontrivial) local minima $u \in W_0^{1,2}(\mathcal{D})$. These are solutions of the eigenvalue problem* (6.27) *or its equivalent weak form* (6.28), *which admit a countable set of solutions $\{(u_n, \lambda_n) : n \in \mathbb{N}\}$ with $0 < \lambda_1 \leq \lambda_2 \leq \cdots$ and $\lambda_n \to \infty$ (the eigenvalues counted with their multiplicity*[4]*) with $(u_n)_{n\in\mathbb{N}}$ constituting an (orthogonal) basis for $L^2(\mathcal{D})$, and $(\frac{u_n}{\sqrt{\lambda_n}})_{n\in\mathbb{N}}$ being a basis for the Hilbert space $W_0^{1,2}(\mathcal{D})$, equipped with the inner product $\langle u, v\rangle_{W_0^{1,2}(\mathcal{D})} = \int_{\mathcal{D}} \nabla u(x) \cdot \nabla v(x) dx$.*

Moreover, the global minimum is attained at the eigenfunction u_1 corresponding to the first eigenvalue λ_1, which is simple and strictly positive, $\lambda_1 > 0$, and characterized in terms of the Rayleigh–Courant–Fisher variational formula (6.29), *while $u_1 > 0$ on $\mathcal{D}$.*

Proof. We are only interested in nontrivial solutions. So, to ease notation, we will replace $W_0^{1,2}(\mathcal{D}) \setminus \{0\}$ with $W_0^{1,2}(\mathcal{D})$ in problem (6.26), assuming that $u \neq 0$. The proof is broken up into 4 steps.

1. The connection between the two problems has already been established. Any local minimum will satisfy the first-order condition of either (6.25) or (6.26).

3 i. e., with sufficiently regular boundary $\partial\mathcal{D}$ for example piecewise C^1. Simple connectedeness and regularity are required for λ_1 to be simple and $u_1 > 0$ (Thm. 8.5.1 [12]).

4 Each eigenvalue is recorded in the sequence as many times as the dimension of the corresponding eigenspace $\mathbf{N}(\lambda I - \mathsf{A})$, which is finite.

Concerning the first problem, one may use the methodology of Lagrange multipliers, according to which, upon rewriting problem (6.25) as $\min_{u\in W_0^{1,2}(\mathcal{D})} F(u)$ subject to $F_1(u) = 0$, for $F(u) = \|\nabla u\|^2_{L^2(\mathcal{D})}$, $F_1(u) = \|u\|^2_{L^2(\mathcal{D})}$, for any local minimum u, there exists a Lagrange multiplier $\lambda \in \mathbb{R}$ such that $DF(u) + \lambda DF_1(u) = 0$. A simple calculation reduces this first-order condition to the eigenvalue problem (6.27).

In order to calculate the first-order conditions for the Rayleigh functional, we perturb around a local minimum u, and expanding $\epsilon \mapsto \phi(\epsilon) = F_R(u + \epsilon v)$ around $\epsilon = 0$ (exercise), we obtain the first-order condition $-\Delta u = F_R(u)u$, which upon calling $\lambda = F_R(u)$ reduces again to the eigenvalue problem (6.27).

Concerning the existence of a global minimum for problem (6.25), a standard application of the direct method suffices, using the convexity of the constraint set and the Poincaré inequality, tracing more or less the steps we followed in Proposition 6.2.3.

2. Concerning the eigenvalue problem (6.27), we use the observation that λ is an eigenvalue of the Laplacian if and only if λ^{-1} is an eigenvalue of $(-\Delta)^{-1} : L^2(\mathcal{D}) \to L^2(\mathcal{D})$, which being a self-adjoint positive definite and compact operator admits a countable set of eigenvalues and eigenfunctions $\{(\mu_n, u_n) : n \in \mathbb{N}\}$ such that $\mu_1 > \mu_2 > \cdots > \mu_n \to 0$ (if counted without multiplicities) with the eigenfunctions $(u_n)_{n\in\mathbb{N}}$ forming a basis for the Hilbert space $L^2(\mathcal{D})$, assumed without loss of generality to be normalized (see Theorems 1.3.12 and 1.3.13). Upon defining $\lambda_n = \mu_n^{-1}$, we obtain the required result (in its form without counting multiplicities). One can easily see, using the weak form (6.28), that $\|\nabla u_n\|^2_{L^2(\mathcal{D})} = \lambda_n \|u_n\|^2 = \lambda_n$. So, upon defining $w_n = \frac{u_n}{\sqrt{\lambda_n}}$, we have $\|w_n\|_{W_0^{1,2}} = 1$. Orthogonality follows by the weak form of the eigenvalue problem. For orthogonality of u_n, u_m, $n \neq m$ in $L^2(\mathcal{D})$, it suffices to first use u_m as a test function in the weak form for u_n, then use u_n as a test function in the weak form for u_m, and use symmetry to deduce that $(\lambda_n - \lambda_m)\langle u_n, u_m\rangle_{L^2(\mathcal{D})} = 0$, so that $\lambda_n \neq \lambda_m$ implies that $\langle u_n, u_m\rangle_{L^2(\mathcal{D})} = 0$. A similar argument leads to the orthogonality of $(w_n)_{n\in\mathbb{N}}$, i. e., $\langle w_n, w_m\rangle_{W_0^{1,2}(\mathcal{D})} = \delta_{n,m}$. We now show that $\operatorname{span}(w_n : n \in \mathbb{N})$ is dense in $W_0^{1,2}(\mathcal{D})$, or, equivalently, if $v \in W_0^{1,2}(\mathcal{D})$ is such that $\langle v, w_n\rangle_{W_0^{1,2}(\mathcal{D})} = 0$ for every $n \in \mathbb{N}$, then $v = 0$. Indeed, upon integration by parts, and using the fact that u_n are eigenfunctions of $-\Delta$, we have that $\langle v, w_n\rangle_{W_0^{1,2}(\mathcal{D})} = \sqrt{\lambda_n}\langle v, u_n\rangle_{L^2(\mathcal{D})}$. So, since $\lambda_n > 0$, $\langle v, w_n\rangle_{W_0^{1,2}(\mathcal{D})} = 0$ for every $n \in \mathbb{N}$ is equivalent to $\langle v, u_n\rangle_{L^2(\mathcal{D})} = 0$ for every $n \in \mathbb{N}$, which by the fact that $(u_n)_{n\in\mathbb{N}}$ is a basis for $L^2(\mathcal{D})$ guarantees that $v = 0$ as required. Extending the distinct sequence of eigenvalues by counting multiplicities and reorganizing the above bases of eigenfunctions to a new basis as in Example 1.3.14, we obtain the required result.

3. We now show the variational formula (6.29). There are several ways around this. The Lagrange multiplier formulation of the problem leads directly to this result. Another way to see this is the following: Since for any eigenfunction u with eigenvalue λ, it holds that $F_R(u) = \lambda$, and it is immediate to see that $\lambda_1 \geq \inf_{u\in W_0^{1,2}(\mathcal{D})} F_R(u)$, so we only need to establish that $\lambda_1 = F_R(u_1) \leq \inf_{u\in W_0^{1,2}(\mathcal{D})} F_R(u)$, i. e., that for every $u \in W_0^{1,2}(\mathcal{D})$, $u \neq 0$,

it holds that $F_R(u) \geq \lambda_1$. To this end, expand any $u \in W_0^{1,2}(\mathcal{D})$ in the orthonormal basis $(w_n)_{n\in\mathbb{N}}$ of $W_0^{1,2}(\mathcal{D})$, as

$$u = \sum_{n=1}^{\infty} \langle u, w_n\rangle_{W_0^{1,2}(\mathcal{D})} w_n = \sum_{n=1}^{\infty} \langle \nabla u, \nabla w_n\rangle_{L^2(\mathcal{D})} = \sum_{n=1}^{\infty} \langle u, u_n\rangle_{L^2(\mathcal{D})} u_n,$$

where we used integration by parts, the definition of w_n, and the fact that u_n are eigenfunctions of the Laplacian. Using this expansion, we calculate

$$\begin{aligned}\int_{\mathcal{D}} |\nabla u(x)|^2 dx &= \langle \nabla u, \nabla u\rangle_{L^2(\mathcal{D})}\\ &= \sum_{m=1}^{\infty}\sum_{n=1}^{\infty} \langle u, u_n\rangle_{L^2(\mathcal{D})} \langle u, u_m\rangle_{L^2(\mathcal{D})} \langle \nabla u_n, \nabla u_m\rangle_{L^2(\mathcal{D})} \qquad (6.30)\\ &= \sum_{n=1}^{\infty} \lambda_n \langle u, u_n\rangle^2_{L^2(\mathcal{D})},\end{aligned}$$

where we used the orthogonality of $(u_n)_{n\in\mathbb{N}}$ in $L^2(\mathcal{D})$, and

$$\|u\|^2_{L^2(\mathcal{D})} = \sum_{n=1}^{\infty} \langle u, u_n\rangle^2_{L^2(\mathcal{D})}. \qquad (6.31)$$

Since $\lambda_n \geq \lambda_1$ for every $n \in \mathbb{N}$, it clearly holds that

$$\int_{\mathcal{D}} |\nabla u(x)|^2 dx = \sum_{n=1}^{\infty} \lambda_n \langle u, u_n\rangle^2_{L^2(\mathcal{D})} \geq \lambda_1 \sum_{n=1}^{\infty} \langle u, u_n\rangle^2_{L^2(\mathcal{D})} = \lambda_1 \|u\|^2_{L^2(\mathcal{D})},$$

so $F_R(u) \geq \lambda_1$ for every $u \in W_0^{1,2}(\mathcal{D})$ as required.

In fact, these steps can be used to show that for any $u \in W_0^{1,2}(\mathcal{D})$ with $\|u\|_{L^2(\mathcal{D})} = 1$,

$$-\Delta u = \lambda_1 u, \quad \text{if and only if} \quad \|\nabla u\|^2_{L^2(\mathcal{D})} = \lambda_1. \qquad (6.32)$$

Indeed, assuming that $\|\nabla u\|^2_{L^2(\mathcal{D})} = \lambda_1$, then (6.30) implies that $\lambda_1 = \sum_{n=1}^{\infty} \lambda_n \langle u, u_n\rangle^2_{L^2(\mathcal{D})}$, while assuming that $\|u\|_{L^2(\mathcal{D})} = 1$, (6.31) implies that $\sum_{n=1}^{\infty} \langle u, u_n\rangle^2_{L^2(\mathcal{D})} = 1$. Combining these $\sum_{n=1}^{\infty} (\lambda_n - \lambda_1)\langle u, u_n\rangle^2_{L^2(\mathcal{D})} = 0$ and as $\lambda_n > \lambda_1$ this is an infinite sum of positive terms, hence, $\langle u, u_n\rangle_{L^2(\mathcal{D})} = 0$ for all n such that $\lambda_n > \lambda_1$. Let m be the multiplicity of λ_1 (which is finite). Then, based on our previous observation, we see that $u \in W_0^{1,2}(\mathcal{D})$ such that $\|u\|_{L^2(\mathcal{D})} = 1$ and $\|\nabla u\|^2_{L^2(\mathcal{D})} = \lambda_1$, admits a finite expansion of the form $u = \sum_{n=1}^{m} \langle u, u_n\rangle_{L^2(\mathcal{D})} u_n$, where u_n, $n = 1, \ldots, m$ solve $-\Delta u_n = \lambda_1 u_n$. By the linearity of the problem, u solves $-\Delta u = \lambda_1 u$.

4. Finally, we prove that $\lambda_1 > 0$ and simple, and that $u_1 > 0$ in $\mathcal{D}$.

Consider any nontrivial weak solution u of $-\Delta u = \lambda_1 u$ in $W_0^{1,2}(\mathcal{D})$. We will show that either $u > 0$ in $\mathcal{D}$ or $u < 0$ in $\mathcal{D}$. Consider $u^+ = \max(u, 0)$ and $u^- = \max(-u, 0)$, and note that $u^{\pm} \in W_0^{1,2}(\mathcal{D})$, while $\langle u^+, u^- \rangle_{L^2(\mathcal{D})} = \langle \nabla u^+, \nabla u^- \rangle_{L^2(\mathcal{D})} = 0$, so $\|u\|_{L^2(\mathcal{D})}^2 = \|u^+\|_{L^2(\mathcal{D})}^2 + \|u^-\|_{L^2(\mathcal{D})}^2$ and $\|\nabla u\|_{L^2(\mathcal{D})}^2 = \|\nabla u^+\|_{L^2(\mathcal{D})}^2 + \|\nabla u^-\|_{L^2(\mathcal{D})}^2$. By the variational representation (6.29) for any $w \in W_0^{1,2}(\mathcal{D})$, $w \neq 0$ it holds that $\|\nabla w\|_{L^2(\mathcal{D})}^2 \geq \lambda_1 \|w\|_{L^2(\mathcal{D})}^2$. Setting $w = u^{\pm}$ in this, we obtain that

$$\|\nabla u^{\pm}\|_{L^2(\mathcal{D})}^2 \geq \lambda_1 \|u^{\pm}\|_{L^2(\mathcal{D})}^2. \tag{6.33}$$

We therefore have that for any $u \in W_0^{1,2}(\mathcal{D})$ such that $\|u\|_{L^2(\mathcal{D})}^2 = 1$, and $-\Delta u = \lambda_1 u$, it holds that $\|\nabla u\|_{L^2(\mathcal{D})}^2 = \lambda_1$, so using the above observations,

$$\begin{aligned} \lambda_1 = \|\nabla u\|_{L^2(\mathcal{D})}^2 &= \|\nabla u^+\|_{L^2(\mathcal{D})}^2 + \|\nabla u^-\|_{L^2(\mathcal{D})}^2 \\ &\geq \lambda_1 (\|u^+\|_{L^2(\mathcal{D})}^2 + \|u^-\|_{L^2(\mathcal{D})}^2) = \lambda_1. \end{aligned} \tag{6.34}$$

Hence, combining (6.33) with (6.34), we conclude that

$$\|\nabla u^{\pm}\|_{L^2(\mathcal{D})}^2 = \|u^{\pm}\|_{L^2(\mathcal{D})}^2. \tag{6.35}$$

Combining (6.35) with the observation (6.32) towards the end of step 3, we conclude that $u^{\pm}$ are solutions of $-\Delta u^{\pm} = \lambda_1 u^{\pm}$ with Dirichlet boundary conditions. We may show using a body of arguments, which constitute the regularity theory for the solutions of the Laplacian eigenvalue problem, that any solution of the Laplacian eigenvalue problem in $W_0^{1,2}(\mathcal{D})$, subject to sufficient regularity of the domain $\mathcal{D}$ (assumed here) is in fact a function[5] $u \in W_0^{1,2}(\mathcal{D}) \cap C(\mathcal{D})$. Since $u^{\pm}$ satisfy $-\Delta u^{\pm} = \lambda_1 u^{\pm} \geq 0$, the strong maximum principle for the Laplacian (Theorem 6.2.7) indicates that either $u^{\pm} > 0$ in $\mathcal{D}$ or $u^{\pm} = 0$ (identically) in $\mathcal{D}$, therefore, since u is nontrivial, either $u > 0$ in $\mathcal{D}$ or $u < 0$ in $\mathcal{D}$.

To show that λ_1 is a simple eigenvalue, assume on the contrary that there exist two nontrivial solutions u, $\hat{u}$ of $-\Delta u = \lambda_1 u$ in $W_0^{1,2}(\mathcal{D})$. By the above result, $\int_{\mathcal{D}} \hat{u}(x)dx \neq 0$ so that there exists $\mu \in \mathbb{R}$ such that $\int_{\mathcal{D}}(u(x) - \mu\hat{u}(x))dx = 0$. By the linearity of the problem, $w := u - \mu\hat{u}$ is also a solution of $-\Delta w = \lambda_1 w$, so it also enjoys the property that either $u - \mu\hat{u} > 0$ or $u - \mu\hat{u} < 0$ in $\mathcal{D}$, which combined with the fact that $\int_{\mathcal{D}}(u(x) - \mu\hat{u}(x))dx = 0$, leads to the conclusion that $u = \mu\hat{u}$ in $\mathcal{D}$, therefore, the eigenvalue λ_1 is simple. □

Example 6.2.13 (Variational formulation of higher eigenvalues). A variational formulation exists not for only the first eigenvalue of the Laplacian but also for the other eigenvalues. Define the finite dimensional subspace $V_n = \text{span}(u_1, \dots, u_n)$ and its orthogonal complement $V_n^{\perp}$. By an argument similar as the one used in Proposition 6.2.12, we see that for any $u \in V_n^{\perp}$ it holds that

5 There are certain ways round this result. One way is to argue as follows: Since $u \in W_0^{1,2}(\mathcal{D})$, we have that $-\Delta u = \lambda u \in W_0^{1,2}(\mathcal{D}) \subset L^2(\mathcal{D})$, hence, by Proposition 6.2.9, $u \in W^{2,2}(\mathcal{D})$ and iterating this scheme, we end up with $u \in C^{\infty}(\mathcal{D})$.

$$\int_{\mathcal{D}} |\nabla u(x)|^2 dx = \sum_{i=n+1}^{\infty} \lambda_i \langle u, u_i \rangle_{L^2(\mathcal{D})}^2 \geq \lambda_{n+1} \sum_{i=n+1}^{\infty} \langle u, u_i \rangle_{L^2(\mathcal{D})}^2 = \lambda_{n+1} \|u\|_{L^2(\mathcal{D})}^2,$$

hence,

$$\lambda_{n+1} = \min\{F_R(u) : u \in V_n^{\perp}, u \neq 0\}. \tag{6.36}$$

This extends the variational formulation to other eigenvalues than the first as well as the definition of eigenvalues of the Laplacian in terms of constrained optimization problems on the surface of the unit ball of $W_0^{1,2}(\mathcal{D})$. ◁

We close our discussion of the eigenvalue problem for the Laplace operator with the celebrated Courant–Fisher minimax formula.

Theorem 6.2.14 (Courant–Fisher minimax formula). *Let E_n be the class of all n-dimensional linear subspaces of $W_0^{1,2}(\mathcal{D})$. Then,*

$$\lambda_n = \min_{E \in E_n} \max_{u \in E,\, u \neq 0} F_R(u). \tag{6.37}$$

Proof. Let $V_n = \text{span}(u_1, \dots, u_n) \in E_n$. Before starting with the proof, we note that the sup and the inf are attained in the above formulae by choosing $E = V_n$ and $u = u_n$.

To show the formula, we work as follows: For any subspace $E \subset W_0^{1,2}(\mathcal{D})$, $E \in E_n$ it holds that $E \cap V_{n-1}^{\perp} \neq \{0\}$. Indeed, consider the linear mapping $\mathsf{T} := \mathsf{P}_{V_{n-1}} : E \to V_{n-1}$, and recall that for any finite dimensional linear map it holds that $\dim(E) = \dim(\mathbf{N}(\mathsf{T})) + \dim(\mathbf{R}(\mathsf{T}))$. By definition $\dim(E) = n$, with $\mathbf{R}(\mathsf{T}) \subset V_{n-1}$, we have that $\dim(\mathbf{R}(\mathsf{T})) \leq n-1$, hence, $\dim(\mathbf{N}(\mathsf{T})) \geq n - (n-1) = 1$, so there exists a nonzero element $w \in E$ such that $\mathsf{P}_{V_{n-1}} w = 0$, i. e. $w \in E \cap V_{n-1}^{\perp}$. By (6.36),

$$\lambda_n \leq F_R(w) \leq \max\{F_R(u) : u \in E, u \neq 0\},$$

and since E is an arbitrary n-dimensional subspace of $W_0^{1,2}(\mathcal{D})$, we have that

$$\lambda_n \leq F_R(w) \leq \min_{E \in E_n} \max_{u \in E,\, u \neq 0} F_R(u). \tag{6.38}$$

On the other hand, one may choose $E = V_n$ so that for any $u \in E = V_n$, using similar arguments as in Proposition 6.2.12, we have that

$$F_R(u) = \frac{\sum_{i=1}^n \lambda_i \langle u, u_i \rangle_{L^2(\mathcal{D})}^2}{\sum_{i=1}^n \langle u, u_i \rangle_{L^2(\mathcal{D})}^2} \leq \lambda_n,$$

since $\lambda_1 \leq \cdots \leq \lambda_n$. The above holds for any $u \in E = V_n$, hence,

$$\max_{u \in V_n,\, u \neq 0} F_R(u) \leq \lambda_n.$$

Since the above choice for E is simply one choice for $E \in E_n$, it is easy to conclude that

$$\min_{E\in E_n} \max_{u\in V_n,\, u\neq 0} F_R(u) \leq \lambda_n. \tag{6.39}$$

Combining (6.38) with (6.39), we obtain the required result. □

Other similar minimax or maximin formulae for the eigenvalues can be obtained (see, e. g., [12]), which may further be extended to more general operators than the Laplacian or discrete versions of the Laplacian appearing in graph theory and find applications in various fields, such as for instance image processing etc. They also allow us to obtain important insight on the properties of the eigenfunctions and in many cases lead to approximation algorithms (either analytical or numerical). We provide some examples.

Example 6.2.15 (Comparison of eigenvalues in different domains). Consider two different domains $\mathcal{D}, \mathcal{D}' \subset \mathbb{R}^d$ such that $\mathcal{D} \subset \mathcal{D}'$. Then for every eigenvalue of the Laplacian with Dirichlet boundary conditions, it holds that $\lambda_n(\mathcal{D}') \leq \lambda_n(\mathcal{D})$.

Since $\mathcal{D} \subset \mathcal{D}'$, any function $u \in W_0^{1,2}(\mathcal{D})$ can be extended to a function $u' \in W_0^{1,2}(\mathcal{D}')$ by defining $u' = u\mathbf{1}_{\mathcal{D}} + \mathbf{0}\mathbf{1}_{\mathcal{D}'\setminus\mathcal{D}}$. Clearly, $\|u\|_{L^2(\mathcal{D})} = \|u'\|_{L^2(\mathcal{D}')}$ and $\|\nabla u\|_{L^2(\mathcal{D})} = \|\nabla u'\|_{L^2(\mathcal{D}')}$ so that $F_R(u,\mathcal{D}) = F_R(u',\mathcal{D}')$, where, by $F_R(v,A)$, we denote the Rayleigh quotient keeping track of both the function v and the domain A. Naturally, the set of all n-dimensional subspaces of $W_0^{1,2}(\mathcal{D})$ is a subset of the corresponding set concerning $W_0^{1,2}(\mathcal{D}')$, and the stated result comes by an application of the variational formula (6.37), by noting that the minimum, when the formula is applied for $\mathcal{D}$, is taken over a smaller set than when applied for $\mathcal{D}'$. ◁

Example 6.2.16 (Faber–Krahn inequality). The Faber–Krahn inequality is an important result stating that among all bounded measurable sets $\mathcal{D} \subset \mathbb{R}^d$, of the same measure, the ball has the smallest possible first eigenvalue λ_1 for the Dirichlet Laplacian, i. e., $\lambda_1(\mathcal{D}) \geq \lambda_1(\mathcal{D}^*)$ with equality if and only if $\mathcal{D} = \mathcal{D}^*$, where for any set $\mathcal{D}$, we denote by $\mathcal{D}^*$ the ball in $\mathbb{R}^d$ with the same volume as $\mathcal{D}$.

The proof of the above result is based on the variational representation formula (6.29), upon using the important concept of spherical rearrangement. Given any function $u : \mathcal{D} \to \mathbb{R}$, we define its spherical (or Schwarz) rearrangement $u^* : \mathcal{D}^* \to \mathbb{R}$ as the spherically symmetric radially nonincreasing function of $|x|$, constructed by arranging the level sets of u in balls of the same volume, $u^*(x) := \sup\{y : x \in L_y^*\}$, where, by L_y, we denote the super level set $L_y = \{x \in \mathcal{D} : u(x) \geq y\}$. Two important results are that $\|u\|_{L^2(\mathcal{D})} = \|u^*\|_{L^2(\mathcal{D}^*)}$ and the Polya–Szego[6] inequality $\|\nabla u\|_{L^2(\mathcal{D})} \geq \|\nabla u^*\|_{L^2(\mathcal{D}^*)}$. Using these, we see that $F_R(u^*, \mathcal{D}^*) \leq F_R(u,\mathcal{D})$ for every function $u : \mathcal{D} \to \mathbb{R}$ and every

6 This is an important result, with deep connections with isoperimetric inequalities; for details, see, e. g., [86] and references therein.

domain $\mathcal{D} \subset \mathbb{R}^d$, and applying the variational representation formula (6.29), we obtain the required result. ◁

As we will see, in this chapter and the following one, these very interesting properties of the Laplacian are not restricted to the particular operator, but rather are general properties of a wider class of variational problems related to self adjoint operators.

6.2.4 The Poisson equation and its connection with Dirichlet-like functionals for other boundary conditions

The connection of the Poisson equation with homogeneous Dirichlet (i. e., $u = 0$ on $\partial\mathcal{D}$ in the sense of traces) with the minimizers of the Dirichlet functional can be generalized for other types of boundary conditions. We will not expand too much into this direction here, but we will offer some initial guidelines for the interested reader. We will assume that $\mathcal{D}$ is a bounded and Lipschitz domain.

One interesting extension of variational methods is to the study of functionals F of the form

$$F(u) = \frac{1}{2}\int_{\mathcal{D}} |\nabla u|^2 dx + \frac{1}{2}\int_{\mathcal{D}} c|u|^2 dx - \int_{\mathcal{D}} fu dx, \quad F : W_0^{1,2}(\mathcal{D}) \to \mathbb{R},$$

where $c \in L^\infty(\mathcal{D})$ is a known function. Using methods similar to the ones studied in the previous section, we can show that the problem of minimizing F is related to the Poisson-like equation

$$\begin{aligned} -\Delta u + cu &= f, \quad \text{in } \mathcal{D}, \\ u &= 0, \quad \text{on } \partial\mathcal{D}, \end{aligned} \tag{6.40}$$

understood in the weak sense, i. e.,

$$\int_{\mathcal{D}} \nabla u \cdot \nabla v dx + \int_{\mathcal{D}} cuv dx = \int_{\mathcal{D}} fv dx, \quad \forall v \in W_0^{1,2}(\mathcal{D}).$$

If $c > 0$, then using techniques similar to the ones employed in the previous section, we can extend the results we have obtained for the Poisson equation to the extended Poisson equation (6.40). The condition $c > 0$, allows us to ensure the coercivity of functional F, easily. From the modeling point of view, the term c can be interpreted as a potential.

One popular type of boundary conditions are nonhomogeneous Dirichlet boundary conditions, in which u is specified on the boundary $\partial\mathcal{D}$, and set equal to a given function $g : \mathcal{D} \to \mathbb{R}$. Assuming regularity of $\mathcal{D}$, this case can be reduced to the homogeneous Dirichlet boundary conditions case, using the trace theorem (Proposition 1.5.17(iii)). In particular, if $g \in H^{1/2}(\partial\mathcal{D})$, then there exists $w \in W^{1,2}(\mathcal{D})$ such that $\gamma_0(w) = g$ (where γ_0 is the trace operator). We will then restate the Poisson equation

$$\begin{aligned} -\Delta u + cu &= f, \quad \text{in } \mathcal{D}, \\ u &= g, \quad \text{on } \partial\mathcal{D}, \end{aligned} \tag{6.41}$$

in terms of the new unknown $u_0 = u - w \in W_0^{1,2}(\mathcal{D})$. Then (6.41) becomes

$$\begin{aligned} -\Delta u_0 + cu_0 &= \tilde{f}, \quad \text{in } \mathcal{D}, \\ u &= 0, \quad \text{on } \partial\mathcal{D}, \end{aligned} \tag{6.42}$$

where $\tilde{f} = f - cw + \Delta w$. Note that (6.42) is well defined in the weak sense since $w \in W^{1,2}(\mathcal{D})$, hence $\tilde{f} \in W^{-1,2}(\mathcal{D})$. This requires extending Proposition 6.2.3, for functionals including such terms (see e. g., [12, 51]). This is equivalent to considering the original functional F, as defined on $W_0^{1,2}(\mathcal{D}) + w$, rather than on $W_0^{1,2}(\mathcal{D})$. The theory concerning solvability, regularity, etc., can be modified accordingly.

Another interesting version of the Poisson equation is when Neumann boundary conditions are imposed, i. e., where the normal derivative $\frac{\partial u}{\partial n} = \nabla u \cdot n$ is determined at the boundary (assumed regular and where n is the outward normal vector). In general, the normal derivative is to be understood in the sense of trace, i. e., $\gamma_1(u)$ (see Definition 1.5.16). A common formulation is in terms of homogeneous Neumann boundary conditions, leading to the Poisson-type equation

$$\begin{aligned} -\Delta u + cu &= f, \quad \text{in } \mathcal{D}, \\ \frac{\partial u}{\partial n} &= 0, \quad \text{on } \partial\mathcal{D}. \end{aligned} \tag{6.43}$$

The connection with Proposition 6.2.3 can be made through the weak formulation of (6.43), which now admits the form

$$\int_{\mathcal{D}} \nabla u \cdot \nabla v dx + \int_{\mathcal{D}} cuv dx = \langle f, v \rangle, \quad \forall\, v \in W^{1,2}(\mathcal{D}), \tag{6.44}$$

where now $\langle \cdot, \cdot \rangle$ can be interpreted either as the inner product in $L^2(\mathcal{D})$ if $f \in L^2(\mathcal{D})$ or as the duality pairing between $W^{1,2}(\mathcal{D})$ and its dual space $(W^{1,2}(\mathcal{D}))^*$, if $f \in (W^{1,2}(\mathcal{D}))^*$. The weak formulation (6.44) can be justified using the Green's formula in Proposition 1.5.19(iii) for a suitably smooth test function and using the density of smooth functions in the Sobolev space $W^{1,2}(\mathcal{D})$. In the case where $c = 0$, since for any solution u, the function $u + M$, where M is an arbitrary constant remains a solution, an extra constraint will have to be adopted to specify the solution uniquely. A typical choice for $c = 0$ would be to adopt, e. g., the constraint $\int_{\mathcal{D}} u dx = 0$, i. e., specify the mean of the solution in the domain.

Finally, another popular set of boundary conditions are the so called Robin boundary conditions, leading to the problem

$$\begin{aligned} -\Delta u + cu &= f, \quad \text{in } \mathcal{D}, \\ \frac{\partial u}{\partial n} + hu &= g, \quad \text{on } \partial\mathcal{D}, \end{aligned} \tag{6.45}$$

where $h \in L^\infty(\partial\mathcal{D})$, $h \geq 0$, $g \in L^2(\partial\mathcal{D})$ are given functions specifying the boundary data. The Robin boundary conditions can be understood from the physical point of view as a flux of the quantity described by u on the boundary, which also depends on the local concentration u there. If $h = 0$, this type of boundary condition reduces to Neumann boundary conditions (homogeneous when $g = 0$). The connection with variational problems can be made in terms of the weak form of this equation,

$$\int_{\mathcal{D}} \nabla u \cdot \nabla v dx + \int_{\mathcal{D}} cuv dx + \int_{\partial\mathcal{D}} huv ds = \langle f, v\rangle + \int_{\partial\mathcal{D}} gv ds, \quad \forall\, v \in W^{1,2}(\mathcal{D}), \tag{6.46}$$

with $\langle\cdot,\cdot\rangle$ interpreted as above. The weak formulation (6.46) can be justified using the Green's formula in Proposition 1.5.19(iii) for a suitably smooth test function and using the density of smooth functions in the Sobolev space $W^{1,2}(\mathcal{D})$.

Further extensions to problems in the spirit of (6.40) with variable coefficients (that may or may not be interpreted in terms of minimization problems) will be studied in Chapter 7, Section 7.6.1, in terms of the celebrated Lax–Milgram lemma (see Theorem 7.3.7). The extension to other types of boundary conditions (e. g., Neumann or Robin) is also feasible for the case of variable coefficients. In this book, we will consider mainly Dirichlet-type boundary conditions, but all of the techniques introduced are quite easily extended to Neumann or Robin boundary conditions. For a detailed treatment of the Neumann and Robin problem, the interested reader can consult, e. g., [51].

6.3 Semicontinuity of integral functionals

Before studying the general problem of minimization of integral functionals, using the direct method, we must first establish some semicontinuity properties. In this section, we collect some fundamental results in this direction, concerning the lower semicontinuity of integral functionals $F : X = W^{1,p}(\mathcal{D}) \to \mathbb{R}$ of the form

$$F(u) := \int_{\mathcal{D}} f(x, u(x), \nabla u(x)) dx, \tag{6.47}$$

where $\mathcal{D} \subset \mathbb{R}^d$ is an open set of bounded measure, u is a function from $\mathcal{D}$ to $\mathbb{R}$, $\nabla u = (\frac{\partial}{\partial x_1} u, \dots, \frac{\partial}{\partial x_d} u)$ its gradient, and $f : \mathcal{D} \times \mathbb{R} \times \mathbb{R}^d \to \mathbb{R}$ is a given function. These results highlight the deep connection of lower semicontinuity for F with the convexity of the function $z \mapsto f(x,y,z)$. It is interesting to note that the functional F can also be defined when u is a vector valued function $u : \mathcal{D} \to \mathbb{R}^m$, in which case, ∇u must be replaced by the derivative matrix $Du \in \mathbb{R}^{m\times d}$ (or its transpose) and $f : \mathcal{D} \times \mathbb{R} \times \mathbb{R}^{m\times d} \to \mathbb{R}$. The semicontinuity problem can be treated in this general case with modification of the techniques developed in this section. For an excellent account as well as more detailed results on the subject, the reader may consult [80] or [61]. For simplicity we limit to the case $p \in (1, \infty)$.

Our starting point will be the study of semicontinuity of a related integral functional in Lebesgue spaces, and in particular of the functional $F_o : L^p(\mathcal{D}) \times L^q(\mathcal{D}; \mathbb{R}^s) \to \mathbb{R}$, defined by

$$F_o(u, v) = \int_{\mathcal{D}} f(u(x), v(x))dx, \tag{6.48}$$

for every $u \in L^p(\mathcal{D})$, $v \in L^q(\mathcal{D}; R^s)$ for appropriate p, q, s (not considering, unless explicitly stated, $p = 1, \infty$ or $q = 1, \infty$, even though some of the results can be extended for these cases). Clearly if $u \in W^{1,p}(\mathcal{D})$, then setting $v = \nabla u$, and $s = d$, we have that $F_o(u, v) = F(u)$, where F is the functional defined in (6.47). Hence, establishing semicontinuity results for the auxiliary functional F_o, defined in (6.48), in Lebesgue spaces can lead us to semicontinuity results in Sobolev spaces for the original functional F, defined in (6.47). Of course, some care will have to be taken as to the topologies with which $W^{1,p}(\mathcal{D})$ and $L^p(\mathcal{D}) \times L^q(\mathcal{D}; \mathbb{R}^s)$ are endowed for F and F_o, respectively, to be lower semicontinuous, but, as we shall see shortly, these subtleties can be handled with the use of the Rellich–Kondrachov compact embedding theorem for Sobolev spaces into a suitable Lebesgue space.

Following [80], we will study first the sequential lower semicontinuity of general functionals of the form (6.48), with respect to strong and weak convergence in Lebesgue spaces, and then apply this general result in providing weak lower semicontinuity results for functionals of the form (6.47) in Sobolev spaces. These are classic results in the calculus of variations, in the legacy of De Giorgi (see, e. g., [63]) revisited by many authors in subsequent years (for a detailed bibliographical exposition, see [61]). Our exposition follows closely the proof of Theorem 7.5 in [80] (see also Theorem 3.26 in [61]).

6.3.1 Semicontinuity in Lebesgue spaces

We start by considering weak lower semicontinuity of integral functionals in Lebesgue spaces. Before moving to the more general problem (6.48), we will first consider the problem of semicontinuity of integral functionals $F_L : L^p(\mathcal{D}; \mathbb{R}^m) \to \mathbb{R}$, defined by

$$F_L(v) := \int_{\mathcal{D}} f(x, v(x))dx, \tag{6.49}$$

where $f : \mathcal{D} \times \mathbb{R}^m \to \mathbb{R} \cup \{+\infty\}$ is a suitable function, and $v : \mathcal{D} \to \mathbb{R}^m$ is a vector valued function such that $v \in L^p(\mathcal{D}; \mathbb{R}^m)$.

We impose the following conditions on the function f:

Assumption 6.3.1. Consider the function $f : \mathcal{D} \times \mathbb{R}^m \to \mathbb{R} \cup \{+\infty\}$ with the following properties:

(i) $x \mapsto f(x,z)$ is measurable for every $z \in \mathbb{R}^m$.

(ii) $z \mapsto f(x,z)$ is lower semicontinuous and convex in $\mathbb{R}^m$, a. e. in $x \in \mathcal{D}$.

(iii) There exist $\alpha \in L^{p^*}(\mathcal{D};\mathbb{R}^m)$, where $\frac{1}{p} + \frac{1}{p^*} = 1$, $p \in (1,\infty)$, and $\beta \in L^1(\mathcal{D})$ such that

$$f(x,z) \geq \alpha(x) \cdot z + \beta(x), \quad \text{a. e. } x \in \mathcal{D},\ \forall\, z \in \mathbb{R}^m.$$

A general semicontinuity result is given in the following theorem [61, 80].

Theorem 6.3.2 (Tonelli). *If Assumption* 6.3.1 *holds, the functional F_L, defined in* (6.49), *is weakly sequentially lower semicontinuous in $L^p(\mathcal{D};\mathbb{R}^m)$, $p \in (1,\infty)$.*

Proof. Let us assume first that $f \geq 0$. We would like to show that if $v_n \rightharpoonup v$ in $L^p(\mathcal{D};\mathbb{R}^m)$, then, $L := \liminf_n F_L(v_n) \geq F_L(v)$. If $f \geq 0$, our basic tool would be the Fatou lemma, which requires a subsequence of the original sequence converging a. e. However, the convergence $v_n \rightharpoonup v$ in $L^p(\mathcal{D};\mathbb{R}^m)$ is weak, and in general, we cannot guarantee the existence of a subsequence that converges a. e. To overcome this difficulty, we recall Mazur's lemma (Proposition 1.2.17), according to which, we may construct a new sequence $(w_n)_{n\in\mathbb{N}}$ out of the original one with the property that $w_n \to v$ in $L^p(\mathcal{D};\mathbb{R}^m)$, and apply the above reasoning upon extraction of a further a. e. convergent subsequence. The proof is broken up into 3 steps.

1. Assume that $f \geq 0$. We will first show an auxiliary strong lower semicontinuity result. Let $v_n \to v$ in $L^p(\mathcal{D};\mathbb{R}^m)$. Select a subsequence $(v_{n_k})_{k\in\mathbb{N}}$ such that $\lim_k F_L(v_{n_k}) = L := \liminf_n F_L(v_n)$. We work along this subsequence and select a further subsequence $(v_{n_{k_\ell}})_{\ell\in\mathbb{N}}$ such that $v_{n_{k_\ell}} \to v$ a. e. We may now use the lower semicontinuity of f, which yields that $\liminf_\ell f(x, v_{n_{k_\ell}}(x)) \geq f(x, v(x))$ a. e. and Fatou's lemma to obtain

$$L = \liminf_\ell \int_{\mathcal{D}} f(x, v_{n_{k_\ell}}(x))dx \geq \int_{\mathcal{D}} \liminf_\ell f(x, v_{n_{k_\ell}}(x))dx \geq \int_{\mathcal{D}} f(x, v(x))dx,$$

which is the required strong lower semicontinuity result. Note that for the strong lower semicontinuity, we do not require convexity of f.

2. We now let $v_n \rightharpoonup v$. We first consider a subsequence $(v_{n_\ell})_{\ell\in\mathbb{N}}$ such that $L := \liminf_n F_L(v_n) = \lim_\ell F_L(v_{n_\ell})$. On account of that, for any $\epsilon > 0$, there exists $M(\epsilon) \in \mathbb{N}$ such that

$$L \leq F_L(v_{n_\ell}) \leq L + \epsilon, \quad \forall\, \ell \geq M(\epsilon). \tag{6.50}$$

Fix $\epsilon > 0$, and apply Mazur's lemma to the sequence $(v_{n_\ell})_{\ell\geq M(\epsilon)}$, to construct a new sequence $(w_\ell)_{\ell\in\mathbb{N}}$ (or rather $\ell > M(\epsilon)$) such that $w_\ell \to v$ in $L^p(\mathcal{D};\mathbb{R}^m)$. The new sequence consists of convex combinations of the elements in the tail of the original sequence, which implies that for every ℓ, there exists $K(\ell) \geq M(\epsilon)$ and $\lambda(\ell,i) > 0$ with $\sum_{i=M(\epsilon)}^{K(\ell)} \lambda(\ell,i) = 1$, such that if $w_\ell = \sum_{i=M(\epsilon)}^{K(\ell)} \lambda(\ell,i) v_{n_i}$, then $w_\ell \to v$ in $L^p(\mathcal{D};\mathbb{R}^m)$. Since $z \mapsto f(x,z)$ is convex, for every ℓ it holds that

$$f(x, w_\ell(x)) = f\left(x, \sum_{i=M(\epsilon)}^{K(\ell)} \lambda(\ell,i) v_{n_i}\right) \leq \sum_{i=M(\epsilon)}^{K(\ell)} \lambda(\ell,i) f(x, v_{n_i}). \tag{6.51}$$

As $w_\ell \to v$ in $L^p(\mathcal{D};\mathbb{R}^m)$, we can extract a subsequence, denoted the same for simplicity, such that $w_\ell \to v$ a. e. Clearly, (6.51) holds for the chosen subsequence, so passing (6.51) to this subsequence and integrating over $\mathcal{D}$ using Fatou's lemma, we obtain that

$$F_L(w_\ell) \leq \sum_{i=M(\epsilon)}^{K(\ell)} \lambda(\ell,i) F_L(v_{n_i}). \tag{6.52}$$

Since all $i \geq M(\epsilon)$, by (6.50), $F_L(v_{n_i}) \leq L + \epsilon$, therefore (6.52) implies that $F_L(w_\ell) \leq L + \epsilon$ for all ℓ, hence, $\liminf_\ell F_L(w_\ell) \leq L + \epsilon$. Since $w_\ell \to u$ in $L^p(\mathcal{D};\mathbb{R}^m)$, by the strong lower semicontinuity result we have shown in step 1, it holds that $F_L(v) \leq \liminf_\ell F_L(w_\ell)$. Therefore, $F_L(v) \leq L + \epsilon$ for every $\epsilon > 0$, so taking the limit as $\epsilon \to 0$, leads to the desired result $F_L(v) \leq L$.

3. If f does not satisfy the property $f \geq 0$, then we cannot apply Fatou's lemma, which was crucial in our argument. However, by Assumption 6.3.1(iii), we can consider the function $\bar{f}$, defined by $\bar{f}(x,z) = f(x,z) - \alpha(x)\cdot z - \beta(x) \geq 0$. By our results so far, the functional $\bar{F_L}$, defined by $\bar{F_L}(v) = \int_{\mathcal{D}} \bar{f}(x, v(x))dx$, is weakly sequentially lower semicontinuous in $L^p(\mathcal{D};\mathbb{R}^m)$. We can express F_L as $F_L = \bar{F_L} + F_1$, where $F_1(v) = \int_{\mathcal{D}} (\alpha(x)\cdot v(x) + \beta(x))dx$. If $v_n \rightharpoonup v$ in $L^p(\mathcal{D};\mathbb{R}^m)$, clearly, $\int_{\mathcal{D}} \alpha(x)\cdot v_n(x)dx \to \int_{\mathcal{D}} \alpha(x)\cdot v(x)dx$ (by the definition of weak convergence and since $\alpha \in L^{p^*}(\mathcal{D};\mathbb{R}^m)$) so that $F_1(v_n) \to F_1(v)$ and F_1 is weakly continuous. Therefore, F_L is weakly lower semicontinuous as the sum of a weakly lower semicontinuous and a continuous functional. □

Convexity played a crucial role in the proof of weak lower semicontinuity of the integral functional F when using the Mazur lemma to pass from weak convergence to strong convergence and from that to a. e. convergence. One could think that convexity was imposed as a condition to facilitate our analysis, however, its role is much more fundamental, as is shown in the following proposition [61, 80].

Theorem 6.3.3 (Tonelli). *Consider the integral functional $F_L : L^p(\mathcal{D};\mathbb{R}^m) \to \mathbb{R}\cup\{+\infty\}$, $p \in (1,\infty)$, defined by $F_L(v) = \int_{\mathcal{D}} f(v(x))dx$, for some $f : \mathbb{R}^m \to \mathbb{R}$, lower semicontinuous. If F_L is weakly sequentially lower semicontinuous in $L^p(\mathcal{D};\mathbb{R}^m)$, then f is a convex function.*

Proof. Assume that F_L is weakly sequentially lower semicontinuous. We will show the convexity of $z \mapsto f(z)$. Fix any $z_1, z_2 \in \mathbb{R}^m$, $t \in (0,1)$, and $h \in S_d$, where S_d is the unit sphere of $\mathbb{R}^d$. Define the sequence of functions $(w_{t,n})_{n\in\mathbb{N}}$ by $w_{t,n}(x) = z_2 + \mathfrak{w}_t(n\,(h\cdot x))(z_1 - z_2)$ for every $x \in \mathbb{R}^d$, where $\mathfrak{w}_t : \mathbb{R} \to \mathbb{R}$ is the periodic extension (with period 1) of the indicator function $\mathbf{1}_{[0,t]}$, of the interval $[0,t] \subset [0,1]$. Using the sequence $(w_{t,n})_{n\in\mathbb{N}}$, we construct the sequence of functions $(\mathfrak{w}_{t,n})_{n\in\mathbb{N}}$, defined by $\mathfrak{w}_{t,n}(x) := \mathfrak{w}_t(n\,(h\cdot x)\,)$, for every $x \in \mathbb{R}^d$. As can be shown by an application of the Riemann–Lebesgue lemma (see Lemma 6.10.1 in Section 6.10.1), for every $t \in [0,1]$, it holds that $\mathfrak{w}_{t,n} \overset{\star}{\rightharpoonup} \psi_t$ in $L^\infty(\mathcal{D};\mathbb{R}^m)$, where ψ_t is the constant function $\psi_t(x) = t$ for every $x \in \mathbb{R}^d$. Hence, for every $t \in [0,1]$, it holds that $w_{t,n} \overset{\star}{\rightharpoonup} \bar{w}_t$ in $L^\infty(\mathcal{D};\mathbb{R}^m)$, where $\bar{w}_t$ is the constant function $\bar{w}_t(x) = z_2 + t(z_1 - z_2)$ a. e. $x \in \mathcal{D}$, and therefore $w_{t,n} \rightharpoonup \bar{w}_t$ in $L^p(\mathcal{D};\mathbb{R}^m)$.

Since F_L is weakly sequentially lower semicontinuous, we have that

$$F_L(\bar{w}_t) \le \liminf_n F_L(w_{t,n}) = \liminf_n \int_{\mathcal{D}} f(z_2 + \mathfrak{w}_t(n\,(h\cdot x)\,)(z_1 - z_2)\,)dx. \tag{6.53}$$

Clearly,

$$F_L(\bar{w}_t) = |\mathcal{D}|\,f(z_2 + t(z_1 - z_2)\,) = |\mathcal{D}|\,f(tz_1 + (1-t)z_2\,),$$

where $|\mathcal{D}|$ is the Lebesgue measure of $\mathcal{D}$, so that (6.53) yields

$$|\mathcal{D}|\,f(tz_1 + (1-t)z_2\,) \le \liminf_n \int_{\mathcal{D}} f(z_2 + \mathfrak{w}_t(n\,(h\cdot x)\,)(z_1 - z_2)\,)dx. \tag{6.54}$$

We now calculate the right-hand side of (6.54). First of all, since $\mathfrak{w}_t(n\,(h\cdot x)\,)$ takes only two values 0 and 1, it can be seen that

$$f(z_2 + \mathfrak{w}_t(n\,(h\cdot x)\,)(z_1 - z_2)\,) = \mathfrak{w}_t(n\,(h\cdot x)\,)f(z_1) + (1 - \mathfrak{w}_t(n\,(h\cdot x)\,))f(z_2),$$

and since $\mathfrak{w}_{t,n}(x) := \mathfrak{w}_t(n(h\cdot x)\,)$ has the property $\mathfrak{w}_{t,n} \rightharpoonup \psi_t$ (where ψ_t is the constant function $\psi_t(x) = t$, for every $x \in \mathbb{R}^d$) we see that

$$\begin{aligned}
&\int_{\mathcal{D}} f(z_2 + \mathfrak{w}_t(n\,(h\cdot x)\,)(z_1 - z_2)\,)dx \\
&\quad = \int_{\mathcal{D}} (\mathfrak{w}_t(n\,(h\cdot x)\,)f(z_1) + (1 - \mathfrak{w}_t(n\,(h\cdot x)\,))f(z_2)\,)dx \\
&\quad \to |\mathcal{D}|\,(tf(z_1) + (1-t)f(z_2)\,).
\end{aligned}$$

Substituting the above in (6.54) yields

$$f(\,t\,z_1 + (1-t)\,z_2\,) \le t\,f(z_1) + (1-t)f(z_2),$$

which is the required convexity of f. □

Remark 6.3.4. The theorem is also valid when $f : \mathcal{D}\times\mathbb{R}^m \to \mathbb{R}$ has explicit space dependence, in which case, weak lower semicontinuity implies the convexity of the function $z \mapsto f(x,z)$ a. e. in $\mathcal{D}$. For a very detailed account of lower semicontinuity and well posedness results for integral functionals in Lebesgue spaces and their connection with convexity as well boundeness properties of f, the reader can consult [80] or [61].

We now provide a more general result, concerning functionals of two variables u and v of the general form (6.48) involving strong convergence with respect to one variable and weak convergence with respect to the other. Concerning the function f, we make the following assumption:

Assumption 6.3.5. Let $f : \mathcal{D} \times \mathbb{R}^s \times \mathbb{R}^m \to \mathbb{R} \cup \{+\infty\}$ and $p, q \in (1, \infty)$ such that
(i) f is Carathéodory, i. e., $(y, z) \mapsto f(x, y, z)$ continuous in $\mathbb{R}^s \times \mathbb{R}^m$, a. e. $x \in \mathcal{D}$ and $x \mapsto f(x, y, z)$ is measurable for every $(y, z) \in \mathbb{R}^s \times \mathbb{R}^m$.
(ii) There exist functions $\alpha \in L^{p^*}(\mathcal{D}; \mathbb{R}^m)$, $\beta \in L^1(\mathcal{D})$ and a constant $\gamma \in \mathbb{R}$ such that

$$f(x, y, z) \geq \alpha(x) \cdot z + \beta(x) + c|y|^q, \quad \text{a. e. } x \in \mathcal{D},\ \forall y \in \mathbb{R}^s,\ \forall z \in \mathbb{R}^m.$$

(iii) The function $z \mapsto f(x, y, z)$ is convex a.e in $\mathcal{D}$ and for every $y \in \mathbb{R}^s$.

A general semicontinuity result is given in the following theorem [61, 80].

Theorem 6.3.6. *If Assumption* 6.3.5 *holds, the functional* $F_o : L^q(\mathcal{D}; \mathbb{R}^s) \times L^p(\mathcal{D}; \mathbb{R}^m) \to \mathbb{R}$, *defined by*

$$F_o(u, v) := \int_{\mathcal{D}} f(x, u(x), v(x))dx, \tag{6.55}$$

is lower sequentially semicontinuous with respect to strong convergence in $L^q(\mathcal{D}; \mathbb{R}^s)$ *and weak convergence in* $L^p(\mathcal{D}; \mathbb{R}^m)$.

Proof. We only sketch the proof, which is based upon the observation that if $u_n \to u$ in $L^q(\mathcal{D}; \mathbb{R}^s)$, then the functional $\bar{F}_o : L^p(\mathcal{D}; \mathbb{R}^m) \to \mathbb{R} \cup \{+\infty\}$, defined by freezing the u coordinate at the limit, $v \mapsto \bar{F}_o(v) := F_o(u, v)$, the latter is weakly lower semicontinuous, so if $v_n \rightharpoonup v$ in $L^p(\mathcal{D}; \mathbb{R}^m)$, then $\liminf_n \bar{F}_o(v_n) \geq \bar{F}_o(v) = F_o(v, u)$. Then, for any $n \in \mathbb{N}$,

$$\begin{aligned} F_o(v_n, u_n) &= F_o(v_n, u) + F_o(v_n, u_n) - F_o(v_n, u) \\ &= \bar{F}_o(v_n) + F_o(v_n, u_n) - F_o(v_n, u), \end{aligned}$$

with the remainder term being as small as we wish by letting n become large, so the lower semicontinuity follows by the lower semicontinuity of $\bar{F}_o$. The detailed argument, which requires a careful control of the remainder term, involves certain technicalities of a measure theoretic nature (see [80]). □

Remark 6.3.7. Theorem 6.3.6 will still hold if in Assumption 6.3.5 instead of assuming that f is Carathéodory, we assume that f is a normal integrand, i. e., that $x \mapsto f(x, y, z)$ is measurable for every $(y, z) \in \mathbb{R}^m \times \mathbb{R}^s$ and $(y, z) \mapsto f(x, y, z)$ is a. e. in $\mathcal{D}$ lower semicontinuous (rather than continuous) at the expense of some technical issues in the proof.

The reader could consult, e. g., [80] for such versions of the lower semicontinuity result. Convexity of the function $z \mapsto f(x,y,z)$ is fundamental for the validity of lower semicontinuity result. In fact, an analogue of Theorem 6.3.3 can be shown for the more general class of integral functionals treated in Theorem 6.3.6 both in the case where f is Carathéodory (see, e. g., Theorem 3.15 in [61]) and in the case where f is a normal integrand (see, e. g., Theorem 7.5 in [80]).

Example 6.3.8. The semicontinuity result of Theorem 6.3.6 is interesting in its own right. Consider, for example, the integral functional $F : L^q(\mathcal{D};\mathbb{R}^s) \to \mathbb{R}\cup\{+\infty\}$, defined by $F(u) := \int_{\mathcal{D}} f(x,u(x),(\mathsf{K}u)(x))dx$, where f is a function satisfying Assumption 6.3.5, and $\mathsf{K} : L^p(\mathcal{D};\mathbb{R}^m) \to L^q(\mathcal{D};\mathbb{R}^s)$ is a compact operator. Then, the functional F is weakly lower semicontinuous. Such functionals are called nonlocal functionals and find interesting applications, e. g., in imaging. ◁

6.3.2 Semicontinuity in Sobolev spaces

As an important application of Theorem 6.3.2 (or of its more general form Theorem 6.3.6), we may provide a useful result concerning lower semicontinuity of integral functionals in Sobolev spaces.

Consider a function $f : \mathcal{D}\times\mathbb{R}\times\mathbb{R}^d$ and a function $u : \mathcal{D} \to \mathbb{R}$ such that $u \in W^{1,p}(\mathcal{D})$. Define the functional $F : W^{1,p}(\mathcal{D}) \to \mathbb{R}$ by

$$F(u) := \int_{\mathcal{D}} f(x,u(x),\nabla u(x))dx. \tag{6.56}$$

We make the following assumption on the function f:

Assumption 6.3.9. Let $f : \mathcal{D}\times\mathbb{R}\times\mathbb{R}^d \to \mathbb{R}\cup\{+\infty\}$ satisfying the following:

(i) f is Carathéodory, i. e., the function $x \mapsto f(x,y,z)$ is measurable for every $(y,z) \in \mathbb{R}\times\mathbb{R}^d$, while the function $(y,z) \mapsto f(x,y,z)$ is continuous in $\mathbb{R}\times\mathbb{R}^d$, a. e. $x \in \mathcal{D}$.

(ii) There exist functions $a \in L^{p^*}(\mathcal{D};\mathbb{R}^d)$, $\beta \in L^1(\mathcal{D})$, and a constant $\gamma \in \mathbb{R}$ such that

$$f(x,y,z) \geq a(x)\cdot z + \beta(x) + c|y|^q, \quad \text{a. e. } x \in \mathcal{D},\ \forall y \in \mathbb{R},\ \forall z \in \mathbb{R}^d,$$

where $q \in (1,\mathfrak{s}_p) = (1,\frac{dp}{d-p})$ if $p < d$ and $q \in (1,\infty)$ if $p \geq d$.

(iii) The function $z \mapsto f(x,y,z)$ is convex a. e. in $\mathcal{D}$ and for every $y \in \mathbb{R}$.

A general semicontinuity result is given in the following theorem [61, 80].

Theorem 6.3.10. *Assume that f satisfies Assumption 6.3.9. Then, the functional F, defined in (6.56), is sequentially weakly lower semicontinuous in $W^{1,p}(\mathcal{D})$.*

Proof. Let $(u_n)_{n\in\mathbb{N}} \subset W^{1,p}(\mathcal{D})$ such that $u_n \rightharpoonup u$ in $W^{1,p}(\mathcal{D})$. This implies that $u_n \rightharpoonup u$ in $L^p(\mathcal{D})$ and $v_n := \nabla u_n \rightharpoonup \nabla u$ in $L^p(\mathcal{D})$. By the compact Sobolev embedding theorem (see Theorem 1.5.11), there exists a subsequence $(u_{n_k})_{k\in\mathbb{N}}$ such that $u_{n_k} \to u$ in $L^q(\mathcal{D})$ for any $q \in [1, \frac{dp}{d-p})$ if $p < d$ and $q \in [1, \infty)$ if $p \geq d$ (see also Remark 1.5.12). Assumption 6.3.9 allows us to use Theorem 6.3.6, from which we deduce the sequential weak lower semicontinuity of the functional F. □

Remark 6.3.11. The continuity hypothesis in Assumption 6.3.9(i) may be replaced by lower semicontinuity (see also Remark 6.3.7). Moreover, extension for $q = 1$ is possible.

6.4 A general problem from the calculus of variations

We now consider the problem of minimization of integral functionals $F : W^{1,p}(\mathcal{D}) \to \mathbb{R} \cup \{+\infty\}$ of the form

$$F(u) := \int_{\mathcal{D}} f(x, u(x), \nabla u(x))dx, \tag{6.57}$$

using the direct method of Weierstrass (Theorem 2.2.5). Our approach follows [61].

Assumption 6.4.1. Let $f : \mathcal{D} \times \mathbb{R} \times \mathbb{R}^d \to \mathbb{R} \cup \{+\infty\}$ satisfying the following:

(i) the growth condition

$$f(x, y, z) \leq a_1|z|^p + \beta_1(x) + c_1|y|^r, \quad \text{a.e. } x \in \mathcal{D},\ \forall\, (y, z) \in \mathbb{R} \times \mathbb{R}^d,$$

where $\beta_1 \in L^1(\mathcal{D})$, $a_1 > 0$, $c_1 > 0$ and $1 \leq r < \mathfrak{s}_p = \frac{dp}{d-p}$ if $1 < p < d$, (any r if $p \geq q$)

(ii) the coercivity condition

$$f(x, y, z) \geq a_2|z|^p + \beta_2(x) + c_2|y|^q, \quad \text{a.e. } x \in \mathcal{D},\ \forall\, (y, z) \in \mathbb{R} \times \mathbb{R}^d,$$

where $\beta_2 \in L^1(\mathcal{D})$, $a_2 > 0$, $c_2 \in \mathbb{R}$, and $p > q \geq 1$.

A general semicontinuity result is given in the following theorem [61, 80].

Theorem 6.4.2. *Let Assumptions* 6.3.9 *and* 6.4.1 *hold and assume that* $p \in (1, \infty)$. *Then, for any* $v \in W^{1,p}(\mathcal{D})$, *the problem*

$$\inf\{F(u) \,:\, u \in v + W_0^{1,p}(\mathcal{D})\}, \tag{6.58}$$

where F is defined as in (6.57), *attains its minimum. If the function* $(y, z) \mapsto f(x, y, z)$ *is strictly convex, this minimum is unique.*

Proof. We will use the direct method of the calculus of variations (see the abstract presentation of the Weierstrass theorem 2.2.5). Let $m = \inf_{u\in v+W_0^{1,p}(\mathcal{D})} F(u)$. By Assumption 6.4.1(i), for any $u \in W^{1,p}(\mathcal{D})$, it holds that $F(u) < \infty$, therefore $m < \infty$. This follows by the Sobolev embedding theorem, according to which if $u \in W^{1,p}(\mathcal{D})$, then $u \in L^{\mathfrak{s}_p}(\mathcal{D})$.

Furthermore, by Assumption 6.4.1(ii), for any $u \in W^{1,p}(\mathcal{D})$, it holds that[7] $F(u) > -\infty$, therefore $m > -\infty$.

We now consider a minimizing sequence, i. e., a sequence $(u_n)_{n\in\mathbb{N}} \subset W^{1,p}(\mathcal{D})$, such that $\lim_n F(u_n) = m$. By the coercivity condition, Assumption 6.4.1(i), this sequence is uniformly bounded in $W^{1,p}(\mathcal{D})$. This is immediate to see in the case where $v = 0$, i. e., when $u \in W_0^{1,p}(\mathcal{D})$, by an application of the Lebesgue embedding theorem and the Poincaré inequality: Since $(u_n)_{n\in\mathbb{N}} \subset W_0^{1,p}(\mathcal{D})$ is a minimizing sequence, by Assumption 6.4.1(i) for any $\epsilon > 0$, there exists N such that for all $n > N$,

$$\begin{aligned} &a_2\|\nabla u_n\|^p_{L^p(\mathcal{D};\mathbb{R}^d)} - |c_2|\,\|u_n\|^q_{L^q(\mathcal{D})} + \|\beta_2\|_{L^1(\mathcal{D})} \\ &\quad\le a_2\|\nabla u_n\|^p_{L^p(\mathcal{D};\mathbb{R}^d)} + c_2\|u_n\|^q_{L^q(\mathcal{D})} + \|\beta_2\|_{L^1(\mathcal{D})} \le F(u_n) < m + \epsilon. \end{aligned}$$

Since $1 \le q < p$ by the Lebesgue embedding $\|u_n\|_{L^q(\mathcal{D})} \le c\|u_n\|_{L^p(\mathcal{D})}$ so that

$$a_2\|\nabla u_n\|^p_{L^p(\mathcal{D};\mathbb{R}^d)} - |c_2|c\,\|u_n\|^q_{L^p(\mathcal{D})} + \|\beta_2\|_{L^1(\mathcal{D})} \le m + \epsilon,$$

where c is a generic constant, which may vary from line to line. Recall the Poincaré inequality (Theorem 1.5.13) according to which for every $u_n \in W_0^{1,p}(\mathcal{D})$, $\|\nabla u_n\|_{L^p(\mathcal{D};\mathbb{R}^d)} \ge \mathbf{c}_{\mathcal{P}}\|u_n\|_{L^p(\mathcal{D})}$, where $\mathbf{c}_{\mathcal{P}}$ is the Poincaré constant, so substituting in the above, we obtain

$$a_2\|\nabla u_n\|^p_{L^p(\mathcal{D};\mathbb{R}^d)} - |c_2|cc_{\mathcal{P}}^{-q}\|\nabla u_n\|^p_{L^p(\mathcal{D};\mathbb{R}^d)} + \|\beta_2\|_{L^1(\mathcal{D})} \le m + \epsilon.$$

Since $1 \le q < p$, this implies the existence of a constant c such that $\|\nabla u_n\|_{L^p(\mathcal{D};\mathbb{R}^d)} < c$ for all $n > N$, hence, the sequence $(\nabla u_n)_{n\in\mathbb{N}}$ is uniformly bounded in $L^p(\mathcal{D};\mathbb{R}^d)$. As a further consequence of the Poincaré inequality, we know that $\|\nabla u\|_{L^p(\mathcal{D};\mathbb{R}^d)}$ is an equivalent norm of $W_0^{1,p}(\mathcal{D})$, wherefore $(u_n)_{n\in\mathbb{N}} \subset W_0^{1,p}(\mathcal{D})$ is uniformly bounded and by the reflexivity of $W_0^{1,p}(\mathcal{D})$ (since $p > 1$) there exists a weakly convergent subsequence $(u_{n_k})_{k\in\mathbb{N}} \subset (u_n)_{n\in\mathbb{N}}$ such that $u_{n_k} \rightharpoonup u_o$ for some $u_o \in W_0^{1,p}(\mathcal{D})$. By Assumption 6.3.9, we may apply Theorem 6.3.10 to infer the lower semicontinuity of F in $W_0^{1,p}(\mathcal{D})$, therefore, following the abstract arguments of the Weierstrass theorem, u_o is a minimizer.

If $v \neq 0$, then the minimizing sequence can be expressed as $u_n = v + \bar{u}_n$, where $(\bar{u}_n)_{n\in\mathbb{N}} \subset W_0^{1,p}(\mathcal{D})$, and our arguments can be applied to $\bar{u}_n$. We leave the details as an exercise.

Assume now that $(y,z) \mapsto f(x,y,z)$ is strictly convex, and assume the existence of two minimizers u_o and v_o. Then $w = \frac{1}{2}(u_o + v_o)$ is also a minimizer (since by convexity $F(w) \le \frac{1}{2}(F(u_o) + F(v_o)) = m$), and this yields

---

7 To see this, we need to use the Sobolev embeddings. For instance if $p < d$, then by the Nirenberg–Gagliardo inequality $\|\nabla u\|_{L^p(\mathcal{D};\mathbb{R}^d)} \ge c\|u\|_{L^{\mathfrak{s}_p}(\mathcal{D})}$ for $\mathfrak{s}_p = \frac{dp}{d-p}$. Since $p > q \ge 1$, it holds that $q < \mathfrak{s}_p$, and combining the Nirenberg–Gagliardo inequality with the Lebesgue embedding, we get that $\|\nabla u\|_{L^p(\mathcal{D};\mathbb{R}^d)} \ge c\|u\|_{L^q(\mathcal{D})}$. Using this inequality along with the boundedness Assumption 6.4.1(ii), we can see that F is bounded below.

$$
\begin{aligned}
\int_{\mathcal{D}} \Big\{ & \frac{1}{2} f(x, u_o(x), \nabla u_o(x)) + \frac{1}{2} f(x, v_o(x), \nabla v_o(x)) \\
& - f\Big(x, \frac{1}{2}(u_o(x) + v_o(x)), \frac{1}{2}\nabla u_o(x) + \frac{1}{2}\nabla v_o(x)\Big)\Big\} dx = 0,
\end{aligned} \tag{6.59}
$$

and since by convexity of $(y, z) \mapsto f(x, y, z)$, a. e. $x \in \mathcal{D}$, it holds that

$$
\begin{aligned}
& \frac{1}{2} f(x, u_o(x), \nabla u_o(x)) + \frac{1}{2} f(x, v_o(x), \nabla v_o(x)) \\
& \quad - f\Big(x, \frac{1}{2}(u_o(x) + v_o(x)), \frac{1}{2}\nabla u_o(x) + \frac{1}{2}\nabla v_o(x)\Big) \geq 0, \quad \text{a. e. } x \in \mathcal{D}.
\end{aligned} \tag{6.60}
$$

By combining (6.59) and (6.60), it holds that

$$
\begin{aligned}
& \frac{1}{2} f(x, u_o(x), \nabla u_o(x)) + \frac{1}{2} f(x, v_o(x), \nabla v_o(x)) \\
& \quad - f\Big(x, \frac{1}{2}(u_o(x) + v_o(x)), \frac{1}{2}\nabla u_o(x) + \frac{1}{2}\nabla v_o(x)\Big) = 0, \quad \text{a. e. } x \in \mathcal{D}.
\end{aligned}
$$

But then, the strict convexity of f implies that $u_o(x) = v_o(x)$, $\nabla u_o(x) = \nabla v_o(x)$, a. e. $x \in \mathcal{D}$, hence, $u_o = v_o$ as elements of $v + W_0^{1,p}(\mathcal{D})$. □

Remark 6.4.3. The assumption $p \in (1, \infty)$ is important. If, for example, $p = 1$, then, even though, the gradients of the minimizing sequence are bounded in $L^1(\mathcal{D}; \mathbb{R}^d)$, this does not guarantee the existence of a weak limit as an element of $L^1(\mathcal{D}; \mathbb{R}^d)$, but rather as a Radon measure. This case requires the use the functional setting of BV spaces.

6.5 Differentiable functionals and connection with nonlinear PDEs: the Euler–Lagrange equation

Under certain conditions, we will show that the functional F, defined in (6.56), is differentiable, therefore, the first-order condition leads to a nonlinear PDE, which is the Euler–Lagrange equation for F. It is important to realize that this is true under certain restrictions on f. One may construct examples in which this is not true, even in one spatial dimension (see [16]).

Assumption 6.5.1 (Growth condition). The function $f : \mathcal{D} \times \mathbb{R} \times \mathbb{R}^d \to \mathbb{R}$ satisfies the following conditions:

(i) f is differentiable with respect to y and z and the functions $f_1 := \frac{\partial}{\partial y} f$, $f_2 := \frac{\partial}{\partial z} f = \nabla_z f$ are Carathéodory functions.

(ii) There exist functions $a_i \in L^{p^*}(\mathcal{D})$ and constants $c_i > 0$, $i = 1, 2$ such that the partial derivatives of f with respect to the y and z, denoted by f_1 and f_2, respectively, satisfy the growth conditions

$$|f_i(x,y,z)| < a_i(x) + c_i(|y|^{p-1} + |z|^{p-1}), \quad i = 1, 2.$$

The connection of the minimization problem for the functional F with the Euler–Lagrange PDE is given in the following theorem [61, 80]

Theorem 6.5.2. *Suppose Assumption* 6.5.1 *holds,* $p \in (1, \infty)$. *The minimizer* u_o *of the functional* F *defined in* (6.56) *satisfies the Euler–Lagrange equation*

$$\begin{aligned}\int_{\mathcal{D}} \Big(&\frac{\partial}{\partial y} f(x, u_o(x), \nabla u_o(x)) v(x) \\ &+ \frac{\partial}{\partial z} f(x, u_o(x), \nabla u_o(x)) \cdot \nabla v(x) \Big) dx = 0, \quad \forall\, v \in W_0^{1,p}(\mathcal{D}),\end{aligned} \tag{6.61}$$

which is interpreted as the weak form of the elliptic differential equation

$$-\operatorname{div}\Big(\frac{\partial}{\partial z} f(x, u_o(x), \nabla u_o(x)) \cdot \nabla v(x) \Big) + \frac{\partial}{\partial y} f(x, u_o(x), \nabla u(x)) = 0. \tag{6.62}$$

If furthermore, the function $(y,z) \mapsto f(x,y,z)$ *is convex, then any solution of the Euler–Lagrange equation* (6.61) *is a minimizer of* F.

Proof. The strategy of proof is to show that F is Gâteaux differentiable and recognize the Euler–Lagrange equation as the first-order condition for the minimum.

The key to the existence of the Gâteaux derivative is the existence of the limit

$$\begin{aligned} L &:= \lim_{\epsilon \to 0} \frac{1}{\epsilon}(F(u + \epsilon v) - F(u)) \\ &= \lim_{\epsilon \to 0} \frac{1}{\epsilon} \int_{\mathcal{D}} (f(x, u(x) + \epsilon v(x), \nabla u(x) + \epsilon \nabla v(x)) - f(x, u(x), \nabla u(x)))dx. \end{aligned}$$

Since f is differentiable, we have that

$$\begin{aligned} &f(x, u(x) + \epsilon v(x), \nabla u(x) + \epsilon \nabla v(x)) - f(x, u(x), \nabla u(x)) \\ &= \int_0^{\epsilon} \frac{d}{ds} f(x, u(x) + s\, v(x), \nabla u(x) + s\, \nabla v(x)) ds \\ &= \int_0^1 \frac{d}{dt} f(x, u(x) + \epsilon\, t\, v(x), \nabla u(x) + \epsilon\, t\, \nabla v(x)) dt \end{aligned}$$

a. e. in $\mathcal{D}$, where the last equality holds by the simple change of variable of integration $s = \epsilon t$. By the properties of the function f,

$$\begin{aligned}
&\frac{d}{dt} f(x, u(x) + \epsilon\, t\, v(x), \nabla u(x) + \epsilon\, t\, \nabla v(x)) \\
&\quad = \frac{\partial}{\partial y} f(x, u(x) + \epsilon\, t\, v(x), \nabla u(x) + \epsilon\, t\, \nabla v(x)) \epsilon v(x) \\
&\qquad + \frac{\partial}{\partial z} f(x, u(x) + \epsilon\, t\, v(x), \nabla u(x) + \epsilon\, t\, \nabla v(x)) \cdot \epsilon \nabla v(x),
\end{aligned}$$

so substituting all these into the above, we obtain

$$\frac{1}{\epsilon}(F(u + \epsilon v) - F(u)) = \int_{\mathcal{D}} w_\epsilon(x) dx, \tag{6.63}$$

where

$$\begin{aligned}
w_\epsilon(x) := \int_0^1 \Big(&\frac{\partial}{\partial y} f(x, u(x) + \epsilon\, t\, v(x), \nabla u(x) + \epsilon\, t\, \nabla v(x)) v(x) \\
&+ \frac{\partial}{\partial z} f(x, u(x) + \epsilon\, t\, v(x), \nabla u(x) + \epsilon\, t\, \nabla v(x)) \cdot \nabla v(x) \Big) dt.
\end{aligned}$$

Clearly, by Assumption 6.5.1(i), $w_\epsilon(x) \to w$ a. e. in $\mathcal{D}$, where

$$w(x) := \frac{\partial}{\partial y} f(x, u(x), \nabla u(x)) v(x) + \frac{\partial}{\partial z} f(x, u(x), \nabla u(x)) \cdot \nabla v(x),$$

so if we are allowed to invoke the Lebesgue dominated convergence theorem, (6.63) will lead us to the result that

$$\begin{aligned}
L &= \int_{\mathcal{D}} w(x) dx \\
&= \int_{\mathcal{D}} \Big(\frac{\partial}{\partial y} f(x, u(x), \nabla u(x)) v(x) + \frac{\partial}{\partial z} f(x, u(x), \nabla u(x)) \cdot \nabla v(x) \Big) dx.
\end{aligned}$$

It thus remains to check the existence of a function $h \in L^1(\mathcal{D})$ such that $|w_\epsilon(x)| \leq h(x)$, a. e. in $\mathcal{D}$. To this end, we change variables of integration once more and express

$$\begin{aligned}
w_\epsilon(x) = \frac{1}{\epsilon} \int_0^\epsilon \Big(&\frac{\partial}{\partial y} f(x, u(x) + s v(x), \nabla u(x) + s \nabla v(x)) v(x) \\
&+ \frac{\partial}{\partial z} f(x, u(x) + s v(x), \nabla u(x) + s \nabla v(x)) \nabla v(x) \Big) ds,
\end{aligned}$$

which in turn implies upon using the growth conditions of Assumption 6.5.1(ii) that a. e. $x \in \mathcal{D}$,

$$
\begin{aligned}
|w_\epsilon(x)| &\leq \frac{1}{\epsilon}\int_0^\epsilon (a_1(x) + c_1(|u(x) + sv(x)|^{p-1} + |\nabla u(x) + s\nabla v(x)|^{p-1}))|v|(x)ds \\
&\quad + \frac{1}{\epsilon}\int_0^\epsilon (a_2(x) + c_2(|u(x) + sv(x)|^{p-1} + |\nabla u(x) + s\nabla v(x)|^{p-1}))|\nabla v(x)|ds \\
&\leq c + c_1(\,|u(x)| + |v(x)|)^{p-1} + (|\nabla u(x) + \nabla v(x)|)^{p-1})\,|v(x)| \\
&\quad + c_2(\,|u(x) + |v(x)|^{p-1} + (|\nabla u(x) + \nabla v(x)|)^{p-1})\,|\nabla v(x)| =: h(x).
\end{aligned}
$$

The above estimate is obtained by majorization of the integrand above. We observe that since $v \in W_0^{1,p}(\mathcal{D})$, both v and $|\nabla v|$ are in L^p, and by standard use of Hölder' s inequality, we may obtain that $h \in L^1(\mathcal{D})$.

We thus conclude that for every $v \in W_0^{1,p}(\mathcal{D})$,

$$
\begin{aligned}
DF(u; v) &:= \langle DF(u), v\rangle \\
&= \int_{\mathcal{D}} \Big(\frac{\partial}{\partial y} f(x, u(x), \nabla u(x))v(x) + \frac{\partial}{\partial z} f(x, u(x), \nabla u(x)) \cdot \nabla v(x)\Big)dx,
\end{aligned}
$$

where $DF(u)$ is now understood as an element of $(W_0^{1,p}(\mathcal{D}))^* = W^{-1,p^*}(\mathcal{D})$.

Therefore, if u_o is a minimizer, then it must hold that

$$
\begin{aligned}
0 &= \langle DF(u_o), v\rangle \\
&= \int_{\mathcal{D}} \Big(\frac{\partial}{\partial y} f(x, u_o(x), \nabla u_o(x))v(x) + \frac{\partial}{\partial z} f(x, u_o(x), \nabla u_o(x)) \cdot \nabla v(x)\Big)dx
\end{aligned}
$$

for all $v \in W_0^{1,p}(\mathcal{D})$, which is (6.61). A further integration by parts leads to the weak form (6.62) of the Euler–Lagrange equation.

Assume now that $(y, z) \mapsto f(x, y, z)$ is convex and consider any solution u_o of the Euler–Lagrange equation (6.61). By the convexity and differentiability of f for any u, we have that

$$
\begin{aligned}
f(x, u(x), \nabla u(x)) - f(x, u_o(x), \nabla u_o(x)) &\geq \frac{\partial}{\partial y} f(x, u_o(x), \nabla u_o(x))(u(x) - u_o(x)) \\
&\quad + \frac{\partial}{\partial z} f(x, u_o(x), \nabla u_o(x)) \cdot \nabla(u - u_o)(x).
\end{aligned}
$$

We integrate over all $x \in \mathcal{D}$, and using $u - u_o \in W_0^{1,p}$ as a test function in the Euler–Lagrange equation (6.61), we conclude that $F(u) - F(u_o) \geq 0$, hence, u_o is a minimizer for F. □

Remark 6.5.3. Using more sophisticated techniques, one may obtain related results under more relaxed assumptions. We will not enter into details here, but we refer the interested reader to the detailed discussion in Section 3.4.2 in [61].

6.6 Regularity results in the calculus of variations

We now consider the problem of regularity of minimizers of integral functionals. As we will see under certain conditions on the function f, the minimizer of the integral functional F, is not just in $W^{1,p}(\mathcal{D})$ but enjoys further regularity properties. We have already seen that in Section 6.2.2, within the simpler framework of the Poisson equation, focusing not so much on the properties of the corresponding functional but rather on the related PDE corresponding to the first-order condition. In this section, we consider a more general problem concerning regularity of minimizers, focusing on the properties of the functional, which is minimized. We present an introduction to a beautiful theory, initiated by De Giorgi and taken up and further developed by many important researchers, which, through some delicate analytical estimates, allows us to establish Hölder continuity of the minimizer. Then, using an important technique, the difference quotient technique introduced by Nirenberg, which approximates derivatives of minimizers in terms of difference quotients, combined with the Euler–Lagrange equation, we may obtain results on the higher differentiability of the minimizer (e. g., show that $u \in W^{2,p}_{\text{loc}}(\mathcal{D})$). The problem of regularity is a long standing, highly interesting, and important part of the calculus of variations. As a matter of fact, Hilbert stressed its importance in the formulation of his 19th problem. This short introduction is only skimming the surface of this deep subject; we refer the reader to the monograph of Giusti (see [83]) as well as the recent lecture notes of Beck (see [22]), on which the presentation of this section is based.

The proofs are long and technical, so, we summarize the main strategy here in case one wishes to skip the next section, which contains the details, and move directly to Section 6.6.2.

The key point to proving the Hölder continuity of minimizers is the observation (see Proposition 6.6.7) that minimizers u satisfy an important class of inequalities (related to estimates of $\|\nabla(u-k)^{\pm}\|^p_{L^p(B(x_0,r))}$ from above in terms of $\|(u-k)^{\pm}\|^p_{L^p(B(x_0,R))}$, $r < R$) called Caccioppoli-type inequalities:

$$\int_{A^+(k,x_0,r)} |\nabla u(x)|^p dx \le c_0\Big((R-r)^{-p} \int_{A^+(k,x_0,R)} (u(x)-k)^p dx + |A^+(k,x_0,R)| \Big),$$

$$\int_{A^-(k,x_0,r)} |\nabla u(x)|^p dx \le c_0\Big((R-r)^{-p} \int_{A^-(k,x_0,R)} (k-u(x))^p dx + |A^-(k,x_0,R)| \Big),$$

where $A^{+}(k, x_0, R) := \{x \in B(x_0, R) : u(x) > k\}$, and $A^{-}(k, x_0, R) := \{x \in B(x_0, R) : u(x) < k\}$ are the local (within the ball $B(x_0, R)$) super- and sub-level sets for u, respectively, at level k, $r < R$, and using the notation $|A|$ for the Lebesgue measure of any set A. It turns out that if a function u satisfies Caccioppoli-type inequalities for all levels k, (the class of such functions is called the De Giorgi class), then, it is locally Hölder continuous with an appropriate exponent α (see Theorem 6.6.3). This can be proved by noting that the Caccioppoli inequalities for all levels k (or for all levels above a critical one) provide an important estimate for the oscillation of the function u,

$$\operatorname{osc}(x_0, R) := \sup_{x \in B(x_0,R)} u(x) - \inf_{x \in B(x_0,R)} u(x),$$

of the form

$$\operatorname{osc}\left(x_0, \frac{\rho}{4}\right) \leq c\frac{\rho}{4} + \lambda \operatorname{osc}(x_0, \rho), \quad \forall \rho \text{ such that } 0 < \rho \leq R_0,$$

which eventually leads to an estimate of the form

$$\operatorname{osc}(x_0, r) \leq c\left[\left(\frac{r}{R}\right)^{\alpha} + r^{\alpha}\right]$$

for appropriate $c > 0$ and $\alpha \in (0, 1)$. This last estimate guarantees the local Hölder continuity of u, with exponent α (see Theorem 1.9.18).

6.6.1 The De Giorgi class

We begin by introducing an important class of functions, the De Giorgi class [22, 83], whose properties play a crucial role in the regularity theory in the calculus of variations.

Definition 6.6.1 (The De Giorgi class). Consider a function $u \in W^{1,p}(\mathcal{D})$, $1 \leq p < \infty$, and for any ball $B(x_0, R) \subset \mathcal{D}$, define the super level set

$$A^{+}(k, x_0, R) := \{x \in B(x_0, R) : u(x) > k\},$$

and the sub level set

$$A^{-}(k, x_0, R) := \{x \in B(x_0, R) : u(x) < k\}$$

using the notation $|A|$ for the Lebesgue measure of any measurable set A. For $r < R$, we introduce the Caccioppoli type inequalities:

$$
\begin{aligned}
(C+) \quad & \int\limits_{A^+(k,x_0,r)} |\nabla u(x)|^p dx \leq c_0\Big((R-r)^{-p} \int\limits_{A^+(k,x_0,R)} (u(x)-k)^p dx + |A^+(k,x_0,R)|\Big),\\
(C-) \quad & \int\limits_{A^-(k,x_0,r)} |\nabla u(x)|^p dx \leq c_0\Big((R-r)^{-p} \int\limits_{A^-(k,x_0,R)} (k-u(x))^p dx + |A^-(k,x_0,R)|\Big),
\end{aligned}
\tag{6.64}
$$

and define as

$$
\begin{aligned}
DG_p^{\pm}(k_0,R_0,\mathcal{D}) := \{u \in W^{1,p}_{\mathrm{loc}}(\mathcal{D}) \,:\, \exists c_0, k_0, R_0 \text{ such that } \forall\, B(x_0,r) \subset\subset B(x_0,R) \subset \mathcal{D},\\
\text{with } R \leq R_0, \text{ and } \forall k \geq k_0, \text{ inequalities (6.64)}(C\pm) \text{ hold}\}.
\end{aligned}
$$

(i) The set $DG_p(k_0,R_0,\mathcal{D}) = DG_p^+(k_0,R_0,\mathcal{D}) \cap DG_p^-(k_0,R_0,\mathcal{D}) \subset W^{1,p}_{\mathrm{loc}}(\mathcal{D})$ is called the De Giorgi class at level k_0.
(ii) If a function u satisfies the Caccioppoli inequalities (6.64) for all $k \in \mathbb{R}$, i. e., $u \in \bigcap_{k_0\in\mathbb{R}} DG_p(k_0,R_0,\mathcal{D}) =: DG_p(R_0,\mathcal{D})$, we say that u belongs to the De Giorgi class.
(iii) If in the Caccioppoli inequalities (6.64) the second term on the right-hand side is missing, the corresponding De Giorgi class is called homogeneous and is denoted by $DGH_p(\mathcal{D})$.

The motivation for defining the De Giorgi class arises as minimizers of integral functionals (under growth conditions) satisfy Caccioppoli inequalities, and therefore belong to the De Giorgi class; for example minimizers of $F(u) = \int_{\mathcal{D}} f(x,u(x),\nabla u(x))dx$ if the Carathéodory function f satisfies the growth condition $|z|^p \leq f(x,y,z) \leq L(1+|z|^p)$ for every $y \in \mathbb{R}$, $z \in \mathbb{R}^d$, and a. e. $x \in \mathcal{D}$ for a suitable $L > 0$ (see Section 6.6.2). On the other hand, functions in DG_p (minimizers or not) satisfy Hölder continuity properties.

Remark 6.6.2. The De Giorgi class is often defined in terms of more general Caccioppoli type inequalities. For instance in [83], the Caccioppoli inequalities are stated as

$$
\begin{aligned}
&\int\limits_{A^+(k,r)} |\nabla u(x)|^p dx\\
&\quad \leq c_0\Big(\frac{c}{(R-r)^p} \int\limits_{A^+(k,r)} (u(x)-k)^p dx + (c_{\mathrm{NH}} + k^p R^{-d\epsilon})|A^+(k,R)|^{1-\frac{p}{d}+\epsilon}\Big),
\end{aligned}
\tag{6.65}
$$

for $0 < \epsilon \leq \frac{p}{d}$, and c_0, c_{NH} positive constants. The reason for generalizing the De Giorgi class as above is because minimizers or quasi-minimizers for functionals satisfying more general growth conditions than $|z|^p \leq f(x,y,z) \leq L(1+|z|^p)$, for example,

$$
L_0|z|^p - a(x)|y|^q - b(x) \leq f(x,y,z) \leq L_1|z|^p + a(x)|y|^q + b(x) \tag{6.66}
$$

for suitable exponent q and positive functions a, b, can be shown to satisfy Caccioppoli type inequalities of the above form (see Remark 6.6.9 below as well as Section 6.10.4 in the appendix of the chapter). Importantly, if $a = 0$ in (6.66), then we may take $c_{NH} = 0$ in the generalized Caccioppoli inequality (6.65). The results concerning local boundedness for functions in the De Giorgi class can be extended for the above generalization using very similar arguments (see [83] for detailed proofs). In particular, using similar techniques as the one used in Theorem 6.6.3, it can be shown (see Theorem 7.2 in [83]) that any function in $DG_p^+(\mathcal{D})$ is locally bounded from above in $\mathcal{D}$ and for any $x_0 \in \mathcal{D}$ and R sufficiently small, with an estimate given in terms of cubes $Q(x_0, R)$ rather than balls $B(x_0, R)$ as

$$\sup_{Q(x_0,\frac{R}{2})} u \le c_2\left(\left(\fint_{Q(x_0,R)} (u^+)^p dx\right)^{\frac{1}{p}} + k_0 + c_{NH}R^{\frac{d\epsilon}{p}}\right), \tag{6.67}$$

with c_{NH} being the same constant as in (6.65), where $\fint_A$ denotes the mean value of a function over A. In fact, under the same assumptions (see Theorem 7.3 in [83] or Theorem 1 in [67]) estimate (6.67) may be generalized for any $\rho < R$ (with R sufficiently small) and any $q > 0$, as

$$\sup_{Q(x_0,\rho)} u \le c_3\left(\left(\frac{1}{(R-\rho)^d} \int_{Q(x_0,R)} (u^+)^q dx\right)^{\frac{1}{q}} + k_0 + c_{NH}R^{\frac{d\epsilon}{p}}\right). \tag{6.68}$$

An important result of De Giorgi (see, e. g., [22] or [83]) is that membership of functions in the De Giorgi classes $DG_p^{\pm}(k_0, R_0, \mathcal{D})$ implies boundedness for the function above and below, respectively, while membership of functions in $DG_p(R_0, \mathcal{D})$ implies local Hölder continuity for this function. This result is stated in the following theorem ([22] or [83]).

Theorem 6.6.3 (De Giorgi). *Let $u \in DG_p(k_0, R_0, \mathcal{D})$.*
(i) *Then, for every ball $B(x_0, R) \subset \mathcal{D}$, with $R \le R_0$, there exists a constant c_{DG} (depending only on d, p and the constant c_0) such that*

$$\begin{aligned}
&- k_0 - c_{DG}\left(R + R^{-d\frac{p+1}{p}}|A^-(-k_0, x_0, R)|\left(\int_{A^-(-k_0,x_0,R)} (-u(x) - k_0)^p dx\right)^{1/p}\right)\\
&\le \inf_{x\in B(x_0,\frac{R}{2})} u(x)\\
&\le \sup_{x\in B(x_0,\frac{R}{2})} u(x)\\
&\le k_0 + c_{DG}\left(R + R^{-d\frac{p+1}{p}}|A^+(k_0, x_0, R)|\left(\int_{A^+(k_0,x_0,R)} (+u(x) - k_0)^p dx\right)^{1/p}\right).
\end{aligned}$$

(ii) *If $p > 1$ and u satisfies the Caccioppoli inequalities* (6.64) *for all $k \in \mathbb{R}$ and $R_0 = 1$, then u is locally Hölder continuous in $\mathcal{D}$, i. e., $u \in C^{0,\alpha}(\mathcal{D})$ for some $\alpha > 0$ depending on d, p, and c_0.*

Proof. See Section 6.10.2. □

Remark 6.6.4. The estimate for the upper bound in Theorem 6.6.3 can be given in terms of the mean value integral of the excess function on $B(x_0, R)$. This can be seen by expressing

$$\begin{aligned}
&R^{-d\frac{p+1}{p}} |A^+(k_0, x_0, R)| \Big(\int_{A^+(k_0,x_0,R)} (u(x) - k_0)^p dx \Big)^{1/p} \\
&= \Big(R^{-d} \int_{A^+(k_0,x_0,R)} (u(x) - k_0)^p dx \Big)^{1/p} (R^{-d}|A^+(k_0, x_0, R)|) \\
&\leq c_1 \Big(\fint_{B(x_0,R)} ((u(x) - k_0)^+)^p dx \Big)^{1/p}
\end{aligned}$$

for an appropriate constant $c_2 > 0$, independent of R, since $A^+(k_0, x_0, R) \subset B(x_0, R)$ and $|B(x_0, R)| = \omega_d R^d$, where ω_d is the volume of the unit ball in $\mathbb{R}^d$.

In fact, as shown by Di Benedetto and Trudinger (see Theorem 3 in [67]) positive functions that belong to the De Giorgi class satisfy a Harnack type inequality. We present the result in its form for functions satisfying the generalized Caccioppoli inequalities (6.65).

Theorem 6.6.5 (Harnack inequality [67]). *If $u \geq 0$ and $u \in DG_p(\mathcal{D})$, (with $c_{\mathrm{NH}} = 0$ and $k_0 = 0$), then for every $\sigma \in (0, 1)$ there exists a constant $c_H = c_H(p, d, \sigma, c_0)$ such that*[8]

$$\sup_{B(0,\sigma R)} u \leq c_H \inf_{B(0,\sigma R)} u. \tag{6.69}$$

Proof. See [67]; see also Section 6.10.3 for a sketch. □

The Harnack inequality is an important result, which provides useful information for solutions (or subsolutions) of elliptic type PDEs. From this one may derive the maximum principle[9] as well as other useful qualitative and quantitative properties of solutions (or sub- and supersolutions). While the Harnack inequality can be derived from

---

8 In general, it satisfies the generalized Harnack inequality $\sup_{B(x_0,\sigma R)} u \leq c_H(\inf_{B(x_0,\sigma R)} u + c_{\mathrm{NH}} R^{\frac{de}{p}})$.

9 This can be seen as follows: Assume u is in the homogeneous De Giorgi class, so it is continuous. Let its minimum value be 0 without loss of generality. The set $A_o := \{x \in \mathcal{D} : u(x) = 0\}$ is closed (by continuity). Furthermore, using the Harnack inequality (6.69), we can see that is also open. If $\mathcal{D}$ is connected, then $A_o = \mathcal{D}$. Hence, $u = 0$ over the whole of $\mathcal{D}$ (see, e. g., [83]).

the PDE directly, it is useful to note that the important contribution of the approach of [67] is that it derives the Harnack inequality directly from the variational integral and not using at all the corresponding Euler–Lagrange equation. Hence, this approach is general and does not require any additional smoothness properties on f.

6.6.2 Hölder continuity of minimizers

We now show that minimizers belong to the De Giorgi class, and hence, by Theorem 6.6.3(ii), enjoy Hölder continuity properties.

Definition 6.6.6 (Q-minimizers and minimizers). Consider any open subset $\mathcal{D}' \subset \mathcal{D}$ and $Q \geq 1$. A function $u \in W^{1,p}(\mathcal{D})$ such that $F(u;\mathcal{D}') \leq QF(u+\phi;\mathcal{D}')$ for any $\mathcal{D}' \subset \mathcal{D}$ and any $\phi \in W_0^{1,p}(\mathcal{D})$ is called a Q-minimizer for F. If the above property holds for every $\phi \leq 0$ (resp. $\phi \geq 0$), it is called a sub (resp. super) Q-minimizer. When $Q = 1$, we recover the standard notion of minimizer (or sub and super minimizing resp.).

The notion of Q-minimizers and sub/superminimizers is related to solutions and sub-/supersolutions of PDEs, a connection that can be made using the Euler–Lagrange equation for the variational problem (if this exists).

A very important step towards the regularity of solutions to variational problems is provided in the following proposition (see e. g., [83]).

Proposition 6.6.7. *Assume that* $F : W^{1,p}(\mathcal{D}) \to \mathbb{R}$ *satisfies the growth conditions*

$$|z|^p \leq f(x,u,z) \leq L(1+|z|^p) \tag{6.70}$$

for some $L > 0$.

(i) *If* $u \in W^{1,p}(\mathcal{D})$ *is a sub Q-minimizer for F, there exists* $c = c(p,L,Q)$ *such that for every* $k \in \mathbb{R}$ *and every pair* $B(x_0,r) \subset\subset B(x_0,R)$ *we have that u satisfies the Caccioppoli type inequality*

$$\int_{A^+(k,x_0,r)} |\nabla u(x)|^p dx \leq c\Big((R-r)^{-p} \int_{A^+(k,x_0,R)} (u(x)-k)^p dx + |A^+(k,x_0,R)|\Big). \tag{6.71}$$

(ii) *If* $u \in W^{1,p}(\mathcal{D})$ *is a super Q-minimizer for F, there exists* $c' = c'(p,L,Q)$ *such that for every* $k \in \mathbb{R}$ *and every pair* $B(x_0,r) \subset\subset B(x_0,R)$ *we have that u satisfies the Caccioppoli type inequality*

$$\int_{A^-(k,x_0,r)} |\nabla u(x)|^p dx \leq c'\Big((R-r)^{-p} \int_{A^-(k,x_0,R)} (k-u(x))^p dx + |A^-(k,x_0,R)|\Big). \tag{6.72}$$

Proof. (i) The proof proceeds in 3 steps.

1. For any r_1, r_2 such that $r \le r_1 < r_2 \le R$ and $k \in \mathbb{R}$, we will show that any sub Q-minimizer of F, for a suitable $\theta < 1$, satisfies the estimate

$$\|\nabla u\|^p_{L^p(A^+(k,x_0,r_1))} \le \theta\|\nabla u\|^p_{L^p(A^+(k,x_0,r_2))} + \|(u-k)\|^p_{L^p(A^+(k,x_0,r_2))} + |A^+(k,x_0,r_2)|. \tag{6.73}$$

Consider any r_1, r_2 such that $r \le r_1 < r_2 \le R$, and choose a cut off function $\psi \in C_c^\infty(B(x_0,r_2))$ such that $0 \le \psi \le 1$ and $\psi \equiv 1$ in $B(x_0,r_1)$ with $|\nabla\psi| \le \frac{2}{r_2-r_1}$. Since u is a sub Q-minimizer, we choose as test function $\phi = -\psi(u-k)^+ \le 0$ and $F(u;A^+(k,x_0,r_2)) \le Q\,F(u+\phi;A^+(k,x_0,r_2))$, where the set[10] $A^+(k,x_0,r_2)$ is the intersection of the k superlevel set for u with the ball $B(x_0,r_2)$. To simplify the notation in the intermediate calculations, we will use the notation $A_i := A^+(k,x_0,r_i)$, $i = 1,2$ and note that $A_1 \subset A_2$.

Using the growth condition for the functional, we see that

$$\begin{aligned}\int_{A_2} |\nabla u(x)|^p dx &\overset{(6.70)}{\le} \int_{A_2} f(x,u(x),\nabla u(x))dx \\ &\le Q\int_{A_2} f(x,u(x)+\phi(x),\nabla(u+\phi)(x))dx \\ &\overset{(6.70)}{\le} QL\int_{A_2}(1+|\nabla(u+\phi)(x)|^p)dx.\end{aligned} \tag{6.74}$$

Since $\nabla(u-k)^+ = \nabla u\mathbf{1}_{\{u>k\}}$, by a standard algebraic inequality and the Leibniz rule, using the properties of the cutoff function ψ, we have that

$$|\nabla(u+\phi)(x)|^p \le c[(1-\psi(x))^p|\nabla u(x)|^p + (r_2-r_1)^{-p}(u(x)-k)^p], \quad \text{on } A_2,$$

for an appropriate constant c. Since $\psi \equiv 1$ on $B(x_0,r_1)$, we have that $1-\psi \equiv 0$ on this set. Hence, integrating the above inequality over $A_2 := A^+(k,x_0,r_2)$, we see that the first term contributes to the integral only over $A^+(k,x_0,r_2) \cap (B(x_0,r_2) \setminus B(x_0,r_1)) = A^+(k,x_0,r_2) \setminus A^+(k,x_0,r_1) = A_2 \setminus A_1$. This observation leads to the estimate

$$\begin{aligned}&\int_{A_2} |\nabla(u+\phi)(x)|^p dx \\ &\le c\Big[\int_{A_2\setminus A_1} |\nabla u(x)|^p dx + \int_{A_2}(1+(r_2-r_1)^{-p}(u(x)-k)^p)dx\Big] \\ &\le c\Big[\int_{A_2\setminus A_1} |\nabla u(x)|^p dx + (r_2-r_1)^{-p}\int_{A_2}(u(x)-k)^p)dx + |A_2|\Big].\end{aligned} \tag{6.75}$$

10 This is not necessarily an open set, but we may approximate it by an open set.

Combining (6.74) with (6.75) along with the trivial observation that $A_1 := A^+(k, x_0, r_1) \subset A_2 := A^+(k, x_0, r_2)$, since $r_1 < r_2$, we obtain the estimate

$$\int_{A_1} |\nabla u|^p dx \le c\Big[\int_{A_2 \setminus A_1} |\nabla u(x)|^p dx + (r_2 - r_1)^{-p} \int_{A_2} (u(x) - k)^p) dx + |A_2| \Big]. \tag{6.76}$$

Note that (6.76) is very close to (6.73), but not quite so. The fact that on the right-hand side of (6.76), we have the integral of $|\nabla u|^p$ on $A_2 \setminus A_1 := A^+(k, x_0, r_2) \setminus A^+(k, x_0, r_1)$ rather than on $A_2 := A^+(k, x_0, r_2)$ (as it appears on the left-hand side) is annoying. As the integral on the right-hand side is on an annulus type region (i. e., on $A_2 := A^+(k, x_0, r_2)$ with a hole $A_1 := A^+(k, x_0, r_1)$), we employ the so called hole-filling technique of Wilder, which consists of adding the term $c \int_{A_1} |\nabla u|^p dx$ on both sides of (6.76). We obtain upon dividing by $1 + c$ and redefining the constant c (using the same notation) and setting $\theta = \frac{c}{c+1} < 1$, that

$$\int_{A_1} |\nabla u(x)|^p dx \le \theta \int_{A_2} |\nabla u(x)|^p dx + (r_2 - r_1)^{-p} \int_{A_2} (u(x) - k)^p) dx + |A_2|, \tag{6.77}$$

where we also used the elementary estimate $\frac{c}{c+1} < c$. This is (6.73).

2. We will show that the Caccioppoli type inequality

$$\|\nabla u\|^p_{L^p(A^+(k,x_0,r))} \le c\big((R - r)^{-p} \|u - k\|^p_{L^p(A^+(k,x_0,R))} + |A^+(k, x_0, R)|\big), \tag{6.78}$$

holds by iterating appropriately (6.73) (equiv. (6.77)).

We define the real valued functions

$$\begin{aligned}
\varphi(\rho) &:= \int_{A^+(k,x_0,\rho)} |\nabla u(x)|^p dx, \\
\widehat{\varphi}_1(\rho) &:= \int_{A^+(k,x_0,\rho)} (u(x) - k)^p) dx, \\
\widehat{\varphi}_2(\rho) &:= |A^+(k, x_0, \rho)|, \\
\widehat{\varphi}(\rho_1, \rho_2) &:= \rho_1^{-p} \widehat{\varphi}_1(\rho_2) + \widehat{\varphi}_2(\rho_2),
\end{aligned} \tag{6.79}$$

and note that φ is nondecreasing, while $\widehat{\varphi}$ is nondecreasing in the second variable.

Since r_1, r_2 in step 1 were arbitrary, we pick any ρ, ρ' such that $r \le \rho \le \rho' \le R$, set $r_1 = \rho$ and $r_2 = \rho'$, and we redress the estimate (6.73) (equiv. (6.77)) using the functions defined in (6.79) as

$$\varphi(\rho) \le \theta\varphi(\rho') + \widehat{\varphi}(\rho' - \rho, \rho'), \quad \forall\, r \le \rho \le \rho' \le R. \tag{6.80}$$

We claim that (6.80) may be iterated to yield

$$\varphi(r) \le c\widehat{\varphi}(r,R) \tag{6.81}$$

for an appropriate constant c (depending on θ), which is the required estimate (6.78).

3. It remains to prove that (6.80) implies (6.81).

To see this, consider the increasing sequence $\rho_n = r + (1-\lambda^n)(R-r)$, $n = 0, 1, \ldots$, for an appropriate $0 < \lambda < 1$, such that $\lambda^{-p}\theta < 1$, and note that $\rho_n - \rho_{n-1} = \lambda^{n-1}(1-\lambda)(R-r)$. We write the inequality choosing each time the pair $\rho = \rho_n$, $\rho' = \rho_{n+1}$ to get

$$\varphi(\rho_n) \le \theta\varphi(\rho_{n+1}) + KM^{n-1}\widehat{\varphi}_1(\rho_{n+1}) + \widehat{\varphi}_2(\rho_{n+1}), \quad n = 0, \ldots, N-1,$$

where

$$M := \lambda^{-p}, \quad K := (1-\lambda)^{-p}(R-r)^{-p},$$

and upon multiplying each one by θ^n and adding, we obtain that

$$\begin{aligned}
\varphi(r) &\le \theta^N\varphi(\rho_N) + KM^{-1}\sum_{m=0}^{N-1}(\theta M)^m\widehat{\varphi}_1(\rho_{m+1}) + \sum_{m=0}^{N-1}\theta^m\widehat{\varphi}_2(\rho_{m+1}) \\
&\le \theta^N\varphi(\rho_N) + KM^{-1}\left(\sum_{m=0}^{N-1}(\theta M)^m\right)\widehat{\varphi}_1(R) + \left(\sum_{m=0}^{N-1}\theta^m\right)\widehat{\varphi}_2(R) \\
&\le \theta^N\varphi(\rho_N) + Kc_1\widehat{\varphi}_1(R) + c_2\widehat{\varphi}_2(R),
\end{aligned}$$

where, for the second estimate, we used the fact that $\widehat{\varphi}_i(\rho(m)) \le \widehat{\varphi}_i(R)$ for every $m = 1, \ldots, N-1$, $i = 1, 2$ (since $\widehat{\varphi}_i$ are nondecreasing and ρ_m is increasing), and we have set

$$\begin{aligned}
c_1 &:= M^{-1}\sum_{m=0}^{\infty}(\theta M)^m = \frac{M^{-1}}{1-\theta M}, \\
c_2 &:= \sum_{m=0}^{\infty}\theta^m = \frac{1}{1-\theta}.
\end{aligned}$$

Passing to the limit as $N \to \infty$ and choosing $c = \max(c_1, c_2)$, we deduce (6.81). The proof is complete.

(ii) If u is a super Q-minimizer, then $-u$ is a sub Q-minimizer, and the proof follows by (i) upon replacing u with $-u$ and k by $-k$. □

We are now ready to conclude the local Hölder continuity of minimizers.

Theorem 6.6.8 (Hölder continuity of Q-minimizers). *Assume that $F : W^{1,p}(\mathcal{D}) \to \mathbb{R}$ satisfies the growth conditions*

$$|z|^p \le f(x,u,z) \le L(1+|z|^p) \tag{6.82}$$

for some $L > 0$. Then, any Q-minimizer is locally Hölder continuous.

Proof. Any Q-minimizer is in $DG_p(R_0, \mathcal{D})$ by Proposition 6.6.7. Then by Theorem 6.6.3 u is locally Hölder continuous with a suitable exponent α. □

Remark 6.6.9. The result of Proposition 6.6.7 can be generalized in order to deal with more general structure conditions (see, e. g., [83]). Assume that f is a Carathéodory function satisfying the structure condition

$$L_0|z|^p - b(x)|y|^q - a(x) \leq f(x,y,z) \leq L_1|z|^p + b(x)|y|^q + a(x), \tag{6.83}$$

where $1 < p \leq q < \mathfrak{s}_p = \frac{pd}{d-p}$ and a, b are positive functions in $L^s(\mathcal{D})$ and L^r, respectively, with $s > \frac{d}{p}$ and $r > \frac{\mathfrak{s}_p}{\mathfrak{s}_p - q}$. Then, for every $x_0 \in \mathcal{D}$, there exists a R_0 depending on u and b such that for every $0 < r < R < \min(R_0, \operatorname{dist}(x_0, \partial\mathcal{D}))$, and every $k \geq 0$, we have the Caccioppoli-like inequality

$$\int_{A^+(k,r)} |\nabla u(x)|^p dx \leq \frac{c}{(R-r)^p} \int_{A^+(k,r)} (u(x) - k)^p dx + c(\|a\|_{L^s(\mathcal{D})} + k^p R^{-d\epsilon})|A^+(k,R)|^{1-\frac{p}{d}+\epsilon},$$

where $\epsilon > 0$ is such that $\frac{1}{s} = \frac{p}{d} - \epsilon$ and $\frac{1}{r} = 1 - \frac{q}{\mathfrak{s}_p} - \epsilon$. Note that if $s = r = \infty$, then we may choose $\epsilon = \frac{p}{d}$, in which case $k^p R^{-d\epsilon}|A^+(k,R)|^{1-\frac{p}{n}+\epsilon} = (\frac{k}{R})^p |A^+(k,R)|$, leading to a Caccioppoli inequality very similar to (6.71). This is a local result, as the radius of the ball in which it holds depends on the data of the problem near x_0. The proof of this result follows along the same lines as in Proposition 6.6.7, with a little extra complications (see Section 6.10.4 for a simplified case or [83] for the result in its full generality). One may also note that if local boundedness for the minimizer has been shown (see Remark 6.6.2), then the structure condition (6.83) can be replaced by a local structure condition of the form $|z|^p - a_c(x) \leq f(x,y,z) \leq L|z|^p + a_c(x)$, with the positive function a_c depending on the local upper bound. Therefore, by Proposition 6.6.7, a quasi-minimizer will satisfy Caccioppoli type inequalities of the form (6.64). Moreover, Q-minimizers for functionals satisfying the structure condition (6.83) satisfy Harnack type inequalities, in particular an inequality of the form given in Theorem 6.6.5, as long as $a = 0$.

6.6.3 Further regularity

Under more restrictive assumptions on the data of the problem, one may show that the minimizer enjoys further regularity properties.

In order to proceed, we will need to define the difference and the translation operators in the direction i, denoted by $\Delta_{h,i}$ and $\tau_{h,i}$, respectively.

Definition 6.6.10 (Difference and translation operators). For any function $u : \mathcal{D} \to \mathbb{R}$, any direction vector e_i, and $h \in \mathbb{R}$, we may define

(i) The translation operator $\tau_{h,i}u(x) := u(x + he_i)$,
(ii) The partial difference operator $\Delta_{h,i}(x) := \frac{u(x+he_i)-u(x)}{h} = \frac{\tau_{h,i}u-u}{h}(x)$.

Strictly speaking, the above operators are well defined for functions on $\mathcal{D}_h := \{x \in \mathcal{D} : \text{dist}(x, \partial\mathcal{D}) > |h|\}$.

The idea behind introducing the partial difference operator is that in the limit as $h \to 0$, under certain conditions, may approximate the partial derivatives of a function, in terms of this operator. In fact, as will be shown shortly, uniform L^p bounds for the $\Delta_{h,i}u$ for some $p > 1$ guarantee membership of $u \in W^{1,p}$. John Nirenberg used this idea in producing a theory for the regularity of elliptic problems. Here we adapt some of these ideas, following mainly the approach of [76] to discuss further regularity results for the solution of variational problems.

We start by collecting a number of useful results concerning the difference operator.

Proposition 6.6.11 (Properties of the difference operator).
(i) *The Leibnitz property* $\Delta_{h,i}(u_1u_2) = (\tau_{h,i}u_1)(\Delta_{h,i}u_2) + (\Delta_{h,i}u_1)u_2$ *is satisfied for any* u_1, u_2.
(ii) *Any function* $u \in L^1(\mathcal{D})$ *satisfies the integration by parts formula*

$$\int_{\mathcal{D}} \phi(x)\Delta_{h,i}u(x)dx = -\int_{\mathcal{D}} u(x)\Delta_{-h,i}\phi(x)dx, \quad \forall\, \phi \in C_c^1(\mathcal{D}),\ |h| < \text{dist}(\text{supp}\,\phi, \partial\mathcal{D}).$$

(iii) *If* $u \in W^{1,p}(\mathcal{D})$, *then* $\Delta_{h,i}u \in W^{1,p}(\mathcal{D})$ *and* $\frac{\partial}{\partial x_j}(\Delta_{h,i}u) = \Delta_{h,i}(\frac{\partial}{\partial x_j}u)$.
(iv) *Consider* $u \in L^p_{\text{loc}}(\mathcal{D})$ *with* $1 < p \le \infty$, *and let* p^* *be the conjugate exponent* $\frac{1}{p} + \frac{1}{p^*} = 1$. *Then for any* $i \in \{1, \dots, n\}$, *the partial derivative* $\frac{\partial u}{\partial x_i} \in L^p_{\text{loc}}(\mathcal{D})$, *if and only if for every* $\mathcal{D}' \subset\subset \mathcal{D}$ *there exists a constant* c *(depending on* $\mathcal{D}'$*) such that*

$$\left|\int_{\mathcal{D}'} (\Delta_{h,i}u)(x)\phi(x)dx\right| \le c\|\phi\|_{L^{p^*}(\mathcal{D}')}, \quad \forall\, \phi \in C_c^1(\mathcal{D}').$$

Proof. Claims (i)–(iii) follow by elementary algebraic manipulations, integration by parts using the translation invariance, and the fact that $W^{1,p}$ is a vector space, respectively, and are left as exercises to the reader. We only sketch the proof of (iv). If $\frac{\partial u}{\partial x_i} \in L^p_{\text{loc}}(\mathcal{D})$, then we express $(\tau_{h,i}u - u)(x) = \int_0^1 \frac{\partial u}{\partial x_i}(x + he_it)dt$ and estimate $\|\tau_{h,i}u - u\|_{L^p(\mathcal{D}')} \le |h|\,\|\frac{\partial u}{\partial x_i}\|_{L^p(\mathcal{D}')}$. We then estimate $|\int_{\mathcal{D}'}(\Delta_{h,i}u)(x)\phi(x)dx|$ with the use of the Hölder inequality. For the reverse implication, fix $\mathcal{D}' \subset\subset \mathcal{D}$, and note that for $\phi \in C_c^1(\mathcal{D}')$ we have that $\lim_{h\to 0}\Delta_{-h,i}\phi(x) = \frac{\partial}{\partial x_i}\phi(x)$, so using dominated convergence and the integration by parts formula

$$\left|\int_{\mathcal{D}'} u(x)\frac{\partial}{\partial x_i}\phi(x)dx\right| = \left|\lim_{h\to 0}\int_{\mathcal{D}'} u(x)(\Delta_{-h,i}\phi)(x)\right|$$

$$= \left| - \lim_{h \to 0} \int_{\mathcal{D}'} (\Delta_{h,i} u)(x) \phi(x) \right| \leq c \|\phi\|_{L^{p^*}(\mathcal{D}')},$$

with the last estimate following by duality since we have assumed that $\Delta_{h,i} u \in L^p(\mathcal{D}')$. □

Statement (iv) in Proposition 6.6.11 is crucial and will be used both in the direct and in the converse direction. In particular, it is important to stress that if $u \in W^{1,p}(\mathcal{D}_{|h|})$, then for every $i \in \{1, \dots, d\}$,

$$\|\Delta_{h,i} u\|_{L^p(\mathcal{D})} \leq \|\nabla u\|_{L^p(\mathcal{D}_{|h|})},$$

where $\mathcal{D}_{|h|}$ is the $|h|$-neighborhood of $\mathcal{D}$, as well as that for small enough h ($h < h_0$ with h_0 depending on the d and $\mathrm{dist}(\mathcal{D}', \partial\mathcal{D})$), we have that $\|\Delta_{h,i} u\|_{L^p(\mathcal{D}')} \leq c \|\frac{\partial u}{\partial x_i}\|_{L^p(\mathcal{D})}$. On the converse side,

$$\text{if } \forall\, \mathcal{D}' \subset\subset \mathcal{D}\; \exists\, c \text{ such that } \forall\, h < \mathrm{dist}(\mathcal{D}', \mathcal{D}), \text{ we have } \|\Delta_{h,i} u\|_{L^p(\mathcal{D}')} \leq c,$$

$$\text{then } \frac{\partial u}{\partial x_i} \in L^p(\mathcal{D}) \text{ and } \left\| \frac{\partial u}{\partial x_i} \right\|_{L^p(\mathcal{D})} \leq c.$$

We present here a simple example that illustrates how the use of the partial difference operator as a test function may lead to $W^{2,2}_{\mathrm{loc}}(\mathcal{D})$ estimates for the solution of a variational problem, or even more general elliptic problems. Starting from this estimate, one may employ more sophisticated techniques and obtain Hölder continuity of the derivatives of the solution.

Following Evans (see [76]), let $A = (A_1, \dots, A_d) : \mathbb{R}^d \to \mathbb{R}^d$ be a C^1 function and consider the nonlinear eliptic system

$$\begin{aligned} -\operatorname{div} \cdot (A(\nabla u)) &= g, && \text{in } \mathcal{D}, \\ u &= 0 && \text{on } \partial\mathcal{D}. \end{aligned} \tag{6.84}$$

In the special case, where $A_i = \frac{\partial f}{\partial z_i}$, $i = 1, \dots, d$, for some C^2 scalar valued function $f : \mathbb{R}^d \to \mathbb{R}$, the system (6.84) can be recognized as the Euler–Lagrange equation for the minimization of the functional $F(u) := \int_{\mathcal{D}} f(\nabla u) dx$.

We will adopt the following assumptions on the vector field A:

Assumption 6.6.12. The vector field $A : \mathbb{R}^d \to \mathbb{R}^d$ satisfies the following:

(i) There exists $c_0 > 0$ such that for any $z \in \mathbb{R}^d$, it holds that $\sum_{i,j=1}^d \frac{\partial A_i}{\partial z_j}(z) \xi_i \xi_j \geq c_0 |\xi|^2$ for every $\xi = (\xi_1, \dots, \xi_d) \in \mathbb{R}^d$.

(ii) There exists a constant c_1 such that for every $i \in \{1, \dots, d\}$ and $z \in \mathbb{R}^d$, it holds that $|\frac{\partial A_i}{\partial z_j}(z)| < c_1$.

The above assumptions in the special case where $A = \nabla f$ (variational case) may be expressed in terms of the Hessian of the function f and can be related to the convexity properties of the function f. Furthermore, for the purposes of this section, we have made rather heavy assumptions on A. Similar results, as the ones stated here can be proved under weaker assumptions, using more elaborate and technical arguments. The next proposition provides a regularity result in the spirit of [76].

Proposition 6.6.13. *Under Assumption* 6.6.12*, on the vector field* A*, any solution* $u \in W_0^{1,2}(\mathcal{D})$ *of* (6.84) *is also in* $W_{\mathrm{loc}}^{2,2}(\mathcal{D})$*, and in particular, for any* $\mathcal{D}' \subset\subset \mathcal{D}$*, it holds that*

$$\|D^2 u\|_{L^2(\mathcal{D}')} \leq c(\|g\|_{L^2(\mathcal{D})} + \|\nabla u\|_{L^2(\mathcal{D})})$$

for an appropriate constant c *(depending on* $\mathcal{D}'$*).*

Proof. Our aim is to show that for any open set $\mathcal{D}' \subset\subset \mathcal{D}$, it holds that

$$\int_{\mathcal{D}'} |D^2 v(x)|^2 dx \leq c\Big(\int_{\mathcal{D}} g(x)^2 dx + \int_{\mathcal{D}} |\nabla u(x)|^2 dx\Big),$$

which implies that $u \in W_{\mathrm{loc}}^{2,2}(\mathcal{D})$. The proof follows in 3 steps.

1. Fix an open set $\mathcal{D}' \subset\subset \mathcal{D}$, and choose an open set $\mathcal{D}''$ such that $\mathcal{D}' \subset\subset \mathcal{D}'' \subset\subset \mathcal{D}$. We also consider a cutoff function $\psi \in C_c^\infty(\mathcal{D}'')$ such that $\psi \equiv 1$ on $\mathcal{D}'$ and $0 \leq \psi \leq 1$. For a suitably small h, we fix $i \in \{1, \dots, d\}$ and consider as test function for the weak formulation of the problem $\phi = -\Delta_{-h,i}(\psi^2 \Delta_{h,i} u)$. Using the integration by parts formula and the properties of the difference operator, the weak form yields

$$\sum_{j=1}^{d} \int_{\mathcal{D}} \Delta_{h,i}(A_j(\nabla u))(x) \frac{\partial}{\partial x_j}(\psi^2 \Delta_{h,i} u)(x) dx = -\int_{\mathcal{D}} g(x)\, \Delta_{-h,i}(\psi^2 \Delta_{h,i} u)(x) dx. \tag{6.85}$$

Using the more convenient notation $v = \nabla u$ (resp. componentwise $v_j = \frac{\partial u}{\partial x_j}$) and the fact that the difference and the partial derivative operator commute, (6.85) can be expressed as

$$\begin{aligned}
&\sum_{j=1}^{d} \int_{\mathcal{D}} \Big[\psi(x)^2 \Delta_{h,i}(A_j(\nabla u))(x) \Delta_{h,i} v_j(x) \\
&\qquad + 2\psi(x) \frac{\partial \psi}{\partial x_j}(x) \Delta_{h,i}(A_j(\nabla u))(x) \Delta_{h,i} u(x)\Big] dx \\
&\quad = -\int_{\mathcal{D}} g(x)\, \Delta_{-h,i}(\psi^2 \Delta_{h,i} u)(x) dx.
\end{aligned}$$

Since A_j are differentiable, we have for any $j = 1, \dots, d$ that for every pair $w_\ell = (w_{\ell,1}, \dots, w_{\ell,d}) \in \mathbb{R}^d$, $\ell = 1, 2$,

$$\begin{aligned}
A_j(w_2) - A_j(w_1) &= \int_0^1 \frac{d}{ds} A_j(sw_2 + (1-s)w_1)ds \\
&= \int_0^1 \sum_{k=1}^d \frac{\partial}{\partial z_k} A_j(sw_2 + (1-s)w_1)(w_{2,k} - w_{1,k})ds \\
&= \left(\int_0^1 D_z A_j(sw_2 + (1-s)w_1)ds \right) \cdot (w_2 - w_1).
\end{aligned}$$

We apply the above for the choice $w_2 = \nabla u(x + he_i)$ and $w_1 = \nabla u(x)$ for any $x \in \mathcal{D}$, and recalling the definition of $\Delta_{h,i}(A_j(\nabla u)) = \Delta_{h,i}(A_j(v))$, we obtain

$$\begin{aligned}
\Delta_{h,i}(A_j(\nabla u)) = \Delta_{h,i}(A_j(v)) &= \frac{1}{h}(A_j(v(x + he_i)) - A_j(v(x)) \\
&= \sum_{k=1}^d \left(\int_0^1 \frac{\partial}{\partial z_k} A_j(s\nabla u(x + he_i) + (1-s)\nabla u(x))ds \right) \frac{1}{h}(v_k(x + he_i) - v_k(x)) \\
&= \sum_{k=1}^d a_{jk}(x)\Delta_{h,i} v_k(x),
\end{aligned}$$

where

$$a_{jk}(x) = \int_0^1 \frac{\partial}{\partial z_k} A_j(s\nabla u(x + he_i) + (1-s)\nabla u(x))ds.$$

By the assumption on A_j, we have that the matrix $\mathbf{A} = (a_{jk})_{j,k=1,\dots,d}$ satisfies the uniform ellipticity condition

$$\sum_{j,k=1}^d a_{jk} z_j z_k \geq c_0 |z|^2,$$

for some $c_0 > 0$.

We substitute the above in the weak formulation of the equation to get

$$\begin{aligned}
&\int_{\mathcal{D}} \sum_{j,k=1}^d \psi(x)^2 a_{jk}(x)\Delta_{h,i} v_k(x)\Delta_{h,i} v_j(x)dx \\
&\qquad + 2\int_{\mathcal{D}} \sum_{j,k=1}^d \psi(x)\frac{\partial \psi}{\partial x_j}(x) a_{jk}(x)\Delta_{h,i} v_k(x)\Delta_{h,i} u(x)dx \\
&\quad = -\int_{\mathcal{D}} g(x)\,\Delta_{-h,i}(\psi^2 \Delta_{h,i} u)(x)dx,
\end{aligned}$$

which leads to the following inequality:

$$\begin{aligned}\int_{\mathcal{D}} \sum_{j,k=1}^{d} \psi(x)^2 a_{jk}(x)\Delta_{h,i}v_k(x)\Delta_{h,i}v_j(x)dx \\ \leq 2\left|\int_{\mathcal{D}} \sum_{j,k=1}^{d} \psi(x)\frac{\partial\psi}{\partial x_j}(x)a_{jk}(x)\Delta_{h,i}v_k(x)\Delta_{h,i}u(x)dx\right| \\ + \left|\int_{\mathcal{D}} g(x)\,\Delta_{-h,i}(\psi^2\Delta_{h,i}u(x))dx\right|.\end{aligned} \tag{6.86}$$

2. We now estimate each of the above terms separately as follows:
For the first one,

$$\begin{aligned}c_0 \int_{\mathcal{D}''} \psi(x)^2|\Delta_{h,i}v(x)|^2 dx &\leq c_0 \int_{\mathcal{D}} \psi(x)^2|\Delta_{h,i}v(x)|^2 dx \\ &\leq \int_{\mathcal{D}} \sum_{j,k=1}^{d} \psi(x)^2 a_{jk}(x)\Delta_{h,i}v_k(x)\Delta_{h,i}v_j(x)dx,\end{aligned} \tag{6.87}$$

where, starting from the right to the left, we have used the ellipticity condition, based on the fact that $\mathcal{D}'' \subset \mathcal{D}$.

For the second one, we estimate using the Cauchy–Schwarz inequality (weighted by an arbitrary constant ϵ_1) as

$$\begin{aligned}&\left|\int_{\mathcal{D}} \sum_{j,k=1}^{d} \psi(x)\frac{\partial\psi}{\partial x_j}(x)a_{jk}(x)\Delta_{h,i}v_k(x)\Delta_{h,i}u(x)dx\right| \\ &\leq c_1 \int_{\mathcal{D}''} \psi(x)|\Delta_{h,i}v(x)|\,|\Delta_{h,i}u(x)|dx \\ &\leq c_1\epsilon_1 \int_{\mathcal{D}''} \psi(x)^2|\Delta_{h,i}v(x)|^2 dx + \frac{c_1}{4\epsilon_1}\int_{\mathcal{D}''} |\Delta_{h,i}u(x)|^2 dx.\end{aligned} \tag{6.88}$$

For the third one, by similar arguments (weighting by an arbitrary constant ϵ_2),

$$\left|\int_{\mathcal{D}} g(x)\,\Delta_{-h,i}(\psi^2\Delta_{h,i}u)(x)dx\right| \leq \frac{1}{4\epsilon_2}\int_{\mathcal{D}} g(x)^2 dx + \epsilon_2 \int_{\mathcal{D}} |\Delta_{-h,i}(\psi^2\Delta_{h,i}u)(x)|^2 dx. \tag{6.89}$$

We must somehow estimate the second integral, which consists of the differences of u. We use the notation $w = \Delta_{h,i}u$ and note that since $\psi^2\Delta_{h,i}u \in W^{1,2}(\mathcal{D})$, we have that

$$\int_{\mathcal{D}} |\Delta_{-h,i}(\psi^2\Delta_{h,i}u)(x)|^2 dx = \int_{\mathcal{D}} |\Delta_{-h,i}(\psi^2 w)(x)|^2 dx \leq c' \int_{\mathcal{D}_{|h|}} |\nabla(\psi^2 w)(x)|^2 dx$$

$$\le c'' \int_{\mathcal{D}_{|h|}} \psi(x)^4 |\nabla w(x)|^2 dx + c'' \int_{\mathcal{D}_{|h|}} 2\psi(x)^2 |\nabla \psi(x)|^2 w(x)^2 dx$$
$$\le c'' \int_{\mathcal{D}''} \psi^2 |\nabla w|^2 dx + c''' \int_{\mathcal{D}_{''}} w^2 dx$$
$$= c'' \int_{\mathcal{D}''} \psi(x)^2 |\Delta_{h,i} v(x)|^2 dx + c''' \int_{\mathcal{D}''} w(x)^2 dx,$$

where, for the first integral, we used the fact that $\psi^4 \le \psi^2$ (since $0 \le \psi \le 1$), for both the fact that $\operatorname{supp}\psi = \mathcal{D}'' \subset \mathcal{D}$, and finally the observation that $\nabla w = \nabla \Delta_{h,i} u = \Delta_{h,i}(\nabla u) = \Delta_{h,i} v$, as the difference operator and the gradient operator commute. Substituting the above estimate in (6.89), we have for the third term that

$$\begin{aligned} &\left| \int_{\mathcal{D}} g(x)\, \Delta_{-h,i}(\psi^2 \Delta_{h,i} u)(x) dx \right| \\ &\quad \le \frac{c}{\epsilon_2} \int_{\mathcal{D}} g(x)^2 dx + \epsilon_2 c'' \int_{\mathcal{D}''} \psi(x)^2 |\Delta_{h,i} v(x)|^2 dx + \epsilon_2 c''' \int_{\mathcal{D}''} |\Delta_{h,i} u(x)|^2 dx, \end{aligned} \tag{6.90}$$

upon substituting $w = \Delta_{h,i} u$ once more.

3. We use (6.87), (6.88), and (6.90) in the weak form estimate (6.86) to obtain upon combining like terms that

$$(c_0 - \epsilon_1 c_1 - \epsilon_2 c'') \int_{\mathcal{D}''} \psi(x)^2 |\Delta_{h,i} v(x)|^2 dx$$
$$\le \left(\frac{c_1}{4\epsilon_1} + \epsilon_2 c''' \right) \int_{\mathcal{D}''} |\Delta_{h,i} u(x)|^2 dx + \frac{1}{4\epsilon_2} \int_{\mathcal{D}} g(x)^2 dx.$$

Choosing ϵ_1, ϵ_2 so that $c_0 - \epsilon_1 c_1 - \epsilon_2 c'' \ge \frac{c_0}{2} > 0$, dividing by $\frac{c_0}{2}$, and redefining the constant c, we end up with

$$\begin{aligned} \int_{\mathcal{D}''} \psi(x)^2 |\Delta_{h,i} v(x)|^2 dx &\le c\left(\int_{\mathcal{D}} g(x)^2 dx + \int_{\mathcal{D}''} |\Delta_{h,i} u(x)|^2 dx \right) \\ &\le c\left(\int_{\mathcal{D}} g(x)^2 dx + \int_{\mathcal{D}} |\nabla u(x)|^2 dx \right), \end{aligned} \tag{6.91}$$

where for the last estimate, we used Proposition 6.6.11. Since $\mathcal{D}' \subset \mathcal{D}''$ and $\psi \equiv 1$ on $\mathcal{D}'$, we have that

$$\int_{\mathcal{D}'} |\Delta_{h,i} v(x)|^2 dx \le \int_{\mathcal{D}''} \psi(x)^2 |\Delta_{h,i} v(x)|^2 dx \le c\left(\int_{\mathcal{D}} g(x)^2 dx + \int_{\mathcal{D}} |\nabla u(x)|^2 dx \right),$$

and by the arbitrary nature of $\mathcal{D}''$ and the fact that inequality (6.91) holds for every h sufficiently small, we conclude using once more Proposition 6.6.11 that

$$\int_{\mathcal{D}'} |D^2 v(x)|^2 dx \leq \int_{\mathcal{D}''} \psi(x)^2 |\Delta_{h,i} v(x)|^2 dx \leq c\Big(\int_{\mathcal{D}} g(x)^2 dx + \int_{\mathcal{D}} |\nabla u(x)|^2 dx\Big),$$

which implies that $u \in W^{2,2}_{\text{loc}}(\mathcal{D})$. □

Since the constant c in the above estimate may depend on $\mathcal{D}'$, the estimates in Proposition 6.6.13 are interior estimates and cannot be continued up to the boundary without further complications. We will present such a result for a more general problem in the next chapter, in Section 7.6.2. Furthermore, the method can be extended for systems of elliptic equations.

6.7 A semilinear elliptic problem and its variational formulation

Let $\mathcal{D} \subset \mathbb{R}^d$ be a bounded domain, and consider the semilinear elliptic problem

$$\begin{aligned} -\Delta u &= f(x,u), \quad \text{in } \mathcal{D}, \\ u &= 0 \quad \text{on } \partial\mathcal{D}, \end{aligned} \tag{6.92}$$

where $f : \mathcal{D} \times \mathbb{R} \to \mathbb{R}$ is a continuous function, which, under extra conditions on the function f (that will be discussed in detail shortly), is associated with the functional $F : X \to \mathbb{R}$,

$$F(u) = \frac{1}{2} \int_{\mathcal{D}} |\nabla u(x)|^2 dx - \int_{\mathcal{D}} \Big(\int_0^{u(x)} f(x,s) ds \Big) dx, \tag{6.93}$$

where X is an appropriately chosen Banach space, a possible choice for which being $X := W^{1,2}_0(\mathcal{D})$.

Note that this is a special case of the general class of functionals studied in Sections 6.4 and 6.5. Using the slightly modified notation $F(u) = \int_{\mathcal{D}} \bar{f}(x, u(x), \nabla u(x)) dx$ for the general functional of the abovementioned sections, we have that $\bar{f}(x,y,z) = \frac{1}{2}|z|^2 + \int_0^y f(x,s) ds$. The general theory, developed in Sections 6.4 and 6.5, naturally applies to this specific case. However, this simpler special and separable form, which often appears in various important applications, will serve as a nice example of how the general conditions may be sharpened, and on how the specific form of the functional may inherit to the corresponding Euler–Lagrange equation certain rather special properties, not necessarily valid in the general case.

6.7.1 The case where sub- and supersolutions exist

We start our study of semilinear problems considering first the case, where the nonlinear function is such that a weak sub- and supersolution exists. This allows the treatment of (6.92) without growth conditions on the nonlinearity f, as well as obtaining detailed and localized information on its solution.

Definition 6.7.1 (Weak sub and supersolutions). Let

$$C^+ := \{v \in W_0^{1,2}(\mathcal{D}) \, : \, v(x) \geq 0 \text{ a. e.}\}.$$

(i) The function $\underline{u} \in W^{1,2}(\mathcal{D})$ is a weak subsolution of (6.92) if it is a weak solution of the inequality

$$\begin{aligned} -\Delta \underline{u} - f(x, \underline{u}) \leq 0, &\quad \text{in } \mathcal{D}, \\ \underline{u} \leq 0, &\quad \text{on } \partial\mathcal{D}, \end{aligned} \tag{6.94}$$

or equivalently that

$$\int_{\mathcal{D}} \nabla \underline{u}(x) \cdot \nabla v(x) dx - \int_{\mathcal{D}} f(x, \underline{u}(x)) v(x) dx \leq 0, \quad \forall \, v \in C^+.$$

(ii) The function $\bar{u} \in W^{1,2}(\mathcal{D})$ is a weak supersolution of (6.92) if it is a weak solution of the inequality

$$\begin{aligned} -\Delta \bar{u} - f(x, \bar{u}) \geq 0, &\quad \text{in } \mathcal{D}, \\ \bar{u} \geq 0, &\quad \text{on } \partial\mathcal{D}, \end{aligned} \tag{6.95}$$

or equivalently that

$$\int_{\mathcal{D}} \nabla \bar{u}(x) \cdot \nabla v(x) dx - \int_{\mathcal{D}} f(x, \bar{u}(x)) v(x) dx \geq 0, \quad \forall \, v \in C^+.$$

We now show (following [121]) how the information of existence of a sub- and a supersolution for (6.92), in the weak sense, can be combined with the calculus of variations in order to provide existence of weak solutions.

Proposition 6.7.2. *Assume that $f : \mathcal{D} \times \mathbb{R} \to \mathbb{R}$ is continuous and such that a (weak) subsolution $\underline{u}$ and a (weak) supersolution $\bar{u}$ exist for* (6.92) *(in the sense of Definition* 6.7.1*), with the property that $\underline{u} \leq \bar{u}$ a. e. in $\mathcal{D}$. Then, there exists a weak solution $u \in W_0^{1,2}(\mathcal{D})$ of* (6.92) *with the property $\underline{u} \leq u \leq \bar{u}$ a. e. in $\mathcal{D}$.*

Proof. Consider the modified function

$$\tilde{f}(x,s) := \begin{cases} f(x,\underline{u}(x)) & \text{if } s \le \underline{u}(x),\ x \in \overline{\mathcal{D}}, \\ f(x,s) & \text{if } \bar{u}(x) \le s \le \underline{u}(x),\ x \in \overline{\mathcal{D}}, \\ f(x,\bar{u}(x)) & \text{if } s \ge \bar{u}(x),\ x \in \overline{\mathcal{D}}, \end{cases}$$

and its primitive $\tilde{\Phi}(x,s) = \int_0^s \tilde{f}(x,\sigma)d\sigma$. It is easy to observe that the function $\tilde{\Phi}$ is sublinear.

We define the functional $\tilde{F} : X := W_0^{1,2}(\mathcal{D})$ by

$$\tilde{F}(u) = \frac{1}{2}\int_{\mathcal{D}} |\nabla u(x)|^2\, dx - \int_{\mathcal{D}} \tilde{\Phi}(x,u(x))\, dx.$$

This functional is easily seen to be sequentially weakly lower semicontinuous, and coercive by the sublinearity of $\tilde{\Phi}$. Note that the weak semicontinuity comes from the convexity of the functional in ∇u (see Theorem 6.3.10). Therefore, by a standard application of the direct method of the calculus of variations, a minimizer $u_o \in X := W_0^{1,2}(\mathcal{D})$ exists. We cannot at this point say anything about uniqueness of the minimizer since there are no conditions on f that guarantee strict convexity. It is also possible to show (see Section 6.7.2 for a more general case) that the minimizer u_o satisfies an Euler–Lagrange equation of the form (6.92), with f replaced by $\tilde{f}$.

We claim that the minimizer u_o has the property $\underline{u} \le u_o \le \bar{u}$.

One way to prove this claim is by using the Euler–Lagrange equation for the functional $\tilde{F}$, which is the weak form of the semilinear elliptic equation

$$\begin{aligned} -\Delta u_o - \tilde{f}(x,u_o) = 0, &\quad \text{in } \mathcal{D}, \\ u_o = 0 &\quad \text{on } \partial\mathcal{D}. \end{aligned} \tag{6.96}$$

Since $\underline{u}$ is a weak subsolution of (6.92), it satisfies the inequality (6.94), and from that, we observe that the function $v = \underline{u} - u_o$ satisfies the inequality[11]

$$-\Delta v - (f(x,\underline{u}) - \tilde{f}(x,u_o)) = -\Delta v - (f(x,v+u_o) - \tilde{f}(x,u_o)) \le 0.$$

We multiply by $v^+ = (\underline{u} - u_o)^+$ and integrate over $\mathcal{D}$ to obtain

$$\int_{\mathcal{D}} |\nabla v^+(x)|^2\, dx - \int_{\mathcal{D}} (f(x,\underline{u}(x)) - \tilde{f}(x,u_o(x)))v^+(x)\, dx \le 0, \tag{6.97}$$

where we have used the fact that

11 Weakly and for a positive test function.

$$-\int_{\mathcal{D}} \Delta v(x) v^+(x)\, dx = \int_{\mathcal{D}} |\nabla v^+(x)|^2\, dx.$$

By the definition of $\tilde{f}$, we note that if $v^+ \neq 0$, i. e., when we consider an $x \in \mathcal{D}$ such that $\underline{u}(x) \geq u_o(x)$, then $f(x, \underline{u}(x)) = \tilde{f}(x, u_o(x))$, so the contribution of the subset $\mathcal{D}_+ := \{x \in \mathcal{D} : \underline{u}(x) \geq u_o(x)\}$ to the second integral in (6.97) is zero. Obviously, the contribution to the same integral of the subset $\mathcal{D}_- := \{x \in \mathcal{D} : \underline{u}(x) \leq u_o(x)\}$ is also vanishing since $v^+(x) = 0$ for every $x \in \mathcal{D}_-$. Therefore, (6.97) becomes

$$\int_{\mathcal{D}} |\nabla v^+(x)|^2\, dx \leq 0,$$

which, by the positivity of this term, immediately yields that

$$\int_{\mathcal{D}} |\nabla v^+(x)|^2\, dx = 0,$$

so $\nabla v^+(x) = 0$, a. e., $x \in \mathcal{D}$. That means that $v^+ = c$ (constant) a. e. By the properties of u_o and $\underline{u}$, we see that $v^+ = 0$ on $\partial\mathcal{D}$, therefore, $v^+ = 0$ a. e. in $\mathcal{D}$. That implies that $\underline{u} \leq u_o$ a. e. in $\mathcal{D}$.

To obtain the other inequality, repeat the same steps, this time using (6.95) and multiplying with $w^+ = (u_o - \bar{u})^+$.

We conclude the proof by noting that since $\underline{u} \leq u_o \leq \bar{u}$, by the definition of $\tilde{f}$, it follows that $\tilde{f}(x, u_o) = f(x, u_o)$, therefore, (6.96) coincides with (6.92). □

The method of super- and sub-solutions, when applicable, is an important method allowing us to obtain sharp *a priori* estimates on the solutions of variational problems and the associated Euler–Lagrange PDE; it will be revisited in Chapter 7 (see Section 7.6.3).

The above scheme may be combined with an appropriate fixed point scheme, which allows us to obtain more information on solutions of (6.92). The following theorem (see [49]) shows how the existence of sub- and supersolutions guarantees the existence of solutions for (6.92):

Theorem 6.7.3. *Suppose that $f(x,u) = f_L(u) + f_0(x)$ with f_L Lipchitz and $f_0 \in (W_0^{1,2}(\mathcal{D}))^* = W^{-1,2}(\mathcal{D})$ and that there exist a sub- and supersolution to* (6.92), *$\underline{u}$ and $\bar{u}$, respectively, such that $\underline{u} \leq \bar{u}$. Then there exist two solutions $\underline{w}$ and $\overline{w}$ to* (6.92) *such that*

$$\underline{u} \leq \underline{w} \leq \overline{w} \leq \bar{u}.$$

Furthermore, $\underline{w}$ and $\overline{w}$ are the minimal and maximal (respectively) solutions between $\underline{u}$ and $\bar{u}$, linked to the sense that any other solution u of (6.92) *satisfies*

$$\underline{u} \leq \underline{w} \leq u \leq \overline{w} \leq \bar{u}.$$

Proof. Let L be the Lipschitz constant of f_L, fix $\rho > L$, and for a given $v \in L^2(\mathcal{D})$, consider the weak solution $u \in W_0^{1,2}(\mathcal{D})$ to the problem

$$-\Delta u + \rho u = \rho v + f_L(v) + f_0. \tag{6.98}$$

By the standard arguments of the calculus of variations, this problem admits a unique solution, and let $\mathsf{T}_\rho : L^2(\mathcal{D}) \to L^2(\mathcal{D})$ be the map defined in terms of the solution of this problem, by $\mathsf{T}_\rho(v) = u$. Clearly, a fixed point of the map T_ρ is a solution to (6.92). The continuity of $\mathsf{T}_\rho : L^2(\mathcal{D}) \to L^2(\mathcal{D})$ follows from the fact that upon considering (6.98) with a given right-hand side $g \in (W_0^{1,2}(\mathcal{D}))^\star$, by taking the weak form of the equation (6.98) using as test function v, then using the Poincaré inequality on one side and the Cauchy–Schwarz inequality on the other, and dividing by the norm of v to obtain that $\|v\|_{W_0^{1,2}(\mathcal{D})} \le c\|g\|_{(W_0^{1,2}(\mathcal{D}))^\star}$ for an appropriate constant c, and combining this observation with the Lipschitz property of f_L. It is important for the proof to note that for the choice of ρ as $\rho > L$, this map is also monotone, i. e., for any $v_1 \ge v_2$, it follows that $\mathsf{T}_\rho(v_1) \ge \mathsf{T}_\rho(v_2)$. This essentially follows from the comparison principle for elliptic problems. Indeed, set $f_i = \rho v_i + f_L(v_i) + f_0$, $i = 1, 2$, and observe that for any $x \in \mathcal{D}$

$$\begin{aligned} f_1(x) - f_2(x) &= \rho(v_1(x) - v_2(x)) + (f_L(v_1(x)) - f_L(v_2(x))) \\ &\ge \rho|v_1(x) - v_2(x)| - |f_L(v_1(x)) - f_L(v_2(x))| \ge (\rho - L)|v_1(x) - v_2(x)| \ge 0, \end{aligned}$$

where we used the Lipschitz continuity of f_L and the fact that $\rho > L$. We then see that $u_i = \mathsf{T}_\rho(v_i)$ solve in $W_0^{1,2}(\mathcal{D})$ the elliptic problem

$$-\Delta u_i + \rho u_i = f_i, \quad i = 1, 2,$$

and since $f_1 \ge f_2$, the comparison principle (see Proposition 6.2.6 and Example 6.2.8) yields that $u_1 \ge u_2$; hence, the monotonicity of T_ρ.

We now construct the solutions $\underline{w}$ and $\overline{w}$ using the following iterative schemes: Consider the sequence $(\underline{u}_n)_{n\in\mathbb{N}}$ (defined as $\underline{u}_n = \mathsf{T}_\rho^n \underline{u}$) and the sequence $(\bar{u}_n)_{n\in\mathbb{N}}$, defined as $\bar{u}_n = \mathsf{T}_\rho^n \bar{u}$, $n = 0, 1, \dots$, with the convention that $\mathsf{T}_\rho^0 = I$, the identity map. It is our aim to show that the sequences $(\underline{u}_n)_{n\in\mathbb{N}}$ and $(\bar{u}_n)_{n\in\mathbb{N}}$ are convergent, and the respective limits are the solutions $\underline{w}$ and $\overline{w}$ we are seeking.

We claim that, by construction, the sequence $(\underline{u}_n)_{n\in\mathbb{N}}$ is increasing. Note that for the first two terms, it holds that $\underline{u}_0 = \underline{u} \le \underline{u}_1 = \mathsf{T}_\rho \underline{u}$. Indeed, $\underline{u}_1 = \mathsf{T}_\rho \underline{u}$ is the solution to the elliptic problem (we omit the explicit statement of the boundary conditions, which are always homogeneous Dirichlet)

$$-\Delta \underline{u}_1 + \rho \underline{u}_1 = \rho \underline{u}_0 + f_L(\underline{u}_0) + f_0 = \rho \underline{u} + f_L(\underline{u}) + f_0. \tag{6.99}$$

Since $\underline{u}$ is a subsolution, we have, by rearranging, that

$$-\Delta \underline{u} \le f_L(\underline{u}) + f_0,$$

which combined with (6.99) yields

$$-\Delta \underline{u}_1 + \rho \underline{u}_1 \geq \rho \underline{u}_0 + f_L(\underline{u}_0),$$

and by the comparison principle (see Proposition 6.2.6 and Example 6.2.8), we see that $\underline{u}_1 \geq \underline{u}_0$. Now recall that T_ρ is monotone, therefore, applying T_ρ recursively, yields that $\underline{u}_{n-1} \leq \underline{u}_n$ for any $n \in \mathbb{N}$, so $(\underline{u}_n)_{n\in\mathbb{N}}$ is an increasing sequence.

We also claim that, by construction, the sequence $(\bar{u}_n)_{n\in\mathbb{N}}$ is decreasing. The proof is similar, but we need to use instead that the sequence is constructed by applying the monotone operator T_ρ to a supersolution, $\underline{u}$ rather than a subsolution. The details are left to the reader.

On the other hand, since by assumption, $\underline{u} \leq \bar{u}$, we have that $\underline{u}_0 \leq \bar{u}_0$ and applying iteratively the monotone operator T_ρ leads to the conclusion that $\underline{u}_n \leq \bar{u}_n$ for every $n \in \mathbb{N}$.

The above considerations allow us to conclude that

$$\underline{u} \leq \underline{u}_{n-1} \leq \underline{u}_n \leq \bar{u}_n \leq \bar{u}_{n-1} \leq \bar{u}, \quad \forall n \in \mathbb{N}.$$

Since $(\underline{u}_n)_{n\in\mathbb{N}}$ is an increasing sequence bounded above by $\bar{u}$, it converges pointwise to some function $\underline{w}$, which by a straightforward application of the Lebesgue dominated convergence theorem is such that $\underline{u}_n \to \underline{w}$ in $L^2(\mathcal{D})$. Similarly, since $(\bar{u}_n)_{n\in\mathbb{N}}$ is a decreasing sequence bounded below by $\underline{u}$, there exists a function $\overline{w}$ such that $\bar{u}_n \to \overline{w}$ in $L^2(\mathcal{D})$. Since $\underline{u}_{n+1} = \mathsf{T}_\rho(\underline{u}_n)$ for any n, taking the limit as $n \to \infty$ in both sides, and using the continuity of T_ρ in $L^2(\mathcal{D})$, we conclude that $\underline{w} = \mathsf{T}_\rho(\underline{w})$, so $\underline{w}$ is a fixed point for T_ρ, hence, a solution for (6.92). Similarly, by taking the limit as $n \to \infty$ in $\bar{u}_{n+1} = \mathsf{T}_\rho(\bar{u}_n)$, we conclude that $\overline{w} = \mathsf{T}_\rho(\overline{w})$, so $\overline{w}$ is a fixed point for T_ρ, hence, a solution for (6.92).

For the minimal and maximal nature of $\underline{w}$ and $\overline{w}$, respectively, let u be any solution of (6.92) such that $\underline{u} \leq u \leq \bar{u}$. Apply the monotone operator T_ρ for n times, and taking into account that u as a solution of (6.92) is a fixed point of T_ρ, we see that $\underline{u}_n \leq u \leq \bar{u}_n$ for any $n \in \mathbb{N}$, and passing to the limit as $n \to \infty$, we obtain the required result. □

Example 6.7.4. Consider the logistic equation $-\Delta u = u(\lambda - u)$ with homogeneous Dirichlet boundary conditions. Let λ_1 be the first eigenvalue of $-\Delta$ with homogeneous Dirichlet boundary conditions and $\phi_1 > 0$ the corresponding eigenfunction (see Prop. 6.2.12). If $\lambda > \lambda_1$, then this equation admits a solution $u \in W_0^{1,2}(\mathcal{D})$ satisfying $0 < u < \lambda$. This can be shown by noting that $\overline{u} = \lambda$ is a supersolution, whereas for $\epsilon > 0$ sufficiently small, $\underline{u} = \epsilon\phi_1$ is a subsolution such that $\underline{u} \leq \overline{u}$, where $\phi_1 > 0$ is the eigenfunction of the Laplacian $-\Delta$ corresponding to λ_1. This follows easily by the properties of ϕ_1, which is an L^∞ function. Then, a straightforward application of Theorem 6.7.3 yields the required result. ◁

6.7.2 Growth conditions on the nonlinearity

We now consider the general case, where we do not have information concerning sub- or supersolutions, but we impose instead appropriate growth conditions on the nonlinear function f. The results in this section are in the spirit of [15] or [79].

Assumption 6.7.5. The Carathéodory function $f : \mathcal{D} \times \mathbb{R} \times \mathbb{R}$ satisfies the:
(i) Growth condition: $|f(x,s)| \leq c|s|^r + \alpha_1(x)$ for every $(x,s) \in \mathcal{D} \times \mathbb{R}$, where $0 \leq r < \mathfrak{s}_2 - 1 = \frac{d+2}{d-2}$ and $\alpha_1 \in L^{r^*}(\mathcal{D})$, where $\mathfrak{s}_2 = \frac{2d}{d-2}$ if $d \geq 3$. In the cases $d = 1, 2$ no growth constraint on f is required.
(ii) Nonresonance condition: There exists $\nu < \lambda_1$ such that $\limsup_{|s|\to\infty} \frac{f(x,s)}{s} \leq \nu$, uniformly in $x \in \mathcal{D}$, where $\lambda_1 > 0$ is the first eigenvalue of the Dirichlet Laplacian.

As we will see, Assumption 6.7.5(i) guarantees the differentiability of the functional F, Assumption 6.7.5(ii) that it is bounded below and coercive, so the semilinear equation (6.92) corresponds to the first-order condition for the minimization problem of the functional F, which is well posed.

Proposition 6.7.6. *If f satisfies Assumption* 6.7.5, *then the semilinear equation* (6.92) *admits a solution in $u \in W_0^{1,2}(\mathcal{D})$.*

Proof. As mentioned above, the strategy of the proof is to show that $F : X := W_0^{1,2}(\mathcal{D}) \to \mathbb{R}$ is differentiable, coercive, and sequentially weakly lower semicontinuous. Hence, the minimization problem is well posed, and the minimum satisfies the first-order condition, which is (6.92). The proof is broken up into 4 steps.

1. F is $C^1(X;\mathbb{R})$ (i. e., continuously Fréchet differentiable).

As F consists of the Dirichlet functional, for which we have already established this result (see Example 2.1.10), plus perturbation by the integral functional $F_\Phi(u) = \int_{\mathcal{D}} \Phi(x,u(x))dx$, where $\Phi(x,s) := \int_0^s f(x,r)dr$, we only have to check the continuous Fréchet differentiability for the latter. This requires the growth condition of Assumption 6.7.5(i).

We claim that the Gâteaux derivative of F_Φ can be expressed as

$$\int_{\mathcal{D}} (DF_\Phi(u))(x)v(x)\,dx = \int_{\mathcal{D}} f(x,u(x))v(x)\,dx, \quad \forall\, v \in W_0^{1,2}(\mathcal{D}),$$

or in more compact notation, using the duality pairing $\langle\cdot,\cdot\rangle$ between $(W_0^{1,2}(\mathcal{D}))^* = W^{-1,2}(\mathcal{D})$ and $W_0^{1,2}(\mathcal{D})$, as

$$\langle DF_\Phi(u)), v\rangle = \langle f(\cdot,u), v\rangle.$$

Indeed, for any $\epsilon > 0$, consider

$$\frac{1}{\epsilon}(F_\Phi(u+\epsilon v)-F_\Phi(u))=\int_{\mathcal{D}}\frac{1}{\epsilon}(\Phi(x,u(x)+\epsilon v(x))-\Phi(x,u(x)))\,dx,$$

and considering x as fixed, define the real valued function $\psi:\mathbb{R}\to\mathbb{R}$ by $t\mapsto\psi(t):=\Phi(x,u(x)+tv(x))$, which, by the definition of Φ, is a C^1 function and such that $\psi'(t)=f(x,u(x)+tv(x))v(x)$. Furthermore, for fixed $x\in\mathcal{D}$, consider the sequence of functions $\{w_\epsilon:\epsilon>0\}$ defined by

$$w_\epsilon(x):=\frac{1}{\epsilon}(\Phi(x,u(x)+\epsilon v(x))-\Phi(x,u(x)))=\frac{1}{\epsilon}(\psi(\epsilon)-\psi(0)).$$

By the properties of the function ψ, we have that for each $x\in\mathcal{D}$ (fixed), it holds that $w_\epsilon(x)\to\psi'(0)=f(x,u(x))v(x)$, as $\epsilon\to0^+$. Therefore, the sequence of functions $w_\epsilon\to w$ a. e. in $\mathcal{D}$, as $\epsilon\to0^+$, where the function $w:\mathcal{D}\to\mathbb{R}$ is defined by $w(x):=f(x,u(x))v(x)$. We thus conclude that

$$\frac{1}{\epsilon}(F_\Phi(u+\epsilon v)-F_\Phi(u))=\int_{\mathcal{D}}w_\epsilon(x)\,dx.$$

The remaining step is passing this a. e. convergence inside an integral, which can be accomplished using the Lebesgue dominated convergence theorem. This requires establishing that $|w_\epsilon|\le h$ for every $\epsilon>0$, where $h\in L^1(\mathcal{D})$. To this end, we note that for every $x\in\mathcal{D}$,

$$w_\epsilon(x)=\frac{1}{\epsilon}(\psi(\epsilon)-\psi(0))=\frac{1}{\epsilon}\int_0^\epsilon\frac{d\psi}{dt}(t)dt=\frac{1}{\epsilon}\int_0^\epsilon f(x,u(x)+tv(x))v(x)dt,$$

so

$$\begin{aligned}|w_\epsilon(x)|&\le\frac{1}{\epsilon}\int_0^\epsilon|f(x,u(x)+tv(x))|\,|v(x)|dt\\&\le\frac{1}{\epsilon}\int_0^\epsilon(c|u(x)+tv(x)|^r+a_1(x))|v(x)|dt\\&\le(c|u(x)+v(x)|^r+a_1(x))|v(x)|=:h(x),\end{aligned}$$

where we have first used the growth condition on f, and then the observation that since for $r>0$ the function $t\mapsto|u(x)+tv(x)|^r$ is increasing, we may majorise the integral by using the upper estimate $|u(x)+tv(x)|^r<|u(x)+v(x)|^r$. The latter estimate is of course independent of t, therefore the upper bound of the integral is obtained trivially. All that is left to do is to check whether $h\in L^1(\mathcal{D})$. Clearly, this requires careful choice of the exponent r, a task in which the fact that $u,v\in W_0^{1,2}(\mathcal{D})$ and the Sobolev embedding theorem proves very helpful. We observe that the growth condition in conjunction with the Sobolev embedding is sufficient for our claim to hold.

We will allow ourselves a little discussion around this last point (consider this as some sort of reverse engineering that motivates our choice of assumptions; readers more familiar with such calculations may skip this paragraph). Our immediate reaction is to say that since $u, v \in W_0^{1,2}(\mathcal{D})$, then they are definitely $L^2(\mathcal{D})$ functions, therefore, $r = 1$ would work. However, we can do much better than that recalling that since $u, v \in W_0^{1,2}(\mathcal{D})$, they in fact enjoy much better integrability properties than square integrability as a consequence of the Sobolev embedding theorem (see Theorem 1.5.11) according to which $W_0^{1,2}(\mathcal{D}) \hookrightarrow L^p(\mathcal{D})$ for appropriate values of p ($1 \le p < \mathfrak{s}_2$, where $\mathfrak{s}_2 = \frac{2d}{d-2}$ for $d > 3$ and $\mathfrak{s}_2 = \infty$ for $d = 1, 2$). That tells us that we may consider $u, v \in L^p(\mathcal{D})$ for some $p > 2$, thus allowing for h to be $L^1(\mathcal{D})$, even for values of r greater than 1. Since the highest value of p we are allowed to consider when applying the Sobolev embedding theorem depends on the dimension, we leave it as p for the time being in order to obtain an estimate for r. This can be done through the use of Hölder's inequality. Consider for example the term $|u|^r v$, where both $u, v \in L^p(\mathcal{D})$. In order that $|u|^r v \in L^1(\mathcal{D})$, we estimate

$$\big\| |u|^r v \big\|_{L^1(\mathcal{D})} = \int_{\mathcal{D}} |u(x)|^r v(x)\, dx \le \left\{ \int_{\mathcal{D}} |u(x)|^{r\ell}\, dx \right\}^{1/\ell} \left\{ \int_{\mathcal{D}} |v(x)|^{\ell^*}\, dx \right\}^{1/\ell^*}.$$

To keep the right-hand side bounded, we need to make the choice $\ell^* = \frac{\ell}{\ell-1} = p$ and $r\ell = p$, which leads to the conclusion that the largest value that may be assigned to r is $r = p-1$, where p is the largest p allowed in order to ensure that $W_0^{1,2}(\mathcal{D}) \hookrightarrow L^p(\mathcal{D})$. This can be found by Theorem 1.5.11 to be $p = \mathfrak{s}_2 = \frac{2d}{d-2}$ if $d \ge 3$ and $p = \infty$ (thus leading to unrestricted growth for f) if $d = 1, 2$. A similar argument allows us to estimate the term $a_1 v$ and find out the exact integrability conditions that must be imposed on the function a_1, given that $v \in L^p(\mathcal{D})$.

We have so far proved the claim that F_Φ is Gâteaux differentiable. In order to prove that it is Fréchet differentiable, by Proposition 2.1.14, it suffices to show that for u fixed, the mapping $u \mapsto DF_\Phi(u)$ considered as a mapping $W_0^{1,2}(\mathcal{D}) \to (W_0^{1,2}(\mathcal{D}))^*$ is continuous, i. e., if we consider a sequence $(u_n)_{n\in\mathbb{N}} \subset W_0^{1,2}(\mathcal{D})$ such that $u_n \to u$ in $W_0^{1,2}(\mathcal{D})$, then $DF_\Phi(u_n) \to DF_\Phi(u)$ in $(W_0^{1,2}(\mathcal{D}))^*$.

For a sequence $(u_n)_{n\in\mathbb{N}}$ such that $u_n \to u$ in $W_0^{1,2}(\mathcal{D})$, we have, by the Sobolev embedding theorem, that $u_n \to u$ in $L^s(\mathcal{D})$ for $s \in [1, \mathfrak{s}_2]$ (note that at this point we do not require compactness of the embedding, so $s = \mathfrak{s}_2$ is allowed). The best possible choice for s is to take $s = \mathfrak{s}_2$ (since if $u_n \to u$ in $L^{\mathfrak{s}_2}(\mathcal{D})$, then this convergence holds for any $L^{s'}(\mathcal{D})$ with $s' < \mathfrak{s}_2$). We may also choose a subsequence $(u_{n_k})_{k\in\mathbb{N}}$ such that $u_{n_k} \to u$ a. e. in $\mathcal{D}$ and $|u_{n_k}(x)| \le w(x)$ for some $w \in L^{\mathfrak{s}_2}(\mathcal{D})$. We will work along this subsequence, which we relabel as $(u_k)_{k\in\mathbb{N}}$ for convenience, and note that, by an application of Hölder's inequality,

$$|\langle DF_\Phi(u_k) - DF_\Phi(u), v\rangle| = \left|\int_{\mathcal{D}} (f(x,u_k(x)) - f(x,u(x)))v(x)dx\right| \le \|f(\cdot,u_k) - f(\cdot,u)\|_{L^{s^*}(\mathcal{D})}^{1/s^*}\|v\|_{L^s(\mathcal{D})}^{1/s}. \tag{6.100}$$

We claim that $\|f(\cdot,u_k) - f(\cdot,u)\|_{L^{s^*}(\mathcal{D})} \to 0$, as $k \to \infty$. Define the sequence of functions $(w_k)_{k\in\mathbb{N}}$, by $w_k(x) = f(x,u_k(x)) - f(x,u(x))$, and note that $w_k \to 0$ a. e. in $\mathcal{D}$. Our claim reduces to $\|w_k\|_{L^{s^*}(\mathcal{D})} \to 0$ as $k \to \infty$, which will follow by a straighforward application of Lebesgue's dominated convergence theorem if $|w_k(x)|^{s^*} < h(x)$, a. e. in $\mathcal{D}$ for some $h \in L^1(\mathcal{D})$. By the growth condition for f in Assumption 6.7.5(i), we see that

$$|w_k(x)|^{s^*} \le \left|2a_1(x) + c\,|u_k(x)|^r + c\,|u(x)|^r\right|^{s^*} \le \left|2a_1(x) + c\,|w(x)|^r + c\,|u(x)|^r\right|^{s^*}$$
$$\le c'\left(a_1(x)^{s^*} + |w(x)|^{r\,s^*} + |w(x)|^{r\,s^*}\right),$$

for some $c' > 0$. If we define h by $h(x) := c'(a_1(x)^{s^*} + |w(x)|^{r\,s^*} + |u(x)|^{r\,s^*})$ for every x, then, by the proper choice of r, we can guarantee that $h \in L^1(\mathcal{D})$. Since $w, u \in L^s(\mathcal{D})$, we can see that $|w|^{r\,s^*}, |v|^{r\,s^*} \in L^1(\mathcal{D})$, as long as $r\,s^* = s$, or equivalently $r = \frac{s}{s^*} = s - 1 = \mathfrak{s}_2 - 1$. Similar arguments lead us to the conclusion that if $a_1 \in L^{r_1}(\mathcal{D})$, then r_1 must be chosen so that $s^* = r_1$, i. e., $r_1 = \frac{s}{s-1} = \frac{\mathfrak{s}_2}{\mathfrak{s}_2-1}$, in order for $a_1^{s^*} \in L^1(\mathcal{D})$. Hence, under these assumptions, we may guarantee that $\|w_k\|_{L^{s^*}(\mathcal{D})} \to 0$, as $k \to \infty$.

Therefore, dividing (6.100) by $\|v\|_{W_0^{1,2}(\mathcal{D})}$ (and noting that by the Sobolev embedding $\|v\|_{L^s(\mathcal{D})} \le c_{\mathfrak{s}}\|v\|_{W_0^{1,2}(\mathcal{D})}$ for some constant $c_{\mathfrak{s}}$ depending on the domain $\mathcal{D}$ and the dimension), we conclude that upon taking the supremum over all $\|v\|_{W_0^{1,2}(\mathcal{D})} \le 1$, that $\|DF_\Phi(u_k) - DF_\Phi(u)\|_{(W_0^{1,2}(\mathcal{D}))^*} \to 0$, as $k \to \infty$.

Note, that we have shown the required result along the selected subsequence $(u_k)_{k\in\mathbb{N}}$, and in order to guarantee continuity, we must show that this holds for the whole sequence. This can be achieved by resorting to the Urysohn property for the strong convergence (see, e. g., Remark 1.1.52) for the sequence $(DF_\Phi(u_n))_{n\in\mathbb{N}} \subset (W_0^{1,2}(\mathcal{D}))^*$ in the standard fashion. This concludes the proof that F_Φ is Fréchet differentiable at u. Note that the same steps also lead to the conclusion that F_Φ is C^1.

2. F is sequentially weakly lower semicontinuous on X.

This follows directly by Tonelli's general result Theorem 6.3.10. An alternative way to see it without the use of this general result is to note that $F := F_D + F_\Phi$, i. e., it is a perturbation of the Dirichlet functional (that we proved its weak lower semicontinuity in Proposition 6.2.1), plus perturbation by the integral functional $F_\Phi(u) = \int_{\mathcal{D}} \Phi(x,u(x))dx$, where $\Phi(x,s) := \int_0^s f(x,r)dr$. The weak continuity of the functional $F_\Phi : X = W_0^{1,2}(\mathcal{D}) \to \mathbb{R}$ follows again by an application of the Sobolev embedding theorem. Consider a sequence $(u_n)_{n\in\mathbb{N}} \in X$ such that $u_n \rightharpoonup u$ in $X = W_0^{1,2}(\mathcal{D})$. By the compact embedding $W_0^{1,2}(\mathcal{D}) \overset{c}{\hookrightarrow} L^p(\mathcal{D})$, where $p = \mathfrak{s}_2$, we see that there exists a subsequence $(u_{n_k})_{k\in\mathbb{N}}$ such that $u_{n_k} \to u$ in $L^p(\mathcal{D})$ and a further subsequence $(u_{n_{k_\ell}})_{\ell\in\mathbb{N}}$ such that $u_{n_{k_\ell}} \to u$ a. e. in $\mathcal{D}$. By the growth condition on f, we have the growth condition $|\Phi(x,s)| \leq c|s|^{r+1} + \alpha_2$ on the primitive of f. Consider now the Nemitskii operator Φ_N, defined by $(\Phi_N(u))(x) = \Phi(x,u(x))$, for every $x \in \mathcal{D}$, which from the theory of the Nemitskii operator maps $L^p(\mathcal{D})$ into $L^{\frac{p}{r+1}}(\mathcal{D})$ continuously (see, e. g., [110] or [144]). Therefore, $\Phi_N(u_{n_{k_\ell}}) \to \Phi_N(u)$ in $L^{\frac{p}{r+1}}(\mathcal{D})$ and by the Lebesgue embedding theorem also in $L^1(\mathcal{D})$. This implies $F_\Phi(u_{n_{k_\ell}}) \to F_\Phi(u)$ (see, e. g., [57]).

3. The functional $F : X := W_0^{1,2}(\mathcal{D}) \to \mathbb{R}$ is bounded below and coercive. This requires the no resonance condition of Assumption 6.7.5(ii).

This condition on f implies that there exists a constant c such that $\Phi(x,s) < c + \frac{\lambda}{2}s^2$ for every $s \in \mathbb{R}$ and a. e. in $\mathcal{D}$. Then,

$$F(u) \geq \frac{1}{2}\int_{\mathcal{D}} |\nabla u(x)|^2 dx - c\,|\mathcal{D}| - \frac{\lambda}{2}\int_{\mathcal{D}} u(x)^2 dx, \tag{6.101}$$

where we use the simplified notation $|\mathcal{D}| = \mu_{\mathcal{L}_d}(\mathcal{D})$. By the variational characterization of eigenvalues of the Laplacian (see Proposition 6.2.12, and in particular (6.29)), we have that

$$\lambda_1 = \min_{u\in W_0^{1,2}(\mathcal{D})} \frac{\int_{\mathcal{D}} |\nabla u(x)|^2 dx}{\int_{\mathcal{D}} u(x)^2 dx}.$$

Hence, for any $u \in W_0^{1,2}(\mathcal{D})$, we have the Poincaré's inequality

$$\lambda_1 \int_{\mathcal{D}} u(x)^2 dx \leq \int_{\mathcal{D}} |\nabla u(x)|^2 dx, \tag{6.102}$$

and by the choice of $\lambda \leq \lambda_1$, we have that

$$\frac{1}{2}\int_{\mathcal{D}} |\nabla u(x)|^2 dx - \frac{\lambda}{2}\int_{\mathcal{D}} u(x)^2 dx \geq 0,$$

which when combined with (6.101), leads to the result that

$$F(u) \geq -c\,|\mathcal{D}|.$$

Hence, the functional is bounded below.

For coercivity, we need to assume that $\lambda < \lambda_1$. By Poincaré's inequality (6.102), the estimate (6.101) yields

$$F(u) \geq \frac{1}{2}\left(1 - \frac{\lambda}{\lambda_1}\right)\int_{\mathcal{D}} |\nabla u(x)|^2 dx - c\,|\mathcal{D}|,$$

and coercivity follows.

4. Combining the results of steps 1,2, and 3 above, the claim follows. □

Example 6.7.7 (Nontrivial solutions). If the function f is such that $f(x,0) = 0$ a. e. in $\mathcal{D}$, then the semilinear problem (6.92) always admits the trivial solution $u = 0$, which in this case is definitely a local minimizer of the functional F. Under what conditions may Proposition 6.7.6 yield minimizers (hence, solutions of (6.92)) other than the trivial solution?

That clearly depends on the rate by which $f(x,s)$ approaches 0, as $s \to 0$. A condition that will ensure that a nontrivial solution will exist is to assume that $\limsup_{s\to 0^+} \frac{f(s)}{s} > \lambda_1$. One way to see why this condition is sufficient, is to note that this condition guarantees that there exist $c > \lambda_1$ and $\delta > 0$ such that, as long as $s \in [0,\delta]$, the function f satisfies the lower bound $f(s) > c\,s$ for $s \in [0,\delta]$, which in turn implies that its primitive Φ satisfies $\Phi(s) > \frac{c}{2}s^2$, for such s. It is then possible to choose $\epsilon > 0$ small enough so that the function $v : \mathcal{D} \to \mathbb{R}$, defined by $v(x) = \epsilon\phi_1(x)$, where ϕ_1 is the eigenfuction corresponding to the eigenvalue λ_1, satisfies $v(x) \in [0,\delta]$ a. e. in $\mathcal{D}$ (we have shown that $\phi_1 > 0$ and $\phi_1 \in L^\infty(\mathcal{D})$). That means $\int_\mathcal{D} \Phi(v(x))dx \geq \frac{c}{2}\int_\mathcal{D} v(x)^2 dx$ and, using that, we conclude by the properties of c that

$$F(v) \leq \frac{\epsilon^2}{2}(\lambda_1 - c)\int_\mathcal{D} \phi_1(x)^2 dx < 0 = F(0).$$

This means that $u = 0$ is not a global minimum, and therefore the minimizers provided by Proposition 6.7.6 cannot be the trivial solution of (6.92). An alternative (but equivalent) way of seeing that is to note that $v = \epsilon\phi_1$ for $\epsilon > 0$ small enough is a subsolution of (6.92); the result may follow by an application of Proposition 6.7.2. ◁

Example 6.7.8 (Uniqueness). There is no information provided by Proposition 6.7.6 regarding the uniqueness of the solution. This requires extra conditions on f. Typically, such conditions are monotonicity conditions, e. g., the extra assumption that $(f(s_1) - f(s_2))(s_1 - s_2) \leq 0$ for every $s_1, s_2 \in \mathbb{R}$. Uniqueness can be easily checked by assuming two solutions u_1, u_2 of the PDE, deriving the corresponding PDE for their difference $w = u_1 - u_2$, and then using w as test function in the weak form of this PDE (details are left as an exercise). Possibility of multiple solutions in semilinear elliptic problems, and the study of the qualitative behavior of solutions as one or more parameters of the system vary is a very active field of study in nonlinear analysis called bifurcation theory. ◁

Example 6.7.9. Certain variants of the above theorem may be easily be constructed, in which a linear part can be included, at the cost of having to handle in certain cases a more complicated eigenvalue problem. For example, one may consider nonlinearities of the form $f(x,s) = -a(x)s + f_0(x,s)$, where $a(x) \in L^\infty(\mathcal{D})$, $a(x) \geq 0$, a. e. in $\mathcal{D}$. Then, we may treat the eigenvalue problem for the perturbed operator $-\Delta + a(x)I$ on $W_0^{1,2}(\mathcal{D})$, whose first eigenvalue $\lambda_{1,a} > 0$ satisfies the modified variational representation $\lambda_{1,a} = \inf_{v \in W_0^{1,2}(\mathcal{D})} F_{R,a}(v)$, where

$$F_{R,a}(v) := \frac{\int_{\mathcal{D}} |\nabla u(x)|^2 \, dx + \int_{\mathcal{D}} a(x) u(x)^2 \, dx}{\int_{\mathcal{D}} u(x)^2 \, dx}.$$

Then, an extension of Proposition 6.7.6, where now Assumption 6.7.5 is posed for f_0, with λ_1 replaced by $\lambda_{1,a}$, can be formulated. The coercivity estimates can be simplified working in the equivalent norm $\| \cdot \|_1$, defined by $\|u\|_1^2 = F_D(u) + \langle au, u \rangle_{L^2(\mathcal{D})}$. The details are left to the reader (see also [15]). ◁

6.7.3 Regularity for semilinear problems

We close our study of the semilinear problem (6.92) with a brief sojourn on the regularity properties of its weak solutions. It will be shown that under certain conditions on the nonlinearity, the weak solutions can be more regular than $W_0^{1,2}(\mathcal{D})$ functions, in particular they can be continuous functions. Our approach, the bootstrap method, builds on the regularity properties of the Laplacian and the linear Poisson equation (see [42]).

Assumption 6.7.10. The function f is Carathéodory and satisfies the growth condition

$$|f(x,s)| \leq c(1 + |s|^p), \quad \text{a. e. } x \in \mathcal{D}, \ \forall s \in \mathbb{R},$$

(a) for some $p \geq 1$ if $d \leq 2$ or (b) for $p < \frac{4}{d-2}$ if $d \geq 3$.

Proposition 6.7.11. *Let Assumption* 6.7.10 *hold. If* $u \in W_0^{1,2}(\mathcal{D})$ *satisfies* (6.92) *and furthermore is such that* $f(\cdot, u(\cdot)) \in (W_0^{1,2}(\mathcal{D}))^*$, *then* $u \in L^\infty(\mathcal{D}) \cap C(\mathcal{D})$.

Proof. Let us consider the case $d > 2$. Since by Proposition 6.2.11, we have regularity results for the linear equation $-\Delta u + \lambda u = f_0$ for some f_0 known and $\lambda > 0$. Our aim is to break up (6.92) into sub-problems, which are of this form, and then build the regularity result on that. To this end, let us consider the auxiliary problems

$$-\Delta u_1 + u_1 = g_1(t, u), \quad u_1 \in W_0^{1,2}(\mathcal{D}), \tag{6.103}$$

and

$$-\Delta u_2 + u_2 = g_2(t, u), \quad u_2 \in W_0^{1,2}(\mathcal{D}), \tag{6.104}$$

where the functions g_1, g_2 are chosen so as to satisfy $|g_1(x,s)| \leq c$, $|g_2(x,s)| \leq c|s|^p$, and $g(x,s) = g_1(x,s) + g_2(x,s) - s$, a. e. $x \in \mathcal{D}$ and for all $s \in \mathbb{R}$. By adding (6.103) and (6.104), we see that $u = u_1 + u_2$. If we can show the required regularity for u_1 and u_2 separately, then we have the stated result.

Since $|g_1(x,s)| \le c$, and the domain is bounded $g_1(t,u) \in L^2(\mathcal{D})$, hence, any weak solution u_1 of (6.103) satisfies $u_1 \in L^\infty \cap C(\mathcal{D})$, by Proposition 6.2.11. Thus, it remains to show that $u_2 \in L^\infty \cap C(\mathcal{D})$.

Since $u \in W_0^{1,2}(\mathcal{D})$ by the Sobolev embedding, we have that $u \in L^q(\mathcal{D})$ for $q \le \frac{2d}{d-2}$, and since $|g_2(x,s)| \le c|s|^p$, it follows that $g_2(x,u) \in L^{\frac{q}{p}}(\mathcal{D})$. If $\frac{q}{p} > \frac{d}{2}$, we may immediately apply Proposition 6.2.11 for (6.104) to yield $u_2 \in L^\infty(\mathcal{D}) \cap C(\mathcal{D})$. Choosing $q = \frac{2d}{d-2}$, we have the required condition on p. □

By more refined estimates, one can improve the range of values of p, for which the stated regularity result holds. For more details the reader may consult e. g. [42].

6.8 A variational formulation of the p-Laplacian

6.8.1 The p-Laplacian Poisson equation

In this section, we are going to apply the general formulation of the calculus of variations to study the problem of the minimization of the functional $F : W_0^{1,p}(\mathcal{D}) \to \mathbb{R}$, where $\mathcal{D}$ is a bounded domain, defined by

$$F(u) := \frac{1}{p}\int_{\mathcal{D}} |\nabla u(x)|^p \, dx - \int_{\mathcal{D}} f(x)u(x)\, dx, \quad \forall u \in W_0^{1,p}(\mathcal{D}),\ p \in (1,\infty), \tag{6.105}$$

and its connection with the quasilinear elliptic partial differential equation

$$\begin{aligned} -\nabla \cdot (|\nabla u|^{p-2}\nabla u) &= f, \quad \text{in } \mathcal{D}, \\ u &= 0 \quad \text{on } \partial\mathcal{D}, \end{aligned} \tag{6.106}$$

with the divergence operator $\nabla\cdot = \operatorname{div}$ (defined in (1.8)) interpreted in the weak sense.

Recalling that $(W_0^{1,p}(\mathcal{D}))^* = W^{-1,p^*}(\mathcal{D})$ with $\frac{1}{p} + \frac{1}{p^*} = 1$, and denoting by $\langle \cdot,\cdot \rangle$ the duality pairing between $W^{-1,p^*}(\mathcal{D})$ and $W_0^{1,p}(\mathcal{D})$, one may define the nonlinear operator $\mathsf{A} := (-\Delta_p) : W_0^{1,p}(\mathcal{D}) \to W^{-1,p^*}(\mathcal{D})$ by

$$\langle \mathsf{A}(u), v\rangle := \langle -\Delta_p u, v\rangle := \int_{\mathcal{D}} |\nabla u(s)|^{p-2} \nabla u(x)\cdot \nabla v(x) dx, \quad \forall\, u,v \in W_0^{1,p}(\mathcal{D}), \tag{6.107}$$

which is called the p-Laplace operator. For $p = 2$, this is a linear operator that coincides with the Laplacian.

Proposition 6.8.1. *Let $p \in (1,\infty)$. For every $f \in W^{-1,p^*}(\mathcal{D})$, there exists a unique solution of the quasilinear elliptic problem* (6.106). *This solution is the minimizer of the functional $F : X \to \mathbb{R}$ defined by* (6.105).

Proof. Consider the Banach spaces $X := W_0^{1,p}(\mathcal{D})$, its dual $X^\star := W^{-1,p^\star}(\mathcal{D})$, and let $\langle\cdot,\cdot\rangle$ be the duality pairing among them. The proof consists in showing the minimization problem for the functional F is well posed on the Banach space $X := W_0^{1,p}(\mathcal{D})$, and identifying (6.106) as the first-order condition for this problem. We proceed in 4 steps.

1. F is strictly convex, weakly sequentially lower semicontinuous and coercive.

The convexity of the functional $F : X = W^{1,p}(\mathcal{D}) \to \mathbb{R}$ follows by the convexity of the function $g : \mathbb{R}^d \to \mathbb{R}$, defined by $g(z) = \frac{1}{p}|z|^p$, for $p > 1$. Strict convexity follows by its resemblance with the L^p norm of ∇u (see Theorem 2.6.5). The weak lower sequential semicontinuity of F follows by a direct application of Theorem 6.3.10 by taking into account the convexity of the function $\varphi(z) = \frac{1}{p}|z|^p$. Coercivity of F on $W_0^{1,p}(\mathcal{D})$ follows by the use of Poincaré inequality. Indeed observe that

$$F(u) \geq \frac{1}{p}\int_{\mathcal{D}} |\nabla u(x)|^p\,dx - \int_{\mathcal{D}} |f(x)|\,|u(x)|\,dx \geq \frac{1}{p}\|\nabla u\|^p_{L^p(\mathcal{D})} - c\|u\|_{L^p(\mathcal{D})},$$

where for the last term we used a Hölder estimate for the term $\|fu\|_{L^1(\mathcal{D})}$ (the constant c depends on $\|f\|_{L^{p^\star}(\mathcal{D})}$). Using the Poincaré inequality and in particular Proposition 1.5.14, which guarantees that $\|\nabla u\|_{L^p(\mathcal{D})}$ is an equivalent norm for $X = W_0^{1,p}(\mathcal{D})$, along with the fact that $p > 1$, leads us to the coercivity result.

2. By step 1 and using the standard arguments of the direct method, the minimization problem for F is well posed and admits a unique minimizer.

3. F is Fréchet differentiable on $X = W_0^{1,p}(\mathcal{D})$ with derivative

$$\langle DF(u), v\rangle = \int_{\mathcal{D}} |\nabla u(x)|^{p-2}\nabla u(x)\cdot\nabla v(x)\,dx - \int_{\mathcal{D}} f(x)v(x)\,dx, \quad \forall\, v \in W_0^{1,p}(\mathcal{D}).$$

To show this, take any $v \in X$, and consider the difference

$$\frac{1}{\epsilon}(F(u+\epsilon v) - F(u)) = \int_{\mathcal{D}} \frac{1}{\epsilon p}\left(|\nabla u(x) + \epsilon\nabla v(x)|^p - |\nabla v(x)|^p\right)dx - \int_{\mathcal{D}} f(x)v(x)\,dx. \quad (6.108)$$

Consider next the real valued function $t \mapsto \phi(t) := \frac{1}{p}|\nabla u(x) + t\nabla v(x)|^p$, which is differentiable with derivative $\frac{d\phi}{dt}(t) = |\nabla u(x) + t\nabla v(x)|^{p-2}(\nabla u(x) + t\nabla v(x))\cdot\nabla v(x)$ for any fixed x. That means, the family of functions $(w_\epsilon)_{\epsilon>0}$, defined by $w_\epsilon := \frac{1}{\epsilon p}(|\nabla u + \epsilon\nabla v|^p - |\nabla v|^p)$, converges a. e. in $\mathcal{D}$ to the function $w := |\nabla u|^{p-2}\nabla u\cdot\nabla v$ as $\epsilon \to 0$.[12] Using the above family, we express (6.108) as

$$\frac{1}{\epsilon}(F(u+\epsilon v) - F(u)) = \int_{\mathcal{D}} w_\epsilon(x)\,dx - \int_{\mathcal{D}} f(x)v(x)\,dx,$$

12 Another way to think of it is setting $\epsilon = \frac{1}{n}$, $n \in \mathbb{N}$ and obtaining the corresponding sequence $(\bar{w}_n)_{n\in\mathbb{N}}$, where $\bar{w}_n = w_{\frac{1}{n}}$.

and since $w_\epsilon \to |\nabla u|^{p-2}\nabla u \cdot \nabla v$ a. e. in $\mathcal{D}$ as $\epsilon \to 0$, if we can exchange the limit with the integral over $\mathcal{D}$, we can identify the Gâteaux derivative in the form above.

For this step, we need to invoke a Lebesgue dominated convergence argument, i. e., we need to ensure that $|w_\epsilon| < h$ for every $\epsilon > 0$, where $h \in L^1(\mathcal{D})$. Note that for every $x \in \mathcal{D}$,

$$w_\epsilon(x) = \frac{1}{\epsilon p}\left(|\nabla u(x) + \epsilon \nabla v(x)|^p - |\nabla v(x)|^p\right) = \frac{1}{\epsilon}(\phi(\epsilon) - \phi(0)) = \frac{1}{\epsilon}\int_0^\epsilon \phi'(t)dt$$
$$= \frac{1}{\epsilon}\int_0^\epsilon |\nabla u(x) + t\nabla v(x)|^{p-2}(\nabla u(x) + t\nabla v(x)) \cdot \nabla v(x)dt,$$

therefore,

$$|w_\epsilon(x)| \le \frac{1}{\epsilon}\int_0^\epsilon \left||\nabla u(x) + t\nabla v(x)|^{p-2}(\nabla u(x) + t\nabla v(x)) \cdot \nabla v(x)\right|dt$$
$$= \frac{1}{\epsilon}\int_0^\epsilon |\nabla u(x) + t\nabla v(x)|^{p-1}|\nabla v(x)|dt \le \left(|\nabla u(x)| + |\nabla v(x)|\right)^{p-1}|\nabla v(x)| =: h(x).$$

This last estimate follows easily from the fact that for every $\xi, \xi' \in \mathbb{R}^d$, we have, by the triangle inequality, that $|\xi + \xi'| \le |\xi| + |\xi'|$, so since $p > 1$, it follows that $|\xi + \xi'|^{p-1} \le (|\xi| + |\xi'|)^{p-1}$. Apply this for $\xi = \nabla u(x)$ and $\xi' = t\nabla v(x)$, and since $t \in (0, 1)$, majorize the resulting expression by setting $t = 1$, obtaining like that an upper bound for the integral. Since $u, v \in W_0^{1,p}(\mathcal{D})$, we have that $\nabla u, \nabla v \in L^p(\mathcal{D})$. Therefore, $(|\nabla u| + |\nabla v|)^{p-1} \in L^{p^*}(\mathcal{D})$, and by Hölder's inequality $h := (|\nabla u| + |\nabla v|)^{p-1}|\nabla v| \in L^1(\mathcal{D})$. We have then established the fact that $|w_\epsilon| \le h$, where $h \in L^1(\mathcal{D})$, and the proof is concluded. To show Fréchet differentiability, we will use Proposition 2.1.14 along with the elementary (but not so easy to prove!) inequality $|\,|z_1|^{p-2}z_1 - |z_2|^{p-2}z_2| \le c\,[|z_1| + |z_2|]^{p-2}|z_1 - z_2|$ for any $z_1, z_2 \in \mathbb{R}^d$, applied to $z_1 = \nabla u_n(x)$, $z_2 = \nabla u(x)$, for any sequence $(u_n)_{n\in\mathbb{N}}$ such that $u_n \to u$ in $W^{1,p}(\mathcal{D})$ (the details are left as exercise).

4. By step 3, we recognize the p-Laplace Poisson equation as the first-order condition for the minimization of F, which shows existence of solutions of (6.106). To show uniqueness of solutions, note that the existence of a weak minimizer is not enough as critical points are not necessarily minima. We may use the elementary[13] (and easier to prove!) inequality

$$c_1\,[|z_1| + |z_2|]^{p-2}|z_1 - z_2|^2 < \left(|z_1|^{p-2}z_1 - |z_2|^{p-2}z_2\right) \cdot (z_1 - z_2)$$

13 see, e. g., [49].

for some constant $c_1 > 0$. Assume two solutions $u_1, u_2 \in W_0^{1,p}(\mathcal{D})$ for (6.106), subtract, and then take the weak form for the difference using as test function $u_1 - u_2$. By applying the above inequality for $z_i = \nabla u_i$, a. e. in $\mathcal{D}$, we guarantee that $u_1 = u_2$, and the proof is complete. The uniqueness is also related to the strict monotonicity of the p-Laplace operator (see Example 9.3.3). □

6.8.2 A quasilinear nonlinear elliptic equation involving the p-Laplacian

In this section, we are going to apply the general formulation of the calculus of variations to study the problem of the minimization of the functional $F : W_0^{1,p}(\mathcal{D}) \to \mathbb{R}$, where $\mathcal{D}$ is a bounded domain, defined by

$$\begin{aligned} F(u) &:= \frac{1}{p}\int_{\mathcal{D}} |\nabla u(x)|^p \, dx - \int_{\mathcal{D}} \Phi(x, u(x))\, dx, \quad p \in (1,\infty), \\ \Phi(x,s) &= \int_0^s f(x,\sigma)d\sigma, \end{aligned} \tag{6.109}$$

where $f : \mathcal{D} \times \mathbb{R} \to \mathbb{R}$ is a given nonlinear function, and its connection with the quasilinear elliptic partial differential equation

$$\begin{aligned} -\nabla \cdot (|\nabla u|^{p-2}\nabla u) &= f(x,u), \quad \text{in } \mathcal{D}, \\ u &= 0 \quad \text{on } \partial\mathcal{D}. \end{aligned} \tag{6.110}$$

This problem may be considered as a generalization of the p-Laplacian Poisson equation we have studied in Section 6.8.1. In the particular case, where $p = 2$, the problem reduces to a semilinear elliptic equation of the form studied in Section 6.7.

Example 6.8.2 (The case $f(x,s) = -\mu|s|^{p-2}s + f_0(x)$, $\mu \geq 0$). We start our study of the general class of equation (6.110) with the particular but very interesting case, where $f(x,s) = -\mu|s|^{p-2}s + f_0(x)$, where $\mu > 0$ is a constant and $f_0 : \mathcal{D} \to \mathbb{R}$ a given function. One can easily extend the result of Proposition 6.8.1 for this choice in the case where $\mu \geq 0$. Moreover, the solution is unique, as one can see by assuming the existence of two solutions u_1, u_2, taking the difference $w = u_1 - u_2$, the corresponding PDE for w and using $v = w$ as test function in its weak form, to obtain

$$\begin{aligned} &\int_{\mathcal{D}} (|\nabla u_1|^{p-2}\nabla u_1 - |\nabla u_2|^{p-2}\nabla u_2)\cdot(\nabla u_1 - \nabla u_2)dx \\ &\qquad + \mu\int_{\mathcal{D}} (|u_1|^{p-2}u_1 - |u_2|^{p-2})(u_1 - u_2)dx = 0. \end{aligned} \tag{6.111}$$

Using the inequality

$$(|\xi_1|^{p-2}\xi_1 - |\xi_2|^{p-2}\xi_2) \cdot (\xi_1 - \xi_2) \geq (|\xi_1|^{p-1} - |\xi_2|^{p-1})(|\xi_1| - \xi_2|) \geq 0,$$

which is valid for any $\xi_1, \xi_2 \in \mathbb{R}^d$ (but also for any real numbers $\xi_1, \xi_2 \in \mathbb{R}$), in (6.111) we obtain, since $\lambda \geq 0$, that $u_1 = u_2$ and $\nabla u_1 = \nabla u_2$ a. e. in $\mathcal{D}$, hence the uniqueness of the solution.

The quasilinear equation in this case enjoys also a weak maximum principle stating that if $f \geq 0$, then $u \geq 0$ in $\mathcal{D}$. This can be shown by taking the weak form of the equation with $u^- = \max(-u, 0)$, and recalling that $\nabla u^- = -\nabla u \mathbf{1}_{u\leq 0}$, we obtain that

$$\int_{\mathcal{D}} |\nabla u^-|^p dx + \mu \int_{\mathcal{D}} |u^-|^p dx = -\int_{\mathcal{D}} f u^- dx.$$

Since the left-hand side is positive and the right-hand side is negative, we conclude that $|\nabla u^-|^p + \mu |u^-|^p = 0$, which leads to the conclusion that $u^- = 0$, hence, $u \geq 0$ in $\mathcal{D}$. A similar argument yields a comparison principle, according to which if u_1, u_2 are two solutions of (6.110) for $f_{0,1} \geq f_{0,2}$, then $u_1 \geq u_2$. ◁

Example 6.8.3 (Nonlinear eigenvalue problems). The nonlinearity treated in Example 6.8.2 becomes particularly interesting if $\mu < 0$ and $f_0 = 0$. In this case, expressing $\mu = -\lambda$, for $\lambda > 0$, we are led to a nonlinear eigenvalue problem

$$\begin{aligned} -\Delta_p u &= \lambda |u|^{p-2} u, \quad \text{in } \mathcal{D}, \\ u &= 0, \quad \text{on } \partial\mathcal{D}. \end{aligned} \tag{6.112}$$

In the special case, where $p = 2$, this reduces to the standard eigenvalue problem for the Laplacian, which was studied in Section 6.2.3. Problem (6.112) has been heavily studied. Here we will be only interested in the smaller $\lambda > 0$, for which problem (6.112) admits a nontrivial solution. This value, denoted by $\lambda_{1,p} > 0$, is called the first eigenvalue of the p-Laplacian operator, and the corresponding solution $\phi_{1,p}$, the first eigenfunction. Importantly, if $\mathcal{D}$ is bounded simply connected and of sufficiently smooth boundary, $\phi_{1,p}$ does not vanish anywhere on $\mathcal{D}$ and can be chosen so that $\phi_{1,p} > 0$. One may further show that $\lambda_{1,p}$ satisfies a variational principle in terms of a generalized Rayleigh quotient

$$\lambda_{1,p} = \inf_{W_0^{1,p}(\mathcal{D})\setminus\{0\}} \frac{\|\nabla u\|^p_{L^p(\mathcal{D})}}{\|u\|^p_{L^p(\mathcal{D})}}.$$

This infimum is attained, a fact that can be shown, using the direct method on the generalized Rayleigh–Fisher functional $F_{R,p}(u) := \frac{\|\nabla u\|^p_{L^p(\mathcal{D})}}{\|u\|^p_{L^p(\mathcal{D})}}$, or rather, noting (by homogeneity) the equivalence of the above problem with the problem of minimizing $\|\nabla u\|^p_{L^p(\mathcal{D})}$ on $\{u \in W_0^{1,p}(\mathcal{D}) : \|u\|_{L^p(\mathcal{D})} = 1\}$. The strict positivity of $\phi_{1,p}$ is a more delicate result requiring extensions of the maximum principle (or the Harnack inequality) for the p-Laplacian operator (see Remark 6.6.9, see also [8] or [100]).

Moreover, any function u such that $\|u\|_{L^p(\mathcal{D})} = 1$, for which $\|\nabla u\|_{L^p(\mathcal{D})} = \lambda_{1,p}$, necessarily satisfies (6.112) for $\lambda = \lambda_{1,p}$ and is equal to $u = \pm\phi_{1,p}$. This follows by the uniqueness property for positive solutions of (6.112) on the unit ball of $L^p(\mathcal{D})$. An easy proof of this remark has been given in [23]. Consider any two positive solutions u_1, u_2 of (6.112) on the unit ball of $L^p(\mathcal{D})$, and define $v = \frac{1}{2}(u_1^p + u_2^p)$, and $w = v^{1/p}$. One can easily see that w is on the unit ball itself. A straightforward, yet rather tedious, calculation yields that upon defining $h = \frac{u_1^p}{u_1^p + u_2^p}$,

$$|\nabla w|^p = v\big|h\,\nabla \ln u_1 + (1-h)\,\nabla \ln u_2\big|^p \leq v\big[h\,|\nabla \ln u_1|^p + (1-h)\nabla \ln u_2|^p\big]$$
$$= \frac{1}{2}(|\nabla u_1|^p + |\nabla u_2|^p),$$

where we used the convexity of the function $\xi \mapsto |\xi|^p$. Upon integration, we have that

$$\int_{\mathcal{D}} |\nabla w|^p dx \leq \frac{1}{2}\Big(\int_{\mathcal{D}} |\nabla u_1|^p dx + \int_{\mathcal{D}} |\nabla u_2|^p dx\Big) = \lambda_1. \tag{6.113}$$

Since w is admissible, it must be a minimizer of $u \mapsto \int_{\mathcal{D}} |\nabla u|^p dx$, on the unit ball of $L^p(\mathcal{D})$, therefore, (6.113) implies that $|\nabla w|^p = \frac{1}{2}|\nabla u_1|^p + \frac{1}{2}|\nabla u_2|^p$ a. e. in $\mathcal{D}$, which in turn implies that $|h\,\nabla \ln u_1 + (1-h)\,\nabla \ln u_2|^p = v[h\,|\nabla \ln u_1|^p + (1-h)\nabla \ln u_2|^p]$. Therefore, $\nabla \ln u_1 = \nabla \ln u_2$ a. e. so that $u_1 = c u_2$ for some constant $c > 0$. ◁

We will impose the following conditions on f:

Assumption 6.8.4. $f : \mathcal{D} \times \mathbb{R} \to \mathbb{R}$ is a Carathéodory function such that satisfies
(i) The growth condition $|f(x,s)| \leq c|s|^{r-1} + a(x)$ a. e. in $x \in \mathcal{D}$, for $1 < r < \mathfrak{s}_p$, for every $s \in \mathbb{R}$, where $\mathfrak{s}_p = \frac{dp}{d-p}$ of $p < d$ and $\mathfrak{s}_p = \infty$ if $p \geq d$ and $a \in L^{p^*}(\mathcal{D})$, with $\frac{1}{p} + \frac{1}{p^*} = 1$.
(ii) The asymptotic condition $\limsup_{|s|\to\infty} \frac{pF(x,s)}{|s|^p} < \lambda_{1,p}$ uniformly in x, where $\lambda_{1,p} > 0$ is defined as $\lambda_{1,p} = \inf_{W_0^{1,p}(\mathcal{D})\setminus\{0\}} \frac{\|\nabla u\|^p_{L^p(\mathcal{D})}}{\|u\|^p_{L^p(\mathcal{D})}}$.

The following proposition [15] provides an existence result for the quasilinear equation (6.110):

Proposition 6.8.5. *Let the function f satisfy Assumption* 6.8.4. *The functional F, defined in* (6.109), *admits a minimum, which corresponds to a solution of the nonlinear PDE problem* (6.110).

Proof. We will first show the existence of a minimum for F, and then associate it through the use of the Euler–Lagrange equation with a solution of (6.110). The existence of the minimum uses the direct method of the calculus of variations, which requires sequential weak lower semicontinuity for F and coercivity. It is convenient to express Φ as $\Phi(x,s) = \frac{\lambda_{1,p}}{p}|s|^p + \Phi_1(x,s)$ with the function Φ_1 satisfying $\limsup_{s\to\pm\infty} \frac{\Phi_1(x,s)}{|s|^p} < 0$ uniformly in $x \in \mathcal{D}$. The proof proceeds in 3 steps.

1. Sequential weak lower semicontinuity follows easily (the first part of the functional enjoys this property by convexity as already seen in the proof of Proposition 6.8.1, and the second part enjoys this property as a result of the compactness of the embedding $W_0^{1,p}(\mathcal{D}) \hookrightarrow L^r(\mathcal{D})$ for $1 \le r < \frac{dp}{d-p} =: \mathfrak{s}_p$).

2. We now check the coercivity of the functional. Suppose that it is not. Then, there exists a sequence $(u_n)_{n\in\mathbb{N}} \subset W_0^{1,p}(\mathcal{D})$ such that $F(u_n) < c$ for some $c > 0$, while $\|u_n\|_{W_0^{1,p}(\mathcal{D})} \to \infty$. Note that by the form of the functional, if such a sequence exists, it must also hold that (at least for some subsequence) $|u_n| \to \infty$ a. e. Consider the normalized sequence $(w_n)_{n\in\mathbb{N}}$, defined by $w_n(x) = \frac{u_n}{\|u_n\|_{W_0^{1,p}(\mathcal{D})}}$, which is clearly bounded in $W_0^{1,p}(\mathcal{D})$, hence, by reflexivity there exists $w_0 \in W_0^{1,p}(\mathcal{D})$ with $\|w_0\|_{W_0^{1,p}(\mathcal{D})} \le 1$ such that $w_n \rightharpoonup w_0$ in $W_0^{1,p}(\mathcal{D})$, while, by the Rellich–Kondrachov compact embedding (up to a subsequence), $w_n \to w_0$ in $L^p(\mathcal{D})$. We now divide $F(u_n)$ by $\|u_n\|^p_{W_0^{1,p}(\mathcal{D})}$, and we obtain

$$\frac{1}{\|u_n\|^p_{W_0^{1,p}(\mathcal{D})}} F(u_n) = \frac{1}{p}\int_{\mathcal{D}} |\nabla w_n(x)|^p dx - \frac{\lambda_{1,p}}{p}\int_{\mathcal{D}} |w_n(x)|^p dx - \int_{\mathcal{D}} \frac{\Phi_1(x, u_n(x))}{|u_n(x)|^p} |w_n(x)|^p dx,$$

where we have trivially expressed $\Phi(x, u_n(x)) = \frac{\Phi(x,u_n(x))}{|u_n(x)|^p} |u_n(x)|^p$. Rearranging so as to bring $\frac{1}{p}\int_{\mathcal{D}} |\nabla w_n(x)|^p dx$ to the left-hand side and keeping in mind the properties of Φ_1 and the sequence $(w_n)_{n\in\mathbb{N}}$, after taking the limit superior, we obtain that

$$\limsup_{n\to\infty} \int_{\mathcal{D}} |\nabla w_n(x)|^p dx \le \lambda_{1,p} \int_{\mathcal{D}} |w(x)|^p dx. \tag{6.114}$$

By the sequential weak lower semicontinuity of the norm in $W_0^{1,p}(\mathcal{D})$,

$$\int_{\mathcal{D}} |\nabla w_o(x)|^p dx \le \liminf_{n\to\infty} \int_{\mathcal{D}} |\nabla w_n(x)|^p dx \le \limsup_{n\to\infty} \int_{\mathcal{D}} |\nabla w_n(x)|^p dx \le \lambda_{1,p} \int_{\mathcal{D}} |w_0(x)|^p dx,$$

whereas by the Poincaré inequality (or in fact by the definition of $\lambda_{1,p}$), we also have

$$\lambda_{1,p} \int_{\mathcal{D}} |w(x)|^p dx \le \int_{\mathcal{D}} |\nabla w_o(x)|^p dx.$$

Combining the above two, we have that

$$\int_{\mathcal{D}} |\nabla w_o(x)|^p dx = \lambda_{1,p} \int_{\mathcal{D}} |w_0(x)|^p dx.$$

This implies that $w_0 = \pm\phi_{1,p} > 0$, where $\phi_{1,p} > 0$ is the first eigenfunction of the p-Laplacian. Without loss of generality, let us assume that $w_0 = \phi_{1,p} > 0$. Moreover, by the uniform convexity of $W_0^{1,p}(\mathcal{D})$, the sequence $(w_n)_{n\in\mathbb{N}}$ also satisfies $w_n \to w_0$ in $W_0^{1,p}(\mathcal{D})$.

To reach a contradiction reconsider $F(u_n) < c$, and note that by the definition of $\lambda_{1,p}$ it holds that

$$\frac{1}{p}\int_{\mathcal{D}} |\nabla u_n(x)|^p dx - \frac{\lambda_{1,p}}{p}\int_{\mathcal{D}} |u_n(x)|^p dx \geq 0,$$

so $F(u_n) < c$ implies that

$$\int_{\mathcal{D}} \Phi_1(x, u_n(x) dx \geq -c,$$

and dividing again by $\|u_n\|^p_{W_0^{1,2}(\mathcal{D})} \to \infty$, we obtain using a similar rearrangement that

$$\int_{\mathcal{D}} \frac{\Phi_1(x, u_n(x)}{|u_n(x)|^p} \frac{|u_n(x)|^p}{\|u_n\|^p_{W_0^{1,2}(\mathcal{D})}} dx = \int_{\mathcal{D}} \frac{\Phi_1(x, u_n(x)}{|u_n(x)|^p} |w_n(x)|^p dx \geq -\frac{c}{\|u_n\|^p_{W_0^{1,p}(\mathcal{D})}},$$

and passing to the limit noting that $w_0 = \phi_{1,p} > 0$, we see that the left-hand side is negative, while the right-hand side converges to 0, which leads to a contradiction.

3. We now show that the minimizer is a solution of the related quasilinear PDE. Under Assumption 6.8.4(i), the functional $F : W_0^{1,p}(\mathcal{D}) \to \mathbb{R}$ is Fréchet differentiable, and its Fréchet derivative $DF : W_0^{1,p}(\mathcal{D}) \to (W_0^{1,p}(\mathcal{D}))^* = W^{1,-p^*}(\mathcal{D})$ satisfies

$$\langle DF(u), v\rangle = \int_{\mathcal{D}} \left(|\nabla u(x)|^{p-2}\nabla u(x)\cdot\nabla v(x) - f(x, u(x))v(x)\right) dx, \quad \forall\, v \in W_0^{1,p\cdot}(\mathcal{D}).$$

Indeed, the functional F consists of two parts: a) $F_0(u) = \frac{1}{p}\|\nabla u\|^p_{W_0^{1,p}(\mathcal{D})}$, which was studied in the previous section, where it was shown that it enjoys the stated properties (see the proof of Proposition 6.8.1), and b) the second part $F_1(u) := \int_{\mathcal{D}} \Phi(x, u)\, dx$, which deserves further consideration. With arguments similar as to those used in Proposition 6.7.6, we see that $DF_1(u) = f(\cdot, u)$, which can be expressed as $DF_1(u) = \mathsf{N}_f(u)$, where N_f is the Nemitskii operator generated by the function f, defined by $(\mathsf{N}_f(u))(x) = f(x, u(x))$ a. e. in $\mathcal{D}$. The continuity of the Nemitskii operator $\mathsf{N}_f : W_0^{1,p}(\mathcal{D}) \to (W_0^{1,p}(\mathcal{D}))^*$ follows from the compactness of the embedding $W_0^{1,p}(\mathcal{D}) \hookrightarrow L^r(\mathcal{D})$ for $1 \leq r \leq \frac{dp}{d-p} =: \mathfrak{s}_p$ (see Theorem 1.5.11) and repeating almost verbatim the arguments in the proof of Proposition 6.7.6. The proof is now complete. □

Remark 6.8.6. One may generalize the nonresonance condition of Assumption 6.8.4(ii) in various ways (see, e. g., [9]). For example, we may assume that Assumption 6.8.4(ii) holds for sets of strictly positive measure. Another set of conditions are the so called

Landesman–Lazer type conditions. Assume that the nonlinearity Φ is of the more general form $\Phi(x,s) = \frac{\lambda_{1,p}}{p}|s|^p + \Phi_1(x,s) + h(x)$, and define $G^{\pm}(x) := \limsup_{s\to\pm\infty} \frac{\Phi_1(x,s)}{s}$ (the uniform limit in $x \in \mathcal{D}$). Then, under the assumption

$$\int_{\mathcal{D}} G^{+}(x)\phi_{1,p}(x)dx \le \int_{\mathcal{D}} h(x)\phi_{1,p}(x)dx \le \int_{\mathcal{D}} G^{-}(x)\phi_{1,p}(x)dx, \tag{6.115}$$

problem (6.110) admits a solution. The proof proceeds along the same lines as the proof of Proposition 6.8.5, and is left as an exercise (see also [9]). Condition (6.115) is called a Landesman–Lazer type condition. These conditions will also play an important role when trying to identify solutions of (6.110), which are not minimizers of the functional (6.109) (see Section 8.4, see also [10]).

6.9 Γ-convergence and the calculus of variations

Γ-convergence is an important tool in the calculus of variations. This is natural, as the calculus of variations is associated with problems related to the minimization of integral functionals, and Γ-convergence is a very convenient and natural tool for the study of the limiting behavior of sequences of minimization problems. In this section, we provide some examples of this connection and refer the reader to more specialized texts (e. g., [62] or [30]) for more advanced applications.

6.9.1 A first result in homogenization

A short introduction to the problem of homogenization

Homogenization problems is an important class of problems, both from the theoretical and the practical view point (see [25, 26, 50, 102, 108, 125, 135]). While this is a large class of rather diverse problems, spanning a range of applications from fluid mechanics, electromagnetics, material science, or even financial economics (see, e. g., [17, 81, 138, 139]), we restrict ourselves here for simplicity to the concrete problem of homogenization of elliptic problems, which are directly related to variational problems, and for which the use of Γ-convergence techniques can be directly employed for an interesting viewpoint to the problem.

For any vector field $w = (w_1, \dots, w_d)$, we define the divergence operator $\operatorname{div} w = \sum_{i=1}^{d} \frac{\partial w_i}{\partial x_i}$, interpreted in the weak sense (see (1.8)). We then consider the family of elliptic problems

$$\begin{aligned} -\operatorname{div}\left(a\left(\frac{x}{\epsilon}\right)u_\epsilon(x)\right) &= f, \quad x \in \mathcal{D}, \\ u_\epsilon(x) &= 0, \quad x \in \partial\mathcal{D}, \end{aligned} \tag{6.116}$$

where $\mathcal{D} \subset \mathbb{R}^d$ is a bounded domain, assumed regular for simplicity, and $a = (a_{ij})_{i,j=1,\dots,d}$ is a symmetric matrix with elements depending on $x \in \mathcal{D}$, satisfying the positive definite condition of existence of a constant $c_0 > 0$ such that

$$\sum_{i,j=1}^{d} a_{ij}(x)\xi_i\xi_j \geq c_0 \sum_{i=1}^{d} |\xi_i|^2, \quad \forall\, \xi = (\xi_1, \dots, \xi_d) \in \mathbb{R}^d, \text{ a. e. } x \in \mathcal{D}. \tag{6.117}$$

For simplicity, we will also assume that the coefficients of a are bounded above and below uniformly in x, i. e., that there exists a constant $\beta > 0$ such that $|a(x)\xi| \leq \beta|\xi|$ for all $\xi \in \mathbb{R}^d$ and all $x \in \mathcal{D}$ (possibly a. e. on $\mathcal{D}$). We favor the notation div over $\nabla\cdot$ for the divergence operator in this section, as we believe it makes it easier for the reader to follow the various integration by parts arguments that will be used in the sequel.

The problem of homogenization deals with the case where the functions a_{ij} are periodic with periodicity Y, where Y is an appropriate subset of $\mathbb{R}^d$. A very common case (which is what we will assume here from now on) is the case where $Y = (0, L_1) \times \cdots \times (0, L_d)$ for $L_i > 0$, $i = 1, \dots, d$. For such a choice for Y, a function $g : \mathbb{R}^d \to \mathbb{R}$ is called Y-periodic if

$$g(x + kL_ie_i) = g(x), \quad \text{a. e. on } \mathbb{R}^d,\ \forall\, k \in \mathbb{Z},\ i = 1, \dots, d, \tag{6.118}$$

where $\{e_1, \dots, e_d\}$ is the canonical basis of $\mathbb{R}^d$. Our second assumption concerning the coefficients matrix a of the problem (6.116) is that the functions a_{ij} (defined on the whole of $\mathbb{R}^d$) are Y-periodic in the above sense. From the mathematical point of view periodic functions enjoy interesting asymptotic properties, one of the most important being that if g is Y-periodic, then the sequence of functions $(g_\epsilon)_{\epsilon>0}$, defined by $g_\epsilon(\frac{x}{\epsilon})$, converges (in the appropriate weak sense) to the average of the function g over its fundamental domain Y, i. e.,

$$g_\epsilon \rightharpoonup \fint_Y g(y)dy := \frac{1}{|Y|}\int_Y g(y)dy, \tag{6.119}$$

as $\epsilon \to 0$, in an appropriate Lebesgue space $L^p(\mathbb{R}^d)$ (see Theorem 6.10.2 in Section 6.10.1 for details; see also [50] Theorem 2.6). Other types of structures are also possible, including random structures (see, e. g., [17, 26]).

The practical implications of this periodicity assumption is that the problem presents a microstructure, which consists of a fast oscillating periodic structure. Depending on the concrete application related to the PDE (6.116), this can model various situations, which often arise in practice. For example, in a heat conduction context, (6.116) can be used to study the equilibrium temperature distribution in a medium that presents a periodic microstructure in its heat conduction properties, a situation commonly encountered in composite media in nature.

An important question that arises is whether in the limit as $\epsilon \to 0$, we may replace problem (6.116) with a simpler problem, ideally with constant coefficients of the form

$$\begin{aligned} -\operatorname{div}(\bar{a}u(x)) &= f, \quad x \in \mathcal{D}, \\ u(x) &= 0, \quad x \in \partial\mathcal{D}, \end{aligned} \tag{6.120}$$

where $\bar{a}$ is a constant coefficient $d \times d$ matrix. This problem will capture the overall, "averaged" effects of the periodic microstructure of the coefficients a, on the solution of the problem. Such a reduced approximation, if and when it exists, comes with several benefits, both conceptual and practical. On the conceptual level, being able to provide a formula for $\bar{a}$ in terms of the original rapidly oscillating coefficients a allows us to understand the effects on the macroscale of the rapid variability on the microscale of the coefficients. This has important consequences from the point of view of applications; e. g., in material science such understanding often leads to better understanding of the properties of the material, or even to construction procedures for new materials with desired properties. It also has important consequence from the point of view of numerical analysis, as the direct numerical treatment of problem (6.116) for small enough $\epsilon > 0$ is very difficult, or even impossible by at least current computer capabilities, whereas, even if this was not the case, prone to numerical instabilities. Being able to study the approximate asymptotic problem (6.120) instead, with sufficient understanding of the errors in involved when replacing (6.116) with (6.120), is a great simplifying step, which is often employed in numerical treatment of problem (6.116) (see, e. g., [26]). The important property (6.119) is expected to play a crucial role in the connection between a and the asymptotic coefficient $\bar{a}$, but certain surprises await us in this respect! As a quick spoiler, we mention here that $\bar{a} \neq \fint_Y a(y)dy$, contrary to what intuition may lead us to expect!

Connection between homogenization and Γ-convergence

The main aim of this section is to connect the problem of homogenization of problem (6.116) with Γ-convergence. This is an important theme in this literature, in fact one of the original motivations of Γ-convergence was along these lines. A very nice and detailed cover of such themes can be found in [62] (see also [25] and [30]) on which our presentation is based.

Our first step in this direction is to connect problems (6.116) and (6.120) with appropriate optimization problems. We consider (6.116) (as the case for (6.120) is similar, and even simpler).

Arguments similar to those employed in Section 6.5 can be used to show that the weak solution of the PDE problem (6.117), defined in terms of finding $u_\epsilon \in W_0^{1,2}(\mathcal{D})$,

$$\int_{\mathcal{D}} a\left(\frac{x}{\epsilon}\right)\nabla u_\epsilon(x) \cdot \nabla v(x)dx = \int_{\mathcal{D}} f(x)v(x)dx, \quad \forall\, v \in W_0^{1,2}(\mathcal{D}). \tag{6.121}$$

The weak solution $u_\epsilon \in W_0^{1,2}(\mathcal{D})$ is in fact the minimizer of the quadratic functional $\hat{F}_\epsilon : W_0^{1,2}(\mathcal{D}) \times W_0^{1,2}(\mathcal{D}) \to \mathbb{R}$, defined by

$$\hat{F}_\epsilon(u) := \int_{\mathcal{D}} a\left(\frac{x}{\epsilon}\right)\nabla u(x) \cdot \nabla u(x)dx - \int_{\mathcal{D}} 2f(x)u(x)dx. \tag{6.122}$$

An alternative convenient formulation is to consider $\hat{F}_\epsilon$, as defined on the whole of $L^2(\mathcal{D})$, by defining it as in (6.122) in $W_0^{1,2}(\mathcal{D})$ and assigning to it the value $+\infty$ in $L^2(\mathcal{D}) \setminus W_0^{1,2}(\mathcal{D})$. The connection of $\hat{F}_\epsilon$ with the PDE (6.117) can be seen by a simple calculation of the derivative of $\hat{F}_\epsilon$, and then taking the first-order condition for the minimization, with the factor 2 on the second term coming from the symmetry of the matrix a. The well posedness of the variational problem of minimizing (6.122) follows by a straightforward generalization of the arguments in Section 6.4. A key point comes from assumption (6.117), which allows us to obtain a coercivity estimate for F_ϵ based on the observation that

$$c_0' \|u\|^2_{W_0^{1,2}(\mathcal{D})} \leq c_0 \|\nabla u\|_{L^2(\mathcal{D})} \leq \int_{\mathcal{D}} a\left(\frac{x}{\epsilon}\right)\nabla u(x) \cdot \nabla u(x)dx \tag{6.123}$$

for $c_0' > 0$, where the far LHS estimate comes from the Poincaré inequality (Theorem 1.5.13). This estimate guarantees that a minimizing sequence for F_ϵ is bounded in $W_0^{1,2}(\mathcal{D})$, hence, by weak compactness arguments, we can complete the steps of the direct method of the calculus of variations. The connection of the minimization problem (6.122) with the weak form (6.121) of the PDE (6.116) follows using similar steps as in Section 6.5.

Remark 6.9.1. Note that we loosely refer to a family of functionals $(F_\epsilon)_{\epsilon>0}$ as a sequence and consider its limit as $\epsilon \to 0$, keeping in mind that we take any sequence $(\epsilon_n)_{n\in\mathbb{N}}$, such that $\epsilon_n \to 0$, and consider the corresponding sequence $(F_n)_{n\in\mathbb{N}}$.

We may prove the following result (see, e. g., [25]):

Proposition 6.9.2. *Consider the family of functionals $(F_\epsilon)_{\epsilon>0}$ defined on $W_0^{1,2}(\mathcal{D})$ as*

$$F_\epsilon(u) = \int_{\mathcal{D}} a\left(\frac{x}{\epsilon}\right)|\nabla u(x)|^2 dx,$$

where a is a Y-periodic function. Define the averaged symmetric matrix $\bar{a} \in \mathbb{R}^{d\times d}$ by its action on the standard basis $\{e_1, \dots, e_d\}$ of $\mathbb{R}^d$ in terms of

$$\bar{a}e_i := \fint_Y a(y)(e_i + \nabla_y \phi_i(y))dy := \frac{1}{|Y|}\int_Y a(y)(e_i + \nabla_y \phi_i(y))dy, \quad i = 1, \dots, d, \tag{6.124}$$

where ϕ_i is the solution of the “cell” problems

$$
\begin{aligned}
&-\operatorname{div}_y[a(y)(e_i+\nabla_y\phi_i(y))]=0, \quad y\in Y,\ i=1,\dots,d,\\
&\phi_i,\ \nabla\phi_j \quad Y\text{-periodic},\\
&\fint_Y \phi_i(y)dy := \frac{1}{|Y|}\int_Y \phi_i(y)dy = 0, \quad i=1,\dots,d,
\end{aligned}
\tag{6.125}
$$

and define the functional F on $W_0^{1,2}(\mathcal{D})$ by

$$F(u)=\int_{\mathcal{D}} \bar{a}|\nabla u(x)|^2 dx.$$

Then, $F_\epsilon \xrightarrow{\Gamma} F$ in the weak topology of $W_0^{1,2}(\mathcal{D})$.

Proof. The Banach space $X := W_0^{1,2}(\mathcal{D})$ satisfies the assumptions of Prop. 2.8.14, so we can prove the Γ-convergence result using the sequential characterization, replacing the strong convergence by weak convergence.

The demanding part of the proof is the identification of a recovery sequence $(\bar{u}_\epsilon)_{\epsilon>0}$ for any $u \in W_0^{1,2}(\mathcal{D})$. To this end, for any $u \in W_0^{1,2}(\mathcal{D})$, consider the solution $\bar{u}_\epsilon \in W_0^{1,2}(\mathcal{D})$ of the problem

$$-\operatorname{div}\left(a\left(\frac{x}{\epsilon}\right)\nabla\bar{u}_\epsilon\right) = -\operatorname{div}(\bar{a}\nabla u), \tag{6.126}$$

which can be stated in the weak formulation

$$\int_{\mathcal{D}} a\left(\frac{x}{\epsilon}\right)\nabla\bar{u}_\epsilon\cdot\nabla v dx = \int_{\mathcal{D}} \bar{a}\nabla u\cdot\nabla v dx, \quad \forall\, v\in W_0^{1,2}(\mathcal{D}). \tag{6.127}$$

The existence of $\bar{u}_\epsilon$ follows by variational considerations, or the Lax–Milgram theorem (see Section 7.6). We claim that $(\bar{u}_\epsilon)_{\epsilon>0}$, as above is a recovery sequence.

Before proving this claim (in step 2), we prove that $\bar{u}_\epsilon \rightharpoonup u$ in $W_0^{1,2}(\mathcal{D})$ (see step 1). This calls for some preparatory results, which are presented in step 0 below.

0. We first note the existence of the limits

$$
\begin{aligned}
\bar{u}_\epsilon &\rightharpoonup \bar{u}, \quad \text{in } W_0^{1,2}(\mathcal{D}),\\
\bar{u}_\epsilon &\to \bar{u}, \quad \text{in } L^2(\mathcal{D}),\\
a\left(\frac{x}{\epsilon}\right)\nabla(\bar{u}_\epsilon) &\rightharpoonup w \quad \text{in } (L^2(\mathcal{D}))^d,
\end{aligned}
\tag{6.128}
$$

for some $\bar{u} \in W_0^{1,2}(\mathcal{D})$, to be identified later on as u, and some $w \in (L^2(\mathcal{D}))^d$. The convergences in (6.128) come from weak compactness arguments, after obtaining uniform in ϵ bounds in the corresponding function spaces for the above sequences. Note that the strong convergence in $L^2(\mathcal{D})$ holds up to a subsequence and is justified by the compact embedding of $W_0^{1,2}(\mathcal{D})$ in $L^2(\mathcal{D})$. Passing to the limit as $\epsilon \to 0$ in (6.127) and using (6.128), we obtain the important result:

$$\int_{\mathcal{D}} w \cdot \nabla v dx = \int_{\mathcal{D}} \bar{a}\nabla u \cdot \nabla v dx, \quad \forall\, v \in W_0^{1,2}(\mathcal{D}). \tag{6.129}$$

Identifying the limit $\bar{u} = u$ (see part 1) is the most demanding part of the proof. Here, we choose a straightforward approach that works for this quadratic functional, which requires the following preparation:

The strategy is to take the weak form of (6.126) with a suitable test function (a fast oscillating test function) that will help us identify the above limits (see [25, 135]). These are test functions of the form $v_\epsilon^{(i)}\psi$, with $\psi \in C_c^\infty(\mathcal{D})$ and $v_\epsilon^{(i)}$ chosen, as

$$v_\epsilon^{(i)}(x) = x_i + \epsilon\phi_i\left(\frac{x}{\epsilon}\right), \quad i = 1, \ldots, d, \tag{6.130}$$

for ϕ_i to be determined shortly. We will use the notation $y = \frac{x}{\epsilon}$ and select $y \mapsto \phi_i(y)$ so that the functions $v_\epsilon^{(i)}$ satisfy (the weak form of) the equation

$$-\operatorname{div}\left(a\left(\frac{x}{\epsilon}\right)\nabla v_\epsilon^{(i)}(x)\right) = 0, \quad i = 1, \ldots, d. \tag{6.131}$$

By straightforward calculation (and noting that $\operatorname{div} := \operatorname{div}_x = \frac{1}{\epsilon}\operatorname{div}_y$), we see that $v_\epsilon^{(i)}$ satisfies (6.131) if ϕ_i is chosen as the (weak) solution of

$$-\operatorname{div}_y(a(y)\nabla\phi_i(y)) = \operatorname{div}_y(a(y)e_i), \quad \text{in } \mathbb{R}^d,\ i = 1, \ldots, d, \tag{6.132}$$

with periodic boundary conditions in Y. This set of equations are the so called cell equations (6.125). Note that the weak formulation of (6.131), using as test function $v = \bar{u}_\epsilon\psi$, yields

$$\begin{aligned} 0 &= \int a\left(\frac{x}{\epsilon}\right)\nabla v_\epsilon^{(i)} \cdot \nabla(\bar{u}^\epsilon\psi)dx \\ \Longrightarrow &\int a\left(\frac{x}{\epsilon}\right)\nabla v_\epsilon^{(i)} \cdot \nabla\bar{u}^\epsilon\, \psi dx = -\int a\left(\frac{x}{\epsilon}\right)\nabla v_\epsilon^{(i)} \cdot \nabla\psi\, \bar{u}^\epsilon dx. \end{aligned} \tag{6.133}$$

Note that in the above, the fact that $\psi \in C_c^\infty(\mathcal{D})$ allows us to take the integration either on the whole of $\mathbb{R}^d$. denoted by $\int$, or on $\mathcal{D}$. denoted by $\int_{\mathcal{D}}$.

We also collect some important properties of the family of test functions $(v_\epsilon^{(i)})_{\epsilon>0}$:

We define the functions $w_i^{(0)}$ in terms of $w_i^{(0)}(x) = x_i$, for which $\nabla w_i^{(0)}(x) = e_i$, $i = 1, \ldots, d$. We first claim that

$$\begin{aligned} v_\epsilon^{(i)} &\rightharpoonup w_i^{(0)}, \quad \text{in } L^2(\mathcal{D}), \\ \nabla v_\epsilon^{(i)} &\rightharpoonup e_i, \quad \text{in } L^2(\mathcal{D}). \end{aligned} \tag{6.134}$$

These results follow by Theorem 6.10.2 (see the appendix) applied to the sequence of functions $(v_\epsilon^{(i)})_{\epsilon>0}$. To obtain the first of these results, we have to take into account the properties of ϕ (and in particular that the average over Y of ϕ_i vanishes), while for the second one we need the fact that $\nabla v_\epsilon^{(i)} = e_i + \nabla_y \phi_i(\frac{x}{\epsilon})$ (and ϕ_i is periodic). Then, a straightforward application of Theorem 6.10.2 yields that $v_\epsilon^{(i)} \rightharpoonup w_0^{(i)}$ in $W^{1,2}(\mathcal{D})$, hence, by the compact embedding of $W^{1,2}(\mathcal{D})$ in $L^2(\mathcal{D})$, we conclude the existence of a subsequence (see also Remark 1.5.12) such that $v_\epsilon^{(i)} \to w_0^{(i)}$ (strong) in $L^2(\mathcal{D})$. We conclude that

$$\begin{aligned} v_\epsilon^{(i)} &\rightharpoonup x_i, && \text{in } W^{1,2}(\mathcal{D}), \\ v_\epsilon^{(i)} &\to x_i, && \text{in } L^2(\mathcal{D}), \quad \text{(up to a subsequence)}, \\ \nabla v_\epsilon^{(i)} &\rightharpoonup e_i, && \text{in } L^2(\mathcal{D}), \end{aligned} \tag{6.135}$$

where (abusing notation), by x_i, we mean the function $w_0^{(i)}$.

We finally consider the asymptotic behavior of the term $a(\frac{x}{\epsilon})\nabla v_\epsilon^{(i)}$ that will be needed in what follows. We have (using Theorem 6.10.2 and (6.124)) that

$$a\Big(\frac{\cdot}{\epsilon}\Big)\nabla v_\epsilon^{(i)} = a\Big(\frac{\cdot}{\epsilon}\Big)\Big(e_i + \nabla_y \phi_i\Big(\frac{\cdot}{\epsilon}\Big)\Big) \rightharpoonup \frac{1}{|Y|}\int_Y a(y)(e_i + \nabla_y \phi_i(y))dy =: \bar{a} e_i. \tag{6.136}$$

1. We now show that $\bar{u} = u$. Take the weak formulation of (6.126) using as test function $v = v_\epsilon^{(i)}\psi$. By (6.131) and its consequence (6.133), using the convention for integrals adopted here, we see that

$$\begin{aligned} \int a\Big(\frac{x}{\epsilon}\Big)\nabla \bar{u}_\epsilon \cdot \nabla\psi\, v_\epsilon^{(i)} dx - \int a\Big(\frac{x}{\epsilon}\Big)\nabla v_\epsilon^{(i)} \cdot \nabla\psi\, \bar{u}_\epsilon dx = \int \bar{a}\nabla u \cdot \nabla v_\epsilon^{(i)} \psi dx \\ + \int \bar{a}\nabla u \cdot \nabla\psi\, v_\epsilon^{(i)} dx. \end{aligned} \tag{6.137}$$

Passing to the limit as $\epsilon \to 0$ in (6.137) yields

$$\int w \cdot \nabla\psi\, x_i dx - \int \bar{a} e_i \cdot \nabla\psi\, \bar{u} dx = \int \bar{a}\nabla u \cdot e_i\, \psi dx + \int \bar{a}\nabla u \cdot \nabla\psi\, x_i dx. \tag{6.138}$$

In the above, for the first term on the LHS, we used the convergences (6.135) (and in particular that up to a chosen subsequence $v_\epsilon^{(i)} \to x_i$ strong in $L^2(\mathcal{D})$, a fact that is required since the convergence $a(\frac{x}{\epsilon})\nabla \bar{u}_\epsilon \rightharpoonup w$ is weak in $L^2(\mathcal{D})$), while for the second term we used (6.136) along with the fact that for a chosen subsequence $\bar{u}_\epsilon \to \bar{u}$ in $L^2(\mathcal{D})$ (strong). We also need to recall Remark 1.1.52. The convergence of the RHS is straightforward.

Noting that $\nabla(\psi x_i) = e_i\psi + \nabla\psi\, x_i$, we may rearrange (6.138) as

$$\int w \cdot (\nabla(\psi\, x_i) - e_i\psi)dx - \int \bar{a} e_i \cdot \nabla\psi\, \bar{u} dx = \int \bar{a}\nabla u \cdot \nabla(\psi x_i)dx,$$

which is rearranged as

$$\int w \cdot e_i \psi dx + \int \bar{a} e_i \cdot \nabla \psi \bar{u} dx = \int w \cdot \nabla(\psi x_i) dx - \int \bar{a} \nabla u \cdot \nabla(\psi x_i) dx. \tag{6.139}$$

Applying (6.129) using the test function $v = x_i \psi$, we immediately see that the RHS of (6.139) vanishes, yielding

$$\int w \cdot e_i \, \psi dx + \int \bar{a} e_i \cdot \nabla \psi \bar{u} dx = 0, \quad \forall \, \psi \in C_c^\infty(\mathcal{D}). \tag{6.140}$$

This condition will allow us to deduce that $\bar{u} = u$. In order to do that, we rearrange (6.139) in more convenient form by noting that

$$0 = \int \bar{a} e_i \cdot \nabla(\bar{u} \psi) dx = \int \bar{a} e_i \cdot \nabla \bar{u} \, \psi dx + \int \bar{a} e_i \cdot \nabla \psi \, \bar{u} dx, \tag{6.141}$$

where the LHS vanishes by a simple integration by parts argument, taking into account that $\bar{u}\psi$ vanishes on the boundary and that $\nabla(\bar{a} e_i) = 0$, since $\bar{a}$ is a constant matrix. Substituting (6.141) in (6.140), we obtain (using also the symmetry of $\bar{a}$),

$$\int w \cdot e_i \, \psi dx = \int \bar{a} e_i \cdot \nabla \bar{u} \, \psi dx = \int_{\mathcal{D}} e_i \cdot \bar{a} \nabla \bar{u} \, \psi dx, \quad \forall \, \psi \in C_c^\infty(\mathcal{D}), \tag{6.142}$$

from which we can deduce that $w = \bar{a} \nabla \bar{u}$.

Upon this identification, we return to (6.129), which now yields

$$\int_D w \cdot \nabla v dx = \int_D \bar{a} \nabla \bar{u} \cdot \nabla v dx = \int_D \bar{a} \nabla u \cdot \nabla v, \quad \forall \, v \in W_0^{1,2}(\mathcal{D}), \tag{6.143}$$

which by the symmetry of the matrix $\bar{a}$ implies that

$$\int_D \bar{a} \nabla v \cdot \nabla \bar{u} dx = \int_D \bar{a} \nabla v \cdot \nabla u dx, \quad \forall \, v \in W_0^{1,2}(\mathcal{D}). \tag{6.144}$$

Upon integration by parts, (6.144) yields

$$\int_D \bar{u}(-\operatorname{div}(\bar{a} \nabla v)) dx = \int_D u(-\operatorname{div}(\bar{a} \nabla v)) dx, \quad \forall \, v \in W_0^{1,2}(\mathcal{D}), \tag{6.145}$$

which can be interpreted as

$$\int_D \bar{u} v^* dx = \int_D u v^* dx, \quad \forall \, v^* \in (W_0^{1,2}(\mathcal{D}))^*, \tag{6.146}$$

which implies the required result $\bar{u} = u$.[14]

2. We prove that $(\bar{u}_\epsilon)_{\epsilon>0}$, defined by (6.126), is a recovery sequence for $(F_\epsilon)_{\epsilon>0}$. Indeed, multiplying (6.126) by $\bar{u}_\epsilon$ and integrating by parts (using the boundary conditions), we obtain

$$\int_{\mathcal{D}} \nabla\bar{u}_\epsilon(x)\cdot a\Big(\frac{x}{\epsilon}\Big)\nabla\bar{u}_\epsilon(x)dx = \int_{\mathcal{D}} \nabla\bar{u}_\epsilon(x)\cdot\bar{a}\nabla u(x)dx \to \int_{\mathcal{D}} \nabla u(x)\cdot\bar{a}\nabla u(x)dx, \tag{6.147}$$

where, for the final step, we used the fact that $\nabla\bar{u}_\epsilon \rightharpoonup \nabla u$ in $L^2(\mathcal{D})$ (established in step 1). Relation (6.147) shows that $(u_\epsilon)_{\epsilon>0}$ is a recovery sequence.

3. We now show the liminf inequality. Consider any $u_\epsilon \rightharpoonup u$ in $W_0^{1,2}(\mathcal{D})$. By the ellipticity condition (6.117), we have (removing, for notational simplicity, explicit x dependence – except for the a term) that

$$\begin{aligned}0 &\le \int_{\mathcal{D}} \nabla(u_\epsilon-\bar{u}_\epsilon)\cdot a\Big(\frac{x}{\epsilon}\Big)\nabla(u_\epsilon-\bar{u}_\epsilon)\\ &= \int_{\mathcal{D}} \nabla u_\epsilon\cdot a\Big(\frac{x}{\epsilon}\Big)\nabla u_\epsilon dx - 2\int_{\mathcal{D}} \nabla u_\epsilon\cdot a\Big(\frac{x}{\epsilon}\Big)\nabla\bar{u}_\epsilon dx + \int_{\mathcal{D}} \nabla\bar{u}_\epsilon\cdot a\Big(\frac{x}{\epsilon}\Big)\nabla\bar{u}_\epsilon dx.\end{aligned} \tag{6.148}$$

Consider the second term in the second line of (6.148)

$$\begin{aligned}\int_{\mathcal{D}} \nabla u_\epsilon\cdot a\Big(\frac{x}{\epsilon}\Big)\nabla\bar{u}_\epsilon dx &= -\int_{\mathcal{D}} \operatorname{div}\Big(a\Big(\frac{x}{\epsilon}\Big)\nabla\bar{u}_\epsilon\Big)u_\epsilon dx\\ &\overset{(6.126)}{=} -\int_{\mathcal{D}} \operatorname{div}(\bar{a}\nabla u)u_\epsilon dx = \int_{\mathcal{D}} \nabla u_\epsilon\cdot\bar{a}\nabla u dx \to F(u),\end{aligned}$$

where the last claim follows by $\nabla u_\epsilon \rightharpoonup \nabla u$ in $L^2(\mathcal{D})$.

Similarly for the third term in the second line of (6.148),

$$\begin{aligned}\int_{\mathcal{D}} \nabla\bar{u}_\epsilon\cdot a\Big(\frac{x}{\epsilon}\Big)\nabla\bar{u}_\epsilon dx &= -\int_{\mathcal{D}} \operatorname{div}\Big(a\Big(\frac{x}{\epsilon}\Big)\nabla\bar{u}_\epsilon\Big)\bar{u}_\epsilon dx\\ &\overset{(6.126)}{=} -\int_{\mathcal{D}} \operatorname{div}(\bar{a}\nabla u)\bar{u}_\epsilon dx = \int_{\mathcal{D}} \nabla\bar{u}_\epsilon\cdot\bar{a}\nabla u dx \to F(u),\end{aligned}$$

where the last claim follows by $\nabla\bar{u}_\epsilon \rightharpoonup \nabla u$ in $L^2(\mathcal{D})$.

Taking the liminf on (6.148) and using the two observations above, we obtain the liminf inequality for Γ-convergence. □

14 For this last step, recall that for any $v^* \in W_0^{1,2}(\mathcal{D})^*$ there exists a unique $v \in W_0^{1,2}(\mathcal{D})$ such that $-\operatorname{div}(\bar{a}\nabla v) = v^*$ with the equality interpreted in $W_0^{1,2}(\mathcal{D})$. For any $u^* \in W_0^{1,2}(\mathcal{D})^*$ find the corresponding $u \in W_0^{1,2}(\mathcal{D})$, and then calculate $\int_{\mathcal{D}}(\bar{u}-u)u^*\,dx = -\int_{\mathcal{D}}(\bar{u}-u)\operatorname{div}(\bar{a}\nabla u)dx$, which is equal to 0 by (6.145).

We now note that the original sequence of functionals $(\hat{F}_\epsilon)_{\epsilon>0}$, whose minimizers are associated with the weak solutions of (6.116) is of the form $\hat{F}_\epsilon = F_\epsilon + G$, where G is the functional $u \mapsto G(u) = -2\int_{\mathcal{D}} fu dx$. Interpreting the family $(F_\epsilon)_{\epsilon>0}$ as a sequence of functionals $(F_n)_{n\in\mathbb{N}}$ upon choosing an appropriate sequence $(\epsilon_n)_{n\in\mathbb{N}}$, $\epsilon_n \to 0$, we see that the Γ-convergence result in Proposition 6.9.2 implies that

$$\hat{F}_\epsilon := F_\epsilon + G \xrightarrow{\Gamma} \hat{F} := F + G,$$

as $\epsilon \to 0$ since the functional G is continuous. Moreover, by condition (6.117) (see also (6.123)), we see that the family $(\hat{F}_\epsilon)_{\epsilon>0}$ is bounded below by the lower semicontinuous coercive functional $F_0 + G$, where $u \mapsto F_0(u) = c_0 \int_{\mathcal{D}} |\nabla u|^2$ (with c_0 as in (6.117)). Hence, the family $(\hat{F}_\epsilon)_{\epsilon>0}$ is equicoercive (see Proposition 2.8.13), and by the fundamental theorem of Γ-convergence (Theorem 2.8.11), we can deduce the convergence of minimizers of $(\hat{F}_\epsilon)_{\epsilon>0}$ to the minimizer of $\hat{F}$. This in turn provides the required answer to our original problem of homogenization, yields a positive answer to the question of whether the solutions $(u_\epsilon)_{\epsilon>0}$ of the family of problems (6.116) converges in the appropriate sense to the solution u of problem (6.120) for some constant matrix $\bar{a}$, whose exact form can be identified in terms of the space dependent coefficients a. In particular, the convergence is in the weak topology of $W_0^{1,2}(\mathcal{D})$, and the exact form of $\bar{a}$ is provided by (6.124).

Remark 6.9.3. The well posedness of problem (6.125) that defines the fundamental auxiliary functions ϕ_i that are needed for the construction of the oscillating test functions can arise from general variational arguments or their extensions related to the celebrated Lax–Milgram lemma (that will be introduced in the following chapter, see Section 7.3). The choice of boundary conditions and fixing the average of the solution to 0 is related to the well posedness (and the uniqueness) of the solution. For more details, the reader can consult Sec. 4.7 in [50]. See also Example 6.9.4 for an explicitly solved example.

Example 6.9.4. The 1D case can be computed explicitly in closed form. For simplicity, assume that $Y = (0,1)$. The cell equation (6.125) reads

$$-\frac{d}{dy}\left(a(y)\left(1 + \frac{d}{dy}\phi\right)\right) = 0, \tag{6.149}$$

which is readily integrated to yield (up to an arbitrary constant C),

$$1 + \frac{d}{dy}\phi(y) = Ca^{-1}(y) \implies \phi(1) - \phi(0) = C\int_0^1 a^{-1}(y)dy - 1,$$

and from the condition on the periodicity of ϕ, we obtain that

$$C = \frac{1}{\int_0^1 a^{-1}(y)dy} = \left(\fint_Y \frac{1}{a(y)}dy\right)^{-1}.$$

Note that in order to obtain ϕ, we must integrate (6.149) once more, thus introducing one more constant that will be determined by the condition that ϕ is chosen to have zero average.

We now use the representation (6.124) for the averaged (homogenized) coefficient $\bar{a}$, which yields

$$\bar{a} = \fint_Y a(y)\left(1 + \frac{d\phi}{dy}\right)dy = C = \left(\fint_Y \frac{1}{a(y)}dy\right)^{-1}.$$

It is interesting to note that $\bar{a}$ is the inverse of the harmonic mean of a, rather than its mean, as our intuition may probably mislead us. This 1D case can be also treated directly, without resort to the general theory presented here, which of course is indispensable for the case of dimension higher than 1. ◁

The above theory can be extended to more general functionals, which are non-quadratic, and for which the corresponding Euler–Lagrange equations are nonlinear. The treatment is more involved and more technical. We refer the reader to e. g. [25] for a simplified version of such results in 1D and to the references in the beginning of the section for more general results. Moreover, other approaches to homogenization problems, bypassing the use or explicit mention of Γ-convergences have been proposed and applied successfully to various problems—in fact, many of the initial rigorous approaches to this problem were not in the spirit of Γ-concergence (see, e. g., [25, 26, 50] etc).

6.9.2 Γ-convergence for the *p*-Laplacian

Another interesting, and relatively easy to handle, application of Γ-convergence in the calculus of variations is in the behavior of the p-Laplacian as the exponent p varies. As a quick flavor of the application of Γ-convergence in this field, consider the family of functionals $(F_p)_{p>1}$, $F_p : W_0^{1,p}(\mathcal{D}) \to \mathbb{R}$,

$$F_p(u) := \int_{\mathcal{D}} |\nabla u(x)|^p \, dx, \tag{6.150}$$

which are conveniently redefined as

$$F_p(u) := \begin{cases} \int_{\mathcal{D}} |\nabla u(x)|^p \, dx, & \text{if } u \in W_0^{1,p}(\mathcal{D}), \\ +\infty, & \text{if } u \in L^1(\mathcal{D}) \setminus W_0^{1,p}(\mathcal{D}). \end{cases} \tag{6.151}$$

For the sake of simplicity, we avoid here dividing by $1/p$.

An interesting question that arises is the behavior of the family $(F_p)_{p>1}$, as well as the minimizers of relevant families of functionals, such as, e. g., $F_p + G$, as the exponent $p \to p_0$, where p_0 is an appropriate exponent. This question is also related to the solutions of the p-Laplacian Poisson equation, or certain eigenvalues of the p-Laplacian (called variational eigenvalues) as p-varies.

The following result, obtained in [109] is an interesting step in this direction:

Proposition 6.9.5 ([109]). *Let $\mathcal{D} \subset \mathbb{R}^d$, a bounded domain with Lipschitz boundary, and consider the family of functionals $(F_p)_{p>1}$, where F_p is defined as in (6.151). Then, $F_p \xrightarrow{\Gamma} F_{p_0}$, in $L^{p_0}(\mathcal{D})$, as $p \to p_0 \in (1, \infty)$.*

Proof. The proof requires different steps for $p \to p_0^+$ and $p \to p_0^-$. Here, we only provide the case $p \to p_0^+$ in detail and refer the reader to [109] for the general case.

Let $p \to p_0^+$. By the Lebesgue embeddings, we have that $L^p(\mathcal{D}) \subset L^{p_0}(\mathcal{D})$ for all $p > p_0$ (see Theorem 1.4.2). On account of that, we can work directly on the larger space $L^{p_0}(\mathcal{D})$, irrespective of the value of p. Moreover, when using the notation $p \to p_0^+$, in order to be compliant with the notation we have used here for Γ-convergence, we will have in mind any sequence $p_n \to p_0^+$, $p_n > p_0$ for all $n \in \mathbb{N}$, and a corresponding sequence of functions $u_{p_n} \to u$ in $L^{p_0}(\mathcal{D})$, where even if $u_{p_n} \in L^{p_n}(\mathcal{D})$, it is considered (by the above arguments) as $u_{p_n} \in L^{p_0}(\mathcal{D})$.

We will show the result by checking the liminf and the limsup inequality for Γ-convergence.

(i) The liminf inequality: For the liminf inequality, it suffices to consider a sequence $u_{p_n} \to u$ in $L^{p_0}(\mathcal{D})$ such that F_{p_n} is bounded for all p_n. This implies that $u_{p_n} \in W_0^{1,p_n}(\mathcal{D})$, and by the same arguments as above, we can consider $u_{p_n} \in W_0^{1,p_0}(\mathcal{D})$ for all p_n. The uniform boundedness of the sequence in $W_0^{1,p_0}(\mathcal{D})$, by weak compactness arguments (valid for the range of p_n and p_0 we are interested in), imply the existence of a subsequence (denoted the same for simplicity) such that $u_{p_n} \rightharpoonup u$ in $W_0^{1,p_0}(\mathcal{D})$. By the Poincaré inequality, $\|\nabla u\|_{L^{p_0}(\mathcal{D})}$ is an equivalent norm for $W_0^{1,p_0}(\mathcal{D})$. Hence, by the weak lower semicontinuity of the norm, we have that

$$\begin{aligned}
\int_{\mathcal{D}} |\nabla u_{p_0}|^{p_0} dx &\le \liminf_n \int_{\mathcal{D}} |\nabla u_{p_n}|^{p_0} dx \overset{(1)}{\le} \liminf_n \Big(\int_{\mathcal{D}} |\nabla u_{p_n}|^{p_n} dx \Big)^{\frac{p_0}{p_n}} |\mathcal{D}|^{\frac{p_n - p_0}{p_n}} \\
&\le \liminf_n \Big(\int_{\mathcal{D}} |\nabla u_{p_n}|^{p_n} dx \Big)^{\frac{p_0}{p_n}} \limsup_n |\mathcal{D}|^{\frac{p_n - p_0}{p_n}} \\
&= \liminf_n \Big(\int_{\mathcal{D}} |\nabla u_{p_n}|^{p_n} dx \Big)^{\frac{p_0}{p_n}},
\end{aligned} \tag{6.152}$$

where for (1) we used Hölder's inequality.[15] By estimate (6.152) and the fact that the choice of the sequence $p_n \to p_0$ is arbitrary, we conclude the liminf inequality.

15 Expressing $|\nabla u_{p_n}|^{p_0} = |\nabla u_{p_n}|^{p_0} 1$, and then applying Hölder for the choice $\bar{p} = p_n/p_0$ and conjugate exponent $\bar{q} = \frac{p_0}{p_n - p_0}$.

(ii) The limsup inequality: We will construct a recovery sequence. As before, consider an arbitrary sequence $p_n \to p_0$ and any $u \in W_0^{1,p_0}(\mathcal{D})$. If $u = 0$, then the constant sequence $u_n = 0$ can be used as a recovery sequence. We consider the case $u \neq 0$. By the density of $C_c^\infty(\mathcal{D})$ in $W_0^{1,p_0}(\mathcal{D})$, we can find a sequence $(u_n)_{n\in\mathbb{N}} \subset C_c^\infty(\mathcal{D})$ such that $u_n \to u$ in $W^{1,p_0}(\mathcal{D})$. We will show that such a sequence can be used as a recovery sequence.

To show that, note first that since $u_n \to u$ in $W^{1,p_0}(\mathcal{D})$, we have that $\nabla u_n \to \nabla u$ in $L^{p_0}(\mathcal{D})$. Moreover, since the sequence consists of smooth functions, ∇u_n is bounded in $L^\infty(\mathcal{D})$, and we can write $\|\nabla u_n\|_{L^{p_0}(\mathcal{D})} = \|\nabla u_n\|_{L^\infty(\mathcal{D})} \|\frac{\nabla u_n}{\|\nabla u_n\|_{L^\infty(\mathcal{D})}}\|_{L^{p_0}(\mathcal{D})}$. To simplify notation, we will denote by $c_n := \|\nabla u_n\|_{L^\infty(\mathcal{D})} > 0$ and $w_n := \frac{\nabla u_n}{\|\nabla u_n\|_{L^\infty(\mathcal{D})}}$. Note that $|w_n| \le 1$. Finally, we observe that $\liminf_n c_n =: c > 0$. For this last one, if not, then the liminf inequality (we just proved) for the family of functionals F_p, along with the use of the Hölder inequality, leads to the result that $u = 0$, which is a contradiction.[16]

We are now ready to prove the limsup inequality. By the strong convergence of $u_n \to u$ in $W^{1,p_0}(\mathcal{D})$, we have that

$$\|\nabla u\|_{L^{p_0}(\mathcal{D})} = \lim_n \|\nabla u_n\|_{L^{p_0}(\mathcal{D})} = \lim_n \big(c_n \|w_n\|_{L^{p_0}(\mathcal{D})}\big). \tag{6.153}$$

Since $|w_n| \le 1$, and $p_n > p_0$, we see that $|w_n|^{p_n} \le |w_n|^{p_0}$. So, upon integrating over $\mathcal{D}$ and then taking the $1/p_0$ power, we obtain

$$c_n\Big(\int_{\mathcal{D}} |w_n|^{p_n} dx\Big)^{1/p_0} \le c_n\Big(\int_{\mathcal{D}} |w_n|^{p_0} dx\Big)^{1/p_0} = c_n\|w_n\|_{L^{p_0}(\mathcal{D})},$$

so that after taking on both sides the limsup, we obtain that

$$\begin{aligned}
\lim_n \big(c_n\|w_n\|_{L^{p_0}(\mathcal{D})}\big) &\ge \limsup_n c_n\Big(\int_{\mathcal{D}} |w_n|^{p_n} dx\Big)^{1/p_0} \\
&= \limsup_n c_n^{\frac{p_0-p_n}{p_n}} \Big(\int_{\mathcal{D}} |\nabla u_n|^{p_n} dx\Big)^{1/p_0} \\
&\ge \liminf_n c_n^{\frac{p_0-p_n}{p_n}} \limsup_n \Big(\int_{\mathcal{D}} |\nabla u_n|^{p_n} dx\Big)^{1/p_0} \\
&= \limsup_n \Big(\Big(\int_{\mathcal{D}} |\nabla u_n|^{p_n} dx\Big)^{1/p_n}\Big)^{p_n/p_0} \\
&= \limsup_n \Big(\int_{\mathcal{D}} |\nabla u_n|^{p_n} dx\Big)^{1/p_n},
\end{aligned} \tag{6.154}$$

16 Indeed, the liminf inequality yields $F_{p_0}(u) = \|\nabla u\|_{L^{p_0}(\mathcal{D})} \le \liminf_n \|\nabla u_n\|_{L^{p_n}(\mathcal{D})} \le \liminf_n \|\nabla u_n\|_{L^\infty(\mathcal{D})} |\mathcal{D}|^{1/p_n} = 0$, from which $u = 0$ follows.

where in the above we used the fact that $\liminf_n c_n = c > 0$. Combining (6.153) and (6.154), we obtain the desired limsup inequality. □

The Γ-convergence result established in Proposition 6.9.5, obtained in [109], along with the equicoercivity of the family of functionals $(F_p)_{p>1}$ allows us to obtain interesting results on the convergence of minimizers of functionals related to F_p, and hence to interesting results concerning PDE problems related to such functionals in terms of the Euler–Lagrange equations. Such results are related, e. g., with the continuity of eigenvalues of the p-Laplacian (see problem (6.112)) or solutions of quasilinear PDEs involving the p-Laplacian (see problem (6.106)) with respect to p. Details are left to the interested reader (see also [109]).

6.10 Appendix

6.10.1 A version of the Riemann–Lebesgue lemma

Lemma 6.10.1. *Let $u \in L^\infty_{\mathrm{loc}}(\mathbb{R})$ be a periodic function of period 1. For every $\epsilon > 0$, define the family of functions $\{u_\epsilon : \epsilon > 0\}$, by $u_\epsilon(x) = u(\frac{x}{\epsilon})$ for every $x \in \mathbb{R}$. Then, $u_\epsilon \overset{\star}{\rightharpoonup} \bar{u} := \int_{(0,1)} u(z)dz$ in $L^\infty(I)$, for every $I \subset \mathbb{R}$ bounded and measurable.*

Proof. Since $u \in L^\infty_{\mathrm{loc}}(\mathbb{R})$, the sequence $(u_\epsilon)_{\epsilon>0}$ is bounded. Hence, there exists a subsequence (which we do not relabel) such that $u_\epsilon \overset{\star}{\rightharpoonup} v$, as $\epsilon \to 0^+$, where v is some function in $L^\infty(\mathbb{R})$. We need to identify the limit, and in particular show that $v(x) = \bar{u}$, a. e. in $(0, 1)$. To this end, consider any Lebesgue point $x_0 \in (0, 1)$, and any interval $I_\delta = (x_0 - \frac{\delta}{2}, x_0 + \frac{\delta}{2})$. We note that

$$\left|\int_{I_\delta} (v(x) - \bar{u})dx\right| = \lim_{\epsilon\to 0^+}\left|\int_{I_\delta}\left(u\left(\frac{x}{\epsilon}\right) - \bar{u}\right)dx\right| = \lim_{\epsilon\to 0^+}\left|\epsilon \int_{\frac{1}{\epsilon}I_\delta} (u(y) - \bar{u})dy\right|, \tag{6.155}$$

where we first used the fact that $u_\epsilon \overset{\star}{\rightharpoonup} v$, as $\epsilon \to 0^+$, and then we changed variables in the integral. Let $k(\epsilon) = [\frac{\delta}{\epsilon}]$ (so that $\frac{\delta}{\epsilon} - 1 < k(\epsilon) \le \frac{\delta}{\epsilon}$), and we express the interval I_δ as $\frac{1}{\epsilon}I_\delta = \bigcup_{i=1}^{k(\epsilon)} I_i \cup A_\epsilon$, where $I_i = (x_i - \frac{1}{2}, x_i + \frac{1}{2})$ are unit intervals centered at some points x_i, $i = 1, \ldots, k(\epsilon)$, and A_ϵ is a remainder set of Lebesgue measure $|A_\epsilon| \le \frac{\delta}{\epsilon} - k(\epsilon) \le \frac{\delta}{\epsilon} - (\frac{\delta}{\epsilon} - 1) = 1$. In other words, we have broken up the interval $\frac{1}{\epsilon}I_\delta$, which is centered at a Lebesgue point x_0 and has length $\frac{\delta}{\epsilon}$ into $k(\epsilon)$ intervals of unit length, centered at points x_i, plus a small remainder A_ϵ of Lebesgue measure less than 1. Using the additivity of the integral, we see that

$$\begin{aligned}\int_{\frac{1}{\epsilon}I_\delta}(u(y)-\bar{u})dy &= \sum_{i=1}^{k(\epsilon)}\int_{I_i}(u(y)-\bar{u})dy+\int_{A_\epsilon}(u(y)-\bar{u})dy \\ &= k(\epsilon)\int_{I_1}(u(y)-\bar{u})dy+\int_{A_\epsilon}(u(y)-\bar{u})dy=\int_{A_\epsilon}(u(y)-\bar{u})dy,\end{aligned} \tag{6.156}$$

where we used the 1-periodicity of u and the definition of $\bar{u}$, and we may obtain the estimate

$$\left|\int_{A_\epsilon}(u(y)-\bar{u})dy\right| \leq 2\|u\|_{L^\infty(0,1)}|A_\epsilon| \leq 2\|u\|_{L^\infty(0,1)}. \tag{6.157}$$

Substituting (6.156) in (6.155) and using estimate (6.157), we obtain

$$\left|\int_{I_\delta}(v(x)-\bar{u})dx\right| = \lim_{\epsilon\to 0^+}\left|\epsilon\int_{A_\epsilon}(u(y)-\bar{u})dy\right| = 0. \tag{6.158}$$

Recall that we have chosen $x_0 \in (0,1)$ to be a Lebesgue point for v so that $\lim_{\delta\to 0^+}\frac{1}{\delta}\int_{I_\delta}v(x)dx = v(x_0)$. This clearly implies that $\lim_{\delta\to 0^+}\frac{1}{\delta}\int_{I_\delta}(v(x)-\bar{u})dx = v(x_0)-\bar{u}$. On the other hand, (6.158) yields that $\lim_{\delta\to 0^+}\frac{1}{\delta}\int_{I_\delta}(v(x)-\bar{u})dx = 0$, so $v(x_0) = \bar{u}$. Since x_0 is a Lebesgue point, this property holds a. e. □

The above Lemma, which finds important applications in fields like, e. g., homogenization theory, holds in any dimension as well as for the weak convergence in L^p. For details and the general proof, the reader may consult, e. g., [80] Lemma 2.85, or [50] Theorem 2.6.

Theorem 6.10.2 ([50]). *Let f be a Y-periodic function in $L^p(Y)$, and define the sequence of functions $(f_\epsilon)_{\epsilon>0}$, setting $f_\epsilon(x) = f(\frac{x}{\epsilon})$ a. e. in $\mathbb{R}^d$. Then,*

$$f_\epsilon \rightharpoonup ⨍_Y f(y)dy := \frac{1}{|Y|}\int_Y f(y)dy,$$

with the convergence being
- *weak in $L^p(D)$ for any $D \subset \mathbb{R}^d$, open and bounded, if $p \in [1,\infty)$*
- *weak* in $L^\infty(\mathbb{R}^d)$ is $p = \infty$.*

6.10.2 Proof of Theorem 6.6.3

(i) The idea of the proof is based upon the observation that a function is bounded above in some ball $B(x_0,\rho)$ if the measure of the super level set $|A^+(k,x_0,\rho)|$ vanishes for some k^* sufficiently large. More precisely, to use the Caccioppoli type inequalities in conjunction with a technical inequality due to Stampacchia, in order to guarantee that $|A^+(k, x_0, \rho)|$ vanishes for k large enough. The proof is broken up into 4 steps.

1. We first provide the technical inequality upon which the proof will be based. This is a Stampacchia-like inequality (see [132]), which states that if $\phi : [k_0, \infty) \times [R_1, R_2] \to \mathbb{R}_+$, is nonincreasing and nondecreasing in the first and second variable, respectively, and satisfies an inequality of the form

$$\begin{gathered}\phi(k_2, r_1) \leq c_S (k_2 - k_1)^{-\alpha} [(r_2 - r_1)^{-\beta} + (k_2 - k_1)^{-\beta}] \phi(k_1, r_2)^{\theta}, \\ \text{for every } r_1 < r_2, \text{ and } k_1 < k_2,\end{gathered} \tag{6.159}$$

with $\alpha > 0$, $\beta > 0$ and $\theta > 1$, then there exists k^* such that $\phi(k^*, R_1) = 0$. This k^* can be expressed as $k^* = k_0 + \bar{k}$ with

$$\bar{k}^{\alpha} = c'_S (R_2 - R_1)^{-\beta} \phi(k_0, R_2)^{\theta - 1} + (R_2 - R_1)^{\alpha}, \tag{6.160}$$

where $c'_S = 2^{\frac{\theta(\alpha+\beta)+\theta-1}{\theta-1}} c_S$.

To show that (6.159) implies that $\phi(k^\star) = 0$ for $k^\star = k_0 + \bar{k}$ with $\bar{k}$ as in (6.160), consider the sequences $\widehat{k}_n = \bar{k}(1 - 2^{-n})$ and $\widehat{r}_n = r + 2^{-n}(R_2 - R_1)$, apply the inequality (6.159) for the choice $k_1 = \widehat{k}_{n-1}, k_2 = \widehat{k}_n, r_1 = \widehat{r}_n, r_2 = \widehat{r}_{n-1}$, and use this to prove inductively that $\phi(\widehat{k}_n, \widehat{r}_n) \leq 2^{-n\frac{\alpha+\beta}{\theta-1}} \phi(\widehat{k}_0, \widehat{r}_0)$ for any $n \in \mathbb{N}$. Then use the monotonicity properties of ϕ to see that $\phi(k_0 + \bar{k}, r) \leq \phi(\widehat{k}_n, \widehat{r}_n)$, and pass to the limit as $n \to \infty$.

2. We now assume that $u \in DG_p(k_0, R_0, \mathcal{D})$, and we shall show that upon fixing a radius $R \leq R_0$ such that $B(x_0, R) \subset \mathcal{D}$, and a $k \geq k_0$, and for any r'_1, r'_3 (not to be confused with the above sequence!) with the property $\frac{R}{2} < r'_1 < r'_2 := \frac{r'_1 + r'_3}{2} < r'_3 < R$, for any $k_0 \leq m < k$, we have the simultaneous estimates,

$$|A^+(k, x_0, r'_1)| \leq (k - m)^{-p} \int_{A^+(m, x_0, r'_3)} (u(x) - m)^p dx, \tag{6.161}$$

and

$$\begin{aligned}&\int_{A^+(k, x_0, r'_1)} (u(x) - k)^p dx \\ &\quad \leq c |A^+(m, x_0, r'_3)|^{\frac{p}{d}} [(r'_3 - r'_1)^{-p} + (k - m)^{-p}] \int_{A^+(m, x_0, r'_3)} (u(x) - m)^p dx.\end{aligned} \tag{6.162}$$

Note that these are Caccioppoli type estimates, connecting the L^p norms of the excess function over superlevel sets at different levels (and involving intersections with balls of different radii). We will simplify that notation by setting $A_i(k) := A^+(k, x_0, r'_i)$, $i = 1, 2, 3$, and clearly

$$A_1(k) \subset A_2(k) \subset A_3(k), \quad \forall\, k > k_0. \tag{6.163}$$

We will use a cutoff function technique. Consider a cutoff function

$$\psi \in C_c^\infty(B(x_0, r_2'), [0,1]) \;:\; \psi \equiv 1 \text{ on } B(x_0, r_1') \text{ and } |\nabla \psi| \le \frac{4}{r_3' - r_1'}. \tag{6.164}$$

Assume for simplicity[17] that $p < d$. By the properties of ψ, we have that

$$\begin{aligned}
\int_{A_1(k)} (u(x)-k)^p dx &\overset{(6.164)}{=} \int_{A_1(k)} \psi(x)^p (u(x)-k)^p dx \\
&\overset{(6.163)}{\le} \int_{A_2(k)} \psi(x)^p (u(x)-k)^p dx \\
&\le |A_2(k)|^{\frac{\mathfrak{s}_p - p}{p}} \Big(\int_{A_2(k)} \psi(x)^{\mathfrak{s}_p} (u(x)-k)^{\mathfrak{s}_p} dx \Big)^{\frac{p}{\mathfrak{s}_p}} \\
&= |A_2(k)|^{\frac{p}{d}} \Big(\int_{A_2(k)} \psi^{\mathfrak{s}_p} (u(x)-k)^{\mathfrak{s}_p} dx \Big)^{\frac{p}{\mathfrak{s}_p}},
\end{aligned} \tag{6.165}$$

where we also used the Hölder inequality to create the $L^{\mathfrak{s}_p}$ norm for $v = \psi(u-k)$, where $\mathfrak{s}_p = \frac{dp}{d-p}$ is the critical exponent in the Sobolev embedding (recall the Gagliardo–Niremberg–Sobolev inequalities). This last step is done in an attempt to bring into the game the gradient of the norm and involve in this way the Caccioppoli inequalities.

We will continue our estimates using the Sobolev inequality, for $v = \psi u$, raised to the power p. Bearing in mind that on $A_2(k)$, it holds that $\nabla v = \psi \nabla u + (u-k)\nabla \psi$, and we have that

$$\begin{aligned}
&\Big(\int_{A_2(k)} \psi(x)^{\mathfrak{s}_p} (u(x)-k)^{\mathfrak{s}_p} dx \Big)^{p/\mathfrak{s}_p} \\
&\quad \le c_S \int_{A_2(k)} |\nabla(\psi(u-k))(x)|^p dx \\
&\quad = c_S \int_{A_2(k)} |\psi(x)\nabla u(x) + (u(x)-k)\nabla\psi(x)|^p dx \\
&\quad \le c_1 \int_{A_2(k)} (\psi(x)^p |\nabla u(x)|^p + (u(x)-k)^p |\nabla \psi(x)|^p) dx \\
&\quad \overset{(6.164)}{\le} c_2 \Big(\int_{A_2(k)} |\nabla u(x)|^p dx + 4^p (r_3' - r_1')^{-p} \int_{A_2(k)} (u(x)-k)^p dx \Big),
\end{aligned} \tag{6.166}$$

17 If $p > d$, the continuity of minimizers follows from the Sobolev embedding theorem. The case $p = d$ requires some proof, which is along the lines of the case $p > d$ presented here if not a little simpler.

where we also used the properties of the cutoff function ψ. Combining (6.165) with (6.166), we have that

$$\int\limits_{A_1(k)} (u(x)-k)^p dx \le c_3 \, |A_2(k)|^{\frac{d}{p}} \Big(\int\limits_{A_2(k)} |\nabla u(x)|^p dx + 4^p (r_3' - r_1')^{-p} \int\limits_{A_2(k)} (u(x)-k)^p dx \Big). \tag{6.167}$$

If $u \in DG_p^+(\mathcal{D})$, then we may further estimate $\int_{A_2(k)} |Du(x)|^p dx$ by the L^p norm of $u-k$ on the larger domain $A_3(k)$ and the measure of this set. Using the relevant Caccioppoli inequality (6.64) for the choice $r = r_2'$, $R = r_3'$ and bearing in mind that $r_3' - r_2' = \frac{1}{2}(r_3' - r_1')$, we have that

$$\begin{aligned}\int\limits_{A_2(k)} |\nabla u(x)|^p dx &\le c_0 \Big((r_3' - r_2')^{-p} \int\limits_{A_3(k)} (u(x)-k)^p dx + |A_3(k)| \Big) \\ &= c_0 \Big(2^p (r_3' - r_1')^{-p} \int\limits_{A_3(k)} (u(x)-k)^p dx + |A_3(k)| \Big).\end{aligned} \tag{6.168}$$

By (6.168) and (6.167) (as $\int_{A_2(k)} (u(x)-k)^p dx \le \int_{A_3(k)} (u(x)-k)^p dx$ since $A_2(k) \subset A_3(k)$ and $u > k$ on these sets) leads to the estimate

$$\int\limits_{A_1(k)} (u(x)-k)^p dx \le c_4 |A_2(k)|^{\frac{d}{p}} \Big((r_3' - r_1')^{-p} \int\limits_{A_3(k)} (u(x)-k)^p dx + |A_3(k)| \Big), \tag{6.169}$$

for an appropriate constant c_4.

So far, we only have estimates for the same level k, but varying the radius of the ball with which we take the intersection of the superlevel set. We now try to assess the effect of trying to vary the level at which the super level set is taken in the above estimate. Consider a lower level $m < k$, ($m \in [k_0, k)$) and the corresponding super level sets $A_3(m) = \{x \in B(x_0, r_3') \, : \, u(x) > m\}$. We first estimate $|A_3(k)|$ in terms of relevant quantities on $A_3(m)$. Clearly, $u(x) > k$ implies that $u(x) > m$ so that $A_3(k) \subset A_3(m)$. This leads to the estimate $\int_{A_3(k)} (u(x) - m)^p dx \le \int_{A_3(m)} (u(x) - m)^p dx$, and multiplying both sides with $(k-m)^{-p}$, while on $A_3(k)$ we have that $u(x) > k$ so that $\frac{u(x)-m}{k-m} > 1$, we easily conclude that

$$
\begin{aligned}
|A_1(k)| \le |A_3(k)| &\le (k-m)^{-p} \int_{A_3(k)} (u(x)-m)^p dx \\
&\le (k-m)^{-p} \int_{A_3(m)} (u(x)-m)^p dx,
\end{aligned}
\tag{6.170}
$$

where the first inequality arises trivially from the fact that as $r_1' < r_3'$, $A_1(k) \subset A_3(k)$. Note that (6.170) is the first of the simultaneous estimates, i. e., (6.161).

Our second estimate is that

$$
\int_{A_3(k)} (u(x)-k)^p dx \le \int_{A_3(k)} (u(x)-m)^p dx \le \int_{A_3(m)} (u(x)-m)^p dx, \tag{6.171}
$$

where we first used the fact that on $A_3(k)$, $0 \le u-k \le u-m$, and then that $A_3(k) \subset A_3(m)$.

Using (6.170) in (6.169) to estimate the term $|A_3(k)|$ inside the bracket, we obtain the symmetric form for (6.169) as

$$
\begin{aligned}
&\int_{A_1(k)} (u(x)-k)^p dx \\
&\quad \le c_4 |A_3(m)|^{\frac{p}{d}} \left[(r_3'-r_1')^{-p} + (k-m)^{-p}\right] \int_{A_3(m)} (u(x)-m)^p dx,
\end{aligned}
\tag{6.172}
$$

which is the second of the simultaneous estimates, i. e., (6.162).

3. We will now try to combine the simultaneous estimates (6.161)–(6.162) into a single Stampacchia-like inequality, which will guarantee that $|A(k, x_0, \rho)|$ vanishes for k large enough.

Since $r_1' < r_3'$ and $m < k$ are arbitrary, we relabel them as $r_1 < r_2$ and $k_1 < k_2$, respectively, and defining the functions

$$
\begin{aligned}
&\phi_1(k,r) := \int_{A^+(k,x_0,r)} (u(x)-k)^p dx, \quad \phi_2(k,r) = |A^+(k,x_0,r)|, \\
&G(s_1,s_2;p) := s_1^{-p} + s_2^{-p},
\end{aligned}
$$

we express the simultaneous estimates (6.161)–(6.162) as

$$
\begin{aligned}
&\phi_1(k_2,r_1) \le c\, G(r_2-r_1, k_2-k_1; p)\phi_1(k_1,r_2)\phi_2(k_1,r_2)^{\frac{p}{d}}, \\
&\phi_2(k_2,r_1) \le G(0, k_2-k_1; p)\phi_1(k_1,r_2).
\end{aligned}
\tag{6.173}
$$

We raise the first one to the power q_1 and the second to the power q_2 (for q_1, q_2 to be specified shortly), and multiply to get

$$
\begin{aligned}
&\phi_1(k_2,r_1)^{q_1}\phi_2(k_2,r_1)^{q_2} \\
&\quad \le c_5\, G(r_2-r_1, k_2-k_1; p)^{q_1} G(0, k_2-k_1; p)^{q_2} \phi_1(k_1,r_2)^{q_1+q_2}\phi_2(k_1,r_2)^{\frac{q_1 p}{d}}.
\end{aligned}
\tag{6.174}
$$

We define the function

$$\phi(k,r) = \phi_1(k,r)^{q_1}\phi_2(k,r)^{q_2},$$

and note that if ϕ vanishes for values of k above a certain threshold, then $|A^+(k,x_0,r)|$ also vanishes, which is in fact the type of estimate we seek.

We now see that if q_1, q_2 are chosen so that there exists a $\theta > 1$ such that

$$q_1 + q_2 = \theta q_1, \quad \text{and} \quad \frac{q_1 p}{d} = \theta q_2, \tag{6.175}$$

then inequality (6.174) becomes

$$\phi(k_2,r_1) \le c_5 G(0,k_2-k_1;p)^{q_2} G(r_2-r_1,k_2-k_1;p)^{q_1}\phi(k_1,r_2)^{\theta}, \quad \theta > 1. \tag{6.176}$$

Noting further that $(x+y)^r \le c_r(x^r+y^r)$ for any $x,y \ge 0$ and an appropriate constant c_r, inequality (6.176) becomes (for an appropriate constant c_6)

$$\phi(k_2,r_1) \le c_6 G(0,k_2-k_1;pq_2) G(r_2-r_1,k_2-k_1;pq_1)\phi(k_1,r_2)^{\theta}, \quad \theta > 1, \tag{6.177}$$

which is in the general form of the Stampacchia inequality (6.159).

In order to proceed further, we express the equations (6.175) in terms of $q = \frac{q_1}{q_2}$ as

$$\theta = 1 + \frac{1}{q}, \quad \text{and} \quad \theta = \frac{p}{d}q,$$

which shows that we may find an infinity of such choices as long as q satisfies the above algebraic system, the solution of which yields

$$\theta = \frac{1}{2} + \sqrt{\frac{1}{4} + \frac{p}{d}} > 1, \quad \theta = \frac{p}{d}q,$$

where we note that $\theta(\theta-1) = \frac{p}{d}$. To simplify the resulting inequality, we choose $q_2 = 1$ so that $q_1 = q$, which brings (6.177) to the form

$$\phi(k_2,r_1) \le c'\,(k_2-k_1)^{-p}((r_2-r_1)^{-pq} + (k_2-k_1)^{-pq})\,\phi(k_1,r_2)^{\theta}, \quad \theta > 1, \tag{6.178}$$

which for $c' = c_6$ is in the exact form (6.159) for $R_1 = \frac{R}{2}$, $R_2 = R$ and the exponents $\alpha = p$, $\beta = pq$ and $\theta > 1$, as given above. We therefore have that $\phi(k_0+\bar{k},R_1) = \phi(k_0+\bar{k},\frac{R}{2}) = 0$ for $\bar{k}$ satisfying

$$\bar{k}^p = c'\left(\frac{R}{2}\right)^{-pq}\phi(k_0,R)^{\theta-1} + \left(\frac{R}{2}\right)^p,$$

which implies that for an appropriate constant c_7,

$$\begin{aligned}
\bar{k} &\le c_7(R^{-q}\phi(k_0,R)^{\frac{(\theta-1)}{p}} + R) \\
&= c_7\Big(R^{-q}|A^+(k_0,x_0,R)|^{\frac{\theta-1}{p}}\Big(\int_{A^+(k_0,x_0,R)} (u(x)-k_0)^p\Big)^{\frac{(\theta-1)q}{p}} + R\Big) \\
&= c_7\Big(R^{-q}|A^+(k_0,x_0,R)|^{\frac{\theta-1}{p}}\Big(\int_{A^+(k_0,x_0,R)} (u(x)-k_0)^p\Big)^{\frac{1}{p}} + R\Big) \\
&= c_7\Big(\Big(\frac{|A^+(k_0,x_0,R)|}{R^d}\Big)^{\frac{\theta-1}{p}}\Big(\frac{1}{R^d}\int_{A^+(k_0,x_0,R)} (u(x)-k_0)^p\Big)^{\frac{1}{p}} + R\Big),
\end{aligned}$$

where we used the fact that $\frac{(\theta-1)q}{p} = \frac{d(\theta-1)\theta}{p^2} = \frac{1}{p}$. This leads to the stated upper bound for $c_{DG} = c_7$.

4. Since $u \in DG_p(k_0, R_0, \mathcal{D})$, we also have that $-u \in DG_p(-k_0, R_0, \mathcal{D})$ so that working with $-u$ in the place of u, we also obtain the stated lower bound.

(ii) To show the local Hölder continuity of u, we first note that since $u \in DG_p(R_0, \mathcal{D})$ from (i), we have that the stated bounds hold for every $k_0 \in \mathbb{R}$. Hence, $\|u\|_{L^\infty} \le c\|u\|_{L^p} + c'$, i. e., we have an L^∞ bound from an L^p bound (recall that we assume that $u \in W^{1,p}(\mathcal{D})$). We therefore have to estimate the Hölder norm for $C^{0,\alpha}$ for an appropriate choice of $\alpha > 0$. For notational convenience, we redefine all constants, except c_0 and c_{DG}.

We introduce the notion of the oscillation of a locally bounded function u over a ball $B(x_0, R) \subset \mathcal{D}$, as

$$\begin{aligned}
&\operatorname{osc}(x_0,R) = M(x_0,R) - m(x_0,R), \quad \text{where} \\
&M(x_0,R) := \sup_{x\in B(x_0,R)} u(x), \quad \text{and} \quad m(x_0,R) := \inf_{x\in B(x_0,R)} u(x).
\end{aligned} \tag{6.179}$$

The oscillation is a quantity that may provide important information on the local continuity properties of u. In fact, if for any $r < R < R_0$

$$\operatorname{osc}(x_0,r) \le \hat{c}\Big[\Big(\frac{r}{R}\Big)^\alpha + r^\alpha\Big], \tag{6.180}$$

for an appropriate constant $\hat{c}$, then u is locally Hölder continuous with Hölder exponent α (see Theorem 1.9.18). Our aim is to show that for a function in the De Giorgi class, there exists an $\alpha > 0$ such that (6.180) holds, hence, u is locally Hölder continuous. We break the proof in 4 steps.

1. By assumption, we have that $u \in DG_p^+(R_0,\mathcal{D}) = \bigcap_{k_0\in\mathbb{R}} DG_p(k_0,R_0,\mathcal{D})$. We now consider some $R > 0$ such that $B(x_0, 2R) \subset\subset \mathcal{D}$ with $2R \le R_0$ and define

$$\bar{k} = \frac{1}{2}(M(x_0,2R) + m(x_0,2R)).$$

Clearly, $u \in DG_p^+(\bar{k}, R_0, \mathcal{D})$. Without loss of generality, we may assume that $A^+(\bar{k}, x_0, R)$ has the property[18] that

$$|A^+(\bar{k}, x_0, R)| \leq \frac{1}{2}|B(x_0, R)| = \frac{1}{2}\omega_d R^d, \tag{6.181}$$

where ω_d is the volume of the unit ball in $\mathbb{R}^d$.

We now consider the sequence

$$k_n = M(x_0, 2R) - 2^{-n-1} \operatorname{osc}(x_0, 2R), \quad n = 0, 1, \dots, \tag{6.182}$$

which clearly satisfies $k_0 = \bar{k}$ and $k_n \uparrow M(x_0, 2R)$, and as by assumption $u \in DG_p^+(k_n, R_0, \mathcal{D})$ for every $n \in \mathbb{N}$, apply the L^∞ bound (upper bound) for every k_n to get a series of estimates on the supremum of u in the ball $B(x_0, \frac{R}{2})$, in terms of L^p bounds in $B(x_0, R)$. Indeed, applying part (i) of the present theorem, we have that for every $n \in \mathbb{N}$,

$$\begin{aligned} M\left(x_0, \frac{R}{2}\right) &= \sup_{B(x_0, \frac{R}{2})} u(x) \\ &\leq k_n + c_{\mathrm{DG}}\left(\left(\frac{|A^+(k_n, x_0, R)|}{R^d}\right)^{\frac{\theta-1}{p}} \left(\frac{1}{R^d} \int_{A^+(k_n, x_0, R)} (u(x) - k_n)^p dx\right)^{\frac{1}{p}} + R\right). \end{aligned} \tag{6.183}$$

We estimate

$$\begin{aligned} \int_{A^+(k_n, x_0, R)} (u(x) - k_n)^p dx &\leq \Big[\sup_{A^+(k_n, x_0, R)} (u(x) - k_n)\Big]^p |A^+(k_n, x_0, R)| \\ &\leq \Big[\sup_{B(x_0, 2R)} (u(x) - k_n)\Big]^p |A^+(k_n, x_0, R)| \\ &= (M(x_0, 2R) - k_n)^p |A^+(k_n, x_0, R)|, \end{aligned}$$

and use (6.203) to enhance inequality (6.183) in order to obtain

$$\begin{aligned} M\left(x_0, \frac{R}{2}\right) &= \sup_{B(x_0, \frac{R}{2})} u(x) \\ &\leq k_n + c_{\mathrm{DG}}\left[\left(\frac{|A^+(k_n, x_0, R)|}{R^d}\right)^{\frac{\theta}{p}} (M(x_0, 2R) - k_n) + R\right], \quad \forall n \in \mathbb{N}. \end{aligned} \tag{6.184}$$

18 To see why we can make the above assumption, we have for any $k \in \mathbb{R}$ that $|A^+(k, x_0, R)| + |A^-(k, x_0, R)| = |B(x_0, R)|$, so either $|A^+(k, x_0, R)| \leq \frac{1}{2}|B(x_0, R)|$, or $|A^-(k, x_0, R)| \leq \frac{1}{2}|B(x_0, R)|$. If the first inequality holds, we are ok. If the second hold that our assumption is valid for $-u$ (which by assumption satisfies $-u \in DG_p^+(R_0, \mathcal{D}) = \bigcap_{k_0 \in \mathbb{R}} DG_p(-k_0, R_0, \mathcal{D})$, we work with $-u$ instead. The choice of 1/2 is indicative, one may choose any $\beta < 1$.

2. We now require some decay estimates for the sequence of the terms $(\frac{|A^+(k_n,x_0,R)|}{R^d})^{\frac{\theta}{p}}$ that ensure that there exists an $n^* \in \mathbb{N}$ (to be specified shortly in step 4, see (6.215)), such that for all $n \geq n^*$ we may guarantee that

$$\left(\frac{|A^+(k_n,x_0,R)|}{R^d}\right)^{\frac{\theta}{p}} < \frac{1}{2c_{DG}}. \tag{6.185}$$

We defer the proof of this claim until step 4. Accepting estimate (6.185) for the time being, we proceed with the bounds (6.184) fixing $n = n^*$ to get that

$$\begin{aligned} M\left(x_0, \frac{R}{2}\right) &\leq k_{n^*} + \frac{1}{2}(M(x_0,2R) - k_{n^*}) + c_{DG}R \\ &= M(x_0,2R) - 2^{-(n^*+2)} \operatorname{osc}(x_0,2R) + c_{DG}R, \end{aligned} \tag{6.186}$$

where we used the definition of the sequence k_n (see (6.182)). Clearly,

$$m\left(x_0, \frac{R}{2}\right) = \inf_{B(x_0,\frac{R}{2})} u \geq \inf_{B(x_0,2R)} u = M(x_0,2R), \tag{6.187}$$

so multiplying (6.187) by -1 and adding the result to (6.186), we obtain an estimate for the oscillation in terms of

$$\operatorname{osc}\left(x_0, \frac{R}{2}\right) \leq c_{DG}R + (1 - 2^{-(n^*+2)}) \operatorname{osc}(x_0,2R). \tag{6.188}$$

Recall that R was arbitrary and such that $2R \leq R_0$, so we may rephrase our estimate as (upon setting $\rho = R/2$),

$$\operatorname{osc}\left(x_0, \frac{\rho}{4}\right) \leq c\frac{\rho}{4} + \lambda \operatorname{osc}(x_0,\rho), \quad \forall \rho \in (0,R_0], \text{ where } \frac{1}{4} < \lambda < 1, \tag{6.189}$$

where we have redefined the constant $c = 8c_{DG}$ and set $\lambda = 1 - 2^{-(n^*+2)} < 1$.

3. We claim that the estimate (6.189) guarantees that for any $r < R < R_0$,

$$\operatorname{osc}(x_0,r) \leq \widehat{c}\left[\left(\frac{r}{R}\right)^{\alpha} + r^{\alpha}\right], \tag{6.190}$$

for an appropriate constant $\widehat{c}$, which is an estimate of the form (6.180), which, as stated in the beginning of the proof, is a crucial estimate that guarantees the local Hölder continuity of u with Hölder exponent α.

To prove the claim (6.190), consider the nonnegative (and nondecreasing) function ϕ defined by $\phi(r) = \operatorname{osc}(x_0,r)$ for any $r \in [0,R_0]$, and express (6.189) as

$$\phi\left(\frac{\rho}{4}\right) \leq c\frac{\rho}{4} + \lambda\phi(\rho), \quad \text{with } \frac{1}{4} < \lambda < 1, \ \forall \rho \in (0,R_0). \tag{6.191}$$

Fix any pair r, R such that $0 < r < R < R_0$ and

$$\text{choose } N \quad \text{such that} \quad \frac{R}{4^{N+1}} < r < \frac{R}{4^N}. \tag{6.192}$$

We set $\rho_n = \frac{R}{4^{n+1}}$, $n = 0, \dots, N-1$, in (6.191) to get

$$\phi\Big(\frac{R}{4^{n+1}}\Big) \le c\frac{R}{4^{n+1}} + \lambda\phi\Big(\frac{R}{4^n}\Big), \quad \text{with } \frac{1}{4} < \lambda < 1,\ n = 0, \dots, N-1. \tag{6.193}$$

We start iterating (6.193) backwards from $n = N-1$ to $n - 0$, which yields

$$\begin{aligned} \phi\Big(\frac{R}{4^N}\Big) &\le \lambda^N\phi(R) + cR\Big(\frac{\lambda^{N-1}}{4} + \frac{\lambda^{N-2}}{4^2} + \dots + \frac{\lambda}{4^{N-1}} + \frac{1}{4^N}\Big) \\ &= \lambda^N\phi(R) + cR\lambda^N\sum_{i=1}^{N}\Big(\frac{1}{4\lambda}\Big)^i \le \lambda^N\,(\phi(R) + c\,c_1\,R), \end{aligned} \tag{6.194}$$

where $c_1 = \sum_{i=0}^{\infty}(\frac{1}{4\lambda})^i < \infty$ as $4\lambda > 1$. From the choice of N and the monotonicity of the function ϕ, from (6.194) we conclude that

$$\phi(r) \le \phi\Big(\frac{R}{4^N}\Big) \le \lambda^N\,(\phi(R) + cc_3R). \tag{6.195}$$

Recalling that $\phi(r) = \operatorname{osc}(x_0, r)$, we observe that (6.195) is very close to our claim (6.190) if we could connect λ^N with an appropriate power of $\frac{r}{R}$.

Indeed, by the choice of N in (6.192), we have that $\frac{1}{4^{N+1}} < \frac{r}{R} < \frac{1}{4^N}$. We express $\lambda = (\frac{1}{4})^\alpha$ (for $\alpha = -\frac{\ln 4}{\ln\lambda}$) and rewrite

$$\lambda^N = \lambda^{-1}\lambda^{N+1} = \lambda^{-1}\Big(\frac{1}{4^{N+1}}\Big)^\alpha \overset{(6.192)}{<} \lambda^{-1}\Big(\frac{r}{R}\Big)^\alpha. \tag{6.196}$$

Combining (6.196) with (6.195), we deduce the required estimate (6.190).

4. It remains to prove claim (6.185).

Choose any levels k_1', k_2' such that $\bar{k} \le k_1' < k_2' < M(x_0, 2R)$, and define the cutoff function

$$v = (k_2' - k_1')\mathbf{1}_{\{u \ge k_2\}'} + (u - k_1')\mathbf{1}_{\{k_1' < u < k_2'\}}, \tag{6.197}$$

which by construction vanishes on $B(x_0, R) \setminus A^+(k_1', x_0, R)$. Note that $|B(x_0, R) \setminus A^+(k_1', x_0, R) \ge \frac{1}{2}|B(x_0, R)|$, as by $\bar{k} \le k_1'$, we clearly have $B(x_0, R) \setminus A^+(\bar{k}, x_0, R) \subset B(x_0, R) \setminus A^+(k_1', x_0, R)$, and we have already established (recall (6.181)) that $|B(x_0, R) \setminus A^+(\bar{k}, x_0, R)| \ge \frac{1}{2}|B(x_0, R)|$.

The function $v \in W^{1,p}(\mathcal{D})$ and $\nabla v = \nabla u \mathbf{1}_{\{k_1' < u < k_2'\}}$ hence

$$\int_{B(x_0,R)} |\nabla v(x)| dx = \int_{A^+(k_1',x_0,R) \setminus A^+(k_2',x_0,R)} |\nabla u(x)| dx. \tag{6.198}$$

We estimate separately the left- and the right-hand side of (6.198).

To estimate the left-hand side of (6.198), we will use the Sobolev embedding $W^{1,1}(B(x_0,R)) \hookrightarrow L^{\mathfrak{s}_1}(B(x_0,R))$ for $\mathfrak{s}_1 = \frac{d}{d-1}$ for the function v on $B(x_0,R)$, which gives (upon denoting by c_S the constant in the Sobolev inequality) that

$$\begin{aligned} \int_{B(x_0,R)} |\nabla v(x)| dx &\geq c_S \Big(\int_{B(x_0,R)} |v(x)|^{\mathfrak{s}_1} dx \Big)^{\frac{1}{\mathfrak{s}_1}} \\ &\geq c_S \Big(\int_{A^+(k_2',x_0,R)} |v(x)|^{\mathfrak{s}_1} dx \Big)^{\frac{1}{\mathfrak{s}_1}} \overset{(6.197)}{=} c_S \, |k_2' - k_1'| \, |A^+(k_2',x_0,R)|^{\frac{1}{\mathfrak{s}_1}}, \end{aligned} \tag{6.199}$$

where, for the estimate in the second line, we used the trivial observation that $A^+(k_2, x_0, R) \subset B(x_0, R)$.

We proceed further by estimating the right-hand side of (6.198) using Hölder's inequality as

$$\begin{aligned} &\int_{A^+(k_1',x_0,R) \setminus A^+(k_2',x_0,R)} |\nabla u(x)| dx \\ &\leq |A^+(k_1',x_0,R) \setminus A^+(k_2',x_0,R)|^{\frac{1}{p^*}} \Big(\int_{A^+(k_1',x_0,R) \setminus A^+(k_2',x_0,R)} |\nabla u(x)|^p dx \Big)^{\frac{1}{p}} \\ &\leq |A^+(k_1',x_0,R) \setminus A^+(k_2',x_0,R)|^{\frac{1}{p^*}} \Big(\int_{A^+(k_1',x_0,R)} |\nabla u(x)|^p dx \Big)^{\frac{1}{p}}, \end{aligned} \tag{6.200}$$

where $\frac{1}{p} + \frac{1}{p^*} = 1$, and for the last estimate, we used the trivial estimate arising from $A^+(k_1', x_0, R) \setminus A^+(k_2', x_0, R) \subset A^+(k_1', x_0, R)$. Importantly, we may use the Caccioppoli inequality (6.64) to estimate the last integral. Indeed, since $u \in DG_p(\mathcal{D})$, using the Caccioppoli inequality (6.64) for the level k_1' and the radii $r_1 = R$, $r_2 = 2R$, we have that

$$\int_{A^+(k_1',x_0,R)} |\nabla u(x)|^p dx \leq c_0 \Big(R^{-p} \int_{A^+(k_1',x_0,2R)} (u(x) - k_1')^p \Big) dx + |A^+(k_1', x_0, 2R)|). \tag{6.201}$$

We first combine (6.198)–(6.200) and raise to the power p to get

$$\begin{aligned} c_S^p\,|k_2'-k_1'|^p\,|A^+(k_2',x_0,R)|^{\frac{p}{s_1}} &\le |A^+(k_1',x_0,R)\setminus A^+(k_2',x_0,R)|^{\frac{p}{p^*}} \int\limits_{A^+(k_1',x_0,R)} |\nabla u(x)|^p dx \\ &\overset{(6.201)}{\le} c_0\big(|A^+(k_1',x_0,R)|-|A^+(k_2',x_0,R)|\big)^{p-1} \\ &\quad\times\Big(R^{-p}\int\limits_{A^+(k_1',x_0,2R)} (u(x)-k_1')^p dx + |A^+(k_1',x_0,2R)|\Big), \end{aligned} \tag{6.202}$$

where for the last estimate, we used the facts that $\frac{p}{p^*}=p-1$, $|A^+(k_1',x_0,R)\setminus A^+(k_2',x_0,R)| = |A^+(k_1',x_0,R)|-|A^+(k_2',x_0,R)|$ and the Caccioppoli inequality (6.64) for the choice $r_1=R$, $r_2=2R$ and the level k_1' (which holds since $u\in DG_p(\mathcal{D})$). We proceed by further estimating the right-hand side of (6.202) as follows: We first note that since $A^+(k_1',x_0,2R)\subset B(x_0,2R)$, it holds that

$$|A^+(k_1',x_0,2R)| \le |B(x_0,2R)| = (2R)^d\omega_d, \tag{6.203}$$

where ω_d is the volume of the unit ball in $\mathbb{R}^d$, and moreover,

$$\begin{aligned} \int\limits_{A^+(k_1',x_0,2R)} (u(x)-k_1')^p dx &\le |A^+(k_1',x_0,2R)|\,(M(x_0,2R)-k_1')^p \\ &\le 2^d\omega_d R^d (M(x_0,2R)-k_1')^p, \end{aligned} \tag{6.204}$$

by a similar argument. Combining the estimates (6.203)–(6.204) with (6.202), and then raising to the power $\frac{1}{p-1}$ and using the definition of s_1, we obtain that

$$\begin{aligned} &|k_2'-k_1'|^{\frac{p}{p-1}}\,|A^+(k_2',x_0,R)|^{\frac{p(d-1)}{(p-1)d}} \\ &\quad\le c_2\big(|A^+(k_1',x_0,R)|-|A^+(k_2',x_0,R)|\big)\big(R^{d-p}(M(x_0,2R)-k_1')^p+R^d\big)^{\frac{1}{p-1}} \\ &\quad\le c_3\big(|A^+(k_1',x_0,R)|-|A^+(k_2',x_0,R)|\big)\big(R^{\frac{d-p}{p-1}}(M(x_0,2R)-k_1')^{\frac{p}{p-1}}+R^{\frac{d}{p-1}}\big), \end{aligned} \tag{6.205}$$

for appropriate constants c_2, c_3.

Estimate (6.205) holds for any levels $k_1'<k_2'$. We choose the increasing sequence

$$k_n = M(x_0,2R) - 2^{-n-1}\operatorname{osc}(x_0,2R), \quad n=0,1,\dots,$$

for which $k_0=\bar{k}$, and apply (6.205) recursively for the choices $k_1'=k_{n-1}$ and $k_2'=k_n$, where upon observing that

$$k_n - k_{n-1} = 2^{-n-1}\operatorname{osc}(x_0,2R), \quad \text{and} \quad M(x_0,2R)-k_{n-1} = 2^{-n-1}\operatorname{osc}(x_0,2R),$$

so (after some algebraic rearrangement), for every n,

$$\begin{aligned}|A^+(k_n, x_0, R)|^{\frac{p(d-1)}{(p-1)d}} \\ \leq c_3 R^{\frac{d-p}{p-1}} \left(1 + \left(\frac{2^{n+1}R}{\operatorname{osc}(x_0, 2R)}\right)^{\frac{p}{p-1}}\right)[|A^+(k_{n-1}, x_0, R)| - |A^+(k_n, x_0, R)|].\end{aligned} \tag{6.206}$$

We now fix an n^* (to be specified shortly in (6.215). Since $\{k_n : n \in \mathbb{N}\}$ is increasing, for every $n \leq n^*$, we have $k_n \leq k_{n^*}$, therefore, $u > k_{n^*}$ implies that $u > k_n$, hence, $A^+(k_{n^*}, x_0, R) \subset A^+(k_n, x_0, R)$. So it holds that

$$|A^+(k_{n^*}, x_0, R)| \leq |A^+(k_n, x_0, R)|, \quad n = 1, \dots, n^*. \tag{6.207}$$

Moreover,

$$\left(\frac{2^{n+1}R}{\operatorname{osc}(x_0, 2R)}\right)^{\frac{p}{p-1}} \leq \left(\frac{2^{n^*+1}R}{\operatorname{osc}(x_0, 2R)}\right)^{\frac{p}{p-1}}, \quad n = 1, \dots, n^*. \tag{6.208}$$

Combining (6.207)–(6.208) with (6.206), we have that for all $n = 1, \dots, n^*$,

$$\begin{aligned}|A^+(k_{n^*}, x_0, R)|^{\frac{p(d-1)}{(p-1)d}} \\ \leq c_3 R^{\frac{d-p}{p-1}} \left(1 + \left(\frac{2^{n^*+1}R}{\operatorname{osc}(x_0, 2R)}\right)^{\frac{p}{p-1}}\right)[|A^+(k_{n-1}, x_0, R)| - |A^+(k_n, x_0, R)|].\end{aligned} \tag{6.209}$$

Adding (6.209) for $n = 1, \dots, n^*$ and noting that $k_0 = \bar{k}$, we have

$$\begin{aligned}n^*|A^+(k_{n^*}, x_0, R)|^{\frac{p(d-1)}{(p-1)d}} \\ \leq c_3 R^{\frac{d-p}{p-1}} \left(1 + \left(\frac{2^{n^*+1}R}{\operatorname{osc}(x_0, 2R)}\right)^{\frac{p}{p-1}}\right)[|A^+(\bar{k}, x_0, R)| - |A^+(k_{n^*}, x_0, R)|].\end{aligned} \tag{6.210}$$

We further estimate

$$|A^+(\bar{k}, x_0, R)| - |A^+(k_{n^*}, x_0, R)| \leq |A^+(\bar{k}, x_0, R)| \overset{(6.181)}{\leq} \frac{1}{2}B(x_0, R) = \frac{1}{2}\omega_d R^d \tag{6.211}$$

and upon combining (6.210) and (6.211), we get

$$n^*|A^+(k_{n^*}, x_0, R)|^{\frac{p(d-1)}{(p-1)d}} \leq c_3 \frac{\omega_d}{2} R^{\frac{p(d-1)}{p-1}} \left(1 + \left(\frac{2^{n^*+1}R}{\operatorname{osc}(x_0, 2R)}\right)^{\frac{p}{p-1}}\right). \tag{6.212}$$

Raising to the power $\gamma = \frac{\theta}{p}\frac{(p-1)d}{p(d-1)}$, and rearranging,

$$\left(\frac{|A^+(k_{n^*},x_0,R)|}{R^d}\right)^{\frac{\theta}{p}} \le c_6\left[1+\left(\frac{2^{n^*+1}R}{\mathrm{osc}(x_0,2R)}\right)^{\frac{p}{p-1}}\right]^{\gamma}\left(\frac{1}{n^*}\right)^{\gamma} \le c_4\left[1+\left(\frac{2^{n^*+1}R}{\mathrm{osc}(x_0,2R)}\right)^{\frac{p}{p-1}}\right]^{\gamma}, \tag{6.213}$$

as long as $n^* \ge 1$, where $c_4 = (c_3\frac{\omega_d}{2})^{\gamma}$.

Since $(k_n)_{n\in\mathbb{N}}$ is increasing for every $n > n^*$, we have that $k_n \ge k_{n^*}$, hence, $A^+(k_n,x_0,R) \subset A^+(k_{n^*},x_0,R)$, i. e., $|A^+(k_n,x_0,R)| \le |A^+(k_{n^*},x_0,R)|$, which, combined with (6.213), leads to the decay estimate

$$\left(\frac{|A^+(k_n,x_0,R)|}{R^d}\right)^{\frac{\theta}{p}} \le c_4\left[1+\left(\frac{2^{n^*+1}R}{\mathrm{osc}(x_0,2R)}\right)^{\frac{p}{p-1}}\right]^{\gamma}, \quad \forall n > n^* \ge 1. \tag{6.214}$$

This is essentially (6.185) as long as we manage to control the term in the square bracket, which still involves n^*. To this end, we choose $n^* \ge 1$ such that

$$2^{n^*}R \le c_5\ \mathrm{osc}(x_0,2R), \tag{6.215}$$

for a constant c_5 such that $(1+(2c_5)^{\frac{p}{p-1}})^{\gamma} = \frac{1}{2c_{DG}}$. If such an $n^* \ge 1$ exists, then (6.214) reduces to the crucial claim (6.185), and we are done. If (6.215) is not satisfied for any $n^* \ge 1$, then this implies that there exists a constant c' such that $\mathrm{osc}(x_0,2R) \le c'R$, and since R is arbitrary the Hölder continuity of u follows by Theorem 1.9.18.

Remark 6.10.3. If the function u satisfies the homogeneous Caccioppoli inequalities instead, then the oscillation criterion (6.190) (or the related criterion (6.191)) simplifies to $\mathrm{osc}(x_0,r) \le \hat{c}(\frac{r}{R})^{\alpha}$ (or the related criterion $\mathrm{osc}_{Q(x,\rho)}\, u \le c\,(\frac{\rho}{R})^{\alpha}\,\mathrm{osc}_{Q(x,R)}\, u$).

6.10.3 Proof of Theorem 6.6.5

We only sketch the main steps of the proof here. We will work in terms of cubes $Q(x_0,R)$ rather than balls $B(x_0,R)$ noting that the boundeness and Hölder continuity results of functions in De Giorgi classes can be generalized if in the defining Caccioppoli inequalities balls are replaced by cubes. Our basic aim is to find a lower bound for $\inf_{Q(x_0,R)} u$, which combined with the upper bounds already obtained and the Hölder continuity property for minimizers, will yield the Harnack inequality. The estimates are local, and upon translations and scaling we may choose $x_0 = 0$ and $R = 1$ at certain points in order to simplify the arguments and notation. Our approach follows [83].

1. The first step is based on a lower estimate of the infimum of a positive function in $DG_p(\mathcal{D})$. For every $z_0 \in \mathcal{D}$, $R > 0$ and $T > 1/2$, chosen so that the resulting cubes $Q(z_0,R)$ and $Q(z_0,TR)$ are in $\mathcal{D}$, and a level θ such that $|A^-(\theta,R)| \le \gamma|Q(z_0,R)|$ for some $\gamma > 0$, there exists a constant $c = c(\gamma,T,R,z_0) > 0$ such that $\inf_{Q(z_0,TR)} u \ge c(\gamma,T,R)\theta$.

This is an interesting and useful result, which allows us to obtain lower bounds on the infimum of a positive function in the De Giorgi class on a larger cube by estimates on the measure of the sublevel sets at a particular level θ. We can see that it suffices to show this result for $R = 1$ and $z_0 = 0$. We will use the simplified notation $Q(R)$ for the resulting cubes.

1(a). We start with a useful observation: Assume that for some $\theta > 0$ and $\gamma > 0$ we have that $|A^-(\theta,1)| \leq \gamma|Q(1)|$, and consider any $T > 1/2$. Then, the condition $|A^-(\theta,1)| \leq \gamma|Q(1)|$ implies that[19] $|A^-(\theta,2T)| \leq (1 - \frac{1-\gamma}{(2T)^d})|Q(2T)|$. This means that if the stated condition holds for a level θ on a cube of size 1, then it also holds for the same level (but a different constant γ) for a cube of size $2T$ for any $T > 1/2$. Consider $R' = 2T$ as a new cube side. Then, the condition is a condition of the form $|A^-(\theta,R')| \leq (1 - \frac{1-\gamma}{(2T)^d})|Q(R')|$, which we wish to connect with a result on the $\inf_{Q(R'/2)} u$, and upon rescaling $R' = 1$, the result we seek must essentially connect a condition of the form $|A^-(\theta,1)| \leq \gamma|Q(1)|$ with a lower bound of the form $\inf_{Q(\frac{1}{2})} u \geq \lambda\theta$.

1(b). It therefore suffices to show the following: If $u \geq 0$ on $Q(2)$ and as long as for some $\gamma \in (0,1)$, $\theta > 0$, it holds that $|A^-(\theta,1)| \leq \gamma|Q(1)|$, then $\inf_{Q(\frac{1}{2})} u \geq \lambda(\gamma)\theta$ for some constant $\lambda(\gamma) > 0$. Note that there is no restriction on the size of γ in this claim. This is Lemma 7.5 in [83].

To prove this claim we first note that for a positive function $u \in DG_p^-(\mathcal{D})$, and for any $k_1 < k_2 < \theta$, it holds that

$$\begin{aligned}(k_2 - k_1)|A^-(k_1,\rho)|^{\frac{d-1}{d}} &\leq c|A^-(k_2,\rho) \setminus A^-(k_1,\rho)|^{\frac{p-1}{p}}\Big(\int_{A^-(k_2,\rho)} |\nabla u|^p\Big)^{1/p} \\ &\leq c|A^-(k_2,\rho) - A^-(k_1,\rho)|^{\frac{p-1}{p}} k_1(R-\rho)^{-1}|A^-(k_2,R)|^{1/p}.\end{aligned} \tag{6.216}$$

To show (6.216), we need to work as follows: The first two inequalities in (6.216) follow by a Poincaré type inequality (see Example 1.5.15) applied to the cutoff function $v(x) = 0\mathbf{1}_{u\geq k_2} + (k-u)\mathbf{1}_{k_1<u<k_2} + (k_2-k_1)\mathbf{1}_{u\leq k_1}$ defined on $Q(\rho)$ for some $\rho \in [\frac{1}{2},1]$ by noting that (a) under the stated condition on $|A^-(\theta,1)|$, the measure of the set $A_o \subset Q(\rho)$, where v vanishes, is positive and (b) $\nabla v \neq 0$ only on $A^-(k_2,\rho) \setminus A^-(k_1,\rho)$, with $\nabla v = -\nabla u$ on this set leads to the estimate

$$\Big(\int_{Q(\rho)} v^{\frac{d}{d-1}} dx\Big)^{\frac{d-1}{d}} \leq c_1 \int_{A^-(k_2,\rho)\setminus A^-(k_1,\rho)} |\nabla v| dx. \tag{6.217}$$

Since $v \geq 0$, we may estimate the integral on the left-hand side from below by

19 To see this, since $A^+(\theta,1) \subset A^+(\theta,2T)$, estimate $|A^+(\theta,2R')| \geq |A^+(\theta,1)| = |Q| - |A^-(\theta,1)| \geq (1-\gamma)|Q| \geq \frac{1-\gamma}{(2R')^d}|Q(2R')|$ since $R' > \frac{1}{2}$ and $|Q(2R')| = (2R')^d|Q|$. Then, take the complement.

$$\left(\int_{Q(\rho)} v^{\frac{d}{d-1}}dx\right)^{\frac{d-1}{d}} \geq \left(\int_{A^-(k_1,\rho)} v^{\frac{d}{d-1}}dx\right)^{\frac{d-1}{d}} \geq (k_2-k_1)|A^-(k_1,\rho)|^{\frac{d-1}{d}}, \tag{6.218}$$

and the right-hand side in terms of the Hölder inequality as

$$\begin{aligned}\int_{A^-(k_2,\rho)\setminus A^-(k_1,\rho)} |\nabla v|dx &\leq \left|A^-(k_2,\rho)-A^-(k_1,\rho)\right|^{\frac{p-1}{p}}\left(\int_{A^-(k_2,\rho)\setminus A^-(k_1,\rho)} |\nabla v|^p dx\right)^{\frac{1}{p}} \\ &\leq \left|A^-(k_2,\rho)-A^-(k_1,\rho)\right|^{\frac{p-1}{p}}\left(\int_{A^-(k_2,\rho)} |\nabla v|^p dx\right)^{\frac{1}{p}}.\end{aligned} \tag{6.219}$$

Combining (6.217), (6.218), (6.219), we obtain the first two inequalities in (6.216). To obtain the third inequality in (6.216), we now recall the relevant Caccioppoli inequality that u satisfies, which is

$$\int_{A^-(k_2,\rho)} |\nabla u|^p dx \leq \frac{c}{(R-\rho)^p}\int_{A^-(k_2,R)} (k_2-u)^p dx + c\left|A^-(k_2,R)\right| \leq \frac{c_2}{(R-\rho)^p}\left|A^-(k_2,R)\right|. \tag{6.220}$$

We raise to the power $\frac{1}{p}$, and this provides the third inequality in (6.216).

Having obtained the estimate (6.216), we may proceed towards the proof of claim 1(b) as follows: Set $\rho = 1$, $R = 2$ in (6.216), and raise to the power $\frac{p}{p-1}$ to obtain

$$(k_2-k_1)^{\frac{p}{p-1}}\left|A^-(k_1,1)\right|^{\frac{p(d-1)}{d(p-1)}} \leq c k_2^{\frac{p}{p-1}}\left|A^-(k_2,1)-A^-(k_1,1)\right|.$$

We write this result along the sequence $k_i' = 2^{-i}\theta$, setting $k_2 = k_i'$, $k_1 = k_{i+1}'$ for every $i = 0,\dots,n^\star-1$ for some $n^\star$ to be determined shortly. Adding the results and noting that since $A^-(k_{i+1}',1) \subset A^-(k_i',1)$, we have that $|A^-(k_{i+1}',1)| \leq |A^-(k_i',1)|$ so that $|A^-(k_{n^\star}',1)| \leq A^-(k_i',1)|$ for every $i = 0,\dots,n^\star-1$, hence

$$n^\star\left|A^-(k_{n^\star}',1)\right| \leq c'\left(\left|A^-(k_0',1)\right| - \left|A^-(k_{n^\star}',1)\right|\right) \leq c'\left|A^-(k_0',1)\right|.$$

This provides an estimate for $|A^-(k_{n^\star}',1)| = |A^-(2^{-n^\star}\theta,1)| \leq \gamma_0|Q(1)|$, as long as $n^\star$ is chosen sufficiently large. This estimate provides us with a level, i. e., $2^{-n^\star}\theta$, such that the sublevel set for the given level has measure satisfying

$$\left|A^-(2^{-n^\star},1)\right| \leq \gamma(n^\star)\left|Q(1)\right|, \tag{6.221}$$

with $\gamma(n^\star)$ inversely proportional to $n^\star$. It is important to note that upon choosing $n^\star$ sufficiently large, $\gamma(n^\star)$ can be made as small as we wish. Hence, it suffices to show the claim for an appropriately sufficiently small γ.

1(c). We now fix any level $\theta' > 0$, and assume that it is chosen so that $|A^-(\theta',1)| \leq \gamma'|Q(1)|$. We will show that as long as $\gamma' > 0$ is small enough, then a positive u satisfies

$\inf_{Q(\frac{1}{2})} u \geq \frac{\theta'}{2}$. Our starting point is (6.216), which we now treat inside the unit cube (i. e., $\rho < R < 1$), and we further estimate as

$$(k_2 - k_1)|A^-(k_1,\rho)|^{\frac{d-1}{d}} \leq c|A^-(k_2,\rho) \setminus A^-(k_1,\rho)|^{\frac{p-1}{p}} \frac{k}{(R-\rho)}|A^-(k_2,R)|^{\frac{1}{p}}$$
$$\leq c|A^-(k_2,\rho)|^{\frac{p-1}{p}} \frac{k}{(R-\rho)}|A^-(k_2,R)|^{\frac{1}{p}} \leq \frac{c}{(R-\rho)}|A^-(k_2,R)|,$$

where we used the obvious inequality $|A^-(k_2,\rho)| \leq |A^-(k_2,R)|$. Rearranging the last inequality, we obtain

$$|A^-(k_1,\rho)| \leq c((k_2-k_1)(R-\rho))^{-\frac{d}{d-1}}|A^-(k_2,R)|^{\frac{d}{d-1}}. \tag{6.222}$$

This is true for any $\rho, R \in [0,1]$. We now consider the sequences $r_i = \frac{1}{2}(1+2^{-i})$ and $k_i' = \frac{\theta'}{2}(1+2^{-i})$ and apply (6.222) consecutively for $\rho = r_{i+1}$, $R = r_i$ and $k_2 = k_i'$, $k_1 = k_{i+1}'$ for every $i \in N$ to obtain the iterative scheme

$$|A^-(k_{i+1}',r_{i+1})| \leq 4^{\frac{2d}{d-1}}(4^{\frac{d}{d-1}})^i|A^-(k_i',r_i)|^{\frac{d}{d-1}}. \tag{6.223}$$

Noting that $\frac{d}{d-1} > 1$, and since $4^{\frac{d}{d-1}} > 1$, setting $\alpha = \frac{1}{d-1}$, $c_0 = 4^{\frac{2d}{d-1}}$ and $c_1 = 4^{\frac{d}{d-1}}$, the inequality (6.223) assumes the form $|A^-(k_{i+1}',r_{i+1})| \leq c_0 c_1^i |A^-(k_i',r_i)|^{1+\alpha}$. We will show that as long as $|A^-(k_0',r_0)| \leq c_0^{-1/\alpha}c_1^{-\frac{1}{\alpha^2}}$, then $|A^-(k_i',r_i)| \leq c_1^{-i/\alpha}|A^-(k_0',r_0)|$, so $|A^-(k_i',r_i)| \to 0$. This can be proved easily by induction. To guarantee the condition on the initial value of the sequence, since we know that $|A^-(\theta',1)| \leq \gamma'|Q(1)|$, it suffices to have $\gamma' \leq c_0^{-1/\alpha}c_1^{-\frac{1}{\alpha^2}}$. As long a γ' is sufficiently small (as given above), then since $|A^-(k_i',r_i)| \to 0$, and because $k_i' \downarrow \frac{\theta'}{2}$ and $r_i \downarrow \frac{1}{2}$, this result implies that $u \geq \frac{\theta'}{2}$ on $Q(1/2)$, hence the required result. By (6.221), we apply this result for the level $\theta' = 2^{-n^\star}\theta$, choosing $n^\star$ sufficiently large so that $\gamma(n^\star)$ satisfies the smallness condition for γ' above, so the bound $\inf_{Q(1/2)} u \geq \frac{1}{2}\theta'$, which provides us with the required lower bound $\inf_{Q(1/2)} u \geq \frac{1}{2}2^{-n^\star}\theta$.

2. We will now show that

$$u(x_0) \leq c \inf_{Q(x_0,R)} u. \tag{6.224}$$

It suffices to show that $w(x) := \frac{u(x)}{u(x_0)} \geq c > 0$ in $Q(x_0,R)$.

Clearly w is in the same De Giorgi class as u so that for every $x \in \mathcal{D}$ and $\rho < R < \frac{1}{2}\operatorname{dist}(x,\partial\mathcal{D})$, it satisfies the oscillation criterion (see (6.190) and Remark 6.10.3)

$$\operatorname{osc}_{Q(x,\rho)} w \leq c\left(\frac{\rho}{R}\right)^\alpha \operatorname{osc}_{Q(x,R)} w \leq c\left(\frac{\rho}{R}\right)^\alpha \|w\|_{L^\infty(Q(x,R))}. \tag{6.225}$$

We now consider the function w in a cube centered at x_0 of varying side $Q(x_0,\tau)$, and fixing a lower bound K for $\|w\|_{L^\infty(Q(x_0,\tau))}$, we will look for the largest value of τ such that

$\|w\|_{L^\infty(Q(x_0,\tau))} \geq K$. Clearly, as $K \to \infty$, it must hold that $\tau \to 0$. We will adopt the scaling $K = K(\tau) := (1-\tau)^{-\delta}$ for $\tau \in [0,1)$ and some $\delta > 0$ to be chosen shortly, and note that there must exist a $\tau_0 \in [0,1)$, the largest value of τ with the property we seek, i. e. such that $\|w\|_{L^\infty(Q(x_0,\tau_0))} \geq (1-\tau_0)^{-\delta}$. For any $\tau > \tau_0$, it must hold that $\|w\|_{L^\infty(x_0,\tau)} < (1-\tau)^{-\delta}$.

By the continuity of w, there exists a $\hat{x} \in \overline{Q(x_0,\tau_0)}$ such that $w(\hat{x}) = \|w\|_{L^\infty(Q(x_0,\tau_0))}$. As $Q(\hat{x}, \frac{1-\tau_0}{2}) \subset Q(x_0, \frac{1+\tau_0}{2})$ and $\frac{1+\tau_0}{2} > \tau_0$, we have that

$$\|w\|_{L^\infty(Q(\hat{x},\frac{1-\tau_0}{2}))} \leq \|w\|_{L^\infty(Q(\hat{x},\frac{1+\tau_0}{2}))} < K\left(\frac{1+\tau_0}{2}\right) = 2^\delta(1-\tau_0)^{-\delta}. \tag{6.226}$$

We apply the oscillation criterion (6.225) for the cubes $Q(\hat{x},\rho)$ and $Q(\hat{x},R)$, for $R = \frac{1-\tau_0}{2}$ and $\rho = \epsilon^\alpha R$, for $\epsilon < 1$ to be chosen shortly. This, combined with the estimate (6.226), yields that

$$\operatorname{osc}_{Q(\hat{x},\frac{1-\tau_0}{2}\epsilon)} w \leq c\epsilon^\alpha 2^\delta (1-\tau_0)^{-\delta},$$

and since for every $x \in Q(\hat{x}, \frac{1-\tau_0}{2}\epsilon)$, it clearly holds that

$$w(x) \geq \inf_{Q(\hat{x},\frac{1-\tau_0}{2}\epsilon)} w = \sup_{Q(\hat{x},\frac{1-\tau_0}{2}\epsilon)} w - \operatorname{osc}_{Q(\hat{x},\frac{1-\tau_0}{2}\epsilon)} w \geq (1-\tau_0)^{-\delta}(1-c2^\delta\epsilon^\alpha)$$

by the choice of $\hat{x}$. Choosing ϵ appropriately (e. g., such that $1-c2^\delta\epsilon^\alpha = \frac{1}{2}$), we obtain that

$$w(x) \geq \frac{1}{2}(1-\tau_0)^{-\delta}, \quad \text{on } Q\left(\hat{x}, \frac{1-\tau_0}{2}\epsilon\right).$$

That means $|A^-(\theta_0,R_0)| = 0 \leq \gamma|Q(R_0)|$ for $\theta_0 = \frac{1}{2}(1-\tau_0)^{-\delta}$ and $R_0 = \frac{1-\tau_0}{2}\epsilon$ so that we may apply the result of step 1 for the choice $T = 2$, and $\gamma = 0$ to obtain that

$$w(x) \geq \mu\theta_0 = \mu\frac{1}{2}(1-\tau_0)^{-\delta}, \quad \text{on } Q(\hat{x},2R_0) = Q(\hat{x},(1-\tau_0)\epsilon).$$

But setting $\theta_1 = \mu\frac{1}{2}(1-\tau_0)^{-\delta}$ and $R_1 = (1-\tau_0)\epsilon$, this implies that $|A^-(\theta_1,R_1)| = 0 \leq \gamma|Q(R_1)|$, so using once more the result of step 1 for the choice $T = 2$, we obtain that

$$w(x) \geq \mu\theta_1 = \mu^2\frac{1}{2}(1-\tau_0)^{-\delta}, \quad \text{on } Q(\hat{x},2R_1) = Q(\hat{x},2(1-\tau_0)\epsilon).$$

We will keep iterating the above argument up to level $n^\star$ so that we obtain the estimate

$$w(x) \geq \mu^{n^\star}\frac{1}{2}(1-\tau_0)^{-\delta}, \quad \text{on } Q(\hat{x},2R_1) = Q(\hat{x},2^{n^\star-1}(1-\tau_0)\epsilon).$$

We will choose $n^\star$ so that the corresponding cube $Q(\hat{x},2^{n^\star-1}(1-\tau_0)\epsilon)$ contains the cube $Q(x_0,1)$. For this choice, we see that there exists a constant $c > 0$ such that $w(x) \geq c > 0$, therefore $u(x_0) \leq c \inf_{Q(x_0,1)} u(x)$. Clearly, upon scaling, this result is true for any $Q(x_0, R)$, for an appropriate choice of $c > 0$.

To conclude the proof, let us choose a cube $Q(x_1, \rho) \subset \mathcal{D}$ for a suitable $x_1 \in \mathcal{D}$ and $\rho > 0$ and consider a point $x_0 \in \overline{Q(x_1, \rho)}$ such that $u(x_0) = \sup_{Q(x_1,\rho)} u$. Since (6.224) holds for any R, we choose an R sufficiently larger than ρ, i. e., $R = 3\rho$, so that

$$u(x_0) = \sup_{Q(x_1,\rho)} u \leq c \inf_{Q(x_0,R)} u \leq c \inf_{Q(x_1,\rho)} u(x),$$

which is the required result on $Q(x_1, \rho)$. The general result follows by a covering argument.

Remark 6.10.4. The proof presented here follows the alternative approach of Giusti, [83], which uses the assumption of Hölder continuity, which follows as long as Caccioppoli type inequalities of the form (6.64) hold (see also Remark 6.6.2). The proof of Di Benedetto and Trudinger in [67] uses a slightly different approach, using the upper bound (6.68) together with a lower bound derived by the estimates of step 1 of the proof of Theorem 6.6.5, combined with a covering result due to Krylov and Safonov (see [97], see also Proposition 7.2 in [83]), according to which, for positive functions in the De Giorgi class $DG_p(\mathcal{D})$, there exists a $q > 0$ such that

$$\inf_{Q(x_0,\frac{R}{2})} u \geq c\left(\left(\fint_{Q(x_0,R)} u^q dx\right)^{\frac{1}{q}} - c_2 R^{\alpha}\right),$$

with c_2 as in Remark 6.6.9.

6.10.4 Proof of generalized Caccioppoli estimates

In this section, we provide a proof of the generalized Caccioppoli estimates of Remark 6.6.9. For simplicity, we provide the proof in the case where $q = p$, $s = r = \infty$, and $\epsilon = \frac{p}{d}$, referring to Theorem 7.1 in [83] for the general case. In particular, using the same arguments and bringing the extra terms resulting from the modified growth condition on the left-hand side to the right, (6.74) can be modified as

$$\int_{A_2} |\nabla u(x)|^p dx \leq c\left(\int_{A_2} |\nabla(u+\phi)(x)|^p dx + \int_{A_2} b(x)|(u+\phi)(x)|^p dx + \int_{A_2} b(x)|u(x)|^p dx + \int_{A_2} a(x)dx\right), \tag{6.227}$$

where we have chosen the constant $c > 0$ large enough.[20] Working as above, by the properties of ϕ and ψ, this reduces to

20 Larger than all individual constants that appear.

$$\int_{A_2} |\nabla u(x)|^p dx \le c_1\Big(\int_{A_2} |(1-\psi)^p \nabla u(x)|^p dx + (r_2 - r_1)^{-p} \int_{A_2} (u(x) - k)^p dx \\ + \int_{A_2} b(x)|(u+\phi)(x)|^p dx + \int_{A_2} b(x)|u(x)|^p dx + \int_{A_2} a(x) dx \Big), \tag{6.228}$$

which now has 3 extra terms that we must control. In order to do this, we add $\int_{A_2} b(x)|u(x)|^p dx$ on both sides of this inequality, and trying to obtain a term of the form $\int_{A_2} b(x)(1-\psi)^2 |u(x)|^p dx$ out of this addition on the right-hand side, we express u on A_2 as $u = (1-\psi)u + \psi((u-k)^+ + k)$, noting that for an appropriate constant $c(p) > 0$, we have (dropping the explicit dependence on x for convenience) that $|u|^p \le c(p)((1-\psi)^p |u|^p + \psi^p |(u-k)^+|^p + \psi^p k^p$. We see that the first contribution in the above estimate, combined with the similar contributions that arise from the third and fourth terms on the right-hand side of (6.228) upon expressing u on A_2 as $u + \phi = (1-\psi)u + k\psi$, and using the estimate $|u+\phi|^p \le c(p)((1-\psi)^p|u|^p + k^p\psi^p)$, will provide the term $\int_{A_2} b(x)(1-\psi)^2|u(x)|^p dx$ we seek for on the right-hand side. Combining the above, and using an appropriate constant c_2, we have an estimate of the form

$$\int_{A_2} |\nabla u(x)|^p dx + \int_{A_2} b|u(x)|^p dx \le c_2\Big(\int_{A_2} (1-\psi)^p |\nabla u(x)|^p dx + \int_{A_2} (1-\psi)^p |u(x)|^p dx \\ + (r_2 - r_1)^{-p} \int_{A_2} (u(x) - k)^p dx + \int_{A_2} b\psi^p |(u-k)^p dx \\ + k^p \int_{A_2} b\psi^p dx + \int_{A_2} a(x) dx \Big).$$

The last two terms can be estimated using $|A_2|$ and the relevant Lebesgue norm of a and b as

$$k^p \int_{A_2} b\psi^p dx + \int_{A_2} a(x) dx \le (\|b\|_{L^\infty(B(R))} k^p + \|a\|_{L^\infty(B(R))})|A_2|.$$

The term that is still problematic is $\int_{A_2} b\psi^p |(u-k)^p dx$. As we would like to connect this term with estimates of the gradient of u, our strategy is to first use the Hölder inequality to create the relevant norm for which the Sobolev inequality holds, and then apply the Sobolev inequality in order to introduce the gradient into the game. To this end, we estimate

$$\int_{A_2} b\psi^p |(u-k)^p dx \le \Big(\int_{A_2} \psi^{s_p} |(u-k)^+|^{s_p} dx \Big)^{\frac{p}{s_p}} \|b\|_{L^\infty(B(R))} |A_2|^{\frac{p}{d}} \\ \le c_R \int_{A_2} |\nabla(\psi(u-k)^+)|^p dx$$

$$\leq c'_R \Big(\int_{A_2} |\nabla u|^p dx + (r_2 - r_1)^{-p} \int_{A_2} ((u-k)^+)^p dx \Big),$$

where, for the third inequality, we used the Sobolev inequality, and then our standard estimate for $\int_{A_2} |\nabla \psi (u-k)^+|^p dx$. The constants c_R and c'_R depend on R, and are proportional to

$$c_R = \|b\|_{L^\infty(B(R))} |A_2|^{\frac{p}{d}} \leq \|b\|_{L^\infty(B(R))} |B(R)|^{\frac{p}{d}}.$$

Choosing R sufficiently small (depending on the data of the problem), we may take $c'_R < \frac{1}{2}$, and hence bring the term $\int_{A_2} |\nabla u|^p dx$ to the left-hand side with a positive sign. Combining the above, and observing that $1 - \psi = \mathbf{1}_{A_2 \setminus A_1}$, so the integral of any function multiplied by $(1-\psi)$ reduces to the integral over $A_2 \setminus A_1$, we have for a new constant $c_3 > 0$ that

$$\begin{aligned}
&\int_{A_2} |\nabla u(x)|^p dx + \int_{A_2} b |u(x)|^p dx \\
&\quad \leq c_3 \Big(\int_{A_2 \setminus A_1} |\nabla u(x)|^p dx + \int_{A_2 \setminus A_1} |u(x)|^p dx \\
&\qquad + (r_2 - r_1)^{-p} \int_{A_2} (u(x) - k)^p dx + (\|b\|_{L^\infty(B(R))} k^p + \|a\|_{L^\infty(B(R))}) |A_2| \Big),
\end{aligned}$$

which is trivially enhanced by noting that $A_1 \subset A_2$ as

$$\begin{aligned}
&\int_{A_1} |\nabla u(x)|^p dx + \int_{A_1} b |u(x)|^p dx \\
&\quad \leq c_3 \Big(\int_{A_2 \setminus A_1} |\nabla u(x)|^p dx + \int_{A_2 \setminus A_1} |u(x)|^p dx \\
&\qquad + (r_2 - r_1)^{-p} \int_{A_2} (u(x) - k)^p dx + (\|b\|_{L^\infty(B(R))} k^p + \|a\|_{L^\infty(B(R))}) |A_2| \Big).
\end{aligned}$$

We can now use the hole filling technique for the positive quantity $|\nabla u|^p + b|u|^p$, adding the integral of this quantity over A_1 multiplied by $c_3 > 0$ on both sides to get, upon rearranging the same inequality but now with a new constant, $c_4 < 1$. The proof may now proceed as in step 2 of Proposition 6.6.7.

6.10.5 The Nemitskii operator

The Nemitskii operator is a useful tool in the treatment of variational problems (see e. g., [110, 128]).

Let $\mathcal{D} \subset \mathbb{R}^d$ be a bounded open set, and consider the function $f : \mathcal{D} \times \mathbb{R} \to \mathbb{R}$. The Nemitskii operator related to f is the operator $u \mapsto \mathsf{N}_f(u)$, with $(\mathsf{N}_f(u))(x) = f(x, u(x))$, understood as a mapping between appropriate $L^p(\mathcal{D})$ spaces.

Assumption 6.10.5. The function $f : \mathcal{D} \times \mathbb{R} \to \mathbb{R}$ satisfies the following properties:
(i) f is a Carathéodory function, i. e., $x \mapsto f(x,s)$ is measurable for all $s \in \mathbb{R}$, and $s \mapsto f(x,s)$ is continuous a. e. $x \in \mathcal{D}$.
(ii) f satisfies the bounds: There exist $\alpha > 0$, $c_0, c > 0$, such that

$$|f(x,s)| \le c_0 + c|s|^\alpha, \quad \text{a. e. } x \in \mathcal{D},\ \forall\, s \in \mathbb{R}.$$

Proposition 6.10.6. *Suppose that $\mathcal{D} \subset \mathbb{R}^d$ is open and bounded, and that f satisfies Assumption 6.10.5. Then the Nemitskii operator $\mathsf{N}_f : L^p(\mathcal{D}) \to L^{p/\alpha}(\mathcal{D})$, $1 < p, q < \infty$, is bounded and continuous.*

Proof. Let $(u_n)_{n\in\mathbb{N}} \subset L^p(\mathcal{D})$ such that $u_n \to u$ in $L^p(\mathcal{D})$, i. e., $\|u_n - u\|_{L^p(\mathcal{D})} \to 0$. Then there exists a subsequence (denoted the same for simplicity) such that $u_n(x) \to u(x)$ a. a. $x \in \mathcal{D}$, i. e., $u_n \to u$ a. e. in $\mathcal{D}$, and such that $|u_n| \le h$ a. e. in $\mathcal{D}$, for some $h \in L^p(\mathcal{D})$ (see Theorem 1.4.9(iv)). Then, $f(x, u_n(x)) \to f(x, u(x))$, and by the bounds on f, it is easy to see that $|f(x, u_n(x))| \le c_0 + c|u_n(x)|^\alpha \le c_0 + c|h|^\alpha \in L^{p/\alpha}(\mathcal{D})$. This observation allows us the use of the Lebesgue dominated convergence theorem. By continuity of f, we have that $f(x, u_n(x)) - f(x, u(x)) \to 0$ so that $\|f(\cdot, u_n) - f(\cdot, u)\|_{L^{p/\alpha}(\mathcal{D})} \to 0$. But this implies that $\|\mathsf{N}_f(u_n) - \mathsf{N}_f(u)\|_{L^{p/\alpha}(\mathcal{D})} \to 0$.

This result is so far achieved for the selected subsequence. However, by the above arguments, any sequence $(u_n)_{n\in\mathbb{N}} \subset L^p(\mathcal{D})$ such that $u_n \to u$ in $L^p(\mathcal{D})$ has a further subsequence such that $\mathsf{N}_f(u_n) \to \mathsf{N}_f(u)$ in $L^{p/\alpha}(\mathcal{D})$, so we obtain the stated result (recall the Urysohn property, Remark 1.1.52). □

By appropriate use of the Sobolev embedding theorems, we may consider Nemitskii operators as mappings $\mathsf{N}_f : W^{1,p}(\mathcal{D}) \to L^{\hat{q}}(\mathcal{D})$ for appropriate choice of $\hat{q}$.

Proposition 6.10.7. *Suppose that $\mathcal{D} \subset \mathbb{R}^d$ open and bounded, and that f satisfies Assumption 6.10.5. For any $p \in \mathbb{N}$, let $p_c = \mathsf{s}_p$ be the maximal exponent required for the compact embedding $W^{1,p}(\mathcal{D}) \overset{c}{\hookrightarrow} L^{p_c-\epsilon}(\mathcal{D})$ for any $\epsilon \in (0, p_c - 1]$ to hold (Theorem 1.5.11). Then $\mathsf{N}_f : L^p(\mathcal{D}) \to L^{(p_c-\epsilon)/\alpha}(\mathcal{D})$, bounded and continuous.*

Proof. Consider a sequence $(u_n)_{n\in\mathbb{N}} \subset W^{1,p}(\mathcal{D})$ such that $u_n \rightharpoonup u$ in $W^{1,p}(\mathcal{D})$. By the Rellich–Kondrachov embedding $W^{1,p}(\mathcal{D}) \overset{c}{\hookrightarrow} L^{p_c-\epsilon}(\mathcal{D})$, hence, there exists a subsequence (denoted the same for simplicity) such that $u_n \to u$ in $L^{p_c-\epsilon}(\mathcal{D})$ and a further subsequence (denoted the same again) such that $u_n(x) \to u(x)$ a. a. $x \in \mathcal{D}$, and $|u_n| \le h$ for $h \in L^{p_c-\epsilon}(\mathcal{D})$. Using the above arguments, we may see that $\mathsf{N}_f(u_n) \to \mathsf{N}_f(u)$ in $L^{(p_c-\epsilon)/\alpha}(\mathcal{D})$. □

Depending on the growth condition on f, we may interpret N_f as an operator $\mathsf{N}_f : W^{1,p}(\mathcal{D}) \to Y$, where Y is an appropriately chosen function space. As an example of this construction, we provide the following:

Example 6.10.8 (The Nemitskii operator as $\mathsf{N}_f : W_0^{1,p}(\mathcal{D}) \to (W_0^{1,p}(\mathcal{D}))^*$). If f satisfies Assumption 6.10.5 for $a = p_c - \epsilon - 1$, for $p_c = s_p$, then the Nemitskii operator can be considered as an operator $\hat{\mathsf{N}}_f : W_0^{1,p}(\mathcal{D}) \to (W_0^{1,p}(\mathcal{D}))^*$, which is completely continuous (i. e., weak to strong continuous).

Consider the case where $a = p_c - \epsilon - 1$. Then, we note that $p_c/a = (p_c - \epsilon)^*$, the conjugate exponent of $p_c - \epsilon$. Set $X = W_0^{1,p}(\mathcal{D})$ for simplicity of notation and $r = p_c - \epsilon$. By Proposition 6.10.7, interpreted as above, we have that

$$\|\mathsf{N}_f(u_n) \to \mathsf{N}_f(u)\|_{L^{r^*}(\mathcal{D})} = \|\mathsf{N}_f(u_n) \to \mathsf{N}_f(u)\|_{(L^r(\mathcal{D}))^*} \to 0 \tag{6.229}$$

for any $u_n \rightharpoonup u$ in $W_0^{1,p}(\mathcal{D})$. We now interpret the Nemitskii operator $\hat{\mathsf{N}}_f$ as a variational operator $\hat{\mathsf{N}}_f : X \to X^*$, defined by

$$\langle \hat{\mathsf{N}}_f(u), v\rangle = \int_{\mathcal{D}} f(x, u(x))v(x)dx, \quad \forall\, v \in X,$$

where $\langle\cdot,\cdot\rangle$ is the duality pairing between X and X^*. Then, for any $v \in X = W_0^{1,p}(\mathcal{D}) \subset L^r(\mathcal{D})$, we have

$$\begin{aligned}
|\langle \hat{\mathsf{N}}_f(u_n), -\hat{\mathsf{N}}_f(u), v\rangle| &\le \int_{\mathcal{D}} |f(x, u_n(x)) - f(x, u(x))|\, |v(x)| dx \\
&\le \|f(\cdot, u_n) - f(\cdot, u)\|_{(L^r(\mathcal{D}))^*} \|v\|_{L^r(\mathcal{D})} \\
&\le \|f(\cdot, u_n) - f(\cdot, u)\|_{L^{r^*}(\mathcal{D})} \|v\|_{W_0^{1,p}(\mathcal{D})},
\end{aligned} \tag{6.230}$$

where, on the second line, we first used the Hölder inequality interpreting v as an element of $L^r(D)$, and then reinterpreted it as an element of $X = W_0^{1,p}(\mathcal{D})$ exploiting the Sobolev embedding $X \subset L^r(\mathcal{D})$. Since (6.230) holds for any $v \in X$, recalling the definition of the dual norm, we obtain that

$$\|\hat{\mathsf{N}}_f(u_n), -\hat{\mathsf{N}}_f(u)\|_{X^*} \le c\|f(\cdot, u_n) - f(\cdot, u)\|_{L^{r^*}(\mathcal{D})} \to 0, \tag{6.231}$$

where we used (6.229) (c is a constant related to the embedding $X \subset L^r(\mathcal{D})$). Hence, by (6.231), we conclude that

$$\hat{\mathsf{N}}_f(u_n) \to \hat{\mathsf{N}}_f(u) \quad \text{in } X^* = (W_0^{1,p}(\mathcal{D}))^* = W^{-1,p}(\mathcal{D}).$$

Hence, we conclude that the operator $\mathsf{N} : W_0^{1,p}(\mathcal{D}) \to (W_0^{1,p}(\mathcal{D}))^*$ is weak to strong continuous, or, as is usually called, completely continuous.

7 Variational inequalities

The aim of this chapter is the study of variational inequalities. These are important counterparts to PDEs and involve inequalities rather than equalities. One way to derive variational inequalities is to study minimization problems for certain functionals on closed and convex sets rather than on a linear subspace. As a first example in a long series of related applications one may consider the so called obstacle problem, which consists of the equilibration of an elastic membrane over an obstacle, which results to a differential inequality involving the Laplacian and the given obstacle. In order to set the ideas, we start our treatment of variational inequalities with this problem, establish its connections with free boundary value problems, and consider certain regularity issues. We then study a general class of variational inequalities, not necessarily related to minimization problems, and present a general theoretical framework developed by Lions, Stampacchia, Lax, and Milgram for their treatment. Certain approximation methods for problems of this type, e. g., the penalization method or internal approximation schemes, which may give rise to numerical algorithms are presented. The chapter closes with application in a general class of variational inequalities involving elliptic operators. Importantly, this framework allows also the treatment of a general class of PDEs, uniformly elliptic boundary value problems, a direction which is pursued here as well. Variational inequalities find important applications in various fields, including decision theory, engineering and mechanics, mathematical finance and management science (see, e. g., [24, 38, 139]). Many monographs have been dedicated to the theoretical and numerical aspects of variational inequalities (see, e. g., [24, 38, 94, 127] or [136]).

7.1 Motivation

We will first provide perspective regarding the material of this chapter through some examples.

Example 7.1.1 (Minimization in $\mathbb{R}^d$). When treating the minimization problem of a C^1 function on an interval $[a, b]$, then the minimum x_o is located either at an interior point or at its end points, and it is easy to check that in any case it holds that $\frac{df}{dx}(x_o)\,(x - x_o) \geq 0$ for every $x \in [a, b]$.

Now, let C be a closed, convex, and bounded subset in $\mathbb{R}^d$, consider $f : C \to \mathbb{R}$ continuously differentiable, and denote by $\nabla f := Df$ its gradient. Consider an $x_o \in C$ such that $f(x_o) = \min_{x \in C} f(x)$. Given any $z \in C$, set $z_t = (1-t)x_o + tz$, $t \in [0,1]$. Consider the function $\phi : [0,1] \to \mathbb{R}$, defined by $\phi(t) = f((1-t)x_o + tz)$. Clearly, ϕ has a minimum at $t_o = 0$. Since ϕ is differentiable, and denoting $D\phi =: \frac{d\phi}{dt}$, by our previous observation, it holds that $\phi'(0)t \geq 0$ for every $t \in [0,1]$, or equivalently $\frac{d\phi}{dt}(0) = 0$. But, $\frac{d\phi}{dt}(0) = \nabla f(x_o) \cdot (z - x_o)$, where $\cdot$ denotes the inner product in $\mathbb{R}^d$. Hence, $\nabla f(x_o) \cdot (z - x_o) \geq 0$ for every $z \in C$. ◁

https://doi.org/10.1515/9783111333298-007

Example 7.1.2 (Laplace equation). Recall (see Example 6.1.1) that the minimization of the Dirichlet functional $F : W_0^{1,2}(\mathcal{D}) \to \mathbb{R}$, defined by $F(u) = \frac{1}{2}\int_{\mathcal{D}} |\nabla u|^2\,dx - \int_{\mathcal{D}} fu\,dx$ over the whole of $W_0^{1,2}(\mathcal{D})$, leads to a first-order condition, which can be expressed in the form of the Laplace equation $-\Delta u = f$ on $\mathcal{D}$, complemented with homogeneous Dirichlet boundary conditions on $\mathcal{D}$. Motivated by the previous examples and in particular by Example 7.1.1, it would be interesting to ask the question: What would happen if instead of minimizing the functional F over the whole of $W_0^{1,2}(\mathcal{D})$ we minimized over a closed convex and bounded set $C \subset W_0^{1,2}(\mathcal{D})$ instead? A useful example of such a set is $C_0 := \{u \in W_0^{1,2}(\mathcal{D}) : u(x) \geq \phi(x) \text{ a. e. } x \in \mathcal{D}\}$ for a suitable choice of a given function ϕ with suitable regularity. As already seen in Proposition 2.4.5, for an abstract formulation of this problem, the first-order condition will be of the form $\langle DF(u_o), u - u_o\rangle \geq 0$ for every $u \in C_0$, which, recalling that $\langle DF(u), v\rangle = \langle -\Delta u - f, v\rangle$ for every $v \in X$, allows us to interpret the first-order condition for the problem as a differential inequality

$$\text{find } u_o \in C_0 : \langle -\Delta u_o - f, u - u_o\rangle \geq 0, \quad \forall\, v \in C_0, \tag{7.1}$$

instead of a differential equation. Such inequalities are called variational inequalities, and can be expressed in an equivalent form as free boundary value problems. ◁

In all the above examples, which we used as motivation, variational inequalities were introduced as first-order conditions for minimization problems in convex subsets of a Banach space. This is of course not the general case, as one may obtain interesting examples of variational inequalities in a more general context, not necessarily related to minimization problems. The basic aim of this chapter, therefore, is to develop a theory for the treatment of problems of the general form

$$\text{find } \mathsf{x}_o \in C : \langle \mathsf{A}(\mathsf{x}_o), \mathsf{x} - \mathsf{x}_o\rangle \geq 0, \quad \text{for all } \mathsf{x} \in C, \tag{7.2}$$

where $C \subset X$ is a closed convex subset of a Banach space X and $\mathsf{A} : X \to X^\star$ is an operator, possibly nonlinear and not related to a derivative or subdifferential, so problem (7.2) is not necessarily the first-order condition of an optimization problem. This requires the development of an elegant theory initiated by Lax–Milgram–Lions and Stampacchia, which has also led to important developments to the theory of monotone type operators. Note furthermore that in the case where $C = X$ (or a linear subspace), the variational inequality (7.2) reduces to an equation.[1]

1 For any $\mathsf{z} \in X$, one may choose the test function $\mathsf{x} \in X$ so that $\mathsf{x} - \mathsf{x}_o = \pm\mathsf{z}$, so that (7.2) yields $\langle \mathsf{A}(\mathsf{x}_o), \mathsf{z}\rangle = 0$ for all $\mathsf{z} \in X$, which leads to the equation $\mathsf{A}(\mathsf{x}_o) = 0$.

7.2 Warm up: free boundary value problems for the Laplacian

To provide a feeling for variational inequalities, we first start with a simple problem related to Example 7.1.2 above and in particular the inequality

$$\text{find } u_o \in C : \langle -\Delta u_o, u - u_o \rangle \geq 0, \quad \text{for all } u \in C, \tag{7.3}$$

for

$$C = \{u \in W^{1,2}(\mathcal{D}) : u = f_0, \text{ on } \partial\mathcal{D}, \ u \geq \phi \text{ a. e. in } \mathcal{D}\},$$

where $\mathcal{D}$ is a bounded domain of sufficiently smooth boundary, ϕ is a given function (called the obstacle), f_0 is a given function specifying the boundary data and the LHS of (7.3) is interpreted as $\langle -\Delta u_o, u - u_o \rangle = \int_{\mathcal{D}} \nabla u_o \cdot \nabla (u - u_o) dx$. Clearly, the equality $u = f_0$ on $\partial\mathcal{D}$ is to be understood in the sense of traces, i. e., that $\gamma_0 u = f_0$ (see Definition 1.5.16, see also [1]). Furthermore, if one may find a sufficiently smooth function f_1 defined on the whole of $\mathcal{D}$ such that $\gamma_0 f_1 = f_0$, then, setting $\bar{u} = u - f_1$, $\bar{\phi} = \phi - f_1$ and $f = \Delta f_1$, we see that problem (7.3) reduces to problem (7.1), i.e,

$$\text{find } u_o \in C_0, : \langle -\Delta u_o - f, u - u_o \rangle \geq 0, \quad \text{for all } u \in C_0, \tag{7.4}$$

for

$$C_0 = \{u \in W_0^{1,2}(\mathcal{D}) : u \geq \phi \text{ a. e. in } \mathcal{D}\},$$

and the above choice of f. Naturally, one has to be very careful on the properties of the extension function f_1 (i. e., membership in the right function space) as well as to what sense is Δf_1 defined so that the appropriate weak form for (7.4) is used.

Problem (7.3) is called an obstacle problem and may be considered as modeling an elastic membrane whose shape is described by the function u without any external force exerted on it, apart from the fact that it has to lie over an obstacle, which is described by the function ϕ. Naturally, even though, no external forces are applied to the membrane, the underlying obstacle exerts some force on it, and this may serve as a physical understanding of the connection between (7.3) and (7.4). Obstacle problems find interesting applications in other fields apart from mechanics, as, for example, in decision theory, mathematical finance etc.

By the symmetry of the operator $A = -\Delta$ both variational inequalities (7.3) and (7.4) can be understood as the first-order condition for some minimization problem, and in particular

$$\min_{u \in C} F_{\mathcal{D}}(u), \quad F_{\mathcal{D}}(u) := \frac{1}{2} \int_{\mathcal{D}} |\nabla u|^2 dx, \tag{7.5}$$

and

$$\min_{u \in C_0} F_0(u), \quad F_0(u) := \frac{1}{2}\int_{\mathcal{D}} |\nabla u|^2 dx - \int_{\mathcal{D}} f u dx, \tag{7.6}$$

respectively. For simplicity in this section, we may assume that the obstacle ϕ is a smooth function.

By the connection of problem (7.3) (resp. (7.4)) with the minimization problem (7.5) (resp. (7.6)) using techniques from the calculus of variations (such as those developed in the previous chapter), we may show some solvability results for these variational inequalities. Furthermore, one may show, concerning problem (7.3), that if its solution is sufficiently regular, it satisfies $-\Delta u_o \geq 0$ with

$$\begin{aligned} -\Delta u_o &= 0, \quad \text{where } u_o > \phi, \\ -\Delta u_o &> 0 \quad \text{where } u_o = \phi, \\ u_o &\geq \phi, \quad u_o|_{\partial\mathcal{D}} = f_0, \end{aligned} \tag{7.7}$$

so that, upon defining the coincidence set

$$\mathcal{C} := \{x \in \mathcal{D} : u(x) = \phi(x)\},$$

the solution is in general a superharmonic function,[2] which satisfies the Laplace equation on the complement of $\mathcal{C}$. It is interesting to note that, apart from the fixed boundary $\partial\mathcal{D}$, where the solution is prescribed and equal to the known function f_0, the coincidence set defines another boundary (on which the solution coincides with the obstacle function ϕ), which is not known *a priori* but rather depends on the solution of the problem. For that reason, problems of this type are often called *free boundary value problems.* Note that (7.7) can be rephrased in the more compact complementarity form[3]

$$u_o \geq \phi, \quad -\Delta u_o \geq 0, \quad (u_o - \phi)\Delta u_o = 0, \quad u_o|_{\partial\mathcal{D}} = f_0.$$

Similarly, for problem (7.3), we may have an analogous specification for the solution with 0 replaced by f on the first two equations in (7.7) and $u_o|_{\partial\mathcal{D}} = 0$. The next proposition (see, e. g., [78] and references therein) collects some fundamental properties of the free boundary problem for the Laplacian.

Proposition 7.2.1. *Assume that the bounded domain $\mathcal{D}$ as well as the functions $\phi : \overline{\mathcal{D}} \to \mathbb{R}$ and $f_0 : \partial\mathcal{D} \to \mathbb{R}$ are of class C^1, while $\phi \leq f_0$ on $\partial\mathcal{D}$. Then,*
(i) *The variational inequality* (7.3) *admits a unique solution $u_o \in C$.*
(ii) *The solution u_o also satisfies* (7.7).

2 i. e., a function such that $\Delta u \leq 0$.

3 If $u_o > \phi$, then $(u_o - \phi)\Delta u_o = 0$ implies that $\Delta u_o = 0$. If $-\Delta u_o > 0$, then $(u_o - \phi)\Delta u_o = 0$ implies that $u_o = \phi$. An alternative formulation could be as $\min(-\Delta u_o, u_o - \phi) = 0$.

Proof. (i) For this symmetric problem, the solvability of (7.3) follows from its connection with the variational problem (7.5), so we will use the direct method for the calculus of variations, essentially following the standard procedure in Theorem 6.4.2, which we briefly repeat here to refresh our memory. We sketch it in 3 steps.

1. By standard arguments, we select a minimizing sequence $(u_n)_{n\in\mathbb{N}} \subset C$. Since $F_{\mathcal{D}}(u_n) \to m := \inf\{F_{\mathcal{D}}(u) : u \in C\}$, by the form of $F_{\mathcal{D}}$, we have that $\|\nabla u_n\|_{L^2(\mathcal{D})} \leq c$ for some $c > 0$ and n sufficiently large. Note that since we are not in $W_0^{1,2}(\mathcal{D})$, hence, $\|\nabla u\|_{L^2(\mathcal{D})}$ does not constitute an equivalent norm, we need to establish boundedness of $\|u_n\|_{W^{1,2}(\mathcal{D})}$ from the above estimate, before we are allowed to proceed in the selection of a weakly convergent subsequence and passage to the limit.

To this end, we need to modify $(u_n)_{n\in\mathbb{N}}$ to a new sequence $(\hat{u}_n)_{n\in\mathbb{N}} \subset W_0^{1,2}(\mathcal{D})$. Let $\widehat{f}_0 \in C^1(\overline{\mathcal{D}})$, be an extension of f_0, and define $\widehat{\phi} := \max\{\phi, \widehat{f}_0\}$. Then, by the assumptions imposed on f_0 and ϕ, setting $\hat{u}_n := u_n - \widehat{\phi}$ we have that $\hat{u}_n \in W_0^{1,2}(\mathcal{D})$ for any $n \in \mathbb{N}$, and therefore Poincaré's inequality holds for any element of the new sequence, hence, $\|u_n - \widehat{\phi}\|_{L^2(\mathcal{D})} \leq \mathbf{c}_{\mathcal{P}}^{-1}\|\nabla u_n - \nabla\widehat{\phi}\|_{L^2(\mathcal{D})}$. Then, by combining the triangle inequality and the Poincaré inequality, we have that

$$\begin{aligned}
\|u_n\|_{W^{1,2}(\mathcal{D})} &= \|u_n\|_{L^2(\mathcal{D})} + \|\nabla u_n\|_{L^2(\mathcal{D})} \\
&\leq \|u_n - \widehat{\phi}\|_{L^2(\mathcal{D})} + \|\widehat{\phi}\|_{L^2(\mathcal{D})} + \|\nabla u_n\|_{L^2(\mathcal{D})} \\
&\leq \mathbf{c}_{\mathcal{P}}^{-1}\|\nabla u_n - \nabla\widehat{\phi}\|_{L^2(\mathcal{D})} + \|\widehat{\phi}\|_{L^2(\mathcal{D})} + \|\nabla u_n\|_{L^2(\mathcal{D})}
\end{aligned}$$

so that the quantity on the left-hand side is estimated by $\|\nabla u_n\|_{L^2(\mathcal{D})}$, which is bounded. Therefore, $\|u_n\|_{W^{1,2}(\mathcal{D})} < c'$ for some $c' > 0$, hence, by reflexivity, there exists a $u_o \in W^{1,2}(\mathcal{D})$ and a subsequence $u_{n_k} \rightharpoonup u_o$ in $W^{1,2}(\mathcal{D})$. If $u_o \in C$, then by the weak lower semicontinuity of $F_{\mathcal{D}}$, we are done.

2. The fact that the limit $u_o \in C$ follows by the convexity of C. We first recall the fact that for convex sets, the weak and the strong closures coincide (see Proposition 1.2.12). To check that C is closed, consider a sequence $(v_n)_{n\in\mathbb{N}} \subset C$ such that $v_n \to v$ in $W^{1,2}(\mathcal{D})$. We will show that $v \in C$. Indeed, since $v_n \to v$ in $W_0^{1,2}(\mathcal{D})$, by the compact embedding $W_0^{1,2}(\mathcal{D}) \overset{c}{\hookrightarrow} L^2(\mathcal{D})$, it follows that there exists a subsequence (denoted the same) $v_n \to v$ in $L^2(\mathcal{D})$, hence, a further subsequence such that $v_{n_k} \to v$ a. e. in $\mathcal{D}$. Since $v_{n_k} \in C$, it follows that $v_{n_k}(x) \geq \phi(x)$, a. e. in $\mathcal{D}$, and passing to the limit in this subsequence, the inequality remains valid, so $v(x) \geq \phi(x)$, a. e. in $\mathcal{D}$. Therefore, $v \in C$, hence, C is closed. Using this observation and the convexity of C we conclude that $u_o \in C$.

3. Steps 1 and 2 complete the existence of a minimizer $u_o \in C$, while derivation of the Euler–Lagrange equation (first-order condition) yields that u_o solves (7.3). Uniqueness follows by a strict convexity argument based on the observation that if two such solutions exist, then their gradients coincide, and by the Poincaré inequality applied to their difference, so do they.

(ii) We prove the above claims in 3 steps.

1. We first prove that $-\Delta u_o \geq 0$ in $\mathcal{D}$. Let $\psi \in C_c^\infty(\mathcal{D})$ be a test function such that $\psi \geq 0$. Then, if $u_o \in C$, it is clear that $v = u_o + \psi \in C$. Setting $u = v$ in the variational inequality (7.3), we obtain $\int_{\mathcal{D}} \nabla u_o(x) \cdot \nabla \psi(x) dx \geq 0$, which is the weak form for the inequality $-\Delta u_o \geq 0$ (as a simple integration by parts argument, plus a density argument of $C_c^\infty(\mathcal{D})$ in $W_0^{1,2}(\mathcal{D})$ shows).

2. We show next that $-\Delta u_o = 0$ in $\{u_o > \phi\} \cap \mathcal{D}$. We claim that $u_o - \phi$ is lower semicontinuous in $\mathcal{D}$ (or equivalently by the continuity of ϕ, that u_o is lower semicontinuous). For the time being, we accept this claim, which will be proved in step 3. Then, we can argue as follows: By the lower semicontinuity of $u_o - \phi$ in $\mathcal{D}$, we have that the set $\mathcal{D}_+ := \{x \in \mathcal{D} : u_o(x) - \phi(x) > 0\}$ is open, so for every $x \in \mathcal{D}_+$, we can find an $r > 0$ with the property $B(x,r) \subset\subset \mathcal{D}_+$ and consider a test function $\psi \in C_c^\infty(B(x,r))$. Therefore, if $\epsilon > 0$ is small enough, we have that both $u_o + \epsilon\psi - \phi > 0$ and $u - \epsilon\psi - \phi > 0$. That means $v_1 = u_o + \epsilon\psi \in C$, so letting $u = v_1$ in (7.4), we get $\int_{\mathcal{D}} \nabla u_o(x) \cdot \nabla\psi(x) dx \geq 0$. Moreover, by the same arguments $v_2 = u_o - \epsilon\psi \in C$, so letting $u = v_2$ into (7.4), we get $\int_{\mathcal{D}} \nabla u_o(x) \cdot \nabla\psi(x) dx \leq 0$. Combining these two (along with the arbitrary nature of the ball $B(x,r) \subset\subset \{u_o > \phi\}$), we conclude that

$$\int_{\mathcal{D}} \nabla u_o(x) \cdot \nabla\psi(x) dx = 0, \quad \forall\, \psi \in C_c^\infty(\mathcal{D} \setminus C),$$

which is the weak form of $-\Delta u_o = 0$ on $\mathcal{D} \setminus C$.

3. We now show that under the assumptions on the data of the problem, u_o is lower semicontinuous. To show this, we may argue as follows:

3(a). We have already shown in step 1 that $-\Delta u_o \geq 0$ in $\mathcal{D}$ in the weak sense, so the mapping

$$r \mapsto \frac{1}{|B(x,r)|} \int_{B(x,r)} u_o(z) dz =: \fint_{B(x,r)} u_o(z) dz =: w(r)$$

is decreasing on $(0, \operatorname{dist}(x, \partial\mathcal{D}))$. This can be easily seen by using the function

$$v_R(x) := R^{2-d}\left[\left(\frac{|x|}{R}\right)^{-(d-2)} + \frac{d-2}{2}\left(\frac{|x|}{R}\right)^2 - \frac{d}{2}\right] \mathbf{1}_{B(0,R)}(|x|),$$

which is a $C^{1,1}$ function, except at the origin, while a simple calculation in spherical coordinates yields $\Delta v_R = \frac{d(d-2)\omega_d}{|B(0,R)|} \mathbf{1}_{B(0,R)}$, where ω_d is the volume of the d-dimensional unit ball ($|B(0,R)| = \omega_d R^d$). We may also note that $\frac{d}{dR} v_R(x) \geq 0$ for every $x \in B(0,R)$; hence, if $R_1 < R_2$, it holds that $v_{R_1}(x) \leq v_{R_2}(x)$ for any $x \in B(0,R_2)$. So $\psi := v_{R_2} - v_{R_1} \geq 0$ is a suitable test function in $C^{1,1}$ for the weak form of $-\Delta u_o \geq 0$. Note that ψ vanishes outside $B(0,R_2)$ and satisfies

$$\Delta\psi = d(d-2)\omega_d \left(\frac{1}{|B(0,R_2)|} \mathbf{1}_{B(0,R_2)} - \frac{1}{|B(0,R_1)|} \mathbf{1}_{B(0,R_1)} \right). \tag{7.8}$$

Since $-\Delta u_o \geq 0$ in the weak sense, using (7.8), we have for the test function $\psi = v_{R_2} - v_{R_1}$ that

$$0 \leq \int_{\mathcal{D}} (-\Delta u_o)(x)\psi(x)dx = \int_{\mathcal{D}} u_o(x)(-\Delta\psi)(x)dx = d(d-2)\omega_d(w(R_1) - w(R_2)),$$

so that the function $R \mapsto w(R) = \fint_{B(0,R)} u_o(x)dx$ is decreasing. By appropriately shifting the origin to any point $x \in \mathcal{D}$, we have that the function $R \mapsto \fint_{B(x,R)} u_o(z)dz$ is decreasing.

3(b). By the monotonicity obtained in step 3(a), the function

$$\bar{u}(x) := \lim_{R\to 0} \fint_{B(x,R)} u_o(z)dz, \quad \forall\, x \in \mathcal{D},$$

is well defined, and again by the same arguments satisfies the lower bound,

$$\fint_{B(x,R)} u_o(z)dz \leq \bar{u}(x) := \lim_{R\to 0} \fint_{B(x,R)} u_o(z)dz, \quad \forall\, R > 0. \tag{7.9}$$

Moreover, $\bar{u}(x) = u_o(x)$ for any Lebesgue point of u_o so that $\bar{u} = u_o$, a. e. in $\mathcal{D}$.

3(c). We will show that, by construction, the function $\bar{u}$ is lower semicontinuous, hence, since $\bar{u} = u_o$, a. e. in $\mathcal{D}$, u_o has a lower semicontinuous version. We will identify u_o with $\bar{u}$ so that it is pointwise defined at every point. Then, (7.9) is a version of the mean value property for super harmonic functions.

To check the lower semicontinuity of $\bar{u}$, consider any sequence $(x_n)_{n\in\mathbb{N}} \subset \mathcal{D}$ such that $x_n \to x$, and note that for any fixed radius $R > 0$, it holds that

$$\begin{aligned}\fint_{B(x_n,R)} u_o(z)dz &= \frac{1}{|B(x_n,R)|}\int_{\mathcal{D}} u_o(z)\mathbf{1}_{B(x_n,R)}(z)dz\\ &= \frac{1}{\omega_d R^d}\int_{\mathcal{D}} u_o(z)\mathbf{1}_{B(x_n,R)}(z)dz,\end{aligned}$$

so noting that $u_o\mathbf{1}_{B(x_n,R)} \to u_o\mathbf{1}_{B(x,R)}$ a. e., and passing to the limit as $n \to \infty$ using the Lebesgue dominated convergence theorem, we get

$$\begin{aligned}\lim_{n\to\infty} \fint_{B(x_n,R)} u_o(z)dz &= \frac{1}{\omega_d R^d}\lim_{n\to\infty}\int_{\mathcal{D}} u_o(z)\mathbf{1}_{B(x_n,R)}(z)dz\\ &= \frac{1}{\omega_d R^d}\int_{\mathcal{D}} u_o(z)\mathbf{1}_{B(x,R)}(z)dz = \fint_{B(x,R)} u_o(z)dz = \bar{u}(x),\end{aligned} \tag{7.10}$$

since for any choice of centers x, x_n, it holds that $|B(x_n,R)| = |B(x,R)| = \omega_d R^d$. Furthermore, applying (7.9) for $x = x_n$ for every $n \in \mathbb{N}$, then taking the limit inferior for the resulting inequality and using (7.10) for the LHS, we get that

$$\bar{u}(x) = \lim_{n\to\infty} \fint_{B(x_n,R)} u_o(z)dz \le \liminf_{n\to\infty} \bar{u}(x_n),$$

which is the required lower semicontinuity result. The proof is complete. □

7.3 The Lax–Milgram–Stampacchia theory

We now present a first approach towards variational inequalities, which is essentially centered or inspired by results related to minimization of quadratic functionals over convex and closed subsets of Hilbert spaces, but importantly leads to results for variational inequality problems of a more general nature, not directly related to minimization problems. These considerations lead to a celebrated theory, pioneered by Lax, Milgram, Lions, and Stampacchia (see, e. g., [38] or [94]). In this chapter, we will focus on the Hilbert space case. The more general problem of variational inequalities in Banach spaces will be treated later on in Chapter 9, after the machinery of monotone operators has been introduced.

Throughout this chapter, let $X := H$ be a Hilbert space, with dual $X^\star$ not necessarily identified with X. Let $\langle\cdot,\cdot\rangle_X$ be the inner product in X, $\|\cdot\| := \|\cdot\|_X$ its norm defined by $\|x\|^2 = \langle x, x\rangle_X$, and let us denote by $\langle\cdot,\cdot\rangle := \langle\cdot,\cdot\rangle_{X^\star,X}$ the duality pairing between $X^\star$ and X. Recall that in this context, we may define a bijection

$$\begin{aligned} &i_X : X \to X^\star, \quad \text{such that} \quad \forall x^\star \in X^\star, \\ &\langle x^\star, z\rangle = \langle i_X^{-1}(x), z\rangle_X, \quad \forall z \in X, \text{with } \|x^\star\|_{X^\star} = \|i_X^{-1}(x^\star)\|_X. \end{aligned} \tag{7.11}$$

The inverse of this map $i_X^{-1} : X^\star \to X$ is the map associated with the Riesz–Fréchet representation theorem (see Theorem 1.1.14 and discussion below). Clearly, $\langle i_X(x_1), x_2\rangle = \langle x_1, x_2\rangle_X$ for every $x_1, x_2 \in X$. We recall further that the dual space $X^\star$ can be turned into a Hilbert space with norm $\|x^\star\|_{X^\star}^2 = \langle i_X^{-1}(x^\star), i_X^{-1}(x^\star)\rangle_X = \|i_X^{-1}(x^\star)\|^2$. Importantly, we do not necessarily consider i_X as the identity map (i. e., we do not identify $x^\star = i_X(x)$ with x), so X^* is not necessarily identified by X (see Remark 1.1.15).

The Lax–Milgram–Stampacchia theory deals with variational inequalities in Hilbert space, involving a bilinear mapping $a : X \times X \to \mathbb{R}$, and in particular problems of the form:

Given a closed convex subset C of the Hilbert space X and $x^\star \in X^\star$, find $x_o \in C$ such that

$$a(x_o, x - x_o) \ge \langle x^\star, x - x_o\rangle, \quad \forall x \in C.$$

If a was symmetric (i. e., such that $a(x, z) = a(z, x)$ for every $x, z \in X$), we may associate the bilinear form a with the Gâteaux derivative of the quadratic functional $F : X \to \mathbb{R}$, defined by $F(x) = a(x, x)$ (see Example 2.1.8), connect the variational inequality with a minimization problem over C, and by employing the general theory of convex minimization

(Theorem 2.4.5) obtain well posedness results. However, the Lax–Milgram–Stampacchia theory goes beyond that, it allows us to treat variational inequalities, even in the non-symmetric case, which is not directly related to minimization problems.

Before presenting the Lax–Milgram–Stampacchia theory, we need to introduce the notion of continuous and coercive bilinear forms.

Definition 7.3.1 (Continuous and coercive bilinear form). Let $X = H$ be a Hilbert space. The bilinear form $\mathfrak{a} : X \times X \to \mathbb{R}$ is called
(i) continuous, if there exists a constant $c > 0$ such that $|\mathfrak{a}(x_1, x_2)| \leq c \, \|x_1\| \, \|x_2\|$, for every $x_1, x_2 \in X$.
(ii) coercive, if there exists a constant $c_o > 0$ such that $\mathfrak{a}(x, x) \geq c_o \, \|x\|^2$ for every $x \in X$.

Remark 7.3.2. If a continuous and coercive bilinear form $\mathfrak{a}$ is also symmetric[4] it can be used to define a new inner product on X by $\langle x_1, x_2 \rangle_* := \mathfrak{a}(x_1, x_2)$ for every $x_1, x_2 \in X$. We may then renorm X using the norm defined by this new inner product (see [33]).

A bilinear form, which is continuous, can be used to define a linear and bounded operator.

Proposition 7.3.3. *Let X be a Hilbert space and $\mathfrak{a} : X \times X \to \mathbb{R}$ a continuous and coercive bilinear form. The form $\mathfrak{a}$ defines an operator $A : X \to X^\star$, by the relation*

$$\langle Ax, z \rangle := \mathfrak{a}(x, z), \quad \forall \, z \in X,$$

that is linear, bounded, and satisfies the monotonicity property

$$\langle Ax_1 - Ax_2, x_1 - x_2 \rangle \geq c_o \, \|x_1 - x_2\|^2. \tag{7.12}$$

Proof. By the continuity of the bilinear form $\mathfrak{a}$, for any fixed $x \in X$, fixed, $\mathfrak{a}(x, \cdot)$ is a functional hence, by the Riesz representation, there exists a unique $x_o \in X^\star$ such that $\mathfrak{a}(x, z) = \langle x_o, z \rangle$ for all $z \in X$. We will denote $x_o = Ax$, and this will define an operator $A : X \to X^\star$. The operator A is bounded, since

$$\big|\langle Ax, z \rangle\big| = \big|\mathfrak{a}(x, z)\big| \leq c \, \|x\| \, \|z\|, \quad \forall \, x \in X, \; z \in X,$$

which leads to $\|A\|_{\mathcal{L}(X, X^\star)} \leq c$. The monotonicity property (7.12) follows by the linearity of the operator A and the coercivity property of the bilinear form. □

Example 7.3.4. A concrete example for the above abstract formulation is the variational definition of the Laplacian for functions defined on a bounded domain $\mathcal{D}$, which are not C^2 but rather only weakly differentiable once. In this case, we consider $X = W_0^{1,2}(\mathcal{D})$ and $X^* = W^{-1,2}(\mathcal{D})$, both of which are Hilbert spaces. In fact $X \subset X_0 = L^2(\mathcal{D})$, but its

4 i. e., $\mathfrak{a}(x_1, x_2) = \mathfrak{a}(x_2, x_1)$ for every $x_1, x_2 \in X$.

inner product does not coincide with that of X_0. The Laplacian can then be defined in this setting in terms of the bilinear form $\mathfrak{a} : X \times X \to \mathbb{R}$, with $\mathfrak{a}(u,v) = \int_{\mathcal{D}} \nabla u(x) \nabla v(x) dx$, by using an integration by parts argument. Clearly $\mathsf{A} = -\Delta$ can no longer be defined as an operator $\mathsf{A} : X \to X$, but rather has to be interpreted as an operator $\mathsf{A} : X \to X^*$. ◁

We are now ready to present the celebrated Lax–Milgram–Stampacchia theory of variational inequalities (see, e. g., [48] or [94]).

Theorem 7.3.5 (Lax–Milgram–Stampacchia I). *Let $C \subset X$ be a closed and convex subset of the Hilbert space X, not necessarily identified with its dual X^*, and $\mathfrak{a} : X \times X \to \mathbb{R}$ a continuous and coercive bilinear form.*

(i) *For every $x^\star \in X^*$, there exists a unique $x_o \in C$ such that*

$$\mathfrak{a}(x_o, x - x_o) \geq \langle x^\star, x - x_o \rangle, \quad \forall x \in C, \tag{7.13}$$

or equivalently, using Proposition 7.3.3

$$\langle \mathsf{A} x_o, x - x_o \rangle \geq \langle x^\star, x - x_o \rangle, \quad \forall x \in C. \tag{7.14}$$

(ii) *If furthermore, $\mathfrak{a}$ is symmetric, then x_o is the unique minimizer on C of the functional $\Phi : X \to \mathbb{R}$, defined by*

$$\Phi(x) = \frac{1}{2}\mathfrak{a}(x,x) - \langle x^\star, x \rangle.$$

Proof. (i) We will denote by $\langle \cdot, \cdot \rangle_X$ the inner product in X and by $\langle \cdot, \cdot \rangle$ the duality pairing between X^* and X. Let $P_C : X \to C$ be the projection operator from X to the closed convex $C \subset X$, defined by $x_o := P_C(x) = \arg\min_{z \in C} \|x - z\|$. This map is nonexpansive (pseudo-contraction) and is characterized by the variational inequality $\langle x - x_o, z - x_o \rangle_X \leq 0$ for every $z \in C$ or its equivalent formulation $x - x_o \in N_C(x_o)$ (see Theorem 2.5.3).

Note that (7.14) can be expressed in the equivalent form

$$\langle \rho(x^\star - \mathsf{A} x_o), x - x_o \rangle = \langle \rho j^{-1}(x^\star - \mathsf{A} x_o), x - x_o \rangle_X \leq 0, \quad \forall x \in C,$$

where we used the map $i_X : X \to X^*$ defined in (7.11), simplifying notation to $j := i_X$. Fix $\rho > 0$ (to be determined shortly) and upon multiplying the variational inequality with ρ and observing that trivially $\rho j^{-1}(x^\star - \mathsf{A} x_o) = \rho j^{-1}(x^\star - \mathsf{A} x_o) + x_o - x_o$ the above inequality can be expressed as

$$\langle \rho j^{-1}(x^\star - \mathsf{A} x_o) + x_o - x_o, x - x_o \rangle_X \leq 0, \quad \forall x \in C. \tag{7.15}$$

However, recalling the characterization of projection in terms of variational inequalities, it can be seen that (7.15) is equivalent to

$$x_o = P_C(\rho j^{-1}(x^\star - \mathsf{A} x_o) + x_o) \tag{7.16}$$

for some $\rho > 0$. It thus suffices to show that (7.16) holds for some $\rho > 0$.

We define the family of maps $\mathsf{T}_\rho : C \to C$ by $\mathsf{T}_\rho(\mathsf{x}) = \mathsf{P}_C(\rho j^{-1}(\mathsf{x}^\star - \mathsf{A}\mathsf{x}) + \mathsf{x})$ for every $\mathsf{x} \in C$, and we claim that for the proper choice of ρ the map T_ρ is a strict contraction, so by the Banach fixed point theorem (see Theorem 3.1.2) the claim follows. To this end, consider $\mathsf{x}_1, \mathsf{x}_2 \in X$, and observe that

$$\begin{aligned}
&\|\mathsf{T}_\rho(\mathsf{x}_1) - \mathsf{T}_\rho(\mathsf{x}_2)\| \\
&\quad = \|\mathsf{P}_C(\rho j^{-1}(\mathsf{x}^\star - \mathsf{A}\mathsf{x}_1) + \mathsf{x}_1) - \mathsf{P}_C(\rho j^{-1}(\mathsf{x}^\star - \mathsf{A}\mathsf{x}_2) + \mathsf{x}_2)\| \\
&\quad \le \|\rho j^{-1}(\mathsf{x}^\star - \mathsf{A}\mathsf{x}_1) + \mathsf{x}_1 - \rho j^{-1}(\mathsf{x}^\star - \mathsf{A}\mathsf{x}_2) - \mathsf{x}_2\| \\
&\quad = \|\mathsf{x}_1 - \mathsf{x}_2 - \rho j^{-1}(\mathsf{A}\mathsf{x}_1 - \mathsf{A}\mathsf{x}_2)\|,
\end{aligned} \tag{7.17}$$

where we used the properties of projection operators and the mapping $j := i_X$. Moreover, using the fact that X is a Hilbert space

$$\begin{aligned}
&\|\mathsf{x}_1 - \mathsf{x}_2 - \rho j^{-1}(\mathsf{A}\mathsf{x}_1 - \mathsf{A}\mathsf{x}_2)\|^2 \\
&\quad = \langle \mathsf{x}_1 - \mathsf{x}_2 - \rho j^{-1}(\mathsf{A}\mathsf{x}_1 - \mathsf{A}\mathsf{x}_2), \mathsf{x}_1 - \mathsf{x}_2 - \rho j^{-1}(\mathsf{A}\mathsf{x}_1 - \mathsf{A}\mathsf{x}_2)\rangle_X \\
&\quad = \|\mathsf{x}_1 - \mathsf{x}_2\|^2 + \rho^2\|j^{-1}(\mathsf{A}\mathsf{x}_1 - \mathsf{A}\mathsf{x}_2)\|^2 - 2\rho\langle \mathsf{x}_1 - \mathsf{x}_2, j^{-1}(\mathsf{A}\mathsf{x}_1 - \mathsf{A}\mathsf{x}_2)\rangle_X.
\end{aligned} \tag{7.18}$$

By the linearity of A, setting $\mathsf{z} = \mathsf{x}_1 - \mathsf{x}_2$ to simplify notation, we see that

$$\|j^{-1}(\mathsf{A}\mathsf{x}_1 - \mathsf{A}\mathsf{x}_2)\|^2 = \|\mathsf{A}\mathsf{x}_1 - \mathsf{A}\mathsf{x}_2\|_{X^\star}^2 = \|\mathsf{A}\mathsf{z}\|_{X^\star}^2 \le c\,\|\mathsf{z}\|^2,$$

since the operator A is bounded, and

$$(\mathsf{x}_1 - \mathsf{x}_2, j^{-1}(\mathsf{A}\mathsf{x}_1 - \mathsf{A}\mathsf{x}_2)) = \langle \mathsf{A}\mathsf{z}, \mathsf{z}\rangle = a(\mathsf{z}, \mathsf{z}) \ge c_o\|\mathsf{z}\|^2,$$

where we used the definition of the operator A, and the coercivity of the bilinear form a (Assumption 7.3.1(ii)). Therefore, (7.18) can be estimated as

$$\|\mathsf{x}_1 - \mathsf{x}_2 - \rho j^{-1}(\mathsf{A}\mathsf{x}_1 - \mathsf{A}\mathsf{x}_2)\|^2 \le (1 + c\rho^2 - 2\rho c_o)\|\mathsf{x}_1 - \mathsf{x}_2\|^2,$$

and choosing $\rho > 0$ such that $0 < \varrho^2 := 1 + c^2\rho^2 - 2\rho c_o < 1$, we have

$$\|\mathsf{x}_1 - \mathsf{x}_2 - \rho j^{-1}(\mathsf{A}\mathsf{x}_1 - \mathsf{A}\mathsf{x}_2)\| < \varrho\|\mathsf{x}_1 - \mathsf{x}_2\|,$$

which when combined with (7.17) leads to

$$\|\mathsf{T}_\rho(\mathsf{x}_1) - \mathsf{T}_\rho(\mathsf{x}_2)\| < \varrho\|\mathsf{x}_1 - \mathsf{x}_2\|,$$

therefore, T_ρ is a strict contraction when $\rho \in (0, \frac{2c_o}{c^2})$.

(ii) Even though, this follows directly by the abstract results of Theorem 2.4.5, upon noting that in this case the variational inequality coincides with the first-order condition for the minimization problem in question, we provide an elementary proof, which highlights the importance of symmetry.[5]

By the definition of Φ, and using the symmetry of a, we see that for all $x \in C$,

$$\begin{aligned}\Phi(x) &= \Phi(x_o + (x - x_o))\\ &= \frac{1}{2}a(x_o + (x - x_o), x_o + (x - x_o)) - \langle x^\star, x_o + (x - x_o)\rangle\\ &= \frac{1}{2}a(x_o, x_o) - \langle x^\star, x_o\rangle + a(x - x_o, x_o) + \frac{1}{2}a(x - x_o, x - x_o) - \langle x^\star, x - x_o\rangle\\ &= \Phi(x_o) + \frac{1}{2}a(x - x_o, x - x_o) + a(x - x_o, x_o) - \langle x^\star, x - x_o\rangle.\end{aligned} \tag{7.19}$$

By coercivity of a it follows that $a(x - x_o, x - x_o) \geq 0$. Now, let x_o be the unique solution of (7.13) (equivalently (7.14)). The symmetry of a implies that $a(x - x_o, x_o) - \langle x^\star, x - x_o\rangle \geq 0$ for any $x \in C$, therefore, (7.19) implies that $\Phi(x) \geq \Phi(x_o)$ for every $x \in C$ so that x_o is the minimizer of Φ in C. □

Remark 7.3.6 (Solution of VIs as the zero of an operator equation). An important observation, arising in the course of the proof of the above theorem, is that the solution of the variational inequality is related to a fixed point of an appropriate operator, $T_\rho : C \to C$, i. e., its reformulation as

$$x_o = T_\rho(x_o) \quad \text{where } x \mapsto T_\rho(x) := P_C(-\rho F(x) + x), \tag{7.20}$$

using the simplified notation $F(x) = -i_X^{-1}(x^\star - Ax)$. An alternative way to express (7.20) is in terms of the normal cone of C. Recall that $z_0 = P_C(z)$ if and only if $z - z_0 \in N_C(z_0)$. Applying that to $z_0 = x_o$ and $z = -\rho F(x_o) + x_o$, and rearranging, we conclude that the solution to the variational inequality satisfies

$$0 \in \rho F(x_o) + N_C(x_o), \quad \rho > 0, \tag{7.21}$$

i. e., it is characterized by the zero of the operator $M = \rho F + N_C$. Note that as defined, the operator F, satisfies monotonicity properties (since A does), and this is important for both formulations (7.20) and (7.21). In fact, these formulations, go beyond the limited case considered in Theorem 7.3.5 and can be extended (under appropriate modifications of course) for nonlinear operators and also for the more general case of Banach spaces. These generalizations will be considered in Chapter 9, later on in this book, and will depend crucially on monotonicity properties of the operators involved. Moreover, reformulations (7.20) and (7.21) are well suited for the numerical solution of variational inequalities.

5 A nice alternative way to handle the symmetric case is to use Remark 7.3.2 (see [33]).

In the special case where $C = X$, we recover the well known Lax–Milgram theorem [33, 76], which forms the basis for the study of linear PDEs.

Theorem 7.3.7 (Lax–Milgram). *Let X be a Hilbert space and $a : X \times X \to \mathbb{R}$ a continuous and coercive bilinear form. Then for any $x^\star \in X^\star$, there exists a unique solution $x_o \in X$ to the equation*

$$a(x_o, x) = \langle x^\star, x\rangle, \quad \forall x \in X.$$

Proof. We will use the Lax–Milgram–Stampacchia theorem 7.3.5 in the special case where $C = X$. Let $x_o \in X$ be the unique solution of (7.14), and consider any $z \in X$. Then, since we now work on the whole of the vector space and not only on a closed convex subset, $x_1 = x_o + z \in X$ and $x_2 = x_o - z \in X$. Applying (7.14) with $x = x_1$, we see that $a(x_o, z) \geq \langle x^\star, z\rangle$. Applying (7.14) once more with $x = x_2$, we see that we see that $a(x_o, z) \leq \langle x^\star, z\rangle$. Combining these two inequalities, and relabeling the arbitrary z as x, we obtain the desired result. Furthermore, in the case of symmetry, this solution has minimizing properties. □

The following equivalent form of the variational inequality (7.14) is often useful [94, 136]:

Theorem 7.3.8 (Minty). *The variational inequality*

$$\text{find } x_o \in C : a(x_o, x - x_o) = \langle Ax_o, x - x_o\rangle \geq \langle x^\star, x - x_o\rangle, \quad \forall x \in C, \tag{7.22}$$

is equivalent to

$$\text{find } x_o \in C : a(x, x - x_o) = \langle Ax, x - x_o\rangle \geq \langle x^\star, x - x_o\rangle, \quad \forall x \in C. \tag{7.23}$$

Proof. Suppose $x_o \in C$ solves (7.22). Then, for any $x \in C$

$$a(x, x - x_o) = a(x - x_o, x - x_o) + a(x_o, x - x_o) \geq a(x_o, x - x_o) \geq \langle x^\star, x - x_o\rangle,$$

where we used first the linearity of a with respect to the first argument and then the coercivity to deduce that $a(x - x_o, x - x_o) > 0$. Therefore, a solution of (7.22) is a solution of (7.23).

Conversely, suppose now that $x_o \in C$ solves (7.23). For any $x \in C$ and $\epsilon \in (0, 1)$, define $z = x_o + \epsilon(x - x_o) \in C$, and apply (7.23) for the pair $(x_o, z) \in C \times C$. This yields,

$$a(x_o + \epsilon(x - x_o), \epsilon(x - x_o)) \geq \langle x^\star, \epsilon(x - x_o)\rangle, \quad \forall x \in C$$

which in the limit as $\epsilon \to 0^+$ leads to (7.22). □

Note that Minty's reformulation (7.23) differs from the original formulation of the variational inequality (7.14), in that while in the original problem (7.14) the operator A acts on the solution x_o of the inequality, in the reformulated version (7.23) it acts on any element $x \in C$. This is important, since in (7.23), we may choose such elements of C so that Ax has desirable properties, thus leading to convenient schemes for the study of qualitative properties of the solution of the variational inequality, or even numerical schemes.

The reformulation by Minty allows us to obtain a saddle point reformulation of variational inequalities (see e. g.,[136]).

Proposition 7.3.9. *Consider the map $L : X \times X \to \mathbb{R}$ defined by*

$$L(x_1, x_2) = a(x_1, x_1 - x_2) - \langle x^\star, x_1 - x_2 \rangle, \quad \forall\, x_1, x_2 \in X.$$

Then, $x_o \in C$ is a solution of the variational inequality

$$a(x_o, x - x_o) \geq \langle x^\star, x - x_o \rangle, \quad \forall\, x \in C, \tag{7.24}$$

if and only if (x_o, x_o) is a saddle point for L over $C \times C$ in the sense that

$$L(x_o, x) \leq L(x_o, x_o) \leq L(z, x_o), \quad \forall\, x, z \in C. \tag{7.25}$$

Proof. Suppose that $x_o \in C$ is a solution of (7.24). Then, setting $x_1 = x_o$ and $x_2 = x$, we see that (7.24) implies that $L(x_o, x) \leq 0$ for every $x \in C$. On the other hand, if $x_o \in C$ is a solution of (7.24), by Minty's trick, we also have that $a(x, x - x_o) \geq \langle x^\star, x - x_o \rangle$ for every $x \in C$, which, setting $x_1 = x$ and $x_2 = x_o$, leads to $L(x, x_o) \geq 0$ for every $x \in C$. Since $L(x_o, x_o) = 0$, (7.25) is immediate.

For the converse, assume that $0 = L(x_o, x_o) \leq L(x, x_o)$ for every $x \in C$. By the definition of L, this is equivalent to $0 \leq a(x, x - x_o) - \langle x^\star, x - x_o \rangle$ for every $x \in C$, which is Minty's form for (7.24). Similarly, if $L(x_o, x) \leq L(x_o, x_o) = 0$ for every $x \in C$, by the definition of L, we obtain (7.24). □

As can be seen from the proof of the Lions–Stampacchia theorem 7.3.5, linearity of the operator $A : X \to X^\star$ played a small role in the actual existence proof, which was based on the Banach contraction theorem. We may then extend the validity of this theorem to a class of nonlinear operators, satisfying a similar coercivity condition.

Assumption 7.3.10. The (possibly nonlinear) operator $A : X \to X^\star$ satisfies the following properties:

(i) A is Lipschitz, i. e., $\|A(x_1) - A(x_2)\|_{X^\star} \leq c\|x_1 - x_2\|_X$ for every $x_1, x_2 \in X$,

(ii) A is coercive (strongly monotone), i. e., there exists $c_o > 0$ such that

$$\langle A(x_1) - A(x_2), x_1 - x_2 \rangle \geq c_o \|x_1 - x_2\|^2, \quad \forall\, x_1, x_2 \in X.$$

We then have the following [94]:

Theorem 7.3.11 (Lions–Stampacchia II). *Let X be a Hilbert space and $C \subset X$ be a closed convex set. If the (nonlinear) operator $\mathsf{A} : X \to X^\star$ satisfies Assumption 7.3.10, then the variational inequality*

$$\text{find } x_o \in C : \langle \mathsf{A}(x_o), x - x_o \rangle \geq \langle x^\star, x - x_o \rangle, \quad \forall x \in C, \tag{7.26}$$

admits a unique solution for every $x^\star \in X^\star$. Moreover, the solution map $x^\star \mapsto x_o$ is Lipschitz with constant $\frac{1}{c_o}$.

Proof. The proof follows very closely that of Theorem 7.3.5. By exactly the same steps, we observe that a solution of (7.26) is a fixed point of the map $T_\rho : C \to C$ by $T_\rho(x) = P_C(\rho i_X^{-1}(x^\star - \mathsf{A}(x)) + x)$ for every $x \in C$, for some $\rho > 0$, while (7.17) and (7.18) are formally the same, so

$$\begin{aligned}\|T_\rho(x_1) - T_\rho(x_2)\|^2 &\leq \|x_1 - x_2\|^2 + \|\mathsf{A}(x_1) - \mathsf{A}(x_2)\|^2 - 2\rho\langle \mathsf{A}(x_1) - \mathsf{A}(x_2), x_1 - x_2\rangle \\ &\leq (1 + \rho^2 c^2 - 2\rho c_o)\|x_1 - x_2\|^2,\end{aligned}$$

where we used Lipschitz continuity for the second term and coercivity (strong monotonicity) for the third term. From the above estimate, we can find $\rho > 0$ such that T_ρ is a contraction, and the proof proceeds as in Theorem 7.3.5. □

Remark 7.3.12. We now revisit Remark 7.3.6 for the case of variational inequalities generated by nonlinear operators. The reformulation of the variational inequality (7.26) as the zero of an appropriate operator that was introduced in Remark 7.3.6 generalizes the nonlinear case as well in terms of the operator equation

$$0 \in \rho F(x_o) + N_C(x_o), \quad \rho > 0. \tag{7.27}$$

The monotonicity of the nonlinear operator $F = -i_X^{-1}(x^\star - \mathsf{A})$, inherited by the monotonicity of A, as well as the properties of the normal cone operator N_C, guarantee the solvability of the operator equation (7.27), turning the equivalent formulation (7.27) to a very powerful tool for both analytic and numerical considerations. We shall return to the properties of equation (7.27) in Chapter 9, where we introduce the concept and study the surjectivity properties of monotone operators.

Example 7.3.13. Minty's trick can be reformulated for equations. Consider the (possibly nonlinear) continuous operator $\mathsf{A} : X \to X^\star$, and the variational inequality $\langle \mathsf{A}(z) - x^\star, z - x\rangle \geq 0$ for every $z \in X$. Then, $\mathsf{A}(x) = x^\star$ in $X^\star$. Conversely, if A satisfies the monotonicity property $\langle \mathsf{A}(x_1) - \mathsf{A}(x_2), x_1 - x_2\rangle \geq 0$ for every $x_1, x_2 \in X$ and $\mathsf{A}(x) = x^\star$ in $X^\star$, then $\langle \mathsf{A}(z) - x^\star, z - x\rangle \geq 0$ for every $z \in X$.

To show the first direction set $z = z' + tx \in X$ for arbitrary $z' \in X$ and $t > 0$, substitute into the variational inequality and pass to the limit as $t \to 0^+$ using continuity to obtain $\langle \mathsf{A}(x), z'\rangle \geq 0$. Repeat the same with $-z'$ to obtain $\langle \mathsf{A}(x), z'\rangle \leq 0$ so that $\langle \mathsf{A}(x), z'\rangle = 0$ for every $z' \in X$, and the result follows. For the reverse direction, by monotonicity $\langle A(z) - A(x), z - x\rangle \geq 0$ for every $z \in X$, so since $A(x) = x^\star$, we get the variational inequality. $\triangleleft$

There are many examples where the abstract setting of the Lax–Milgram–Stampacchia theory can be applied. Its application to elliptic PDEs, variational inequalities, and free boundary value problems will be studied in Section 7.6.

7.4 Variational inequalities of the second kind

An interesting class of variational inequalities are variational inequalities of the form,

> Given a closed convex subset C of a Hilbert space $X = H$, a proper lower semicontinuous convex function $\varphi : C \to \mathbb{R} \cup \{+\infty\}$ and an element $x^\star \in X^\star = H^\star$,
>
> $$\text{find } x_o \in C : \mathfrak{a}(x_o, x - x_o) + \varphi(x) - \varphi(x_o) \geq \langle x^\star, x - x_o\rangle, \quad \forall x \in C.$$

Variational inequalities of this general type are called variational inequalities of the second kind and are related to quasivariational inequalities [24]. In the special case where $\varphi = 0$, we obtain the familiar case of a variational inequality,[6] treated in Section 7.3 in terms of the Lax–Milgram–Stampacchia theory. In the case where the bilinear form $\mathfrak{a}$ is symmetric, this variational inequality can be interpreted as the first-order condition for the minimization on C of a convex functional, therefore, the existence of a solution to the variational inequality can be obtained by the standard theory for convex optimization presented in Section 2.4. However, this problem is well posed for the general case where $\mathfrak{a}$ is a continuous and coercive bilinear form, not necessarily symmetric, by an appropriate extension of the Lax–Milgram–Stampacchia Theorem 7.3.5. This is displayed in the following theorem (see e. g., [24, 38, 136]):

Theorem 7.4.1. *Let X be a Hilbert space, $C \subset X$ closed and convex, $\mathfrak{a} : X \times X \to \mathbb{R}$ a continuous and coercive bilinear form, and $\varphi : C \to \mathbb{R} \cup \{+\infty\}$ a proper convex and lower semicontinuous function.*

(i) *For any $x^\star \in X^\star$, there exists a unique $x_o \in C$ such that*

$$\mathfrak{a}(x_o, x - x_o) + \varphi(x) - \varphi(x_o) \geq \langle x^\star, x - x_o\rangle, \quad \forall x \in C. \tag{7.28}$$

(ii) *If furthermore, $\mathfrak{a}$ is symmetric, then x_o is the minimizer in C of the convex functional $F : X \to \mathbb{R} \cup \{+\infty\}$ defined by*

$$F(x) = \frac{1}{2}\mathfrak{a}(x, x) + \varphi(x) - \langle x^\star, x\rangle, \quad \forall x \in X.$$

6 Sometimes called variational inequality of the first kind.

Proof. As $X^\star$ is not necessarily identified with X, the setting is exactly the same as in Section 7.3.

(i) Define the function $\varphi_C : X \to \mathbb{R} \cup \{+\infty\}$ by $\varphi_C(x) = \varphi(x)(1 + I_C(x))$, which is equal to φ for any $x \in C$ and equal to $+\infty$ for any $x \notin C$. It can be seen that the variational inequality (7.28) is equivalent to the following formulation:

$$\text{Find } x_o \in C \ : \ a(x_o, x - x_o) + \varphi_C(x) - \varphi_C(x_o) \geq \langle x^\star, x - x_o \rangle, \quad \forall x \in X,$$

where now we assume that x takes values over the whole Hilbert space X rather than simply on C (as is the original formulation of the problem). We further use the definition of the operator A to express the variational inequality as the following:

$$\text{Find } x_o \in C \ : \ \langle Ax_o, x - x_o \rangle + \varphi_C(x) - \varphi_C(x_o) \geq \langle x^\star, x - x_o \rangle, \quad \forall x \in X,$$

which, using the map $i_X : X \to X^\star$, subsequently denoted for simplicity as $j := i_X$, is further expressed as

$$\text{Find } x_o \in C \ : \ \langle j^{-1}(Ax_o - x^\star), x - x_o \rangle_X + \varphi_C(x) - \varphi_C(x_o) \geq 0, \quad \forall x \in X.$$

We fix a $\rho > 0$ (to be determined shortly), multiply the variational inequality with ρ, and observing the trivial fact that $\rho j^{-1}(Ax_o - x^\star) = x_o - (x_o - \rho j^{-1}(Ax_o - x^\star))$, we obtain the equivalent form for the variational inequality as

$$\langle x_o - (x_o - \rho j^{-1}(Ax_o - x^\star)), x - x_o \rangle_X + \rho\varphi_C(x) - \rho\varphi_C(x_o) \geq 0, \quad \forall x \in X. \tag{7.29}$$

The function $\rho\varphi_C : X \to \mathbb{R} \cup \{+\infty\}$ is convex, proper, and lower semicontinuous, so the proximity operator $\text{prox}_{\rho\varphi_C}$ is well defined. Recall (see Proposition 4.4.2(i), and in particular inequality (4.31)[7]), that if $z_0 = \text{prox}_{\rho\varphi_C} x$, then, z_0 satisfies the variational inequality

$$\langle z_0 - x, z - z_0 \rangle_X + \rho\varphi_C(z) - \rho\varphi_C(z_0) \geq 0, \quad \forall z \in X. \tag{7.30}$$

Comparing (7.30) with (7.29) (and in particular setting $z \to x$, $z_0 \to x_o + \rho j^{-1}(Ax_o - x^\star)$ in (7.30)), we see that the solution of the variational inequality should be such that

$$x_o = \text{prox}_{\rho\varphi_C}(x_o - \rho j^{-1}(Ax_o - x^\star)). \tag{7.31}$$

Note that if this is true, then we automatically have that $x_o \in C$ by the definition of the function[8] φ_C. It thus suffices to show that (7.31) holds for some $\rho > 0$.

---

7 Appropriately modified for the case where $X^\star$ is not identified with X.

8 Which is infinite if its argument is not in C and by the fact that $\text{prox}_{\rho\varphi_C}$ minimizes a perturbation of this function.

We therefore consider the family of maps $T_\rho : C \to C$ defined by

$$T_\rho(x) := \text{prox}_{\rho\varphi_C}(x - \rho j^{-1}(Ax - x^\star)), \quad \text{for every } x \in C.$$

We will show that there exists $\rho > 0$ such that T_ρ has a fixed point by using Banach's contraction theorem. So, by (7.31), this fixed point is the solution we seek. To show that T_ρ is a strict contraction for the proper choice of $\rho > 0$, we need to use the fact that the proximity operator $\text{prox}_{\rho\varphi_C}$ associated with the convex proper and lower semicontinuous function $\rho\varphi_C$ is a nonexpansive operator (see Proposition 4.4.2(ii)). From this point onwards, the proof follows almost verbatim the steps of the proof of Theorem 7.3.5 by simply replacing the projection operator P_C by the proximity operator $\text{prox}_{\rho\varphi_C}$.

(ii) Suppose that a is symmetric and $x_o \in C$ satisfies (7.28). Then, for any $x \in C$, we see that

$$\begin{aligned} F(x) - F(x_o) &= \frac{1}{2}a(x,x) - \frac{1}{2}a(x_o,x_o) + \varphi(x) - \varphi(x_o) - \langle x^\star, x - x_o\rangle \\ &= a(x_o, x - x_o) + \frac{1}{2}a(x_o - x, x_o - x) + +\varphi(x) - \varphi(x_o) - \langle x^\star, x - x_o\rangle \geq 0, \end{aligned}$$

where we have explicitly used the symmetry property for a, the fact that $a(x_o-x, x_o-x) \geq 0$ (by the coercivity), and the fact that $x_o \in C$ satisfies (7.28). Therefore, $F(x) \geq F(x_o)$ for every $x \in C$, hence, x_o is the minimizer of F in C.

For the converse, assume that a is symmetric, and $x_o \in C$ is the minimizer of F on C. By the convexity of C for any $x \in C$, and any $t \in (0,1)$, we have that $z = (1-t)x_o + tx \in C$ so that since x_o is the minimizer of F, it holds that $F(x_o) \leq F(z)$. Using the definition of F and the fact that a is bilinear and symmetric yields

$$\begin{aligned} &\frac{1}{2}t^2 a(x_o - x, x_o - x) + t\, a(x_o, x - x_o) + \varphi((1-t)x_o + tx) - \varphi(x_o) \\ &\quad \geq t\,\langle x^\star, x - x_o\rangle. \end{aligned} \tag{7.32}$$

By the convexity of φ, we have that

$$\varphi((1-t)x_o + tx) - \varphi(x_o) \leq t\varphi(x) - t\varphi(x_o),$$

which when combined with (7.32) yields the inequality

$$\frac{1}{2}t^2 a(x_o - x, x_o - x) + t\, a(x_o, x - x_o) + t(\varphi(x) - \varphi(x_o)) \geq t\,\langle x^\star, x - x_o\rangle.$$

Dividing by t and passing to the limit as $t \to 0^+$, we obtain (7.28). □

A Minty type equivalent reformulation (similar to the one presented in Theorem 7.3.8 for variational inequalities of the first type) is also valid for variational inequalities of the second type.

Theorem 7.4.2 (Minty). *The variational inequality* (7.28) *is equivalent to solving for* $x_o \in C$ *such that*

$$a(x, x - x_o) + \varphi(x) - \varphi(x_o) \geq \langle x^\star, x - x_o \rangle, \quad \forall x \in C,$$

or in terms of the operator A *as*

$$\langle Ax, x - x_o \rangle + \varphi(x) - \varphi(x_o) \geq \langle x^\star, x - x_o \rangle, \quad \forall x \in C.$$

Proof. The proof follows closely the proof of Theorem 7.3.8. For the converse, we need to take into account the convexity of the function φ. The details are left to the reader. □

7.5 Approximation methods and numerical techniques

7.5.1 The penalization method

The penalization method [24, 136] is a useful method when dealing with variational inequalities. The essence of the method is trying to introduce the constraints involved in the variational inequality into the operator defining the variational inequality, and then reduce the problem to one on the whole space X rather on the constraint set C.

To fix ideas, let X be a Hilbert space, $C \subset X$ a nonempty closed convex set, $a : X \times X \to \mathbb{R}$ a bilinear continuous and coercive form, and consider the variational inequality

$$x_o \in C : a(x_o, x - x_o) \geq \langle x^\star, x - x_o \rangle, \quad \forall x \in C. \tag{7.33}$$

We next introduce a lower semicontinuous convex function $\varphi : X \to [0, \infty]$ with the property that $\varphi(x) = 0$ if and only if $x \in C$. An example of such a function is the indicator function of C, i. e., $\varphi_o = I_C$. However, note that while this can be used for theoretical arguments of existence, for practical purposes (e. g., numerical treatment) it may not be such a good choice. Using a family of lower semicontinuous convex real valued positive functions $(\varphi_\epsilon)_{\epsilon>0}$ such that $\varphi_\epsilon \to \varphi_o$ for $\epsilon \to 0^+$, and choosing $\varphi_\epsilon = \frac{1}{\epsilon}\varphi$, we may define a family of variational inequalities of the form

$$x_{o,\epsilon} \in X : a(x_{o,\epsilon}, x - x_{o,\epsilon}) + \frac{1}{\epsilon}\varphi(x) - \frac{1}{\epsilon}\varphi(x_{o,\epsilon}) \geq \langle x^\star, x - x_{o,\epsilon} \rangle, \quad \forall x \in X. \tag{7.34}$$

An alternative formulation (choosing β so that the inequality $\phi(x) - \phi(z) \geq \langle \beta(z), z - x \rangle$ holds for all $x, z \in X$) is

$$x_{o,\epsilon} \in X : a(x_{o,\epsilon}, x) + \frac{1}{\epsilon}\langle \beta(x_{o,\epsilon}), x \rangle = \langle x^\star, x \rangle, \quad \forall x \in X, \tag{7.35}$$

for a suitable penalty function β, which must satisfy a monotonicity property. The family of variational inequalities (7.34) or equations (7.35) is called the penalized version of (7.33), and the function φ is called the penalty function. Note the important difference between (7.33) and (7.34) (or (7.35)): the former is solved over the closed convex subset C (modeling the constraints), whereas the latter are solved over the whole space X, turning them into equations rather than inequalities. The solution is however effectively constrained in C, through the introduction of the penalty terms in (7.34) (or (7.35)). The unconstrained problem is definitely easier to handle than the original constrained version. The important contribution of the penalization method is the fact that it can be shown that, as $\epsilon \to 0$, the family of solutions $(\mathsf{x}_{o,\epsilon})_{\epsilon>0}$, of (7.34) (or (7.35)), has a limit, which in fact is the solution x_o of the original problem (7.33). The penalty method may be used for theoretical purposes (e. g., showing existence or regularity of solutions) [24] or for numerical approximation [136]. Before providing the convergence result for the penalty method, we provide some examples. It is worth noting that the choice of penalty function may not be unique, so one has the liberty of choosing the particular penalty function that better suits the problem at hand.

Example 7.5.1. The choice of penalty terms is not unique. A possible penalty term may be $\frac{1}{\epsilon}\beta(\mathsf{x}) = \frac{1}{\epsilon}(\mathsf{x} - P_C(\mathsf{x}))$, where P_C is the projection operator to the closed convex set $C \subset X$. Note the monotonicity property of the penalty term, i. e., that $\langle \beta(\mathsf{x}_1) - \beta(\mathsf{x}_2), \mathsf{x}_1 - \mathsf{x}_2 \rangle \geq 0$ for every $\mathsf{x}_1, \mathsf{x}_2 \in X$, which in this example follows by the properties of the projection operator. ◁

Example 7.5.2. Let $\mathcal{D} \subset \mathbb{R}^d$ a bounded and smooth domain, $X = W_0^{1,2}(\mathcal{D})$, and identify elements of X with functions $u : \mathcal{D} \to \mathbb{R}$, and elements of $X^\star = W^{-1,2}(\mathcal{D})$ with functions $f : \mathcal{D} \to \mathbb{R}$. Consider the bilinear form $\mathfrak{a}$ related to the elliptic operator $\mathsf{A} = -\operatorname{div}(\mathbf{A}\nabla) + a \cdot \nabla + a_0 I$, where $\mathbf{A} = (a_{ij})_{i,j=1,\dots,d}$, $a = (a_1, \dots, a_d)$ with $a_{ij}, a_i, a_0 \in L^\infty(\mathcal{D})$, $i, j = 1, \dots, d$ given functions, and the closed convex sets $C_1 = \{u \in W_0^{1,2}(\mathcal{D}) : u \geq \psi \text{ a. e.}\}$ and $C_2 = \{u \in W_0^{1,2}(\mathcal{D}) : u \leq \psi \text{ a. e.}\}$ for given obstacle ψ. Then, one may check by direct substitution in the inequality defining the projection that $P_{C_1}(u) = \sup\{u, \psi\}$, hence, $u - P_{C_1}(u) = -(\psi - u)^+$, whereas $P_{C_2}(u) = \inf\{u, \psi\}$, hence, $u - P_{C_2}(u) = (u - \psi)^+$.

This leads to the possible penalization schemes for the variational inequalities

$$
\begin{aligned}
&\text{find } u \in C_1 : \langle \mathsf{A}u, v - u \rangle \geq \langle f, v - u \rangle, \quad \forall\, v \in C_1, \\
&\langle \mathsf{A}u_\epsilon, v \rangle - \frac{1}{\epsilon}\langle (\psi - u_\epsilon)^+, v \rangle = \langle f, v \rangle, \quad \forall v \in X, \quad (P_\epsilon)
\end{aligned}
\tag{7.36}
$$

and

$$
\begin{aligned}
&\text{find } u \in C_2 : \langle \mathsf{A}u, v - u \rangle \geq \langle f, v - u \rangle, \quad \forall\, v \in C_2, \\
&\langle \mathsf{A}u_\epsilon, v \rangle + \frac{1}{\epsilon}\langle (u_\epsilon - \psi)^+, v \rangle = \langle f, v \rangle, \quad \forall v \in X, \quad (P'_\epsilon)
\end{aligned}
\tag{7.37}
$$

in the limit $\epsilon \to 0^+$. Indeed, one may informally see that in this limit (P_ϵ) will make sense if $(\psi - u)^+ = 0$ so that $u \geq \psi$, while (P'_ϵ) will make sense if $(u - \psi)^+ = 0$ so that $u \leq \psi$. Smooth approximations of the above functions are also possible. ◁

The following proposition (see, e. g., [24] or [136]) proves the convergence of such approximation schemes:

Proposition 7.5.3. *Assume that the bilinear form* $\mathfrak{a} : X \times X \to R$ *is continuous and coercive, and that the penalty function* $\varphi : X \to [0, \infty]$ *satisfies* $\varphi(\mathrm{x}) = 0$ *if and only if* $\mathrm{x} \in C$. *Let* $\mathrm{x}_{o,\epsilon}$ *be the solution of* (7.34) *and* x_o *the solution of* (7.33). *Then,* $\lim_{\epsilon\to 0} \mathrm{x}_{o,\epsilon} = \mathrm{x}_o$ *in* X *and* $\lim_{\epsilon\to 0} \frac{1}{\epsilon}\varphi(\mathrm{x}_{o,\epsilon}) = 0$. *Similarly for scheme* (7.35) *for* β *monotone vanishing only on* C.

Proof. We work in terms of (7.34), the case of (7.35) being similar. For each $\epsilon > 0$ fixed, the penalized problem (7.34) is a variational inequality of the second kind, and the existence of a unique solution $\mathrm{x}_{o,\epsilon}$ is guaranteed, e. g., by Theorem 7.4.1, considering the case $C = X$. Alternatively, one can show existence directly by using finite dimensional approximations (see Example 7.5.4 for (7.35)). In order to pass to the limit, as $\epsilon \to 0^+$, we need some a priori estimates on the solution. Since for any $\mathrm{x} \in X$, it holds that $\frac{1}{\epsilon}\varphi(\mathrm{x}) \geq 0$ with equality if and only if $\mathrm{x} \in C$, choosing $\mathrm{x} \in C$ as test element in (7.34), the contribution of $\frac{1}{\epsilon}\varphi(\mathrm{x})$ vanishes, and we obtain upon rearrangement

$$\mathfrak{a}(\mathrm{x}_{o,\epsilon}, \mathrm{x}_{o,\epsilon}) + \frac{1}{\epsilon}\varphi(\mathrm{x}_{o,\epsilon}) \leq \mathfrak{a}(\mathrm{x}_{o,\epsilon}, \mathrm{x}) - \langle \mathrm{x}^\star, \mathrm{x} - \mathrm{x}_{o,\epsilon} \rangle, \quad \forall \mathrm{x} \in C. \tag{7.38}$$

Since $\frac{1}{\epsilon}\varphi(\mathrm{x}_{o,\epsilon}) \geq 0$ and $\mathfrak{a}$ is coercive, we estimate the left-hand side of (7.38) as

$$c_o \|\mathrm{x}_{o,\epsilon}\|^2 \leq \mathfrak{a}(\mathrm{x}_{o,\epsilon}, \mathrm{x}_{o,\epsilon}) \leq \mathfrak{a}(\mathrm{x}_{o,\epsilon}, \mathrm{x}_{o,\epsilon}) + \frac{1}{\epsilon}\varphi(\mathrm{x}_{o,\epsilon}).$$

On the other hand, by continuity, the right-hand side of (7.38) is estimated by

$$\begin{aligned}\mathfrak{a}(\mathrm{x}_{o,\epsilon}, \mathrm{x}) - \langle \mathrm{x}^\star, \mathrm{x} - \mathrm{x}_{o,\epsilon} \rangle &\leq |\mathfrak{a}(\mathrm{x}_{o,\epsilon}, \mathrm{x}) - \langle \mathrm{x}^\star, \mathrm{x} - \mathrm{x}_{o,\epsilon} \rangle| \\ &\leq c\|\mathrm{x}_{o,\epsilon}\| \, \|\mathrm{x}\| + \|\mathrm{x}^\star\|_{X^\star} \|\mathrm{x}\| \leq c_1\|\mathrm{x}_{o,\epsilon}\| + c_2,\end{aligned}$$

for some finite constants c_1, c_2 related to $\|\mathrm{x}\|$ and $\|\mathrm{x}^\star\|_{X^\star}$, but independent of ϵ. Combining the above estimates with (7.38), we obtain the a priori estimate

$$c_o \|\mathrm{x}_{o,\epsilon}\|^2 \leq c_1\|\mathrm{x}_{o,\epsilon}\| + c_2, \quad \forall \epsilon > 0,$$

which leads to a uniform bound $\|\mathrm{x}_{o,\epsilon}\| \leq c$ for all $\epsilon > 0$. That implies the existence of a subsequence of $(\mathrm{x}_{o,\epsilon})_{\epsilon>0}$ (denoted the same for simplicity) and $\bar{\mathrm{x}} \in X$ such that $\mathrm{x}_{o,\epsilon} \rightharpoonup \bar{\mathrm{x}}$ as $\epsilon \to 0^+$. We will show that this convergence is in fact strong, but before doing so, we first need to consider the behavior of the penalty term in the limit. For that, consider (7.38) again, but now estimate the left-hand side, since $\mathfrak{a}(\mathrm{x}_{o,\epsilon}, \mathrm{x}_{o,\epsilon}) \geq 0$, as

$$\frac{1}{\epsilon}\varphi(\mathrm{x}_{o,\epsilon}) \leq \mathfrak{a}(\mathrm{x}_{o,\epsilon}, \mathrm{x}_{o,\epsilon}) + \frac{1}{\epsilon}\varphi(\mathrm{x}_{o,\epsilon}),$$

and by a similar estimation of the right-hand side, we obtain

$$\frac{1}{\epsilon}\varphi(x_{o,\epsilon}) \le c_1\|x_{o,\epsilon}\| + c_2 \le c_3, \quad \forall\, \epsilon > 0,$$

where c_3 is independent of ϵ. For the last inequality, we used the fact that $\|x_{o,\epsilon}\|$ is uniformly bounded (which was our previous estimate). This implies upon multiplying with ϵ that

$$\varphi(x_{o,\epsilon}) \le c_3\epsilon, \quad \forall\, \epsilon > 0,$$

and since φ is a positive function, passing to the limit (along the chosen subsequence) as $\epsilon \to 0^+$ and using the weak lower semicontinuity of φ, we conclude that $\varphi(\bar{x}) = 0$, therefore $\bar{x} \in C$.

By the positivity of φ, (7.38) implies that

$$a(x_{o,\epsilon}, x_{o,\epsilon}) \le a(x_{o,\epsilon}, x) - \langle x^\star, x - x_{o,\epsilon}\rangle, \quad \forall\, x \in C,\ \forall\, \epsilon > 0,$$

so taking the limit superior as $\epsilon \to 0^+$ along the chosen subsequence and using the fact that $x_{o,\epsilon} \rightharpoonup \bar{x}$, we see that

$$\limsup_{\epsilon\to 0^+} a(x_{o,\epsilon}, x_{o,\epsilon}) \le a(\bar{x}, x) - \langle x^\star, x - \bar{x}\rangle, \quad \forall\, x \in C.$$

Note that the weak convergence for $x_{o,\epsilon}$ does not guarantee the existence of limit for $a(x_{o,\epsilon}, x_{o,\epsilon})$, and this is the need of taking the limit superior.

By the coercivity of a, it holds that $a(x_{o,\epsilon} - \bar{x}, x_{o,\epsilon} - \bar{x}) \ge 0$ for any $\epsilon > 0$, which when rearranged, using the fact that a is bilinear, yields

$$a(x_{o,\epsilon}, \bar{x}) + a(\bar{x}, x_{o,\epsilon}) - a(\bar{x}, \bar{x}) \le a(x_{o,\epsilon}, x_{o,\epsilon}), \quad \forall\, \epsilon > 0,$$

which upon taking the limit inferior as $\epsilon \to 0^+$ along the chosen subsequence and using the fact that $x_{o,\epsilon} \rightharpoonup \bar{x}$ to handle the linear terms leads to

$$a(\bar{x}, \bar{x}) + a(\bar{x}, \bar{x}) - a(\bar{x}, \bar{x}) = a(\bar{x}, \bar{x}) \le \liminf_{\epsilon\to 0^+} a(x_{o,\epsilon}, x_{o,\epsilon}).$$

Since $\liminf_{\epsilon\to 0^+} a(x_{o,\epsilon}, x_{o,\epsilon}) \le \limsup_{\epsilon\to 0^+} a(x_{o,\epsilon}, x_{o,\epsilon})$ combining the above inequalities leads to the conclusion that

$$a(\bar{x}, \bar{x}) \le a(\bar{x}, x) - \langle x^\star, x - \bar{x}\rangle, \quad \forall\, x \in C,$$

therefore, $\bar{x} \in C$ is a solution of (7.33), and by the uniqueness of solutions to problem (7.33), we conclude that $\bar{x} = x_o$. Therefore, we have obtained that $x_{o,\epsilon} \rightharpoonup x_o$ in X. By the standard argument (based on the Urysohn property, see Remark 1.1.52), we have convergence for the whole sequence and not just for the chosen subsequence.

It remains to show that the convergence is strong. To this end, note that by the positivity of φ and the coercivity of a, it holds that

$$0 \le c_o \|x_{o,\epsilon} - x_o\|^2 + \frac{1}{\epsilon}\varphi(x_{o,\epsilon}) \le a(x_{o,\epsilon} - x_o, x_{o,\epsilon} - x_o) + \frac{1}{\epsilon}\varphi(x_{o,\epsilon}).$$

By the fact that a is bilinear,

$$a(x_{o,\epsilon} - x_o, x_{o,\epsilon} - x_o) = a(x_{o,\epsilon}, x_{o,\epsilon}) - a(x_o, x_{o,\epsilon}) - a(x_{o,\epsilon}, x_o) + a(x_o, x_o),$$

and since $x_{o,\epsilon}$ solves the penalized problem (7.38), setting $x = x_o$ as test element, adding the terms $-a(x_o, x_{o,\epsilon})$ and $a(x_o, x_o)$ we obtain that

$$\begin{aligned} &a(x_{o,\epsilon}, x_{o,\epsilon}) - a(x_o, x_{o,\epsilon}) - a(x_{o,\epsilon}, x_o) + a(x_o, x_o) + \frac{1}{\epsilon}\varphi(x_{o,\epsilon}) \\ &\quad \le -\langle x^\star, x_o - x_{o,\epsilon}\rangle - a(x_o, x_{o,\epsilon}) + a(x_o, x_o). \end{aligned}$$

Combining the above,

$$\begin{aligned} 0 &\le c_o \|x_{o,\epsilon} - x_o\|^2 + \frac{1}{\epsilon}\varphi(x_{o,\epsilon}) \\ &\le -\langle x^\star, x_o - x_{o,\epsilon}\rangle - a(x_o, x_{o,\epsilon}) + a(x_o, x_o). \quad \forall \epsilon > 0, \end{aligned}$$

and as above, passing first to the limit inferior, and then to the limit superior as $\epsilon \to 0^+$, and keeping in mind that $x_{o,\epsilon} \rightharpoonup x_o$, we see that

$$0 \le \liminf_{\epsilon \to 0^+}\left(c_o \|x_{o,\epsilon} - x_o\|^2 + \frac{1}{\epsilon}\varphi(x_{o,\epsilon})\right) \le \limsup_{\epsilon \to 0^+}\left(c_o \|x_{o,\epsilon} - x_o\|^2 + \frac{1}{\epsilon}\varphi(x_{o,\epsilon})\right) \le 0$$

so that

$$\lim_{\epsilon \to 0^+}\left(c_o \|x_{o,\epsilon} - x_o\|^2 + \frac{1}{\epsilon}\varphi(x_{o,\epsilon})\right) = 0,$$

and since $\frac{1}{\epsilon}\varphi(x_{o,\epsilon}) \ge 0$ (note that in face this term could be omitted earlier on because of positivity), we conclude that $\|x_{o,\epsilon} - x_o\| \to 0$, therefore $x_{o,\epsilon} \to x$. □

Example 7.5.4 (An existence proof based on penalization). As already mentioned, penalization may be used to provide an alternative existence proof for the variational inequality (7.33) (see, e. g., [24]). To this end, instead of using e. g., Theorem 7.4.1 to show the existence for the penalized problem (7.34) or (7.35), we may do so directly by using a finite dimensional approximation instead, whose solvability is guaranteed by the Brouwer fixed point theorem, and passing to the limit as the dimensions go to infinity (Galerkin method). For concreteness, let us consider (7.35), and to be more precise, the situation encountered in Example 7.5.2 and in particular (7.36), following [24]. Consider a basis $(e_m)_{m\in\mathbb{N}}$ for X, let $X_m = \text{span}\{e_k,\ k = 1, \dots, m\}$, and consider the approximate (finite dimensional) problems

$$a(u_m, v_m) + \frac{1}{\epsilon}\langle \beta(u_m), v_m\rangle = \langle f, v_m\rangle, \quad \forall\, v_m \in X_m, \tag{7.39}$$

with the explicit dependence of u_m on ϵ, omitted for easing the notation. Then by an application of the Brouwer fixed point theorem (and in particular Proposition 3.2.13) problem (7.39) admits, for every $\epsilon > 0$ fixed, a unique $u_m \in X_m$, for any $m \in \mathbb{N}$. The family of solutions $(u_m)_{m\in\mathbb{N}}$ is uniformly bounded in $m \in \mathbb{N}$, as can be shown by using as test function $v_m = u_m - v_o$ for any $v_o \in C$ in (7.39), where by noting that $\beta(v_o) = 0$, (7.39) yields

$$\mathfrak{a}(u_m, u_m - v_o) + \frac{1}{\epsilon}\langle \beta(u_m) - \beta(v_o), u_m - v_o\rangle = \langle f, u_m - v_o\rangle, \tag{7.40}$$

which by the monotonicity of the penalty term β provides the inequality

$$\mathfrak{a}(u_m - v_o, u_m - v_o) \le \langle f, u_m - v_o\rangle - \mathfrak{a}(v_o, u_m - v_o),$$

where, by the coercivity and continuity of $\mathfrak{a}$, we get that $\alpha\|u_m - v_o\|^2 \le (\|f\|_{X^*} + c\|v_o\|)\|u_m - v_o\|$, from which we have that $\|u_m - v_o\| \le c$, and therefore $\|u_m\| \le c'$ for some c, c' independent of m. Then, by reflexivity, we may extract a subsequence (denoted the same) and $u_\epsilon \in X$ such that $u_m \rightharpoonup u_\epsilon$ in $X = W_0^{1,2}(\mathcal{D})$, as $m \to \infty$, and by the compact embedding of $W_0^{1,2}(\mathcal{D})$ in $L^2(\mathcal{D})$, a further subsequence (still denoting the same) such that $u_m \to u_\epsilon$ in $L^2(\mathcal{D})$. Passing to the limit, as $m \to \infty$, we may show that x_ϵ satisfies the penalized form of the equation. To pass to the limit, we first note that, by (7.40), and since we have already established the fact that x_m is bounded, we see that $\langle \frac{1}{\epsilon}\beta(u_m), u_m - v_o\rangle \le c$ for some c independent of m. Recalling the special form of $\beta(u) = -(\psi - u)^+$, we write

$$\begin{aligned}\|(\psi - u_m)^+\|^2_{L^2(\mathcal{D})} &= \langle (\psi - u_m)^+, (\psi - u_m)\rangle = \langle -(\psi - u_m)^+, (u_m - \psi)\rangle = \langle \beta(u_m), (u_m - \psi)\rangle \\ &= \langle \beta(u_m), (u_m - v_0)\rangle + \langle \beta(u_m), (v_o - \psi)\rangle \\ &\le \langle \beta(u_m), u_m - v_o\rangle \le c',\end{aligned}$$

as $\langle \beta(u_m), v_o - \psi\rangle = -\langle (\psi - u)^+, v_o - \psi\rangle \le 0$, since $v_o \ge \psi$. This implies that $\frac{1}{\epsilon}\|\beta(u_m)\| < c''$ with c'' independent of m (but not ϵ), so again, by reflexivity, there exists $\bar{\beta} \in L^2(\mathcal{D})$ such that $\beta(u_m) \rightharpoonup \bar{\beta}$ as $m \to \infty$. With this information at hand, we now take an arbitrary $v \in X$ and consider a $(v_m)_{m\in\mathbb{N}}$, with $v_m \in X_m$ such that $v_m \to v$ in X. Then for any $m \in \mathbb{N}$, we take v_m as test function in (7.39) so that taking $m \to \infty$, for $\epsilon > 0$ fixed, we have

$$\mathfrak{a}(u_\epsilon, v) + \frac{1}{\epsilon}\langle \bar{\beta}, v\rangle = \langle f, v\rangle,$$

so if we can show that $\bar{\beta} = \beta(u_\epsilon)$, we are done. But this can be seen by passing to the subsequence for which $u_m \to u_\epsilon$ in $L^2(\mathcal{D})$. For more general types of penalty functions, one could obtain this result by monotonicity arguments, which will be discussed in detail in Chapter 9. In fact, one could obtain an existence result for a general penalized equation in terms of finite dimensional approximations using the arguments leading to the celebrated Browder–Minty surjectivity theorem for monotone operators (see Theorem 9.3.8). ◁

7.5.2 Internal approximation schemes

In this section, we consider the approximation of variational inequalities of the first kind of the type

$$\text{find } x_o \in C : \langle Ax_o, x - x_o\rangle \geq \langle x^\star x - x_o\rangle, \quad \forall x \in C, \tag{7.41}$$

where $C \subset X$ is a closed convex set, $A : X \to X^\star$ is a linear operator associated with a continuous and coercive bilinear form $a : X \times X \to \mathbb{R}$, and $x^\star \in X^*$ is given. We assume for simplicity that X is a Hilbert space. By the standard Lax–Milgram–Stampacchia theory, this admits a unique solution $x_o \in C$.

A key element in the approximation scheme is a family of closed subspaces $\{X_h : h > 0\}$, $X_h \subset X$, and closed convex sets $\{C_h : h > 0\}$, $C_h \subset X_h$, parameterized with $h > 0$, where h is some parameter converging to 0. Usually these approximations are chosen to be finite dimensional, for instance in a Hilbert space with an orthonormal basis, we may choose as this approximation the sequence of finite dimensional projections to the basis (Fourier expansion). We then replace the variational inequality (7.41) with its approximate version for any[9] $h > 0$,

$$\text{find } x_h \in C_h : \langle Ax_h, x - x_h\rangle \geq \langle x^\star, x - x_h\rangle, \quad \forall x \in C_h. \tag{7.42}$$

By the standard Lax–Milgram–Stampacchia theory, the approximation (7.42) admits a unique solution x_h for every $h > 0$. We can then construct the family of solutions $(x_h)_{h>0}$ (also seen as a sequence by selecting $h = \frac{1}{n}$). Our aim here is, assuming that we have a solution for (7.42), to show that under certain conditions, $x_h \to x_o$ as $h \to 0$.

We need the following assumption on the family of approximations (see [136]):

Assumption 7.5.5 (The family $\{C_h : h > 0\}$). The family of approximations $\{C_h : h > 0\}$ is such that

(i) For any $(x_h)_{h>0} \subset X$, with $x_h \in C_h$ for every $h > 0$, if $x_h \rightharpoonup x$, then[10] $x \in C$.
(ii) For every $z \in C$, there exists $\hat{z}_h \in C_h$ such that $\hat{z}_h \to z$ (strongly) in X, as $h \to 0$.

The following proposition (see [136]) shows convergence of the approximation scheme (7.42) to solutions of the variational inequality (7.41):

Proposition 7.5.6. *Let $a : X \times X \to \mathbb{R}$ be a continuous and coercive bilinear form and $x^\star \in X^*$ given. Consider also a family of approximations $\{C_h : h > 0\}$ for the closed*

9 Note that $x^\star \in X^\star$ can be considered as generating a linear form $\mathfrak{L}$ by $\mathfrak{L}(x) = \langle x^\star, x\rangle$ for any $x \in X$. In general, one may also approximate the linear form as well by finite dimensional approximation, for example, use $x_h^\star$ in the duals of finite dimensional spaces so that $x_h^\star$ converges in some appropriate sense to $x^\star$. For simplicity we omit this procedure here.

10 This always holds if $C_h \subset C$ for every $h > 0$.

convex set $C \subset X$ that satisfies Assumption 7.5.5. Then, for the solution x_h of (7.42), we have that $x_h \to x_o$, as $h \to 0$.

Proof. The solution of (7.42) satisfies the uniform bound $\|x_h\| < c'$, $h > 0$, for some c' independent of h. Indeed, using the bilinear form version of (7.42), we have that $a(x_h, x_h) \leq a(x_h, x) - \langle x^\star, x - x_h \rangle$ for every $x \in C_h$, which by the coercivity and continuity of a yields the estimate

$$c_o\|x_h\|^2 \leq c\|x\|\,\|x_h\| + c_1(\|x\| + \|x_h\|), \quad \forall x \in C_h.$$

Choose an arbitrary $z \in C$, and consider the approximating sequence $\hat{z}_h \in C_h$ (which exists by Assumption 7.5.5(ii)) as test function $x = \hat{z}_h$. Since $\hat{z}_h \to z$, we have that $\hat{z}_h$ is uniformly bounded, i. e., $\|\hat{z}_h\| < c_1$ for some constant independent of $h > 0$, hence, $c_o\|x_h\|^2 \leq c_3\|x\| + c_4$ for appropriate constants c_3, c_4 independent of $h > 0$, and this implies that $\|x_h\| < c'$ for every $h > 0$.

By the uniform boundedness of $(x_h)_{h>0}$ and reflexivity, there exists a subsequence $(x_{h_n})_{n\in\mathbb{N}}$, $h_n \to 0$, and a $\bar{x} \in X$ such that $x_{h_n} \rightharpoonup \bar{x}$. By Assumption 7.5.5(i), we have that $\bar{x} \in C$. We will show that in fact, $\bar{x} = x_o$ is the solution of (7.41). Moving along the subsequence $(h_n)_{n\in\mathbb{N}}$ in (7.42) and adopting the simpler notation $x_n := x_{h_n}$ and $C_n := C_{h_n}$, we have that for every $n \in \mathbb{N}$,

$$a(x_n, z - x_n) \geq \langle x^\star, z - x_n \rangle, \quad \forall z \in C_n. \tag{7.43}$$

Choose an arbitrary $z \in C$, and consider the approximating sequence $\hat{z}_h \in C_h$, $\hat{z}_h \to z$ (which exists by Assumption 7.5.5(ii)), pass to the subsequence $(h_n)_{n\in\mathbb{N}}$, adopt the notation $z_n := \hat{z}_{h_n}$, and then for every n, use as test function $z = z_n \in C_n$ in (7.43) to obtain

$$a(x_n, z_n - x_n) \geq \langle x^\star, z_n - x_n \rangle, \quad n \in \mathbb{N}. \tag{7.44}$$

We rearrange (7.44) as

$$a(x_n, x_n) \leq a(x_n, z_n) - \langle x^\star, z_n - x_n \rangle, \quad \forall n \in \mathbb{N}, \tag{7.45}$$

and taking the limit inferior on both sides of the above inequality, keeping in mind that $\langle x^\star, z_n - x_n \rangle \to \langle x^\star, z - \bar{x} \rangle$ and $a(x_n, z_n) \to a(\bar{x}, z)$, since $x_n \rightharpoonup \bar{x}$ and $z_n \to z$, we have

$$\liminf_{n\to\infty} a(x_n, x_n) \leq a(\bar{x}, z) + \langle x^\star, z - \bar{x} \rangle. \tag{7.46}$$

We claim that

$$a(\bar{x}, \bar{x}) \leq \liminf_{n\to\infty} a(x_n, x_n). \tag{7.47}$$

Assuming the claim to hold for the time being and combining (7.47) with (7.46) yields $a(\bar{x}, \bar{x}) \leq a(\bar{x}, z) + \langle x^\star, z - \bar{x} \rangle$, which upon rearrangement becomes

$$a(\bar{x}, z - \bar{x}) \geq \langle x^*, z - \bar{x} \rangle,$$

and since $z \in C$ was arbitrary, we conclude (by the uniqueness of the solution of (7.41)) that $\bar{x} = x_o$. Performing the same steps for any subsequence of $\{x_h : h > 0\}$ and using the Urysohn property (Remark 1.1.52), we conclude that the whole sequence is $x_h \rightharpoonup x_o$.

To complete weak convergence, it remains to prove claim (7.47). By coercivity

$$a(x_n - \bar{x}, x_n - \bar{x}) \geq c_o \|x_n - \bar{x}\|^2 \geq 0$$

for every $n \in \mathbb{N}$, so upon rearrangement

$$a(x_n, \bar{x}) + a(\bar{x}, x_n) - a(\bar{x}, \bar{x}) \leq a(x_n, x_n), \quad \forall n \in \mathbb{N}.$$

Hence, taking the limit inferior on both sides and recalling that $x_n \rightharpoonup \bar{x}$, we obtain the claim.

To show the strong convergence, we will use coercivity once more to estimate (upon replacing $\bar{x} = x_o$ now that the equality has been established)

$$c_o \|x_n - x_o\|^2 \leq a(x_n - x_o, x_n - x_o) = a(x_n, x_n) - a(x_n, x_o) - a(x_o, x_n) + a(x_o, x_o),$$

whereupon taking limit superior on both sides and recalling that $x_n \rightharpoonup x_o$, we have that

$$c_o \limsup_{n\to\infty} \|x_n - x_o\|^2 \leq \limsup_{n\to\infty} a(x_n, x_n) - a(x_o, x_o). \tag{7.48}$$

Since x_n solves (7.43) and $x_o \in C$, by Assumption 7.5.5(ii), we may choose a sequence $\widehat{x}_{o,n} \to x_o$, $\widehat{x}_{o,n} \in C_n$, set $z = \widehat{x}_{o,n}$ as test function in (7.43) to obtain the analogue for (7.44). Rearranging to bind $a(x_n, x_n)$ from above and noting that $x_n \rightharpoonup x_o$, and $\widehat{x}_{o,n} \to x_o$ so as to handle the term $a(x_n, \widehat{x}_{o,n})$, and then taking the limit superior we obtain

$$\limsup_{n\to\infty} a(x_n, x_n) \leq a(x_o, x_o).$$

Combining this result with (7.48) yields $\limsup_n \|x_n - x_o\|^2 \leq 0$ so that $x_n \to x_o$, and then by standard arguments using the Urysohn property, we conclude that $x_h \to x_o$. □

7.5.3 Operator methods

A broad class of solution methods for the solution of variational inequalities is based on their reformulation in terms of the equivalent problem of finding the zero of an appropriate operator [21]. This approach is quite general; we will sketch it here for a general class of variational inequalities (VI) of the form

$$\text{find } x_o \in C : \langle F(x_o), x - x_o \rangle \geq 0, \quad \forall x \in C \subset X, \tag{7.49}$$

where X is a Hilbert space, $C \subset X$ is a closed convex set, and $F : X \to X^*$ is a nonlinear operator satisfying the monotonicity property

$$\langle F(x_1) - F(x_2), x_1 - x_2 \rangle \geq 0, \quad \forall x_1, x_2 \in X.$$

Clearly, (7.49) encompasses the VIs considered in the Lions–Stampacchia theorem (both in the linear and the nonlinear case). Note that we do not necessarily identify X with its dual here.

Working, almost verbatim, as in the proof of the Lions–Stampacchia theorem (see Theorems 7.3.5 and 7.3.11), we can show that the VI (7.49) is equivalent to the operator equations

$$0 \in \rho i_X^{-1} F(x_o) + N_C(x_o) \iff x_o = P_C(x_o - \rho i_X^{-1} F(x_o)), \quad \rho > 0, \tag{7.50}$$

where i_X is the Riesz isomorphism connecting X to its dual X^*.[11]

The equivalent formulation (7.50) can be used to produce a wide class of powerful numerical methods for the solution of VIs of the form (7.49). These methods will be better motivated (and their scope will be greatly extended) once we study the properties of monotone operators in Chapter 9. But, at this point, we stress that essentially they can be considered as extensions of the numerical methods for optimization introduced in Section 4.5, involving the subdifferential operator. This is not a mere coincidence, as these methods largely build upon monotonicity methods.

The first type of numerical method for solving VI (7.49) is an iterative method relying on the fixed point equation in (7.50) (the second equivalence), i. e.,

$$x_{n+1} = P_C(x_n - \rho i_X^{-1} F(x_n)), \quad \rho > 0. \tag{7.51}$$

The iterative scheme (7.51) leads to the projection algorithm (or Goldstein–Levitin–Polyak scheme [84, 116]). For suitable choice of ρ, the operator $T_\rho = P_C(I - \rho i_X^{-1} F)$ can be shown to be a contraction (see the proof of Theorems 7.3.5 and 7.3.11), so the convergence of the iterative scheme given in (7.51) follows by the arguments used in the proof of the Banach fixed point theorem (see Theorem 3.1.2).

An extension to the basic scheme (7.51) is in terms of the proximal algorithm, which is based upon the first relation in (7.50). Upon defining the operator $M = \rho i_X^{-1} F + N_C$, this becomes $0 \in M(x_o)$. This will be approximated by the fixed point scheme

$$x_n - x_{n+1} \in \gamma M(x_{n+1}), \quad \gamma > 0, \tag{7.52}$$

which, if convergent to some limit x_o leads to a solution of $0 \in M(x_o)$. The iterative scheme (7.52) resembles the proximal scheme for the solution of optimization problems

11 For any $x^* \in X^*$ we denote by $i_X^{-1}(x^*) = x_o \in X$, the element of X is as in Theorem 1.1.14.

introduced in Section 4.5.1; indeed it is identical to that if $M = \partial\phi$ for some proper convex lower semicontinuous function ϕ. Recalling the analysis of the proximal method (see Proposition 4.5.2), we see that it relies mainly on monotonicity properties of the subdifferential operator, hence, in principle, it may also be valid for the more general operator $M = \rho i_X^{-1} F + \mathrm{N}_C$, as long as it enjoys similar monotonicity type properties as the subdifferential. In Chapter 9, we will see that this is indeed the case. For the particular case where $M = \rho i_X^{-1} F + \mathrm{N}_C$ (and setting without loss of generality $\gamma = 1$), the fixed point scheme (7.52) can be seen to reduce (using Theorem 2.5.3) to

$$(\mathrm{x}_n - \rho i_X^{-1} F(\mathrm{x}_{n+1})) - \mathrm{x}_{n+1} \in \mathrm{N}_C(\mathrm{x}_{n+1}) \iff \mathrm{x}_{n+1} = P_C(\mathrm{x}_n - \rho i_X^{-1} F(\mathrm{x}_{n+1})). \tag{7.53}$$

The scheme (7.53) is called the proximal scheme for variational inequalities and can be shown to converge, using arguments very similar to the ones used in the proof of Proposition 4.5.2. Clearly, a version of scheme (7.53) with variable ρ for each iteration can be used.

To conclude this section, we introduce another type of numerical iterative schemes inspired by (7.50), which use the concept of operator splitting [21]. By this, we mean expressing the operator $F_\gamma := \gamma i_X^{-1} F + \mathrm{N}_C$ as $F_\gamma = F_1 + F_2$, where $F_1 = \gamma i_X^{-1} F$ and $F_2 = \mathrm{N}_C$, and look for x, z such that $\mathrm{x} - \mathrm{z} \in F_1(\mathrm{x})$ and $\mathrm{z} - \mathrm{x} \in F_2(\mathrm{x})$, turning the previous inclusions into iterative schemes for achieving a fixed point. Schemes characterized as the above are called splitting schemes and can be considered as extensions of the forward-backward algorithms introduced in Section 4.5.2, where we studied the special case of such schemes when $F_i = \partial\varphi_i$, $i = 1, 2$, where φ_i are proper convex lower semicontinuous functions. Similar splitting schemes can be devised, e. g., the forward-backward or the Douglas–Ratchford scheme. The proof of convergence of these schemes proceeds along the same lines as the proof of the relevant results in Section 4.5.2 (these proofs were based mainly on monotonicity properties of the relevant operators and not on the specific definition of the subdifferential). We shall return to these types of algorithms in Chapter 9, in the context of finding the zeros of maximal monotone operators.

All the methods presented in this section can be combined with methods of approximating infinite dimensional variational inequalities by finite dimensional methods, such as those presented in Section 7.5.2.

7.6 Application: boundary and free boundary value problems

7.6.1 An important class of bilinear forms

Let $\mathcal{D} \subset \mathbb{R}^d$, a bounded domain, with sufficiently smooth boundary $\partial\mathcal{D}$, and consider the functions $a_{ij}, a_i, a_0 \in L^\infty(\mathcal{D})$, $i, j = 1, \dots, d$. For any $u, v : \mathcal{D} \to \mathbb{R}$ such that $u, v \in X := W_0^{1,2}(\mathcal{D})$, define the bilinear form $a : X \times X \to \mathbb{R}$, by

$$\mathfrak{a}(u,v) := \int_{\mathcal{D}} \left(\sum_{i,j=1}^{d} a_{ij}(x) \frac{\partial u}{\partial x_i}(x) \frac{\partial v}{\partial x_j}(x) + \sum_{i=1}^{d} a_i(x) \frac{\partial u}{\partial x_i}(x) v(x) + a_0(x)\, u(x)\, v(x) \right) dx. \tag{7.54}$$

A simple integration by parts exercise shows that the operator $\mathsf{A} : X \to X^{\star}$, defined by this bilinear form through $\langle \mathsf{A}u, v\rangle = \mathfrak{a}(u,v)$ for every $u, v \in X = W_0^{1,2}(\mathcal{D})$, admits the form of the second-order differential operator

$$\mathsf{A}u = -\sum_{i,j=1}^{d} \frac{\partial}{\partial x_i}\left(a_{ij} \frac{\partial u}{\partial x_j} \right) + \sum_{i=1}^{d} a_i \frac{\partial u}{\partial x_i} + a_0 u, \tag{7.55}$$

or in more compact form

$$\mathsf{A}u = -\operatorname{div}(\mathbf{A}\nabla u) + a \cdot \nabla u + a_0 u, \tag{7.56}$$

upon defining the matrix and the vector valued functions $\mathbf{A} = (a_{ij})_{i,j=1,\dots,d}$, and $a = (a_1, \dots, a_d)$, respectively. As anticipated (building upon the intuition developed in Section 6.2.4), these bilinear forms will be associated with the weak formulation of partial differential equations of the form $\mathsf{A}u = f$ for some given f, with homogeneous Dirichlet boundary conditions on ∂D (because of our choice of the function space $X = W_0^{1,2}(\mathcal{D})$ in the definition of the bilinear form $\mathfrak{a}$).

If $a_i = 0$ and $a_{ij} = a_{ji}$ for every $i, j = 1, \dots, d$, the bilinear form $\mathfrak{a}$ is symmetric and the operator A is related to the Gâteaux derivative of the functional $F : X \to \mathbb{R}$, defined by $F(u) = \frac{1}{2}\mathfrak{a}(u,u)$. In the special case where the matrix $\mathbf{A} = I_{d\times d}$, $a_i = 0$ for every $i = 1, \dots, d$ and $a_0 = 0$, the operator $\mathsf{A} = -\Delta$, where Δ is the Laplacian.

The following will be standing assumptions in the remaining part of this chapter:

Assumption 7.6.1. The functions $a_{ij}, a_i, a_0 \in L^{\infty}(\mathcal{D})$ for $i, j = 1, \dots, d$.

Assumption 7.6.2 (Uniform ellipticity). The matrix valued function $\mathbf{A} = (a_{ij})_{i,j=1,\dots,d}$ satisfies the uniform ellipticity condition, i. e., there exists a constant $c_{\mathrm{E}} > 0$ such that

$$\sum_{i,j=1}^{d} a_{ij}(x) z_i z_j \geq c_{\mathrm{E}} |z|^2, \quad \text{a. e. } x \in \mathcal{D},\ \forall\, z \in \mathbb{R}^d.$$

The corresponding operator A is called uniformly elliptic.[12]

We will assume in what follows that $a_{ij} = a_{ji}$ so that the uniform ellipticity assumptions can be easily checked in terms of the spectral properties of $\mathbf{A}$. Some of the above assumptions may be too restrictive and can be relaxed at the expense of having to employ more sophisticated arguments.

12 The constant c_{E} plays the role of the coercivity constant c_o used in the abstract formulation in Section 7.3.

The following proposition (see e. g., [51, 76]) collects some useful facts for the bilinear form $\mathfrak{a}$.

Proposition 7.6.3. *Let Assumption 7.6.1 hold, and suppose that $\mathbf{A} = (a_{ij})_{i,j=1,\dots,d}$ satisfies the uniform ellipticity condition (Assumption 7.6.2). Moreover, assume that*

$$\|a\|_{L^\infty(\mathcal{D})} \leq \frac{c_E}{4\sqrt{c_p}}, \quad \|a_0\|_{L^\infty(\mathcal{D})} \leq \frac{c_E}{4c_p}, \tag{7.57}$$

with $c_p = \mathbf{c}_{\mathcal{P}}^{-2}$, where $\mathbf{c}_{\mathcal{P}}$ is the constant appearing in Poincaré's inequality (see Theorem 1.5.13). Then, the bilinear form $\mathfrak{a} : X \times X \to \mathbb{R}$, with $X = W_0^{1,2}(\mathcal{D})$ is continuous and coercive.

Proof. Continuity follows by a simple application of the Hölder inequality, since

$$\begin{aligned}|\mathfrak{a}(u,v)| \leq {} & \max_{i,j} \|a_{ij}\|_{L^\infty(\mathcal{D})} \|\nabla u\|_{L^2(\mathcal{D})} \|\nabla v\|_{L^2(\mathcal{D})} \\ & + \max_i \|a_i\|_{L^\infty(\mathcal{D})} \|\nabla u\|_{L^2(\mathcal{D})} \|v\|_{L^2(\mathcal{D})} + \|a_0\|_{L^\infty(\mathcal{D})} \|u\|_{L^2(\mathcal{D})} \|v\|_{L^2(\mathcal{D})},\end{aligned}$$

from which it easily follows that there exists a $c_1 > 0$ such that

$$|\mathfrak{a}(u,v)| \leq c_1 \|u\|_{W_0^{1,2}(\mathcal{D})} \|v\|_{W_0^{1,2}(\mathcal{D})}.$$

Note that continuity holds over the whole of $W^{1,2}(\mathcal{D})$ and not only on $W_0^{1,2}(\mathcal{D})$.

To show the coercivity property, we first consider the bilinear form $\mathfrak{a}_0 : X \times X \to \mathbb{R}$, defined by

$$\mathfrak{a}_0(u,v) = \int_{\mathcal{D}} \sum_{i,j=1}^d a_{ij}(x) \frac{\partial u}{\partial x_i}(x) \frac{\partial v}{\partial x_j}(x) dx.$$

Clearly,

$$\mathfrak{a}(u,v) = \mathfrak{a}_0(u,v) + \int_{\mathcal{D}} \left(\sum_{i=1}^d a_i(x) \frac{\partial u}{\partial x_i}(x) v(x) + a_0(x)\, u(x)\, v(x) \right) dx. \tag{7.58}$$

By the uniform ellipticity Assumption 7.6.2, we see that

$$\mathfrak{a}_0(u,u) \geq c_E \|\nabla u\|^2_{L^2(\mathcal{D})},$$

so $\mathfrak{a}_0(u,u)$ always keeps a positive sign. We now estimate, using (7.58) and Hölder's inequality,

$$\begin{aligned}\mathfrak{a}(u,u) &\geq \mathfrak{a}_0(u,u) - \|a\|_{L^\infty(\mathcal{D})} \|\nabla u\|_{L^2(\mathcal{D})} \|u\|_{L^2(\mathcal{D})} - \|a_0\|_{L^\infty(\mathcal{D})} \|u\|^2_{L^2(\mathcal{D})} \\ &\geq c_E \|\nabla u\|^2_{L^2(\mathcal{D})} - \|a\|_{L^\infty(\mathcal{D})} \|\nabla u\|_{L^2(\mathcal{D})} \|u\|_{L^2(\mathcal{D})} - \|a_0\|_{L^\infty(\mathcal{D})} \|u\|^2_{L^2(\mathcal{D})}.\end{aligned} \tag{7.59}$$

If $a = a_0 = 0$, then by the Poincaré inequality (which in fact guarantees that $\|\nabla u\|_{L^2(\mathcal{D})}$ is an equivalent norm for $W_0^{1,2}(\mathcal{D})$). We conclude that $\mathfrak{a}$ is coercive on $X = W_0^{1,2}(\mathcal{D})$. If $a \neq 0$ and $a_0 \neq 0$, we need to subordinate the resulting terms to the leading contribution of $\mathfrak{a}_0(u,u)$ so that coercivity is guaranteed. To this end, estimate

$$\|\nabla u\|_{L^2(\mathcal{D})}\|u\|_{L^2(\mathcal{D})} \leq \epsilon\|\nabla u\|^2_{L^2(\mathcal{D})} + \frac{1}{4\epsilon}\|u\|^2_{L^2(\mathcal{D})}$$

so that

$$\begin{aligned} &-\|a\|_{L^\infty(\mathcal{D})}\|\nabla u\|_{L^2(\mathcal{D})}\|u\|_{L^2(\mathcal{D})} - \|a_0\|_{L^\infty(\mathcal{D})}\|u\|^2_{L^2(\mathcal{D})} \\ &\geq -\epsilon\|a\|_{L^\infty(\mathcal{D})}\|\nabla u\|^2_{L^2(\mathcal{D})} - \left(\frac{\|a\|_{L^\infty(\mathcal{D})}}{4\epsilon} + \|a_0\|_{L^\infty(\mathcal{D})}\right)\|u\|^2_{L^2(\mathcal{D})} \\ &\geq -\epsilon\|a\|_{L^\infty(\mathcal{D})}\|\nabla u\|^2_{L^2(\mathcal{D})} - \left(\frac{\|a\|_{L^\infty(\mathcal{D})}}{4\epsilon} + \|a_0\|_{L^\infty(\mathcal{D})}\right)c_p\|\nabla u\|^2_{L^2(\mathcal{D})}, \end{aligned} \tag{7.60}$$

where, in the last estimate, we used the Poincaré inequality (see Theorem 1.5.13) and $c_p := \mathbf{c}_{\mathcal{P}}^{-2}$, where, by $\mathbf{c}_{\mathcal{P}}$, we denote the Poincaré constant. Suppose that ϵ is chosen such that $c_p \frac{\|a\|_{L^\infty(\mathcal{D})}}{4\epsilon} \leq \frac{c_E}{8}$, this leading to the condition $\|b\|_{L^\infty(\mathcal{D})} \leq \epsilon\frac{c_E}{2c_p}$. Then, $\epsilon\|a\|_{L^\infty(\mathcal{D})} \leq \epsilon^2\frac{c_E}{2c_p}$, and we now choose ϵ such that $\epsilon^2\frac{c_E}{2c_p} = \frac{c_E}{8}$, i. e., $\epsilon = \frac{\sqrt{c_p}}{2}$. This choice leads to a condition for $\|a\|_{L^\infty(\mathcal{D})}$ of the form $\|a\|_{L^\infty(\mathcal{D})} \leq \frac{c_E}{4\sqrt{c_p}}$. Suppose, furthermore, that $\|a_0\|_{L^\infty(\mathcal{D})}$ is such that $\|a_0\|_{L^\infty(\mathcal{D})}c_p \leq \frac{c_E}{4}$, i. e., that $\|a_0\|_{L^\infty(\mathcal{D})} \leq \frac{c_E}{4c_p}$. Then, we may estimate the right-hand side of (7.60) as

$$\begin{aligned} &-\|a\|_{L^\infty(\mathcal{D})}\|\nabla u\|_{L^2(\mathcal{D})}\|u\|_{L^2(\mathcal{D})} - \|a_0\|_{L^\infty(\mathcal{D})}\|u\|^2_{L^2(\mathcal{D})} \\ &\geq -\frac{c_E}{8}\|\nabla u\|^2_{L^2(\mathcal{D})} - \frac{c_E}{8}\|\nabla u\|^2_{L^2(\mathcal{D})} - -\frac{c_E}{4}\|\nabla u\|^2_{L^2(\mathcal{D})} = -\frac{c_E}{2}\|\nabla u\|^2_{L^2(\mathcal{D})}, \end{aligned}$$

which when combined with (7.59) leads to

$$\mathfrak{a}(u,u) \geq \frac{c_E}{2}\|\nabla u\|^2_{L^2(\mathcal{D})},$$

and using the fact that $\|\nabla u\|_{L^2(\mathcal{D})}$ is an equivalent norm for $W_0^{1,2}(\mathcal{D})$ (see Example 1.5.14), we are led to the required coercivity result. □

7.6.2 Boundary value problems

Within the framework of Section 7.6.1 (and using the same notation), we first consider the application of the Lax–Milgram theorem 7.3.7 to the solvability of linear boundary value problems of the form

$$-\operatorname{div}(\mathbf{A}\nabla u) + a \cdot \nabla u + a_0 u = f, \quad \text{in } \mathcal{D},$$
$$u = 0 \quad \text{on } \partial\mathcal{D}. \tag{7.61}$$

We consider the Hilbert spaces $X := W_0^{1,2}(\mathcal{D})$, $X^\star = W^{-1,2}(\mathcal{D})$, and upon defining the elliptic operator $\mathsf{A} : X \to X^\star$, associated with the bilinear form $\mathfrak{a} : X \times X \to \mathbb{R}$ defined in (7.54), by $\mathsf{A} = -\operatorname{div}(\mathbf{A}\nabla) + a \cdot \nabla + a_0 I$, equation (7.61) can be expressed in the abstract form $\mathsf{A}u = f$, assuming $f \in X^\star = W^{-1,2}(\mathcal{D})$.

A variational formulation of equation (7.61), leading to the so called concept of weak solutions for the boundary value problem (7.61), will prove very useful in the treatment of this equation.

Definition 7.6.4 (Weak solutions). A function $u \in X := W_0^{1,2}(\mathcal{D})$ is a weak solution of the boundary value problem (7.61), for a given $f \in X^\star = W^{-1,2}(\mathcal{D})$, if,

$$\mathfrak{a}(u, v) = \langle f, v \rangle, \quad \forall\, v \in X = W_0^{1,2}(\mathcal{D}), \tag{7.62}$$

where $\mathfrak{a} : X \times X \to \mathbb{R}$ is the bilinear form defined in (7.54), and $\langle \cdot, \cdot \rangle$ is the duality pairing between $W_0^{1,2}(\mathcal{D})$ and $W^{-1,2}(\mathcal{D})$.

Remark 7.6.5 (Other boundary conditions). The boundary value problem (7.61) can be considered (subject to regularity assumptions on $\partial\mathcal{D}$) with other boundary conditions, apart from homogeneous Dirichlet boundary conditions, by appropriately modifying the weak form. This can be done using the extension of the Green formula (see Proposition 1.5.19) and working similarly as in Section 6.2.4. For example, if nonhomogeneous Dirichlet boundary conditions of the form $u = g$ on $\partial\mathcal{D}$ are considered with (7.61), then the simple transformation $u_0 = u - g$ (as long as $g \in H^{1/2}(\mathcal{D})$) can be used to transform the problem to a version with homogeneous Dirichlet boundary conditions, upon suitable modification of f. If on the other hand Neumann or Robin boundary conditions are to be used, then the bilinear form $\mathfrak{a}$ and the corresponding weak form will have to be defined on $X = W^{1,2}(\mathcal{D})$, rather than on $W_0^{1,2}(\mathcal{D})$. For instance if the Robin boundary condition $\frac{\partial u}{\partial n} + hu = g$ on $\partial\mathcal{D}$ is to be employed (where $h \in L^\infty(\partial\mathcal{D})$, $h \geq 0$, $g \in L^2(\partial\mathcal{D})$, are given functions), then under the assumption $a = 0$, the weak formulation has to be modified by setting $X = W^{1,2}(\mathcal{D})$ and by using the modified bilinear form

$$\mathfrak{a}_R(u, v) = \mathfrak{a}(u, v) + \int_{\mathcal{D}} huv ds,$$

and replacing the term $\langle f, v \rangle$ in (7.62) by $\langle f, v \rangle + \int_{\partial\mathcal{D}} gn ds$. This leads to the weak formulation of (7.61) (with $a = 0$) as

$$\mathfrak{a}_R(u, v) = \langle f, v \rangle + \int_{\partial\mathcal{D}} gn ds, \quad \forall\, v \in W^{1,2}(\mathcal{D}).$$

We will not consider these problems in detail here, but their treatment requires (in most cases) straightforward extensions of the methods presented here for the Dirichlet problem. We refer the interested reader to, e. g., [51].

In certain special cases, in which (7.61) could be identified as the Euler–Lagrange equation for the minimization of an integral functional, the existence of weak solutions has been studied using techniques from the calculus of variations in Chapter 6. However, the Lax–Milgram theory (developed in Section 7.3 and in particular Theorem 7.3.7) can be used (see, e. g., [76]) to study more general cases.

Proposition 7.6.6. *Let Assumption 7.6.1 hold and* $\mathbf{A} = (a_{ij})_{i,j=1,\dots,d}$ *satisfy the uniform ellipticity condition of Assumption 7.6.2. Moreover, assume that* a, a_0 *satisfy condition (7.57). Then, for any* $f \in (W_0^{1,2}(\mathcal{D}))^* = W^{-1,2}(\mathcal{D})$, *the boundary value problem (7.61) admits a unique weak solution* $u \in W_0^{1,2}(\mathcal{D})$, *satisfying the bound*

$$\|u\|_{W_0^{1,2}(\mathcal{D})} \leq c\|f\|_{(W_0^{1,2}(\mathcal{D}))^*}$$

for some constant $c > 0$, *which depends on the ellipticity constant* c_E *and the domain* $\mathcal{D}$. *If furthermore,* $f \in L^2(\mathcal{D})$, *the operator* $\mathsf{T} : L^2(\mathcal{D}) \to L^2(\mathcal{D})$ *defined by* $\mathsf{T}f = u$, *where* u *is the solution of (7.61), is a compact operator.*

Proof. Let $X = W_0^{1,2}(\mathcal{D})$ and $X^\star = W^{-1,2}(\mathcal{D})$, which are Hilbert spaces. Since a, a_0 satisfy condition (7.57) by Proposition 7.6.3, the bilinear form $\mathfrak{a}$ is continuous and coercive on $X = W_0^{1,2}(\mathcal{D})$. Then, by a straightforward application of the Lax–Milgram theorem 7.3.7, we conclude the existence of a unique solution of the boundary value problem (7.61). The bound for the solution follows by the coercivity of the bilinear form, according to which setting $v = u$ in the weak form yields

$$c_E\|u\|^2_{W_0^{1,2}(\mathcal{D})} \leq \mathfrak{a}(u,u) = \langle f, u\rangle \leq \|f\|_{(W_0^{1,2}(\mathcal{D}))^*}\|u\|_{W_0^{1,2}(\mathcal{D})},$$

and the result follows upon dividing both sides with $\|u\|_{W_0^{1,2}(\mathcal{D})}$. The compactness of the solution operator T follows from the compact embedding of $W_0^{1,2}(\mathcal{D}) \overset{c}{\hookrightarrow} L^2(\mathcal{D})$ (see Theorem 1.5.11). □

Of course, the conditions of Proposition 7.6.6 can be rather restrictive. One may treat the solvability of the boundary value problem (7.61), even when a, a_0 do not satisfy condition (7.57) using a combination of the Lax–Milgram theorem 7.3.7 and the Fredholm theory of compact operators (see Theorem 1.3.9). The idea (see, e. g., [76]) is based on the observation that the bilinear form $\mathfrak{a}(u,v) = \langle \mathsf{A}u, v\rangle_{L^2(\mathcal{D})} + \gamma\langle u, v\rangle_{L^2(\mathcal{D})}$ is coercive for large enough values of γ, so the corresponding operator $\mathsf{A} + \gamma I$ is invertible, and by the compact embedding $W_0^{1,2}(\mathcal{D}) \overset{c}{\hookrightarrow} L^2(\mathcal{D})$ its inverse $\mathsf{T} = (\mathsf{A} + \gamma I)^{-1} : L^2(\mathcal{D}) \to L^2(\mathcal{D})$ is compact. We may then treat the original problem in terms of a perturbation of T and use the well developed theory of the solvability of equations involving compact operators to proceed. The following theorem [76] is based on this strategy.

Theorem 7.6.7. *Let Assumption* 7.6.1 *hold and* $\mathbf{A} = (a_{ij})_{i,j=1,\dots,d}$ *satisfy the uniform ellipticity condition of Assumption* 7.6.2. *Then, either problem* (7.61) *admits a unique solution for any* $f \in L^2(\mathcal{D})$ *or the homogeneous problem* $\mathsf{A}w = 0$ *admits a nontrivial weak solution in* $W_0^{1,2}(\mathcal{D})$.

Proof. It is easy to see that for large enough $\gamma > 0$, the boundary value problem

$$\begin{aligned} \mathsf{A}u + \gamma u &= f, \quad \text{in } \mathcal{D}, \\ u &= 0 \quad \text{on } \partial\mathcal{D}, \end{aligned} \tag{7.63}$$

admits a unique weak solution $u \in W_0^{1,2}(\mathcal{D})$ for every $f \in W_0^{1,2}(\mathcal{D})$. This follows by a straightforward application of the Lax–Milgram theorem to the bilinear, continuous, and coercive form $\mathfrak{a}_\gamma : X\times X \to \mathbb{R}$ defined by $\mathfrak{a}_\gamma(u,v) = \mathfrak{a}(u,v)+\gamma\langle u,v\rangle$, for every $u, v \in X$, which is associated to the linear operator $\mathsf{A}_\gamma := \mathsf{A} + \gamma I : X \to X^\star$. One may therefore define the solution mapping $f \mapsto u$, which in fact coincides with the operator $(\mathsf{A}+\gamma I)^{-1} : (W_0^{1,2}(\mathcal{D}))^* \to W_0^{1,2}(\mathcal{D})$, which, by the Lax–Milgram theorem, is linear and continuous. If we restrict the data of the problem to $f \in L^2(\mathcal{D})$, by the Rellich–Kondrachov compact embedding theorem, we see that the solution mapping $(\mathsf{A} + \gamma I)^{-1} : L^2(\mathcal{D}) \to L^2(\mathcal{D})$ is linear and compact.

We may then define the operator $\mathsf{A}_\gamma = \mathsf{A}+\gamma I$, and express the original problem (7.61) as $\mathsf{A}_\gamma u = \gamma u + f$. By the properties of the solution map, we may express the boundary value problem (7.61) in operator form as

$$u = \mathsf{T}_\gamma u + f_\gamma, \tag{7.64}$$

where $\mathsf{T}_\gamma = \gamma A_\gamma^{-1}$ and $f_\gamma := \mathsf{A}_\gamma^{-1} f$ is given. By the compactness of the operator $\mathsf{T}_\gamma : L^2(\mathcal{D}) \to L^2(\mathcal{D})$, equation (7.64) is a linear Fredholm operator equation of the second type, which can then be treated by the standard theory of such equations. In particular, a straightforward application of the Fredholm alternative (Theorem 1.3.9 applied to the compact operator T_γ) yields that either (7.61) has a unique solution for any $f \in L^2(\mathcal{D})$, or the eigenvalue problem $\mathsf{A}u = 0$ with homogeneous Dirichlet conditions, admits a weak solution. □

Remark 7.6.8. The adjoint operator $\mathsf{A}^\star$ plays an important role in the study of linear problems of the form (7.61). In particular, one may easily see that if $\mathsf{A}^\star w = 0$ admits other solutions in $W_0^{1,2}(\mathcal{D})$ apart from the trivial, then $\mathsf{A}u = f$ admits a solution if and only if $\langle f, w\rangle = 0$ for any solution w of $\mathsf{A}^\star w = 0$. This simple observation finds interesting applications in, e. g., homogenization theory [50]). Moreover, if $\mathsf{A}w = 0$ admits nontrivial solutions, or equivalently $\mathbf{N}(\mathsf{A}) \neq \{0\}$, then by the Fredholm alternative (Theorem 1.3.9 applied to the compact operator T_γ), one may conclude that $\dim(\mathbf{N}(\mathsf{A})) = \dim(\mathbf{N}(\mathsf{A}^*) < \infty$. For more details, see, e. g., [76].

We close this section with two results (see e. g., [76]) which generalize the spectral theory for the Laplace operator developed in Section 6.2.3.

Theorem 7.6.9. *Let Assumption* 7.6.1 *hold and* $\mathbf{A} = (a_{ij})_{i,j=1,\dots,d}$ *satisfy the uniform ellipticity condition of Assumption* 7.6.2. *Consider the eigenvalue problem* $\mathsf{A}u = \lambda u$ *in* $W_0^{1,2}(\mathcal{D})$.

(i) *In general, the set of eigenvalues is an at most countable set, which, if infinite, is of the form* $(\lambda_n)_{n\in\mathbb{N}}$ *with* $\lambda_n \to \infty$, *as* $n \to \infty$.

(ii) *If the bilinear form* $\mathfrak{a}$ *is symmetric, then all eigenvalues are real, and satisfy* $0 < \lambda_1 \leq \lambda_2 \leq \lambda_3 \leq \cdots$, *with* $\lambda_n \to \infty$ *as* $n \to \infty$, *while the first eigenvalue admits a variational representation as*

$$\lambda_1 = \min\{\mathfrak{a}(u,u) \; : \; u \in W_0^{1,2}(\mathcal{D}), \; \|u\|_{L^2(\mathcal{D})} = 1\},$$

with the minimum attained for a function u_1, *satisfying the eigenvalue problem for* $\lambda = \lambda_1$.

Moreover, there exists an orthornormal basis $(u_n)_{n\in\mathbb{N}}$ *of* $L^2(\mathcal{D})$ *consisting of eigenfunctions, i. e., solutions of the problem* $\mathsf{A}u_n = \lambda_n u_n$, $u_n \in W_0^{1,2}(\mathcal{D})$, $n \in \mathbb{N}$.

Proof. (i) As above choose $\gamma \in \mathbb{R}$ large enough so that the bilinear form $\mathfrak{a}_\gamma(u,v) = \mathfrak{a}(u,v) + \gamma\langle u,v\rangle_{L^2(\mathcal{D})}$ is coercive in $W_0^{1,2}(\mathcal{D})$, and express the eigenvalue problem in the equivalent form $(\mathsf{A} + \gamma I)u = (\gamma + \lambda)u$, which, using the inverse of $\mathsf{A}_\gamma := \mathsf{A} + \gamma I$, can be expressed as $u = \frac{\gamma+\lambda}{\gamma}\mathsf{T}_\gamma u$, where $\mathsf{T}_\gamma = \gamma(\mathsf{A} + \gamma I)^{-1}$. The operator $\mathsf{T}_\gamma : L^2(\mathcal{D}) \to L^2(\mathcal{D})$ is linear, bounded, and compact, by our standard argument based on the Rellich–Kondrachov–Sobolev embedding. We therefore conclude that λ is an eigenvalue of A if $\frac{\gamma}{\gamma+\lambda}$ is an eigenvalue of the linear compact operator T_γ. By Theorem 1.3.12, the eigenvalues of T_γ either form a finite set or a countable set $(\mu_n)_{n\in\mathbb{N}}$ with $\mu_n \to 0$, as $n \to \infty$, from which the result follows.

(ii) The proof follows similar arguments as the ones used for the study of the eigenvalues of the Laplacian, only that here we can resort directly to the Lax–Milgram lemma for the bilinear form $\mathfrak{a}$ associated with the operator A, combined with the compact Sobolev embeddings. The result then follows by the abstract theory for the eigenvalue problem for compact operators, as in the case of the Laplacian. The variational representation for λ_1 follows by repeating the steps in Proposition 6.2.12 for the bilinear form $\mathfrak{a}$ instead of the Dirichlet integral $\langle\nabla u, \nabla u\rangle_{L^2(\mathcal{D})}$. The existence of the basis follows by Theorem 1.3.13. □

Apart from spectral theory, many of the properties that we have seen for the Laplacian and the Poisson equation hold for more general elliptic equations of the form studied in this section. For instance, under restrictive coefficients on the data of the problem, one may obtain analogues of the maximum principle or the method of sub- and supersolutions. For lack of space, we do not present these extensions here, but we will present such results for the more general case of variational inequalities (which reduces to elliptic equations in the special case $C = X$) in Section 7.6.3.

Working similarly as in Proposition 6.6.13, we may obtain higher regularity for the weak solutions of (7.61). The following proposition (see e. g., [76]) displays such a result.

Proposition 7.6.10. *Let Assumption* 7.6.1 *hold and* $\mathbf{A} = (a_{ij})_{i,j=1,\dots,d}$ *satisfy the uniform ellipticity condition of Assumption* 7.6.2. *Assume moreover, that* $a_{ij} \in C^1(\mathcal{D})$, $i,j = 1,\dots,d$, *and* $f \in L^2(\mathcal{D})$.

If $u \in W_0^{1,2}(\mathcal{D})$ *is a solution of* (7.61), *then* $u \in W_{\text{loc}}^{2,2}(\mathcal{D})$. *The same result follows if* $u \in W^{1,2}(\mathcal{D})$ *and* $\mathrm{A}u \le f$ *in the sense of distributions.*

Proof. The proof follows essentially in the same fashion as for Proposition 6.6.13 by noting that (using the same notation) $A_j(\nabla u) = \sum_{i=1}^d a_{ij}(x)\frac{\partial u}{\partial x_i}, j = 1,\dots,d$, is linear, but there is an explicit spatial dependence on the coefficients a_{ij}, which will affect the action of the difference operator. The lower-order terms of the operator A will be moved to the right-hand side, replacing the term g in Proposition 6.6.13 by $f - a \cdot \nabla u - a_0 u$. These modifications lead to a weak form:

$$
\begin{aligned}
&\sum_{j,k=1}^{d} \int_{\mathcal{D}} (\tau_{h,i}a_{jk})(x)\Big(\Delta_{h,i}\frac{\partial u}{\partial x_j}\Big)(x)\Big(\Delta_{h,i}\frac{\partial u}{\partial x_k}\Big)(x)\psi(x)^2 dx \\
&\quad + \sum_{j,k=1}^{d} \int_{\mathcal{D}} \frac{\partial \psi^2}{\partial x_k}(x)(\tau_{h,i}a_{jk})(x)\Big(\Delta_{h,i}\frac{\partial u}{\partial x_j}\Big)(x)(\Delta_{h,i}u)(x)dx \\
&\quad + \sum_{j,k=1}^{d} \int_{\mathcal{D}} \psi(x)^2(\Delta_{h,i}a_{jk})(x)\frac{\partial u}{\partial x_j}(x)\Big(\Delta_{h,i}\frac{\partial u}{\partial x_k}\Big)(x)dx \\
&\quad + \sum_{j,k=1}^{d} \int_{\mathcal{D}} \frac{\partial \psi^2}{\partial x_k}(x)(\Delta_{h,i}a_{jk})(x)\frac{\partial u}{\partial x_j}(x)(\Delta_{h,i}u)(x)dx \\
&= \int_{\mathcal{D}} \Big(f(x) - \sum_{j=1}^{d} a_j(x)\frac{\partial u}{\partial x_j}(x) - a_0(x)u(x)\Big)(-\Delta_{-h,i}(\psi^2\Delta_{h,i}u)(x))dx.
\end{aligned}
$$

The estimates proceed as in Proposition 6.6.13 with the first two terms being essentially the same; the next two can be bounded by noting that $a_{jk} \in C^1(\mathcal{D})$ so that the norms of the term $\Delta_{h,i}a_{jk}$ are bounded, whereas the right-hand side can be handled by (discrete) integration by parts and the fact that $a_j, a_0 \in L^\infty(\mathcal{D})$, so the norms of the resulting terms $\Delta_{h,i}a_j$ are bounded. Details are left to the reader (see also [76]). □

For weak solutions under the extra assumption that $a_{ij} \in C^1(\overline{\mathcal{D}})$, $i,j = 1,\dots d$, and if the domain satisfies extra regularity properties, we may prove regularity up to the boundary, i. e., that $u \in W^{2,2}(\mathcal{D})$.

Definition 7.6.11 (C^2 domains). A domain $\mathcal{D} \subset \mathbb{R}^d$ is said to be of class C^2 if for every $x_o \in \partial\mathcal{D}$ there exists a neighborhood $N(x_o) \subset \overline{\mathcal{D}}$ and a C^2-diffeomorphism $g : \overline{B^+} \to \overline{N(x_o)}$, where $B^+ = \{x \in \mathbb{R}^d \,:\, |x| < 1,\ x_d > 0\}$ is a half ball.

Since a domain of class C^2 can be mapped locally through a diffeomorphism g to a half ball, if we manage to show the required estimates concerning regularity on half balls, then we may transfer these estimates to any compact domain. This is essentially the strategy of the proof of global regularity results. The following proposition [76]) displays a result along these lines.

Proposition 7.6.12. *Let $\mathcal{D}$ be a bounded domain with $\partial\mathcal{D}$ of class C^2. Let Assumption* 7.6.1 *hold, and* $\mathbf{A} = (a_{ij})_{i,j=1,\dots,d}$ *satisfy the uniform ellipticity condition of Assumption* 7.6.2. *Moreover, assume that $a_{ij} \in C^1(\overline{\mathcal{D}})$, $i,j = 1,\dots,d$ and $f \in L^2(\mathcal{D})$.*

If $u \in W_0^{1,2}(\mathcal{D})$ is a weak solution of (7.61)*, then $u \in W^{2,2}(\mathcal{D})$, and there is a constant $c > 0$ depending on $\mathcal{D}$ and $\|a_{ij}\|_{C^1(\overline{\mathcal{D}})}$, $i,j = 1,\dots,d$, such that $\|u\|_{W^{2,2}(\mathcal{D})} \leq c\|f\|_{L^2(\mathcal{D})}$.*

Proof. Our aim is to show that for any $x_o \in \partial\mathcal{D}$ we can obtain local estimates of the form established in Proposition 7.6.10. The proof proceeds in 4 steps.

1. Since we consider $\mathcal{D}$ to be compact, we can reduce our argument to a finite number of neighborhoods so that it suffices to obtain the relevant estimate for any neighborhood $N(x_o)$ of a point $x_o \in \partial\mathcal{D}$.

2. By step 1, consider any neighborhood $N(x_o)$ of a point $x_o \in \partial\mathcal{D}$, assumed without loss of generality to be $x_o = 0$, and a diffeomorphism $g : B^+ \to N(x_o)$, which is essentially a change of variables $z \to x$ that transforms the half ball B^+ to the neighborhood $N(x_o)$.

It will be convenient to adopt the convention of expressing the transformation g as $z = (z_1,\dots,z_d) \mapsto g(z) = x = (x_1,\dots,x_d)$, so $z = g^{-1}(x)$, and expressing its inverse in coordinate form as $g^{-1} = (\mathfrak{z}_1,\dots,\mathfrak{z}_d)$ so that $w_{ki} := \frac{\partial z_k}{\partial x_i} = \frac{\partial \mathfrak{z}_k}{\partial x_i}(x)$.

Recall the change of variables formula from multivariate calculus:

$$\int_{N(x_o)} \Psi(x)dx = \int_{g^{-1}(N(x_o))=B^+} \Psi(g(z))|\det(Dg(z))|dz$$

for any suitable $\Psi : N(x_o) \to \mathbb{R}$. Since we deal with weak formulations of elliptic problems, we will apply this formula to

$$\Psi = \sum_{i,j=1}^{d} a_{ij}\frac{\partial u}{\partial x_i}\frac{\partial \phi}{\partial x_j} + \sum_{i=1}^{d} a_i\frac{\partial u}{\partial x_i}\phi + a_0\phi - f\phi$$

for a test function ϕ localized in $N(x_o)$, so the right-hand side of the above will provide the weak form in the new variables z, which in fact will be a weak form on the half ball B^+. After some algebra involving elementary arguments based on the chain rule, and taking care in properly transforming the partial derivatives $\frac{\partial}{\partial x_i}$, we conclude that in the new variables z, the elliptic equation becomes an elliptic equation in B^+ of the form

$$\int_{B^+}\left[\sum_{i,j=1}^{d} \widehat{a}_{ij}(z)\frac{\partial \widehat{u}}{\partial z_i}(z)\frac{\partial \widehat{\phi}}{\partial z_j}(z) + \sum_{i=1}^{d}\widehat{a}_i(z)\frac{\partial \widehat{u}}{\partial z_i}(z)\widehat{\phi}(z) + \widehat{a_0}(z)\widehat{u}(z)\widehat{\phi}(z)\right]dz = \int_{B^+}\widehat{f}(z)\widehat{\phi}(z)dz, \quad (7.65)$$

where $\widehat{\phi}$ is a test function localized on B^+, and

$$\widehat{u}(z) = u(g(z)), \quad J(z) = |\det(Dg(z))|,$$

$$\hat{a}_{ij}(z) = J(z) \sum_{k,\ell} a_{k\ell}(g(z)) \frac{\partial z_i}{\partial x_k} \frac{\partial z_j}{\partial x_\ell},$$

$$\hat{a}_i(z) = J(z) \sum_k a_k(g(z)) \frac{\partial z_i}{\partial x_k},$$

$$\widehat{a_0}(z) = J(z) a_0(g(z), \quad \hat{f}(z) = J(z) f(g(z)).$$

One can see that the matrix $\widehat{\mathbf{A}} := (\hat{a}_{ij})_{i,j=1}^d$ satisfies a uniform ellipticity condition so that the transformed system is an elliptic equation on B^+. Indeed, observing that for any $\xi' = (\xi'_1, \dots, \xi'_d) \in \mathbb{R}^d$, we can write, upon defining $w_{ik} = \sqrt{J} \frac{\partial z_i}{\partial x_k}$, that

$$\sum_{i,j=1}^d \hat{a}_{ij} \xi'_i \xi'_j = \sum_{k,\ell=1}^d \sum_{i,j=1}^d a_{k\ell} w_{ik} w_{j\ell} \xi'_i \xi'_j = \sum_{k,\ell=1}^d a_{k\ell} \xi_k \xi_\ell \geq c_{\mathrm{E}} |\xi|^2, \tag{7.66}$$

where $\xi_k = \sum_{i=1}^d w_{ik} \xi'_i$, and for the last inequality, we used the uniform ellipticity for the matrix $\mathbf{A} = (a_{k\ell})_{k,\ell=1}^d$. Note that in vector notation $\xi = Dg^{-1}\xi'$, and since $g \circ g^{-1} = \mathrm{Id}$, the identity transformation, it holds that $DgDg^{-1} = I_{d\times d}$, hence, there exists a constant c' such that $|\xi| \geq c'|\xi'|$. This, combined with (7.66), leads to $\sum_{i,j=1}^d a'_{ij} \xi'_i \xi'_j \geq c'_{\mathrm{E}} |\xi'|^2$, which is the required uniform ellipticity for $\widehat{\mathbf{A}} = (\hat{a}_{ij})_{i,j=1}^d$.

Moreover, $\hat{u} \in W^{2,2}(B^+)$ if and only if $u \in W^{2,2}(N(x_o))$. So it suffices to establish the required regularity for elliptic equations on half balls.

3. We now consider the elliptic equation (7.65) on half balls B^+. As stated before, if $\hat{u} \in W^{1,2}(B^+)$ and $\hat{\psi} \in C_c^\infty(B^+)$, choosing any $k \in \{1, \dots, d-1\}$, it holds that $\hat{v} = \Delta_{-h,k}(\psi'^2 \Delta_{h,k} \hat{u}) \in W_0^{1,2}(B^+)$ so that it can be used as a test function in the weak form (7.65). Then, by essentially repeating the same steps as in Proposition 6.6.13, we can show that for all the partial derivatives $\frac{\partial^2 \hat{u}}{\partial z_\ell z_k}$, $k, \ell = 1, \dots, d$, $k + \ell < 2d$, it holds that upon defining U' to be the original half ball and V', one with half the radius,

$$\sum_{\substack{k+\ell<2d \\ k,\ell=1}}^d \left\| \frac{\partial^2 u}{\partial z_\ell \partial z_k} \right\|_{L^2(V)} \leq c(\|\hat{f}\|_{L^2(U')} + \|\hat{u}\|_{W^{1,2}(U')}). \tag{7.67}$$

The only second partial derivative that we cannot control using this type of test function is $\frac{\partial^2 \hat{u}}{\partial z_d^2}$, as the above choice was based on the use of translations parallel to the flat boundary. For this derivative we need special treatment.

To this end, we rewrite the equation (in the weak form) as

$$a_{dd} \frac{\partial^2 \hat{u}}{\partial z_d^2} = - \sum_{\substack{k+\ell<2d \\ k,\ell=1}}^d \hat{a}_{ij} \frac{\partial^2 \hat{u}}{\partial z_i \partial z_j} + \sum_{i=1}^d \left(\hat{a}_i - \sum_{j=1}^d \frac{\partial \hat{a}_{ij}}{\partial z_j} \right) \frac{\partial \hat{u}}{\partial z_i} + \widehat{a_0} \hat{u} - \hat{f},$$

and by the uniform ellipticity condition (upon choosing $\xi = e_n = (0, \dots, 0, 1)$), we have that $\hat{a}_{dd} > c_{\mathrm{E}} > 0$ so that

$$\left|\frac{\partial^2 \widehat{u}}{\partial z_d^2}\right| \le c\left(\sum_{\substack{k+\ell<2d\\ k,\ell=1}}^{d}\left|\frac{\partial^2 \widehat{u}}{\partial z_i \partial z_j}\right| + |\nabla \widehat{u}| + |\widehat{u}| + |\widehat{f}|\right) \quad \text{a. e.,}$$

which upon integrating provides the missing term in (7.67) and leads to the required estimate

$$\|\widehat{u}\|_{W^{2,2}(V')} \le c(\|\widehat{f}\|_{L^2(U')} + \|\widehat{u}\|_{W^{1,2}(U')}).$$

4. We consider any point $x_o \in \partial\mathcal{D}$ and a neighborhood $N(x_o)$. By using the properties of the boundary and possibly a change of variables, we may assume that

$$\mathcal{D} \cap B(x_o, r) = \{x \in B(x_o, r) \ : \ x_d > \gamma(x_1, \dots, x_{d-1})\},$$

for some $r > 0$ and a function $\gamma : \mathbb{R}^{d-1} \to \mathbb{R}$, which is C^2. Denoting by g^{-1} the transformation that takes $\mathcal{D}$ to a half ball, we may assume (without loss of generality) that $g^{-1}(x_o) = 0$, and we choose U' to be a half ball of radius $\rho > 0$ such that $U \in g^{-1}(\mathcal{D} \cap N(x_o))$, i. e., $U' = \{z \in \mathbb{R}^d \ : \ |z| < \rho, \ z_d > 0\}$. We also set $V' = \{z \in \mathbb{R}^d \ : \ |z| < \frac{\rho}{2}, \ z_d > 0\}$. Then, the estimates of step 3 are valid for V' and U'. Undoing the transformation, this estimate is transferred to a similar local estimate for the solution of the elliptic problem u in the original domain $\mathcal{D}$, and a covering argument based on step 1 can be used to complete the proof. □

Remark 7.6.13. The result of Proposition 7.6.12 can be generalized for non-homogeneous Dirichlet data of sufficient smoothness, i. e., if there exists $g \in W^{2,2}(\mathcal{D})$ such that the weak solution of (7.61) satisfies $u - g \in W_0^{1,2}(\mathcal{D})$, with the estimate for the solution modified as $\|u\|_{W^{2,2}(\mathcal{D})} \le c(\|f\|_{L^2(\mathcal{D})} + \|g\|_{W^{2,2}(\mathcal{D})})$.

7.6.3 Free boundary value problems

We now turn our attention to applications of the Lax–Milgram–Stampacchia theorem 7.3.5. We will consider the application of this theorem to the general class of bilinear forms $\mathfrak{a}$, defined in (7.54), and the corresponding operator A, and consider the problem

$$\begin{aligned} &\text{find } u \in C \ : \ \langle \mathrm{A}u, v - u\rangle \ge \langle f, v - u\rangle, \quad \forall\, v \in C, \\ &C := \{v \in W_0^{1,2}(\mathcal{D}) \ : \ v(x) \ge \psi(x) \text{ a. e. } x \in \mathcal{D}\}, \end{aligned} \tag{7.68}$$

where $\psi : \mathcal{D} \to \mathbb{R}$ is a suitable given function. This is a generalization of the problem treated in Section 7.2 for the special case of the Laplacian.

Theorem 7.6.14. *Let Assumption* 7.6.1 *hold and* $\mathbf{A} = (a_{ij})_{i,j=1,\dots,d}$ *satisfy the uniform ellipticity condition of Assumption* 7.6.2. *Moreover, assume that* a, a_0 *satisfy condition* (7.57). *Then, for any* $f \in W^{-1,2}(\mathcal{D})$, *the variational inequality* (7.68) *admits a unique solution.*

Proof. Let $X := W_0^{1,2}(\mathcal{D})$ and $X^\star = W^{-1,2}(\mathcal{D})$. By the assumptions of the data of the problem applying Proposition 7.6.3, we see that the corresponding bilinear for $\mathfrak{a} : X \times X \to \mathbb{R}$ is continuous and coercive. Furthermore, C is a closed convex subset of X (see step 2 in the proof of Proposition 7.2.1). Therefore, by a straightforward application of Theorem 7.3.5, we conclude. □

If the obstacle ψ and the function f satisfy certain smoothness conditions, then, working similarly as in Proposition 7.2.1, we may show that the solution of the variational inequality (7.68) satisfies the differential inequalities of the form

$$\begin{aligned} \mathsf{A}u &= f, \quad \text{if } u > \psi, \\ \mathsf{A}u &> f, \quad \text{if } u = \psi. \end{aligned} \tag{7.69}$$

An alternative formulation is the so called complementarity form

$$u \geq \psi, \quad \mathsf{A}u - f \geq 0, \quad \text{and} \quad \langle \mathsf{A}u - f, u - \psi \rangle = 0, \tag{7.70}$$

where by $\mathsf{A}u - f \geq 0$ we mean that $\langle \mathsf{A}u - f, w \rangle \geq 0$ for every $w \in W_0^{1,2}(\mathcal{D})$ such that $w \geq 0$. This is the complementarity form of the variational inequality, which is equivalent to the original formulation. The equivalence can be seen by choosing appropriate test functions, e. g., $v = u + w$ for arbitrary $w \geq 0$, $v = \psi$ and $v = 2u - \psi$.

Problems of the general type (7.69) are called free boundary value problems or obstacle problems; they find important applications in mechanics, image processing, the theory of stochastic processes, and mathematical finance. Unlike the boundary value problems treated in the previous section, where the values of the unknown function u are specified on a known domain (e. g., for the homogeneous Dirichlet problem $u = 0$ on $\partial\mathcal{D}$, with the set $\partial\mathcal{D}$ being *a priori* specified), here the set $C \subset \mathcal{D}$, on which $u = \phi$ is unspecified and will only be determined after problem (7.69) is solved, therefore leading to an unknown (free) boundary.

Remark 7.6.15. The result of Theorem 7.6.14 can be generalized in various ways. For example, we may consider the double obstacle problem by using the convex set $C = \{u \in W^{1,2}(\mathcal{D}) : \psi_1(x) \leq u(x) \leq \psi_2(x), \text{ a. e. } x \in \mathcal{D}\}$, for suitable obstacle function ψ_1, ψ_2. Another possible generalization is to include nonhomogeneous boundary data $g : \partial\mathcal{D} \to \mathbb{R}$, $g \in H^{1/2}(\partial\mathcal{D})$ (see Proposition 1.5.17) satisfying $\psi \leq g$ on $\partial\mathcal{D}$. In this case, we need to work in $W^{1,2}(\mathcal{D})$ and consider the closed convex set $C \subset W^{1,2}(\mathcal{D})$, defined by $C = \{u \in W^{1,2}(\mathcal{D}) : u(x) \geq \psi(x), \text{ a. e. } x \in \mathcal{D},\ u(x) = g(x),\ x \in \partial\mathcal{D}\}$.

Remark 7.6.16. If the bilinear form $\mathfrak{a}$ is noncoercive, then we may resort to a perturbation type argument similar to that used in Theorem 7.6.7, combined with the Leray–Schauder alternative (Theorem 3.3.5). In particular, assuming for simplicity that we consider homogeneous Dirichlet boundary conditions, we may replace $\mathfrak{a}$ with $\mathfrak{a}_\gamma$, defined as $\mathfrak{a}_\gamma(u, v) = \mathfrak{a}(u, v) + \gamma\langle u, v \rangle$, for every $u, v \in W_0^{1,2}(\mathcal{D})$, with $\gamma > 0$ chosen so that $\mathfrak{a}_\gamma$ is coercive. Then, for any $w \in L^2(\mathcal{D})$, we consider the variational inequality $\mathfrak{a}_\gamma(u, v - u) \geq$

$\langle f, v-u\rangle + \gamma\langle w, v-u\rangle$ for every $v \in C$, and define the map $w \mapsto \mathsf{T}_\gamma w := u$, where u is the solution of the above variational inequality. This is a continuous and compact map from $L^2(\mathcal{D})$ onto itself, a fixed point of which is a solution of the original inequality. The existence of the fixed point can be obtained by the Leray–Schauder alternative. This requires to show that for any $\lambda \in (0,1)$ any solution of the equation $w = \lambda \mathsf{T}_\gamma w$ admits an *a priori* bound. By the definition of T_γ, we can see that any solution of $w = \lambda\mathsf{T}_\gamma w$ will satisfy the variational inequality $\mathfrak{a}_{\gamma(1-\lambda)}(\mathsf{T}_\gamma w, v - \mathsf{T}_\gamma w) \geq \langle f, v - \mathsf{T}_\gamma w\rangle$, or equiv. $\mathfrak{a}_{\gamma(1-\lambda)}(\widehat{w}, v-\widehat{w}) \geq \langle f, v-\widehat{w}\rangle$ for any $v \in C$. Hence, deriving the required bounds requires some rather delicate *a priori* estimates on a family of related variational inequalities, which in turn require certain restrictive conditions on the data of the problem (see [127]).

The solution of variational inequalities associated with the bilinear form (7.54) enjoy important and useful qualitative properties, which are related to comparison and maximum principles, under certain restrictive assumptions on the coefficients of the operator A. Without striving for full generality, we will restrict ourselves to the following simple case:

Consider the variational inequality

$$\begin{aligned}
&\text{find } u \in C : \langle \mathsf{A}u, v-u\rangle \geq \langle f, v-u\rangle, \quad \forall v \in C,\\
&C = \{u \in W^{1,2}(\mathcal{D}) : u(x) \geq \psi(x), \text{ a.e. } x \in \mathcal{D},\ u(x) = g(x),\ x \in \partial\mathcal{D}\},
\end{aligned} \tag{7.71}$$

for given obstacle $\psi : \overline{\mathcal{D}} \to \mathbb{R}$ and boundary data $g : \partial\mathcal{D} \to \mathbb{R}$, $g \in H^{1/2}(\partial\mathcal{D})$ satisfying $\psi \leq g$ on $\mathcal{D}$. We will also use the notation C_i for the set defined in (7.71) for suitable obstacle and boundary data ψ_i, g_i, and denote by u_i the solution of (7.71) in C_i with f replaced by f_i, $i = 1, 2$.

We are now ready to state and prove a comparison result and a maximum principle for the variational inequality (7.71).

Proposition 7.6.17 (Comparison result and maximum principle). *Let Assumption* 7.6.1 *hold and* $\mathbf{A} = (a_{ij})_{i,j=1,\dots d}$ *satisfy the uniform ellipticity condition of Assumption* 7.6.2. *Moreover, assume that* $a_i = a_0 = 0$, $i = 1, \dots, d$. *Consider suitable obstacle, boundary data, and forcing terms* ψ, ψ_i, g, g_i, f, f_i, *respectively, satisfying* $\psi \leq g$, $\psi_i \leq g_i$, *and let* u, u_i *be the corresponding solutions to the variational inequalities* (7.71), $i = 1, 2$.

(i) *Suppose that* $\psi_1 \geq \psi_2$, $g_1 \geq g_2$ *and* $f_1 \geq f_2$. *Then,* $u_1 \geq u_2$.

(ii) *u satisfies the bounds*

$$\begin{aligned}
&u \geq m := \min\Big(0, \inf_{\partial\mathcal{D}} g\Big) \quad \text{a. e. in } \mathcal{D} \quad \text{if } f \geq 0 \quad \text{and}\\
&u \leq M := \max\Big(0, \sup_{\partial\mathcal{D}} g, \sup_{\mathcal{D}} \psi\Big) \quad \text{a. e. in } \mathcal{D} \quad \text{if } f \leq 0.
\end{aligned}$$

Proof. Note that under the stated assumptions, by Proposition 7.6.3, the corresponding bilinear form is continuous and coercive in $W_0^{1,2}(\mathcal{D})$.

(i) Since $\psi_1 \geq \psi_2$ and $g_1 \geq g_2$, the sets C_1, C_2 are such that

$$\begin{aligned} &u_1 \vee u_2 := \max(u_1, u_2) \in C_1, \quad \text{and} \\ &u_1 \wedge u_2 := \min(u_1, u_2) \in C_2, \quad \forall\, u_1 \in C_1,\ u_2 \in C_2. \end{aligned} \tag{7.72}$$

This is elementary to check, but for the convenience of the reader, the proof is presented in Lemma 7.7.1, in the appendix of the chapter.

For the variational inequality (7.71) for u_1, use as test function $v = \max(u_1, u_2) \in C_1$ (by (7.72)). By the identity $\max(u_1, u_2) = u_1 + (u_2 - u_1)^+$, this yields

$$\langle \mathsf{A}u_1, (u_2 - u_1)^+ \rangle \geq \langle f_1, (u_2 - u_1)^+ \rangle. \tag{7.73}$$

For the variational inequality (7.71) for u_2, use as test function $v = \min(u_1, u_2) \in C_2$ (by (7.72)). By the identity $\min(u_1, u_2) = u_2 - (u_2 - u_1)^+$, this yields

$$\langle \mathsf{A}u_2, (u_2 - u_1)^+ \rangle \leq \langle f_2, (u_2 - u_1)^+ \rangle. \tag{7.74}$$

Multiplying (7.73) by −1, rearranging, and summing with (7.74), we obtain that

$$\langle \mathsf{A}u_2 - \mathsf{A}u_1, (u_2 - u_1)^+ \rangle + \langle f_1 - f_2, (u_2 - u_1)^+ \rangle \leq 0,$$

which, since $f_1 \geq f_2$, leads to

$$\langle \mathsf{A}u_2 - \mathsf{A}u_1, (u_2 - u_1)^+ \rangle \leq 0. \tag{7.75}$$

Since $g_1 \geq g_2$, and $u_i \in C_i$, $i = 1, 2$, it is clear the $(u_2 - u_1)^+$ vanishes on $\partial\mathcal{D}$, so $(u_2 - u_1)^+ \in W_0^{1,2}(\mathcal{D})$. This, combined with (7.75), leads to $(u_2 - u_1)^+ = 0$. Indeed, a quick calculation yields that

$$\begin{aligned} 0 \geq \langle \mathsf{A}u_2 - \mathsf{A}u_1, (u_2 - u_1)^+ \rangle &= \int_{\mathcal{D}} \mathbf{A}\nabla(u_2 - u_1) \cdot \nabla(u_2 - u_1)^+ dx \\ &= \int_{\mathcal{D}} \mathbf{A}\nabla(u_2 - u_1)^+ \cdot \nabla(u_2 - u_1)^+ dx \geq c_{\mathrm{E}} \|\nabla(u_2 - u_1)^+\|^2_{L^2(\mathcal{D})} \geq c_{\mathrm{E}} \mathbf{c}_{\mathcal{P}}^2 \|(u_2 - u_1)^+\|^2_{L^2(\mathcal{D})}, \end{aligned}$$

where we used subsequently the uniform ellipticity and the Poincaré inequality so that $(u_2 - u_1)^+ = 0$, and $u_1 \geq u_2$ follows.

(ii) For the weak maximum principle, let us first consider the case $f \geq 0$. Let $m = \min(0, \inf_{\partial\mathcal{D}} g)$, and consider as test function $v = \max(u, m) = u + (m - u)^+$. Since $v \geq u$, it is easy to check that $v \in C$, so it is an acceptable test function. The weak form of the variational inequality (7.71) for this choice yields

$$\int_{\mathcal{D}} \mathbf{A}\nabla u \cdot \nabla(m - u)^+ dx \geq \int_{\mathcal{D}} f(m - u)^+ dx. \tag{7.76}$$

Noting that

$$\int_{\mathcal{D}} \mathbf{A}\nabla u \cdot \nabla(m-u)^+ dx = -\int_{\mathcal{D}} \mathbf{A}\nabla(m-u)\cdot\nabla(m-u)^+ dx$$
$$= -\int_{\mathcal{D}} \mathbf{A}\nabla(m-u)^+ \cdot \nabla(m-u)^+ dx,$$

and substituting this in (7.76), we obtain

$$-\int_{\mathcal{D}} \mathbf{A}\nabla(m-u)^+ \cdot \nabla(m-u)^+ dx \geq \int_{\mathcal{D}} f(m-u)^+ dx \geq 0, \tag{7.77}$$

where, for the last inequality, we used the fact that $f \geq 0$. Therefore, combining (7.77) and the uniform ellipticity condition for $\mathbf{A}$, we obtain

$$c_{\mathrm{E}} \|\nabla(m-u)^+\|^2_{L^2(\mathcal{D};\mathbb{R}^d)} \leq \int_{\mathcal{D}} \mathbf{A}\nabla(m-u)^+ \cdot \nabla(m-u)^+ dx \leq 0,$$

which leads to $(m-u)^+$ being a constant a. e. in $\mathcal{D}$, and since it vanishes on $\mathcal{D}$ (by the choice of m), it holds that $(m-u)^+ = 0$ a. e. in $\mathcal{D}$, therefore, $u \geq m$ a. e. in $\mathcal{D}$.

If $f \leq 0$, let $M = \max(0, \sup_{\partial\mathcal{D}} g, \sup_{\mathcal{D}} \psi)$, consider as test function $v = \min(u, M) = u - (u-M)^+$, and proceed accordingly. □

Remark 7.6.18. In the case where $\psi_i = -\infty$, $i = 1, 2$ or $\psi = -\infty$, we are essentially working with $C_i = X$ or $C = X$, and the above comparison results and maximum principles apply to the corresponding elliptic equation.

The above comparison principle motivates the useful concept of super- and subsolutions for variational inequalities [40].

Definition 7.6.19 (Super- and subsolutions).

(i) An element $\bar{u} \in X = W^{1,2}(\mathcal{D})$ is called a supersolution of the variational inequality (7.71) if $\bar{u} \geq \psi$ in $\mathcal{D}$, $\bar{u} \geq g$ on $\partial\mathcal{D}$ and $\mathrm{A}\bar{u} - f \geq 0$ (meaning that $\langle \mathrm{A}\bar{u} - f, w\rangle \geq 0$ for every $w \in W^{1,2}_0(\mathcal{D})$, $w \geq 0$).

(ii) An element $\underline{u} \in X = W^{1,2}(\mathcal{D})$ is called a subsolution of the variational inequality (7.71) if $\underline{u} \geq \psi$ in $\mathcal{D}$, $\underline{u} \leq g$ on $\partial\mathcal{D}$ and $\mathrm{A}\underline{u} - f \leq 0$ (meaning that $\langle \mathrm{A}\bar{u} - f, w\rangle \leq 0$ for every $w \in W^{1,2}_0(\mathcal{D})$, $w \geq 0$).

Proposition 7.6.20. *Let Assumption* 7.6.1 *hold and* $\mathbf{A} = (a_{ij})_{i,j=1,\dots,d}$ *satisfy the uniform ellipticity condition of Assumption* 7.6.2. *Moreover, assume that* $a_i = a_0 = 0$, $i = 1, \dots, d$.

(i) *Let* u *and* $\bar{u}$ *be any solution and any supersolution of* (7.71), *respectively. Then,* $u \leq \bar{u}$.

(ii) *Let* u *and* $\underline{u}$ *be any solution and any subsolution of* (7.71), *respectively. Then,* $\underline{u} \leq u$.

Proof. (i) Since $u \ge \psi$ and $\bar{u} \ge \psi$, it also holds that $\min(u, \bar{u}) \ge \psi$ so that we may use it as a test function. Setting $v = \min(u, \bar{u}) = u - (u - \bar{u})^+$ in (7.71), we obtain

$$\langle \mathsf{A}u - f, (u - \bar{u})^+ \rangle \le 0. \tag{7.78}$$

Since $\bar{u}$ is a supersolution, $\mathsf{A}\bar{u} - f \ge 0$, and since $(u - \bar{u})^+ \ge 0$,

$$\langle \mathsf{A}\bar{u} - f, (u - \bar{u})^+ \rangle \ge 0. \tag{7.79}$$

Combining (7.78) and (7.79), we obtain that

$$\langle \mathsf{A}u - \mathsf{A}\bar{u}, (u - \bar{u})^+ \rangle \le 0,$$

therefore, $(u - \bar{u})^+ = 0$, and $u \le \bar{u}$.

(ii) Use the test function $v = u \vee \underline{u} = \max(u, \underline{u}) = u + (\underline{u} - u)^+$, and proceed as above. □

Example 7.6.21. The *a priori* bounds provided by Proposition 7.6.20 can be refined since if any two supersolutions $\bar{u}_1, \bar{u}_2$ are given, $\bar{u} = \bar{u}_1 \wedge \bar{u}_2 = \min(\bar{u}_1, \bar{u}_2)$ is also a supersolution, and hence, we may apply Proposition 7.6.20 for $\bar{u}$. Clearly, this procedure can be repeated for a finite number of iterations, so if $\bar{u}_i$, $i = 1, \dots, n$, are supersolutions, then $\bar{u} = \min(\bar{u}_1, \dots, \bar{u}_n)$ is also a supersolution. A similar construction is possible with subsolutions; if $\underline{u}_1, \underline{u}_2$ are two subsolutions, then $\underline{u} = \underline{u}_1 \vee \underline{u}_2 = \max(\underline{u}_1, \underline{u}_2)$ is also a subsolution.

To check this claim, we reason as follows: Clearly, since $\bar{u}_i \ge \psi$ on $\mathcal{D}$ and $\bar{u}_i \ge g$ on $\partial\mathcal{D}$, $i = 1, 2$, it holds that $\bar{u} \ge \psi$ on $\mathcal{D}$ and $\bar{u} \ge g$ on $\partial\mathcal{D}$, so it only remains to check that $\mathsf{A}\bar{u} - f \ge 0$. To this end, consider the closed convex set $C' := \{u \in W^{1,2}(\mathcal{D}) : u \ge \bar{u} \text{ in } \mathcal{D},\ u = g \text{ on } \partial\mathcal{D}\}$ and the corresponding variational inequality

$$\text{find } u \in C' : \langle \mathsf{A}u, v - u \rangle \ge \langle f, v - u \rangle, \quad \forall\, v \in C', \tag{7.80}$$

which admits the equivalent complementarity form

$$u \in C' \quad \mathsf{A}u - f \ge 0, \quad \langle \mathsf{A}u - f, u - \bar{u} \rangle = 0. \tag{7.81}$$

By the Lions–Stampacchia theory, (7.80) admits a unique solution $u \in C'$, hence, $u \ge \bar{u}$. On the other hand, $\bar{u}_i$ are supersolutions of (7.80), since $\bar{u}_i \ge \bar{u}$, by Proposition 7.6.20, $u \le \bar{u}_i$, $i = 1, 2$, hence, $u \le \bar{u}_1 \wedge \bar{u}_2 = \bar{u}$. By the uniqueness of u since $u \ge \bar{u}$ and $u \le \bar{u}$, we have that $u = \bar{u}$. That, by the complementarity form (7.81), leads to $\mathsf{A}\bar{u} - f \ge 0$ so that $\bar{u}$ is a supersolution of (7.71) ◁

One may generalize the above constructions to more general problems involving more general possibly nonlinear operators (see, e. g., [40]).

We close our treatment of free boundary value problems with a regularity result. To avoid unnecessary technicalities, let us consider problem (7.71) under the extra assumptions that $a_i = a_0 = g = 0$, $i = 1, \dots, d$, i. e.,

$$\begin{aligned}&\text{find } u \in C : \langle \mathsf{A}u, v-u\rangle \geq \langle f, v-u\rangle, \quad \forall\, v \in C,\\ &C = \{u \in W_0^{1,2}(\mathcal{D}) : u(x) \geq \psi(x), \text{ a. e. } x \in \mathcal{D}\},\end{aligned} \tag{7.82}$$

for given obstacle $\psi : \overline{\mathcal{D}} \to \mathbb{R}$ such that $\psi \leq 0$ on $\partial\mathcal{D}$.

The following proposition (see [24]) provides a regularity result for variational inequalities.

Proposition 7.6.22. *Let Assumption* 7.6.1 *hold and* $\mathbf{A} = (a_{ij})_{i,j=1,\dots,d}$ *satisfy the uniform ellipticity condition of Assumption* 7.6.2. *Moreover, assume that* $a_i = a_0 = 0$, $i = 1,\dots,d$.

Under the additional assumptions that $a_{ij} \in C^1(\overline{\mathcal{D}})$, $i,j = 1,\dots,d$, $f \in L^2(\mathcal{D})$ *and* $\mathsf{A}\psi \in L^2(\mathcal{D})$, *the solution* u *of the variational inequality* (7.82) *satisfies* $u \in W^{2,2}(\mathcal{D})$.

Proof. We sketch the main argument of the proof.

First of all, we note that under the stated assumptions, by Proposition 7.6.3, the corresponding bilinear form is continuous and coercive in $W_0^{1,2}(\mathcal{D})$. The proof uses a penalization argument, the regularity results for boundary value problems we have developed in Section 7.6.2, and a passage to the limit, which requires some delicate estimates.

The penalized version (in the sense of Section 7.5.1) of (7.82) is equation

$$\mathfrak{a}(u_\epsilon, v) + \int_{\mathcal{D}} \frac{1}{\epsilon}(u_\epsilon - \mathsf{P}_C(u_\epsilon))v\,dx = \int_{\mathcal{D}} fv\,dx, \quad \forall\, v \in W_0^{1,2}(\mathcal{D}),\ \epsilon > 0. \tag{7.83}$$

Recall (see Example 7.5.2) that for the specific C, we consider here $\mathsf{P}_C(u) = \max(u,\psi)$ so that $u - \mathsf{P}_C(u) = -(\psi - u)^+ \in W_0^{1,2}(\mathcal{D})$ (since $\psi \leq 0$ on $\partial\mathcal{D}$).

We proceed in 4 steps.

1. We will first obtain a uniform bound in ϵ for the term $\frac{1}{\epsilon}\|(\psi - u_\epsilon)^+\|_{L^2(\mathcal{D})}$.

We will use as test function $v = u_\epsilon - \mathsf{P}_C(u_\epsilon) = -(\psi - u_\epsilon)^+ \in W_0^{1,2}(\mathcal{D})$ in (7.83). We have the important estimate:

$$\begin{aligned}\mathfrak{a}(u_\epsilon, u_\epsilon - \mathsf{P}_C(u_\epsilon)) &= \mathfrak{a}(-u_\epsilon, (\psi - u_\epsilon)^+)\\ &= \mathfrak{a}(\psi - u_\epsilon, (\psi - u_\epsilon)^+) - \mathfrak{a}(\psi, (\psi - u_\epsilon)^+)\\ &= \mathfrak{a}((\psi - u_\epsilon)^+, (\psi - u_\epsilon)^+) - \mathfrak{a}(\psi, (\psi - u_\epsilon)^+)\\ &= \mathfrak{a}((\psi - u_\epsilon)^+, (\psi - u_\epsilon)^+) - \langle \mathsf{A}\psi, (\psi - u_\epsilon)^+\rangle,\end{aligned} \tag{7.84}$$

where we used the fact that $\nabla(\psi - u_\epsilon)^+ = \nabla(\psi - u_\epsilon)\,\mathbf{1}_{\psi - u_\epsilon > 0}$.

Combining (7.83) (for $v = u_\epsilon - \mathsf{P}_C(u_\epsilon)$) with (7.84), we have that

$$\mathfrak{a}((\psi - u_\epsilon)^+, (\psi - u_\epsilon)^+) - \langle \mathsf{A}\psi, (\psi - u_\epsilon)^+\rangle + \frac{1}{\epsilon}\|(\psi - u_\epsilon)^+\|^2_{L^2(\mathcal{D})} = -\int_{\mathcal{D}} f(\psi - u_\epsilon)^+ dx. \tag{7.85}$$

Since $\mathfrak{a}((\psi - u_\epsilon)^+, (\psi - u_\epsilon)^+) \geq 0$ by coercivity, (7.85) implies that for all $\epsilon > 0$, we have

$$\frac{1}{\epsilon}\|(\psi - u_\epsilon)^+\|^2_{L^2(\mathcal{D})} \leq \langle \mathsf{A}\psi, (\psi - u_\epsilon)^+\rangle - \int_{\mathcal{D}} f(\psi - u_\epsilon)^+ dx$$

$$\leq (\|A\psi\|_{L^2(\mathcal{D})} + \|f\|_{L^2(\mathcal{D})})\|(\psi - u_\epsilon)^+\|_{L^2(\mathcal{D})},$$

which upon dividing yields the bound $\frac{1}{\epsilon}\|(\psi - u_\epsilon)^+\|_{L^2(\mathcal{D})} \leq (\|A\psi\|_{L^2(\mathcal{D})} + \|f\|_{L^2(\mathcal{D})})$.

2. We now return to (7.83), which in operator form implies that $Au_\epsilon - \frac{1}{\epsilon}(\psi - u_\epsilon)^+ = f$, and solving with respect to Au_ϵ, since both $\frac{1}{\epsilon}(\psi - u_\epsilon)^+$ and f are bounded in $L^2(\mathcal{D})$, so is Au_ϵ (with a bound independent of ϵ). Hence, $Au_\epsilon \in L^2(\mathcal{D})$ so that, by arguments similar to those employed in Propositions 7.6.10 and 7.6.12, we have that $u_\epsilon \in W^{2,2}(\mathcal{D})$ (uniformly bounded in ϵ). By reflexivity, there exists a weakly convergent subsequence $u_\epsilon \rightharpoonup \hat{u}$ in $W^{2,2}(\mathcal{D})$.

3. Working as in Proposition 7.5.3 or Example 7.5.4, we can show that $u_\epsilon \to u$ in $W_0^{1,2}(\mathcal{D})$ as $\epsilon \to 0$, with $u = P_C(u) \in C$. In particular, we express $u - u_\epsilon = (u - \psi) - (\psi - u_\epsilon)^- + (\psi - u_\epsilon)^+$ so that defining $w_\epsilon := (u - \psi) - (\psi - u_\epsilon)^-$, we have $u - u_\epsilon = w_\epsilon + (\psi - u_\epsilon)^+$. In what follows, c_i, $i = 1, \dots, 5$, is used to denote positive constants, which we do not bother to specify. We have already established a bound of the form $\|(\psi - u_\epsilon)^+\|_{L^2(\mathcal{D})} < c_1\epsilon$ in step 1 above. Inserting this estimate in (7.85), we have that $\mathfrak{a}((\psi - u_\epsilon)^+, (\psi - u_\epsilon)^+) \leq c_2\epsilon$, and by coercivity of $\mathfrak{a}$, we conclude that $\|(\psi - u_\epsilon)^+\|_{W_0^{1,2}(\mathcal{D})} \leq c_3\sqrt{\epsilon}$. Since $u - u_\epsilon = w_\epsilon + (\psi - u_\epsilon)^+$, if we show that $\|w_\epsilon\|_{W_0^{1,2}(\mathcal{D})} \leq c_4\sqrt{\epsilon}$, then we have that $\|u - u_\epsilon\|_{W_0^{1,2}(\mathcal{D})} \leq c_5\sqrt{\epsilon}$, hence, the stated convergence result in $W_0^{1,2}(\mathcal{D})$. To show the bound for w_ϵ, use as test function $-w_\epsilon$ in (7.83) and $v = \psi + (\psi - u_\epsilon)^-$ (so that $v - u = -w_\epsilon$) as test function in (7.82). We add the resulting inequalities to obtain

$$\mathfrak{a}(u - u_\epsilon, w_\epsilon) + \frac{1}{\epsilon}\langle(\psi - u_\epsilon)^+, w_\epsilon\rangle \leq 0. \tag{7.86}$$

Clearly, by the definition of w_ϵ, we have that $\langle(\psi - u_\epsilon)^+, w_\epsilon\rangle = \langle(\psi - u_\epsilon)^+, u - \psi\rangle \geq 0$ (since $u \in C$), hence, (7.86) implies that $\mathfrak{a}(u - u_\epsilon, w_\epsilon) \leq 0$. This is equivalent with $\mathfrak{a}(w_\epsilon + (\psi - u_\epsilon)^+, w_\epsilon) \leq 0$, which is rearranged as $\mathfrak{a}(w_\epsilon, w_\epsilon) \leq -\mathfrak{a}((\psi - u_\epsilon)^+, w_\epsilon)$, and using the facts that $\mathfrak{a}$ is continuous and coercive, we have that

$$c_E\|w_\epsilon\|^2_{W_0^{1,2}(\mathcal{D})} \leq c\|(\psi - u_\epsilon)^+\|_{W_0^{1,2}(\mathcal{D})}\|w_\epsilon\|_{W_0^{1,2}(\mathcal{D})},$$

from which follows the required bound on $\|w_\epsilon\|_{W_0^{1,2}(\mathcal{D})}$.

4. Combining steps 2 and 3, we pass to the limit, and since $u_\epsilon \rightharpoonup \hat{u}$ in $W^{2,2}(\mathcal{D})$ and $u_\epsilon \to u$ in $W_0^{1,2}(\mathcal{D})$, where u is a solution of (7.82), we obtain the stated regularity result. □

Remark 7.6.23. From estimate (7.84), and using the elementary inequality $A\psi \leq (A\psi)^+$, may further estimate $\mathfrak{a}(u_\epsilon, u_\epsilon - P_C(u_\epsilon)) \geq \mathfrak{a}((\psi - u_\epsilon)^+, (\psi - u_\epsilon)^+) + \langle(A\psi)^+, u_\epsilon - P_C(u_\epsilon)\rangle$. Proceeding with the proof, using this modified estimate instead of (7.84), we see that it is sufficient to assume that $(A\psi)^+ \in L^2(\mathcal{D})$. Furthermore, one could also include the terms a_i, a_0, $i = 1, \dots, d$, as long as coercivity of $\mathfrak{a}$ holds with a little further elaboration.

7.6.4 Semilinear variational inequalities

The above results may be extended to semilinear elliptic variational inequalities as well. A convenient class of such problems is the class of problems for which the nonlinearity is such that the resulting nonlinear operator A enjoys the continuity and monotonicity properties of Assumption 7.3.10.

Given a matrix valued function $\mathbf{A} \in L^\infty(\mathcal{D}; \mathbb{R}^{d\times d})$ and a function $f : \mathbb{R} \to \mathbb{R}$, consider the semilinear operator $\mathsf{A} : X \to X^\star$ defined by

$$\langle \mathsf{A}(u), v\rangle = \int_{\mathcal{D}} \mathbf{A}(x)\nabla u(x)\cdot \nabla v(x)dx + \int_{\mathcal{D}} a_0(x)u(x)v(x)dx - \int_{\mathcal{D}} f(u(x))v(x)dx$$

and the associated semilinear variational inequality

$$\begin{aligned} &u \in C : \langle \mathsf{A}(u), v-u\rangle \geq 0, \quad \forall\, v \in C,\\ &C = \{v \in W_0^{1,2}(\mathcal{D}) : v(x) \geq \psi(x), \text{ a. e. } x \in \mathcal{D}\}. \end{aligned} \tag{7.87}$$

As the linear part of the operator, under the standard conditions of the uniform ellipticity Assumption 7.6.2, definitely satisfies the properties of Assumption 7.3.10, it remains to choose f in such a way that it does not upset monotonicity and continuity. To simplify the exposition, we focus on the case of homogeneous Dirichlet conditions. The following proposition (see e. g., [94]) provides a solvability result for such variational inequalities.

Proposition 7.6.24. *Assume that $f : \mathbb{R} \to \mathbb{R}$ is Lipschitz and nonincreasing. Then the semilinear variational inequality* (7.87) *admits a unique solution.*

Proof. We simply need to check whether A satisfies the conditions of Assumption 7.3.10. We break the operator into two contributions: $\mathsf{A} = \mathsf{A}_0 + \mathsf{A}_1$, where A_0 is the elliptic operator associated with the bilinear form (i. e., when we omit the term related to the nonlinear term f), and A_1 is the Nemitsky operator related to the function f. The linear part A_0 is (trivially) Lipschitz continuous, so we simply have to check the nonlinear part A_1. Recalling the definition of the norm of the dual space $X^\star$, we consider any $v, u, w \in X = W_0^{1,2}(\mathcal{D})$. By the Sobolev embedding theorem $v, u, w \in L^2(\mathcal{D})$, hence,

$$\begin{aligned} |\langle \mathsf{A}_1(u) - \mathsf{A}_1(w), v\rangle| &\leq \int_{\mathcal{D}} |f(u(x)) - f(w(x))|\,|v(x)|dx \\ &\leq L\|u-w\|_{L^2(\mathcal{D})}\|v\|_{L^2(\mathcal{D})} \leq Lc\|u-w\|_{W_0^{1,2}(\mathcal{D})}\,\|v\|_{W_0^{1,2}(\mathcal{D})}, \end{aligned}$$

for some appropriate constant $c > 0$ (related to the best constant in the Poincaré inequality). This leads to

$$\|\mathsf{A}_1(u) - \mathsf{A}_1(w)\|_{(W_0^{1,2}(\mathcal{D}))^\star} = \sup_{\|v\|_{W_0^{1,2}(\mathcal{D})}=1} |\langle \mathsf{A}_1(u) - \mathsf{A}_1(w), v\rangle| \leq Lc\,\|u-w\|_{W_0^{1,2}(\mathcal{D})},$$

which is the required Lipschitz continuity result.

For the monotonicity, it suffices to note that

$$\langle \mathsf{A}_1(u) - \mathsf{A}_1(w), u - w\rangle = -\int_{\mathcal{D}} (f(u(x)) - f(w(x))(u(x) - w(x))dx \geq 0,$$

since $f(u(x)) - f(w(x))(u(x) - w(x)) \leq 0$, a. e. $x \in \mathcal{D}$ by the fact that f is nonincreasing. Therefore,

$$\begin{aligned}\langle \mathsf{A}(u) - \mathsf{A}(w), u - w\rangle &= \langle \mathsf{A}_0 u - \mathsf{A}_0 w, u - w\rangle - \langle \mathsf{A}_1(u) - \mathsf{A}_1(w), u - w\rangle \\ &\geq \langle \mathsf{A}_0 u - \mathsf{A}_0 w, u - w\rangle \geq c_{\mathrm{E}} \|u - w\|^2_{W_0^{1,2}(\mathcal{D})},\end{aligned}$$

which is the strict monotonicity condition.

Since A satisfies the conditions of Assumption 7.3.10, applying the Lions–Stampacchia theorem 7.3.11, we obtain the stated result. □

Example 7.6.25. The method of sub- and supersolutions can be extended to the semilinear case to provide bounds for the solutions of (7.87). In particular, if $a_0 > c_{\mathrm{E}} > 0$, we may show that the solution of (7.87) satisfies

$$m := \min\left(0, \frac{f(0)}{c_{\mathrm{E}}}\right) \leq u \leq M := \max\left(\sup_{x\in\mathcal{D}} \psi, \frac{f(0)}{c_{\mathrm{E}}}\right).$$

To show this, use $v = \max(u, m)$ and $v = \min(u, M) = u - (u - M)^+$, respectively, which are both in $W_0^{1,2}(\mathcal{D})$, as test functions in (7.87), working essentially as in the proof of Proposition 7.6.17 along with the monotonicity of f. To illustrate the technique, consider the upper bound. Using the test function $v = \min(u, M) = u - (u - M)^+$ in the weak formulation, we obtain

$$\begin{aligned}\int_{\mathcal{D}} \mathbf{A}\nabla u \nabla (u - M)^+ dx &\leq \int_{\mathcal{D}} (f(u) - a_0 u)(u - M)^+ dx \\ &= \int_{\{u>M\}} (f(u) - a_0 u)(u - M) dx \leq (f(0) - c_{\mathrm{E}} M)|\mathcal{D}| \leq 0,\end{aligned}$$

where we used the fact that since $0 < M < u$, we have $c_{\mathrm{E}} M < c_{\mathrm{E}} u \leq a_0 u$, and since f is not increasing, that $f(u) \leq f(0)$, whereas by the Poincaré inequality, there exists a constant $c_p > 0$ such that $c_p \|(u - M)^+\|_{L^2(\mathcal{D})} \leq \int_{\mathcal{D}} \mathbf{A}\nabla u \nabla (u - M)^+ dx$, so $(u - M)^+ = 0$ a. e.

The special case where $\psi = -\infty$ generalizes the results of Section 6.7 on semilinear elliptic equations to operators more general than the Laplacian. ◁

7.7 Appendix

7.7.1 An elementary lemma

Lemma 7.7.1. *Consider the functions* $\psi_i : \bar{\mathcal{D}} \to \mathbb{R}$, $g_i : \partial\mathcal{D} \to \mathbb{R}$, *such that* $\psi_i \le g_i$ *on* $\partial\mathcal{D}$, $i = 1, 2$, *and the sets*

$$C_i = \{u \in W^{1,2}(\mathcal{D}) : u(x) \ge \psi_i(x),\ a.e.\, x \in \mathcal{D},\ u(x) = g_i(x),\ x \in \partial\mathcal{D}\}, \quad i = 1, 2.$$

If $\psi_1 \ge \psi_2$ *and* $g_1 \ge g_2$, *then*

$$u_1 \vee u_2 := \max(u_1, u_2) \in C_1, \quad \textit{and} \quad u_1 \wedge u_2 := \min(u_1, u_2) \in C_2, \quad \forall\, u_1 \in C_1,\ u_2 \in C_2.$$

Proof. Consider any $u_i \in C_i$, $i = 1, 2$. Letting $w = \max(u_1, u_2)$, we claim that $w \in C_1$.

For any $x \in \mathcal{D}$, either $u_1(x) \ge u_2(x)$ or $u_1(x) \le u_2(x)$. If $x \in \mathcal{D}$ such that $u_1(x) \ge u_2(x)$, then $w(x) = u_1(x)$ and since $u_1 \in C_1$, we conclude that $w(x) \ge \psi_1$. If, on the other hand, $x \in \mathcal{D}$ such that $u_1(x) \le u_2(x)$, then $w(x) = u_2(x)$. Recalling the fact that $u_1 \in C_1$, clearly, $\psi_1(x) \le u_1(x) \le u_2(x) = w(x)$. Therefore, in any case $w \ge \psi_1$.

For any $x \in \partial\mathcal{D}$, either $u_1(x) \ge u_2(x)$ or $u_1(x) \le u_2(x)$. In the first case, $w(x) = \max(u_1(x), u_2(x)) = u_1(x) = g_1(x)$, where the last equality comes from the fact that $u_1 \in C_1$. In the second case, we have on the one hand that

$$w(x) = \max\big(u_1(x), u_2(x)\big) = u_2(x) = g_2(x) \le g_1(x), \tag{7.88}$$

where we used the fact that $u_2 \in C_2$, and that $g_2 \le g_1$. On the other hand, since $u_1 \in C_1$, we have that $g_1(x) = u_1(x) \le u_2(x) = w(x)$, hence, $w(x) \ge g_1(x)$. Combining this with (7.88), we get that $w(x) = g_1(x)$.

We conclude that $w \ge \psi_1$ a. e. on $\mathcal{D}$, and $w = g_1$ a. e. on $\partial\mathcal{D}$, therefore $w \in C_1$. The other claim follows in a similar fashion. □

8 Critical point theory

In this chapter, we focus on a deeper study of critical points for nonlinear functionals in Banach spaces, i. e., on points $x \in X$ such that the Fréchet derivative of the functional vanishes. Such points may correspond to minima, maxima, or saddle points of the functional. Since the Fréchet derivative of a functional can be interpreted as a nonlinear operator equation, the answer to whether a given functional admits a critical point or not has important applications to questions related to the solvability of certain nonlinear operator equations. Furthermore, if the functionals in question are integral functionals, then the study of their critical points is related to the study of nonlinear PDEs. There is a well developed theory for the study of critical points of nonlinear functionals, with important results, such as the mountain pass or the saddle point theorems, which provide conditions for the existence of critical points for a wide class of nonlinear functionals (including integral functionals). In this chapter, we provide a brief introduction to this theory, focusing on applications to integral functionals and nonlinear PDEs. Critical point theory is an important part of nonlinear analysis, and there are several books or lecture notes devoted to it (see, e. g., [79, 91, 120] or [133] upon which our presentation is based).

8.1 Motivation

The general aim of this methodology is to identify a critical point for a given functional $F : X \to \mathbb{R}$. To obtain a geometric intuition, assume that the functional can be understood as a depiction of a surface with level sets $\{x \in X \, : \, F(x) = c\}$. Consider two points on the surface, $(x_0, F(x_0))$ and $(x_1, F(x_1))$, as well as paths on the surface connecting them in the form $(\gamma(\cdot), F(\gamma(\cdot)))$, where $\gamma : [0,1] \to X$ is a continuous curve such that $\gamma(0) = x_0$ and $\gamma(1) = x_1$. The function $F(\gamma(\cdot))$ measures the elevation of the path on the surface from the "plane" X, whereas $\max_{t\in[0,1]} F(\gamma(t))$ is the maximal elevation from the plane along a given path γ. In a more picturesque mode, consider the functional F as representing the surface of a mountain and a hiker wishing to go on a walk on the mountain, from point $(x_0, F(x_0))$ to $(x_1, F(x_1))$; then $\max_{t\in[0,1]} F(\gamma(t))$ is the highest point on the mountain the hiker has reached during the hike. Now consider scanning all possible paths γ between these two points (collected in a set of paths Γ), looking for the one such path that will minimize the maximal elevation, i. e., searching for a solution to the problem $\inf_{\gamma\in\Gamma} \max_{t\in[0,1]} F(\gamma(t))$. Quoting [119], such a path will be the optimal path between the two points since the energy required by the hiker would be minimal. Roughly speaking, if the initial and final points are two low points in the valleys of the mountain, then this path must definitely (unless the mountain is flat) pass through the ridge of the mountain, hence, will pass through a critical point of the functional F.

This intuitive result is in fact the basis of the majority of results presented in this chapter, which, not surprisingly, go under the general name of mountain pass theorems.

https://doi.org/10.1515/9783111333298-008

Such results were obtained in the 1940s and 1950s in finite dimensions by a number of authors, such as for instance Morse or Courant. We will not enter into details for the finite dimensional case but mention a result by Courant that essentially states that a coercive function $f \in C^1(\mathbb{R}^d;\mathbb{R})$ that possesses two distinct strict relative minima x_0, x_1 possesses a third critical point x_2 (distinct from the other two and not a relative minimizer), which in fact can be characterized by $f(x_2) = \inf_{A\in K} \max_{x\in A} f(x)$, where

$$K = \{A \subset \mathbb{R}^d \;:\; A \text{ compact and connected},\; x_0, x_1 \in A\}.$$

Note that the critical point is characterised via a minimax approach.

8.2 The mountain pass and the saddle point theorems

8.2.1 The mountain pass theorem

The mountain pass theorem allows us to show the existence of a critical point z_0 for a functional that we know admits a strict local minimum at some point x_0, i. e., if there exists some $\epsilon > 0$ such that $F(x_0) < F(x)$ for all $x \in B(x_0, \epsilon)$. There are various formulations and variants of the mountain pass theorem, here we present its fundamental version as first introduced by Ambrosetti and Rabinowitz [7] (see also [79, 91, 120]).

Before stating the theorem for the convenience of the reader, we recall the Palais–Smale conditions (see Section 3.6.3).

Definition 8.2.1 (Palais–Smale conditions). A functional[1] $F \in C^1(X;\mathbb{R})$ satisfies the Palais–Smale condition if any sequence $(x_n)_{n\in\mathbb{N}} \subset X$ such that $F(x_n)$ is bounded and $DF(x_n) \to 0$ has a (strongly) convergent subsequence.

Theorem 8.2.2 (Mountain Pass (Ambrosetti–Rabinowitz)). *Let X be a Banach space, and consider $F \in C^1(X;\mathbb{R})$ satisfying the Palais–Smale condition. Assume there exists an open neighborhood[2] U of x_0 and a point $x_1 \notin \overline{U}$ such that*

$$\max(F(x_0), F(x_1)) < c_0 \leq \inf_{x\in\partial U} F(x). \tag{8.1}$$

Consider the set of paths connecting the points $x_0, x_1 \in X$,

$$\Gamma = \{\gamma \in C([0,1];X) \;:\; \gamma(0) = x_0,\; \gamma(1) = x_1\}.$$

Then,

1 i. e., Frechet differentiable with continuous derivative.

2 For example, $U = B_X(x_0, \rho)$ for some $\rho \in (0, \|x_1 - x_0\|)$.

$$c := \inf_{\Gamma} \max_{t \in [0,1]} F(\gamma(t)) \geq c_0$$

is a critical value for the functional F, i. e., there exists $z_0 \in X$ such that $DF(z_0) = 0$ and $F(z_0) = c$.

Proof. The set of paths $Y = C([0,1];X)$, endowed with the norm $\|\gamma\|_Y = \max_{t\in[0,1]} \|\gamma(t)\|_X$ is a Banach space. Then, the set $\Gamma \subset Y$ can be turned into a complete metric space with the metric $d(\gamma_1, \gamma_2) = \|\gamma_1 - \gamma_2\|_Y$. Consider the functional $\Psi : \Gamma \to \mathbb{R}$, defined by $\Psi(\gamma) = \max_{t\in[0,1]} F(\gamma(t))$, which by the assumptions is lower semicontinuous (see also Example 2.2.3). For any $\gamma \in \Gamma$, clearly, $\Psi(\gamma) = \max_{t\in[0,1]} F(\gamma(t)) \geq \max(F(x_0), F(x_1))$ so that taking the infimum over all paths $c := \inf_\Gamma \Psi \geq \max(F(x_0), F(x_1))$, and the functional Ψ is bounded below.

We now apply the Ekeland variational principle (Theorem 3.6.1) to Ψ. For every $\epsilon > 0$, there exists an element of $\Gamma \subset Y$; let us denote it by γ_ϵ, with the properties

$$\begin{aligned} &\Psi(\gamma_\epsilon) \leq c + \epsilon, \\ &\Psi(\gamma) \geq \Psi(\gamma_\epsilon) - \epsilon\|\gamma - \gamma_\epsilon\|_Y, \quad \forall \gamma \in \Gamma. \end{aligned} \tag{8.2}$$

By standard arguments, for the function $t \mapsto F(\gamma_\epsilon(t))$, there exists at least a point in $[0,1]$, where it achieves its maximum, and let $M_\epsilon = \arg\max_{t\in[0,1]}(F \circ \gamma_\epsilon)(t)$ be the set of all such points (depending on ϵ). Clearly, $\Psi(\gamma_\epsilon) = F(\gamma_\epsilon(t_\epsilon))$ for some $t_\epsilon \in M_\epsilon$. Choosing $\gamma(t) = \gamma_\epsilon(t) + \delta h\phi(t)$ for arbitrary $\delta > 0$, $h \in X$ and for some smooth function ϕ such that $\phi(0) = \phi(1) = 0$ with $\phi(t) \equiv 1$ for $t \in [k, 1-k]$ for some suitable $k > 0$, so that ϕ is sufficiently localized around t_ϵ. Using the continuity properties of F and DF and working in a similar fashion as in Proposition 3.6.4, we conclude from the second condition of (8.2) that $\|DF(\gamma_\epsilon(t_\epsilon))\|_{X^\star} < \epsilon$. We then use the Palais–Smale condition and pass to an appropriate subsequence, which converges to a critical point. □

Remark 8.2.3. The mountain pass theorem has originally been proved by a different technique, which uses the so called deformation lemma (see [91]).

Example 8.2.4 (Mountain pass geometry). An example where the mountain pass theorem can be used is if there is a local minimum at x_0 and another (distant) point x_1 such that $F(x_0) > F(x_1)$. Then, the theorem guarantees the existence of a critical point for the functional. Since $c \geq c_0 > \max\{F(x_0), F(x_1)\}$, this point is different from the points x_0 and x_1. ◁

Remark 8.2.5. The claim of Theorem 8.2.2 remains valid, even if the strict inequality in (8.1) is replaced by the weaker condition $\inf_{x:\|x\|=\rho} F(x)\} \geq \max\{F(x_0), F(x_1)\}$, but the proof requires a more careful consideration of the case of equality (the strict inequality case reduces to Theorem 8.2.2, for details, see, e. g., [79]).

8.2.2 Generalizations of the mountain pass theorem

The classical mountain pass theorem admits many important generalizations in various directions. One such direction is, by replacing the paths $\gamma : [0,1] \to X$, along which we monitor the value of the functional $F : X \to \mathbb{R}$, with more general mappings of the form $\gamma : K \to X$, where K is a general compact metric space. In this setup, we replace the set of two points $\{\gamma(0), \gamma(1)\}$, on which we have information concerning the value of the functional, by information on the value of the functional on $\gamma(K_0) \subset X$ for a closed subset $K_0 \subset K$. Then, extending the arguments used in the proof of Theorem 8.2.2, following [79] or [91], we have an interesting generalization of the mountain pass theorem. Alternative proofs of the generalized mountain pass theorem may be found. An interesting proof, which still uses the Ekeland variational principle, but bypasses the machinery of convex analysis, was proposed in [34]; it is the one presented below.

Theorem 8.2.6. *Let X be a Banach space, $X^\star$ its dual, $\langle\cdot,\cdot\rangle$ the duality pairing between X and $X^\star$, and consider $F \in C^1(X;\mathbb{R})$. Let K be a compact metric space and $K_0 \subset K$ be a closed subset of K. Finally, for a given function $\gamma_0 \in C(K_0;X)$, let*

$$\Gamma := \{\gamma \in C(K;X) \,:\, \gamma = \gamma_0 \text{ on } K_0\}, \quad \text{and} \quad c = \inf_{\gamma\in\Gamma}\max_{s\in K} F(\gamma(s)).$$

Suppose that

$$\max_{s\in K} F(\gamma(s)) > \max_{s\in K_0} F(\gamma(s)), \quad \forall\, \gamma \in \Gamma. \tag{8.3}$$

Then, there exists a sequence $(x_n)_{n\in\mathbb{N}}$ such that $F(x_n) \to c$ and $\|DF(x_n)\|_{X^\star} \to 0$. If in addition F satisfies the Palais–Smale condition, then c is a critical value.

Proof. We sketch the proof, which uses a perturbation argument and the Ekeland variational principle to construct a sequence of approximate critical points, which converge to the required critical point. We proceed in 5 steps.

1. For arbitrary $\epsilon > 0$, we define the perturbed functional $F_\epsilon : \Gamma \times K \to \mathbb{R} \cup \{+\infty\}$ by

$$(\gamma, s) \mapsto F_\epsilon(\gamma, s) := F(\gamma(s)) + \epsilon d(s), \quad \text{where } d(s) = \min\{\operatorname{dist}(s, K_0), 1\},$$

and the functional $\Psi_\epsilon : \Gamma \to \mathbb{R} \cup \{+\infty\}$ by

$$\gamma \mapsto \Psi_\epsilon(\gamma) := \max_{s\in K} F_\epsilon(\gamma(s), s).$$

It can be seen that by the stated assumptions Ψ_ϵ satisfies the conditions of the Ekeland variational principle.

2. We apply the Ekeland variatonal principle to Ψ_ϵ to guarantee the existence of a $\gamma_\epsilon \in \Gamma$ such that

$$
\begin{aligned}
\Psi_\epsilon(\gamma) - \Psi_\epsilon(\gamma_\epsilon) + \epsilon d(\gamma_\epsilon, \gamma) \geq 0, \quad \forall \gamma \in \Gamma, \\
c_\epsilon \leq \Psi_\epsilon(\gamma_\epsilon) \leq c_\epsilon + \epsilon,
\end{aligned} \tag{8.4}
$$

where $c_\epsilon = \inf_{\gamma \in \Gamma} \Psi_\epsilon(\gamma)$. Since clearly $c \leq c_\epsilon \leq c + \epsilon$, the second inequality becomes $c \leq \Psi_\epsilon(\gamma_\epsilon) \leq c + 2\epsilon$.

3. We now define $K^{(\epsilon)}_{\max}(\gamma)$ to be the set of maximizers of the perturbed functional $F_\epsilon \circ \gamma$ for a given $\gamma \in \Gamma$, i. e.,

$$
K^{(\epsilon)}_{\max}(\gamma) := \{s \in K \,:\, F_\epsilon(\gamma(s), s) = \Psi_\epsilon(s)\} = \arg\max_{s \in K} F_\epsilon(\gamma(s), s).
$$

By assumption $K^{(\epsilon)}_{\max}(\gamma) \subset K \setminus K_0$. We claim the existence of an $s_\epsilon \in K^{(\epsilon)}_{\max}(\gamma_\epsilon)$ such that $\gamma_\epsilon(s_\epsilon)$ is almost a critical point. In particular, we claim that

$$
\exists\, s_\epsilon \in K^{(\epsilon)}_{\max}(\gamma_\epsilon) \,:\, \|DF(\gamma_\epsilon(s_\epsilon))\|_{X^\star} \leq 2\epsilon. \tag{8.5}
$$

Recalling that ϵ is arbitrary, we make the choice $\epsilon = \frac{1}{n}$, $n \in \mathbb{N}$, to construct a sequence $(\mathrm{x}_n)_{n \in \mathbb{N}}$ with $\mathrm{x}_n = \gamma_{1/n}(s_{1/n})$, for which $F(\mathrm{x}_n) \to c$ and $\|DF(\mathrm{x}_n)\|_{X^\star} \to 0$. If the Palais–Smale condition holds, then c is a critical value for F.

4. The proof of claim (8.5) follows from another claim, that of the existence of a pseudo-gradient vector field for the continuous map $f := DF \circ \gamma : K \to X^\star$ for any fixed $\epsilon > 0$. More precisely, we claim that for any $\epsilon > 0$, there exists a locally Lipschitz map $v_\epsilon : K \to X$ with the property that

$$
\|v_\epsilon(s)\|_X \leq 1, \quad \text{and} \quad \langle f(s), v_\epsilon(s)\rangle \geq \|f(s)\|_{X^\star} - \epsilon, \quad \forall\, s \in K. \tag{8.6}
$$

Assume for the time being that the claim in (8.6) holds. For any $\delta > 0$, sufficiently small, consider

$$
\gamma_{\delta,\epsilon} \in \Gamma \,:\, \gamma_{\delta,\epsilon}(s) = \gamma_\epsilon(s) - \delta \phi_\epsilon(s) v_\epsilon(s),
$$

where v_ϵ is the pseudo-gradient vector field of (8.6) and

$$
\phi_\epsilon \in C(K; [0,1]) \,:\, \phi_\epsilon \equiv 1 \text{ on } K^{(\epsilon)}_{\max}(\gamma_\epsilon), \text{ and } \phi_\epsilon \equiv 0 \text{ on } K_0.
$$

Setting $\gamma = \gamma_{\delta,\epsilon}$ in (8.4), and then calculating the result at a point $s_{\delta,\epsilon} \in \arg\max_{s \in K} \Psi_\epsilon(\gamma_{\delta,\epsilon}) \neq \emptyset$ so that $\Psi_\epsilon(\gamma_{\delta,\epsilon}(s_{\delta,\epsilon})) = F(\gamma_{\delta,\epsilon}(s_{\delta,\epsilon}))$, we obtain that

$$
F(\gamma_\epsilon(s_{\delta,\epsilon}) - \delta \phi_\epsilon(s_{\delta,\epsilon}) v_\epsilon(s_{\delta,\epsilon})) + \epsilon d(s_{\delta,\epsilon}) - \Psi_\epsilon(\gamma_\epsilon) + \epsilon\delta \geq 0. \tag{8.7}
$$

Using the fact that $F \in C^1(K; \mathbb{R})$, we have

$$
F(\gamma_\epsilon(s_{\delta,\epsilon}) - \delta \phi_\epsilon(s_{\delta,\epsilon}) v_\epsilon(s_{\delta,\epsilon})) = F(\gamma_\epsilon(s_{\delta,\epsilon})) - \langle DF(\gamma_\epsilon(s_{\delta,\epsilon})), \delta \phi_\epsilon(s_{\delta,\epsilon}) v_\epsilon(s_{\delta,\epsilon})\rangle + o(\delta),
$$

while, clearly,

$$F(\gamma_\epsilon(s_{\delta,\epsilon})) + \epsilon d(s_{\delta,\epsilon}) = F_\epsilon(\gamma_\epsilon(s_{\delta,\epsilon}), s_{\delta,\epsilon}) \leq \max_{s\in K} F_\epsilon(\gamma_\epsilon(s), s) = \Psi_\epsilon(\gamma_\epsilon).$$

Combining these, we conclude, by (8.7), that

$$\langle DF(\gamma_\epsilon(s_{\delta,\epsilon})), \delta\phi_\epsilon(s_{\delta,\epsilon})v_\epsilon(s_{\delta,\epsilon})\rangle \leq \epsilon + o(1). \tag{8.8}$$

Recall that K is compact. We choose a subsequence $\delta_n \to 0$ such that $s_{\delta_n,\epsilon} \to s_\epsilon$ for some $s_\epsilon \in K^{(\epsilon)}_{\max}$ and pass to the limit in (8.8) using the properties (8.6) of the pseudo-gradient vector field v_ϵ to prove our initial claim (8.5).

5. It remains to prove that a pseudo-gradient vector field can be constructed. This can be done via a partition of unity approach as follows: Consider any continuous map $f : K \to X^\star$, and fix any $s \in K$. By the definition of $\|f(s)\|_{X^\star} = \sup\{\langle f(s), x\rangle : \|x\|_X \leq 1\}$ for any $\epsilon > 0$, we may find $x_\epsilon(s) \in X$, with $\|x_\epsilon(s)\|_X \leq 1$ such that $\langle f(s), x_\epsilon(s)\rangle \geq \|f(s)\|_{X^\star} - \epsilon$. By the continuity of f, there exists a neighborhood $N(s)$ of $s \in K$ such that this inequality holds in the whole neighborhood, i. e., $\langle f(s'), x_\epsilon(s)\rangle \geq \|f(s')\|_{X^\star} - \epsilon$ for every $s' \in N(s)$. Consider the open cover $\mathscr{C} = \{N(s) : s \in K\}$ of K, which, by compactness, admits a locally finite refinement $\{N_i : i = 1,\dots M\}$ such that for any $i = 1,\dots,M$ there exists a set $\{s_i : i = 1,\dots,M\} \subset K$ with the corresponding neighborhoods $N(s_i)$ having the property $N_i \subset N(s_i)$. For each (fixed) $s_i \in K$, we can work as above and obtain a local pseudo-gradient vector x_i, which works for every $s' \in N_i$. Then, we may construct the pseudo-gradient vector field by patching up these local vectors, as follows: Define the functions $\rho_i : K \to [0,1]$ by $\rho_i(s) = \frac{d_i(s)}{\sum_{j=1}^M d_j(s)}$, where $d_i(s) = \operatorname{dist}(s, K \setminus N_i)$, which are locally Lipshitz continuous and the vector field $v : K \to X$ by $v(s) = \sum_{i=1}^M \rho_i(s)x_i$, which essentially assigns to any s the vector x_i, as long as $s \in N_i$. This mapping has the desired properties. □

8.2.3 The saddle point theorem

An interesting extension of the mountain pass theorem is the saddle point theorem [79, 91, 120].

Theorem 8.2.7 (Saddle Point). *Let X be a Banach space, and $F \in C^1(X;\mathbb{R})$ a functional satisfying the Palais–Smale condition. Let $X_1 \subset X$ be a finite dimensional subspace and X_2 the complement of X_1, i. e., $X = X_1 \oplus X_2$. Assume that there exists $r > 0$, such that*

$$\max_{x\in\partial X_1(r)} F(x) \leq a < b \leq \inf_{x\in X_2} F(x), \tag{8.9}$$

where $X_1(r) := X_1 \cap B(0,r)$ and $\partial X_1(r) = \{x \in X_1,\ \|x\| = r\}$. Let

$$\Gamma = \{\gamma \in C(\overline{X_1(r)}; X) : \gamma(x) = x,\ \forall x \in \partial X_1(r)\},$$

and define

$$c := \inf_{\gamma\in\Gamma} \sup_{\mathrm{x}\in\overline{X_1(r)}} F(\gamma(\mathrm{x})).$$

Then, $c > -\infty$ and is a critical value of the functional F.

Proof. Let $K = \overline{X_1(r)}$ and $K_0 = \partial X_1(r)$. These are compact sets, since X_1 is finite dimensional. To apply the general minimax Theorem 8.2.6, we need to show that

$$\max_{t\in K} F(\gamma(t)) > \max_{t\in K_0} F(\gamma(t)), \quad \forall \gamma \in \Gamma.$$

Note that now t is interpreted as $\mathrm{x} \in X_1$, i. e., as a vector in $\mathbb{R}^m$, where $m = \dim(X_1)$. With our choice for K and K_0, we have that

$$\max_{t\in K_0} F(\gamma(t)) = \max_{\mathrm{x}\in\partial X_1(r)} F(\gamma(\mathrm{x})) \le a.$$

It suffices to prove that there exists $\mathrm{x}_0 \in \overline{X_1(r)}$ such that $\gamma(\mathrm{x}_0) \in X_2$, since by assumption $\inf_{X_2} F \ge b$, and evidently $\max_{t\in K} F(\gamma(t)) \ge \inf_{X_2} F \ge b$.

Let P_{X_1} be the projection onto X_1 (a linear map). In order to show that there exists $\mathrm{x}_0 \in \overline{X_1(r)}$ such that $\gamma(\mathrm{x}_0) \in X_2$, it is equivalent to show that $\mathsf{P}_{X_1}\gamma : \overline{X_1(r)} \to X_1$ has a zero.[3] Note also that the map $\mathsf{P}_{X_1}\gamma|_{\partial X_1(r)} = I$ is the identity map, since by construction of Γ, $\gamma(\mathrm{x}) = \mathrm{x}$ for $\mathrm{x} \in \partial X_1(r)$, therefore, $\langle \mathsf{P}_{X_1}\gamma(x), x\rangle \ge 0$ for all $\|x\| = r$. Then, by Brouwer's fixed point theorem (and in particular Corollary 3.2.6), there exists a point x_0, $\|x_0\| \le r$ such that $\mathsf{P}_{X_1}\gamma(x_0) = 0$. □

8.3 Applications in semilinear elliptic problems

The general theory presented above finds important applications in PDEs. We present here a selection of applications (inspired by [79, 91, 120, 133]).

As usual, let $\mathcal{D} \subset \mathbb{R}^d$ be a sufficiently smooth bounded domain, and consider the nonlinear elliptic PDE

$$\begin{aligned} -\Delta u(x) &= \lambda u + f(x,u), \quad x \in \mathcal{D} \\ u(x) &= 0 \quad x \in \partial\mathcal{D}, \end{aligned} \tag{8.10}$$

where $\lambda \in \mathbb{R}$ is an appropriate parameter, and $f : \mathcal{D} \times \mathbb{R} \to \mathbb{R}$ is a given function. As already seen in Chapter 6 and in particular Section 6.7, under certain conditions, equation (8.10) can be understood as the Euler–Lagrange equation of the functional $F : X := W_0^{1,2}(\mathcal{D}) \to \mathbb{R}$, where

3 Recall that $\mathrm{y} \in X_2$ is equivalent to $\mathsf{P}_{X_1}\mathrm{y} = 0$.

$$F(u) = \int_{\mathcal{D}} \Big(\frac{1}{2} |\nabla u(x)|^2 - \frac{\lambda}{2} u(x)^2 - \Phi(x, u(x)) \Big) \, dx,$$
$$\Phi(x, u) = \int_0^u f(x, s) ds. \tag{8.11}$$

Therefore, the problem of solvability for equation (8.10) can be understood as equivalent to searching for critical points for the functional F.

Using the direct method of the calculus of variations under appropriate conditions on f, we have shown in Section 6.7 the existence of minima for F, which in turn are obtained in terms of the Euler–Lagrange equation as solutions of the PDE (8.10). However, not all solutions to (8.10) can be characterized as minima of F. Therefore, in this chapter, having obtained some familiarity with the study of critical points for functionals, we will take a complementary route, and try to show how we can obtain a critical point of the functional (8.11), thus providing a solvability result for (8.10). The general tools developed in the previous section will provide valuable help in this direction, thus leading to a better understanding of solutions of equations of the form (8.10). We will show that by varying the conditions on the nonlinearity f different tools, such as the mountain pass or the saddle point theorem, will become handy. As we will see, the value of the parameter $\lambda \in \mathbb{R}$ will play an important role in our study and in particular its value compared with the eigenvalues of the Laplacian operator $-\Delta$, in the afore mentioned Sobolev space setting. Recalling the discussion in Section 6.2.3, this problem admits a set of solutions $(\lambda_n)_{n\in\mathbb{N}}$, $0 < \lambda_1 \leq \lambda_2 \leq \cdots \leq \lambda_n \leq \cdots$ and $\lambda_n \to \infty$ as $n \to \infty$, with corresponding eigenfunctions $(u_n)_{n\in\mathbb{N}}$. Of special interest is the comparison of λ with the first eigenvalue of the Laplacian on $\mathcal{D}$ with Dirichlet boundary conditions, $\lambda_1 > 0$. The reader should note that the importance of the spectrum of the Laplacian in the treatment of (8.10) arises from the fact that if $\lambda = \lambda_n$ for some $n \in \mathbb{N}$, then the operator $\mathsf{A} = -\Delta - \lambda_n I$ is not boundedly invertible, so the approach of expressing (8.10) in terms of the compact operator $\mathsf{T} = (-\Delta - \lambda I)^{-1}$ and using a fixed point scheme to construct solutions (as done for instance in Section 7.6) cannot be applied at least in a straightforward fashion. Problems of the type (8.10), when $\lambda = \lambda_n$ for some $n \in \mathbb{N}$, will be called resonant problems.

8.3.1 Superlinear growth at infinity

We first study the case, where f is a nonlinear function, satisfying a superlinear growth condition at infinity and displaying slower than linear growth at 0 (so that for small values of the solution the linear problem is a good approximation to the semilinear problem). These conditions are formulated in the assumption below (in the case where the dimension of the domain is $d \geq 3$).

Assumption 8.3.1.

(i) The function $f : \mathcal{D}\times\mathbb{R} \to \mathbb{R}$ is a Carathéodory function, which for every $(x, s) \in \mathcal{D}\times\mathbb{R}$ satisfies the growth condition $|f(\cdot, s)| \le c|s|^r + a_1$, a. e. in $\mathcal{D}$, where $1 < r < \mathfrak{s}_2 - 1$ with $\mathfrak{s}_2 = \frac{2d}{d-2}$, $d \ge 3$, the critical Sobolev exponent for the embedding $W_0^{1,2}(\mathcal{D}) \hookrightarrow L^2(\mathcal{D})$ to hold.[4]

(ii) For small s the function f has the following behavior: $\lim_{s\to 0} \frac{f(x,s)}{|s|} = 0$ uniformly in $x \in \mathcal{D}$.

(iii) For large s the function f is superquadratic: There exist $a > 2$ and $R > 0$ such that $0 < a\Phi(x, s) \le sf(x, s)$ for $|s| \ge R$, uniformly in $x \in \overline{\mathcal{D}}$.

An example of a function satisfying this assumption is $f(x, s) = s^{\frac{d+2}{d-2}-\epsilon}$, $\epsilon > 0$, $d \ge 3$.[5] An existence result is given in the following proposition [79, 120].

Proposition 8.3.2. *Under Assumption* 8.3.1, *if* $\lambda < \lambda_1$, *where* $\lambda_1 > 0$ *is the first eigenvalue of the Dirichlet Laplacian on* $\mathcal{D}$, *the nonlinear elliptic problem* (8.10) *admits a positive nontrivial solution.*

Proof. We will apply the mountain pass theorem to show that F admits a critical point, which is the solution we seek. For that, we need to check that F satisfies the conditions for the mountain pass theorem to hold. In particular, we need to show that F is C^1, satisfies the Palais–Smale condition, as well as the existence of a $u_1 \in X$ and a $\rho > 0$ such that $0 = \max\{F(0), F(u_1)\} < \inf_{u\in X,\ \|u\|=\rho} F(u)$. In particular, we will show that 0 is a strict local minimum for F (see Example 8.2.4). Then an application of the mountain pass theorem (see Theorem 8.2.2) will guarantee the existence of a critical point, different than 0, hence, a nontrivial solution.

We proceed in 4 steps, leaving the verification of the Palais–Smale condition last.

1. On account of the growth assumption 8.3.1(i), the functional $F : W_0^{1,2}(\mathcal{D}) \to \mathbb{R}$ is C^1 (see the proof of Proposition 6.7.6).

2. We now identify $u = 0$ as a strict local minimum of F. To see this, we first note that the conditions on the nonlinearity f imply that for every $\epsilon > 0$, there exists δ such that $|f(x, s)| < \epsilon|s|$ for $|s| < \delta$, which upon integration implies that $|\Phi(x, s)| \le \frac{1}{2}\epsilon|s|^2$ for $|s| < \delta$, which provides information on Φ for small s. In particular, this estimate yields $F(0) = 0$. For the large s behavior, we use Assumption 8.3.1(i) to remark that $|\Phi(x, s)| < c'|s|^{r+1}$ for some $c' > 0$, and for $|s| > \delta$. Therefore, for any $s \in \mathbb{R}$, it holds that $|\Phi(x, s)| < \frac{1}{2}\epsilon|s|^2 + c'|s|^{r+1}$. Using this estimate for Φ allows us to provide a lower bound for the functional F as follows:

$$F(u) \ge \frac{1}{2}\|\nabla u\|^2_{L^2(\mathcal{D})} - \frac{\lambda + \epsilon}{2}\|u\|^2_{L^2(\mathcal{D})} - c'\|u\|^{r+1}_{L^{r+1}(\mathcal{D})}.$$

---

4 See Theorem 1.5.11(i).

5 We note that In the case where $d = 1, 2$, the growth condition in Assumption 8.3.1(i) is not needed as long as $r > 1$.

Note that the second and the third term are subordinate to the first term. Recall the Poincaré inequality according to which $\|\nabla u\|^2_{L^2(\mathcal{D})} \geq \lambda_1 \|u\|^2_{L^2(\mathcal{D})}$, where $\lambda_1 > 0$ is the first eigenvalue of the operator $-\Delta$ with homogeneous Dirichlet boundary conditions on $\mathcal{D}$, and the Sobolev embedding according to which for the choice of r, there exists c_1 such that $\|u\|_{L^{r+1}(\mathcal{D})} \leq c_1 \|\nabla u\|_{L^2(\mathcal{D})}$. This leads to a lower bound for F of the form

$$F(u) \geq \frac{1}{2}\left(1 - \frac{\lambda + \epsilon}{\lambda_1}\right)\|\nabla u\|^2_{L^2(\mathcal{D})} - c_2 \|\nabla u\|^{r+1}_{L^2(\mathcal{D})}. \tag{8.12}$$

Since $\lambda < \lambda_1$, $\epsilon > 0$ is arbitrarily small, and $r > 1$, there exists a $\rho > 0$ such that for any u satisfying $\|\nabla u\|_{L^2(\mathcal{D})} < \rho$, $F(u) > 0 = F(0)$, hence, 0 is a strict local minimum.

3. We now try to show the existence of a $u_1 \in X$ such that $F(u_1) < F(0)$. It suffices to show that for any $u_0 \neq 0$, it holds that $F(\rho u_0) \to -\infty$, as $\rho \to \infty$, thus (moving along this radial direction) guaranteeing the existence of a u_1 with the required properties.

Indeed, by Assumption 8.3.1(iii),[6] we see that for large enough $|s| > R$, $\Phi(x, s) \geq |s|^a + c_3$, for some constant c_3. On the other hand, by the growth Assumption 8.3.1(i) and upon integrating, we see that for any s, it holds that $\Phi(x, s) \geq \frac{c}{r}|s|^{r+1} + a_1 s$, so for any $s \in \mathbb{R}$, there exist two constants, $c_4 > 0$ and c_5, such that $\Phi(x, s) \geq c_4 |s|^a + c_5$. Using this estimate, we obtain the upper bound for the functional

$$F(u) \leq \frac{1}{2}\|\nabla u\|^2_{L^2(\mathcal{D})} - \frac{\lambda}{2}\|u\|^2_{L^2(\mathcal{D})} - c_4 \|u\|^a_{L^a(\mathcal{D})} - c_5 |\mathcal{D}|. \tag{8.13}$$

Let us choose a u_0 such that $\|\nabla u_0\|_{L^2(\mathcal{D})} = 1$. By the Poincaré inequality $\|u_0\|_{L^2(\mathcal{D})} \leq c' \|\nabla u_0\|_{L^2(\mathcal{D})}$, and set $\beta := \|u_0\|_{L^a(\mathcal{D})}$ and $\gamma := \|u_0\|_{L^2(\mathcal{D})} < \infty$. Now consider $u = \rho u_0$. The upper bound for F implies that

$$F(\rho u_0) \leq \frac{1}{2}\rho^2 - \frac{\lambda}{2}\rho^2 \gamma^2 - c_4 \rho^a \beta^a - c_5 |\mathcal{D}| \to -\infty, \quad \text{as } \rho \to \infty,$$

since $a > 2$, as claimed.

By the estimates above, we have the existence of a $u_1 \in X$ and a $\rho > 0$ such that $0 = \max\{F(0), F(u_1)\} < \inf_{u \in X,\, \|u\| = \rho} F(u)$. If F satisfies the Palais–Smale condition, then an application of the mountain pass theorem (see Theorem 8.2.2) will guarantee the existence of a critical point, different than 0, hence, a nontrivial solution.

4. It therefore remains to check the Palais–Smale condition. Consider a Palais–Smale sequence $(u_n)_{n \in \mathbb{N}}$. We first note that $\|u_n\|_X = \|\nabla u_n\|_{L^2(\mathcal{D})}$ is bounded for any n. Indeed, assume that the sequence u_n satisfies $|F(u_n)\| \leq c'$ and $DF(u_n) \to 0$. Then, it holds that

$$|\langle DF(u_n), u_n \rangle| \leq \|DF(u_n)\|_{X^*} \|u_n\|_X \leq \epsilon \|u_n\|_X \tag{8.14}$$

for large enough n. Therefore,

6 Rewrite Assumption 8.3.1(iii), as $0 < a\Phi \leq s\Phi'$ and integrate.

$$
\begin{aligned}
-\epsilon\|\nabla u_n\|_{L^2(\mathcal{D})} &\le \langle DF(u_n), u_n\rangle \\
&= \|\nabla u_n\|^2_{L^2(\mathcal{D})} - \lambda\|u_n\|^2_{L^2(\mathcal{D})} - \int_{\mathcal{D}} f(x, u_n(x))\, u_n(x)dx \le \epsilon\|\nabla u_n\|_{L^2(\mathcal{D})},
\end{aligned}
$$

so that using the lower bound, we obtain an estimate for the term involving f,

$$
-\int_{\mathcal{D}} f(x, u_n(x))\, u_n(x)dx \ge -\|\nabla u_n\|^2_{L^2(\mathcal{D})} - \epsilon\|\nabla u_n\|_{L^2(\mathcal{D})} + \lambda\|u_n\|^2_{L^2(\mathcal{D})}. \tag{8.15}
$$

On the other hand,

$$
F(u_n) = \frac{1}{2}\|\nabla u_n\|^2_{L^2(\mathcal{D})} - \frac{\lambda}{2}\|u_n\|^2_{L^2(\mathcal{D})} - \int_{\mathcal{D}} \Phi(x, u_n(x))dx \le c'.
$$

We rearrange

$$
\begin{aligned}
F(u_n) &= \frac{1}{2}\|\nabla u_n\|^2_{L^2(\mathcal{D})} - \frac{\lambda}{2}\|u_n\|^2_{L^2(\mathcal{D})} \\
&\quad - \int_{\mathcal{D}} \Big(\Phi(x, u_n(x)) - \frac{1}{a}f(x, u_n(x))u_n(x)\Big)dx - \frac{1}{a}\int_{\mathcal{D}} f(x, u_n(x))u_n(x)dx \\
&\ge \Big(\frac{1}{2} - \frac{1}{a}\Big)\|u_n\|^2_X - \lambda\Big(\frac{1}{2} - \frac{1}{a}\Big)\|u_n\|^2_{L^2(\mathcal{D})} - \frac{\epsilon}{a}\|u_n\|_X \\
&\quad - \int_{\mathcal{D}} \Big(\Phi(x, u_n(x)) - \frac{1}{a}f(x, u_n(x))u_n(x)\Big)dx,
\end{aligned} \tag{8.16}
$$

where we used the estimate (8.15) and $\|u_n\|_X = \|\nabla u_n\|_{L^2(\mathcal{D})}$. Recalling Assumption 8.3.1(iii), upon defining for any fixed $n \in \mathbb{N}$,

$$
\mathcal{D}_{n,+} = \{x \in \mathcal{D} : |u_n(x)| > R\}, \quad \mathcal{D}_{n,-} = \{x \in \mathcal{D} : |u_n(x)| \le R\},
$$

we see that, upon breaking the integral over $\mathcal{D}$ to a sum over the integral on $\mathcal{D}_{n,+}$ and $\mathcal{D}_{n,-}$,

$$
\begin{aligned}
&-\int_{\mathcal{D}} \Big(\Phi(x, u_n(x)) - \frac{1}{a}f(x, u_n(x))u_n(x)\Big)dx \\
&\quad \ge -\int_{\mathcal{D}_{n,-}} \Big(\Phi(x, u_n(x)) - \frac{1}{a}f(x, u_n(x))u_n(x)\Big)dx \ge c_0,
\end{aligned}
$$

for some constant c_0 independent of n, where, for the first estimate, we used Assumption 8.3.1(iii) to eliminate the integral on $\mathcal{D}_{n,+}$, and for the second we used the fact that on $\mathcal{D}_{n,-}$, $|u_n| < R$ (which is independent of n), and therefore both f and Φ are also bounded by a bound independent of n. Using this last estimate in (8.16), along with the facts that $F(u_n) \leq c'$, we see that

$$c' \geq \left(\frac{1}{2} - \frac{1}{a}\right)\|u_n\|_X^2 - \lambda\left(\frac{1}{2} - \frac{1}{a}\right)\|u_n\|_{L^2(\mathcal{D})}^2 - \epsilon\|u_n\|_X + c_0.$$

Since $a > 2$, this implies using the Poincaré inequality that, if $\lambda > 0$,

$$c' \geq \left(\frac{1}{2} - \frac{1}{a}\right)\left(1 - \frac{\lambda}{\lambda_1}\right)\|u_n\|_X^2 - \epsilon\|u_n\|_X + c_0,$$

or if $\lambda < 0$ that

$$c' \geq \left(\frac{1}{a} - \frac{1}{2}\right)\|u_n\|_X^2 - \epsilon\|u_n\|_X + c_0.$$

So, in any case, since $\epsilon > 0$ is arbitrary, $a > 2$, and $\lambda < \lambda_1$, we conclude the existence of a constant c_1 (independent of n) such that $\|u_n\|_X < c_1$. By the reflexivity of X, there exists a weakly converging in X subsequence $u_{n_k} \rightharpoonup \bar{u}$ for some $\bar{u} \in X$. By the compact embedding of $X = W_0^{1,2}(\mathcal{D})$ into $L^{r+1}(\mathcal{D})$, there exists a further subsequence (denoted the same; see also Remark 1.5.12) so that $u_{n_k} \to \bar{u}$ in $L^{r+1}(\mathcal{D})$. This, taking into account the growth conditions on f, implies that the operator $\mathsf{K} : X \to X^\star$, defined by $u \mapsto f(x, u)$, is compact. We also note that $DF(u) = \mathsf{L}u + \mathsf{K}u$, with $\mathsf{L} = -\Delta - \lambda I : X \to X^\star$ being boundedly invertible for $\lambda < \lambda_1$. Consider a Palais–Smale sequence $(u_n)_{n\in\mathbb{N}}$, i. e., $\mathsf{L}u_n + \mathsf{K}u_n \to 0$. This is expressed as $u_n + \mathsf{L}^{-1}\mathsf{K}(u_n) \to 0$ so that since $\mathsf{L}^{-1}\mathsf{K}$ is compact and $(u_n)_{n\in\mathbb{N}}$ is bounded, then $(\mathsf{L}^{-1}\mathsf{K}(u_n))_{n\in\mathbb{N}}$ has a convergent subsequence (denoted the same), and since $u_n + \mathsf{L}^{-1}\mathsf{K}(u_n) \to 0$, we conclude that $(u_n)_{n\in\mathbb{N}}$ has a convergent subsequence as well, hence, the Palais–Smale property holds. □

8.3.2 Nonresonant semilinear problems with asymptotic linear growth at infinity and the saddle point theorem

We now consider the elliptic problem (8.10) but dropping the condition of superlinear growth for the nonlinearity at infinity, assuming instead that f has asymptotically linear growth at infinity (with respect to s). To simplify the exposition and the arguments, we assume a slightly modified version of (8.10) in the form

$$\begin{aligned} -\Delta u &= \lambda u + f_0(u) + h, \quad \text{in } \mathcal{D}, \\ u &= 0, \quad \text{on } \partial\mathcal{D}, \end{aligned} \tag{8.17}$$

for $u \in W_0^{1,2}(\mathcal{D})$, and given $\lambda \in \mathbb{R}$, and h a function on $\mathcal{D}$. We first focus on $\lambda \neq \lambda_n$, where $(\lambda_n)_{n\in\mathbb{N}}$ are the eigenvalues of the Dirichlet Laplacian. We assume that $d \geq 3$.

We impose the following conditions on f_0 and h:

Assumption 8.3.3.
(i) The function $f_0 : \mathbb{R} \to \mathbb{R}$ satisfies the growth condition $|f_0(s)| \leq c|s|^r + a_1$ for every $s \in \mathbb{R}$, where $1 < r < \mathfrak{s}_2 - 1$ with $\mathfrak{s}_2 = \frac{2d}{d-2}$, $d \geq 3$, the critical Sobolev exponent for the embedding $W_0^{1,2}(\mathcal{D}) \hookrightarrow L^2(\mathcal{D})$ to hold,[7] and $h \in L^{r^*}(\mathcal{D})$.
(ii) $\lim_{s\to\pm\infty} \frac{f(x,s)}{s} = \gamma_\pm$, for some $\gamma_\pm \in \mathbb{R}$.

In the cases where $d = 1, 2$, the growth condition in Assumption 8.3.3 is not required. The following proposition [79, 120] provides an existence result.

Proposition 8.3.4. *Let f_0, h satisfy Assumption 8.3.3, and assume that $\lambda_k < \lambda + \gamma_\pm < \lambda_{k+1}$ for some $k \geq 1$, where $(\lambda_n)_{n\in\mathbb{N}}$ are the eigenvalues of the Dirichlet Laplacian on $\mathcal{D}$. Then, (8.17) has a nontrivial weak solution for every $h \in L^{r^*}(\mathcal{D})$.*

Proof. We will look for weak solutions of (8.17), which will be identified as critical points of the integral functional $F : X = W_0^{1,2}(\mathcal{D}) \to \mathbb{R}$, defined in (8.11), for $f(x,s) = f_0(s) + h(x)$. We will use the saddle point Theorem 8.2.7, identifying a splitting of $X = X_1 \oplus X_2$, with X_1 finite dimensional such that condition (8.9) of this theorem is satisfied. Since $\lambda_k < \lambda + \gamma_\pm < \lambda_{k+1}$, we choose

$$X_1 := \operatorname{span}\{\phi_1, \dots, \phi_k\}, \quad X_2 := \overline{\operatorname{span}\{\phi_i \,:\, i \geq k+1\}}, \tag{8.18}$$

where ϕ_i are the eigenfunctions of the Dirichlet Laplacian normalized so that

$$\|\nabla \phi_n\|^2_{L^2(\mathcal{D})} = 1, \quad \text{and,} \quad \|\phi_n\|^2_{L^2(\mathcal{D})} = \frac{1}{\lambda_n}, \qquad n \in \mathbb{N}. \tag{8.19}$$

These are orthogonal with respect to the inner product in $L^2(\mathcal{D})$ (see Proposition 6.2.12), so $X = X_1 \oplus X_2$. We intend to show that this choice actually satisfies (8.9). We proceed in 6 steps, leaving the verification of the Palais–Smale condition last.

1. By the growth condition in Assumption 8.3.3(i), this functional is $C^1(X;\mathbb{R})$ (see the proof of Proposition 6.7.6).

2. Since $\lambda_k < \lambda + \gamma_\pm < \lambda_k$, let us choose μ_1, μ_2 such that $\lambda_k < \mu_1 < \lambda + \gamma_\pm < \mu_2 < \lambda_{k+1}$, so by Assumption 8.3.3(ii), there exists $R > 0$ such that $\mu_1 < \lambda + \frac{f_0(s)}{s} < \mu_2$ for $|s| > R$,. Upon integration $\frac{1}{2}\mu_1 s^2 + c_1' \leq \frac{\lambda}{2}s^2 + \Phi_0(s) \leq \frac{1}{2}\mu_2 s^2 + c_2'$ for every $s \in \mathbb{R}$, where Φ_0 is the antiderivative of f_0, and c_1', c_2' are two appropriate constants.

3. For any $u \in \sum_{n\geq 1} c_i \phi_i$, $\|u\|_X^2 = \sum_{i\geq 0} c_i^2$ and $\|u\|^2_{L^2(\mathcal{D})} = \sum_{i\geq 1} \frac{1}{\lambda_i} c_i^2$. For any $v \in X_1$, i. e., $v = \sum_{i=1}^k c_i\phi_i$ for some $c_i \in \mathbb{R}$, so by linearity $-\Delta v = \sum_{i=1}^k \lambda_i c_i \phi_i$. Multiplying by v, integrating by parts and using the orthogonality of the ϕ_i in $L^2(\mathcal{D})$, we conclude that

$$\int_{\mathcal{D}} |\nabla v(x)|^2 dx = \sum_{i=1}^k \lambda_i c_i^2 \|\phi_i\|^2_{L^2(\mathcal{D})} \leq \lambda_k \|v\|^2_{L^2(\mathcal{D})}, \tag{8.20}$$

7 See Theorem 1.5.11(i).

where for the last estimate we used the fact that $\lambda_i \leq \lambda_k$ for $i = 1, \dots, k$. With this estimate at hand, using the bounds for $\frac{\lambda}{2}s^2 + \Phi_0(s)$ from step 2, we see that for every $v \in X_1$,

$$\begin{aligned} F(v) &\leq \frac{1}{2}\|\nabla v\|^2_{L^2(\mathcal{D})} - \frac{1}{2}\mu_1\|v\|^2_{L^2(\mathcal{D})} - c_1'|\mathcal{D}| + \|h\|_{L^2(\mathcal{D})}\|v\|_{L^2(\mathcal{D})} \\ &\leq \frac{1}{2}\left(1 - \frac{\mu_1}{\lambda_k}\right)\|\nabla v\|^2_{L^2(\mathcal{D})} - c_1'|\mathcal{D}| + \|h\|_{L^2(\mathcal{D})}\|u\|_{L^2(\mathcal{D})} \\ &\leq \frac{1}{2}\left(1 - \frac{\mu_1}{\lambda_k}\right)\|\nabla v\|^2_{L^2(\mathcal{D})} - c_1'|\mathcal{D}| + \frac{1}{\lambda_1^{1/2}}\|h\|_{L^2(\mathcal{D})}\|\nabla v\|_{L^2(\mathcal{D})}, \end{aligned}$$

where, we first used (8.20) and then the Poincaré inequality $\|\nabla v\|^2_{L^2(\mathcal{D})} \geq \lambda_1\|v\|^2_{L^2(\mathcal{D})}$, which is valid for any $v \in W_0^{1,2}(\mathcal{D})$. Since $\mu_1 > \lambda_k$, the above estimate implies that $F(v) \to -\infty$ for any $v \in X_1$ with $\|v\|_X = \|\nabla v\|_{L^2(\mathcal{D})} \to \infty$.

4. Now consider any $w \in X_2$. This can be expressed as $w = \sum_{i=k+1}^{\infty} c_i\phi_i$ and, by similar arguments as above, we can see that for any $w \in X_2$, it holds that $\|\nabla w\|^2_{L^2(\mathcal{D})} \geq \lambda_{k+1}\|w\|^2_{L^2(\mathcal{D})}$. Using that, we then estimate F from below as

$$\begin{aligned} F(w) &\geq \frac{1}{2}\|\nabla w\|^2_{L^2(\mathcal{D})} - \frac{1}{2}\mu_2\|w\|^2_{L^2(\mathcal{D})} - c_2'|\mathcal{D}| - \|h\|_{L^2(\mathcal{D})}\|w\|_{L^2(\mathcal{D})} \\ &\geq \frac{1}{2}\left(1 - \frac{\mu_2}{\lambda_{k+1}}\right)\|\nabla w\|^2_{L^2(\mathcal{D})} - c_2'|\mathcal{D}| - \|h\|_{L^2(\mathcal{D})}\|w\|_{L^2(\mathcal{D})} \\ &\geq \frac{1}{2}\left(1 - \frac{\mu_2}{\lambda_{k+1}}\right)\|\nabla w\|^2_{L^2(\mathcal{D})} - c_2'|\mathcal{D}| - \frac{1}{\lambda_1^{1/2}}\|h\|_{L^2(\mathcal{D})}\|\nabla w\|_{L^2(\mathcal{D})}, \end{aligned}$$

where for the last estimate we used the Poincaré inequality. Since $\mu_2 < \lambda_{k+1}$, the above estimate implies that $F(w) \to \infty$ for any $w \in X_2$ with $\|w\|_X = \|\nabla w\|_{L^2(\mathcal{D})} \to \infty$, and also provides a finite lower bound for $F(w)$ for any $w \in X_2$.

5. The fact that $F(v) \to -\infty$ for any $v \in X_1$ with $\|v\|_X = \|\nabla v\|_{L^2(\mathcal{D})} \to \infty$, while $F(w)$ has a finite lower bound for any $w \in X_2$, implies the existence of a $\rho > 0$ such that $\max_{v\in X_1,\ \|v\|=\rho} F(v) < \inf_{w\in X_2} F(w)$, so that, if the Palais–Smale condition holds, we may apply the saddle point Theorem 8.2.7 and conclude the existence of a nontrivial critical point for F, which corresponds to the solution we seek.

6. It only remains to check the validity of the Palais–Smale condition. By similar arguments as in the proof of Proposition 8.3.2 (Step 4), it suffices to check that any Palais–Smale sequence is bounded in $X = W_0^{1,2}(\mathcal{D})$. We argue by contradiction. Consider any Palais–Smale sequence $(u_n)_{n\in\mathbb{N}}$, i. e., a sequence such that $F(u_n) < c'$ and $\|DF(u_n)\|_{X^*} \to 0$, while at the same time $\|u_n\| \to \infty$ as $n \to \infty$. Define $\psi_n = \frac{u_n}{\|u_n\|_X}$, which satisfies $\|\psi_n\|_X = 1$ for every n. Since $\|DF(u_n)\|_{X^*} \to 0$, we have that for any $\epsilon > 0$ and any $\phi \in X = W_0^{1,2}(\mathcal{D})$ (see (8.14) and the following analysis), we have,

$$-\epsilon\|\phi\|_X \leq \int_{\mathcal{D}} (\nabla u_n(x)\cdot\nabla\phi(x) - \lambda u_n(x)\phi(x) - f_0(u_n(x))\phi(x) - h(x)\phi(x))dx \leq \epsilon\|\phi\|_X, \quad (8.21)$$

for n large enough. Dividing through by $\|u_n\|_X$ leads to the conclusion that

$$-\epsilon\frac{\|\phi\|_X}{\|u_n\|_X} \leq \int_{\mathcal{D}}\Big(\nabla\psi_n(x)\cdot\nabla\phi(x) - \lambda\psi_n(x)\phi(x) - \frac{1}{\|u_n\|_X}f_0(u_n(x))\phi(x) - h(x)\phi(x)\Big)dx \leq \epsilon\frac{\|\phi\|_X}{\|u_n\|_X},$$

so

$$\lim_{n\to\infty}\int_{\mathcal{D}}\Big(\nabla\psi_n(x)\cdot\nabla\phi(x) - \lambda\psi_n(x)\phi(x) - \frac{1}{\|u_n\|_X}f_0(u_n(x))\phi(x) - \frac{1}{\|u_n\|_X}h(x)\phi(x)\Big)dx = 0. \tag{8.22}$$

Since $\|\psi_n\|_X = 1$ for every n, the sequence $(\psi_n)_{n\in\mathbb{N}}$ is uniformly bounded in X with respect n, so by reflexivity of X, there exists a subsequence (denoted the same for simplicity) such that $\psi_n \rightharpoonup \psi$ in X, by the compact embedding of $X = W_0^{1,2}(\mathcal{D})$, a further subsequence such that $\psi_n \to \psi$ in $L^2(\mathcal{D})$ and a further subsequence such that $\psi_n \to \psi$ a. e. in $\mathcal{D}$. We express the term $\frac{1}{\|u_n\|_X}f_0(u_n(x)) = \frac{f_0(u_n(x))}{u_n(x)}\psi_n(x)$ and note that in the limit, as $n \to \infty$, we have that[8] $\frac{1}{\|u_n\|_X}f_0(u_n) \to \gamma_+\psi^+ - \gamma^-\psi^-$ a. e. in $\mathcal{D}$ (using the convention $\psi^- = -\max(-\psi, 0)$). By the Lebesgue dominated convergence theorem passing to the limit in (8.22), we conclude that

$$\int_{\mathcal{D}}\Big(\nabla\psi(x)\cdot\nabla\phi(x) - (\lambda\psi(x) + \gamma_+\psi^+(x) - \gamma_-\psi^-(x))\phi(x)\Big)dx = 0 \quad \forall\,\phi \in W_0^{1,2}(\mathcal{D}),$$

which, by $\psi = \psi^+ - \psi^-$, is recognized as the weak form for the elliptic problem

$$\begin{aligned} -\Delta\psi &= (\lambda + \gamma_+)\psi^+ - (\lambda + \gamma_-)\psi^- \quad \text{in } \mathcal{D}, \\ \psi &= 0, \quad \text{on } \partial\mathcal{D}. \end{aligned}$$

We claim that since $\lambda_k < \lambda + \gamma_\pm < \lambda_{k+1}$, this implies that $\psi = 0$. This can be seen as follows: The elliptic problems $-\Delta u = (\lambda + \gamma_\pm)u$ admit only the trivial solution $u = 0$. The function m, defined by $m(x, \psi(x)) = (\lambda + \gamma_+)\psi^+(x) + (\lambda + \gamma_-)\psi^-(x)$, satisfies the inequalities

$$m \geq f_1 = (\lambda + \gamma_+)\psi, \text{ if } \gamma_+ \geq \gamma_-, \quad \text{and} \quad m \leq f_2 = (\lambda + \gamma_-)\psi, \text{ if } \gamma_+ < \gamma_-,$$

8 To see this, simply note that since $\|u_n\|_X \to \infty$, either $u_n(x) \to \infty$, in which case $\psi_n(x) \geq 0$ and $\frac{1}{\|u_n\|_X}f_0(u_n(x)) \to \gamma_+\psi^+(x)$ or $u_n(x) \to -\infty$, in which case $\psi_n(x) \leq 0$ and $\frac{1}{\|u_n\|_X}f_0(u_n(x)) \to -\gamma_-\psi^-(x)$, $\psi^- = \max(-\psi, 0)$.

a. e. in $\mathcal{D}$, so by the comparison principle for elliptic problems (see, e. g., Proposition 6.2.6), applied for the RHSs m,f_1 to obtain $\psi \geq 0$, and m,f_2 to obtain $\psi \leq 0$, we get $\psi = 0$.

We now set $\phi = \psi_n$ in (8.21) and dividing by $\|u_n\|_X$, we obtain that

$$\left|\int_{\mathcal{D}} \left(|\nabla \psi_n(x)|^2 - \lambda \psi_n(x)^2 - \frac{1}{\|u_n\|_X} f_0(u_n(x))\psi_n(x) - \frac{1}{\|u_n\|_X} h(x)\psi_n(x) \right) dx\right| \leq \epsilon \frac{\|\psi_n\|_X}{\|u_n\|_X},$$

which, recalling that $\|\nabla \psi_n\|_{L^2(\mathcal{D})} = \|\psi_n\|_X = 1$, implies that

$$\left|1 - \int_{\mathcal{D}} \left(\lambda \psi_n(x)^2 + \frac{1}{\|u_n\|_X} f_0(u_n(x))\psi_n(x) + \frac{1}{\|u_n\|_X} h(x)\psi_n(x) \right) dx\right| \leq \epsilon \frac{\|\psi_n\|_X}{\|u_n\|_X},$$

which is further rearranged as

$$\left|1 - \int_{\mathcal{D}} \left[\left(\lambda + \frac{f_0(u_n(x))}{u_n(x)} \right) \psi_n(x)^2 + \frac{1}{\|u_n\|_X} h(x)\psi_n(x) \right] dx\right| \leq \epsilon \frac{\|\psi_n\|_X}{\|u_n\|_X}.$$

Passing to the limit, as $n \to \infty$, and keeping in mind that $\lambda + \frac{f_0(u_n(x)}{u_n(x))} \to \lambda + \gamma_\pm$ a. e. depending on whether $u_n(x) \to +\infty$ or $u_n(x) \to -\infty$, while $\psi_n(x) \to 0$ a. e., we conclude that $1 \leq 0$, which is a contradiction. Hence, any Palais–Smale sequence must be bounded. This concludes the proof. □

8.3.3 Resonant semilinear problems and the saddle point theorem

We now turn our attention to the resonant case $\lambda = \lambda_k$. We know, even in the linear case, that resonant problems may not have a solution, unless additional conditions hold (see Section 7.6.2). As a quick illustration of that, consider the linear problem

$$\begin{aligned} -\Delta u(x) &= \lambda_k u(x) + h(x), \quad \text{in } \mathcal{D}, \\ u(x) &= 0, \quad \text{on } \partial\mathcal{D}, \end{aligned}$$

where λ_k is an eigenvalue of the operator $-\Delta$ with homogeneous Dirichlet boundary conditions, and h is a known function. As one can see very easily, by multiplying with the corresponding eigenfunction ϕ_k and integrating over all $\mathcal{D}$, the above problem admits a solution only for these functions h that satisfy the compatibility condition $\int_{\mathcal{D}} h(x)\phi_k(x)dx = 0$, or put otherwise are orthogonal in terms of the standard inner product of $L^2(\mathcal{D})$ to the k eigenfunction ϕ_k. Such compatibility conditions are expected to be needed also for the nonlinear resonant case, and in fact they do, as we will see in the following discussion and the proposition that follows.

Consider again the elliptic problem (8.17) in the resonant case $\lambda = \lambda_k$,

$$\begin{aligned} -\Delta u &= \lambda_k u + f(x,u), \quad x \in \mathcal{D}, \\ u &= 0, \quad x \in \partial\mathcal{D}, \end{aligned} \tag{8.23}$$

where λ_k is an eigenvalue of the Laplacian on $\mathcal{D}$ with homogeneous Dirichlet boundary conditions.

We need to impose the following assumptions on f:

Assumption 8.3.5. We assume that $f : \mathcal{D} \times \mathbb{R} \to \mathbb{R}$ satisfies the following properties:

(i) f is continuous, and there exists a constant $c_1' \geq 0$ such that $|f(x,s)| \leq c_1'$ for every $x \in \mathcal{D}$, $s \in \mathbb{R}$.

(ii) The condition $\int_{\mathcal{D}} \Phi(x,s)dx \to \pm\infty$, as $|s| \to \infty$, where $\Phi(x,s) = \int_0^s f(x,\sigma)d\sigma$, holds.

This case is treated in the following proposition [120]:

Proposition 8.3.6. *Under Assumption* 8.3.5, *the resonant problem* (8.23) *admits a nontrivial weak solution.*

Proof. The proof uses the saddle point Theorem 8.2.7. It follows upon the steps of the proof of Proposition 8.3.4, and we use the same notation as there. We will also use the same splitting as in (8.18), i. e.,

$$X_1 := \operatorname{span}\{\phi_1, \ldots, \phi_k\}, \quad X_2 := \overline{\operatorname{span}\{\phi_i : i \geq k+1\}},$$

where ϕ_i are the eigenfunctions of the Dirichlet Laplacian, normalized as in (8.19).

We proceed in 3 steps leaving the verification of the Palais–Smale condition last.

1. It is straightforward to show that F is bounded below on X_2. Indeed, for any $w = \sum_{i=k+1}^{\infty} c_i\phi_i \in X_2$, we have, by linearity, that $-\Delta w - \lambda_k w = \sum_{i=k+1}^{\infty}(\lambda_i - \lambda_k)c_i\phi_i$, and multiplying with w and integrating by parts, using the orthogonality of the eigenfunctions ϕ_i in $L^2(\mathcal{D})$, we obtain that

$$\begin{aligned} \int_{\mathcal{D}} (|\nabla w(x)|^2 - \lambda_k w(x)^2)dx &= \sum_{i=k+1}^{\infty} c_i^2(\lambda_i - \lambda_k)\int_{\mathcal{D}} \phi_i(x)^2 dx \\ &= \sum_{i=k+1}^{\infty}\left(1 - \frac{\lambda_k}{\lambda_i}\right)c_i^2 \geq \left(1 - \frac{\lambda_k}{\lambda_{k+1}}\right)\sum_{i=k+1}^{\infty} c_i^2 \\ &= \left(1 - \frac{\lambda_k}{\lambda_{k+1}}\right)\|w\|_X^2, \end{aligned} \tag{8.24}$$

where, for the second equality, we used the fact that[9] $\|\phi_i\|^2_{L^2(\mathcal{D})} = \frac{1}{\lambda_i}$, and then the fact that $\lambda_i \geq \lambda_{k+1}$ for all[10] $i = k+1, \dots$. We now estimate $F(w)$ as

$$F(w) = \int_{\mathcal{D}} \left(\frac{1}{2} |\nabla w(x)|^2 - \frac{\lambda_k}{2} w(x)^2 - \Phi(x, w(x)) \right) dx$$
$$\geq \frac{1}{2}\left(1 - \frac{\lambda_k}{\lambda_{k+1}}\right) \|w\|_X^2 - \int_{\mathcal{D}} |\Phi(x, w(x))| dx.$$

By Assumption 8.3.5(i), there exists $c_1' = \sup_{x \in \mathcal{D}, s \in \mathbb{R}} f(x,s)$ such that $|\Phi(x,s)| < c_1'|s|$ for every $x \in \mathcal{D}$ and $s \in \mathbb{R}$, so $\int_{\mathcal{D}} |\Phi(x, w(x))| dx \leq c_1' \|w\|_{L^1(\mathcal{D})}$, therefore obtaining the lower bound,

$$F(w) \geq \frac{1}{2}\left(1 - \frac{\lambda_k}{\lambda_{k+1}}\right) \|w\|_X^2 - c_1' \|w\|_{L^1(\mathcal{D})}. \tag{8.25}$$

By a combination of the Hölder and Poincaré inequalities

$$\|w\|_{L_1 \mathcal{D}} \leq |\mathcal{D}|^{1/2} \|w\|_{L^2(\mathcal{D})} \leq \frac{|\mathcal{D}|^{1/2}}{\lambda_1^{1/2}} \|\nabla w\|_{L^2(\mathcal{D})} = \frac{|\mathcal{D}|^{1/2}}{\lambda_1^{1/2}} \|w\|_X,$$

which combined with (8.25), we obtain the lower bound

$$F(w) \geq \frac{1}{2}\left(1 - \frac{\lambda_k}{\lambda_{k+1}}\right) \|w\|_X^2 - c_2' \|w\|_X, \quad \forall\, w \in X_2,$$

for an appropriate constant c_2'. This finite lower bound guarantees that $\inf_{w \in X_2} F(w) > -\infty$.

2. We now show that $F(v) \to -\infty$ for $v \in X_1$ such that $\|v\|_X \to \infty$. As before, for any $v = \sum_{i=1}^k c_i \phi_i \in X_1$, we have that

$$\int_{\mathcal{D}} (|\nabla v(x)|^2 - \lambda_k v(x)^2) dx = \sum_{i=1}^k (\lambda_i - \lambda_k) c_i^2 \int_{\mathcal{D}} \phi_i(x)^2 dx = \sum_{i=1}^k \left(1 - \frac{\lambda_k}{\lambda_i}\right) c_i^2.$$

For any of the eigenvalues λ_i, for which $\lambda_i < \lambda_k$, this term can be estimated above by a negative quadratic term in $\|v\|_X$, but a problem arises for the term $i = k$ (or if the eigenvalue λ_k is not simple with all the other terms i such that $\lambda_i = \lambda_k$). We therefore

9 Since the ϕ_i are normalized so that $\|\nabla \phi_i\|_{L^2(\mathcal{D})} = \|\phi_i\|_X = 1$ and $-\Delta \phi_i = \lambda_i \phi_i$, by multiplying by ϕ_i and integrating by parts, it follows easily that $\|\phi_i\|^2_{L^2(\mathcal{D})} = \frac{1}{\lambda_i}$.

10 For the last part, we used the observation that for the eigenfunctions ϕ_i, the fact that $\int_{\mathcal{D}} \phi_i(x)\phi_j(x) dx = \delta_{ij}$ implies also that $\int_{\mathcal{D}} \nabla \phi_i(x) \cdot \nabla \phi_j(x) dx = \delta_{ij}$, by a simple integration by parts argument.

decompose X_1 into two orthogonal components: $X_1^{(k)} = \text{span}\{\phi_i \,:\, \lambda_i = \lambda_k\}$ and $X_1^{(-,k)} = \text{span}\{\phi_i \,:\, \lambda_i < \lambda_k\}$, and express any $v \in X_1$ as $v = v_k + v_{-,k}$ with $v_k \in X_1^{(k)}$ and $v_{-,k} \in X_1^{(-,k)}$. With this decomposition in mind,

$$\int_{\mathcal{D}} (|\nabla v(x)|^2 - \lambda_k v(x)^2)dx = \sum_{i=1}^{k} \left(1 - \frac{\lambda_k}{\lambda_i}\right) c_i^2$$
$$\leq \left(1 - \frac{\lambda_k}{\lambda_{k-1}}\right) \sum_{i=1}^{k} c_i^2 = \left(1 - \frac{\lambda_k}{\lambda_{k-1}}\right) \|v_{-,k}\|_X^2,$$

with all the sums taken over indices such that $\lambda_i < \lambda_k$, so $\lambda_i \leq \lambda_{k-1}$. We therefore have that

$$F(v) = \frac{1}{2} \int_{\mathcal{D}} (|\nabla v(x)|^2 - \lambda_k v(x)^2)dx - \int_{\mathcal{D}} \Phi(x, v(x))dx$$
$$\leq \frac{1}{2}\left(1 - \frac{\lambda_k}{\lambda_{k-1}}\right) \|v_{-,k}\|_X^2 - \int_{\mathcal{D}} \Phi(x, v(x))dx \leq - \int_{\mathcal{D}} \Phi(x, v(x))dx, \quad \forall\, v \in X_1,$$

and by Assumption 8.3.5(ii), $F(v) \to -\infty$, as $\|v\|_X \to \infty$.

3. It remains to prove that F satisfies the Palais–Smale condition, for which it is enough to show that any Palais–Smale sequence is bounded in $X = W_0^{1,2}(\mathcal{D})$ (see proof of Proposition 8.3.2 (Step 4)). Consider any Palais–Smale sequence $(u_n)_{n\in\mathbb{N}}$ and for any n, decompose $u_n = u_{-,n} + u_n^0 + u_{+,n}$, where $u_{-,n} \in X_1^{(-,k)}$, $u_n^0 \in X_1^{(k)}$, and $u_{+,n} \in X_2$. Since $(u_n)_{n\in\mathbb{N}}$ is a Palais–Smale sequence, it holds that $DF(u_n) \to 0$ in $X^\star$, therefore (see (8.14) and the following analysis) for any $\epsilon > 0$ and $\phi \in X$, there exists N such that for all $n > N$, and it holds that

$$|\langle DF(u_n), \phi\rangle| = \left|\int_{\mathcal{D}} (\nabla u_n(x) \cdot \nabla\phi(x) - \lambda_k u_n(x)\phi(x) - f(x, u_n(x))\phi(x))dx\right| < \epsilon\|\phi\|_X. \quad (8.26)$$

To prove the boundedness of the Palais–Smale sequence, we will have to choose ϕ properly in order to show the boundedness of the norm of each component separately. We initially choose $\phi = -u_{-,n}$. We first estimate each term in (8.26) separately. By orthogonality and similar arguments as those employed above, noting that for eigenfunctions in $X_1^{(-,k)}$, $\lambda_i \leq \lambda_{k-1}$

$$\int_{\mathcal{D}} (\nabla u_n(x) \cdot (-\nabla u_{-,n}(x)) - \lambda_k u_n(x)(-u_{-,n}(x)\,)dx$$
$$= + \int_{\mathcal{D}} \left(-(-\Delta)u_{-,n}(x)u_{-,n}(x) + \lambda_k u_{-,n}(x)^2\right)dx \geq (\lambda_k - \lambda_{k-1})\|u_{-,n}\|_{L^2(\mathcal{D})}^2,$$

while using Assumption 8.3.5(i)

$$-\int_{\mathcal{D}} f(x,u_n(x))(-u_{-,n}(x))dx \geq -\int_{\mathcal{D}} |f(x,u_n(x))u_{-,n}(x)|dx \geq -c_1'\|u_n\|_{L^1(\mathcal{D})}$$
$$\geq -c_1'|\mathcal{D}|^{1/2}\|u_n\|_{L^2(\mathcal{D})},$$

where, in the last estimate, we used Hölder's inequality. Inserting these estimates in (8.26), we conclude that

$$(\lambda_k - \lambda_{k-1})\|u_{-,n}\|^2_{L^2(\mathcal{D})} - c_1'|\mathcal{D}|^{1/2}\|u_n\|_{L^2(\mathcal{D})} \leq \langle DF(u_n),(-u_{-,n})\rangle < \epsilon\|u_{-,n}\|_X,$$

which, since ϵ is arbitrary, proves the boundeness of $\|u_{-,n}\|^2_{L^2(\mathcal{D})}$. However, since $u_{-,n} \in X_1^{(k)}$, which is a finite dimensional space, all norms are equivalent,[11] and the boundedness of $\|u_{-,n}\|_X$ is concluded.

We next set $\phi = u_{+,n}$. Using orthogonality and the estimate (8.25), which holds for any $w \in X_2$, we obtain the estimate

$$\begin{aligned}\langle DF(u_n),u_{+,n}\rangle &= \int_{\mathcal{D}} (\nabla u_{+,n}(x)\cdot\nabla u_{+,n}(x) - \lambda_k u_{+,n}(x))dx - \int_{\mathcal{D}} f(x,u_n(x))u_{+,n}(x)dx\\
&\geq \left(1-\frac{\lambda_k}{\lambda_{k+1}}\right)\|u_{+,n}\|_X^2 - \int_{\mathcal{D}} |f(x,u_n(x)||u_{+,n}|dx\\
&\geq \left(1-\frac{\lambda_k}{\lambda_{k+1}}\right)\|u_{+,n}\|_X^2 - c_1'\|u_{+,n}\|_{L^1(\mathcal{D})}\\
&\geq \left(1-\frac{\lambda_k}{\lambda_{k+1}}\right)\|u_{+,n}\|_X^2 - \frac{1}{\lambda_1^{1/2}}c_1'|\mathcal{D}|^{1/2}\|u_{+,n}\|_X,\end{aligned}$$

where, for the final step, we used a combination of the Hölder and Poincaré inequality. Combining the above estimate with (8.26), we conclude that

$$\left(1-\frac{\lambda_k}{\lambda_{k+1}}\right)\|u_{+,n}\|_X^2 - c_1'\left(\frac{|\mathcal{D}|}{\lambda_1}\right)^{1/2}\|u_{+,n}\|_X \leq \langle DF(u_n),(u_{+,n})\rangle < \epsilon\|u_{+,n}\|_X,$$

from which the boundedness of $\|u_{+,n}\|_X$ follows.

It remains to verify that $\|u_n^0\|_X$ is bounded. For this term, the above strategy cannot be used, as the quadratic term of F in $\|u_n^0\|_X$ will vanish since $u_n^0 \in X_1^{(k)}$. We will thus use the condition $|F(u_n)| < c'$, which holds since $(u_n)_{n\in\mathbb{N}}$ is a Palais–Smale sequence. Using the decomposition $u_n = u_{-,n} + u_n^0 + u_{+,n}$, we obtain that

$$F(u_n) = \frac{1}{2}(\|u_{+,n}\|_X^2 + \|u_{-,n}\|_X^2 - \lambda_k(\|u_{+,n}\|^2_{L^2(\mathcal{D})} + \|u_{-,n}\|^2_{L^2(\mathcal{D})}) - \int_{\mathcal{D}} \Phi(x,u_n(x))dx$$

---

11 For any $u_{-,n} = \sum_{i=1}^{k-1} c_i\phi_i$, it holds that $\|u_{-,n}\|_X^2 = \|\nabla u_{-,n}\|^2_{L^2(\mathcal{D})} = \sum_{i=1}^{k-1} c_i^2$, while $\|u_{-,n}\|^2_{L^2(\mathcal{D})} = \sum_{i=1}^{k-1} c_i^2\|\phi_i\|^2_{L^2(\mathcal{D})} = \sum_{i=1}^{k-1}\frac{1}{\lambda_i}c_i^2$, from which it easily follows that $\frac{1}{\lambda_k}\|v_n^-\|_X^2 \leq \|u_{-,n}\|^2_{L^2(\mathcal{D})} \leq \frac{1}{\lambda_1}\|v_n^-\|_X^2$.

$$= \frac{1}{2}(\|u_{+,n}\|_X^2 + \|u_{-,n}\|_X^2 - \lambda_k(\|u_{+,n}\|_{L^2(\mathcal{D})}^2 + \|u_{-,n}\|_{L^2(\mathcal{D})}^2)$$
$$- \int_{\mathcal{D}} (\Phi(x, u_n(x)) - \Phi(x, u_n^0(x)))dx - \int_{\mathcal{D}} \Phi(x, u_n^0(x))dx.$$

Note that

$$\Phi(x, u_n(x)) - \Phi(x, u_n^0(x)) = \int_0^{u_n(x)} f(x,s)ds - \int_0^{u_n^0(x)} f(x,s)ds = \int_{u_n^0(x)}^{u_n(x)} f(x,s)ds,$$

so by Assumption 8.3.5(i),

$$|\Phi(x, u_n(x)) - \Phi(x, u_n^0(x))| < c_1'|u_n(x) - u_n^0(x)| = c_1'|(u_{+,n} + u_{-,n})(x)|, \tag{8.27}$$

and its integral over $\mathcal{D}$ is bounded by $\|u_{+,n} + u_{-,n}\|_{L^1(\mathcal{D})}$. This allows us to estimate:

$$|F(u_n)| = \Big|\frac{1}{2}(\|u_{+,n}\|_X^2 + \|u_{-,n}\|_X^2) - \lambda_k(\|u_{+,n}\|_{L^2(\mathcal{D})}^2 + \|u_{-,n}\|_{L^2(\mathcal{D})}^2)$$
$$- \int_{\mathcal{D}} [\Phi(x, u_n(x)) - \Phi(x, u_n^0(x))]dx - \int_{\mathcal{D}} \Phi(x, u_n^0(x))dx\Big|$$
$$\geq -\Big|\frac{1}{2}(\|u_{+,n}\|_X^2 + \|u_{-,n}\|_X^2) - \lambda_k(\|u_{+,n}\|_{L^2(\mathcal{D})}^2 + \|u_{-,n}\|_{L^2(\mathcal{D})}^2)$$
$$- \int_{\mathcal{D}} [\Phi(x, u_n(x)) - \Phi(x, u_n^0(x))]dx\Big| + \Big|\int_{\mathcal{D}} \Phi(x, u_n^0(x))dx\Big|.$$

where we used the reverse triangle inequality. All terms on the RHS of the above estimate, but the last, depend on $u_{+,n}$, $u_{-,n}$, hence are bounded. This observation, combined with the fact that $|F(u_n)| < c'$, for some constant c', leads (upon rearrangement) to the conclusion that there exists a constant c_3' such that $\int_{\mathcal{D}} \Phi(x, u_n^0(x))dx < c_3'$, which by Assumption 8.3.5(ii) leads to the boundedness of $\|u_n^0\|_X$. The proof is complete. □

8.4 Applications in quasilinear elliptic problems

In this section, we present applications of critical point theory to the functional

$$F(u) := \frac{1}{p}\int_{\mathcal{D}} |\nabla u(x)|^p\, dx - \int_{\mathcal{D}} \Phi(x, u(x))\, dx, \quad p \in (1, \infty), \tag{8.28}$$

and its connection with the quasilinear elliptic partial differential equation

$$\begin{aligned} -\nabla \cdot (|\nabla u|^{p-2}\nabla u) &= f(x,u), \quad \text{in } \mathcal{D}, \\ u &= 0 \quad \text{on } \partial\mathcal{D}, \end{aligned} \tag{8.29}$$

where $\mathcal{D}$ is a bounded sufficiently smooth domain. The existence of minimizers to this functional under appropriate conditions on the nonlinearity f was studied in Section 6.8.2 using the direct method of the calculus of variations. The condition that guaranteed the coercivity of the functional F, and hence the existence of a minimizer, was essentially Assumption 6.8.4, according to which $\limsup_{|s|\to\infty} \frac{p\Phi(x,s)}{|s|^p} < \lambda_{1,p}$, where $\lambda_{1,p} > 0$ is the first eigenvalue of the p-Laplacian operator, defined as

$$\lambda_{1,p} = \inf_{u\in W_0^{1,p}(\mathcal{D})\setminus\{0\}} \frac{\|\nabla u\|^p_{L^p(\mathcal{D})}}{\|u\|^p_{L^p(\mathcal{D})}}, \tag{8.30}$$

which is clearly related to the best constant in the relevant Poincaré inequality. The infimum is attained for a function $\phi_{1,p} > 0$ called the first eigenfunction of the p-Laplacian operator (see, e. g., [8]). Recall that $\lambda_{1,p}$ was also related to the existence of nontrivial solutions to the quasilinear problem

$$\begin{aligned} -\nabla\cdot(|\nabla u|^{p-2}\nabla u) &= \lambda|u|^{p-2}u, \quad \text{in } \mathcal{D}, \\ u &= 0 \quad \text{on } \partial\mathcal{D}, \end{aligned} \tag{8.31}$$

with $\lambda_{1,p} > 0$ being the smallest value of λ, for which this equation admits such a solution.

We will see that by relaxing this assumption, we may obtain other critical points of the functional F using the mountain pass or the saddle point theorem.

8.4.1 The p-Laplacian and the mountain pass theorem

We will see that by relaxing Assumption 6.8.4 on f, we may obtain other critical points of the functional F related to solutions of the quasilinear elliptic equation (8.29) using the mountain pass theorem. Our approach follows [64].

We will impose the following conditions on f:

Assumption 8.4.1. $f : \mathcal{D}\times\mathbb{R}\to\mathbb{R}$ is a Carathéodory function such that it satisfies:

(i) The growth condition $|f(\cdot,s)| \le c|s|^r + a_1$, a. e. in $\mathcal{D}$, for $p-1 < r < \mathfrak{s}_p - 1$, where $\mathfrak{s}_p = \frac{dp}{d-p}$ if $d > p$ and $p-1 < r < \infty$ if $d \le p$.

(ii) The asymptotic condition that there exists a $\theta > p$ and a $R > 0$ such that

$$0 < \theta\Phi(x,s) \le sf(x,s), \quad x\in\overline{\mathcal{D}},\ |s| \ge R.$$

(iii) The condition for small values of s that $\limsup_{s\to 0}\frac{pf(x,s)}{|s|^{p-2}s} \le \mu < \lambda_{1,p}$ uniformly in $x\in\mathcal{D}$, where $\lambda_{1,p} > 0$, is defined as in (8.30).

Using the mountain pass theorem, we will show the existence of a positive solution for (8.29), where the positivity arises from the asymptotic positivity of Φ, while the remainder of Assumption 8.4.1(ii) is needed in order to guarantee the validity of the Palais–

Smale property for F. At the same time, this condition makes the functional unbounded from below, so the critical point that will be obtained cannot correspond to a minimum. On the other hand, Assumption 8.4.1(iii) will guarantee the existence of a $\rho > 0$ and an $\alpha > 0$ such that $F(u) \geq \alpha$ for every $u \in W_0^{1,p}(\mathcal{D})$ such that $\|u\|_{W_0^{1,p}(\mathcal{D})} = \rho$, a condition which is essential for the application of the mountain pass theorem.

We then have the following proposition [64]:

Proposition 8.4.2. *Under Assumption* 8.4.1, *problem* (8.29) *admits a nontrivial solution. If furthermore* $f(x,0) = 0$, *it is a positive solution* $u \geq 0$.

Proof. Concerning the properties of the functional F, the reader may consult Section 6.8 in order to show that $F \in C^1(X;\mathbb{R})$ for $X = W_0^{1,p}(\mathcal{D})$. We will concentrate in showing that this functional has the right structure for the mountain pass theorem to apply. The proof follows in 5 steps.

1. We will show first that F is unbounded below. By integrating Assumption 8.4.1(ii), with respect to s, we conclude that $\Phi(x,s) \geq \gamma s^\theta$ for $s \geq R$ and all x (for a suitable $R > 0$ and $\gamma > 0$), which allows us to get control of the growth rate of the functional when u takes large values, whereas the growth condition (Assumption 8.4.1(i)) provides sufficient information when u takes values below R. We therefore have, using the notation $\mathcal{D}_+ = \{x \in \mathcal{D} : u(x) \geq R\}$ and $\mathcal{D}_- = \mathcal{D} \setminus \mathcal{D}_+$, that

$$\begin{aligned} F(u) &= \frac{1}{p}\|u\|^p_{W_0^{1,p}(\mathcal{D})} - \int_{\mathcal{D}} \Phi(x,u(x))\,dx \\ &= \frac{1}{p}\|u\|^p_{W_0^{1,p}(\mathcal{D})} - \int_{D_+} \Phi(x,u(x))dx - \int_{D_-} \Phi(x,u(x))dx \\ &\leq \frac{1}{p}\|u\|^p_{W_0^{1,p}(\mathcal{D})} - \gamma \int_{D_+} |u(x)|^\theta dx + c_1, \end{aligned} \tag{8.32}$$

where c_1 is a bound for last integral.[12] Since we are interested in showing that F is unbounded below, consider any $u \in W_0^{1,p}(\mathcal{D})$ (bounded), and take $w = \varrho u$, for $\varrho \to \infty$. Substituting that into (8.32), we see that

$$F(\varrho u) \leq \frac{\varrho^p}{p}\|u\|^p_{W_0^{1,p}(\mathcal{D})} - \varrho^\theta \gamma \int_{D_+} |u(x)|^\theta dx - c_1,$$

which, since $\theta > p$, shows that $F(\varrho u) \to -\infty$, as $\varrho \to \infty$.

12 On $\{x : u(x) < R\}$, by the growth condition, the nonlinearity is bounded by a constant, and so the term can be estimated as above.

2. We now show the existence of $\rho > 0$ and $\alpha > 0$ such that $F(u) \geq \alpha$ for every $u \in W_0^{1,p}(\mathcal{D})$ such that $\|u\|_{W_0^{1,p}(\mathcal{D})} = \rho$. Combining Assumption 8.4.1(iii) with the growth condition, we have that there exist $\mu \in (0, \lambda_{1,p})$ and $c_2 > 0$ such that

$$\Phi(x,s) \leq \frac{\mu}{p}|s|^p + c_2|s|^{q_1}, \quad x \in \mathcal{D},\ s \in \mathbb{R}, \tag{8.33}$$

for $q_1 \in \{\max(p, r+1), \mathfrak{s}_p\}$. Indeed, by Assumption 8.4.1(iii), there exists $\mu \in (0, \lambda_{1,p})$ such that $\limsup_{s\to 0} \frac{pf(x,s)}{|s|^{p-2}s} < \mu$ uniformly in $x \in \mathcal{D}$, and a δ (depending on μ) such that $\frac{p\Phi(x,s)}{|s|^{p-2}s} \leq \mu$ for $|s| < \delta$, and integrating (assume without loss of generality that $f(x,0) = 0$), we obtain that $\Phi(x,s) \leq \frac{\mu}{p}|s|^p$ for $|s| < \delta$. On the other hand, integrating the growth condition, we obtain that $|\Phi(x,s)| \leq c_3(|s|^q + 1)$ for all $s \in \mathbb{R}$, and the claim easily follows. Having obtained (8.33), we estimate F as follows:

$$\begin{aligned} F(u) &= \frac{1}{p}\|u\|^p_{W_0^{1,p}(\mathcal{D})} - \int_{\mathcal{D}} \Phi(x,u(x))\,dx \\ &\geq \frac{1}{p}\|u\|^p_{W_0^{1,p}(\mathcal{D})} - \frac{\mu}{p}\|u\|^p_{L^p(\mathcal{D})} - c_2\|u\|^{q_1}_{L^{q_1}(\mathcal{D})}. \end{aligned} \tag{8.34}$$

By the choice of $q_1 < \mathfrak{s}_p$, we have that $W_0^{1,p}(\mathcal{D}) \hookrightarrow L^{q_1}(\mathcal{D})$ so that there exists a constant c_3 such that $\|u\|_{L^{q_1}(\mathcal{D})}\| \leq c_3\|u\|_{W_0^{1,p}(\mathcal{D})}$, whereas by the definition of $\lambda_{1,p}$ (or the Poincaré inequality in fact) gives us that $\lambda_{1,p}\|u\|^p_{L^p(\mathcal{D})} \leq \|u\|^p_{W_0^{1,p}(\mathcal{D})}$. Using these estimates in (8.34), we obtain

$$F(u) \geq \|u\|^p_{W_0^{1,p}(\mathcal{D})}\left(\frac{1}{p}\left(1 - \frac{\mu}{\lambda_{1,p}}\right) - c_4\|u\|^{q_1-p}_{W_0^{1,p}(\mathcal{D})}\right).$$

Setting $\rho > 0$ such that $1 - c_4\rho^{q_1-p} > 0$, and since $\mu < \lambda_{1,p}$, we can see that there exists $\alpha > 0$, for which our claim follows.

3. It remains to verify that F satisfies the Palais–Smale condition. This is done in two substeps.

3(a). We first show that any Palais–Smale sequence is bounded, i. e., for any sequence $(u_n)_{n\in\mathbb{N}}$ such that $\|F(u_n)\|_X < c'$ for some $c' > 0$ and $DF(u_n) \to 0$ in $X^\star$, it holds that[13] $\|u_n\|_X \leq c$ for an appropriate constant $c > 0$. We express $\Phi(x,s) = \Phi(x,s) - \frac{1}{\theta}f(x,s)s + \frac{1}{\theta}f(x,s)s$ in order to employ the asymptotic growth condition of Assumption 8.4.1(ii) on f. Using the notation $\Phi_0(x,s) := \Phi(x,s) - \frac{1}{\theta}f(x,s)s$, we express $\Phi(x,s) = \Phi_0(x,s) + \frac{1}{\theta}f(x,s)s$, and noting that by Assumption 8.4.1(ii) $\Phi_0(x,s) \leq 0$ for $|s| \geq R$, so upon setting $\mathcal{D}_{n,+} := \{x \in \mathcal{D} : |u_n(x)| \geq R\}$ with $\mathcal{D}_{n,-} = \mathcal{D} \setminus \mathcal{D}_{n,+}$, we have that

13 $DF(u_n) \to 0$ implies that $\|DF(u_n)\|_{X^\star} = \sup\{|\langle DF(u_n), v\rangle| : \|v\|_X \leq 1\} \to 0$, and the claim follows from that, choosing $v = \frac{u_n}{\|u_n\|_X}$.

$$
\begin{aligned}
\int_{\mathcal{D}} \Phi_0(x, u_n(x))dx &= \int_{\mathcal{D}_{n,+}} \Phi_0(x, u_n(x))dx + \int_{\mathcal{D}_{n,-}} \Phi_0(x, u_n(x))dx \\
&\leq \int_{\mathcal{D}_{n,-}} \Phi_0(x, u_n(x))dx \leq c_1,
\end{aligned}
\tag{8.35}
$$

with the last estimate arising from the growth condition Assumption 8.4.1(i). Since our information on $F(u_n)$ and $DF(u_n)$ guarantee that $F(u_n)$ is bounded, while $|\langle DF(u_n), u_n\rangle| \leq \epsilon\|u_n\|_X$ for any $\epsilon > 0$ for large enough n (since $DF(u_n) \to 0$ in $X^\star$), we may try a linear combination of $F(u_n)$ and $\langle DF(u_n), u_n\rangle$ so as to form the integral of Φ_0, whose sign we may control. Combining the above, and using the definitions of F and DF and Φ_0,

$$
\begin{aligned}
F(u_n) - \frac{1}{\theta}\langle DF(u_n), u_n\rangle &= \left(\frac{1}{p} - \frac{1}{\theta}\right)\int_{\mathcal{D}} |\nabla u_n(x)|^p dx - \int_{\mathcal{D}} \Phi_0(u_n(x))dx \\
&\geq \left(\frac{1}{p} - \frac{1}{\theta}\right)\int_{\mathcal{D}} |\nabla u_n(x)|^p dx - c_1,
\end{aligned}
$$

using (8.35), so by rearranging and using the boundedness assumptions provided, we get an estimate of the form $c_2\|u_n\|_X^2 \leq \epsilon\|u_n\|_X + c_3$ with $c_2 = \frac{1}{p} - \frac{1}{\theta} > 0$, from which we conclude that $(u_n)_{n\in\mathbb{N}}$ is bounded.

3(b). Having established that a Palais–Smale sequence is bounded by the reflexivity of $X = W_0^{1,p}(\mathcal{D})$ ($p > 1$ and finite), there exists a subsequence $(u_{n_k})_{k\in\mathbb{N}}$ and a $u \in X$ such that $u_{n_k} \rightharpoonup u$ in X. We will show that this convergence is strong, i. e., $u_n \to u$ in X, thus establishing the Palais–Smale property.

We first observe that upon defining

$$
\varphi(u) = \frac{1}{p}\|u\|_X^p = \frac{1}{p}\|\nabla u\|_{L^p(\mathcal{D})}^p,
$$

and noting that $\langle D\varphi(u), v\rangle = \int_{\mathcal{D}} |\nabla u(x)|^{p-2}\nabla u(x) \cdot \nabla v(x)dx$ for every $v \in X$, that

$$
\begin{aligned}
&\langle DF(u_{n_k}), u_{n_k} - u\rangle + \int_{\mathcal{D}} f(x, u_{n_k}(x))(u_{n_k}(x) - u(x))dx \\
&= \langle D\varphi(u_{n_k}), \nabla u_{n_k} - \nabla u\rangle.
\end{aligned}
$$

Since $DF(u_{n_k}) \to 0$ (strong) and $u_{n_k} - u$ is bounded, we have that $\langle DF(u_{n_k}), u_{n_k} - u\rangle \to 0$. By the compact embedding of $X = W_0^{1,p}(\mathcal{D}) \overset{c}{\hookrightarrow} L^{r+1}(\mathcal{D})$, there exists a subsequence of $(u_{n_k})_{k\in\mathbb{N}}$ (denoted the same) such that $u_{n_k} \to u$ in $L^q(\mathcal{D})$, for $q = r + 1$, hence, moving to this subsequence and taking also into account the growth condition, which provides a uniform bound for $f(\cdot, u_{n_k}(\cdot))$ in $L^{q^*}(\mathcal{D})$, we have $\int_{\mathcal{D}} f(x, u_{n_k}(x))(u_{n_k}(x) - u(x))dx \to 0$. This leads to the fact that $\langle D\varphi(u_{n_k}), \nabla u_{n_k} - \nabla u\rangle \to 0$, which, combined with the fact that $u_{n_k} \rightharpoonup u$ in X, will lead us to the conclusion that $u_{n_k} \to u$ in X. To see the last claim, note that φ is a convex functional, hence, $\varphi(u) \geq \varphi(u_{n_k}) + \langle D\varphi(u_{n_k}), u - u_{n_k}\rangle$, and taking the limit superior (combined with the fact that the last term converges to 0), we have that $\varphi(u) \geq \limsup_k \varphi(u_{n_k})$. On the other hand, by the weak lower semicontinuity of φ (it is in fact a norm), we have that $\varphi(u) \leq \liminf_k \varphi(u_{n_k})$, so

$$\liminf_k \varphi(u_{n_k}) = \limsup_k \varphi(u_{n_k}) = \lim_k \varphi(u_{n_k}) = \varphi(u).$$

Hence, $u_{n_k} \rightharpoonup u$, and $\|u_{n_k}\|_X \to \|u\|_X$, and by the uniform convexity of $X = W_0^{1,p}(\mathcal{D})$ (see Example 2.6.10) and the Radon–Riesz property (see Example 2.6.13), we conclude that $u_{n_k} \to u$ in X. Using the standard trick, we may show the result for the whole sequence.

4. Using the results of steps 1, 2, and 3, we therefore conclude the existence of a critical point u for F by an application of the mountain pass theorem.

5. It remains to show that if $f(x, 0) = 0$, then this critical point satisfies $u \geq 0$. To see that, define the function f_+ by $f_+(x, s) = f(x, \frac{s+|s|}{2})$. It is easy to see that $f_+(x, s) = 0$ for $s \leq 0$, while $f_+(x, s) = f(x, s)$ for $s > 0$. It is also easy to see that all the assumptions on f are still true for f_+, so repeating all the above, we may find a critical point u_+ for the functional F_+, which is obtained by exchanging f with f_+. This critical point is a weak solution of the quasilinear PDE:

$$\begin{aligned} -\Delta_p u_+ &= f_+(x, u_+), \quad \text{in } \mathcal{D}, \\ u_+ &= 0 \quad \text{on } \partial\mathcal{D}. \end{aligned}$$

Using $v = \max(-u_+, 0) \in W_0^{1,p}(\mathcal{D})$ as a test function, defining $\mathcal{D}_- = \{x \in \mathcal{D} : u_+ < 0\}$, we obtain that $-\int_{\mathcal{D}_-} |\nabla u_+(x)|^p dx = -\int_{\mathcal{D}_-} f_+(x, u_+(x))v(x)dx \geq 0$ so that $\nabla u_+(x) = 0$ a. e. in $\mathcal{D}_-$. Since, $\nabla v(x) = -\nabla u_+(x)$ for $x \in \mathcal{D}_-$ and $\nabla v(x) = 0$ for $x \in \mathcal{D} \setminus \mathcal{D}_-$, we see that $\nabla u_+(x) = 0$ a. e. in $\mathcal{D}_-$ implies that $\nabla v(x) = 0$ a. e. in $\mathcal{D}$. Therefore, $\|v\|_{W_0^{1,p}(\mathcal{D})} = 0$ and $v = 0$ a.e in $\mathcal{D}$, which in turn leads to the conclusion that $u \geq 0$ a. e. in $\mathcal{D}$. □

Remark 8.4.3. The condition in Assumption 8.4.1(iii) can be generalized as $\limsup_{s\to 0} \frac{pf(x,s)}{|s|^{p-2}s} \leq b(x) < \lambda_{1,p}$, where $b \in L^\infty(\mathcal{D})$ is a function such that $|\{x \in \mathcal{D} : b(x) < \lambda_{1,p}\}| > 0$, where $\lambda_{1,p} > 0$ is defined as in (8.30).

Remark 8.4.4. In the case where $f(x, 0) = 0$, one may also find a negative solution $u < 0$ by defining the function f_- as $f_-(x, s) = f(x, \frac{s-|s|}{2})$, which has the property $f_-(x, s) = 0$ for $s \geq 0$. Then applying Proposition 8.4.2 for the functional F_- in which f is replaced by f_-, one may find a nontrivial critical point, which is a weak solution for problem (8.29) with f replaced by f_-. The negativity of u is shown similarly as in the proof of Proposition 8.4.2 by using as test function $v = \max(u, 0)$.

8.4.2 Resonant problems for the p-Laplacian and the saddle point theorem

We now consider problem (8.29) but for the case where $\Phi(x,s) = \frac{\lambda_{1,p}}{p}|s|^p + \Phi_0(x,s)$ (or equivalently $f(x,s) = \lambda_{1,p}|s|^{p-2}s + f_0(x,s)$, where $\lambda_{1,p} > 0$ is the first eigenvalue of the p-Laplacian (given in (8.30)). This choice leads to the quasilinear problem

$$\begin{aligned} -\nabla\cdot(|\nabla u|^{p-2}\nabla u) &= \lambda_{1,p}|u|^{p-2}u + f_0(x,u), \quad \text{in } \mathcal{D}, \\ u &= 0 \quad \text{on } \partial\mathcal{D}, \end{aligned} \tag{8.36}$$

which is called resonant since if $f_0 = 0$ it reduces to the eigenvalue problem (8.31) that admits nontrivial solutions, which are scalar multiples of the eigenfunction $\phi_{1,p} > 0$ and, in analogy with the standard Laplacian operator case ($p = 2$), leads to noninvertibility of the operator $\mathrm{A}(u) := -\Delta_p u - \lambda_{1,p}|u|^{p-2}u$. An interesting question is whether problem (8.36) continues to admit nontrivial solutions when $f_0 \neq 0$. This problem can be treated using the saddle point theorem. Our approach follows [10].

We impose the following assumption on f_0 (whose antiderivative will be denoted by Φ_0):

Assumption 8.4.5. The Carathéodory function $f_0 : \mathcal{D}\times\mathbb{R}\to\mathbb{R}$ satisfies the following:
(i) f_0 is bounded and satisfies $\lim_{s\to\pm\infty} f(x,s) = \gamma_\pm(x)$, a. e.; for $\gamma_\pm \in L^\infty(\mathcal{D})$.
(ii) $\int_{\mathcal{D}} \gamma_+(x)\phi_{1,p}(x)dx < 0 < \int_{\mathcal{D}} \gamma_-(x)\phi_{1,p}(x)dx$ (or $\int_{\mathcal{D}} \gamma_-(x)\phi_{1,p}(x)dx < 0 < \int_{\mathcal{D}} \gamma_+(x)\phi_{1,p}(x)dx$).

The condition of Assumption 8.4.5(ii) is a Landesman–Lazer type condition.

We then have the following proposition [10]:

Proposition 8.4.6. *Under Assumption* 8.4.5 *problem* (8.36) *admits a nontrivial solution.*

Proof. Let $X_1 = \mathrm{span}(\phi_1)$, which is an one-dimensional space, and X_2 its orthogonal complement, so that $W_0^{1,p}(\mathcal{D}) = X_1 \oplus X_2$. In order to show that F satisfies the necessary geometry to apply the saddle point Theorem 8.2.7, we need to show that F is unbounded below on X_1, while it is bounded below on X_2. The fact that $F \in C^1(X;\mathbb{R})$ for $X = W_0^{1,p}(\mathcal{D})$ has already been established. The proof proceeds in 4 steps.

1. To show that F is unbounded below on X_1, consider any $u = \lambda\phi_{1,p}$, and note that since $\phi_{1,p}$ satisfies $\lambda_{1,p}\|\phi_{1,p}\|^p_{L^p(\mathcal{D})} = \|\phi_{1,p}\|^p_{W_0^{1,p}(\mathcal{D})}$, it holds that

$$\begin{aligned} F(\lambda\phi_{1,p}) &= \frac{1}{p}\|\lambda\phi_{1,p}\|^p_{W_0^{1,p}(\mathcal{D})} - \frac{\lambda_{1,p}}{p}\|\lambda\phi_{1,p}\|^p_{L^p(\mathcal{D})} - \int_{\mathcal{D}} \Phi_0(x,\lambda\phi_{1,p}(x))dx \\ &= -\int_{\mathcal{D}} \Phi_0(x,\lambda\phi_{1,p}(x))dx = -\lambda\int_{\mathcal{D}} \frac{\Phi_0(x,\lambda\phi_{1,p}(x))}{\lambda\phi_{1,p}(x)}\phi_{1,p}(x)dx. \end{aligned}$$

Since $\phi_{1,p} > 0$, as $\lambda \to \infty$, we have Assumption 8.4.5(i) that $\frac{\Phi_0(x,\lambda\phi_{1,p}(x))}{\lambda\phi_{1,p}(x))} \to \gamma_+(x)$ so that using Lebesgue's dominated convergence $\int_{\mathcal{D}} \frac{\Phi_0(x,\lambda\phi_{1,p}(x))}{\lambda\phi_{1,p}(x)}\phi_{1,p}(x)dx \to \int_{\mathcal{D}} \gamma_+(x)\phi_{1,p}(x)dx$ and, by Assumption 8.4.5(ii), $F(\lambda\phi_{1,p}) \to -\infty$, as $\lambda \to \infty$ (with a similar reasoning if the alternative condition holds).

2. We now consider $u \in X_2$. By the definition of $\lambda_{1,p}$ (see (8.30)) for any $u \in X_2$, there exists a $\lambda > \lambda_{1,p}$ such that $\lambda\|u\|^p_{L^p(\mathcal{D})} < \|u\|^p_{W_0^{1,p}(\mathcal{D})}$. We then see that

$$\begin{aligned} F(u) &= \frac{1}{p}\|u\|^p_{W_0^{1,p}(\mathcal{D})} - \frac{\lambda_{1,p}}{p}\|u\|^p_{L^p(\mathcal{D})} - \int_{\mathcal{D}} \Phi_0(x,u(x))dx \\ &\geq \frac{1}{p}\Big(1 - \frac{\lambda_{1,p}}{\lambda}\Big)\|u\|^p_{W_0^{1,p}(\mathcal{D})} - \int_{\mathcal{D}} \Phi_0(x,u(x))dx \\ &\geq \frac{1}{p}\Big(1 - \frac{\lambda_{1,p}}{\lambda}\Big)\|u\|^p_{W_0^{1,p}(\mathcal{D})} - c_1'\|u\|_{L^1(\mathcal{D})}, \end{aligned}$$

where we used Assumption 8.4.5(i), and c_1' is an appropriate constant. By the Hölder inequality $\|u\|_{L^1(\mathcal{D})} \leq c_2'\|u\|_{L^p(\mathcal{D})}$ and using once more the Poincaré inequality, we obtain the lower bound

$$F(u) \geq \frac{1}{p}\Big(1 - \frac{\lambda_{1,p}}{\lambda}\Big)\|u\|^p_{W_0^{1,p}(\mathcal{D})} - c_3'\|u\|_{W_0^{1,p}(\mathcal{D})}, \quad \forall\, u \in X_2,$$

from which we conclude that $\min_{u\in X_2} F(u) > -\infty$.

3. It remains to show that F satisfies the Palais–Smale condition. As in the proof of Proposition 8.4.2, it suffices to show that any Palais–Smale sequence is bounded in $X = W_0^{1,p}(\mathcal{D})$.

3(a). We will show that if $(u_n)_{n\in\mathbb{N}}$ is a Palais–Smale sequence, then it is bounded. We may either follow the route in step 3(a) in Proposition 8.4.2 or the following alternative route by contradiction. Assume not, and consider a Palais–Smale sequence for F, $(u_n)_{n\in\mathbb{N}}$ such that $\|u_n\|_{W_0^{1,p}(\mathcal{D})} \to \infty$. Define $(w_n)_{n\in\mathbb{N}}$ with $w_n = \frac{u_n}{\|u_n\|_{W_0^{1,p}(\mathcal{D})}}$, which is bounded and by the reflexivity of $W_0^{1,p}(\mathcal{D})$; there exists a $w_0 \in W_0^{1,p}(\mathcal{D})$ and a subsequence (denoted the same) such that $w_n \rightharpoonup w_0$ in $W_0^{1,p}(\mathcal{D})$, while by the compact embedding $W_0^{1,p}(\mathcal{D}) \hookrightarrow L^p(\mathcal{D})$, there exists a further subsequence (still denoted the same) such that $w_n \to w_0$ in $L^p(\mathcal{D})$. We will show that since $(u_n)_{n\in\mathbb{N}}$ is a Palais–Smale sequence, $w_n \to w_0$ in $W_0^{1,p}(\mathcal{D})$. Indeed, since $F(u_n) < c$, dividing by $\|u_n\|^p_{W_0^{1,p}(\mathcal{D})}$, we have that

$$\frac{1}{p}\|w_n\|^p_{W_0^{1,p}(\mathcal{D})} - \frac{\lambda_{1,p}}{p}\|w_n\|^p_{L^p(\mathcal{D})} - \int_{\mathcal{D}} \frac{\Phi(x,u_n)}{\|u_n\|^p_{W_0^{1,p}(\mathcal{D})}}dx \leq \frac{c}{\|u_n\|^p_{W_0^{1,p}(\mathcal{D})}}. \tag{8.37}$$

Note that on account of Assumption 8.4.5 (i), the asymptotic behavior of $\Phi(x,u)$ is $\gamma_+(x)u(x)$ for $u(x) \to \infty$ and $\gamma_-(x)u(x)$ for $u(x) \to -\infty$; it is straightforward to see that

$\int_{\mathcal{D}} \frac{\Phi(x,u_n)}{\|u_n\|^p_{W_0^{1,p}(\mathcal{D})}} dx \to 0$, as $\|u_n\|_{W_0^{1,p}(\mathcal{D})} \to \infty$. Furthermore, since $w_n \to w_0$ in $L^p(\mathcal{D})$, it holds that $\|w_n\|^p_{L^p(\mathcal{D})} \to \|w_0\|^p_{L^p(\mathcal{D})}$. Taking the limit superior in (8.37), we conclude that

$$\limsup_n \|w_n\|^p_{W_0^{1,p}(\mathcal{D})} \le \lambda_{1,p}\|w_0\|^p_{L^p(\mathcal{D})}. \tag{8.38}$$

On the other hand, by the weak lower semicontinuity of the norm, it holds that $\|w_0\|^p_{W_0^{1,p}(\mathcal{D})} \le \liminf_n \|w_n\|^p_{W_0^{1,p}(\mathcal{D})}$, which combined with the Poincaré inequality (or equivalently the definition of $\lambda_{1,p}$), yields $\lambda_{1,p}\|w_0\|^p_{L^p(\mathcal{D})} \le \|w_0\|^p_{W_0^{1,p}(\mathcal{D})}$ and leads to the conclusion that

$$\lambda_{1,p}\|w_0\|^p_{L^p(\mathcal{D})} \le \liminf_n \|w_n\|^p_{W_0^{1,p}(\mathcal{D})}. \tag{8.39}$$

Combining (8.38) with (8.39), we conclude the fact that
(a) $\lim_n \|w_n\|^p_{W_0^{1,p}(\mathcal{D})} = \|w_0\|^p_{W_0^{1,p}(\mathcal{D})}$, while
(b) $\lambda_{1,p}\|w_0\|^p_{L^p(\mathcal{D})} = \|w_0\|^p_{W_0^{1,p}(\mathcal{D})}$.

By the uniform convexity of $W_0^{1,p}(\mathcal{D})$ (see Example 2.6.10) and the Radon–Riesz property (Example 2.6.13), we conclude from (a) that $w_n \to w$ (strong) in $W_0^{1,p}(\mathcal{D})$. Fact (b) implies that w_0 is a minimizer of problem (8.30) so that w_0 is a multiple of $\phi_{1,p}$ (either $\phi_{1,p}$ or $-\phi_{1,p}$ since it is of norm 1). Let us assume without loss of generality that $w_0 = \phi_{1,p} > 0$. Hence, $w_n \to \phi_{1,p}$ in $W_0^{1,p}(\mathcal{D})$.

We now consider the condition that $DF(u_n) \to 0$ in $(W_0^{1,p}(\mathcal{D}))^*$, which allows us to conclude that for any $\epsilon > 0$ there exists an $N \in \mathbb{N}$ such that for any $n > N$,

$$-\epsilon\|u_n\|_{W_0^{1,p}(\mathcal{D})} < \|u_n\|^p_{W_0^{1,p}(\mathcal{D})} - \lambda_{1,p}\|u_n\|^p_{L^p(\mathcal{D})} - \int_{\mathcal{D}} f(x,u_n(x))u_n(x)dx < -\epsilon\|u_n\|_{W_0^{1,p}(\mathcal{D})}, \tag{8.40}$$

while $|F(u_n)| < c$ gives that

$$-c \le \frac{1}{p}\|u_n\|^p_{W_0^{1,p}(\mathcal{D})} - \frac{\lambda_{1,p}}{p}\|u_n\|^p_{L^p(\mathcal{D})} - \int_{\mathcal{D}} \Phi_0(x,u_n(x))dx \le c. \tag{8.41}$$

Multiplying (8.41) by $-p$ and adding to (8.42), we can get rid of the terms involving $\|u_n\|^p_{W_0^{1,p}(\mathcal{D})}$ and $\|u_n\|^p_{L^p(\mathcal{D})}$ and conclude that

$$\left|\int_{\mathcal{D}} (f_0(x,u_n(x))u_n(x) - p\Phi_0(x,u_n(x)))dx\right| \le c\,p + \epsilon\|u_n\|^p_{W_0^{1,p}(\mathcal{D})},$$

which upon dividing with $\|u_n\|_{W_0^{1,p}(\mathcal{D})}$ leads to

$$\left|\int_{\mathcal{D}}\left(f_0(x,u_n(x))w_n(x) - p\frac{\Phi_0(x,u_n(x))}{u_n(x)}w_n(x)\right)dx\right| \leq \frac{c}{\|u_n\|_{W_0^{1,p}(\mathcal{D})}} + \epsilon. \tag{8.42}$$

Since $w_n \to \phi_1 > 0$, we have that $u_n \to \infty$ so that $f(x,u_n(x)) \to \gamma_\infty(x)$ and $\frac{\Phi_0(x,u_n(x))}{u_n(x)} \to \gamma_+(x)$. Therefore, passing to the limit in (8.42) and using the Lebesgue dominated convergence theorem, we conclude that $\int_{\mathcal{D}} \gamma_+(x)\phi_1(x)dx = 0$, which is in contradiction with Assumption 8.4.5(ii). Therefore, $(u_n)_{n\in\mathbb{N}}$ is bounded. If $w_0 \to -\phi_1$, then we proceed similarly.

3(b) To show that any Palais–Smale sequence converges, we proceed exactly as in step 3(b) of Proposition 8.4.2, using boundedness of $(u_n)_{n\in\mathbb{N}}$ to guarantee the existence of a weak limit, and then uniform convexity of $W_0^{1,p}(\mathcal{D})$ to show that this limit is strong. The details are left to the reader.

4. Using steps 1, 2, and 3 using the saddle point theorem, we conclude the proof. □

9 Monotone type operators

This chapter is dedicated to the theory and applications of monotone type operators. We will introduce various notions of monotonicity (i. e., monotonicity, maximal monotonicity and pseudomonotonicity) and study the remarkable properties of these operators in particular with respect to surjectivity. The theory of monotone type operators plays an important role in applications. In closing the chapter, we try to present a variety of applications, including quasilinear elliptic PDEs, variational inequalities with nonlinear operators, as well as gradient flows in Hilbert spaces or evolution equations in Gelfand triples. Monotone type operators and their applications are covered in many books, see, e. g., [18, 32, 107, 110] or [128], upon which our approach is based.

Unless explicitly stated otherwise, throughout this chapter, all Hilbert spaces are identified with their duals ($H = H^\star$).

9.1 Motivation

The motivation for monotone operators is closely connected with the theory of convexity. Recall for instance that a Gâteaux differentiable function is convex if and only if $\langle D\varphi(\mathsf{x}_2) - D\varphi(\mathsf{x}_1), \mathsf{x}_2 - \mathsf{x}_1\rangle \geq 0$ holds for any $\mathsf{x}_1, \mathsf{x}_2 \in C$ for a suitable convex subset (see Theorem 2.3.23). Recall also that a similar property holds, even for nonsmooth convex functions in terms of the subdifferential, since for any $\mathsf{x}_i^\star \in \partial\varphi(\mathsf{x}_i)$, $i = 1, 2$, we have that $\langle \mathsf{x}_1^\star - \mathsf{x}_2^\star, \mathsf{x}_1 - \mathsf{x}_2\rangle \geq 0$ (see Proposition 4.2.1). These properties have proved to be extremely useful in a number of applications, e. g., in the convergence of iterative schemes for the minimization of convex functions, either smooth or not (see, e. g., Section 4.5).

A similar property was also encountered when considering linear operators related to coercive bilinear forms in Chapter 7, and has played a crucial role in the development of the Lions–Stampacchia and Lax–Milgram theory for the treatment of variational inequalities and linear elliptic equations, respectively. Important linear or nonlinear operators, single valued or multivalued, whether related to derivatives of convex functionals or not, enjoy the property $\langle \mathsf{A}(\mathsf{x}_1) - \mathsf{A}(\mathsf{x}_2), \mathsf{x}_1 - \mathsf{x}_2\rangle \geq 0$. Two examples are the Laplacian and the p-Laplacian, and we have already seen, in Chapters 6 and 7, how this property can be used appropriately in order to obtain important conclusions concerning solvability of operator equations, properties of eigenvalues, or approximation results for these operators.

It is the aim of this chapter to present the elegant theory of monotone operators, as well as its various extensions, and how monotonicity may be used to provide important and detailed information concerning properties of such operators, which turn out to be natural extensions of the properties that the Gâteaux derivatives of convex functions enjoy, and also how this abstract and general theory may be used in applications in order to provide solvability results for nonlinear operator equations, which are related to nonlinear PDEs, variational inequalities, or evolution equations.

https://doi.org/10.1515/9783111333298-009

9.2 Boundedness and continuity notions for nonlinear operators

In this section, we introduce (or refresh) some notions of boundedness and continuity for nonlinear operators that will be important for what follows.

Definition 9.2.1. An operator $A : \mathbf{D}(A) \subset X \to X^\star$ is called

(i) Locally bounded at $x \in X$ if there exists a neighborhood $N(x)$ of x such that the set $A(N(x)) = \{A(y) : y \in N(x) \cap \mathbf{D}(A)\}$ is bounded in $X^\star$.

(ii) Bounded if it maps bounded sets of X to bounded sets of $X^\star$.

Remark 9.2.2. If A were a linear operator, then continuity is equivalent to boundedness. However, this is not true for nonlinear operators in general.

Definition 9.2.3 (Modes of continuity). Let X be a Banach space and $X^\star$ its dual. An operator $A : \mathbf{D}(A) \subset X \to X^\star$ is called

(i) Hemicontinuous at $x \in \operatorname{int}(\mathbf{D}(A))$ if for every $z \in X$ with $x + tz \in \mathbf{D}(A)$ for $t \in [0, t_0)$, $t_0 > 0$, we have $A(x + tz) \overset{\star}{\rightharpoonup} A(x)$ in $X^\star$, as $t \to 0^+$.

(ii) Demicontinuous (strong to weak continuous) at $x \in \mathbf{D}(A)$ if for every $(x_n)_{n\in\mathbb{N}} \subset \mathbf{D}(A)$ with $x_n \to x$, we have $A(x_n) \overset{\star}{\rightharpoonup} A(x)$ in $X^\star$.

(iii) Continuous (strong to strong continuous) if $\|A(x) - A(z)\|_{X^\star} \to 0$, as long as $\|x - z\|_X \to 0$.

(iv) Completely continuous (weak to strong continuous) if $x \rightharpoonup z$ in X implies that $A(x) \to A(z)$ in $X^\star$.

The ordering above is from weaker to stronger notion in the sense that

Completely continuous $\Longrightarrow$ Continuous $\Longrightarrow$ Demicontinuous $\Longrightarrow$ Hemicontinuous.

Definition 9.2.4. An operator $A : \mathbf{D}(A) \subset X \to X^\star$ is called coercive if $\frac{\langle A(x), x\rangle}{\|x\|} \to \infty$, as $\|x\| \to \infty$.

The linear operators, which are generated by bilinear coercive forms (see Section 7.3), are coercive operators. An example of a nonlinear coercive operator is the p-Laplace operator (see Example 9.3.3).

The above definitions for multivalued operators follow accordingly, replacing $A(x)$ by $x^* \in A(x)$.

9.3 Monotone operators

9.3.1 Monotone operators, definitions, and examples

Let X be a Banach space, $X^\star$ its topological dual, and by $\langle\cdot,\cdot\rangle$ we denote the duality pairing between X and $X^\star$.

Definition 9.3.1 (Monotone operators). Let $\mathsf{A} : \mathbf{D}(\mathsf{A}) \subset X \to 2^{X^\star}$ be a (possibly multivalued) operator,

(i) A is called monotone if

$$\langle x_1^\star - x_2^\star, x_1 - x_2 \rangle \geq 0, \quad \forall\, x_i^\star \in \mathsf{A}(x_i),\ x_i \in \mathbf{D}(\mathsf{A}) \subset X,\ i = 1, 2. \tag{9.1}$$

(ii) A is called strictly monotone if it is monotone and $\langle x_1^\star - x_2^\star, x_1 - x_2 \rangle = 0$ for some $(x_i, x_i^\star) \in \mathbf{Gr}(\mathsf{A})$, $i = 1, 2$, implies $x_1 = x_2$.

If A is single valued, then the monotonicity condition (9.1) simplifies to

$$\langle \mathsf{A}(x_1) - \mathsf{A}(x_1), x_1 - x_2 \rangle \geq 0, \quad \forall\, x_1, x_2 \in \mathbf{D}(\mathsf{A}).$$

Increasing functions and their corresponding Nemitskii operators can be considered as monotone operators. The same holds for linear operators defined by coercive bilinear forms. Also, according to Proposition 4.2.1, the subdifferential of a convex function is a monotone operator, hence, also the Gâteaux derivative of a differentiable convex function. As a final example, we mention the duality map $\mathsf{J} : X \to 2^{X^\star}$, which is a monotone (possibly multivalued) operator, as the subdifferential of the norm (see Proposition 4.2.17).

Example 9.3.2. Let H be a Hilbert space. A (possibly multivalued) operator $\mathsf{A} : \mathbf{D}(\mathsf{A}) \subset H \to 2^H$ is monotone if and only if

$$\|x_1 - x_2\| \leq \|(x_1 - x_2) + \lambda(z_1 - z_2)\|, \quad \forall\, x_i \in \mathbf{D}(\mathsf{A}),\ z_i \in \mathsf{A}(x_i),\ \lambda > 0,\ i = 1, 2. \tag{9.2}$$

Inequality (9.2) implies that for each $\lambda > 0$, the operator $(I + \lambda\mathsf{A})^{-1}$ is defined on $\mathbf{R}(I + \lambda\mathsf{A})$.

Indeed, the direct assertion is immediate by expanding $\|(x_1 - x_2) + \lambda(z_1 - z_2)\|^2$ in terms of the inner product and using monotonicity, while for the converse, note that (9.2) implies that $2\lambda\langle z_1 - z_2, x_1 - x_2 \rangle + \lambda^2 \|z_1 - z_2\|^2 \geq 0$, so dividing by λ and passing to the limit $\lambda \to 0^+$ yields the monotonicity of A. ◁

Example 9.3.3. Let $\mathcal{D} \subset \mathbb{R}^d$ be a bounded domain with sufficiently smooth boundary. The p-Laplace operator $\mathsf{A} = -\Delta_p : W_0^{1,p}(\mathcal{D}) \to W^{-1,p^\star}(\mathcal{D})$, defined by

$$\langle \mathsf{A}(u), v \rangle := \langle -\Delta_p u, v \rangle := \int_{\mathcal{D}} |\nabla u(x)|^{p-2} \nabla u(x) \cdot \nabla v(x)\, dx, \quad \forall\, u, v \in W_0^{1,p}(\mathcal{D}),$$

where $p \in (1, \infty)$ (see Section 6.8.1), is strictly monotone. The special case $p = 2$ corresponds to a linear operator, which is easily identified with the Laplacian.

We will equip $W_0^{1,p}(\mathcal{D})$ with the equivalent norm $\|u\|_{1,p} = (\sum_{i=1}^d \|D_i u\|_{L^p(\mathcal{D})}^p)^{\frac{1}{p}}$, where, by $D_i u$, we denote the partial derivative as $D_i u = \frac{\partial u}{\partial x_i}$, $i = 1, \ldots, d$. Noting that

$$
\begin{aligned}
\langle \mathsf{A}(u_1), u_2 \rangle &= \sum_{i=1}^{d} \langle |D_i u_1|^{p-2} D_i u_1, D_i u_2 \rangle \leq \sum_{i=1}^{d} \|D_i u_1\|_{L^p(\mathcal{D})}^{p-1} \|D_i u_2\|_{L^p(\mathcal{D})} \\
&\leq \left(\sum_{i=1}^{d} \|D_i u_1\|_{L^p(\mathcal{D})}^{p} \right)^{1/p^*} \left(\|D_i u_2\|_{L^p(\mathcal{D})}^{p} \right)^{1/p} \leq \|u_1\|_{1,p}^{p-1} \|u_2\|_{1,p},
\end{aligned}
$$

we obtain that

$$
\begin{aligned}
&\langle \mathsf{A}(u_1) - \mathsf{A}(u_2), u_1 - u_2 \rangle \\
&\quad = \langle \mathsf{A}(u_1), u_1 \rangle - \langle \mathsf{A}(u_1), u_2 \rangle - \langle \mathsf{A}(u_2), u_1 \rangle + \langle \mathsf{A}(u_2), u_2 \rangle \\
&\quad \geq \|u_1\|_{1,p}^{p} - \|u_1\|_{1,p}^{p-1} \|u_2\|_{1,p} - \|u_2\|_{1,p}^{p-1} \|u_1\|_{1,p} + \|u_2\|_{1,p}^{p} \\
&\quad = (\|u_1\|_{1,p}^{p-1} - \|u_2\|_{1,p}^{p-1})(\|u_1\|_{1,p} - \|u_2\|_{1,p}) \geq 0.
\end{aligned}
$$

The strict monotonicity of A follows from the elementary inequality

$$
(|a|^{p-2} a - |b|^{p-2} b) \cdot (a - b) \geq c\, |a - b|^{p}
$$

for all $a, b \in \mathbb{R}^d$ and fixed $p \geq 2$, $c > 0$ (which follows by the convexity of the p-th power of the Euclidean norm $|\cdot|$ in $\mathbb{R}^d$). We thus have

$$
\begin{aligned}
\langle \mathsf{A}(u_1) - \mathsf{A}(u_2), u_1 - u_2 \rangle &= \langle \mathsf{A}(u_1), u_1 - u_2 \rangle - \langle \mathsf{A}(u_2), u_1 - u_2 \rangle \\
&= \int_{\mathcal{D}} \sum_{i=1}^{d} (|D_i u_1|^{p-2} D_i u_1 - |D_i u_2|^{p-2} D_i u_2) \cdot (D_i u_1 - D_i u_2) dx \\
&\geq c \int_{\mathcal{D}} \sum_{i=1}^{d} |D_i u_1 - D_i u_2|^{p} dx = c\, \|u_1 - u_2\|_{1,p}^{p},
\end{aligned}
$$

which guarantees the strict monotonicity of A. ◁

In this section, for reasons of simplicity and clarity of exposition, we will focus on the case of single valued monotone operators. The general case where A is multivalued will be treated in detail in the general context of maximal monotone and pseudomonotone operators in Sections 9.4 and 9.5, respectively.

9.3.2 Local boundedness of monotone operators

A fundamental property of monotone operators is local boundedness (see, e. g., [3]).

Theorem 9.3.4. *Let* $\mathsf{A} : \mathbf{D}(\mathsf{A}) \subset X \to 2^{X^*}$ *be a monotone operator. Then,* A *is locally bounded on* $\operatorname{int}(\mathbf{D}(\mathsf{A}))$.

Proof. For simplicity, let A be single valued. Assume that the assertion of the theorem is not true. Then there exists $x_o \in \operatorname{int}(\mathbf{D}(\mathsf{A}))$ and a sequence $(x_n)_{n\in\mathbb{N}} \subset X$ such that $x_n \to x_o$

in X, but $\|A(x_n)\|_{X^\star} \to \infty$. We define the sequences $(a_n)_{n\in\mathbb{N}}, (b_n)_{n\in\mathbb{N}} \subset \mathbb{R}$ with $a_n := \|x_n - x\| \to 0$ and $b_n := \|A(x_n)\|_{X^\star} \to \infty$, as well as the sequence $(c_n)_{n\in\mathbb{N}}$ with $c_n := \max(a_n^{1/2}, b_n^{-1})$, which satisfies the properties $c_n \to 0$, and of course $b_n c_n \geq 1$ and $a_n \leq c_n^2$.

Consider now $z \in X$ arbitrary, and let $z_n = x_o + c_n z \in \mathbf{D}(A)$ for large enough n (since $x_o \in \operatorname{int}(\mathbf{D}(A))$ and $c_n \to 0$). Choose $r > 0$ such that $\bar{z} = x_o + rz \in \mathbf{D}(A)$ with $\|A(\bar{z})\|_{X^\star} < \infty$. Using the monotonicity of A, we have that

$$\langle A(z_n) - A(\bar{z}), z_n - \bar{z}\rangle = (c_n - r)\langle A(z_n) - A(\bar{z}), z\rangle \geq 0,$$

and since $c_n \to 0$, choosing n large enough, we may guarantee that $c_n - r < 0$ so that $\langle A(z_n), z\rangle \leq \langle A(\bar{z}), z\rangle$, which by the properties of $\bar{z}$, allows us by the use of the Banach–Steinhaus uniform bound principle (Theorem 1.1.7) to conclude that there exists $c > 0$ independent of n such that $\|A(z_n)\|_{X^\star} < c$ for all n sufficiently large.

We now apply the monotonicity property once more to the points x_n and z_n. This gives $0 \leq \langle A(x_n) - A(z_n), x_n - z_n\rangle$, which, by the definition of z_n, upon rearrangement yields the bound

$$\begin{aligned} c_n\langle A(x_n), z\rangle &\leq \langle A(x_n), x_n - x_o\rangle - \langle A(z_n), (x_n - x_o) - c_n z\rangle \\ &\leq b_n a_n + c\,(a_n + c_n\|z\|) \leq b_n c_n^2 + c(\,c_n^2 + c_n\|z\|). \end{aligned}$$

Dividing by $b_n\, c_n^2$, we obtain that

$$\frac{1}{b_n c_n}\langle A(x_n), z\rangle \leq 1 + c\left(\frac{1}{b_n} + \frac{1}{b_n c_n}\|z\|\right) \leq 1 + \frac{c}{b_n} + \|z\| \leq c',$$

for some appropriate constant $c' > 0$, uniformly in n, (we used the fact $b_n c_n \geq 1$ for all $n \in \mathbb{N}$ and that $b_n^{-1} \to 0$). Applying the Banach–Steinhaus principle once more, we have that $\frac{1}{b_n c_n}\|A(x_n)\|_{X^\star} < c''$ for an appropriate constant $c'' > 0$, uniformly in n. But $\frac{1}{b_n c_n}\|A(x_n)\|_{X^\star} = \frac{1}{c_n}$, so we conclude that $\frac{1}{c_n} < c''$ uniformly in n, which is in contradiction with the assumption that $c_n \to 0$. □

Remark 9.3.5. Note that A may be unbounded at $\partial\mathbf{D}(A)$. For example, let $(x_n)_{n\in\mathbb{N}} \subset \mathbb{R}^d$, $d \geq 2$, be a sequence such that $|x_n| = 1$ for every $n \in \mathbb{N}$ and consider the mapping $A : \overline{B}(0,1) \subset \mathbb{R}^d \to \mathbb{R}^d$, defined by

$$A(x) = \begin{cases} x, & \text{for } x \in \overline{B}(0,1),\ x \neq x_n, \\ (n+1)x_n, & \text{for } x = x_n, \end{cases}$$

which is a monotone operator with $\mathbf{D}(A) = \overline{B}(0,1)$, but is unbounded on $\partial\mathbf{D}(A)$.

9.3.3 Hemicontinuity and demicontinuity

It is clear that in general, demicontinuity implies hemicontinuity. In the following result we will see that for monotone operators, the converse is also true:

Proposition 9.3.6. *Let X be a reflexive Banach space. Any monotone and hemicontinuous operator* $\mathsf{A} : \mathbf{D}(\mathsf{A}) \subset X \to X^\star$ *is demicontinuous on* $\operatorname{int}(\mathbf{D}(\mathsf{A}))$.

Proof. Suppose that A is hemicontinuous on $\operatorname{int}(\mathbf{D}(\mathsf{A}))$. Let $x \in \operatorname{int}(\mathbf{D}(\mathsf{A}))$ and $(x_n)_{n\in\mathbb{N}} \subset \operatorname{int}(\mathbf{D}(\mathsf{A}))$ be such that $x_n \to x$ in X. From Theorem 9.3.4, we know that A is locally bounded on $\operatorname{int}(\mathbf{D}(\mathsf{A}))$. So we may assume that the sequence $(\mathsf{A}(x_n))_{n\in\mathbb{N}}$ is bounded in $X^\star$ and, by reflexivity, there exists a subsequence $(\mathsf{A}(x_{n_k}))_{k\in\mathbb{N}}$ of $(\mathsf{A}(x_n))_{n\in\mathbb{N}}$ and a $x^\star \in X^\star$ such that $\mathsf{A}(x_{n_k}) \rightharpoonup x^\star$ in $X^\star$. We have

$$\langle \mathsf{A}(x_{n_k}) - \mathsf{A}(z), x_{n_k} - z\rangle \geq 0, \quad \forall z \in \mathbf{D}(\mathsf{A}),$$

and passing to the limit, we obtain

$$\langle x^\star - \mathsf{A}(z), x - z\rangle \geq 0, \quad \forall z \in \mathbf{D}(\mathsf{A}). \tag{9.3}$$

Since $\operatorname{int}(\mathbf{D}(\mathsf{A}))$ is an open set, for any $z \in X$ there exists $t_o > 0$, depending on z, such that $x_t = x + t z \in \operatorname{int}(\mathbf{D}(\mathsf{A}))$ for all t with $0 < t \leq t_o$. Set $z = x_t$ in (9.3) to get

$$\langle x^\star - \mathsf{A}(x + t z), z\rangle \leq 0.$$

Letting $t \to 0^+$, we obtain by the hemicontinuity of A that

$$\langle x^\star - \mathsf{A}(x), z\rangle \leq 0, \quad \forall z \in X,$$

so $\mathsf{A}(x) = x^\star$, i. e., $\mathsf{A}(x_{n_k}) \rightharpoonup \mathsf{A}(x)$, and using the Urysohn property (see Remark 1.1.52), we conclude that the whole sequence $\mathsf{A}(x_n) \rightharpoonup \mathsf{A}(x)$, hence, A is demicontinuous. □

The properties of monotonicity and hemicontinuity have the following interesting implication (often called Minty's trick):

Lemma 9.3.7 (Minty). *If* $\mathsf{A} : \mathbf{D}(\mathsf{A}) \subset X \to X^\star$, *with* $\mathbf{D}(\mathsf{A}) = X$ *(X not necessarily reflexive), is monotone and hemicontinuous, and* $(x_o, x_o^\star) \in X \times X^\star$ *such that*

$$\langle x_o^\star - \mathsf{A}(x), x_o - x\rangle \geq 0, \quad \forall x \in X, \tag{9.4}$$

then $x_o^\star = \mathsf{A}(x_o)$.

Proof. For any $t > 0$, let $x = x_o + tz$, where $z \in X$. Then, it follows from (9.4) that

$$\langle x_o^\star - \mathsf{A}(x_o + tz), z\rangle \leq 0.$$

Taking the limit as $t \to 0^+$, hemicontinuity gives $\langle x_o^\star - A(x_o), z\rangle \le 0$ for all $z \in X$, and replacing z by $-z$, we obtain $x_o^\star = A(x_o)$. □

9.3.4 Surjectivity of monotone operators and the Minty–Browder theory

In this section, we consider the question of surjectivity of monotone operators $A : \mathbf{D}(A) \subset X \to X^\star$ in the case where X is a separable reflexive Banach space. Even though, we will consider more general results concerning the surjectivity of a more general class of operators (see Section 9.4.4). We include this discussion here as a good point to revisit the Faedo–Galerkin method, which is an important approximation method (already encountered in Chapter 7) in a more general and abstract framework. Our approach follows [123].

The following important theorem holds for coercive and monotone operators, and guarantees the solvability of the operator equation $A(x) = f$ for any $f \in X^\star$.

Theorem 9.3.8 (Minty–Browder). *Let X be a reflexive and separable Banach space, and $A : \mathbf{D}(A) \subset X \to X^\star$ be a monotone, hemicontinuous, and coercive operator, with $\mathbf{D}(A) = X$. Then $\mathbf{R}(A) = X^\star$. Moreover, for any $x^\star \in X^\star$, the solution set $S(x^\star) := \{x \in X \,:\, A(x) = x^\star\}$ is closed, bounded, and convex. If, furthermore, A is strictly monotone, then for any $x^\star \in X^\star$, the solution is unique.*

Proof. The proof is broken up into 5 steps.

1. Since X is separable, there exists an increasing sequence[1] $(X_n)_{n\in\mathbb{N}}$ of finite dimensional subspaces of X with $\overline{\bigcup_{n=1}^\infty X_n} = X$. For each n, fixed, let $j_n : X_n \to X$ be the injection mapping of X_n into X, let $j_n^\star : X^\star \to X_n^\star$ be its adjoint map, and define $A_n : X_n \to X_n^\star$ by $A_n = j_n^* A j_n$.

Since A is coercive, and $\langle A_n(x), x\rangle = \langle A(x), x\rangle$ for every $x \in X_n$, we conclude that A_n is also coercive. On the other hand, by the hemicontinuity of A, we conclude that A_n is continuous. By a straightforward application of Brouwer's fixed point theorem (see Proposition 3.2.13), we have that $\mathbf{R}(A_n) = X_n^\star$. Then, for any given $x^\star \in X^\star$ and any $n \in \mathbb{N}$, there exists $x_n \in X_n$ such that

$$A_n(x_n) = j_n^* x^\star, \tag{9.5}$$

which can be interpreted as

$$\langle A(x_n), z_n\rangle = \langle x^\star, z_n\rangle, \quad \forall\, z_n \in X_n. \tag{9.6}$$

1 Since X is separable, there exists a sequence $(z_n)_{n\in\mathbb{N}}$ whose span is dense in X. Then the sequence $(X_n)_{n\in\mathbb{N}}$ can be constructed by setting $X_n := \operatorname{span}\{z_1, \dots, z_n\}$, which is a finite dimensional subspace of X. If necessary, we may proceed with the bases of the finite dimensional spaces X_n. Note that in general $\dim(X_n) \neq n$.

Coercivity plays an important role here, as it allows us to restrict attention to bounded closed sets, which are compact in X_n by finite dimensionality. Note that (9.6) can be interpreted in terms of the canonical inclusion operator $j_n : X_n \to X$, or rather its adjoint $I_n^* : X^* \to X_n^* \simeq X_n$, which allows us to consider the restriction of $x^\star$ on X_n, ($x_n^\star := x^\star |_{X_n} = j_n^* x^\star$). Using that, (9.6) can be understood as the equation $j_n^*(A(x_n) - x^\star) = 0$.

2. We now show that the sequence of approximate solutions $(x_n)_{n\in\mathbb{N}}$ is bounded. Indeed,

$$\frac{\langle A(x_n), x_n\rangle}{\|x_n\|} = \frac{\langle A_n(x_n), x_n\rangle}{\|x_n\|} = \frac{\langle j_n^* x^\star, x_n\rangle}{\|x_n\|} = \frac{\langle x^\star, x_n\rangle}{\|x_n\|} \le \|x^\star\|_{X^\star},$$

and since A is coercive, it follows that $\|x_n\|$ is bounded, i. e., there exists $c_1 > 0$ such that $\|x_n\| \le c_1$ for every $n \in \mathbb{N}$.

We now show that the sequence $(A(x_n))_{n\in\mathbb{N}}$ is bounded as well. Indeed, since A is monotone, by Theorem 9.3.4, it is locally bounded, therefore,

$$\exists\, r, c > 0 \quad \text{if } \|x\| \le r \quad \text{then } \|A(x)\|_{X^\star} \le c. \tag{9.7}$$

By the monotonicity of A, we have

$$\langle A(x_n) - A(x), x_n - x\rangle \ge 0. \tag{9.8}$$

By (9.5), we have $\langle A(x_n), x_n\rangle = \langle x^\star, x_n\rangle$, which in turn implies that

$$|\langle A(x_n), x_n\rangle| \le \|x^\star\|_{X^\star} \|x_n\| \le c_1 \|x^\star\|_{X^\star}. \tag{9.9}$$

We have that[2] using (9.8),

$$\begin{aligned}\|A(x_n)\|_{X^\star} &= \sup_{x\in X,\, \|x\|=r} \frac{1}{r}\langle A(x_n), x\rangle \\ &\le \sup_{x\in X,\, \|x\|=r} \frac{1}{r}(\langle A(x), x\rangle + \langle A(x_n), x_n\rangle - \langle A(x), x_n\rangle) \\ &\le \frac{1}{r}(c\, r + c_1\|x^\star\|_{X^\star} + c\, c_1) < \infty,\end{aligned}$$

and using (9.7) and (9.7), we conclude that $(A(x_n))_{n\in\mathbb{N}}$ is bounded.

3. By the reflexivity of X, we may assume that there exists a subsequence of $(x_n)_{n\in\mathbb{N}}$, denoted the same for simplicity, and a $x_o \in X$, such that $x_n \rightharpoonup x_o$ in X. The proof will be complete if we show that $A(x_o) = x^\star$.

2 Express $\langle A(x_n), x\rangle = \langle A(x_n), x_n\rangle + \langle A(x), x - x_n\rangle + \langle A(x_n) - A(x), x - x_n\rangle \le \langle A(x_n), x_n\rangle + \langle A(x), x - x_n\rangle$, where we used (9.8).

We assert that $A(x_n) \rightharpoonup x^\star$. By the density of $\bigcup_{n\in\mathbb{N}} X_n$ in X suffices to show that

$$\langle A(x_n), z\rangle \to \langle x^\star, z\rangle, \quad \forall\, z \in \bigcup_{n\in\mathbb{N}} X_n. \tag{9.10}$$

Suppose not. Then, there exists $\bar{z} \in \bigcup_{n\in\mathbb{N}} X_n$, an $\epsilon > 0$, and a subsequence $(x_{n_k})_{k\in\mathbb{N}}$ such that

$$\big|\langle A(x_{n_k}), \bar{z}\rangle - \langle x^\star, \bar{z}\rangle\big| \geq \epsilon, \quad k \in \mathbb{N}. \tag{9.11}$$

Clearly, $\bar{z} \in \bigcup_{n\in\mathbb{N}} X_n$ implies the existence of some $N \in \mathbb{N}$ such that $\bar{z} \in X_N$ so that $\bar{z} \in X_n$ for all $n \geq N$. Pick K such that $n_K \geq N$. Then $n_k \geq N$ for all $k \geq K$. Since $\bar{z} \in X_n$ for all $n \geq N$, the above choice implies that $\bar{z} \in X_{n_k}$ for all $k \geq K$. Applying (9.6) for this choice of test function, we have that

$$\langle A(x_{n_k}), \bar{z}\rangle = \langle x^\star, \bar{z}\rangle, \tag{9.12}$$

which contradicts (9.11). Hence, (9.10) holds for any $z \in \bigcup_{n\in\mathbb{N}} X_n$ and by density for any $z \in X$.

Finally, we have (working along the subsequence of $(x_n)_{n\in\mathbb{N}}$ such that $x_n \rightharpoonup x_o$),

$$\langle A(x_n), x_n\rangle = \langle A_n(x_n), x_n\rangle = \langle j_n^* x^\star, x_n\rangle = \langle x^\star, x_n\rangle \to \langle x^\star, x_o\rangle. \tag{9.13}$$

By the monotonicity of A, we have

$$\langle A(x) - A(x_n), x - x_n\rangle \geq 0, \quad \forall\, x \in X,$$

and for $n \to \infty$, it follows from (9.13) that

$$\langle A(x) - x^\star, x - x_o\rangle \geq 0, \quad \forall\, x \in X,$$

and letting $x = x_o + t\,z$, for arbitrary $z \in X$, we obtain,

$$\langle A(x_o + t\,z) - x^\star, z\rangle \geq 0, \quad \forall\, z \in X,\ \forall\, t > 0.$$

Passing to the limit, as $t \to 0^+$, and using the hemicontinuity of A, we conclude that $A(x_o) = x^\star$. To pass from the selected subsequence to the whole sequence, we use the Urysohn property (see Remark 1.1.52).

4. Consider any $x^\star \in X^\star$. We have just proved that $S(x^\star) \neq \emptyset$. Since A is coercive, $S(x^\star)$ is bounded. We now prove that it is also convex. We let $x_1, x_2 \in S(x^\star)$, i. e., $A(x_i) = x^\star$, $i = 1, 2$, set $x_t = t x_1 + (1-t) x_2$, $t \in [0, 1]$. Then,

$$\langle x^\star - A(z), x_t - z\rangle = \langle x^\star - A(z), t(x_1 - z)\rangle + \langle x^\star - A(z), (1-t)(x_2 - z)\rangle$$
$$= t\langle A(x_1) - A(z), x_1 - z\rangle + (1-t)\langle A(x_2) - A(z), x_2 - z\rangle \geq 0, \quad \forall z \in X,$$

by the monotonicity of A, so by Lemma 9.3.7, we have $A(x_t) = x^\star$, therefore $x_t \in S(x^\star)$, hence, $S(x^\star)$ is convex.

We now show that $S(x^\star)$ is closed.[3] Fix any $x^\star \in X^\star$, and let $(x_n)_{n\in\mathbb{N}} \subset S(x^\star)$ (i. e., $A(x_n) = x^\star$) with $x_n \to x$. Then, for each $z \in X$, (using first the fact that $x_n \to x$, and then the fact that $x_n \in S(x^\star)$), we have that

$$\langle x^\star - A(z), x - z\rangle = \lim_{n\to\infty}\langle x^\star - A(z), x_n - z\rangle = \lim_{n\to\infty}\langle A(x_n) - A(z), x_n - z\rangle \geq 0,$$

where for the last inequality we have used the monotonicity of the operator A. Therefore, for each $z \in X$, it holds that $\langle x^\star - A(z), x - z\rangle \geq 0$, which in turn, by Lemma 9.3.7, implies that $A(x) = x^\star$. Therefore, $x \in S(x^\star)$, and $S(x^\star)$ is closed.

5. Under the extra assumption that A is strictly monotone, let $x_1 \neq x_2$ be two solutions of $A(x) = x^\star$. Then, $A(x_1) = x^\star = A(x_2)$ and

$$0 < \langle A(x_1) - A(x_2), x_1 - x_2\rangle = 0,$$

which is a contradiction, hence we have uniqueness. □

Example 9.3.9. The surjectivity results provided by the Lions–Stampacchia and Lax–Milgram theorems in Chapter 7 can be considered as special cases of the Minty–Browder surjectivity result. For instance, if $a : X\times X \to \mathbb{R}$ is a bilinear coercive form, then the linear operator $A : X \to X^\star$ defined by $\langle Ax, z\rangle = a(x, z)$ for every $z \in X$, is a coercive operator, satisfying the conditions required by Theorem 9.3.8, so the operator equation $Ax = z^\star$ admits a unique solution for every $z^\star \in X^\star$. Important partial differential equations, such as the Poisson equation, can be solved using such abstract techniques, extending the results obtained in Chapters 6 and 7). ◁

Example 9.3.10. Let $\mathcal{D} \subset \mathbb{R}^d$ be a bounded domain of sufficiently smooth boundary, and consider the p-Laplace Poisson equation $-\Delta_p u = f$ for a given $f \in L^{p^\star}(\mathcal{D})$, with $\frac{1}{p} + \frac{1}{p^\star} = 1$, $p > 1$. Then, the nonlinear operator $A := -\Delta_p : W_0^{1,p}(\mathcal{D}) \to W_0^{-1,p^\star}(\mathcal{D})$ is continuous, coercive, bounded, and monotone (see Section 6.8 and Example 9.3.3), hence, by Theorem 9.3.8, $\mathbf{R}(A) = X^\star$, so the equation $-\Delta_p u = f$ admits a unique solution in $W_0^{1,p}(\mathcal{D})$ for every $f \in W^{-1,p^\star}(\mathcal{D})$. The solvability of this equation was treated using the direct method of the calculus of variations in Section 6.8. ◁

3 Since $S(x^\star)$ is convex, we do not have to worry about discriminating between strong and weak closedness.

9.4 Maximal monotone operators

9.4.1 Maximal monotone operators definitions and examples

Definition 9.4.1. A monotone operator $\mathsf{A} : \mathbf{D}(\mathsf{A}) \subset X \to 2^{X^\star}$ is called maximal monotone if it has the following property:

$$\mathbf{Gr}(\mathsf{A}) \subset \mathbf{Gr}(\mathsf{A}') \text{ with } \mathsf{A}' \text{ monotone implies that } \mathsf{A} = \mathsf{A}'.$$

$\mathbf{Gr}(\mathsf{A})$ is called a maximal monotone graph, or we say that A generates a maximal monotone graph.

In other words, an operator is maximal monotone if it has no proper monotone extension, i. e., any monotone extension of A coincides with itself; for any monotone operator A', $\mathbf{Gr}(\mathsf{A}) \subset \mathbf{Gr}(\mathsf{A}')$ implies $\mathsf{A}' = \mathsf{A}$. An alternative way to put it is that A is maximal monotone if its graph is not properly contained in the graph of any other monotone operator, or equivalently, the graph of A is maximal with respect to inclusion among graphs of monotone operators. As an illustrative example of that, consider the set valued map $\phi : \mathbb{R} \to 2^{\mathbb{R}}$ defined by $\phi(x) = 0\,\mathbf{1}_{\{x<0\}} + A\mathbf{1}_{\{x=0\}} + 1\,\mathbf{1}_{\{x>0\}}$, which is maximal monotone if and only if $A = [0,1]$ (see [115]).

The following characterization of maximal monotonicity is simple, but often very helpful when trying to determine whether an element of $(\mathsf{x},\mathsf{x}^\star) \in X \times X^\star$ belongs to the graph of a maximal monotone operator A or not:

Proposition 9.4.2. *A monotone operator* $\mathsf{A} : \mathbf{D}(\mathsf{A}) \subset X \to 2^{X^\star}$ *is maximal monotone if and only if it satisfies the property*

$$\langle \mathsf{x}^\star - \mathsf{z}^\star, \mathsf{x} - \mathsf{z}\rangle \geq 0, \quad \forall\, (\mathsf{x},\mathsf{x}^\star) \in \mathbf{Gr}(\mathsf{A}) \textit{ implies } (\mathsf{z},\mathsf{z}^\star) \in \mathbf{Gr}(\mathsf{A}). \tag{9.14}$$

Proof. Suppose that A is maximal monotone but (9.14) does not hold, so that there exists $(\mathsf{z},\mathsf{z}^\star) \notin \mathbf{Gr}(\mathsf{A})$ with the property $\langle \mathsf{x}^\star - \mathsf{z}^\star, \mathsf{x} - \mathsf{z}\rangle \geq 0, \forall\, (\mathsf{x},\mathsf{x}^\star) \in \mathbf{Gr}(\mathsf{A})$. Then $\mathbf{Gr}(\mathsf{A}) \cup (\mathsf{z},\mathsf{z}^\star)$ is a monotone graph with the property that $\mathbf{Gr}(\mathsf{A}) \subset \mathbf{Gr}(\mathsf{A}) \cup (\mathsf{z},\mathsf{z}^\star)$, a fact that contradicts the maximal monotonicity of A.

For the converse, suppose that (9.14) holds, but A is not maximal monotone. Then there must be a monotone graph G', which is a proper extension of $\mathbf{Gr}(\mathsf{A})$. If $(\mathsf{z},\mathsf{z}^\star) \in G' \setminus \mathbf{Gr}(\mathsf{A})$, then $(\mathsf{z},\mathsf{z}^\star) \notin \mathbf{Gr}(\mathsf{A})$ (i. e., $\mathsf{z} \notin \mathbf{D}(\mathsf{A})$ and $\mathsf{z}^\star \notin \mathsf{A}(\mathsf{z})$), but by monotonicity of G', $\langle \mathsf{x}^\star - \mathsf{z}^\star, \mathsf{x} - \mathsf{z}\rangle \geq 0, \forall\, (\mathsf{x},\mathsf{x}^\star) \in \mathbf{Gr}(\mathsf{A})$. This contradicts the hypothesis that (9.14) holds. □

We will show shortly that many operators we have encountered so far enjoy the maximal monotonicity property, such as for instance the subdifferential of convex functions, linear operators $\mathsf{A} : X \to X^\star$ such that $\langle \mathsf{A}\mathsf{x}, \mathsf{x}\rangle \geq 0$ for every $\mathsf{x} \in X$, or the p-Laplace operator.

Example 9.4.3. It is obvious that if $\mathsf{A} : \mathbf{D}(\mathsf{A}) \subset X \to 2^{X^\star}$ is a maximal monotone operator, then for each $\lambda > 0$, $\lambda\,\mathsf{A}$ is also maximal monotone. ◁

Example 9.4.4. Let X be a reflexive Banach space. An operator A defines a maximal monotone graph on $X \times X^\star$ if and only A^{-1} defines a maximal monotone graph on $X^\star \times X$.

The result follows from the fact that the inverse operator A^{-1} is the operator that has the inverse graph of A, i. e., $x \in \mathsf{A}^{-1}(x^\star)$ if and only if $x^\star \in \mathsf{A}(x)$. ◁

9.4.2 Properties of maximal monotone operators

The next proposition (see e. g., [107]) links maximal monotonicity of the operator A with properties of the set $\mathsf{A}(x)$.

Proposition 9.4.5. *Let $\mathsf{A} : \mathbf{D}(\mathsf{A}) \subset X \to 2^{X^\star}$ be a maximal monotone operator. Then,*

(i) *For each $x \in \mathbf{D}(\mathsf{A})$, $\mathsf{A}(x)$ is a convex and weak* closed subset of $X^\star$ (if X is reflexive, it is weakly closed).*
(ii) *If $x \in \operatorname{int}(\mathbf{D}(\mathsf{A}))$, then $\mathsf{A}(x)$ is weak* compact (if X is reflexive, it is weakly compact).*
(iii) *The graph of A is closed in $X \times X^\star_{w^\star}$ and in $X_w \times X^\star$. This property is sometimes referred to as demiclosedness.*
(iv) *Let X be reflexive, and consider a sequence $((x_n, x_n^\star))_{n\in\mathbb{N}} \subset \mathbf{Gr}(\mathsf{A})$ such that $x_n \rightharpoonup x$, $x_n^\star \rightharpoonup x^\star$ and $\limsup_n \langle x_n^\star - x^\star, x_n - x\rangle \le 0$. Then $(x, x^\star) \in \mathbf{Gr}(\mathsf{A})$ and $\langle x_n^\star, x_n\rangle \to \langle x^\star, x\rangle$.*
(v) *Let X be reflexive, and consider a sequence $((x_n, x_n^\star))_{n\in\mathbb{N}} \subset \mathbf{Gr}(\mathsf{A})$ such that $x_n \rightharpoonup x$, $x_n^\star \rightharpoonup x^\star$ and $\limsup_{n,m} \langle x_n^\star - x_m^\star, x_n - x_m\rangle \le 0$. Then $(x, x^\star) \in \mathbf{Gr}(\mathsf{A})$ and $\langle x_n^\star, x_n\rangle \to \langle x^\star, x\rangle$.*

Proof. (i) Let $x_1^\star, x_2^\star \in \mathsf{A}(x)$, and set $x_t^\star = (1-t)x_1^\star + t x_2^\star$, $t \in [0,1]$. Then, for any $(z, z^\star) \in \mathbf{Gr}(\mathsf{A})$ we have

$$\langle x_t^\star - z^\star, x - z\rangle = (1-t)\langle x_1^\star - z^\star, x - z\rangle + t\langle x_2^\star - z^\star, x - z\rangle \ge 0.$$

By maximal monotonicity of A (see Proposition 9.4.2), we get that $x_t^\star \in \mathsf{A}(x)$, hence, $\mathsf{A}(x)$ is a convex set.

We now prove that $\mathsf{A}(x)$ is weak* closed. To this end, consider a net $\{x_\alpha^\star : \alpha \in \mathcal{I}\} \subseteq \mathsf{A}(x)$ that converges weak* (respectively weakly if X is reflexive) to $x^\star$ in $X^\star$. We then have

$$\langle x_\alpha^\star - z^\star, x - z\rangle \ge 0, \quad \forall (z, z^\star) \in \mathbf{Gr}(\mathsf{A}),$$

and passing to the limit, we obtain that

$$\langle x^\star - z^\star, x - z\rangle \ge 0, \quad \forall (z, z^\star) \in \mathbf{Gr}(\mathsf{A}),$$

and so invoking Proposition 9.4.2, we deduce that $x^\star \in \mathsf{A}(x)$, hence the weak* (respectively weak) closedness of $\mathsf{A}(x)$.

(ii) Since A is maximal monotone, it is also monotone, hence, by Theorem 9.3.4, we have that it is locally bounded at each interior point of $\mathbf{D}(A)$. Then, using the weak* compactness result (Alaoglu, Theorem 1.1.36(iii)) or the Eberlein–Šmulian theorem (Theorem 1.1.59) in the case where X is reflexive, we obtain the stated result.

(iii) Consider a net $\{(x_\alpha, x_\alpha^\star) : \alpha \in \mathcal{I}\} \subset \mathbf{Gr}(A)$ such that $x_\alpha \to x$ in X and $x_\alpha^\star \xrightarrow{\star} x^\star$ in $X^\star$. For any $(z, z^\star) \in \mathbf{Gr}(A)$ it holds that $\langle x_\alpha^\star - z^\star, x_\alpha - z\rangle \geq 0$, and passing to the limit, $\langle x^\star - z^\star, x - z\rangle \geq 0$ so that, by application of Proposition 9.4.2, we deduce that $(x, x^\star) \in \mathbf{Gr}(A)$, hence, the $X \times X^\star_{w^\star}$ closedness of the graph of A. For the other property, consider a net $(x_\alpha, x_\alpha^\star) \in \mathbf{Gr}(A)$ such that $x_\alpha \rightharpoonup x$ in X and $x_\alpha^\star \to x^\star$ in $X^\star$, and proceed accordingly.

(iv) Since $(x_n, x_n^\star) \in \mathbf{Gr}(A)$ we have

$$\langle z^\star - x_n^\star, z - x_n\rangle \geq 0, \quad \forall (z, z^\star) \in \mathbf{Gr}(A), \forall n \in \mathbb{N}.$$

Taking the limit superior and using the property that $\limsup_n \langle x_n^\star, x_n\rangle \leq \langle x^\star, x\rangle$, we obtain

$$\langle z^\star - x^\star, z - x\rangle \geq 0, \quad \forall (z, z^\star) \in \mathbf{Gr}(A),$$

hence, by Proposition 9.4.2, we deduce that $(x^\star, x) \in \mathbf{Gr}(A)$. Now, by the monotonicity of A, we have

$$\langle x^\star - x_n^\star, x - x_n\rangle \geq 0, \quad \forall n \in \mathbb{N}.$$

Taking the limit inferior and using the facts that $x_n \rightharpoonup x$ and $x_n^\star \rightharpoonup x^\star$, we deduce that $\liminf_n \langle x_n^\star, x_n\rangle \geq \langle x^\star, x\rangle$. This, combined with the relevant inequality for the limsup, implies that $\langle x_n^\star, x_n\rangle \to \langle x^\star, x\rangle$.

(v) Using the monotonicity of A, we have that $(x_n, x_n^\star), (x_m, x_m^\star) \in \mathbf{Gr}(A)$ implies $\langle x_n^\star - x_m^\star, x_n - x_m\rangle \geq 0$ for every $n, m \in \mathbb{N}$, so combined with the hypothesis, we obtain that

$$\lim_{n,m\to\infty} \langle x_n^\star - x_m^\star, x_n - x_m\rangle = 0. \tag{9.15}$$

To conclude the proof, we take a subsequence $((x_{n_k}, x_{n_k}^\star))_{k\in\mathbb{N}}$ such that $\langle x_{n_k}^\star, x_{n_k}\rangle \to \limsup_n \langle x_n^\star, x_n\rangle = \mu$ and work along this subsequence using maximal monotonicity, repeating the steps of (v). In particular, let $(n_k)_{k\in\mathbb{N}}$ be a subsequence of $(n)_{n\in\mathbb{N}}$ such that $\langle x_{n_k}^\star, x_{n_k}\rangle \to \mu$. Then, from (9.15), we have that

$$0 = \lim_{n_k\to\infty}\Big[\lim_{n_\ell\to\infty} \langle x_{n_k}^\star - x_{n_\ell}^\star, x_{n_k} - x_{n_\ell}\rangle\Big] = 2\mu - 2\langle x^\star, x\rangle.$$

Hence, $\mu = \langle x^\star, x\rangle = \lim_n \langle x_n^\star, x_n\rangle$. Since A is monotone, this implies that

$$\langle z^\star - x^\star, z - x \rangle \geq 0, \quad \forall (z, z^\star) \in \mathbf{Gr}(A).$$

By the maximal monotonicity of A, $(x, x^\star) \in \mathbf{Gr}(A)$. □

Remark 9.4.6. If $\mathbf{D}(A) = X$, and X is reflexive, then, from Proposition 9.4.5, $A(x)$ is a weakly compact and convex subset of $X^\star$.

We know that in general weak convergence in X and $X^\star$ does not imply continuity of the duality pairing between the two spaces, i. e., if $x_n \rightharpoonup x$ in X and $x_n^\star \rightharpoonup x^\star$, then it is not necessarily true that $\langle x_n^\star, x_n \rangle \to \langle x^\star, x \rangle$. However, if $x_n^\star \in A(x_n)$ where A is a maximal monotone operator, and $\limsup_n \langle x_n^\star, x_n - x \rangle \leq 0$, then the continuity property of the duality pairing holds.

Theorem 9.4.7. *Let $A : \mathbf{D}(A) \subset X \to 2^{X^\star}$ be a maximal monotone operator with $\mathbf{D}(A) = X$. Then A is weak* upper semicontinuous from X into $X^\star$ (if X is reflexive weak upper semicontinuous).*

Proof. For simplicity, we only prove the reflexive case. Consider any $x \in X$. Assume that for a given open weak neighborhood V of $A(x)$, there exists a sequence $(x_n)_{n\in\mathbb{N}} \subset X$ with $x_n \to x$ and $x_n^\star \in A(x_n)$ such that $x_n^\star \notin V$ for all n. Since, by Theorem 9.3.4, A is locally bounded at the point x, the sequence $(x_n^\star)_{n\in\mathbb{N}}$ is bounded in $X^\star$, so there exists a subsequence $(x_{n_k}^\star)_{k\in\mathbb{N}}$ converging weakly to an element $x^\star$.

We have

$$\langle x_{n_k}^\star - z^\star, x_{n_k} - z \rangle \geq 0, \quad \forall (z, z^\star) \in \mathbf{Gr}(A),$$

so passing to the limit, we obtain

$$\langle x^\star - z^\star, x - z \rangle \geq 0, \quad \forall (z, z^\star) \in \mathbf{Gr}(A).$$

Since A is maximal monotone, $x^\star \in A(x)$. But, on the other hand, since V is weakly open, $X^\star \setminus V$ is weakly closed, and this implies that $x^\star \notin V$, which is a contradiction. □

9.4.3 Criteria for maximal monotonicity

The next result (see e. g., [107]) provides conditions under which a monotone operator is maximal monotone.

Theorem 9.4.8. *Let X be a reflexive Banach space. Let $A : \mathbf{D}(A) \subset X \to 2^{X^\star}$ be a monotone operator with $\mathbf{D}(A) = X$ such that for each $x \in X$, $A(x)$ is a nonempty, weakly closed and convex subset of $X^\star$. Suppose that for every $x, z \in X$ the mapping $t \mapsto A(x + tz)$ is upper semicontinuous[4] from $[0, 1]$ to $X_w^\star$. Then A is maximal monotone.*

---

4 in the sense of Definition 1.8.2, recall also Proposition 1.8.6.

Proof. We will use Proposition 9.4.2. Suppose that for some $(z, z^\star) \in X \times X^\star$ it holds that

$$\langle x^\star - z^\star, x - z\rangle \geq 0, \quad \forall (x, x^\star) \in \mathbf{Gr}(A).$$

We must show that $z^\star \in A(z)$. Suppose that $z^\star \notin A(z)$. Since, by assumption, $A(z)$ is a weakly closed, and convex subset of $X^\star$, from the second separation theorem, we can find $z_0 \in X$ such that

$$\langle z^\star, z_0\rangle > \langle z_0^\star, z_0\rangle, \quad \forall z_0^\star \in A(z).$$

Let $x_t = z + t z_0$, $t \in [0, 1]$, and set

$$U = \{z_0^\star \in X^\star \ : \ \langle z^\star, z_0\rangle > \langle z_0^\star, z_0\rangle\}.$$

Clearly, U is a weak neighborhood of $A(z)$. Since $x_t \to z$, as $t \to 0^+$, and $t \mapsto A(x + tz)$ is upper semicontinuous, for small enough t we have that $A(x_t) \subseteq U$. Let $x_t^\star \in A(x_t)$. Then for small t, we obtain that

$$0 \leq \langle x_t^\star - z^\star, x_t - z\rangle = \langle x_t^\star - z^\star, z_0\rangle < 0,$$

which is a contradiction. So, $z^\star \in A(z)$, and thus, A is maximal monotone. Note that in passing to the limit, we have used the local boundeness theorem for monotone operators (Theorem 9.3.4) as well Alaoglu's weak* compactness result (Theorem 1.1.36(iii)). □

Theorem 9.4.8 has the following interesting implication:

Corollary 9.4.9. *Let X be a reflexive Banach space. If $A : \mathbf{D}(A) \subset X \to X^\star$ is monotone, hemicontinuous, and $\mathbf{D}(A) = X$, then A is maximal monotone.*

Example 9.4.10. The maximal monotonicity of the p-Laplacian can be proved using Corollary 9.4.9. Recall that the p-Laplacian can be defined in variational form by

$$\langle A(u), v\rangle = \int_{\mathcal{D}} |\nabla u(x)|^{p-2} \nabla u(x) \cdot \nabla v(x) dx,$$

from which follows that A is demicontinuous (hence, also hemicontinuous). Since A is monotone (see Example 9.3.3), by Corollary 9.4.9, A is maximal monotone. ◁

For linear operators defined on the whole of X, monotonicity and maximal monotonicity are equivalent.

Proposition 9.4.11. *Let X be a reflexive Banach space. A linear operator $A : \mathbf{D}(A) \subset X \to X^\star$, with $\mathbf{D}(A) = X$, is maximal monotone if and only if it is monotone.*

Proof. This follows from a direct application of Proposition 9.4.2. To this end, we take any fixed $(z, z^\star) \in X \times X^\star$ such that $\langle A(x) - z^\star, x - z\rangle \geq 0$ for every $x \in X$, and we will prove that $z^\star = A(z)$. For arbitrary $z_0 \in X$, and $t > 0$, set $x = z + tz_o$ in the above inequality. This yields $\langle A(z + tz_o) - z^\star, (z + tz_o) - z\rangle \geq 0$ or equivalently $\langle A(z) - z^\star, z_0\rangle + t\langle A(z_0), z_0\rangle \geq 0$, where we have used the linearity of A. Letting $t \to 0$, we obtain that $\langle A(z) - z^\star, z_0\rangle \geq 0$ for every $z_0 \in X$. Repeating the steps with z_0 replaced by $-z_0$ allows us to conclude that $z^\star = A(z)$. Alternatively, we could use directly Corollary 9.4.9. □

9.4.4 Surjectivity results

Maximal monotone operators enjoy some very interesting surjectivity results (see, e. g., [18, 107, 110]), which make them very useful in applications.

The following theorem [18, 107, 110] provides a very flexible surjectivity result.

Theorem 9.4.12. *Let X be a reflexive Banach space, $A : \mathbf{D}(A) \subset X \to 2^{X^\star}$ a maximal monotone operator, and $B : \mathbf{D}(B) \subset X \to X^\star$ a monotone, hemicontinuous, bounded, and coercive operator with $\mathbf{D}(B) = X$. Then $\mathbf{R}(A + B) = X^\star$.*

Proof. Since monotonicity is a property that is invariant under translations, we may assume[5] that $(0, 0) \in \mathbf{Gr}(A)$. We must show that given any $x^\star \in X^\star$, the inclusion $x^\star \in A(x) + B(x)$ has a solution $x \in \mathbf{D}(A) \subset X$. Since A is maximal monotone, this problem is equivalent to the following: given $x^\star \in X^\star$, find $x \in \mathbf{D}(A)$ such that

$$\langle x^\star - B(x) - z^\star, x - z\rangle \geq 0, \quad \forall\,(z, z^\star) \in \mathbf{Gr}(A). \tag{9.16}$$

First, we establish an a priori bound for the solutions of this problem. Let $x \in \mathbf{D}(A)$ be a solution of (9.16). Since $(0, 0) \in \mathbf{Gr}(A)$, we obtain $\langle x^\star - B(x), x\rangle \geq 0$, which leads to $\langle B(x), x\rangle \leq \|x^\star\|_{X^\star}\|x\|$. Since B is coercive, we conclude that there exists $c > 0$ such that $\|x\| \leq c$.

Let $\mathcal{X}_F$ be the family of all finite dimensional subspaces X_F of X, ordered by inclusion. For each $X_F \in \mathcal{X}_F$, let $j_F : X_F \to X$ be the inclusion map and $j_F^\star : X^\star \to X_F^\star$ its adjoint projection map. Let $A_F = j_F^\star A j_F$ and $B_F = j_F^\star B j_F$. Then, clearly A_F is monotone, and B_F is continuous since B is hemicontinuous and $\dim(X_F) < \infty$. Also, let $K_F = X_F \cap \overline{B}(0, c)$. Then, if $x_F^\star = j_F^\star x^\star$, from the Debrunner–Flor theorem 3.5.1, there exists $x_F \in K_F$ such that

$$\langle x^\star - B(x_F) - z^\star, x_F - z\rangle \geq 0, \quad \forall\,(z, z^\star) \in \mathbf{Gr}(A),\ z \in X_F. \tag{9.17}$$

Note that $\langle B_F(x_F), x_F - z\rangle = \langle B(x_F), x_F - z\rangle$, $\langle j_F^\star z^\star, x_F - z\rangle = \langle z^\star, x_F - z\rangle$ and $\langle j_F^\star x^\star, x_F - z\rangle = \langle x^\star, x_F - z\rangle$.

Since B is bounded, there exists $c_1 > 0$ such that $\|B(x_F)\| \leq c_1$ for all $x_F \in X_F$ and $X_F \in \mathcal{X}_F$. Then, for any $X_F \in \mathcal{X}_F$, we consider the set

5 Otherwise for some $(x_0, x_0^\star) \in \mathbf{Gr}(A)$, we consider the operator $z \mapsto A_1(z) = A(z + x_0) - x_0^\star$ and $z \mapsto B_1(z) = B(z + x_0) - x_0^\star$ for all $z \in \mathbf{D}(A) - x_0$.

$$U(X_F) = \bigcup \{(x_{F'}, B(x_{F'})) \in X \times X^\star \ : \ x_{F'} \text{ solves (9.17) on } X_{F'}, \text{ with } \mathcal{X}_F \ni X_{F'} \supseteq X_F\}.$$

It is easy to see, that the family $\{\overline{U(X_F)}^w \ : \ X_F \in \mathcal{X}_F\}$ has the finite intersection property.[6] Since for all $X_F \in \mathcal{X}_F$ it holds that $U(X_F) \subseteq \bar{B}(0,c) \times \bar{B}(0,c_1)$ and the latter is weakly compact in the reflexive space $X \times X^\star$, we deduce that there exists $(\bar{x}, \bar{x}^\star) \in \bigcap_{X_F \in \mathcal{X}_F} \overline{U(X_F)}^w$, where, by $\overline{A}^w$, we denote the weak closure of a set A. From Proposition 1.1.62, we know that for any fixed $X_F \in \mathcal{X}_F$, we can find a sequence $((\bar{x}_n, B(\bar{x}_n)))_{n\in\mathbb{N}} \subset U(X_F)$ such that $(\bar{x}_n, B(\bar{x}_n)) \rightharpoonup (\bar{x}, \bar{x}^\star)$ in $X \times X^\star$, i. e., $\bar{x}_n \rightharpoonup \bar{x}$ and $B(\bar{x}_n) \rightharpoonup \bar{x}^\star$. We will show that $\bar{x}$ is the required solution.

Now, we show that there exists a $(z_0, z_0^\star) \in \mathbf{Gr}(A)$ such that

$$\langle x^\star - \bar{x}^\star - z_0^\star, x - z_0 \rangle \le 0. \tag{9.18}$$

Suppose not. Then

$$\langle x^\star - \bar{x}^\star - z^\star, \bar{x} - z \rangle > 0, \quad \forall\, (z, z^\star) \in \mathbf{Gr}(A).$$

This implies that $x^\star - \bar{x}^\star \in A(\bar{x})$ since A is maximal monotone. Then if $z^\star = x^\star - \bar{x}^\star$ and $z = \bar{x}$, we have $\langle x^\star - \bar{x}^\star - z^\star, \bar{x} - z \rangle = 0$, which is a contradiction.

From (9.17), we obtain

$$\langle x^\star - B(\bar{x}_n) - z^\star, \bar{x}_n - z \rangle \ge 0, \quad \forall\, (z, z^\star) \in \mathbf{Gr}(A), \quad z \in X_F. \tag{9.19}$$

We now consider $X_{F_0} \in \mathcal{X}_F$ such that $z_0 \in X_{F_0}$, where the fixed element z_0 satisfies (9.18) (we note that $\bigcup_{X_F \in \mathcal{X}_F} X_F = X$), and restrict the above for every $z \in X_{F_0}$, so taking the limes superior, we obtain

$$\limsup_n \langle B(\bar{x}_n), \bar{x}_n \rangle \le \langle \bar{x}^\star, z \rangle + \langle x^\star - z^\star, \bar{x} - z \rangle, \quad \forall\, (z, z^\star) \in \mathbf{Gr}(A).$$

Setting $z = z_0$ and $z^\star = z_0^\star$, we obtain

$$\limsup_n \langle B(\bar{x}_n), \bar{x}_n \rangle \le \langle \bar{x}^\star, z_0 \rangle + \langle x^\star - z_0^\star, \bar{x} - z_0 \rangle.$$

We claim that

$$\langle \bar{x}^\star, z_0 \rangle + \langle x^\star - z_0^\star, \bar{x} - z_0 \rangle \le \langle \bar{x}^\star, \bar{x} \rangle.$$

Suppose not; then $\langle x^\star - \bar{x}^\star - z_0^\star, \bar{x} - z_0 \rangle > 0$, which contradicts (9.18).

6 Consider any $X_{F_1}, X_{F_2} \in \mathcal{X}_F$, then for any $X_{F_3} \in \mathcal{X}_F$ such that $X_{F_1} \cup X_{F_2} \subset X_{F_3}$, it holds that $U(X_{F_3}) \subset U(X_{F_1}) \cap U(X_{F_2})$.

So, $\limsup_n \langle \mathsf{B}(\bar{\mathsf{x}}_n), \bar{\mathsf{x}}_n\rangle \le \langle \bar{\mathsf{x}}^\star, \bar{\mathsf{x}}\rangle$, and since B is maximal monotone, by Proposition 9.4.5(iv), $(\bar{\mathsf{x}}, \bar{\mathsf{x}}^\star) \in \mathrm{Gr}(\mathsf{B})$ and $\langle \mathsf{B}(\bar{\mathsf{x}}_n), \bar{\mathsf{x}}_n\rangle \to \langle \bar{\mathsf{x}}^\star, \bar{\mathsf{x}}\rangle$. It then follows by (9.19) that

$$\langle \mathsf{x}^\star - \mathsf{B}(\bar{\mathsf{x}}) - \mathsf{z}^\star, \bar{\mathsf{x}} - \mathsf{z}\rangle \ge 0, \quad \forall\, (\mathsf{z}, \mathsf{z}^\star) \in \mathbf{Gr}(\mathsf{A}),$$

so that $\bar{\mathsf{x}}$ has the required property. This completes the proof. □

For the remaining part of this section, we assume that $\mathsf{J} : X \to X^\star$ is the duality map corresponding to a uniformly convex renorming of both X and $X^\star$ (see Theorem 2.6.16).

From Theorem 9.4.12, we obtain an important characterization [18, 107, 110] of maximal monotone operators.

Theorem 9.4.13. *Let X be a reflexive Banach space, and let $\mathsf{A} : \mathbf{D}(\mathsf{A}) \subset X \to 2^{X^\star}$ be a monotone operator. Then* A *is maximal monotone if and only if* $\mathbf{R}(\mathsf{A}+\lambda\mathsf{J}) = X^\star$ *for all* $\lambda > 0$.

Proof. Let A be maximal monotone. The operator J is continuous, (strictly) monotone, coercive, and bounded (see Theorem 2.6.17). Then we may apply Theorem 9.4.12 with $\mathsf{B} = \lambda\mathsf{J}$ for any $\lambda > 0$ and infer that $\mathbf{R}(\mathsf{A} + \lambda\mathsf{J}) = X^\star$.

Conversely, assume that for any $\lambda > 0$, $\mathbf{R}(\mathsf{A} + \lambda\mathsf{J}) = X^\star$, and suppose that A is not maximal monotone. Then, by Proposition 9.4.2, there exists $(\mathsf{z}_0, \mathsf{z}_0^\star) \in (X \times X^\star) \setminus \mathbf{Gr}(\mathsf{A})$ such that

$$\langle \mathsf{z}^\star - \mathsf{z}_0^\star, \mathsf{z} - \mathsf{z}_o\rangle \ge 0, \quad \forall\, (\mathsf{z}, \mathsf{z}^\star) \in \mathbf{Gr}(\mathsf{A}). \tag{9.20}$$

Since, $\mathbf{R}(\mathsf{A} + \lambda\mathsf{J}) = X^\star$ for all $\lambda > 0$, there exists $(\mathsf{x}_o, \mathsf{x}_o^\star) \in \mathbf{Gr}(\mathsf{A})$ such that

$$\lambda\mathsf{J}(\mathsf{x}_o) + \mathsf{x}_o^\star = \lambda\mathsf{J}(\mathsf{z}_0) + \mathsf{z}_0^\star. \tag{9.21}$$

We set $(\mathsf{z}, \mathsf{z}^\star) = (\mathsf{x}_o, \mathsf{x}_o^\star)$ in (9.20) so that

$$0 \le \langle \mathsf{x}_o^\star - \mathsf{z}_0^\star, \mathsf{x}_o - \mathsf{z}_0\rangle \overset{(9.21)}{=} \lambda\langle \mathsf{J}(\mathsf{z}_0) - \mathsf{J}(\mathsf{x}_o), \mathsf{x}_o - \mathsf{z}_0\rangle \le 0,$$

where the last inequality follows by the monotonicity of J. Hence, $0 = \langle \mathsf{J}(\mathsf{z}_0) - \mathsf{J}(\mathsf{x}_o), \mathsf{z}_0 - \mathsf{x}_o\rangle$, and since J is strictly monotone $\mathsf{x}_o = \mathsf{z}_0$. Combining this with (9.21), we see that $\lambda\mathsf{J}(\mathsf{z}_0) - \lambda\mathsf{J}(\mathsf{x}_o) = \mathsf{x}_o^\star - \mathsf{z}_0^\star = 0$, so $\mathsf{x}_o^\star = \mathsf{z}_0^\star$. But since $\mathsf{x}_o^\star \in \mathsf{A}(\mathsf{x}_o) = \mathsf{A}(\mathsf{z}_0)$ and (by assumption) $\mathsf{z}_0^\star \notin \mathsf{A}(\mathsf{z}_0)$, we reach a contradiction. □

Example 9.4.14. Let $\mathcal{D} \subset \mathbb{R}^d$ be open, and let $\mathsf{A} : \mathbb{R} \to 2^{\mathbb{R}}$ be a maximal monotone operator with $(0, 0) \in \mathbf{Gr}(\mathsf{A})$. Let $\widehat{\mathsf{A}} : L^2(\mathcal{D}) \to 2^{L^2(\mathcal{D})}$ be the realization of A on the Hilbert space $L^2(\mathcal{D})$ defined by $\widehat{\mathsf{A}}(u) = \{v \in L^2(\mathcal{D}) \;:\; v(x) \in \mathsf{A}(u(x)) \text{ a. e.}\}$. We claim that $\widehat{\mathsf{A}}$ is maximal monotone.

Since the monotonicity of $\widehat{\mathsf{A}}$ is obvious, it suffices to show that for any $\lambda > 0$, $\mathbf{R}(\lambda I + \widehat{\mathsf{A}}) = L^2(\mathcal{D})$. Let $v \in L^2(\mathcal{D})$, so there exists $u(x)$, unique such that $v(x) \in \lambda u(x) + \mathsf{A}u(x)$, a. e. It then follows (see Example 9.3.2) that $u(x) = (\lambda I + \mathsf{A})^{-1}v(x)$, a. e. Observe that u is measurable. Also, since $(0, 0) \in \mathbf{Gr}(\mathsf{A})$, and $(\lambda I + \mathsf{A})^{-1}$ is nonexpansive, we have that

$|u(x)| = |(\lambda I + \mathsf{A})^{-1}v(x)| \le |v(x)|$ a. e. So $u \in L^2(\mathcal{D})$, and we have established the maximality of $\widehat{\mathsf{A}}$. If $\mathcal{D}$ is bounded, then we can drop the hypothesis that $(0,0) \in \mathbf{Gr}(\mathsf{A})$ since $L^\infty(\mathcal{D}) \subset L^2(\mathcal{D})$. ◁

We are now ready to prove the main surjectivity result [18, 107, 110] on maximal monotone operators.

Theorem 9.4.15. *Let X be a reflexive Banach space and $\mathsf{A} : \mathbf{D}(\mathsf{A}) \subset X \to 2^{X^\star}$ a maximal monotone and coercive operator. Then $\mathbf{R}(\mathsf{A}) = X^\star$.*

Proof. Since monotonicity is a property, which is invariant under translations, it is sufficient to prove that $0 \in \mathbf{R}(\mathsf{A})$. Let $(\epsilon_n)_{n\in\mathbb{N}}$ be a sequence of positive numbers such that $\epsilon_n \to 0$, as $n \to \infty$. By Theorem 9.4.13, there exists $(\mathsf{x}_n, \mathsf{x}_n^\star) \in \mathbf{Gr}(\mathsf{A})$ such that

$$\mathsf{x}_n^\star + \epsilon_n \mathsf{J}(\mathsf{x}_n) = 0. \tag{9.22}$$

Since A is coercive, there exists $c > 0$ such that $\|\mathsf{x}_n\| \le c$. By the reflexivity of X, there exists $\mathsf{x}_o \in X$ such that $\mathsf{x}_n \rightharpoonup \mathsf{x}_o$ in X (up to subsequences). On the other hand, by (9.22), we have

$$\|\mathsf{x}_n^\star\|_{X^\star} = \epsilon_n \,\|\mathsf{J}(\mathsf{x}_n)\|_{X^\star} = \epsilon_n \,\|\mathsf{x}_n\| \le \epsilon_n \, c \to 0, \quad \text{as } n \to \infty,$$

i. e., $\mathsf{x}_n^\star \to 0$ in $X^\star$. By the monotonicity of A, we have

$$\langle \mathsf{z}^\star - \mathsf{x}_n^\star, \mathsf{z} - \mathsf{x}_n \rangle \ge 0, \quad \forall\, (\mathsf{z}, \mathsf{z}^\star) \in \mathbf{Gr}(\mathsf{A}),$$

and taking the limit as $n \to \infty$ implies that

$$\langle \mathsf{z}^\star, \mathsf{z} - \mathsf{x}_o \rangle \ge 0, \quad \forall\, (\mathsf{z}, \mathsf{z}^\star) \in \mathbf{Gr}(\mathsf{A}).$$

Hence, since A is maximal monotone, we conclude that $0 \in \mathsf{A}(\mathsf{x}_o)$. The proof is complete. □

Theorem 9.4.16 (Browder). *Let X be a reflexive Banach space and $\mathsf{A} : \mathbf{D}(\mathsf{A}) \subset X \to 2^{X^\star}$ be a maximal monotone operator. Then, $\mathbf{R}(\mathsf{A}) = X^\star$ if and only if A^{-1} is locally bounded.*

Proof. If A is maximal monotone, so is A^{-1} (see Example 9.4.4), hence, by Theorem 9.3.4, A^{-1} is locally bounded on $X^\star$.

Conversely, assume that A^{-1} is locally bounded on $X^\star$. It suffices to show that $\mathbf{R}(\mathsf{A})$ is both open and closed in $X^\star$.

We first show that $\mathbf{R}(\mathsf{A})$ is closed in $X^\star$. Let $(\mathsf{x}_n^\star)_{n\in\mathbb{N}} \subset X^\star$ be a sequence such that $\mathsf{x}_n^\star \in \mathsf{A}(\mathsf{x}_n)$ and $\mathsf{x}_n^\star \to \mathsf{x}^\star$ in $X^\star$. Since A^{-1} is locally bounded, $(\mathsf{x}_n)_{n\in\mathbb{N}}$ is a bounded sequence. By the reflexivity of X, there is a subsequence (denote the same for simplicity) such that $\mathsf{x}_n \rightharpoonup \mathsf{x}$. Hence, $(\mathsf{x}, \mathsf{x}^\star) \in \mathbf{Gr}(\mathsf{A})$ (see Proposition 9.4.5(iii)).

We show next that $\mathbf{R}(\mathsf{A})$ is open in $X^\star$. Let $\mathsf{x}^\star \in R(\mathsf{A})$, i. e., $\mathsf{x}^\star \in \mathsf{A}(\mathsf{x})$ for some $\mathsf{x} \in \mathbf{D}(\mathsf{A})$. Since the maximal monotonicity remains invariant under a translations, we may without loss of generality assume that $x = 0$. Let $r > 0$ such that A^{-1} is bounded on $\overline{B}_{X^\star}(x^\star, r)$. We claim that

$$\text{if } z^\star \in \overline{B}_{X^\star}\left(x^\star, \frac{r}{2}\right), \quad \text{then } z^\star \in \mathbf{R}(A), \tag{9.23}$$

from which it follows directly that $\mathbf{R}(A)$ is open.

To show this, we work as follows: By Theorem 9.4.13, there exists for any $\lambda > 0$ a solution $x_\lambda \in \mathbf{D}(A)$ of the equation

$$\lambda J(x_\lambda) + x_\lambda^\star = z^\star, \quad x_\lambda^\star \in A(x_\lambda). \tag{9.24}$$

Since $(0, x^\star) \in (x, x^\star) \in \mathbf{Gr}(A)$, by the monotonicity of A, we have

$$\langle z^\star - \lambda J(x_\lambda) - x^\star, x_\lambda - 0 \rangle \geq 0,$$

which implies that $\|z^\star - x^\star\|_{X^\star} \|x_\lambda\| - \lambda \|x_\lambda\|^2 \geq 0$, hence,

$$\lambda \|x_\lambda\| \leq \|z^\star - x^\star\|_{X^\star} < \frac{r}{2}, \quad \forall \lambda > 0.$$

From (9.24), by rearranging and taking the norm, we have

$$\|z^\star - x_\lambda^\star\|_{X^\star} = \lambda \|x_\lambda\| < \frac{r}{2}, \tag{9.25}$$

and thus from (9.24) and (9.25), we have

$$\|x_\lambda^\star - x^\star\|_{X^\star} \leq \|x_\lambda^\star - z^\star\|_{X^\star} + \|z^\star - x^\star\|_{X^\star} < r, \quad \forall \lambda > 0.$$

Since A^{-1} is bounded on $\overline{B}_{X^\star}(x^\star, r)$, the set of solutions $\{x_\lambda \in A^{-1}(x_\lambda^\star) : \lambda > 0\}$ remains bounded and

$$\|z^\star - x_\lambda^\star\| = \lambda \|x_\lambda\| \to 0, \quad \text{as } \lambda \to 0.$$

Hence, since $\mathbf{R}(A)$ is closed in $X^\star$, we have that $z^\star \in \mathbf{R}(A)$, which proves claim (9.23). The proof is complete. □

Remark 9.4.17. A simple consequence of Theorem 9.4.16 is that if A is maximal monotone and $\mathbf{D}(A)$ is bounded, then A is surjective.

9.4.5 Maximal monotonicity of the subdifferential and the duality map

The subdifferential and the duality map are important examples of maximal monotone operators.

Maximal monotonicity of the subdifferential

The maximal monotonicity of the subdifferential was shown in [126]. In reflexive Banach spaces, the coercivity of the operator $\partial\varphi$, under extra conditions, is equivalent to coercivity of φ (Theorem 3.2.41 [107]). The proof of maximal monotonicity we present here is an alternative proof, due to [130] (see also [6]).

Theorem 9.4.18 (Rockafellar). *Let X be a Banach space and $\varphi : X \to \mathbb{R} \cup \{+\infty\}$ be a lower semicontinuous proper convex function. Then the subdifferential $\partial\varphi : X \to 2^{X^\star}$ is a maximal monotone operator.*

Proof. Consider any $(x, x^\star) \in X \times X^\star$, fixed, such that

$$\langle z^\star - x^\star, z - x\rangle \geq 0, \quad \forall (z, z^\star) \in \mathbf{Gr}(\partial\varphi). \tag{9.26}$$

If we show that $(x, x^\star) \in \mathbf{Gr}(\partial\varphi)$, then, by Proposition 9.4.2, $\partial\varphi$ is a maximal monotone operator.

Define $\varphi_x : X \to \mathbb{R} \cup \{+\infty\}$ by $z \mapsto \varphi_x(z) = \varphi(z + x)$, and consider the perturbed function $\psi_x : X \to \mathbb{R} \cup \{+\infty\}$, defined by $z \mapsto \psi_x(z) := \varphi_x(z) + \frac{1}{2}\|z\|^2$. Since the Legendre–Fenchel conjugate of φ_x, $\varphi_x^\star$ is a proper function, there exists $x_o^\star \in X^\star$ for which $\varphi_x^\star(x_o^\star)$ is finite. We claim that for every $z \in X$, $\langle x^\star, z\rangle - \psi_x(z) < \infty$ so that $\psi_x^\star(x^\star) < \infty$. Indeed,

$$\langle x^\star, z\rangle - \psi_x(z) = \langle x^\star, z\rangle - \varphi(z + x) - \frac{1}{2}\|z\|^2 = \langle x^\star, z\rangle - \varphi_x(z) - \frac{1}{2}\|z\|^2. \tag{9.27}$$

By the definition of $\varphi_x^\star$, it holds that $\varphi_x^\star(x_o^\star) \geq \langle x_o^\star, z\rangle - \varphi_x(z)$, so by (9.27)

$$\begin{aligned}\langle x^\star, z\rangle - \psi_x(x) &\leq \langle x^\star, z\rangle - \langle x_o^\star, z\rangle + \varphi_x^\star(x_o^\star) - \frac{1}{2}\|z\|^2 \\ &= \varphi_x^\star(x_o^\star) + \langle x^\star - x_o^\star, z\rangle - \frac{1}{2}\|z\|^2.\end{aligned} \tag{9.28}$$

On the other hand, $\langle x^\star - x_o^\star, z\rangle - \frac{1}{2}\|z\|^2 \leq \frac{1}{2}\|x^\star - x_o^\star\|^2$, as can be easily seen by noting that the function $\bar{\phi} : X \to \mathbb{R}\cup\{+\infty\}$ defined by $\bar{\phi}(z) = \frac{1}{2}\|z\|^2$ has Legendre–Fenchel transform $\bar{\phi}^\star(z^\star) = \frac{1}{2}\|z^\star\|^2$, and we then use the definition of the Legendre–Fenchel transform to note that $\frac{1}{2}\|z^\star\|^2 \geq \langle z^\star, z\rangle - \frac{1}{2}\|z\|^2$ for every $z^\star \in X^\star$ and every $z \in X$, applying that for the choice $z^\star = x^\star - x_o^\star$. Substituting that in (9.28) yields the estimate

$$\langle x^\star, z\rangle - \psi_x(x) \leq \varphi_x^\star(x_o^\star) + \frac{1}{2}\|x^\star - x_o^\star\|^2 < \infty.$$

Having guaranteed that $\psi_x^\star(x^\star) < \infty$, and since by definition $\psi_x^\star(x^\star) = \sup_{x\in X}(\langle x^\star, x\rangle - \psi_x(x))$ given any $\epsilon_n = \frac{1}{n^2}$, $n \in \mathbb{N}$, we may find $x_n \in X$ such that $\psi_x^\star(x^\star) - \epsilon_n \leq (\langle x^\star, x_n\rangle - \psi_x(x_n))$. Combining the above

$$\psi_x(x) - \psi_x(x_n) + \frac{1}{n^2} - \langle x^\star, x\rangle \geq -\psi_x^\star(x^\star) - \psi_x(x_n) + \frac{1}{n^2} \geq -\langle x^\star, x\rangle, \quad \forall x \in X. \tag{9.29}$$

Clearly (recall Definition 4.2.23) this means that $x^\star$ belongs to the approximate subdifferential of ψ_x, in fact that $x^\star \in \partial_{\epsilon_n}\psi_x(x_n)$. We may thus apply the Brøndsted–Rockafellar approximation theorem 4.2.25 to guarantee the existence of a sequence $((z_n, z_n^\star))_{n\in\mathbb{N}} \subset X \times X^\star$ such that

$$z_n^\star \in \partial\psi_x(z_n), \quad \|z_n - x_n\| < \frac{1}{n}, \quad \text{and} \quad \|z_n^\star - x^\star\| < \frac{1}{n}. \tag{9.30}$$

However, by the definition of ψ_x and y, $\partial\psi_x(z) = \partial\varphi(z + x) + J(z)$. Combining that with $z_n^\star \in \partial\psi_x(z_n)$ yields the existence of a sequence $\bar{z}_n^\star \in J(z_n)$ such that $z_n^\star - \bar{z}_n^\star \in \partial\varphi(z_n + x)$. We now apply (9.26) for the choice $x_o = z_n + x$ and $x_o^\star = z_n^\star - \bar{z}_n^\star$ and obtain that $\langle z_n^\star - \bar{z}_n^\star - x^\star, z_n + x - x\rangle \geq 0$, so

$$\langle z_n^\star - x^\star, z_n\rangle \geq \langle \bar{z}_n^\star, z_n\rangle. \tag{9.31}$$

Since $\bar{z}_n^\star \in J(z_n)$, it holds that $\langle \bar{z}_n^\star, z_n\rangle = \|z_n\|^2$, from the definition of the duality map, so (9.31) yields

$$\|z_n\|^2 = \langle \bar{z}_n^\star, z_n\rangle \leq \langle z_n^\star - x^\star, z_n\rangle = |\langle z_n^\star - x^\star, z_n\rangle| \leq \|z_n^\star - x^\star\| \, \|z_n\|,$$

and dividing by $\|z_n\|$ provides the estimate

$$\|z_n\| \leq \|z_n^\star - x^\star\| < \frac{1}{n},$$

where in the last inequality we have used the third condition in (9.30). But, then using the triangle inequality on the second condition of (9.30) yields that

$$\|x_n\| = \|z_n + (x_n - z_n)\| \leq \frac{1}{n} + \|z_n\| < \frac{2}{n},$$

so $x_n \to 0$ as $n \to \infty$. We may now pass to the limit as $n \to \infty$ in (9.29), taking into account the lower semicontinuity of ψ_x to obtain that $\psi_x(0) + \psi_x^\star(x^\star) \leq 0$, so $x^\star \in \partial\psi_x(0) = \partial\varphi(x) + J(0) = \partial\varphi(x)$. The proof is complete. □

Remark 9.4.19. If X is reflexive, then a simpler proof of the maximal monotonicity of the subdifferential based on Proposition 9.4.13 can be made. The same applies for the density of the domain of the subdifferential of φ in the domain of φ. To illustrate this alternative proof, let $x \in \text{dom}(\varphi)$ and x_λ be the solution of the equation $0 \in J(x_\lambda - x) + \lambda\partial\varphi(x_\lambda)$. Taking the duality pairing of that with $x_\lambda - x$, we obtain that

$$\|x_\lambda - x\|^2 = -\lambda\langle\partial\varphi(x_\lambda), x_\lambda - x\rangle \leq \lambda(\varphi(x) - \varphi(x_\lambda)).$$

Since φ is bounded from below by an affine function (see Proposition 2.3.19), this inequality implies that $\lim_{\lambda\to 0} x_\lambda = x$. Since $x_\lambda \in \mathbf{D}(\partial\varphi)$ and x is arbitrary in $\text{dom}\,\varphi$, we conclude that $\overline{\mathbf{D}(\partial\varphi)} = \overline{\text{dom}\,\varphi}$.

Maximal monotonicity of the duality map

Example 9.4.20. The duality map $\mathsf{J} : X \to 2^{X^\star}$ is a maximal monotone operator since $\mathsf{J}(\mathsf{x}) = \partial\varphi(\mathsf{x})$, where $\varphi(\mathsf{x}) = \frac{1}{2}\|\mathsf{x}\|^2$ (see Proposition 4.2.17). The surjectivity properties of J follow from the surjectivity properties of maximal monotone operators. Similar results follow for the generalized duality maps J_p, for the same reason. ◁

9.4.6 Yosida approximations

The Yosida approximation (see e. g., [18, 107, 110]) is a very useful concept with important applications. In a nutshell, the Yosida approximation is an one parameter family of single valued operators, which approximates (in a sense to become precise shortly) a possibly multivalued maximal monotone operator. The theory of Yosida approximations plays an important role in the development of the theory of nonlinear semigroups and in the theory of regularization.

Throughout this section, let X be a reflexive Banach space. By Theorem 2.6.16, we may assume that X and $X^\star$ are both locally uniformly convex.[7] Let $\mathsf{A} : \mathbf{D}(\mathsf{A}) \subset X \to 2^{X^\star}$ be a maximal monotone operator. By Theorem 9.4.13 the inclusion

$$0 \in \mathsf{J}(\mathsf{x}_\lambda - \mathsf{x}) + \lambda \mathsf{A}(\mathsf{x}_\lambda)$$

has a solution $\mathsf{x}_\lambda \in \mathbf{D}(\mathsf{A})$, i. e., there exists $\mathsf{x}_\lambda^\star \in \mathsf{A}(\mathsf{x}_\lambda)$ such that $\mathsf{J}(\mathsf{x}_\lambda - \mathsf{x}) + \lambda \mathsf{x}_\lambda^\star = 0$, for any fixed $\mathsf{x} \in X$ and $\lambda > 0$.

We claim that this solution is unique. Indeed, suppose that $\mathsf{x}_{i,\lambda} \in \mathbf{D}(\mathsf{A})$, $i = 1, 2$, satisfy

$$\mathsf{J}(\mathsf{x}_{i,\lambda} - \mathsf{x}) + \lambda\, \mathsf{x}_{i,\lambda}^\star = 0, \quad \mathsf{x}_{i,\lambda}^\star \in \mathsf{A}(\mathsf{x}_{i,\lambda}),\ i = 1, 2.$$

Then, by the monotonicity of A,

$$0 \le \lambda\langle \mathsf{x}_{1,\lambda}^\star - \mathsf{x}_{2,\lambda}^\star, \mathsf{x}_{1,\lambda} - \mathsf{x}_{2,\lambda}\rangle = \langle \mathsf{J}(\mathsf{x}_{2,\lambda} - \mathsf{x}) - \mathsf{J}(\mathsf{x}_{1,\lambda} - \mathsf{x}), (\mathsf{x}_{1,\lambda} - \mathsf{x}) - (\mathsf{x}_{2,\lambda} - \mathsf{x})\rangle \le 0,$$

with the last inequality arising by the monotonicity of J. Since J is strictly monotone (see Theorem 2.6.17), we conclude that $\mathsf{x}_{1,\lambda} = \mathsf{x}_{2,\lambda}$. A similar argument implies that the corresponding $\mathsf{x}_\lambda^\star \in \mathsf{A}(\mathsf{x}_\lambda)$ for which $\mathsf{J}(\mathsf{x}_\lambda - \mathsf{x}) + \lambda \mathsf{x}_\lambda^\star = 0$ is also unique. Therefore, given $\lambda > 0$, to each $\mathsf{x} \in X$ we may associate a unique element $\mathsf{x}_\lambda \in X$ and a unique element $\mathsf{x}_\lambda^\star \in \mathsf{A}(\mathsf{x})$ such that $\mathsf{J}(\mathsf{x}_\lambda - \mathsf{x}) + \lambda \mathsf{x}_\lambda^\star = 0$.

The above observations lead to the following definition:

Definition 9.4.21 (Resolvent operator and Yosida approximation). Let X be a reflexive Banach space, $X^\star$ its dual, $\langle\cdot,\cdot\rangle$ the duality pairing among them (assuming further a

7 Recall also that a locally uniformly convex Banach space is also strictly convex (see Section 2.6.1.3), hence, by Theorem 2.6.17, the duality map J is single valued.

renorming such that both X and its dual $X^\star$ are locally uniformly convex), $\mathsf{J} : X \to X^\star$ the (single valued) duality map, and $\mathsf{A} : \mathbf{D}(\mathsf{A}) \subset X \to 2^{X^\star}$ a maximal monotone (possibly multivalued) operator.

For each $\mathsf{x} \in X$ and $\lambda > 0$, consider the solution $\mathsf{x}_\lambda \in X$ of the operator inclusion

$$0 \in \mathsf{J}(\mathsf{x}_\lambda - \mathsf{x}) + \lambda \mathsf{A}(\mathsf{x}_\lambda). \tag{9.32}$$

(i) The family of (single valued) operators $\mathsf{R}_\lambda : X \to \mathbf{D}(\mathsf{A})$, defined for each $\lambda > 0$ by

$$R(\lambda, \mathsf{A}) = \mathsf{R}_\lambda(\mathsf{x}) := \mathsf{x}_\lambda,$$

is called the family of resolvent operators of A.

(ii) The family of (single valued) operators $\mathsf{A}_\lambda : X \to X^\star$, defined by

$$\mathsf{A}_\lambda(\mathsf{x}) := -\frac{1}{\lambda}(\mathsf{J}(\mathsf{x}_\lambda - \mathsf{x})), \quad \lambda > 0,$$

is called the family of Yosida approximations of A. From the definition, it is clear that $\mathsf{A}_\lambda(\mathsf{x}) \in \mathsf{A}(\mathsf{x}_\lambda) = \mathsf{A}(\mathsf{R}_\lambda(\mathsf{x}))$.

(iii) For any $\mathsf{x} \in X$, we define the element of minimal norm $\mathsf{A}^0(\mathsf{x}) \in X^\star$ as the element with the property

$$\|\mathsf{A}^0(\mathsf{x})\|_{X^\star} = \mathfrak{m}(\mathsf{A}(\mathsf{x})) := \inf_{\mathsf{x}^\star \in \mathsf{A}(\mathsf{x})} \|\mathsf{x}^\star\|_{X^\star}.$$

By the properties of X and $X^\star$, the element $\mathsf{A}^0(\mathsf{x}) \in X^\star$ is well defined and unique.

Example 9.4.22 (Resolvent operators in Hilbert space). In the special case where $X = H$ is a Hilbert space, since the duality mapping J coincides with the identity I, the resolvent operators can be identified to the family of operators $\mathsf{R}_\lambda : H \to H$, defined by $\mathsf{R}_\lambda := (I + \lambda \mathsf{A})^{-1}$. ◁

Example 9.4.23 (Yosida approximations of the subdifferential). Let X and $X^\star$ be as in Definition 9.4.21, consider $\varphi : X \to \mathbb{R} \cup \{+\infty\}$ a proper lower semicontinuous convex function, and let $\mathsf{A} := \partial\varphi : X \to 2^{X^\star}$. Then for any $\lambda > 0$, the resolvent operator for the subdifferential R_λ is related to a regularization problem for φ, in the sense that for any $\mathsf{x} \in X$, it holds that $\mathsf{R}_\lambda(\mathsf{x}) = \arg\min_{\mathsf{z} \in X} \varphi_{\mathsf{x},\lambda}(\mathsf{z})$, where $\varphi_{\mathsf{x},\lambda}(\mathsf{z}) := \frac{1}{2\lambda}\|\mathsf{x} - \mathsf{z}\|^2 + \varphi(\mathsf{z})$, and the function $\mathsf{x} \mapsto \varphi_\lambda(\mathsf{x}) := \inf_{\mathsf{z} \in X} \varphi_{\mathsf{x},\lambda}(\mathsf{z})$ can be considered as the generalization of the notion of the Moreau–Yosida regularization (we have studied in Hilbert space in Section 4.4.2) for Banach spaces. In the special case where $\varphi = I_C$, where $C \subset X$ is a closed convex set, recalling that $\partial I_C = N_C$ (see Example 4.1.6), we may consider R_λ as a generalization of the projection operator.[8] ◁

8 If $X = H$ is a Hilbert space, then, for any $\mathsf{x} \in X$, we can see that $\mathsf{R}_\lambda(\mathsf{x})$ solves the variational inequality $\langle \mathsf{x} - \mathsf{R}_\lambda(\mathsf{x}), \mathsf{z} - \mathsf{R}_\lambda(\mathsf{x}) \rangle \le 0$ for every $\mathsf{z} \in C$, which characterizes the projection P_C (see Proposition 2.5.3).

The resolvent family has the following useful properties [18, 107, 110]:

Proposition 9.4.24 (Properties of resolvent family). *Let X be a reflexive Banach space (assuming further a renorming such that both X and its dual $X^\star$ are locally uniformly convex) and $\mathsf{A} : \mathbf{D}(\mathsf{A}) \subset X \to 2^{X^\star}$ a maximal monotone operator. The operators $\mathsf{R}_\lambda : X \to \mathbf{D}(\mathsf{A})$ are well defined for every $\lambda > 0$, single valued, bounded, and satisfy the property*

$$\lim_{\lambda \to 0} \mathsf{R}_\lambda(x) = x, \quad \forall x \in \overline{\mathrm{conv}(\mathbf{D}(\mathsf{A}))}. \tag{9.33}$$

Proof. We have already established the fact that A_λ is single valued.

We now show that A_λ is bounded for any $\lambda > 0$. We note that $\mathsf{A}_\lambda(x) \in \mathsf{A}(\mathsf{R}_\lambda(x))$. For any $(z, z^\star) \in \mathbf{Gr}(\mathsf{A})$, by the monotonicity of A, we have

$$\begin{aligned} \langle z^\star, \mathsf{R}_\lambda(x) - z\rangle &\le \langle \mathsf{A}_\lambda(x), \mathsf{R}_\lambda(x) - z\rangle = -\frac{1}{\lambda}\langle \mathsf{J}(\mathsf{R}_\lambda(x) - x), \mathsf{R}_\lambda(x) - z\rangle \\ &= -\frac{1}{\lambda}\langle \mathsf{J}(\mathsf{R}_\lambda(x) - x), \mathsf{R}_\lambda(x) - x\rangle - \frac{1}{\lambda}\langle \mathsf{J}(\mathsf{R}_\lambda(x) - x), x - z\rangle. \end{aligned}$$

Hence,

$$\|\mathsf{R}_\lambda(x) - x\|^2 \le -\lambda\langle z^\star, \mathsf{R}_\lambda(x) - z\rangle - \langle \mathsf{J}(\mathsf{R}_\lambda(x)) - x), x - z\rangle, \tag{9.34}$$

from which using the reverse triangle inequality, we have $\|\mathsf{R}_\lambda(x)\|^2 \le c_1\|\mathsf{R}_\lambda(x)\| + c_2$ for suitable $c_1, c_2 > 0$ depending on x, z, $z^\star$, λ, which implies that $\mathsf{R}_\lambda(x)$ maps bounded sets into bounded sets, so it is bounded.

To show (9.33), consider $(\lambda_n)_{n\in\mathbb{N}} \subset \mathbb{R}$ such that $\lambda_n \to 0$, and $\mathsf{J}(\mathsf{R}_{\lambda_n}(x) - x) \rightharpoonup z_o^\star$ in $X^\star$. From (9.34), we have that

$$\limsup_{n\to\infty} \|\mathsf{R}_{\lambda_n}(x) - x\|^2 \le \langle z_o^\star, z - x\rangle, \quad \forall z \in \mathbf{D}(\mathsf{A}),$$

and this holds for every $z \in \overline{\mathrm{conv}\,\mathbf{D}(\mathsf{A})}$. In particular, this inequality remains valid for $z = x$, which implies the desired result. □

The resolvent family and its properties can be used to show important properties for maximal monotone operators. The following proposition illustrates this point:

Proposition 9.4.25. *Let X be a reflexive Banach space (assuming further a renorming such that both X and its dual $X^\star$ are locally uniformly convex), and let $\mathsf{A} : \mathbf{D}(\mathsf{A}) \subset X \to 2^{X^\star}$ be a maximal monotone operator. Then, both $\overline{\mathbf{D}(\mathsf{A})}$ and $\overline{\mathbf{R}(\mathsf{A})}$ are convex.*

Proof. By Proposition 9.4.24, for every $x \in \overline{\mathrm{conv}\,\mathbf{D}(\mathsf{A})}$, we have $\mathsf{R}_\lambda(x) \to x$, as $\lambda \to 0^+$. Since $\mathsf{R}_\lambda(x) \in \mathbf{D}(\mathsf{A})$, we have $x \in \overline{\mathbf{D}(\mathsf{A})}$. Hence, $\overline{\mathbf{D}(\mathsf{A})} = \overline{\mathrm{conv}\,\mathbf{D}(\mathsf{A})}$, which proves the convexity of $\overline{\mathbf{D}(\mathsf{A})}$. Since $\mathbf{R}(\mathsf{A}) = \mathbf{D}(\mathsf{A}^{-1})$ and A^{-1} is also maximal monotone, we conclude that $\overline{\mathbf{R}(\mathsf{A})}$ is also convex. □

The following theorem (see, e. g., [20, 107]) summarizes the properties of the Yosida approximation:

Theorem 9.4.26 (Properties of Yosida approximation). *Let X be a reflexive Banach space (assuming further a renorming such that both X and its dual $X^\star$ are locally uniformly convex), $A : \mathbf{D}(A) \subset X \to 2^{X^\star}$ a maximal monotone operator, and $\lambda > 0$. Then,*

(i) *$A_\lambda : X \to X^\star$ is single valued with $\mathbf{D}(A_\lambda) = X$, monotone, bounded, and demicontinuous, hence maximal monotone.*[9]

(ii) *It holds that*

$$\|A_\lambda(x)\|_{X^\star} \le m(A(x)) := \inf_{x^\star \in A(x)} \|x^\star\|_{X^\star}, \quad \forall\, x \in \mathbf{D}(A). \tag{9.35}$$

If $A^0(x) \in X^\star$ is the element of minimal norm, (i. e., $\|A^0(x)\|_{X^\star} = m(A(x))$), then

$$A_\lambda(x) \to A^0(x), \quad \text{in } X^\star, \text{ as } \lambda \to 0. \tag{9.36}$$

(iii) *Consider arbitrary sequences $(\lambda_n)_{n\in\mathbb{N}} \subset \mathbb{R}$, and $(x_n)_{n\in\mathbb{N}} \subset X$ with the properties*

$$\begin{aligned} &\lambda_n \to 0, \quad x_n \rightharpoonup x \quad \text{in } X, \quad A_{\lambda_n}(x_n) \rightharpoonup x^\star \quad \text{in } X^\star, \\ &\text{and} \quad \limsup_{n,m} \langle A_{\lambda_n}(x_n) - A_{\lambda_m}(x_m), x_n - x_m \rangle \le 0. \end{aligned} \tag{9.37}$$

Then,

$$(x, x^\star) \in \mathbf{Gr}(A), \quad \text{and} \quad \lim_{n,m} \langle A_{\lambda_n}(x_n) - A_{\lambda_m}(x_m), x_n - x_m \rangle = 0. \tag{9.38}$$

Proof. (i) We consider each of the claimed properties separately.

(a) We have already seen that A_λ is single valued, and that $\mathbf{D}(A_\lambda) = X$ in the introduction of the section.

(b) For the monotonicity of A_λ, observe that

$$\begin{aligned} \langle A_\lambda(x_1) - A_\lambda(x_2), x_1 - x_2 \rangle &= \langle A_\lambda(x_1) - A_\lambda(x_2), R_\lambda(x_1) - R_\lambda(x_2) \rangle \\ &\quad + \langle A_\lambda(x_1) - A_\lambda(x_2), (x_1 - R_\lambda(x_1)) - (x_2 - R_\lambda(x_2)) \rangle. \end{aligned} \tag{9.39}$$

We also notice that $A_\lambda(x_i) \in A(R_\lambda(x_i))$ for any $x_i \in X$, $i = 1, 2$, and since A is monotone, this implies that

$$\langle A_\lambda(x_1) - A_\lambda(x_2), R_\lambda(x_1) - R_\lambda(x_2) \rangle \ge 0. \tag{9.40}$$

The second term on the right-hand side of (9.39), by the definition of A_λ, is identified as

9 See Proposition 9.4.9.

$$\langle A_\lambda(x_1) - A_\lambda(x_2), (x_1 - R_\lambda(x_1)) - (x_2 - R_\lambda(x_2))\rangle$$
$$= \frac{1}{\lambda}\langle J(x_1 - R_\lambda(x_1)) - J(x_2 - R_\lambda(x_2)), (x_1 - R_\lambda(x_1)) - (x_2 - R_\lambda(x_2))\rangle \ge 0, \tag{9.41}$$

by the monotonicity of the duality mapping J. Therefore, combining (9.39) with (9.40) and (9.41) yields

$$\langle A_\lambda(x_1) - A_\lambda(x_2), x_1 - x_2\rangle \ge 0,$$

hence, A_λ is monotone.

(c) Concerning boundedness, by Proposition 9.4.24, R_λ is bounded so that by definition A_λ is also bounded.

(d) Now we prove that A_λ is demicontinuous. To this end, let $(x_n)_{n\in\mathbb{N}}$ be an arbitrary sequence in $\mathbf{D}(A_\lambda) = X$ with $x_n \to x$, and we must show that $A_\lambda(x_n) \rightharpoonup A_\lambda(x)$ (since X is reflexive).

Set $z_n = R_\lambda(x_n)$ and $z_n^\star = A_\lambda(x_n)$. Then, $\lambda z_n^\star + J(z_n - x_n) = 0$. So we have that

$$\langle J(z_n - x_n) - J(z_m - x_m), x_m - x_n\rangle$$
$$= \langle J(z_n - x_n) - J(z_m - x_m), (z_n - x_n) - (z_m - x_m)\rangle + \lambda\langle z_n^\star - z_m^\star, z_n - z_m\rangle.$$

Since $\langle J(z_n - x_n) - J(z_m - x_m), x_m - x_n\rangle \to 0$, as $n, m \to \infty$, and since both summands in the right-hand side of the above inequality are nonnegative (by the monotonicity of J and A_λ), we have

$$\lim_{n,m\to\infty}\langle z_n^\star - z_m^\star, z_n - z_m\rangle = 0,$$

and

$$\lim_{n,m\to\infty}\langle J(z_n - x_n) - J(z_m - x_m), (z_n - x_n) - (z_m - x_m)\rangle = 0.$$

Since R_λ, A_λ, and J are bounded, we may assume by reflexivity that (up to subsequences) $z_n \rightharpoonup z_0$ in X, $z_n^\star \rightharpoonup z_0^\star$ in $X^\star$ and $J(z_n - x_n) \rightharpoonup x_o^\star$ in $X^\star$, for some $z_0 \in X$, $z_0^\star, x_o^\star \in X^\star$. Applying Proposition 9.4.5(vi), we have that $(z_0, z_0^\star) \in \mathbf{Gr}(A)$, $J(z_0 - x) = x_o^\star$ and $\lambda z_0^\star + J(z_0 - x) = 0$. Hence, $z_0 \in R_\lambda(x)$ and $z_0^\star = A_\lambda(x)$. By the Urysohn property (see Remark 1.1.52), this holds for the whole sequence. We therefore conclude that if $x_n \to x$, then $A_\lambda(x_n) \rightharpoonup A_\lambda(x)$, i. e., A_λ is demicontinuous.

(ii) Consider any $(x, x^\star) \in \mathbf{Gr}(A)$, and recall that $A_\lambda(x) \in A(R_\lambda(x))$. Apply the monotonicity property of A for the pairs $(x, x^\star) \in \mathbf{Gr}(A)$ and $(x_\lambda, A_\lambda(x)) \in \mathbf{Gr}(A)$, recalling that $R_\lambda(x) = x_\lambda$. This yields

$$0 \le \langle x^\star - A_\lambda(x), x - R_\lambda(x)\rangle = \langle x^\star + \lambda^{-1}J(x_\lambda - x), x - x_\lambda\rangle$$
$$= \langle x^\star, x - x_\lambda\rangle - \lambda^{-1}\langle J(x_\lambda - x), x_\lambda - x\rangle = \langle x^\star, x - x_\lambda\rangle - \lambda^{-1}\|x_\lambda - x\|^2$$
$$\le \|x^\star\|_{X^\star}\|x - x_\lambda\| - \lambda^{-1}\|x_\lambda - x\|^2.$$

Note that for the first equality, we have used the definition of $A_\lambda(x) = -\lambda^{-1}J(x_\lambda - x)$. We therefore obtain

$$0 \leq \|x^\star\|_{X^\star} \|x - x_\lambda\| - \lambda^{-1}\|x_\lambda - x\|^2,$$

which upon dividing by $\|x - x_\lambda\|$ yields

$$\lambda^{-1}\|x_\lambda - x\| \leq \|x^\star\|_{X^\star}, \tag{9.42}$$

and recalling the definition of A_λ, according to which $A_\lambda(x) = \lambda^{-1}J(x - x_\lambda)$, we obtain the inequality

$$\|A_\lambda(x)\| \leq \|x^\star\|_{X^\star}, \quad \forall\,(x, x^\star) \in \mathbf{Gr}(A). \tag{9.43}$$

Since, (9.43) holds for all $x^\star \in A(x)$, (9.35) follows by taking the infimum in (9.43) over all $x^\star \in A(x)$.

The element $A^0(x) \in X^\star$ is well defined and unique (since $X^\star$ is local uniformly convex it is also strictly convex by (2.25), and then we use Theorem 2.6.5). Since $\|A_\lambda(x)\|_{X^\star} \leq \mathfrak{m}(A(x))$ for every $\lambda > 0$, if we take a sequence $\lambda_n \to 0$, by the reflexivity of $X^\star$, there exists $z_0^\star \in X^\star$ such that $A_{\lambda_n}(x) \rightharpoonup z_0^\star$ in $X^\star$. By Proposition 9.4.24, $R_{\lambda_n}(x) \to x$, and since $A_{\lambda_n}(x) \in A(R_{\lambda_n}(x))$ by the closedness properties of maximal monotone operators (see Proposition 9.4.5(iii)), we have that $z_0^\star \in A(x)$. By the definition of $\mathfrak{m}(A(x))$, since $z_0^\star \in A(x)$, we have that $\mathfrak{m}(A(x)) \leq \|z_0^\star\|_{X^\star}$. On the other hand, since $A_{\lambda_n}(x) \rightharpoonup z_0^\star$ in $X^\star$, by the weak lower semicontinuity of the norm,[10] we have that $\liminf_n \|A_{\lambda_n}(x)\| \geq \|z_0^\star\|_{X^\star}$, and since $\|A_{\lambda_n}(x)\|_{X^\star} \leq \mathfrak{m}(A(x))$ for every n, $\liminf_n \|A_{\lambda_n}(x)\|_{X^\star} \leq \mathfrak{m}(A(x))$ so that $\|z_0^\star\|_{X^\star} \leq \mathfrak{m}(A(x))$. This leads us to conclude that

$$\|z_0^\star\|_{X^\star} = \mathfrak{m}(A(x)) = \|A^0(x)\|_{X^\star}.$$

Therefore, the limit $z_0^\star$ is an element of minimal norm, and by the uniqueness of $A^0(x)$, we conclude that $z_0^\star = A^0(x)$. Furthermore, a similar argument shows that $\limsup_n \|A_{\lambda_n}(x)\|_{X^\star} \leq \|z_0^\star\|_{X^\star}$, while the weak lower semicontinuity of the norm implies $\liminf_n \|A_{\lambda_n}(x)\|_{X^\star} \geq \|z_0^\star\|_{X^\star}$ so that $\lim_n \|A_{\lambda_n}(x)\|_{X^\star} = \|z_0^\star\|_{X^\star}$. Since $A_\lambda(x) \rightharpoonup A^0(x)$ and $\|A_\lambda(x)\|_{X^\star} \to \|A^0(x)\|_{X^\star}$, the local uniform convexity of $X^\star$ leads us to the conclusion $A_\lambda(x) \to A^0(x)$ (see Example 2.6.13).

(iii) Recall that $A_\lambda(x) \in A(R_\lambda(x)) = A(x_\lambda)$ for any $\lambda > 0$. Consider two sequences $(\lambda_n)_{n\in\mathbb{N}} \subset \mathbb{R}$, $(x_n)_{n\in\mathbb{N}} \subset X$, satisfying property (9.37). Then,

10 See Proposition 1.1.54(ii).

$$\begin{aligned}&\langle A_{\lambda_n}(x_n) - A_{\lambda_m}(x_m), x_n - x_m\rangle\\&\quad= \langle A_{\lambda_n}(x_n) - A_{\lambda_m}(x_m), R_{\lambda_n}(x_n) - R_{\lambda_m}(x_m)\rangle\\&\qquad+ \langle A_{\lambda_n}(x_n) - A_{\lambda_m}(x_m), (x_n - R_{\lambda_n}(x_n)) - (x_m - R_{\lambda_m}(x_m))\rangle\\&\quad\geq \langle A_{\lambda_n}(x_n) - A_{\lambda_m}(x_m), (x_n - R_{\lambda_n}(x_n)) - (x_m - R_{\lambda_m}(x_m))\rangle,\end{aligned} \tag{9.44}$$

since $\langle A_{\lambda_n}(x_n) - A_{\lambda_m}(x_m), R_{\lambda_n}(x_n) - R_{\lambda_m}(x_m)\rangle \geq 0$ (by the monotonicity of A and since $A_{\lambda_n}(x_n) \in A(R_{\lambda_n}(x_n))$, $A_{\lambda_m}(x_m) \in A(R_{\lambda_m}(x_m))$). We now use the definition of $A_\lambda(x)$, and substituting $A_{\lambda_n}(x_n)$ and $A_{\lambda_m}(x_\mu)$ for their equals in (9.44), we obtain the inequality

$$\begin{aligned}&\langle A_{\lambda_n}(x_n) - A_{\lambda_m}(x_m), x_n - x_m\rangle\\&\quad\geq \langle \lambda_n^{-1}J(x_n - R_{\lambda_n}(x_n)) - \lambda_m^{-1}J(x_m - R_{\lambda_m}(x_m)), (x_n - R_{\lambda_n}(x_n)) - (x_m - R_{\lambda_m}(x_m))\rangle.\end{aligned} \tag{9.45}$$

By the fact that the sequences chosen satisfy property (9.37), inequality (9.45) yields

$$\lim_{n,m\to\infty} \langle A_{\lambda_n}(x_n) - A_{\lambda_m}(x_m), x_n - x_m\rangle = 0,$$

and

$$\lim_{n,m\to\infty} \langle A_{\lambda_n}(x_n) - A_{\lambda_m}(x_m), R_{\lambda_n}(x_n) - R_{\lambda_m}(x_m)\rangle = 0.$$

Applying Proposition 9.4.5(vi), we conclude that $(x, x^\star) \in \mathbf{Gr}(A)$. Hence, (9.38) holds. □

Remark 9.4.27. Let X be a reflexive Banach space (assuming a renorming so that X and $X^\star$ are both locally uniformly convex). Let $A : \mathbf{D}(A) \subset X \to 2^{X^\star}$ and $B : \mathbf{D}(B) \subset X \to 2^{X^\star}$ be two maximal monotone operators satisfying $\mathbf{D}(A) \cap \mathbf{D}(B) \neq \emptyset$. Let $\lambda > 0$, and consider B_λ, the Yosida approximation of B. Given $z^\star \in X^\star$, we claim that the operator inclusion $z^\star \in A(x_\lambda) + B_\lambda(x_\lambda) + J(x_\lambda)$ has a unique solution $x_\lambda \in \mathbf{D}(A)$.

Indeed, note that the operator $B_\lambda + J$ is monotone, demicontinuous, coercive, and bounded with $\mathbf{D}(B_\lambda + J) = X$. By Theorem 9.4.12, we have $\mathbf{R}(A + B_\lambda + J) = X^\star$. The uniqueness of the solution x_λ follows from the monotonicity of A and B_λ and the strict monotonicity of J. So we have $z^\star = x_\lambda^\star + B_\lambda(x_\lambda) + J(x_\lambda)$, $x_\lambda^\star \in A(x_\lambda)$.

Proposition 9.4.28. *Let X be a reflexive Banach space (assuming a renorming such that X and $X^\star$ are locally uniformly convex) and $A : \mathbf{D}(A) \subset X \to 2^{X^\star}$, $B : \mathbf{D}(B) \subset X \to 2^{X^\star}$ be maximal monotone operators, and $\mathbf{D}(A) \cap \mathbf{D}(B) \neq \emptyset$. Then, $\mathbf{R}(A + B + J) = X^\star$ if and only if $\|B_\lambda(x_\lambda)\|_{X^\star} \leq c$ for some $c > 0$, $\lambda_0 > 0$, and every $\lambda \in (0, \lambda_0]$.*

Proof. We will denote by B_λ and R_λ^B the Yosida approximation and the resolvent, respectively, of the operator B.

Let $z^\star \in \mathbf{R}(A + B + J)$. Then, there exist $(x, x_1^\star) \in \mathbf{Gr}(A)$ and $(x, x_2^\star) \in \mathbf{Gr}(B)$ such that $z^\star = x_1^\star + x_2^\star + J(x)$. Let $x_\lambda \in \mathbf{D}(A)$ be the unique solution of

$$z^\star = x_\lambda^\star + B_\lambda(x_\lambda) + J(x_\lambda), \quad x_\lambda^\star \in A(x_\lambda). \tag{9.46}$$

Using the monotonicity of J and A, we have

$$\begin{aligned} 0 &\le \langle J(x_\lambda) - J(x), x_\lambda - x\rangle = \langle x_1^* - x_\lambda^*, x_\lambda - x\rangle + \langle x_2^* - B_\lambda(x_\lambda), x_\lambda - x\rangle \\ &\le \langle x_2^* - B_\lambda(x_\lambda, x_\lambda - x\rangle. \end{aligned} \tag{9.47}$$

Recall that $B_\lambda(x_\lambda) = \frac{1}{\lambda}J(x_\lambda - R_\lambda^B(x_\lambda))$, where, by R_λ^B, we denote the resolvent of B, so $\lambda J^{-1}(B_\lambda(x_\lambda)) = x_\lambda - R_\lambda^B(x_\lambda)$, which implies that $x_\lambda = R_\lambda^B(x_\lambda) + \lambda J^{-1}(B_\lambda(x_\lambda))$. Putting this in (9.47) yields

$$0 \le \langle x_2^* - B_\lambda(x_\lambda), R_\lambda^B(x_\lambda) - x\rangle + \langle x_2^* - B_\lambda(x_\lambda), \lambda J^{-1}(B_\lambda(x_\lambda))\rangle. \tag{9.48}$$

Since $B_\lambda(x_\lambda) \in B(R_\lambda^B(x_\lambda))$, and because B is monotone, the first term on the right-hand side of (9.48) is nonpositive, so

$$0 \le \langle x_2^* - B_\lambda(x_\lambda), \lambda J^{-1}(B_\lambda(x_\lambda))\rangle,$$

which implies that $\lambda\|B_\lambda(x_\lambda)\|^2 \le \langle x_2^*, \lambda J^{-1}(B_\lambda(x_\lambda))\rangle$, from which it follows that $\|B_\lambda(x_\lambda)\|_{X^*} \le \|x_2^*\|_{X^*}, \lambda > 0$.

For the converse, let $z^* \in X^*$, arbitrary. We assume now that $\|B_\lambda(x_\lambda)\|_{X^*} \le c$ for some $c > 0$ and all $\lambda \in (0, \lambda_0]$. Let $(z_0, z_0^*) \in \mathbf{Gr}(A)$, and multiply equation (9.46) by $x_\lambda - z_0$. Using the monotonicity of A, we have

$$\begin{aligned} \|x_\lambda\|^2 &= \langle z_0^*, x_\lambda - z_0\rangle + \langle J(x_\lambda), z_0\rangle - \langle x_\lambda^*, x_\lambda - z_0\rangle - \langle B_\lambda(x_\lambda), x_\lambda - z_0\rangle \\ &\le \langle z_0^*, x_\lambda - z_0\rangle + \langle J(x_\lambda), z_0\rangle - \langle z_0^*, x_\lambda - z_0\rangle - \langle B_\lambda(x_\lambda), x_\lambda - z_o\rangle. \end{aligned}$$

We now use the boundedness hypothesis for $\{B_\lambda(x_\lambda)\}_{\lambda\in(0,\lambda_0]}$, and we have $\|x_\lambda\|^2 \le c_1\|x_\lambda\| + c_2$ for some $c_1, c_2 \ge 0$ and every $\lambda \in (0, \lambda_0]$, so $\{x_\lambda : \lambda \in (0, \lambda_0]\}$ is bounded. Then, from (9.46), it follows that $\{x_\lambda^* : \lambda \in (0, \lambda_0]\}$ is bounded, so we can find a sequence $\lambda_n \to 0^+$ such that $x_{\lambda_n} \rightharpoonup x$ in X and $x_{\lambda_n}^* \rightharpoonup x_1^*$, $B_{\lambda_n}(x_{\lambda_n}) \rightharpoonup x_2^*$, $J(x_{\lambda_n}) \rightharpoonup x_3^*$ in X^*. From (9.46), we have

$$\begin{aligned} 0 = &\langle x_{\lambda_n}^* + J(x_{\lambda_n}) - x_{\lambda_m}^* - J(x_{\lambda_m}), x_{\lambda_n} - x_{\lambda_m}\rangle \\ &+ \langle B_{\lambda_n}(x_{\lambda_n}) - B_{\lambda_m}(x_{\lambda_m}), x_{\lambda_n} - x_{\lambda_m}\rangle. \end{aligned} \tag{9.49}$$

Since A + J is monotone, we have

$$\langle B_{\lambda_n}(x_{\lambda_n}) - B_{\lambda_m}(x_{\lambda_m}), x_{\lambda_n} - x_{\lambda_m}\rangle \le 0,$$

and so

$$\limsup_{n,m\to\infty}\langle B_{\lambda_n}(x_{\lambda_n}) - B_{\lambda_m}(x_{\lambda_m}), x_{\lambda_n} - x_{\lambda_m}\rangle \le 0.$$

From Proposition 9.4.5(vi), we have that $(x, x_2^*) \in \mathbf{Gr}(B)$ and

$$\lim_{n,m\to\infty} \langle B_{\lambda_n}(x_{\lambda_n}) - B_{\lambda_m}(x_{\lambda_m}), x_{\lambda_n} - x_{\lambda_m} \rangle = 0.$$

Therefore, from (9.49), it follows that

$$\lim_{n,m\to\infty} \langle x^\star_{\lambda_n} + J(x_{\lambda_n}) - x^\star_{\lambda_m} - J(x_{\lambda_m}), x_{\lambda_n} - x_{\lambda_m} \rangle = 0,$$

and because A is monotone,

$$\limsup_{n,m\to\infty} \langle J(x_{\lambda_n}) - J(x_{\lambda_m}), x_{\lambda_n} - x_{\lambda_m} \rangle \leq 0.$$

Since J is maximal monotone with $\mathbf{D}(J) = X$, we can apply again Proposition 9.4.5(v) to obtain that $x^\star_3 = J(x)$. Also, $\lim_{n,m\to\infty} \langle J(x_{\lambda_n}) - J(x_{\lambda_m}), x_{\lambda_n} - x_{\lambda_m} \rangle = 0$, and so $\lim_{n,m\to\infty} \langle x^\star_{\lambda_n} - x^\star_{\lambda_m}, x_{\lambda_n} - x_{\lambda_m} \rangle = 0$. Hence, another application of Proposition 9.4.5(vi) yields that $x^\star_1 \in A(x)$. Then, $z^\star - x^\star_{\lambda_n} - J(x_{\lambda_n}) \rightharpoonup z^\star - x^\star_1 - J(x) = x^\star_2$ in $X^\star$, so finally $z^\star \in \mathbf{R}(A + B + J)$. □

We close our treatment of the Yosida approximation for maximal monotone operators by an important example, the Yosida approximation for the subdifferential. This extends the analysis for the Moreau–Yosida approximation and the proximity operator in Hilbert spaces (see Section 4.4) to the more general case of reflexive Banach spaces. The following theorem [18, 107, 110] is a generalization of Proposition 4.4.6 for reflexive Banach spaces:

Theorem 9.4.29. *Let X be a reflexive Banach space (assuming a renorming such that X and $X^\star$ are locally uniformly convex), $\varphi : X \to \mathbb{R} \cup \{+\infty\}$ be a proper convex lower semicontinuous function, and $A = \partial\varphi$, and consider its Yosida approximation, defined by*

$$\varphi_\lambda(x) := \inf_{z\in X} \left\{ \frac{1}{2\lambda} \|x - z\|^2 + \varphi(z) \right\}, \quad \lambda > 0.$$

For every $\lambda > 0$, the function φ_λ is convex, finite, and Gâteaux differentiable on X and $A_\lambda = \partial\varphi_\lambda = D\varphi_\lambda$. If $X = H$ is a Hilbert space, then φ_λ is Fréchet differentiable. In addition,
(i) $\varphi_\lambda(x) = \frac{1}{2\lambda}\|x - R_\lambda(x)\|^2 + \varphi(R_\lambda(x))$, $\forall \lambda > 0$, and $x \in X$.
(ii) $\varphi(R_\lambda(x)) \leq \varphi_\lambda(x) \leq \varphi(x)$, $\forall \lambda > 0$, and $x \in X$.
(iii) $\lim_{\lambda\to 0^+} \varphi_\lambda(x) = \varphi(x)$, $\forall x \in X$.

Proof. (i) It is a simple consequence of Proposition 4.2.17 and Theorem 4.2.12. For every $x \in X$, fixed, define $\phi_x : X \to \mathbb{R} \cup \{+\infty\}$ by $\phi_x(z) = \frac{1}{2\lambda}\|x - z\|^2 + \varphi(z)$, and note that $\partial\phi_x(z) = \frac{1}{\lambda}J(z - x) + \partial\varphi(z)$. This implies (see Proposition 4.3.1) that every solution z_0 of $0 \in \frac{1}{\lambda}J(z - x) + \partial\varphi(z)$ is a minimizer of φ_x, and since $z_0 = R_\lambda(x)$, we obtain (i).

(ii) This is immediate from (i) and the definition of φ_λ.

(ii) We need to consider two cases. If $x \in \operatorname{dom}\varphi$, then by Proposition 9.4.24, $\lim_{\lambda\to 0^+} R_\lambda(x) = x$. Using (ii) and the lower semicontinuity of φ, we obtain that $\varphi_\lambda(x) \to \varphi(x)$, as $\lambda \to 0$.

Now, suppose that $x \notin \operatorname{dom}\varphi$. We must show that $\varphi_\lambda(x) \to \infty$, as $\lambda \to 0$. If this is not the case, then there exist $\lambda_n > 0$ and $c > 0$ such that

$$\varphi_{\lambda_n}(x) = \varphi(R_{\lambda_n}(x)) + \frac{1}{2\lambda_n}\|x - R_{\lambda_n}(x)\|^2 \leq c, \quad \forall n \geq 1.$$

It then follows that

$$\varphi(R_{\lambda_n}(x)) \leq c \quad \text{and} \quad \|x - R_{\lambda_n}(x)\| \to 0 \quad \text{as } n \to \infty.$$

From these facts and the lower semicontinuity of φ, it follows that $\varphi(x) \leq c$ which is a contradiction. Therefore,

$$\lim_{\lambda \to 0^+} \varphi_\lambda(x) = \varphi(x), \quad \forall x \in X.$$

We shall now prove the Gâteaux differentiability of φ_λ for every $x \in X$. From (i), it is easy to see that

$$0 \leq \varphi_\lambda(z) - \varphi_\lambda(x) - \langle A_\lambda(x), z - x\rangle \leq \langle A_\lambda(z) - A_\lambda(x), z - x\rangle, \quad \forall \lambda > 0,\ x, z \in X, \tag{9.50}$$

where the left inequality follows by the fact that $A_\lambda(x) = \partial\varphi_\lambda(x)$. For arbitrary $z_0 \in X$, setting $z = x + tz_o$, $t > 0$ in the above, and dividing by t, we obtain,

$$\lim_{t \to 0} \frac{\varphi_\lambda(x + tz_o) - \varphi_\lambda(x)}{t} = \langle A_\lambda(x), z_0\rangle, \quad \forall z_0 \in X,$$

because A_λ is demicontinuous. Hence, φ_λ is Gâteaux differentiable at any $x \in X$, and $\partial\varphi_\lambda(x) = A_\lambda(x) = D\phi_\lambda(x)$. □

In the special case where $X = H$ a Hilbert space, the resolvent operator and the Yosida approximation enjoy supplementary properties [18, 107, 110].

Proposition 9.4.30. *Let $X = H$ be a Hilbert space and $A : \mathbf{D}(A) \subset H \to 2^H$ a maximal monotone operator. Then,*
(i) *the resolvent $R_\lambda = (I + \lambda A)^{-1}$ is nonexpansive on H and,*
(ii) *the Yosida approximation A_λ is Lipschitz with Lipschitz constant $\frac{1}{\lambda}$.*

Proof. (i) The fact that the resolvent can be expressed in the form $R_\lambda = (I + \lambda A)^{-1}$ has already been discussed in Example 9.4.22. Consider $(x_i, x_i^\star) \in \mathbf{Gr}(A)$, $i = 1, 2$. Exploiting the inner product structure of H for any $\lambda > 0$,

$$\begin{aligned} \|x_1 - x_2 + \lambda(x_1^\star - x_2^\star)\|^2 &= \|x_1 - x_2\|^2 + \lambda\langle x_1^\star - x_2^\star, x_1 - x_2\rangle + \lambda^2\|x_1^\star - x_2^\star\|^2 \\ &\geq \|x_1 - x_2\|^2, \end{aligned} \tag{9.51}$$

since the monotonicity of A yields $\langle x_1^\star - x_2^\star, x_1 - x_2\rangle \geq 0$. This implies

$$\|x_1 - x_2\| \le \|x_1 - x_2 + \lambda(x_1^\star - x_2^\star)\|,$$

which is the nonexpansive property for R_λ. Note that the converse of (i) is also true since taking the limit of (9.51) as $\lambda \to 0$ yields the monotonicity of A.

(ii) Set $x_{i,\lambda} = (I + \lambda A)^{-1} x_i$, $i = 1, 2$, so that

$$x_1 - x_2 \in x_{1,\lambda} - x_{1,\lambda} + \lambda(A(x_{1,\lambda}) - A(x_{2,\lambda})),$$

and taking the inner product in H by $A(x_{1,\lambda}) - A(x_{2,\lambda})$ yields the required result. □

The case of monotone type operators, defined as $A : \mathbf{D}(A) \subset X \to 2^X$ (rather than in general as $A : \mathbf{D}(A) \subset X \to 2^{X^\star}$), finds important application in the theory of evolution problems. We will consider their properties in detail in Chapter 10.

9.4.7 Sum of maximal monotone operators

The following remarkable theorem (see, e. g., [20, 107]), addresses the maximal monotonicity of sums of operators.

Theorem 9.4.31 (Rockafellar). *Let X be a reflexive Banach space (assuming a renorming such that X and $X^\star$ are locally uniformly convex) and $A : \mathbf{D}(A) \subset X \to 2^{X^\star}$, $B : \mathbf{D}(B) \subset X \to 2^{X^\star}$ maximal monotone operators. Suppose that* $\operatorname{int}(\mathbf{D}(A)) \cap \mathbf{D}(B) \neq \emptyset$. *Then, $A + B$ is a maximal monotone operator.*

Proof. Without loss of generality,[11] we may assume that $0 \in \operatorname{int}(\mathbf{D}(A)) \cap \mathbf{D}(B)$, $0 \in A(0)$, $0 \in B(0)$. For $x^\star \in X^\star$ and $\lambda > 0$, we consider the equation

$$x^\star \in J(x_\lambda) + A(x_\lambda) + B_\lambda(x_\lambda), \tag{9.52}$$

where B_λ is the Yosida approximation of B. Since B_λ is demicontinuous and bounded, it follows from Remark 9.4.27 that (9.52) has a solution $x_\lambda \in \mathbf{D}(A)$ for any $\lambda > 0$. Since A and B_λ are monotone, upon taking the duality pairing of (9.52) by x_λ, we see that

$$\|x_\lambda\| \le \|x^\star\|_{X^\star}, \quad \forall \lambda > 0.$$

Since $0 \in \operatorname{int}(\mathbf{D}(A))$, it follows from Theorem 9.3.4 that A is locally bounded at 0. Hence, there exist $\rho > 0$ and $c > 0$ such that

$$\|x^\star\| \le c, \quad \forall x^\star \in \bigcup \{A(x), \|x\| \le \rho\}.$$

11 Indeed, if $x_0 \in \operatorname{int}(\mathbf{D}(A)) \cap \mathbf{D}(B)$, then consider $A_1(x) = A(x + x_0) - A(x_0)$ and $B_1(x) = B(x + x_0) - B(x_0)$.

Now, for $\lambda > 0$, define $z_\lambda = \frac{\rho}{2} J^{-1}(x_\lambda^*)\|x_\lambda^*\|_{X^*}^{-1}$, where $x_\lambda^* = x^* - B_\lambda(x_\lambda) - J(x_\lambda) \in A(x_\lambda)$, so that $\|z_\lambda\| = \frac{\rho}{2} < \rho$, hence $z_\lambda \in \mathbf{D}(A)$. Let $(z_\lambda, z_\lambda^*) \in \mathbf{Gr}(A)$. Then, $\|z_\lambda^*\|_{X^*} < c$.

Using the monotonicity of A, we obtain

$$0 \le \langle x_\lambda^* - z_\lambda^*, x_\lambda - z_\lambda \rangle = \langle x_\lambda^*, x_\lambda \rangle + \langle z_\lambda^*, z_\lambda \rangle - \langle x_\lambda^*, z_\lambda \rangle - \langle z_\lambda^*, x_\lambda \rangle, \quad \forall z_\lambda^* \in A(x_\lambda),$$

which implies that

$$\frac{\rho}{2}\|x_\lambda^*\|_{X^*} = \langle x_\lambda^*, z_\lambda \rangle \le c\frac{\rho}{2} + c\|x^*\|_{X^*} + \|x^*\|_{X^*}^2.$$

We have thus shown that $\{x_\lambda : \lambda > 0\}$, $\{B_\lambda x_\lambda : \lambda > 0\}$, and $\{x_\lambda^* : \lambda > 0\}$ are bounded sets of X and X^*, respectively. Since X is reflexive, we may assume that

$$x_\lambda \rightharpoonup x_0, \quad J(x_\lambda) \rightharpoonup x_0^*, \quad B_\lambda(x_\lambda) \rightharpoonup x_1^*, \quad x_\lambda^* \rightharpoonup x_2^*, \quad \text{as } \lambda \to 0.$$

Since A is monotone, equality

$$J(x_\lambda) + B_\lambda(x_\lambda) + x_\lambda^* = x^*, \quad x_\lambda^* \in A(x_\lambda),$$

implies that

$$0 = \langle J(x_\lambda) + x_\lambda^* - J(x_\mu) - x_\mu^*, x_\lambda - x_\mu \rangle + \langle B_\lambda(x_\lambda) - B_\mu(x_\mu), x_\lambda - x_\mu \rangle. \tag{9.53}$$

Therefore,

$$\limsup_{\lambda,\mu\to 0} \langle B_\lambda(x_\lambda) - B_\mu(x_\mu), x_\lambda - x_\mu \rangle \le 0.$$

It then follows that $(x_0, x_1^*) \in \mathbf{Gr}(B)$ and

$$\lim_{\lambda,\mu\to 0} \langle B_\lambda(x_\lambda) - B_\mu(x_\mu), x_\lambda - x_\mu \rangle = 0.$$

By (9.53), we have that

$$\lim_{\lambda,\mu\to 0} \langle J(x_\lambda) + x_\lambda^* - J(x_\mu) - x_\mu^*, x_\lambda - x_\mu \rangle = 0.$$

Since $J + A$ is maximal monotone, by Proposition 9.4.5(vi), $(x_0, x_0^* + x_2^*) \in \mathbf{Gr}(A + J)$. Passing to the limit as $\lambda \to 0$ in (9.52), we obtain

$$x_0^* + x_2^* + x_1^* = x^*,$$

where $x_2^* \in A(x_0)$, $x_1^* \in B(x_0)$, i. e., $x^* \in \mathbf{R}(J + A + B)$. It then follows that $\mathbf{R}(J + A + B) = X^*$ so that, by Theorem 9.4.13, $A + B$ is maximal monotone. □

A maximal monotonicity criterion for sums of operators in Hilbert spaces is given by the next theorem [88].

Theorem 9.4.32. *Let H be a Hilbert space, and $\mathsf{A} : \mathbf{D}(\mathsf{A}) \subset H \to 2^H$ and $\mathsf{B} : \mathbf{D}(\mathsf{B}) \subset H \to 2^H$ maximal monotone operators. Assume furthermore that for every $(z, z^\star) \in \mathbf{Gr}(\mathsf{A})$ and $\lambda > 0$ holds $\langle z^\star, \mathsf{B}_\lambda(z)\rangle \geq 0$. Then, $\mathsf{A} + \mathsf{B}$ is maximal monotone.*

Proof. Let $x^\star \in H$, and let $x_\lambda \in \mathbf{D}(\mathsf{A})$ be the unique solution of the regularized problem $x_\lambda + x_\lambda^\star + \mathsf{B}_\lambda(x_\lambda) = x^\star$ with $(x_\lambda, x_\lambda^\star) \in \mathbf{Gr}(\mathsf{A})$. According to Proposition 9.4.28, it suffices to prove that $\{\mathsf{B}_\lambda(x_\lambda) : \lambda > 0\}$ is bounded in H.

Take the inner product of the above equation with $\mathsf{B}_\lambda(x_\lambda)$, and use the hypothesis to get

$$\langle \mathsf{B}_\lambda(x_\lambda), x_\lambda\rangle + \|\mathsf{B}_\lambda(x_\lambda)\|^2 \leq \|x^\star\| \, \|\mathsf{B}_\lambda(x_\lambda)\|.$$

Let $(z, z^\star) \in \mathbf{Gr}(\mathsf{A})$. By monotonicity, we have

$$0 \leq \langle x_\lambda^\star - z^\star, x_\lambda - z\rangle = \langle x^\star - x_\lambda - \mathsf{B}_\lambda(x_\lambda) - z^\star, x_\lambda - z\rangle,$$

which upon rearrangement yields

$$\begin{aligned}\|x_\lambda\|^2 + \langle \mathsf{B}_\lambda(x_\lambda), x_\lambda\rangle &\leq \langle x^\star - z^\star, x_\lambda - z\rangle + \langle x_\lambda + \mathsf{B}_\lambda(x_\lambda), z\rangle \\ &\leq c_1 + c_2\|x_\lambda\| + c_3\|\mathsf{B}_\lambda(x_\lambda)\|,\end{aligned}$$

where $c_1, c_2, c_3 > 0$ are independent of $\lambda > 0$.

Let $x' \in \mathbf{D}(\mathsf{B})$. By the monotonicity of B_λ, we have $0 \leq \langle \mathsf{B}_\lambda(x_\lambda) - \mathsf{B}_\lambda(x'), x_\lambda - x'\rangle$, so

$$\langle \mathsf{B}_\lambda(x'), x_\lambda - x'\rangle + \langle \mathsf{B}_\lambda(x_\lambda), x'\rangle \leq \langle \mathsf{B}_\lambda(x_\lambda), x_\lambda\rangle,$$

from which it follows that

$$-c_4 - c_5\|x_\lambda\| - c_6\|\mathsf{B}_\lambda(x_\lambda)\| \leq \langle \mathsf{B}_\lambda(x_\lambda), x_\lambda\rangle,$$

where $c_4, c_5, c_6 > 0$ are independent of $\lambda > 0$. Therefore,

$$\begin{aligned}\|x_\lambda\|^2 &\leq -\langle \mathsf{B}_\lambda(x_\lambda), x_\lambda\rangle + c_1 + c_2\|x_\lambda\| + c_3\|\mathsf{B}_\lambda(x_\lambda)\| \\ &\leq c_7 + c_8\|x_\lambda\| + c_9\|\mathsf{B}_\lambda(x_\lambda)\|,\end{aligned}$$

where $c_7, c_8, c_9 > 0$ are independent of $\lambda > 0$. Hence, we finally have that

$$\begin{aligned}\|\mathsf{B}_\lambda(x_\lambda)\|^2 &\leq -\mathsf{B}_\lambda(x_\lambda), x_\lambda\rangle + \|x^\star\| \, \|\mathsf{B}_\lambda(x_\lambda)\| \\ &\leq c_4 + c_5\|x_\lambda\| + c_{10}\|\mathsf{B}_\lambda(x_\lambda)\|,\end{aligned}$$

with $c_{10} > 0$ independent of $\lambda > 0$. Since $\|x_\lambda\|$ is bounded, we deduce that the set $\{\mathsf{B}_\lambda(x_\lambda) : \lambda > 0\}$ is bounded, and so $\mathsf{A} + \mathsf{B}$ is maximal monotone. □

9.5 Pseudomonotone operators

Pseudomonotone operators is a general class of monotone type operators playing an important role in nonlinear analysis.

9.5.1 Pseudomonotone operators, definitions, examples, and properties

We will start with the presentation of the theory of pseudomonotone operators in the single valued case.

Definition 9.5.1. Let X be a reflexive Banach space. The operator $\mathsf{A} : X \to X^\star$ is called pseudomonotone for every $(x_n)_{n\in\mathbb{N}} \subset X$ such that

$$\begin{cases} x_n \rightharpoonup x \\ \limsup_{n\to\infty}\langle \mathsf{A}(x_n), x_n - x\rangle \le 0 \end{cases} \tag{9.54}$$
$$\Longrightarrow \langle \mathsf{A}(x), x - z\rangle \le \liminf_{n\to\infty}\langle \mathsf{A}(x_n), x_n - z\rangle, \quad \forall z \in X.$$

Remark 9.5.2. As noted in [82], condition (9.54) may look strange, but it is in fact an easy to check condition, that allows us to guarantee that

$$\begin{cases} x_n \rightharpoonup x, \\ \mathsf{A}(x_n) \rightharpoonup y^\star, \\ \langle \mathsf{A}(x_n), x_n\rangle \to \langle y^\star, x\rangle \end{cases} \Longrightarrow \mathsf{A}(x) = y^\star, \tag{9.55}$$

or even the stronger result

$$\begin{cases} x_n \rightharpoonup x, \\ \mathsf{A}(x_n) \rightharpoonup y^\star, \\ \limsup_{n\to\infty}\langle \mathsf{A}(x_n), x_n\rangle \le \langle y^\star, x\rangle \end{cases} \Longrightarrow \mathsf{A}(x) = y^\star. \tag{9.56}$$

Condition (9.56) (called the M-condition by Brezis [31]) is required for the proof of surjectivity of pseudomonotone operators, using the Faedo–Galerkin approach (see Theorem 9.5.5).

To see that the pseudomonotonicity condition (9.54) implies (9.56) (which in turn implies (9.55)), we assume that (9.54) holds, and consider a sequence $(x_n)_{n\in\mathbb{N}} \subset X$ satisfying the LHS column of condition (9.55). Then,

$$\limsup_{n\to\infty}\langle \mathsf{A}(x_n), x_n - x\rangle = \limsup_{n\to\infty}\langle \mathsf{A}(x_n), x_n\rangle - \lim_{n\to\infty}\langle \mathsf{A}(x_n), x\rangle \le 0,$$

as $(x_n)_{n\in\mathbb{N}}$ satisfies the LHS of (9.56). Hence, by (9.54), for any $z \in X$,

$$\langle A(x), x - z\rangle \leq \liminf_{n\to\infty} \langle A(x_n), x_n - z\rangle. \tag{9.57}$$

However, we note that

$$\begin{aligned}\limsup_{n\to\infty} \langle A(x_n), x_n - z\rangle &= \limsup_{n\to\infty} (\langle A(x_n), x_n - x\rangle + \langle A(x_n), x - z\rangle) \\ &\leq \limsup_{n\to\infty} \langle A(x_n), x_n - x\rangle + \limsup_{n\to\infty} \langle A(x_n), x - z\rangle \\ &= \limsup_{n\to\infty} \langle A(x_n), x_n - x\rangle + \langle y^\star, x - z\rangle \leq 0 + \langle y^\star, x - z\rangle.\end{aligned} \tag{9.58}$$

Combining (9.57) and (9.58), we have that

$$\langle A(x), x - z\rangle \leq \liminf_{n\to\infty} \langle A(x_n), x_n - z\rangle \leq \limsup_{n\to\infty} \langle A(x_n), x_n - z\rangle \leq \langle y^\star, x - z\rangle.$$

Since this holds for any $z \in X$, we take an arbitrary $\bar{z}$ and apply the above for $z = x \pm \bar{z}$. This yields $A(x) = y^\star$.

The following proposition [107, 110] provides useful characterizations and properties of pseudomonotone operators.

Proposition 9.5.3. *Let X be a reflexive Banach space and $A, B : X \to X^\star$ operators. Then the following hold:*
(i) *If A is monotone and hemicontinuous, then A is pseudomonotone.*
(ii) *If A is completely continuous, then A is pseudomonotone.*
(iii) *If A and B are pseudomonotone, then $A + B$ is pseudomonotone.*
(iv) *If A is pseudomonotone and locally bounded, then A is demicontinuous.*

Proof. (i) Let $(x_n)_{n\in\mathbb{N}}$ be a sequence in X such that $x_n \rightharpoonup x$ and $\limsup_n \langle A(x_n), x_n - x\rangle \leq 0$. Since A is monotone it holds that $\langle A(x_n) - A(x), x_n - x\rangle \geq 0$, therefore

$$\liminf_{n\to\infty} \langle A(x_n), x_n - x\rangle \geq \liminf_{n\to\infty} \langle A(x), x_n - x\rangle = 0,$$

so $\lim_n \langle A(x_n), x_n - x\rangle = 0$. For arbitrary $\bar{z} \in X$, we set $z = x + t(\bar{z} - x)$, $t > 0$. By the monotonicity of A,

$$\langle A(x_n) - A(z), x_n - (x + t(\bar{z} - x))\rangle \geq 0,$$

so

$$t\langle A(x_n), x - \bar{z}\rangle \geq -\langle A(x_n), x_n - x\rangle + \langle A(z), x_n - x\rangle + t\langle A(z), x - \bar{z}\rangle.$$

It then follows that for all $\bar{z} \in X$, since $x_n \rightharpoonup x$ and $t > 0$,

$$\liminf_{n\to\infty} \langle A(x_n), x_n - \bar{z}\rangle \geq \langle A(z), x - \bar{z}\rangle.$$

Since A is hemicontinuous, we conclude that $A(z) \rightharpoonup A(x)$ as $t \to 0^+$. Therefore, for all $\bar{z} \in X$, we have

$$\liminf_{n\to\infty} \langle A(x_n), x_n - \bar{z}\rangle \geq \langle A(x), x - \bar{z}\rangle,$$

i. e. A is pseudomonotone.

(ii) Let $(x_n)_{n\in\mathbb{N}}$ be a sequence in X such that $x_n \rightharpoonup x$. Since A is completely continuous, $A(x_n) \to A(x)$ in $X^\star$. Therefore, for every $z \in X$, we have

$$\langle A(x), x - z\rangle = \lim_{n\to\infty} \langle A(x_n), x_n - z\rangle,$$

i. e., A is pseudomonotone.

(iii) Let $(x_n)_{n\in\mathbb{N}}$ be a sequence in X such that $x_n \rightharpoonup x$ and

$$\limsup_{n\to\infty} \langle A(x_n) + B(x_n), x_n - x\rangle \leq 0. \tag{9.59}$$

We claim that

$$\limsup_{n\to\infty} \langle A(x_n), x_n - x\rangle \leq 0, \quad \limsup_{n\to\infty} \langle B(x_n), x_n - x\rangle \leq 0. \tag{9.60}$$

Suppose that $\limsup_n \langle A(x_n), x_n - x\rangle = \alpha > 0$. From (9.59), we have that $\limsup_n \langle B(x_n), x_n - x\rangle \leq -\alpha$. Since B is pseudomonotone for all $z \in X$, we have

$$\langle B(x), x - z\rangle \leq \liminf_{n\to\infty} \langle B(x_n), x_n - z\rangle.$$

For $z = x$, we have that $0 \leq -\alpha < 0$, which is a contradiction. Also, (9.60) holds, and since A and B are pseudomonotone, we have

$$\langle A(x), x - z\rangle \leq \liminf_{n\to\infty} \langle A(x_n), x_n - z\rangle,$$
$$\langle B(x), x - z\rangle \leq \liminf_{n\to\infty} \langle B(x_n), x_n - z\rangle.$$

Adding the above, we get that for all $z \in X$,

$$\langle A(x) + B(x), x - z\rangle \leq \liminf_{n\to\infty} \langle A(x_n) + B(x_n), x_n - z\rangle,$$

i. e., A + B is pseudomonotone.

(iv) Let $(x_n)_{n\in\mathbb{N}}$ be a sequence in X such that $x_n \to x$. Since A is locally bounded, the sequence $(A(x_n))_{n\in\mathbb{N}}$ is bounded. The space X is reflexive, so there exists a subsequence $(A(x_{n_k}))_{k\in\mathbb{N}}$, with $A(x_{n_k}) \rightharpoonup x^\star$, as $k \to \infty$. Therefore, $\lim_k \langle A(x_{n_k}), x_{n_k} - x\rangle = 0$. So, by the pseudomonotonicity of A, we have

$$\langle A(x), x - z\rangle \leq \liminf_{k\to\infty} \langle A(x_{n_k}), x_{n_k} - z\rangle = \langle x^\star, x - z\rangle, \quad \forall z \in X.$$

It then follows that $A(x) = x^\star$. By the standard argument (based on the Urysohn property), we have that $A(x_n) \rightharpoonup A(x)$, i. e., A is demicontinuous. □

Example 9.5.4 (The shift of pseudomonotone maps is pseudomonotone). Let $A : X \to X^*$ be pseudomonotone, fix $x_0 \in X$, and consider the shifted operator $A_{x_0} : X \to X^*$, defined by $x \mapsto A_{x_0}(x) = A(x + x_0)$. Then, A_{x_0} is pseudomonotone.

It is easy to see that A_{x_0} is bounded. To check the remaining condition, let $x_n \rightharpoonup x$ in X, and $\limsup_n \langle A_{x_0}(x_n), x_n - x\rangle \le 0$. By the definition of A_{x_0}, this implies that

$$\limsup_n \big\langle A(x_n + x_0), (x_n + x_0) - (x + x_0)\big\rangle \le 0. \tag{9.61}$$

Consider the sequence $(\bar{x}_n)_{n\in\mathbb{N}}$ with $\bar{x}_n = x_n + x_0$ that satisfies $\bar{x}_n \rightharpoonup \bar{x} := x + x_0$, in terms of which (9.61) becomes $\limsup_n \langle A(\bar{x}_n), \bar{x}_n - \bar{x}\rangle \le 0$. By the pseudomonotonicity of A, this implies that for any $\bar{z}$,

$$\liminf_n \big\langle A(\bar{x}_n), \bar{x}_n - \bar{z}\big\rangle \ge \big\langle A(\bar{x}), \bar{x} - \bar{z}\big\rangle \Longrightarrow \liminf_n \big\langle A_{x_0}(x_n), x_n - z\big\rangle \ge \big\langle A_{x_0}(x), x - z\big\rangle,$$

(the last inequality being valid by applying the first for $\bar{z} = z + x_0$ for arbitrary $z \in X$). ◁

9.5.2 Surjectivity results for pseudomonotone operators

One of the most important features of pseudomonotone operators are their surjectivity properties. We will present here two important surjectivity results: a) one due to Brezis [31] employing the Galerkin approximation method and b) a more abstract approach due to Browder and Hess [37], each one of interest in its own right.

9.5.2.1 Surjectivity of pseudomonotone operators I: the Faedo–Galerkin approximation [31]

The Browder–Minty surjectivity theorem for monotone operators (see Theorem 9.3.8 in Section 9.3.4) can be extended for pseudomonotone operators using the Faedo–Galerkin approach (which is closely related to a frequently used numerical approximation approach to solving operator equations of the form $A(x) = x^\star$, bearing the same name). For the proof of surjectivity of pseudomonotone operators, fitting as reference is that it was first used by Brezis in [31] (see also Ch. 2 in [128], or [82], whose approach we follow here and references therein).

Theorem 9.5.5. *Let X be a reflexive and separable Banach space and $A : X \to X^*$ be a pseudomonotone, bounded, and coercive operator. Then for any $x^\star \in X^*$, there exists a solution to the operator equation $A(x) = x^\star$.*

Proof. This result is a generalization of the corresponding result of Theorem 9.3.8 in Section 9.3.4, following essentially the same path for steps 1 and 2 and differentiating in step 3, as far as the passage to the limit for the sequence of approximate solutions is concerned.

1. This step is the same as step 1 of Theorem 9.3.8. Take a sequence of finite dimensional subspaces $(X_n)_{n\in\mathbb{N}}$ such that $X_n \subset X_{n+1} \subset X$, with $\bigcup_{n\in\mathbb{N}} X_n$ dense in X, and define the sequence of finite dimensional approximations $x_n \in X_n$ by

$$\langle A(x_n), z\rangle = \langle x^\star, z\rangle, \quad \forall z \in X_n, \tag{9.62}$$

which is a finite dimensional equation that can be treated by a fixed point theorem, such as, e. g., the Brouwer fixed point theorem. The necessary demicontinuity property for A is provided by Proposition 9.5.3(iv).

2. Once existence of $x_n \in X_n$ solving (9.62) is guaranteed, coercivity is used once more to obtain a priori bounds for $\|x_n\| := \|x_n\|_X$ independent of n. For instance, using $z = x_n \in X_n$ in (9.62) allows us to get by coercivity that

$$c_0(\|x_n\|)\|x_n\| \le \langle A(x_n), x_n\rangle = \langle x^\star, x_n\rangle \le \|x^\star\|_{X^\star}\|x_n\|,$$

for a suitable function $s \to c_0(s) > 0$, increasing, from which we obtain a uniform bound (i. e., independent of n) for $\|x_n\|$. Note that, in the above, we used the inclusion $X_n \subset X$ to interpret x_n as an element of X.

3. For the final step, we carefully pass to the limit as $n \to \infty$ for the sequence of approximate solutions $(x_n)_{n\in\mathbb{N}}$. The a priori bounds $\|x_n\| < C$, C independent of n and the reflexivity of X guarantee the existence of a subsequence (denoted the same for simplicity) and $x \in X$ such that $x_n \rightharpoonup x$ in X. By the boundedness of A and reflexivity of X^*, we also have a subsequence (denoted the same) and a $y^* \in X^*$ such that $A(x_n) \rightharpoonup y^*$. We claim that $y^* = x^*$ and that $A(x) = x^*$.

By pseudomonotonity (Remark 9.5.2), to prove our claim it suffices to show that

$$\langle A(x_n), z\rangle \to \langle x^\star, z\rangle, \quad \forall z \in \bigcup_{n\in\mathbb{N}} X_n, \tag{9.63}$$

which, by density of $\bigcup_{n\in\mathbb{N}} X_n$ in X, implies the same result for all $z \in X$. Note that (9.63) combined with $A(x_n) \rightharpoonup y^\star$ implies that $y^\star = x^\star$, hence, we may use condition (M) of Remark 9.5.2 to obtain the required result $A(x) = x^*$.

Suppose that (9.63) does not hold. Then, there exists $\bar{z} \in \bigcup_{n\in\mathbb{N}} X_n$, an $\epsilon > 0$ and a subsequence $(x_{n_k})_{k\in\mathbb{N}}$ such that

$$|\langle A(x_{n_k}), \bar{z}\rangle - \langle x^\star, \bar{z}\rangle| \ge \epsilon, \quad k \in \mathbb{N}. \tag{9.64}$$

Clearly, $\bar{z} \in \bigcup_{n\in\mathbb{N}} X_n$ implies the existence of some $N \in \mathbb{N}$ such that $\bar{z} \in X_N$ so that $z \in X_n$ for all $n \ge N$. Pick K such that $n_K \ge N$. Then $n_k \ge N$ for all $k \ge K$. Since $z \in X_n$ for all

$n \geq N$, the above choice implies that $\bar{z} \in X_{n_k}$ for all $k \geq K$. Applying (9.62) for this choice of test function, we have that

$$\langle A(x_{n_k}), \bar{z} \rangle = \langle x^\star, \bar{z} \rangle, \tag{9.65}$$

which contradicts (9.64). Hence, (9.63) holds for any $z \in \bigcup_{n \in \mathbb{N}} X_n$ and by density for any $z \in X$. However, since $A(x_n) \rightharpoonup y^*$, we also have that $\langle A(x_n), z \rangle \to \langle y^*, z \rangle$ for any $z \in X$. Comparing this with (9.63), we see that $y^* = x^\star$, and the proof is complete. □

An alternative proof of this step, by passing Remark 9.5.2 through the use of appropriately chosen sequences of test functions $(z_n)_{n \in \mathbb{N}}$, is presented in Section 9.8 in the Appendix.

9.5.2.2 Surjectivity of pseudomonotone operators II: an abstract approach [37]

This approach, due to Browder and Hess [37], deals with the general case of multivalued maps.

We first define the notion of pseudomonotonicity for multivalued maps.

Definition 9.5.6. Let X be a reflexive Banach space. The operator $A : X \to 2^{X^*}$ is called pseudomonotone if it satisfies the following condition:

- If $(x_n)_{n \in \mathbb{N}} \subset X$ is a sequence such that $x_n \rightharpoonup x$ in X, and if for $x_n^\star \in A(x_n)$ it holds that $\limsup_n \langle x_n^\star, x_n - x \rangle \leq 0$, then for every $z \in X$ there exists $z^\star = z^\star(z) \in A(x)$ with the property

$$\liminf_{n \to \infty} \langle x_n^\star, x_n - z \rangle \geq \langle z^\star(z), x - z \rangle.$$

Next we introduce the concept of coercivity for multivalued maps.

Definition 9.5.7. Let X be a Banach space. The multivalued operator $T : X \to 2^{X^*}$ is called coercive if there exists a real valued function $c : \mathbb{R}_+ \to \mathbb{R}_+$ with $\lim_{r \to \infty} c(r) \to +\infty$ such that $\langle x^\star, x \rangle \geq c(\|x\|)\|x\|$ for all $(x, x^\star) \in \mathbf{Gr}(T)$.

We also need the following proposition, due to Browder and Hess [37], whose proof is based on the theory of the Brouwer degree:

Proposition 9.5.8. *Let X be a finite dimensional Banach space, $T : X \to 2^{X^*}$ a mapping such that for each $x \in X$, $T(x)$ is a nonempty, bounded, closed, convex subset of $X^\star$. Suppose that T is coercive and upper semicontinuous from X to 2^{X^*}. Then, $\mathbf{R}(T) = X^\star$.*

We are now ready to present the surjectivity result of Browder and Hess (Theorem 3 in [37]).

Theorem 9.5.9. *Let X be a reflexive Banach space and $A : X \to 2^{X^*}$ a pseudomonotone and coercive operator satisfying the extra assumptions*

(i) *The set $A(x)$ is nonempty, bounded, closed, and convex for all $x \in X$.*

(ii) A *is upper semicontinuous from every finite dimensional subspace $X_F \subset X$ to the weak topology of $X^\star$ (in the sense of Definition* 1.8.2*)*

Then, A *is surjective, i. e.,* $\mathbf{R}(A) = X^\star$.

Proof. The proof follows along the lines of Theorem 9.4.12. First of all, note that it suffices to show that $0 \in \mathbf{R}(A)$.

We start with a finite dimensional approximation. Let $\mathcal{X}_F$ be the family of all finite dimensional subspaces X_F of X ordered by inclusion, and for any $X_F \in \mathcal{X}_F$, let $j_F : X_F \to X^\star$, the inclusion mapping of X_F into X, and $j_F^\star : X^\star \to X_F^\star$ its dual map, which is the projection of $X^\star$ onto $X_F^\star$, and define the operator $A_F : X_F \to 2^{X_F^\star}$ by $A_F := j_F^\star A j_F$. By property (i) for the operator A, for every $x \in X_F$, $A(x)$ is a nonempty, convex, and closed (hence weakly compact) subset of $X^\star$. Furthermore, $j_F^\star$ is continuous from the weak topology on $X^\star$ to $X_F^\star$.[12] By property (ii), A is upper semicontinuous from X_F to $2^{X^\star}$, $X^\star$ endowed with the weak topology, therefore, A_F is upper semicontinuous from X_F to $2^{X_F^\star}$. Finally, A_F inherits the coercivity property from A since for any pair $(x, x_F^\star) \in \mathbf{Gr}(A_F)$, there exists some $x^\star \in A(x)$ such that $x_F^\star = j_F^\star x^\star$ and

$$\langle x_F^\star, x\rangle_{X_F^\star, X_F} = \langle j_F^\star x^\star, x\rangle_{X_F^\star, X_F} = \langle x^\star, x\rangle \geq c(\|x\|)\|x\|.$$

By Proposition 9.5.8, it follows that $\mathbf{R}(A_F) = X_F^\star$ for all $X_F \in \mathcal{X}_F$.

Hence, for any $X_F \in \mathcal{X}_F$, there exists a $x_F \in X_F$ such that $0 \in A_F(x_F)$, which in turn implies that $0 = j_F^\star x_F^\star$ for some $x_F^\star \in A(x_F)$ with $x_F^\star \in X_F^\star$. The set $\{x_F : X_F \in \mathcal{X}_F\}$ is uniformly bounded by the coercivity of A. Let $c > 0$ be this uniform bound. For any $X_F \in \mathcal{X}_F$, define the set

$$U(X_F) := \bigcup \{x_{F'} : 0 \in A(x_{F'}) \text{ with } \mathcal{X}_F \ni X_{F'} \supset X_F\}.$$

By the uniform boundedness of the sets $\{x_F : X_F \in \mathcal{X}_F\}$, it is seen that $U(X_F)$ is contained in the closed ball $\overline{B}(0, c)$ of X, which is weakly compact. The family $\{\overline{U(X_F)}^w : X_F \in \mathcal{X}_F\}$ has the finite intersection property (see Proposition 1.9.10), therefore, $\bigcap_{X_F \in \mathcal{X}_F} \overline{U(X_F)}^w \neq \emptyset$.

Let $x_o \in \bigcap_{X_F \in \mathcal{X}_F} \overline{U(X_F)}^w$ and $x \in X$, arbitrary. We will show that $0 \in A(x_o)$. Let $X_F \in \mathcal{X}_F$ such that $\{x_o, x\} \subset X_F$. By Proposition 1.1.62, there exists a sequence $\{x_{F_k}\} \subset U(X_F)$ such that $x_{F_k} \rightharpoonup x_o$ in X. For each k, $0 = j_{F_k}^\star x_{F_k}^\star$, so

$$\langle x_{F_k}^\star, x_{F_k} - x_o\rangle = 0,$$

therefore, taking the upper limit yields $\limsup_k \langle x_{F_k}^\star, x_{F_k} - x_o\rangle = 0$. By the pseudomonotonicity of A, there exists $x^\star = x^\star(x) \in A(x_o)$ such that

12 Since $X_F^\star$ is finite dimensional, the weak and the strong topology coincide.

$$0 = \liminf_k \langle x^\star_{F_k}, x_{F_k} - x \rangle \ge \langle x^\star(x), x_o - x \rangle \tag{9.66}$$

Assume now that $0 \notin A(x_o)$. Then, we may apply the separation theorem to the sets $\{0\}$ and $A(x_o)$ and find $\bar{x} \in X$ such that $0 < \inf_{x^\star \in A(x_o)} \langle x^\star, x_o - \bar{x} \rangle \le \langle x^\star_o, x_o - \bar{x} \rangle$, which is in contradiction with (9.66). □

In the proof of the main existence theorem for maximal monotone operators, we have used the fact that a maximal monotone operator is pseudomonotone. So, it is clear that Theorem 9.5.9 can be generalized as follows:

Theorem 9.5.10. *Let X be a reflexive Banach space, $A : X \to 2^{X^\star}$ be a maximal monotone operator and $B : X \to X^\star$ be a pseudomonotone, bounded, and coercive operator with $\mathbf{D}(B) = X$. Then $\mathbf{R}(A + B) = X^\star$.*

Take $A : X \to 2^{X^\star}$ to be the operator $A(x) = 0$ for every $x \in X$. Then from the above theorem, we have the following result:

Corollary 9.5.11. *Let X be a reflexive Banach space and $A : X \to X^\star$ be a pseudomonotone, bounded, and coercive operator with $\mathbf{D}(A) = X$. Then $\mathbf{R}(A) = X^\star$.*

9.6 Applications of monotone type operators

The surjectivity theorems for monotone type operators find important applications in the theory of partial differential equations and variational inequalities.

9.6.1 Quasilinear elliptic equations I: using maximal monotonicity

Let $\mathcal{D} \subseteq \mathbb{R}^d$ be a bounded domain with smooth boundary $\partial\mathcal{D}$. We consider the quasilinear elliptic boundary value problem, which consists, given $\lambda \ge 0$ and $f \in L^{p^\star}(\mathcal{D})$, of finding $u \in X = W^{1,p}_0(\mathcal{D})$ such that (omitting explicit x dependence for simplicity)

$$\begin{aligned} -\operatorname{div}(|\nabla u|^{p-2}\nabla u) + \lambda u &= f \quad \text{on } \mathcal{D},\ \lambda \ge 0, \\ u &= 0 \quad \text{on } \partial\mathcal{D}, \end{aligned} \tag{9.67}$$

or its weak formulation

$$\int_{\mathcal{D}} |\nabla u|^{p-2}\nabla u \cdot \nabla v dx + \lambda \int_{\mathcal{D}} uv dx = \int_{\mathcal{D}} fv dx, \quad \forall v \in X. \tag{9.68}$$

This is an extension of the p-Laplace Poisson equation (see Section 6.8 and Example 9.3.10), involving the perturbed operator $A = -\Delta_p + \lambda I$ for $\lambda \ge 0$.

We will bring (9.68) into an appropriate abstract operator form, which will allow for the use of surjectivity results for maximal monotone operators.

As above, we define the operator $\mathsf{A} : X \to X^\star$ and the functional $b \in X^\star$ by

$$\begin{aligned} \langle \mathsf{A}(u), v\rangle_{X^\star,X} &:= \int_{\mathcal{D}} |\nabla u|^{p-2}\nabla u \cdot \nabla v dx + \lambda \int_{\mathcal{D}} uv dx, \quad \forall\, u, v \in X, \\ \langle b, v\rangle_{X^\star,X} &:= \int_{\mathcal{D}} f v dx, \quad \forall\, v \in X. \end{aligned} \tag{9.69}$$

The following proposition (see e. g., [128]) provides an existence result.

Proposition 9.6.1. *For $p \geq \frac{2d}{d+2}$, $p \in (1,\infty)$, and $f \in L^{p^\star}(\mathcal{D})$, problem* (9.67) *has a unique weak solution.*

Proof. The proof is broken up into 4 steps.

1. For $p \geq \frac{2d}{d+2}$, the operator A and the functional b (see (9.69)) are well defined so that (9.68) is equivalent to the operator equation $\mathsf{A}u = b$ in $X^\star$.

To show the above, we note (using Hölder's inequality) that for any $u, v \in X$,

$$\begin{aligned} |\langle \mathsf{A}(u), v\rangle_{X^\star,X}| &\leq \int_{\mathcal{D}} |\nabla u|^{p-1}|\nabla v| dx + \lambda \int_{\mathcal{D}} |uv| dx \\ &\leq \Big(\int_{\mathcal{D}} |\nabla u|^{(p-1)p^\star} dx\Big)^{1/p^\star}\Big(\int_{\mathcal{D}} |\nabla v|^p dx\Big)^{1/p} + \lambda\Big(\int_{\mathcal{D}} |u|^2 dx\Big)^{1/2}\Big(\int_{\mathcal{D}} |v|^2 dx\Big)^{1/2} \\ &= \|\nabla u\|^{p-1}_{L^p(\mathcal{D})}\|\nabla v\|_{L^p(\mathcal{D})} + \lambda\|u\|_{L^2(\mathcal{D})}\|v\|_{L^2(\mathcal{D})}. \end{aligned}$$

For $p \geq \frac{2d}{d+2}$, we have the embedding $X = W^{1,p}(\mathcal{D}) \hookrightarrow L^2(\mathcal{D})$ (see Theorem 1.5.11) so that for all $v \in X$ it holds that $\|v\|_{L^2(\mathcal{D})} \leq c_1\|v\|_X = c_1\|\nabla v\|_{L^p(\mathcal{D})}$ for an appropriate constant $c_1 > 0$. It then follows that

$$\begin{aligned} |\langle \mathsf{A}(u), v\rangle_{X^\star,X}| &\leq \|\nabla u\|^{p-1}_{L^p(\mathcal{D})}\|\nabla v\|_{L^p(\mathcal{D})} + \lambda\|u\|_{L^2(\mathcal{D})}\|v\|_{L^2(\mathcal{D})} \\ &\leq c(\|\nabla u\|^{p-1}_{L^p(\mathcal{D})} + \lambda\|\nabla u\|_{L^p(\mathcal{D})})\|\nabla v\|_{L^p(\mathcal{D})} \end{aligned}$$

for a suitable constant $c > 0$. Therefore, since $\|\nabla v\|_{L^p(\mathcal{D})} = \|v\|_X$, we have

$$\|\mathsf{A}(u)\|_{X^\star} = \sup_{v\in X,\ \|v\|_X\leq 1} |\langle \mathsf{A}u, v\rangle_{X^\star,X}| \leq c(\|\nabla u\|^{p-1}_{L^p(\mathcal{D})} + \lambda\|\nabla u\|_{L^p(\mathcal{D})}),$$

which implies that $\mathsf{A}(u) \in X^\star$, and the operator $\mathsf{A}; X \to X^\star$ is bounded.

Now, by Hölder's inequality, we have

$$|\langle b, v\rangle_{X^\star,X}| \leq \|f\|_{L^{p^\star}(\mathcal{D})}\|v\|_{L^p(\mathcal{D})} \leq c_2\|f\|_{L^{p^\star}(\mathcal{D})}\|v\|_X,$$

since $X := W^{1,p}(\mathcal{D}) \hookrightarrow L^p(\mathcal{D})$, i.e, $\|v\|_{L^p(\mathcal{D})} \leq c_2\|v\|_X$. Therefore, $\|b\|_{X^\star} \leq c_2\|f\|_{L^p(\mathcal{D})}$ so that $b \in X^\star$.

It then follows that the weak formulation (9.68) is equivalent to the operator equation $\mathsf{A}u = b$ in $X^\star$. Note that in the case that $\lambda = 0$, the restriction $p \geq \frac{2d}{d+2}$ is not necessary.

2. The operator A is continuous and coercive. To show continuity, let $(u_n)_{n\in\mathbb{N}} \subset X$ be a sequence such that $u_n \to u$ in X. Then, $\nabla u_n \to \nabla u$ in $L^p(\mathcal{D})$ so that if $\phi(s) = |s|^{p-2}s$, then $\phi(\nabla u_n) \to \phi(\nabla u)$ in $L^{p^*}(\mathcal{D})$. For a suitable constant $c > 0$,

$$\begin{aligned}\langle \mathsf{A}(u_n) - \mathsf{A}(u), v\rangle_{X^\star,X} &= \int_{\mathcal{D}} (\phi(\nabla u_n) - \phi(\nabla u))\nabla v dx + \lambda \int_{\mathcal{D}} (u_n - u) v dx \\ &\leq \|\phi(\nabla u_n) - \phi(\nabla u)\|_{L^{p^*}(\mathcal{D})} \|\nabla v\|_{L^p(\mathcal{D})} + \lambda \|u_n - u\|_{L^2(\mathcal{D})} \|v\|_{L^2(\mathcal{D})} \\ &\leq \|\phi(\nabla u_n) - \phi(\nabla u)\|_{L^{p^*}(\mathcal{D})} \|\nabla v\|_{X} + c\|u_n - u\|_X \|v\|_X,\end{aligned}$$

where we used the Sobolev embeddings once more, which in turn, since $\|\nabla v\|_{L^p(\mathcal{D})} = \|v\|_X$, implies that

$$\|\mathsf{A}(u_n) - \mathsf{A}(u)\|_{X^\star} \leq \|\phi(\nabla u_n) - \phi(\nabla u)\|_{L^{p^*}(\mathcal{D})} + c\|u_n - u\|_X.$$

Therefore, $\mathsf{A}(u_n) \to \mathsf{A}(u)$, i. e., A is continuous.

To show coercivity, note that

$$\langle \mathsf{A}(u), u\rangle_{X^\star,X} = \|u\|_X^p + \lambda \|u\|^2_{L^2(\mathcal{D})} \geq \|u\|_X^p,$$

which implies that $\frac{\langle \mathsf{A}(u),u\rangle_{X^\star,X}}{\|u\|_X} \geq \|u\|_X^{p-1} \to \infty$, as $\|u\|_X \to \infty$ if $p > 1$.

3. The operator A is strictly monotone. Indeed, (simplifying notation $\langle\cdot,\cdot\rangle_{X^\star,X}$ to $\langle\cdot,\cdot\rangle$ and $\|\cdot\|_X$ to $\|\cdot\|$), we have for every $u_1, u_2 \in X$,

$$\begin{aligned}\langle \mathsf{A}(u_1) - \mathsf{A}(u_2), u_1 - u_2\rangle &= \langle \mathsf{A}(u_1), u_1\rangle + \langle \mathsf{A}(u_2), u_2\rangle - \langle \mathsf{A}(u_1), u_2\rangle - \langle \mathsf{A}(u_2), u_1\rangle \\ &\geq \|u_1\|^p + \|u_2\|^p - \|u_1\|^{p-1}\|u_2\| - \|u_2\|^{p-1}\|u_2\| + \lambda \int_{\mathcal{D}} (u_1 - u_2)^2 dx \\ &= (\|u_1\|^{p-1} - \|u_2\|^{p-1})(\|u_1\| - \|u_2\|) + \lambda \int_{\mathcal{D}} (u_1 - u_2)^2 dx,\end{aligned}$$

and the strict monotonicity of A from the elementary inequality, stating that for some $c > 0$,[13]

$$(a|a|^{p-2} - b|b|^{p-2})(a - b) \geq c|a - b|^p, \quad \forall\, a, b \in \mathbb{R},\ p \geq 2.$$

4. By Corollary 9.4.9, A is maximal monotone. Unique solvability follows by a straightforward application of the surjectivity theorem for maximal monotone operators (Theorem 9.4.15). □

---

13 To prove this inequality, note that if $0 \leq b \leq a$, then $a^{p-1} - b^{p-1} = \int_0^{a-b} (p-1)(t+b)^{p-2} dt \geq \int_0^{a-b} (p-1) t^{p-2} dt = (a-b)^{p-1}$. In the case $b \leq 0 \leq a$, we have $a^{p-1} + |b|^{p-1} \geq c(a + |b|)^{p-1}$ with $c = 2^{2-p}$.

9.6.2 Quasilinear elliptic equations II: using pseudomonotonicity

We now revisit problem (9.67) for more general perturbations of the p-Laplacian than λI for $\lambda \geq 0$. In this case, the resulting operator is not necessarily maximal monotone and an alternative approach using pseudomonotonicity is required.

For $\mathcal{D} \subset \mathbb{R}^d$, a bounded domain with sufficiently smooth boundary, we consider the problem of finding $u \in X = W_0^{1,p}(\mathcal{D})$ such that

$$\begin{aligned} -\operatorname{div}(|\nabla u|^{p-2}\nabla u) + g(u) = f \quad & \text{on } \mathcal{D}, \\ u = 0 \quad & \text{on } \partial\mathcal{D}, \end{aligned} \tag{9.70}$$

for $g : \mathbb{R} \to \mathbb{R}$ and $f \in L^{p^*}(\mathcal{D})$ given. In the special case that $g(u) = \lambda u$, for $\lambda \geq 0$, (9.70) reduces to (9.67) and the approach using maximal monotonicity can be used. However, for $g(u) = \lambda u$, with $\lambda \in \mathbb{R}$, or for even more general nonlinearities g, the approach using maximal monotonicity may not be appropriate. On the other hand, the more general framework of pseudomonotone operators is applicable.

We impose the following conditions on the nonlinearity:

Assumption 9.6.2. The function $g : \mathbb{R} \to \mathbb{R}$ satisfies the following:
(i) Continuity: g is continuous.
(ii) Coercivity: $\inf_{s\in\mathbb{R}} g(s)s > -\infty$.
(iii) Growth condition: $|g(s)| \leq c_0 + c_1|s|^{r-1}$ for $r < p_c$, where $p_c = p_c(p,d)$ is the critical exponent required for the Sobolev embedding theorem to hold.

We define the operator $\mathsf{A} : X \to X^*$ by

$$\begin{aligned} \langle \mathsf{A}(u), v\rangle_{X^*,X} &:= \langle \mathsf{A}_0(u), v\rangle_{X^*,X} + \langle \mathsf{A}_1(u), v\rangle_{X^*,X} \\ &:= \int_{\mathcal{D}} |\nabla u|^{p-2}\nabla u \cdot \nabla v dx + \int_{\mathcal{D}} g(u) v dx, \quad \forall\, u, v \in X, \end{aligned} \tag{9.71}$$

and the functional $b \in X^*$ by

$$\langle b, v\rangle_{X^*,X} := \int_{\mathcal{D}} f v dx, \quad \forall\, v \in X. \tag{9.72}$$

By inspection of (9.71), we see that $\mathsf{A}_0 = \mathsf{L}_p$ (the p- Laplace operator) and $\mathsf{A}_1 = \mathsf{N}_g$ (the Nemitskii operator generated by the function g), both considered as operators from $X = W_0^{1,p}(\mathcal{D})$ to $X^* = W^{-1,p^*}(\mathcal{D})$. For the p-Laplace operator and its variational formulation, see Section 6.8, and for the Nemitskii operator and its various interpretations (see Section 6.10.5 and in particular Example 6.10.8).

In terms of the above, the quasilinear equation (9.70) can be brought into the abstract formulation

$$\mathsf{A}(u) := (\mathsf{A}_0 + \mathsf{A}_1)(u) = b, \quad \text{in } X^*, \tag{9.73}$$

which can be treated in terms of the surjectivity theorems for monotone type operators. As noted above, A_0 is maximal monotone, but A may fail to be so because of the general perturbation A_1, by the Nemitskii operator. However, making use of the stability with respect to summation theorem for pseudomonotone operators, we can still rely on the more relaxed notion of pseudomonotonicity and the corresponding surjectivity properties for the perturbed operator $\mathsf{A}_0 + \mathsf{A}_1$.

The following proposition (see e. g., [128]) addresses existence of solutions.

Proposition 9.6.3. *Let Assumption* 9.6.2 *hold and* $p \in (1, \infty)$. *Then, for every* $f \in X^* = W^{-1,p^*}(\mathcal{D})$, *the quasilinear equation* (9.70) *admits a solution* $u \in X = W_0^{1,p}(\mathcal{D})$.

Proof. The proof proceeds in 3 steps.

1. We first establish the pseudomonotonicity property for A.

We have already established (see Examples 9.3.3 and 9.4.10) that the operator A_0 is strictly monotone, continuous, and bounded. Hence, by Proposition 9.5.3, it is pseudomonotone. The operator $\mathsf{A}_1 = \mathsf{N}_g$, considered as an operator $\mathsf{N}_g : X \to X^*$ under the growth Assumption 9.6.2(ii), is completely continuous and bounded (see Section 6.10.5 and in particular Example 6.10.8), hence, (by Proposition 9.5.3) it is pseudomonotone. Hence, $\mathsf{A} = \mathsf{A}_0 + \mathsf{A}_1$ is pseudomonotone as well (Proposition 9.5.3).

2. We now establish coercivity of A.

The coercivity of A_0 has already been established (see the proof of Proposition 9.6.1, step 2, setting $\lambda = 0$). On the other hand, the contribution of A_1 to A does not affect coercivity. Indeed, there exists $c > 0$ such that

$$\langle \mathsf{A}_1(u), u \rangle = \int_{\mathcal{D}} g(u(x))u(x)dx \geq -c > -\infty,$$

because of the coercivity assumption 9.6.2(ii). Coercivity of A follows by

$$\frac{\langle \mathsf{A}(u), u \rangle}{\|u\|} = \frac{\langle \mathsf{A}_0(u), u \rangle}{\|u\|} - \frac{c}{\|u\|} \geq \|u\|^{p-1} - \frac{c}{\|u\|} \to \infty, \quad \text{as } \|u\| \to \infty.$$

3. Since A is pseudomonotone and coercive, using the surjectivity Theorem 9.5.9, the solvability of (9.73) follows, and hence of (9.70). □

9.6.3 Semilinear elliptic inclusions

Let $\mathcal{D} \subset \mathbb{R}^d$ be a bounded domain with C^2 boundary[14] $\partial\mathcal{D}$, and let $\beta : \mathbb{R} \to 2^{\mathbb{R}}$ be a maximal monotone map. Given $f \in L^2(\mathcal{D})$, we consider the following nonlinear Dirichlet problem:

$$\begin{aligned} -\Delta u + \beta(u) \ni f \quad &\text{in } \mathcal{D}, \\ u = 0, \quad &\text{on } \partial\mathcal{D}. \end{aligned} \tag{9.74}$$

Let $H = L^2(\mathcal{D})$, and $\mathsf{A} : \mathbf{D}(\mathsf{A}) \subset H \to H$, with $\mathbf{D}(\mathsf{A}) = W_0^{1,2}(\mathcal{D}) \cap W^{2,2}(\mathcal{D})$. Also, let B be the realization of β in H, i. e., $\mathsf{B}(u) = \{v \in H : v(x) \in \beta(u(x))\ a.e\}$. We know that B is maximal monotone (see Example 9.4.3). Without loss of generality, we may assume that β is the subdifferential of a proper convex lower semicontinuous function $\varphi : \mathbb{R} \to \mathbb{R} \cup \{+\infty\}$, i. e., $\beta = \partial\varphi$. We will need the resolvent of the maximal monotone operator (graph) β and β_λ, its Yosida approximation (see Definition 9.4.21 and Example 9.4.23). For the ease of the reader, we recall it here: For any $s \in \mathbb{R}$, we have that $\mathsf{R}_\lambda^\beta(s) = R(\lambda, \beta) = y$, where y is the unique solution of the equation $y + \lambda\beta(y) = s$, whereas, $\beta_\lambda(s) = \frac{1}{\lambda}(s - y)$. Let

$$\varphi_\lambda(s) := \inf_{z \in \mathbb{R}} \left\{ \frac{1}{2\lambda} \|s - z\|^2 + \varphi(z) \right\},$$

so, by Theorem 9.4.29, it holds that $\beta_\lambda = \partial\varphi_\lambda$.

We need to introduce the following lemma:

Lemma 9.6.4. *For every $u \in \mathbf{D}(\mathsf{A})$ and $\lambda > 0$, it holds that* $\langle \mathsf{A}u, \mathsf{B}_\lambda(u) \rangle \geq 0$.

Proof. We have that

$$\langle \mathsf{A}u, \mathsf{B}_\lambda(u) \rangle = -\int_{\mathcal{D}} \Delta u(x) \beta_\lambda(u(x)) dx = \int_{\mathcal{D}} |\nabla u(x)|^2 \frac{d}{ds} \beta_\lambda(u(x)) dx.$$

Since $\beta_\lambda(u(x)) = \partial\varphi_\lambda(u(x))$, and because φ is a lower semicontinuous and convex function on $\mathbb{R}$, we have $\frac{d}{ds}\beta_\lambda(u(x)) = \frac{d^2}{ds^2}\phi_\lambda(u(x)) \geq 0$ a. e. Therefore, $\langle \mathsf{A}u, \mathsf{B}_\lambda(u) \rangle \geq 0$ for every $u \in \mathbf{D}(\mathsf{A})$ and $\lambda > 0$. □

The following proposition (see e. g., [18, 19]) deals with existence results for the inclusion.

Proposition 9.6.5. *If $f \in L^2(\mathcal{D})$ and $\beta : \mathbb{R} \to 2^{\mathbb{R}}$ is maximal monotone, then* (9.74) *has a unique solution $u \in W_0^{1,2}(\mathcal{D}) \cap W^{2,2}(\mathcal{D})$.*

14 This hypothesis is needed for reasons related to the regularity of the Poisson equation and the definition of the Laplace operator.

Proof. Let $H = L^2(\mathcal{D})$ and $\mathsf{B} : H \to 2^H$ be the realization of β on H, which is maximal monotone (see Example 9.4.3). Also define $\mathsf{A} : \mathbf{D}(\mathsf{A}) \subset H \to H$ by $\mathsf{A}u = -\Delta u$ with $\mathbf{D}(\mathsf{A}) = W_0^{1,2}(\mathcal{D}) \cap W^{2,2}(\mathcal{D})$. From Lemma 9.6.4, we know that $\langle \mathsf{A}u, \mathsf{B}_\lambda(u)\rangle \geq 0$ for every $u \in \mathbf{D}(\mathsf{A})$ and every $\lambda > 0$. So by Theorem 9.4.32, $\mathsf{A} + \mathsf{B}$ is maximal monotone. Let $\lambda_1 > 0$ be the first eigenvalue of A. Recalling the variational characterization of λ_1 (Rayleigh quotient), we have that $\mathsf{A} - \lambda_1 I$ is monotone. Hence, $\mathsf{A} + \mathsf{B} - \lambda_1 I$ is monotone. But $\mathbf{R}(\mathsf{A} + \mathsf{B} - \lambda_1 I + (1 + \lambda_1)I) = \mathbf{R}(\mathsf{A} + \mathsf{B} + I) = H$, because $\mathsf{A} + \mathsf{B}$ is maximal monotone. Therefore, $\mathsf{A} + \mathsf{B} - \lambda_1 I$ is maximal monotone too, and so $R(\mathsf{A} + \mathsf{B}) = \mathbf{R}(\mathsf{A} + \mathsf{B} - \lambda_1 I + \lambda_1 I) = H$. The uniqueness of the solution is clear. □

Example 9.6.6 (The porous medium equation). Consider the porous medium equation

$$-\Delta(w^\alpha) + g(w) = f, \quad \text{in } \mathcal{D}, \tag{9.75}$$

subject to homogeneous Dirichlet boundary conditions, where $\alpha > 1$. This is a degenerate elliptic equation, since the action of the Laplacian operator is incapacitated on the boundary. However, one may consider the Kirchhoff transformation to a new variable $u = w^\alpha$ and express (9.75) in terms of

$$-\Delta u + \beta(u) = f, \quad \text{in } \mathcal{D}, \tag{9.76}$$

where $\beta(u) = g(u^{1/\alpha})$. If β is maximal monotone, then an application of Proposition 9.6.5 provides solvability of (9.75). ◁

9.6.4 Variational inequalities with monotone type operators

Let X be a reflexive Banach space, $\mathsf{A} : X \to 2^{X^\star}$ a maximal monotone operator, and $C \subset \mathbf{D}(\mathsf{A}) \subset X$ a convex closed set. Given $\mathsf{x}^\star \in X^\star$, we consider the variational inequality of finding $\mathsf{x}_o \in C$ such that there exists $\mathsf{x}_o^\star \in \mathsf{A}(\mathsf{x}_o)$ with the property

$$\langle \mathsf{x}_o^\star - \mathsf{x}^\star, \mathsf{z} - \mathsf{x}_o \rangle \geq 0, \quad \forall\, \mathsf{z} \in C. \tag{9.77}$$

Variational inequalities can be reformulated in terms of inclusions related to monotone type operators and, starting from that, the abstract theory can be applied for their study. The following propositions (in the spirit of e. g., [18, 31, 101] are in this direction.

Proposition 9.6.7. *Suppose that either* $\operatorname{int}(C) \neq \emptyset$ *or* $\operatorname{int}(\mathbf{D}(\mathsf{A})) \cap C \neq \emptyset$. *Then the variational inequality* (9.77) *is equivalent to the inclusion*

$$\mathsf{x}_o \in C : \mathsf{x}^\star \in \mathsf{A}(\mathsf{x}_o) + \partial I_C(\mathsf{x}_o). \tag{9.78}$$

Proof. Assume that x_o solves the inclusion (9.78). Then, there exists $\mathsf{x}_o^\star \in \mathsf{A}(\mathsf{x}_o)$ such that $\mathsf{x}^\star - \mathsf{x}_o^\star \in \partial I_C(\mathsf{x}_o)$, and using the definition of the subdifferential, we see that x_o is indeed a solution of the variational inequality (9.77). Note that for this direction, we do not require the conditions $\operatorname{int}(C) \neq \emptyset$ or $\operatorname{int}(\mathbf{D}(\mathsf{A})) \cap C \neq \emptyset$.

For the converse, assume that x_o and $x_o^\star \in \mathsf{A}(x_o)$ solve the variational inequality (9.77). By monotonicity, we have that for every $z \in C$ and $z^\star \in \mathsf{A}(z)$, it holds that

$$0 \leq \langle z^\star - x_o^\star, z - x_o \rangle = \langle z^\star - x^\star, z - x_o \rangle + \langle x^\star - x_o^\star, z - x_o \rangle,$$

which, using (9.77), implies

$$\langle z^\star - x^\star, z - x_o \rangle \geq 0. \tag{9.79}$$

Consider now any $\bar{z}^\star \in \partial I_C(z)$. By the definition of the subdifferential, since $x_o \in C$, this yields

$$\langle \bar{z}^\star, z - x_o \rangle \geq 0, \tag{9.80}$$

so that adding (9.79) and (9.80), we conclude that

$$\langle z^\star + \bar{z}^\star - x^\star, z - x_o \rangle \geq 0, \quad \forall z \in C,\ z^\star \in \mathsf{A}(z),\ \bar{z}^\star \in I_C(z). \tag{9.81}$$

Since either $\operatorname{int}(C) \neq \emptyset$ or $\operatorname{int}(\mathbf{D}(\mathsf{A})) \cap C \neq \emptyset$, we have, by Theorem 9.4.31, that $\mathsf{B} := \mathsf{A} + \partial I_C$ is a maximal monotone operator so that (9.81) implies that $x^\star \in \mathsf{B}(x_o) = \mathsf{A}(x_o) + I_C(x_o)$. □

Using the restatement of the variational inequality (9.77) in terms of the inclusion (9.78), we may provide an existence result for (9.77).

Proposition 9.6.8. *Let X be a reflexive Banach space. Assume that $\mathsf{A} : \mathbf{D}(\mathsf{A}) \subset X \to 2^{X^\star}$ is maximal monotone and coercive. Suppose, furthermore, that either $\operatorname{int}(C) \neq \emptyset$ or $\operatorname{int}(\mathbf{D}(\mathsf{A})) \cap C \neq \emptyset$. Then the variational inequality (9.77) admits a solution.*

Proof. By Proposition 9.6.7, it suffices to consider the inclusion (9.78). As above, the operator $\mathsf{B} := \mathsf{A} + \partial I_C$ is maximal monotone, and as B inherits the coercivity property from A, the surjectivity Theorem 9.4.15 yields existence. □

Pseudomonotone operators find important applications in variational inequalities. Let X be a reflexive Banach space, $C \subset X$ closed and convex, and $\mathsf{A} : X \to X^\star$ a nonlinear operator. For any $x^\star \in X$, we consider the variational inequality

$$\text{find } x_o \in C : \langle \mathsf{A}(x_o) - x^\star, x - x_o \rangle \geq 0, \quad \forall x \in C. \tag{9.82}$$

We make the following assumptions on the operator A:

Assumption 9.6.9. The operator $\mathsf{A} : X \to X^\star$ satisfies the following conditions:
(i) A is pseudomonotone.
(ii) There exist constants $c_1, c_2 > 0$ such that $\langle \mathsf{A}(x), x \rangle \geq c_1 \|x\|^2 - c_2$ for every $x \in X$.
(iii) There exist constants $c_3, c_4 > 0$ such that $\|\mathsf{A}(x)\|_{X^\star} \leq c_3 + c_4 \|x\|$ for every $x \in X$.

Proposition 9.6.10. *Let X be a reflexive Banach space, $C \subset X$ closed and convex, and $A : X \to X^\star$ satisfy Assumption* 9.6.9. *Then the variational inequality* (9.82) *admits a solution for any $x^\star \in X^\star$.*

Proof. We first observe that (9.82) is equivalent to the operator equation

$$A(x_o) + \partial I_C(x_o) \ni x^\star,$$

hence, it is sufficient to show that the multivalued operator $A + B : X \to 2^{X^\star}$, where $B = \partial I_C$, is surjective. The operator A is pseudomonotone, bounded, and coercive (by Assumption 9.6.9), while B is a maximal monotone operator. Hence, using Theorem 9.5.10 (with the roles of A and B interchanged), we obtain the required solvability result. □

9.7 Numerical schemes for finding zeros of monotone operators

The theory of monotone type operators can be used for the construction of concrete numerical iterative schemes (see, e. g., [21]) for the solution of operator equations (or inclusions) of the form

$$0 \in A(x), \quad A \text{ of monotone type.} \tag{9.83}$$

Example 9.7.1. An important special class of problems, such as (9.83), is the case that $A = A_1 + A_2$, the sum of two monotone type operators. Examples of such equations can be found in, e. g.,

- In the solution of optimization problems of the type $\min_{x\in X}(\varphi_1(x) + \varphi_2(x))$, where φ_i, $i = 1, 2$ are proper convex lower semicontinuous functions (see Section 4.5). In this case, $A = \partial\varphi_1 + \partial\varphi_2$, which is a monotone type operator. Methods such as the proximal method or operator splitting methods were developed and studied in this context there.
- In the solution of variational inequalities (VIs) of the type: Find $x_o \in C$ such that $\langle F(x_o), x - x_o\rangle \geq 0$ for every $x \in C$, where C is a closed convex subset of a Hilbert space X (see Section 7.5.3). Then, $A = \rho i_X^{-1} F + N_C$, where i_X is the Riesz isometry and N_C is the normal cone operator for C (see also Proposition 9.6.7). The proximal method or operator splitting were extended from their minimization context to that of VIs in Section 7.5.3. ◁

Clearly, there is a common thread between the above two examples, even though, for pedagogical reasons, they were presented separately. We will show that these are special cases of general methods for the solution of operator inclusions, such as (9.83). For simplicity, we let X be a Hilbert space.

9.7.1 Proximal methods

The proximal method for solving (9.83) is inspired by the iterative scheme

$$\frac{1}{\gamma}(x_n - x_{n+1}) \in A(x_{n+1}) \iff x_{n+1} = R(\gamma, A)(x_n), \quad \gamma > 0,$$

where, by $R(\gamma, A) = (I + \gamma A)^{-1}$, we denote the resolvent operator for A. If A is maximal monotone, the operator $R(\gamma, A)$ is well defined and enjoys nice properties (see Section 9.4.6). This method, or rather its variable step version

$$x_{n+1} = R(\gamma_n, A)(x_n) = (I + \gamma_n A)^{-1}(x_n), \quad \gamma_n > 0, \tag{9.84}$$

is called the proximal method for the solution of the operator equation (or inclusion) (9.83). It can be easily seen that it is a generalization of the methods presented in Example 9.7.1 (recall the monotonicity properties of the subdifferential or the normal cone operator). The convergence of (9.84) can be covered by Proposition 4.5.2 (simply replacing the proximal operator $\text{prox}_{\lambda_n \varphi}$ by the resolvent operator $R(\gamma_n, A)$).[15] The reader is encouraged to repeat the proof of Proposition 4.5.2 for this case as an exercise.

9.7.2 Splitting methods

In cases where A can be expressed in terms of $A = A_1 + A_2$, with A_i, $i = 1, 2$, enjoying "nice" properties (see, e. g., Example 9.7.1), then a common practice is to "split" the problem (9.83) into two problems of the form

$$\begin{aligned} x - z &\in A_1(x), \\ z - x &\in A_2(x). \end{aligned}$$

We will then treat the above as a system, which will provide the pair (x, z), with x being the solution we are looking for. Among the reasons for doing so is if the resolvent operator for A_1 or A_2 alone is easier to calculate the full resolvent operator for A. Various splitting schemes are possible, with the forward-backward and Douglas–Rachford scheme being among the more popular ones.

9.7.2.1 Splitting: forward-backward scheme

In this splitting, assuming that A_1 is single valued, for any $\gamma > 0$, we express

$$A_1 + A_2 = \gamma^{-1}((I + \gamma A_2) - (I - \gamma A_1)), \quad \gamma > 0,$$

15 Recall also that if $A = \partial\varphi$, then $R(\gamma, A)$ is related to $\text{prox}_{\gamma\varphi}$ (see Example 9.4.23).

and thus split problem (9.83) into the system

$$\begin{cases} z \in (I + \gamma A_2)(x), \\ z = (I - \gamma A_1)(x) \end{cases} \Longleftrightarrow \begin{cases} x = R(\gamma, A_2)(z), \\ z = (I - \gamma A_1)(x). \end{cases}$$

This leads to the iterative scheme

$$x_{n+1} = R(\gamma, A_2)(I - \gamma A_1)(x_n), \quad \gamma > 0,$$

or its relaxed version

$$\begin{aligned} z_n &= (I - \gamma A_1)(x_n), \\ x_{n+1} &= x_n + \lambda_n (R(\gamma, A_2)(z_n) - x_n), \end{aligned}$$

for $\gamma > 0$ and suitable relaxation parameters λ_n. It can easily be checked that the algorithms obtained are generalizations of those referred to in Example 9.7.1 for the particular cases where $A_1 = \partial\varphi_1$, $A_2 = \partial\varphi_2$ and $A_1 = j^{-1}F$, $A_2 = N_C$, analyzed in Sections 4.5.2.1 and 7.5.3, respectively. The convergence of the algorithm can be proved using similar steps as those for the proof of Proposition 4.5.7.

9.7.2.2 Splitting: Douglas–Rachford scheme

The Douglas–Rachford scheme is another popular spliting scheme, using the Cayley reflection operator

$$C(\gamma, A) := 2R(\gamma, A) - I, \quad \gamma > 0.$$

If A is a maximal monotone operator, then it can be easily shown that $C(\gamma, A)$ is nonexpansive. Moreover, it can be shown by direct computation that

$$(I - \gamma A) = C(\gamma, A)(I + \gamma A), \quad \gamma > 0.$$

With this information at hand, we can now use the splitting

$$\gamma A = \gamma(A_1 + A_2) = (I + \gamma A_1) - (I - \gamma A_2) = (I + \gamma A_1) - C(\gamma, A_2)(I + \gamma A_2)$$

so that we decompose $0 \in A(x)$ as

$$\begin{cases} z \in (I + \gamma A_2)(x), \\ y = C(\gamma, A_2)(z), \\ y \in (I + \gamma A_1)(x) \end{cases} \iff \begin{cases} x = R(\gamma, A_2)(z), \\ y \in C(\gamma, A_2)(z), \\ x = R(\gamma, A_1)(y) \end{cases}$$

$$\iff \begin{cases} x = R(\gamma, A_2)(z), \\ R(\gamma, A_2)(z) = R(\gamma, A_1)(C(\gamma, A_2)(z)), \end{cases}$$

$$\iff \begin{cases} x = R(\gamma, A_2)(z), \\ z = (\frac{1}{2}I + \frac{1}{2}C(\gamma, A_1)(C(\gamma, A_2)))(z), \end{cases}$$

where for the last step we reexpressed the Cayley reflection operators in terms of the resolvents (using the definition), leading to $z = C(\gamma, A_1)C(\gamma, A_2)(z)$, which is then expressed in the averaged operator form stated above. The reason for expressing it in this form, which uses the averaged operator $T := \frac{1}{2}I + \frac{1}{2}C(\gamma, A_1)C(\gamma, A_2)$, is that in this form we can make use of the Krasnoselskii–Mann iteration scheme (see Theorem 3.4.9) to approximate the fixed point z of the operator T, and then use $x = R(\gamma, A_2)(z)$ to obtain the required solution.

The above discussion motivates the following iterative scheme:

$$\begin{cases} y_{n+1} = R(\gamma, A_2)(z_n), \\ x_{n+1} = R(\gamma, A_1)(2y_{n+1} - z_n), \\ z_{n+1} = z_n + \lambda_n(x_{n+1} - y_{n+1}), \end{cases} \tag{9.85}$$

for $\gamma > 0$, where $\lambda_n > 0$ is a relaxation parameter. The iterative scheme (9.85) is called the Douglas–Rachford scheme. The reader can easily check that in the special case where $A_i = \partial\varphi_i$, $i = 1, 2$, the scheme (9.85) coincides with the Douglas–Rachford scheme for minimization of the function $\varphi_1 + \varphi_2$, discussed in Section 4.5.2.2. The convergence of the general Douglas–Rachford scheme (9.85) can be obtained using Proposition 4.5.12, simply putting the operators $R(\gamma, A_i)$ in the place of the proximal operators (we leave it as a simple exercise to the reader to check that the proof generalizes to this case essentially with minimal changes).

Example 9.7.2 (Douglas–Rachford scheme for variational inequalities).
The Douglas–Rachford scheme (9.85) can be easily adapted for the solution of variational inequalities. Following Example 9.7.1, we set $A_1 = pi_X^{-1}F$, and $A_2 = N_C$, and recall that for this case $R(\gamma, A_2) = \Pi_C$, where Π_C is the projection operator on the closed convex set C. Using the above information (9.85) reduces to the Douglas–Rachford scheme for the solution of variational inequalities. ◁

9.8 Appendix

We present here an alternative way for step 3 in the proof of Theorem 9.5.5, bypassing Remark 9.5.2 by appropriate choice of test functions in the approximations (9.62). This is simply an equivalent reformulation for equations, but nevertheless we present it here, as it can also be useful for variational inequalities.

Using the density of $\bigcup_{n\in\mathbb{N}} X_n$ in X and the fact that $X_n \subset X_{n+1}$, for any $z \in X$, we first construct a sequence $(z_n)_{n\in\mathbb{N}}$, $z_n \in X_{\ell_n}$ for all $n \geq \mathbb{N}$ (where $(\ell_n)_{n\in\mathbb{N}}$ is an increasing sequence) such that $z_n \to x$ (strong) in X. For the construction of this sequence, we work as follows: Take $\epsilon = \frac{1}{n}$. By density for each $n \in \mathbb{N}$, there exists an $\ell_n \in N$ and a $z_n \in X_{\ell_n}$ such that $\|z_n - z\| < \frac{1}{n}$. Since $X_n \subset X_{n+1}$, we can always choose ℓ_n such that $\ell_n < \ell_{n+1}$.[16] The construction of such sequences will be used for the choice of appropriate test functions in the Galerkin approximation.

Take an arbitrary $z \in X$ and perform the above approximation in terms of the sequence $(z_n)_{n\in\mathbb{N}}$, $z_n \in X_{\ell_n}$. For any $n \in \mathbb{N}$, sufficiently large, e. g., larger than ℓ_1, pick m such that $\ell_m \leq n$ (this is always possible since $\ell_m \uparrow$). We now use as test functions in (9.62), $z_m - x_n \in X_n$, $m \leq n$ (note that $z_m \in X_{\ell_m} \subset X_n$),

$$\langle A(x_n), z_m - x_n\rangle = \langle x^\star, z_m - x_n\rangle. \tag{9.86}$$

We now express

$$\begin{aligned}\langle A(x_n), x_n - x\rangle &= \langle A(x_n), x_n - z_m\rangle + \langle A(x_n), z_m - x\rangle \\ &\overset{(9.86)}{=} \langle x^\star, x_n - z_m\rangle + \langle A(x_n), z_m - x\rangle \\ &\leq \langle x^\star, x_n - z_m\rangle + \|A(x_n)\|_{X^\star}\|z_m - x\| \\ &\leq \langle x^\star, x_n - z_m\rangle + C\|z_m - x\|, \quad m \leq n,\end{aligned} \tag{9.87}$$

where the last estimate follows by the fact that $A(x_n)$ is uniformly bounded in X^*. Fixing m and taking the limsup as $n \to \infty$ in (9.87), we obtain that

$$\limsup_n \langle A(x_n), x_n - x\rangle \leq \langle x^\star, x - z_m\rangle + C\|z_m - x\|, \quad m \in \mathbb{N}, \tag{9.88}$$

where we used the fact that $x_n \rightharpoonup x$. Since (9.88) holds for all $m \in \mathbb{N}$, we pass to the limit, as $m \to \infty$, and using the fact that $z_m \to x$ (strong), we conclude that

$$\limsup_n \langle A(x_n), x_n - x\rangle \leq 0. \tag{9.89}$$

The pseudomonotonicity of A (see Definition 9.5.1) then implies that

16 If $z_n \in X_m$ with $m < \ell_{n-1}$, then, since $X_m \subset X_k$, for all $k \geq m$ we can choose $\ell_n > \ell_{n-1}$ and by $X_m \subset X_{\ell_n}$ consider $z_n \in X_m \subset X_{\ell_n}$ as an element of X_{ℓ_n}.

$$\langle A(x), x - z\rangle \le \liminf_n \langle A(x_n), x_n - z\rangle, \quad \forall z \in X. \tag{9.90}$$

We apply the above for $z \in \bigcup_{n\in\mathbb{N}} X_n$. That implies the existence of an $N \in \mathbb{N}$ such that $z \in X_N$, and since $X_n \subset X_{n+1}$, it is easy to see that $z \in X_n$ for all $n \ge N$. This allows us to use $x_n - z$ as a test function in (9.62) so that

$$\langle A(x_n), x_n - z\rangle = \langle x^\star, x_n - z\rangle, \quad \forall n \ge N. \tag{9.91}$$

Since $x_n \to x$, we have that $\langle x^\star, x_n - z\rangle \to \langle x^\star, x - z\rangle$. Taking the liminf as $n \to \infty$ in (9.91), we obtain

$$\liminf_n \langle A(x_n), x_n - z\rangle = \lim_n \langle x^\star, x_n - z\rangle = \langle x^\star, x - z\rangle. \tag{9.92}$$

Comparing (9.90) and (9.92), we obtain that

$$\langle x^\star, x - z\rangle \ge \langle A(x), x - z\rangle, \quad \forall z \in \bigcup_{n\in\mathbb{N}} X_n. \tag{9.93}$$

Since $\bigcup_{n\in\mathbb{N}} X_n$ is dense in X, (9.93) holds for any $z \in X$,[17] which leads us to the result that $\langle A(x), z\rangle = \langle f, z\rangle$ for any $z \in X$.

17 Express (9.93) as $\langle A(x) - x^\star, x - z\rangle \le 0$ for all $z \in \bigcup_{n\in\mathbb{N}} X_n$. Now, by density, for any $z \in X$ and $\epsilon > 0$, there exists $z_\epsilon \in \bigcup_{n\in\mathbb{N}} X_n$ such that $\|z_\epsilon - z\| \le \epsilon$. Apply (9.93) in its current form for z_ϵ, and pass to the limit, as $\epsilon \to 0$, to obtain the result for any $z \in X$.

10 Evolution problems

10.1 Introduction

This chapter is devoted to the study of evolution problems in Banach spaces. We will focus on differential equations (or inclusions) of the form

$$\begin{aligned} \frac{dx}{dt}(t) &= A(t, x(t)) + f(t), \quad t \in (0, T], \\ x(0) &= x_0, \end{aligned} \tag{10.1}$$

and consider the problem of finding a function $x : [0, T] \to X$, with certain properties (e. g., continuity, differentiability etc.) that satisfies (10.1) for given data of the problem (i. e., the function f, the operator A, and the initial condition x_0).[1]

Evolution problems appear in many fields of applications, such as biology, physics, mechanics, economics, optimization, machine learning, and are studied intensely. In this chapter, we give an introduction to this fascinating field, focusing on two basic strands of the literature, connected with different ways of defining the operator A:

(i) The case where A is defined as $A : \mathbf{D}(A) \subset X \to X$ for a suitable domain of definition $\mathbf{D}(A)$, which is related to the theory of linear and nonlinear semigroups and allows for the treatment of the problem in terms of a family of operators $\{S(t)\}_{t \geq 0}$ in terms of which (under certain conditions) the solution of (10.1) can be constructed, and
(ii) The case where A is more loosely defined as $A : X \to X^*$, leading to a variational construction of solutions in terms of the concept of the evolution triple.

In both cases, we will heavily rely on the concept of monotone type operators and their properties, presented in Chapter 9. At this point, it is useful to state for the convenience of the reader that having expressed our evolution system in the form (10.1), we will require that $A =: A_{\mathcal{D}}$ is dissipative, i. e., that $A_{\mathcal{M}} = -A$ is of monotone type. This will become clear shortly, as the monotonicity of $A_{\mathcal{M}} = -A$ will emerge as a natural implication of the requirement that the solution of (10.1) satisfies some contraction properties. This is an important fact the reader has to bear in mind, which will also be our key strategy of attack to this problem here.[2] For various reasons, in almost half of the literature, problem (10.1) can be found in the equivalent form

$$\begin{aligned} \frac{dx}{dt}(t) + A(t, x(t)) &= f(t), \quad t \in (0, T], \\ x(0) &= x_0, \end{aligned} \tag{10.2}$$

1 We use the notation $t \in (0, T]$ in (10.1) since it may be that x is not continuously differentiable, for example, when weak solutions are considered.

2 Other approaches, not related to monotonicity properties are of course still feasible, but these are outside the scope of the present work.

https://doi.org/10.1515/9783111333298-010

in which case (using the same arguments) its treatment is facilitated if A is of monotone type.

Notation 10.1.1. While we will adopt formulation (10.1) here,[3] to familiarize the reader with the equivalent formulation (10.2), we make the following note:

$$\begin{cases} \frac{dx}{dt} = A_{\mathcal{D}}(t,x) + f \\ x(0) = x_0 \end{cases} \iff \begin{cases} \frac{dx}{dt} + A_{\mathcal{M}}(t,x) = f \\ x(0) = x_0, \end{cases} \tag{10.3}$$

where $A_{\mathcal{M}} = -A_{\mathcal{D}}$ is a monotone type operator. When needed, we will use the notation $A_{\mathcal{D}}$ for the operator A to emphasize that we require dissipativity (as in formulation (10.1)) and the notation $A_{\mathcal{M}}$ for the operator A to emphasize that we require monotonicity (as in formulation (10.2)).

The theory and applications of evolution problems is an important part of nonlinear analysis with a large number of excellent monographs and textbooks dedicated to it. Among these, we cite here the ones that have mostly influenced our approach and presentation in this chapter, i. e., [18, 19, 42, 89, 128].

10.2 Evolution problems and nonlinear semigroups

The theory of nonlinear semigroups is an important methodology for the study of evolution problems in the case of operators $A : \mathbf{D}(A) \subset X \to X$ or $A : \mathbf{D}(A) \subset X \to 2^X$ (see e. g., [18, 19, 89, 128]).

10.2.1 Nonlinear semigroups

Linear semigroups (see Section 1.7) can offer important insights and are important tools in the study of the solutions of linear differential equations in Banach spaces, which can also be extended in certain cases to the study of semilinear differential equations. The success of these techniques for linear problems sets the challenging question of whether one can extend the theory of linear semigroups so that an analogous treatment can apply to nonlinear equations. Of course, there are important issues with this programme, essentially related to the fact that the superposition principle, upon which the variation of constants formula (Duhamel's principle) that was used for the construction of solutions of linear evolution equations in terms of linear semigroups was based, does not hold for nonlinear problems.

We start with a necessary definition.

3 Which we find more natural since, for instance, a student is introduced to differential equations in a first course on the subject using this formulation.

Definition 10.2.1 (Nonlinear semigroup). Let X be a Banach space and $K \subset X$ be a nonempty subset of X.

A strongly continuous nonlinear operator semigroup on K is a family of nonlinear operators $\{S(t) : t \geq 0\}$, $S(t) : K \to K$, for every $t \geq 0$, satisfying

(i) $S(0) = I_K$.
(ii) $S(t_2)S(t_1)x = S(t_2 + t_1)x$ for all $t_1, t_2 \in \mathbb{R}_+$, $x \in K$.
(iii) For every $x \in K$, the mapping $t \mapsto S(t)x$ is continuous from $\mathbb{R}_+$ into X.
(iv) If moreover,

$$\|S(t)x_1 - S(t)x_2\| \leq e^{\omega t}\|x_1 - x_2\|, \quad x_1, x_2 \in K,\ t > 0,$$

the semigroup is called of type ω, whereas if $\omega = 0$, it is called a contraction semigroup.

For simplicity, we will consider here the case where $K = X$.

10.2.2 Interlude: dissipative operators

How much can be said concerning the generalization of the theory of linear semigroups (presented in Section 1.7) for the case where the semigroups are nonlinear? In other words, can we address the question of which nonlinear operators can be considered as generators of a nonlinear semigroup of contractions? This is the subject of the celebrated Crandall–Liggett theorem [58].

Before presenting this result, we must introduce the relevant class of operators that can act as generators of nonlinear semigroups of contractions. This is the class of nonlinear dissipative operators, i. e., operators that satisfy a condition similar to that in Definition 1.7.5, however, properly generalized for the general case that A is a nonlinear (possible multivalued) operator defined on a Banach space X.

Definition 10.2.2 (Dissipative operators). Let X be a Banach space with dual $X^\star$, $J : X \to 2^{X^\star}$ be the duality mapping, and consider the nonlinear (possibly multivalued) operator $A : \mathbf{D}(A) \subset X \to 2^X$.

(i) A is called dissipative if for any $(x_i, y_i) \in \mathbf{Gr}(A)$, $i = 1, 2$,

$$\exists\, x^\star \in J(x_1 - x_2) \quad \text{such that} \quad \langle x^\star, y_1 - y_2\rangle \leq 0 \tag{10.4}$$

or equivalently

$$\|x_1 - x_2 - \lambda(y_1 - y_2)\| \geq \|x_1 - x_2\| \quad \forall \lambda > 0. \tag{10.5}$$

(ii) A is called m-dissipative if it is dissipative and $\mathbf{R}(I - \lambda \mathsf{A}) = X$ for some $\lambda > 0$ (and as shown below in Prop. 10.2.7(ii) for all $\lambda > 0$).

(iii) A is called accretive if $-\mathsf{A}$ is dissipative and m-accretive if $-\mathsf{A}$ is m-dissipative.

Remark 10.2.3. Note that unlike our definitions of nonlinear operators $\mathsf{A} : \mathbf{D}(\mathsf{A}) \subset X \to 2^{X^*}$ in Chapter 9, for this section, which focuses on nonlinear semigroup theory, we restrict our attention to operators $\mathsf{A} : \mathbf{D}(\mathsf{A}) \subset X \to 2^X$. This of course would require modification of the domain of the operator so that its range is included in 2^X. In this respect, the operators encountered here can be considered as special cases of the types of operators studied in Chapter 9.

Remark 10.2.4. It is important to note that the definition of dissipativity resembles the definition of monotonicity for $-\mathsf{A}$ and importantly, as we will see shortly, A being m-dissipative implies that $-\mathsf{A}$ is maximal monotone (in the sense of Chapter 9, but now considering the range of A as subset of X, rather than as subset of X^*). This opens the road to transferring results and constructions for maximal monotone operators (such as for example the resolvent or the Yosida approximation, see Definition 9.4.21) to m-dissipative operators. These must be accordingly modified of course, to account for the sign change and the change of the range).

We provide the relevant definitions:

Definition 10.2.5 (Resolvent and Yosida approximation). Let $\mathsf{A} : \mathbf{D}(\mathsf{A}) \subset X \to 2^X$ be an m-dissipative operator.

(i) The resolvent of A is the family of single valued operators from X to X:

$$(R(\lambda, \mathsf{A}))_{\lambda>0} = (\mathsf{R}_\lambda)_{\lambda>0}, \quad \text{defined by} \quad R(\lambda, \mathsf{A}) = \mathsf{R}_\lambda = (I - \lambda \mathsf{A})^{-1}.$$

(ii) The Yosida approximation of A is the family of single valued operators from X to X:

$$(\mathsf{A}_\lambda)_{\lambda>0}, \quad \text{defined by} \quad \mathsf{A}_\lambda = \frac{1}{\lambda}(\mathsf{R}_\lambda - I).$$

Notation 10.2.6. Note the use of the same notation as in Chapter 9, but with the sign difference due to dissipativity (compare with Definition 9.4.21). Which definition applies will be clear from the context, or indicated. If there is no ambiguity concerning the choice of operator A, we will use the simplified notation R_λ instead of $R(\lambda, \mathsf{A})$ (as already used in Definition 10.2.5).

Before we proceed, we collect here some useful properties of nonlinear dissipative operators (see, e. g., [89]).

Proposition 10.2.7. *The following hold:*

(i) *Condition* (10.4) *is equivalent to condition* (10.5). *Moreover,* $R(\lambda, \mathsf{A}) = \mathsf{R}_\lambda$ *is well defined on* $\mathbf{R}(I - \lambda \mathsf{A})$, *and it holds that*

$$\|R(\lambda,\mathsf{A})\mathsf{x}_1 - R(\lambda,\mathsf{A})\mathsf{x}_2\| := \|\mathsf{R}_\lambda\mathsf{x}_1 - \mathsf{R}_\lambda\mathsf{x}_2\| \le \|\mathsf{x}_1 - \mathsf{x}_2\|, \quad \forall\, \mathsf{x}_1,\mathsf{x}_2 \in \mathbf{R}(I-\lambda\mathsf{A}).$$

Clearly, if A *is m-dissipative, then the above hold for all X.*

(ii) *If* $\mathbf{R}(I-\lambda\mathsf{A}) = X$ *for some* $\lambda > 0$, *then* $\mathbf{R}(I-\lambda\mathsf{A}) = X$ *for all* $\lambda > 0$.

(iii) *If* A *is m-dissipative (equiv.* $-\mathsf{A}$ *is m-accretive), then* A *is maximal dissipative (equiv.* $-\mathsf{A}$ *is maximal monotone).*

(iv) *If* A *is m-dissipative, then* A *is closed, i. e., if* $\mathsf{x}_n \to \mathsf{x}_0$, $\mathsf{y}_n \to \mathsf{y}_0$, *with* $(\mathsf{x}_n,\mathsf{y}_n) \in \mathbf{Gr}(\mathsf{A})$, *for all* $n \in \mathbb{N}$, *then* $(\mathsf{x}_0,\mathsf{y}_0) \in \mathbf{Gr}(\mathsf{A})$.

(v) *If* A *is m-dissipative, then*

$$\forall\, (\mathsf{x}_\lambda)_{\lambda>0} \text{ with } \mathsf{x}_\lambda \to \mathsf{x} \text{ and } \mathsf{y}_\lambda \in \mathsf{A}_\lambda\mathsf{x}_\lambda \to y \text{ as } \lambda \to 0, \text{ we have } (\mathsf{x},\mathsf{y}) \in \mathbf{Gr}(\mathsf{A}).$$

(vi) *For every* $0 \le \mu \le \lambda$, *and* $\mathsf{x} \in \mathbf{D}(R(\lambda,\mathsf{A}))$, *it holds that*

$$R(\lambda,\mathsf{A})\mathsf{x} = R(\mu,\mathsf{A})\left(\frac{\mu}{\lambda}\mathsf{x} + \frac{\lambda-\mu}{\lambda}R(\lambda,\mathsf{A})\mathsf{x}\right). \tag{10.6}$$

(vii) *For* $\mathsf{x} \in \mathbf{D}(R(\lambda,\mathsf{A})) \cap \mathbf{D}(\mathsf{A})$, *we have (compare with Def.* 9.4.21 *where X is reflexive)*

$$\|\mathsf{A}_\lambda\mathsf{x}\| \le |\mathsf{A}\mathsf{x}|_0 := \inf\{\|\mathsf{y}\| \;:\; \mathsf{y} \in \mathsf{A}(\mathsf{x})\},$$
$$\|R(\lambda,\mathsf{A})\mathsf{x} - \mathsf{x}\| \le \lambda|\mathsf{A}\mathsf{x}|_0 := \lambda\inf\{\|\mathsf{y}\| \;:\; \mathsf{y} \in \mathsf{A}(\mathsf{x})\}.$$

Proof. We simplify notation by $\mathsf{R}_\lambda = R(\lambda,\mathsf{A})$ and we may use Ax to indicate any $\mathsf{y}^* \in \mathsf{A}(\mathsf{x})$.

(i) To show that (10.4) implies (10.5), observe that for any $\mathsf{x}^* \in \mathsf{J}(\mathsf{x})$,

$$\begin{aligned}\|(I-\lambda\mathsf{A})\mathsf{x}\| &\ge \frac{\langle \mathsf{x}^*, (I-\lambda\mathsf{A})\mathsf{x}\rangle}{\|\mathsf{x}^*\|_{X^*}} = \frac{\langle \mathsf{x}^*, (I-\lambda\mathsf{A})\mathsf{x}\rangle}{\|\mathsf{x}\|} = \frac{\langle \mathsf{x}^*,\mathsf{x}\rangle - \lambda\langle \mathsf{x}^*,\mathsf{A}\mathsf{x}\rangle}{\|\mathsf{x}\|}\\ &\ge \frac{\langle \mathsf{x}^*,\mathsf{x}\rangle}{\|\mathsf{x}\|} = \|\mathsf{x}\|,\end{aligned}$$

where we used the obvious lower bound for $\|(I-\lambda\mathsf{A})\mathsf{x}\|$ (arising from $|\langle \mathsf{z}^\star,\mathsf{z}\rangle| \le \|\mathsf{z}^\star\|_{X^*}\|\mathsf{z}\|$), and then (10.4) and the fact that since $\mathsf{x}^* \in \mathsf{J}(\mathsf{x})$, it holds that $\langle \mathsf{x}^*,\mathsf{x}\rangle = \|\mathsf{x}\|^2 = \|\mathsf{x}^*\|_{X^*}^2$.

To show that (10.5) implies (10.4), consider any point $\mathsf{z}_\lambda^\star \in \mathsf{J}((I-\lambda\mathsf{A})\mathsf{x})$. Note that $\|\mathsf{z}_\lambda^\star\|_{X^*} = \|(I-\lambda\mathsf{A})\mathsf{x}\|$, while $\langle \mathsf{z}_\lambda^\star, (I-\lambda\mathsf{A})\mathsf{x}\rangle = \|\mathsf{z}_\lambda^\star\|_{X^*}^2 = \|(I-\lambda\mathsf{A})\mathsf{x}\|^2$, so the normalized $\hat{\mathsf{z}}_\lambda^* = \frac{\mathsf{z}_\lambda^\star}{\|\mathsf{z}_\lambda^\star\|_{X^*}}$ satisfies

$$\langle \hat{\mathsf{z}}_\lambda^*, (I-\lambda\mathsf{A})\mathsf{x}\rangle = \|(I-\lambda\mathsf{A})\mathsf{x}\| \ge \|\mathsf{x}\|,$$

where for the last inequality we used (ii). Expanding the LHS, we obtain that

$$\|\mathsf{x}\| \le \langle \hat{\mathsf{z}}_\lambda^*, \mathsf{x}\rangle - \lambda\langle \hat{\mathsf{z}}_\lambda^*, \mathsf{A}\mathsf{x}\rangle, \quad \lambda > 0, \tag{10.7}$$

which since $\langle \hat{\mathsf{z}}_\lambda^*, \mathsf{x}\rangle \le \|\mathsf{x}\|$ leads to the conclusion that

$$\lambda\langle \hat{z}_\lambda^*, Ax\rangle \le 0, \quad \forall\,\lambda > 0. \tag{10.8}$$

We will show that as $\lambda \to 0$, $\hat{z}_\lambda^*$ tends to an element $x^\star \in J(x)$ to conclude. To show this, we rearrange (10.7) to obtain

$$\|x\| \le \langle \hat{z}_\lambda^*, x\rangle + \lambda\langle -\hat{z}_\lambda^*, Ax\rangle \le \langle \hat{z}_\lambda^*, x\rangle + \lambda\|z_\lambda^*\|_{X^*}\|Ax\| = \langle \hat{z}_\lambda^*, x\rangle + \lambda\|Ax\|,$$

(since $\|\hat{z}_\lambda^*\|_{X^*} = 1$), which leads to

$$\|x\| - \lambda\|Ax\| \le \langle \hat{z}_\lambda^*, x\rangle. \tag{10.9}$$

Now, by the fact that $\|\hat{z}_\lambda^*\|_{X^*} = 1$ for each $\lambda > 0$, there exists some $z_0^* \in X^*$ and a subsequence (denoted the same) such that $\hat{z}_\lambda^* \overset{*}{\rightharpoonup} z_0^*$. By the weak lower semicontinuity of the norm, it follows that $\|z_0^*\|_{X^*} \le 1$. Passing to the limit as $\lambda \to 0$ in (10.9), we obtain $\|x\| \le \langle z_0^*, x\rangle$. But, on the other hand, $\langle z_0^*, x\rangle \le \|z_0^*\|_{X^*}\|x\| \le \|x\|$, and combining these two, we obtain $\langle z_0^*, x\rangle = \|x\|$. Moreover, (10.8) implies that $\langle z_\lambda^*, Ax\rangle \le 0$ for all $\lambda > 0$ and passing to the limit as $\lambda \to 0$, we obtain that $\langle z_0^*, Ax\rangle \le 0$. Summing up, we have obtained a point $z_0^* \in X^*$ with the following properties:

$$\|z_0^*\|_{X^*} \le 1, \quad \langle z_0^*, x\rangle = \|x\|, \quad \langle z_0^*, Ax\rangle \le 0.$$

Define $x_0^* = \|x\|z_0^*$. By the above, this satisfies

$$\|x_0^*\|_{X^*} \le \|x\|, \quad \langle x_0^*, x\rangle = \|x\|^2, \quad \langle x_0^*, Ax\rangle \le 0.$$

The second one implies $\|x\|^2 \le \|x_0^*\|_{X^*}\|x\|$, hence $\|x\| \le \|x_0^*\|_{X^*}$, which combined with the first one yields $\|x_0^*\|_{X^*} = \|x\|$. Hence, x_0^* satisfies

$$\langle x_0^*, x\rangle = \|x\|^2 = \|x_0^*\|_{X^*}^2, \quad \langle x_0^*, Ax\rangle \le 0,$$

therefore

$$x_0^* \in J(x), \quad \text{and} \quad \langle x_0^*, Ax\rangle \le 0,$$

i. e., (10.4) is satisfied. The non expansive property for R_λ follows from (10.5).

(ii) Assume the property holds for $\lambda_0 > 0$, i. e., that $\mathbf{R}(I - \lambda_0 A) = X$. We also note that

$$I - \lambda A = \frac{\lambda}{\lambda_0}\left[I - \left(1 - \frac{\lambda_0}{\lambda}\right)R_{\lambda_0}\right](I - \lambda_0 A), \quad \lambda > 0. \tag{10.10}$$

For any $x \in X$ fixed we define the operator $T : X \to X$, acting on any $z \in X$ by

$$Tz = x + \left(1 - \frac{\lambda_0}{\lambda}\right)R_{\lambda_0} z.$$

Since R_{λ_0} is non expansive, T satisfies the bound

$$\|Tz_1 - Tz_2\| \le \left|1 - \frac{\lambda_0}{\lambda}\right| \|z_1 - z_2\|,$$

hence, is a contraction as long as $\lambda > \frac{\lambda_0}{2}$, and by Banach contraction theorem has a fixed point $z_0 \in X$. By the definition of T, this implies that

$$x = \left[I - \left(1 - \frac{\lambda_0}{\lambda}\right)R_{\lambda_0}\right] z_0. \tag{10.11}$$

Since $\mathbf{R}(I - \lambda_0 A) = X$, we have that for any $x \in X$ there exists $y := R_{\lambda_0} x \in \mathbf{D}(A)$ such that $x \in y - \lambda_0 Ay$, i. e., $R_{\lambda_0} x \in \mathbf{D}(A)$ for any $x \in X$. Hence, the same holds for z_0 and let us call $\hat{z}_0 = R_{\lambda_0} z_0 \in \mathbf{D}(A)$, where $z_0 \in \hat{z}_0 - \lambda_0 A\hat{z}_0$. This is interpreted as $z_0 \in (I - \lambda_0 A)\mathbf{D}(A)$, so by (10.11), we obtain that $x \in [I - (1 - \frac{\lambda_0}{\lambda})R_{\lambda_0}](I - \lambda_0 A)\mathbf{D}(A)$. Since $x \in X$ is arbitrary, so is $\frac{\lambda}{\lambda_0} x \in X$, so that any $z \in X$ satisfies $z \in \frac{\lambda}{\lambda_0}[I - (1 - \frac{\lambda_0}{\lambda})R_{\lambda_0}](I - \lambda_0 A)\mathbf{D}(A) = (I - \lambda A)\mathbf{D}(A)$, hence $\mathbf{R}(I - \lambda A) = X$, as long as $\lambda > \frac{\lambda_0}{2}$.

We now repeat the argument once more to obtain that $\mathbf{R}(I - \lambda A) = X$, as long as $\lambda > \frac{1}{2}\frac{\lambda_0}{2} = \frac{\lambda_0}{4}$, and continue likewise to obtain the surjectivity of $I - \lambda A$ for all $\lambda > 0$.

(iii) Let $\hat{A}$ be a dissipative extension of A, and let $(\hat{x}, \hat{y}) \in \mathbf{Gr}(\hat{A})$. Since A is m-dissipative, $\mathbf{R}(I - \lambda A) = X$. Choosing $\hat{x} - \lambda\hat{y} \in X$, there exist $(x_0, y_0) \in \mathbf{Gr}(A)$ such that $x_0 - \lambda y_0 = \hat{x} - \lambda\hat{y}$ or equiv. $(x_0 - \hat{x}) = \lambda(y_0 - \hat{y})$. On the other hand, since $\hat{A}$ is an extension of A, denoted by $A \subset \hat{A}$ and $(x_0, y_0) \in \mathbf{Gr}(A)$, it clearly holds that $(x_0, y_0) \in \hat{A}$. By the dissipative property of $\hat{A}$, we then have for the pairs $(\hat{x}, \hat{y}), (x_0, y_0) \in \mathbf{Gr}(\hat{A})$:

$$\|x_0 - \hat{x}\| \le \|(x_0 - \hat{x}) - \lambda(y_0 - \hat{y})\| = 0,$$

which implies that $x_0 = \hat{x}$, and consequently $y_0 = \hat{y}$. But this implies also the opposite inclusion $\hat{A} \subset A$, hence the maximality property.

(iv) Consider a sequence $(x_n, y_n)_{n \in \mathbb{N}}$, such that $(x_n, y_n) \in \mathbf{Gr}(A)$ for all $n \in \mathbb{N}$ and $x_n \to x_0$ and $y_n \to y$. For any n fixed, we apply (10.5) for (x_n, y_n) and an arbitrary $(\bar{x}, \bar{y}) \in \mathbf{Gr}(A)$, and pass to the limit as $n \to \infty$ to obtain $\|x_0 - \bar{x}\| \le \|x_0 - \lambda\bar{x} - (\bar{x} - \lambda\bar{y})\|$. On the other hand, by m-dissipativity, $\mathbf{R}(I - \lambda A) = X$, hence applying this property to $x_0 - \lambda y_0 \in X$, there exists $(\hat{x}_0, \hat{y}_0) \in \mathbf{Gr}(A)$ such that $x_0 = \lambda y_0 = \hat{x}_0 - \lambda\hat{y}_0$. We now apply the above inequality for the choice $\bar{x} = \hat{x}_0$ and $\bar{y} = \hat{y}_0$ to obtain that $x_0 = \hat{x}_0$ and subsequently $y_0 = \hat{y}_0$. Hence, $(x_0, y_0) \in \mathbf{Gr}(A)$.

(v) To show this, note that $A_\lambda x \in AR_\lambda x$ for every $x \in \mathbf{D}(R_\lambda) = \mathbf{R}(I - \lambda A)$. Indeed, let $x \in \mathbf{R}(I - \lambda A)$ so that $z = R_\lambda x$ is well defined. By the definition of R_λ, there exists $(z, \hat{z}) \in \mathbf{Gr}(A)$ such that $x = z - \lambda\hat{z}$, i. e. $z = x + \lambda\bar{z}$. Then, by the definition of A_λ, we have that $A_\lambda x = \frac{1}{\lambda}(z - x) = \bar{z} \in Az = AR_\lambda x$ proving our claim. We observe next that $A_\lambda x_\lambda$ is bounded for all $\lambda > 0$, as by assumption $A_\lambda x_\lambda \to y$. Since $R_\lambda x_\lambda - x_\lambda = \lambda A_\lambda x_\lambda$, passing to the limit as $\lambda \to 0$, we see that $R_\lambda x_\lambda - x_\lambda \to 0$ hence, $R_\lambda x_\lambda \to x$, as $\lambda \to 0$. Since $A_\lambda x \in AR_\lambda x$ for every $x \in \mathbf{D}(R_\lambda) = \mathbf{R}(I - \lambda A)$, and $\mathbf{R}(I - \lambda A) = X$ for all $\lambda > 0$, because A is m-dissipative, we conclude that $(x, y) \in \mathbf{Gr}(A)$ by the fact that A is closed (see (iv)).

(vi) To show this, consider any $x \in \mathbf{D}(R_\lambda)$. This implies that there exists $(x_0, y_0) \in \mathbf{Gr}(A)$ such that $x = x_0 - \lambda y_0$. By definition $x_0 = R_\lambda x$. Then,

$$\frac{\mu}{\lambda}x + \frac{\lambda - \mu}{\lambda}R_\lambda x = \frac{\mu}{\lambda}x + \frac{\lambda - \mu}{\lambda}x_0 = x_0 - \mu y_0 \in (I - \mu A)x_0,$$

which implies that

$$R_\lambda x = x_0 = R_\mu\Big(\frac{\mu}{\lambda}x + \frac{\lambda - \mu}{\lambda}R_\lambda x\Big).$$

(vii) Consider any $y \in A(x)$ (arbitrary). Clearly, $x - \lambda y \in (I - \lambda A)x$, so that $R_\lambda(x - \lambda y) = (I - \lambda A)^{-1}(x - \lambda y) = x$. Hence, by the definition of A_λ, we have that

$$\begin{aligned} \|A_\lambda x\| &= \lambda^{-1}\|R_\lambda x - x\| = \lambda^{-1}\big\|R_\lambda x - R_\lambda(x - \lambda y)\big\| \\ &\le \lambda^{-1}\big\|x - (x - \lambda y)\big\| = \|y\|, \end{aligned} \tag{10.12}$$

where we also used the fact that R_λ is nonexpansive. Since $y \in A(x)$ is arbitrary, the above holds also for the infimum of the RHS over all $y \in A(x)$, hence,

$$\|A_\lambda x\| \le \lambda|Ax|_0.$$

Another rearrangement of (10.12) immediately yields,

$$\|R_\lambda x - x\| \le \lambda\|y\|,$$

and again, by the fact that $y \in A(x)$ is arbitrary, by taking the infimum over all such y, we obtain

$$\|R_\lambda x - x\| \le \lambda|Ax|_0.$$

The proof is complete. □

Finally, a convenient quantity is the semi-inner product.

Definition 10.2.8. For any $x, z \in X$, we define

$$[x, z]_+ = \inf_{\lambda>0} \frac{\|x + \lambda z\| - \|x\|}{\lambda} = \lim_{\lambda\to 0^+} \frac{\|x + \lambda z\| - \|x\|}{\lambda}.$$

The semi-inner product $[x, z]_+$ can be interpreted as the directional derivative of the norm at point $x \in X$ in the direction z.

In terms of the semi-inner product, dissipativity can be equivalently expressed as

$$[x_1 - x_2, y_1 - y_2]_+ \le 0, \quad \forall\,(x_i, y_i) \in \mathbf{Gr}(A),\ i = 1, 2.$$

10.2.3 Generation of nonlinear semigroups of contractions

We are now ready to extend the theory of generation of linear contraction semigroups to nonlinear semigroups. As in the linear case, covered by the Lumer–Phillips theory, we will see that m-dissipativity plays an important role for semigroup generation in the nonlinear case as well. These results will provide important information concerning the differential equation $\frac{dx}{dt} = A(x)$, in the case where A is an m-dissipative or in general a dissipative operator.

The following is essentially the celebrated Crandall–Liggett theorem [58]; see also [18, 19]:

Theorem 10.2.9 (Crandall–Liggett). *Let* A *be a dissipative operator satisfying the condition*

$$\overline{\mathbf{D}(A)} \subset \mathbf{R}(I - \lambda A), \tag{10.13}$$

for all sufficiently small $\lambda > 0$. *Then*

$$S(t)x := \lim_{n\to\infty} \left(I - \frac{t}{n}A\right)^{-n} x = \lim_{\lambda\to 0^+} (I - \lambda A)^{-[t/\lambda]} x$$

exists for all $x \in \overline{\mathbf{D}(A)}$, *uniformly in* $t \in I$, *where* I *is any compact interval of* $[0, \infty)$, *and defines a nonlinear semigroup of contractions on* $\overline{\mathbf{D}(A)}$. *Moreover, it holds that*

$$\|S(t_2)x - S(t_1)x\| \leq |t_2 - t_1|\, |Ax|_0, \quad \forall\, t_2, t_1 \geq 0,\ x \in \mathbf{D}(A),$$

where $|Ax|_0 = \inf\{\, \|y\| \ :\ y \in A(x)\}$.

Clearly, condition (10.13) *holds true if* A *is m-dissipative.*

Remark 10.2.10 (Connection with the Lumer–Phillips theorem). The Crandall–Ligget theorem is an extension of the Lumer–Phillips theorem (Theorem 1.7.6). Clearly, if A is m-dissipative and linear, a direct application of Theorem 10.2.9 would allow us to extract the assertion of the Lumer–Phillips theorem, as long as we can guarantee that A is also closed and densely defined. That A is closed can be guaranteed by the m-dissipativity (see Proposition 10.2.7(iv)). That it is densely defined can be obtained from the range condition (10.13) and m-dissipativity. We stress of course that the Lumer–Phillips theorem preceded the Crandall–Ligget theorem and the original proof is different. If A is linear, it can be easily established that the corresponding semigroup is a linear semigroup.

Proof of Thm. 10.2.9. The resolvent operator $R(\lambda, A) = R_\lambda = (I - \lambda A)^{-1}$ and its properties play a crucial role in the proof. In the interest of simplified notation, we will use the notation

$$R_\lambda := R(\lambda, A).$$

0. The strategy of the proof is to show that, for fixed $t > 0$, the sequence $((I - \frac{t}{n}\mathsf{A})^{-n}\mathsf{x})_{n\in\mathbb{N}}$ is Cauchy in X, so that it has a limit, through which we will define the semigroup. We will then prove the relevant properties for this sequence, and show that they are preserved in the limit.

We will use the simplified notation $\mu = \frac{t}{n}$ and $\lambda = \frac{t}{m}$, and note that

$$\left(I - \frac{t}{n}\mathsf{A}\right)^{-n}\mathsf{x} = \mathsf{R}_\mu^n\mathsf{x} \quad \text{and} \quad \left(I - \frac{t}{m}\mathsf{A}\right)^{-m}\mathsf{x} = \mathsf{R}_\lambda^m\mathsf{x}.$$

Hence, to show that $((I - \frac{t}{n}\mathsf{A})^{-n}\mathsf{x})_{n\in\mathbb{N}}$ is Cauchy in X, we need to obtain bounds on the quantity $a_{nm} := \|\mathsf{R}_\mu^n\mathsf{x} - \mathsf{R}_\lambda^m\mathsf{x}\|$.

Before proceeding, we note that the range condition (10.13) guarantees that $\overline{\mathbf{D}(\mathsf{A})} \subset \mathbf{D}(\mathsf{R}_\lambda)$, so the following arguments are well posed:

1. The crucial estimate that will be used for the rest of the proof is that for any $\mathsf{x} \in \mathbf{D}(\mathsf{A})$,

$$\begin{aligned}\|\mathsf{R}_\mu^n\mathsf{x} - \mathsf{R}_\lambda^m\mathsf{x}\| \le \{&((n\mu - m\lambda)^2 + n\mu(\lambda-\mu))^{1/2} \\ &+ (m\lambda(\lambda-\mu) + (m\lambda - n\mu)^2)^{1/2}\}|\mathsf{A}\mathsf{x}|_0, \quad \forall\, 0 < \mu \le \lambda,\ m \le n,\end{aligned} \tag{10.14}$$

where $|\mathsf{A}\mathsf{x}|_0 := \inf\{\|\mathsf{y}\| \ : \ \mathsf{y} \in \mathsf{A}(\mathsf{x})\}$.

Estimate (10.14) is based on the nonexpanding property of the operator R_λ (see Proposition 10.2.7), the important identity

$$\mathsf{R}_\lambda\mathsf{x} = \mathsf{R}_\mu\left(\frac{\mu}{\lambda}\mathsf{x} + \frac{\lambda-\mu}{\lambda}\mathsf{R}_\lambda\mathsf{x}\right), \quad \forall\, x \in \mathbf{D}(\mathsf{R}_\lambda),$$

established in Proposition 10.2.7(vi) (see eq. (10.6)) and the linear inequality

$$a_{jk} \le \frac{\mu}{\lambda}a_{j-1,k-1} + \frac{\lambda-\mu}{\lambda}a_{j-1,k}, \quad j \le n,\ k \le m,$$

where $a_{jk} = \|\mathsf{R}_\mu^j\mathsf{x} - \mathsf{R}_\lambda^k\mathsf{x}\|$, and some rather tedious combinatorial arguments, based on an induction procedure and the (generalized) binomial theorem. A complete proof of (10.14) is provided in the appendix of the chapter (Section 10.5.1, Lemma 10.5.1), so as not to disrupt the flow of the main proof.

2. To prove the Cauchy property of the sequence $((I - \frac{t}{n}\mathsf{A})^{-n}\mathsf{x})_{n\in\mathbb{N}}$, in X, we fix an arbitrary t and select

$$\mu = \frac{t}{n}, \quad \lambda = \frac{t}{m}, \quad m \le n.$$

We now apply (10.14) for this choice, noting that the upper bound simplifies considerably to yield

$$\|R^n_{\frac{t}{n}}x - R^m_{\frac{t}{m}}x\| \le 2t\Big(\frac{1}{m} - \frac{1}{n}\Big)^{1/2} |Ax|_0.$$

This shows that $\|R^n_{\frac{t}{n}}x - R^m_{\frac{t}{m}}x\| \to 0$, as $n, m \to \infty$, hence, $((I - \frac{t}{n}A)^{-n}x)_{n\in\mathbb{N}}$ is Cauchy in X. This implies the existence of a limit for this sequence, which we use for defining

$$S(t)x := \lim_{n\to\infty} R^n_{\frac{t}{n}}x = \lim_{n\to\infty}\Big(I - \frac{t}{n}A\Big)^{-n} x.$$

In fact, this works for any $x \in \overline{\mathbf{D}(A)}$, by taking a $(x_n)_{n\in\mathbb{N}} \subset \mathbf{D}(A)$, $x_n \to x$, expressing $R^n_{\frac{t}{n}}x - R^m_{\frac{t}{m}}x = R^n_{\frac{t}{n}}x - R^n_{\frac{t}{n}}x_n + R^n_{\frac{t}{n}}x_n - R^m_{\frac{t}{m}}x_n + R^m_{\frac{t}{m}}x_n - R^m_{\frac{t}{m}}x$, treating the second pair as above, and using the non expansivity of $R_{\frac{t}{n}}$ and $R_{\frac{t}{m}}$ combined with $x_n \to x$ for the other pairs.

3. The nonexpansive property of $S(t)$ follows by the nonexpansive property of the operators R_μ. Indeed, for any $x, z \in \overline{\mathbf{D}(A)}$, we have that

$$\|R^n_{\frac{t}{n}}x - R^n_{\frac{t}{n}}z\| \le \|R^{n-1}_{\frac{t}{n}}x - R^{n-1}_{\frac{t}{n}}z\| \le \cdots \le \|x - z\|,$$

from which, passing to the limit, as $n \to \infty$, we conclude that

$$\|S(t)x - S(t)z\| \le \|x - z\|, \quad \forall\, x, y \in \overline{\mathbf{D}(A)}.$$

4. We now establish the continuity with respect to t of the family of nonlinear operators $\{S(t)\}_{t\ge 0}$. In fact, we will show something stronger, i. e., the Lipschitz continuity of the family. Fix s, t, such that $0 \le t \le s$, and set

$$\mu = \frac{t}{n}, \quad \lambda = \frac{s}{n}, \quad n = m.$$

We now apply (10.14) for this choice, noting the upper bound simplifies considerably to yield

$$\|R^n_{\frac{t}{n}}x - R^n_{\frac{s}{n}}x\| \le \Big(\Big((s-t)^2 + \frac{t(s-t)}{n}\Big)^{1/2} + \Big((s-t)^2 + \frac{s(s-t)}{n}\Big)^{1/2}\Big)|Ax|_0.$$

Passing to the limit, as $n \to \infty$, we obtain

$$\|S(t)x - S(s)x\| \le 2|t - s|\, |Ax|_0,$$

from which the Lipschitz continuity of $t \mapsto S(t)x$ holds for all $x \in \mathbf{D}(A)$.

5. We finally establish the semigroup property $S(t + s) = S(t)S(s)$.

We first note that for any $t > 0$ and any natural number p, it holds that

$$S(t)^p = S(pt). \tag{10.15}$$

Indeed,

$$S(t)^p x = \left(\lim_{n\to\infty} R^n_{\frac{t}{n}}\right)^p x = \lim_{n\to\infty} R^{pn}_{\frac{t}{n}} x.$$

On the other hand,

$$S(pt)x = \lim_{n\to\infty} R^n_{\frac{pt}{n}} = \lim_{n\to\infty} R^{pn}_{\frac{pt}{pn}} = \lim_{n\to\infty} R^{pn}_{\frac{t}{n}}.$$

Combining these two, we obtain (10.15).

We now consider any two $s, t \in \mathbb{Q}$. Let $t = \frac{p_1}{q_1}$, $s = \frac{p_2}{q_2}$. Then,

$$\begin{aligned} S(t+s) &= S\left(\frac{p_1 q_2 + p_2 q_1}{q_1 q_2}\right) = \left(S\left(\frac{1}{q_1 q_2}\right)\right)^{p_1 q_2 + p_2 q_1} \\ &= \left(S\left(\frac{1}{q_1 q_2}\right)\right)^{p_1 q_2} \left(S\left(\frac{1}{q_1 q_2}\right)\right)^{p_2 q_1} \\ &= S\left(\frac{p_1 q_2}{q_1 q_2}\right) S\left(\frac{p_2 q_1}{q_1 q_2}\right) = S\left(\frac{p_1}{q_1}\right) S\left(\frac{q_1}{q_2}\right) = S(t)S(s), \end{aligned} \tag{10.16}$$

where we used (10.15) repeatedly. To complete the proof, consider two sequences in $\mathbb{Q}$, $t_n \to t$ and $s_n \to s$, apply (10.16) to t_n, s_n for any $n \in \mathbb{N}$, and then pass to the limit, as $n \to \infty$, using continuity. □

10.2.4 Nonlinear semigroups and evolution problems: a first encounter

We now address the important question of how is the nonlinear semigroup $\{S(t)\}_{t\geq 0}$ generated by A related to the solution of the Cauchy problem

$$\frac{dx}{dt}(t) = A(x(t)), \quad x(0) = x_0. \tag{10.17}$$

In particular, if we define $x(t) = S(t)x_0$, is $t \mapsto x(t)$ a solution of (10.17) and if so, in which sense? Here we obtain an important difference with the linear case. Unlike the linear case in which it is known (see Theorem 1.7.3) that $x(t)$ as above for $x_0 \in \mathbf{D}(A)$ is a classical solution of (10.17), in the nonlinear case the function $t \mapsto x(t) = S(t)x_0$ is not in general (strongly) differentiable, so $x(t)$ is not necessarily a solution of (10.17) in the classical (or as is usually the terminology here) a strong solution. While it is true that, if a strong solution to (10.17) exists such that $x_0 \in \overline{\mathbf{D}(A)}$, then under the assumptions of Theorem 10.2.9 this is unique and can be represented by $x(t) = S(t)x_0$, $t \geq 0$, where $\{S(t)\}_{t\geq 0}$ is the nonlinear semigroup generated by the Crandall–Ligget theorem (see, e. g., Theorem 5.6 in [89]), in general there is no guarantee that $x(t) = S(t)x_0$ is a strong solution of the Cauchy problem (10.17). That means that having obtained the generation of a strongly continuous semigroup $\{S(t)\}_{t\geq 0}$, by A, this cannot be used without other considerations to guarantee the existence of a strong solution for (10.17) in terms of $x(t) = S(t)x_0$, $t \geq 0$, (extra conditions need to be imposed) as we did for the linear case.

It is rather a solution in a weak sense, called a mild solution or integral solution. To guarantee that $x(t) = S(t)x_0$ is a solution of (10.17), we need to impose extra conditions, such as X reflexive and A closed.

We announce the following result that will be proved (in a more general setting) later (see Theorem 10.2.21) [18].

Theorem 10.2.11. *Let* A *satisfy the assumptions made in Theorem* 10.2.9. *Additionally, let* X *be reflexive and* $x_0 \in \mathbf{D}(A)$. *Then,* $x(t) = S(t)x_0$ *is the strong solution of the Cauchy problem* (10.17).

The obvious question is: What about the case where X is not reflexive? This discussion will require the introduction of a new weaker concept of solutions, called mild or integral solutions, and a rather involved discussion concerning the identification of these solutions with $x(t) = S(t)x_0$. We opt to do this in the general case of the nonhomogeneous Cauchy problem (10.17), and then deduce the connection of the nonlinear semigroup with the Cauchy problem as a special case. However, at this point, we can motivate the relevant notion of generalized solutions.

Consider an interval $[0,T]$ and break it up into n subintervals, each of length h. Clearly, since T is fixed and finite, $n \to \infty$ is equivalent to $h \to 0$. Now, for each $t \in [0,T]$, there is an $i \in \mathbb{N}$ such that $t \in [ih,(i+1)h)$, $i = 0,1,\dots,n = [T/h]$ (where, by $[\cdot]$, we denote the integer part). Then for each t, we set $x_h(t) := x_{h,i}$, where $x_{h,i}$ is a constant. Note that $\frac{dx}{dt}(t)$ can be approximated by $\frac{x_{h,i+1}-x_{h,i}}{h} = \frac{x_{h,i}-x_{h,i+1}}{-h}$, so that we can replace the differential equation by the difference equation, corresponding to the backward Euler scheme,

$$\frac{x_{h,i+1}-x_{h,i}}{h} \in A(x_{h,i+1}) \implies x_{h,i} \in x_{h,i+1} - hA(x_{h,i+1}), \implies x_{h,i+1} = R_h(x_{h,i}), \quad h = \frac{T}{n}, \tag{10.18}$$

with initial condition $x_{h,0} = x_0 = x(0)$. Note that (10.18) is equivalent to the Euler scheme for $\frac{dx_h}{dt} = A_h(x_h)$. Consider the family of piecewise constant functions $(x_h)_{h>0}$, such that

$$\forall\, t \in [0,T], \quad x_h(t) = x_{h,i}, \quad \text{if } t \in \big[ih,(i+1)h\big). \tag{10.19}$$

By induction, we may see that $x_h(t) = x_{h,i} = (I - hA)^{-i}x_{h,0} = (I - hA)^{-i}x_0$. To simplify the argument, assume that $t = ih$ so that the above can be expressed as $x_h(t) = (I - \frac{t}{i}A)^{-i}x_0$. Since t is finite, the limit, as $h \to 0$, corresponds to the limit $i \to \infty$, hence we expect that $\lim_{h\to 0} x_h(t) = \lim_{i\to\infty}(I - \frac{t}{i}A)^{-i}x_0 = S(t)x_0$, which is the nonlinear semigroup constructed in Theorem 10.2.9. This is a hand waving argument that will be turned rigorous in the next section, but indicates the connection of the nonlinear semigroup $\{S(t)\}_{t\geq 0}$, with the limit of the discretized solution as the time step $h \to 0$.

What will be shown in the next section is that the sequence of functions $(x_h)_{h>0}$ converges to a continuous function x, which will be considered as some type of weak solution of the Cauchy problem (10.17), called mild solutions. In fact, by definition, mild solutions will be defined as appropriate continuous limits of the solution of the discretized backward Euler scheme. It is closely connected with a relative notion of weak solutions, introduced by Benilan and called integral solutions. For a general Banach space X, the nonlinear semigroup constructed in terms of the exponential formula in the Crandall–Liggett theorem 10.2.9 is related with a mild or integral solution. Only under extra assumptions on X this solution will be a strong solution (i. e., the semigroup will be strongly differentiable).

We introduce all these notions, and we show these results for the general nonhomogeneous Cauchy problem in Section 10.2.6:

10.2.5 Generation of nonlinear C_0-semigroups

The Crandall–Liggett theorem (see Theorem 10.2.9) can be extended for the generation of more general nonlinear C_0-semigroups (rather than simply contraction semigroups).

Let $\{S(t)\}_{t\geq 0}$ be a nonlinear semigroup of ω-type, i.e, satisfying

$$\|S(t)x_1 - S(t)x_2\| \leq e^{\omega t}\|x_1 - x_2\|, \quad x_1, x_2 \in K,\ t > 0.$$

For the theory concerning the generation of such semigroups, a crucial role is played by the so called ω-dissipative operators.

Definition 10.2.12 (ω-dissipative operator). Let $A : \mathbf{D}(A) \subset X \to X$ be a nonlinear operator.

(i) A is called ω-dissipative if $A_\omega := A - \omega I$ is dissipative.
(ii) A is called ω-m dissipative if $A_\omega := A - \omega I$ is m-dissipative.
(iii) A is called ω-accretive (or ω- m accretive) if $-A_\omega$ is dissipative (or m-dissipative).

The properties of ω-dissipative and ω-m dissipative operators can be easily read off from the corresponding properties of dissipative and m-dissipative operators, by carefully rexpressing them for A_ω. We may also modify accordingly Proposition 10.2.7, especially the statements concerning the resolvent operator $R(\lambda, A) = (I - \lambda A)^{-1}$. Straightforward calculations easily show that A being ω-dissipative is equivalent to

$$\|x_1 - x_2 - \lambda(y_1 - y_2)\| \geq (1 - \omega\lambda)\|x_1 - x_2\|, \quad \forall\,(x_i, y_i) \in \mathbf{Gr}(A),\ i = 1, 2,\ 0 < \lambda < \frac{1}{|\omega|}.$$

Consequently, the resolvent operator $R(\lambda, A) : X \to X$ is well defined and single valued, as long as $\lambda \in (0, \frac{1}{|\omega|})$. Moreover, for such λ, it is Lipschitz, with Lipschitz constant $\frac{1}{1-\lambda\omega}$. Similar properties hold for the Yosida approximation.

The Crandall–Liggett generation theorem (Thm. 10.2.9) can be generalized for ω-dissipative operators as follows [18]:

Theorem 10.2.13. *Let* A *be an ω-dissipative operator on X, $\omega \in \mathbb{R}$, satisfying the range condition, $\overline{\mathbf{D}(A)} \subset \mathbf{R}(I - \lambda A)$, for sufficiently small $\lambda > 0$. Then, there exists a strongly continuous semigroup $\{S(t)\}_{t\geq 0}$ of type ω on $\overline{\mathbf{D}(A)}$, which is given by the exponential formula*

$$S(t)x = \lim_{\lambda \to 0^+} (I - \lambda A)^{-[t/\lambda]} x, \quad \forall\, x \in \overline{\mathbf{D}(A)}.$$

Moreover, for every $x \in \mathbf{D}(A)$ *and* $0 \le t_1 \le t_2$, *it holds that*

$$\|S(t_2)x - S(t_1)x\| \le \begin{cases} (t_2 - t_1)e^{\omega t_2}\, \|Ax\|, & \omega > 0, \\ (t_2 - t_1)e^{\omega t_1}\, \|Ax\|, & \omega < 0. \end{cases}$$

10.2.6 Nonlinear semigroups and evolution problems

We now consider the nonhomogeneous problem

$$\begin{aligned} \frac{dx}{dt} &\in Ax + f, \\ x(0) &= x_0 \end{aligned} \tag{10.20}$$

in a Banach space X for $t \in [0, T]$, or $t \in (0, T]$, where $x_0 \in X$ is an initial condition, A is a suitable nonlinear operator (possibly multivalued), and $f \in L^1(0, T; X)$ is a given forcing term. In the case where $f = 0$, problem (10.20) reduces to the homogeneous Cauchy problem (10.17).

Unlike the case of linear operators, in which the representation formula (1.15) for the solution of (10.20), in terms of the linear semigroup $\{S(t)\}_{t\ge 0} = \{e^{tA}\}_{t\ge 0}$ can be used for the construction of the solution to the inhomogeneous problem, the nonlinear case presents several problems. This is due to the nonlinear nature of the problem, which no longer allows for the use of the principle of superposition in the construction of solutions. On account of that, we may not directly use the nonlinear semigroup $\{S(t)\}_{t\ge 0}$ constructed in the previous section to construct and represent solutions to the nonlinear inhomogeneous. However, we may take a different route, which in fact may be even more useful, as it has direct links to the numerical analysis of differential equations of the type (10.20). This is a time discretization scheme, akin to the backward Euler scheme (10.18), modified to take into account the nonhomogeneous term

$$\frac{x_{h,n+1} - x_{h,n}}{h} \in A(x_{h,n+1}) + f_{h,n}, \tag{10.21}$$

where the subscript h is included to emphasize the important of the choice of the discretization parameter h. As already noted for a fixed interval $[0, T]$, we can choose $h = \frac{T}{n}$ for a suitable n, and then construct the family of piecewise constant functions, as in (10.19). We will see that in the limit, as $h \to 0$, this family attains a limit, which will be interpreted as a type of weak solutions of (10.20), that will be called mild solutions. We shall see that these are closely related to another type of weak solutions called integral solutions. Hence, the construction of solutions to (10.20) will be obtained in terms of this limiting procedure, obtaining thus the existence of mild solutions. To upgrade these mild solutions to strong solutions, we will impose extra conditions on the Banach space X, and the forcing term f, in terms of regularity theorems. This is essentially the road map of the current section.

10.2.6.1 Concepts of solutions: mild, integral, and strong

We start by defining the various notions of solutions that will be considered for (10.20) (see e. g. [18]). To proceed further, we need a generalization of the backward Euler scheme (10.21), involving a more general discretization scheme, with possibly variable time steps (as would also be suitable for a more realistic numerical approximation scheme).

Definition 10.2.14 (ϵ-approximate solution). Consider a partition $\{t_i^{(n)}\ i = 0,\dots,n\} \subset [0,T]$ of the time interval $[0,T]$, such that $0 = t_0^{(n)} < t_1^{(n)} < \cdots < t_n^{(n)} = T(n) \le T$, into n intervals, and a collection $\{f_i^{(n)},\ i = 0,\dots,n\} \subset X$ such that the piecewise constant function

$$f_n : [0,T(n)] \to X \quad \text{with } f_n(t) = \begin{cases} f_i^{(n)}, & t \in (t_{i-1}^{(n)}, t_i^{(n)}], \\ f_n(0) = f_i^{(0)} \end{cases}$$

satisfies $\|f - f_n\|_{L^1(0,T;X)} \to 0$ as $n \to \infty$.

The collection $\{\mathsf{x}_i^{(n)},\ i = 0,\dots,n\} \subset X$, and the corresponding piecewise constant function

$$\mathsf{x}_n : [0,T(n)] \to X \quad \text{with } \mathsf{x}_n(t) = \begin{cases} \mathsf{x}_i^{(n)}, & t \in (t_i^{(n)}, t_{i+1}^{(n)}], \\ \mathsf{x}_n(0) = \mathsf{x}_0^{(n)} \end{cases} \tag{10.22}$$

is called an ϵ-approximate solution of (10.20) if it satisfies

$$\frac{\mathsf{x}_i^{(n)} - \mathsf{x}_{i-1}^{(n)}}{t_i^{(n)} - t_{i-1}^{(n)}} - \mathsf{A}(\mathsf{x}_i^{(n)}) \ni f_i^{(n)}, \quad i = 1,\dots,n,$$

for a choice of n such that

$$\max_{i=1,\dots,n}\{|t_{i+1}^{(n)} - t_i^{(n)}|, T - T(n)\} < \epsilon, \quad \|f_n - f\|_{L^1((0,T);X)} < \epsilon, \quad \|\mathsf{x}_0 - \mathsf{x}_0^{(n)}\| < \epsilon.$$

Note that in the special case where $t_i^{(n)} - t_{i-1}^{(n)} = h = \frac{T}{n}$ for all $i = 1,\dots,n$, we obtain the standard backward Euler discretization scheme with constant step size h. We may then consider the limit as $n \to \infty$ or equivalently $h \to 0$.

By increasing n, the partition $\{t_i^{(n)}\ :\ i = 0,\dots,n\}$ becomes increasingly finer, and our intuition implies that the corresponding piecewise constant function x_n, as defined in (10.22), will approximate the solution x of the corresponding Cauchy problem (10.20). This corresponds to changing the number of intervals n in the partition of $[0,T]$, for

$n = 1, 2, \ldots$, thus creating a sequence $(\mathsf{x}_n)_{n\in\mathbb{N}}$ of piecewise constant functions (each one defined as in (10.22)). The solution to the Cauchy problem (10.20) will then be identified by the limit of $(\mathsf{x}_n)_{n\in\mathbb{N}}$ as $n \to \infty$. Importantly, from the point of view of numerical analysis, the solution to (10.22) can be approximated by x_N, an element of the sequence $(\mathsf{x}_n)_{n\in\mathbb{N}}$, for sufficiently large N. In fact, one could also consider different partitions, e. g., $\{t_i^{(n)},\ i = 0, \ldots, n\}$, $\{s_j^{(m)},\ i = 0, \ldots, m\}$ and different approximations of the forcing term f, and obtain similar results for the corresponding ϵ-approximate solutions $\{\mathsf{x}_i^{(n)},\ i = 0, \ldots, n\}$ and $\{\bar{\mathsf{x}}_j^{(m)},\ i = 0, \ldots, m\}$ or equiv. for the corresponding sequence of piecewise constat functions (see Theorem 10.2.18).

The concept of the mild solution for (10.20) is defined in terms of the ϵ-approximate solution.

Definition 10.2.15 (Mild solution). A mild solution of (10.20) is a continuous function, $\mathsf{x} : [0, T] \to X$ such that for every $\epsilon > 0$ there exists an ϵ-approximate solution x_n such that

$$\|\mathsf{x}(t) - \mathsf{x}_n(t)\| \le \epsilon, \quad \forall\, t \in [0, T], \quad \mathsf{x}_n(0) = \mathsf{x}_0.$$

In some sense, the above definition implies that the mild solution will be obtained as the limit, as $\epsilon \to 0$ (or, loosely speaking, $n \to \infty$) of an ϵ-approximate solution. Hence, showing the existence of a mild solution for the nonlinear nonhomogeneous problem (10.20) requires showing that we can pass to the limit, as $n \to \infty$, in the sequence $(\mathsf{x}_n)_{n\in\mathbb{N}}$.

Closely connected to mild solutions are integral solutions:

Definition 10.2.16 (Integral solutions). An integral solution of (10.20) is a function $\mathsf{x} \in C([0, T]; X)$ such that $\mathsf{x}(0) = \mathsf{x}_0$ and

$$\frac{1}{2}\|\mathsf{x}(t) - \mathsf{z}\|^2 \le \frac{1}{2}\|\mathsf{x}(s) - \mathsf{x}\|^2 + \int_s^t [f(\tau) - \mathsf{A}(\mathsf{z}), \mathsf{x}(\tau) - \mathsf{z}]_+ d\tau,$$
$$\forall\, \mathsf{z} \in \mathbf{D}(\mathsf{A}), \quad 0 \le s \le t \le T,$$

where $[\cdot, \cdot]_+$ denotes the semi-inner product (see Definition 10.2.8).

The connection between mild and integral solutions is not obvious but will be established in Theorem 10.2.18 below.

We finally state the concept of strong solution.

Definition 10.2.17 (Strong solution). A strong solution of (10.20) is a function $\mathsf{x} \in W^{1,1}([0, T]; X) \cap C([0, T]; X)$, such that $\frac{d\mathsf{x}}{dt}(t) - f(t) \in \mathsf{A}(\mathsf{x}(t))$ a. e. $t \in (0, T]$, with $\mathsf{x}(0) = \mathsf{x}_0$.

10.2.6.2 Existence of mild solutions and connection with nonlinear semigroups

The connection of ϵ-approximate solutions to mild solutions, hence providing an existence result for the latter, is given in the next theorem [18, 59].

Theorem 10.2.18. *Let $\mathrm{x}_0 \in \overline{\mathbf{D}(\mathrm{A})}$, let A be a dissipative operator, and assume that for every $\epsilon > 0$, sufficiently small, problem* (10.20) *admits an ϵ-approximate solution. This is guaranteed if the range condition* (10.13) *holds (with λ and ϵ related) or if A is m-dissipative.*

Then, (10.20) *admits a unique mild solution, which can be approximated by an ϵ-approximate solution x_n satisfying the estimate*

$$\|\mathrm{x}(t) - \mathrm{x}_n(t)\| \leq \delta(\epsilon), \quad t \in [0, T - \epsilon],$$

for some continuous function $\epsilon \mapsto \delta(\epsilon)$ such that $\delta(0) = 0$.

Finally, the mild solution satisfies the following stability property: Given any $f \in L^1(0,T;X)$, $\bar{f} \in L^1(0,T;X)$, the corresponding solutions x, $\bar{\mathrm{x}}$ satisfy the estimate

$$\begin{aligned} \|\mathrm{x}(t) - \bar{\mathrm{x}}(t)\| &\leq \|\mathrm{x}(s) - \bar{\mathrm{x}}(s)\| \\ &\quad + \int_s^t [\mathrm{x}(\tau) - \bar{\mathrm{x}}(\tau), f(\tau) - \bar{f}(\tau)]_+ d\tau, \quad 0 \leq s \leq t \leq T, \end{aligned} \tag{10.23}$$

where $[\cdot,\cdot]_+$ denotes the semi-inner product (see Definition 10.2.8*).*

Remark 10.2.19. The stability estimate (10.23) is important for obtaining a number of interesting features for the mild solution, such as uniquencess or Lipschitz continuity. Moreover, x and $\bar{\mathrm{x}}$ can correspond either to different forcing terms $f, \bar{f}$ or to different discretizations of the same forcing term f (a version that is important for numerical analysis considerations).

Proof of Theorem 10.2.18. We follow the approach in [18]. The general strategy for the proof is as follows:

Consider two partitions of $[0,T]$,

$$\begin{aligned} P_n &:= \{t_i^{(n)},\ i = 0,\dots,n\}, \quad \text{with } h_i^{(n)} = t_{i+1}^{(n)} - t_i^{(n)} \\ \bar{P}_m &:= \{s_j^{(m)},\ j = 0,\dots,m\}, \quad \text{with } \bar{h}_j^{(m)} = s_{j+1}^{(m)} - s_j^{(m)}, \end{aligned} \tag{10.24}$$

the discrete approximations of two functions $f, \bar{f}$, in these partitions, respectively,

$$f_n^D := \{f_i^{(n)},\ i = 0,\dots,n\} \quad \text{and} \quad \bar{f}_m^D := \{\bar{f}_j^{(m)},\ j = 0,\dots,m\},$$

and the two dimensional array

$$f_{n,m}^D := \{f_{ij}^{(n,m)},\ i = 0,\dots,n,\ j = 0,\dots,m\}, \quad \text{with } f_{ij}^{(n,m)} = \|f_i^{(n)} - \bar{f}_j^{(m)}\|.$$

Note that even if $f = \bar{f}$, it may still hold $f_n^D \neq \bar{f}_m^D$, as these will correspond to two different discretization schemes for the same function.

Take (10.20) with $f, \bar{f}$, and the corresponding ϵ-approximate solutions

$$\mathrm{x}_n^D := \{\mathrm{x}_i^{(n)},\ i = 0,\dots,n\} \quad \text{and} \quad \bar{\mathrm{x}}_m^D := \{\bar{\mathrm{x}}_j^{(m)},\ j = 0,\dots,m\},$$

to the above partitions, and define the two-dimensional array $e^D_{n,m}$ of their differences:

$$e^D_{n,m} := \{e^{(n,m)}_{ij},\ i = 0,\dots,n,\ j = 0,\dots,m\}, \quad \text{where } e^{(n,m)}_{i,j} = \|\mathsf{x}^{(n)}_i - \bar{\mathsf{x}}^{(m)}_j\|.$$

Finally, for each $n, m \in \mathbb{N}$ fixed, we will define the operator $\mathsf{L}^D_{n,m}$ acting on two-dimensional arrays $a^D_{n,m} = \{a^{(n,m)},\ i = 0,\dots,n,\ j = 0,\dots,m\}$ by

$$(\mathsf{L}^D_{n,m} a^D_{n,m})_{ij} := (1/h^{(n)}_i)(a^{(n,m)}_{i,j} - a^{(n,m)}_{i-1,j}) + (1/\bar{h}^{(m)}_j)(a^{(n,m)}_{i,j} - a^{(n,m)}_{i,j-1}) \tag{10.25}$$

for all i, j compatible with the allowed indices' values (see below for boundary conditions), and note its connection with the differential operator L acting on functions $a : [0,T] \times [0,T] \to \mathbb{R}$, defined by

$$(\mathsf{L}a)(t,s) := \frac{\partial}{\partial t} a(t,s) + \frac{\partial}{\partial s} a(t,s). \tag{10.26}$$

The operator $\mathsf{L}^D_{n,m}$ can be considered as the discretized version of the operator L, a comment that is formal at this point, but will be made fully rigorous in the course of the proof.

In what follows, we will consider the sequences $(\mathsf{x}^D_n)_{n\in\mathbb{N}}$ (or $(\bar{\mathsf{x}}^D_m)_{m\in\mathbb{N}}$), which would correspond to the sequences of ϵ-approximate solutions of (10.20), for different possible partitions of $[0,T]$, i. e., the ones we will obtain by varying $n \in N$ (or $m \in \mathbb{N}$), and try to establish that these sequences do have a limit, as $n \to \infty$ (or $m \to \infty$), which satisfies the definition of a mild solution for (10.20). In doing so, we need to identify each element x^D_n of the sequence $(\mathsf{x}^D_n)_{n\in\mathbb{N}}$ with a piecewise constant function $\mathsf{x}_n : [0,T] \to X$, and interpret $(\mathsf{x}^D_n)_{n\in\mathbb{N}}$ as a sequence of piecewise constant functions $(\mathsf{x}_n)_{n\in\mathbb{N}} \subset L^1((0,T);X)$ that has a limit x in this function space, which will be shown to satisfy all the required properties so that it is a mild solution for (10.20). Similarly for $(\bar{\mathsf{x}}^D_m)_{m\in\mathbb{N}}$. An argument to show the existence of a limit would be to establish the Cauchy property of the sequences $(\mathsf{x}^D_n)_{n\in\mathbb{N}}$ (or $(\bar{\mathsf{x}}^D_m)_{m\in\mathbb{N}}$). This could follow if, assuming $f = \bar{f}$, so that the sequences $(\mathsf{x}^D_n)_{n\in\mathbb{N}}$ and $(\bar{\mathsf{x}}^D_m)_{m\in\mathbb{N}}$ are essentially different approximations of the same function, we could show that the sequence of differences $e^D_{n,m} \to 0$ as $n, m \to \infty$. Such a result requires detailed estimates for the terms $e^{(n,m)}_{ij}$, $i = 0,\dots,n, j = 0,\dots,m$ for each $n, m \in \mathbb{N}$ fixed, and then a careful passage to the limit as $n, m \to \infty$. The required bounds for e^D_{mn} will be obtained upon the observation that they satisfy a difference inequality, defined in terms of the discrete operator $\mathsf{L}^D_{n,m}$ defined in (10.25), which is in turn related to a first-order linear PDE, related to the operator L defined in (10.26), which admits an exact analytic solution, in terms of which the required estimates for the difference inequality are obtained. These steps are presented in detail below.

The proof follows in several steps.

1. We first obtain a stability estimate for ϵ-approximate solutions, which is of interest in its own right. Using the above notation, and fixing $n, m \in \mathbb{N}$, we have that the two-dimensional array $e^D_{n,m}$ satisfies for all $i = 1,\dots,n$ and $j = 1,\dots,m$ the estimate

$$e_{i,j}^{(n,m)} \leq \frac{\bar{h}_j^{(m)}}{h_i^{(n)} + \bar{h}_j^{(m)}} e_{i-1,j}^{(n,m)} + \frac{h_i^{(n)}}{h_i^{(n)} + \bar{h}_j^{(m)}} e_{i,j-1}^{(n,m)} + \frac{h_i^{(n)} \bar{h}_j^{(m)}}{h_i^{(n)} + \bar{h}_j^{(m)}} f_{ij}^{(n,m)}, \tag{10.27}$$

or its equivalent formulation

$$\frac{1}{h_i^{(n)}}\left(e_{i,j}^{(n,m)} - e_{i-1,j}^{(n,m)}\right) + \frac{1}{\bar{h}_j^{(m)}}\left(e_{i,j}^{(n,m)} - e_{i,j-1}^{(n,m)}\right) \leq f_{ij}^{(n,m)}, \tag{10.28}$$

or in terms of the discrete operator $\mathsf{L}_{n,m}^D$ defined in (10.25):

$$(\mathsf{L}_{n,m}^D e_{n,m}^D)_{i,j} \leq (f_{n,m}^D)_{i,j} \iff \mathsf{L}_{n,m}^D e_{n,m}^D \leq f_{n,m}^D, \tag{10.29}$$

where the left inequality holds for all suitable valued of the indices i, j, and the right inequality is a shorthand notation for the left one. We note that (10.27) (or its equivalent formulations (10.28) and (10.29)) still holds, replacing $f_{ij}^{(n,m)}$ by $\hat{f}_{ij}^{(n,m)} = [\mathsf{x}_i^{(n)} - \bar{\mathsf{x}}_j^{(m)}, f_i^{(n)} - \bar{f}_j^{(m)}]_+$.

Estimate (10.27) (or its equivalents) follows by the definition of ϵ-approximate solutions, the properties of the operator A (and in particular dissipativity) and the definition and properties of $[\mathsf{x}, \mathsf{z}]_+$. So as not to disrupt the flow of the arguments, the proof of (10.27) is provided in the appendix of the chapter, Section 10.5.2.

Estimate (10.27) (or its equivalents) has to be complemented with an appropriate boundary condition. This corresponds to specifying the double array $e_{n,m}^D = \{e_{ij}^{(n,m)},\ i = 0, \dots, n,\ j = 0, \dots, m\}$ for $i = 0$ and $j = 0$. By a modification of the arguments that led to (10.27) (see Section 10.5.3 in the appendix of the chapter), we can obtain the estimates:

$$\|\mathsf{x}_i^{(n)} - \mathsf{x}\| \leq \|\mathsf{x}_0^{(n)} - \mathsf{x}\| + \sum_{i_1=0}^{i} h_{i_1}^{(n)}(\|f_{i_1}^{(n)}\| + \|\mathsf{y}\|), \quad \forall\, \mathsf{x} \in \mathbf{D}(\mathsf{A}), \quad \mathsf{y} \in \mathsf{A}(\mathsf{x}). \tag{10.30}$$

A similar estimate can be obtained for $\|\bar{\mathsf{x}}_j^{(m)} - \mathsf{x}\|$, in terms of

$$\|\bar{\mathsf{x}}_j^{(m)} - \mathsf{x}\| \leq \|\bar{\mathsf{x}}_0^{(m)} - \mathsf{x}\| + \sum_{j_1=0}^{j} \bar{h}_{j_1}^{(m)}(\|\bar{f}_{j_1}^{(m)}\| + \|\mathsf{y}\|), \quad \forall\, \mathsf{x} \in \mathbf{D}(\mathsf{A}), \quad \mathsf{y} \in \mathsf{A}(\mathsf{x}). \tag{10.31}$$

Even though x may not be equal to x_0, by the assumption that $\mathsf{x}_0 \in \overline{\mathbf{D}(\mathsf{A})}$, we may approximate x_0 arbitrarily close with an $\mathsf{x} \in \mathbf{D}(\mathsf{A})$, so that (10.30) and (10.31) can be interpreted as inequalities for $\|\mathsf{x}_i^{(n)} - \mathsf{x}_0\| = e_{i,0}^{(n,m)}$ for all $i = 1, \dots, n$ and $\|\bar{\mathsf{x}}_j^{(m)} - \mathsf{x}_0\| = e_{0,j}^{(n,m)}$, for all $j = 1, \dots, m$. These can be interpreted as

$$
\begin{aligned}
e_{i,0}^{(n,m)} &\leq \mathfrak{g}_i, \quad i = 1, \dots, n, \\
e_{0,j}^{(n,m)} &\leq \bar{\mathfrak{g}}_j, \quad j = 1, \dots, m,
\end{aligned} \tag{10.32}
$$

where $\mathfrak{g}_i$ and $\bar{\mathfrak{g}}_j$ are appropriate bounds consistent with (10.30) and (10.31), and are to be considered as boundary conditions. As a final reformulation of (10.32), we "unify" the two arrays, $\{\mathfrak{g}_i,\ i = 1, \dots, n\}$ and $\{\bar{\mathfrak{g}}_j,\ j = 1, \dots, m\}$, into a single double array $\{g_{ij},\ i = 0, \dots, n,\ j = 0, \dots, m\}$ chosen such that

$$
g_{ij} = \begin{cases} \mathfrak{g}_i, & i = 1, \dots, n,\ j = 0, \\ \bar{\mathfrak{g}}_j, & j = 1, \dots, n,\ i = 0, \end{cases}
$$

in terms of which the boundary condition in (10.32) can be expressed as

$$
e_{ij}^{(n,m)} \leq g_{ij}, \quad i = 0, \ \text{or}\ j = 0.
$$

We are now in position to complete our estimates by incorporating in (10.27) (or its equivalents) the appropriate boundary conditions, to conclude that in terms of the operator $\mathsf{L}_{n,m}^D$, $e_{n,m}^D$ satisfies the system of inequalities

$$
\begin{aligned}
&\begin{cases} (\mathsf{L}_{n,m}^D e_{n,m}^D)_{ij} \leq f_{ij}^{(n,m)}, & i = 1, \dots, n,\ j = 1, \dots, m \\ (e_{n,m}^D)_{ij} \leq g_{ij}, & i = 0, \ \text{or}\ j = 0. \end{cases} \\
\Longleftrightarrow\ &\begin{cases} \mathsf{L}_{n,m}^D e_{n,m}^D \leq f_{n,m}^D, & \\ (e_{n,m}^D)_{ij} \leq g_{ij}, & i = 0, \ \text{or}\ j = 0. \end{cases}
\end{aligned} \tag{10.33}
$$

The system (10.33) can be considered as a discrete boundary value problem.

Obtaining a solution to the system of inequalities (10.33) allows us to obtain useful estimates for $e_{ij}^{(n,m)}$, and hence for $e_{n,m}^D$ (for any fixed $n, m \in \mathbb{N}$), which will then be used to establish the behavior of $e_{n,m}^D$ as $n, m \to \infty$, allowing us to establish that the sequence of piecewise constant functions, defined by the discretized solution, is Cauchy, hence convergent. Such estimates are crucial for the proof of the theorem.

However, problem (10.33) is not easy to solve explicitly. The strategy is to replace problem (10.33) by a continuous problem, in terms of a PDE related to the operator L (defined in (10.26)), which is more tractable and yields a solution in semi-closed form, shows that, for sufficiently fine grids, the solution of the continuous problem is arbitrarily close to the solutions of the discrete problem (10.33), and then use the solution of the relevant continuous problem for our estimates. These tasks are performed in the following steps:

2. Our first observation is that, for any $n, m \in \mathbb{N}$ fixed, in order to obtain an upper bound for $e_{n,m}^D = \{e_{ij}^{(n,m)},\ i = 0, \dots, n,\ j = 0, \dots, m\}$, we may use $E_{n,m}^D = \{E_{ij}^{(n,m)},\ i = 0, \dots, n,\ j = 0, \dots, m\}$, which is the solution of the equation

$$\begin{cases} (\mathsf{L}^D_{n,m} E^D_{n,m})_{ij} = f^{(n,m)}_{ij}, & i = 1, \dots, n,\ j = 1, \dots, m \\ (E^D_{n,m})_{ij} = g_{ij}, & i = 0,\ \text{or}\ j = 0. \end{cases}$$
$$\Longleftrightarrow \begin{cases} \mathsf{L}^D_{n,m} E^D_{n,m} = f^D_{n,m}, \\ (E^D_{n,m})_{ij} = g_{ij}, & i = 0,\ \text{or}\ j = 0. \end{cases} \tag{10.34}$$

The system (10.34) can be considered as a discrete boundary value problem, which is identical to (10.33), except for the equality sign.

It is easy to see that for positive initial and boundary data, $E^{(n,m)}_{i,j}$ remains positive, as long as $f^{(n,m)}_{ij} \geq 0$. Since, for any fixed $n, m \in \mathbb{N}$, the error term $e^D_{n,m}$ satisfies (10.33), multiplying this inequality by -1, adding it to (10.34), we obtain that

$$\mathsf{L}^D_{n,m}(E^D_{n,m} - e^D_{n,m}) \geq 0, \tag{10.35}$$

so by the above remark

$$e^{(n,m)}_{ij} \leq E^{(n,m)}_{ij}. \tag{10.36}$$

Therefore, (10.36) provides us with an upper bound to $e^{(n,m)}_{ij}$, which can be used to show the convergence of the ϵ-approximate solution to a mild solution, as long as we can estimate e_{ij}. For this last task, we can use as an approximation for $E^{(n,m)}_{ij}$ the value $E(t,s)$ of a function $E : [0,T] \times [0,T] \to \mathbb{R}$, which is the solution to an appropriate continuous problem, which can be obtained analytically, hence leading to a useful bound for $e^D_{n,m}$.

3. Upon defining the function $F : [0,T] \times [0,T] \to \mathbb{R}$ by $F(t,s) = \|f(t) - \bar{f}(s)\|$, we note that (10.34) can be expressed as the discretized version of the linear first-order PDE

$$\mathsf{L}E(t,s) := \frac{\partial}{\partial t} E(t,s) + \frac{\partial}{\partial s} E(t,s) = F(t,s), \tag{10.37}$$

for the (unknown) function $E : [0,T] \times [0,T] \to \mathbb{R}$.

Indeed, using the two partitions of $[0,T]$ in (10.24), we observe that

$$f^{(n,m)}_{ij} = F(t^{(n)}_i, s^{(m)}_j).$$

Upon defining $E^{(n,m)}_{ij} := E(t^{(n)}_i, s^{(m)}_j)$, we can then approximate the functions $E_t := \frac{\partial E}{\partial t}$ and $E_s := \frac{\partial E}{\partial s}$ in terms of the finite difference approximations

$$\begin{aligned} E^{(n,m)}_{t,ij} &:= \frac{\partial E}{\partial t}(t^{(n)}_i, s^{(m)}_j) \simeq \frac{1}{h^{(n)}_i}(E^{(n,m)}_{ij} - E^{(n,m)}_{i-1,j}) \\ E^{(n,m)}_{s,ij} &:= \frac{\partial E}{\partial s}(t^{(n)}_i, s^{(m)}_j) \simeq \frac{1}{\bar{h}^{(m)}_j}(E^{(n,m)}_{ij} - E^{(n,m)}_{i,j-1}). \end{aligned} \tag{10.38}$$

Plugging these into (10.37), we see that we can (formally) approximate the solution of (10.37) in terms of the solution of the difference equation

$$(\mathsf{L}^D_{n,m}E^{(n,m)})_{ij} := \frac{1}{h_i^{(n)}}(E_{ij}^{(n,m)} - E_{i-1,j}^{(n,m)}) + \frac{1}{\bar{h}_j^{(m)}}(E_{ij}^{(n,m)} - E_{i,j-1}^{(n,m)}) \\ = f_{ij}^{(n,m)}. \tag{10.39}$$

However, (10.39) is nothing but the first equation in system (10.34). This implies that if indeed (10.34) is a good approximation of its continuous version (10.37), we may use the function E calculated on the relevant points $t_i^{(n)}$, $s_j^{(m)}$ as an estimate for $(E^D_{n,m})_{ij} = E_{ij}^{(n,m)}$ and, by (10.36), as an estimate to our quantity of interest $(e^D_{n,m})_{ij} = e_{ij}^{(n,m)}$. What makes this prospect very appealing is the fact that (10.37) admits an analytic solution.

To fully compare (10.34) with its continuous version (10.37), and to obtain the analytic solution to the latter, we need to assign some boundary conditions to (10.37).

The relevant boundary conditions for (10.37) are

$$E(0,s) = \bar{\mathfrak{g}}(s), \quad \forall\, s \in [0,T],$$
$$E(t,0) = \mathfrak{g}(t), \quad \forall\, t \in [0,T],$$

where the functions $\bar{\mathfrak{g}} : [0,T] \to \mathbb{R}$ and $\mathfrak{g} : [0,T] \to \mathbb{R}$ are chosen so that

$$\bar{\mathfrak{g}}(s_m^{(j)}) = \bar{\mathfrak{g}}_j, \quad j = 1,\dots,m,$$
$$\mathfrak{g}(t_n^{(i)}) = \mathfrak{g}_i, \quad i = 1,\dots,n.$$

It turns out that in view of being able to obtain an easy analytic representation of the solution to (10.40), it is better to express these boundary conditions in terms of a single function of one variable $g : \mathbb{R} \to \mathbb{R}$, i. e.,

$$E(t,s) = g(t-s) \quad \text{for } t = 0 \text{ or } s = 0,$$

with $g : [-T,T] \to \mathbb{R}$ chosen such that

$$g(t-s) = g(-s) = \bar{\mathfrak{g}}(s), \quad t = 0, \quad \text{and}$$
$$g(t-s) = g(t) = \mathfrak{g}(t), \quad s = 0.$$

Let us assume for the time being that such a function g has been found. For such a boundary condition, we consider the continuous problem

$$(\mathsf{L}E)(t,s) := \frac{\partial}{\partial t}E(t,s) + \frac{\partial}{\partial s}E(t,s) = F(t,s), \\ E(t,s) = g(t-s), \quad \text{for } t = 0 \text{ or } s = 0. \tag{10.40}$$

It is easy to check by direct differentiation[4] that the solution of (10.40) can be expressed in the form

4 Taking also into account the parametric dependence of the integrands on t and s!

$$E(t,s) = \begin{cases} g(t-s) + \int_0^t F(\tau, s-t+\tau)d\tau, & s \geq t, \\ g(t-s) + \int_0^s F(t-s+\tau, \tau)d\tau, & t \geq s. \end{cases} \tag{10.41}$$

We now determine the appropriate form of the function g. Clearly, the boundary conditions for E, when discretized using the partitions of $[0,T]$, must approximate the boundary conditions of $E^D_{n,m}$, which in turn (see (10.33) and (10.34)) must be the same as the boundary conditions of $e^D_{n,m}$. By the definition of $e^D_{n,m}$, these are related to $\|x_i^{(n)} - x\|$ and $\|\bar{x}_j^{(m)} - x\|$ for any $x \in D(A)$, chosen sufficiently close to the initial condition of the differential equation.

As a result of the above reasoning, estimates (10.30) and (10.31) will give us an idea concerning the relevant boundary condition for the continuous equation. Indeed, using the previous notation, we see that a choice for $\mathfrak{g}$ and $\bar{\mathfrak{g}}$ can be

$$\begin{aligned} \bar{\mathfrak{g}}(s) &\simeq \|\bar{x}_0^{(m)} - x\| + \int_0^s (\|\bar{f}(\tau)\| + \|y\|)d\tau, \\ \mathfrak{g}(t) &\simeq \|x_0^{(n)} - x\| + \int_0^t (\|f(\tau)\| + \|y\|)d\tau, \end{aligned} \tag{10.42}$$

where $x \in \mathbf{D}(A)$ and $y \in A(x)$ are (so far) arbitrary.

Hence, choosing the function g as

$$\begin{aligned} g(t-s) \simeq {} & \|x_0^{(n)} - x\| + \|\bar{x}_0^{(m)} - x\| + \int_0^t (\|f(\tau)\| + \|y\|)d\tau \\ & + \int_0^s (\|\bar{f}(\tau)\| + \|y\|)d\tau, \end{aligned} \tag{10.43}$$

is a possibility for the boundary condition, which is compatible with the above discretization. Note that the above are approximations and not strict equalities. The terms $\|x_0^{(n)} - x\|$ and $\|\bar{x}_0^{(m)} - x\|$ can be made arbitrarily small by appropriate choice of x (as long as $x_0^{(n)}, \bar{x}_0^{(m)} \in \mathbf{D}(A)$), whereas the integral terms are approximations of the corresponding sums. Note that this approximation is not unique, it will be slightly modified to the needs of the proof in the next step.

To conclude this step, we state that using the above considerations we can estimate for any $n, m \in \mathbb{N}$ the discrete error term $e^D_{n,m}$ by the solution (10.41) of the PDE (10.40) for $F(t,s) := \|f(t) - \bar{f}(s)\|$ and g, as in (10.43) (or a suitable modification of that). This upper bound will then be used to show results concerning the discretized solutions, e. g., convergence to a continuous function that can be considered as a mild solution etc.

This approximation is so far a bit sketchy, but it will be made rigorous in step 4 below.

4. Let $\Delta := [0,T] \times [0,T]$, consider any two partitions $\{t_i^{(n)},\ i = 0,\dots,n\} \subset [0,T]$, $\{s_j^{(m)},\ j = 0,\dots,m\} \subset [0,T]$ and the corresponding grid $\Delta_{nm} = \{(t_i^{(n)}, s_j^{(m)}) : i = 0,\dots,n,\ j = 0,\dots,m\}$. Let us keep $n,m \in \mathbb{N}$ fixed for the time being. Any function $u : \Delta \to \mathbb{R}$ can be approximated by its discretized version $u_{n,m}^D = \{u_{ij}^{(n,m)} : i = 0,\dots,n,\ j = 0,\dots,m\}$, where

$$u_{ij}^{(n,m)} = u(t_i^{(n)}, s_j^{(m)}), \quad t \in (t_i^{(n)}, t_{i+1}^{(n)}],\ s \in (s_j^{(m)}, s_{j+1}^{(m)}].$$

We point out here that with every double array $u_{n,m}^D = \{u_{ij}^{(n,m)},\ i = 0,\dots,n, j = 0,\dots,m\}$ we can associate the piecewise continuous function $u_{n,m} : \Delta \to \mathbb{R}$, defined by

$$u_{n,m}(t,s) = u_{ij}^{(n,m)}, \quad (t,s) \in (t_i^{(n)}, t_{i+1}^{(n)}] \times (s_j^{(m)}, s_{j+1}^{(m)}]. \tag{10.44}$$

Moreover, we may define an operator L_{nm} acting on the space of such piecewise continuous functions by $\mathsf{L}_{nm} u_{n,m} = \mathsf{L}_{n,m}^D u_{n,m}^D$ for any $u_{n,m}$ of the form (10.44). To avoid unnecessary complications, and to emphasize that these correspond to discretizations on the grid Δ_{nm} of the corresponding continuous quantities, we may identify the piecewise function $u_{n,m}$ with the array $u_{n,m}^D$, and the operator L_{nm} with $\mathsf{L}_{n,m}^D$ and use the same notation for both:

$$u_{n,m}(t,s) = u_{n,m}^D(t,s), \quad \mathsf{L}_{nm} = \mathsf{L}_{n,m}^D. \tag{10.45}$$

The operator $L_{n,m}^D$ can be interpreted as the discretization of the operator L for the grid Δ_{nm}. Our intuition dictates that for fine grids, i. e., in the limit, as $n,m \to \infty$, the operator $L_{n,m}^D$ approximates in an appropriate sense the continuous operator L, in the sense that solution for the discretized system approaches the solution of the corresponding continuous system. To turn this intuitive result to a rigorous result, providing concrete error bounds for the approximation, we require the introduction of an appropriate norm. We will be interested in comparing the solution E of the PDE

$$\begin{cases} (LE)(t,s) = F(t,s), \\ E(t,s) = g(t-s), \quad t = 0, \text{ or } s = 0, \end{cases} \quad \text{denoted by} \quad E = S^C(g,F), \tag{10.46}$$

with the solution of the discretized problem

$$\begin{cases} L_{n,m}^D E_{n,m}^D = F_{n,m}^D, \\ (E_{n,m}^D)_{ij} = (g_{n,m}^D)_{ij}, \quad i = 0 \text{ or } j = 0, \end{cases} \quad \text{denoted by} \quad E_{n,m}^D = S_{n,m}^D(g_{n,m}^D, F_{n,m}^D), \tag{10.47}$$

where in (10.47) the first equation is assumed to hold for all elements i, j, of $v_{n,m}^D := L_{n,m}^D E_{n,m}^D$ and $F_{n,m}^D$, $v_{ij}^{(n,m)}$ and $F_{ij}^{(n,m)}$, respectively, with $i \neq 0$ and $j \neq 0$. Clearly, (10.47) is the discretization of (10.46) on the grid Δ_{nm}, with the first relation in (10.47) being the discretization of the equation in the domain, and the second being the discretization of the boundary condition.

By the linearity of these problems, it follows that

$$\begin{aligned} S^C(g_1+g_2,F) &= S^C(g_1,F)+S^C(g_2,F), \\ S^C(g,F_1+F_2) &= S^C(g,F_1)+S^C(g,F_2), \end{aligned} \tag{10.48}$$

and similarly for the discrete solution operator.

The result we are after is to show that

$$\begin{aligned} \|E-E_{n,m}\|_{L^\infty(\Delta)} &= \|E-E^D_{n,m}\|_{L^\infty(\Delta)} \\ &= \|S^C(g,F)-S^D_{n,m}(g^D_{n,m},F^D_{n,m})\|_{L^\infty(\Delta)} \to 0, \end{aligned} \tag{10.49}$$

(where we use the construction (10.44) and the identification (10.45)) as long as the data of the discrete problem, $F^D_{n,m}$, $g^D_{n,m}$, are sufficiently close to the corresponding data F, g of the continuous problem, i. e., as the grid Δ_{nm} becomes sufficiently fine, which corresponds to the limit $n,m\to\infty$. This calls for the definition of an appropriate norm.

Motivated by the fact that the functions F that we will focus on are of the form $F(t,s)=\|f(t)-f(s)\|\le\|f(t)\|+\|f(s)\|$ for all $(t,s)\in\Delta$, we define the following norm:

$$\begin{aligned} \|\phi\|_* = \inf\{\|\phi_1\|_{L^1((0,T);X)}+\|\phi_2\|_{L^1((0,T);X)} \ &: \ \phi_1,\phi_2\in L^1((0,T);X), \\ \text{such that } |\phi(t,s)|\le\phi_1(t)+\phi_2(s), \ &\text{a. e. on } \Delta\}. \end{aligned} \tag{10.50}$$

The following estimate is the key result of this step: Using the definitions of the mappings S^C and $S^D_{n,m}$ in (10.46) and (10.47), respectively, it holds that

$$\begin{aligned} &\|S^C(g,F)-S^D_{n,m}(g^D_{n,m},F^D_{n,m})\|_{L^\infty(\Delta)} \to 0 \quad \text{if} \\ &\|F-F^D_{n,m}\|_* \to 0, \quad \text{and} \quad \|g-g^D_{n,m}\|_{L^\infty((-T,T);X)}. \end{aligned} \tag{10.51}$$

The proof of (10.51) is technical, and is provided as a separate proposition which is of interest in its own right (Proposition 10.5.3) in Section 10.5.4 in the appendix of the chapter so as not to disrupt the course of the current proof.

5. We are now in position to show that the sequence $(x_n)_{n\in\mathbb{N}}$ of piecewise continuous functions, defined as in Definition 10.2.14, converges to a continuous function. We will consider two possible partitions of the same interval $[0,T]$ and the corresponding discretizations of the same function $f\in L^1((0,T);X)$,

$$\begin{aligned} f_n(t) &= f_i^{(n)}, \quad t\in(t_i^{(n)},t_{i+1}^{(n)}], \\ \bar f_m(s) &= f_j^{(m)}, \quad s\in(s_j^{(m)},s_{j+1}^{(m)}]. \end{aligned} \tag{10.52}$$

These approximations can be chosen so that

$$\|f_n - f\|_{L^1((0,T);X)} \to 0 \quad \text{as } n \to \infty,$$
$$\|\bar{f}_m - f\|_{L^1((0,T);X)} \to 0 \quad \text{as } m \to \infty,$$

as follows by the approximation of integrals using their discretizations. For this choice of f and $\bar{f}$, the resulting approximations $\{\mathsf{x}_i^{(n)},\ i = 0,\dots,n\}$ and $\{\bar{\mathsf{x}}_j^{(m)},\ j = 0,\dots,m\}$ and sequences of piecewise constant functions, correspond to two different discrete approximations of the solution of the same equation (10.20) for two different partitions, $\{t_i^{(n)},\ i = 0,\dots,n\}$ and $\{s_j^{(m)},\ j = 0,\dots,m\}$ of the interval $[0,T]$, possibly starting at different approximations $\mathsf{x}_0^{(n)} = \mathsf{x}_0^n$, $\bar{\mathsf{x}}_0^{(m)} = \mathsf{x}_0^m$ of the initial condition x_0.

Our aim is to monitor the behavior of these sequences, as $n, m \to \infty$, i. e., in the limit of very fine discretizations. What we expect is that for such fine discretizations, no matter if we choose the partition $\{t_i^{(n)},\ i = 0,\dots,n\}$ or $\{s_j^{(m)},\ j = 0,\dots,m\}$, the corresponding approximations lead to the same result as $n, m \to \infty$, hence converging to a limiting procedure that will be recognized (by definition) as the mild solution to (10.20).

To this end, fix any n, m, and consider two arbitrary partitions $P_n = \{t_i^{(n)},\ i = 0,\dots,n\}$, $P_m = \{s_j^{(m)},\ j = 0,\dots,m\}$ of $[0,T]$, and define the sequence of piecewise constant functions $(\mathsf{x}_n)_{n\in\mathbb{N}}$, whose any two terms, x_n and x_m, are defined as the functions such that for any $t \in [0,T]$,

$$\mathsf{x}_n(t) = \mathsf{x}_i^{(n)}, \quad t \in (t_i^{(n)}, t_{i+1}^{(n)}],$$
$$\mathsf{x}_m(t) = \bar{\mathsf{x}}_j^{(m)}, \quad t \in (s_j^{(m)}, s_{j+1}^{(m)}].$$

Our aim is to show that the sequence $(\mathsf{x}_n)_{n\in\mathbb{N}}$, defined as above, is convergent in $C([0,T];X)$, so that the limit x can be interpreted as the mild solution of (10.20).

Consider the grid $\Delta_{nm} := \{(t_i^{(n)}, s_j^{(m)}),\ i = 0,\dots,n,\ j = 0,\dots,m\} \subset [0,T] \times [0,T]$. Given any $(t,s) \in [0,T] \times [0,T]$, there exists a pair (i,j) such that $(t,s) \in (t_i^{(n)}, t_{i+1}^{(n)}] \times (s_j^{(m)}, s_{j+1}^{(m)}]$. Then, for such (t,s),

$$\|\mathsf{x}_n(t) - \mathsf{x}_m(s)\| = \|\mathsf{x}_i^{(n)} - \bar{\mathsf{x}}_j^{(m)}\| = e_{ij}^{(n,m)}, \tag{10.53}$$

where $e_{n,m}^D = \{e_{ij}^{(n,m)},\ i = 0,\dots,n,\ j = 0,\dots,m\}$ is the solution to the problem (10.33). Monitoring this quantity will be used to show that the sequence $(\mathsf{x}_n)_{n\in\mathbb{N}}$, defined above, is Cauchy, hence convergent.

Following the strategy of step 3, we will estimate $e_{n,m}^D$ in terms of the solution of the discrete problem

$$\begin{aligned} L_{n,m}^D E_{n,m}^D &= F_{n,m}^D, \\ E_{ij}^{(n,m)} &= g_{ij}^{(n,m)}, \quad i = 0 \text{ or } j = 0, \end{aligned} \tag{10.54}$$

where the forcing term $F^D_{n,m}$ and boundary term $g^D_{n,m}$ finite sequences are defined in terms of the piecewise constant functions $F^D_{nm} : \Delta \to \mathbb{R}$, $g^D_{nm} : [-T,T] \to \mathbb{R}$ (recall the construction (10.44) and the identification (10.45)), given by

$$\begin{aligned} F^D_{n,m}(t,s) &= \|f_n(t) - f_m(s)\|, \\ g^D_{nm}(t-s) &= \int_0^{|t-s|} (\|f(\tau)\| + \|y\|)d\tau + \|f - f_n\|_{L^1((0,T);X)} \\ &\quad + \|f - f_m\|_{L^1((0,T);X)} + 2(\|x_0^{(n)} - x\| + \|x_0^{(m)} - x\|). \end{aligned} \tag{10.55}$$

Note that the functions $F^D_{nm} = F^D_{n,m}$ and $g^D_{nm} = g^D_{n,m}$ can be considered as approximations of the functions

$$\begin{aligned} F(t,s) &= \|f(t) - f(s)\|, \\ g(t-s) &= \int_0^{|t-s|} (\|f(\tau)\| + \|y\|)d\tau + 4\|x_0 - x\|, \end{aligned} \tag{10.56}$$

in the norms $\|\cdot\|_*$ and $\|\cdot\|_{L^\infty((-T,T))}$, respectively.

To show that F^D_{nm} approximates F, we work as follows: Observe that for any n, m and $(t,s) \in \Delta$, there exists $i = 1,\dots,n$ and $j = 1,\dots,m$ such that $t \in (t_i^{(n)}, t_{i+1}^{(n)}]$ and $s \in (s_j^{(m)}, s_{j+1}^{(m)}]$. Then,

$$\begin{aligned} |F(t,s) - F^D_{n,m}(t,s)| &= \big|\, \|f(t) - f(s)\| - \|f_n(t) - f_m(s)\| \,\big| \\ &\le \|f(t) - f_n(t) + f_m(s) - f(s)\| \le \|f_n(t) - f(t)\| + \|f_m(s) - f(s)\|, \end{aligned} \tag{10.57}$$

where for the first inequality we used the reverse triangle inequality and for the last the triangle inequality. Hence, $\varphi(t,s) := F(t,s) - F^D_{n,m}(t,s)$ satisfies $|\varphi(t,s)| \le \varphi_1(t) + \varphi_2(s)$ a. e. in Δ, for $\varphi_1(t) = \|f_n(t) - f(t)\|$ and $\varphi_2(s) = \|f_m(s) - f(s)\|$, and since $f_n \to f$ in $L^1((0,T);X)$, as $n \to \infty$ (and of course similarly for f_m), we see by the definition of the norm $\|\cdot\|_*$ that

$$\|F - F^D_{mn}\|_* \to 0, \quad \text{as } n, m \to \infty. \tag{10.58}$$

The fact that

$$\|g - g^D_{nm}\|_{L^\infty((-T,T);X)} \to 0, \quad \text{as } n, m \to \infty \tag{10.59}$$

follows directly from the definition of g and its discretization g^D_{nm}.

For any $n, m \in \mathbb{N}$, we have that

$$\|x_n(t) - x_m(s)\| \le E_{ij}^{(n,m)}, \quad \text{where } E^D_{n,m} = S^D_{n,m}(g^D_{n,m}, F^D_{n,m}),$$

with i, j chosen as above. This inequality is essentially a comparison principle for the discrete operator L^D_{nm}. We now take the limsup in this inequality to obtain that

$$\begin{aligned}\limsup_{n,m\to\infty}\|\mathsf{x}_n(t)-\mathsf{x}_m(s)\| &\le \limsup_{n,m\to\infty} E_{ij}^{(n,m)}\\ &= \limsup_{n,m\to\infty} (S^D(g^D_{nm},F^D_{n,m}))_{ij}\\ &\le (S^C(g,F))(t,s), \quad \forall\,(t,s)\in\Delta,\end{aligned} \tag{10.60}$$

where the second estimate comes from (10.51). We now have an explicit bound for the quantity on the LHS by (10.41).

Setting $t=s$ in (10.60) and noting that $F(t,t)=0$, we obtain, by (10.41), that

$$\limsup_{n,m\to\infty}\|\mathsf{x}_n(t)-\mathsf{x}_m(t)\| \le 4\|\mathsf{x}-\mathsf{x}_0\| \to 0, \tag{10.61}$$

by an appropriate choice of the arbitrary $\mathsf{x}\in \mathbf{D}(\mathsf{A})$ since $\mathsf{x}_0\in\overline{\mathbf{D}(\mathsf{A})}$. This estimate is uniform in $[0,T]$. Therefore, the sequence $(\mathsf{x}_n)_{n\in\mathbb{N}}$ is Cauchy and has a uniform limit $\mathsf{x}:[0,T]\to X$.

It remains to show that the function x is a continuous function. For that, we return to the estimate (10.60) for the choice $t=t+h$ and $s=t$. Applying (10.41) for this choice, we obtain

$$\begin{aligned}\|\mathsf{x}(t+h)-\mathsf{x}(t)\| &\le g(h)+\int_0^t F(\tau+h,\tau)d\tau\\ &= 4\|\mathsf{x}_0-\mathsf{x}\|+\int_0^{|h|}(\|f(\tau)\|+\|\mathsf{y}\|)d\tau+\int_0^t\|f(\tau+h)-f(\tau)\|d\tau,\end{aligned} \tag{10.62}$$

from which follows the continuity of x. □

10.2.6.3 When is a mild solution a strong solution: regularity

Under certain extra conditions, the mild solution satisfies additional regularity. Such an example is provided in he following theorem [18]):

Theorem 10.2.20. *Let the assumptions of Theorem* 10.2.18 *hold. Additionally, assume that X is reflexive,* A *is closed,* $\mathsf{x}_0\in\mathbf{D}(\mathsf{A})$, *and* $f\in W^{1,1}([0,T];X)$. *Then, the mild solution of* (10.20) *is a strong solution, which moreover satisfies* $\mathsf{x}\in W^{1,\infty}([0,T];X)$.

Proof (Sketch). Estimate (10.23) plays a crucial role in the proof. Applying it for f and $\bar f(\cdot)=f(\cdot+h)$ for any $h>0$, and integrating between $s=0$ and t, yields an estimate for $\|\mathsf{x}(t+h)-\mathsf{x}(t)\|$ in terms of $\|f(\cdot+h)-f(\cdot)\|_{L^1((0,T);X)}$, which, since $f\in W^{1,1}([0,T];X)$, is bounded by Ch. In particular, we obtain

$$\|\mathsf{x}(t+h)-\mathsf{x}(t)\| \le \|\mathsf{x}(h)-\mathsf{x}_0\|+\int_0^t\|f(\tau+h)-f(\tau)\|d\tau$$

$$\leq \|x(h) - x_0\| + Ch.$$

We further need an estimate for $\|x(h)-x(0)\|$, which is obtained by using estimate (10.23) once more, this time integrating from $s = 0$ to $t = h$, but using $\bar{f} = -Ax_0$ (which cancels the RHS for the differential equation for $\bar{x}$ in the neighborhood of $t = 0$, thus leading to $\bar{x}(h) = x_0$). This leads to an estimate

$$\|x(h) - x_0\| \leq \int_0^h \|f(\tau) - \bar{f}\| d\tau \leq C'h.$$

Combining the two estimates, we obtain that $\|x(t + h) - x(t)\| \leq Ch$ for an appropriate constant $C > 0$.

This estimate shows that the mild solution $x : [0, T] \to X$ is Lipschitz. But since X is reflexive (hence satisfies the Radon–Nikodym property), x is a. e. differentiable (analogue of Rademacher's theorem). □

10.2.6.4 Connection between nonlinear semigroups and the homogeneous Cauchy problem

In the particular case where $f = 0$, we can apply the above theorem to obtain the strong solution of the homogeneous problem, and thus connect the nonlinear semigroup $\{S(t)\}_{t\geq 0}$ generated by the dissipative operator A with the solution of the homogeneous Cauchy problem [18].

Theorem 10.2.21. *Let the assumptions of Theorem* 10.2.18 *hold.*

(i) *If* $x_0 \in \overline{\mathbf{D}(A)}$, *then* $x(t) = S(t)x_0$ *is the mild solution of the homogeneous Cauchy problem.*

(ii) *If* X *is reflexive and* $x_0 \in \mathbf{D}(A)$, *then* $x(t) = S(t)x_0$ *is the strong solution of the homogeneous Cauchy problem.*

10.3 Evolution problems in evolution triples

The approach of evolution problems using the concept of evolution triples (or Gel'fand triples) is a key approach for the study of problems involving operators $A : \mathbf{D}(A) \subset X \to X^*$ or $A : \mathbf{D}(A) \subset X \to 2^{X^*}$. Throughout this section we assume X is reflexive.

10.3.1 Setting of the evolution problem

We reconsider the Cauchy problem

$$\frac{dx}{dt} = A_{\mathcal{D}}(x) + f, \quad x(0) = x_0, \tag{10.63}$$

in the case where the nonlinear operator $\mathsf{A}_\mathcal{D}$ is no longer considered as an operator $\mathsf{A}_\mathcal{D} : \mathbf{D}(\mathsf{A}) \subset X \to X$, but rather in a more generalized sense as an operator $\mathsf{A}_\mathcal{D} : \mathbf{D}(\mathsf{A}) \subset X \to X^*$. Note that in these two cases the domain of the operator $\mathbf{D}(\mathsf{A}_\mathcal{D})$ is different, with that domain being in principle larger in the second case,[5] thus leading to a different operator with a different image set, i. e., X^* instead of X. We have seen in Section 10.2 that, in the case where $\mathsf{A}_\mathcal{D} : \mathbf{D}(\mathsf{A}_\mathcal{D}) \subset X \to X$, the well posedness of (10.63) requires dissipativity properties for $\mathsf{A}_\mathcal{D}$. We will establish in this section that in the case where $\mathsf{A}_\mathcal{D} : \mathbf{D}(\mathsf{A}_\mathcal{D}) \subset X \to X^*$ (or for the multivalued case), the treatment of problem (10.63) requires monotonicy properties (of the type discussed in Chapter 9) for the operator $\mathsf{A} = -\mathsf{A}_\mathcal{D}$.

Notation 10.3.1. As we are more familiar with monotonicity properties for A rather than $-\mathsf{A}$, it is more convenient (and perhaps pleasing to the eye) to work directly with $-\mathsf{A}_\mathcal{D}$, and express (10.63) in the equivalent form $\frac{dx}{dt} + \mathsf{A}(x) = f$.

Moving away from the rather constrained semigroup context, we are allowed to study the more general nonlinear (and nonautonomous) Cauchy problem

$$\begin{cases} \frac{dx}{dt}(t) = \mathsf{A}_\mathcal{D}(t, x(t)) + f(t), \\ x(0) = x_0, \end{cases} \iff \begin{cases} \frac{dx}{dt}(t) + \mathsf{A}(t, x(t)) = f(t), \\ x(0) = x_0, \end{cases} \tag{10.64}$$

where $T > 0$ is fixed, $\mathsf{A}_\mathcal{D} = -\mathsf{A} : [0, T] \times X \to X^*$ is a given family of nonlinear operators, and $f : I = [0, T] \to X^*$ is a given function, within the convenient setting of evolution (or Gel'fand) triples

$$X \overset{c}{\hookrightarrow} H \hookrightarrow X^*,$$

in the case where $\mathsf{A}(t, \cdot) = -\mathsf{A}_\mathsf{D}(t, \cdot)$ satisfies monotonicity properties. The Cauchy problem consists in, for a given f, finding $x \in W^{1,p}(I; X)$ satisfying (10.64), with the first equation interpreted a. e. on $[0, T]$. Note that the initial condition makes sense if $W^{1,p}(I; X) \hookrightarrow C([0, T]; H)$ continuously (see Proposition 1.6.13).

The standing assumptions on the family of operators $(\mathsf{A}(t, \cdot))_{t \in [0,T]} = (-\mathsf{A}_\mathcal{D}(t, \cdot))_{t \in [0,T]}$ for this section will be the following:

Assumption 10.3.2. The family of nonlinear operators $(\mathsf{A}_\mathcal{D}(t, \cdot))_{t \in [0,T]}$ or equivalently $(\mathsf{A}(t, \cdot))_{t \in [0,T]}$, where $\mathsf{A} = -\mathsf{A}_\mathcal{D}$, satisfies the following hypotheses:
(i) The mapping $t \mapsto \mathsf{A}(t, x)$ is measurable.
(ii) $\|\mathsf{A}(t, x)\|_{X^*} \le a(t) + c\,\|x\|_X^{p-1}$ a. e. on $[0, T]$ with $a \in L^{p^*}((0, T))$, $c > 0$, $p, p^* \in (1, \infty)$ and $\frac{1}{p} + \frac{1}{p^*} = 1$.

5 For example, if $\mathsf{A}_\mathcal{D}$ is a linear operator, it is no longer restrained to be compatible with Definition 1.7.2 for some semigroup.

(iii) There exists $c_1 > 0$ and $\theta \in L^1_+((0,T);\mathbb{R})$ such that

$$\langle \mathsf{A}(t,\mathsf{x}),\mathsf{x}\rangle_{X^\star,X} \geq c_1 \,\|\mathsf{x}\|^p_X - \theta(t), \quad \text{a. e. on } [0,T], \text{ for all } \mathsf{x} \in X.$$

We will study problem (10.64) under two possible extra assumptions on the family of nonlinear operators $(\mathsf{A}(t,\cdot))_{t\in[0,T]} = (-\mathsf{A}_{\mathcal{D}}(t,\cdot))_{t\in[0,T]}$, (a) monotonicity for $\mathsf{A}(t,\cdot)$ and (b) pseudomonotonicity for $\mathsf{A}(t,\cdot)$. Case (a) will be treated in Section 10.3.3, whereas case (b) will be treated in Sections 10.3.4.2 and 10.3.4.1.

10.3.2 Some preliminaries

We will first need some preliminaries in order to set up problem (10.64) appropriately.

We begin by recalling the definition of an evolution triple (or Gel'fand triple) as a triple of spaces $X \hookrightarrow H \hookrightarrow X^\star$ with X a separable and reflexive Banach space with dual $X^\star$, and H, which is a separable Hilbert space identified with its dual, with the embedding $X \hookrightarrow H$ dense and continuous (or even compact for some results). For simplicity, we will use the notation I for the intervals $[0,T]$ and $(0,T)$ interchangeably, depending on the context. In this setting, we may also consider the Sobolev–Bochner space $W^{1,p}(I;X) := W^{1,p,p^*}(I;X,X^\star)$, consisting of functions $\mathsf{x}(\cdot) : I \to X$ such that $\mathsf{x}(\cdot) \in L^p(I;X)$ and $\frac{d\mathsf{x}}{dt}(\cdot) \in L^{p^*}(I;X^\star)$. For more information on Sobolev spaces on evolution triples and their embeddings, we refer to Section 1.6.

We will also consider the Lebesgue–Bochner spaces

$$X_T := L^p((0,T);X), \quad \text{and} \quad X_T^* = L^{p^*}((0,T);X^*),$$

for a suitable $p \in (1,\infty)$ (and $p^* = \frac{p}{p-1}$ its conjugate exponent, following our usual notation). Clearly, X_T^* is the dual of X_T. Moreover, to simplify notation, we set

$$W_T = \left\{\mathsf{x} \in X_T \;:\; \frac{d\mathsf{x}}{dt} \in X_T^*\right\}.$$

Recalling the definition of Sobolev–Bochner spaces (see Definition 1.6.9), we see that

$$W_T = W^{1,p,p^*}((0,T);X,X^*) =: W^{1,p}((0,T);X).$$

As a result of the Aubin–Lions lemma (see Theorem 1.6.13(ii)), we have that $W_T \overset{c}{\hookrightarrow} L^p((0,T);H)$. Moreover, by Theorem 1.6.13(i), as long as we show that problem (10.64) admits solutions $\mathsf{x} \in W^{1,p,p^*}(I;X,X^*) = W_T$, then $t \mapsto \mathsf{x}(t)$ can be considered as a continuous function in H, so $\mathsf{x}(0)$ makes sense in H and the initial condition makes sense.

We first must make sense of the (weak) derivative with respect to time, in an operator sense. Note that, for simplicity, we may consider the special case where $\mathsf{x}(0) = 0$.

The problem for more general initial condition $x(0) = x_o \in H$ can be treated using a translation (see, e. g., [18]).

Recall that $\frac{dx}{dt}$ is defined in terms of an appropriate integration by parts (see Definition 1.6.7). Since we focus on the case where $x(0) = 0$, we will consider the operator $L : D(L) \to X^*$, defined by

$$Lx = \frac{dx}{dt}, \quad D(L) = \left\{x \in X_T \, : \, \frac{dx}{dt} \in X_T^*, \, x(0) = 0\right\},$$
$$\langle Lx, z\rangle = \int_0^T \left\langle \frac{dx}{dt}(t), z(t)\right\rangle dt, \quad \forall z \in X_T.$$

The next thing we need to sort out, is the proper understanding of the family of operators $(A(t,\cdot))_{t\geq 0}$, where for each $t \in [0,T]$, $A(t,\cdot) : X \to X^*$. These must be understood in the sense of operators from X_T to X_T^*. This is done in terms of an appropriate Nemitskii mapping, defined as

$$\widehat{A} : X_T \to X_T^*, \quad (\widehat{A}(x))(t) = A(t, x(t)), \quad t \in [0,T]. \tag{10.65}$$

Using the above constructions, we can express problem (10.64) in the following form:

$$\text{Find } x \in D(L) \, : \, \langle Lx + \widehat{A}(x), z\rangle = \langle f, z\rangle, \quad \forall z \in X_T, \tag{10.66}$$

where $\langle\cdot,\cdot\rangle$ is the duality pairing between X_T and X_T^*. Equation (10.66) is an abstract operator formulation of the evolution problem. If the operator $L + \widehat{A}$ enjoys monotonicity or pseudomonotonicity properties, then this will allow us to use the abstract theory concerning the surjectivity properties of monotone type operators to provide existence results. We require the following [89, 102]:

Proposition 10.3.3. *The linear operator $L : X_T \to X_T^*$ is a maximal monotone operator.*

Proof. Monotonicity is immediate by using the integration by parts formula (see Theorem 1.6.13), linearity, and the boundary conditions, to show that for any $x_1, x_2 \in \mathbf{D}(L)$, we have that

$$\langle Lx_1 - Lx_2, x_1 - x_2\rangle_{X_T^*, X_T} = \frac{1}{2}\|x_1(T) - x_2(T)\|_H^2 - \frac{1}{2}\|x_1(0) - x_2(0)\|_H^2 \geq 0,$$

since $x_1(0) = x_2(0)$.

To show maximality, it suffices to prove that for any $(x, x^\star) \in X_T \times X_T^\star$ such that

$$\langle x^\star - L(z), x - z\rangle_{X_T^\star, X_T} \geq 0, \quad \forall z \in \mathbf{D}(L), \tag{10.67}$$

it holds that $x \in \mathbf{D}(L)$ and $x^\star = L(x)$.

To this end, assume that (10.67) holds and choose $z = \phi\bar{z} \in X_T$, where $\phi \in C_c^\infty((0,T);\mathbb{R})$ and $\bar{z} \in X$. Clearly, $\frac{dz}{dt} = \frac{d\phi}{dt}\bar{z}$ and $z \in \mathbf{D}(L)$, while $z(T) = 0$, so using the integration by parts formula, it is easy to see that $\langle L(z), z\rangle_{X_T^\star, X_T} = 0$, hence, applying (10.67) to the chosen z, we obtain

$$0 \le \langle x^\star, x\rangle_{X_T^\star, X_T} - \int_0^T \left(\langle x^\star(t), \phi(t)\bar{z}\rangle_{X^\star,X} + \left\langle \frac{d\phi}{dt}(t)\bar{z}, x(t)\right\rangle_{X^\star,X}\right) dt, \quad \forall \bar{z} \in X.$$

Since $\bar{z} \in X$ was arbitrary, this inequality holds also for $\lambda\bar{z}$ for all $\lambda \in \mathbb{R}$. Because $\langle x^\star, x\rangle_{X_T^\star, X_T}$ is independent of $\bar{z}$ (sending $\lambda \to \pm\infty$), we obtain

$$\int_0^T \left(\langle x^\star(t), \phi(t)\bar{z}\rangle_{X^\star,X} + \left\langle \frac{d\phi}{dt}(t)\bar{z}, x(t)\right\rangle_{X^\star,X}\right) dt = 0 \quad \forall \phi \in C_c^\infty((0,T);\mathbb{R}),\ \bar{z} \in X,$$

and from the definition of the generalized derivative (see Definition 1.6.7), we obtain that $\frac{dx}{dt} = x^\star$ and $\frac{dx}{dt} \in L^{p^*}(I; X^\star)$, i. e., $x \in W^{1,p}(I; X)$.

It remains to show that $x \in \mathbf{D}(L)$. By the integration by parts formula, from (10.67), we obtain

$$\begin{aligned} 0 &\le \left\langle \frac{dx}{dt} - \frac{dz}{dt}, x - z\right\rangle_{X_T^\star, X_T} \\ &= \frac{1}{2}\left(\|x(T) - z(T)\|_H^2 - \|x(0) - z(0)\|_H^2\right). \end{aligned} \tag{10.68}$$

Now, if $z \in \mathbf{D}(L)$, then $z(0) = 0$, so (10.68) takes the form

$$\|x(T) - z(T)\|_H^2 - \|x(0)\|_H^2 \ge 0. \tag{10.69}$$

Since $X \hookrightarrow H$ is dense, we choose a sequence $(\bar{z}_n)_{n\in\mathbb{N}} \subset X$ such that $\bar{z}_n \to \frac{1}{T}x(T)$ in H. Set $z_n(t) = t\bar{z}_n$ so that $z_n \in \mathbf{D}(L)$ for every n. Setting $z = z_n$ in (10.69) and letting $n \to \infty$, we obtain that $x(0) = 0$, i. e., $x \in \mathbf{D}(L)$.

We therefore conclude that $L : X_T \to X_T^\star$ is a maximal monotone operator. □

We next consider the operator $\widehat{A} : X_T \to X_T^*$, defined in (10.65), and discuss its properties.

Proposition 10.3.4. *Assume that the family of operators satisfies Assumption* 10.3.2. *Then,*

(i) $\widehat{A}$ *is bounded.*
(ii) $\widehat{A}$ *is coercive.*

Proof. (i) By Assumption 10.3.2(i), for each $x \in X$ and $z \in X$, the real function $t \mapsto \langle A(t, x(t)), z\rangle_{X^\star,X}$ is measurable on $[0,T]$. Hence, by Pettis theorem (see Theorem 1.6.2), for each $x \in X$, the function $t \to A(t, x(t))$ is measurable from $[0, T]$ to $X^\star$. By the growth condition (Assumption 10.3.2(ii); take p^* power and integrate on $[0, T]$), we have $\|\widehat{A}(x)\|_{X_T^\star} \le c_1\|a\|_{L^{p^*}((0,T))} + c_2\|x\|_{X_T}^p$ for all $x \in X_T$, and appropriate constants $c_1, c_2 > 0$.

(ii) Coercivity follows by Assumption 10.3.2(iii). For each $x \in X_T$, we have

$$\begin{aligned}\langle \widehat{A}(x), x\rangle_{X_T^\star, X_T} &= \int_0^T \langle A(t, x(t)), x(t)\rangle_{X^\star, X} dt \\ &\ge \int_0^T (c_1\|x(t)\|_X^p - \theta(t))dt \ge c_1\|x\|_{X_T}^p - c_0\end{aligned}$$

for a suitable constant c_0 related to the L^1 norm of θ. This in turn implies, since $p > 1$, that $\frac{1}{\|x\|_{X_T}}\langle \widehat{A}(x), x\rangle_{X_T^\star, X_T} \to \infty$, as $\|x\|_{X_T} \to \infty$, hence the coercivity. □

Concerning monotonicity properties, it is very easy to show that $\widehat{A}$ inherits monotonicity from A. However, pseudomonotonicity is a bit more involved and requires the definition of a new concept: that of pseudomonotonicity with respect to the domain of an appropriate linear operator.

Definition 10.3.5. Let Y be a reflexive Banach space, and $M : \mathbf{D}(M) \subset Y \to Y^\star$ be a linear maximal monotone operator. An operator $K : Y \to Y^\star$ is called M-pseudomonotone, if for any $(y_n)_{n\in\mathbb{N}} \subset \mathbf{D}(M)$ such that (i) $y_n \rightharpoonup y$ in Y, (ii) $M(y_n) \rightharpoonup M(y)$ in $Y^\star$, (iii) $K(y_n) \rightharpoonup y^\star$ in $Y^\star$, and (iv) $\limsup_n \langle K(y_n), y_n - y\rangle_{Y^*,Y} \le 0$, it holds that

$$K(y) = y^\star \quad \text{and} \quad \lim_n \langle K(y_n), y_n\rangle_{Y^*,Y} = \langle K(y), y\rangle_{Y^*,Y}.$$

Monotonicity properties of $\widehat{A}$ are treated in the next proposition [88, 101]:

Proposition 10.3.6. *Let Assumption* 10.3.2 *hold and set* $M = L$ *in Definition* 10.3.5. *Then,*

(i) *If* $x \mapsto A(t, x)$ *is demicontinuous and monotone a. e. in* $[0, T]$, *then* $\widehat{A}$ *is demicontinuous and monotone.*
(ii) *If* $x \mapsto A(t, x)$ *is pseudomonotone a. e. in* $[0, T]$, *then* $\widehat{A}$ *is L-pseudomonotone.*

Proof. (i) Monotonicity is immediate. Indeed, for all $x, z \in X_T$, we have

$$\langle \widehat{A}(x) - \widehat{A}(z), x - z\rangle_{X_T^\star, X_T} = \int_0^T \underbrace{\langle A(t, x(t)) - A(t, z(t)), x(t) - z(t)\rangle_{X^\star, X}}_{\ge 0,\text{ by monotonicity of } x\mapsto A(t,x)} dt \ge 0.$$

To show demicontinuity, let $x_n \to x$ in X_T, as $n \to \infty$. By passing to a subsequence, if necessary, we may assume that $x_n(t) \to x(t)$ in X for any t a. e. on $[0, T]$, as $n \to \infty$. Then, by demicontinuity of $x \mapsto A(t, x)$, given $z \in X_T$

$$\langle A(t, x_n(t)), z(t)\rangle_{X^\star,X} \to \langle A(t, x(t)), z(t)\rangle_{X^\star,X}, \quad \text{a. e. on } [0, T].$$

By Assumption 10.3.2(ii), we apply the dominated convergence theorem, and we obtain that

$$\langle \widehat{A}(x_n), z\rangle_{X_T^\star, X_T} = \int_0^T \langle A(t, x_n(t)), z(t)\rangle_{X^\star, X} dt \to \int_0^T \langle A(t, x(t)), z(t)\rangle_{X^\star, X} dt$$
$$= \langle \widehat{A}(x), z\rangle_{X_T^\star, X_T}$$

as $n \to \infty$. Since $z \in X_T$ is arbitrary, we conclude that $\widehat{A}(x_n) \rightharpoonup \widehat{A}(x)$ in $X_T^\star$ hence, $\widehat{A}$ is demicontinuous.

(ii) We set $M = L = \frac{d}{dt}$ the derivative operator. We will show that $\widehat{A}$ is L-pseudomonotone (see Definition 10.3.5).

Let $(x_n)_{n\in\mathbb{N}} \subset \mathbf{D}(L)$ such that $x_n \rightharpoonup x$ in X_T, $\frac{dx_n}{dt} \rightharpoonup \frac{dx}{dt}$ in $X_T^\star$, $\widehat{A}(x_n) \rightharpoonup x^\star$ in $X_T^\star$ and

$$\limsup_n \langle \widehat{A}(x_n), x_n - x\rangle_{X_T^\star, X_T} = \limsup_n \int_0^T \langle A(t, x_n(t)), x_n(t) - x(t)\rangle_{X^\star, X} dt \le 0.$$

For any $t \in [0, T]$, let $\xi_n(t) = \langle A(t, x_n(t)), x_n(t) - x(t)\rangle_{X^\star, X}$. Since $W^{1,p}((0, T); X)$ is compactly embedded in $L^p((0, T); H)$ (see Theorem 1.6.13, see also [101] p. 58), we may assume that $x_n(t) \to x(t)$ a. e. in H. Let $N \subset [0, T]$ be the Lebesgue null set, outside of which conditions (ii) and (iii) of Assumption 10.3.2 hold. Then, for each $t \in [0, T] \setminus N$,

$$\xi_n(t) \ge c_1\|x_n(t)\|_X^p - \theta(t) \ge (a(t) + c\|x_n(t)\|_X^{p-1})\|x(t)\|_X =: \theta_n(t). \tag{10.70}$$

Let $K = \{t \in [0, T] : \liminf_n \xi_n(t) < 0\}$, which is clearly a measurable set. Suppose that the Lebesgue measure $|K| > 0$. By (10.70), $(x_n(t))_{n\in\mathbb{N}} \subset X$ is bounded for $t \in K \cap ([0, T] \setminus N) \neq \emptyset$. So, we may assume that $x_n(t) \rightharpoonup x(t)$ in X. Fix t, and choose a suitable subsequence $(\xi_{n_k}(t))_{k\in\mathbb{N}}$ such that $\liminf_n \xi_n(t) = \lim_k \xi_{n_k}(t)$. Since A is pseudomonotone, $\langle A(t, x_{n_k}(t)), x_{n_k}(t) - x(t)\rangle_{X^\star, X} \to 0$, which is a contradiction since $t \in K$. So $\mu(K) = 0$, i. e., $\liminf_n \xi_n(t) \ge 0$, a. e. in $[0, T]$ (i. e. $(\xi_n))_{n\in\mathbb{N}}$ bounded below for all but finite terms). So,

$$0 \le \int_0^T \liminf_n \xi_n(t) dt \le \liminf_n \int_0^T \xi_n(t) dt \le \limsup_n \int_0^T \xi_n(t) dt \le 0,$$

(using Fatou's lemma on $\xi_n + c$, with $-c$ the above constant) hence, $\int_0^T \xi_n(t)dt \to 0$.

Write $|\xi_n(t)| = \xi_n^+(t) + \xi_n^-(t) = \xi_n(t) + 2\xi_n^-(t)$. Note that $\xi_n^-(t) \to 0$, a. e. on $[0, T]$. It is clear from (10.70) that $(\theta_n)_{n\in\mathbb{N}} \subset L^1((0, T))$ is uniformly integrable. Since, $0 \le \xi_n^-(t) \le \theta_n^-(t)$, we conclude that $\int_0^T \xi_n^-(t)dt \to 0$. Therefore, $\int_0^T |\xi_n(t)|dt \to 0$. We may assume that $\xi_n(t) \to 0$ a. e. on $[0, T]$.

Since A is pseudomonotone, we have $A(t, x_n(t)) \rightharpoonup A(t, x(t))$ in $X^\star$, and

$$\langle A(t, x_n(t)), x_n(t)\rangle_{X^\star, X} \to \langle A(t, x(t)), x(t)\rangle_{X^\star, X}, \quad \text{a. e. on } [0, T].$$

Integrate in t and use dominated convergence, so that $\widehat{A}(x_n) \rightharpoonup \widehat{A}(x)$ in $X_T^\star$ and

$$\langle \widehat{A}(x_n), x_n \rangle_{X_T^\star, X_T} \to \langle \widehat{A}(x), x \rangle_{X_T^\star, X_T},$$

therefore, $\widehat{A}$ is L-pseudomonotone. □

10.3.3 Evolution problems using evolution triples: monotone operators

We require the following for the nonlinear operators $A(t,\cdot) = -A_{\mathcal{D}}(t,\cdot)$:

Assumption 10.3.7. The family of nonlinear operators $(A(t,\cdot))_{t\in[0,T]}$ satisfies Assumption 10.3.2 and additionally

- The mapping $x \mapsto A(t,x)$ is demicontinuous and monotone a. e. in $[0,T]$.

An existence and uniqueness result concerning problem (10.64) is given in the following theorem [88, 101]:

Theorem 10.3.8. *Suppose that Assumption* 10.3.7 *holds. Then, for every* $x_0 \in H$ *and* $f \in L^{p^*}((0,T);X^\star)$, *the problem* (10.64) *has a unique solution.*

Proof. For simplicity let $x(0) = 0$ (the general case requires a translation, e. g, [18]). The strategy of the proof is to express (10.64) as an abstract operator equation and use the surjectivity theorems for monotone type operators to show solvability. Using the notation $I = (0,T)$, the function space setting is $X_T := L^p(I;X)$, so $X_T^\star := L^{p^*}(I;X^\star)$. We define the linear operator $L : \mathbf{D}(L) \subset X_T \to X_T^\star$, by $Lx = \frac{dx}{dt}$ for every $x \in \mathbf{D}(L) = \{x \in W^{1,p}(I;X) : x(0) = 0\}$, where $W^{1,p}(I;X) := W^{1,p,p^*}(I;X,X^\star)$, as well as the operator $\widehat{A} : X_T \to X_T^\star$ by $\widehat{A}(x)(\cdot) = A(\cdot, x(\cdot))$, and express (10.64) as the operator equation

$$(L + \widehat{A})(x) = f \tag{10.71}$$

for any $f \in L^{p^*}(I;X^\star)$. If $\mathbf{R}(L + \widehat{A}) = X_T^\star$, then the claim follows. We intend to use the surjectivity result of Theorem 9.4.12 to conclude the proof.

The proof follows in 4 steps.

1. First we prove the uniqueness. Let $x_1, x_2 \in W^{1,p}(I;X)$ be two solutions of (10.64). Then, $x_1(0) = x_2(0)$ and $\frac{dx_i}{dt}(t) + A(t, x_i(t)) = f(t)$ a. e. on $[0,T]$, $i = 1,2$, where upon subtracting and taking the duality pairing of the resulting equation with $x_1(t) - x_2(t)$, integrating over $[0,t]$, and then using the integration by parts formula (see Theorem 1.6.13), we obtain by the monotonicity of A

$$\|x_1(t) - x_2(t)\|_H^2 = 2\int_0^t \left\langle \frac{dx_1}{dt}(s) - \frac{dx_2}{dt}(s), x_1(s) - x_2(s) \right\rangle_{X^\star,X} ds$$

$$= -2 \int_0^t \langle A(s, x_1(s)) - A(s, x_2(s)), x_1(s) - x_2(s) \rangle_{X^*, X} ds \le 0,$$

hence $x_1 = x_2$, and the solution of (10.64) is unique.

2. The linear operator $L : X_T \to X_T^*$ is a maximal monotone operator (see Proposition 10.3.3).

3. The operator $\widehat{A}$ is (a) bounded and coercive (see Proposition 10.3.4) and (b) demicontinuous and monotone (see Proposition 10.3.6).

4. Since $L : X_T \to X_T^*$ is maximal monotone (by step 2) and $\widehat{A} : X_T \to X_T^*$ is bounded, monotone, demicontinuous, and coercive (by step 3), by Theorem 9.4.12, $\mathbf{R}(L + \widehat{A}) = X_T^*$, so that (10.71) admits a solution, hence problem (10.64) has a solution. This is unique by step 1. □

10.3.4 Evolution problems using evolution triples: pseudomonotone operators

Let $(X, H, X^\star)$ be an evolution triple with compact embedding $X \overset{c}{\hookrightarrow} H$ (recall that X is reflexive), and assume that the operators $A(t, \cdot) = -A_{\mathcal{D}}(t, \cdot)$ satisfy the conditions of Assumption 10.3.9.

We now replace the monotonicity assumption on the operators family $A(t, \cdot) = -A_{\mathcal{D}}(t, \cdot)$ in problem (10.64) by pseudomonotonicity.

Assumption 10.3.9. The family of nonlinear operators $(A(t, \cdot))_{t \in [0,T]}$ satisfies Assumption 10.3.2 and additionally
- The mapping $x \mapsto A(t, x)$ is pseudomonotone a. e. $t \in [0, T]$.

We will address the Cauchy problem (10.64) for the case where A is a pseudomonotone operators using two approaches. One approach involving a finite dimensional approximation in terms of the Faedo–Galerkin approach (following the exposition in [128]) and a more abstract approach based on the monotonicity properties of the operator L and the alternative formulation (10.66) (based on [88, 101]).

10.3.4.1 Evolution problems for pseudomonotone operators I: the Galerkin method

We first consider problem (10.64) using a less abstract approach, which can also lead to the numerical approximation of the Cauchy problem, the Faedo–Galerkin approach. This is an extension of the corresponding approach for the time independent problem (see Section 9.5.2.1). We will use the same notation as Section 9.5.2.1, consider the evolution triple $X \hookrightarrow H \hookrightarrow X^*$ (with the standard assumptions), and use the notation X_T, X_T^* and H_T for the corresponding spaces containing the vector valued functions $t \mapsto x(t)$. Our approach follows [128]. We will assume X is reflexive and separable and that the embedding $H \overset{c}{\hookrightarrow} X$ is compact.

Our strategy is as follows:

- Use an appropriate collection of finite dimensional spaces $(X_n)_{n\in\mathbb{N}}$, such that $X_n \subset X_{n+1}$, and $\bigcup_{n\in\mathbb{N}} X_n$ dense in X, to obtain, for each $n \in \mathbb{N}$, a finite dimensional approximation x_n, such that $\mathsf{x}_n(t) \in X_n$ is a solution of (10.64) in the finite dimensional space X_n, expressed into the form

$$\left\langle \frac{d\mathsf{x}_n}{dt}(t), \mathsf{z} \right\rangle_H + \langle \mathsf{A}(t, \mathsf{x}_n(t)), \mathsf{z} \rangle_{X^*,X} = \langle f(t), \mathsf{z} \rangle, \quad \text{a. e. } t \in [0,T], \ \forall \mathsf{z} \in X_n,$$
$$\mathsf{x}_n \in L^\infty((0,T); X_n), \quad \frac{d\mathsf{x}_n}{dt} \in L^{p^*}((0,T); X_n), \quad \mathsf{x}_n(0) = \mathsf{x}_{0,n}. \tag{10.72}$$

  Importantly, $(\mathsf{x}_{0,n})_{n\in\mathbb{N}}$ is a sequence such that $\mathsf{x}_{0,n} \in X_{\ell_n}$ for each $n \in \mathbb{N}$, where $(\ell_n)_{n\in\mathbb{N}}$ is an increasing sequence, and $\mathsf{x}_{0,n} \to \mathsf{x}_0$ (strong) in H (whose existence is guaranteed by the density of $\bigcup_{n\in\mathbb{N}} X_n$ in X; see discussion in Section 9.8). Problem (10.72) is essentially a system of ODEs in $\mathbb{R}^{\dim(X_n)}$.[6] Standard results from ODE theory can be used to guarantee that the resulting system is well posed, hence (10.72) makes sense as a differential equation.
- The membership of x_n, for each $n \in \mathbb{N}$, in the function spaces stated in (10.72) is important, as it allows us to obtain bounds for the sequence $(\mathsf{x}_n)_{n\in\mathbb{N}}$ in the corresponding norms that are independent of n. This step, which is crucial for the Galerkin method to work, usually goes under the name a priori estimates for the solution.
- Having guaranteed a priori bounds, use weak compactness arguments to show that the sequence $(\mathsf{x}_n)_{n\in\mathbb{N}}$, constructed above, converges weakly to a function x such that $\mathsf{x}(t) \in X$, and $\frac{d\mathsf{x}}{dt}(t) \in X^*$.
- Show, using monotonicity arguments, that x is a solution of (10.64).

Once the conceptual part is settled, we can now pass to the main result for this section. We start with an assumption concerning the finite dimensional approximation spaces.

Assumption 10.3.10. Consider a family of finite dimensional spaces $(X_n)_{n\in\mathbb{N}}$ such that
(i) $X_n \subset X_{n+1}$, with $\bigcup_{n\in\mathbb{N}} X_n$ dense in X.
(ii) There exists a self adjoint projector $\Pi_n : H \to H$ such that $\Pi_n(X) = X_n$ and $\|P_n|_X\|_{\mathcal{L}(X,X)}$ is bounded independently of n.

Remark 10.3.11. Since X is separable (see Section 1.9.1 Definition 1.9.5 and the discussion following it), the existence of $(X_n)_{n\in\mathbb{N}}$ satisfying $\overline{\bigcup_{n\in\mathbb{N}} X_n} = X$ follows by choosing $X_n = \text{span}(\mathsf{z}_1, \dots, \mathsf{z}_n)$ for a sequence $(\mathsf{z}_n)_{n\in\mathbb{N}}$ dense in X. If a basis $(b_n)_{n\in\mathbb{N}}$ exists for X, then we

---

6 To see this, let $\dim(X_n) = k_n$ and simply take a basis $\{b_1, \dots, b_{k_n}\}$ for the finite dimensional space X_n. Then, expand $\mathsf{x}_n(t)$ on this basis as $\mathsf{x}_n(t) = \sum_{i=1}^{k_n} c_i(t) b_i$, where $c_i(t) \in \mathbb{R}$ are appropriate coefficients, and then use $\mathsf{z} = b_i$, $i = 1, \dots, k_n$, as test functions in (10.72) to obtain a system of ODEs in $\mathbb{R}^{k_n}$ for the coefficients $\{c_1(t), \dots, c_{k_n}(t)\}$.

may construct $X_n = \operatorname{span}(b_1, \dots, b_n)$, in which case, we can guarantee that $\dim(X_n) = n$. Moreover, the projection operators Π_n can be explicitly constructed using a basis for the finite dimensional spaces X_n by projecting any element of H onto this basis (however Assumption 10.3.10 may require special choice for the basis; see [128]).

An existence result in this spirit is given in the following proposition [128]:

Proposition 10.3.12. *Suppose that Assumptions* 10.3.2, 10.3.9, *and* 10.3.10 *hold. Then, for every* $\mathsf{x}_0 \in H$ *and* $f \in L^{p^*}((0,T);X^\star)$ *problem* (10.64) *has a solution* $\mathsf{x} \in L^p((0,T);X) \cap L^\infty((0,T);H)$, $\frac{d\mathsf{x}}{dt} \in L^{p^*}((0,T);X^*)$. *Moreover, this solution can be approximated as the weak limit in the corresponding spaces of (a subsequence) of the sequence of finite dimensional approximations* (10.72).

Proof. The proof consists of 3 parts.

1. A priori estimates for the finite dimensional approximation (10.72). Note that here we start by the equation, and build the membership of the solution to the required function space framework as we proceed.

Existence of solutions such that $\mathsf{x}_n(t) \in X_n$ is obtained by standard ODE arguments (see above), hence we may use $\mathsf{z} = \mathsf{x}_n(t)$ as test function in (10.72). This yields

$$\frac{1}{2}\frac{d}{dt}\|\mathsf{x}_n(t)\|_H^2 + \langle \mathsf{A}(t, \mathsf{x}_n(t)), \mathsf{x}_n(t)\rangle_{X^*,X} = \langle f(t), \mathsf{x}_n(t)\rangle,$$

where we have used the fact that $\mathsf{x}_n \in X_n \subset X \hookrightarrow H$. Using the coercivity properties of A (see Assumption 10.3.2(iii) and setting $\theta(t) = 0$ for simplicity), the above yields the estimate

$$\begin{aligned}\frac{1}{2}\frac{d}{dt}\|\mathsf{x}_n(t)\|_H^2 + c_1\|\mathsf{x}_n(t)\|_X^p &\le \langle f(t), \mathsf{x}_n(t)\rangle \le \|f(t)\|_{X^*}\|\mathsf{x}_n(t)\|_X \\ &\le C(\epsilon)\|f(t)\|_{X^*}^{p^*} + \epsilon\|\mathsf{x}_n(t)\|_X^p,\end{aligned}$$

where, for the final estimate, we used the weighted Hölder inequality for an arbitrary $\epsilon > 0$. Then choosing ϵ sufficiently small (e. g., $\epsilon = c_1/2$), we bring the term $\epsilon\|\mathsf{x}_n(t)\|_X^p$ to the LHS to obtain an estimate of the form

$$\frac{1}{2}\frac{d}{dt}\|\mathsf{x}_n(t)\|_H^2 + c_2\|\mathsf{x}_n(t)\|_X^p \le C(\epsilon)\|f(t)\|_{X^*}^{p^*}, \tag{10.73}$$

for a constant $c_2 > 0$. The actual value of c_2 is not important; what is important is the fact that c_2 is independent of n.

By omitting the $c_2\|\mathsf{x}_n(t)\|_X^p$ term in (10.73) and integrating over $[0,t]$ for any $t \le T$, we obtain

$$\|\mathsf{x}_n(t)\|_H^2 \le \|\mathsf{x}_0\|_H^2 + 2\int_0^t \|f(s)\|_{X^*}^{p^*} ds \le \|\mathsf{x}_0\|_H^2 + 2\int_0^T \|f(s)\|_{X^*}^{p^*} ds,$$

which, by the fact that $f \in L^{p^*}((0,T);X^*)$, leads to a bound for x_n in $L^\infty((0,T);H)$, of the form $\|x_n\|_{L^\infty((0,T);H)} \le C_1$ for C_1 independent of n.

Reinstating the term $c_2\|x_n(t)\|_X^p$ in (10.73) and integrating over $[0,T]$, we obtain after rearranging,

$$c_2 \int_0^T \|x_n(t)\|^p dt \le \frac{1}{2}\|x_0\|_H^2 - \frac{1}{2}\|x_n(T)\|_H^p + \int_0^T \|f(s)\|_{X^*}^{p^*} ds$$
$$\le \frac{1}{2}\|x_0\|_H^2 + \int_0^T \|f(s)\|_{X^*}^{p^*} ds,$$

which leads to a bound $\|x_n\|_{L^p((0,T);X)} \le C_2$, for C_2 independent of n.

We now consider the bounds for the derivative $\frac{dx_n}{dt}$. Consider any $z \in L^p((0,T);X)$. Note that since $\frac{dx_n}{dt}(t) \in X_k$, we have that $\Pi_n \frac{dx_n}{dt}(t) = \frac{dx_n}{dt}(t)$. Clearly, by $X_n \subset X \hookrightarrow H \hookrightarrow X^*$, any element of X_n can be considered as an element of X^* as well. With the above considerations at hand, we have that

$$\left\langle \frac{dx_n}{dt}(t), z(t) \right\rangle_{X^*,X} = \left\langle \Pi_n \frac{dx_n}{dt}(t), z(t) \right\rangle_{X^*,X} = \left\langle \frac{dx_n}{dt}(t), \Pi_n z(t) \right\rangle_{X^*,X}, \tag{10.74}$$

where we used the fact that Π_n is self adjoint. But, $\Pi_n z(t) \in X_n$, so it can be used as a test function for the finite dimensional approximation equation (10.72). Combining this observation with (10.74), we obtain

$$\left\langle \frac{dx_n}{dt}(t), z(t) \right\rangle_{X^*,X} = \langle f(t), \Pi_n z(t)\rangle_{X^*,X} - \langle A(t, x_n(t)), \Pi_n z(t)\rangle_{X^*,X}.$$

We integrate over all $t \in [0,T]$ to obtain

$$\begin{aligned} \left\langle \frac{dx_n}{dt}, z \right\rangle_{X_T^*,X_T} &= \int_0^T \langle f(t) - A(t, x_n(t)), \Pi_n z(t)\rangle_{X^*,X} dt = \langle f - \widehat{A}(x_n), \Pi_n z\rangle_{X_T^*,X_T} \\ &\le (\|f\|_{X_T^*} + \|\widehat{A}(x_n)\|_{X_T^*})\|\Pi_n z\|_{X_T} \le C(\|f\|_{X_T^*} + \|\widehat{A}(x_n)\|_{X_T^*})\|z\|_{X_T}, \end{aligned} \tag{10.75}$$

where we used the operator $\widehat{A}$ defined in (10.65) and the properties of the projection operator Π_n. The term $\|f\|_{X_T^*}$ is bounded by assumption. Moreover, the term $\|\widehat{A}(x_n)\|_{X_T^*}$, also bounded, since the operator A satisfies boundedness properties, and we have already established a bound for x_n in X_T. Therefore, $\|f\|_{X_T^*} + \|\widehat{A}(x_n)\|_{X_T^*}$ is bounded independently of n. Combining all these together, we obtain by (10.75) that

$$\left\langle \frac{dx_n}{dt}, z \right\rangle_{X_T^*,X_T} \le c_3\|z\|_{X_T}, \quad \forall z \in X_T = L^p((0,T);X),$$

for some constant c_3 independent of n, hence (recall the definition of the dual norm) $\|\frac{dx_n}{dt}\|_{X_T^*} \leq c_3$.

Hence, we have the important a priori estimates

$$\|x_n\|_{L^\infty((0,T);H)} \leq c, \quad \|x_n\|_{L^p((0,T);X)} \leq c, \quad \left\|\frac{dx_n}{dt}\right\|_{L^{p^*}((0,T);X^*)} \leq c, \tag{10.76}$$

for c independent of n.

2. We are now ready to pass to the limit, as $n \to \infty$. By weak compactness, (10.76) guarantees the existence of an $x \in L^\infty((0,T);H) \cap L^p((0,T);X)$ and an $x^* \in L^{p^*}((0,T);X^*)$ such that

$$\begin{aligned} x_n &\rightharpoonup x \quad \text{in } L^p((0,T);X), \\ x_n &\overset{\star}{\rightharpoonup} x \quad \text{in } L^\infty((0,T);H), \\ \frac{dx_n}{dt} &\rightharpoonup x^* \quad \text{in } L^{p^*}((0,T);X^*). \end{aligned} \tag{10.77}$$

The fact that the limit for the first two convergences is the same, follows by considering both as convergence in the space of distributions and using arguments related to the uniqueness of the limit. Moreover, $x^* = \frac{dx}{dt}$, a fact that follows by the definition of the weak derivative.[7] Finally, again by weak compactness arguments, $x_n(T) \rightharpoonup \xi$ in H, and this can be identified with $x(T)$ in H.[8]

3. It remains to show that the function $x : [0,T] \to X$ obtained above solves the equation, a task that requires the finite dimensional approximation (10.72) for any $t \in [0,T]$ and some attention since all the convergences involved are weak (which do not

7 This is a pretty standard argument that goes as follows: The weak derivative of $x : [0,T] \to X$ is defined in terms of the integration by parts argument $\langle \frac{dx}{dt}, \phi \rangle = -\langle x, \frac{d\phi}{dt} \rangle$, with the duality pairing considered between any two suitable function spaces X_T and X_T^* and with the equality holding for any suitable test function $\phi : [0,T] \to X$. For any $n \in \mathbb{N}$, this formula yields $\langle \frac{dx_n}{dt}, \phi \rangle = -\langle x_n, \frac{d\phi}{dt} \rangle$. Passing to the limit, this gives $\langle x^*, \phi \rangle = -\langle x, \frac{d\phi}{dt} \rangle = \langle \frac{dx}{dt}, \phi \rangle$, and this allows us to identify $x^* = \frac{dx}{dt}$. Note that we have been deliberately sloppy about the definition of the pairings $\langle \cdot, \cdot \rangle$.

8 To see this, consider any suitable z and use the integration by parts formula (1.13) for any x_n. This gives

$$\langle x_n(T), z(T) \rangle_H - \langle x_n(0), z(0) \rangle_H + \int_0^T \left(\left\langle x_n(\tau), \frac{dz}{dt}(\tau) \right\rangle + \left\langle \frac{dx_n}{dt}(\tau), z(\tau) \right\rangle \right) d\tau.$$

Passing to the limit, as $n \to \infty$ in the above, yields

$$\langle \xi, z(T) \rangle_H - \langle x_0, z(0) \rangle_H + \int_0^T \left(\left\langle x(\tau), \frac{dz}{dt}(\tau) \right\rangle + \left\langle \frac{dx}{dt}(\tau), z(\tau) \right\rangle \right) d\tau.$$

Comparing that with the corresponding integration by parts formula for x, we obtain the required statement.

always behave well with nonlinear operators). Consider any $z \in L^p((0,T);X)$, and for any $t \in [0,T]$ consider $z(t) \in X$. By the density of $\bigcup_{n\in\mathbb{N}} X_n$ in X, this (using similar arguments as above) can be approximated by a $(z_n(t))_{n\in\mathbb{N}} \subset X$, with $z_n(t) \in X_{\ell_n}$ for each $n \in \mathbb{N}$, for a sequence $(\ell_n)_{n\in\mathbb{N}}$, $\ell_n \uparrow$, which will be used to construct a test function $x_n(t) - z_n(t) \in H_{\ell_n}$ for the finite dimensional approximation (10.72). We note that $z_n(t) \to z(t)$ (strong) in X and clearly $x_n(t) - z_n(t) \rightharpoonup x(t) - z(t)$ in X. Our goal is to show that passing to the limit, as $n \to \infty$, (in the appropriate subsequence) (10.72) tends to the weak formulation of the original problem, i. e., in the space $L^p((0,T);X)$.

Working as above (see also Section 9.8), considering the finite dimensional approximation (10.72) for any $t \in [0,T]$ and using the test function $x_n(t) - z_n(t) \in H_{\ell_n}$, the weak formulation (10.72), upon integration over $[0,T]$ yields

$$\begin{aligned}\int_0^T \left\langle \frac{dx_n}{dt}(t), x_n(t) - z_n(t) \right\rangle dt + \int_0^T \langle A(t, x_n(t)), x_n(t) - z_n(t)\rangle dt \\ = \int_0^T \langle f(t), x_n(t) - z_n(t)\rangle dt,\end{aligned}$$

which is rearranged as

$$\begin{aligned}\int_0^T \langle A(t, x_n(t)), x_n(t) - z_n(t)\rangle dt &= \int_0^T \langle f(t), x_n(t) - z_n(t)\rangle dt \\ &\quad - \int_0^T \left\langle \frac{dx_n}{dt}(t), x_n(t) - z_n(t) \right\rangle dt \\ &= \int_0^T \langle f(t), x_n(t) - z_n(t)\rangle dt - \frac{1}{2}\|x_n(T)\|_H^2 + \frac{1}{2}\|x_n(0)\|^2 \\ &\quad + \int_0^T \left\langle \frac{dx_n}{dt}(t), z_n(t) \right\rangle dt.\end{aligned} \tag{10.78}$$

We intend to take the limsup in the above equality. Before doing so, we note that

$$\begin{aligned}\int_0^T \langle f(t), x_n(t) - z_n(t)\rangle dt &\to \int_0^T \langle f(t), x(t) - z(t)\rangle dt, \\ \|x_n(0)\| = \|x_{0,n}\| &\to \|x_0\|, \\ \int_0^T \left\langle \frac{dx_n}{dt}(t), z_n(t) \right\rangle dt &\to \int_0^T \left\langle \frac{dx}{dt}(t), z(t) \right\rangle dt,\end{aligned} \tag{10.79}$$

since (by order of appearance)

$$\mathsf{x}_n - \mathsf{z}_n \rightharpoonup \mathsf{x} - \mathsf{z}, \quad \text{in } L^p((0,T);X),$$
$$\mathsf{x}_{n,0} \to \mathsf{x}_0 \quad \text{in } H,$$
$$\frac{d\mathsf{x}_n}{dt} \rightharpoonup \frac{d\mathsf{x}}{dt} \quad \text{in } L^{p^*}((0,T);X^*), \quad \text{and} \quad \mathsf{z}_n \to \mathsf{z} \quad \text{in } L^p((0,T);X).$$

Note that the reason for choosing $(\mathsf{z}_n)_{n\in\mathbb{N}}$ as in the beginning of step 3 above, is to guarantee the last convergence in (10.79), clearly, because of the weak convergence of $\frac{d\mathsf{x}_n}{dt}$ requires strong convergence of z_n to guarantee the stated result. The results in (10.79) cover all terms in (10.78) but the second, treated by lower semicontinuity of the norm,

$$\limsup_n(-\|\mathsf{x}_n(T)\|_H^2) = -\liminf_n\|\mathsf{x}_n(T)\|_H^2 \le -\|\mathsf{x}(T)\|_H^2. \tag{10.80}$$

We now take the limsup for both sides of (10.78) using (10.79) and (10.80) to obtain

$$\begin{aligned}
&\limsup_n \int_0^T \langle A(t,\mathsf{x}_n(t)), \mathsf{x}_n(t) - \mathsf{z}_n(t)\rangle dt \\
&\le \int_0^T \langle f(t), \mathsf{x}(t) - \mathsf{z}(t)\rangle dt - \frac{1}{2}\|\mathsf{x}(T)\|_H^2 + \frac{1}{2}\|\mathsf{x}_0\|_H^2 + \int_0^T \left\langle \frac{d\mathsf{x}}{dt}(t), \mathsf{z}(t)\right\rangle dt \\
&= \int_0^T \left\langle f(t) - \frac{d\mathsf{x}}{dt}(t), \mathsf{x}(t) - \mathsf{z}(t)\right\rangle dt.
\end{aligned} \tag{10.81}$$

where we used $-\frac{1}{2}\|\mathsf{x}(T)\|_H^2 + \frac{1}{2}\|\mathsf{x}_0\|_H^2 = -\int_0^T \langle \frac{d\mathsf{x}}{dt}(t), \mathsf{x}(t)\rangle dt$. We now estimate

$$\begin{aligned}
&\limsup_n \int_0^T \langle A(t,\mathsf{x}_n(t)), \mathsf{x}_n(t) - \mathsf{z}(t)\rangle dt \\
&= \limsup_n\left(\int_0^T \langle A(t,\mathsf{x}_n(t)), \mathsf{x}_n(t) - \mathsf{z}_n(t)\rangle dt + \int_0^T \langle A(t,\mathsf{x}_n(t)), \mathsf{z}_n(t) - \mathsf{z}(t)\rangle dt\right) \\
&\le \limsup_n(\int_0^T \langle A(t,\mathsf{x}_n(t)), \mathsf{x}_n(t) - \mathsf{z}_n(t)\rangle dt + \underbrace{\lim_n \int_0^T \langle A(t,\mathsf{x}_n(t)), \mathsf{z}_n(t) - \mathsf{z}(t)\rangle dt}_{=0 \text{ since } \mathsf{z}_n \to \mathsf{z} \text{ in } L^p((0,T);X)} \\
&\overset{(10.81)}{\le} \int_0^T \left\langle f(t) - \frac{d\mathsf{x}}{dt}(t), \mathsf{x}(t) - \mathsf{z}(t)\right\rangle dt.
\end{aligned} \tag{10.82}$$

Note that again we used the special choice of $(\mathsf{z}_n)_{n\in\mathbb{N}}$ such that $\mathsf{z}_n \to \mathsf{z}$ in $X_T = L^p((0,T);X)$ to eliminate the underbraced term above.

Note that in terms of the operator $\widehat{A}$ (defined in (10.65)), we see that (10.82) becomes

$$\limsup_n \langle \widehat{A}(x_n), x_n - z\rangle_{X_T^*, X_T} \le \left\langle f - \frac{dx}{dt}, x - z\right\rangle_{X_T^*, X_T}, \quad \forall z \in X_T. \tag{10.83}$$

Choosing $z = x$, we obtain that

$$\limsup_n \langle \widehat{A}(x_n), x_n - x\rangle_{X_T^*, X_T} \le 0.$$

By the pseudomonotonicity of $\widehat{A}$ (see Assumption 10.3.9 and Proposition 10.3.6), this yields that

$$\liminf_n \langle A(x), x - z\rangle_{X_T^*, X_T} \le \langle \widehat{A}(x_n), x_n - z\rangle_{X_T^*, X_T}, \quad \forall z \in X_T. \tag{10.84}$$

Combining (10.84) with (10.83) we get that

$$\begin{aligned} \langle \widehat{A}(x), x - z\rangle_{X_T^*, X_T} &\le \liminf_n \langle \widehat{A}(x_n), x_n - z\rangle_{X_T^*, X_T} \\ &\le \limsup_n \langle \widehat{A}(x_n), x_n - z\rangle_{X_T^*, X_T} \\ &\le \left\langle f - \frac{dx}{dt}, x - z\right\rangle_{X_T^*, X_T}, \quad \forall z \in X_T. \end{aligned} \tag{10.85}$$

Hence,

$$\langle \widehat{A}(x), x - z\rangle_{X_T^*, X_T} \le \left\langle f - \frac{dx}{dt}, x - z\right\rangle_{X_T^*, X_T}, \quad \forall z \in X_T,$$

which leads to

$$\langle \widehat{A}(x), z\rangle_{X_T^*, X_T} = \left\langle f - \frac{dx}{dt}, z\right\rangle_{X_T^*, X_T}, \quad \forall z \in X_T,$$

i. e., x is the solution to the Cauchy problem. □

Remark 10.3.13 (Time discretization or the Rothe method). Another alternative to the Galerkin method is the method of time discretization, or the Rothe method. This is essentially similar to the backward Euler method presented in Section 10.2.6, with the difference that now (10.21) is replaced by

$$\frac{x_{h,n+1} - x_{h,n}}{h} + A(x_{h,n+1}) \ni f_{h,n}, \tag{10.86}$$

(recall that since we are dealing with monotone type operators, we are solving $\frac{dx}{dt} + A(x) = f$ or its multivalued version) and that $A : \mathbf{D}(A) \subset X \to X^*$ and $f : [0, T] \to X^*$ (rather than $A : \mathbf{D}(A) \subset X \to X$ and $f : [0, T] \to X$ as in Section 10.2.6). The underlying idea remains more or less the same, i. e., use (10.86) to construct a sequence of approximations as in (10.19), which can in the appropriate limit can be interpreted as a solution to the Cauchy problem. The existence of the discretized sequence can be proved by using the surjectivity theorems for pseudomonotone operators for the solvability of the operator equation (10.86). A difference with Section 10.2.6 is that now, depending on the properties of f, the solution can reside in X^*. However, more regular forcing terms and initial conditions may allow for more regular solutions residing in X. For a detailed analysis of the Rothe method, see, e. g., [128] Section 8.2.

10.3.4.2 Evolution problems for pseudomonotone operators II: abstract approach

As before, let $(X, H, X^\star)$ be an evolution triple (recall X is reflexive), but we now make the extra assumption that the embedding $X \overset{c}{\hookrightarrow} H$ is compact.

The following surjectivity result is due to Lions (see Theorem 1.1, Chapter 3, p. 316 [101]):

Theorem 10.3.14. *Let X be a reflexive Banach space, $\mathsf{M} : \mathbf{D}(\mathsf{M}) \subset X \to X^\star$, where $\mathbf{D}(\mathsf{M})$ is a dense subspace of X, be a linear maximal monotone operator and $\mathsf{B} : X \to X^\star$ be a demicontinuous, bounded, coercive, and M-pseudomonotone operator, then*

$$\mathbf{R}(\mathsf{M} + \mathsf{B}) = X^\star.$$

Proof. We sketch the proof which uses an interesting regularization argument. The proof proceeds in 3 steps.

1. For any $x^\star \in X^\star$, we approximate $\mathsf{M}x + \mathsf{B}(x) = x^\star$ by

$$\epsilon \mathsf{M}^\star \mathsf{J}^{-1}(\mathsf{M}x_\epsilon) + \mathsf{M}x_\epsilon + \mathsf{B}(x_\epsilon) = x^\star, \quad \epsilon > 0, \tag{10.87}$$

where $\mathsf{J} : X \to X^\star$ is the duality map. We equip $\mathbf{D}(\mathsf{M})$ with the graph norm $\|x\|_{X_0} := \|x\|_X + \|\mathsf{M}x\|_{X^\star}$, and thus turn it into a reflexive Banach space, denoted by X_0, with dual $X_0^\star$. Now, for every $\epsilon > 0$ fixed define the operator $\mathsf{M}_\epsilon : X_0 \to X_0^\star$, by $\mathsf{M}_\epsilon(x) = \epsilon \mathsf{M}^\star \mathsf{J}^{-1}(\mathsf{M}x) + \mathsf{M}x$, which is a bounded hemicontinuous monotone operator. Consider also the restriction of B on X_0, denoted by $\hat{\mathsf{B}} : X_0 \to X_0^\star$, which is clearly pseudomonotone. Then, by Proposition 9.5.3, the operator $\mathsf{M}_\epsilon + \hat{\mathsf{B}} : X_0 \to X_0^\star$ is pseudomonotone, and by Theorem 9.5.9 equation (10.87) admits a solution $x_\epsilon \in X_0$. Furthermore, observe that $\mathsf{J}^{-1}(\mathsf{M}x_\epsilon) \in D(\mathsf{M}^\star)$.

2. We see that $\|x_\epsilon\|_X$ and $\|\mathsf{M}x_\epsilon\|_{X^\star}$ are bounded uniformly in ϵ. For instance to verify the second claim take the duality pairing of (10.87) with $\mathsf{J}^{-1}(\mathsf{M}x_\epsilon) \in D(\mathsf{M}^\star)$, where, by using the monotonicity of $\mathsf{M}^\star$ (which follows by the monotonicity of M) and the definition of the duality map, we obtain

$$\|\mathsf{M}x_\epsilon\|_{X^\star}^2 + \langle \mathsf{B}(x_\epsilon), \mathsf{J}^{-1}(\mathsf{M}x_\epsilon)\rangle_{X^\star,X} \le \langle x^\star, \mathsf{J}^{-1}(\mathsf{M}x_\epsilon)\rangle_{X^\star,X},$$

which (since $\|x_\epsilon\|_X < c_1$ by some $c_1 > 0$ independent of $\epsilon > 0$ and B is bounded) leads to the uniform boundedness of $\|\mathsf{M}x_\epsilon\|_{X^\star}$.

3. By reflexivity and uniform boundedness, there exists a subsequence $\epsilon_n \to 0$, such that upon denoting for simplicity $\mathsf{x}_{\epsilon_n} = \mathsf{x}_n$, we have $\mathsf{x}_n \rightharpoonup \mathsf{x}$ in X, $\mathsf{M}\mathsf{x}_n \rightharpoonup \mathsf{x}^\star$ and $\mathsf{B}(\mathsf{x}_n) \rightharpoonup \mathsf{z}$ in $X^\star$. By the properties of M, we have that $\mathsf{x} \in \mathbf{D}(\mathsf{M})$ and $\mathsf{x}^\star = \mathsf{M}\mathsf{x}$. Moving along this subsequence and taking the duality pairing of (10.87) with $\mathsf{x}_n \rightharpoonup \mathsf{x}$, and using the fact that $\|\mathsf{M}\mathsf{x}_n\|_{X^\star} < c_2$ for some $c_2 > 0$ independent of $n \in \mathbb{N}$ (since the whole sequence $\|\mathsf{M}\mathsf{x}_\epsilon\|_{X^\star} < \epsilon$), we conclude that $\langle \mathsf{B}(\mathsf{x}_n), \mathsf{x}_n - \mathsf{x}\rangle_{X^\star,X} \leq \langle \mathsf{x}^\star, \mathsf{x}_n - \mathsf{x}\rangle_{X^\star,X} + c_2\epsilon_n$, whereupon taking the limit superior, we get that

$$\limsup_{n\to\infty} \langle \mathsf{B}(\mathsf{x}_n), \mathsf{x}_n - \mathsf{x}\rangle \leq 0.$$

By M-pseudomonotonicity of B, we have $\mathsf{B}(\mathsf{x}) = \mathsf{z}$ and $\langle \mathsf{B}(\mathsf{x}_n), \mathsf{x}_n\rangle \to \langle \mathsf{B}(\mathsf{x}), \mathsf{x}\rangle$. Taking the duality pairing of (10.87) (along the chosen subsequence) with $\mathsf{x}_n - \mathsf{z}_0$ for every $\mathsf{z}_0 \in \mathbf{D}(\mathsf{M})$, and using the fact that $\mathsf{M}\mathsf{x}_n$ is uniformly bounded in $X^\star$, we get that

$$\langle \mathsf{x}^\star - \mathsf{M}\mathsf{x} - \mathsf{B}(\mathsf{x}), \mathsf{x} - \mathsf{z}_0\rangle = 0, \quad \forall\, \mathsf{z}_0 \in \mathbf{D}(\mathsf{M}),$$

therefore, by density of $\mathbf{D}(\mathsf{M})$, we have $\mathsf{M}\mathsf{x} + \mathsf{B}(\mathsf{x}) = \mathsf{x}^\star$. □

Theorem 10.3.14 leads to the existence result in the next theorem [88].

Theorem 10.3.15. *Suppose that* A *satisfies the condition of Assumption* 10.3.9. *Then, for every* $\mathsf{x}_0 \in H$ *and* $f \in L^{p^*}((0,T);X^\star)$, *problem* (10.64) *has a solution* $\mathsf{x} \in W^{1,p}((0,T);X)$.

Proof. The strategy of the proof is exactly the same (and uses the same notation) as in Thm. 10.3.8. By this theorem, we know that $\widehat{\mathsf{A}}$ is bounded and coercive. By Thm. 10.3.14 (for $\mathsf{M} = \mathsf{L}$ and $\mathsf{B} = \widehat{\mathsf{A}}$), we get $\mathbf{R}(\mathsf{L} + \widehat{\mathsf{A}}) = X^*$, i. e., problem (10.64) has a solution. □

The results of this section can be applied, for instance, to the heat equation or the generalized heat equation involving the p-Laplace operator (see Section 10.4.6).

10.4 Applications

The general results presented in this chapter can be used in the study of multiple and diverse applications. The selected examples in this section are inspired by e. g., [18, 19, 33, 42, 51, 76, 89, 111, 128], where the reader can also refer to, for more examples.

10.4.1 Semilinear evolution equations

In this section, we examine semilinear evolution equations of the general form

$$\begin{aligned} &\frac{d\mathsf{x}}{dt}(t) = \mathsf{A}\mathsf{x}(t) + f(t,\mathsf{x}(t)), \quad t \in I = [0,T], \\ &\mathsf{x}(0) = \mathsf{x}_0 \in X, \end{aligned} \tag{10.88}$$

where X is a Banach space, $\mathsf{A} : \mathbf{D}(\mathsf{A}) \subset X \to X$ is a linear operator that generates a contraction semigroup (see Section 1.7 and in particular Theorems 1.7.4 and 1.7.6), and $f : [0,T] \times X \to X$ a given nonlinear function. Naturally, the results in this section can be generalized for the case where A is the generator of a C_0-semigroup (i. e., not necessarily a contraction). We impose the following hypotheses on the nonlinearity f:

Assumption 10.4.1. The function $f : [0,T] \times X \to X$ satisfies the following:
(i) For every $\mathsf{x} \in X$, $t \mapsto f(t,\mathsf{x})$ is strongly measurable.
(ii) There exists $k \in L^1((0,T))_+$ such that for a. e. $t \in [0,T]$ and all $\mathsf{x},\mathsf{z} \in X$, we have

$$\begin{aligned} \|f(t,\mathsf{x}) - f(t,\mathsf{z})\| &\le k(t)\|\mathsf{x}-\mathsf{z}\|, \\ \|f(t,0)\| &\le k(t). \end{aligned}$$

Definition 10.4.2. A function $\mathsf{x} \in C([0,T];X)$ is called a mild solution of (10.88) if

$$\mathsf{x}(t) = \mathsf{S}(t)\mathsf{x}_0 + \int_0^t \mathsf{S}(t-\tau)f(\tau,\mathsf{x}(\tau))d\tau, \quad t \in [0,T].$$

A solvability result is shown in the next proposition [33, 89, 111].

Proposition 10.4.3. *If* A *is the infinitesimal generator of a contraction semigroup* $\{\mathsf{S}(t)\}_{t\ge 0}$, f *satisfies the hypotheses of Assumption* 10.4.1 *and* $\mathsf{x}_0 \in X$, *then problem* (10.88) *has a unique mild solution* $\mathsf{x}(\cdot) =: \mathsf{x}(\cdot;\mathsf{x}_0) \in C([0,T];X)$ *satisfying*

$$\begin{aligned} \|\mathsf{x}(t;\mathsf{x}_0)\| &\le \exp\left(\int_0^t k(\tau)d\tau\right)(1+\|\mathsf{x}_0\|), \\ \|\mathsf{x}(t;\mathsf{x}_0) - \mathsf{x}(t;\bar{\mathsf{x}}_0)\| &\le \exp\left(\int_0^t k(\tau)d\tau\right)\|\mathsf{x}_0 - \bar{\mathsf{x}}_0\|. \end{aligned} \tag{10.89}$$

Proof. We introduce the following equivalent norm on $C([0,T];X)$:

$$\|\mathsf{x}\|_1 = \max\left\{\exp\left(-L\int_0^t k(\tau)d\tau\right)\|\mathsf{x}(t)\|,\ t \in [0,T]\right\}$$

for a suitable $L > 1$. We consider the nonlinear operator $F : C([0,T];X) \to C([0,T];X)$, defined by $\mathsf{x} \mapsto F(\mathsf{x})$ with

$$(F(\mathsf{x}))(t) = \mathsf{S}(t)\mathsf{x}_0 + \int_0^t \mathsf{S}(t-\tau)f(\tau,\mathsf{x}(\tau))d\tau,$$

for any $\mathsf{x} \in C([0,T];X)$. Now let $\mathsf{x},\mathsf{z} \in C([0,T];X)$, and calculate

$$\begin{aligned}\|(F(\mathsf{x}))(t) - (F(\mathsf{z}))(t)\| &= \left\| \int_0^t \mathsf{S}(t-\tau)(f(\tau,\mathsf{x}(\tau))) - f(\tau,\mathsf{z}(\tau))d\tau \right\| \\ &\leq \int_0^t k(\tau)\|\mathsf{x}(\tau) - \mathsf{z}(\tau)\|d\tau.\end{aligned} \tag{10.90}$$

Multiplying (10.90) with $\exp(-L\int_0^t k(s)ds) = \exp(-L\int_0^\tau k(s)ds)\exp(-L\int_\tau^t k(s)ds)$,

$$\begin{aligned}&\exp\left(-L\int_0^t k(\tau)d\tau\right)\|(F(\mathsf{x}))(t) - (F(\mathsf{z}))(t)\| \\ &\leq \int_0^t \exp\left(-L\int_\tau^t k(s)ds\right)\exp\left(-L\int_0^\tau k(s)ds\right)k(\tau)\|\mathsf{x}(\tau) - \mathsf{z}(\tau)\|d\tau \\ &\leq \|\mathsf{x} - \mathsf{z}\|_1 \int_0^t \exp\left(-L\int_\tau^t k(s)ds\right)k(\tau)d\tau \leq \frac{1}{L}\|\mathsf{x} - \mathsf{z}\|_1.\end{aligned}$$

Hence, we have

$$\|F(\mathsf{x}) - F(\mathsf{z})\|_1 \leq \frac{1}{L}\|\mathsf{x} - \mathsf{z}\|_1,$$

and since $L > 1$, by Banach's contraction principle, we get that there exists a unique $\mathsf{x}(\cdot;\mathsf{x}_0) \in C([0,T];X)$ such that

$$\mathsf{x}(\cdot;\mathsf{x}_0) = F(\mathsf{x}(\cdot;\mathsf{x}_0)).$$

Clearly, this is the unique mild solution of (10.88).

Expressing $f(\tau,\mathsf{x}(\tau)) = f(\tau,\mathsf{x}(\tau)) - f(\tau,0) + f(\tau,0)$, and using the Lipschitz property,

$$\begin{aligned}\|\mathsf{x}(t;\mathsf{x}_0)\| &\leq \|\mathsf{x}_0\| + \int_0^t \|f(\tau,\mathsf{x}(\tau;\mathsf{x}_0))\|d\tau \\ &\leq \|\mathsf{x}_0\| + \int_0^t (\|f(\tau,0)\| + k(\tau)\|\mathsf{x}(\tau;\mathsf{x}_0)\|)d\tau \\ &\leq \|\mathsf{x}_0\| + \int_0^t k(\tau)(1 + \|\mathsf{x}(\tau;\mathsf{x}_0)\|)d\tau.\end{aligned}$$

Using Gronwall's inequality, we get that

$$\|\mathsf{x}(t;\mathsf{x}_0)\| \leq \exp\left(\int_0^t k(\tau)d\tau\right)(1 + \|\mathsf{x}_0\|).$$

Moreover, we have

$$\begin{aligned}\|\mathsf{x}(t;\mathsf{x}_0)-\mathsf{x}(t;\bar{\mathsf{x}}_0)\| &\leq \|\mathsf{S}(t)(\mathsf{x}_0-\bar{\mathsf{x}}_0)\| + \left\|\int_0^t \mathsf{S}(t-\tau)(f(\tau,\mathsf{x}(\tau;\mathsf{x}_0))-f(\tau,\mathsf{x}(\tau;\bar{\mathsf{x}}_0)))d\tau\right\| \\ &\leq \|\mathsf{x}_0-\bar{\mathsf{x}}_0\| + \int_0^t k(\tau)\|\mathsf{x}(\tau;\mathsf{x}_0)-\mathsf{x}(\tau;\bar{\mathsf{x}}_0)\|d\tau,\end{aligned}$$

where we used the Lipschitz property of f and the contraction property for S. Using Gronwall's inequality again

$$\|\mathsf{x}(t;\mathsf{x}_0)-\mathsf{x}(t;\bar{\mathsf{x}}_0)\| \leq \exp\left(\int_0^t k(\tau)d\tau\right)\|\mathsf{x}_0-\bar{\mathsf{x}}_0\|,$$

and the proof is complete. □

10.4.2 Elliptic operators in $L^2(\mathcal{D})$ and parabolic equations

As a concrete example of the application of the Lumer–Phillips theorem (see Theorem 1.7.6) to semilinear problems, which was sketched in a abstract setting in Section 10.4.1, we consider the case of parabolic problems in bounded domains $\mathcal{D} \subset \mathbb{R}^d$ of the form

$$\begin{aligned}&\frac{\partial u}{\partial t}(t,x) = \sum_{i,j=1}^{d}\frac{\partial}{\partial x_i}\left(a_{ij}(x)\frac{\partial u}{\partial x_j}(t,x)\right) + f(t,u(t,x)), \quad (t,x)\in[0,T]\times\mathcal{D},\\ &u(t,x)=0, \quad (t,x)\in[0,T]\times\partial\mathcal{D},\\ &u(0,x)=u_0(x), \quad x\in\mathcal{D},\end{aligned} \tag{10.91}$$

for a given function $f : [0,T]\times\mathbb{R}\to\mathbb{R}$, satisfying Assumption 10.4.1 (note that in the abstract formulation f can also be interpreted as $f : [0,T]\times X\to X$ for $X = L^2(\mathcal{D})$), and a given initial condition u_0. The case where f is independent of u reduces problem (10.91) to the linear parabolic case (that can be treated directly in terms of Theorem 1.7.3, for appropriate definition of the operator A). Homogeneous Dirichlet boundary conditions are assumed for simplicity, however, other boundary conditions can be chosen. Moreover, generalized solutions, e. g., mild solutions of the system can be considered.

The coefficients $a_{ij} = a_{ji}$ are in principle space-dependent, and satisfy a standard ellipticity condition:

$$\exists\ c>0 \text{ such that } \sum_{i,j=1}^{d} a_{ij}(x)\xi_i\xi_j \geq c|\xi|^2, \quad \forall\,\xi=(\xi_1,\dots,\xi_d)\in\mathbb{R}^d, \text{ a. e. in } \mathcal{D}. \tag{10.92}$$

In the case where $a_{ij} = \delta_{ij}$, (10.91) reduces to the heat equation.

We will write (10.91) in abstract operator form. To make better connection with the theory of elliptic systems discussed in Chapters 6 and 7, we consider the operator A, defined by

$$\mathsf{A}u = -\sum_{i,j=1}^{d} \frac{\partial}{\partial x_i}\left(a_{ij}\frac{\partial u}{\partial x_j}\right), \tag{10.93}$$

where $a_{ij} = a_{ji} \in C^1(\mathcal{D})$ satisfy the uniform ellipticity condition (10.92), with domain

$$\mathbf{D}(\mathsf{A}) = \{u \in W_0^{1,2}(\mathcal{D}) \ : \ \mathsf{A}u \in L^2(\mathcal{D})\}. \tag{10.94}$$

The operator A is closely related to a bilinear form $\mathfrak{a} : W_0^{1,2}(\mathcal{D}) \times W_0^{1,2}(\mathcal{D}) \to \mathbb{R}$, namely

$$\mathfrak{a}(u,v) := \int_{\mathcal{D}} \sum_{i,j=1}^{d} a_{ij}(x)\frac{\partial u}{\partial x_i}(x)\frac{\partial v}{\partial x_j}(x)dx, \tag{10.95}$$

which, by the stated conditions on a_{ij}, is continuous and coercive in $W_0^{1,2}(\mathcal{D})$ (see Section 7.6.1). The operator A is related to the bilinear form $\mathfrak{a}$ in terms of

$$\langle \mathsf{A}u, v\rangle_{L^2(\mathcal{D})} = \mathfrak{a}(u,v), \quad \forall\, v \in L^2(\mathcal{D}), \tag{10.96}$$

and $\mathbf{D}(\mathsf{A}) \subsetneq W_0^{1,2}(\mathcal{D})$ is the set of u such that the above definition is well defined, in the sense that $\mathsf{A}u \in L^2(\mathcal{D})$. We emphasize that, while similar, the operator A defined here is different that the one defined in Chapters 7 and 9 on account of the different domains (see Remark 10.4.5). If $a_{ij} = \delta_{ij}$, $\mathsf{A} = -\Delta$ (the Laplace operator) and $\mathfrak{a}$ reduces to the Dirichlet form.

Being already familiar with the operator A, it is easy to see that it enjoys monotonicity type properties, so we anticipate that $\mathsf{A}_{\mathcal{D}} = -\mathsf{A}$, will be dissipative, hence a good candidate for the generator of a semigroup if defined on the proper domain. Based on this observation, we will work with the operator $\mathsf{A}_{\mathcal{D}} = -\mathsf{A}$, in terms of which the parabolic equation (10.91) becomes

$$\begin{aligned} \frac{du}{dt} &= \mathsf{A}_{\mathcal{D}}u + f, \quad \mathsf{A}_{\mathcal{D}} = -\mathsf{A}, \\ u(0) &= u_0, \end{aligned} \tag{10.97}$$

where the boundary conditions are incorporated into the choice of domain for the operator $\mathsf{A} = -\mathsf{A}_{\mathcal{D}}$ (see (10.94)). Our strategy towards the study of (10.91) is to replace it by its equivalent formulation (10.97), show that $\mathsf{A}_{\mathcal{D}} = -\mathsf{A}$ with $\mathbf{D}(\mathsf{A}_{\mathcal{D}}) = \mathbf{D}(\mathsf{A})$ (as in (10.94)) is the infinitesimal generator of a C_0-semigroup of contractions on $X = L^2(\mathcal{D})$, and then use the general theory developed in Section 10.4.1 for the connection between C_0-semigroups and the solution of differential equations to obtain existence and regularity results for the parabolic equation (10.91).

Remark 10.4.4. We note here that, instead of defining $\mathbf{D}(\mathsf{A})$ in terms of (10.94), we could alternatively use the definition $\mathbf{D}(\mathsf{A}) = W_0^{1,2}(\mathcal{D}) \cap W^{2,2}(\mathcal{D})$. Clearly if $u \in W^{2,2}(\mathcal{D})$ for well-behaved a_{ij}, it is guaranteed that $\mathsf{A}u \in L^2(\mathcal{D})$. The question of whether $\mathsf{A}u \in L^2(\mathcal{D})$ is less straightforward, the positive answer comes from elliptic regularity theory.

Remark 10.4.5. Note that the definition of A as in (10.94) is different than the one used in Section 7.6.1, and used elsewhere (Chapter 9) in this book. This definition was of the operator $\mathsf{A}' : W_0^{1,2}(\mathcal{D}) \to (W_0^{1,2}(\mathcal{D}))^* = W^{-1,2}(\mathcal{D})$, whose domain was the whole of $W_0^{1,2}(\mathcal{D})$, but, in return, the image Au was in the larger space, $(W_0^{1,2}(\mathcal{D}))^* = W^{-1,2}(\mathcal{D})$, and not necessarily restricted to $L^2(\mathcal{D})$.

The crucial abstract result concerning parabolic problems is provided in the following proposition [33, 111]

Proposition 10.4.6. *The operator* $\mathsf{A}_\mathcal{D} = -\mathsf{A} : \mathbf{D}(\mathsf{A}) \subset X \to X$, *defined as in* (10.94) *and* (10.96), *is the generator of a contraction semigroup on* $X = L^2(\mathcal{D})$. *Similarly, if* $\mathbf{D}(\mathsf{A}) = W_0^{1,2}(\mathcal{D}) \cap W^{2,2}(\mathcal{D})$.

Proof. This can be established either by direct application of the Hille–Yosida (see Theorem 1.7.4) or the Lumer–Phillips theorem (see Theorem 1.7.6, or as a special case of the Crandall–Ligget theorem, see Theorem 10.2.9).

We choose the last option. It is easy to see that the operator $\mathsf{A}_\mathcal{D}$ is dissipative. Indeed, being in a Hilbert space setting, a simple integration by parts formula shows, for every $u \in \mathbf{D}(\mathsf{A})$ (with $\mathbf{D}(\mathsf{A})$ as in (10.94)), we have that $\langle u, \mathsf{A}_\mathcal{D} u\rangle \leq 0$ (recall that A is linear).

To prove m-dissipativity, it suffices to check the range condition $\mathbf{R}(I - \lambda \mathsf{A}_\mathcal{D}) = X$ for any $\lambda > 0$, where $X = L^2(\mathcal{D})$. Upon rescaling and redefining f in terms of multiplication by a scalar, this corresponds to looking at the solvability of the operator problem $(\mathsf{A} + \lambda I)u = f$ for any $f \in X = L^2(\mathcal{D})$. This is related to the variational problem $a(u, v) + \lambda\langle u, v\rangle = \langle f, v\rangle$ for any $v \in W_0^{1,2}(\mathcal{D})$, which (from the Lax–Milgram lemma (see Theorem 7.3.7)) is known to have a solution. Since $f \in L^2(\mathcal{D})$, this solution satisfies $Au \in L^2(\mathcal{D})$.[9] Similarly, if $\mathbf{D}(\mathsf{A}) = W_0^{1,2}(\mathcal{D}) \cap W^{2,2}(\mathcal{D})$.

So far, we have established that $\mathsf{A}_\mathcal{D}$ generates a semigroup on $\overline{\mathbf{D}(\mathsf{A}_\mathcal{D})} = \overline{\mathbf{D}(\mathsf{A})}$. To show that this semigroup can be extended to the whole of X, it remains to show the density of $\mathbf{D}(\mathsf{A})$ in $X = L^2(\mathcal{D})$. If $\mathbf{D}(\mathsf{A})$ is chosen to be $W_0^{1,2}(\mathcal{D}) \cap W^{2,2}(\mathcal{D})$, then this follows by standard results on Sobolev spaces. If we choose $\mathbf{D}(\mathsf{A}) = \{u \in W_0^{1,2}(\mathcal{D}) : \mathsf{A}u \in L^2(\mathcal{D})\}$, we can work as follows: Let $v \in L^2(\mathcal{D})$ such that $\langle v, u\rangle = 0$ for all $u \in \mathbf{D}(\mathsf{A})$. Using the same rescaling step as above, we can work with $\lambda I + \mathsf{A}$ in lieu of $I - \lambda \mathsf{A}_\mathcal{D}$. Since $\lambda I + \mathsf{A}$

9 This fact can be seen either by a sharper version of the Lax–Milgram lemma or by elliptic regularity theory if we set $\mathbf{D}(\mathsf{A}) = W_0^{1,2}(\mathcal{D}) \cap W^{2,2}(\mathcal{D})$.

is invertible, there exists $w \in \mathbf{D}(\mathsf{A})$ such that $v = (\lambda I + \mathsf{A})w$. Use $u = w$ in the above to obtain

$$0 = \langle v, w\rangle = \langle(\lambda I + \mathsf{A})w, w\rangle = \lambda\|w\|^2 + \mathfrak{a}(w, w) \geq \lambda\|w\|^2,$$

so $w = 0$, and $v = 0$. This guarantees the density of $\mathbf{D}(\mathsf{A})$. □

We can now apply the abstract results of Section 10.4.1 to obtain solvability results for semilinear parabolic equations:

Theorem 10.4.7. *Under the stated assumptions, for any $u_0 \in L^2(\mathcal{D})$ problem,* (10.91) *admits a unique mild solution $u \in C([0, T]; L^2(\mathcal{D}))$ satisfying the estimates* (10.89).

Proof. The proof follows by combining Propositions 10.4.6 and 10.4.3. □

Clearly if f is independent of u and $f \in L^1((0, T); L^2(D))$, using the above, we obtain a mild solution of the linear parabolic equation. If u_0 and f have better regularity properties, then exploiting the linearity of the operator A, we may obtain classical solutions.

10.4.3 A symmetric hyperbolic semilinear problem

Another concrete example of the application of the Lumer–Phillips theorem (see Theorem 1.7.6) to semilinear problems, which was sketched in a abstract setting in Section 10.4.1, is the study of symmetric hyperbolic semilinear problems.

Consider a symmetric hyperbolic problem related to a wave-like problem,

$$\frac{\partial^2 u}{\partial t^2}(t,x) \underbrace{- \sum_{i,j=1}^{d} \frac{\partial}{\partial x_i}\left(a_{ij}(x)\frac{\partial u}{\partial x_j}(t,x)\right)}_{=:\mathsf{A}u} = f(t, u(t,x)), \tag{10.98}$$

for a given forcing term f satisfying Assumption 10.4.1 (for the choice $X = W_0^{1,2}(\mathcal{D})$) and the operator A as in Section 10.4.2, equation (10.93) (including the smoothness of a_{ij} as well as the ellipticity condition (10.92)). This is complemented with the following boundary and initial conditions:

$$\begin{aligned} &u(t,x) = 0 \quad \text{on } [0,T] \times \partial\mathcal{D}, \\ &u(0,x) = u_0(x), \quad \frac{\partial u}{\partial t}(0,x) = u_1(x), \quad \text{in } \mathcal{D}. \end{aligned}$$

The homogeneous Dirichlet boundary conditions are used for simplicity, other boundary conditions, such as, e. g., inhomogeneous Dirichlet, Neumann, or Robin are possible. Moreover, generalized solutions, e. g., mild solutions of the equation can be considered.

Problem (10.98) can be expressed as a first-order problem, setting $u : [0, T] \to X$ and $w = (w_1, w_2)^T = (u, \frac{du}{dt})^T$, as

$$\frac{d}{dt}\begin{pmatrix} w_1 \\ w_2 \end{pmatrix} = \underbrace{\begin{pmatrix} 0 & I \\ -\mathsf{A} & 0 \end{pmatrix}}_{=:\mathcal{A}_{\mathcal{D}}} \begin{pmatrix} w_1 \\ w_2 \end{pmatrix} + \begin{pmatrix} 0 \\ f \end{pmatrix}, \tag{10.99}$$

or, in compact form, in terms of the operator $\mathcal{A}_{\mathcal{D}} w = (w_2, \mathsf{A}w_1)^T$ and $\mathfrak{f} = (0, f)^T$ as

$$\frac{dw}{dt} = \mathcal{A}_{\mathcal{D}} w + \mathfrak{f}. \tag{10.100}$$

For the boundary conditions considered here, we will choose

$$\begin{aligned} &\mathbf{D}(\mathcal{A}_{\mathcal{D}}) = \mathbf{D}(\mathsf{A}) \times L^2(\mathcal{D}), \quad \text{with} \\ &\mathbf{D}(\mathsf{A}) = \{u \in W_0^{1,2}(\mathcal{D}) \, : \, \mathsf{A}u \in L^2(\mathcal{D})\}, \quad \text{or} \quad \mathbf{D}(\mathsf{A}) = W_0^{1,2}(\mathcal{D}) \cap W^{2,2}(\mathcal{D}). \end{aligned} \tag{10.101}$$

The crucial abstract result concerning hyperbolic problems is provided in the following proposition [33, 111].

Proposition 10.4.8. *Let $\mathcal{X} := W_0^{1,2}(\mathcal{D}) \times L^2(\mathcal{D})$, which is a Hilbert space, when equipped with the norm $\|w\|_{\mathcal{X}} := \mathfrak{a}(w_1, w_1) + \|w_2\|^2_{L^2(\mathcal{D})}$, coming from the inner product $\langle w, \hat{w}\rangle_{\mathcal{X}} = \mathfrak{a}(w_1, \hat{w}_1) + \langle w_2, w_2\rangle_{L^2(\mathcal{D})}$.*

The operator $\mathcal{A}_{\mathcal{D}} : \mathbf{D}(\mathcal{A}_{\mathcal{D}}) \subset \mathcal{X} \to \mathcal{X}$ (defined as in (10.99) *and* (10.101)*) is the generator of a C_0-semigroup on $\mathcal{X}$.*

Proof. Even though we could establish the result using the special case of the Crandall–Liggett theorem for linear operators (as in Proposition 10.4.6), for variety, we use here directly the Hille–Yosida theorem (Theorem 1.7.4) for the operator $\mathcal{A}_{\mathcal{D}}$. For convenience we will use a rescaled version replacing $I - \lambda\mathcal{A}_{\mathcal{D}}$ with $\lambda I - \mathcal{A}_{\mathcal{D}}$.

We consider first the surjectivity of $\mathcal{A}_{\mathcal{D}}^{(\lambda)} := \lambda I - \mathcal{A}_{\mathcal{D}}$. Consider any $\mathfrak{f} \in \mathcal{X}$, and the corresponding operator equation $(\lambda I - \mathcal{A}_{\mathcal{D}})w = \mathfrak{f}$, for any $\mathfrak{f} = (f_1, f_2)^T \in \mathcal{X}$. This reduces to the system

$$\begin{cases} \lambda w_1 - w_2 = f_1 \\ \mathsf{A}w_1 + \lambda w_2 = f_2 \end{cases} \Longrightarrow \begin{cases} w_2 = \lambda w_1 - f_1 \\ \mathsf{A}w_1 + \lambda^2 w_1 = \lambda f_1 + f_2. \end{cases} \tag{10.102}$$

The second equation in (10.102) is essentially identical to the problem we considered for the parabolic problem in Proposition 10.4.6, so either by a more refined version of Lax–Milgram or by elliptic regularity theory, we may obtain for each $f = \lambda f_1 + f_2 \in L^2(\mathcal{D})$ a unique $w_1 \in \mathbf{D}(\mathsf{A})$ that satisfies it. Then, the first equation leads to the corresponding $w_2 \in L^2(\mathcal{D})$. This shows that $\mathbf{R}(\lambda I - \mathcal{A}_{\mathcal{D}}) = \mathcal{X}$.

To obtain the required bound for $(\lambda I - \mathcal{A}_{\mathcal{D}})^{-1}$, we first note that by taking the inner product in $L^2(\mathcal{D})$ of the second equation on the left block of (10.102) with w_2 (and recalling that, by (10.96), the definition of the operator A in this setting is such that $\langle \mathsf{A}w_1, v\rangle_{L^2(\mathcal{D})} = \mathfrak{a}(w_1, v)$ for every $v \in L^2(\mathcal{D})$), after some manipulation, yields

$$\mathfrak{a}(w_1, w_2) + \lambda\langle w_2, w_2\rangle_{L^2(\mathcal{D})} = \langle f_2, w_2\rangle_{L^2(\mathcal{D})}. \tag{10.103}$$

Then,

$$\begin{aligned}\lambda\|w\|_{\mathcal{X}}^2 &= \lambda\big(\mathfrak{a}(w_1,w_1) + \langle w_2,w_2\rangle_{L^2(\mathcal{D})}\big)\\ &\overset{(10.103)}{=} \mathfrak{a}(w_1,\lambda w_1) - \mathfrak{a}(w_1,w_2) + \langle f_2,w_2\rangle_{L^2(\mathcal{D})} = \mathfrak{a}(w_1,\lambda w_1 - w_2) + \langle f_2,w_2\rangle_{L^2(\mathcal{D})}\\ &\overset{(10.102)}{=} \mathfrak{a}(w_1,f_1) + \langle f_2,w_2\rangle_{L^2(\mathcal{D})} = \langle w,\mathfrak{f}\rangle_{\mathcal{X}} \le \|w\|_{\mathcal{X}}\|\mathfrak{f}\|_{\mathcal{X}},\end{aligned}$$

which is turn leads to the required estimate

$$\|w\|_{\mathcal{X}} = \|(\lambda I - \mathcal{A}_{\mathcal{D}})^{-1}\mathfrak{f}\|_{\mathcal{X}} \le \frac{1}{\lambda}\|\mathfrak{f}\|_{\mathcal{X}}.$$

The density of $\mathbf{D}(\mathcal{A}_{\mathcal{D}})$ in $\mathcal{X}$ follows by similar reasoning as in Proposition 10.4.6.

Combining the above, we see that by the Hille–Yosida theorem (rescaled version), $\mathcal{A}_{\mathcal{D}}$ is the generator of a C_0-semigroup of contractions in $\mathcal{X}$. □

We are now ready to discuss the solvability and regularity of the hyperbolic equation (10.98).

Proposition 10.4.9. *Under the stated assumptions, the semilinear hyperbolic equation* (10.98) *for any* $(u_0, \frac{du_0}{dt}) \in \mathcal{X}$ *admits a unique mild solution satisfying the estimates* (10.89).

Proof. We easily see that if f satisfies Assumption 10.4.1 for $X = W_0^{1,2}(\mathcal{D})$, then $\mathfrak{f}$ satisfies this assumption in $\mathcal{X}$. We then apply Proposition 10.4.3 to the first-order problem (10.100), and in combination with Proposition 10.4.8. □

In the case where f is independent of u, (10.98) reduces to a linear symmetric hyperbolic equation. For $w_0 = (u_0, \frac{du_0}{dt}) \in \mathcal{X}$ and $\mathfrak{f} \in L^1((0,T);\mathcal{X})$, (10.98) admits a unique mild solution, whereas for more regular data it admits classical solutions. In this case, we can use Theorem 1.7.3 according to which if $w_0 \in \mathbf{D}(\mathcal{A}_{\mathcal{D}})$, then (10.98) admits a classical solution, i. e., a solution $w \in C([0,\infty);\mathbf{D}(\mathcal{A}_{\mathcal{D}})) \cap C^1((0,\infty);\mathcal{X})$. Recalling the definition of the above function spaces, this implies that $u \in C([0,\infty);W_0^{1,2}(\mathcal{D}))$, $\mathrm{A}u \in C([0,\infty);L^2(\mathcal{D}))$, $\frac{du}{dt} \in L^2(\mathcal{D})$.

10.4.4 Gradient flows in Hilbert spaces

An important class of evolution problems are gradient flows. These are Cauchy problems of the form

$$\frac{d\mathsf{x}}{dt}(t) \in \mathsf{A}_{\mathcal{D}}(\mathsf{x}(t)) := -\partial\varphi(\mathsf{x}(t)), \quad \mathsf{x}(0) = \mathsf{x}_0, \tag{10.104}$$

where $\varphi : X \to \mathbb{R} \cup \{+\infty\}$ is a proper convex lower semicontinuous function. In this section, we will focus on the case where $X = H$ a Hilbert space. Before proceeding to the study of problem (10.104), we present problems of this form by a few examples.

Example 10.4.10. Consider the problem of minimizing the proper convex lower semi-continuous function $\varphi : X \to \mathbb{R} \cup \{+\infty\}$. The minimizer $x_o \in X$ satisfies the first-order condition $0 \in \partial\varphi(x_o)$. A common numerical scheme for the approximation of the minimizer is either the gradient (or subgradient) scheme

$$x_{n+1} - x_n \in -\lambda\partial\phi(x_n), \quad \lambda > 0,$$

or the proximal scheme

$$x_{n+1} - x_n \in -\lambda\partial\phi(x_{n+1}), \quad \lambda > 0.$$

For $\lambda > 0$ small, both can be identified as an approximation of the solution of the gradient flow (10.104): the gradient scheme corresponding to the Euler approximation and the proximal scheme corresponding to the backward Euler approximation for (10.104). ◁

Example 10.4.11. Let $\mathcal{D}$ be a bounded open domain in $\mathbb{R}^d$, $X = W_0^{1,2}(\mathcal{D})$, and consider the $\phi(u) = \int_{\mathcal{D}} |\nabla u|^2 dx$ to be the Dirichlet functional. Then (10.104) corresponds to the heat equation

$$\frac{\partial u}{\partial t} = \Delta u,$$

with homogeneous Dirichlet boundary conditions. ◁

Since $A_{\mathcal{M}} = -A_{\mathcal{D}} = \partial\varphi(\cdot)$ is a monotone type operator, $A_{\mathcal{D}}$ is a dissipative operator, and we anticipate that it generates a nonlinear semigroup so that, by the general theory developed in Section 10.3, problem (10.104) is well posed. Moreover, within the Hilbert space context, and if $X = H$ is identified with its dual, the theory of nonlinear semigroups presented in Section 10.2 can also be applied to problem (10.104).

Here we present an alternative route that allows us to illustrate the importance and use of the Yosida approximations in the construction of solutions of differential equations of the general form

$$\begin{cases} \frac{dx}{dt}(t) \in A_{\mathcal{D}}(x(t)), & t \in [0,T] \\ x(0) = x_0 \end{cases} \iff \begin{cases} \frac{dx}{dt}(t) + A_{\mathcal{M}}(x(t)) \ni 0, & t \in [0,T] \\ x(0) = x_0, \end{cases} \tag{10.105}$$

where $A_{\mathcal{M}} = -A_{\mathcal{D}}$, possibly multivalued operators ($A_{\mathcal{D}}$ dissipative and $A_{\mathcal{M}}$ of monotone type). To this point, we will work with the monotone version of (10.105), denote $A_{\mathcal{M}} = A$ in order to simplify the notation, and replace the original problem $\frac{dx}{dt} + A(x) \ni 0$ with the approximate equation

$$\begin{aligned} \frac{dx_\lambda}{dt}(t) + A_\lambda(x_\lambda(t)) = 0, \quad t \in [0,T],\ \lambda > 0, \\ x(0) = x_0, \end{aligned} \tag{10.106}$$

where $(\mathsf{A}_\lambda)_{\lambda>0}$ is the family of Yosida approximations of the monotone type operator $\mathsf{A} = \mathsf{A}_{\mathcal{M}}$ (defined as in Definition 9.4.21, with $R(\lambda, \mathsf{A}) = (I + \lambda\mathsf{A})^{-1}$). Since A_λ is Lipschitz continuous, it follows that (10.106) is well posed for any $\lambda > 0$, and admits a unique solution $\mathsf{x}_\lambda \in C^1([0,T];H)$.

A reasonable question that arises is the following: Since the Yosida approximation $(\mathsf{A}_\lambda)_{\lambda>0}$ approximates the operator A in the limit as $\lambda \to 0^+$, can we claim that the limit of $(\mathsf{x}_\lambda)_{\lambda>0}$, as $\lambda \to 0^+$, if it exists, is a solution to the original problem (10.105)?

Remark 10.4.12. We define $\mathsf{R}(\lambda,\mathsf{A})$ and A_λ, here as in Definition 9.4.21. If we wanted to work in terms of the formulation $\frac{dx}{dt} \in \mathsf{A}_{\mathcal{D}}(\mathsf{x})$, then the approximation would be modified to $\frac{dx_\lambda}{dt} = \mathsf{A}_{\mathcal{D},\lambda}(\mathsf{x}_\lambda)$, where $(\mathsf{A}_{\mathcal{D},\lambda})_{\lambda>0}$ is the Yosida approximation of the dissipative operator $\mathsf{A}_{\mathcal{D}}$, as in Definition 10.2.5, with $\mathsf{R}(\lambda,\mathsf{A}_{\mathcal{D}}) = (I - \lambda\mathsf{A}_{\mathcal{D}})^{-1}$, since $\mathsf{A}_{\mathcal{D}}$ is dissipative.

The answer to the above question is affirmative, leading to an existence result. This is shown in the following theorem [32, 76].

Theorem 10.4.13. *For each* $\mathsf{x}_0 \in \mathbf{D}(\mathsf{A})$, *there exists a unique function* $\mathsf{x} \in C([0,T];H)$, *with* $\frac{dx}{dt} \in L^\infty((0,T);H)$, *which is a solution of* (10.105).

Proof. We simplify the notation by using $\mathsf{R}_\lambda := \mathsf{R}(\lambda,\mathsf{A})$ (as in Def. 9.4.21). By Proposition 9.4.30, A_λ is Lipschitz continuous, hence, by Theorem 3.1.5, the approximate problem (10.106) admits a unique solution $\mathsf{x}_\lambda \in C^1([0,T];H)$. To pass to the limit, as $\lambda \to 0^+$, and show that this limit is a solution to the equation (10.105), we need to derive some estimates concerning uniform (in λ) boundedness of x_λ and $\frac{dx_\lambda}{dt}$.

The proof proceeds in 4 steps.

1. We claim that for any $\lambda > 0$, it holds that $\|\frac{dx_\lambda}{dt}(t)\| \le \|\mathsf{A}^0(\mathsf{x}_0)\|$ (for A^0 see Def. 9.4.21). To obtain this bound, consider the solution of problems

$$\begin{aligned} \frac{dx_{i,\lambda_i}}{dt}(t) + \mathsf{A}_{\lambda_i}(\mathsf{x}_{i,\lambda_i}(t)) = 0, \quad t \in [0,T], \quad \lambda_i > 0, \\ \mathsf{x}_{i,\lambda_i}(0) = \mathsf{x}_i, \quad i = 1,2. \end{aligned} \tag{10.107}$$

Set $\lambda_1 = \lambda_2 = \lambda$. Subtracting, taking the inner product with $\mathsf{x}_{1,\lambda}(t) - \mathsf{x}_{2,\lambda}(t)$ and using the integration by parts formula and the definition of the subdifferential, we conclude that $\frac{d}{dt}(\|\mathsf{x}_{1,\lambda} - \mathsf{x}_{2,\lambda}\|^2)(t) \le 0$, which leads to the estimate $\|\mathsf{x}_{1,\lambda}(t) - \mathsf{x}_{2,\lambda}(t)\| \le \|\mathsf{x}_1 - \mathsf{x}_2\|$ for $t \in [0,T]$. Since the solutions to the approximate system (10.107) are unique choosing any $h > 0$ and setting $\mathsf{x}_2 = \mathsf{x}_{1,\lambda}(h)$, we have that $\mathsf{x}_{2,\lambda}(t) = \mathsf{x}_{1,\lambda}(t+h)$, so the previous estimate becomes $\|\mathsf{x}_{1,\lambda}(t+h) - \mathsf{x}_{1,\lambda}(t)\| \le \|\mathsf{x}_{1,\lambda}(h) - \mathsf{x}_1\|$ and dividing by h and passing to the limit, as $h \to 0$, we have that

$$\left\|\frac{dx_{1,\lambda}}{dt}(t)\right\| \le \left\|\frac{dx_{1,\lambda}}{dt}(0)\right\| = \|\mathsf{A}_\lambda(\mathsf{x}_1)\| \le \|\mathsf{A}^0(\mathsf{x}_1)\|, \tag{10.108}$$

where we used the properties of the Yosida approximation (see Theorem 9.4.26(ii)). Since $\mathsf{x}_1 \in H$ was arbitrary, the claim follows.

2. We now consider problem (10.107) for $\lambda_1 \neq \lambda_2$ but the same initial condition $x_1 = x_2 = x_0$. Using the notation $x_{i,\lambda_i} = x_{\lambda_i}$, we will show that

$$\|x_{\lambda_1}(t) - x_{\lambda_2}(t)\| \leq \frac{1}{\sqrt{2}}(\lambda_1 + \lambda_2)^{1/2} t^{1/2} \|A^0(x_0)\|, \quad \forall \lambda_1, \lambda_2 > 0,\ t \in [0, T]. \tag{10.109}$$

To prove the claim (10.109), we need to estimate x_{λ_1} and x_{λ_2}. Subtracting (10.107), taking the inner product with the difference $x_{\lambda_1}(t) - x_{\lambda_2}(t)$ for any t and working similarly as above, we obtain

$$\begin{aligned} &\frac{1}{2}\frac{d}{dt}\|x_{\lambda_1}(t) - x_{\lambda_2}(t)\|^2 \\ &\quad + \langle A_{\lambda_1}(x_{\lambda_1}(t)) - A_{\lambda_2}(x_{\lambda_2}(t)), x_{\lambda_1}(t) - x_{\lambda_2}(t)\rangle = 0. \end{aligned} \tag{10.110}$$

Here we may not use the monotonicity properties of A_{λ_1} directly (as the above involves the difference of A_{λ_1} and A_{λ_2}). We will then have to rearrange this duality pairing, using the resolvent operator R_{λ_1}. The definition of the Yosida approximation rearranged as $x_\lambda = R_\lambda(x_\lambda) + \lambda A_\lambda(x_\lambda)$ for every $\lambda > 0$ allows us to express the difference $x_{\lambda_1} - x_{\lambda_2}$ as

$$x_{\lambda_1} - x_{\lambda_2} = R_{\lambda_1}(x_{\lambda_1}) - R_{\lambda_2}(x_{\lambda_2}) + \lambda_1 A_{\lambda_1}(x_{\lambda_1}) - \lambda_2 A_{\lambda_2}(x_{\lambda_2}).$$

Therefore,

$$\begin{aligned} &\langle A_{\lambda_1}(x_{\lambda_1}(t)) - A_{\lambda_2}(x_{\lambda_2}(t)), x_{\lambda_1}(t) - x_{\lambda_2}(t)\rangle \\ &\quad = \langle A_{\lambda_1}(x_{\lambda_1}(t)) - A_{\lambda_2}(x_{\lambda_2}(t)), R_{\lambda_1}(x_{\lambda_1}(t)) - R_{\lambda_2}(x_{\lambda_2}(t))\rangle \\ &\quad + \langle A_{\lambda_1}(x_{\lambda_1}(t)) - A_{\lambda_2}(x_{\lambda_2}(t)), \lambda_1 A_{\lambda_1}(x_{\lambda_1}(t)) - \lambda_2 A_{\lambda_2}(x_{\lambda_2}(t))\rangle. \end{aligned} \tag{10.111}$$

Consider the first term of the RHS. By Theorem 9.4.26(i), we have that for every $\lambda_i > 0$, $x \in H$, it holds that $A_{\lambda_i}(x) \in A(R_{\lambda_i}(x))$, $i = 1, 2$, so that we may estimate this term using the monotonicity of $A = \partial\varphi$ (see Proposition 4.2.1). Applying that for λ_i, $x_{\lambda_i}(t)$, $i = 1, 2$, we obtain immediately that

$$\langle A_{\lambda_1}(x_{\lambda_1}(t)) - A_{\lambda_2}(x_{\lambda_2}(t)), R_{\lambda_1}(x_{\lambda_1}(t)) - R_{\lambda_2}(x_{\lambda_2}(t))\rangle \geq 0.$$

Using the above in (10.111) implies

$$\begin{aligned} &\langle A_{\lambda_1}(x_{\lambda_1}(t)) - A_{\lambda_2}(x_{\lambda_2}(t)), x_{\lambda_1}(t) - x_{\lambda_2}(t)\rangle \\ &\quad \geq \langle A_{\lambda_1}(x_{\lambda_1}(t)) - A_{\lambda_2}(x_{\lambda_2}(t)), \lambda_1 A_{\lambda_1}(x_{\lambda_1}(t)) - \lambda_2 A_{\lambda_2}(x_{\lambda_2}(t))\rangle \\ &\quad = \lambda_1 \|A_{\lambda_1}(x_{\lambda_1}(t))\|^2 + \lambda_2 \|A_{\lambda_2}(x_{\lambda_2}(t))\|^2 - (\lambda_1 + \lambda_2)\langle A_{\lambda_1}(x_{\lambda_1}(t)), A_{\lambda_2}(x_{\lambda_2}(t))\rangle \\ &\quad \geq \lambda_1 \|A_{\lambda_1}(x_{\lambda_1}(t))\|^2 + \lambda_2 \|A_{\lambda_2}(x_{\lambda_2}(t))\|^2 - (\lambda_1 + \lambda_2)|\langle A_{\lambda_1}(x_{\lambda_1}(t)), A_{\lambda_2}(x_{\lambda_2}(t))\rangle| \\ &\quad \geq \lambda_1 \|A_{\lambda_1}(x_{\lambda_1}(t))\|^2 + \lambda_2 \|A_{\lambda_2}(x_{\lambda_2}(t))\|^2 - (\lambda_1 + \lambda_2)\|A_{\lambda_1}(x_{\lambda_1}(t))\| \, \|A_{\lambda_2}(x_{\lambda_2}(t))\| \\ &\quad \geq -\frac{\lambda_2}{4}\|A_{\lambda_1}(x_{\lambda_1}(t))\|^2 - \frac{\lambda_1}{4}\|A_{\lambda_2}(x_{\lambda_2}(t))\|^2 \geq -\frac{\lambda_1 + \lambda_2}{4}\|A^0(x_0)\|^2, \end{aligned}$$

where for the penultimate estimate we have used the algebraic inequality

$$(\lambda_1+\lambda_2)a\,b \le \lambda_1\Big(a^2+\frac{1}{4}b^2\Big)+\lambda_2\Big(b^2+\frac{1}{4}a^2\Big),$$

and the final estimate comes from (10.108) of step 1.

Substituting this estimate in (10.110) yields the differential inequality

$$\frac{1}{2}\frac{d}{dt}\|x_{\lambda_1}(t)-x_{\lambda_2}(t)\|^2 \le \frac{\lambda_1+\lambda_2}{4}\|A^0(x_0)\|^2,$$

and since the right-hand side is independent of t, this can be easily integrated to yield (10.109).

3. Let us consider the family $(x_\lambda)_{\lambda>0}$ obtained as solutions of the approximate problem and try to go to the limit as $\lambda \to 0^+$. By the a priori bounds (10.108) in step 1, we see that for any $T > 0$, there exists a function $z \in L^2((0,T);H)$ such that

$$\frac{dx_\lambda}{dt} \rightharpoonup z, \quad \text{in } L^2((0,T);H) \quad \text{as} \quad \lambda \to 0^+.$$

Furthermore, the limit satisfies the estimate $\|z(t)\| \le \|A^0(x_0)\|$.

The a priori bound (10.109) in step 2 is in fact an equicontinuity result, which by the Ascoli–Arzelá theorem (Thm. 1.9.13) guarantees the existence of a function x such that

$$x_\lambda \to x, \quad \text{uniformly in } C([0,T];H) \quad \text{as} \quad \lambda \to 0^+.$$

A simple argument shows that $z = \frac{dx_\lambda}{dt}$.

4. We now show that $x(t) \in \mathbf{D}(A)$ for a. a. $t \in [0,T]$, and that $\frac{dx}{dt}(t)$ and $x(t)$ satisfy the equation for a. a. $t \in [0,T]$. It is immediate to see that $R_\lambda(x_\lambda) \to x$ uniformly in $C([0,T];H)$ as $\lambda \to 0$. Indeed, by the definition of the resolvent and the Yosida approximation,

$$\|R_\lambda(x_\lambda(t)) - x_\lambda(t)\| = \lambda\|A_\lambda(x_\lambda(t))\| = \lambda\Big\|\frac{dx_\lambda}{dt}(t)\Big\| \le \lambda\,\|A^0(x_0)\|.$$

Therefore,

$$\|R_\lambda(x_\lambda(t)) - x(t)\| \le \|R_\lambda(x_\lambda(t)) - x_\lambda(t)\| + \|x_\lambda(t)-x(t)\| \to 0, \quad \text{as}\,\lambda \to 0^+,$$

with both terms of the right-hand side converging uniformly, and the result follows.

The solution of the approximate problem satisfies equality

$$-\frac{dx_\lambda}{dt}(t) = A_\lambda(x_\lambda(t)) \in A(R_\lambda(x_\lambda(t))).$$

This means that $(-x_\lambda'(t), R_\lambda(x_\lambda(t)) \in \mathbf{Gr}(A)$, and since A is a maximal monotone operator, its graph is weakly closed, so going to the limit as $\lambda \to 0^+$ and recalling that $R_\lambda(x_\lambda) \to x$, we obtain that $(-\frac{dx}{dt}(t), x(t)) \in \mathbf{Gr}(A)$ a. e. in $[0,T]$, therefore $-\frac{dx}{dt}(t) \in A(x(t))$. □

Remark 10.4.14. In fact, the solution can be extended on $[0, \infty)$, see, e. g., [32].

The solution to the evolution problem (10.105) is related to a family of operators $\{\mathsf{S}(t)\}_{t\geq 0}$, with $\mathsf{S}(t) : H \to H$ for every $t \geq 0$, defined for any fixed t by $\mathsf{S}(t)\mathsf{x} := \mathsf{x}(t)$ for every $\mathsf{x} \in H$, where $\mathsf{x}(\cdot)$ is the solution of the evolution problem (10.105) for the initial condition $\mathsf{x}(0) = \mathsf{x}_o \in H$. By the properties of the solution, it can be seen that this family of operators is in fact the nonlinear semigroup studied in Section 10.2.

10.4.5 A quasilinear parabolic problem I: nonlinear semigroup approach

We now consider the quasilinear parabolic problem

$$\begin{aligned} &\frac{\partial u}{\partial t}(t,x) = \Delta_p u(t,x) + f(t,x), \quad (t,x) \in [0,T] \times \mathcal{D}, \\ &u(0,x) = u_0(x), \quad x \in \mathcal{D}, \\ &u(t,x) = 0, \quad x \in [0,T] \times \partial\mathcal{D}, \end{aligned} \tag{10.112}$$

for $\Delta_p u := \nabla \cdot (|\nabla u|^{p-2} \nabla u)$, with $p \in (1, \infty)$, in a bounded domain $\mathcal{D}$, or suitable mild or weak formulations. This is expressed in operator form as

$$\frac{du}{dt} = \mathsf{A}_{\mathcal{D}}(u) + f,$$

where $\mathsf{A}_{\mathcal{D}} = \Delta_p = -\mathsf{A}_{\mathcal{M}}$ is related to the p-Laplace operator studied in Section 6.8 (see (6.107) although the domain will differ).

We have already seen that the operator

$$\mathsf{A}_{\mathcal{M}} = -\Delta_p : W_0^{1,p}(\mathcal{D}) \to (W_0^{1,p}(\mathcal{D}))^* = W^{-1,p^*}(\mathcal{D})$$

is maximal monotone. On the other hand, here, in the interest of trying to fit this operator within the context of nonlinear semigroup theory, by further restricting the domain of $\mathsf{A}_{\mathcal{M}}$ (equiv. $\mathsf{A}_{\mathcal{D}}$) to

$$\mathbf{D}(\mathsf{A}_{\mathcal{D}}) = \{u \in W_0^{1,p}(\mathcal{D}) \; : \; \mathsf{A}(u) \in L^{p^*}(\mathcal{D})\}, \tag{10.113}$$

we can show that the operator $\mathsf{A}_{\mathcal{D}}$ is an m-dissipative operator (resp. $\mathsf{A}_{\mathcal{M}} = -\mathsf{A}_{\mathcal{D}}$ an m-accretive operator). This is the approach considered in this section (see, e. g., [19]). We emphasize that defining $\mathbf{D}(\mathsf{A}_{\mathcal{D}})$ as in (10.113), the operator considered in this section is different than the one in Section 6.8. Some important properties of the operator $\mathsf{A}_{\mathcal{D}}$ are collected in the following proposition [18, 19].

Proposition 10.4.15. *The operator $\mathsf{A}_{\mathcal{D}} = \Delta_p$, defined on $\mathbf{D}(\mathsf{A}_{\mathcal{D}}) = \{u \in W_0^{1,p}(\mathcal{D}) \; : \; \mathsf{A}_{\mathcal{D}} u \in L^p(\mathcal{D})\}$, for $p \in (1, \infty)$, is*
(i) *dissipative on $X = L^p(\mathcal{D})$.*

(ii) *m-dissipative on* $X = L^p(\mathcal{D})$ *if* $p > \mathfrak{s}_p$, *where* $\mathfrak{s}_p$ *is the critical exponent for the embedding* $W_0^{1,p}(\mathcal{D}) \hookrightarrow L^{\mathfrak{s}_p}(\mathcal{D})$ *to hold.*

Proof. (i) To show dissipativity, we simply have to check the relevant condition in Definition 10.2.2.

For simplicity, we only consider the $p > 2$ case here. We denote by $\langle \cdot, \cdot \rangle$ the duality pairing between $L^p(\mathcal{D})$ and $L^{p^*}(\mathcal{D})$. Recall that for any $w \in L^p(\mathcal{D})$, $\mathsf{J}(w) = \|w\|_{L^p(\mathcal{D})}^{2-p} |w|^{p-2} w$. Setting $w = u - v$, we have that

$$\begin{aligned}
&\langle \mathsf{J}(u-v), \Delta_p u - \Delta_p v \rangle \\
&= \|w\|_{L^p(\mathcal{D})}^{2-p} \int_{\mathcal{D}} |w|^{p-2} w[(\nabla \cdot (|\nabla u|^{p-2} \nabla u) - (\nabla \cdot (|\nabla v|^{p-2} \nabla v)] dx \\
&= -\|w\|_{L^p(\mathcal{D})}^{2-p} \int_{\mathcal{D}} \nabla(|w|^{p-2} w) \cdot (|\nabla u|^{p-2} \nabla u - |\nabla v|^{p-2} \nabla v) dx \\
&= -\|w\|_{L^p(\mathcal{D})}^{2-p} \int_{\mathcal{D}} (p-1)|w|^{p-2} \underbrace{(\nabla u - \nabla v) \cdot (|\nabla u|^{p-2} \nabla u - |\nabla v|^{p-2} \nabla v)}_{\geq 0 \text{ by monotonicity}} dx \leq 0.
\end{aligned}$$

In the above, we have skipped some technical steps required to handle the $|w|$ term; these can either be supplied by the reader or simply assume p even. Hence, we obtain the dissipativity of $\mathsf{A}_{\mathcal{D}}$. The cases for the other values of p are similar but with the appropriate form of the duality map.

(ii) For m-dissipativity, we must show (upon rescaling) that $\mathbf{R}(\lambda I - \mathsf{A}_{\mathcal{D}}) = X$, i. e., we require the solvability of the operator equation $(\lambda I - \Delta_p)u = f$ in $u \in X = L^p(\mathcal{D})$ for any $f \in X = L^p(\mathcal{D})$. This follows by the monotonicity properties of the operator $\mathsf{A}_{\mathcal{M}} : W_0^{1,p}(\mathcal{D}) \to (W_0^{1,p}(\mathcal{D}))^*$, where $\mathsf{A}_{\mathcal{M}} = -\mathsf{A}_{\mathcal{D}}$, and the embedding $W_0^{1,p}(\mathcal{D}) \hookrightarrow L^{\mathfrak{s}_p}(\mathcal{D})$. □

Having obtained the m-dissipativity of the operator $\mathsf{A}_{\mathcal{D}}$, we can apply the Crandall–Ligget theorem 10.2.9 to show that $\mathsf{A}_{\mathcal{D}}$ generates a C_0 nonlinear semigroup of contractions on $\mathbf{D}(\mathsf{A}_{\mathcal{D}}) = \{u \in W_0^{1,p}(\mathcal{D}) \ : \ \mathsf{A}_{\mathcal{D}}(u) \in L^p(\mathcal{D})\}$. This can be used to show the existence of solutions to problem (10.112) for $f = 0$ for any $u_0 \in \mathbf{D}(\mathsf{A}_{\mathcal{D}})$. Moreover, we can use Theorem 10.2.18 to study the existence of mild solutions to problem (10.112) for $f \neq 0$ such that $f \in L^1((0,T); L^p(\mathcal{D}))$.

The above results will remain true for the operator $\mathsf{A}(u) = \Delta_p u - c(x,u)$ in the case where $u \mapsto c(x,u)$ enjoys monotonicity properties.

10.4.6 A quasilinear parabolic problem II: pseudomonotonicity approach

We now consider generalizations of problem (10.112) of the form

$$\frac{\partial u}{\partial t} = \Delta_p u + c(x,u) + f, \quad \Delta_p u = \nabla \cdot (|\nabla u|^{p-2} \nabla u) \tag{10.114}$$

subject to suitable initial and boundary conditions, i. e., equation (10.112) perturbed by the term $c(x,u)$, without any assumptions of monotonicity on the function $u \mapsto c(x,u)$. This is of the form $\frac{du}{dt} = \mathsf{A}(u) + f$, for $\mathsf{A} = \Delta_p + c$.

In this case, the nonlinear semigroup approach (that was based on dissipativity of the operator A, or equivalently the accretivity of −A) is no longer feasible, as the perturbation of Δ_p by the term c, if c does not enjoy monotonicity properties, may ruin the dissipativity properties of A. However, pseudomonotonicity, which is a more robust property with respect to perturbations, may still persist. Hence, by a proper definition of the operator A using an evolution triple, we may obtain solvability results (see, e. g., [128]). This approach, allows for a more general definition of the p-Laplace operator than in Section 10.4.5 (and consequently of A), as $\mathsf{A} : W_0^{1,p}(\mathcal{D}) \to (W_0^{1,p}(\mathcal{D}))^*$.

The appropriate evolution triple setting would be

$$W_0^{1,p}(\mathcal{D}) \hookrightarrow L^2(\mathcal{D}) \hookrightarrow (W_0^{1,p}(\mathcal{D}))^* = W^{-1,p^*}(\mathcal{D}),$$

as long as p is such that the first embedding is guaranteed by the Sobolev embedding theorem (and is also compact). Moreover, we require $p \in (1,\infty)$ so that $W_0^{1,p}(\mathcal{D})$ is reflexive. So we can choose p such that $1 < p < \infty$ and $2 < \mathfrak{s}_p$, where $\mathfrak{s}_p$ is the critical exponent corresponding to p. For example, if $p < d$, then $\mathfrak{s}_p = \frac{dp}{d-p}$, and in this case, $2 < \mathfrak{s}_p$ implies that $p > \frac{2d}{2+d}$, so that in this case the choice of p should be such that $p > \max(1, \frac{2d}{2+d})$. If $p \geq d$, where $\mathfrak{s}_p$ can be chosen arbitrarily large, the restriction $2 < \mathfrak{s}_p$ is not important.

Moreover, we know that if c satisfies coercivity, continuity, and appropriate growth conditions (see Assumption 9.6.2), then the operator $\mathsf{A} : W_0^{1,p}(\mathcal{D}) \to (W_0^{1,p}(\mathcal{D}))^* = W^{-1,p^*}(\mathcal{D})$ is pseudomonotone. This allows us to use Theorem 10.3.15 or its Galerkin version Proposition 10.3.12, to show for any $f \in L^{p^*}((0,T); W^{-1,p^*}(\mathcal{D}))$ and any $u_0 \in L^2(\mathcal{D})$ the existence of a solution u of (10.114) such that $u \in L^p((0,T); W_0^{1,p}(\mathcal{D}))$, $\frac{du}{dt} \in L^{p^*}((0,T); W^{-1,p^*}(\mathcal{D}))$, and $u \in C((0,T); L^2(\mathcal{D}))$.

10.5 Appendix

10.5.1 Proof of the fundamental estimate (10.14)

This fundamental estimate is given in the following lemma [58]:

Lemma 10.5.1. *The estimate*

$$\begin{aligned} \|\mathsf{R}_\mu^n \mathsf{x} - \mathsf{R}_\lambda^m \mathsf{x}\| \leq \{&((n\mu - m\lambda)^2 + n\mu(\lambda - \mu))^{1/2} \\ &+ (m\lambda(\lambda - \mu) + (m\lambda - n\mu)^2)^{1/2}\}\|\mathsf{A}\mathsf{x}\|_0, \quad \forall\, 0 < \mu \leq \lambda,\ m \leq n \end{aligned} \tag{10.115}$$

holds for all $\mathsf{x} \in \mathbf{D}(\mathsf{A})$.

Proof. The proof is based on the useful identity

$$R_\lambda x = R_\mu\left(\frac{\mu}{\lambda}x + \frac{\lambda-\mu}{\lambda}R_\lambda x\right), \quad \forall x \in \mathbf{D}(R_\lambda), \tag{10.116}$$

established in Proposition 10.2.7(vi) (see eq. (10.6)).

We apply (10.116) for the choice $x = R_\mu^{k-1}x$. This yields

$$R_\lambda^k x = R_\mu\left(\frac{\mu}{\lambda}R_\lambda^{k-1}x + \frac{\lambda-\mu}{\lambda}R_\lambda^k x\right) \tag{10.117}$$

Note that

$$\begin{aligned}
\|R_\mu^j x - R_\lambda^k x\| &= \left\|R_\mu(R_\mu^{j-1}x) - R_\mu\left(\frac{\mu}{\lambda}R_\lambda^{k-1}x + \frac{\lambda-\mu}{\lambda}R_\lambda^k x\right)\right\| \\
&\le \left\|R_\mu^{j-1}x - \frac{\mu}{\lambda}R_\lambda^{k-1}x - \frac{\lambda-\mu}{\lambda}R_\lambda^k x\right\| \\
&= \left\|\left(\frac{\mu}{\lambda} + \frac{\lambda-\mu}{\lambda}\right)R_\mu^{j-1}x - \frac{\mu}{\lambda}R_\lambda^{k-1}x - \frac{\lambda-\mu}{\lambda}R_\lambda^k x\right\| \\
&\le \frac{\mu}{\lambda}\|R_\mu^{j-1}x - R_\lambda^{k-1}x\| + \frac{\lambda-\mu}{\lambda}\|R_\mu^{j-1}x - R_\lambda^k x\|,
\end{aligned} \tag{10.118}$$

where we used the contraction property of R_μ. Upon defining $a_{jk} = \|R_\mu^j x - R_\lambda^k x\|$, we see that (10.118) leads to the linear inequality

$$a_{jk} \le \frac{\mu}{\lambda}a_{j-1,k-1} + \frac{\lambda-\mu}{\lambda}a_{j-1,k}. \tag{10.119}$$

Iteration of this linear inequality yields

$$\begin{aligned}
a_{n,m} \le &\sum_{j=0}^{m}\binom{n}{j}\left(\frac{\mu}{\lambda}\right)^j\left(\frac{\lambda-\mu}{\lambda}\right)^{m-j}a_{0,m-j} \\
&+ \sum_{j=m}^{n}\binom{j-1}{m-1}\left(\frac{\mu}{\lambda}\right)^m\left(\frac{\lambda-\mu}{\lambda}\right)^{j-m}a_{n-j,0}, \quad m \le n.
\end{aligned} \tag{10.120}$$

This can be verified by a simple, yet tedious, induction argument (more details on that can be found in the proof of Lemma 10.5.2 below).

Note that

$$\begin{aligned}
a_{0,m-j} = \|x - R_\lambda^{m-j}\| &= \left\|\sum_{\ell=0}^{m-j-1}(R_\lambda^{\ell+1}x - R_\lambda^\ell x)\right\| \\
&\le \sum_{\ell=0}^{m-j-1}\|R_\lambda^{\ell+1}x - R_\lambda^\ell x\|.
\end{aligned} \tag{10.121}$$

For any ℓ,

$$\|R_\lambda^{\ell+1}x - R_\lambda^\ell x\| = \|R_\lambda R_\lambda^\ell x - R_\lambda R_\lambda^{\ell-1}x\| \leq \|R_\lambda^\ell x - R_\lambda^{\ell-1}x\|, \tag{10.122}$$

which upon induction yields

$$\|R_\lambda^{\ell+1}x - R_\lambda^\ell x\| \leq \|R_\lambda x - x\|. \tag{10.123}$$

Combining (10.123) with (10.121), we obtain that

$$\begin{aligned} a_{0,m-j} &\leq \sum_{\ell=0}^{m-j-1} \|R_\lambda^{\ell+1}x - R_\lambda^\ell x\| \\ &\leq \sum_{\ell=0}^{m-j-1} \|R_\lambda x - x\| = (m-j)\|R_\lambda x - x\|. \end{aligned} \tag{10.124}$$

Using similar arguments, we obtain

$$a_{n-j,0} \leq (n-j)\|R_\mu x - x\|. \tag{10.125}$$

Substituting these in (10.120), we obtain

$$\begin{aligned} a_{n,m} &\leq \sum_{j=0}^{m} \binom{n}{j}\left(\frac{\mu}{\lambda}\right)^j\left(\frac{\lambda-\mu}{\lambda}\right)^{m-j}(m-j)\|R_\lambda x - x\| \\ &+ \sum_{j=m}^{n} \binom{j-1}{m-1}\left(\frac{\mu}{\lambda}\right)^m\left(\frac{\lambda-\mu}{\lambda}\right)^{j-m}(n-j)\|R_\mu x - x\|. \end{aligned} \tag{10.126}$$

We note further (see Proposition 10.2.7(vii)) that

$$\begin{aligned} \|R_\lambda x - x\| &\leq \lambda|Ax|_0, \\ \|R_\mu x - x\| &\leq \mu|Ax|_0. \end{aligned} \tag{10.127}$$

The above estimates, substituted in (10.126), yield

$$\begin{aligned} a_{n,m} \leq \Bigg\{&\lambda\sum_{j=0}^{m} \binom{n}{j}\left(\frac{\mu}{\lambda}\right)^j\left(\frac{\lambda-\mu}{\lambda}\right)^{m-j}(m-j) \\ &+ \mu\sum_{j=m}^{n} \binom{j-1}{m-1}\left(\frac{\mu}{\lambda}\right)^m\left(\frac{\lambda-\mu}{\lambda}\right)^{j-m}(n-j)\Bigg\}|Ax|_0. \end{aligned} \tag{10.128}$$

To proceed further, we estimate each of the terms above separately.

For the first one, we have

$$
\begin{aligned}
I_1 &= \lambda \sum_{j=0}^{m} \binom{n}{j} \left(\frac{\mu}{\lambda}\right)^{j} \left(\frac{\lambda-\mu}{\lambda}\right)^{m-j} (m-j) \\
&= \lambda \sum_{j=0}^{m} \binom{n}{j}^{1/2} \left(\frac{\mu}{\lambda}\right)^{j/2} \left(\frac{\lambda-\mu}{\lambda}\right)^{(m-j)/2} \binom{n}{j}^{1/2} \left(\frac{\mu}{\lambda}\right)^{j/2} \left(\frac{\lambda-\mu}{\lambda}\right)^{(m-j)/2} (m-j) \\
&\le \lambda \left(\sum_{j=0}^{m} \binom{n}{j} \left(\frac{\mu}{\lambda}\right)^{j} \left(\frac{\lambda-\mu}{\lambda}\right)^{m-j} \right)^{1/2} \left(\sum_{j=0}^{m} \binom{n}{j} \left(\frac{\mu}{\lambda}\right)^{j} \left(\frac{\lambda-\mu}{\lambda}\right)^{m-j} (m-j)^2 \right)^{1/2} \\
&\le \lambda \left(\sum_{j=0}^{n} \binom{n}{j} \left(\frac{\mu}{\lambda}\right)^{j} \left(\frac{\lambda-\mu}{\lambda}\right)^{n-j} \right)^{1/2} \left(\sum_{j=0}^{m} \binom{n}{j} \left(\frac{\mu}{\lambda}\right)^{j} \left(\frac{\lambda-\mu}{\lambda}\right)^{m-j} (m-j)^2 \right)^{1/2} \\
&= \lambda \left(\frac{\mu}{\lambda} + \frac{\lambda-\mu}{\lambda} \right)^{n/2} \left(\sum_{j=0}^{m} \binom{n}{j} \left(\frac{\mu}{\lambda}\right)^{j} \left(\frac{\lambda-\mu}{\lambda}\right)^{m-j} (m-j)^2 \right)^{1/2} \\
&= \lambda \left(\sum_{j=0}^{m} \binom{n}{j} \left(\frac{\mu}{\lambda}\right)^{j} \left(\frac{\lambda-\mu}{\lambda}\right)^{m-j} (m-j)^2 \right)^{1/2},
\end{aligned}
$$

where we first used the Cauchy–Schwarz inequality, then simply added the extra term (remaining in the sum from $m \le n$ to n), and then used the binomial expansion to eliminate the first sum in the product.

We continue the estimation of the first term with the same strategy, adding the remainder terms in the sum from m to n,

$$
\begin{aligned}
I_1 &\le \lambda \left(\sum_{j=0}^{n} \binom{n}{j} \left(\frac{\mu}{\lambda}\right)^{j} \left(\frac{\lambda-\mu}{\lambda}\right)^{n-j} (m-j)^2 \right)^{1/2} \\
&= \lambda \left(m^2 \sum_{j=0}^{n} \binom{n}{j} \left(\frac{\mu}{\lambda}\right)^{j} \left(\frac{\lambda-\mu}{\lambda}\right)^{n-j} - 2m \sum_{j=0}^{n} \binom{n}{j} \left(\frac{\mu}{\lambda}\right)^{j} \left(\frac{\lambda-\mu}{\lambda}\right)^{n-j} j \right. \\
&\quad \left. + \sum_{j=0}^{n} \binom{n}{j} \left(\frac{\mu}{\lambda}\right)^{j} \left(\frac{\lambda-\mu}{\lambda}\right)^{n-j} j^2 \right)^{1/2} \\
&= \lambda \left(m^2 - 2mn\frac{\mu}{\lambda} + \left(\frac{\mu}{\lambda}\right)^2 n^2 - \left(\frac{\mu}{\lambda}\right)^2 n + \left(\frac{\mu}{\lambda}\right) n \right)^{1/2} \\
&= \lambda \left(\left(m - n\frac{\mu}{\lambda} \right)^2 + n \left(\frac{\mu}{\lambda}\right) \left(\frac{\lambda-\mu}{\lambda}\right) \right)^{1/2} = ((\lambda m - n\mu)^2 + n\mu(\lambda-\mu))^{1/2}.
\end{aligned}
\tag{10.129}
$$

In the above, we used the elementary results

$$\sum_{j=0}^{n}\binom{n}{j}a^j b^{n-j} = (a+b)^n,$$
$$\sum_{j=0}^{n}\binom{n}{j}a^j b^{n-j} j = na(a+b)^{n-1},$$
$$\sum_{j=0}^{n}\binom{n}{j}a^j b^{n-j} j^2 = a^2n^2(a+b)^{n-2} - a^2n(a+b)^{n-2} + an(a+b)^{n-1},$$

(which are obtained by the binomial theorem, and then by differentiation once and twice with respect to a), applied for $a = \frac{\mu}{\lambda}$ and $b = 1 - a = \frac{\lambda-\mu}{\lambda}$.

The estimation of the second term requires the generalized binomial theorem according to which

$$\sum_{j=m}^{\infty}\binom{j-1}{m-1}b^{j-m} = \frac{1}{(1-b)^m}, \quad b \in (0,1). \tag{10.130}$$

Moreover, upon differentiating (10.130) with respect to b yields

$$\begin{aligned}\sum_{j=m}^{\infty}\binom{j-1}{m-1}b^{j-m}(j-m) &= \frac{mb}{(1-b)^{m+1}},\\ \sum_{j=m}^{\infty}\binom{j-1}{m-1}b^{j-m}(j-m)^2 &= \frac{mb(mb+1)}{(1-b)^{m+2}}.\end{aligned} \tag{10.131}$$

With this in mind, we estimate

$$\begin{aligned}I_2 &= \mu\sum_{j=m}^{n}\binom{j-1}{m-1}\left(\frac{\mu}{\lambda}\right)^m\left(\frac{\lambda-\mu}{\lambda}\right)^{j-m}(n-j)\\ &\le \mu\left(\sum_{j=m}^{n}\binom{j-1}{m-1}\left(\frac{\mu}{\lambda}\right)^m\left(\frac{\lambda-\mu}{\lambda}\right)^{j-m}\right)^{1/2}\\ &\quad\times\left(\sum_{j=m}^{n}\binom{j-1}{m-1}\left(\frac{\mu}{\lambda}\right)^m\left(\frac{\lambda-\mu}{\lambda}\right)^{j-m}(n-j)^2\right)^{1/2}.\end{aligned} \tag{10.132}$$

To further estimate this term, note that

$$\begin{aligned}\sum_{j=m}^{n}\binom{j-1}{m-1}\left(\frac{\mu}{\lambda}\right)^m\left(\frac{\lambda-\mu}{\lambda}\right)^{j-m} &\le \left(\frac{\mu}{\lambda}\right)^m\sum_{j=m}^{\infty}\binom{j-1}{m-1}\left(\frac{\lambda-\mu}{\lambda}\right)^{j-m}\\ &= \left(\frac{\mu}{\lambda}\right)^m\frac{1}{(1-\frac{\lambda-\mu}{\lambda})^m} = 1,\end{aligned} \tag{10.133}$$

where we used (10.130) for $b = \frac{\lambda-\mu}{\lambda}$. Using that in (10.132), we obtain

$$\begin{aligned} I_2 &\le \mu\left(\sum_{j=m}^{n}\binom{j-1}{m-1}\left(\frac{\mu}{\lambda}\right)^m\left(\frac{\lambda-\mu}{\lambda}\right)^{j-m}(n-j)^2\right)^{1/2} \\ &\le \mu\left(\sum_{j=m}^{\infty}\binom{j-1}{m-1}\left(\frac{\mu}{\lambda}\right)^m\left(\frac{\lambda-\mu}{\lambda}\right)^{j-m}(n-j)^2\right)^{1/2}. \end{aligned} \tag{10.134}$$

We note that

$$\begin{aligned} &\sum_{j=m}^{\infty}\binom{j-1}{m-1}\left(\frac{\mu}{\lambda}\right)^m\left(\frac{\lambda-\mu}{\lambda}\right)^{j-m}(n-j)^2 \\ &\quad\times\sum_{j=m}^{\infty}\binom{j-1}{m-1}\left(\frac{\mu}{\lambda}\right)^m\left(\frac{\lambda-\mu}{\lambda}\right)^{j-m}((n-m)+(m-j))^2 \\ &\quad=\left(\frac{\mu}{\lambda}\right)^m\sum_{j=m}^{\infty}\binom{j-1}{m-1}\left(\frac{\lambda-\mu}{\lambda}\right)^{j-m}((n-m)^2-2(n-m)(m-j)+(m-j)^2) \\ &\quad=(n-m)^2-2(n-m)m\frac{b}{a}+\frac{mb(mb+1)}{a^2}=\left(n-m+\frac{mb}{a}\right)^2+\frac{mb}{a}, \end{aligned} \tag{10.135}$$

where we used (10.130) and (10.131) for the choice $a = \frac{\mu}{\lambda}$ and $b = 1 - a = \frac{\lambda-\mu}{\lambda}$.

Substituting the above in (10.134) yields

$$I_2 \le (m\lambda(\lambda-\mu) + (m\lambda - n\mu)^2)^{1/2}. \tag{10.136}$$

Combining the above the proof is complete. □

Lemma 10.5.2. *Let the double sequence $a_{n,m}$ satisfy the inequality*

$$a_{jk} \le \frac{\mu}{\lambda}a_{j-1,k-1} + \frac{\lambda-\mu}{\lambda}a_{j-1,k}. \tag{10.137}$$

Then,

$$\begin{aligned} a_{n,m} &\le \sum_{j=0}^{m}\binom{n}{j}\left(\frac{\mu}{\lambda}\right)^j\left(\frac{\lambda-\mu}{\lambda}\right)^{m-j}a_{0,m-j} \\ &\quad+\sum_{j=m}^{n}\binom{j-1}{m-1}\left(\frac{\mu}{\lambda}\right)^m\left(\frac{\lambda-\mu}{\lambda}\right)^{j-m}a_{n-j,0}, \quad m \le n, \end{aligned} \tag{10.138}$$

and

$$a_{n,m} \leq \sum_{j=0}^{n} \binom{n}{j} \left(\frac{\mu}{\lambda}\right)^{j} \left(\frac{\lambda-\mu}{\lambda}\right)^{m-j} a_{0,m-j}, \quad n \leq m. \tag{10.139}$$

Proof. The proof is by induction from the set $S_N := \{1,\dots,N\}\times\{1,\dots,N\}$ to the set $S_{N+1} := \{1,\dots,N+1\} \times \{1,\dots,N+1\}$. Note that because of the nature of the iterative scheme, to check the claim, we need to consider both the case $n \leq m$ and the case $m \leq n$. Going from S_N to S_{N+1} for the induction step requires to take into account several separate cases, so the proof is tedious but in any other respect elementary. We only provide a sample of the relevant calculations.

The extension from S_N to S_{N+1} requires the following steps:

1. $(1,N) \to (1,N+1)$;
2. $(1,N+1) \to (2,N+1) \to \cdots \to (N+1,N+1)$;
3. $(N,N) \to (N+1,N+1)$;
4. $(N,1) \to (N+1,1)$;
5. $(N+1,1) \to (N+1,2) \to \cdots \to (N+1,N+1)$.

As a sample of the calculations, we show the last step (step 5), which requires an induction argument to show that the stated estimates hold for $(N+1,m)$ for all $m \leq N+1$. Since we are in the situation $m \leq n$, we can focus on estimate (10.138) only. Consider going from $(N+1,m)$ to $(N+1,m-1)$ to $(N+1,m)$. We have the iterative scheme

$$a_{N+1,m} \leq a a_{N,m-1} + b a_{N,m}, \tag{10.140}$$

where, for simplicity, we have set $a = \frac{\mu}{\lambda}$, $b = 1-a = \frac{\lambda-\mu}{\lambda}$. Since the induction hypothesis holds for N, using the estimate (10.138) and regrouping similar terms, we have from (10.140) that

$$\begin{aligned} a_{N+1,m} \leq & \underbrace{\left(a \sum_{j=0}^{m-1} a^j b^{N-j} \binom{N}{j} a_{0,m-1-j} + b \sum_{j=0}^{m} a^j b^{N-j} \binom{N}{j} a_{0,m-j} \right)}_{I_1} \\ & + \underbrace{\left(a \sum_{j=m-1}^{N} a^{m-1} b^{j-m+1} \binom{j-1}{m-1-1} a_{N-j,0} + b \sum_{j=m}^{N} a^m b^{j-m} \binom{j-1}{m-1} a_{N-j,0} \right)}_{I_2}. \end{aligned} \tag{10.141}$$

We treat each term separately.

$$
\begin{aligned}
I_1 &= \sum_{j=0}^{m-1} a^{j+1} b^{N-j} \binom{N}{j} a_{0,m-1-j} + b \sum_{j=0}^{m} a^j b^{N-j} \binom{N}{j} a_{0,m-j} \\
&= \sum_{j=1}^{m} a^j b^{N-j} \binom{N}{j-1} a_{0,m-j} + b \sum_{j=0}^{m} a^j b^{N-j} \binom{N}{j} a_{0,m-j} \\
&= b \sum_{j=1}^{m} a^j b^{N-j} \left(\binom{N}{j-1} + \binom{N}{j} \right) a_{0,m-j} + b a^0 b^{N-0} \binom{N}{0} a_{0,m} \\
&= b \sum_{j=1}^{m} a^j b^{N-j} \binom{N+1}{j} a_{0,m-j} + b a^0 b^{N-0} \binom{N}{0} a_{0,m} \\
&= b \sum_{j=0}^{m} a^j b^{N-j} \binom{N+1}{j} a_{0,m-j} \\
&= \sum_{j=0}^{m} a^j b^{N+1-j} \binom{N+1}{j} a_{0,m-j},
\end{aligned}
\tag{10.142}
$$

where we used the combinatorial identity

$$
\binom{N}{j-1} + \binom{N}{j} = \binom{N+1}{j}.
$$

For the second term, we have

$$
\begin{aligned}
I_2 &= \sum_{j=m-1}^{N} a^m b^{j-m+1} \binom{j-1}{m-1-1} a_{N-j,0} + \sum_{j=m}^{N} a^m b^{j+1-m} \binom{j-1}{m-1} a_{N-j,0} \\
&= \sum_{j=m}^{N} a^m b^{j-m+1} \left(\binom{j-1}{m-1-1} + \binom{j-1}{m-1} \right) a_{N-j,0} + a^m b^0 \binom{m-2}{m-2} a_{0,N-m+1} \\
&= \sum_{j=m}^{N} a^m b^{j-m+1} \binom{j}{m-1} a_{N-j,0} + a^m a_{0,N-m+1} \\
&= \sum_{j=m+1}^{N+1} a^m b^{j-m} \binom{j-1}{m-1} a_{N+1-j,0} + a^m a_{0,N-m+1} \\
&= \sum_{j=m}^{N+1} a^m b^{j-m} \binom{j-1}{m-1} a_{N+1-j,0},
\end{aligned}
\tag{10.143}
$$

where we used the combinatorial identity

$$
\binom{j-1}{m-2} + \binom{j-1}{m-1} = \binom{j}{m-1}.
$$

Combining (10.141), (10.142), and (10.143), we conclude that the estimate holds for $N+1$.

The other cases proceed similarly. □

10.5.2 Proof of (10.27)

Since $\{x_i^{(n)},\ i = 0, \dots, n\}$, $\{\bar{x}_j^{(m)},\ j = 0, \dots, m\}$ are ϵ-approximate solutions of the equation for the discretizations $\{f_i^{(n)},\ i = 0, \dots, n\}$, $\{\bar{f}_j^{(m)},\ j = 0, \dots, m\}$, respectively, they satisfy

$$\begin{aligned} \frac{x_{i+1}^{(n)} - x_i^{(n)}}{h_i^{(n)}} - A(x_{i+1}^{(n)}) \ni f_{i+1}^{(n)}, \quad i = 1, \dots, n-1, \\ \frac{\bar{x}_{j+1}^{(m)} - \bar{x}_j^{(m)}}{\bar{h}_j^{(m)}} - A(\bar{x}_{j+1}^{(m)}) \ni \bar{f}_{j+1}^{(m)}, \quad j = 1, \dots, m-1. \end{aligned} \tag{10.144}$$

For the sake of lighter notation, we will from now on use the simplified notation $h_i = h_i^{(n)} = t_{i+1}^{(n)} - t_i^{(n)}$, $\bar{h}_j = \bar{h}_j^{(m)} = s_{j+1}^{(m)} - s_j^{(m)}$.

By the dissipativity of A (equiv. accretivity of $-A$), and the definition of $[x, y]_+$, for any $x_i, y_i \in -A(x_i)$, $i = 1, 2$, it holds that

$$0 \le [x_1 - x_2, y_1 - y_2]_+. \tag{10.145}$$

We apply this for the choice $x_1 = x_{i+1}^{(n)}$, $y_1 = f_{i+1}^{(n)} + h_i^{-1}(x_i^{(n)} - x_{i+1}^{(n)})$, and then for the choice $x_1 = \bar{x}_{j+1}^{(m)}$, $y_1 = \bar{f}_{j+1}^{(m)} + \bar{h}_j^{-1}(\bar{x}_j^{(m)} - \bar{x}_{j+1}^{(m)})$. This, along with the property that $[x, y_1 + y_2]_+ \le [x, y]_+ + [x, z]_+$, yields

$$\begin{aligned} 0 \le{} & [x_{i+1}^{(n)} - \bar{x}_{j+1}^{(m)}, f_{i+1}^{(n)} - \bar{f}_{j+1}^{(m)}]_+ + h_i^{-1}[x_{i+1}^{(n)} - \bar{x}_{j+1}^{(m)}, x_i^{(n)} - x_{i+1}^{(n)}]_+ \\ & + \bar{h}_j^{-1}[x_{i+1}^{(n)} - \bar{x}_{j+1}^{(m)}, \bar{x}_{j+1}^{(m)} - \bar{x}_j^{(m)}]_+. \end{aligned} \tag{10.146}$$

We now consider each of the second and the third terms on the RHS of the above separately. For any $x_1, x_2, z \in X$, we have that

$$\begin{aligned} [x_1 - x_2, z - x_1]_+ &= \inf_{\lambda > 0}(\|x_1 - x_2 + \lambda(z - x_1)\| - \|x_1 - x_2\|) \\ &\le \|x_1 - x_2 + (z - x_1)\| - \|x_1 - y\| = \|z - x_2\| - \|x_1 - x_2\|, \end{aligned} \tag{10.147}$$

and

$$\begin{aligned} [x_1 - x_2, x_2 - z]_+ &= \inf_{\lambda > 0}(\|x_1 - x_2 + \lambda(x_2 - z)\| - \|x_1 - x_2\|) \\ &\le \|x_1 - x_2 + (x_2 - z)\| - \|x_1 - x_2\| = \|x_1 - z\| - \|x_1 - x_2\|. \end{aligned} \tag{10.148}$$

Applying (10.147) for $x_1 = x_{i+1}^{(n)}$, $x_2 = \bar{x}_{j+1}^{(m)}$, $z = x_i^{(n)}$ and (10.148) for the same x_1, x_2 and $z = \bar{x}_j^{(m)}$ and combining the results with (10.146), we obtain

$$\begin{aligned} 0 \le{} & [x_{i+1}^{(n)} - \bar{x}_{j+1}^{(m)}, f_{i+1}^{(n)} - \bar{f}_{j+1}^{(m)}]_+ + h_i^{-1}(\|x_i^{(n)} - \bar{x}_{j+1}^{(m)}\| - \|x_{i+1}^{(n)} - \bar{x}_{j+1}^{(m)}\|) \\ & + \bar{h}_j^{-1}(\|x_{i+1}^{(n)} - \bar{x}_j^{(m)}\| - \|x_{i+1}^{(n)} - \bar{x}_{j+1}^{(m)}\|). \end{aligned} \tag{10.149}$$

Finally, since for any $x, z \in X$,

$$[x, z]_+ = \inf_{\lambda>0}(\|x + \lambda z\| - \|x\|) \le \|x + z\| - \|x\| \le \|z\|, \tag{10.150}$$

(where for the last one we used the triangle inequality), we obtain that

$$[x_{i+1}^{(n)} - \bar{x}_{j+1}^{(m)}, f_i^{(n)} - \bar{f}_j^{(m)}]_+ \le \|f_{i+1}^{(n)} - \bar{f}_{j+1}^{(m)}\|. \tag{10.151}$$

Combining (10.149) with (10.151), we obtain

$$0 \le f_{i+1,j+1}^{(n,m)} + h_i^{-1}(e_{i,j+1)}^{(n,m)} - e_{i+1,j+1}^{(n,m)}) + \bar{h}_j^{-1}(e_{i+1,j}^{(n,m)} - e_{i+1,j+1}^{(n,m)}),$$

which is equivalent (upon shifting the indices) to (10.34). We rearrange, bringing $e_{i+1,j+1}^{(n,m)}$ on the LHS, to obtain the estimate

$$e_{i+1,j+1}^{(n,m)} - \alpha e_{i,j+1}^{(n,m)} - \beta e_{i+1,j}^{(n,m)} - \gamma f_{i+1,j+1}^{(n,m)} \le 0, \tag{10.152}$$

with

$$\alpha = \frac{\bar{h}_j}{h_i + \bar{h}_j}, \quad \beta = \frac{h_i}{h_i + \bar{h}_j}, \quad \gamma = \frac{h_i \bar{h}_j}{h_i + \bar{h}_j} \tag{10.153}$$

(note that α, β, γ depend on i and j in general for nonhomogeneous grids, but so as not to clutter the notation, we omit this explicit dependence). Upon redefining the constants and the indices, we obtain the stated result (10.27).

10.5.3 Proof of the boundary estimates (10.30) and (10.31)

We now prove the boundary estimates (10.30) and (10.31).

For the sake of lighter notation, we will from now on use the simplified notation $h_i = h_i^{(n)} = t_{i+1}^{(n)} - t_i^{(n)}$, $\bar{h}_j = \bar{h}_j^{(m)} = s_{j+1}^{(m)} - s_j^{(m)}$.

Recall that the dissipativity of A (equiv. accretivity of −A) implies that for any $(x_i, y_i) \in$ **Gr**$(-A)$, $i = 1, 2$, it holds that

$$\|x_1 - x_2\| \le \|x_1 - x_2 + \delta(y_1 - y_2)\|, \quad \forall\, \delta > 0. \tag{10.154}$$

We now apply (10.154) for the choice (a) $x_1 = x_{i+1}^{(n)}$ and $y_1 = f_{i+1}^{(n)} + h_i^{-1}(x_i^{(n)} - x_{i+1}^{(n)})$ (the fact that $y_1 \in -Ax_1$ for this choice comes from the definition of discretized solution as above), (b) $x_2 = x \in$ **D**(A) and any $y_2 = y \in -A(x)$, and (c) $\delta = h_i$. For this choice, (10.154) yields

$$\begin{aligned} \|x_{i+1}^{(n)} - x\| &\le \|x_{i+1}^{(n)} - x + h_i(f_{i+1}^{(n)} + h_i^{-1}(x_i^{(n)} - x_{i+1}^{(n)})) - y)\| \\ &\le \|x_i^{(n)} - x + h_i(f_{i+1}^{(n)} + y)\| \le \|x_i^{(n)} - x\| + h_i(\|f_{i+1}^{(n)}\| + \|y\|). \end{aligned} \tag{10.155}$$

This provides a telescopic relation, which upon iteration yields

$$\|x_i^{(n)} - x\| \leq \|x_0^{(n)} - x\| + \sum_{i_1=1}^{i} h_{i_1-1}(\|f_{i_1}^{(n)}\| + \|y\|). \tag{10.156}$$

A similar estimate can be obtained for $\|\bar{x}_j^{(m)} - x\|$, in terms of

$$\|\bar{x}_j^{(m)} - x\| \leq \|\bar{x}_0^{(m)} - x\| + \sum_{j_1=1}^{j} \bar{h}_{j_1-1}(\|\bar{f}_{j_1}^{(m)}\| + \|y\|). \tag{10.157}$$

Note that in the above estimates, $x \in \mathbf{D}(A)$ and $y \in -A(x)$ are arbitrary. Since in general the initial condition x_0 may not be an element of $\mathbf{D}(A)$, we may not be able to set $x = x_0$. However, if $x_0 \in \overline{\mathbf{D}(A)}$, then we may always choose an $x \in \mathbf{D}(A)$ arbitrarily close to the initial condition x_0 (or $x_0, \bar{x}_0$).

The above estimates will give us an idea concerning the relevant boundary condition for the continuous equation. Indeed, using the previous notation, we see that a choice for g_1 and g_2 can be

$$\begin{aligned} g_1(s) &\simeq \|\bar{x}_0^{(m)} - x\| + \int_0^s (\|\bar{f}(\tau))\| + \|y\|)d\tau, \\ g_2(t) &\simeq \|x_0^{(n)} - x\| + \int_0^t (\|f(\tau))\| + \|y\|)d\tau. \end{aligned} \tag{10.158}$$

Choosing the function g as in (10.43) i. e.,

$$\begin{aligned} g(t-s) \simeq \|x_0^{(n)} - x\| + \|\bar{x}_0^{(m)} - x\| + \int_0^t (\|f(\tau))\| + \|y\|)d\tau \\ + \int_0^s (\|\bar{f}(\tau))\| + \|y\|)d\tau \end{aligned} \tag{10.159}$$

is a possibility for the boundary condition, which is compatible with the above discretization. Note that the above are approximations and not strict equalities. The terms $\|x_0^{(n)} - x\|$ and $\|\bar{x}_0^{(m)} - x\|$ can be made arbitrarily small by appropriate choice of x (as long as $x_0^{(n)}, \bar{x}_0^{(m)} \in \mathbf{D}(A)$), whereas the integral terms are approximations of the corresponding sums. Note that this approximation is not unique, it will be slightly modified to the needs of the proof in the next step.

10.5.4 Proof of (10.51)

We now prove the estimate (10.51).

For the sake of lighter notation, we will from now on use the simplified notation $h_i = h_i^{(n)} = t_{i+1}^{(n)} - t_i^{(n)}$, $\bar{h}_j = \bar{h}_j^{(m)} = s_{j+1}^{(m)} - s_j^{(m)}$. We present the following proposition [59]

Proposition 10.5.3. *Consider the sequences of piecewise constant functions* $(g_{n,m}^D)_{nm}$ *and* $(F_{n,m}^D)_{nm}$ *(recall the construction in* (10.44) *and the identification* (10.45)*) such that*

$$\|F_{n,m}^D - F\|_* \to 0, \quad \text{and} \quad \|g_{n,m}^D - g\|_{L^\infty((-T,T);X)} \to 0, \quad \text{as } n,m \to \infty.$$

Then

$$\|S_{n,m}^D(g_{n,m}^D, F_{n,m}^D) - S^C(g,F)\|_{L^\infty(\Delta)} \to 0, \quad \text{as } n,m \to \infty.$$

Proof. The proof is technical and proceeds in various steps. To avoid heavy notation, we will omit the explicit n,m-dependence on the grid, e. g., denote $h_i = t_{i+1} - t_i = t_{i+1}^{(n)} - t_i^{(n)}$, $\bar{h}_j = s_{j+1} - s_j = s_{j+1}^{(m)} - s_j^{(m)}$, etc.

1. For any $u \in C^2(\Delta)$ with $u_{tt}, u_{ss} \in L^\infty(\Delta)$, it holds that

$$L_{n,m}^D u = Lu + \mathcal{E}_{n,m}^D, \tag{10.160}$$

where the error $\mathcal{E}_{n,m}^D = \{\mathcal{E}_{ij}^{(n,m)}, i = 0,\dots,n, j = 0,\dots,m\}$, with the simplified notation $k = (n,m)$ has the property $|\mathcal{E}_{ij}^{(n,m)}| \to 0$ as $n,m \to \infty$ for all i,j.

This follows easily from a Taylor expansion argument. For a sufficiently smooth u, it holds that

$$\begin{aligned} \left|\frac{\partial u}{\partial t}(t_i, s_j) - \frac{u_{i,j} - u_{i-1,j}}{h_i}\right| &\le h_i \|u_{tt}\|_{L^\infty(\Delta)}, \\ \left|\frac{\partial u}{\partial s}(t_i, s_j) - \frac{u_{i,j} - u_{i,j-1}}{\bar{h}_j}\right| &\le \bar{h}_j \|u_{ss}\|_{L^\infty(\Delta)}. \end{aligned} \tag{10.161}$$

Hence, we can express

$$\begin{aligned} \frac{\partial u}{\partial t}(t_i, s_j) &= \frac{u_{i,j} - u_{i-1,j}}{h_i} + \mathcal{E}_{ij}^{t;(n,m)} \\ \frac{\partial u}{\partial s}(t_i, s_j) &= \frac{u_{i,j} - u_{i,j-1}}{\bar{h}_j} + \mathcal{E}_{ij}^{s;(n,m)}, \end{aligned} \tag{10.162}$$

where the terms $\mathcal{E}_{ij}^{t;(n,m)}$ and $\mathcal{E}_{ij}^{s;(n,m)}$ correspond to the error induced when approximating the partial derivatives of u with respect to t and s, respectively, by the corresponding finite difference approximation. Using these, we can construct (using the construction in (10.44) and the convention (10.45)) the piecewise continuous functions $\mathcal{E}_{n,m}^t = \mathcal{E}^{t;(n,m)}$

and $\mathcal{E}^{s;(n,m)} = \mathcal{E}^{s}_{n,m}$, which are bounded in the $L^\infty(\Delta)$ norm. Hence, upon addition, (10.162) leads to the result that

$$(Lu)(t_i, s_j) = (L^D_k u_k)_{ij} + \mathcal{E}^{(n,m)}_{ij} \tag{10.163}$$

where $\mathcal{E}^D_{n,m} = (\mathcal{E}^{(n,m)}_{ij},\ i = 0, \dots, n,\ j = 0, \dots, m)$ with

$$\begin{aligned} \mathcal{E}^{(n,m)}_{ij} &= \mathcal{E}^{t;(n,m)}_{ij} + \mathcal{E}^{s;(n,m)}_{ij}, \\ |\mathcal{E}^{(n,m)}_{ij}| &\le h_i \|u_{tt}\|_{L^\infty(\Delta)} + \bar{h}_j \|u_{ss}\|_{L^\infty(\Delta)}. \end{aligned} \tag{10.164}$$

Clearly, as $n, m \to \infty$, it holds that $h_i = t^{(n)}_{i+1} - t^{(n)}_i \to 0$ and $\bar{h}_j = s^{(m)}_{j+1} - s^{(m)}_j \to 0$, so, by the above, we have that $\mathcal{E}^{(n,m)}_{ij} \to 0$ for all i, j.

2. We now provide an a priori estimate for the solution of the continuous problem $E = S^C(g, F)$ in terms of the data g, F. We use the representation (10.41).

For any function $F : \Delta = [0, T] \times [0, T] \to \mathbb{R}$, we define the set

$$\begin{aligned} K(F) := \{(\phi, \psi) \ :\ |F(t, s)| \le \phi(t) + \psi(s) \text{ a. e. in } [0, T] \times [0, T]\} \\ \subset L^1((0, T); X) \times L^1((0, T); X). \end{aligned} \tag{10.165}$$

By definition

$$\|F\|_* = \inf\{\|\phi\|_{L^1((0,T);X)} + \|\psi\|_{L^1((0,T);X)} \ :\ (\phi, \psi) \in K(F)\}. \tag{10.166}$$

Now take any $(\phi, \psi) \in K(F)$. Then it holds that

$$\begin{aligned} \left|\int_0^t F(\tau, s - t + \tau) d\tau\right| &\le \int_0^t |F(\tau, s - t + \tau)| d\tau \\ &\le \int_0^t |\phi(\tau)| d\tau + \int_t |\psi(s - t + \tau)| d\tau \\ &\le \|\phi\|_{L^1((0,T);X)} + \|\psi\|_{L^1((0,T);X)} =: \Lambda(\phi, \psi), \end{aligned} \tag{10.167}$$

in the case where $t \le s$, and similarly for the case $s \le t$. Since (10.167) holds for any $\Lambda(\phi, \psi)$ obtained for any choice $(\phi, \psi) \in K(F)$, it also holds for the infimum of this quantity over $K(F)$, hence,

$$\left|\int_0^t F(\tau, t - s + \tau) d\tau\right| \le \|F\|_*, \quad \forall\, t \le s, \tag{10.168}$$

with the same bound holding in the case $s \le t$ as well. Finally, it is easy to see that

$$|g(t-s)| \le \|g\|_{L^\infty((-T,T))}, \quad \forall\, s,t \in \Delta. \tag{10.169}$$

Combining the representation (10.41) with (10.168) and (10.169), we conclude that

$$\|S^C(g,F)\|_{L^\infty(\Delta)} \le \|g\|_{L^\infty((-T,T))} + \|F\|_*. \tag{10.170}$$

3. We now obtain an a priori estimate, similar to that obtained in step 2 for the continuous problem, for the discrete problem, i. e., for the solution mapping $S^D_{n,m}(g^D_{n,m}, F^D_{n,m})$. This is easier to see by expressing the discrete problem

$$h_i^{-1}(u_{ij} - u_{i-1,j}) + \bar{h}_j^{-1}(u_{ij} - u_{i,j-1}) = f_{ij}, \tag{10.171}$$

to its equivalent form

$$\begin{aligned} u_{ij} &= a_{ij}u_{i-1,j} + a_{ji}u_{i,j-1} + \frac{h_i\bar{h}_j}{h_i+\bar{h}_j}f_{ij} \\ a_{ij} &= \frac{\bar{h}_j}{h_i+\bar{h}_j}, \quad a_{ji} = \frac{h_i}{h_i+\bar{h}_j}, \quad a_i + a_j = 1, \end{aligned} \tag{10.172}$$

for arbitrary f_{ij}. Clearly, for the same boundary conditions, problems (10.171) and (10.172) have the same solution operator $S^D_{n,m}$. The fact that (10.172) displays the structure of a convex combination along with the observation that $a_{ij} < \bar{h}_j$ and $a_{ji} < h_i$ allows us to obtain a useful a priori estimate for the solution of (10.171) in terms of its data.

We start by the observation that for any data, due to the linearity of the problem and the superposition principle

$$\begin{aligned} S^D_{n,m}(g^D_{n,m}, F^D_{n,m}) &= S^D_{n,m}(0, F^D_{n,m}) \\ &\quad + S^D_{n,m}(g^D_{n,m}, 0), \end{aligned} \tag{10.173}$$

where in the above, the first term corresponds to the solution of (10.171) (equiv. (10.172)) with zero boundary data and the second term to the solution of (10.171) (equiv. (10.172)) with zero RHS.

We first try to estimate $S^D_{n,m}(g^D_{n,m}, 0)$. Note that

$$\begin{aligned} u_{i,j} &= a_{i,1}u_{i-1,1} + a_{1,i}u_{i,0} \\ &= a_{i,1}(a_{i-1,1}u_{i-2,1} + a_{1,i-1}u_{i-1,0}) + a_{1,i}u_{i,0} \\ &\le a_{i-1,1}u_{i-2,1} + a_{1,i-1}u_{i-1,0} + a_{1,i}u_{i,0} \\ &= a_{i-1,1}(a_{i-2,1}u_{i-3,1} + a_{1,i-2}u_{i-2,0}) + a_{1,i-1}u_{i-1,0} + a_{1,i}u_{i,0} \\ &\le a_{i-2,1}u_{i-3,1} + a_{1,i-2}u_{i-2,0} + a_{1,i-1}u_{i-1,0} + a_{1,i}u_{i,0} \\ &\le u_{i-3,1} + h_{i-2}g_{i-2,0} + h_{i-1}g_{i-1,0} + h_ig_{i,0}. \end{aligned} \tag{10.174}$$

Based on this structure and iterating, a simple induction step yields

$$u_{i,1} \leq \sum_{\ell=0}^{i} h_\ell g_{\ell,0} \leq \|g\|_{L^1(-T,T)}, \quad i = 1, \ldots, n. \tag{10.175}$$

Similarly for $u_{j,1}$. Having established these bounds, we can work in an analogous fashion for all the other indices. This establishes the bound

$$\|S^D_{n,m}(g^D_{n,m}, 0)\|_{L^\infty(\Delta)} \leq \|g\|_{L^1(-T,T)}. \tag{10.176}$$

We now turn our attention to $S^D_{n,m}(0, F_k)$. Motivated by the continuous case, we choose the relevant bound for the discretized functions as a limit for the discretized solution. We claim that

$$\|S^D_{n,m}(0, F^D_{n,m})\|_{L^\infty(\Delta_{nm})} \leq \|F^D_{n,m}\|_*, \tag{10.177}$$

where all the relevant quantities are for the piecewise constant function $u_{n,m} = u^D_{n,m} := S^D_{n,m}(0, F^D_{n,m})$ and $F_{n,m}$ on the grid Δ_{nm}.

We will show this claim using induction. Consider any pair of piecewise constant functions $\phi^D_n = \{\phi^{(n)}_i\ ;\ i = 1, \ldots, n\}$ and $\psi^D_m = \{\psi^{(m)}_j\ :\ j = 1, \ldots, m\}$ such that $|f^{(n,m)}_{ij}| \leq \phi^{(n)}_i + \psi^{(m)}_j$. By the above, we mean, as usual, that $\phi^D_n(t) = \phi^{(n)}_i$ for $t \in [t^{(n)}_i, t^{(n)}_{i+1})$ and $\psi^D_m(s) = \psi^{(m)}_j$ for $s \in [s^{(m)}_j, s^{(m)}_{j+1})$.

We may now consider the discretized version of the set $K(F)$ defined in (10.165),

$$\begin{aligned} K^D_{n,m}(F^D_{n,m}) &:= \{(\phi^D_n, \psi^D_m)\ :\ |f^{(n,m)}_{ij}| \leq \phi^{(n)}_i + \psi^{(m)}_j,\ i = 1, \ldots, n,\ j = 1, \ldots, m\} \\ &\subset L^1((0,T); X) \times L^1((0,T); X), \end{aligned}$$

along with the discretized version of the norm $\|\cdot\|_*$ defined in (10.166):

$$\|F^D_{n,m}\|_* = \inf\{\|\phi^D_n\|_{L^1((0,T);X)} + \|\psi^D_m\|_{L^1((0,T);X)}\ :\ (\phi^{(n)}_i, \psi^{(m)}_j) \in K^D_{n,m}(F^D_{n,m})\}.$$

Since $\|\phi^D_n\|_{L^1(0,T)} = \sum_{i=1}^n h_i \phi^{(n)}_i$ and $\|\psi^D_m\|_{L^1(0,T)} = \sum_{j=1}^m \bar{h}_j \psi^{(m)}_j$, we see that

$$\|F^D_{n,m}\|_* = \inf\left\{\sum_{i=1}^n h_i \phi^{(n)}_i + \sum_{j=1}^m \bar{h}_j \psi^{(m)}_j\ :\ (\phi^{(n)}_i, \psi^{(m)}_j) \in K^D_{n,m}(F^D_{n,m})\right\}.$$

For any $(\phi^D_n, \psi^D_m) \in K^D_{n,m}(F^D_{n,m})$, it follows that $u^D_{n,m} = S^D_{n,m}(0, F^D_{n,m})$ satisfies the inequality

$$\begin{aligned} u_{ij} &= a_{ij} u_{i-1,j} + a_{ji} u_{i,j-1} + \frac{h_i \bar{h}_j}{h_i + \bar{h}_j} f^{(n,m)}_{ij} \\ &\leq a_{ij} u_{i-1,j} + a_{ji} u_{i,j-1} + \frac{h_i \bar{h}_j}{h_i + \bar{h}_j} (\phi^{(n)}_i + \psi^{(m)}_j). \end{aligned} \tag{10.178}$$

We maintain that for every i, j, it holds that

$$u_{ij} \leq \sum_{\ell_1=1}^{i} h_{\ell_1} \phi_{\ell_1}^{(n)} + \sum_{\ell_2=1}^{j} \bar{h}_{\ell_2} \psi_{\ell_2}^{(m)}. \tag{10.179}$$

Clearly, because of the choice of homogeneous boundary conditions, (10.179) holds for the boundary. We will show by induction that it holds for all (i,j). Some care has to be taken because of the double index, but essentially the argument can proceed as follows: Assume that the claim holds for $(i-1,j)$ and $(i,j-1)$. Set

$$\begin{aligned} S_i &= \sum_{\ell_1=1}^{i} h_{\ell_1} \phi_{\ell_1}^{(n)}, \\ T_j &= \sum_{\ell_2=1}^{j} \bar{h}_{\ell_2} \psi_{\ell_2}^{(m)}. \end{aligned} \tag{10.180}$$

Then, (10.178) yields

$$\begin{aligned} u_{ij} &\leq a_{ij}(S_{i-1} + T_j) + a_{ji}(S_i + T_{j-1}) + \frac{h_i \bar{h}_j}{h_i + \bar{h}_j} (\phi_i^{(n)} + \psi_j^{(m)}) \\ &= a_{ij}(S_{i-1} + T_{j-1} + \bar{h}_j \psi_j^{(m)}) + a_{ji}(S_{i-1} + h_i \phi_i^{(n)} + T_{j-1}) + \frac{h_i \bar{h}_j}{h_i + \bar{h}_j} (\phi_i^{(n)} + \psi_j^{(m)}) \\ &= S_{i-1} + T_{j-1} + \left(a_{ji} h_i + \frac{h_i \bar{h}_j}{h_i + \bar{h}_j} \right) \phi_i^{(n)} + \left(a_{ij} \bar{h}_j + \frac{h_i \bar{h}_j}{h_i + \bar{h}_j} \right) \psi_j^{(m)} \\ &= S_{i-1} + T_{j-1} + h_i \phi_i^{(n)} + \bar{h}_j \psi_j^{(m)} = S_i + T_j, \end{aligned}$$

which is (10.179). In the above, we used the definition of a_{ij} and a_{ji} and the property $a_{ij} + a_{ji} = 1$. Hence, the induction hypothesis holds for i, j. Some care has to be taken when passing, e. g., from $i-1, j-1$ to $i-1, j$, etc., but the steps are similar.

Having established (10.179), using the fact that all quantities are positive, we see that

$$u_{ij} \leq \sum_{i=1}^{n} h_i \phi_i^{(n)} + \sum_{j=1}^{m} \bar{h}_j \psi_j^{(m)}, \quad i = 1, \ldots, n,\ j = 1, \ldots, m,$$

which leads to the estimate

$$\|S_{n,m}^D(0, F_{n,m}^D)\|_{L^\infty(\Delta)} \leq \sum_{i=1}^{n} h_i \phi_i^{(n)} + \sum_{j=1}^{m} \bar{h}_j \psi_j^{(m)}, \tag{10.181}$$

and since (10.181) holds for any $(\phi_n^D, \psi_m^D) \in K_{n,m}^D(F_{n,m}^D)$, it also holds for the infimum of the RHS over $K_{n,m}^D(F_{n,m}^D)$, which by definition coincides with $\|F_{n,m}^D\|_*$. Hence, (10.177) is established.

Combining (10.173) with (10.176) and (10.177) leads to the required estimate

$$\|S^D_{n,m}(g^D_{n,m}, F^D_{n,m})\|_{L^\infty(\Delta)} \le \|g^D_{n,m}\|_{L^\infty(-T,T))} + \|F^D_{n,m}\|_*. \tag{10.182}$$

4. Let g^S, F^S be smooth approximations to g and F chosen so that for any $\delta > 0$,

$$\|g - g^S\|_{L^\infty((-T,T))} < \delta, \quad \|F - F^S\|_* < \delta, \tag{10.183}$$

and consider $E^S := S^C(g^S, F^S)$. We will show that E^S is "close" to $E^D_{n,m} = S^D_{n,m}(g^D_{n,m}, F^D_{n,m})$, as long as $g^D_{n,m}, F^D_{n,m}$ are sufficiently close to g, F. In particular, we will show that

$$\begin{aligned}\|S^C(g^S, F^S) - S^D_{n,m}(g^D_{n,m}, F^D_{n,m})\|_{L^\infty(\Delta)}& \\ \le \|\mathcal{E}^{(n,m)}\|_* + \|g^S - g^D_{n,m}\|_{L^\infty((-T,T))} &+ \|F^S - F^D_{n,m}\|_*.\end{aligned} \tag{10.184}$$

By definition

$$\begin{aligned}LE^S &= F^S,\\ E^S(t,s) &= g^S(t-s), \quad t = 0 \text{ or } s = 0,\end{aligned} \tag{10.185}$$

where the first equation holds for all $(t,s) \in \Delta$. Moreover, since E^S is smooth, (10.160) applied to E^S yields

$$L^D_{n,m}E^S = LE^S + \mathcal{E}^D_{n.m} = F^S + \mathcal{E}^D_{n.m}, \tag{10.186}$$

where the discretization error $\mathcal{E}^D_{nm}$ satisfies

$$|\mathcal{E}^{(n,m)}_{ij}| \le \max\{h_i, \bar{h}_j\}\left(\left\|\frac{\partial E^S}{\partial t^2}\right\|_{L^\infty(\Delta)} + \left\|\frac{\partial E^S}{\partial s^2}\right\|_{L^\infty(\Delta)}\right). \tag{10.187}$$

Note that (10.186) is to be interpreted as

$$\begin{aligned}(L^D_{n,m}E^S)_{ij} &:= (LE^S)(t^{(n)}_i, s^{(m)}_j) + \mathcal{E}^{(n,m)}_{ij}\\ &= F^S(t^{(n)}_i, s^{(m)}_j) + \mathcal{E}^{(n,m)}_{ij}.\end{aligned} \tag{10.188}$$

We will use (10.186) to express E^S in an alternative way in terms of the discrete solution operator $S^D_{n,m}$. Before doing so, we consider $w^D_{n,m} = S^D_{n,m}(0, \mathcal{E}^D_{n,m})$, i. e., the solution of

$$\begin{aligned}L^D_{n,m}w^D_{n,m} &= \mathcal{E}^D_{n,m},\\ (w^D_{n,m})_{ij} &= 0, \quad i = 0, \text{ or } j = 0,\end{aligned} \tag{10.189}$$

with the first equation interpreted as in (10.188). Subtracting (10.186) and (10.189), we obtain

$$
\begin{aligned}
L^D_{n,m}(E^S - w^D_{n,m}) &= F^S,\\
(E^S - w^D_{n,m})_{ij} &= (g^S)_{ij}, \quad i = 0 \text{ or } j = 0,
\end{aligned} \tag{10.190}
$$

where, for a smooth function, e. g., g^S by $(g^S)_{ij}$, we mean $(g^S)_{ij} = g^S(t_i^{(n)}, s_j^{(m)})$.

It follows from (10.190) that $E^S - w^D_{n,m} = S^D_{n,m}(g^S, F^S)$, hence, by the initial definition of E^S, it holds that

$$
S^C(g^S, F^S) - S^D_{n,m}(0, \mathcal{E}^D_{nm}) = S^D_{n,m}(g^S, F^S). \tag{10.191}
$$

By the properties of the solution operator, we have that

$$
\begin{aligned}
S^C(g^S, F^S) &- S^D_{n,m}(g^D_{n,m}, F^D_{n,m})\\
&= S^D_{n,m}(0, \mathcal{E}^D_{n,m}) + S^D_{n,m}(g^S, F^S) - S^D_{n,m}(g^D_{n,m}, F^D_{n,m})\\
&= S^D_{n,m}(0, \mathcal{E}^D_{n,m}) + S^D_{n,m}(g^S - g^D_{n,m}, F^S - F^D_{n,m}).
\end{aligned} \tag{10.192}
$$

We then have

$$
\begin{aligned}
\|S^C(g^S, F^S) &- S^D_{n,m}(g^D_{n,m}, F^D_{n,m})\|_{L^\infty(\Delta)}\\
&\le \|S^D_{n,m}(0, \mathcal{E}^D_{n,m})\|_{L^\infty(\Delta)} + \|S^D_{n,m}(g^S - g^D_{n,m}, F^S - F^D_{n,m})\|_{L^\infty(\Delta)}.
\end{aligned} \tag{10.193}
$$

Using (10.177), we obtain the bound

$$
\|S^D_{n,m}(0, \mathcal{E}^D_{n,m})\|_{L^\infty(\Delta)} \le \|\mathcal{E}^D_{n,m}\|_*, \tag{10.194}
$$

whereas using (10.182), we obtain

$$
\begin{aligned}
\|S^D_{n,m}(g^S &- g^D_{n,m}, F^S - F^D_{n,m})\|_{L^\infty(\Delta)}\\
&\le \|g^S - g^D_{n,m}\|_{L^\infty((-T,T))} + \|F^S - F^D_{n,m}\|_*.
\end{aligned} \tag{10.195}
$$

Combining these two, we obtain (10.184).

5. We now estimate

$$
\begin{aligned}
\|S^D_{n,m}(g^D_{n,m}, F^D_{n,m}) &- S^C(g, F)\|_{L^\infty(\Delta)}\\
&= \|S^D_{n,m}(g^D_{n,m}, F^D_{n,m}) - S^C(g^S, F^S) + S^C(g^S, F^S) - S^C(g, F)\|_{L^\infty(\Delta)}\\
&\le \|S^D_{n,m}(g^D_{n,m}, F^D_{n,m}) - S^C(g^S, F^S)\|_{L^\infty(\Delta)}\\
&\quad + \|S^C(g^S, F^S) - S^C(g, F)\|_{L^\infty(\Delta)}\\
&= \|S^D_{n,m}(g^D_{n,m}, F^D_{n,m}) - S^C(g^S, F^S)\|_{L^\infty(\Delta)}\\
&\quad + \|S^C(g^S - g, F^S - F)\|_*.
\end{aligned}
$$

We will use the estimate (10.184) obtained in step 4 for the first term and estimate (10.170) in step 2 to estimate the second term of the above, respectively. Combining these estimates, we obtain

$$\begin{aligned} &\|S^D_{n,m}(g^D_{n,m}, F^D_{n,m}) - S^C(g, F)\|_{L^\infty(\Delta)} \\ &\quad \leq \|\mathcal{E}^D_{nm}\|_* + \|g^S - g^D_{n,m}\|_{L^\infty((-T,T))} + \|F^S - F^D_{n,m}\|_* \\ &\quad\quad + \|g^S - g\|_{L^\infty((-T,T))} + \|F^S - F\|_*. \end{aligned} \tag{10.196}$$

We now consider each term of estimate (10.196) separately.

By (10.187), it easily follows that

$$\|\mathcal{E}^D_{nm}\|_* \to 0, \quad \text{as } n, m \to \infty,$$

which corresponds to the limit $h_i, \bar{h}_j \to 0$. We then use (10.183) for an arbitrarily small $\delta > 0$, and estimate

$$\begin{aligned} \|g^S - g^D_{n,m}\|_{L^\infty((-T,T))} &\leq \|g^S - g\|_{L^\infty((-T,T))} + \|g - g^D_{n,m}\|_{L^\infty((-T,T))} \\ &\leq \delta + \|g - g^D_{n,m}\|_{L^\infty((-T,T))}, \end{aligned}$$

which, taking $\delta \to 0$ and $n, m \to \infty$, leads us to the result that

$$\|g^S - g^D_{n,m}\|_{L^\infty((-T,T))} \to 0, \quad \text{as } n, m \to \infty,$$

with a similar result for $\|F^S - F^D_{n,m}\|_*$. Finally, by (10.183), the terms $\|g^S - g\|_{L^\infty((-T,T))}$, $\|F^S - F\|_*$ can be chosen arbitrarily small.

Combining the above, we obtain the desired result. □

Bibliography

[1] Robert A Adams and John JF Fournier. *Sobolev spaces*, volume 140. Academic Press, 2003.

[2] Ravi P Agarwal, Maria Meehan, and Donal O'Regan. *Fixed point theory and applications*, volume 141. Cambridge University Press, 2001.

[3] Yakov Alber and Irina Ryazantseva. *Nonlinear ill-posed problems of monotone type*. Springer, 2006.

[4] Fernando Albiac and Nigel John Kalton. *Topics in Banach space theory*, volume 233. Springer, 2016.

[5] Charalampos D. Aliprantis and Kim C. Border. *Infinite dimensional analysis: a hitchhiker's guide*. Springer, 2006.

[6] M Marques Alves and Benar Fux Svaiter. A new proof for maximal monotonicity of subdifferential operators. *Journal of Convex Analysis*, 15(2):345, 2008.

[7] Antonio Ambrosetti and Paul H Rabinowitz. Dual variational methods in critical point theory and applications. *Journal of Functional Analysis*, 14(4):349–381, 1973.

[8] Aomar Anane. Simplicité et isolation de la premiere valeur propre du p-laplacien avec poids. *Comptes Rendus de L'Académie Des Sciences. Série 1, Mathématique*, 305(16):725–728, 1987.

[9] Aomar Anane and Jean-Pierre Gossez. Strongly nonlinear elliptic problems near resonance: a variational approach. *Communications in Partial Differential Equations*, 15(8):1141–1159, 1990.

[10] David Arcoya and Luigi Orsina. Landesman-Lazer conditions and quasilinear elliptic equations. *Nonlinear Analysis: Theory, Methods & Applications*, 28(10):1623–1632, 1997.

[11] Francisco J Aragón Artacho, Jonathan M Borwein, and Matthew K Tam. Douglas–Rachford feasibility methods for matrix completion problems. *ANZIAM Journal*, 55(4):299–326, 2014.

[12] Hedy Attouch, Giuseppe Buttazzo, and Gérard Michaille. *Variational analysis in Sobolev and BV spaces: applications to PDEs and optimization*, volume 17. SIAM, 2014.

[13] Jean P. Aubin and Ivar Ekeland. *Applied nonlinear analysis*. Wiley New York, 1984.

[14] Jean-Pierre Aubin and Hélène Frankowska. *Set-valued analysis*. Springer Science & Business Media, 2009.

[15] Marino Badiale and Enrico Serra. *Semilinear elliptic equations for beginners: existence results via the variational approach*. Springer Science & Business Media, 2010.

[16] John M Ball and Victor J Mizel. One-dimensional variational problems whose minimizers do not satisfy the Euler–Lagrange equation. In *Analysis and Thermomechanics*, pages 285–348. Springer, 1987.

[17] G Barbatis, IG Stratis, and AN Yannacopoulos. Homogenization of random elliptic systems with an application to Maxwell's equations. *Mathematical Models and Methods in Applied Sciences*, 25(7):1365–1388, 2015.

[18] Viorel Barbu. *Nonlinear differential equations of monotone types in Banach spaces*. Springer Science & Business Media, 2010.

[19] Viorel Barbu. *Semigroup approach to nonlinear diffusion equations*. World Scientific, 2022.

[20] Viorel Barbu and Teodor Precupanu. *Convexity and optimization in Banach spaces*. Springer Science & Business Media, 2012.

[21] Heinz H Bauschke and Patrick L Combettes. *Convex analysis and monotone operator theory in Hilbert spaces*. Springer Science & Business Media, 2011.

[22] Lisa Beck. *Elliptic regularity theory, volume 19 of Lecture Notes of the Unione Matematica Italiana*, 2016.

[23] Marino Belloni and Bernd Kawohl. A direct uniqueness proof for equations involving the p-Laplace operator. *Manuscripta Mathematica*, 109(2):229–231, 2002.

[24] Alain Bensoussan and J-L Lions. *Applications of variational inequalities in stochastic control*, volume 12. Elsevier, 2011.

[25] Leonid Berlyand and Volodymyr Rybalko. *Getting acquainted with homogenization and multiscale*. Springer, 2018.

[26] Xavier Blanc and Claude Le Bris. *Homogenization theory for multiscale problems*, volume 21. Springer, 2023.

https://doi.org/10.1515/9783111333298-011

[27] Jonathan M Borwein and D Russell Luke. Duality and convex programming. In *Handbook of mathematical methods in imaging*, pages 229–270. Springer, 2011.
[28] Jonathan M Borwein and Jon D Vanderwerff. *Convex functions: constructions, characterizations and counterexamples*, volume 109. Cambridge University Press Cambridge, 2010.
[29] Jonathan M Borwein and Qiji J Zhu. *Techniques of variational analysis*. Springer, 2005.
[30] Andrea Braides. *Gamma-convergence for beginners*, volume 22. Clarendon Press, 2002.
[31] Haïm Brezis. Equations et inéquations non linéaires dans les espaces vectoriels en dualité. *Annales de L'Institut Fourier*, 18:115–175, 1968.
[32] Haim Brezis. *Operateurs maximaux monotones et semi-groupes de contractions dans les espaces de Hilbert*, volume 50. North Holland, 1973.
[33] Haim Brezis. *Functional analysis, Sobolev spaces and partial differential equations*. Springer, 2011.
[34] Haïm Brezis and Louis Nirenberg. Remarks on finding critical points. *Communications on Pure and Applied Mathematics*, 44(8–9):939–963, 1991.
[35] Arne Brøndsted and Tyrell R Rockafellar. On the subdifferentiability of convex functions. *Proceedings of the American Mathematical Society*, 16(4):605–611, 1965.
[36] Felix E Browder. Nonexpansive nonlinear operators in a Banach space. *Proceedings of the National Academy of Sciences of the United States of America*, 54(4):1041–1044, 1965.
[37] Felix E Browder and Peter Hess. Nonlinear mappings of monotone type in Banach spaces. *Journal of Functional Analysis*, 11(3):251–294, 1972.
[38] Anca Capatina. *Variational inequalities and frictional contact problems*. Springer, 2014.
[39] James Caristi. Fixed point theorems for mappings satisfying inwardness conditions. *Transactions of the American Mathematical Society*, 215:241–251, 1976.
[40] Siegfried Carl, Vy K Le, and Dumitru Motreanu. *Nonsmooth variational problems and their inequalities: comparison principles and applications*. Springer Science & Business Media, 2007.
[41] Nigel L. Carothers. *A short course on Banach space theory*. Cambridge University Press, 2004.
[42] Thierry Cazenave. An introduction to semilinear elliptic equations. *Editora do Instituto de Matemática, Universidade Federal do Rio de Janeiro, Rio de Janeiro*, 164, 2006.
[43] Antonin Chambolle and Pierre-Louis Lions. Image recovery via total variation minimization and related problems. *Numerische Mathematik*, 76(2):167–188, 1997.
[44] Antonin Chambolle and Thomas Pock. A first-order primal-dual algorithm for convex problems with applications to imaging. *Journal of Mathematical Imaging and Vision*, 40(1):120–145, 2011.
[45] Antonin Chambolle and Thomas Pock. An introduction to continuous optimization for imaging. *Acta Numerica*, 25:161–319, 2016.
[46] Antonin Chambolle and Thomas Pock. On the ergodic convergence rates of a first-order primal–dual algorithm. *Mathematical Programming*, 159(1–2):253–287, 2016.
[47] Kung-Ching Chang. *Methods in nonlinear analysis*. Springer Science & Business Media, 2006.
[48] Michel Chipot. *Elements of nonlinear analysis*. Birkhäuser, 2000.
[49] Michel Chipot. *Elliptic equations: an introductory course*. Springer Science & Business Media, 2009.
[50] Doina Cioranescu and Patrizia Donato. *An introduction to homogenization, volume 17 of Oxford Lecture Series in Mathematics and its Applications*, volume 4. The Clarendon Press Oxford University Press, New York, 1999.
[51] Doina Cioranescu, Patrizia Donato, and Marian P Roque. *An introduction to second order partial differential equations: classical and variational solutions*. World Scientific, 2018.
[52] Ioana Cioranescu. *Geometry of Banach spaces, duality mappings and nonlinear problems*, volume 62. Springer Science & Business Media, 1990.
[53] Christian Clason. *Introduction to functional analysis*. Springer Nature, 2020.
[54] Patrick L Combettes and Valérie R Wajs. Signal recovery by proximal forward-backward splitting. *Multiscale Modeling & Simulation*, 4(4):1168–1200, 2005.
[55] Patrick L Combettes and Isao Yamada. Compositions and convex combinations of averaged nonexpansive operators. *Journal of Mathematical Analysis and Applications*, 425(1):55–70, 2015.

[56] John B Conway. *A course in functional analysis*, volume 96. Springer Science & Business Media, 2013.
[57] David G Costa. *An invitation to variational methods in differential equations*. Springer Science & Business Media, 2010.
[58] MG Crandall and TM Liggett. Generation of semi-groups of nonlinear transformations on general Banach spaces. *American Journal of Mathematics*, pages 265–298, 1971.
[59] MG Crandall and LC Evans. On the relation of the operator $\partial/\partial s + \partial/\partial\tau$ to evolution governed by accretive operators. *Israel Journal of Mathematics*, pages 261–278, 1975.
[60] Riccardo Cristoferi. γ-convergence, lecture notes.
[61] Bernard Dacorogna. *Direct methods in the calculus of variations*, volume 78. Springer Verlag, 2008.
[62] Gianni Dal Maso. *An introduction to Γ-convergence*, volume 8. Springer Science & Business Media, 2012.
[63] Ennio De Giorgi. Teoremi di semicontinuita nel calcolo delle variazioni. *Istituto Nazionale di Alta Matematica, Roma*, 1969, 1968.
[64] Pablo De Nápoli and María Cristina Mariani. Mountain pass solutions to equations of p-Laplacian type. *Nonlinear Analysis: Theory, Methods & Applications*, 54(7):1205–1219, 2003.
[65] Gerard Debreu. *Theory of value: an axiomatic analysis of economic equilibrium*, volume 17. Yale University Press, 1959.
[66] Klaus Deimling. *Nonlinear functional analysis*. Dover, 2010.
[67] Emmanuele Di Benedetto and Neil S Trudinger. Harnack inequalities for quasi-minima of variational integrals. *Annales de L'Institut Henri Poincaré. Analyse Non Linéaire*, 1(4):295–308, 1984.
[68] Joseph Diestel. *Geometry of Banach spaces-selected topics, volume 485 of Lecture Notes in Mathematics*. Springer, 1975.
[69] Joseph Diestel and John Jerry Uhl. *Vector measures*, volume 15. American Mathematical Society, 1977.
[70] James Dugundji. *Topology*. Allyn and Bacon, Inc, 1970.
[71] Michael Edelstein. On fixed and periodic points under contractive mappings. *Journal of the London Mathematical Society*, 1(1):74–79, 1962.
[72] Ivar Ekeland. Sur les problèmes variationnels. *CR Seanc. Acad. Sci. Paris*, 275:1057–1059, 1972.
[73] Ivar Ekeland. On the variational principle. *Journal of Mathematical Analysis and Applications*, 47(2):324–353, 1974.
[74] Ivar Ekeland and Roger Temam. *Convex analysis and variational problems*, volume 28. SIAM, 1999.
[75] Klaus-Jochen Engel and Rainer Nagel. *One-parameter semigroups for linear evolution equations*, volume 194. Springer Science & Business Media, 1999.
[76] Lawrence C Evans. *Partial differential equations*. American Mathematical Society, Providence, 1998.
[77] Marián Fabian, Petr Habala, Petr Hájek, Vicente Montesinos, and Václav Zizler. *Banach space theory: the basis for linear and nonlinear analysis*. Springer, 2011.
[78] Alessio Figalli. Free boundary regularity in obstacle problems. *arXiv preprint arXiv:1807.01193*, 2018.
[79] Dario C. Figueiredo. *Lectures on the Ekeland variational principle with applications and detours*. Springer Berlin etc, 1989.
[80] Irene Fonseca and Giovanni Leoni. *Modern methods in the calculus of variations: L^p spaces*. Springer New York, 2007.
[81] Jean-Pierre Fouque, George Papanicolaou, Ronnie Sircar, and Knut Sølna. *Multiscale stochastic volatility for equity, interest rate, and credit derivatives*. Cambridge University Press, 2011.
[82] Jan Francu. Monotone operators. A survey directed to applications to differential equations. *Aplikace Matematiky*, 35(4):257–301, 1990.
[83] Enrico Giusti. *Direct methods in the calculus of variations*. World Scientific, 2003.
[84] AA Goldstein. Convex programming in Hilbert space. *Bulletin of the American Mathematical Society*, 70(6):709–710, 1964.
[85] Christopher Heil. *Introduction to real analysis*, volume 280. Springer, 2019.
[86] Antoine Henrot. *Extremum problems for eigenvalues of elliptic operators*. Springer Science & Business Media, 2006.

[87] Theophil Henry Hildebrandt and Lawrence M Graves. Implicit functions and their differentials in general analysis. *Transactions of the American Mathematical Society*, 29(1):127–153, 1927.

[88] Shouchuan Hu and Nikolaos Socrates Papageorgiou. *Handbook of multivalued analysis*, volumes 1 and 2. Kluwer Dordrecht, 1997, 2000.

[89] Kazufumi Ito and Franz Kappel. *Evolution equations and approximations*, volume 61. World Scientific, 2002.

[90] Kazufumi Ito and Karl Kunisch. *Lagrange multiplier approach to variational problems and applications*, volume 15. SIAM, 2008.

[91] Youssef Jabri. *The Mountain Pass Theorem: variants, generalizations and some applications*, volume 95. Cambridge University Press, 2003.

[92] John L Kelley and MH Stone. *General topology*, volume 233. van Nostrand Princeton, 1955.

[93] S Kesavan. *Topics in functional analysis and applications*. New Age International Limited Publishers, New Delhi, 2003.

[94] David Kinderlehrer and Guido Stampacchia. *An introduction to variational inequalities*. Academic Press, New York, 1981.

[95] William Art Kirk. A fixed point theorem for mappings which do not increase distances. *The American Mathematical Monthly*, 72(9):1004–1006, 1965.

[96] Gottfried Köthe. *Topological vector spaces*. Springer-Verlag, 1969.

[97] Nikolai Vladimirovich Krylov and Mikhail V Safonov. A certain property of solutions of parabolic equations with measurable coefficients. *Izvestiya Rossiiskoi Akademii Nauk. Seriya Matematicheskaya*, 44(1):161–175, 1980.

[98] Giovanni Leoni. *A first course in Sobolev spaces*. American Mathematical Society, 2009.

[99] Elliott H Lieb, Michael Loss, et al. *Analysis*, volume 14. American Mathematical Society Providence, Rhode Island, 2001.

[100] Peter Lindqvist. On the equation. *Proceedings of the American Mathematical Society*, 109(1):157–164, 1990.

[101] Jacques-Louis Lions. *Quelques méthodes de résolution des problemes aux limites non linéaires*, volume 31. Dunod Paris, 1969.

[102] Pierre-Louis Lions and Panagiotis E Souganidis. Homogenization of degenerate second-order pde in periodic and almost periodic environments and applications. *Annales de l'IHP Analyse non linéaire*, 22:667–677, 2005.

[103] Anthony R Lovaglia. Locally uniformly convex Banach spaces. *Transactions of the American Mathematical Society*, 78(1):225–238, 1955.

[104] Robert E Megginson. *Banach space theory*, volume 183. Springer Science & Business Media, 2012.

[105] Ernest Michael. Continuous selections. I. *Annals of Mathematics*, 361–382, 1956.

[106] Jean-Jacques Moreau. Proximité et dualité dans un espace hilbertien. *Bulletin de la Société Mathématique de France*, 93:273–299, 1965.

[107] Nikolaos S Papageorgiou and Sophia Th Kyritsi-Yiallourou. *Handbook of applied analysis*, volume 19. Springer Science & Business Media, 2009.

[108] George Papanicolaou, Alain Bensoussan, and J-L Lions. *Asymptotic analysis for periodic structures*. Elsevier, 1978.

[109] Enea Parini. Continuity of the variational eigenvalues of the p-Laplacian with respect to p. *Bulletin of the Australian Mathematical Society*, 83(3):376–381, 2011.

[110] Dan Pascali and Silviu Sburlan. *Nonlinear mappings of monotone type*. Springer, 1978.

[111] Amnon Pazy. *Semigroups of linear operators and applications to partial differential equations*, volume 44. Springer Science & Business Media, 2012.

[112] Electra V Petracou and Athanasios N Yannacopoulos. Decision theory under risk and applications in social sciences: II. Game theory. In *Mathematical modeling with multidisciplinary applications*, pages 421–447, 2013.

[113] Juan Peypouquet. *Convex optimization in normed spaces: theory, methods and examples*. Springer, 2015.

[114] Robert R Phelps. *Convex functions, monotone operators and differentiability*, volume 1364. Springer, 1993.
[115] Robert R Phelps. Lectures on maximal monotone operators. *Extracta Mathematicae*, 12(3):193–230, 1997.
[116] Boris T Polyak and ES Levitin. Constrained minimization methods. *USSR Computational Mathematics and Mathematical Physics*, 6(5):1–50, 1966.
[117] Lee C Potter and KS Arun. A dual approach to linear inverse problems with convex constraints. *SIAM Journal on Control and Optimization*, 31(4):1080–1092, 1993.
[118] Radu Precup. *Methods in nonlinear integral equations*. Springer Science & Business Media, 2013.
[119] Patrizia Pucci and Vicentiu Radulescu. The impact of the mountain pass theory in nonlinear analysis: a mathematical survey. *Bollettino dell'Unione Matematica Italiana*, 3(3):543–582, 2010.
[120] Paul H Rabinowitz. *Minimax methods in critical point theory with applications to differential equations*, volume 65. American Mathematical Society, 1986.
[121] Vicentiu D Radulescu. *Qualitative analysis of nonlinear elliptic partial differential equations*. Hindawi Publishing Corporation, 2008.
[122] Simeon Reich. Weak convergence theorems for nonexpansive mappings in Banach spaces. *Journal of Mathematical Analysis and Applications*, 67(2):274–276, 1979.
[123] Michael Renardy and Robert C. Rogers. *An introduction to partial differential equations*, volume 13. Springer Verlag, 2004.
[124] Sergey Repin. A posteriori error estimation for variational problems with uniformly convex functionals. *Mathematics of Computation*, 69(230):481–500, 2000.
[125] Gary Francis Roach, Ioannis G Stratis, and Athanasios N Yannacopoulos. *Mathematical analysis of deterministic and stochastic problems in complex media electromagnetics*, volume 42. Princeton University Press, 2012.
[126] R Tyrell Rockafellar et al. On the maximal monotonicity of subdifferential mappings. *Pacific Journal of Mathematics*, 33(1):209–216, 1970.
[127] Jose-Francisco Rodrigues. *Obstacle problems in mathematical physics*, volume 134. Elsevier, 1987.
[128] Tomáš Roubíček. *Nonlinear partial differential equations with applications*, volume 153. Springer Science & Business Media, 2013.
[129] Filippo Santambrogio. *A course in the calculus of variations: optimization, regularity, and modeling*. Springer Nature, 2023.
[130] Stephen Simons. A new proof of the maximal monotonicity of subdifferentials. *Journal of Convex Analysis*, 16(1):165–168, 2009.
[131] Maurice Sion. On general minimax theorems. *Pacific Journal of Mathematics*, 8(1):171–176, 1958.
[132] Guido Stampacchia. Le problème de dirichlet pour les équations elliptiques du second ordre à coefficients discontinus. *Annales de L'Institut Fourier*, 15:189–257, 1965.
[133] Michael Struwe. *Variational methods: applications to nonlinear partial differential equations and Hamiltonian systems*. Springer Verlag, 2008.
[134] W A Sutherland. *Introduction to metric and topological spaces*. Oxford University Press, 1975.
[135] Luc Tartar. *The general theory of homogenization: a personalized introduction*, volume 7. Springer Science & Business Media, 2009.
[136] Raymond Trémolières, J-L Lions, and R Glowinski. *Numerical analysis of variational inequalities*, volume 8. Elsevier, 2011.
[137] S Troyanski. On locally uniformly convex and differentiable norms in certain non-separable Banach spaces. *Studia Mathematica*, 37(2):173–180, 1971.
[138] Andrianos E Tsekrekos and Athanasios N Yannacopoulos. Optimal switching decisions under stochastic volatility with fast mean reversion. *European Journal of Operational Research*, 251(1):148–157, 2016.
[139] Andrianos E Tsekrekos and Athanasios N Yannacopoulos. *Variational inequalities in management science and finance: modelling, analysis, numerics and applications*. Springer Nature, 2025.

[140] Jack Warga. *Optimal control of differential and functional equations*. Academic Press, 2014.
[141] Stephen Willard. *General topology*. Courier Dover Publications, 2004.
[142] James Yeh. *Real analysis: theory of measure and integration*. World Scientific, 2006.
[143] Eberhard Zeidler. *Nonlinear functional analysis vol. 1: fixed-point theorems*. Springer-Verlag Berlin and Heidelberg GmbH & Co. K, 1986.
[144] Eberhard Zeidler. *Nonlinear functional analysis and its applications, part II/B: nonlinear monotone operators*, volume 2. Springer Verlag, 1989.

Index

https://doi.org/10.1515/9783111333298-012

www.ingramcontent.com/pod-product-compliance
Lightning Source LLC
Chambersburg PA
CBHW080348220326
41598CB00030B/4636